KV-449-226

OXFORD SCIENCE PUBLICATIONS

Concise Dictionary of Chemistry

Concise Dictionary of Chemistry

OXFORD NEW YORK
OXFORD UNIVERSITY PRESS
1985

Oxford University Press, Walton Street, Oxford OX2 6DP
Oxford New York Toronto
Delhi Bombay Calcutta Madras Karachi
Kuala Lumpur Singapore Hong Kong Tokyo
Nairobi Dar es Salaam Cape Town
Melbourne Auckland
and associated companies in
Beirut Berlin Ibadan Nicosia

Oxford is a trademark of Oxford University Press

ISBN 0 19 866143 6

Text prepared by
Market House Books Ltd., Aylesbury
Printed in Great Britain by
Cox and Wyman Ltd., Reading

Preface

This dictionary is derived from the *Concise Science Dictionary*, published by Oxford University Press in 1984. It consists of all the entries relating to chemistry in this dictionary, including physical chemistry, as well as many of the terms used in biochemistry.

The more physical aspects of physical chemistry and the physics itself will be found in the *Concise Dictionary of Physics*, which is a companion volume to this dictionary. The *Concise Dictionary of Biology* contains a more thorough coverage of the biophysical entries from the *Concise Science Dictionary* together with the entries relating to biology.

SI units are used throughout this book and its companion volumes.

JD, 1985

Editor

John Daintith B.Sc., Ph.D.

Advisors

B. S. Beckett B.Sc., B.Phil., MA (Ed.)
R. A. Hands B.Sc. Michael Lewis MA

Contributors

Tim Beardsley BA
Lionel Bender B.Sc.
W. M. Clarke B.Sc.
Derek Cooper Ph.D., FRIC
E. K. Duintith B.Sc.
D. E. Edwards B.Sc., M.Sc.
Malcolm Hart B.Sc., M.I.Biol.
Robert S. Hine B.Sc., M.Sc.
Ann Lockwood B.Sc.
J. Valerie Neal B.Sc., Ph.D.
R. A. Prince MA
Jackie Smith BA
Brian Stratton B.Sc., M.Sc.
Elizabeth Tootill B.Sc., M.Sc.
David Eric Ward B.Sc., M.Sc., Ph.D

A

absolute 1. Not dependent on or relative to anything else, e.g. *absolute zero. **2.** Denoting a temperature measured on an *absolute scale*, a scale of temperature based on absolute zero. The usual absolute scale now is that of thermodynamic *temperature; its unit, the kelvin, was formerly called the degree absolute (°A) and is the same size as the degree Celsius. In British engineering practice an absolute scale with Fahrenheit-size degrees has been used: this is the Rankine scale.

absolute alcohol *See* ethanol.

absolute configuration A way of denoting the absolute structure of an optical isomer (*see* optical activity). Two conventions are in use: The D-L convention relates the structure of the molecule to some reference molecule. In the case of sugars and similar compounds, the dextrorotatory form of glyceraldehyde ($HOCH_2CH(OH)CHO$), 2,3-dihydroxypropanal) was used. The rule is as follows. Write the structure of this molecule down with the asymmetric carbon in the centre, the –CHO group at the top, the –OH on the right, the $-CH_2OH$ at the bottom, and the –H on the left. Now imagine that the central carbon atom is at the centre of a tetrahedron with the four groups at the corners and that the –H and –OH come out of the paper and the –CHO and $-CH_2OH$ groups go into the paper. The resulting three-dimensional structure was taken to be that of *d*-glyceraldehyde and called D-glyceraldehyde. Any compound that contains an asymmetric carbon atom having this configuration belongs to the D-series. One having the opposite configuration belongs to the L-series. It is important to note that the prefixes D- and L- do not stand for dextrorotatory and laevorotatory (i.e. they are not the same as *d*- and *l*-). In fact the arbitrary configuration assigned to D-glyceraldehyde is now known to be the correct one for the dextrorotatory form, although this was not known at the time. However, all D-compounds are not dextrorotatory. For instance, the acid obtained by oxidizing the –CHO group of glyceraldehyde is glyceric acid (1,2-dihydroxypropanoic acid). By convention, this belongs to the D-series, but it is in fact laevorotatory; i.e. its name can be written as D-glyceric acid or *l*-glyceric acid. To avoid confusion it is better to use + (for dextrorotatory) and – (for laevorotatory), as in D-(+)-glyceraldehyde and D-(–)-glyceric acid.

The D–L convention can also be used with alpha amino acids (compounds with the $-NH_2$ group on the same carbon as the –COOH group). In this case the molecule is imagined as being viewed along the H–C bond between the hydrogen and the asymmetric carbon atom. If the clockwise order of the other three groups is –**CO**OH, –**R**, $-\mathbf{N}H_2$, the amino acid belongs to the D-series; otherwise it belongs to the L-series. This is known as the *CORN rule*.

The R–S convention is a convention based on priority of groups attached to the chiral carbon atom. The order of priority is I, Br, Cl, SO_3H, $OCOCH_3$, OCH_3, OH, NO_2, NH_2, $COOCH_3$, $CONH_2$, $COCH_3$, CHO, CH_2OH, C_6H_5, C_2H_5, CH_3, H, with hydrogen lowest. The molecule is viewed with the group of lowest priority behind the chiral atom. If the clockwise arrangement of the other three groups is in descending priority, the compound belongs to the R-series; if the descending order is anticlockwise it is in the S-series. D-(+)-glyceraldehyde is R-(+)-glyceraldehyde.

absolute temperature *See* absolute; temperature.

absolute zero Zero of thermodynamic *temperature (0 kelvin) and the lowest temperature theoretically attainable. It is the temperature at which the kinetic energy of

planar formula | structure in 3 dimensions | Fischer projection

D−(+)− glyceraldehyde (2,3-dihydroxypropanal)

D−alanine (R is CH_2 in the CORN rule). The molecule is viewed with H on top

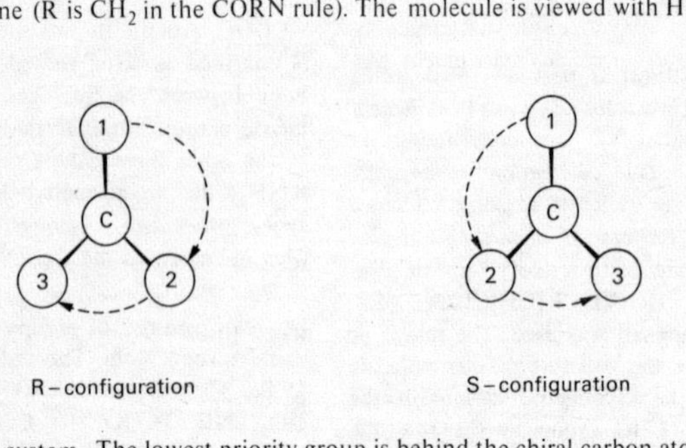

R-configuration

S-configuration

R-S system. The lowest priority group is behind the chiral carbon atom

atoms and molecules is minimal. It is equivalent to −273.15°C or −459.67°F. *See also* zero-point energy; cryogenics.

absorption 1. (in chemistry) The take up of a gas by a solid or liquid, or the take up of a liquid by a solid. Absorption differs from *adsorption in that the absorbed substance permeates the bulk of the absorbing substance. **2.** (in physics) The conversion of the energy of electromagnetic radiation, sound, streams of particles, etc., into other forms of energy on passing through a medium. A beam of light, for instance, passing through a medium, may lose intensity because of two effects: scattering of light out of the beam, and absorption of photons by atoms or molecules in the medium. When a photon is absorbed, there is a transition to an excited state.

absorption indicator A type of indicator used in reactions that involve precipitation. The yellow dye fluorescein is a common example, used for the reaction

$$NaCl(aq) + AgNO_3(aq) \rightarrow AgCl(s) + NaNO_3(aq)$$

As silver nitrate solution is added to the sodium chloride, silver chloride precipitates. As long as Cl^- ions are in excess, they adsorb on the precipitate particles. At the end point, no Cl^- ions are left in solution and negative fluorescein ions are then adsorbed, giving a pink colour to the precipitate.

absorption spectrum *See* spectrum.

abundance 1. The ratio of the total mass of a specified element in the earth's crust to the total mass of the earth's crust, often expressed as a percentage. For example, the abundance of aluminium in the earth's crust is about 8%. **2.** The ratio of the number of atoms of a particular isotope of an element to the total number of atoms of all the isotopes present, often expressed as a percentage. For example, the abundance of uranium-235 in natural uranium is 0.71%. This is the *natural abundance*, i.e. the abundance as found in nature before any enrichment has taken place.

accelerator A substance that increases the rate of a chemical reaction, i.e. a catalyst.

acceptor A compound, molecule, ion, etc., to which electrons are donated in the formation of a coordinate bond.

accumulator (secondary cell; storage battery) A type of *voltaic cell or battery that can be recharged by passing a current through it from an external d.c. supply. The charging current, which is passed in the opposite direction to that in which the cell supplies current, reverses the chemical reactions in the cell. The common types are the *lead–acid accumulator and the *nickel–iron accumulator.

acetaldehyde *See* ethanal.

acetals Organic compounds formed by addition of alcohol molecules to aldehyde molecules. If one molecule of aldehyde (RCHO) reacts with one molecule of alcohol (R′OH) a *hemiacetal* is formed (RCH(OH)OR′). The rings of aldose sugars are hemiacetals. Further reaction with a second alcohol molecule produces a full acetal ($RCH(OR')_2$). The formation of acetals is reversible; acetals can be hydrolysed back to aldehydes in acidic solutions. In synthetic organic chemistry aldehyde groups are often converted into acetal groups to protect them before performing other reactions on different groups in the molecule. *See also* ketals.

$$\text{R–CHO (aldehyde)} + \text{R'OH (alcohol)} \longrightarrow \text{RCH(OR')(OH) (hemiacetal)}$$

$$\text{RCH(OR')(OH)} + \text{R'OH} \longrightarrow \text{RCH(OR')}_2 \text{ (acetal)} + H_2O$$

Formation of acetals

acetamide *See* ethanamide.

acetate *See* ethanoate.

acetic acid *See* ethanoic acid.

acetone *See* propanone.

acetylation *See* acylation.

acetyl chloride *See* ethanoyl chloride.

acetylcholine A substance that is released at some (*cholinergic*) nerve endings. Its function is to pass on a nerve impulse to the next nerve (i.e. at a synapse) or to initiate muscular contraction. Once acetylcholine has been released, it has only a transitory effect because it is rapidly broken down by the enzyme *acetylcholinesterase*.

acetylene *See* ethyne.

acetylenes *See* alkynes.

acetyl group *See* ethanoyl group.

acetylide *See* carbide.

Acheson process An industrial process for the manufacture of graphite by heating coke mixed with clay. The reaction involves the production of silicon carbide, which loses silicon at 4150°C to leave graphite. The process was patented in 1896 by the US inventor Edward Goodrich Acheson (1856–1931).

acid 1. A type of compound that contains hydrogen and dissociates in water to produce positive hydrogen ions. The reaction, for an acid HX, is commonly written:

$$HX \rightleftharpoons H^+ + X^-$$

In fact, the hydrogen ion (the proton) is solvated, and the complete reaction is:

$$HX + H_2O \rightleftharpoons H_3O^+ + X^-$$

The ion H_3O^+ is the *oxonium ion* (or *hydroxonium ion* or *hydronium ion*). This definition of acids comes from the *Arrhenius theory*. Such acids tend to be corrosive substances with a sharp taste, which turn litmus red and give colour changes with other *indicators. They are referred to as *protonic acids* and are classified into *strong acids*, which are almost completely dissociated in water (e.g. sulphuric acid and hydrochloric acid), and *weak acids*, which are only partially dissociated (e.g. ethanoic acid and hydrogen sulphide). The strength of an acid depends on the extent to which it dissociates, and is measured by its *dissociation constant. *See also* base.

2. In the *Lowry–Brønsted theory* of acids and bases (1923), the definition was extended to one in which an acid is a proton donor, and a base is a proton acceptor. For example, in

$$HCN + H_2O \rightleftharpoons H_3O^+ + CN^-$$

the HCN is an acid, in that it donates a proton to H_2O. The H_2O is acting as a base in accepting a proton. Similarly, in the reverse reaction H_3O^+ is an acid and CN^- a base. In such reactions, two species related by loss or gain of a proton are said to be *conjugate*. Thus, in the reaction above HCN is the *conjugate acid* of the base CN^-, and CN^- is the *conjugate base* of the acid HCN. Similarly, H_3O^+ is the conjugate acid of the base H_2O. An equilibrium, such as that above, is a competition for protons between an acid and its conjugate base. A strong acid has a weak conjugate base, and vice versa. Under this definition water can act as both acid and base. Thus in

$$NH_3 + H_2O \rightleftharpoons NH_4^+ + OH^-$$

the H_2O is the conjugate acid of OH^-. The definition also extends the idea of acid–base reaction to solvents other than water. For instance, liquid ammonia, like water, has a high dielectric constant and is a good ionizing solvent. Equilibria of the type

$$NH_3 + Na^+Cl^- \rightleftharpoons Na^+NH_2^- + HCl$$

can be studied, in which NH_3 and HCl are acids and NH_2^- and Cl^- are their conjugate bases.

3. A further extension of the idea of acids and bases was made in the *Lewis theory* (G. N. Lewis, 1923). In this, a *Lewis acid* is a compound or atom that can accept a pair of electrons and a *Lewis base* is one that can donate an electron pair. This definition encompasses 'traditional' acid–base reactions. In

$$HCl + NaOH \rightarrow NaCl + H_2O$$

the reaction is essentially

$$H^+ + :OH^- \rightarrow H:OH$$

i.e. donation of an electron pair by OH^-. But it also includes reactions that do not involve ions, e.g.

$$H_3N: + BCl_3 \rightarrow H_3NBCl_3$$

in which NH_3 is the base (donor) and BCl_3 the acid (acceptor). The Lewis theory establishes a relationship between acid–base reactions and *oxidation–reduction reactions.

acid anhydrides (acyl anhydrides) Compounds that react with water to form an acid. For example, carbon dioxide reacts with water to give carbonic acid:

$$CO_2(g) + H_2O(aq) \rightleftharpoons H_2CO_3(aq)$$

A particular group of acid anhydrides are anhydrides of carboxylic acids. They have a general formula of the type R.CO.O.CO.R′, where R and R′ are alkyl or aryl groups.

R–C=O, HO–C=O ; HO–C=O, R′–C=O —($-H_2O$)→ R–C=O, O, R′–C=O

carboxylic acids   acid anhydride

Formation of a carboxylic acid anhydride

For example, the compound ethanoic anhydride ($CH_3.CO.O.CO.CH_3$) is the acid anhydride of ethanoic (acetic) acid. Organic acid anhydrides can be produced by dehydrating acids (or mixtures of acids). They are usually made by reacting an acyl halide with the sodium salt of the acid. They react readily with water, alcohols, phenols, and amines and are used in *acylation reactions.

acid dissociation constant *See* dissociation.

acid dye *See* dye.

acid halides *See* acyl halides.

acidic 1. Describing a compound that is an acid. **2.** Describing a solution that has an excess of hydrogen ions. **3.** Describing a compound that forms an acid when dissolved in water. Carbon dioxide, for example, is an acidic oxide.

acidic hydrogen (acid hydrogen) A hydrogen atom in an *acid that forms a positive ion when the acid dissociates. For instance, in methanoic acid

$$HCOOH \rightleftharpoons H^+ + HCOO^-$$

the hydrogen atom on the carboxylate group is the acidic hydrogen (the one bound directly to the carbon atom does not dissociate).

acidimetry Volumetric analysis using standard solutions of acids to determine the amount of base present.

acidity constant *See* dissociation.

acid rain *See* pollution.

acid salt A salt of a polybasic acid (i.e. an acid having two or more acidic hydrogens) in which not all the hydrogen atoms have been replaced by positive ions. For example, the dibasic acid carbonic acid (H_2CO_3) forms acid salts (hydrogencarbonates) containing the ion HCO_3^-. Some salts of monobasic acids are also known as acid salts. For instance, the compound potassium hydrogendifluoride, KHF_2, contains the ion $[F...H-F]^-$, in which there is hydrogen bonding between the fluoride ion F^- and a hydrogen fluoride molecule.

acid value A measure of the amount of free acid present in a fat, equal to the number of milligrams of potassium hydroxide needed to neutralize this acid. Fresh fats contain glycerides of fatty acids and very little free acid, but the glycerides decompose slowly with time and the acid value increases.

Acrilan A tradename for a synthetic fibre. *See* acrylic resins.

acrylate *See* propenoate.

acrylic acid *See* propenoic acid.

acrylic resins Synthetic resins made by polymerizing esters or other derivatives of acrylic acid (propenoic acid). Examples are poly(propenonitrile) (e.g. *Acrilan*), and poly-(methyl 2-methylpropenoate) (polymethyl methacrylate, e.g. *Perspex*).

acrylonitrile *See* propenonitrile.

actinides *See* actinoids.

actinium Symbol Ac. A silvery radioactive metallic element belonging to group IIIB of the periodic table; a.n. 89; mass number of most stable isotope 227 (half-life 21.7 years); m.p. 1050 ± 50°C; b.p. 3300°C (estimated). Actinium–227 occurs in natural uranium to an extent of about 0.715%. Actinium–228 (half-life 6.13 hours) also occurs in nature. There are 22 other artificial isotopes, all radioactive and all with very short half-lives. Its chemistry is similar to that of lanthanum. It has no uses and was discovered by A. Debierne in 1899.

actinium series *See* radioactive series.

actinoid contraction A smooth decrease in atomic or ionic radius with increasing proton number found in the *actinoids.

actinoids (actinides) A series of elements in the *periodic table, generally considered to range in atomic number from thorium (90) to lawrencium (103) inclusive. The actinoids all have two outer *s*-electrons (a $7s^2$ configuration), follow actinium, and are classified together by the fact that increasing proton number corresponds to filling of the $5f$ level. In fact, because the $5f$ and $6d$ levels are close in energy the filling of the $5f$ orbitals is not smooth. The outer electron configurations are as follows:
89 actinium (Ac) $6d^17s^2$
90 thorium (Th) $6d^27s^2$
91 protactinium (Pa) $5f^26d^17s^2$
92 uranium (Ur) $5f^36d7s^2$
93 neptunium (Np) $5f^57s^2$ (or $5f^46d^17s^2$)
94 plutonium (Pu) $5f^67s^2$
95 americium (Am) $5f^77s^2$
96 curium (Cm) $5f^76d^1s^2$
97 berkelium (Bk) $5f^86d7s^2$ (or $5f^97s^2$)
98 californium (Cf) $5f^{10}7s^2$
99 einsteinium (Es) $5f^{11}7s^2$
100 fermium (Fm) $5f^{12}7s^2$
101 mendelevium (Md) $5f^{13}7s^2$
102 nobelium (Nb) $5f^{14}7s^2$
103 lawrencium (Lw) $5f^{14}6d^1s^2$

The first four members (Ac to Ur) occur naturally. All are radioactive and this makes investigation difficult because of self-heating, short lifetimes, safety precautions, etc. Like the *lanthanoids, the actinoids show a smooth decrease in atomic and ionic radius with increasing proton number. The lighter members of the series (up to americium) have *f*-electrons that can participate in bonding, unlike the lanthanoids. Consequently, these elements resemble the transition metals in forming coordination complexes and displaying variable valency. As a result of increased nuclear charge, the heavier members (curium to lawrencium) tend not to use their inner *f*-electrons in forming bonds and resemble the lanthanoids in forming compounds containing the M^{3+} ion. The reason for this is pulling of these inner electrons towards the centre of the atom by the increased nuclear charge. Note that actinium itself does not have a $5f$ electron, but it is usually classified with the actinoids because of its chemical similarities. *See also* transition elements.

actinometer Any of various instruments for measuring the intensity of electromagnetic radiation. Recent actinometers use the *photoelectric effect but earlier instruments depended either on the fluorescence produced by the radiation on a screen or on the amount of chemical change induced in some suitable substance.

action spectrum A graphical plot of the efficiency of electromagnetic radiation in producing a photochemical reaction against the wavelength of the radiation used. For example, the action spectrum for photosynthesis using light shows a peak in the region 670–700 nm. This corresponds to a maximum absorption in the absorption spectrum of chlorophylls in this region.

activated alumina *See* aluminium hydroxide.

activated charcoal *See* charcoal.

activated complex The association of atoms of highest energy formed in the *transition state of a chemical reaction.

activation analysis An analytical technique that can be used to detect most elements when present in a sample in milligram quantities (or less). In *neutron activation analysis* the sample is exposed to a flux of thermal neutrons in a nuclear reactor. Some of these neutrons are captured by nuclides in the sample to form nuclides of the same atomic number but a higher mass number. These newly formed nuclides emit gamma radiation, which can be used to identify the element present by means of a gamma-ray spectrometer. Activation analysis has also been employed using charged particles, such as protons or alpha particles.

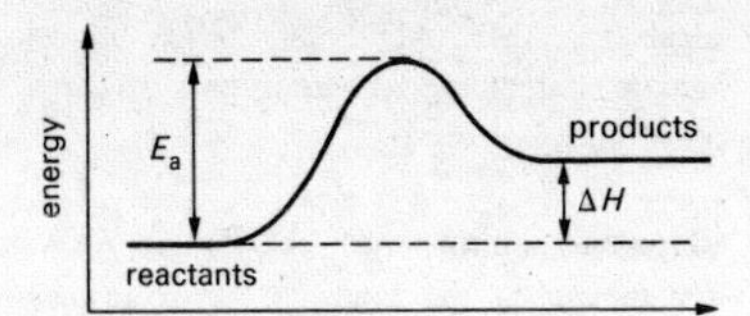

Reaction profile
(for an endothermic reaction)

activation energy The minimum energy required for a chemical reaction to take place. In a reaction, the reactant molecules come together and chemical bonds are stretched, broken, and formed in producing the products. During this process the energy of the system increases to a maximum, then decreases to the energy of the products. The activation energy is the difference between the maximum energy and the energy of the reactants; i.e. it is the energy barrier that has to be overcome for the reaction to proceed. The activation energy determines the way in which the rate of the reaction varies with temperature (*see* Arrhenius equation). It is usual to express activation energies in joules per mole of reactants.

active mass *See* mass action.

active site **1.** A site on the surface of a catalyst at which activity occurs. **2.** The site on the surface of an *enzyme molecule that binds the substrate molecule. The properties of an active site are determined by the three-dimensional arrangement of the polypeptide chains of the enzyme and their constituent amino acids. These govern the nature of the interaction that takes place and hence the degree of substrate specificity and susceptibility to *inhibition.

activity **1.** Symbol *a*. A thermodynamic function used in place of concentration in equilibrium constants for reactions involving nonideal gases and solutions. For example, in a reaction

$$A \rightleftharpoons B + C$$

the true equilibrium constant is given by

$$K = a_B a_C / a_A$$

where a_A, a_B, and a_C are the activities of the components, which function as concentrations (or pressures) corrected for nonideal behaviour. *Activity coefficients* (symbol γ) are defined for gases by $\gamma = a/p$ (where p is pressure) and for solutions by $\gamma = aX$ (where X is the mole fraction). Thus, the equilibrium constant of a gas reaction has the form

$$K_p = \gamma_B p_B \gamma_C p_C / \gamma_A p_A$$

The equilibrium constant of a reaction in solution is

$$K_c = \gamma_B X_B \gamma_C X_C / \gamma_A X_A$$

The activity coefficients thus act as correction factors for the pressures or concentrations. *See also* fugacity.

2. Symbol *A*. The number of atoms of a radioactive substance that disintegrate per unit time. The *specific activity* (*a*) is the activity per unit mass of a pure radioisotope. *See* radiation units.

acyclic Describing a compound that does not have a ring in its molecules.

acyl anhydrides *See* acid anhydrides.

acylation The process of introducing an acyl group (RCO–) into a compound. The usual method is to react an alcohol with an acyl halide or a carboxylic acid anhydride; e.g.

$$RCOCl + R'OH \rightarrow RCOOR' + HCl$$

The introduction of an acetyl group (CH_3CO-) is *acetylation*, a process used for protecting –OH groups in organic synthesis.

acylglycerol *See* glyceride.

acyl group A group of the type RCO–, where R is an organic group. An example is the acetyl group CH_3CO-.

R
C=O
X

Acyl halide: X is a halogen atom

acyl halides (acid halides) Organic compounds containing the group –CO.X, where X is a halogen atom. Acyl chlorides, for instance, have the general formula RCOCl. The group RCO– is the *acyl group*. In systematic chemical nomenclature acyl-halide names end in the suffix *-oyl*; for example, ethanoyl chloride, CH_3COCl. Acyl halides react readily with water, alcohols, phenols, and amines and are used in *acylation reactions. They are made by replacing the –OH group in a carboxylic acid by a halogen using a halogenating agent such as PCl_5.

addition polymerization *See* polymerization.

addition reaction A chemical reaction in which one molecule adds to another. Addition reactions occur with unsaturated compounds containing double or triple bonds, and may be *electrophilic or *nucleophilic. An example of electrophilic addition is the reaction of hydrogen chloride with an alkene, e.g.

$$HCl + CH_2{:}CH_2 \rightarrow CH_3CH_2Cl$$

An example of nucleophilic addition is the addition of hydrogen cyanide across the carbonyl bond in aldehydes to form *cyanohydrins. *Addition–elimination* reactions are ones in which the addition is followed by elimination of another molecule (*see* condensation reaction).

adduct A compound formed by an addition reaction. The term is used particularly for compounds formed by coordination between a Lewis acid (acceptor) and a Lewis base (donor). *See* acid.

adenine A *purine derivative. It is one of the major component bases of *nucleotides and the nucleic acids *DNA and *RNA.

adenosine A nucleoside comprising one adenine molecule linked to a D-ribose sugar molecule. The phosphate-ester derivatives of adenosine, AMP, ADP, and *ATP, are of fundamental biological importance as carriers of chemical energy.

adenosine diphosphate (ADP) *See* ATP.

adenosine monophosphate (AMP) *See* ATP.

adenosine triphosphate *See* ATP.

adhesive A substance used for joining surfaces together. Adhesives are generally colloidal solutions, which set to gels. There are many types including animal glues (based on collagen), vegetable mucilages, and synthetic resins (e.g. *epoxy resins).

adiabatic demagnetization A technique for cooling a paramagnetic salt, such as potassium chrome alum, to a temperature near *absolute zero. The salt is placed between the poles of an electromagnet and the heat produced during magnetization is removed by liquid helium. The salt is then isolated thermally from the surroundings and the field is switched off; the salt is demagnetized adiabatically and its temperature falls. This is because the demagnetized state, being less ordered, involves more energy than the magnetized state. The extra energy can come only from the internal, or thermal, energy of the substance.

adiabatic process Any process that occurs without heat entering or leaving a system. In general, an adiabatic change involves a fall or rise in temperature of the system. For example, if a gas expands under adiabatic conditions, its temperature falls (work is done against the retreating walls of the container). The *adiabatic equation* describes the relationship between the pressure (p) of an ideal gas and its volume (V), i.e. $pV^{\gamma} = K$, where γ is the ratio of the principal specific *heat capacities of the gas and K is a constant.

adipic acid *See* hexanedioic acid.

ADP *See* ATP.

adrenaline (epinephrine) A hormone, produced by the medulla of the adrenal glands, that increases heart activity, improves the power and prolongs the action of muscles, and increases the rate and depth of breathing to prepare the body for 'fright, flight, or fight'. At the same time it inhibits digestion and excretion.

adsorbate A substance that is adsorbed on a surface.

adsorption The formation of a layer of gas on the surface of a solid or, less frequently, of a liquid. There are two types depending on the nature of the forces involved. In *chemisorption* a single layer of molecules, atoms, or ions is attached to the adsorbent surface by chemical bonds. In *physisorption* adsorbed molecules are held by the weaker *van der Waals' forces.

aerosol A colloidal dispersion of a solid or liquid in a gas. The commonly used aerosol sprays contain an inert propellant liquefied under pressure. Halogenated alkanes, such as dichlorodifluoromethane, are commonly used in aerosol cans. This use has been criticized on the grounds that these compounds persist in the atmosphere and may eventually (it is claimed) affect the *ozone layer.

agar An extract of certain species of red seaweeds that is used as a gelling agent in microbiological culture media, foodstuffs, medicines, and cosmetic creams and jellies. *Nutrient agar* consists of a broth made from beef extract or blood that is gelled with agar and used for the cultivation of bacteria, fungi, and some algae.

agate A variety of *chalcedony that forms in rock cavities and has a pattern of concentrically arranged bands or layers that lie parallel to the cavity walls. These layers are frequently alternating tones of brownish-red. *Moss agate* does not show the same banding and is a milky chalcedony containing moss-like or dendritic patterns formed by inclusions of manganese and iron oxides. Agates are used in jewellery and for ornamental purposes.

air *See* earth's atmosphere.

alabaster *See* gypsum.

alanine *See* amino acid.

albumin (albumen) One of a group of globular proteins that are soluble in water but form insoluble coagulates when heated. Albumins occur in egg white, blood, milk, and plants. Serum albumins, which constitute about 55% of blood plasma protein, help regulate the osmotic pressure and hence plasma volume. They also bind and transport fatty acids. α-lactalbumin is one of the proteins in milk.

alcohols Organic compounds that contain the –OH group. In systematic chemical nomenclature alcohol names end in the suffix *-ol*. Examples are methanol, CH_3OH, and ethanol, C_2H_5OH. *Primary alcohols* have two hydrogen atoms on the carbon joined to the –OH group (i.e. they contain the

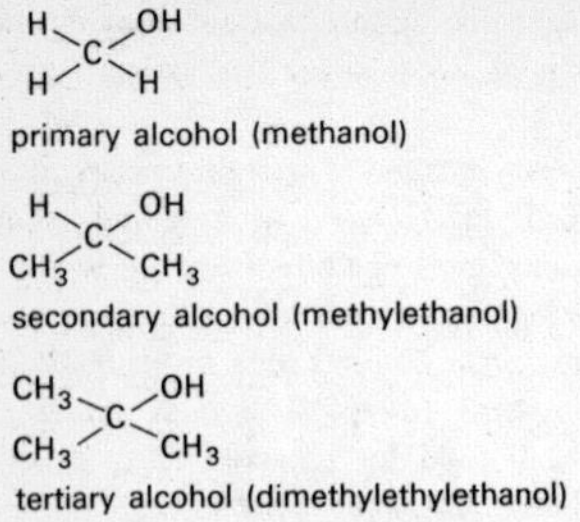

H, OH, C, H, H
primary alcohol (methanol)

H, OH, C, CH_3, CH_3
secondary alcohol (methylethanol)

CH_3, OH, C, CH_3, CH_3
tertiary alcohol (dimethylethylethanol)

Examples of alcohols

group $-CH_2-OH$); *secondary alcohols* have one hydrogen on this carbon (the other two bonds being to carbon atoms, as in $(CH_3)_2CHOH$); *tertiary alcohols* have no hydrogen on this carbon (as in $(CH_3)_3COH$). Primary and secondary alcohols differ in their reactions with potassium dichromate(VI) in sulphuric acid as the following schemes indicate:

primary alcohol → aldehyde → carboxylic acid

secondary alcohol → ketone

tertiary alcohol – no reaction

Other characteristics of alcohols are reaction with acids to give *esters and dehydration to give *alkenes or *ethers. Alcohols that have two –OH groups in their molecules are *diols* (or *dihydric alcohols*), those with three are *triols* (or *trihydric alcohols*), etc.

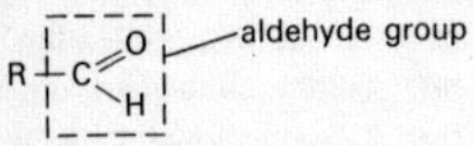

R–C(=O)–H — aldehyde group

Aldehyde structure

aldehydes Chemical compounds that contain the group –CHO (the *aldehyde group*; i.e. a carbonyl group (C=O) with a hydrogen atom bound to the carbon atom). In systematic chemical nomenclature, aldehyde names end with the suffix *-al*. Examples of aldehydes are methanal (formaldehyde), HCOH, and ethanal (acetaldehyde), CH_3CHO. Aldehydes are formed by oxidation of primary *alcohols; further oxidation yields carboxylic acids. They are reducing agents and tests for aldehydes include *Fehling's test and *Tollen's reagent. Aldehydes have certain characteristic addition and condensation reactions. With sodium hydrogensulphate(IV) they form addition compounds of the type $[RCOH(SO_3)H]^-Na^+$. Formerly these were known as *bisulphite addition compounds*. They also form addition compounds with hydrogen cyanide to give *cyanohydrins and with alcohols to give *acetals and undergo condensation reactions to yield *oximes, *hydrazones, and *semicarbazones. Aldehydes readily polymerize. *See also* ketones.

aldohexose *See* monosaccharide.

aldol *See* aldol reaction.

aldol reaction A reaction of aldehydes of the type

$$2CH_3RCHO \rightarrow CH_3RCH(OH)CH_2RCHO$$

where R represents a hydrocarbon group. The resulting compound is a hydroxyaldehyde (i.e. an aldehyde–alcohol, or *aldol*), containing an alcohol (–OH) group on one carbon atom and an aldehyde group (–CHO) on another carbon atom. The reaction occurs in the presence of hydroxide ions (i.e. it is base-catalysed) and the first step is formation of a negative (carbanion) $^-CH_2RCHO$, which reacts with a molecule of the aldehyde.

aldose *See* monosaccharide.

aldosterone A hormone produced by the adrenal glands that controls excretion of sodium by the kidneys and thereby maintains the balance of salt and water in the body fluids.

alicyclic compound A compound that contains a ring of atoms and is aliphatic. Cyclohexane, C_6H_{12}, is an example.

aliphatic compounds Organic compounds that are *alkanes, *alkenes, or *alkynes or their derivatives. The term is used to denote compounds that do not have the special stability of *aromatic compounds. All noncyclic organic compounds are aliphatic. Cyclic aliphatic compounds are said to be *alicyclic*.

alkali A *base that dissolves in water to give hydroxide ions.

alkali metals (group I elements) The elements of the first group of the *periodic table (group IA): lithium (Li), sodium (Na), potassium (K), rubidium (Rb), caesium (Cs), and francium (Fr). All have a characteristic electron configuration that is a noble gas structure with one outer *s*-electron. They are typical metals (in the chemical sense) and readily lose their outer electron to form stable M^+ ions with noble-gas configurations. All are highly reactive, with the reactivity (i.e. metallic character) increasing down the group. There is a decrease in ionization energy from lithium (520 $kJ\,mol^{-1}$) to caesium (380 $kJ\,mol^{-1}$). The second ionization energies are much higher and divalent ions are not formed. Other properties also change down the group. Thus, there is an increase in atomic and ionic radius, an increase in density, and a decrease in melting and boiling point. The standard electrode potentials are low and negative, although they do not show a regular trend because they depend both on ionization energy (which decreases down the group) and the hydration energy of the ions (which increases).

All the elements react with water (lithium slowly; the others violently) and tarnish rapidly in air. They can all be made to react with chlorine, bromine, sulphur, and hydrogen. The hydroxides of the alkali metals are strongly alkaline (hence the name) and do not decompose on heating. The salts are generally soluble. The carbonates do not decompose on heating, except at very high temperatures. The nitrates (except for lithium) decompose to give the nitrite and oxygen:

$$2MNO_3(s) \rightarrow 2MNO_2(s) + O_2(g)$$

Lithium nitrate decomposes to the oxide. In fact lithium shows a number of dissimilarities to the other members of group I and in many ways resembles magnesium (*see* diagonal relationship). In general, the stability of salts of oxo acids increases down the group (i.e. with increasing size of the M^+ ion). This trend occurs because the smaller cations (at the top of the group) tend to polarize the oxo anion more effectively than the larger cations at the bottom of the group.

alkalimetry Volumetric analysis using standard solutions of alkali to determine the amount of acid present.

alkaline **1.** Describing an alkali. **2.** Describing a solution that has an excess of hydroxide ions (i.e. a pH greater than 7).

alkaline-earth metals (group II elements) The elements of the second group of the *periodic table (group IIA): beryllium (Be), magnesium (Mg), calcium (Ca), strontium (Sr), and barium (Ba). The elements are sometimes referred to as the 'alkaline earths', although strictly the 'earths' are the oxides of the elements. All have a characteristic electron configuration that is a noble-gas structure with two outer *s*-electrons. They are typical metals (in the chemical sense) and readily lose both outer electrons to form stable M^{2+} ions; i.e. they are strong reducing agents. All are reactive, with the reactivity increasing down the group. There is a decrease in both first and second ionization energies down the group. Although there is a significant difference between the first and second ionization energies of each element, compounds containing univalent ions are not known. This is because the divalent ions have a smaller size and larger charge, leading to higher hydration energies (in solution) or lattice energies (in solids). Consequently, the overall energy charge favours the formation of divalent compounds. The third ionization energies are much higher than the second ionization energies, and trivalent compounds (containing M^{3+}) are unknown.

Beryllium, the first member of the group, has anomalous properties because of the small size of the ion; its atomic radius (0.112 nm) is much less than that of magnesium (0.16 nm). From magnesium to radium there is a fairly regular increase in atomic and ionic radius. Other regular changes take place in moving down the group from magnesium. Thus, the density and melting and boiling points all increase. Beryllium, on the other hand, has higher boiling and melting points than calcium and its density lies between those of calcium and strontium. The standard electrode potentials are negative and show a regular small decrease from magnesium to barium. In some ways beryllium resembles aluminium (*see* diagonal relationship).

All the metals are rather less reactive than the alkali metals. They react with water and oxygen (beryllium and magnesium form a protective surface film) and can be made to react with chlorine, bromine, sulphur, and hydrogen. The oxides and hydroxides of the metals show the increasing ionic character in moving down the group: beryllium hydroxide is amphoteric, magnesium hydroxide is only very slightly soluble in water and is weakly basic, calcium hydroxide is sparingly soluble and distinctly basic, strontium and barium hydroxides are quite soluble and basic. The hydroxides decompose on heating to give the oxide and water:

$$M(OH)_2(s) \rightarrow MO(s) + H_2O(g)$$

The carbonates also decompose on heating to the oxide and carbon dioxide:

$$MCO_3(s) \rightarrow MO(s) + CO_2(g)$$

The nitrates decompose to give the oxide:

$$2M(NO_3)_2(s) \rightarrow 2MO(s) + 4NO_2(g) + O_2(g)$$

As with the *alkali metals, the stability of salts of oxo acids increases down the group. In general, salts of the alkaline-earth elements are soluble if the anion has a single charge (e.g. nitrates, chlorides). Most salts with a doubly charged anion (e.g. carbonates, sulphates) are insoluble. The solubilities of salts of a particular acid tend to decrease down the group. (Solubilities of hydroxides increase for larger cations.)

alkaloid One of a group of nitrogenous organic compounds derived from plants and having diverse pharmacological properties. Alkaloids include morphine, cocaine, atropine, quinine, and caffeine, most of which are used in medicine as analgesics (pain relievers) or anaesthetics. Some alkaloids are poisonous, e.g. strychnine and coniine, and colchicine inhibits cell division.

alkanes (paraffins) Saturated hydrocarbons with the general formula C_nH_{2n+2}. In systematic chemical nomenclature alkane names end in the suffix *-ane*. They form a *homologous series (the *alkane series*) methane (CH_4), ethane (C_2H_6), propane (C_3H_8), butane (C_4H_{10}), pentane (C_5H_{12}), etc. The lower members of the series are gases; the high-molecular weight alkanes are waxy solids. Alkanes are present in natural gas and petroleum. They can be made by heating the sodium salt of a carboxylic acid with soda lime:

$$RCOO^-Na^+ + Na^+OH^- \rightarrow Na_2CO_3 + RH$$

Other methods include the *Wurtz reaction and *Kolbé's method. Generally the alkanes are fairly unreactive. They form haloalkanes with halogens when irradiated with ultraviolet radiation.

alkenes (olefines; olefins) Unsaturated hydrocarbons that contain one or more double carbon–carbon bonds in their molecules. In systematic chemical nomenclature alkene names end in the suffix *-ene*. Alkenes that have only one double bond form a ho-

but-1-ene

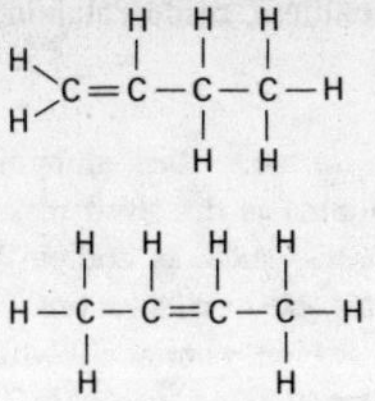

but-2-ene

Butene isomers

mologous series (the *alkene series*) starting ethene (ethylene), $CH_2{:}CH_2$, propene, $CH_3CH{:}CH_2$, etc. The general formula is C_nH_{2n}. Higher members of the series show isomerism depending on position of the double bond; for example, butene (C_4H_8) has two isomers, which are (1) but-1-ene ($C_2H_5CH{:}CH_2$) and (2) but-2-ene ($CH_3CH{:}CHCH_3$). Alkenes can be made by dehydration of alcohols (passing the vapour over hot pumice):

$$RCH_2CH_2OH - H_2O \rightarrow RCH{:}CH_2$$

An alternative method is the removal of a hydrogen atom and halogen atom from a haloalkane by potassium hydroxide in hot alcoholic solution:

$$RCH_2CH_2Cl + KOH \rightarrow KCl + H_2O + RCH{:}CH_2$$

Alkenes typically undergo *addition reactions to the double bond. *See also* hydrogenation; oxo reaction; ozonolysis; Ziegler process.

alkoxides Compounds formed by reaction of alcohols with sodium or potassium metal. Alkoxides are saltlike compounds containing the ion $R{-}O^-$.

alkylbenzenes Organic compounds that have an alkyl group bound to a benzene ring. The simplest example is methylbenzene (toluene), $CH_3C_6H_5$. Alkyl benzenes can be made by the *Friedel–Crafts reaction.

alkyl group A group obtained by removing a hydrogen atom from an alkane, e.g. methyl group, CH_3-, derived from methane.

alkyl halides *See* haloalkanes.

alkynes (acetylenes) Unsaturated hydrocarbons that contain one or more triple carbon–carbon bonds in their molecules. In systematic chemical nomenclature alkyne names end in the suffix *-yne*. Alkynes that have only one triple bond form a *homologous series: ethyne (acetylene), $CH{\equiv}CH$, propyne, $CH_3CH{\equiv}CH$, etc. They are made by the action of potassium hydroxide in alcohol solution on haloalkanes containing halogen atoms on adjacent carbon atoms; for example:

$$RCHClCH_2Cl + 2KOH \rightarrow 2KCl + 2H_2O + RCH{\equiv}CH$$

Like *alkenes, alkynes undergo addition reactions.

allotropy The existence of elements in two or more different forms (*allotropes*). In the case of oxygen, there are two forms: 'normal' dioxygen (O_2) and ozone, or trioxygen (O_3). These two allotropes have different molecular configurations. More commonly, allotropy occurs because of different crystal structures in the solid, and is particularly prevalent in groups IV, V, and VI of the periodic table. In some cases, the allotropes are stable over a temperature range, with a definite transition point at which one changes into the other. For instance, tin has two allotropes: white (metallic) tin stable above 13.2°C and grey (nonmetallic) tin stable below 13.2°C. This form of allotropy is called *enantiotropy*. Carbon also has two allotropes – diamond and graphite – although graphite is the stable form at all temperatures. This form of allotropy, in which there is no transition temperature at which the two are in equilibrium, is called *monotropy*. *See also* polymorphism.

allowed bands *See* energy bands.

alloy A material consisting of two or more metals (e.g. brass is an alloy of copper and zinc) or a metal and a nonmetal (e.g. steel is an alloy of iron and carbon, sometimes with other metals included). Alloys may be compounds, *solid solutions, or mixtures of the components.

alloy steels *See* steel.

allyl group Formerly, the organic group $CH_2{:}CHCH_2-$. It was used in names such as *allyl alcohol* ($CH_2{:}CHCH_2OH$, prop-2-en-1-ol).

Alnico A tradename for a series of alloys, containing iron, aluminium, nickel, cobalt, and copper, used to make permanent magnets.

alpha-iron *See* iron.

alpha-naphthol test A biochemical test to detect the presence of carbohydrates in solution, also known as *Molisch's test* (after the Austrian chemist H. Molisch (1856–1937), who devised it). A small amount of alcoholic alpha-naphthol is mixed with the test solution and concentrated sulphuric acid is poured slowly down the side of the test tube. A positive reaction is indicated by the formation of a violet ring at the junction of the two liquids.

alpha particle A helium nucleus emitted by a larger nucleus during the course of the type of radioactive decay known as *alpha decay*. As a helium nucleus consists of two protons and two neutrons bound together as a stable entity the loss of an alpha particle involves a decrease in *nucleon number of 4 and decrease of 2 in the *atomic number, e.g. the decay of a uranium–238 nucleus into a thorium–234 nucleus. A stream of alpha particles is known as an *alpha-ray* or *alpha-radiation*.

alum *See* aluminium potassium sulphate; alums.

alumina *See* aluminium oxide; aluminium hydroxide.

aluminate A salt formed when aluminium hydroxide or γ-alumina is dissolved in solutions of strong bases, such as sodium hydroxide. Aluminates exist in solutions containing the aluminate ion, commonly written $[Al(OH)_4]^-$. In fact the ion probably is a complex hydrated ion and can be regarded as formed from a hydrated Al^{3+} ion by removal of four hydrogen ions:

$$[Al(H_2O)_6]^{3+} + 4OH^- \rightarrow 4H_2O + [Al(OH)_4(H_2O)_2]^-$$

Other aluminates and polyaluminates, such as $[Al(OH)_6]^{3-}$ and $[(HO)_3AlOAl(OH)_3]^{2-}$, are also present. *See also* aluminium hydroxide.

aluminium Symbol Al. A silvery-white lustrous metallic element belonging to *group III of the periodic table; a.n. 13; r.a.m. 26.98; r.d. 2.7; m.p. 660°C; b.p. 2467°C. The metal itself is highly reactive but is protected by a thin transparent layer of the oxide, which forms quickly in air. Aluminium and its oxide are amphoteric. The metal is extracted from purified bauxite (Al_2O_3) by electrolysis; the main process uses a *Hall–Heroult cell but other electrolytic methods are under development, including conversion of bauxite with chlorine and electrolysis of the molten chloride. Pure aluminium is soft and ductile but its strength can be increased by work-hardening. A large number of alloys are manufactured; alloying elements include copper, manganese, silicon, zinc, and magnesium. Its lightness, strength (when alloyed), corrosion resistance, and electrical conductivity (62% of that of copper) make it suitable for a variety of uses, including vehicle and aircraft construction, building (window and door frames), and overhead power cables. Although it is the third most abundant element in the earth's crust (8.1% by weight) it was not isolated until 1825 by H. C. Oersted (1777–1851).

aluminium acetate *See* aluminium ethanoate.

aluminium chloride A whitish solid, $AlCl_3$, which fumes in moist air and reacts violently with water (to give hydrogen chloride). It is known as the anhydrous salt (hexagonal; r.d. 2.44 (fused solid); m.p. 190°C (2.5 atm.); sublimes at 178°C) or the hexahydrate $AlCl_3.6H_2O$ (rhombic; r.d. 2.398; loses water at 100°C), both of which are deliquescent. Aluminium chloride may be prepared by passing hydrogen chloride or chlorine over hot aluminium or (industrially) by passing chlorine over heated aluminium oxide and carbon. The chloride ion is polarized by the small positive aluminium ion and the bonding in the solid is intermediate between covalent and ionic. In the liquid and vapour phases dimer molecules exist, Al_2Cl_6, in which there are chlorine bridges making coordinate bonds to aluminium atoms. The $AlCl_3$ molecule can also form compounds with other molecules that donate pairs of electrons (e.g. amines or hydrogen sulphide); i.e. it acts as a Lewis *acid. At high temperatures the Al_2Cl_6 molecules in the vapour dissociate to (planar) $AlCl_3$ molecules. Aluminium chloride is used commercially as a catalyst in the cracking of oils. It is also a catalyst in certain other organic reactions, especially the Friedel–Crafts reaction.

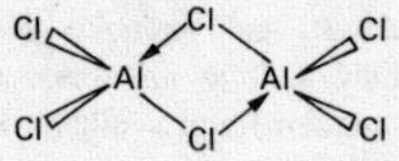

Structure of aluminium trichloride dimer

aluminium ethanoate (aluminium acetate) A white solid, $Al(OOCCH_3)_3$, which decomposes on heating, is very slightly soluble in cold water, and decomposes in warm water. The normal salt, $Al(OOCCH_3)_3$, can only be made in the absence of water (e.g. ethanoic anhydride and aluminium chloride at 180°C); in water it forms the basic salts $Al(OH)(OOCCH_3)_2$ and $Al_2(OH)_2(OOCCH_3)_4$. The reaction of aluminium hydroxide with ethanoic acid gives these basic salts directly. The compound is used extensively in dyeing as a mordant, particularly in combination with aluminium sulphate (known as *red liquor*); in the paper and board industry for sizing and hardening; and in tanning. It was previously used as an antiseptic and astringent.

aluminium hydroxide A white crystalline compound, $Al(OH)_3$; r.d. 2.42–2.52. The compound occurs naturally as the mineral *gibbsite* (monoclinic). In the laboratory it can be prepared by precipitation from solutions of aluminium salts. Such solutions contain the hexaquoaluminium(III) ion with six water molecules coordinated, $[Al(H_2O)_6]^{3+}$. In neutral solution this ionizes:

$$[Al(H_2O)_6]^{3+} \rightleftharpoons H^+ + [Al(H_2O)_5OH]^{2+}$$

The presence of a weak base such as S^{2-} or $CO_3^{\ 2-}$ (by bubbling hydrogen sulphide or carbon dioxide through the solution) causes further ionization with precipitation of aluminium hydroxide

$$[Al(H_2O)_6]^{3+}(aq) \rightarrow Al(H_2O)_3(OH)_3(s) + 3H^+(aq)$$

The substance contains coordinated water molecules and is more correctly termed *hydrated aluminium hydroxide*. In addition, the precipitate has water molecules trapped in it and has a characteristic gelatinous form. The substance is amphoteric. In strong bases the *aluminate ion is produced by loss of a further proton:

$$Al(H_2O)_3(OH)_3(s) + OH^-(aq) \rightleftharpoons [Al(H_2O)_2(OH)_4]^-(aq) + H_2O(l)$$

On heating, the hydroxide transforms to a mixed oxide hydroxide, AlO.OH (rhombic; r.d. 3.01). This substance occurs naturally as *diaspore* and *boehmite*. Above 450°C it transforms to γ-alumina.

In practice various substances can be produced that are mixed crystalline forms of $Al(OH)_3$, AlO.OH, and aluminium oxide (Al_2O_3) with water molecules. These are known as *hydrated alumina*. Heating the hydrated hydroxide causes loss of water, and produces various *activated aluminas*, which differ in porosity, number of remaining –OH groups, and particle size. These are used as catalysts (particularly for organic dehydration reactions), as catalyst supports, and in chromatography. Gelatinous freshly precipitated aluminium hydroxide was formerly widely used as a mordant for dyeing and calico printing because of its ability to form insoluble coloured *lakes with vegetable dyes. *See also* aluminium oxide.

aluminium oxide (alumina) A white or colourless oxide of aluminium occurring in two main forms. The stable form α-alumina (r.d. 3.97; m.p. 2020°C; b.p. 2980 $\pm$ 60°C) has colourless hexagonal or rhombic crystals; γ-alumina (r.d. 3.5–3.9) transforms to the α-form on heating and is a white microcrystalline solid. The compound occurs naturally as *corundum* or *emery* in the α-form with a hexagonal-close-packed structure of oxide ions with aluminium ions in the octahedral interstices. The gemstones ruby and sapphire are aluminium oxide coloured by minute traces of chromium and cobalt respectively. A number of other forms of aluminium oxide have been described (β-, δ-, and ζ-alumina) but these contain alkali-metal ions. There is also a short-lived spectroscopic suboxide AlO. The highly protective film of oxide formed on the surface of aluminium metal is yet another structural variation, being a defective rock-salt form (every third Al missing).

Pure aluminium oxide is obtained by dissolving the ore bauxite in sodium hydroxide solution; impurities such as iron oxides remain insoluble because they are not amphoteric. The hydrated oxide is precipitated by seeding with material from a previous batch and this is then roasted at 1150–1200°C to give pure α-alumina, or at 500–800°C to give γ-alumina. The bonding in aluminium hydroxide is not purely ionic due to polarization of the oxide ion. Although the compound might be expected to be amphoteric, α-alumina is weakly acidic, dissolving in alkalis to give solutions containing aluminate ions; it is resistant to acid attack. In contrast γ-alumina is typically amphoteric dissolving both in acids to give aluminium salts and in bases to give aluminates. α-alumina is one of the hardest materials known (silicon carbide and diamond are harder) and is widely used as an abrasive in both natural (corundum) and synthetic forms. Its refractory nature makes alumina brick an ideal material for furnace linings and alumina is also used in cements for high-temperature conditions. *See also* aluminium hydroxide.

aluminium potassium sulphate (potash alum; alum) A white or colourless crystalline compound, $Al_2(SO_4)_3.K_2SO_4.24H_2O$; r.d. 1.757; loses $18H_2O$ at 92.5°C; becomes anhydrous at 200°C. It forms cubic or octahedral crystals that are soluble in cold water, very soluble in hot water, and insoluble in ethanol and acetone. The compound occurs naturally as the mineral *kalinite*. It is a double salt and can be prepared by recrystallization from a solution containing equimolar quantities of potassium sulphate and aluminium sulphate. It is used as a mordant for dyeing and in the tanning and finishing of leather goods (for white leather). *See also* alums.

aluminium sulphate A white or colourless crystalline compound, $Al_2(SO_4)_3$, known as the anhydrous compound (r.d. 2.71; decomposes at 770°C) or as the hydrate $Al_2(SO)_3.18H_2O$ (monoclinic; r.d. 1.69; loses water at 86.5°C). The anhydrous salt is soluble in water and slightly soluble in ethanol; the hydrate is very soluble in water and insoluble in ethanol. The compound occurs naturally in the rare mineral *alunogenite* ($Al_2(SO)_3.18H_2O$). It may be prepared by dissolving aluminium hydroxide or china clays (aluminosilicates) in sulphuric acid. It decomposes on heating to sulphur dioxide, sulphur trioxide, and aluminium oxide. Its solutions are acidic because of hydrolysis.

Aluminium sulphate is commercially one of the most important aluminium compounds; it is used in sewage treatment (as a flocculating agent) and in the purification of drinking water, the paper industry, and in the preparation of mordants. It is also a fire-proofing agent. Aluminium sulphate is often wrongly called *alum* in these industries.

aluminium trimethyl *See* trimethylaluminium.

alums A group of double salts with the formula $A_2SO_4.B_2(SO_4)_3.24H_2O$, where A is a monovalent metal and B a trivalent metal. The original example contains potassium and aluminium (called *potash alum* or simply *alum*); its formula is often written $AlK(SO_4)_2.12H_2O$ (aluminium potassium sulphate-12-water). *Ammonium alum* is $AlNH_4(SO_4)_2.12H_2O$, *chrome alum* is $KCr(SO_4)_2.12H_2O$ (*see* potassium chromium sulphate), etc. The alums are isomorphous and can be made by dissolving equivalent amounts of the two salts in water and recrystallizing. *See also* aluminium sulphate.

alunogenite A mineral form of hydrated *aluminium sulphate, $Al_2(SO_4)_3.18H_2O$.

amalgam An alloy of mercury with one or more other metals. Most metals form amalgams (iron and platinum are exceptions), which may be liquid or solid. Some contain definite intermetallic compounds, such as $NaHg_2$.

americium Symbol Am. A radioactive metallic transuranic element belonging to the *actinoids; a.n. 95; mass number of most stable isotope 243 (half-life 7.95×10^3 years); r.d. 13.67 (20°C); m.p. 994 ± 4°C; b.p. 2607°C. Ten isotopes are known. The element was discovered by G. T. Seaborg and associates in 1945, who obtained it by bombarding uranium-238 with alpha particles.

amethyst The purple variety of the mineral *quartz. It is found chiefly in Brazil, the Urals (Soviet Union), Arizona (USA), and Uruguay. The colour is due to impurities, especially iron oxide. It is used as a gemstone.

amides Organic compounds containing the group $-CO.NH_2$ (the *amide group*). Amides are volatile solids; examples are ethanamide, CH_3CONH_2, and propanamide, $C_2H_5CONH_2$. They are made by heating the ammonium salt of the corresponding carboxylic acid.

amination A chemical reaction in which an amino group ($-NH_2$) is introduced into a molecule. Examples of amination reaction include the reaction of halogenated hydrocarbons with ammonia (high pressure and temperature) and the reduction of nitro compounds and nitriles.

amines Organic compounds derived by replacing one or more of the hydrogen atoms in ammonia by organic groups. *Primary amines* have one hydrogen replaced, e.g. methylamine, CH_3NH_2. They contain the functional group $-NH_2$ (the *amino group*). *Secondary amines* have two hydrogens replaced, e.g. methylethylamine, $CH_3(C_2H_5)NH$. The group $=NH$ is the *imino group*. *Tertiary amines* have all three replaced, e.g. trimethylamine, $(CH_3)_3N$. Amines are produced by the decomposition of organic matter. They can be made by reducing nitro compounds or amides.

amine salts Salts similar to ammonium salts in which the hydrogen atoms attached to the nitrogen are replaced by one or more organic groups. Amines readily form salts by reaction with acids, gaining a proton to form a positive ammonium ion, They are named as if they were substituted derivatives of ammonium compounds; for example, dimethylamine ($(CH_3)_2NH$) will react with hydrogen chloride to give dimethylammonium chloride, which is an ionic compound $[(CH_3)_2NH_2]^+Cl^-$. When the amine has a common nonsystematic name the suffix *-ium* can be used; for example, phenylamine (aniline) would give $[C_6H_5NH_3]^+Cl^-$, known as anilinium chloride. Formerly, such compounds were sometimes called *hydrochlorides*, e.g. aniline hydrochloride with the formula $C_6H_5NH_2.HCl$.

Salts formed by amines are crystalline substances that are readily soluble in water. Many insoluble *alkaloids (e.g. quinine and atropine) are used medicinally in the form of soluble salts ('hydrochlorides'). If alkali (sodium hydroxide) is added to solutions of such salts the free amine is liberated.

If all four hydrogen atoms of an ammonium salt are replaced by organic groups a *quaternary ammonium compound* is formed. Such compounds are made by reacting tertiary amines with halogen compounds; for example, trimethylamine ($(CH_3)_3N$) with chloromethane (CH_3Cl) gives tetramethylammonium chloride, $(CH_3)_4N^+Cl^-$. Salts of this type do not liberate the free amine when alkali is added, and quaternary hydroxides (such as $(CH_3)_4N^+OH^-$) can be isolated. Such compounds are strong alkalis, comparable to sodium hydroxide.

amino acid Any of a group of water-soluble organic compounds that possess both a carboxyl (–COOH) and an amino (–NH_2) group attached to the α-carbon atom. Amino acids can be represented by the general formula R–CH(NH_2)COOH. R may be hydrogen or an organic group and determines the properties of any particular amino acid. Through the formation of peptide bonds, amino acids join together to form short chains (*peptides) or much longer chains (*polypeptides). Proteins are composed of various proportions of about 20 commonly occurring amino acids (see table). The sequence of these amino acids in the protein polypeptides determines the shape, properties, and hence biological role of the protein. Some amino acids that never occur in proteins are nevertheless important, e.g. ornithine and citrulline, which are intermediates in the urea cycle.

Plants and many microorganisms can synthesize amino acids from simple inorganic compounds, but animals rely on adequate supplies in their diet. The *essential amino acids must be present in the diet whereas others can be manufactured from them.

aminobenzene *See* phenylamine.

amino group *See* amines.

ammine A coordination *complex in which the ligands are ammonia molecules. An example of an ammine is the tetraamminecopper(II) ion $[Cu(NH_3)_4]^{2+}$.

ammonia A colourless gas, NH_3, with a strong pungent odour; r.d. 0.59 (relative to air); m.p. –74°C; b.p. –30.9°C. It is very soluble in water and soluble in alcohol. The compound may be prepared in the laboratory by the reaction of ammonium salts with bases such as calcium hydroxide, or by the hydrolysis of a nitride. Industrially it is made by the *Haber process and over 80 million tonnes per year are used either directly or in combination. Major uses are the manufacture of nitric acid, ammonium nitrate, ammonium phosphate, and urea (the last three as fertilizers), explosives, dyestuffs and resins.

Liquid ammonia has some similarity to water as it is hydrogen bonded and has a moderate dielectric constant, which permits it to act as an ionizing solvent. It is weakly self-ionized to give ammonium ions, NH_4^+ and amide ions, NH_2^-. It also dissolves electropositive metals to give blue solutions, which are believed to contain solvated electrons. Ammonia is extremely soluble in water giving basic solutions that contain solvated NH_3 molecules and small amounts of the ions NH_4^+ and OH^-. The combustion of ammonia in air yields nitrogen and

amino acid	abbreviation	formula
alanine	ala	$CH_3-CH(NH_2)-COOH$
*arginine	arg	$H_2N-C(=NH)-NH-CH_2-CH_2-CH_2-CH(NH_2)-COOH$
asparagine	asn	$H_2N-C(=O)-CH_2-CH(NH_2)-COOH$
aspartic acid	asp	$HOOC-CH_2-CH(NH_2)-COOH$
cysteine	cys	$HS-CH_2-CH(NH_2)-COOH$
glutamic acid	glu	$HOOC-CH_2-CH_2-CH(NH_2)-COOH$
glutamine	gln	$H_2N(O=)C-CH_2-CH_2-CH(NH_2)-COOH$
glycine	gly	$H-CH(NH_2)-COOH$
*histidine	his	$HC=C(-NH-CH=N-)-CH_2-CH(NH_2)-COOH$ (imidazole ring: HC=C, N, NH, CH)
*isoleucine	ile	$CH_3-CH_2-CH(CH_3)-CH(NH_2)-COOH$
*leucine	leu	$(H_3C)_2CH-CH_2-CH(NH_2)-COOH$

*lysine	lys	$H_3N-CH_2-CH_2-CH_2-CH_2-C(H)(NH_2)-COOH$
*methionine	met	$CH_3-S-CH_2-CH_2-C(H)(NH_2)-COOH$
*phenylalanine	phe	$C_6H_5-CH_2-C(H)(NH_2)-COOH$
proline	pro	H_2C-CH_2, H_2C, $CH-COOH$, N, H (ring) → H, OH, $C-CH_2$, H_2C, $CH-COOH$, N, H (ring) 4-hydroxyproline
serine	ser	$HO-CH_2-C(H)(NH_2)-COOH$
*threonine	thr	$CH_3-CH(OH)-C(H)(NH_2)-COOH$
*tryptophan	trp	C_6H_4(ring)$-C(=CH-N)-CH_2-C(H)(NH_2)-COOH$
tyrosine	tyr	$HO-C_6H_4-CH_2-C(H)(NH_2)-COOH$
*valine	val	$(H_3C)(H_3C)CH-C(H)(NH_2)-COOH$

*an essential amino acid

The amino acids occurring in proteins

water. In the presence of catalysts NO, NO_2, and water are formed; this last reaction is the basis for the industrial production of nitric acid. Ammonia is a good proton acceptor (i.e. it is a base) and gives rise to a series of ammonium salts, e.g. NH_3 + HCl → $NH_4{}^+$ + Cl^-. It is also a reducing agent. The participation of ammonia in the *nitrogen cycle is a most important natural process. Nitrogen-fixing bacteria are able to achieve similar reactions to those of the Haber process, but under normal conditions of temperature and pressure. These release ammonium ions, which are converted by nitrifying bacteria into nitrite and nitrate ions.

ammoniacal Describing a solution in which the solvent is aqueous ammonia.

ammonia–soda process *See* Solvay process.

ammonium alum *See* alums.

ammonium carbonate A colourless or white crystalline solid, $(NH_4)_2CO_3$, usually encountered as the monohydrate. It is very soluble in cold water. The compound decomposes slowly to give ammonia, water, and carbon dioxide. Commercial 'ammonium carbonate' is a double salt of ammonium hydrogencarbonate and ammonium aminomethanoate (ammonium carbamate), $NH_4HCO_3.NH_2COONH_4$. This material is manufactured by heating a mixture of ammonium chloride and calcium carbonate and recovering the product as a sublimed solid. It readily releases ammonia and is the basis of sal volatile. It is also used in dyeing and wool preparation and in baking powders.

ammonium chloride (sal ammoniac) A white or colourless cubic solid, NH_4Cl; r.d. 1.53; sublimes at 340°C. It is very soluble in water and slightly soluble in ethanol but insoluble in ether. It may be prepared by fractional crystallization from a solution containing ammonium sulphate and sodium chloride or ammonium carbonate and calcium chloride. Pure samples may be made directly by the gas-phase reaction of ammonia and hydrogen chloride. Because of its ease of preparation it can be manufactured industrially alongside any plant that uses or produces ammonia. The compound is used in dry cells, metal finishing, and in the preparation of cotton for dyeing and printing.

ammonium ion The monovalent cation NH_4^+. It may be regarded as the product of the reaction of ammonia (a Lewis base) with a hydrogen ion. The ion has tetrahedral symmetry. The chemical properties of ammonium salts are frequently very similar to those of equivalent alkali-metal salts.

ammonium nitrate A colourless crystalline solid, NH_4NO_3; r.d. 1.72; m.p. 169.6°C; b.p. 210°C. It is very soluble in water and soluble in ethanol. The crystals are rhombic when obtained below 32°C and monoclinic above 32°C. It may be readily prepared in the laboratory by the reaction of nitric acid with aqueous ammonia. Industrially, it is manufactured by the same reaction using ammonia gas. Vast quantities of ammonium nitrate are used as fertilizers (over 20 million tonnes per year) and it is also a component of some explosives.

ammonium sulphate A white rhombic solid, $(NH_4)_2SO_4$; r.d. 1.67; decomposes at 235°C. It is very soluble in water and insoluble in ethanol. It occurs naturally as the mineral *mascagnite*. Ammonium sulphate was formerly manufactured from the 'ammoniacal liquors' produced during coal-gas manufacture but is now produced by the direct reaction between ammonia gas and sulphuric acid. It is decomposed by heating to release ammonia (and ammonium hydrogensulphate) and eventually water, sulphur dioxide, and ammonia. Vast quantities of ammonium sulphate are used as fertilizers.

amorphous Describing a solid that is not crystalline; i.e. one that has no long-range order in its lattice. Many powders that are described as 'amorphous' in fact are composed of microscopic crystals, as can be demonstrated by X-ray diffraction. *Glasses are examples of true amorphous solids.

amount of substance Symbol *n*. A measure of the number of entities present in a substance. The specified entity may be an atom, molecule, ion, electron, photon, etc., or any specified group of such entities. The amount of substance of an element, for example, is proportional to the number of atoms present. For all entities, the constant of proportionality is the *Avogadro constant. The SI unit of amount of substance is the *mole.

AMP *See* ATP; cyclic AMP.

ampere Symbol A. The SI unit of electric current. The constant current that, maintained in two straight parallel infinite conductors of negligible cross section placed one metre apart in a vacuum, would produce a force between the conductors of 2×10^{-7} N m^{-1}. This definition replaced the earlier international ampere defined as the current required to deposit 0.001 118 00 gram of silver from a solution of silver nitrate in one second. The unit is named after A. M. Ampère (1775–1836).

ampere-hour A practical unit of electric charge equal to the charge flowing in one hour through a conductor passing one ampere. It is equal to 3600 coulombs.

ampere-turn The SI unit of magnetomotive force equal to the magnetomotive force produced when a current of one ampere flows through one turn of a magnetizing coil.

amphiboles A large group of rock-forming metasilicate minerals. They have a structure of silicate tetrahedra linked to form double endless chains, in contrast to the single chains of the *pyroxenes, to which they are closely related. They are present in many igneous and metamorphic rocks. The amphiboles show a wide range of compositional variation but conform to the general formula: $X_{2-3}Y_5Z_8O_{22}(OH)_2$, where X = Ca, Na, K, Mg, or Fe^{2+}; Y = Mg, Fe^{2+}, Fe^{3+}, Al, Ti, or Mn; and Z = Si or Al. The hydroxyl ions may be replaced by F, Cl, or O. Most amphiboles are monoclinic, including cummingtonite, $(Mg,Fe^{2+})_7(Si_8O_{22})(OH)_2$; tremolite, $Ca_2Mg_5(Si_8O_{22})(OH,F)_2$; actinolite, $Ca_2(Mg,Fe^{2+})_5(Si_8O_{22})(OH,F)_2$; *hornblende, $NaCa_2(Mg,Fe^{2+},Fe^{3+},Al)_5((Si,Al)_8O_{22})(OH,F)_2$; edenite, $NaCa_2(Mg,Fe^{2+})_5(Si_7AlO_{22})(OH,F)_2$; and riebeckite, $Na_2,Fe_3^{2+}(Si_8O_{22})(OH,F)_2$. Anthophyllite, $(Mg,Fe^{2+})_7(Si_8O_{22})(OH,F)_2$, and gedrite, $(Mg,Fe^{2+})_6Al(Si,Al)_8O_{22})(OH,F)_2$, are orthorhombic amphiboles.

ampholyte A substance that can act as either an acid, in the presence of a strong base, or a base, when in the presence of a strong acid.

ampholyte ion *See* zwitterion.

amphoteric Describing a compound that can act as both an acid and a base (in the traditional sense of the term). For instance, aluminium hydroxide is amphoteric: as a base $Al(OH)_3$ it reacts with acids to form aluminium salts; as an acid H_3AlO_3 it reacts with alkalis to give *aluminates. Oxides of metals are typically basic and oxides of nonmetals tend to be acidic. The existence of amphoteric oxides is sometimes regarded as evidence that an element is a *metalloid.

a.m.u. *See* atomic mass unit.

amylase (diastase) Any of a group of closely related enzymes that degrade starch, glycogen, and other polysaccharides. Plants contain both α- and β-amylases; animals possess only α-amylases, found in pancreatic juice and also (in humans and some other species) in saliva. Amylases cleave the long polysaccharide chains, producing a mixture of glucose and maltose.

amyl group Formerly, any of several isomeric groups with the formula $C_5H_{11}-$.

amylopectin A *polysaccharide comprising highly branched chains of glucose molecules.

It is one of the constituents (the other being amylose) of *starch.

amylose A *polysaccharide consisting of linear chains of between 100 and 1000 linked glucose molecules. Amylose is a constituent of *starch. In water, amylose reacts with iodine to give a characteristic blue colour.

anabolism The metabolic synthesis of proteins, fats, and other constituents of living organisms from molecules or simple precursors. This process requires energy in the form of ATP. *See* metabolism. *Compare* catabolism.

analysis The determination of the components in a chemical sample. *Qualitative analysis* involves determining the nature of a pure unknown compound or the compounds present in a mixture. Various chemical tests exist for different elements or types of compound, and systematic analytical procedures can be used for mixtures. *Quantitative analysis* involves measuring the proportions of known components in a mixture. Chemical techniques for this fall into two main classes: *volumetric analysis and *gravimetric analysis. In addition, there are numerous physical methods of qualitative and quantitative analysis, including spectroscopic techniques, mass spectrometry, polarography, chromatography, activation analysis, etc.

anglesite A mineral form of *lead(II) sulphate, $PbSO_4$.

angstrom Symbol Å. A unit of length equal to 10^{-10} metre. It was formerly used to measure wavelengths and intermolecular distances but has now been replaced by the nanometre. 1 Å = 0.1 nanometre. The unit is named after the Swedish pioneer of spectroscopy A. J. Ångstrom (1814–74).

anhydride A compound that produces a given compound on reaction with water. For instance, sulphur trioxide is the (acid) anhydride of sulphuric acid

$$SO_3 + H_2O \rightarrow H_2SO_4$$

See also acid anhydrides.

anhydrite An important rock-forming anhydrous mineral form of calcium sulphate, $CaSO_4$. It is chemically similar to *gypsum but is harder and heavier and crystallizes in the rhombic form (gypsum is monoclinic). Under natural conditions anhydrite slowly hydrates to form gypsum. It occurs chiefly in white and greyish granular masses and is often found in the caprock of certain salt domes. It is used as a raw material in the chemical industry and in the manufacture of cement and fertilizers.

anhydrous Denoting a chemical compound lacking water: applied particularly to salts lacking their water of crystallization.

aniline *See* phenylamine.

anilinium ion The ion $C_6H_5NH_3^+$, derived from *phenylamine.

animal charcoal *See* charcoal.

animal starch *See* glycogen.

anion A negatively charged *ion, i.e. an ion that is attracted to the *anode in *electrolysis. *Compare* cation.

anionic detergent *See* detergent.

anionic resin *See* ion exchange.

anisotropic Denoting a medium in which certain physical properties are different in different directions. Wood, for instance, is an anisotropic material: its strength along the grain differs from that perpendicular to the grain. Single crystals that are not cubic are anisotropic with respect to some physical properties, such as the transmission of electromagnetic radiation. *Compare* isotropic.

annealing A form of heat treatment applied to a metal to soften it, relieve internal stresses and instabilities, and make it easier to work or machine. It consists of heating the metal to a specified temperature for a specified time, both of which depend on the metal involved, and then allowing it to cool slowly. It is applied to both ferrous and nonferrous metals and a similar process can be applied to other materials, such as glass.

anode A positive electrode. In *electrolysis anions are attracted to the anode. In an electronic vacuum tube it attracts electrons from the *cathode and it is therefore from the anode that electrons flow out of the device. In these instances the anode is made positive by external means; however in a *voltaic cell the anode is the electrode that spontaneously becomes positive and therefore attracts electrons to it from the external circuit.

anode sludge *See* electrolytic refining.

anodizing A method of coating objects made of aluminium with a protective oxide film, by making them the anode in an electrolytic bath containing an oxidizing electrolyte. Anodizing can also be used to produce a decorative finish by formation of an oxide layer that can absorb a coloured dye.

anthocyanin One of a group of *flavonoid pigments. Anthocyanins occur in various plant organs and are responsible for many of the blue, red, and purple colours in plants (particularly in flowers).

anthracene A white crystalline solid, $C_{14}H_{10}$; r.d. 1.25; m.p. 286°C; b.p. 379.8°C. It is an aromatic hydrocarbon with three fused rings, and is obtained by the distillation of crude oils. The main use is in the manufacture of dyes.

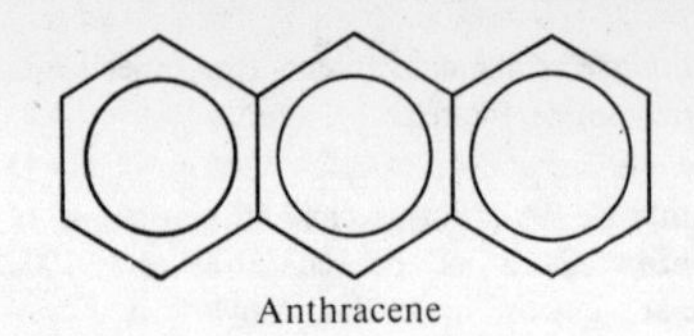

Anthracene

anthracite *See* coal.

antibiotics Substances obtained from microorganisms, especially moulds, that destroy or inhibit the growth of other microorganisms, particularly disease-producing bacteria and fungi. Common antibiotics include penicillin, streptomycin, and tetracyclines. They are used to treat various infections but tend to weaken the body's natural defence mechanisms and can cause allergies. Overuse of antibiotics can lead to the development of resistant strains of microorganisms.

antigorite *See* serpentine.

antimony Symbol Sb. An element belonging to *group VA of the periodic table; a.n. 51; r.a.m. 121.75; r.d. 6.73; m.p. 630.5°C; b.p. 1380°C. Antimony has several allotropes. The stable form is a bluish-white metal. Yellow antimony and black antimony are unstable nonmetallic allotropes made at low temperatures. The main source is stibnite (Sb_2S_3), from which antimony is extracted by reduction with iron metal or by roasting (to give the oxide) followed by reduction with carbon and sodium carbonate. The main use of the metal is as an alloying agent in lead-accumulator plates, type metals, bearing alloys, solders, Britannia metal, and pewter. It is also an agent for producing pearlitic cast iron. Its compounds are used in flame-proofing, paints, ceramics, enamels, glass dyestuffs, and rubber technology. The element will burn in air but is unaffected by water or dilute acids. It is attacked by oxidizing acids and by halogens. It was first reported by Tholden in 1450.

apatite A complex mineral form of *calcium phosphate, $Ca_5(PO_4)_3(OH,F,Cl)$; the commonest of the phosphate minerals. It has a hexagonal structure and occurs widely as an accessory mineral in igneous rocks (e.g. pegmatite) and often in regional and contact metamorphic rocks, especially limestone. Large deposits occur in the Kola Peninsula, USSR. It is used in the production of fertilizers and is a major source of phosphorus. The enamel of teeth is composed chiefly of apatite.

aqua regia A mixture of concentrated nitric acid and concentrated hydrochloric acid in the ratio 1:3 respectively. It is a very powerful oxidizing mixture and will dissolve all metals (except silver, which forms an insoluble chloride) including such noble metals as gold and platinum, hence its name ('royal water'). Nitrosyl chloride (NOCl) is believed to be one of the active constituents.

aqueous Describing a solution in water.

aragonite A rock-forming anhydrous mineral form of calcium carbonate, $CaCO_3$. It is much less stable than *calcite, the commoner form of calcium carbonate, from which it may be distinguished by its greater hardness and specific gravity. Over time aragonite undergoes recrystallization to calcite. Aragonite occurs in cavities in limestone, as a deposit in limestone caverns, as a precipitate around hot springs and geysers, and in high-pressure low-temperature metamorphic rocks; it is also found in the shells of a number of molluscs and corals and is the main constituent of pearls. It is white or colourless when pure but the presence of impurities may tint it grey, blue, green, or pink.

arenes Aromatic hydrocarbons, such as benzene, toluene, and naphthalene.

argentic compounds Compounds of silver in its higher (+2) oxidation state; e.g. argentic oxide is silver(II) oxide (AgO).

argentite A sulphide ore of silver, Ag_2S. It crystallizes in the cubic system but most commonly occurs in massive form. It is dull grey-black in colour but bright when first cut and occurs in veins associated with other silver minerals. Important deposits occur in Mexico, Peru, Chile, Bolivia, Norway, and Czechoslovakia.

argentous compounds Compounds of silver in its lower (+1) oxidation state; e.g. argentous chloride is silver(I) chloride.

arginine *See* amino acid.

argon Symbol Ar. A monatomic noble gas present in air (0.93%); a.n. 18; r.a.m. 39.948; d. 0.00178 g cm^{-3}; m.p. −189°C; b.p. −185°C. Argon is separated from liquid air by fractional distillation. It is slightly soluble in water, colourless, and has no smell. Its uses include inert atmospheres in welding and special-metal manufacture (Ti and Zr), and (when mixed with 20% nitrogen) in gas-filled electric-light bulbs. The element is inert and has no true compounds. Lord Rayleigh and Sir William Ramsey identified argon in 1894.

aromatic compound An organic compound that contains a benzene ring in its molecules or that has chemical properties similar to benzene. Aromatic compounds are unsaturated compounds, yet they do not easily partake in addition reactions. Instead they undergo electrophilic substitution.

Benzene, the archetypal aromatic compound, has an hexagonal ring of carbon atoms and the classical formula (the Kekulé structure) would have alternating double and single bonds. In fact all the bonds in benzene are the same length intermediate between double and single C–C bonds. The properties arise because the electrons in the

π-orbitals are delocalized over the ring, giving an extra stabilization energy of 150 kJ mol^{-1} over the energy of a Kekulé structure. The condition for such delocalization is that a compound should have a planar ring with ($4n + 2$) pi electrons – this is known as the *Huckel rule*. Aromatic behaviour is also found in heterocyclic compounds such as pyridine. *See also* non-benzenoid aromatic; pseudoaromatic.

aromaticity The property characteristic of *aromatic compounds.

Arrhenius equation An equation of the form

$$k = A\exp(-E_A/RT)$$

where k is the rate constant of a given reaction and E_A the *activation energy. A is a constant for a given reaction, called the *pre-exponential factor*. Often the equation is written in logarithmic form

$$\ln k = \ln A - E_A/RT$$

A graph of $\ln k$ against $1/T$ is a straight line with a gradient $-E_A/R$ and an intercept on the $\ln k$ axis of $\ln A$.

Arrhenius theory *See* acid.

arsenate(III) *See* arsenic(III) oxide.

arsenate(V) *See* arsenic(V) oxide.

arsenic Symbol As. A metalloid element of *group V of the periodic table; a.n. 33; r.a.m. 74.92; r.d. 5.7; sublimes at 613°C. It has three allotropes – yellow, black, and grey. The grey metallic form is the stable and most common one. Over 150 minerals contain arsenic but the main sources are as impurities in sulphide ores and in the minerals orpiment (As_2S_3) and realgar (As_4S_4). Ores are roasted in air to form arsenic oxide and then reduced by hydrogen or carbon to metallic arsenic. Arsenic compounds are used in insecticides and as doping agents in semiconductors. The element is included in some lead-based alloys to promote hardening. Confusion can arise because As_4O_6 is often sold as white arsenic. Arsenic compounds are accumulative poisons. The element will react with halogens, concentrated oxidizing acids, and hot alkalis. Albertus Magnus is believed to have been the first to isolate the element in 1250.

arsenic acid *See* arsenic(V) oxide.

arsenic(III) acid *See* arsenic(III) oxide.

arsenic hydride *See* arsine.

arsenic(III) oxide (arsenic trioxide; arsenious oxide; white arsenic) A white or colourless compound, As_4O_6, existing in three solid forms. The commonest has cubic or octahedral crystals (r.d. 3.85; sublimes at 193°C) and is soluble in water, ethanol, and alkali solutions. It occurs naturally as *arsenolite*. A vitreous form can be prepared by slow condensation of the vapour (r.d. 3.74); its solubility in cold water is more than double that of the cubic form. The third modification, which occurs naturally as *claudetite*, has monoclinic crystals (r.d. 4.15). Arsenic(III) oxide is obtained commercially as a byproduct from the smelting of nonferrous sulphide ores; it may be produced in the laboratory by burning elemental arsenic in air. The structure of the molecule is similar to that of P_4O_6, with a tetrahedral arrangement of As atoms edge linked by oxygen bridges. Arsenic(III) oxide is acidic; its solutions were formerly called *arsenious acid* (technically, *arsenic(III) acid*). It forms *arsenate(III)* salts (formerly called *arsenites*). Arsenic(III) oxide is extremely toxic and is used as a poison for vermin; trace doses are used for a variety of medicinal purposes. It is also used for producing opalescent glasses and enamels.

arsenic(V) oxide (arsenic oxide) A white amorphous deliquescent solid, As_2O_5; r.d. 4.32; decomposes at 315°C. It is soluble in water and ethanol. Arsenic(V) oxide cannot be obtained by direct combination of arsenic and oxygen; it is usually prepared by the reaction of arsenic with nitric acid followed by dehydration of the arsenic acid thus formed. It readily loses oxygen on heating to give arsenic(III) oxide. Arsenic(V) oxide is acidic, dissolving in water to give arsenic(V) acid (formerly called *arsenic acid*), H_3AsO_4; the acid is tribasic and slightly weaker than phosphoric acid and should be visualized as $(HO)_3AsO$. It gives *arsenate (V)* salts (formerly called *arsenates*).

arsenic trioxide *See* arsenic(III) oxide.

arsenious acid *See* arsenic(III) oxide.

arsenious oxide *See* arsenic(III) oxide.

arsenite *See* arsenic(III) oxide.

arsenolite A mineral form of *arsenic(III) oxide, As_4O_6.

arsine (arsenic hydride) A colourless gas, AsH_3; m.p. −116.3°C; b.p. −55°C. It is soluble in water, chloroform, and benzene. Liquid arsine has a relative density of 1.69. Arsine is produced by the reaction of mineral acids with arsenides of electropositive metals or by the reduction of many arsenic compounds using nascent hydrogen. It is extremely poisonous and, like the hydrides of the heavier members of group V, is readily decomposed at elevated temperatures (around 260–300°C). Like ammonia and phosphine, arsine has a pyramidal structure.

Arsine gas has a very important commercial application in the production of modern microelectronic components. It is used in a dilute gas mixture with an inert gas and its ready thermal decomposition is exploited to enable other growing crystals to be doped with minute traces of arsenic to give *n*-type semiconductors.

artinite A mineral form of basic *magnesium carbonate, $MgCO_3.Mg(OH)_2.3H_2O$.

aryl group A group obtained by removing a hydrogen atom from an aromatic compound, e.g. phenyl group, C_6H_5-, derived from benzene.

asbestos Any one of a group of fibrous amphibole minerals (amosite, crocidolite (blue asbestos), tremolite, anthophyllite, and actinolite) or the fibrous serpentine mineral chrysotile. Asbestos has widespread commercial uses because of its resistance to heat, chemical inertness, and high electrical resistance. The fibres may be spun and woven into fireproof cloth for use in protective clothing, curtains, brake linings, etc., or moulded into blocks. Exposure to large amounts of asbestos can cause the respiratory disease asbestosis. Canada and the USSR are the largest producers of asbestos; others include South Africa, Zimbabwe, and China.

ascorbic acid *See* vitamin C.

asparagine *See* amino acid.

aspartic acid *See* amino acid.

astatine Symbol At. A radioactive *halogen element; a.n. 85; r.a.m. 211; m.p. 302°C; b.p. 377°C. It occurs naturally by radioactive decay from uranium and thorium isotopes. Astatine forms at least 20 isotopes, the most stable astatine–210 has a half-life of 8.3 hours. It can also be produced by alpha bombardment of bismuth–200. Astatine is stated to be more metallic than iodine; at least 5 oxidation states are known in aqueous solutions. It will form interhalogen compounds, such as AtI and AtCl. The existence of At_2 has not yet been established. The element was synthesized by nuclear bombardment in 1940 by D. R. Corson, K. R. MacKenzie, and E. Segré at the University of California.

asymmetric atom *See* optical activity.

atactic polymer *See* polymer.

atmolysis The separation of a mixture of gases by means of their different rates of diffusion. Usually, separation is effected by allowing the gases to diffuse through the walls of a porous partition or membrane.

atmosphere 1. (atm.) A unit of pressure equal to 101 325 pascals. This is equal to 760.0 mmHg. The actual *atmospheric pressure fluctuates around this value. The unit is usually used for expressing pressures well in excess of standard atmospheric pressure, e.g. in high-pressure chemical processes. **2.** *See* earth's atmosphere.

atmospheric pressure The pressure exerted by the weight of the air above it at any point on the earth's surface. At sea level the atmosphere will support a column of mercury about 760 mm high. This decreases with increasing altitude. The standard value for the atmospheric pressure at sea level in SI units is 101 325 pascals.

atom The smallest part of an element that can exist chemically. Atoms consist of a small dense nucleus of protons and neutrons surrounded by moving electrons. The number of electrons equals the number of protons so the overall charge is zero. The electrons are considered to move in circular or elliptical orbits (*see* Bohr theory) or, more accurately, in regions of space around the nucleus (*see* orbital).

The *electronic structure* of an atom refers to the way in which the electrons are arranged about the nucleus, and in particular the *energy levels that they occupy. Each electron can be characterized by a set of four quantum numbers, as follows:

(1) The *principal quantum number* n gives the main energy level and has values 1, 2, 3, etc. (the higher the number, the further the electron from the nucleus). Traditionally, these levels, or the orbits corresponding to them, are referred to as *shells* and given letters K, L, M, etc. The K-shell is the one nearest the nucleus.

(2) The *orbital quantum number* l, which governs the angular momentum of the electron. The possible values of l are $(n-1)$, $(n-2)$, ... 1, 0. Thus, in the first shell ($n = 1$) the electrons can only have angular momentum zero ($l = 0$). In the second shell ($n = 2$), the values of l can be 1 or 0, giving rise to two *subshells* of slightly different energy. In the third shell ($n = 3$) there are three subshells, with $l = 2$, 1, or 0. The subshells are denoted by letters $s(l = 0)$, $p(l = 1)$, $d(l = 2)$, $f(l = 3)$. The orbital quantum number is sometimes called the *azimuthal quantum number*.

(3) The *magnetic quantum number* m, which governs the energies of electrons in an external magnetic field. This can take values of $+l$, $+(l-1)$, ... 1, 0, -1, ... $-(l-1)$, $-l$. In an s-subshell (i.e. $l = 0$) the value of $m = 0$. In a p-subshell ($l = 1$), m can have values $+1$, 0, and -1; i.e. there are three p-orbitals in the p-subshell, usually designated p_x, p_y, and p_z. Under normal circumstances, these all have the same energy level.

(4) The *spin quantum number* m_s, which gives the spin of the individual electrons and can have the values $+\frac{1}{2}$ or $-\frac{1}{2}$.

According to the *Pauli exclusion principle, no two electrons in the atom can have the same set of quantum numbers. The numbers define the *quantum state* of the electron, and explain how the electronic structures of atoms occur.

atomicity The number of atoms in a given molecule. For example, oxygen (O_2) has an atomicity of 2, ozone (O_3) an atomicity of 3, benzene (C_6H_6) an atomicity of 12, etc.

atomic mass unit (a.m.u.) A unit of mass used to express *relative atomic masses. It is equal to 1/12 of the mass of an atom of the isotope carbon–12 and is equal to $1.660\,33 \times 10^{-27}$ kg. This unit superseded both the physical and chemical mass units based on oxygen–16 and is sometimes called the *unified mass unit* or the *dalton*.

atomic number (proton number) Symbol *Z*. The number of protons in the nucleus of an atom. The atomic number is equal to the number of electrons orbiting the nucleus in a neutral atom.

atomic orbital *See* orbital.

atomic volume The relative atomic mass of an element divided by its density.

atomic weight *See* relative atomic mass.

ATP (adenosine triphosphate) A nucleotide that is of fundamental importance as a carrier of chemical energy in all living organisms. It consists of adenine linked to D-ribose (i.e. adenosine); the D-ribose component bears three phosphate groups, linearly linked together by covalent bonds. These bonds can undergo hydrolysis to yield either a molecule of *ADP* (*adenosine diphosphate*) and inorganic phosphate or a molecule of *AMP* (*adenosine monophosphate*) and pyrophosphate. Both these reactions yield a large amount of energy (about 30.6 kJ mol^{-1}) that is used to bring about such biological processes as muscle contraction, the active transport of ions and molecules across cell membranes, and the synthesis of biomolecules. The reactions bringing about these processes often involve the enzyme-catalysed transfer of the phosphate group to intermediate substrates. Most ATP-mediated reactions require Mg^{2+} ions as *cofactors.

ATP is regenerated by the rephosphorylation of AMP and ADP using the chemical energy obtained from the oxidation of food. This takes place during glycolysis and the Krebs cycle but, most significantly, is also a result of the reduction–oxidation reactions of the electron transport chain, which ultimately reduces molecular oxygen to water (oxidative phosphorylation).

atropine A poisonous crystalline alkaloid, $C_{17}H_{23}NO_3$; m.p. 114–16°C. It can be extracted from deadly nightshade and other solanaceous plants and is used in medicine to treat colic, to reduce secretions, and to dilate the pupil of the eye.

atto- Symbol *a*. A prefix used in the metric system to denote 10^{-18}. For example, 10^{-18} second = 1 attosecond (as).

Aufbau principle A principle that gives the order in which orbitals are filled in successive elements in the periodic table. The order of filling is 1*s*, 2*s*, 2*p*, 3*s*, 3*p*, 4*s*, 3*d*, 4*p*, 5*s*, 4*d*, 5*p*, 6*s*, 4*f*, 5*d*, 6*p*, 7*s*, 5*f*, 6*d*. *See* atom.

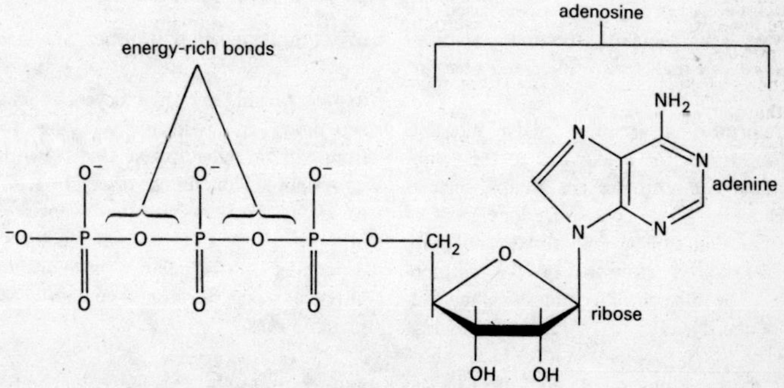

ATP

Auger effect The ejection of an electron from an atom without the emission of an X-or gamma-ray photon, as a result of the de-excitation of an excited electron within the atom. This type of transition occurs in the X-ray region of the emission spectrum. The kinetic energy of the ejected electron, called an *Auger electron*, is equal to the energy of the corresponding X-ray photon minus the binding energy of the Auger electron. The effect was discovered by Pierre Auger (1899–) in 1925.

auric compounds Compounds of gold in its higher (+3) oxidation state; e.g. auric chloride is gold(III) chloride ($AuCl_3$).

aurous compounds Compounds of gold in its lower (+1) oxidation state; e.g. aurous chloride is gold(I) chloride (AuCl).

austenite *See* steel.

autocatalysis *Catalysis in which one of the products of the reaction is a catalyst for the reaction. Reactions in which autocatalysis occurs have a characteristic S-shaped curve for reaction rate against time – the reaction starts slowly and increases as the amount of catalyst builds up, falling off again as the products are used up.

autoclave A strong steel vessel used for carrying out chemical reactions, sterilizations, etc., at high temperature and pressure.

auxochrome A group in a dye molecule that influences the colour due to the *chromophore. Auxochromes are groups, such as $-OH$ and $-NH_2$, containing lone pairs of electrons that can be delocalized along with the delocalized electrons of the chromophore. The auxochrome intensifies the colour of the dye. Formerly, the term was also used of such groups as $-SO_2O^-$, which make the molecule soluble and affect its application.

Avogadro constant Symbol N_A or L. The number of atoms or molecules in one *mole of substance. It has the value $6.022\,52 \times 10^{23}$. Formerly it was called *Avogadro's number.*

Avogadro's law Equal volumes of all gases contain equal numbers of molecules at the same pressure and temperature. The law, often called *Avogadro's hypothesis*, is true only for ideal gases. It was first proposed in 1811 by Count Amadeo Avogadro (1776–1856).

azeotrope (azeotropic mixture; constant-boiling mixture) A mixture of two liquids that boils at constant composition; i.e. the composition of the vapour is the same as that of the liquid. Azeotropes occur because of deviations in Raoult's law leading to a maximum or minimum in the *boiling-point–composition diagram. When the mixture is boiled, the vapour initially has a higher proportion of one component than is present in the liquid, so the proportion of this in the liquid falls with time. Eventually, the maximum and minimum point is reached, at which the two liquids distil together without change in composition. The composition of an azeotrope depends on the pressure.

azides Compounds containing the ion N_3^- or the group $-N_3$.

azimuthal quantum number *See* atom.

azine An organic heterocyclic compound containing a six-membered ring formed from carbon and nitrogen atoms. Pyridine is an example containing one nitrogen atom (C_5H_5N). *Diazines* have two nitrogen atoms in the ring (e.g. $C_4H_4N_2$), and isomers exist depending on the relative positions of the nitrogen atoms. *Triazines* contain three nitrogen atoms.

azo compounds Organic compounds containing the group $-N{=}N-$ linking two other groups. They can be formed by reaction of a diazonium ion with a benzene ring.

azo dye *See* dyes.

azoimide *See* hydrogen azide.

azurite A secondary mineral consisting of hydrated basic copper carbonate, $Cu_3(OH)_2(CO_3)_2$, in monoclinic crystalline form. It is generally formed in the upper zone of copper ore deposits and often occurs with *malachite. Its intense azure-blue colour made it formerly important as a pigment. It is a minor ore of copper and is used as a gemstone.

B

Babbit metal Any of a group of related alloys used for making bearings. They consist of tin containing antimony (about 10%) and copper (1–2%), and often lead. The original alloy was invented in 1839 by the US inventor Isaac Babbit (1799–1862).

Babo's law The vapour pressure of a liquid is decreased when a solute is added, the amount of the decrease being proportional to the amount of solute dissolved. The law was discovered in 1847 by the German chemist Lambert Babo (1818–99). *See also* Raoult's law.

back donation A form of chemical bonding in which a *ligand forms a sigma bond to an atom or ion by donating a pair of electrons, and the central atom donates electrons back by overlap of its *d*-orbitals with empty *p*- or *d*-orbitals on the ligand.

back e.m.f. An electromotive force that opposes the main current flow in a circuit. For example, in an electric cell, *polarization causes a back e.m.f. to be set up by chemical means.

background radiation Low intensity *ionizing radiation present on the surface of the earth and in the atmosphere as a result of *cosmic radiation and the presence of radioisotopes in the earth's rocks, soil, and atmosphere. The radioisotopes are either natural or the result of nuclear fallout or waste gas from power stations. Background counts must be taken into account when measuring the radiation produced by a specified source.

bacteriocidal Capable of killing bacteria. Common bacteriocides are some antibiotics, antiseptics, and disinfectants.

Bakelite A tradename for certain phenol–formaldehyde resins, first introduced in 1909 by the Belgian–American chemist Leo Hendrik Baekeland (1863–1944).

baking soda *See* sodium hydrogencarbonate.

balance An accurate weighing device. The simple *beam balance* consists of two pans suspended from a centrally pivoted beam. Known masses are placed on one pan and the substance or body to be weighed is placed in the other. When the beam is exactly horizontal the two masses are equal. An accurate laboratory balance weighs to the nearest hundredth of a milligram. Specially designed balances can be accurate to a millionth of a milligram. More modern *substitution balances* use the substitution principle. In this calibrated weights are removed from the single lever arm to bring the single pan suspended from it into equilibrium with a fixed counter weight. The substitution balance is more accurate than the two-pan device and enables weighing to be carried out more rapidly. In automatic electronic balances, mass is determined not by mechanical deflection but by electronically controlled compensation of an electric force. A scanner monitors the displacement of the pan support generating a current proportional to the displacement. This current flows through a coil forcing the pan support to return to its original position by means of a magnetic force. The signal generated enables the mass to be read from a digital display. The mass of the empty container can be stored in the balance's computer memory and automatically deducted from the mass of the container plus its contents.

Balmer series *See* hydrogen spectrum.

banana bond Informal name for the type of electron-deficient bond holding the B–H–B bridges in *boranes and similar compounds.

band spectrum *See* spectrum.

band theory *See* energy bands.

bar A c.g.s. unit of pressure equal to 10^6 dynes per square centimetre or 10^5 pascals (approximately 750 mmHg or 0.987 atmosphere). The *millibar* (100 Pa) is commonly used in meteorology.

Barfoed's test A biochemical test to detect monosaccharide (reducing) sugars in solution, devised by the Swedish physician C. T. Barfoed (1815–99). *Barfoed's reagent*, a mixture of ethanoic (acetic) acid and copper(II) acetate, is added to the test solution and boiled. If any reducing sugars are present a red precipitate of copper(II) oxide is formed. The reaction will be negative in the presence of disaccharide sugars as they are weaker reducing agents.

barite *See* barytes.

barium Symbol Ba. A silvery-white reactive element belonging to *group II of the periodic table; a.n. 54; r.a.m. 137.33; r.d. 3.51; m.p. 725°C; b.p. 1640°C. It occurs as the minerals barytes ($BaSO_4$) and witherite. Extraction is by high-temperature reduction of barium oxide with aluminium or silicon in a vacuum, or by electrolysis of fused barium chloride. The metal is used as a getter in vacuum systems. It oxidizes readily in air and reacts with ethanol and water. Soluble barium compounds are extremely poisonous. It was first identified in 1774 by Karl Scheele, and was extracted by Humphry Davy in 1808.

barium bicarbonate *See* barium hydrogencarbonate.

barium carbonate A white insoluble compound, $BaCO_3$; r.d. 4.43. It decomposes on heating to give barium oxide and carbon dioxide:

$$BaCO_3(s) \rightarrow BaO(s) + CO_2(g)$$

The compound occurs naturally as the mineral *witherite* and can bc prepared by adding an alkaline solution of a carbonate to a solution of a barium salt. It is used as a raw material for making other barium salts, as a flux for ceramics, and as a raw material in the manufacture of certain types of optical glass.

barium chloride A white compound, $BaCl_2$. The anhydrous compound has two crystalline forms: an α form (monoclinic; r.d. 3.856), which transforms at 962°C to a β form (cubic; r.d. 3.917; m.p. 963°C; b.p. 1560°C). There is also a dihydrate, $BaCl_2.2H_2O$ (cubic; r.d. 3.1), which loses water at 113°C. It is prepared by dissolving barium carbonate (witherite) in hydrochloric acid and crystallizing out the dihydrate. The compound is used in the extraction of barium by electrolysis.

barium hydrogencarbonate (barium bicarbonate) A compound, $Ba(HCO_3)_2$, which is only stable in solution. It can be formed by the action of carbon dioxide on a suspension of barium carbonate in cold water:

$$BaCO_3(s) + CO_2(g) + H_2O(l) \rightarrow Ba(HCO_3)_2(aq)$$

On heating, this reaction is reversed.

barium hydroxide (baryta) A white solid, $Ba(OH)_2$, sparingly soluble in water. The common form is the octahydrate, $Ba(OH)_2.8H_2O$; monoclinic; r.d. 2.18; m.p. 78°C. It can be produced by adding water to barium monoxide or by the action of sodium hydroxide on soluble barium compounds and is used as a weak alkali in volumetric analysis.

barium oxide A white or yellowish solid, BaO, obtained by heating barium in oxygen or by the thermal decomposition of barium carbonate or nitrate; cubic; r.d. 5.72; m.p. 1920°C; b.p. 2000°C. When barium oxide is heated in oxygen the peroxide, BaO_2, is formed in a reversible reaction that was once used as a method for obtaining oxygen (the *Brin process*). Barium oxide is now used in the manufacture of lubricating-oil additives.

barium peroxide A dense off-white solid, BaO_2, prepared by carefully heating *barium oxide in oxygen; r.d. 4.96; m.p. 450°C. It is used as a bleaching agent. With acids, hydrogen peroxide is formed and the reaction is used in the laboratory preparation of hydrogen peroxide.

barium sulphate An insoluble white solid, $BaSO_4$, that occurs naturally as the mineral *barytes (or *heavy spar*) and can be prepared as a precipitate by adding sulphuric acid to barium chloride solution; r.d. 4.50; m.p. 1580°C. The rhombic form changes to a monoclinic form at 1149°C. It is used as a raw material for making other barium salts, as a pigment extender in surface coating materials (called *blanc fixe*), and in the glass and rubber industries. Barium compounds are opaque to X-rays, and a suspension of the sulphate in water is used in medicine to provide a contrast medium for X-rays of the stomach and intestine. Although barium compounds are extremely poisonous, the sulphate is safe to use because it is very insoluble.

baryta *See* barium hydroxide.

barytes (barite) An orthorhombic mineral form of *barium sulphate, $BaSO_4$; the chief ore of barium. It is usually white but may also be yellow, grey, or brown. Large deposits occur in Andalusia, Spain, and in the USA.

basalt A fine-grained basic igneous rock. It is composed chiefly of calcium-rich plagioclase feldspar and pyroxene; other minerals present may be olivine, magnetite, and apatite. Basalt is the commonest type of lava.

base A compound that reacts with a protonic acid to give water (and a salt). The definition comes from the Arrhenius theory of acids and bases. Typically, bases are metal oxides, hydroxides, or compounds (such as ammonia) that give hydroxide ions in aqueous solution. Thus, a base may be either: (1) An insoluble oxide or hydroxide that reacts with an acid, e.g.

$$CuO(s) + 2HCl(aq) \rightarrow CuCl_2(aq) + H_2O(l)$$

Here the reaction involves hydrogen ions from the acid

$$CuO(s) + 2H^+(aq) \rightarrow H_2O(l) + Cu^{2+}(aq)$$

(2) A soluble hydroxide, in which case the solution contains hydroxide ions. The reaction with acids is a reaction between hydrogen ions and hydroxide ions:

$$H^+ + OH^- \rightarrow H_2O$$

(3) A compound that dissolves in water to produce hydroxide ions. For example, ammonia reacts as follows:

$$NH_3(g) + H_2O(l) \rightleftharpoons NH_4^+(aq) + {}^-OH$$

Similar reactions occur with organic *amines (*see also* nitrogenous base; amine salts). A base that dissolves in water to give hydroxide ions is called an *alkali*. Ammonia and sodium hydroxide are common examples.

The original Arrhenius definition of a base has been extended by the Lowry–Brønsted theory and by the Lewis theory. *See* acid.

base dissociation constant *See* dissociation.

base metal A common relatively inexpensive metal, such as iron or lead, that corrodes, oxidizes, or tarnishes on exposure to air, moisture, or heat, as distinguished from precious metals, such as gold and silver.

base unit A unit that is defined arbitrarily rather than being defined by simple combinations of other units. For example, the ampere is a base unit in the SI system defined in terms of the force produced between two current-carrying conductors, whereas the coulomb is a *derived unit*, defined as the quantity of charge transferred by one ampere in one second.

basic 1. Describing a compound that is a base. **2.** Describing a solution containing an excess of hydroxide ions; alkaline.

basic dye *See* dyes.

basicity constant *See* dissociation.

basic-oxygen process (BOP process) A high-speed method of making high-grade steel. It originated in the *Linnz–Donnewitz (L–D) process*. Molten pig iron and scrap are charged into à tilting furnace, similar to the Bessemer furnace except that it has no tuyeres. The charge is converted to steel by blowing high-pressure oxygen onto the surface of the metal through a water-cooled lance. The excess heat produced enables up to 30% of scrap to be incorporated into the charge. The process has largely replaced the Bessemer and open-hearth processes.

basic salt A compound that can be regarded as being formed by replacing some of the oxide or hydroxide ions in a base by other negative ions. Basic salts are thus mixed salt–oxides (e.g. bismuth(III) chloride oxide, BiOCl) or salt–hydroxides (e.g. lead (II) chloride hydroxide, Pb(OH)Cl).

basic slag *Slag formed from a basic flux (e.g. calcium oxide) in a blast furnace. The basic flux is used to remove acid impurities in the ore and contains calcium silicate, phosphate, and sulphide. If the phosphorus content is high the slag can be used as a fertilizer.

basic stains *See* staining.

battery A number of electric cells joined together. The common car battery, or *accumulator, usually consists of six secondary cells connected in series to give a total e.m.f. of 12 volts. A torch battery is usually a dry version of the *Leclanché primary cell, two of which are often connected in series. Batteries may also have cells connected in parallel, in which case they have the same e.m.f. as a single cell, but their capacity is increased, i.e. they will provide more total charge. The capacity of a battery is usually specified in ampere-hours, the ability to supply 1 A for 1 hr, or the equivalent.

bauxite The chief ore of aluminium, consisting of hydrous aluminium oxides and aluminous laterite. It is a claylike amorphous material formed by the weathering of silicate rocks under tropical conditions. The chief producers are Australia, Guinea, Jamaica, USSR, Brazil, and Surinam.

beam balance *See* balance.

Beckmann thermometer A thermometer for measuring small changes of temperature. It consists of a mercury-in-glass thermometer with a scale covering only 5 or 6°C calibrated in hundredths of a degree. It has two

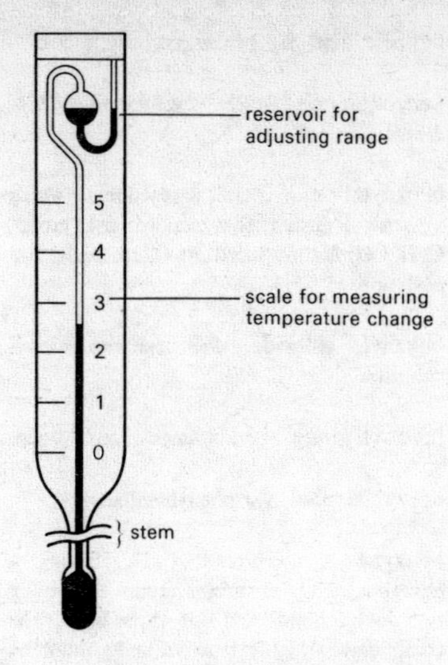

Beckmann thermometer

mercury bulbs, the range of temperature to be measured is varied by running mercury from the upper bulb into the larger lower bulb. It is used particularly for measuring *depression of freezing point or *elevation of boiling point of liquids when solute is added, in order to find relative molecular masses. The instrument was invented by the German chemist E. O. Beckmann (1853–1923).

becquerel Symbol Bq. The SI unit of activity (*see* radiation units). The unit is named after the discoverer of radioactivity A. H. Becquerel (1852–1908).

beet sugar *See* sucrose.

bel Ten *decibels.

bell metal A type of *bronze used in casting bells. It consists of 60–85% copper alloyed with tin, often with some zinc and lead included.

Benedict's test A biochemical test to detect reducing sugars in solution, devised by the US chemist S. R. Benedict (1884–1936). *Benedict's reagent* – a mixture of copper(II) sulphate and a filtered mixture of hydrated sodium citrate and hydrated sodium carbonate – is added to the test solution and boiled. A high concentration of reducing sugars induces the formation of a red precipitate; a lower concentration produces a yellow precipitate. Benedict's test is a more sensitive alternative to *Fehling's test.

beneficiation (ore dressing) The separation of an ore into the valuable components and the waste material (gangue). This may be achieved by a number of processes, including crushing, grinding, magnetic separation, froth flotation, etc. The dressed ore, consisting of a high proportion of valuable components, is then ready for smelting or some other refining process.

benzaldehyde *See* benzenecarbaldehyde.

Kekulé structures

Dewar structures (3 in all)

Benzene

benzene A colourless liquid hydrocarbon, C_6H_6; r.d. 0.88; m.p. 5.5°C; b.p. 80.1°C. It is now made from gasoline from petroleum by catalytic reforming (formerly obtained from coal tar). Benzene is the archetypal *aromatic compound. It has an unsaturated molecule, yet will not readily undergo addition reactions. On the other hand, it does undergo substitution reactions in which hydrogen atoms are replaced by other atoms or groups. This behaviour occurs because of delocalization of *p*-electrons over the benzene ring, and all the C–C bonds in benzene are equivalent and intermediate in length between single and double bonds. *See also* Kekulé structure.

benzenecarbaldehyde (benzaldehyde) A yellowish volatile oily liquid, C_6H_5CHO; r.d. 1.04; m.p. −26°C; b.p. 178.1°C. The compound occurs in almond kernels and has an almond-like smell. It is made from methylbenzene (by conversion to dichloromethyl benzene, $C_6H_5CHCl_2$, followed by hydrolysis). Benzenecarbaldehyde is used in flavourings, perfumery, and the dyestuffs industry.

benzenecarbonyl chloride (benzoyl chloride) A colourless liquid, C_6H_5COCl; r.d. 1.21; m.p. 0°C; b.p. 197.2°C. It is an *acyl halide, used to introduce benzenecarbonyl groups into molecules. *See* acylation.

benzenecarbonyl group (benzoyl group) The organic group C_6H_5CO-.

benzenecarboxylate (benzoate) A salt or ester of benzenecarboxylic acid.

benzenecarboxylic acid (benzoic acid) A white crystalline compound, C_6H_5COOH; r.d. 1.27; m.p. 122.4°C; b.p. 249°C. It occurs naturally in some plants and is used as a food preservative. Benzenecarboxylic acid has a carboxyl group bound directly to a benzene ring. It is a weak carboxylic acid ($K_a = 6.4 \times 10^{-5}$ at 25°C), which is slightly soluble in water. It also undergoes substitution reactions on the benzene ring.

benzene-1,4-diol (hydroquinone; quinol) A white crystalline solid, $C_6H_4(OH)_2$; r.d. 1.33; m.p. 170°C; b.p. 285°C. It is used in making dyes. *See also* quinhydrone electrode.

benzene hexachloride (BHC) A crystalline substance, $C_6H_6Cl_6$, made by adding chlorine to benzene. It is used as a pesticide and, like *DDT, concern has been expressed at its environmental effects.

benzoate *See* benzenecarboxylate.

benzoic acid *See* benzenecarboxylic acid.

benzoquinone *See* cyclohexadiene-1,4-dione.

benzoylation A chemical reaction in which a benzoyl group (benzenecarbonyl group, C_6H_5CO) is introduced into a molecule. *See* acylation.

benzoyl chloride *See* benzenecarbonyl chloride.

benzoyl group *See* benzenecarbonyl group.

benzyl alcohol *See* phenylmethanol.

benzyne A compound, C_6H_4, having a hexagonal ring of carbon atoms containing two double bonds and one triple bond. The compound is highly reactive and cannot be isolated, although benzyne and its derivatives occur as short-lived intermediates in organic reactions.

Bergius process A process for making hydrocarbon mixtures (for fuels) from coal by heating powdered coal mixed with tar and iron(III) oxide catalyst at 450°C under hydrogen at a pressure of about 200 atmospheres. In later developments of the process, the coal was suspended in liquid hydrocarbons and other catalysts were used. The process was developed by the German chemist Friederich Bergius (1884–1949) during World War I as a source of motor fuel.

berkelium Symbol Bk. A radioactive metallic transuranic element belonging to the *actinoids; a.n. 97; mass number of the most stable isotope 247 (half-life 1.4×10^3 years); r.d. (calculated) 14. There are eight known isotopes. It was first produced by G. T. Seborg and associates in 1949 by bombarding americium-241 with alpha particles.

Berthollide compound *See* nonstoichiometric compound.

beryl A hexagonal mineral form of beryllium aluminium silicate, $Be_3Al_2Si_6O_{18}$; the chief ore of beryllium. It may be green, blue, yellow, or white and has long been used as a gemstone. Beryl occurs throughout the world in granite and pegmatites. *Emerald, the green gem variety, occurs more rarely and is of great value. Important sources of beryl are found in Brazil, Madagascar, and the USA.

beryllate A compound formed in solution when beryllium metal, or the oxide or hydroxide, dissolves in strong alkali. The reaction (for the metal) is often written

$$Be + 2OH^-(aq) \rightarrow BeO_2^{2-}(aq) + H_2(g)$$

The ion BeO_2^{2-} is the beryllate ion. In fact, as with the *aluminates, the ions present are probably hydroxy ions of the type $Be(OH)_4^{2-}$ (the *tetrahydroxoberyllate(II)* ion) together with polymeric ions.

beryllia *See* beryllium oxide.

beryllium Symbol Be. A grey metallic element of *group II of the periodic table; a.n. 4; r.a.m. 9.012; r.d. 1.85; m.p. 1285°C; b.p. 2970°C. Beryllium occurs as beryl ($3BeO.Al_2O_3.6SiO_2$) and chrysoberyl ($BeO.Al_2O_3$). The metal is extracted from a fused mixture of BeF_2/NaF by electrolysis or by magnesium reduction of BeF_2. It is used to manufacture Be–Cu alloys, which are used in nuclear reactors as reflectors and moderators because of their low absorption cross section. Beryllium oxide is used in ceramics and in nuclear reactors. Beryllium and its compounds are toxic and can cause serious lung diseases and dermatitis. The metal is resistant to oxidation by air because of the formation of an oxide layer, but will react with dilute hydrochloric and sulphuric acids. Beryllium compounds show high covalent character. The element was isolated independently by F. Wohler and A. A. Bussy in 1828.

beryllium hydroxide A white crystalline compound, $Be(OH)_2$, precipitated from solutions of beryllium salts by adding alkali. Like the oxide, it is amphoteric and dissolves in excess alkali to give *beryllates.

beryllium oxide (beryllia) An insoluble solid compound, BeO; hexagonal; r.d. 3.01; m.p. 2550°C; b.p. 4120°C. It occurs naturally as *bromellite*, and can be made by burning beryllium in oxygen or by the decomposition of beryllium carbonate or hydroxide. It is an important amphoteric oxide, reacting with acids to form salts and with alkalis to form compounds known as *beryllates. Beryllium oxide is used in the production of beryllium and beryllium–copper refractories, transistors, and integrated circuits.

Bessemer process A process for converting *pig iron from a *blast furnace into *steel. The molten pig iron is loaded into a refractory-lined tilting furnace (*Bessemer converter*) at about 1250°C. Air is blown into the furnace from the base and *spiegel is added to introduce the correct amount of carbon. Impurities (especially silicon, phosphorus, and manganese) are removed by the converter lining to form a slag. Finally the furnace is tilted so that the molten steel can be poured off. In the modern VLN (very low nitrogen) version of this process, oxygen and steam are blown into the furnace in place of air to minimize the absorption of nitrogen from the air by the steel. The process is named after the British engineer Sir Henry Bessemer (1813–98), who announced it in 1856. *See also* basic-oxygen process.

beta decay A type of radioactive decay in which an unstable atomic nucleus changes into a nucleus of the same mass number but different proton number. The change involves the conversion of a neutron into a proton with the emission of an electron and an antineutrino ($n \rightarrow p + e^- + \bar{\nu}$) or of a proton into a neutron with the emission of a positron and a neutrino ($p \rightarrow n + e^+ + \nu$). An example is the decay of carbon-14:

$$^{14}_{6}C \rightarrow \ ^{14}_{7}N + e^- + \bar{\nu}$$

The electrons or positrons emitted are called *beta particles* and streams of beta particles are known as *beta radiation.*

beta-iron A nonmagnetic allotrope of iron that exists between 768°C and 900°C.

beta particle *See* beta decay.

BHC *See* benzene hexachloride.

bicarbonate *See* hydrogencarbonate.

bicarbonate of soda *See* sodium hydrogencarbonate.

bimolecular reaction A step in a chemical reaction that involves two molecules. *See* molecularity.

binary Describing a compound or alloy formed from two elements.

binary acid An *acid in which the acidic hydrogen atom(s) are bound directly to an atom other than oxygen. Examples are hydrogen chloride (HCl) and hydrogen sulphide (H_2S). Such compounds are sometimes called *hydracids. Compare* oxo acid.

biochemical oxygen demand (BOD) The amount of oxygen taken up by microorganisms that decompose organic waste matter in water. It is therefore used as a measure of the amount of certain types of organic pollutant in water. BOD is calculated by keeping a sample of water containing a known amount of oxygen for five days at 20°C. The oxygen content is measured again after this time. A high BOD indicates the presence of a large number of microorganisms, which suggests a high level of pollution.

biochemistry The study of the chemistry of living organisms, especially the structure and function of their chemical components (principally proteins, carbohydrates, lipids, and nucleic acids). Biochemistry has advanced rapidly with the development, from the mid-20th century, of such techniques as chromatography, X-ray diffraction, radioisotopic labelling, and electron microscopy. Using these techniques to separate and analyse biologically important molecules, the steps of the metabolic pathways in which they are involved (e.g. glycolysis) have been determined. This has provided some knowledge of how organisms obtain and store energy, how they manufacture and degrade their biomolecules, and how they sense and respond to their environment.

bioluminescence *See* luminescence.

biosynthesis The production of molecules by a living cell, which is the essential feature of *anabolism.

biotin A vitamin in the *vitamin B complex. It is the *coenzyme for various enzymes that catalyse the incorporation of carbon dioxide into various compounds. Adequate amounts are normally produced by the intestinal bacteria; other sources include cereals, vegetables, milk, and liver.

biotite An important rock-forming silicate mineral, a member of the *mica group of minerals, in common with which it has a sheetlike crystal structure. It is usually black, dark brown, or green in colour.

birefringence *See* double refraction.

Birkeland–Eyde process A process for the fixation of nitrogen by passing air through an electric arc to produce nitrogen oxides. It was introduced in 1903 by the Norwegian chemists Kristian Birkeland (1867–1913) and Samuel Eyde (1866–1940).

The process is economic only if cheap hydroelectricity is available.

bismuth Symbol Bi. A white crystalline metal with a pinkish tinge belonging to *group V of the periodic table; a.n. 83; r.a.m. 208.98; r.d. 9.78; m.p. 271.3°C; b.p. 1560°C. The most important ores are bismuthinite (Bi_2S_3) and bismite (Bi_2O_3). Peru, Japan, Mexico, Bolivia, and Canada are major producers. The metal is extracted by carbon reduction of its oxide. Bismuth is the most diamagnetic of all metals and its thermal conductivity is lower than any metal except mercury. The metal has a high electrical resistance and a high Hall effect when placed in magnetic fields. It is used to make low-melting-point casting alloys with tin and cadmium. These alloys expand on solidification to give clear replication of intricate features. It is also used to make thermally activated safety devices for fire-detection and sprinkler systems. More recent applications include its use as a catalyst for making acrylic fibres, as a constituent of malleable iron, as a carrier of uranium–235 fuel in nuclear reactors, and as a specialized thermocouple material. Bismuth compounds (when lead-free) are used for cosmetics and medical preparations. It is attacked by oxidizing acids, steam (at high temperatures), and by moist halogens. It burns in air with a blue flame to produce yellow oxide fumes. C. G. Junine first demonstrated that it was different from lead in 1753.

bisulphate *See* hydrogensulphate.

bisulphite *See* hydrogensulphite; aldehydes.

bittern The solution of salts remaining when sodium chloride is crystallized from sea water.

bitumen *See* petroleum.

bituminous coal *See* coal.

bituminous sand *See* oil sand.

biuret test A biochemical test to detect proteins in solution, named after the substance *biuret* ($H_2NCONHCONH_2$), which is formed when urea is heated. Sodium hydroxide is mixed with the test solution and drops of 1% copper(II) sulphate solution are then added slowly. A positive result is indicated by a violet ring, caused by the reaction of *peptide bonds in the proteins or peptides. Such a result will not occur in the presence of free amino acids.

bivalent (*or* **divalent**) Having a valency of two.

blackdamp (choke damp) Air left depleted in oxygen following the explosion of firedamp in a mine.

black lead *See* carbon.

blanc fixe *See* barium sulphate.

blast furnace A furnace for smelting iron ores, such as haematite (Fe_2O_3) or magnetite (Fe_3O_4), to make *pig iron. The furnace is a tall refractory-lined cylindrical structure that is charged at the top with the dressed ore (*see* beneficiation), coke, and a flux, usually limestone. The conversion of the iron oxides to metallic iron is a reduction process in which carbon monoxide and hydrogen are the reducing agents. The overall reaction can be summarized thus:

$$Fe_3O_4 + 2CO + 2H_2 \rightarrow 3Fe + 2CO_2 + 2H_2O$$

The CO is obtained within the furnace by blasting the coke with hot air from a ring of tuyeres about two-thirds of the way down the furnace. The reaction producing the CO is:

$$2C + O_2 \rightarrow 2CO$$

In most blast furnaces hydrocarbons (oil, gas, tar, etc.) are added to the blast to provide a source of hydrogen. In the modern *direct-reduction process* the CO and H_2 may be produced separately so that the reduction process can proceed at a lower temperature. The pig iron produced by a blast furnace contains about 4% carbon and further refining is usually required to produce steel or cast iron.

blasting gelatin A high explosive made from nitroglycerine and gun cotton (cellulose nitrate).

bleaching powder A white solid regarded as a mixture of calcium chlorate(I), calcium chloride, and calcium hydroxide. It is prepared on a large scale by passing chlorine gas through a solution of calcium hydroxide. Bleaching powder is sold on the basis of available chlorine, which is liberated when it is treated with a dilute acid. It is used for bleaching paper pulps and fabrics and for sterilizing water.

blende A naturally occurring metal sulphide, e.g. zinc blende ZnS.

block *See* periodic table.

block copolymer *See* polymer.

blue vitriol *See* copper(II) sulphate.

boat conformation *See* conformation.

BOD *See* biochemical oxygen demand.

body-centred cubic (b.c.c.) *See* cubic crystal.

boehmite A mineral form of a mixed aluminium oxide and hydroxide, AlO.OH. It is named after the German scientist J. Böhm. *See* aluminium hydroxide.

Bohr theory The theory published in 1913 by the Danish physicist Niels Bohr (1885–1962) to explain the line spectrum of hydrogen. He assumed that a single electron of mass m travelled in a circular orbit of radius r, at a velocity v, around a positively charged nucleus. The *angular momentum of the electron would then be mvr. Bohr proposed that electrons could only occupy orbits in which this angular momentum had certain fixed values, $h/2\pi$, $2h/2\pi$, $3h/2\pi$, ... $nh/2\pi$, where h is the Planck constant. This means that the angular momentum is quantized, i.e. can only have certain values, each of which is a multiple of n. Each permitted value of n is associated with an orbit of different radius and Bohr assumed that when the atom emitted or absorbed radiation of frequency ν, the electron jumped from one orbit to another; the energy emitted or absorbed by each jump is equal to $h\nu$. This theory gave good results in predicting the series of lines observed in the *hydrogen spectrum but not with any other spectrum. The idea of quantized values of angular momentum was later explained by the wave nature of the electron. Each orbit has to have a whole number of wavelengths around it; i.e. $n\lambda = 2\pi r$, where λ is the wavelength and n a whole number. The wavelength of a particle is given by h/mv, so $nh/mv = 2\pi r$, which leads to $mvr = nh/2\pi$. Modern atomic theory does not allow subatomic particles to be treated in the same way as large objects, and Bohr's reasoning is somewhat discredited. However, the idea of quantized angular momentum has been retained.

boiling point (b.p.) The temperature at which the saturated vapour pressure of a liquid equals the external atmospheric pressure. As a consequence, bubbles form in the liquid and the temperature remains constant until all the liquid has evaporated. As the boiling point of a liquid depends on the external atmospheric pressure, boiling points are usually quoted for standard atmospheric pressure (760 mmHg = 101 325 Pa).

boiling-point–composition diagram A graph showing how the boiling point and vapour composition of a mixture of two liquids depends on the composition of the mixture. The abscissa shows the range of compositions from 100% A at one end to 100% B at the other. The diagram has two curves: the lower one gives the boiling points (at a fixed pressure) for the different compositions. The upper one is plotted by taking the composition of vapour at each temperature on the boiling-point curve. The two curves would coincide for an ideal mixture, but generally they are different because of deviations from *Raoult's law. In some cases, they may show a maximum or minimum and coincide at some intermediate composition, explaining the formation of *azeotropes.

Boltzmann constant Symbol k. The ratio of the universal gas constant (R) to the Avogadro constant (N_A). It may be thought of therefore as the gas constant per molecule:

$$k = R/N_A = 1.380\,622 \times 10^{-23}\ \mathrm{J\,K^{-1}}$$

It is named after the Austrian physicist Ludwig Boltzmann (1844–1906).

bomb calorimeter An apparatus used for measuring heats of combustion (e.g. calorific values of fuels and foods). It consists of a strong container in which the sample is sealed with excess oxygen and ignited electrically. The heat of combustion at constant volume can be calculated from the resulting rise in temperature.

bond *See* chemical bond.

bond energy An amount of energy associated with a bond in a chemical compound. It is obtained from the heat of atomization. For instance, in methane the bond energy of the C–H bond is one quarter of the enthalpy of the process

$$CH_4(g) \rightarrow C(g) + 4H(g)$$

Bond energies can be calculated from the standard enthalpy of formation of the compound and from the enthalpies of atomization of the elements. Energies calculated in this way are called *average bond energies* or *bond–energy terms*. They depend to some extent on the molecule chosen; the C–H bond energy in methane will differ slightly from that in ethane. The *bond dissociation energy* is a different measurement, being the energy required to break a particular bond; e.g. the energy for the process:

$$CH_4(g) \rightarrow CH_3\cdot(g) + H\cdot(g)$$

bonding orbital *See* orbital.

bone black *See* charcoal.

Diborane, the simplest of the boranes

borane (boron hydride) Any of a group of compounds of boron and hydrogen, many of which can be prepared by the action of acid on magnesium boride (MgB_2). Others are made by pyrolysis of the products of this reaction in the presence of hydrogen and other reagents. They are all volatile, reactive, and oxidize readily in air, some explosively so. The boranes are a remarkable group of compounds in that their structures cannot be described using the conventional two-electron covalent bond model (*see* electron-deficient compound). The simplest example is *diborane* (B_2H_6). Other boranes include B_4H_{10}, B_5H_9, B_5H_{11}, B_6H_{10}, and $B_{10}H_4$. The larger borane molecules have open or closed polyhedra of boron atoms. In addition, there is a wide range of borane derivatives containing atoms of other elements, such as carbon and phosphorus.

borate Any of a wide range of ionic compounds that have negative ions containing boron and oxygen. Lithium borate, for example, contains the simple anion $B(OH)_4^-$. Most borates, however, are inorganic polymers with rings, chains, or other networks based on the planar BO_3 group or the tetrahedral $BO_3(OH)$ group. 'Hydrated'

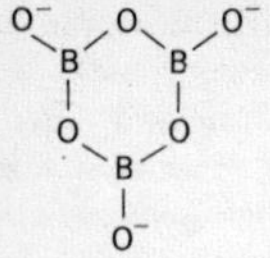

$B_3O_6^{3-}$ as in $Na_3B_3O_6$

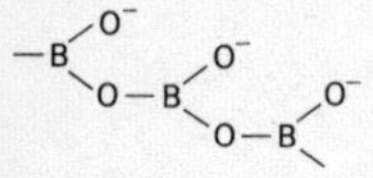

$(BO_2)_n^{n-}$ as in CaB_2O_4

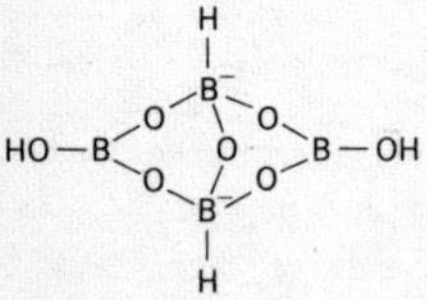

$[B_4O_5(OH)_4]^{2-}$
as in borax $Na_2B_4O_7.10H_2O$

Structure of some typical borate ions

borates are ones containing –OH groups; many examples occur naturally. Anhydrous borates, which contain BO_3 groups, can be made by melting together boric acid and metal oxides.

borax (disodium tetraborate-10-water) A colourless monoclinic solid, $Na_2B_4O_7.10H_2O$, soluble in water and very slightly soluble in ethanol; monoclinic; r.d. 1.73; loses $8H_2O$ at 75°C; loses $10H_2O$ at 320°C. The formula gives a misleading impression of the structure. The compound contains the ion $[B_4O_5(OH)_4]^{2-}$ (*see* borate). Attempts to recrystallize this compound above 60.8°C yield the pentahydrate. The main sources are the borate minerals *kernite* ($Na_2B_4O_7.4H_2O$) and *tincal* ($Na_2B_4O_7.10H_2O$). The ores are purified by carefully controlled dissolution and recrystallization. On treatment with mineral acids borax gives boric acid.

Borax is a very important substance in the glass and ceramics industries as a raw material for making borosilicates. It is also important as a metallurgical flux because of the ability of molten borates to dissolve metal oxides. In solution it partially hydrolyses to boric acid and can thus act as a buffer. For this reason it is used as a laundry pre-soak. It is used medicinally as a mild alkaline antiseptic and astringent for the skin and mucous membranes.

Disodium tetraborate is the source of many industrially important boron compounds, such as barium borate (fungicidal paints), zinc borate (fire-retardant additive in plastics), and boron phosphate (heterogeneous acid catalyst in the petrochemicals industry).

borax–bead test A simple laboratory test for certain metal ions in salts. A small amount of the salt is mixed with borax and a molten bead formed on the end of a piece of platinum wire. Certain metals can be identified by the colour of the bead produced in the oxidizing and reducing parts of a Bunsen flame. For example, iron gives a bead that is red when hot and yellow when cold in the oxidizing flame and a green bead in the reducing flame.

borazon *See* boron nitride.

boric acid Any of a number of acids containing boron and oxygen. Used without qualification the term applies to the compound H_3BO_3 (which is also called *orthoboric acid* or, technically, *trioxoboric (III) acid*). This is a white or colourless solid that is soluble in water and ethanol; triclinic; r.d. 1.435; m.p. 169°C. It occurs naturally in the condensate from volcanic steam vents (suffioni). Commercially, it is made by treating borate minerals (e.g. kernite, $Na_2B_4O_7.4H_2O$) with sulphuric acid followed by recrystallization.

In the solid there is considerable hydrogen bonding between H_3BO_3 molecules resulting in a layer structure, which accounts for the easy cleavage of the crystals. H_3BO_3 molecules also exist in dilute solutions but in more concentrated solutions polymeric acids and ions are formed (e.g. $H_4B_2O_7$; *pyroboric acid* or *tetrahydroxomonoxodiboric (III) acid*). The compound is a very weak acid but also acts as a Lewis *acid in accepting hydroxide ions:

$$B(OH)_3 + H_2O \rightleftharpoons B(OH)_4{}^- + H^+$$

If solid boric acid is heated it loses water and transforms to another acid at 300°C. This is given the formula HBO_2 but is in fact a polymer $(HBO_2)_n$. It is called *metaboric acid* or, technically, *polydioxoboric (III) acid.*

Boric acid is used in the manufacture of glass (borosilicate glass), glazes and enamels, leather, paper, adhesives, and explosives. It is widely used (particularly in the USA) in detergents, and because of the ability of fused boric acid to dissolve other metal oxides it is used as a flux in brazing and welding. Because of its mild antiseptic properties it is used in the pharmaceutical industry and as a food preservative.

boride A compound of boron with a metal. Most metals form at least one boride of the type MB, MB_2, MB_4, MB_6, or MB_{12}. The compounds have a variety of structures; in particular, the hexaborides contain clusters of B_6 atoms. The borides are all hard high-melting materials with metal-like conductivity. They can be made by direct combination of the elements at high temperatures (over 2000°C) or, more usually, by high-temperature reduction of a mixture of the metal oxide and boron oxide using carbon or aluminium. Chemically, they are stable to nonoxidizing acids but are attacked by strong oxidizing agents and by strong alkalis. Magnesium boride (MgB_2) is unusual in that it can be hydrolysed to boranes. Industrially, metal borides are used as refractory materials. The most important are CrB, CrB_2, TiB_2, and ZnB_2. Generally, they are fabricated using high-temperature powder metallurgy, in which the article is produced in a graphite die at over 2000°C and at very high pressure. Items are pressed as near to final shape as possible as machining requires diamond cutters and is extremely expensive.

Born–Haber cycle A cycle of reactions used for calculating the lattice energies of ionic crystalline solids. For a compound MX, the lattice energy is the enthalpy of the reaction

$$M^+(g) + X^-(g) \rightarrow M^+X^-(s) \; \Delta H_L$$

The standard enthalpy of formation of the ionic solid is the enthalpy of the reaction

$$M(s) + \tfrac{1}{2}X_2(g) \rightarrow M^+X^-(s) \; \Delta H_f$$

The cycle involves equating this enthalpy (which can be measured) to the sum of the enthalpies of a number of steps proceeding from the elements to the ionic solid. The steps are:

(1) Atomization of the metal:

$$M(s) \rightarrow M(g) \; \Delta H_1$$

(2) Atomization of the nonmetal:

$$\tfrac{1}{2}X_2(g) \rightarrow X(g) \; \Delta H_2$$

(3) Ionization of the metal:

$$M(g) \rightarrow M^+(g) + e \; \Delta H_3$$

This is obtained from the ionization potential.

(4) Ionization of the nonmetal:

$$X(g) + e \rightarrow X^-(g) \; \Delta H_4$$

This is the electron affinity.

(5) Formation of the ionic solids:

$$M^+(g) + X^-(g) \rightarrow M^+X^-(s) \; \Delta H_L$$

Equating the enthalpies gives:

from which ΔH_L can be found. It is named after the German physicist Max Born (1882–1970) and the chemist Fritz Haber (1868–1934).

bornite An important ore of copper composed of a mixed copper–iron sulphide, Cu_5FeS_4. Freshly exposed surfaces of the mineral are a metallic reddish-brown but a purplish iridescent tarnish soon develops – hence it is popularly known as *peacock ore*. Bornite is mined in Chile, Peru, Bolivia, Mexico, and the USA.

boron Symbol B. An element of *group III of the periodic table; a.n. 5; r.a.m. 10.81; r.d. 2.35; m.p. 2079°C; b.p. 2550°C. It forms two allotropes; amorphous boron is a brown powder but metallic boron is black. The metallic form is very hard (9.3 on Mohs' scale) and is a poor electrical conductor at room temperature. At least three crystalline forms are possible; two are rhombohedral and the other tetragonal. The element is never found free in nature. It occurs as orthoboric acid in volcanic springs in Tuscany, as borates in kernite ($Na_2B_4O_7.4H_2O$), and as colemanite ($Ca_2B_6O_{11}.5H_2O$) in California. Samples usually contain isotopes in the ratio of 19.78% boron-10 to 80.22% boron-11. Extraction is achieved by vapour-phase reduction of boron trichloride with hydrogen on electrically heated filaments. Amorphous boron can be obtained by reducing the trioxide with magnesium powder. Boron when heated reacts with oxygen, halogens, oxidizing acids, and hot alkalis. It is used in semiconductors and in filaments for specialized aerospace applications. Amorphous boron is used in flares, giving a green coloration. The isotope boron-10 is used in nuclear reactor control rods and shields. The element was discovered in 1808 by Sir Humphry Davy and by J. L. Gay-Lussac and L. J. Thenard.

boron carbide A black solid, B_4C, soluble only in fused alkali; it is extremely hard, over 9½ on Mohs' scale; rhombohedral; r.d. 2.52; m.p. 2350°C; b.p. >3500°C. Boron carbide is manufactured by the reduction of boric oxide with petroleum coke in an electric furnace. It is used largely as an abrasive, but objects can also be fabricated using high-temperature powder metallurgy. Boron nitride is also used as a neutron absorber because of its high proportion of boron-10.

boron hydride *See* borane.

boron nitride A solid, BN, insoluble in cold water and slowly decomposed by hot water; r.d. 2.25 (hexagonal); sublimes above 3000°C. Boron nitride is manufactured by heating boron oxide to 800°C on an acid-soluble carrier, such as calcium phosphate, in the presence of nitrogen or ammonia. It is isoelectronic with carbon and, like carbon, it has a very hard cubic form (*borazon*) and a softer hexagonal form; unlike graphite this is a nonconductor. It is used in the electrical industries where its high thermal conductivity and high resistance are of especial value.

boron trichloride A colourless fuming liquid, BCl_3, which reacts with water to give hydrogen chloride and boric acid; r.d. 1.349; m.p. −107°C; b.p. 12.5°C. Boron trichloride is prepared industrially by the exothermic chlorination of boron carbide at above 700°C, followed by fractional distillation. An alternative, but more expensive, laboratory method is the reaction of dry chlorine with boron at high temperature. Boron trichloride is a Lewis *acid, forming stable addition compounds with such donors as ammonia and the amines and is used in the laboratory to promote reactions that liberate these donors. The compound is important industrially as a source of pure boron (reduction with hydrogen) for the electronics industry. It is also used for the preparation of boranes by reaction with metal hydrides.

borosilicate Any of a large number of substances in which BO_3 and SiO_4 units are linked to form networks with a wide range of structures. Borosilicate glasses are particularly important; the addition of boron to the silicate network enables the glass to be fused at lower temperatures than pure silica and also extends the plastic range of the glass. Thus such glasses as Pyrex have a wider range of applications than soda glasses (narrow plastic range, higher thermal expansion) or silica (much higher melting point). Borosilicates are also used in glazes and enamels and in the production of glass wools.

Bosch process *See* Haber process.

Boyle's law The volume (V) of a given mass of gas at a constant temperature is inversely proportional to its pressure (p), i.e. pV = constant. This is true only for an *ideal gas. This law was discovered in 1662 by the Irish physicist Robert Boyle (1627–91). On the continent of Europe it is known as *Mariotte's law* after E. Mariotte (1620–84), who discovered it independently in 1676. *See also* gas laws.

Brackett series *See* hydrogen spectrum.

branched chain *See* chain.

brass A group of alloys consisting of copper and zinc. A typical yellow brass might contain about 67% copper and 33% zinc.

bridge An atom joining two other atoms in a molecule. *See* aluminium chloride; borane.

Brin process A process formerly used for making oxygen by heating barium oxide in air to form the peroxide and then heating the peroxide at higher temperature (>800°C) to produce oxygen

$$2BaO_2 \rightarrow 2BaO + O_2$$

Britannia metal A silvery alloy consisting of 80–90% tin, 5–15% antimony, and sometimes small percentages of copper, lead, and zinc. It is used in bearings and some domestic articles.

British thermal unit (Btu) The Imperial unit of heat, being originally the heat required to raise the temperature of 1lb of water by 1°F. 1 Btu is now defined as 1055.06 joules.

bromate A salt or ester of a bromic acid.

bromic(I) acid (hypobromous acid) A yellow liquid, HBrO. It is a weak acid but a strong oxidizing agent.

bromic(V) acid A colourless liquid, $HBrO_3$, made by adding sulphuric acid to barium bromate. It is a strong acid.

bromide *See* halide.

bromination A chemical reaction in which a bromine atom is introduced into a molecule. *See also* halogenation.

bromine Symbol Br. A *halogen element; a.n. 35; r.a.m. 79.909; r.d. 3.13; m.p. −7.2°C; b.p. 58.78°C. It is a red volatile liquid at room temperature, having a red-brown vapour. Bromine is obtained from brines in the USA (displacement with chlorine); a small amount is obtained from sea water in Anglesey. Large quantities are used to make 1,2-dibromoethane as a petrol additive. It is also used in the manufacture of many other compounds. Chemically, it is intermediate in reactivity between chlorine and iodine. It forms compounds in which it has oxidation states of 1, 3, 5, or 7. The liquid is harmful to human tissue and the vapour irritates the eyes and throat. The element was discovered in 1826 by Antoine Balard.

bromoethane (ethyl bromide) A colourless flammable liquid, C_2H_5Br; r.d. 1.43; m.p. −119°C; b.p. 38.4°C. It is a typical *haloalkane, which can be prepared from ethene and hydrogen bromide. Bromoethane is used as a refrigerant.

bromoform *See* tribromomethane; haloforms.

bromomethane (methyl bromide) A colourless volatile nonflammable liquid, CH_3Br; r.d. 1.73; m.p. −93°C; b.p. 4.5°C. It is a typical *haloalkane.

bromothymol blue An acid–base *indicator that is yellow in acid solutions and blue in alkaline solutions. It changes colour over the pH range 6–8.

bronze Any of a group of alloys of copper and tin, sometimes with lead and zinc present. The amount of tin varies from 1% to 30%. The alloy is hard and easily cast and extensively used in bearings, valves, and other machine parts. Various improved bronzes are produced by adding other elements; for instance, *phosphor bronzes* contain up to 1% phosphorus. In addition certain alloys of copper and metals other than tin are called bronzes – *aluminium bronze* is a mixture of copper and aluminium. Other special bronzes include *bell metal and *gun metal.

Brownian movement The continuous random movement of microscopic solid particles (of about 1 micrometre in diameter) when suspended in a fluid medium. First observed by the botanist Robert Brown (1773–1858) in 1827 when studying pollen particles, it was originally thought to be the manifestation of some vital force. It was later recognized to be a consequence of bombardment of the particles by the continually moving molecules of the liquid. The smaller the particles the more extensive is the motion. The effect is also visible in particles of smoke suspended in a still gas.

brown-ring test A test for ionic nitrates. The sample is dissolved and iron(II) sulphate solution added in a test tube. Concentrated sulphuric acid is then added slowly so that it forms a separate layer. A brown ring (of $Fe(NO)SO_4$) at the junction of the liquids indicates a positive result.

brucite A mineral form of *magnesium hydroxide, $Mg(OH)_2$.

Buchner funnel A type of funnel with an internal perforated tray on which a flat circular filter paper can be placed, used for filtering by suction. It is named after the German chemist Eduard Buchner (1860–1917).

buffer A solution that resists change in pH when an acid or alkali is added or when the solution is diluted. Acidic buffers consist of a weak acid with a salt of the acid. The salt provides the negative ion A^-, which is the conjugate base of the acid HA. An example is carbonic acid and sodium hydrogencarbonate. Basic buffers have a weak base and a salt of the base (to provide the conjugate acid). An example is ammonia solution with ammonium chloride.

In an acidic buffer, for example, molecules HA and ions A^- are present. When acid is added most of the extra protons are removed by the base:

$$A^- + H^+ \rightarrow HA$$

When base is added, most of the extra hydroxide ions are removed by reaction with undissociated acid:

$$OH^- + HA \rightarrow A^- + H_2O$$

Thus, the addition of acid or base changes the pH very little. The hydrogen-ion concentration in a buffer is given by the expression

$$K_a = [H^+] = [A^-]/[HA]$$

i.e. it depends on the ratio of conjugate base to acid. As this is not altered by dilution, the hydrogen-ion concentration for a buffer does not change much during dilution.

In the laboratory, buffers are used to prepare solutions of known stable pH. Natural buffers occur in living organisms, where the biochemical reactions are very sensitive to change in pH. The main natural buffers are H_2CO_3/HCO_3^- and $H_2PO_4^-/HPO_4^{2-}$. Buffer solutions are also used in medicine (e.g. in intravenous injections), in agriculture, and in many industrial processes (e.g. dyeing, fermentation processes, and the food industry).

bumping Violent boiling of a liquid caused by superheating so that bubbles form at a pressure above atmospheric pressure. It can be prevented by putting pieces of porous pot in the liquid to enable bubbles of vapour to form at the normal boiling point.

Bunsen burner A laboratory gas burner having a vertical metal tube into which the gas is led, with a hole in the side of the base of the tube to admit air. The amount of air can be regulated by a sleeve on the tube. When no air is admitted the flame is luminous and smoky. With air, it has a faintly visible hot outer part (the oxidizing part) and an inner blue cone where combustion is incomplete (the cooler reducing part of the flame). The device is named after the German chemist Robert Bunsen (1811–99), who used a similar device (without a regulating sleeve) in 1855.

Bunsen cell A *primary cell consisting of a zinc cathode immersed in dilute sulphuric acid and a carbon anode immersed in concentrated nitric acid. The electrolytes are separated by a porous pot. The cell gives an e.m.f. of about 1.9 volts.

burette A graduated glass tube with a tap at one end leading to a fine outlet tube, used for delivering known volumes of a liquid (e.g. in titration).

buta-1,3-diene (butadiene) A colourless gaseous hydrocarbon, $CH_2{:}CHCH{:}CH_2$; m.p. 109°C; b.p. −4.5°C. It is made by catalytic dehydrogenation of butane (from petroleum or natural gas) and polymerized in the production of synthetic rubbers. The compound is a conjugated *diene in which the electrons in the pi orbitals are partially delocalized over the whole molecule. It can have trans and cis forms, the latter taking part in *Diels–Alder reactions.

butanal (butyraldehyde) A colourless flammable liquid aldehyde, C_3H_9CHO; r.d. 0.8; m.p. −109°C; b.p. 75.7°C.

butane A gaseous hydrocarbon, C_4H_{10}; d. 0.58 g cm^{-3}; m.p. −135°C; b.p. 0°C. Butane is obtained from petroleum (from refinery gas or by cracking higher hydrocarbons). The fourth member of the *alkane series, it has a straight chain of carbon atoms and is isomeric with 2-methylpropane ($CH_3CH(CH_3)CH_3$, formerly called *isobutane*). It can easily be liquefied under pressure and is supplied in cylinders for use as a fuel gas. It is also a raw material for making buta-1,3-diene (for synthetic rubber).

butanedioic acid (succinic acid) A colourless crystalline fatty acid, $(CH_2)_2(COOH)_2$; r.d. 1.5; m.p. 185°C; b.p. 235°C. A weak carboxylic acid occurring in certain plants, it is produced by fermentation of sugar or ammonium tartrate and used as a sequestrant and in making dyes.

butanoic acid (butyric acid) A colourless liquid water-soluble acid, C_3H_7COOH; r.d. 0.96; b.p. 163°C. It is a weak acid (K_a = 1.5×10^{-5} mol dm^{-3} at 25°C) with a rancid odour. Its esters are present in butter and in human perspiration. The acid is used to make esters for flavourings and perfumery.

butanol Either of two aliphatic alcohols with the formula C_4H_9OH. *Butan-1-ol*, $CH_3(CH_2)_3OH$, is a primary alcohol; r.d. 0.81; m.p. 89.5°C; b.p. 117.3°C. *Butan-2-ol*, $CH_3CH(OH)C_2H_5$, is a secondary alcohol; r.d. 0.81; m.p. −114.7°C; b.p. 100°C. Both are colourless volatile liquids obtained from butane and used as solvents.

butanone (methyl ethyl ketone) A colourless flammable water-soluble liquid, $CH_3COC_2H_5$; r.d. 0.8; m.p. −86.4°C; b.p. 77.6°C. It can be made by the catalytic oxidation of butane and is used as a solvent.

butenedioic acid Either of two isomers with the formula HCOOHC:CHCOOH. Both compounds can be regarded as derivatives of ethene in which a hydrogen atom on each carbon has been replaced by a –COOH group. The compounds show cis–trans isomerism. The trans form is *fumaric acid* (r.d. 1.64; sublimes at 200°C) and the cis form is *maleic acid* (r.d. 1.59; m.p. 130°C). Both are colourless crystalline compounds used in making synthetic resins. The cis form is rather less stable than the trans form and converts to the trans form at 120°C. Unlike the trans form it can eliminate water on heating to form a cyclic anhydride containing a –CO.O.CO– group (*maleic anhydride*). Fumaric acid is an intermediate in the Krebs cycle.

butyl group The organic group $CH_3(CH_2)_3$–.

butyl rubber A type of synthetic rubber obtained by copolymerizing 2-methylpropene (CH_2:C(CH_3)CH_3; isobutylene) and methylbuta-1,3-diene (CH_2:C(CH_3)CH:CH_2, isoprene). Only small amounts of isoprene (about 2 mole %) are used. The rubber can be vulcanized. Large amounts were once used for tyre inner tubes.

butyraldehyde *See* butanal.

butyric acid *See* butanoic acid.

by-product A compound formed during a chemical reaction at the same time as the main product. Commercially useful by-products are obtained from a number of industrial processes. For example, calcium chloride is a by-product of the *Solvay process for making sodium carbonate. Propanone is a by-product in the manufacture of *phenol.

C

cadmium Symbol Cd. A soft bluish metal belonging to *group IIB of the periodic table; a.n. 48; r.a.m. 112.41; r.d. 8.65; m.p. 320.9°C; b.p. 765°C. The element's name is derived from the ancient name for calamine, zinc carbonate $ZnCO_3$, and it is usually found associated with zinc ores, such as sphalerite (ZnS), but does occur as the mineral greenockite (CdS). Cadmium is usually produced as an associate product when zinc, copper, and lead ores are reduced. Cadmium is used in low-melting-point alloys to make solders, in Ni–Cd batteries, in bearing alloys, and in electroplating (over 50%). Cadmium compounds are used as phosphorescent coatings in TV tubes. Cadmium and its compounds are extremely toxic at low concentrations; great care is essential where solders are used or where fumes are emitted. It has similar chemical properties to zinc but shows a greater tendency towards complex formation. The element was discovered in 1817 by F. Stromeyer.

cadmium cell *See* Weston cell.

cadmium sulphide A water-insoluble compound, CdS; r.d. 4.82. It occurs naturally as the mineral *greenockite* and is used as a pigment and in semiconductors and fluorescent materials.

caesium Symbol Cs. A soft silvery-white metallic element belonging to *group I of the periodic table; a.n. 55; r.a.m. 132.905; r.d. 1.88; m.p. 28.4°C; b.p. 678°C. It occurs in small amounts in a number of minerals, the main source being carnallite (KCl.$MgCl_2$.$6H_2O$). It is obtained by electrolysis of molten caesium cyanide. The natural isotope is caesium–133. There are 15 other radioactive isotopes. Caesium–137 (half-life 33 years) is used as a gamma source. As the heaviest alkali metal, caesium has the lowest ionization potential of all elements, hence its use in photoelectric cells, etc.

calcination The formation of a calcium carbonate deposit from hard water. *See* hardness of water.

calcinite A mineral form of *potassium hydrogencarbonate, $KHCO_3$.

calcite One of the most common and widespread minerals, consisting of crystalline calcium carbonate, $CaCO_3$. Calcite crystallizes in the rhombohedral system; it is usually colourless or white and has a hardness of 3 on the Mohs' scale. It has the property of double refraction, which is apparent in Iceland spar – the transparent variety of calcite. It is an important rock-forming mineral and is a major constituent in limestones, marbles, and carbonatites.

calcium Symbol Ca. A soft grey metallic element belonging to *group II of the periodic table; a.n. 20; r.a.m. 40.08; r.d. 1.55; m.p. 840°C; b.p. 1484°C. Calcium compounds are common in the earth's crust; e.g. limestone and marble ($CaCO_3$), gypsum ($CaSO_4.2H_2O$), and fluorite (CaF_2). The element is extracted by electrolysis of fused calcium chloride and is used as a getter in vacuum systems and a deoxidizer in producing nonferrous alloys. It is also used as a reducing agent in the extraction of such metals as thorium, zirconium, and uranium.

Calcium is an essential element for living organisms, being required for normal growth and development.

calcium acetylide *See* calcium dicarbide.

calcium bicarbonate *See* calcium hydrogencarbonate.

calcium carbide *See* calcium dicarbide.

calcium carbonate A white solid, $CaCO_3$, which is only sparingly soluble in water. Calcium carbonate decomposes on heating to give *calcium oxide (quicklime) and carbon dioxide. It occurs naturally as the minerals *calcite (rhombohedral; r.d. 2.71) and *aragonite (rhombic; r.d. 2.93). Rocks containing calcium carbonate dissolve slowly in acidified rainwater (containing dissolved CO_2) to cause temporary hardness. In the laboratory, calcium carbonate is precipitated from *limewater by carbon dioxide. Calcium carbonate is used in making lime (calcium oxide) and is the main raw material for the *Solvay process.

calcium chloride A white deliquescent compound, $CaCl_2$, which is soluble in water; r.d. 2.15; m.p. 772°C; b.p. 7600°C. There are a number of hydrated forms, including the monohydrate, $CaCl_2.H_2O$, the dihydrate, $CaCl_2.2H_2O$ (r.d. 0.84), and the hexahydrate, $CaCl_2.6H_2O$ (trigonal; r.d. 1.71; the hexahydrate loses $4H_2O$ at 30°C and the remaining $2H_2O$ at 200°C). Large quantities of it are formed as a byproduct of the *Solvay process and it can be prepared by dissolving calcium carbonate or calcium oxide in hydrochloric acid. Crystals of the anhydrous salt can only be obtained if the hydrated salt is heated in a stream of hydrogen chloride. Solid calcium chloride is used in mines and on roads to reduce dust problems, whilst the molten salt is the electrolyte in the extraction of calcium. An aqueous solution of calcium chloride is used in refrigeration plants.

calcium cyanamide A colourless solid, $CaCN_2$, which sublimes at 1150°C. It is prepared by heating calcium dicarbide at 800°C in a stream of nitrogen:

$$CaC_2(s) + N_2(g) \rightarrow CaCN_2(s) + C(s)$$

The reaction has been used as a method of fixing nitrogen in countries in which cheap electricity is available to make the calcium dicarbide (the *cyanamide process*). Calcium cyanamide can be used as a fertilizer because it reacts with water to give ammonia and calcium carbonate:

$$CaCN_2(s) + 3H_2O(l) \rightarrow CaCO_3(s) + 2NH_3(g)$$

It is also used in the production of melamine, urea, and certain cyanide salts.

calcium dicarbide (calcium acetylide; calcium carbide; carbide) A colourless solid compound, CaC_2; tetragonal; r.d. 2.22; m.p. 450°C; b.p. 2300°C. In countries in which electricity is cheap it is manufactured by heating calcium oxide with either coke or ethyne at temperatures above 2000°C in an electric arc furnace. The crystals consist of Ca^{2+} and C_2^{-} ions arranged in a similar way to the ions in sodium chloride. When water is added to calcium dicarbide, the important organic raw material ethyne (acetylene) is produced:

$$CaC_2(s) + 2H_2O(l) \rightarrow Ca(OH)_2(s) + C_2H_2(g)$$

calcium fluoride A white crystalline solid, CaF_2; r.d. 3.2; m.p. 1360°C; b.p. 2500°C. It occurs naturally as the mineral *fluorite (or fluorspar) and is the main source of fluorine. The *calcium fluoride structure* (*fluorite structure*) is a crystal structure in which the calcium ions are each surrounded by eight fluoride ions arranged at the corners of a cube. Each fluoride ion is surrounded by four calcium ions at the corners of a tetrahedron.

calcium hydrogencarbonate (calcium bicarbonate) A compound, $Ca(HCO_3)_2$, that is stable only in solution and is formed when water containing carbon dioxide dissolves calcium carbonate:

$$CaCO_3(s) + H_2O(l) + CO_2(g) \rightarrow Ca(HCO_3)_2(aq)$$

It is the cause of temporary *hardness in water, because the calcium ions react with soap to give scum. Calcium hydrogencarbonate is unstable when heated and decomposes to give solid calcium carbonate. This explains why temporary hardness is removed by boiling and the formation of 'scale' in kettles and boilers.

calcium hydroxide (slaked lime) A white solid, $Ca(OH)_2$, which dissolves sparingly in water (*see* limewater); hexagonal; r.d. 2.24. It is manufactured by adding water to calcium oxide, a process that evolves much heat and is known as slaking. It is used as a cheap alkali to neutralize the acidity in certain soils and in the manufacture of mortar, whitewash, bleaching powder, and glass.

calcium nitrate A white deliquescent compound, $Ca(NO_3)_2$, that is very soluble in water; cubic; r.d. 2.50; m.p. 561°C. It can be prepared by neutralizing nitric acid with calcium carbonate and crystallizing it from solution as the tetrahydrate $Ca(NO_3)_2.4H_2O$, which exists in two monoclinic crystalline forms (α, r.d. 1.9; β, r.d. 1.82). There is also a trihydrate, $Ca(NO_3)_2.3H_2O$. The anhydrous salt can be obtained from the hydrate by heating but it decomposes on strong heating to give the oxide, nitrogen dioxide, and oxygen. Calcium nitrate is sometimes used as a nitrogenous fertilizer.

calcium octadecanoate (calcium stearate) An insoluble white salt, $Ca(CH_3(CH_2)_{16}COO)_2$, which is formed when soap is mixed with water containing calcium ions and is the scum produced in hard-water regions.

calcium oxide (quicklime) A white solid compound, CaO, formed by heating calcium in oxygen or by the thermal decomposition of calcium carbonate; cubic; r.d. 3.35; m.p. 2600°C; b.p. 2850°C. On a large scale, calcium carbonate in the form of limestone is heated in a tall tower (lime kiln) to a temperature above 550°C:

$$CaCO_3(s) \leftrightharpoons CaO(s) + CO_2(g)$$

Although the reaction is reversible, the carbon dioxide is carried away by the upward current through the kiln and all the limestone decomposes. Calcium oxide is used to make calcium hydroxide, as a cheap alkali for treating acid soil, and in extractive metallurgy to produce a slag with the impurities (especially sand) present in metal ores.

calcium phosphate(V) A white insoluble powder, $Ca_3(PO_4)_2$; r.d. 3.14. It is found naturally in the mineral *apatite ($Ca_5(PO_4)_3(OH,F,Cl)$) and as rock phosphate. It is also the main constituent of animal bones. Calcium phosphate can be prepared by mixing solutions containing calcium ions and hydrogenphosphate ions in the presence of an alkali:

$$HPO_4^{2-} + OH^- \rightarrow PO_4^{3-} + H_2O$$

$$3Ca^{2+} + 2PO_4^{3-} \rightarrow Ca_3(PO_4)_2$$

It is used extensively as a fertilizer. The compound was formerly called *calcium orthophosphate* (*see* phosphate).

calcium stearate *See* calcium octadecanoate.

calcium sulphate A white solid compound, $CaSO_4$; r.d. 2.96; 1450°C. It occurs naturally as the mineral *anhydrite, which has a rhombic structure, transforming to a monoclinic form at 200°C. More commonly, it is found as the dihydrate, *gypsum, $CaSO_4.2H_2O$ (monoclinic; r.d. 2.32). When heated, gypsum loses water at 128°C to give the hemihydrate, $2CaSO_4.H_2O$, better known as *plaster of Paris. Calcium sulphate is sparingly soluble in water and is a cause of permanent *hardness of water. It is used in the manufacture of certain paints, ceramics, and paper. The naturally occurring forms are used in the manufacture of sulphuric acid.

Calgon Tradename for a water-softening agent. *See* hardness of water.

caliche A mixture of salts found in deposits between gravel beds in the Atacama and Tarapaca regions of Chile. They vary from 4 m to 15 cm thick and were formed by periodic leaching of soluble salts during wet geological epochs, followed by drying out of inland seas in dry periods. They are economically important as a source of nitrates. A typical composition is $NaNO_3$ 17.6%, NaCl 16.1%, Na_2SO_4 6.5%, $CaSO_4$ 5.5%, $MgSO_4$ 3.0%, KNO_3 1.3%, $Na_2B_4O_7$ 0.94%, $KClO_3$ 0.23%, $NaIO_3$ 0.11%, sand and gravel to 100%.

californium Symbol Cf. A radioactive metallic transuranic element belonging to the *actinoids; a.n. 98; mass number of the most stable isotope 251 (half-life about 700 years). Nine isotopes are known; californium–252 is an intense neutron source, which makes it useful in neutron *activation analysis and potentially useful as a radiation source in medicine. The element was first produced by G. T. Seaborg and associates in 1950.

calomel *See* mercury(I) chloride.

calomel half cell (calomel electrode) A type of half cell in which the electrode is mercury coated with calomel (HgCl) and the electrolyte is a solution of potassium chloride and saturated calomel. The standard electrode potential is −0.2415 volt (25°C). In the calomel half cell the reactions are

$$HgCl(s) \rightleftharpoons Hg^+(aq) + Cl^-(aq)$$

$$Hg^+(aq) + e \rightleftharpoons Hg(s)$$

The overall reaction is

$$HgCl(s) + e \rightleftharpoons Hg(s) + Cl^-(aq)$$

This is equivalent to a $Cl_2(g)|Cl^-(aq)$ half cell in which the pressure is the dissociation pressure of HgCl.

calorie The quantity of heat required to raise the temperature of 1 gram of water by 1°C (1 K). The calorie, a c.g.s. unit, is now largely replaced by the *joule, an *SI unit. 1 calorie = 4.186 8 joules.

Calorie (kilogram calorie; kilocalorie) 1000 calories. This unit is still in limited use in estimating the energy value of foods, but is obsolescent.

calorific value The heat per unit mass produced by complete combustion of a given substance. Calorific values are used to express the energy values of fuels; usually these are expressed in megajoules per kilogram ($MJ\,kg^{-1}$). They are also used to measure the energy content of foodstuffs; i.e. the energy produced when the food is oxidized in the body. The units here are kilojoules per gram ($kJ\,g^{-1}$), although Calories (kilocalories) are often still used in nontechnical contexts. Calorific values are measured using a *bomb calorimeter.

calx A metal oxide formed by heating an ore in air.

camphor A white crystalline cyclic ketone, $C_{10}H_{16}O$; r.d. 0.99; m.p. 179°C; b.p. 204°C. It was formerly obtained from the wood of the Formosan camphor tree, but can now be synthesized. The compound has a characteristic odour associated with its use in mothballs. It is a plasticizer in celluloid.

Canada balsam A yellow-tinted resin used for mounting specimens in optical microscopy. It has similar optical properties to glass.

candela Symbol Cd. The *SI unit of luminous intensity equal to the luminous intensity in the perpendicular direction of the *black-body radiation from an area of 1/600 000 square metre at the temperature of freezing platinum under a pressure of 101 325 pascals.

cane sugar *See* sucrose.

Cannizzaro reaction A reaction of aldehydes to give carboxylic acids and alcohols. It occurs in the presence of strong bases with aldehydes that do not have alpha hydrogen atoms. For example, benzenecarbaldehyde gives benzenecarboxylic acid and benzyl alcohol:

$$2C_6H_5CHO \rightarrow C_6H_5COOH + C_6H_5CH_2OH$$

Aldehydes that have alpha hydrogen atoms undergo the *aldol reaction instead. The Cannizzaro reaction is an example of a *disproportionation. It was discovered in 1853 by the Italian chemist Stanislao Cannizzaro (1826–1910).

canonical form One of the possible structures of a molecule that together form a *resonance hybrid.

capric acid *See* decanoic acid.

caproic acid *See* hexanoic acid.

caprolactam A white crystalline substance, $C_6H_{11}NO$; r.d. 1.02; m.p. 170°C; b.p. 150°C. It is a *lactam containing the –NH.CO– group with five CH_2 groups making up the rest of the seven-membered ring. Caprolactam is used in making *nylon.

caprylic acid *See* octanoic acid.

carat 1. A measure of fineness (purity) of gold. Pure gold is described as 24-carat gold. 14-carat gold contains 14 parts in 24 of gold, the remainder usually being copper. **2.** A unit of mass equal to 0.200 gram, used to measure the masses of diamonds and other gemstones.

carbamide *See* urea.

carbanion An organic ion with a negative charge on a carbon atom; i.e. an ion of the type R_3C^-. Carbanions are intermediates in certain types of organic reaction (e.g. the *aldol reaction).

carbene A species of the type R_2C:, in which the carbon atom has two electrons that do not form bonds. *Methylene*, :CH_2, is the simplest example. Carbenes are highly reactive and exist only as transient intermediates in certain organic reactions. They attack double bonds to give cyclopropane derivatives. They also cause insertion reactions, in which the carbene group is inserted between the carbon and hydrogen atoms of a C–H bond:

$$C{-}H + {:}CR_2 \rightarrow C{-}CR_2{-}H$$

carbide Any of various compounds of carbon with metals or other more electropositive elements. True carbides contain the ion C^{4-} as in Al_4C_3. These are saltlike compounds giving methane on hydrolysis, and were formerly called *methanides*. Compounds containing the ion $C_2^{\ 2-}$ are also saltlike and are known as *dicarbides*. They yield ethyne (acetylene) on hydrolysis and were formerly called *acetylides*. The above types of compound are ionic but have partially covalent bond character, but boron and silicon form true covalent carbides, with giant molecular structures. In addition, the transition metals form a range of interstitial carbides in which the carbon atoms occupy interstitial positions in the metal lattice. These substances are generally hard materials with metallic conductivity. Some transition metals (e.g. Cr, Mn, Fe, Co, and Ni) have atomic radii that are too small to allow individual carbon atoms in the interstitial holes. These form carbides in which the metal lattice is distorted and chains of carbon atoms exist (e.g. Cr_3C_2, Fe_3C). Such compounds are intermediate in character between interstitial carbides and ionic carbides. They give mixtures of hydrocarbons on hydrolysis with water or acids.

carbocyclic *See* cyclic.

carbohydrate One of a group of organic compounds based on the general formula $C_x(H_2O)_y$. The simplest carbohydrates are the *sugars (saccharides), including glucose and sucrose. *Polysaccharides are carbohydrates of much greater molecular weight and complexity; examples are starch, glycogen, and cellulose. Carbohydrates perform many vital roles in living organisms. Sugars, notably glucose, and their derivatives are essential intermediates in the conversion of food to energy. Starch and other polysaccharides serve as energy stores in plants. Cellulose, lignin, and others form the supporting cell walls and woody tissue of plants. Chitin is a structural polysaccharide found in the body shells of many invertebrate animals.

carbolic acid *See* phenol.

carbon Symbol C. A nonmetallic element belonging to *group IV of the periodic table; a.n. 6; r.a.m. 12.011; m.p. ~3550°C; b.p. ~4289°C. Carbon has two main allotropic forms (*see* allotropy).

*Diamond (r.d. 3.52) occurs naturally and small amounts can be produced synthetically. It is extremely hard and has highly refractive crystals. The hardness of diamond results from the covalent crystal structure, in which each carbon atom is linked by covalent bonds to four others situated at the corners of a tetrahedron. The C–C bond length is 0.154 nm and the bond angle is 109.5°.

Graphite (r.d. 2.25), the other allotrope, is a soft black slippery substance (sometimes called *black lead* or *plumbago*). It occurs naturally and can also be made by the *Acheson process. In graphite the carbon atoms are arranged in layers, in which each carbon atom is surrounded by three others to which it is bound by single or double bonds. The layers are held together by much weaker van der Waals' forces. The carbon–carbon bond length in the layers is 0.142 nm and the layers are 0.34 nm apart. Graphite is a good conductor of heat and electricity. It has a variety of uses including electrical contacts, high-temperature equipment, and as a solid lubricant. Graphite mixed with clay is the 'lead' in pencils (hence its alternative name). There are also several amorphous forms of carbon, such as *carbon black and *charcoal.

There are two stable isotopes of carbon (proton numbers 12 and 13) and four radioactive ones (10, 11, 14, 15). Carbon–14 is used in *carbon dating. Chemically, carbon is unique in its ability to form many compounds containing chains and rings of carbon atoms.

carbonate A salt of carbonic acid containing the carbonate ion, CO_3^{2-}. The free ion has a plane triangular structure. Metal carbonates may be ionic or may contain covalent metal–carbonate bonds (complex carbonates) via one or two oxygen atoms. The carbonates of the alkali metals are all soluble but other carbonates are insoluble; they all react with mineral acids to release carbon dioxide.

carbonate minerals A group of common rock-forming minerals containing the anion CO_3^{2-} as the fundamental unit in their structure. The most important carbonate minerals are *calcite, *dolomite, and *magnesite. *See also* aragonite.

carbonation The solution of carbon dioxide in a liquid under pressure.

carbon bisulphide *See* carbon disulphide.

carbon black A fine carbon powder made by burning hydrocarbons in insufficient air. It is used as a pigment and a filler (e.g. for rubber).

carbon cycle One of the major cycles of chemical elements in the environment. Carbon (as carbon dioxide) is taken up from the atmosphere and incorporated into the tissues of plants in photosynthesis. It may then pass into the bodies of animals as the plants are eaten. During the respiration of plants, animals, and organisms that bring about decomposition, carbon dioxide is returned to the atmosphere. The combustion of fossil fuels (e.g. coal and peat) also releases carbon dioxide into the atmosphere.

carbon dating (radiocarbon dating) A method of estimating the ages of archaeological specimens of biological origin. As a result of cosmic radiation a small number of atmospheric nitrogen nuclei are continuously being transformed by neutron bombardment into radioactive nuclei of carbon–14:

$$^{14}_{7}N + n \rightarrow {}^{14}_{6}C + p$$

Some of these radiocarbon atoms find their way into living trees and other plants in the form of carbon dioxide, as a result of photosynthesis. When the tree is cut down photosynthesis stops and the ratio of radiocarbon atoms to stable carbon atoms begins to fall as the radiocarbon decays. The ratio $^{14}C/^{12}C$ in the specimen can be measured and enables the time that has elapsed since the tree was cut down to be calculated. The method has been shown to give consistent results for specimens up to some 40 000 years old, though its accuracy depends upon assumptions concerning the past intensity of the cosmic radiation. The technique was developed by Willard F. Libby (1908–80) and his coworkers in 1946–47.

carbon dioxide A colourless odourless gas, CO_2, soluble in water, ethanol, and acetone; d. 1.977 g dm^{-3} (0°C); m.p. −56.6°C; b.p. −78.5°C. It occurs in the atmosphere (0.03% by volume) but has a short residence time in this phase as it is both consumed by plants during photosynthesis and produced by respiration and by combustion. It is readily prepared in the laboratory by the action of dilute acids on metal carbonates or of heat on heavy-metal carbonates. Carbon dioxide is a byproduct from the manufacture of lime and from fermentation processes.

Carbon dioxide has a small liquid range and liquid carbon dioxide is produced only at high pressures. The molecule CO_2 is linear with each oxygen making a double bond to the carbon. Chemically, it is unreactive and will not support combustion. It dissolves in water to give *carbonic acid.

Large quantities of solid carbon dioxide (*dry ice*) are used in processes requiring large-scale refrigeration. It is also used in

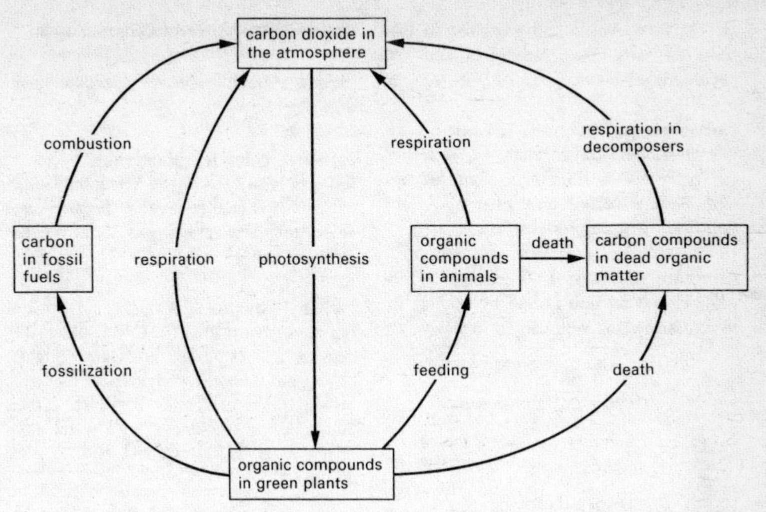

The carbon cycle in nature

fire extinguishers as a desirable alternative to water for most fires, and as a constituent of medical gases as it promotes exhalation. It is also used in carbonated drinks.

The level of carbon dioxide in the atmosphere has been the subject of much environmental controversy as it is argued that extensive burning of fossil fuels will increase the overall CO_2 concentration and then, by the greenhouse effect, increase atmospheric temperatures and cause climatic modification. The full significance of all the factors remains to be established.

carbon disulphide (carbon bisulphide) A colourless highly refractive liquid, CS_2, slightly soluble in water and soluble in ethanol and ether; r.d. 1.261; m.p. −110°C; b.p. 46.3°C. Pure carbon disulphide has an ethereal odour but the commercial product is contaminated with a variety of other sulphur compounds and has a most unpleasant smell. It was previously manufactured by heating a mixture of wood, sulphur, and charcoal; modern processes use natural gas and sulphur. Carbon disulphide is an excellent solvent for oils, waxes, rubber, sulphur, and phosphorus, but its use is decreasing because of its high toxicity and its flammability. It is used for the preparation of xanthates in the manufacture of viscose yarns.

carbon fibres Fibres of carbon in which the carbon has an oriented crystal structure. Carbon fibres are made by heating textile fibres and are used in strong composite materials for use at high temperatures.

carbonic acid A dibasic acid, H_2CO_3, formed in solution when carbon dioxide is dissolved in water:

$$CO_2(aq) + H_2O(l) \rightleftharpoons H_2CO_3(aq)$$

The acid is in equilibrium with dissolved carbon dioxide, and also dissociates as follows:

$$H_2CO_3 \rightleftharpoons H^+ + HCO_3^-$$
$$K_a = 4.5 \times 10^{-7}\ \text{mol dm}^{-3}$$
$$HCO_3^- \rightleftharpoons CO_3^{2-} + H^+$$
$$K_a = 4.8 \times 10^{-11}\ \text{mol dm}^{-3}$$

The pure acid cannot be isolated, although it can be produced in ether solution at $-30°C$. Carbonic acid gives rise to two series of salts: the *carbonates and the *hydrogencarbonates.

carbonium ion An organic ion with a positive charge on a carbon atom; i.e. an ion of the type R_3C^+. Carbonium ions are intermediates in certain types of organic reaction (e.g. *Williamson's synthesis).

carbonize (carburize) To change an organic compound into carbon by heating, or to coat something with carbon in this way.

C≡O	carbon monoxide
O=C=O	carbon dioxide
O=C=C=C=O	tricarbon dioxide (carbon suboxide)

Oxides of carbon

carbon monoxide A colourless odourless gas, CO, sparingly soluble in water and soluble in ethanol and benzene; d. 1.25 g dm^{-3} (0°C); m.p. $-199°C$; b.p. $-191.5°C$. It is flammable and highly toxic. In the laboratory it can be made by the dehydration of methanoic acid (formic acid) using concentrated sulphuric acid. Industrially it is produced by the oxidation of natural gas (methane) or (formerly) by the water-gas reaction. It is formed by the incomplete combustion of carbon and is present in car-exhaust gases.

It is a neutral oxide, which burns in air to give carbon dioxide, and is a good reducing agent, used in a number of metallurgical processes. It has the interesting chemical property of forming a range of transition metal carbonyls, e.g. $Ni(CO)_4$. Carbon monoxide is able to use vacant *p*-orbitals in bonding with metals; the stabilization of low oxidation states, including the zero state, is a consequence of this. This also accounts for its toxicity, which is due to the binding of the CO to the iron in haemoglobin, thereby blocking the uptake of oxygen.

carbon suboxide *See* tricarbon dioxide.

carbon tetrachloride *See* tetrachloromethane.

carbonyl chloride (phosgene) A colourless gas, $COCl_2$, with an odour of freshly cut hay. It is used in organic chemistry as a chlorinating agent, and was formerly used as a war gas.

carbonyl compound A compound containing the carbonyl group >C=O. Aldehydes, ketones, and carboxylic acids are examples of organic carbonyl compounds. Inorganic carbonyls are complexes in which carbon monoxide has coordinated to a metal atom or ion, as in *nickel carbonyl, $Ni(CO)_4$. *See also* ligand.

carbonyl group The group >C=O, found in aldehydes, ketones, carboxylic acids, amides, etc., and in inorganic carbonyl complexes (*see* carbonyl compound).

carborundum *See* silicon carbide.

carboxyl group The organic group $-COOH$, present in *carboxylic acids.

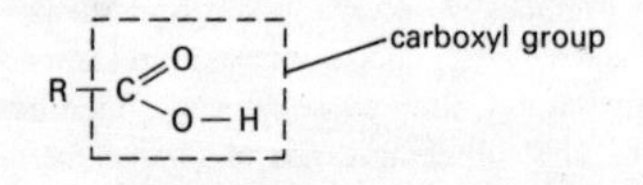

Carboxylic acid structure

carboxylic acids Organic compounds containing the group $-CO.OH$ (the *carboxyl group*; i.e. a carbonyl group attached to a hydroxyl group). In systematic chemical nomenclature carboxylic-acid names end in the suffix *-oic*, e.g. ethanoic acid, CH_3COOH. They are generally weak acids. Many long-chain carboxylic acids occur naturally as esters in fats and oils and are therefore also known as *fatty acids. *See also* glycerides.

carburize *See* carbonize.

carbylamine reaction *See* isocyanide test.

carcinogen Any agent that produces cancer, e.g. tobacco smoke, certain industrial chemicals, and ionizing radiation (such as X-rays and ultraviolet rays).

Carius method A method of determining the amount of sulphur and halogens in an organic compound, by heating the compound in a sealed tube with silver nitrate in concentrated nitric acid. The compound is decomposed and silver sulphide and halides are precipitated, separated, and weighed.

carnallite A mineral consisting of a hydrated mixed chloride of potassium and magnesium, $KCl.MgCl_2.6H_2O$.

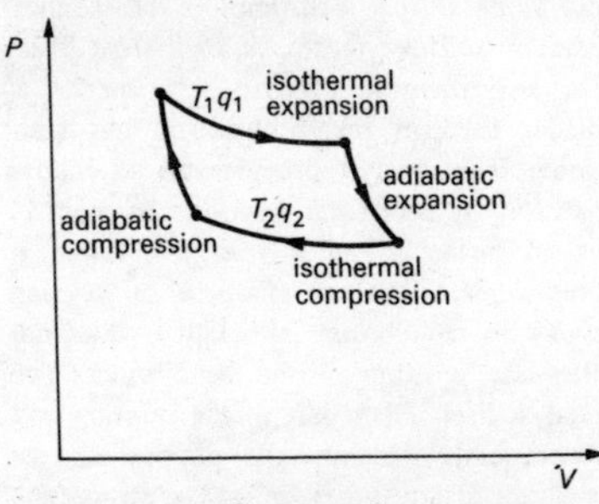

Carnot cycle

Carnot cycle The most efficient cycle of operations for a reversible heat engine. Published in 1824 by the French physicist N. L. S. Carnot (1746–1832), it consists of four operations on the working substance in the engine:

a. Isothermal expansion at thermodynamic temperature T_1 with heat Q_1 taken in.
b. Adiabatic expansion with a fall of temperature to T_2.
c. Isothermal compression at temperature T_2 with heat Q_2 given out.
d. Adiabatic compression with a rise of temperature back to T_1.

According to the *Carnot principle*, the efficiency of any reversible heat engine depends only on the temperature range through which it works, rather than the properties of the working substances. In any reversible engine, the efficiency (η) is the ratio of the work done (W) to the heat input (Q_1), i.e. $\eta = W/Q_1$. As, according to the first law of *thermodynamics, $W = Q_1 - Q_2$, it follows that $\eta = (Q_1 - Q_2)/Q_1$. As the system is reversible, $Q_1/Q_2 = T_1/T_2$, therefore $\eta = (T_1 - T_2)/T_1$. For maximum efficiency T_1 should be as high as possible and T_2 as low as possible.

carnotite A radioactive mineral consisting of hydrated uranium potassium vanadate, $K_2(UO_2)_2(VO_4)_2.nH_2O$. It varies in colour from bright yellow to lemon- or greenish-yellow. It is a source of uranium, radium, and vanadium. The chief occurrences are in the Colorado Plateau, USA; Radium Hill, Australia; and Katanga, Zaïre.

Caro's acid *See* peroxosulphuric(VI) acid.

carotene A member of a class of *carotenoid pigments. Examples are β-carotene and lycopene, which colour carrot roots and ripe tomato fruits respectively. α- and β-carotene yield vitamin A when they are broken down during animal digestion.

carotenoid Any of a group of yellow, orange, red, or brown plant pigments chemically related to terpenes. Carotenoids are responsible for the characteristic colour of many plant organs, such as ripe tomatoes, carrots, and autumn leaves. They also function in the light reactions of photosynthesis.

carrier gas The gas that carries the sample in *gas chromatography.

cascade liquefier An apparatus for liquefying a gas of low *critical temperature. Another gas, already below its critical temperature, is liquified and evaporated at a reduced pressure in order to cool the first gas to below its critical temperature. In practice a series of steps is often used, each step enabling the critical temperature of the next gas to be reached.

cascade process Any process that takes place in a number of steps, usually because the single step is too inefficient to produce the desired result. For example, in various uranium-enrichment processes the separation of the desired isotope is only poorly achieved in a single stage; to achieve better separation the process has to be repeated a number of times, in a series, with the enriched fraction of one stage being fed to the succeeding stage for further enrichment. Another example of cascade process is that operating in a *cascade liquefier.

case hardening The hardening of the surface layer of steel, used for tools and certain mechanical components. The commonest method is to carburize the surface layer by heating the metal in a hydrocarbon or by dipping the red hot metal into molten sodium cyanide. Diffusion of nitrogen into the surface layer to form nitrides is also used.

casein One of a group of phosphate-containing proteins (phosphoproteins) found in milk. Caseins are easily digested by the enzymes of young mammals and represent a major source of phosphorus.

cassiterite A yellow, brown, or black form of tin(IV) oxide, SnO_2, that forms tetragonal, often twinned, crystals; the principal ore of tin. It occurs in hydrothermal veins and metasomatic deposits associated with acid igneous rocks and in alluvial (placer) deposits. The chief producers are Malaysia, Indonesia, Zaîre, and Nigeria.

cast iron A group of iron alloys containing 1.8 to 4.5% of carbon. It is usually cast into specific shapes ready for machining, heat treatment, or assembly. It is sometimes produced direct from the *blast furnace or it may be made from remelted *pig iron.

catabolism The metabolic breakdown of large molecules in living organisms to smaller ones, with the release of energy. Respiration is an example of a catabolic series of reactions. *See* metabolism. *Compare* anabolism.

catalysis The process of changing the rate of a chemical reaction by use of a *catalyst.

catalyst A substance that increases the rate of a chemical reaction without itself undergoing any permanent chemical change (*see also* inhibition). Catalysts that have the same phase as the reactants are *homogeneous catalysts* (e.g. *enzymes in biochemical reactions). Those that have a different phase are *hetereogeneous catalysts* (e.g. metals or oxides used in many industrial gas reactions). The catalyst provides an alternative pathway by which the reaction can proceed, in which the activation energy is lower. It thus increases the rate at which the reaction comes to equilibrium, although it does not alter the position of the equilibrium. The catalyst itself takes part in the reaction and consequently may undergo physical change (e.g. conversion into powder). In certain circumstances, very small quantities of catalyst can speed up very large reactions. Some catalysts are also highly specific in the type of reaction they catalyse, particularly in biochemical reactions.

catalytic cracking *See* cracking.

cataphoresis *See* electrophoresis.

catechol *See* 1,2-dihydroxybenzene.

catecholamine Any of a class of amines (including *dopamine, *adrenaline, and *noradrenaline) that function as neurotransmitters and/or hormones.

catenation The formation of chains of atoms in chemical compounds.

cathetometer A telescope or microscope fitted with crosswires in the eyepiece and mounted so that it can slide along a graduated scale. Cathetometers are used for accurate measurement of lengths without mechanical contact. The microscope type is often called a *travelling microscope*.

cathode A negative electrode. In *electrolysis cations are attracted to the cathode. In vacuum electronic devices electrons are emitted by the cathode and flow to the *anode. It is therefore from the cathode that electrons flow into these devices. However, in a primary or secondary cell the cathode is the electrode that spontaneously becomes negative during discharge, and from which therefore electrons emerge.

cathodic protection *See* sacrificial protection.

cation A positively charged ion, i.e. an ion that is attracted to the cathode in *electrolysis. *Compare* anion.

cationic detergent *See* detergent.

cationic dye *See* dyes.

cationic resin *See* ion exchange.

caustic Describing a substance that is strongly alkaline (e.g. caustic soda).

caustic potash *See* potassium hydroxide.

caustic soda *See* sodium hydroxide.

celestine A mineral form of strontium sulphate, $SrSO_4$.

cell 1. A system in which two electrodes are in contact with an electrolyte. The electrodes are metal or carbon plates or rods or, in some cases, liquid metals (e.g. mercury). In an *electrolytic cell a current from an outside source is passed through the electrolyte to produce chemical change (*see* electrolysis). In a *voltaic cell, spontaneous reactions between the electrodes and electrolyte(s) produce a potential difference between the two electrodes.

Voltaic cells can be regarded as made up of two *half cells, each composed of an electrode in contact with an electrolyte. For instance, a zinc rod dipped in zinc sulphate solution is a $Zn|Zn^{2+}$ half cell. In such a system zinc atoms dissolve as zinc ions, leaving a negative charge on the electrode

$$Zn(s) \rightarrow Zn^{2+}(aq) + 2e$$

The solution of zinc continues until the charge build-up is sufficient to prevent further ionization. There is then a potential difference between the zinc rod and its solution. This cannot be measured directly, since measurement would involve making contact with the electrolyte, thereby introducing another half cell (*see* electrode potential). A rod of copper in copper sulphate solution comprises another half cell. In this case the spontaneous reaction is one in which copper ions in solution take electrons from the electrode and are deposited on the electrode as copper atoms. In this case, the copper acquires a positive charge.

The two half cells can be connected by using a porous pot for the liquid junction (as in the *Daniell cell) or by using a *salt bridge. The resulting cell can then supply current if the electrodes are connected through an external circuit. The cell is written

$$Zn(s)|Zn^{2+}(aq)|Cu^{2+}(aq)|Cu \quad E = 1.10 \text{ V}$$

Here, E is the e.m.f. of the cell equal to the potential of the right-hand electrode minus that of the left-hand electrode for zero current. Note that 'right' and 'left' refer to the cell as written. Thus, the cell could be written

$$Cu(s)|Cu^{2+}(aq)|Zn^{2+}(aq)|Zn(s) \quad E = -1.10 \text{ V}$$

The overall reaction for the cell is

$Zn(s) + Cu^{2+}(aq) \rightarrow Cu(s) + Zn^{2+}(aq)$
This is the direction in which the cell reaction occurs for a positive e.m.f.

The cell above is a simple example of a *chemical cell*; i.e. one in which the e.m.f. is produced by a chemical difference. *Concentration cells* are cells in which the e.m.f. is caused by a difference of concentration. This may be a difference in concentration of the electrolyte in the two half cells. Alternatively, it may be an electrode concentration difference (e.g. different concentrations of metal in an amalgam, or different pressures of gas in two gas electrodes). Cells are also classified into cells *without transport* (having a single electrolyte) and *with transport* (having a liquid junction across which ions are transferred). Various types of voltaic cell exist, used as sources of current, standards of potential, and experimental set-ups for studying electrochemical reactions.
2. *See* photoelectric cell.

celluloid A transparent highly flammable substance made from cellulose nitrate with a camphor plasticizer. It was formerly widely used as a thermoplastic material, especially for film, (a use now discontinued owing to the inflammability of celluloid).

cellulose A polysaccharide that consists of a long unbranched chain of glucose units. It is the main constituent of the cell walls of all higher plants, many algae, and some fungi and is responsible for providing the rigidity of the cell wall. The fibrous nature of extracted cellulose has led to its use in the textile industry for the production of cotton, artificial silk, etc.

cellulose acetate *See* cellulose ethanoate.

cellulose ethanoate (cellulose acetate) A compound prepared by treating cellulose (cotton linters or wood pulp) with a mixture of ethanoic anhydride, ethanoic acid, and concentrated sulphuric acid. Cellulose in the cotton is ethanoylated and when the resulting solution is treated with water, cellulose ethanoate forms as a flocculent white mass. It is used in lacquers, nonshatterable glass, varnishes, and as a fibre (*see also* rayon).

cellulose nitrate A highly flammable material made by treating cellulose (wood pulp) with concentrated nitric acid. Despite the alternative name *nitrocellulose*, the compound is in fact an ester (containing $CONO_2$ groups), not a nitro compound (which would contain $C\text{-}NO_2$). It is used in explosives (as *guncotton*) and celluloid.

Celsius scale A *temperature scale in which the fixed points are the temperatures at standard pressure of ice in equilibrium with water (0°C) and water in equilibrium with steam (100°C). The scale, between these two temperatures, is divided in 100 degrees. The degree Celsius (°C) is equal in magnitude to the *kelvin. This scale was formerly known as the *centigrade scale*; the name was officially changed in 1948 to avoid confusion with a hundredth part of a grade. It is named after the Swedish astronomer Anders Celsius (1701–44), who devised the inverted form of this scale (ice point 100°, steam point 0°) in 1742.

cement Any of various substances used for bonding or setting to a hard material. Portland cement is a mixture of calcium silicates and aluminates made by heating limestone ($CaCO_3$) with clay (containing aluminosilicates) in a kiln. The product is ground to a fine powder. When mixed with water it sets in a few hours and then hardens over a longer period of time due to the formation of hydrated aluminates and silicates.

cementation Any metallurgical process in which the surface of a metal is impregnated by some other substance, especially an obsolete process for making steel by heating bars of wrought iron to red heat for several days in a bed of charcoal. *See also* case hardening.

cementite *See* steel.

centi- Symbol c. A prefix used in the metric system to denote one hundredth. For example, 0.01 metre = 1 centimetre (cm).

centigrade scale *See* Celsius scale.

centrifugal pump *See* pump.

centrifuge A device in which solid or liquid particles of different densities are separated by rotating them in a tube in a horizontal circle. The denser particles tend to move along the length of the tube to a greater radius of rotation, displacing the lighter particles to the other end.

ceramics Inorganic materials, such as pottery, enamels, and refractories. Ceramics are metal silicates, oxides, nitrides, etc.

cerium Symbol Ce. A silvery metallic element belonging to the *lanthanoids; a.n. 58; r.a.m. 140.12; r.d. 6.77 (20°C); m.p. 798°C; b.p. 3433°C. It occurs in allanite, bastnasite, cerite, and monazite. Four isotopes occur naturally: cerium–136, –138, –140, and –142; fifteen radioisotopes have been identified. Cerium is used in mischmetal, a rare-earth metal containing 25% cerium, for use in lighter flints. The oxide is used in the glass industry. It was discovered by M. H. Klaproth in 1803.

cermet A composite material consisting of a ceramic in combination with a sintered metal, used when a high resistance to temperature, corrosion, and abrasion is needed.

cerussite An ore of lead consisting of lead carbonate, $PbCO_3$. It is usually of secondary origin, formed by the weathering of *galena. Pure cerussite is white but the mineral may be grey due to the presence of impurities. It forms well-shaped orthorhombic crystals. It occurs in the USA, Spain, and SW Africa.

cetane *See* hexadecane.

cetane number A number that provides a measure of the ignition characteristics of a Diesel fuel when it is burnt in a standard Diesel engine. It is the percentage of cetane (hexadecane) in a mixture of cetane and 1-methylnaphthalene that has the same ignition characteristics as the fuel being tested. *Compare* octane number.

c.g.s. units A system of *units based on the centimetre, gram, and second. Derived from the metric system, it was badly adapted to use with thermal quantities (based on the inconsistently defined *calorie) and with electrical quantities (in which two systems, based respectively on unit permittivity and unit permeability of free space, were used). For scientific purposes c.g.s. units have now been replaced by *SI units.

chain A line of atoms of the same type in a molecule. In a *straight chain* the atoms are attached only to single atoms, not to groups. Propane, for instance, is a straight-chain alkane, $CH_3CH_2CH_3$, with a chain of three carbon atoms. A *branched chain* is one in which there are side groups attached to the chain. Thus, 3-ethyloctane, $CH_3CH_2CH(C_2H_5)C_5H_{11}$, is a branched-chain alkane in which there is a *side chain* (C_2H_5) attached to the third carbon atom. A *closed chain* is a *ring of atoms in a molecule; otherwise the molecule has an *open chain*.

chain reaction A reaction that is self sustaining as a result of the products of one step initiating a subsequent step. Chemical chain reactions usually involve free radicals as intermediates. An example is the reaction of chlorine with hydrogen initiated by ultraviolet radiation. A chlorine molecule is first split into atoms:

$$Cl_2 \rightarrow Cl\cdot + Cl\cdot$$

These react with hydrogen as follows

$$Cl\cdot + H_2 \rightarrow HCl + H\cdot$$
$$H\cdot + Cl_2 \rightarrow HCl + Cl\cdot \text{ etc.}$$

Combustion and explosion reactions involve similar free-radical chain reactions.

chair conformation *See* conformation.

chalcedony A mineral consisting of a microcrystalline variety of *quartz. It occurs in several forms, including a large number of semiprecious gemstones; for example, sard, carnelian, jasper, onyx, chrysoprase, agate, and tiger's-eye.

chalcogens *See* group VI elements.

chalconides Binary compounds formed between metals and group VI elements; i.e. oxides, sulphides, selenides, and tellurides.

chalcopyrite (copper pyrites) A brassy yellow mineral consisting of a mixed copper–iron sulphide, $CuFeS_2$, crystallizing in the tetragonal system; the principal ore of copper. It is similar in appearance to pyrite and gold. It crystallizes in igneous rocks and hydrothermal veins associated with the upper parts of acid igneous intrusions. Chalcopyrite is the most widespread of the copper ores, occurring, for example, in Cornwall (UK), Sudbury (Canada), Chile, Tasmania (Australia), and Rio Tinto (Spain).

chalk A very fine-grained white rock composed of the skeletal remains of microscopic sea creatures, such as plankton, and consisting largely of *calcium carbonate ($CaCO_3$). It is used in toothpaste and cosmetics. It should not be confused with blackboard 'chalk', which is made from calcium sulphate.

change of phase (change of state) A change of matter in one physical *phase (solid, liquid, or gas) into another. The change is invariably accompanied by the evolution or absorption of energy, even if it takes place at constant temperature (*see* latent heat).

charcoal A porous form of carbon produced by the destructive distillation of organic material. Charcoal from wood is used as a fuel. All forms of charcoal are porous and are used for adsorbing gases and purifying and clarifying liquids. There are several types depending on the source. Charcoal from coconut shells is a particularly good gas adsorbent. *Animal charcoal* (or *bone black*) is made by heating bones and dissolving out the calcium phosphates and other mineral salts with acid. It is used in sugar refining. *Activated charcoal* is charcoal that has been activated for adsorption by steaming or by heating in a vacuum.

charge A property of some *elementary particles that gives rise to an interaction between them and consequently to the host of material phenomena described as electrical. Charge occurs in nature in two forms, conventionally described as *positive* and *negative* in order to distinguish between the two kinds of interaction between particles. Two particles that have similar charges (both negative or both positive) interact by repelling each other; two particles that have dissimilar charges (one positive, one negative) interact by attracting each other.

The natural unit of negative charge is the charge on an *electron, which is equal but opposite in effect to the positive charge on the proton. Large-scale matter that consists of equal numbers of electrons and protons is electrically neutral. If there is an excess of electrons the body is negatively charged; an excess of protons results in a positive charge. A flow of charged particles, especially a flow of electrons, constitutes an electric current. Charge is measured in coulombs, the charge on an electron being 1.602×10^{-19} coulombs.

Charles' law The volume of a fixed mass of gas at constant pressure expands by a constant fraction of its volume at 0°C for each Celsius degree or kelvin its temperature is raised. For any *ideal gas the fraction is approximately 1/273. This can be expressed by the equation $V = V_0(1 + t/273)$, where V_0 is the volume at 0°C and V is its volume at t°C. This is equivalent to the statement that the volume of a fixed mass of gas at constant pressure is proportional to its thermodynamic temperature, $V = kT$, where k is a constant. The law resulted from experiments begun around 1787 by the French scientist J. A. C. Charles (1746–1823) but was properly established only by the more accurate results published in 1802 by the French scientist Joseph Gay-Lussac (1778–1850). Thus the law is also known as *Gay-Lussac's law*. An equation similar to that given above applies to pressures for ideal gases: $p = p_0(1 + t/273)$, a relationship known as *Charles' law of pressures*. *See also* gas laws.

cheddite Any of a group of high explosives made from nitro compounds mixed with sodium or potassium chlorate.

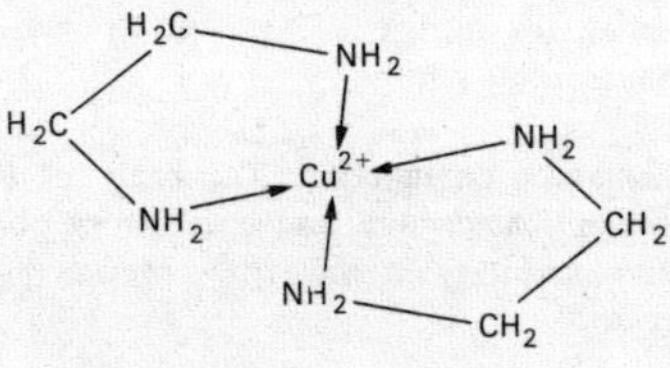

Chelate formed by coordination of two molecules of $H_2N(CH_2)_2NH_2$

chelate An inorganic complex in which a *ligand is coordinated to a metal ion at two (or more) points, so that there is a ring of atoms including the metal. The process is known as *chelation*. The bidentate ligand diaminoethane forms chelates with many ions. *See also* sequestration.

chemical bond A strong force of attraction holding atoms together in a molecule or crystal. Typically chemical bonds have energies of about 1000 kJ mol^{-1} and are distinguished from the much weaker forces between molecules (*see* van der Waals' forces). There are various types.

Ionic (or *electrovalent*) bonds can be formed by transfer of electrons. For instance, the calcium atom has an electron configuration of [Ar]$4s^2$, i.e. it has two electrons in its outer shell. The chlorine atom is [Ne]$3s^2 3p^5$, with seven outer electrons. If the calcium atom transfers two electrons, one to each chlorine atom, it becomes a Ca^{2+} ion with the stable configuration of an inert gas [Ar]. At the same time each chlorine, having gained one electron, becomes a Cl^- ion, also with an inert-gas configuration [Ar]. The bonding in calcium chloride is the electrostatic attraction between the ions.

Covalent bonds are formed by sharing of valence electrons rather than by transfer. For instance, hydrogen atoms have one outer electron ($1s^1$). In the hydrogen molecule, H_2, each atom contributes 1 electron to the bond. Consequently, each hydrogen atom has control of 2 electrons – one of its own and the second from the other atom – giving it the electron configuration of an inert gas [He]. In the water molecule, H_2O, the oxygen atom, with six outer electrons, gains control of an extra two electrons supplied by the two hydrogen atoms. This gives it the configuration [Ne]. Similarly, each hydrogen atom gains control of an extra electron from the oxygen, and has the [He] electron configuration.

A particular type of covalent bond is one in which one of the atoms supplies both the electrons. These are known as *coordinate* (*semipolar* or *dative*) bonds, and written A→B, where the direction of the arrow denotes the direction in which electrons are donated.

Covalent or coordinate bonds in which one pair of electrons is shared are *electron-pair bonds* and are known as *single bonds*. Atoms can also share two pairs of electrons to form *double bonds* or three pairs in *triple bonds*. *See* orbital.

In a compound such as sodium chloride, Na^+Cl^-, there is probably complete transfer of electrons in forming the ionic bond (the bond is said to be *heteropolar*). Alternatively, in the hydrogen molecule H–H, the pair of electrons is equally shared between the two atoms (the bond is *homopolar*). Between these two extremes, there is a whole range of *intermediate bonds*, which have both ionic and covalent contributions. Thus, in hydrogen chloride, H–Cl, the bonding is predominantly covalent with one pair of electrons shared between the two atoms. However, the chlorine atom is more electronegative than the hydrogen and has more control over the electron pair; i.e. the molecule is polarized with a positive charge on the hydrogen and a negative charge on the chlorine. *See also* hydrogen bond; metallic bond.

chemical cell *See* cell.

chemical combination The combination of elements to give compounds. There are three laws of chemical combination.

(1) The *law of constant composition* states that the proportions of the elements in a compound are always the same, no matter how the compound is made. It is also called the *law of constant proportions* or *definite proportions*.

(2) The *law of multiple proportions* states that when two elements A and B combine to form more than one compound, then the masses of B that combine with a fixed mass of A are in simple ratio to one another. For example, carbon forms two oxides. In one, 12 grams of carbon is combined with 16 grams of oxygen (CO); in the other 12 g of carbon is combined with 32 grams of oxygen (CO_2). The oxygen masses combining with a fixed mass of carbon are in the ratio 16:32, i.e. 1:2.

(3) The *law of equivalent proportions* states that if two elements A and B each form a compound with a third element C, then a compound of A and B will contain A and B in the relative proportions in which they react with C. For example, sulphur and carbon both form compounds with hydrogen. In methane 12 g of carbon react with 4 g of hydrogen. In hydrogen sulphide, 32 g of sulphur react with 2 g of hydrogen (i.e. 64 g of S for 4 g of hydrogen). Sulphur and carbon form a compound in which the C:S ratio is 12:64 (i.e. CS_2). The law is sometimes called the law of *reciprocal proportions*.

chemical dating An absolute *dating technique that depends on measuring the chemical composition of a specimen. Chemical dating can be used when the specimen is known to undergo slow chemical change at a known rate. For instance, phosphate in buried bones is slowly replaced by fluoride ions from the ground water. Measurement of the proportion of fluorine present gives a rough estimate of the time that the bones have been in the ground. Another, more accurate, method depends on the fact that amino acids in living organisms are L- optical isomers. After death, these racemize and the age of bones can be estimated by measuring the relative amounts of D- and L-amino acids present.

chemical engineering The study of the design, manufacture, and operation of plant and machinery in industrial chemical processes.

chemical equation A way of denoting a chemical reaction using the symbols for the participating particles (atoms, molecules, ions, etc.); for example,

$$xA + yB \rightarrow zC + wD$$

The single arrow is used for an irreversible reaction; double arrows ($\rightleftharpoons$) are used for reversible reactions. When reactions involve different phases it is usual to put the phase in brackets after the symbol (s = solid; l = liquid; g = gas; aq = aqueous). The numbers x, y, z, and w, showing the relative numbers of molecules reacting, are called the *stoichiometric coefficients*. The sum of the coefficients of the reactants minus the sum of the coefficients of the products (x +

$y - z - w$ in the example) is the *stoichiometric sum.* If this is zero the equation is balanced. Sometimes a generalized chemical equation is considered

$$\nu_1 A$$

In this case the reaction can be written $\Sigma\nu_i A_i = 0$, where the convention is that stoichiometric coefficients are positive for reactants and negative for products. The stoichiometric sum is $\Sigma\nu_i$.

chemical equilibrium A reversible chemical reaction in which the concentrations of reactants and products are not changing with time because the system is in thermodynamic equilibrium. For example, the reversible reaction

$$3H_2 + N_2 \leftrightharpoons 2NH_3$$

is in chemical equilibrium when the rate of the *forward reaction*

$$3H_2 + N_2 \rightarrow 2NH_3$$

is equal to the rate of the *back reaction*

$$2NH_3 \rightarrow 3H_2 + N_2$$

See also equilibrium constant.

chemical equivalent *See* equivalent weight.

chemical fossil Any of various organic compounds found in ancient geological strata that appear to be biological in origin and are assumed to indicate that life existed when the rocks were formed. The presence of chemical fossils in Precambrian strata indicates that life existed over 3000 million years ago.

chemical potential Symbol: μ. For a given component in a mixture, the coefficient $\partial G/\partial n$, where G is the Gibbs free energy and n the amount of substance of the component. The chemical potential is the change in Gibbs free energy with respect to change in amount of the component, with pressure, temperature, and amounts of other components being constant. Components are in equilibrium if their chemical potentials are equal.

chemical reaction A change in which one or more chemical elements or compounds (the *reactants*) form new compounds (the *products*). All reactions are to some extent *reversible*; i.e. the products can also react to give the original reactants. However, in many cases the extent of this back reaction is negligibly small, and the reaction is regarded as *irreversible.*

chemiluminescence *See* luminescence.

chemisorption *See* adsorption.

chemistry The study of the elements and the compounds they form. Chemistry is mainly concerned with effects that depend on the outer electrons in atoms. *See* biochemistry; geochemistry; inorganic chemistry; organic chemistry; physical chemistry.

chert *See* flint.

Chile saltpetre A commercial mineral largely composed of *sodium nitrate from the caliche deposits in Chile. Before the ammonia-oxidation process for nitrates most imported Chilean saltpetre was used by the chemical industry; its principal use today is as an agricultural source of nitrogen.

china clay *See* kaolin.

Chinese white *See* zinc oxide.

chirality The property of existing in left- and right-handed structural forms. *See* optical activity.

chitin A *polysaccharide comprising chains of N-acetyl-D-glucosamine, a derivative of glucose. Chitin is structurally very similar to cellulose and serves to strengthen the supporting structures of various invertebrates. It also occurs in fungi.

chloral *See* trichloroethanal.

chloral hydrate *See* 2,2,2-trichloroethanediol.

chlorates Salts of the chloric acids; i.e. salts containing the ions ClO^- (chlorate(I) or *hypochlorite*), ClO_2^- (chlorate(III) or *chlorite*), ClO_3^- (chlorate(V)), or ClO_4^- (chlorate(VII) or *perchlorate*). When used without specification of an oxidation state the term 'chlorate' refers to a chlorate(V) salt.

chloric acid Any of the oxoacids of chlorine: *chloric(I) acid, *chloric(III) acid, *chloric(V) acid, and *chloric(VII) acid. The term is commonly used without specification of the oxidation state of chlorine to mean chloric(V) acid, $HClO_3$.

chloric(I) acid (hypochlorous acid) A liquid acid that is stable only in solution, HOCl. It may be prepared by the reaction of chlorine with an agitated suspension of mercury(II) oxide. Because the disproportionation of the ion ClO^- is slow at low temperatures chloric(I) acid may be produced, along with chloride ions by the reaction of chlorine with water at 0°C. At higher temperatures disproportionation to the chlorate(V) ion, ClO_3^-, takes place. Chloric(I) acid is a very weak acid but is a mild oxidizing agent and is widely used as a bleaching agent.

chloric(III) acid (chlorous acid) A pale-yellow acid known only in solution, $HClO_2$. It is formed by the reaction of chlorine dioxide and water and is a weak acid and an oxidizing agent.

chloric(V) acid (chloric acid) A colourless unstable liquid, $HClO_3$; r.d. 1.2; m.p. <-20°C; decomposes at 40°C. It is best prepared by the reaction of barium chlorate with sulphuric acid although chloric(V) acid is also formed by the disproportionation of chloric(I) acid in hot solutions. It is both a strong acid and a powerful oxidizing agent; hot solutions of the acid or its salts have been known to detonate in contact with readily oxidized organic material.

chloric(VII) acid (perchloric acid) An unstable liquid acid, $HClO_4$; r.d. 1.76; m.p. −112°C; b.p. 39°C (50 mmHg); explodes at about 90°C at atmospheric pressure. There is also a monohydrate (r.d. 1.88 (solid), 1.77 (liquid); m.p. 48°C; explodes at about 110°C) and a dihydrate (r.d. 1.65; m.p. −17.8°C; b.p. 200°C). Commercial chloric (VII) acid is a water azeotrope, which is 72.5% $HClO_4$, boiling at 203°C. The anhydrous acid may be prepared by vacuum distillation of the concentrated acid in the presence of magnesium perchlorate as a dehydrating agent. Chloric(VII) acid is both a strong acid and a strong oxidizing agent. It is widely used to decompose organic materials prior to analysis, e.g. samples of animal or vegetable matter requiring heavy-metal analysis.

chloride *See* halide.

chlorination **1.** A chemical reaction in which a chlorine atom is introduced into a compound. *See* halogenation. **2.** The treatment of water with chlorine to disinfect it.

chlorine Symbol Cl. A *halogen element; a.n. 17; r.a.m. 35.453; d. 3.214 g dm^{-3}; m.p. −100.98°C; b.p. −34.6°C. It is a poisonous greenish-yellow gas and occurs widely in nature as sodium chloride in seawater and as halite (NaCl), carnallite ($KCl.MgCl_2.6H_2O$), and sylvite (KCl). It is manufactured by the electrolysis of brine and also obtained in the *Downs process for making sodium. It has many applications, including the chlorination of drinking water, bleaching, and the manufacture of a large number of organic chemicals.

It reacts directly with many elements and compounds and is a strong oxidizing agent. Chlorine compounds contain the element in the 1, 3, 5, and 7 oxidation states. It was discovered by Karl Scheele in 1774 and Humphry Davy confirmed it as an element in 1810.

chlorine dioxide A yellowish-red explosive gas, ClO_2; d. 3.09 g dm^{-3}; m.p. −59.5°C; b.p. 9.9°C. It is soluble in cold water but decomposed by hot water to give chloric (VII) acid, chlorine, and oxygen. Because of its high reactivity, chlorine dioxide is best prepared by the reaction of sodium chlorate and moist oxalic acid at 90°−100°C, as the product is then diluted by liberated carbon dioxide. Commercially the gas is produced by the reaction of sulphuric acid containing chloride ions with sulphur dioxide. Chlorine dioxide is widely used as a bleach in flour milling and in wood pulping and also finds application in water purification.

chlorine monoxide *See* dichlorine oxide.

chlorite 1. *See* chlorates. **2.** A group of layered silicate minerals, usually green or white in colour, that are similar to the micas in structure and crystallize in the monoclinic system. Chlorites are composed of complex silicates of aluminium, magnesium, and iron in combination with water, with the formula $(Mg,Al,Fe)_{12}(Si,Al)_8O_{20}(OH)_{16}$. They are most common in low-grade metamorphic rocks and also occur as secondary minerals in igneous rocks as alteration products of pyroxenes, amphiboles, and micas. The term is derived from *chloros*, the Greek word for green.

chloroacetic acids *See* chloroethanoic acids.

chlorobenzene A colourless highly inflammable liquid, C_6H_5Cl; r.d. 1.106; m.p. −45.43°C; b.p. 131.85°C. It is prepared by the direct chlorination of benzene using a halogen carrier (*see* Friedel–Crafts reaction), or manufactured by the *Raschig process. It is used mainly as an industrial solvent.

2-chlorobuta-1,3-diene (chloroprene) A colourless liquid chlorinated diene, $CH_2{:}CClCH{:}CH_2$; r.d. 0.96; b.p. 59°C. It is polymerized to make synthetic rubbers (e.g. neoprene).

chloroethane (ethyl chloride) A colourless flammable gas, C_2H_5Cl; m.p. −45°C; b.p. 132°C. It is made by reaction of ethene and hydrogen chloride and used in making lead tetraethyl for petrol.

chloroethanoic acids (chloroacetic acids) Three acids in which hydrogen atoms in the methyl group of ethanoic acid have been replaced by chlorine atoms. They are: *monochloroethanoic acid* ($CH_2ClCOOH$); *dichloroethanoic acid* ($CHCl_2COOH$); *trichloroethanoic acid* (CCl_3COOH). The presence of chlorine atoms in the methyl group causes electron withdrawal from the COOH group and makes the chloroethanoic acids stronger acids than ethanoic acid itself. The K_a values (in moles dm^{-3} at 25°C) are

CH_3COOH 1.7×10^{-5}

$CH_2ClCOOH$ 1.3×10^{-3}

$CHCl_2COOH$ 5.0×10^{-2}

CCl_3COOH 2.3×10^{-1}

chloroethene (vinyl chloride) A gaseous compound, $CH_2{:}CHCl$; r.d. 0.911; m.p. −153.8°C; b.p. −13.37°C. It is made by chlorinating ethene to give dichloroethane, then removing HCl:

$$C_2H_4 + Cl_2 \rightarrow CH_2ClCH_2Cl \rightarrow CH_2CHCl$$

The compound is used in making PVC.

chloroform *See* trichloromethane.

chloromethane (methyl chloride) A colourless flammable gas, CH_3Cl; r.d. 0.916; m.p. −97°C; b.p. −24°C. It is a *haloalkane, made by direct chlorination of methane and used as a local anaesthetic and refrigerant.

chlorophyll The pigment responsible for the green colour of most plants. The chlorophyll molecule is the principal site of light absorption in the light reactions of photosynthesis. It is a magnesium-containing *porphyrin, chemically related to cytochrome and haemoglobin.

chloroplatinic acid A reddish crystalline compound, H_2PtCl_6, made by dissolving platinum in aqua regia.

chloroprene *See* 2-chlorobuta-1,3-diene.

chlorosulphanes *See* disulphur dichloride.

chlorous acid *See* chloric(III) acid.

choke damp *See* blackdamp.

cholecalciferol *See* vitamin D.

cholesteric crystal *See* liquid crystal.

cholesterol A *sterol occurring widely in animal tissues and also in some higher plants and algae. It can exist as a free sterol or esterified with a long-chain fatty acid. Cholesterol is absorbed through the intestine or manufactured in the liver. It serves principally as a constituent of blood plasma lipoproteins and of the lipid–protein complexes that form cell membranes. It is also important as a precursor of various steroids, especially the bile acids, sex hormones, and adrenocorticoid hormones. The derivative 7-dehydrocholesterol is converted to vitamin D_3 by the action of sunlight on skin. Increased levels of dietary and blood cholesterol have been associated with *atherosclerosis*, a condition in which lipids accumulate on the inner walls of arteries and eventually obstruct blood flow.

choline An amino alcohol, $CH_2OHCH_2N(CH_3)_3OH$. It occurs widely in living organisms as a constituent of certain types of phospholipids – the lecithins and sphingomyelins. It is sometimes classified as a member of the vitamin B complex.

chromate A salt containing the ion CrO_4^{2-}.

chromatogram A record obtained by chromatography. The term is applied to the developed records of *paper chromatography and *thin-layer chromatography and also to the graphical record produced in *gas chromatography.

chromatography A technique for analysing or separating mixtures of gases, liquids, or dissolved substances. The original technique (invented by the Russian botanist Mikhail Tsvet in 1906) is a good example of *column chromatography*. A vertical glass tube is packed with an adsorbing material, such as alumina. The sample is poured into the column and continuously washed through with a solvent (a process known as *elution*). Different components of the sample are adsorbed to different extents and move down the column at different rates. In Tsvet's original application, plant pigments were used and these separated into coloured bands in passing down the column (hence the name chromatography). The usual method is to collect the liquid (the *eluate*) as it passes out from the column in fractions.

In general, all types of chromatography involve two distinct phases – the *stationary phase* (the adsorbent material in the column in the example above) and the *moving phase* (the solution in the example). The separation depends on competition for molecules of sample between the moving phase and the stationary phase. The form of column chromatography above is an example of *adsorption chromatography*, in which the sample molecules are adsorbed on the alumina. In *partition chromatography*, a liquid (e.g. water) is first absorbed by the stationary phase and the moving phase is an immiscible liquid. The separation is then by *partition between the two liquids. In ion-exchange chromatography (*see* ion exchange), the process involves competition between different ions for ionic sites on the stationary phase. *Gel filtration is another chromatographic technique in which the size of the sample molecules is important.

See also gas chromatography; paper chromatography; thin-layer chromatography.

chrome alum *See* potassium chromium sulphate.

chrome iron ore A mixed iron–chromium oxide, $FeO.Cr_2O_3$, used to make ferrochromium for chromium steels.

chrome red A basic lead chromate, $PbO.PbCrO_4$, used as a red pigment.

chrome yellow Lead chromate, $PbCrO_4$, used as a pigment.

chromic acid A hypothetical acid, H_2CrO_4, known only in chromate salts.

chromic anhydride *See* chromium(VI) oxide.

chromic compounds Compounds containing chromium in a higher (+3 or +6) oxidation state; e.g. chromic oxide is chromium (VI) oxide (CrO_3).

chromite A spinel mineral, $FeCr_2O_4$; the principal ore of chromium. It is black with a metallic lustre and usually occurs in massive form. It is a common constituent of peridotites and serpentines. The chief producing countries are Turkey, South Africa, the USSR, the Philippines, and Zimbabwe.

chromium Symbol Cr. A hard silvery *transition element; a.n. 24; r.a.m. 52.00; r.d. 7.19; m.p. 1900°C; b.p. 2640°C. The main ore is chromite ($FeCr_2O_4$). The metal is extracted by heating chromite with sodium chromate, from which chromium can be obtained by electrolysis. Alternatively, chromite can be heated with carbon in an electric furnace to give ferrochrome, which is used in making alloy steels. The metal is also used as a shiny decorative electroplated coating and in the manufacture of certain chromium compounds.

At normal temperatures the metal is corrosion-resistant. It reacts with dilute hydrochloric and sulphuric acids to give chromium(II) salts. These readily oxidize to the more stable chromium(III) salts. Chromium also forms compounds with the +6 oxidation state, as in chromates, which contain the CrO_4^{2-} ion. The element was discovered in 1797 by Vauquelin.

chromium(II) oxide A black insoluble powder, CrO. Chromium(II) oxide is prepared by oxidizing chromium amalgam with air. At high temperatures hydrogen reduces it to the metal.

chromium(III) oxide A green crystalline water-insoluble salt, Cr_2O_3; r.d. 5.21; m.p. 2266°C; b.p. 4000°C. It is obtained by heating chromium in a stream of oxygen or by heating ammonium dichromate. The industrial preparation is by reduction of sodium dichromate with carbon. Chromium(III) oxide is amphoteric, dissolving in acids to give chromium(III) ions and in concentrated solutions of alkalis to give *chromites*. It is used as a green pigment in glass, porcelain, and oil paint.

chromium(IV) oxide (chromium dioxide) A black insoluble powder, CrO_2; m.p. 300°C. It is prepared by the action of oxygen on chromium(VI) oxide or chromium (III) oxide at 420–450°C and 200–300 atmospheres. The compound is unstable.

chromium(VI) oxide (chromium trioxide; chromic anhydride) A red compound, CrO_3; rhombic; r.d. 2.70; m.p. 196°C. It can be made by careful addition of concentrated sulphuric acid to an ice-cooled strong aqueous solution of sodium dichromate with stirring. The mixture is then filtered through sintered glass, washed with nitric acid, then dried out 120°C in a desiccator.

Chromium(VI) oxide is an extremely powerful oxidizing agent, especially to organic matter; it immediately inflames ethanol. It is an acidic oxide and dissolves in water to form 'chromic acid', a powerful oxidizing agent and cleansing fluid for glassware. At 400°C, chromium(VI) oxide loses oxygen to give chromium(III) oxide.

chromium potassium sulphate A red crystalline solid, $K_2SO_4.Cr_2(SO_4)_3.24H_2O$; r.d. 1.91. It is used as a mordant *See also* alums.

chromium steel Any of a group of *stainless steels containing 8–25% of chromium. A typical chromium steel might contain 18% of chromium, 8% of nickel, and 0.15% of carbon. Chromium steels are highly resistant to corrosion and are used for cutlery, chemical plant, ball bearings, etc.

chromophore A group causing coloration in a *dye. Chromophores are generally groups of atoms having delocalized electrons.

chromous compounds Compounds containing chromium in its lower (+2) oxidation state; e.g. chromous chloride is chromium(II) chloride ($CrCl_2$).

chromyl chloride (chromium oxychloride) A dark red liquid, CrO_2Cl_2; r.d. 1.911; m.p. −96.5°C; b.p. 117°C. It is evolved as a dark-red vapour on addition of concentrated sulphuric acid to a mixture of solid potassium dichromate and sodium chloride; it condenses to a dark-red covalent liquid, which is immediately hydrolysed by solutions of alkalis to give the yellow chromate. Since bromides and iodides do not give analogous compounds this is a specific test for chloride ions. The compound is a powerful oxidizing agent, exploding on contact with phosphorus and inflaming sulphur and many organic compounds.

chrysotile *See* serpentine.

cinnabar A bright red mineral form of mercury(II) sulphide, HgS, crystallizing in the hexagonal system; the principal ore of mercury. It is deposited in veins and impregnations near recent volcanic rocks and hot springs. The chief sources include Spain, Italy, and Yugoslavia.

cinnamic acid (3-phenylpropenoic acid) A white crystalline aromatic *carboxylic acid, $C_6H_5CH{:}CHCOOH$; r.d. 1.28 (trans isomer); m.p. 133°C; b.p. 300°C. Esters of cinnamic acid occur in some essential oils.

circular polarization *See* polarization of light.

cis–trans isomerism *See* isomerism.

citrate A salt or ester of citric acid.

citric acid A white crystalline hydroxycarboxylic acid, $HOOCCH_2C(OH)(COOH)CH_2COOH$; r.d. 1.67; m.p. 153°C. It is present in citrus fruits and is an intermediate in the *Krebs cycle in plant and animal cells.

citric-acid cycle *See* Krebs cycle.

Claisen condensation A reaction of esters in which two molecules of the ester react to give a keto ester, e.g.

$$2CH_3COOR \rightarrow CH_3COCH_2COOR + ROH$$

The reaction is catalysed by sodium ethoxide, the mechanism being similar to that of the *aldol reaction.

Clark cell A type of *voltaic cell consisting of an anode made of zinc amalgam and a cathode of mercury both immersed in a saturated solution of zinc sulphate. The Clark cell was formerly used as a standard of e.m.f.; the e.m.f. at 15°C is 1.4345 volts. It is named after the British scientist Hosiah Clark (*d.* 1898).

Clark process *See* hardness of water.

clathrate A solid mixture in which small molecules of one compound or element are trapped in holes in the crystal lattice of another substance. Clathrates are sometimes called *enclosure compounds*, but they are not true compounds (the molecules are not held by chemical bonds). Quinol and ice both form clathrates with such substances as sulphur dioxide and xenon.

Claude process A process for liquefying air on a commercial basis. Air under pressure is used as the working substance in a piston engine, where it does external work and cools adiabatically. This cool air is fed to a counter-current heat exchanger, where it reduces the temperature of the next intake of high-pressure air. The same air is re-compressed and used again, and after several cycles eventually liquefies. The process was perfected in 1902 by the French scientist Georges Claude (1870–1960).

claudetite A mineral form of *arsenic(III) oxide, As_4O_6.

clay A fine-grained deposit consisting chiefly of *clay minerals. It is characteristically plastic and virtually impermeable when wet and cracks when it dries out. In geology the size of the constituent particles is usually taken to be less than 1/256 mm. In soil science clay is regarded as a soil with particles less than 0.002 mm in size.

clay minerals Very small particles, chiefly hydrous silicates of aluminium, sometimes with magnesium and/or iron substituting for all or part of the aluminium, that are the major constituents of clay materials. The particles are essentially crystalline (either platy or fibrous) with a layered structure, but may be amorphous or metalloidal. The clay minerals are responsible for the plastic properties of clay; the particles have the property of being able to hold water. The chief groups of clay minerals are: *kaolinite*, $Al_4Si_4O_{10}(OH)_8$, the chief constituent of *kaolin; *halloysite*, $Al_4Si_4(OH)_8O_{10}.4H_2O$; *illite*, $KAl_4(Si,Al)_8O_{18}.2H_2O$; *montmorillonite*, $(Na,Ca)_{0.33}(Al,Mg)_2Si_4O_{10}(OH)_2.nH_2O$, formed chiefly through alteration of volcanic ash; *vermiculite*, $(Mg,Fe,Al)_3(Al,Si)_4O_{10}(OH)_2.4H_2O$, used as an insulating material and potting soil.

cleavage The splitting of a crystal along planes of atoms in the lattice.

closed chain *See* chain; ring.

close packing The packing of spheres so as to occupy the minimum amount of space. In a single plane, each sphere is surrounded by six close neighbours in a hexagonal arrangement. The spheres in the second plane fit into depressions in the first layer, and so on. Each sphere has 12 other touching spheres. There are two types of close packing. In *hexagonal close packing* the spheres in the third layer are directly over those in the first, etc., and the arrangement of planes is ABAB.... In *cubic close packing* the spheres in the third layer occupy a different set of depressions than those in the first. The arrangement is ABCABC.... *See also* cubic crystal.

Clusius column A device for separating isotopes by thermal diffusion. One form consists of a vertical column some 30 metres high with a heated electric wire running along its axis. The lighter isotopes in a gaseous mixture of isotopes diffuse faster than the heavier isotopes. Heated by the axial wire, and assisted by natural convection, the lighter atoms are carried to the top of the column, where a fraction rich in lighter isotopes can be removed for further enrichment.

coagulation The process in which colloidal particles come together to form larger masses. Coagulation can be brought about by adding ions to neutralize the charges stabilizing the colloid. Ions with a high charge are particularly effective (e.g. alum, containing Al^{3+}, is used in styptics to coagulate blood). Another example of ionic coagulation is in the formation of river deltas, which occurs when colloidal silt particles in rivers are coagulated by ions in sea water. Heating is another way of coagulating certain colloids (e.g. boiling an egg coagulates the albumin).

coal A brown or black carbonaceous deposit derived from the accumulation and alteration of ancient vegetation, which originated largely in swamps or other moist environments. As the vegetation decomposed it formed layers of peat, which were subsequently buried (for example, by marine sediments following a rise in sea level or subsidence of the land). Under the increased pressure and resulting higher temperatures the peat was transformed into coal. Two types of coal are recognized: *humic* (or *woody*) *coals*, derived from plant remains; and *sapropelic coals*, which are derived from algae, spores, and finely divided plant material.

As the processes of coalification (i.e. the transformation resulting from the high temperatures and pressures) continue, there is a progressive transformation of the deposit: the proportion of carbon relative to oxygen rise and volatile substances and water are driven out. The various stages in this process are referred to as the *ranks* of the coal. In ascending order, the main ranks of coal are: *lignite* (or *brown coal*), which is soft, brown, and has a high moisture content; *subbituminous coal*, which is used chiefly by generating stations; *bituminous coal*, which is the most abundant rank of coal; *semibituminous coal*; *semianthracite coal*, which has a fixed carbon content of between 86% and 92%; and *anthracite coal*, which is hard and black with a fixed carbon content of between 92% and 98%.

Most deposits of coal were formed during the Carboniferous and Permian periods. More recent periods of coal formation occurred during the early Jurassic and Tertiary periods. Coal deposits occur in all the major continents; the leading producers include the USA, China, USSR, Poland, UK, South Africa, India, Australia, and West Germany. Coal is used as a fuel and in the chemical industry; by-products include coke and coal tar.

coal gas A fuel gas produced by the destructive distillation of coal. In the late-19th and early-20th centuries coal gas was a major source of energy and was made by heating coal in the absence of air in local gas works. Typically, it contained hydrogen (50%), methane (35%), and carbon monoxide (8%). By-products of the process were *coal tar and coke. The use of this type of gas declined with the increasing availability of natural gas, although since the early 1970s interest has developed in using coal in making *SNG.

coal tar A tar obtained from the destructive distillation of coal. Formerly, coal tar was obtained as a by-product in manufacturing *coal gas. Now it is produced in making coke for steel making. The crude tar contains a large number of organic compounds, such as benzene, naphthalene, methylbenzene, phenols, etc., which can be obtained by distillation. The residue is *pitch*. At one time coal tar was the major source of organic chemicals, most of which are now derived from petroleum and natural gas.

cobalt Symbol Co. A light-grey *transition element; a.n. 27; r.a.m. 58.933; r.d. 8.9; m.p. 1495°C; b.p. 2870°C. Cobalt is ferromagnetic below its Curie point of 1150°C. Small amounts of metallic cobalt are present in meteorites but it is usually extracted from ore deposits worked in Canada, Morocco, and Zaïre. It is present in the minerals cobaltite, smaltite, and erythrite but also associated with copper and nickel as sulphides and arsenides. Cobalt ores are usually roasted to the oxide and then reduced with carbon or water gas. Cobalt is usually alloyed for use. Alnico is a well-known magnetic alloy and cobalt is also used to make stainless steels and in high-strength alloys that are resistant to oxidation at high temperatures (for turbine blades and cutting tools).

The metal is oxidized by hot air and also reacts with carbon, phosphorus, sulphur, and dilute mineral acids. Cobalt salts, usual oxidation states II and III, are used to give a brilliant blue colour in glass, tiles, and pottery. Anhydrous cobalt(II) chloride paper is used as a qualitative test for water and as a heat-sensitive ink. Small amounts of cobalt salts are essential in a balanced diet for mammals (*see* essential element). Artificially produced cobalt-60 is an important radioactive tracer and cancer-treatment agent. The element was discovered by G. Brandt in 1737.

cobalt(II) oxide A pink solid, CoO; cubic; r.d. 6.45; m.p. 1795°C. The addition of potassium hydroxide to a solution of cobalt(II) nitrate gives a bluish-violet precipitate, which on boiling is converted to pink impure cobalt(II) hydroxide. On heating this in the absence of air, cobalt(II) oxide is formed. The compound is readily oxidized in air to form tricobalt tetroxide, Co_3O_4, and is readily reduced by hydrogen to the metal.

cobalt(III) oxide (cobalt sesquioxide) A black grey insoluble solid, Co_2O_3; hexagonal or rhombic; r.d. 5.18; decomposes at 895°C. It is produced by the ignition of cobalt nitrate; the product however never has the composition corresponding exactly to cobalt(III) oxide. On heating it readily forms Co_3O_4, which contains both Co(II) and Co(III), and is easily reduced to the metal by hydrogen. Cobalt(III) oxide dissolves in strong acid to give unstable brown solutions of trivalent cobalt salts. With dilute acids cobalt(II) salts are formed.

cobalt steel Any of a group of *alloy steels containing 5–12% of cobalt, 14–20% of tungsten, usually with 4% of chromium and 1–2% of vanadium. They are very hard but somewhat brittle. Their main use is in high-speed tools.

coenzyme An organic nonprotein molecule that associates with an enzyme molecule in catalysing biochemical reactions. Coenzymes usually participate in the substrate–enzyme interaction by donating or accepting certain chemical groups. Many vitamins are precursors of coenzymes. *See also* cofactor.

coenzyme A (CoA) A complex organic compound that acts in conjunction with enzymes involved in various biochemical reactions, notably the oxidation of pyruvate via the *Krebs cycle and fatty-acid oxidation and synthesis. It comprises principally the B vitamin pantothenic acid, the nucleotide adenine, and a ribose-phosphate group.

coenzyme Q (ubiquinone) Any of a group of related quinone-derived compounds that serve as electron carriers in the *electron transport chain reactions of cellular respiration. Coenzyme Q molecules have side chains of different lengths in different types of organisms but function in similar ways.

cofactor A nonprotein component essential for the normal catalytic activity of an enzyme. Cofactors may be organic molecules (*coenzymes) or inorganic ions. They may activate the enzyme by altering its shape or they may actually participate in the chemical reaction.

coherent units A system of *units of measurement in which derived units are obtained by multiplying or dividing base units without the use of numerical factors. *SI units form a coherent system; for example the unit of force is the newton, which is equal to 1 kilogram metre per second squared ($kg\,m\,s^{-2}$), the kilogram, metre, and second all being base units of the system.

coinage metals A group of three malleable ductile metals forming subgroup IB of the *periodic table: copper (Cu), silver (Ag), and gold (Au). Their outer electronic configurations have the form $nd^{10}(n+1)s^1$. Although this is similar to that of alkali metals, the coinage metals all have much higher ionization energies and higher (and positive) standard electrode potentials. Thus, they are much more difficult to oxidize and are more resistant to corrosion. In addition, the fact that they have *d*-electrons makes them show variable valency (Cu^{I}, Cu^{II}, and Cu^{III}; Ag^{I} and Ag^{II}; $Au^{I}$and Au^{III}) and form a wide range of coordination compounds. They are generally classified with the *transition elements.

coke A form of carbon made by the destructive distillation of coal. Coke is used for blast-furnaces and other metallurgical and chemical processes requiring a source of carbon. Lower-grade cokes, made by heating the coal to a lower temperature, are used as smokeless fuels for domestic heating.

colchicine An *alkaloid derived from the autumn crocus, *Colchicum autumnale*. It inhibits cell division. Colchicine is used in genetics, cytology, and plant-breeding research, and also in cancer therapy to inhibit cell division.

collagen An insoluble fibrous protein found extensively in the connective tissue of skin, tendons, and bone. The polypeptide chains of collagen (containing the amino acids glycine and proline predominantly) form triple-stranded helical coils that are bound together to form fibrils, which have great strength and limited elasticity. Collagen accounts for over 30% of the total body protein of mammals.

colligative properties Properties that depend on the concentration of particles (molecules, ions, etc.) present in a solution, and not on the nature of the particles. Examples of colligative properties are osmotic pressure (*see* osmosis), *lowering of vapour pressure, *depression of freezing point, and *elevation of boiling point.

collodion A thin film of cellulose nitrate made by dissolving the cellulose nitrate in ethanol or ethoxyethane, coating the surface, and evaporating the solvent.

colloids Colloids were originally defined by Thomas Graham in 1861 as substances, such as starch or gelatin, which will not diffuse through a membrane. He distinguished them from *crystalloids* (e.g. inorganic salts), which would pass through membranes. Later it was recognized that colloids were distinguished from true solutions by the presence of particles that were too small to be observed with a normal microscope yet were much larger than normal molecules. Colloids are now regarded as systems in which there are two or more phases, with one (the *dispersed phase*) distributed in the other (the *continuous phase*). Moreover, at least one of the phases has small dimensions (in the range 10^{-9}–10^{-6} m). Colloids are classified in various ways.

Sols are dispersions of small solid particles in a liquid. The particles may be macromolecules or may be clusters of small molecules. *Lyophobic sols* are those in which there is no affinity between the dispersed phase and the liquid. An example is silver chloride dispersed in water. In such colloids the solid particles have a surface charge, which tends to stop them coming together. Lyophobic sols are inherently unstable and in time the particles aggregate and form a precipitate. *Lyophilic sols*, on the other hand, are more like true solutions in which the solute molecules are large and have an affinity for the solvent. Starch in water is an example of such a system. *Association colloids* are systems in which the dispersed phase consists of clusters of molecules that have lyophobic and lyophilic parts. Soap in water is an association colloid (*see* micelle).

Emulsions are colloidal systems in which the dispersed and continuous phases are both liquids, e.g. oil-in-water or water-in-oil. Such systems require an emulsifying agent to stabilize the dispersed particles.

Gels are colloids in which both dispersed and continuous phases have a three-dimensional network throughout the material, so that it forms a jelly-like mass. Gelatin is a common example. One component may sometimes be removed (e.g. by heating) to leave a rigid gel (e.g. silica gel).

Other types of colloid include *aerosols* (dispersions of liquid or solid particles in a gas, as in a mist or smoke) and foams (dispersions of gases in liquids or solids).

colorimetric analysis Quantitative analysis of solutions by estimating their colour, e.g. by comparing it with the colours of standard solutions.

columbium A former name for the element *niobium.

column chromatography *See* chromatography.

combustion A chemical reaction in which a substance reacts rapidly with oxygen with the production of heat and light. Such reactions are often free-radical chain reactions, which can usually be summarized as the oxidation of carbon to form its oxides and the oxidation of hydrogen to form water. *See also* flame.

common salt *See* sodium chloride.

complex A compound in which molecules or ions form coordinate bonds to a metal atom or ion. The complex may be a positive ion (e.g. $[Cu(H_2O)_6]^{2+}$), a negative ion (e.g. $Fe[(CN)_6]^{3-}$), or a neutral molecule (e.g. $PtCl_2(NH_3)_2$). The formation of such coordination complexes is typical behaviour of transition metals. The complexes formed are often coloured and have unpaired electrons (i.e. are paramagnetic). *See also* ligand; chelate.

complexometric analysis A type of volumetric analysis in which the reaction involves the formation of an inorganic *complex.

component A distinct chemical species in a mixture. If there are no reactions taking place, the number of components is the number of separate chemical species. A mixture of water and ethanol, for instance, has two components (but is a single phase). A mixture of ice and water has two phases but one component (H_2O). If an equilibrium reaction occurs, the number of components is taken to be the number of chemical species minus the number of reactions. Thus, in

$$H_2 + I_2 \rightleftharpoons 2HI$$

there are two components. *See also* phase rule.

compound A substance formed by the combination of elements in fixed proportions. The formation of a compound involves a chemical reaction; i.e. there is a change in the configuration of the valence electrons of the atoms. Compounds, unlike mixtures, cannot be separated by physical means. *See also* molecule.

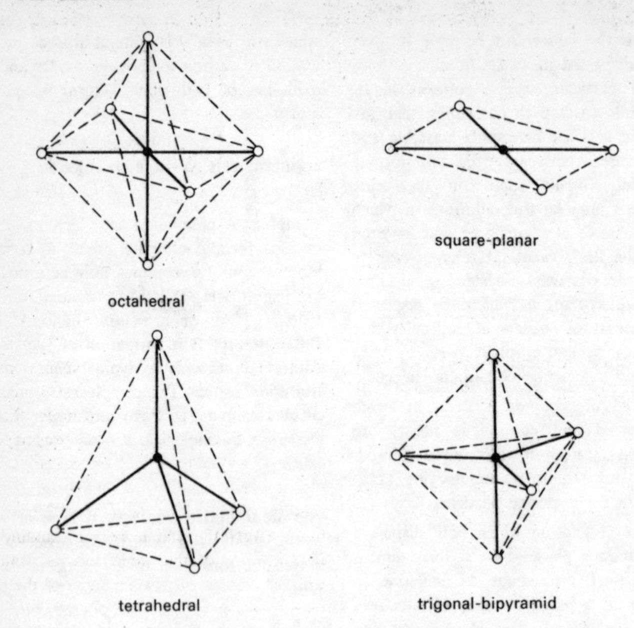

Some common shapes of coordination complexes

concentrated Describing a solution that has a relatively high concentration of solute.

concentration The quantity of dissolved substance per unit quantity of solvent in a solution. Concentration is measured in various ways. The amount of substance dissolved per unit volume (symbol *c*) has units of mol dm^{-3} or mol l^{-1}. It is now called 'concentration' (formerly *molarity*). The *mass concentration* (symbol ρ) is the mass of solute per unit volume of solvent. It has units of kg dm^{-3}, g cm^{-3}, etc. The *molal concentration* (or *molality*; symbol *m*) is the amount of substance per unit mass of solvent, commonly given in units of mol kg^{-1}. *See also* mole fraction.

concentration cell *See* cell.

condensation The change of a vapour or gas into a liquid. The change of phase is accompanied by the evolution of heat (*see* latent heat).

condensation polymerization *See* polymer.

condensation pump *See* diffusion pump.

condensation reaction A chemical reaction in which two molecules combine to form a larger molecule with elimination of a small molecule (e.g. H_2O). *See* aldehydes; ketones.

condenser A device used to cool a vapour to cause it to condense to a liquid. *See* Liebig condenser.

conductiometric titration A type of titration in which the electrical conductivity of the reaction mixture is continuously monitored as one reactant is added. The equivalence point is the point at which this undergoes a sudden change. The method is used for titrating coloured solutions, which cannot be used with normal indicators.

conduction band *See* energy bands.

conductivity water *See* distilled water.

Condy's fluid A mixture of calcium and potassium permanganates (manganate(VII)) used as an antiseptic.

configuration 1. The arrangement of atoms or groups in a molecule. **2.** The arrangement of electrons about the nucleus of an *atom.

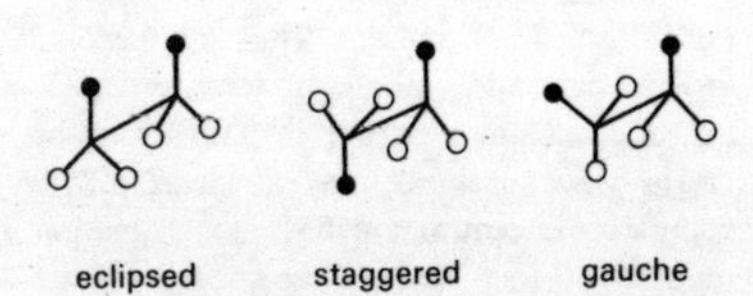

Conformation for rotation about a single bond

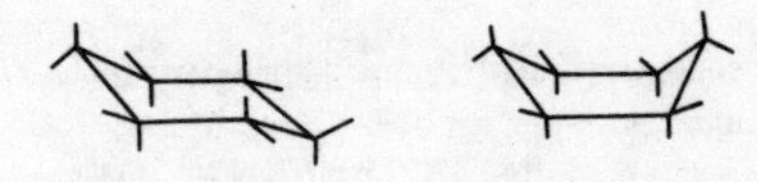

Conformations of cyclohexane ring

conformation Any of the large number of possible shapes of a molecule resulting from rotation of one part of the molecule about a single bond. See illustration.

congeners Elements that belong to the same group in the periodic table.

conjugate acid *or* **base** *See* acids.

conjugated Describing double or triple bonds in a molecule that are separated by one single bond. For example, the organic compound buta-1,3-diene, $H_2C=CH-CH=CH_2$, has conjugated double bonds. In such molecules, there is some delocalization of electrons in the pi orbitals between the carbon atoms linked by the single bond.

conservation law A law stating that the total magnitude of a certain physical property of a system, such as its mass, energy, or charge, remain unchanged even though there may be exchanges of that property between components of the system. For example, imagine a table with a bottle of salt solution (NaCl), a bottle of silver nitrate solution ($AgNO_3$), and a beaker standing on it. The mass of this table and its contents will not change even when some of the contents of the bottles are poured into the beaker. As a result of the reaction between the chemicals two new substances (silver chloride and sodium nitrate) will appear in the beaker:

$$NaCl + AgNO_3 \rightarrow AgCl + NaNO_3,$$

but the total mass of the table and its contents will not change. This *conservation of mass* is a law of wide and general applicability, which is true for the universe as a whole, provided that the universe can be considered a closed system (nothing escaping from it, nothing being added to it). According to Einstein's mass–energy relationship, every quantity of energy (E) has a mass (m), which is given by E/c^2, where c is the speed of light. Therefore if mass is conserved, the law of *conservation of energy* must be of equally wide application.

consolute temperature The temperature at which two partially miscible liquids become fully miscible as the temperature is increased.

constantan An alloy having an electrical resistance that varies only very slightly with temperature (over a limited range around normal room temperatures). It consists of copper (50–60%) and nickel (40–50%) and is used in resistance wire, thermocouples, etc.

constant-boiling mixture *See* azeotrope.

constant proportions *See* chemical combination.

contact process A process for making sulphuric acid from sulphur dioxide (SO_2), which is made by burning sulphur or by roasting sulphide ores. A mixture of sulphur dioxide and air is passed over a hot catalyst

$$2SO_2 + O_2 \rightarrow 2SO_3$$

The reaction is exothermic and the conditions are controlled to keep the temperature at an optimum 450°C. Formerly, platinum catalysts were used but vanadium–vanadium oxide catalysts are now mainly employed (although less efficient, they are less susceptible to poisoning). The sulphur trioxide is dissolved in sulphuric acid

$$H_2SO_4 + SO_3 \rightarrow H_2S_2O_7$$

and the oleum is then diluted.

continuous phase *See* colloid.

continuous spectrum *See* spectrum.

convection A process by which heat is transferred from one part of a fluid to another by movement of the fluid itself. In *natural convection* the movement occurs as a result of gravity; the hot part of the fluid expands, becomes less dense, and is displaced by the colder denser part of the fluid as this drops below it. This is the process that occurs in most domestic hot-water systems between the boiler and the hot-water cylinder. A natural convection current is set up transferring the hot water from the boiler up to the cylinder (always placed above the boiler) so that the cold water from the cylinder can move down into the boiler to be heated. In some modern systems, where small-bore pipes are used or it is inconvenient to place the cylinder above the boiler, the circulation between boiler and hot-water cylinder relies upon a pump. This is an example of *forced convection*, where hot fluid is transferred from one region to another by a pump or fan.

converter The reaction vessel in the *Bessemer process or some similar steel-making process.

coordinate bond *See* chemical bond.

coordination compound A compound in which coordinate bonds are formed (*see* chemical bond). The term is used especially for inorganic *complexes.

coordination number The number of groups, molecules, atoms, or ions surrounding a given atom or ion in a complex or crystal. For instance, in a square-planar complex the central ion has a coordination number of four. In a close-packed crystal (*see* close packing) the coordination number is twelve.

copolymer *See* polymer.

copper Symbol Cu. A red-brown *transition element; a.n. 29; r.a.m. 63.546; r.d. 8.93; m.p. 1083.4°C; b.p. 2582°C. Copper has been extracted for thousands of years; it was known to the Romans as cuprum, a name linked with the island of Cyprus. The metal is malleable and ductile and an excellent conductor of heat and electricity. Copper-containing minerals include cuprite (Cu_2O), azurite ($2CuCO_3.Cu(OH)_2$), chalcopyrite ($CuFeS_2$), and malachite ($CuCO_3.Cu(OH)_2$). Native copper appears in isolated pockets in some parts of the world. The large mines in the USA, Chile, Canada, Zambia, Zaîre, and Peru extract ores containing sulphides, oxides, and carbonates. They are usually worked by smelting, leaching, and electrolysis. Copper metal is used to make electric cables and wires. Its alloys, brass (copper–zinc) and bronze (copper–tin), are used extensively.

Water does not attack copper but in moist atmospheres it slowly forms a characteristic green surface layer (patina). The metal will not react with dilute sulphuric or hydrochloric acids, but with nitric acid oxides of nitrogen are formed. Copper compounds contain the element in the +1 and +2 oxidation states. Copper(I) compounds are mostly white (the oxide is red). Copper (II) salts are blue in solution. The metal also forms a large number of coordination complexes.

copperas *See* iron(II) sulphate.

copper(I) chloride A white solid compound, CuCl; cubic; r.d. 4.14; m.p. 430°C; b.p. 1490°C. It is obtained by boiling a solution containing copper(II) chloride, excess copper turnings, and hydrochloric acid. Copper(I) is present as the $[CuCl_2]^-$ complex ion. On pouring the solution into air-free distilled water copper(I) chloride precipitates. It must be kept free of air and moisture since it oxidizes to copper(II) chloride under those conditions.

Copper(I) chloride is essentially covalent and its structure is similar to that of diamond; i.e. each copper atom is surrounded tetrahedrally by four chlorine atoms and vice versa. In the vapour phase, dimeric and trimeric species are present. Copper(I) chloride is used in conjunction with ammonium chloride as a catalyst in the dimerization of ethyne to but-1-ene-3-yne (vinyl acetylene), which is used in the production of synthetic rubber. In the laboratory a mixture of copper(I) chloride and hydrochloric acid is used for converting benzene diazonium chloride to chlorobenzene – the Sandmeyer reaction.

copper(II) chloride A brown-yellow powder, $CuCl_2$; r.d. 3.386; m.p. 620°C. It exists as a blue-green dihydrate (rhombic; r.d. 2.54; loses H_2O at 100°C). The anhydrous solid is obtained by passing chlorine over heated copper. It is predominantly covalent and adopts a layer structure in which each copper atom is surrounded by four chlorine atoms at a distance of 0.23 and two more at a distance of 0.295. A concentrated aqueous solution is dark brown in colour due to the presence of complex ions such as $[CuCl_4]^{2-}$. On dilution the colour changes to green and then blue because of successive replacement of chloride ions by water molecules, the final colour being that of the $[Cu(H_2O)_6]^{2+}$ ion. The dihydrate can be obtained by crystallizing the solution.

copper glance A mineral form of copper(I) sulphide, Cu_2S.

copper(II) nitrate A blue deliquescent solid, $Cu(NO_3)_2.3H_2O$; r.d. 2.32; m.p. 114.5°C. It may be obtained by reacting either copper(II) oxide or copper(II) carbonate with dilute nitric acid and crystallizing the resulting solution. Other hydrates containing 6 or 9 molecules of water are known. On heating it readily decomposes to give copper(II) oxide, nitrogen dioxide, and oxygen. The anhydrous form can be obtained by reacting copper with a solution of nitrogen dioxide in ethyl ethanoate. It sublimes on heating suggesting that it is appreciably covalent.

copper(I) oxide A red insoluble solid, Cu_2O; r.d. 6.0; m.p. 1235°C. It is obtained by reduction of an alkaline solution of copper(II) sulphate. Since the addition of alkalis to a solution of copper(II) salt results in the precipitation of copper(II) hydroxide the copper(II) ions are complexed with tartrate ions; under such conditions the concentration of copper(II) ions is so low that the solubility product of copper(II) hydroxide is not exceeded.

When copper(I) oxide reacts with dilute sulphuric acid a solution of copper(II) sulphate and a deposit of copper results, i.e. disproportionation occurs.

$Cu_2O + 2H^+ \rightarrow Cu^{2+} + Cu + H_2O$

When dissolved in concentrated hydrochloric acid the $[CuCl_2]^-$ complex ion is formed. Copper(I) oxide is used in the manufacture of rectifiers and the production of red glass.

copper(II) oxide A black insoluble solid, CuO; monoclinic; r.d. 6.3; m.p. 1326°C. It is obtained by heating either copper(II) carbonate or copper(II) nitrate. It decomposes on heating above 800°C to copper(I) oxide and oxygen. Copper(II) oxide reacts readily with mineral acids on warming, with the formation of copper(II) salts; it is also readily reduced to copper on heating in a stream of hydrogen. Copper(II) oxide is soluble in dilute acids forming blue solutions of cupric salts.

copper pyrites *See* chalcopyrite.

copper(II) sulphate A blue crystalline solid, $CuSO_4.5H_2O$; triclinic; r.d. 2.284. The pentahydrate loses $4H_2O$ at 110°C and the fifth H_2O at 150°C to form the white anhydrous compound (rhombic; r.d. 3.6; decomposes above 200°C). The pentahydrate is prepared either by reacting copper(II) oxide or copper(II) carbonate with dilute sulphuric acid; the solution is heated to saturation and the blue pentahydrate crystallizes out on cooling (a few drops of dilute sulphuric acid are generally added to prevent hydrolysis). It is obtained on an industrial scale by forcing air through a hot mixture of copper and dilute sulphuric acid. In the pentahydrate each copper(II) ion is surrounded by four water molecules at the corner of a square, the fifth and sixth octahedral positions are occupied by oxygen atoms from the sulphate anions, and the fifth water molecule is held in place by hydrogen bonding. Copper(II) sulphate has many industrial uses, including the preparation of the Bordeaux mixture (a fungicide) and the preparation of other copper compounds. It is also used in electroplating and textile dying and as a timber preservative. The anhydrous form is used in the detection of traces of moisture.

Copper(II) sulphate pentahydrate is also known as *blue vitriol.*

cordite An explosive mixture of cellulose nitrate and nitroglycerin, with added plasticizers and stabilizers, used as a propellant for guns.

CORN rule *See* absolute configuration.

corrosion Chemical or electrochemical attack on the surface of a metal. *See also* electrolytic corrosion; rusting.

corundum A mineral form of aluminium oxide, Al_2O_3. It crystallizes in the trigonal system and occurs as well-developed hexagonal crystals. It is colourless and transparent when pure but the presence of other elements gives rise to a variety of colours. *Ruby is a red variety containing chromium; *sapphire is a blue variety containing iron and titanium. Corundum occurs as a rock-forming mineral in both metamorphic and igneous rocks. It is chemically resistant to weathering processes and so also occurs in alluvial (placer) deposits. The second hardest mineral after diamond (it has a hardness of 9 on the Mohs' scale), it is used as an abrasive.

coulomb Symbol C. The *SI unit of electric charge. It is equal to the charge transferred by a current of one ampere in one second. The unit is named after Charles de Coulomb (1736–1806), a French physicist.

coupling A type of chemical reaction in which two molecules join together; for example, the formation of an *azo compound by coupling of a diazonium ion with a benzene ring.

covalent bond *See* chemical bond.

covalent crystal A crystal in which the atoms are held together by covalent bonds. Covalent crystals are sometimes called *macromolecular* or *giant-molecular crystals*. They are hard high-melting substances. Examples are diamond and boron nitride.

covalent radius An effective radius assigned to an atom in a covalent compound. In the case of a simple diatomic molecule, the covalent radius is half the distance between the nuclei. Thus, in Cl_2 the internuclear distance is 0.198 nm so the covalent radius is taken to be 0.099 nm. Covalent radii can also be calculated for multiple bonds; for instance, in the case of carbon the values are 0.077 nm for single bonds, 0.0665 nm for double bonds, and 0.0605 nm for triple bonds. The values of different covalent radii can sometimes be added to give internuclear distances. For example, the length of the bond in interhalogens (e.g. ClBr) is nearly equal to the sum of the covalent radii of the halogens involved. This, however, is not always true because of other effects (e.g. ionic contributions to the bonding).

cracking The process of breaking down chemical compounds by heat. The term is applied particularly to the cracking of hydrocarbons in the kerosine fraction obtained from *petroleum refining to give smaller hydrocarbons and alkenes. It is an important process, both as a source of branched-chain hydrocarbons suitable for gasoline (for motor fuel) and as a source of ethene and other alkenes. *Catalytic cracking* is a similar process in which a catalyst is used to lower the temperature required and to modify the products obtained.

cream of tartar *See* potassium hydrogentartrate.

creosote 1. (wood creosote) An almost colourless liquid mixture of phenols obtained by distilling tar obtained by the destructive distillation of wood. It is used medically as an antiseptic and expectorant. **2. (coal-tar creosote)** A dark liquid mixture of phenols and cresols obtained by distilling coal tar. It is used for preserving timber.

cresols *See* methylphenols.

cristobalite A mineral form of *silicon(IV) oxide, SiO_2.

critical pressure The pressure of a fluid in its *critical state; i.e. when it is at its critical temperature and critical volume.

critical state The state of a fluid in which the liquid and gas phases both have the same density. The fluid is then at its critical temperature, critical pressure, and critical volume.

critical temperature 1. The temperature above which a gas cannot be liquefied by an increase of pressure. *See also* critical state. **2.** *See* transition point.

critical volume The volume of a fixed mass of a fluid in its *critical state; i.e. when it is at its critical temperature and critical pressure. The *critical specific volume* is its volume per unit mass in this state: in the past this has often been called the critical volume.

cross linkage A short side chain of atoms linking two longer chains in a polymeric material.

crucible A dish or other vessel in which substances can be heated to a high temperature.

crude oil *See* petroleum.

cryogenic pump A *vacuum pump in which pressure is reduced by condensing gases on surfaces maintained at about 20 K by means of liquid hydrogen or at 4 K by means of liquid helium. Pressures down to 10^{-8} mmHg (10^{-6} Pa) can be maintained; if they are used in conjunction with a *diffusion pump, pressures as low as 10^{-15} mmHg (10^{-13} Pa) can be reached.

cryohydrate A eutectic mixture of ice and some other substance (e.g. an ionic salt) obtained by freezing a solution.

cryolite A rare mineral form of sodium aluminofluoride, Na_3AlF_6, which crystallizes in the monoclinic system. It is usually white but may also be colourless. The only important occurrence of the mineral is in Greenland. It is used chiefly as a flux in the production of aluminium from bauxite.

cryoscopic constant *See* depression of freezing point.

cryostat A vessel enabling a sample to be maintained at a very low temperature. The *Dewar flask is the most satisfactory vessel for controlling heat leaking in by radiation, conduction, or convection. Cryostats usually consist of two or more Dewar flasks nesting in each other. For example, a liquid nitrogen bath is often used to cool a Dewar flask containing a liquid helium bath.

crystal A solid with a regular polyhedral shape. All crystals of the same substance grow so that they have the same angles between their faces. However, they may not have the same external appearance because different faces can grow at different rates, depending on the conditions. The external form of the crystal is referred to as the *crystal habit*. The atoms, ions, or molecules forming the crystal have a regular arrangement and this is the *crystal structure*.

crystal-field theory A theory of the electronic structures of inorganic *complexes, in which the complex is assumed to consist of a central metal atom or ion surrounded by ligands that are negative ions. For example, the complex $[PtCl_4]^{2-}$ is thought of as a Pt^{2+} ion surrounded by four Cl^- ions at the corners of a square. The presence of these ions affects the energies of the *d*-orbitals, causing a splitting of energy levels. The theory can be used to explain the spectra of complexes and their magnetic properties. *Ligand-field theory* is a development of crystal-field theory in which the overlap of orbitals is taken into account.

crystal habit *See* crystal.

crystal lattice The regular pattern of atoms, ions, or molecules in a crystalline substance. A crystal lattice can be regarded as produced by repeated translations of a *unit cell* of the lattice. *See also* crystal system.

crystalline Having the regular internal arrangement of atoms, ions, or molecules characteristic of crystals. Crystalline materials need not necessarily exist as crystals; all metals, for example, are crystalline although they are not usually seen as regular geometric crystals.

crystallite A small crystal, e.g. one of the small crystals forming part of a microcrystalline substance.

crystallization The process of forming crystals from a liquid or gas.

crystallography The study of crystal form and structure. *See also* X-ray crystallography.

crystalloids *See* colloids.

crystal structure *See* crystal.

crystal system A method of classifying crystalline substances on the basis of their unit cell. There are seven crystal systems. If the cell is a parallelopiped with sides *a*, *b*, and *c* and if α is the angle between *b* and *c*, β the angle between *a* and *c*, and γ the angle between *a* and *b*, the systems are:
(1) *cubic* $a=b=c$ and $\alpha=\beta=\gamma=90°$
(2) *tetragonal* $a=b=c$ and $\alpha=\beta=\gamma=90°$
(3) *rhombic* (or *orthorhombic*) $a=b=c$ and $\alpha=\beta=\gamma=90°$
(4) *hexagonal* $a=b=c$ and $\alpha=\beta=\gamma=90°$
(5) *trigonal* $a=b=c$ and $\alpha=\beta=\gamma=90°$
(6) *monoclinic* $a=b=c$ and $\alpha=\gamma=90°=\beta$
(7) *triclinic* $a=b=c$ and $\alpha=\beta=\gamma$

CS gas The vapour from a white solid, $C_6H_4(Cl)CH{:}C(CN)_2$, causing tears and choking, used in 'crowd control'.

cubic close packing *See* close packing.

cubic crystal A crystal in which the unit cell is a cube (*see* crystal system). There are three possible packings for cubic crystals: *simple cubic*, *face-centred cubic*, and *body-centred cubic*. See illustration.

cumene process An industrial process for making phenol from benzene. A mixture of benzene vapour and propene is passed over a phosphoric acid catalyst at 250°C and high pressure

$$C_6H_6 + CH_3CH{:}CH_2 \rightarrow C_6H_5CH(CH_3)_2$$

The product is called *cumene*, and it can be oxidized in air to a peroxide, $C_6H_5C(CH_3)_2O_2H$. This reacts with dilute acid to give phenol (C_6H_5OH) and propanone (acetone, CH_3OCH_3), which is a valuable by-product.

cupellation A method of separating noble metals (e.g. gold or silver) from base metals (e.g. lead) by melting the mixture with a blast of hot air in a shallow porous dish (the *cupel*). The base metals are oxidized, the oxide being carried away by the blast of air or absorbed by the porous container.

cuprammonium ion The tetraamminecopper(II) ion $[Cu(NH_3)_4]^{2+}$. *See* ammine.

cupric compounds Compounds containing copper in its higher (+2) oxidation state; e.g. cupric chloride is copper(II) chloride ($CuCl_2$).

cuprite A red mineral cubic form of copper (I) oxide, Cu_2O; an important ore of copper. It occurs where deposits of copper have been subjected to oxidation. The mineral has been mined as a copper ore in Chile, Zaîre, Bolivia, Australia, the USSR, and the USA.

cuprous compounds Compounds containing copper in its lower (+1) oxidation state; e.g. cuprous chloride is copper(I) chloride (CuCl).

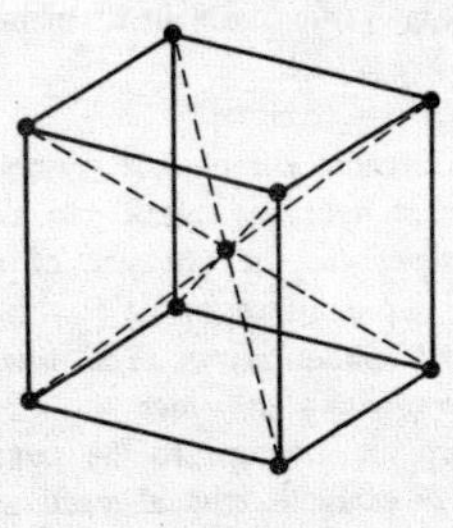
body-centred

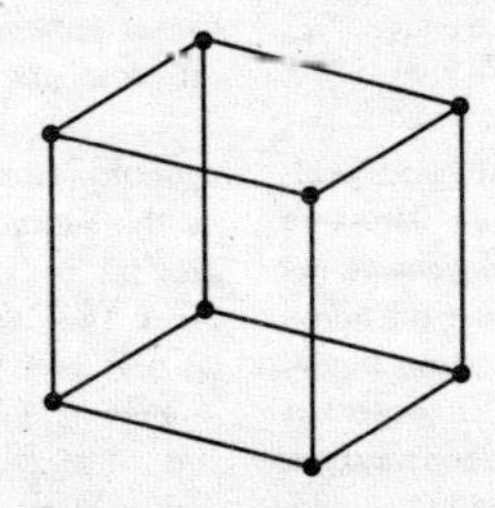
simple cubic

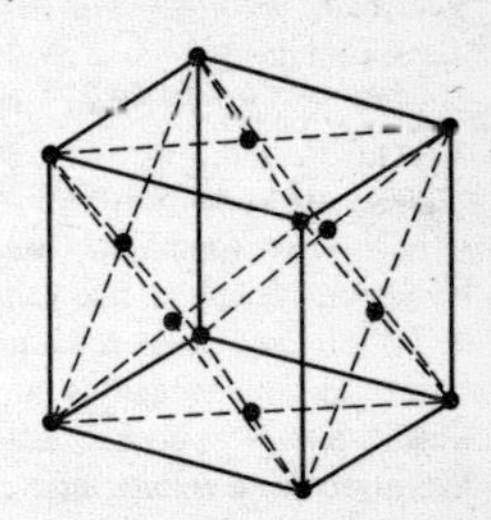
face-centred

Cubic crystal structures

curie The former unit of *activity (*see* radiation units). It is named after the Polish-born French physicist Marie Curie (1867–1934).

Curie point (Curie temperature) The temperature at which a ferromagnetic substance loses its ferromagnetism and becomes only paramagnetic. For iron the Curie point is 760°C and for nickel 356°C.

curium Symbol Cm. A radioactive metallic transuranic element belonging to the *actinoids; a.n. 96; mass number of the most stable isotope 247 (half-life 1.64×10^7 years); r.d. (calculated) 13.51; m.p. 1340±40°C. There are nine known isotopes. The element was first identified by G. T. Seaborg and associates in 1944 and first produced by L. B. Werner and I. Perlman in 1947 by bombarding americium–241 with neutrons.

cyanamide 1. An inorganic salt containing the ion CN_2^{2-}. *See* calcium cyanamide. **2.** A colourless crystalline solid, H_2NCN, made by the action of carbon dioxide on hot sodamide. It is a weakly acidic compound (the parent acid of cyanamide salts) that is soluble in water and ethanol. It is hydrolysed to urea in acidic solutions.

cyanamide process *See* calcium cyanamide.

cyanate *See* cyanic acid.

cyanic acid An unstable explosive acid, HOCN. The compound has the structure H–O–C≡N, and is also called *fulminic acid*. Its salts and esters are *cyanates* (or *fulminates*). The compound is a volatile liquid, which readily polymerizes. In water it hydrolyses to ammonia and carbon dioxide. It is isomeric with another acid, H–N=C=O, which is known as *isocyanic acid*. Its salts and esters are *isocyanates*.

cyanide 1. An inorganic salt containing the cyanide ion CN^-. Cyanides are extremely poisonous because of the ability of the CN^- ion to coordinate with the iron in haemoglobin, thereby blocking the uptake of oxygen by the blood. **2.** A metal coordination complex formed with cyanide ions.

cyanide process A method of extracting gold by dissolving it in potassium cyanide (to form the complex ion $[Au(CN)_2]^-$). The ion can be reduced back to gold with zinc.

cyanine dyes A class of dyes that contain a –CH= group linking two nitrogen-containing heterocyclic rings. They are used as sensitizers in photography.

cyanocobalamin *See* vitamin B complex.

cyanogen A colourless gas, $(CN)_2$, with a pungent odour; soluble in water, ethanol, and ether; d. 2.335 g dm^{-3}; m.p. –27.9°C; b.p. –20.7°C. The compound is very toxic. It may be prepared in the laboratory by heating mercury(II) cyanide; industrially it is made by gas-phase oxidation of hydrogen cyanide using air over a silver catalyst, chlorine over activated silicon(IV) oxide, or nitrogen dioxide over a copper(II) salt. Cyanogen is an important intermediate in the preparation of various fertilizers and is also used as a stabilizer in making nitrocellulose. It is an example of a *pseudohalogen.

cyano group The group –CN in a chemical compound. *See* nitrile.

cyanohydrins Organic compounds formed by the addition of hydrogen cyanide to aldehydes or ketones (in the presence of a base). The first step is attack by a CN^- ion on the carbonyl carbon atom. The final product is a compound in which a –CN and –OH group are attached to the same carbon atom. For example, ethanal reacts as follows

$$CH_3CHO + HCN \rightarrow$$

$$CH_3CH(OH)(CN)$$

The product is 2-hydroxypropanonitrile. Cyanohydrins of this type can be oxidized to α-hydroxy carboxylic acids.

cyanuric acid A white crystalline water-soluble trimer of cyanic acid, $(HNCO)_3$. It is a cyclic compound having a six-membered ring made of alternating imide (NH) and carbonyl (CO) groups.

cyclamates Salts of the acid, $C_6H_{11}.NH.SO_3H$, where $C_6H_{11}-$is a cyclohexyl group. Sodium and calcium cyclamates were formerly used as sweetening agents in soft drinks, etc, until their use was banned when they were suspected of causing cancer.

cyclic Describing a compound that has a ring of atoms in its molecules. In *homocyclic* compounds all the atoms in the ring are the same type, e.g. benzene (C_6H_6) and cyclohexane (C_6H_{12}). These two examples are also examples of *carbocyclic* compounds; i.e. the rings are of carbon atoms. If different atoms occur in the ring, as in pyridine (C_5H_5N), the compound is said to be *heterocyclic.*

cyclic AMP A derivative of *ATP that is widespread in animal cells as an intermediate messenger in many biochemical reactions induced by hormones. Upon reaching their target cells, the hormones activate the enzyme that catalyses cyclic AMP production. Cyclic AMP ultimately activates the enzymes of the reaction induced by the hormone concerned. Cyclic AMP is also involved in controlling gene expression and cell division, in immune responses, and in nervous transmission.

cyclization The formation of a cyclic compound from an open-chain compound. *See* ring.

cyclo- Prefix designating a cyclic compound, e.g. a cycloalkane or a cyclosilicate.

cycloalkanes Cyclic saturated hydrocarbons containing a ring of carbon atoms joined by single bonds. They have the general formula C_nH_{2n}, for example cyclohexane, C_6H_{12}, etc. In general they behave like the *alkanes but are rather less reactive.

cyclohexadiene-1,4-dione (benzoquinone; quinone) A yellow solid, $C_6H_4O_2$; r.d. 1.3; m.p. 116°C. It has a six-membered ring of carbon atoms with two opposite carbon atoms linked to oxygen atoms (C=O) and the other two pairs of carbon atoms linked by double bonds (HC=CH). The compound is used in making dyes. *See also* quinhydrone electrode.

cyclohexane A colourless liquid *cycloalkane, C_6H_{12}; r.d. 0.78; m.p. 6.5°C; b.p. 81°C. It occurs in petroleum and is made by passing benzene and hydrogen under pressure over a heated Raney nickel catalyst at 150°C, or by the reduction of cyclohexanone. It is used as a solvent and paint remover and can be oxidized using hot concentrated nitric acid to hexanedioic acid (adipic acid). The cyclohexane ring is not planar and can adopt boat and chain *conformations; in formulae it is represented by a single hexagon.

cyclonite (RDX) A highly explosive nitro compound, $(CH_2N.NO_2)_3$. It has a cyclic structure with a six-membered ring of alternating CH_2 groups and nitrogen atoms, with each nitrogen being attached to a NO_2 group. It is made by nitrating hexamine, $C_6H_{12}N_4$, which is obtained from ammonia and methanal. Cyclonite is a very powerful explosive used mainly for military purposes.

cyclopentadiene A colourless liquid cyclic *alkene, C_5H_6; r.d. 0.8047; b.p. 42.5°C. It is prepared as a by-product during the fractional distillation of crude benzene from coal tar. It undergoes condensation reactions with ketones to give highly coloured compounds (fulvenes) and readily undergoes polymerization at room temperature to give the dimer, dicyclopentadiene. The compound itself is not *aromatic because it does not have the required number of pi electrons. However, removal of a hydrogen atom produces the stable *cyclopentadienyl ion*, $C_5H_5^-$, which does have aromatic properties. In particular, the ring can coordinate to positive ions in such compounds as *ferrocene.

cyclopentadienyl ion *See* cyclopentadiene.

cysteine *See* amino acid.

cystine A molecule resulting from the oxidation reaction between the sulphydryl (–SH) groups of two cysteine molecules (*see* amino acid). This often occurs between adjacent cysteine residues in polypeptides. The resultant *disulphide bonds* (–S–S–) are important in stabilizing the structure of protein molecules.

cytidine A nucleoside comprising one cytosine molecule linked to a D-ribose sugar molecule. The derived nucleotides, cytidine mono-, di-, and triphosphate (CMP, CDP, and CTP respectively) participate in various biochemical reactions, notably in phospholipid synthesis.

cytochrome Any of a group of proteins, each with an iron-containing *haem group, that form part of the electron transport chain in mitochondria and chloroplasts. Electrons are transferred by reversible changes in the iron atom between the reduced Fe(II) and oxidized Fe(III) states.

cytosine A *pyrimidine derivative. It is one of the principal component bases of *nucleotides and the nucleic acids *DNA and *RNA.

D

dalton *See* atomic mass unit.

Dalton's atomic theory A theory of *chemical combination, first stated by the British chemist John Dalton (1766–1844) in 1803. It involves the following postulates:

(1) Elements consist of indivisible small particles (atoms).

(2) All atoms of the same element are identical; different elements have different types of atom.

(3) Atoms can neither be created nor destroyed.

(4) 'Compound elements' (i.e. compounds) are formed when atoms of different elements join in simple ratios to form 'compound atoms' (i.e. molecules).

Dalton also proposed symbols for atoms of different elements (later replaced by the present notation using letters).

Dalton's law The total pressure of a mixture of gases or vapours is equal to the sum of the partial pressures of its components, i.e. the sum of the pressures that each component would exert if it were present alone and occupied the same volume as the mixture of gases. Strictly speaking, the principle is true only for ideal gases.

Daniell cell A type of primary *voltaic cell with a copper positive electrode and a negative electrode of a zinc amalgam. The zinc-amalgam electrode is placed in an electrolyte of dilute sulphuric acid or zinc sulphate solution in a porous pot, which stands in a solution of copper sulphate in which the copper electrode is immersed. While the reaction takes place ions move through the porous pot, but when it is not in use the cell should be dismantled to prevent the diffusion of one electrolyte into the other. The e.m.f. of the cell is 1.08 volts with sulphuric acid and 1.10 volts with zinc sulphate. It was invented in 1836 by the British chemist John Daniell (1790–1845).

dating techniques Methods of estimating the age of rocks, palaeontological specimens, archaeological sites, etc. *Relative dating techniques* date specimens in relation to one another; for example, *stratigraphy is used to establish the succession of fossils. *Absolute* (or *chronometric*) *techniques* give an absolute estimate of the age and fall into two main groups. The first depends on the existence of something that develops at a seasonally varying rate, as in *dendrochronology and *varve dating. The other uses some measurable change that occurs at a known rate, as in *chemical dating, *radioactive* (or *radiometric*) *dating* (*see* carbon dating; fission-track dating; potassium–argon dating; rubidium–strontium dating; uranium–lead dating), and thermoluminescence.

dative bond *See* chemical bond.

daughter 1. A nuclide produced by radioactive *decay of some other nuclide (the *parent*). **2.** An ion or free radical produced by dissociation or reaction of some other (*parent*) ion or radical.

Davy lamp An oil-burning miner's safety lamp invented by Sir Humphry Davy (1778–1829) in 1816 when investigating firedamp (methane) explosions in coal mines. The lamp has a metal gauze surrounding the flame, which cools the hot gases by conduction and prevents ignition of gas outside the gauze. If firedamp is present it burns within the gauze cage, and lamps of this type are still used for testing for gas.

***d*-block elements** The block of elements in the *periodic table consisting of scandium, yttrium, and lanthanum together with the three periods of transition elements: titanium to zinc, zirconium to cadmium, and hafnium to mercury. These elements all have two outer *s*-electrons and have *d*-electrons in their penultimate shell; i.e. an outer electron configuration of the form $(n-1)d^x ns^2$, where x is 1 to 10. *See also* transition elements.

DDT Dichlorodiphenyltrichloroethane; a colourless organic crystalline compound, $(ClC_6H_4)_2CH(CCl_3)$, made by the reaction of trichloromethanal with chlorobenzene. DDT is the best known of a number of chlorine-containing pesticides used extensively in agriculture in the 1940s and 50s. The compound is stable, accumulates in the soil, and concentrates in fatty tissue, reaching dangerous levels in carnivores high in the food chain. Restrictions are now placed on the use of DDT and similar pesticides.

Deacon process A former process for making chlorine by oxidizing hydrogen chloride in air at 450°C using a copper chloride catalyst. It was patented in 1870 by Henry Deacon (1822–76).

deamination The removal of an amino group ($-NH_2$) from a compound. Enzymatic deamination occurs in the liver and is important in amino-acid metabolism, especially in their degradation and subsequent oxidation. The amino group is removed as ammonia and excreted, either unchanged or as urea or uric acid.

de Broglie wavelength The wavelength of the wave associated with a moving particle. The wavelength (λ) is given by $\lambda = h/mv$, where h is the Planck constant, m is the mass of the particle, and v its velocity. The *de Broglie wave* was first suggested by the French physicist Louis de Broglie (1892–) in 1924 on the grounds that electromagnetic waves can be treated as particles (photons) and one could therefore expect particles to behave in some circumstances like waves. The subsequent observation of electron diffraction substantiated this argument and the de Broglie wave became the basis of *wave mechanics.

debye A unit of electric dipole moment in the electrostatic system, used to express dipole moments of molecules. It is the dipole moment produced by two charges of opposite sign, each of 1 statcoulomb and placed 10^{-18} cm apart, and has the value 3.335 64 $\times$ 10^{-30} coulomb metre.

deca- Symbol da. A prefix used in the metric system to denote ten times. For example, 10 coulombs = 1 decacoulomb (daC).

decahydrate A crystalline hydrate containing ten molecules of water per molecule of compound.

decanoic acid (capric acid) A white crystalline straight-chain saturated *carboxylic acid, $CH_3(CH_2)_8COOH$; m.p. 315°C. Its esters are used in perfumes and flavourings.

decay The spontaneous transformation of one radioactive nuclide into a daughter nuclide, which may be radioactive or may not, with the emission of one or more particles or photons. The decay of N_0 nuclides to give N nuclides after time t is given by $N = N_0 \exp(-\gamma t)$, where γ is called the *decay constant* or the *disintegration constant*. The reciprocal of the decay constant is the *mean life*. The time required for half the original nuclides to decay (i.e. $N = \frac{1}{2}N_0$) is called the *half-life* of the nuclide. The same terms are applied to elementary particles that spontaneously transform into other particles. For example, a free neutron decays into a proton and an electron.

deci- Symbol d. A prefix used in the metric system to denote one tenth. For example, 0.1 coulomb = 1 decicoulomb (dC).

decomposition Chemical reaction in which a compound breaks down into simpler compounds or into elements.

decrepitation A crackling noise produced when certain crystals are heated, caused by changes in structure resulting from loss of water of crystallization.

defect A discontinuity in a crystal lattice. A *point defect* consists either of a missing atom or ion creating a *vacancy* in the lattice (a vacancy is sometimes called a *Schottky defect*) or an extra atom or ion between two normal lattice points creating an *interstitial*. A *Frenkel defect* consists of a vacancy in which the missing atom or ion has moved to an interstitial position. If more than one adjacent point defect occurs in a crystal there may be a slip along a surface causing a *line defect* (or *dislocation*). Defects are caused by strain or, in some cases, by irradiation. All crystalline solids contain an equilibrium number of point defects above absolute zero; this number increases with temperature. The existence of defects in crystals is important in the conducting properties of semiconductors.

definite proportions *See* chemical combination.

degradation A type of organic chemical reaction in which a compound is converted into a simpler compound in stages.

degree A division on a *temperature scale.

degrees of freedom 1. The number of independent parameters required to specify the configuration of a system. This concept is applied in the *kinetic theory to specify the number of independent ways in which an atom or molecule can take up energy. There are however various sets of parameters that may be chosen, and the details of the consequent theory vary with the choice. For example, in a monatomic gas each atom may be allotted three degrees of freedom, corresponding to the three coordinates in space required to specify its position. The mean energy per atom for each degree of freedom is the same, according to the principle of the *equipartition of energy, and is equal to $kT/2$ for each degree of freedom (where k is the *Boltzmann constant and T is the thermodynamic temperature). Thus for a monatomic gas the total molar energy is $3LkT/2$, where L is the Avogadro constant (the number of atoms per mole). As $k = R/L$, where R is the molar gas constant, the total molar energy is $3RT/2$.

In a diatomic gas the two atoms require six coordinates between them, giving six degrees of freedom. Commonly these are interpreted as six independent ways of storing energy: on this basis the molecule has three degrees of freedom for different directions of translational motion, and in addition there are two degrees of freedom for rotation of the molecular axis and one vibrational degree of freedom along the bond between the atoms. The rotational degrees of freedom each contribute their share, $kT/2$, to the total energy; similarly the vibrational degree of freedom has an equal share of kinetic energy and must on average have as much potential energy. The total energy per molecule for a diatomic gas is therefore $3kT/2$ (for translational energy of the whole molecule) plus $2kT/2$ (for rotational energy of each atom) plus $2kT/2$ (for vibrational energy), i.e. a total of $7kT/2$.

2. The least number of independent variables required to define the state of a system in the *phase rule. In this sense a gas has two degrees of freedom (e.g. temperature and pressure).

dehydration 1. Removal of water from a substance. **2.** A chemical reaction in which a compound loses hydrogen and oxygen in the ratio 2:1. For instance, ethanol passed over hot pumice undergoes dehydration to ethene:

$$C_2H_5OH - H_2O \rightarrow CH_2{:}CH_2$$

Substances such as concentrated sulphuric acid, which can remove H_2O in this way, are known as *dehydrating agents*. For example, with sulphuric acid, methanoic acid gives carbon monoxide:

$$HCOOH - H_2O \rightarrow CO$$

dehydrogenase Any enzyme that catalyses the removal of hydrogen atoms in biological reactions. Dehydrogenases occur in many biochemical pathways but are particularly important in driving the electron-transport-chain reactions of cell respiration. They work in conjunction with the hydrogen-accepting coenzymes NAD and FAD.

deliquescence The absorption of water from the atmosphere by a hygroscopic solid to such an extent that a concentrated solution of the solid eventually forms.

delocalization In certain chemical compounds the valence electrons cannot be regarded as restricted to definite bonds between the atoms but are 'spread' over several atoms in the molecule. Such electrons are said to be *delocalized*. Delocalization occurs particularly when the compound contains alternating (conjugated) double or triple bonds, the delocalized electrons being those in the pi *orbitals. The molecule is then more stable than it would be if the electrons were localized, an effect accounting for the properties of benzene and other aromatic compounds. Another example is in the ions of carboxylic acids, containing the carboxylate group $-COO^-$. In terms of a simple model of chemical bonding, this group would have the carbon joined to one oxygen by a double bond (i.e. $C=O$) and the other joined to O^- by a single bond ($C-O^-$). In fact, the two $C-O$ bonds are identical because the extra electron on the O^- and the electrons in the pi bond of $C=O$ are delocalized over the three atoms.

delta-iron *See* iron.

denature 1. To add a poisonous or unpleasant substance to ethanol to make it unsuitable for human consumption (*see* methylated spirits). **2.** To produce a structural change in a protein or nucleic acid that results in the reduction or loss of its biological properties. Denaturation is caused by heat, chemicals, and extremes of pH. The differences between raw and boiled eggs are largely a result of denaturation. **3.** To add another isotope to a fissile material to make it unsuitable for use in a nuclear weapon.

dendrite A crystal that has branched in growth into two parts. Crystals that grow in this way (*dendritic growth*) have a branching treelike appearance.

dendrochronology An absolute *dating technique using the growth rings of trees. It depends on the fact that trees in the same locality show a characteristic pattern of growth rings resulting from climatic conditions. Thus it is possible to assign a definite date for each growth ring in living trees, and to use the ring patterns to date fossil trees or specimens of wood (e.g. used for buildings or objects on archaeological sites) with lifespans that overlap those of living trees. The bristlecone pine (*Pinus aristata*), which lives for up to 5000 years, has been used to date specimens over 8000 years old. Fossil specimens accurately dated by dendrochronology have been used to make corrections to the *carbon-dating technique. Dendrochronology is also helpful in studying past climatic conditions. Analysis of trace elements in sections of rings can also provide information on past atmospheric pollution.

denitrification A chemical process in which nitrates in the soil are reduced to molecular nitrogen, which is released into the atmosphere. This process is effected by the bacterium *Pseudomonas denitrificans*, which uses nitrates as a source of energy for other chemical reactions in a manner similar to respiration in other organisms. *Compare* nitrification. *See* nitrogen cycle.

density The mass of a substance per unit of volume. In *SI units it is measured in kg m^{-3}. *See also* relative density; vapour density.

deoxyribonucleic acid *See* DNA.

depolarization The prevention of *polarization in a *primary cell. For example, maganese(IV) oxide (the *depolarizer*) is placed around the positive electrode of a *Leclanché cell to oxidize the hydrogen released at this electrode.

depression of freezing point The reduction in the freezing point of a pure liquid when another substance is dissolved in it. It is a *colligative property – i.e. the lowering of the freezing point is proportional to the number of dissolved particles (molecules or ions), and does not depend on their nature. It is given by $\Delta t = K_f C_m$, where C_m is the molar concentration of dissolved solute and K_f is a constant (the *cryoscopic constant*) for the solvent used. Measurements of freezing-point depression (using a Beckmann thermometer) can be used for finding relative molecular masses of unknown substances.

derivative A compound that is derived from some other compound and usually maintains its general structure, e.g. trichloromethane (chloroform) is a derivative of methane.

derived unit *See* base unit.

desalination The removal of salt from sea water for irrigation of the land or to provide drinking water. The process is normally only economic if a cheap source of energy, such as the waste heat from a nuclear power station, can be used. Desalination using solar energy has the greatest economic potential since shortage of fresh water is most acute in hot regions. The methods employed include evaporation, often under reduced pressure (flash evaporation); freezing (pure ice forms from freezing brine); *reverse osmosis; *electrodialysis; and *ion exchange.

desiccator A container for drying substances or for keeping them free from moisture. Simple laboratory desiccators are glass vessels containing a drying agent, such as silica gel. They can be evacuated through a tap in the lid.

desorption The removal of adsorbed atoms, molecules, or ions from a surface.

destructive distillation The process of heating complex organic substances in the absence of air so that they break down into a mixture of volatile products, which are condensed and collected. At one time the destructive distillation of coal (to give coke, coal tar, and coal gas) was the principal source of industrial organic chemicals.

detergent A substance added to water to improve its cleaning properties. Although water is a powerful solvent for many compounds, it will not dissolve grease and natural oils. Detergents are compounds that cause such nonpolar substances to go into solution in water. *Soap is the original example, owing its action to the presence of ions formed from long-chain fatty acids (e.g. the octadecanoate (stearate) ion, $CH_3(CH_2)_{16}COO^-$). These have two parts: a nonpolar part (the hydrocarbon chain), which attaches to the grease; and a polar part (the $-COO^-$ group), which is attracted to the water. A disadvantage of soap is that it forms a scum with hard water (*see* hardness of water) and is relatively expensive to make. Various synthetic ('soapless') detergents have been developed from petrochemicals. The commonest, used in washing powders, is sodium dodecylbenzenesulphonate, which contains $CH_3(CH_2)_{11}C_6H_4SO_2O^-$ ions. This, like soap, is an example of an *anionic detergent*, i.e. one in which the active part is a negative ion. *Cationic detergents* have a long hydrocarbon chain connected to a positive ion. Usually they are amine salts, as in $CH_3(CH_2)_{15}N(CH_3)_3{}^+Br^-$, in which the polar part is the $-N(CH_3)_3{}^+$ group. *Nonionic detergents* have nonionic polar groups of the type $-C_2H_4-O-C_2H_4-OH$, which form hydrogen bonds with the water. Synthetic detergents are also used as wetting agents, emulsifiers, and stabilizers for foam.

deuterated compound A compound in which some or all of the hydrogen-1 atoms have been replaced by deuterium atoms.

deuterium (heavy hydrogen) Symbol D. The isotope of hydrogen that has a mass number 2 (r.a.m. 2.0144). Its nucleus contains one proton and one neutron. The abundance of deuterium in natural hydrogen is about 0.015%. It is present in water as the oxide, deuterium oxide (*see* heavy water), from which it is usually obtained by electrolysis or fractional distillation. Its chemical behaviour is almost identical to hydrogen although deuterium compounds tend to react rather more slowly than the corresponding hydrogen compounds. Its physical properties are slightly different from those of hydrogen, e.g. b.p. 23.6 K (hydrogen 20.4 K).

deuterium oxide *See* heavy water.

devitrification Loss of the amorphous nature of glass as a result of crystallization.

Dewar flask A vessel for storing hot or cold liquids so that they maintain their temperature independently of the surroundings. Heat transfer to the surroundings is reduced to a minimum: the walls of the vessel consist of two thin layers of glass (or, in large vessels, steel) separated by a vacuum to reduce conduction and convection; the inner surface of a glass vessel is silvered to reduce radiation; and the vessel is stoppered to prevent evaporation. It was devised around 1872 by the British physicist Sir James Dewar (1842–1923) and is also known by its first trade name *Thermos flask*. *See also* cryostat.

Dewar structure A proposed structure of *benzene, having a hexagonal ring of six carbon atoms with two opposite atoms joined by a long single bond across the ring and with two double C–C bonds, one on each side of the hexagon. Dewar structures contribute to the resonance hybrid of benzene.

dextrorotatory Denoting a chemical compound that rotates the plane of polarization of plane-polarized light to the right (clockwise as observed by someone facing the oncoming radiation). *See* optical activity.

dextrose *See* glucose.

***d*-form** *See* optical activity.

diagonal relationship A relationship within the periodic table by which certain elements in the second period have a close chemical similarity to their diagonal neighbours in the next group of the third period. This is particularly noticeable with the following pairs.

Lithium and magnesium:

(1) both form chlorides and bromides that hydrolyse slowly and are soluble in ethanol;

(2) both form colourless or slightly coloured crystalline nitrides by direct reaction with nitrogen at high temperatures;

(3) both burn in air to give the normal oxide only;

(4) both form carbonates that decompose on heating.

Beryllium and aluminium:

(1) both form highly refractory oxides with polymorphs;

(2) both form crystalline nitrides that are hydrolysed in water;

(3) addition of hydroxide ion to solutions of the salts gives an amphoteric hydroxide, which is soluble in excess hydroxide giving beryllate or aluminate ions $[Be(OH)_4]^{2-}$ and $[Al(OH)_4]^-$;

(4) both form covalent halides and covalent alkyl compounds that display bridging structures;

(5) both metals dissolve in alkalis.

Boron and silicon:

(1) both display semiconductor properties;

(2) both form hydrides that are unstable in air and chlorides that hydrolyse in moist air;

(3) both form acidic oxides with covalent crystal structures, which are readily incorporated along with other oxides into a wide range of glassy materials.

The reason for this relationship is a combination of the trends to increase size down a group and to decrease size along a period, and a similar, but reversed, effect in electronegativity, i.e. decrease down a group and increase along a period.

dialysis A method by which large molecules (such as starch or protein) and small molecules (such as glucose or amino acids) in solution may be separated by selective diffusion through a semipermeable membrane. For example, if a mixed solution of starch and glucose is placed in a closed container made of a semipermeable substance (such as Cellophane), which is then immersed in a beaker of water, the smaller glucose molecules will pass through the membrane into the water while the starch molecules remain behind. The cell membranes of living organisms are semipermeable, and dialysis takes place naturally in the kidneys for the excretion of nitrogenous waste. An artificial kidney (*dialyser*) utilizes the principle of dialysis by taking over the functions of diseased kidneys.

1,6-diaminohexane (hexamethylenediamine) A solid colourless amine, $H_2N(CH_2)_6NH_2$; m.p. 41°C; b.p. 204°C. It is made by oxidizing cyclohexane to hexanedioic acid, reacting this with ammonia to give the ammonium salt, and dehydrating the salt to give hexanedionitrile ($NC(CH_2)_6CN$). This is reduced with hydrogen to the diamine. The compound is used, with hexanedioic acid, in the production of *nylon 6,6.

diamond The hardest known mineral (with a hardness of 10 on Mohs' scale). It is an allotropic form of pure *carbon that has crystallized in the cubic system, usually as octahedra or cubes, under great pressure. Diamond crystals may be colourless and transparent or yellow, brown, or black. They are highly prized as gemstones but also have extensive uses in industry, mainly for cutting and grinding tools. Diamonds occur in ancient volcanic pipes of kimberlite; the most important deposits are in South Africa but others are found in Tanzania, the USA, USSR, and Australia. Diamonds also occur in river deposits that have been derived from weathered kimberlite, notably in Brazil, Zaîre, Sierra Leone, and India. Industrial diamonds are increasingly being produced synthetically.

diaspore A mineral form of a mixed aluminium oxide and hydroxide, AlO.OH. *See* aluminium hydroxide.

diastase *See* amylase.

diastereoisomers Stereoisomers that are not identical and yet not mirror images. For instance, the *d*-form of tartaric acid and the meso form constitute a pair of diastereoisomers. *See* optical activity.

diatomic molecule A molecule formed from two atoms (e.g. H_2 or HCl).

diazine *See* azine.

diazo compounds Organic compounds containing two linked nitrogen compounds. The term includes *azo compounds, diazonium compounds, and also such compounds as diazomethane, CH_2N_2.

Structure of diazonium ion $C_6H_5N_2^+$

diazonium salts Unstable salts containing the ion $C_6H_5N_2^+$ (the *diazonium ion*). They are formed by *diazotization reactions.

diazotization The formation of a *diazonium salt by reaction of an aromatic amine with nitrous acid at low temperature (below 5°C). The nitrous acid is produced in the reaction mixture from sodium nitrite and hydrochloric acid:

$$ArNH_2 + NaNO_2 + HCl \rightarrow$$
$$ArN^+N + Cl^- + Na^+ + OH^- + H_2O$$

dibasic acid An *acid that has two acidic hydrogen atoms in its molecules. Sulphuric (H_2SO_4) and carbonic (H_2CO_3) acids are common examples.

1,2-dibromoethane A colourless liquid *haloalkane, $BrCH_2CH_2Br$; r.d. 2.2; m.p. 9°C; b.p. 121°C. It is made by addition of bromine to ethene and used as an additive in petrol to remove lead as the volatile lead bromide.

dicarbide *See* carbide.

dicarboxylic acid A *carboxylic acid having two carboxyl groups in its molecules. In systematic chemical nomenclature, dicarboxylic acids are denoted by the suffix *-dioic*; e.g. hexanedioic acid, $HOOC(CH_2)_4COOH$.

dichlorine oxide (chlorine monoxide) A strongly oxidizing orange gas, Cl_2O, made by oxidation of chlorine using mercury(II) oxide. It is the acid anhydride of chloric(I) acid.

dichloroethanoic acid *See* chloroethanoic acids.

2,4-dichlorophenoxyacetic acid *See* 2,4-D.

dichroism The property of some crystals, such as tourmaline, of selectively absorbing light vibrations in one plane while allowing light vibrations at right angles to this plane to pass through. Polaroid is a synthetic dichroic material. *See* polarization.

dichromate(VI) A salt containing the ion $Cr_2O_7^{-}$. Solutions containing dichromate (VI) ions are strongly oxidizing.

Diels–Alder reaction A type of chemical reaction in which a compound containing two double bonds separated by a single bond (i.e. a conjugated *diene) adds to a suitable compound containing one double bond (known as the *dienophile*) to give a ring compound. In the dienophile, the double bond must have a carbonyl group on each side. It is named after the German chemists Otto Diels (1876–1954) and Kurt Alder (1902–58), who discovered it in 1928.

diene An *alkene that has two double bonds in its molecule. If the two bonds are separated by one single bond, as in buta-1,3-diene CH_2:CHCH:CH_2, the compound is a *conjugated diene*.

dienophile *See* Diels–Alder reaction.

diethyl ether *See* ethoxyethane.

diffusion 1. The process by which different substances mix as a result of the random motions of their component atoms, molecules, and ions. In gases, all the components are perfectly miscible with each other and mixing ultimately becomes nearly uniform, though slightly affected by gravity (*see also* Graham's law). The diffusion of a solute through a solvent to produce a solution of uniform concentration is slower, but otherwise very similar to the process of gaseous diffusion. In solids, however, diffusion occurs very slowly at normal temperatures. **2.** The passage of elementary particles through matter when there is a high probability of scattering and a low probability of capture.

diffusion pump (condensation pump) A *vacuum pump in which oil or mercury vapour is diffused through a jet, which entrains the gas molecules from the container in which the pressure is to be reduced. The diffused vapour and entrained gas molecules are condensed on the cooled walls of the pump. Pressures down to 10^{-7} Pa can be reached by sophisticated forms of the diffusion pump.

dihedral (dihedron) An angle formed by the intersection of two planes (e.g. two faces of a polyhedron). The *dihedral angle* is the angle formed by taking a point on the line of intersection and drawing two lines from this point, one in each plane, perpendicular to the line of intersection.

dihydrate A crystalline hydrate containing two molecules of water per molecule of compound.

dihydric alcohol *See* diol.

1,2-dihydroxybenzene (catechol) A colourless crystalline phenol, $C_6H_4(OH)_2$; r.d. 1.4; m.p. 105°C; b.p. 240°C. It is used as a photographic developer.

2,3-dihydroxybutanedioic acid *See* tartaric acid.

dilead(II) lead(IV) oxide A red amorphous powder, Pb_3O_4; r.d. 9.1; decomposes at 500°C to lead(II) oxide. It is prepared by heating lead(II) oxide to 400°C and has the unusual property of being black when hot and red-orange when cold. The compound is nonstoichiometric, generally containing less oxygen than implied by the formula. It is largely covalent and has Pb(IV)O_6 octahedral groups linked together by Pb(II) atoms, each joined to three oxygen atoms. It is used in glass making but its use in the paint industry has largely been discontinued because of the toxicity of lead. Dilead(II) lead(IV) oxide is commonly called *red lead* or, more accurately, *red lead oxide*.

dilute Describing a solution that has a relatively low concentration of solute.

dilution The volume of solvent in which a given amount of solute is dissolved.

dilution law *See* Ostwald's dilution law.

dimer An association of two identical molecules linked together. The molecules may react to form a larger molecule, as in the formation of dinitrogen tetroxide (N_2O_4) from nitrogen dioxide (NO_2), or the formation of an *aluminium chloride dimer (Al_2Cl_6) in the vapour. Alternatively, they may be held by hydrogen bonds. For example, carboxylic acids form dimers in organic solvents, in which hydrogen bonds exist between the O of the C=O group and the H of the –O–H group.

dimethylbenzenes (xylenes) Three compounds with the formula $(CH_3)_2C_6H_4$, each having two methyl groups substituted on the benzene ring. 1,2-dimethylbenzene is *o*-xylene, etc. A mixture of the isomers (b.p. 135–145°C) is obtained from petroleum.

dimorphism *See* polymorphism.

dinitrogen oxide (nitrous oxide) A colourless gas, N_2O, d. 1.97 g dm^{-3}; m.p. –90.8°C; b.p. –88.5°C. It is soluble in water, ethanol, and sulphuric acid. It may be prepared by the controlled heating of ammonium nitrate (chloride free) to 250°C and passing the gas produced through solutions of iron(II) sulphate to remove impurities of nitrogen monoxide. It is relatively unreactive, being inert to halogens, alkali metals, and ozone at normal temperatures. It is decomposed on heating above 520°C to nitrogen and oxygen and will support the combustion of many compounds. Dinitrogen oxide is used as an anaesthetic gas ('laughing gas') and as an aerosol propellant.

dinitrogen tetroxide A colourless to pale yellow liquid or a brown gas, N_2O_4; r.d. 1.45 (liquid); m.p. –11.2°C; b.p. 21.2°C. It dissolves in water with reaction to give a mixture of nitric acid and nitrous acid. It may be readily prepared in the laboratory by the reaction of copper with concentrated nitric acid; mixed nitrogen oxides containing dinitrogen oxide may also be produced by heating metal nitrates. The solid compound is wholly N_2O_4 and the liquid is about 99% N_2O_4 at the boiling point; N_2O_4 is diamagnetic. In the gas phase it dissociates to give *nitrogen dioxide*

$$N_2O_4 \rightleftharpoons 2NO_2$$

Because of the unpaired electron this is paramagnetic and brown. Liquid N_2O_4 has been widely studied as a nonaqueous solvent system (self-ionizes to NO^+ and NO_3^-). Dinitrogen tetroxide, along with other nitrogen oxides, is a product of combustion engines and is thought to be involved in the depletion of stratospheric ozone.

diol (dihydric alcohol) An *alcohol containing two hydroxyl groups per molecule.

dioxan A colourless toxic liquid, $C_4H_8O_2$; r.d. 1.03; m.p. 11°C; b.p. 101.5°C. The molecule has a six-membered ring containing four CH_2 groups and two oxygen atoms at opposite corners. It can be made from ethane-1,2-diol and is used as a solvent.

dioxonitric(III) acid *See* nitrous acid.

dioxygenyl compounds Compounds containing the positive ion O_2^+, as in dioxygenyl hexafluoroplatinate O_2PtF_6 – an orange solid that sublimes in vacuum at 100°C. Other ionic compounds of the type

$O_2^+[MF_6]^-$ can be prepared, where M is P, As, or Sb.

diphosphane (diphosphine) A yellow liquid, P_2H_4, which is spontaneously flammable in air. It is obtained by hydrolysis of calcium phosphide. Many of the references to the spontaneous flammability of phosphine (PH_3) are in fact due to traces of P_2H_4 as impurities.

diphosphine *See* diphosphane.

direct dye *See* dyes.

disaccharide A sugar consisting of two linked *monosaccharide molecules. For example, sucrose comprises one glucose molecule and one fructose molecule bonded together.

disilane *See* silane.

dislocation *See* defect.

disodium hydrogenphosphate(V) (disodium orthophosphate) A colourless crystalline solid, Na_2HPO_4, soluble in water and insoluble in ethanol. It is known as the dihydrate (r.d. 2.066), heptahydrate (r.d. 1.68), and dodecahydrate (r.d. 1.52). It may be prepared by titrating phosphoric acid with sodium hydroxide to an alkaline end point (phenolphthalein) and is used in treating boiler feed water and in the textile industry.

disodium orthophosphate *See* disodium hydrogenphosphate(V).

disodium tetraborate-10-water *See* borax.

disperse dye *See* dye.

disperse phase *See* colloids.

displacement reaction *See* substitution reaction.

disproportionation A type of chemical reaction in which the same compound is simultaneously reduced and oxidized. For example, copper(I) chloride disproportionates thus:

$$2CuCl \rightarrow Cu + CuCl_2$$

The reaction involves oxidation of one molecule

$$Cu^{I} \rightarrow Cu^{II} + e$$

and reduction of the other

$$Cu^{I} + e \rightarrow Cu$$

The reaction of halogens with hydroxide ions is another example of a disproportionation reaction, for example

$$Cl_2(g) + 2OH^-(aq) \rightleftharpoons Cl^-(aq) + ClO^-(aq) + H_2O(l)$$

dissociation The breakdown of a molecule, ion, etc., into smaller molecules, ions, etc. An example of dissociation is the reversible reaction of hydrogen iodide at high temperatures

$$2HI(g) \rightleftharpoons H_2(g) + I_2(g)$$

The *equilibrium constant of a reversible dissociation is called the *dissociation constant*. The term 'dissociation' is also applied to ionization reactions of *acids and *bases in water; for example

$$HCN + H_2O \rightleftharpoons H_3O^+ + CN^-$$

which is often regarded as a straightforward dissociation into ions

$$HCN \rightleftharpoons H^+ + CN^-$$

The equilibrium constant of such a dissociation is called the *acid dissociation constant* or *acidity constant*, given by

$$K_a = [H^+][A^-]/[HA]$$

for an acid HA (the concentration of water $[H_2O]$ can be taken as constant). K_a is a measure of the strength of the acid. Similarly, for a nitrogenous base B, the equilibrium

$$B + H_2O \rightleftharpoons BH^+ + OH^-$$

is also a dissociation; with the *base dissociation constant*, or *basicity constant*, given by

$$K_b = [BH^+][OH^-]/[B]$$

For a hydroxide MOH,

$$K_b = [M^+][OH^-]/[MOH]$$

dissociation pressure When a solid compound dissociates to give one or more gaseous products, the dissociation pressure is the pressure of gas in equilibrium with the solid at a given temperature. For example, when calcium carbonate is maintained at a constant high temperature in a closed container, the dissociation pressure at that temperature is the pressure of carbon dioxide from the equilibrium

$$CaCO_3(s) \rightleftharpoons CaO(s) + CO_2(g)$$

distillation The process of boiling a liquid and condensing and collecting the vapour. The liquid collected is the *distillate*. It is used to purify liquids and to separate liquid mixtures (*see* fractional distillation; steam distillation). *See also* destructive distillation.

distilled water Water purified by distillation so as to free it from dissolved salts and other compounds. Distilled water in equilibrium with the carbon dioxide in the air has a conductivity of about 0.8×10^{-6} siemens cm^{-1}. Repeated distillation in a vacuum can bring the conductivity down to 0.043×10^{-6} siemens cm^{-1} at 18°C (sometimes called *conductivity water*). The limiting conductivity is due to self ionization: $H_2O \rightleftharpoons H^+ + OH^-$.

disulphur dichloride (sulphur monochloride) An orange-red liquid, S_2Cl_2, which is readily hydrolysed by water and is soluble in benzene and ether; r.d. 1.678; m.p. −80°C; b.p. 136°C. It may be prepared by passing chlorine over molten sulphur; in the presence of iodine or metal chlorides *sulphur dichloride*, SCl_2, is also formed. In the vapour phase S_2Cl_2 molecules have Cl–S–S–Cl chains. The compound is used as a solvent for sulphur and can form higher *chlorosulphanes* of the type $Cl-(S)_n-Cl$ ($n < 100$), which are of great value in *vulcanization processes.

disulphuric(VI) acid (pyrosulphuric acid) A colourless hygroscopic crystalline solid, $H_2S_2O_7$; r.d. 1.9; m.p. 35°C. It is commonly encountered mixed with sulphuric acid as it is formed by dissolving sulphur trioxide in concentrated sulphuric acid. The resulting fuming liquid, called *oleum* or *Nordhausen sulphuric acid*, is produced during the *contact process and is also widely used in the *sulphonation of organic compounds. *See also* sulphuric acid.

dithionate A salt of dithionic acid, containing the ion $S_2O_6^{\ 2-}$, usually formed by the oxidation of a sulphite using manganese(IV) oxide. The ion has neither pronounced oxidizing nor reducing properties.

dithionic acid An acid, $H_2S_2O_6$, known in the form of its salts (dithionates).

dithionite *See* sulphinate.

dithionous acid *See* sulphinic acid.

divalent (bivalent) Having a valency of two.

***dl*-form** *See* optical activity; racemic mixture.

D-lines Two close lines in the yellow region of the visible spectrum of sodium, having wavelengths 589.0 and 589.6 nm. As they are prominent and easily recognized they are used as a standard in spectroscopy.

DNA (deoxyribonucleic acid) The genetic material of most living organisms, which is a major constituent of the chromosomes within the cell nucleus and plays a central role in the determination of hereditary characteristics by controlling protein synthesis in cells. DNA is a nucleic acid composed of two chains of *nucleotides in which the sugar is *deoxyribose* and the bases are *adenine, *cytosine, *guanine, and *thymine (*compare* RNA). The two chains are wound round each other and linked together by hydrogen bonds between specific complementary bases

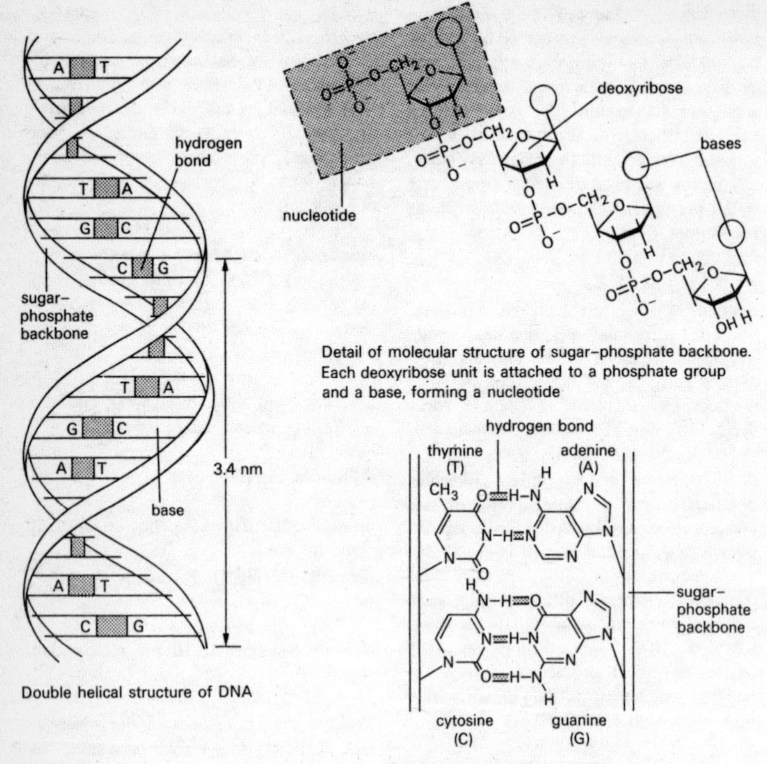

Molecular structure of DNA

to form a spiral ladder-shaped molecule (*double helix*).

When the cell divides, its DNA also replicates in such a way that each of the two daughter molecules is identical to the parent molecule. The hydrogen bonds between the complementary bases on the two strands of the parent molecule break and the strands unwind. Using as building bricks nucleotides present in the nucleus, each strand directs the synthesis of a new one complementary to itself. Replication is initiated, controlled, and stopped by polymerases.

Döbereiner's triads A set of triads of chemically similar elements noted by J. W. Döbereiner in 1817. Even with the inaccurate atomic mass data of the day it was observed that when each triad was arranged in order of increasing atomic mass, then the mass of the central member was approximately the average of the values for the other two. The chemical and physical properties were similarly related. The triads are now recognized as consecutive members of the groups of the periodic table. Examples are: lithium, sodium, and potassium; calcium, strontium, and barium; and chlorine, bromine, and iodine.

dodecanoic acid (lauric acid) A white crystalline *fatty acid, $CH_3(CH_2)_{10}COOH$; r.d. 0.8; m.p. 44°C; b.p. 225°C. Glycerides of the acid are present in natural fats and oils (e.g. coconut and palm-kernel oil).

dodecene A straight-chain alkene, $CH_3(CH_2)_9CH:CH_2$, obtained from petroleum and used in making *dodecylbenzene.

dodecylbenzene A hydrocarbon, $CH_3(CH_2)_{11}C_6H_5$, manufactured by a Friedel–Crafts reaction between dodecene ($CH_3(CH_2)_9CH:CH_2$) and benzene. It can be sulphonated, and the sodium salt of the sulphonic acid is the basis of common *detergents.

dolomite A carbonate mineral consisting of a mixed calcium–magnesium carbonate, $CaCO_3.MgCO_3$, crystallizing in the rhombohedral system. It is usually white or colourless. The term is also used to denote a rock with a high ratio of magnesium to calcium carbonate. *See* limestone.

donor An ion or molecule that provides a pair of electrons in forming a coordinate bond.

dopa (dihydroxyphenylalanine) A derivative of the amino acid tyrosine. It is found in particularly high levels in the adrenal glands and is a precursor in the synthesis of *dopamine, *noradrenaline, and *adrenaline. The laevorotatory form, *L-dopa*, is administered in the treatment of Parkinson's disease, in which brain levels of dopamine are reduced.

dopamine A *catecholamine that is a precursor in the synthesis of *noradrenaline and *adrenaline. It is also believed to function as a neurotransmitter in the brain.

***d*-orbital** *See* orbital.

double bond *See* chemical bond.

double decomposition (metathesis) A chemical reaction involving exchange of radicals, e.g.

$$AgNO_3(aq) + KCl(aq) \rightarrow KNO_3(aq) + AgCl(s)$$

double refraction The property, possessed by certain crystals (notably calcite), of forming two refracted rays from a single incident ray. The *ordinary ray* obeys the normal laws of refraction. The other refracted ray, called the *extraordinary ray*, follows different laws. The light in the ordinary ray is polarized at right angles to the light in the extraordinary ray. Along an optic axis the ordinary and extraordinary rays travel with the same speed. Some crystals, such as calcite, quartz, and tourmaline, have only one optic axis; they are *uniaxial crystals*. Others, such as mica and selenite, have two optic axes; they are *biaxial crystals*. The phenomenon is also known as *birefringence* and the double-refracting crystal as a *birefringent crystal*. *See also* polarization.

double salt A crystalline salt in which there are two different anions and/or cations. An example is the mineral dolomite, $CaCO_3.MgCO_3$, which contains a regular arrangement of Ca^{2+} and Mg^{2+} ions in its crystal lattice. *Alums are double sulphates. Double salts only exist in the solid; when dissolved they act as a mixture of the two separate salts. *Double oxides* are similar.

doublet A pair of associated lines in certain spectra, e.g. the two lines that make up the sodium D-lines.

Downs process A process for extracting sodium by the electrolysis of molten sodium chloride. The *Downs cell* has a central graphite anode surrounded by a cylindrical steel cathode. Chlorine released is led away through a hood over the anode. Molten sodium is formed at the cathode and collected through another hood around the top of the cathode cylinder (it is less dense than the sodium chloride). The two hoods and electrodes are separated by a coaxial cylindrical steel gauze. A small amount of calcium chloride is added to the sodium chloride to lower its melting point. The sodium chloride is melted electrically and kept molten by the current through the cell. More sodium chloride is added as the electrolysis proceeds.

dry cell A primary or secondary cell in which the electrolytes are in the form of a paste. Many torch, radio, and calculator batteries are *Leclanché cells in which the electrolyte is an ammonium chloride paste and the container is the negative zinc electrode (with an outer plastic wrapping).

dry ice Solid carbon dioxide used as a refrigerant. It is convenient because it sublimes at −78°C (195 K) at standard pressure rather than melting.

drying oil A natural oil, such as linseed oil, that hardens on exposure to the air. Drying oils contain unsaturated fatty acids, such as linoleic and linolenic acids, which polymerize on oxidation. They are used in paints, varnishes, etc.

D-series *See* absolute configuration.

Dulong and Petit's law For a solid element the product of the relative atomic mass and the specific heat capacity is a constant equal to about 25 J mol^{-1} K^{-1}. Formulated in these terms in 1819 by the French scientists Pierre Dulong (1785–1838) and A. T. Petit (1791–1820), the law in modern terms states: the molar heat capacity of a solid element is approximately equal to $3R$, where R is the *gas constant. The law is only approximate but applies with fair accuracy at normal temperatures to elements with a simple crystal structure.

Dumas' method 1. A method of finding the amount of nitrogen in an organic compound. The sample is weighed, mixed with copper(II) oxide, and heated in a tube. Any nitrogen present in the compound is converted into oxides of nitrogen, which are led over hot copper to reduce them to nitrogen gas. This is collected and the volume measured, from which the mass of nitrogen in a known mass of sample can be found. **2.** A method of finding the relative molecular masses of volatile liquids by weighing. A thin-glass bulb with a long narrow neck is used. This is weighed full of air at known temperature, then a small amount of sample is introduced and the bulb heated (in a bath) so that the liquid is vaporized and the air is driven out. The tip of the neck is sealed and the bulb cooled and weighed at known (room) temperature. The volume of the bulb is found by filling it with water and weighing again. If the density of air is known, the mass of vapour in a known volume can be calculated.

The techniques are named after the French chemist Jean Baptiste André Dumas (1800–84).

duplet A pair of electrons in a covalent chemical bond.

dyes Substances used to impart colour to textiles, leather, paper, etc. Compounds used for dyeing (*dyestuffs*) are generally organic compounds containing conjugated double bonds. The group producing the colour is the *chromophore; other noncoloured groups that influence or intensify the colour are called *auxochromes. Dyes can be classified according to the chemical structure of the dye molecule. For example, *azo dyes* contain the $-N{=}N-$ group (*see* azo compounds). In practice, they are classified according to the way in which the dye is applied or is held on the substrate.

Acid dyes are compounds in which the chromophore is part of a negative ion (usually an organic sulphonate RSO_2O^-). They can be used for protein fibres (e.g. wool and silk) and for polyamide and acrylic fibres. Originally, they were applied from an acidic bath. *Metallized dyes* are forms of acid dyes in which the negative ion contains a chelated metal atom. *Basic dyes* have chromophores that are part of a positive ion (usually an amine salt or ionized imino group). They are used for acrylic fibres and also for wool and silk, although they have only moderate fastness with these materials.

Direct dyes are dyes that have a high affinity for cotton, rayon, and other cellulose fibres. They are applied directly from a neutral bath containing sodium chloride or sodium sulphate. Like acid dyes, they are usually sulphonic acid salts but are distinguished by their greater substantivity (affinity for the substrate), hence the alternative name *substantive dyes*.

Vat dyes are insoluble substances used for cotton dyeing. They usually contain keto groups, C=O, which are reduced to C–OH groups, rendering the dye soluble (the *leuco form* of the dye). The dye is applied in this form, then oxidized by air or oxidizing agents to precipitate the pigment in the fibres. Indigo and anthroquinone dyes are examples of vat dyes. *Sulphur dyes* are dyes applied by this technique using sodium sulphide solution to reduce and dissolve the dye. Sulphur dyes are used for cellulose fibres.

Disperse dyes are insoluble dyes applied in the form of a fine dispersion in water. They are used for cellulose acetate and other synthetic fibres.

Reactive dyes are compounds that contain groups capable of reacting with the substrate to form covalent bonds. They have high substantivity and are used particularly for cellulose fibres.

dynamic equilibrium *See* equilibrium.

dynamite Any of a class of high explosives based on nitroglycerin. The original form, invented in 1867 by Alfred Nobel, consisted of nitroglycerin absorbed in kieselguhr. Modern dynamites, which are used for blasting, contain sodium or ammonium nitrate sensitized with nitroglycerin and use other absorbers (e.g. wood pulp).

dysprosium Symbol Dy. A soft silvery metallic element belonging to the *lanthanoids; a.n. 66; r.a.m. 162.50; r.d. 8.551 (20°C); m.p. 1412°C; b.p. 2567°C. It occurs in apatite, gadolinite, and xenotime, from which it is extracted by an ion-exchange process. There are seven natural isotopes and twelve artificial isotopes have been identified. It finds limited use in some alloys as a neutron absorber, particularly in nuclear technology. It was discovered by François Lecoq de Boisbaudran in 1886.

dystectic mixture A mixture of substances that has a constant maximum melting point.

E

earth The planet that orbits the sun between the planets Venus and Mars. The earth consists of three layers: the gaseous atmosphere (*see* earth's atmosphere), the liquid *hydrosphere*, and the solid *lithosphere*. The solid part of the earth also consists of three layers: the *crust* with a mean thickness of about 32 km under the land and 10 km under the seas; the *mantle*, which extends some 2900 km below the crust; and the *core*, part of which is believed to be liquid. The composition of the crust is: oxygen 47%, silicon 28%, aluminium 8%, iron 4.5%, calcium 3.5%, sodium and potassium 2.5% each, and magnesium 2.2%. Hydrogen, carbon, phosphorus, and sulphur are all present to an extent of less than 1%.

earth's atmosphere The gas that surrounds the earth. The composition of dry air at sea level is: nitrogen 78.08%, oxygen 20.95%, argon 0.93%, carbon dioxide 0.03%, neon 0.0018%, helium 0.0005%, krypton 0.0001%, and xenon 0.00001%. In addition to water vapour, air in some localities contains sulphur compounds, hydrogen peroxide, hydrocarbons, and dust particles.

ebonite *See* vulcanite.

ebullioscopic constant *See* elevation of boiling point.

echelon A form of interferometer consisting of a stack of glass plates arranged stepwise with a constant offset. It gives a high resolution and is used in spectroscopy to study hyperfine line structure.

eclipsed conformation *See* conformation.

Edison cell *See* nickel–iron accumulator.

EDTA Ethylenediaminetetracetic acid,

$$(HOOCCH_2)_2N(CH_2)_2N(CH_2COOH)_2$$

A compound used as a chelating agent in inorganic chemistry.

effective temperature *See* luminosity.

effervescence The formation of gas bubbles in a liquid by chemical reaction.

efflorescence The process in which a crystalline hydrate loses water, forming a powdery deposit on the crystals.

effusion The flow of a gas through a small aperture. The relative rates at which gases effuse, under the same conditions, is approximately inversely proportional to the square roots of their densities.

eigenfunction An allowed *wave function of a system in quantum mechanics. The associated energies are *eigenvalues*.

Einstein equation 1. The mass–energy relationship announced by Einstein in 1905 in the form $E = mc^2$, where E is a quantity of energy, m its mass, and c is the speed of light. It presents the concept that energy possesses mass. **2.** The relationship $E_{max} = hf - W$, where E_{max} is the maximum kinetic energy of electrons emitted in the photoemissive effect, h is the Planck constant, f the frequency of the incident radiation, and W the *work function of the emitter. This is also written $E_{max} = hf - \phi e$, where e is the electronic charge and ϕ a potential difference, also called the work function. (Sometimes W and ϕ are distinguished as *work function energy* and *work function potential*.) The equation can also be applied to photoemission from gases, when it has the form: $E = hf - I$, where I is the ionization potential of the gas.

einsteinium Symbol Es. A radioactive metallic transuranic element belonging to the *actinoids; a.n. 99; mass number of the most stable isotope 254 (half-life 270 days). Eleven isotopes are known. The element was first identified by A. Ghiorso and associates in debris from the first hydrogen bomb explosion in 1952. Microgram quantities of the element did not become available until 1961.

elastomer A natural or synthetic rubber or rubberoid material, which has the ability to undergo deformation under the influence of a force and regain its original shape once the force has been removed.

electret A permanently electrified substance or body that has opposite charges at its extremities. Electrets resemble permanent magnets in many ways. An electret can be made by cooling certain waxes in a strong electric field.

electric-arc furnace A furnace used in melting metals to make alloys, especially in steel manufacture, in which the heat source is an electric arc. In the direct-arc furnace, such as the Héroult furnace, an arc is formed between the metal and an electrode. In the indirect-arc furnace, such as the Stassano furnace, the arc is formed between two electrodes and the heat is radiated onto the metal.

electrochemical cell *See* cell.

electrochemical equivalent Symbol z. The mass of a given element liberated from a solution of its ions in electrolysis by one coulomb of charge. *See* Faraday's laws (of electrolysis).

electrochemical series *See* electromotive series.

electrochemistry The study of chemical properties and reactions involving ions in solution, including electrolysis and electric cells.

electrochromatography *See* electrophoresis.

electrode 1. A conductor that emits or collects electrons in a cell, thermionic valve, semiconductor device, etc. The *anode* is the positive electrode and the *cathode* is the negative electrode. **2.** *See* half cell.

electrodeposition The process of depositing one metal on another by electrolysis, as in *electroforming and *electroplating.

electrode potential The potential difference produced between the electrode and the solution in a *half cell. It is not possible to measure this directly since any measurement involves completing the circuit with the electrolyte, thereby introducing another half cell. *Standard electrode potentials* E are defined by measuring the potential relative to a standard *hydrogen half cell using 1.0 molar solution at 25°C. The convention is to designate the cell so that the oxidized form is written first. For example,

$$Pt(s)|H_2(g)H^+(aq)|Zn^{2+}(aq)|Zn(s)$$

The e.m.f. of this cell is −0.76 volt (i.e. the zinc electrode is negative). Thus the standard electrode potential of the $Zn^{2+}|Zn$ half cell is −0.76 V. Electrode potentials are also called *reduction potentials*. *See also* electromotive series.

electrodialysis A method of obtaining pure water from water containing a salt, as in *desalination. The water to be purified is fed into a cell containing two electrodes. Between the electrodes is placed an array of *semipermeable membranes alternately semipermeable to positive ions and negative ions. The ions tend to segregate between alternate pairs of membranes, leaving pure water in the other gaps between membranes. In this way, the feed water is separated into two streams: one of pure water and the other of more concentrated solution.

electroforming A method of forming intricate metal articles or parts by *electrodeposition of the metal on a removable conductive mould.

electroluminescence *See* luminescence.

electrolysis The production of a chemical reaction by passing an electric current through an electrolyte. In electrolysis, positive ions migrate to the cathode and negative ions to the anode. The reactions occurring depend on electron transfer at the electrodes and are therefore redox reactions. At the anode, negative ions in solution may lose electrons to form neutral species. Alternatively, atoms of the electrode can lose electrons and go into solution as positive ions. In either case the reaction is an oxidation. At the cathode, positive ions in solution can gain electrons to form neutral species. Thus cathode reactions are reductions.

electrolyte A liquid that conducts electricity as a result of the presence of positive or negative ions. Electrolytes are molten ionic compounds or solutions containing ions, i.e. solutions of ionic salts or of compounds that ionize in solution. Liquid metals, in which the conduction is by free electrons, are not usually regarded as electrolytes.

electrolytic cell A cell in which electrolysis occurs; i.e. one in which current is passed through the electrolyte from an external source.

electrolytic corrosion Corrosion that occurs through an electrochemical reaction. *See* rusting.

electrolytic gas The highly explosive gas formed by the electrolysis of water. It consists of two parts hydrogen and one part oxygen by volume.

electrolytic refining The purification of metals by electrolysis. It is commonly applied to copper. A large piece of impure copper is used as the anode with a thin strip of pure copper as the cathode. Copper (II) sulphate solution is the electrolyte. Copper dissolves at the anode: $Cu \rightarrow Cu^{2+} + 2e$, and is deposited at the cathode. The net result is transfer of pure copper from anode to cathode. Gold and silver in the impure copper form a so-called *anode sludge* at the bottom of the cell, which is recovered.

electromagnetic spectrum The range of wavelengths over which electromagnetic radiation extends. The longest waves (10^5–10^{-3} metres) are radio waves, the next longest (10^{-3}–10^{-6} m) are infrared waves, then comes the narrow band (4–7×10^{-7} m) of visible radiation, followed by ultraviolet waves (10^{-7}–10^{-9} m), X-rays (10^{-9}–10^{-11} m), and gamma rays (10^{-11}–10^{-14} m).

electrometallurgy The uses of electrical processes in the separation of metals from their ores, the refining of metals, or the forming or plating of metals.

electromotive force (e.m.f.) The greatest potential difference that can be generated by a particular source of electric current. In practice this may be observable only when the source is not supplying current, because of its internal resistance.

electromotive series (electrochemical series) A series of chemical elements arranged in order of their *electrode potentials. The hydrogen electrode ($H^+ + e \rightarrow \frac{1}{2}H_2$) is taken as having zero electrode potential. Elements that have a greater tendency than hydrogen to lose electrons to their solution are taken as *electropositive*; those that gain electrons from their solution are below hydrogen in the series and are called *electronegative*. The series shows the order in which metals replace one another from their salts; electropositive metals will replace hydrogen from acids. The chief metals and hydrogen, placed in order in the series, are: potassium, calcium, sodium, magnesium, aluminium, zinc, cadmium, iron, nickel, tin, lead, hydrogen, copper, mercury, silver, platinum, gold.

electron An *elementary particle with a rest mass of $9.109\,558 \times 10^{-31}$ kg and a negative charge of $1.602\,192 \times 10^{-19}$ coulomb. Electrons are present in all atoms in groupings called shells around the nucleus; when they are detached from the atom they are called *free electrons*. The antiparticle of the electron is the *positron*.

electron affinity Symbol *A*. The energy change occurring when an atom or molecule gains an electron to form a negative ion. For an atom or molecule X, it is the energy released for the electron-attachment reaction

$$X(g) + e \rightarrow X^-(g)$$

Often this is measured in electronvolts. Alternatively, the molar enthalpy change, ΔH, can be used.

electron capture **1.** The formation of a negative ion by an atom or molecule when it acquires an extra free electron. **2.** A radioactive transformation in which a nucleus acquires an electron from an inner orbit of the atom, thereby transforming, initially, into a nucleus with the same mass number but an atomic number one less than that of the original nucleus (capture of the electron transforms a proton into a neutron). This type of capture is accompanied by emission of an X-ray photon or Auger electron as the vacancy in the inner orbit is filled by an outer electron.

electron-deficient compound A compound in which there are fewer electrons forming the chemical bonds than required in normal electron-pair bonds. *See* borane.

electron diffraction Diffraction of a beam of electrons by atoms or molecules. The fact that electrons can be diffracted in a similar way to light and X-rays shows that particles can act as waves (*see* de Broglie wavelength). An electron (mass m, charge e) accelerated through a potential difference V acquires a kinetic energy $mv^2/2 = eV$, where v is the velocity of the electron. Thus, the momentum (p) of the electron is $\sqrt{(2eVm)}$. As the de Broglie wavelength (λ) of an electron is given by h/p, where h is the Planck constant, then $\lambda = h/\sqrt{(2eVm)}$. For an accelerating voltage of 3600 V, the wavelength of the electron beam is 0.02 nanometre, some 3×10^4 times shorter than visible radiation.

Electrons then, like X-rays, show diffraction effects with molecules and crystals in which the interatomic spacing is comparable to the wavelength of the beam. They have the advantage that their wavelength can be set by adjusting the voltage. Unlike X-rays they have very low penetrating power. The first observation of electron diffraction was by George Thomson (1892–1975) in 1927, in an experiment in which he passed a beam of electrons in a vacuum through a very thin gold foil onto a photographic plate. Concentric circles were produced by diffraction of electrons by the lattice. The same year Clinton J. Davisson (1881–1958) and Lester Germer (1896–1971) performed a classic experiment in which they obtained diffraction patterns by glancing an electron beam off the surface of a nickel crystal. Both experiments were important verifications of de Broglie's theory and the new quantum theory.

Electron diffraction, because of the low penetration, cannot easily be used to investigate crystal structure. It is, however, employed to measure bond lengths and angles of molecules in gases. Moreover, it is extensively used in the study of solid surfaces and absorption. The main techniques are low-energy electron diffraction (*LEED*) in which the electron beam is reflected onto a fluorescent screen, and high-energy electron diffraction (*HEED*) used either with reflection or transmission in investigating thin films.

electronegative Describing elements that tend to gain electrons and form negative ions. The halogens are typical electronegative elements. For example, in hydrogen chloride, the chlorine atom is more electronegative than the hydrogen and the molecule is polar, with negative charge on the chlorine atom. There are various ways of assigning values for the *electronegativity* of an element. *Mulliken electronegativities* are calculated from $E = (I + A)/2$, where I is ionization potential and A is electron affinity. More commonly, *Pauling electronegativities* are used. These are based on bond dissociation energies using a scale in which fluorine, the most electronegative element, has a value 4. Some other values on this scale are B 2, C 2.5, N 3.0, O 3.5, Si 1.8, P 2.1, S 2.5, Cl 3.0, Br 2.8.

electron microscope A form of microscope that uses a beam of electrons instead of a beam of light (as in the optical microscope) to form a large image of a very small object. In optical microscopes the resolution is limited by the wavelength of the light. High-energy electrons, however, can be associated with a considerably shorter wavelength than light; for example, electrons accelerated to an energy of 10^5 electronvolts have a wavelength of 0.04 nanometre (*see* de Broglie wavelength) enabling a resolution of 0.2–0.5 nm to be achieved. The *transmission electron microscope* has an electron beam, sharply focused by electron lenses, passing through a very thin metallized specimen (less than 50 nanometres thick) onto a fluorescent screen, where a visual image is formed. This image can be photographed. The *scanning electron microscope* can be used with thicker specimens and forms a perspective image, although the resolution and magnification are lower. In this type of instrument a beam of primary electrons scans the specimen and those that are reflected, together with any secondary electrons emitted, are collected. This current is used to modulate a separate electron beam in a TV monitor, which scans the screen at the same frequency, consequently building up a picture of the specimen. The resolution is limited to about 10–20 nm.

electron probe microanalysis (EPM) A method of analysing a very small quantity of a substance (as little as 10^{-13} gram). The method consists of directing a very finely focused beam of electrons on to the sample to produce the characteristic X-ray spectrum of the elements present. It can be used quantitatively for elements with atomic numbers in excess of 11.

electron-spin resonance (ESR) A spectroscopic method of locating electrons within the molecules of a paramagnetic substance (*see* magnetism) in order to provide information regarding its bonds and structure. The spin of an unpaired electron is associated with a magnetic moment that is able to align itself in one of two ways with an applied external magnetic field. These two alignments correspond to different energy levels, with a statistical probability, at normal temperatures, that there will be slightly more in the lower state than in the higher. By applying microwave radiation to the sample a transition to the higher state can be achieved. The precise energy difference between the two states of an electron depends on the surrounding electrons in the atom or molecule. In this way the position of unpaired electrons can be investigated. The technique is used particularly in studying free radicals and paramagnetic substances such as inorganic complexes. *See also* nuclear magnetic resonance.

electron transport chain (respiratory chain) A sequence of biochemical oxidation-reduction reactions that forms the final stage of aerobic respiration. It results in the transfer of electrons or hydrogen atoms derived from the *Krebs cycle to molecular oxygen, with the formation of water. At the same time it conserves energy from good or light in the form of ATP. The chain comprises a series of electron carriers that undergo reversible oxidation-reduction reactions, accepting electrons and then donating them to the next carrier in the chain. In the mitochondria, NADH and $FADH_2$, generated by the Krebs cycle, transfer their electrons to a chain comprising flavin mononucleotide (FMN), *coenzyme Q, and a series of *cytochromes. This process is coupled to the formation of ATP at three sites along the chain. The ATP is then carried across the mitochondrial membrane in exchange for ADP. An electron transport chain also occurs in the light reaction of photosynthesis.

electronvolt Symbol eV. A unit of energy equal to the work done on an electron in moving it through a potential difference of one volt. It is used as a measure of particle energies although it is not an *SI unit. 1 eV = 1.602×10^{-19} joule.

electrophile An ion or molecule that is electron deficient and can accept electrons. Electrophiles are often reducing agents and Lewis *acids. They are either positive ions (e.g. NO_2^+) or molecules that have a positive charge on a particular atom (e.g. SO_3, which has an electron-deficient sulphur atom). In organic reactions they tend to attack negatively charged parts of a molecule. *Compare* nucleophile.

electrophilic addition An *addition reaction in which the first step is attack by an electrophile (e.g. a positive ion) on an electron-rich part of the molecule. An example is addition to the double bonds in alkenes.

electrophilic substitution A *substitution reaction in which the first step is attack by an electrophile. Electrophilic substitution is a feature of reactions of benzene (and its compounds) in which a positive ion approaches the delocalized pi electrons on the benzene ring.

electrophoresis (cataphoresis) A technique for the analysis and separation of colloids, based on the movement of charged colloidal particles in an electric field. There are various experimental methods. In one the sample is placed in a U-tube and a buffer solution added to each arm, so that there are sharp boundaries between buffer and sample. An electrode is placed in each arm, a voltage applied, and the motion of the boundaries under the influence of the field is observed. The rate of migration of the particles depends on the field, the charge on the particles, and on other factors, such as the size and shape of the particles. More simply, electrophoresis can be carried out using an adsorbent, such as a strip of filter paper, soaked in a buffer with two electrodes making contact. The sample is placed between the electrodes and a voltage applied. Different components of the mixture migrate at different rates, so the sample separates into zones. The components can be identified by the rate at which they move. This technique has also been called *electrochromatography*.

Electrophoresis is used extensively in studying mixtures of proteins, nucleic acids, carbohydrates, enzymes, etc. In clinical medicine it is used for determining the protein content of body fluids.

electroplating A method of plating one metal with another by *electrodeposition. The articles to be plated are made the cathode of an electrolytic cell and a rod or bar of the plating metal is made the anode. Electroplating is used for covering metal with a decorative, more expensive, or corrosion-resistant layer of another metal.

electropositive Describing elements that tend to lose electrons and form positive ions. The alkali metals are typical electropositive elements.

electrovalent bond *See* chemical bond.

electrum **1.** An alloy of gold and silver containing 55–88% of gold. **2.** A *German silver alloy containing 52% copper, 26% nickel, and 22% zinc.

element A substance that cannot be decomposed into simpler substances. In an element, all the atoms have the same number of protons or electrons, although the number of neutrons may vary. There are 92 naturally occurring elements. *See also* periodic table; transuranic element.

elevation of boiling point An increase in the boiling point of a liquid when a solid is dissolved in it. The elevation is proportional to the number of particles dissolved (molecules or ions) and is given by $\Delta t = k_B C$, where C is the molal concentration of solute. The constant k_B is the *ebullioscopic constant* of the solvent and if this is known, the molecular weight of the solute can be calculated from the measured value of Δt. The elevation is measured by a Beckmann thermometer. *See also* colligative property.

elimination reaction A reaction in which a molecule decomposes to two molecules, one smaller than the other.

Elinvar Trade name for a nickel–chromium steel containing about 36% nickel, 12% chromium, and smaller proportions of tungsten and manganese. Its elasticity does not vary with temperature and it is therefore used to make hairsprings for watches.

elliptical polarization *See* polarization of light.

eluate *See* chromatography; elution.

eluent *See* chromatography; elution.

elution The process of removing an adsorbed material (*adsorbate*) from an adsorbent by washing it in a liquid (*eluent*). The solution consisting of the adsorbate dissolved in the eluent is the *eluate*. Elution is the process used to wash components of a mixture through a *chromatography column.

elutriation The process of suspending finely divided particles in an upward flowing stream of air or water to wash and separate them into sized fractions.

emanation The former name for the gas radon, of which there are three isotopes: Rn–222 (radium emanation), Rn–220 (thoron emanation), and Rn–219 (actinium emanation).

emerald The green gem variety of *beryl: one of the most highly prized gemstones. The finest specimens occur in the Muzo mines, Colombia. Other occurrences include the Ural Mountains, the Transvaal in South Africa, and Kaligunan in India. Emeralds can also be successfully synthesized.

emery A rock composed of corundum (natural aluminium oxide, Al_2O_3) with magnetite, haematite, or spinel. It occurs on the island of Naxos (Greece) and in Turkey. Emery is used as an abrasive and polishing material and in the manufacture of certain concrete floors.

e.m.f. *See* electromotive force.

emission spectrum *See* spectrum.

empirical Denoting a result that is obtained by experiment or observation rather than from theory.

empirical formula *See* formula.

emulsion A *colloid in which small particles of one liquid are dispersed in another liquid. Usually emulsions involve a dispersion of water in an oil or a dispersion of oil in water, and are stabilized by an *emulsifier*. Commonly emulsifiers are substances, such as *detergents, that have lyophobic and lyophilic parts in their molecules.

enantiomers *See* optical activity.

enantiomorphism *See* optical activity.

endothermic Denoting a chemical reaction that takes heat from its surroundings. *Compare* exothermic.

end point The point in a titration at which reaction is complete as shown by the *indicator.

energy A measure of a system's ability to do work. Like work itself, it is measured in joules. Energy is conveniently classified into two forms: *potential energy* is the energy stored in a body or system as a consequence of its position, shape, or state (this includes gravitational energy, electrical energy, nuclear energy, and chemical energy); *kinetic energy* is energy of motion and is usually defined as the work that will be done by the body possessing the energy when it is brought to rest. For a body of mass m having a speed v, the kinetic energy is $mv^2/2$. The rotational kinetic energy of a body having an angular velocity ω is $I\omega^2/2$, where I is its moment of inertia.

The *internal energy of a body is the sum of the potential energy and the kinetic energy of its component atoms and molecules.

energy band A range of energies that electrons can have in a solid. In a single atom, electrons exist in discrete *energy levels. In a crystal, in which large numbers of atoms are held closely together in a lattice, electrons are influenced by a number of adjacent nuclei and the sharply defined levels of the atoms become bands of allowed energy; this approach to energy levels in solids is often known as the *band theory*. Each band represents a large number of allowed quantum states. Between the bands are *forbidden bands*. The outermost electrons of the atoms (i.e. the ones responsible for chemical bonding) form the *valence band* of the solid. This is the band, of those occupied, that has the highest energy.

The band structure of solids accounts for their electrical properties. In order to move through the solid, the electrons have to change from one quantum state to another. This can only occur if there are empty quantum states close to the electrons. In general, if the valence band is full, electrons cannot change to new quantum states in the same band. For conduction to occur, the electrons have to be in an unfilled band – the *conduction band*. Metals are good conductors either because the valence band and the conduction band are only half-filled or because the conduction band overlaps with the valence band; in either case vacant states are available. In insulators the conduction band and valence band are separated by a wide forbidden band and electrons do not have enough energy to 'jump' from one to the other.

In intrinsic semiconductors the forbidden gap is narrow and, at normal temperatures, electrons at the top of the valence band can move by thermal agitation into the conduction band (at absolute zero, a semiconductor would act as an insulator). Doped semiconductors have extra bands in the forbidden gap.

energy level A definite fixed energy that a molecule, atom, electron, or nucleus can have. In an atom, for example, the atom

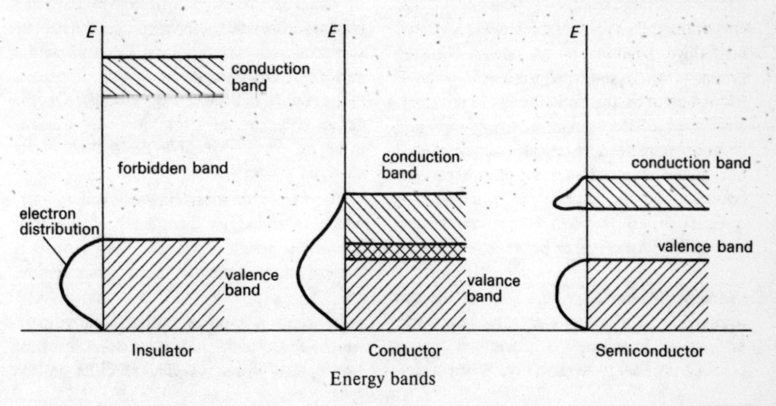

Energy bands

has a fixed energy corresponding to the *orbitals in which its electrons move around the nucleus. The atom can accept a quantum of energy to become an excited atom (*see* excitation) if that extra energy will raise an electron to a permitted orbital. Between the *ground state*, which is the lowest possible energy level for a particular system, and the first excited state there are no permissible energy levels. According to the *quantum theory, only certain energy levels are possible. An atom passes from one energy level to the next without passing through fractions of that energy transition. These levels are usually described by the energies associated with the individual electrons in the atoms, which are always lower than an arbitrary level for a free electron. The energy levels of molecules also involve quantized vibrational and rotational motion.

Engel's salt *See* potassium carbonate.

enols Compounds containing the group $-CH{=}C(OH)-$ in their molecules. *See also* keto–enol tautomerism.

enrichment The process of increasing the abundance of a specified isotope in a mixture of isotopes. It is usually applied to an increase in the proportion of U–235, or the addition of Pu–239 to natural uranium for use in a nuclear reactor or weapon.

enthalpy Symbol H. A thermodynamic property of a system defined by $H = U + pV$, where H is the enthalpy, U is the internal energy of the system, p its pressure, and V its volume. In a chemical reaction carried out in the atmosphere the pressure remains constant and the enthalpy of reaction, ΔH, is equal to $\Delta U + p\Delta V$. For an exothermic reaction ΔH is taken to be negative.

entropy Symbol S. A measure of the unavailability of a system's energy to do work; an increase in entropy is accompanied by a decrease in energy availability. When a system undergoes a reversible change the entropy (S) changes by an amount equal to the energy (Q) absorbed by the system divided by the thermodynamic temperature (T) at which the energy is absorbed, i.e. $\Delta S = \Delta Q/T$. However, all real processes are to a certain extent irreversible changes and in any closed system an irreversible change is always accompanied by an increase in entropy.

In a wider sense entropy can be interpreted as a measure of a system's disorder; the higher the entropy the greater the disorder. As any real change to a closed system tends towards higher entropy, and therefore higher disorder, it follows that the entropy of the universe (if it can be considered a closed system) is increasing and its available energy is decreasing. This increase in the entropy of the universe is one way of stating the second law of *thermodynamics.

enzyme A protein that acts as a *catalyst in biochemical reactions. Each enzyme is specific to a particular reaction or group of similar reactions. Many require the association of certain nonprotein *cofactors in order to function. The molecule undergoing reaction (the *substrate*) binds to a specific *active site on the enzyme molecule to form a short-lived intermediate: this greatly increases (by a factor of up to 10^{20}) the rate at which the reaction proceeds to form the product. Enzyme activity is influenced by substrate concentration and by temperature and pH, which must lie within a certain range. Other molecules may compete for the active site, causing *inhibition of the enzyme or even irreversible destruction of its catalytic properties.

The names of most enzymes end in *-ase*, which is added to the names of the substrates on which they act. Thus *lactase* is the enzyme that acts to break down lactose.

epimerism A type of optical isomerism in which a molecule has two chiral centres; two optical isomers (*epimers*) differ in the arrangement about one of these centres. *See also* optical activity.

epinephrine *See* adrenaline.

epitaxy (epitaxial growth) Growth of a layer of one substance on a single crystal of another, such that the crystal structure in the layer is the same as that in the substrate. It is used in making semiconductor devices.

EPM *See* electron probe microanalysis.

$R_1R_2C{-}CR_3R_4$ (with O bridging the two C atoms)

The functional group in epoxides

epoxides Compounds that contain oxygen atoms in their molecules as part of a three-membered ring. Epoxides are thus *cyclic ethers.*

epoxyethane (ethylene oxide) A colourless flammable gas, C_2H_4O; m.p. −111°C; b.p. 11°C. It is a cyclic ether (*see* epoxides), made by the catalytic oxidation of ethene. It can be hydrolysed to ethane-1,2-diol and also polymerizes to:
... $-O-C_2H_4-O-C_2H_4-$..., which is used for lowering the viscosity of water (e.g. in fire fighting).

epoxy resins Synthetic resins produced by copolymerizing epoxide compounds with phenols. They contain −O− linkages and epoxide groups and are usually viscous liquids. They can be hardened by addition of agents, such as polyamines, that form cross-linkages. Alternatively, catalysts may be used to induce further polymerization of the resin. Epoxy resins are used in electrical equipment and in the chemical industry (because of resistance to chemical attack). They are also used as adhesives.

epsomite A mineral form of *magnesium sulphate heptahydrate, $MgSO_4.7H_2O$.

Epsom salt *See* magnesium sulphate.

equation of state An equation that relates the pressure p, volume V, and thermodynamic temperature T of an amount of substance n. The simplest is the ideal *gas law:

$$pV = nRT,$$

where R is the universal gas constant. Applying only to ideal gases, this equation takes no account of the volume occupied by the gas molecules (according to this law if the pressure is infinitely great the volume becomes zero), nor does it take into account any forces between molecules. A more accurate equation of state would therefore be

$$(p + k)(V - nb) = nRT,$$

where k is a factor that reflects the decreased pressure on the walls of the container as a result of the attractive forces between particles, and nb is the volume occupied by the particles themselves when the pressure is infinitely high. In the *van der Waals equation of state,* proposed by the Dutch physicist J. D. van der Waals (1837–1923),

$$k = n^2a/V^2,$$

where a is a constant. This equation more accurately reflects the behaviour of real gases; several others have done better but are more complicated.

equilibrium A state in which a system has its energy distributed in the statistically most probable manner; a state of a system in which forces, influences, reactions, etc., balance each other out so that there is no net change. A body is said to be in *thermal equilibrium* if no net heat exchange is taking place within it or between it and its surroundings. A system is in *chemical equilibrium when a reaction and its reverse are proceeding at equal rates (*see also* equilibrium constant). These are examples of *dynamic equilibrium,* in which activity in one sense or direction is in aggregate balanced by comparable reverse activity.

equilibrium constant For a reversible reaction of the type

$$xA + yB \rightleftharpoons zC + wD$$

chemical equilibrium occurs when the rate of the forward reaction equals the rate of the back reaction, so that the concentrations of products and reactants reach steady-state values. It can be shown that at equilibrium the ratio of concentrations

$$[C]^z[D]^w/[A]^x[B]^y$$

is a constant for a given reaction and fixed temperature, called the equilibrium constant K_c (where the c indicates concentrations have been used). Note that, by convention, the products on the right-hand side of the reaction are used on the top line of the expression for equilibrium constant. This form of the equilibrium constant was originally introduced in 1863 by C. M. Guldberg and P. Waage using the law of *mass action. They derived the expression by taking the rate of the forward reaction

$$k_f[A]^x[B]^y$$

and that of the back reaction

$$k_b[C]^z[D]^w$$

Since the two rates are equal at equilibrium, the equilibrium constant K_c is the ratio of the rate constants k_f/k_b. The principle that the expression is a constant is known as the *equilibrium law* or *law of chemical equilibrium*.

The equilibrium constant shows the *position* of equilibrium. A low value of K_c indicates that [C] and [D] are small compared to [A]and [B]; i.e. that the back reaction predominates. It also indicates how the equilibrium shifts if concentration changes. For example, if [A] is increased (by adding A) the equilibrium shifts towards the right so that [C] and [D] increase, and K_c remains constant.

For gas reactions, partial pressures are used rather than concentrations. The symbol K_p is then used. Thus, in the example above

$$K_p = p_C{}^z p_D{}^w / p_A{}^x p_B{}^y$$

It can be shown that, for a given reaction $K_p = K_c(RT)^{\Delta\nu}$, where $\Delta\nu$ is the difference in stoichiometric coefficients for the reaction (i.e. $z + w - x - y$). Note that the units of K_p and K_c depend on the numbers of molecules appearing in the stoichiometric equation. The value of the equilibrium constant depends on the temperature. If the forward reaction is exothermic, the equilibrium constant decreases as the temperature rises; if endothermic it increases (*see also* van't Hoff's isochore).

The expression for the equilibrium constant can also be obtained by thermodynamics; it can be shown that the standard equilibrium constant K is given by exp $(-\Delta G\ /RT)$, where ΔG is the standard Gibbs free energy change for the complete reaction. Strictly, the expressions above for equilibrium constants are true only for ideal gases (pressure) or infinite dilution (concentration). For accurate work *activities are used.

equilibrium law *See* equilibrium constant.

equipartition of energy The theory, proposed by Ludwig Boltzmann (1844–1906) and given some theoretical support by James Clerk Maxwell (1831–79), that the energy of gas molecules in a large sample under thermal *equilibrium is equally divided among their available *degrees of freedom, the average energy for each degree of freedom being $kT/2$, where k is the *Boltzmann constant and T is the thermodynamic temperature. The proposition is not generally true if quantum considerations are important, but is frequently a good approximation.

equivalence point The point in a titration at which reaction is complete. *See* indicator.

equivalent proportions *See* chemical combination.

equivalent weight The mass of an element or compound that could combine with or displace one gram of hydrogen (or eight grams of oxygen or 35.5 grams of chlorine) in a chemical reaction. The equivalent weight represents the 'combining power' of the substance. For an element it is the relative atomic mass divided by the valency. For a compound it depends on the reaction considered.

erbium Symbol Er. A soft silvery metallic element belonging to the *lanthanoids; a.n. 68; r.a.m. 167.26; r.d. 9.066 (20°C); m.p. 1529°C; b.p. 2868°C. It occurs in apatite, gadolinite, and xenotine from certain sources. There are six natural isotopes, which are stable, and twelve artificial isotopes are known. It has been used in alloys for nuclear technology as it is a neutron absorber; it is being investigated for other potential uses. It was discovered by C. G. Mosander in 1843.

ergocalciferol *See* vitamin D.

ergosterol A *sterol occurring in fungi, bacteria, algae, and higher plants. It is converted into vitamin D_2 by the action of ultraviolet light.

ESCA *See* photoelectron spectroscopy.

ESR *See* electron-spin resonance.

essential amino acid An *amino acid that an organism is unable to synthesize in sufficient quantities. It must therefore be present in the diet. In man the essential amino acids are arginine, histidine, lysine, threonine, methionine, isoleucine, leucine, valine, phenylalanine, and tryptophan. These are required for protein synthesis and deficiency leads to retarded growth and other symptoms. Most of the amino acids required by man are also essential for all other multicellular animals and for most protozoans.

essential element Any of a number of elements required by living organisms to ensure normal growth, development, and maintenance. Apart from the elements found in organic compounds (i.e. carbon, hydrogen, oxygen, and nitrogen), plants, animals, and microorganisms all require a range of elements in inorganic forms in varying amounts, depending on the type of organism. The *major elements*, present in tissues in relatively large amounts (greater than 0.005%), are calcium, phosphorus, potassium, sodium, chlorine, sulphur, and magnesium. The *trace elements* occur at much lower concentrations and thus requirements are much less. The most important are iron, manganese, zinc, copper, iodine, cobalt, selenium, molybdenum, chromium, and silicon. Each element may fulfil one or more of a variety of metabolic roles.

essential fatty acids *Fatty acids that must normally be present in the diet of certain animals, including man. Essential fatty acids, which include *linoleic and *linolenic acids, all possess double bonds at the same two positions along their hydrocarbon chain and so can act as precursors of *prostaglandins. Deficiency of essential fatty acids can cause dermatosis, weight loss, irregular oestrus, etc. An adult human requires 2–10 g linoleic acid or its equivalent per day.

essential oil A natural oil with a distinctive scent secreted by the glands of certain aromatic plants. *Terpenes are the main constituents. Essential oils are extracted from plants by steam distillation, extraction with cold neutral fats or solvents (e.g. alcohol), or pressing and used in perfumes, flavourings, and medicine. Examples are citrus oils, flower oils (e.g. rose, jasmine), and oil of cloves.

esterification A reaction of an alcohol with an acid to produce an ester and water; e.g.

$$CH_3OH + C_6H_5COOH \rightleftharpoons CH_3OOCC_6H_5 + H_2O$$

The reaction is an equilibrium and is slow under normal conditions, but can be speeded up by addition of a strong acid catalyst. The ester can often be distilled off so that the reaction can proceed to completion. The reverse reaction is ester hydrolysis or *saponification. *See also* labelling.

esters Organic compounds formed by reaction between alcohols and acids. Esters formed from carboxylic acids have the general formula RCOOR′. Examples are ethyl ethanoate, $CH_3COOC_2H_5$, and methyl propanoate, $C_2H_5COOCH_3$. Esters containing simple hydrocarbon groups are volatile fragrant substances used as flavourings in the food industry. Triesters, molecules containing three ester groups, occur in nature as oils and fats. *See also* glycerides.

ethanal (acetaldehyde) A colourless highly flammable liquid aldehyde, CH_3CHO; r.d. 0.78; m.p. −124.6°C; b.p. 20.8°C. It is made from ethene by the *Wacker process and used as a starting material for making many organic compounds. The compound polymerizes if dilute acid is added to give *ethanal trimer* (or *paraldehyde*), which contains a six-membered ring of alternating carbon and oxygen atoms with a hydrogen atom and a methyl group attached to each carbon atom. It is used as a drug for inducing sleep. Addition of dilute acid below 0°C gives *ethanal tetramer* (or *metaldehyde*), which has a similar structure to the trimer but with an eight-membered ring. It is used as a solid fuel in portable stoves and in slug pellets.

ethanamide (acetamide) A colourless solid crystallizing in the form of long white crystals with a characteristic smell of mice, CH_3CONH_2; r.d. 1.159; m.p. 82.3°C; b.p. 221.25°C. It is made by the dehydration of ammonium ethanoate or by the action of ammonia on ethanoyl chloride, ethanoic anhydride, or ethyl ethanoate.

ethane A colourless flammable gaseous hydrocarbon, C_2H_6; m.p. −183°C; b.p. −89°C. It is the second member of the *alkane series of hydrocarbons and occurs in natural gas.

ethanedioic acid *See* oxalic acid.

ethane-1,2-diol (ethylene glycol; glycol) A colourless viscous hygroscopic liquid, CH_2OHCH_2OH; m.p. −13°C; b.p. 197°C. It is made by hydrolysis of epoxyethane (from ethene) and used as an antifreeze and a raw material for making *polyesters (e.g. Terylene).

ethanoate (acetate) A salt or ester of ethanoic acid (acetic acid).

ethanoic acid (acetic acid) A clear viscous liquid or glassy solid *carboxylic acid, CH_3COOH, with a characteristically sharp odour of vinegar; r.d. 1.049; m.p. 16.7°C; b.p. 118.5°C. The pure compound is called *glacial ethanoic acid.* It is manufactured by the oxidation of ethanol or by the oxidation of butane in the presence of dissolved manganese(II) or cobalt(II) ethanoates at 200°C, and is used in making ethanoic anhydride for producing cellulose ethanoates. It is also used in making ethenyl ethanoate (for polyvinylacetate). The compound is formed by the fermentation of alcohol and is present in vinegar, which is made by fermenting beer or wine. 'Vinegar' made from ethanoic acid with added colouring matter is called 'nonbrewed condiment'.

ethanol (ethyl alcohol) A colourless water-soluble *alcohol, C_2H_5OH; r.d. 0.61

$$CH_3{-}O{-}H + H{-}O{-}C({=}O){-}C_2H_5 \rightleftharpoons CH_3{-}O{-}C({=}O){-}C_2H_5 + H_2O$$

methanol + ethanoic acid ⇌ methyl ethanoate + water

Ester formation

(0°C); m.p. −169°C; b.p. −102°C. It is the active principle in intoxicating drinks, in which it is produced by fermentation of sugar using yeast

$$C_6H_{12}O_6 \rightarrow 2C_2H_5OH + 2CO_2$$

The ethanol produced kills the yeast and fermentation alone cannot produce ethanol solutions containing more than 15% ethanol by volume. Distillation can produce a constant-boiling mixture containing 95.6% ethanol and 4.4% water. Pure ethanol (*absolute alcohol*) is made by removing this water by means of drying agents.

The main industrial use of ethanol is as a solvent although at one time it was a major starting point for making other chemicals. For this it was produced by fermentation of molasses. Now ethene has replaced ethanol as a raw material and industrial ethanol is made by hydrolysis of ethene.

ethanoyl chloride (acetyl chloride) A colourless liquid acyl chloride (*see* acyl halides), CH_3COCl, with a pungent smell; r.d. 1.104; m.p. −112.15°C; b.p. 55°C. It is made by reacting ethanoic acid with a halogenating agent such as phosphorus(III) chloride, phosphorus(V) chloride, or sulphur dichloride oxide and is used to introduce ethanoyl groups into organic compounds containing −OH, $-NH_2$, and −SH groups. *See* acylation.

ethanoyl group (acetyl group) The organic group CH_3CO-.

ethene (ethylene) A colourless flammable gaseous hydrocarbon, C_2H_4; m.p. −169°C; b.p. −102°C. It is the first member of the *alkene series of hydrocarbons. It is made by cracking hydrocarbons from petroleum and is now a major raw material for making other organic chemicals (e.g. ethanal, ethanol, ethane-1,2-diol). It can be polymerized to *polyethene. It occurs naturally in plants, in which it acts as a *growth substance promoting the ripening of fruits.

ethenyl ethanoate (vinyl acetate) An unsaturated organic ester, $CH_2{:}CHOOCCH_3$; r.d. 0.9; m.p. −100°C; b.p. 73°C. It is made by catalytic reaction of ethanoic acid and ethene and used to make polyvinylacetate.

ether *See* ethoxyethane; ethers.

ethers Organic compounds containing the group −O− in their molecules. Examples are dimethyl ether, CH_3OCH_3, and diethyl ether, $C_2H_5OC_2H_5$ (*see* ethoxyethane). They are volatile highly flammable compounds made by dehydrating alcohols using sulphuric acid.

ethoxyethane (diethyl ether; ether) A colourless flammable volatile *ether, $C_2H_5OC_2H_5$; r.d. 0.71; m.p. −116°C; b.p. 34.5°C. It can be made by *Williamson's synthesis. It is an anaesthetic and useful organic solvent.

ethyl acetate *See* ethyl ethanoate.

ethyl alcohol *See* ethanol.

ethylamine A colourless flammable volatile liquid, $C_2H_5NH_2$; r.d. 0.69; m.p. −81°C; b.p. 16.6°C. It is a primary amine made by reacting chloroethane with ammonia and used in making dyes.

ethylbenzene A colourless flammable liquid, $C_6H_5C_2H_5$; r.d. 0.8; m.p. −95°C, b.p. 136°C. It is made from ethene and benzene by a *Friedel–Crafts reaction and is used in making phenylethene (for polystyrene).

ethyl bromide *See* bromoethane.

ethylene *See* ethene.

ethylene glycol *See* ethane-1,2-diol.

ethylene oxide *See* epoxyethane.

ethyl ethanoate (ethyl acetate) A colourless flammable liquid ester, $C_2H_5OOCCH_3$; r.d. 0.69; m.p. −81°C; b.p. 16.6°C. It is used as a solvent and in flavourings and perfumery.

ethyl group The organic group C_2H_5–.

ethyl iodide *See* iodoethane.

ethyne (acetylene) A colourless unstable gas, C_2H_2, with a characteristic sweet odour; r.d. 0.618; m.p. −83.25°C; b.p. −79.85°C. It is the simplest member of the *alkyne series of unsaturated hydrocarbons, and is prepared by the action of water on calcium dicarbide or by adding alcoholic potassium hydroxide to 1,2-dibromoethane. It can be manufactured by heating methane to 1500°C in the presence of a catalyst. It is used in oxyacetylene welding and in the manufacture of ethanal and ethanoic acid. Ethyne can be polymerized easily at high temperatures to give a range of products. The inorganic saltlike dicarbides contain the ion C_2^{2-}, although ethyne itself is a neutral compound (i.e. not a protonic acid).

eudiometer An apparatus for measuring changes in volume of gases during chemical reactions. A simple example is a graduated glass tube sealed at one end and inverted in mercury. Wires passing into the tube allow the gas mixture to be sparked to initiate the reaction between gases in the tube.

europium Symbol Eu. A soft silvery metallic element belonging to the *lanthanoids; a.n. 63; r.a.m. 151.96; r.d. 5.245 (20°C); m.p. 822°C; b.p. 1529°C. It occurs in small quantities in bastanite and monazite. Two stable isotopes occur naturally: europium-151 and europium-153, both of which are neutron absorbers. Experimental europium alloys have been tried for nuclear-reactor parts but until recently the metal has not been available in sufficient quantities. Additional uses are being researched since the metal became available in larger quantities. It was discovered by Sir William Crookes in 1889.

eutectic mixture A solid solution consisting of two or more substances and having the lowest freezing point of any possible mixture of these components. The minimum freezing point for a set of components is called the *eutectic point*. Low-melting-point alloys are usually eutectic mixtures.

evaporation The change of state of a liquid into a vapour at a temperature below the boiling point of the liquid. Evaporation occurs at the surface of a liquid, some of those molecules with the highest kinetic energies escaping into the gas phase. The result is a fall in the average kinetic energy of the molecules of the liquid and consequently a fall in its temperature.

exa- Symbol E. A prefix used in the metric system to denote 10^{18} times. For example, 10^{18} metres = 1 exametre (Em).

excitation A process in which a nucleus, electron, atom, ion, or molecule acquires energy that raises it to a quantum state (*excited state*) higher than that of its *ground state. The difference between the energy in the ground state and that in the excited state is called the *excitation energy*. *See* energy level.

exclusion principle *See* Pauli exclusion principle.

exothermic Denoting a chemical reaction that releases heat into its surroundings. *Compare* endothermic.

explosives Substances that undergo rapid chemical reaction evolving heat and causing a sudden increase in pressure. The volume of gas produced by an explosive is great compared with the volume of the original substance. Examples of explosive substances are gunpowder, cellulose nitrate, TNT, nitroglycerin, and cyclonite.

extender An inert substance added to a product (paint, rubber, washing powder, etc.) to dilute it (for economy) or to modify its physical properties.

extraordinary ray *See* double refraction.

F

face-centred cubic (f.c.c.) *See* cubic crystal.

FAD (flavin adenine dinucleotide) A *coenzyme important in various biochemical reactions. It comprises a phosphorylated vitamin B_2 (riboflavin) molecule linked to the nucleotide adenine monophosphate (AMP). FAD is usually tightly bound to the enzyme forming a *flavoprotein*. It functions as a hydrogen acceptor in dehydrogenation reactions, being reduced to $FADH_2$. This in turn is oxidized to FAD by the *electron transport chain, thereby generating ATP (two molecules of ATP per molecule of $FADH_2$).

Fahrenheit scale A temperature scale in which (by modern definition) the temperature of boiling water is taken as 212 degrees and the temperature of melting ice as 32 degrees. It was invented in 1714 by the German scientist G. D. Fahrenheit (1686–1736), who set the zero at the lowest temperature he knew how to obtain in the laboratory (by mixing ice and common salt) and took his own body temperature as 96°F. The scale is no longer in scientific use. To convert to the *Celsius scale the formula is $C = 5(F - 32)/9$.

Fajans' rules Rules indicating the extent to which an ionic bond has covalent character caused by polarization of the ions. Covalent character is more likely if:

(1) the charge of the ions is high;

(2) the positive ion is small or the negative ion is large;

(3) the positive ion has an outer electron configuration that is not a noble-gas configuration.

The rules were introduced by the Polish–American chemist Kasimir Fajans (1887–).

fall-out 1. (*or* **radioactive fall-out**) Radioactive particles deposited from the atmosphere either from a nuclear explosion or from a nuclear accident. *Local fall-out*, within 250 km of an explosion, falls within a few hours of the explosion. *Tropospheric fall-out* consists of fine particles deposited all round the earth in the approximate latitude of the explosion within about one week. *Stratospheric fall-out* may fall anywhere on earth over a period of years. The most dangerous radioactive isotopes in fall-out are the fission fragments iodine–131 and strontium–90. Both can be taken up by grazing animals and passed on to human populations in milk, milk products, and meat. Iodine–131 accumulates in the thyroid gland and strontium–90 accumulates in bones. **2.** (*or* **chemical fall-out**) Hazardous chemicals discharged into and subsequently released from the atmosphere, especially by factory chimneys.

farad Symbol F. The SI unit of capacitance, being the capacitance of a capacitor that, if charged with one coulomb, has a potential difference of one volt between its plates. $1\ F = 1\ C\ V^{-1}$. The farad itself is too large for most applications; the practical unit is the microfarad (10^{-6} F). The unit is named after Michael Faraday (1791–1867).

Faraday constant Symbol *F*. The electric charge carried by one mole of electrons or singly ionized ions, i.e. the product of the *Avogadro constant and the charge on an electron (disregarding sign). It has the value $9.648\,670 \times 10^4$ coulombs per mole. This number of coulombs is sometimes treated as a unit of electric charge called the *faraday*.

Faraday's laws Two laws describing electrolysis:
(1) The amount of chemical change during electrolysis is proportional to the charge passed.
(2) The charge required to deposit or liberate a mass m is given by $Q = Fmz/M$, where F is the Faraday constant, z the charge of the ion, and M the relative ionic mass.

These are the modern forms of the laws. Originally, they were stated by Faraday in a different form:
(1) The amount of chemical change produced is proportional to the quantity of electricity passed.
(2) The amount of chemical change produced in different substances by a fixed quantity of electricity is proportional to the electrochemical equivalent of the substance.

fat A mixture of lipids, chiefly *triglycerides, that is solid at normal body temperatures. Fats occur widely in plants and animals as a means of storing food energy, having twice the calorific value of carbohydrates. In mammals, fat is deposited in a layer beneath the skin (subcutaneous fat) and deep within the body as a specialized adipose tissue.

Fats derived from plants and fish generally have a greater proportion of unsaturated *fatty acids than those from mammals. Their melting points thus tend to be lower, causing a softer consistency at room temperatures. Highly unsaturated fats are liquid at room temperatures and are therefore more properly called *oils.

fatty acid An organic compound consisting of a hydrocarbon chain and a terminal carboxyl group (*see* carboxylic acids). Chain length ranges from one hydrogen atom (methanoic, or formic, acid, HCOOH) to nearly 30 carbon atoms. Ethanoic (acetic), propanoic (propionic), and butanoic (butyric) acids are important in metabolism. Long-chain fatty acids (more than 8–10 carbon atoms) most commonly occur as constituents of certain lipids, notably glycerides, phospholipids, sterols, and waxes, in which they are esterified with alcohols. These long-chain fatty acids generally have an even number of carbon atoms; unbranched chains predominate over branched chains. They may be saturated (e.g. *palmitic (hexadecanoic) acid and *stearic (octadecanoic) acid) or unsaturated, with one double bond (e.g. *oleic (cis-octodec-9-enoic) acid) or two or more double bonds, in which case they are called *polyunsaturated fatty acids* (e.g. *linoleic acid and *linolenic acid). *See also* essential fatty acids.

The physical properties of fatty acids are determined by chain length, degree of unsaturation, and chain branching. Short-chain acids are pungent liquids, soluble in water. As chain length increases, melting points are raised and water-solubility decreases. Unsaturation and chain branching tend to lower melting points.

***f*-block elements** The block of elements in the *periodic table consisting of the lanthanoid series (from cerium to lutetium) and the actinoid series (from thorium to lawrencium). They are characterized by having two *s*-electrons in their outer shell (n) and *f*-electrons in their inner ($n-1$) shell.

f.c.c. Face-centred cubic. *See* cubic crystal.

Fehling's test A chemical test to detect reducing sugars and aldehydes in solution, devised by the German chemist H. C. von Fehling (1812–85). *Fehling's solution* consists of Fehling's A (copper(II) sulphate solution) and Fehling's B (alkaline sodium tartrate 2,3-dihydroxybutanedioate solution), equal amounts of which are added to the test solution. After boiling, a positive result is indicated by the formation of a brick-red precipitate of copper(I) oxide. Methanal, being a strong reducing agent, also produces copper metal; ketones do not react.

feldspars A group of silicate minerals, the most abundant minerals in the earth's crust. They have a structure in which $(Si,Al)O_4$ tetrahedra are linked together with potassium, sodium, and calcium and very occasionally barium ions occupying the large spaces in the framework. The chemical composition of feldspars may be expressed as combinations of the four components:
anorthite (An), $CaAl_2Si_2O_8$;
albite (Ab), $NaAlSi_3O_8$;
orthoclase (Or), $KAlSi_3O_8$; and
celsian (Ce), $BaAl_2Si_2O_8$.
The feldspars are subdivided into two groups: the *alkali feldspars* (including microcline, orthoclase, and sanidine), in which potassium is dominant with a smaller proportion of sodium and negligible calcium; and the *plagioclase feldspars*, which vary in composition in a series that ranges from pure sodium feldspar (albite) through to pure calcium feldspar (anorthite) with negligible potassium. Feldspars form colourless, white, or pink crystals with a hardness of 6 on the Mohs' scale.

feldspathoids A group of alkali aluminosilicate minerals that are similar in chemical composition to the *feldspars but are relatively deficient in silica and richer in alkalis. The structure consists of a framework of $(Si,Al)O_4$ tetrahedra with aluminium and silicon atoms at their centres. The feldspathoids occur chiefly with feldspars but do not coexist with free quartz (SiO_2) as they react with silica to yield feldspars. The chief varieties of feldspathoids are:
nepheline, $KNa_3(AlSiO_4)_4$;
leucite, $KAlSi_2O_6$;
analcime, $NaAlSi_2O_6.H_2O$;
cancrinite, $Na_8(AlSiO_4)_6(HCO_3)_2$;
and the sodalite subgroup comprising
sodalite, $3(NaAlSiO_4).NaCl$;
nosean, $3(NaAlSiO_4).Na_2SO_4$;
haüyne, $3(NaAlSiO_4).CaSO_4$;
and lazurite, $(Na,Ca)_8(Al,Si)_{12}O_{24}(S,SO_4)$ (*see* *lapis lazuli*).

femto- Symbol f. A prefix used in the metric system to denote 10^{-15}. For example, 10^{-15} second = 1 femtosecond (fs).

fermentation A form of anaerobic respiration occurring in certain microorganisms, e.g. yeasts. It comprises a series of biochemical reactions by which sugar is converted to ethanol and carbon dioxide. Fermentation is the basis of the baking, wine, and beer industries.

fermi A unit of length formerly used in nuclear physics. It is equal to 10^{-15} metre. In SI units this is equal to 1 femtometre (fm). It was named after the Italian-born US physicist Enrico Fermi (1901–54).

Fermi–Dirac statistics *See* quantum statistics.

Fermi level The energy level in a solid at which the probability of finding an electron is 1/2. The Fermi level in conductors lies in the conduction band (*see* energy bands), in insulators it lies in the valence band, and in semiconductors it falls in the gap between the conduction band and the valence band. At absolute zero all the electrons would occupy energy levels up to the Fermi level and no higher levels would be occupied.

fermium Symbol Fm. A radioactive metallic transuranic element belonging to the *actinoids; a.n. 100; mass number of the most stable isotope 257 (half-life 10 days). Ten isotopes are known. The element was first identified by A. Ghiorso and associates in debris from the first hydrogen-bomb explosion in 1952.

ferric compounds Compounds of iron in its +3 oxidation state; e.g. ferric chloride is iron(III) chloride, $FeCl_3$.

ferricyanide A compound containing the complex ion $[Fe(CN)_6]^{3-}$, i.e. the hexacyanoferrate(III) ion.

ferrite 1. A member of a class of mixed oxides $MO.Fe_2O_3$, where M is a metal such as cobalt, manganese, nickel, or zinc. The ferrites are ceramic materials that show either ferrimagnetism or ferromagnetism, but are not electrical conductors. For this reason they are used in high-frequency circuits as magnetic cores. **2.** *See* steel.

ferroalloys Alloys of iron with other elements made by smelting mixtures of iron ore and the metal ore; e.g. ferrochromium, ferrovanadium, ferromanganese, ferrosilicon, etc. They are used in making alloy *steels.

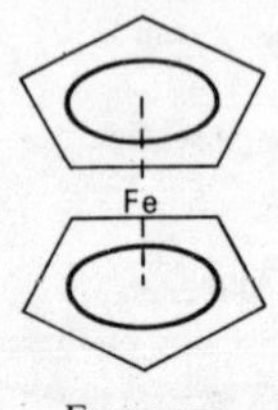

Ferrocene

ferrocene An orange-red crystalline solid, $Fe(C_5H_5)_2$; m.p. 173°C. It can be made by adding the ionic compound $Na^+C_5H_5^-$ (cyclopentadienyl sodium, made from sodium and cyclopentadiene) to iron(III) chloride. In ferrocene, the two rings are parallel, with the iron ion sandwiched between them (hence the name *sandwich compound*). The bonding is between pi orbitals on the rings and *d*-orbitals on the Fe^{2+} ion. The compound can undergo electrophilic substitution on the C_5H_5 rings (they have some aromatic character). It can also be oxidized to the blue ion $(C_5H_5)_2Fe^+$. Ferrocene is the first of a class of similar complexes called *metallocenes.* Its systematic name is *di-π-cyclopentadienyl iron(II).*

ferrocyanide A compound containing the complex ion $[Fe(CN)_6]^{4-}$, i.e. the hexacyanoferrate(II) ion.

ferroelectric materials Ceramic dielectrics, such as Rochelle salt and barium titanate, that have a domain structure making them analogous to ferromagnetic materials. They exhibit hysteresis and usually the piezoelectric effect.

ferrosoferric oxide *See* tri-iron tetroxide.

ferrous compounds Compounds of iron in its +2 oxidation state; e.g. ferrous chloride is iron(II) chloride, $FeCl_2$.

filler A solid inert material added to a synthetic resin or rubber, either to change its physical properties or simply to dilute it for economy.

film badge A lapel badge containing masked photographic film worn by personnel who could be exposed to ionizing radiation. The film is developed to indicate the extent that the wearer has been exposed to harmful radiation.

filter A device for separating solid particles from a liquid or gas. The simplest laboratory filter for liquids is a funnel in which a cone of paper (*filter paper*) is placed. Special containers with a porous base of sintered glass are also used. *See also* Gooch crucible.

filter pump A simple laboratory vacuum pump in which air is removed from a system by a jet of water forced through a narrow nozzle. The lowest pressure possible is the vapour pressure of water.

filtrate The clear liquid obtained by filtration.

filtration The process of separating solid particles using a filter. In vacuum filtration, the liquid is drawn through the filter by a vacuum pump.

fine chemicals Chemicals produced industrially in relatively small quantities and with a high purity; e.g. dyes and drugs.

fineness of gold A measure of the purity of a gold alloy, defined as the parts of gold in 1000 parts of the alloy by mass. Gold with a fineness of 750 contains 75% gold, i.e. 18 *carat gold.

fine structure Closely spaced spectral lines arising from transitions between energy levels that are split by the vibrational or rotational motion of a molecule or by electron spin. They are visible only at high resolution. *Hyperfine structure*, visible only at very high resolution, results from the influence of the atomic nucleus on the allowed energy levels of the atom.

firedamp Methane formed in coal mines.

first-order reaction *See* order.

Fischer–Tropsch process An industrial method of making hydrocarbon fuels from carbon monoxide and hydrogen. The process was invented in 1933 and used by Germany in World War II to produce motor fuel. Hydrogen and carbon monoxide are mixed in the ratio 2:1 (water gas was used with added hydrogen) and passed at 200°C over a nickel or cobalt catalyst. The resulting hydrocarbon mixture can be separated into a higher-boiling fraction for Diesel engines and a lower-boiling gasoline fraction. The gasoline fraction contains a high proportion of straight-chain hydrocarbons and has to be reformed for use in motor fuel. Alcohols, aldehydes, and ketones are also present. The process is also used in the manufacture of SNG from coal. It is named after the German chemist Franz Fischer (1852–1932) and the Czech Hans Tropsch (1839–1935).

fission-track dating A method of estimating the age of glass and other mineral objects by observing the tracks made in them by the fission fragments of the uranium nuclei that they contain. By irradiating the objects with neutrons to induce fission and comparing the density and number of the tracks before and after irradiation it is possible to estimate the time that has elapsed since the object solidified.

Fittig reaction *See* Wurtz reaction.

fixation *See* nitrogen fixation.

fixed point A temperature that can be accurately reproduced to enable it to be used as the basis of a *temperature scale.

flame A hot luminous mixture of gases undergoing combustion. The chemical reactions in a flame are mainly free-radical chain reactions and the light comes from fluorescence of excited molecules or ions or from incandescence of small solid particles (e.g. carbon).

flame test A simple test for metals, in which a small amount of the sample (usually moistened with hydrochloric acid) is placed on the end of a platinum wire and held in a Bunsen flame. Certain metals can be detected by the colour produced: barium (green), calcium (brick red), lithium (crimson), potassium (lilac), sodium (yellow), strontium (red).

flash photolysis A technique for studying free-radical reactions in gases. The apparatus used typically consists of a long glass or quartz tube holding the gas, with a lamp outside the tube suitable for producing an intense flash of light. This dissociates molecules in the sample creating free radicals, which can be detected spectroscopically by a beam of light passed down the axis of the tube. It is possible to focus the spectrometer on an absorption line for a particular product and measure its change in intensity with time using an oscilloscope. In this way the kinetics of very fast free-radical gas reactions can be studied.

flash point The temperature at which the vapour above a volatile liquid forms a combustible mixture with air. At the flash point the application of a naked flame gives a momentary flash rather than sustained combustion, for which the temperature is too low.

flavin adenine dinucleotide *See* FAD.

flavonoid One of a group of naturally occurring phenolic compounds many of which are plant pigments. They include the anthocyanins, flavonols, and flavones. Patterns of flavonoid distribution have been used in taxonomic studies of plant species.

flavoprotein *See* FAD.

flint (chert) Very hard dense nodules of microcrystalline quartz and chalcedony found in chalk and limestone.

flocculation The process of aggregating into larger clumps.

flocculent Aggregated in woolly masses; used to describe precipitates.

fluidization A technique used in some industrial processes in which solid particles suspended in a stream of gas are treated as if they were in the liquid state. Fluidization is useful for transporting powders, such as coal dust. *Fluidized beds*, in which solid particles are suspended in an upward stream, are extensively used in the chemical industry, particularly in catalytic reactions where the powdered catalyst has a high surface area.

fluorescein A yellowish-red dye that produces yellow solutions with a green fluorescence. It is used in tracing water flow and as an *absorption indicator.

fluorescence *See* luminescence.

fluoridation The process of adding very small amounts of fluorine salts (e.g. sodium fluoride, NaF) to drinking water to prevent tooth decay.

fluoride *See* halide.

fluorination A chemical reaction in which a fluorine atom is introduced into a molecule. *See* halogenation.

fluorine Symbol F. A poisonous pale yellow gaseous element belonging to group VII of the periodic table (the *halogens); a.n. 9; r.a.m. 18.9984; d. 1.7 $g\,dm^{-3}$; m.p. −219.62°C; b.p. −188.1°C. The main mineral sources are *fluorite (CaF_2) and *cryolite (Na_3AlF). The element is obtained by electrolysis of a molten mixture of potassium fluoride and hydrogen fluoride. It is used in the synthesis of organic fluorine compounds. Chemically, it is the most reactive and electronegative of all elements. It is a highly dangerous compound, causing severe chemical burns on contact with the skin. The element was identified by Scheele in 1771 and first isolated by Moissan in 1886.

fluorite (fluorspar) A mineral form of calcium fluoride, CaF_2, crystallizing in the cubic system. It is variable in colour; the most common fluorites are green and purple (blue john), but other forms are white, yellow, or brown. Fluorite is used chiefly as a flux material in the smelting of iron and steel; it is also used as a source of fluorine and hydrofluoric acid and in the ceramic and optical-glass industries.

fluorite structure *See* calcium fluoride structure.

fluorocarbons Compounds obtained by replacing the hydrogen atoms of hydrocarbons by fluorine atoms. Their high stability to temperature makes them suitable for a variety of uses, including aerosol propellants, oils, polymers, etc. They are often known as *freons*. There has been some concern that their use in aerosols may cause depletion of the ozone layer.

foam A dispersion of bubbles in a liquid. Foams can be stabilized by *surfactants. Solid foams (e.g. expanded polystyrene or foam rubber) are made by foaming the liquid and allowing it to set. *See also* colloid.

folacin *See* folic acid.

folic acid (folacin) A vitamin of the *vitamin B complex. In its active form, tetrahydrofolic acid, it is a *coenzyme in various reactions involved in the metabolism of amino acids, purines, and pyrimidines. It is synthesized by intestinal bacteria and is widespread in food, especially green leafy vegetables. Deficiency causes poor growth and nutritional anaemia.

fool's gold *See* pyrite.

forbidden band *See* energy bands.

forced convection *See* convection.

formaldehyde *See* methanal.

formalin A colourless solution of methanal (formaldehyde) in water with methanol as a stabilizer; r.d. 1.075–1.085. When kept at temperatures below 25°C a white polymer of methanal separates out. It is used as a disinfectant and preservative for biological specimens.

formate *See* methanoate.

formic acid *See* methanoic acid.

formula A way of representing a chemical compound using symbols for the atoms present. Subscripts are used for the numbers of atoms. The *molecular formula* simply gives the types and numbers of atoms present. For example, the molecular formula of ethanoic acid is $C_2H_4O_2$. The *empirical formula* gives the atoms in their simplest ratio; for ethanoic acid it is CH_2O. The *structural formula* gives an indication of the way the atoms are arranged. Commonly, this is done by dividing the formula into groups; ethanoic acid can be written $CH_3.CO.OH$ (or more usually simply CH_3COOH). Structural formulae can also show the arrangement of atoms or groups in space.

formula weight The relative molecular mass of a compound as calculated from its molecular formula.

formyl group The group HCO–.

Fortin barometer *See* barometer.

fossil fuel Coal, oil, and natural gas, the fuels used by man as a source of energy. They are formed from the remains of living organisms and all have a high carbon or hydrogen content. Their value as fuels relies on the exothermic oxidation of carbon to form carbon dioxide ($C + O_2 \rightarrow CO_2$) and the oxidation of hydrogen to form water ($H_2 + \frac{1}{2}O_2 \rightarrow H_2O$).

fraction *See* fractional distillation.

fractional crystallization A method of separating a mixture of soluble solids by dissolving them in a suitable hot solvent and then lowering the temperature slowly. The least soluble component will crystallize out first, leaving the other components in solution. By controlling the temperature, it is sometimes possible to remove each component in turn.

fractional distillation (fractionation) The separation of a mixture of liquids by distillation. Effective separation can be achieved by using a long vertical column (*fractionating column*) attached to the distillation vessel and filled with glass beads. Vapour from the liquid rises up the column until it condenses and runs back into the vessel. The rising vapour in the column flows over the descending liquid, and eventually a steady state is reached in which there is a decreasing temperature gradient up the column. The vapour in the column has more volatile components towards the top and less volatile components at the bottom. Various *fractions* of the mixture can be drawn off at points on the column. Industrially, fractional distillation is performed in large towers containing many perforated trays. It is used extensively in petroleum refining.

fractionating column *See* fractional distillation.

fractionation *See* fractional distillation.

francium Symbol Fr. A radioactive element belonging to *group I of the periodic table; a.n. 87; r.d. 2.4; m.p. 27±1°C; b.p. 677±1°C. The element is found in uranium and thorium ores. All 22 known isotopes are radioactive, the most stable being francium-223. The existence of francium was confirmed by Marguerite Perey in 1939.

Frasch process A method of obtaining sulphur from underground deposits using a tube consisting of three concentric pipes. Superheated steam is passed down the outer pipe to melt the sulphur, which is forced up through the middle pipe by compressed air fed through the inner tube. The steam in the outer casing keeps the sulphur molten in the pipe.

Fraunhofer lines Dark lines in the solar spectrum that result from the absorption by elements in the solar chromosphere of some of the wavelengths of the visible radiation emitted by the hot interior of the sun.

free electron *See* electron.

free energy A measure of a system's ability to do work. The *Gibbs free energy* (or *Gibbs function*), G, is defined by $G = H - TS$, where G is the energy liberated or absorbed in a reversible process at constant pressure and constant temperature (T), H is the *enthalpy and S the *entropy of the system. Changes in Gibbs free energy, ΔG, are useful in indicating the conditions under which a chemical reaction will occur. If ΔG is positive the reaction will only occur if energy is supplied to force it away from the equilibrium position (i.e. when $\Delta G = 0$). If ΔG is negative the reaction will proceed spontaneously to equilibrium.

The *Helmholtz free energy* (or *Helmholtz function*), F, is defined by $F = U - TS$, where U is the *internal energy. For a reversible isothermal process, ΔF represents the useful work available.

free radical An atom or group of atoms with an unpaired valence electron. Free radicals can be produced by photolysis or pyrolysis in which a bond is broken without forming ions (*see* homolytic fission). Because of their free valency, most free radicals are extremely reactive. *See also* chain reaction.

freeze drying A process used in dehydrating food, blood plasma, and other heat-sensitive substances. The product is deep-frozen and the ice trapped in it is removed by reducing the pressure and causing it to sublime. The water vapour is then removed, leaving an undamaged dry product.

freezing mixture A mixture of components that produces a low temperature. For example, a mixture of ice and sodium chloride gives a temperature of −20°C.

freons *See* fluorocarbons.

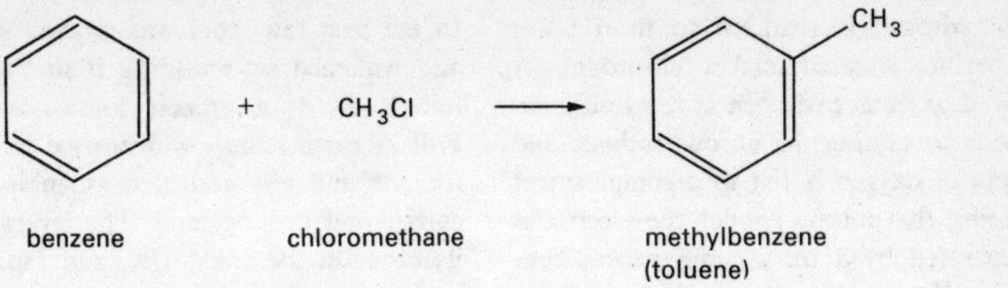

Friedel-Crafts methylation

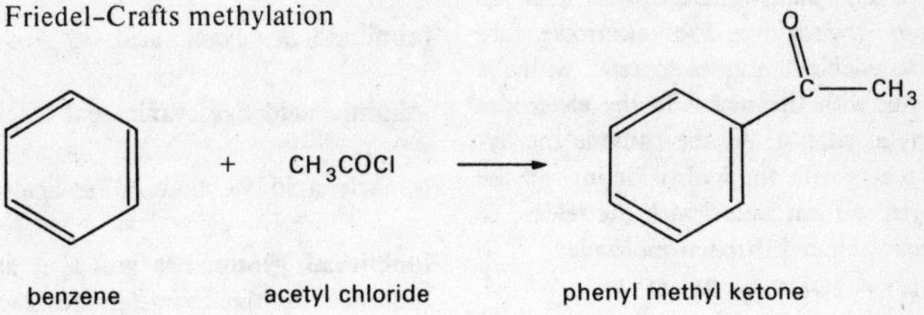

Friedel-Crafts acetylation

Friedel–Crafts reaction A type of reaction in which an alkyl group (from a haloalkane) or an acyl group (from an acyl halide) is substituted on a benzene ring. The product is an alkylbenzene (for alkyl substitution) or an alkyl aryl ketone (for acyl substitution). The reactions occur at high temperature (about 100°C) with an aluminium chloride catalyst. The catalyst acts as an electron acceptor for a lone pair on the halide atom. This polarizes the haloalkane or acyl halide, producing a positive charge on the alkyl or acyl group. The mechanism is then electrophilic substitution. Alcohols and alkenes can also undergo Friedel–Crafts reactions. The reaction is named after the French chemist Charles Friedel (1832–99) and the US chemist James M. Craft (1839–1917).

froth flotation A method of separating mixtures of solids, used industrially for separating ores from the unwanted gangue. The mixture is ground to a powder and water and a frothing agent added. Air is blown through the water. With a suitable frothing agent, the bubbles adhere only to particles of ore and carry them to the surface, leaving the gangue particles at the bottom.

fructose (fruit sugar; laevulose) A simple sugar, $C_6H_{12}O_6$, stereoisomeric with glucose. (Although natural fructose is the D-form, it is in fact laevorotatory.) Fructose occurs in green plants, fruits, and honey and tastes sweeter than sucrose (cane sugar), of which it is a constituent. Derivatives of fructose are important in the energy metabolism of living organisms. Some polysaccharide derivatives (fructans) are carbohydrate energy stores in certain plants.

fruit sugar *See* fructose.

fuel A substance that is oxidized or otherwise changed in a furnace or heat engine to release useful heat or energy. For this purpose wood, vegetable oil, and animal products have largely been replaced by *fossil fuels since the 18th century.

The limited supply of fossil fuels and the expense of extracting them from the earth has encouraged the development of nuclear fuels to produce electricity (*see* nuclear energy).

fuel cell A cell in which the chemical energy of a fuel is converted directly into electrical energy. The simplest fuel cell is one in which hydrogen is oxidized to form water over porous sintered nickel electrodes. A supply of gaseous hydrogen is fed to a compartment containing the porous cathode and a supply of oxygen is fed to a compartment containing the porous anode; the electrodes are separated by a third compartment containing a hot alkaline electrolyte, such as potassium hydroxide. The electrodes are porous to enable the gases to react with the electrolyte, with the nickel in the electrodes acting as a catalyst. At the cathode the hydrogen reacts with the hydroxide ions in the electrolyte to form water, with the release of two electrons per hydrogen molecule:

$$H_2 + 2OH^- \rightarrow 2H_2O + 2e^-$$

At the anode, the oxygen reacts with the water, taking up electrons, to form hydroxide ions:

$$\frac{1}{2}O_2 + H_2O + 2e^- \rightarrow 2OH^-$$

The electrons flow from the cathode to the anode through an external circuit as an electric current. The device is a more efficient converter of electric energy than a heat engine, but it is bulky and requires a continuous supply of gaseous fuels. Their use to power electric vehicles is being actively explored.

fugacity Symbol *f*. A thermodynamic function used in place of partial pressure in reactions involving real gases and mixtures. For a component of a mixture, it is defined by $d\mu = RTd(\ln f)$, where μ is the chemical potential. It has the same units as pressure and the fugacity of a gas is equal to the pressure if the gas is ideal. The fugacity of a liquid or solid is the fugacity of the vapour with which it is in equilibrium. The ratio of the fugacity to the fugacity in some standard state is the *activity. For a gas, the standard state is chosen to be the state at which the fugacity is 1. The activity then equals the fugacity.

fuller's earth A naturally occurring clay material (chiefly montmorillonite) that has the property of decolorizing oil and grease. In the past raw wool was cleaned of grease and whitened by kneading it in water with fuller's earth; a process known as *fulling*. Fuller's earth is now widely used to decolorize fats and oils and also as an insecticide carrier and drilling mud. The largest deposits occur in the USA, UK, and Japan.

fulminate *See* cyanic acid.

fulminic acid *See* cyanic acid.

fumaric acid *See* butenedioic acid.

functional group The group of atoms responsible for the characteristic reactions of a compound. The functional group is –OH for alcohols, –CHO for aldehydes, –COOH for carboxylic acids, etc.

fundamental constants (universal constants) Those parameters that do not change throughout the universe. The charge on an electron, the speed of light in free space, the Planck constant, the gravitational constant, the electric constant, and the magnetic constant are all thought to be examples.

fundamental units A set of independently defined *units of measurement that forms the basis of a system of units. Such a set requires three mechanical units (usually of length, mass, and time) and one electrical unit; it has also been found convenient to treat certain other quantities as fundamental, even though they are not strictly independent. In the metric system the centimetre–gram–second (c.g.s.) system was replaced by the metre–kilogram–second (m.k.s.) system; the latter has now been adapted to provide the basis for *SI units. In British Imperial units the foot–pound–second (f.p.s.) system was formerly used.

furan A colourless liquid compound, C_4H_4O; r.d. 0.94; m.p. –86°C; b.p. 31.4°C.

It has a five-membered ring consisting of four CH_2 groups and one oxygen atom.

furanose A *sugar having a five-membered ring containing four carbon atoms and one oxygen atom.

fused ring *See* ring.

fusible alloys Alloys that melt at low temperature (around 100°C). They have a number of uses, including constant-temperature baths, pipe bending, and automatic sprinklers to provide a spray of water to prevent fires from spreading. Fusible alloys are usually *eutectic mixtures of bismuth, lead, tin, and cadmium. Wood's metal and Lipowitz's alloy are examples of alloys that melt at about 70°C.

fusion 1. Melting. **2.** *See* nuclear fusion.

G

gadolinium Symbol Gd. A soft silvery metallic element belonging to the *lanthanoids; a.n. 64; r.a.m. 157.25; r.d. 7.901 (20°C); m.p. 1312°C; b.p. 3273°C. It occurs in gadolinite, xenotime, monazite, and residues from uranium ores. There are seven stable natural isotopes and eleven artificial isotopes are known. Two of the natural isotopes, gadolinium–155 and gadolinium–157, are the best neutron absorbers of all the elements. The metal has found limited applications in nuclear technology and in ferromagnetic alloys (with cobalt, copper, iron, and cerium). It was discovered by J. C. G. Marignac in 1880.

galactose A simple sugar, $C_6H_{12}O_6$, stereoisomeric with glucose, that occurs naturally as one of the products of the enzymic digestion of milk sugar (lactose) and as a constituent of gum arabic.

galena A mineral form of lead(II) sulphide, PbS, crystallizing in the cubic system; the chief ore of lead. It usually occurs as grey metallic cubes, frequently in association with silver, arsenic, copper, zinc, and antimony. Important deposits occur in Australia (at Broken Hill), Germany, the USA (especially in Missouri, Kansas, and Oklahoma), and the UK.

gallium Symbol Ga. A soft silvery metallic element belonging to group IIIA of the periodic table; a.n. 31; r.a.m. 69.72; r.d. 5.90 (20°C); m.p. 29.78°C; b.p. 2403°C. It occurs in zinc blende, bauxite, and kaolin, from which it can be extracted by fractional electrolysis. It also occurs in gallite, $CuGaS_2$, to an extent of 1%; although bauxite only contains 0.01% this is the only commercial source. The two stable isotopes are gallium–69 and gallium–71; there are eight radioactive isotopes, all with short half-lives. The metal has only a few minor uses (e.g. as an activator in luminous paints) , but gallium arsenide is extensively used as a semiconductor in many applications. Gallium corrodes most other metals because it rapidly diffuses into their lattices. Most gallium(I) and some gallium(II) compounds are unstable. The element was first identified by François Lecoq de Boisbaudran in 1875.

galvanized iron Iron or steel that has been coated with a layer of zinc to protect it from corrosion. Corrugated mild-steel sheets for roofing and mild-steel sheets for dustbins, etc., are usually galvanized by dipping them in molten zinc. The formation of a brittle zinc–iron alloy is prevented by the addition of small quantities of aluminium or magnesium. Wire is often galvanized by a cold electrolytic process as no alloy forms in this process. Galvanizing is an effective method of protecting steel because even if the surface is scratched, the zinc still protects the underlying metal. *See* sacrificial protection.

gamma-iron *See* iron.

gamma radiation Electromagnetic radiation emitted by excited atomic nuclei during the process of passing to a lower excitation state. Gamma radiation ranges in energy from about 10^{-15} to 10^{-11} joule (10 keV to 100 MeV) corresponding to a wavelength range of about 10^{-10} to 10^{-14} metre. A common source of gamma radiation is cobalt–60, the decay process of which is:

$$^{60}_{27}Co \rightarrow ^{60}_{28}Ni \rightarrow Ni$$

The de-excitation of nickel–60 is accompanied by the emission of a gamma-ray photon having an energy of 1.87×10^{-13} J (1.17 MeV).

gangue Rock and other waste material present in an ore.

garnet Any of a group of silicate minerals that conform to the general formula $A_3B_2(SiO_4)_3$. The elements representing A may include magnesium, calcium, manganese, and iron(II); those representing B may include aluminium, iron(III), chromium, or titanium. Six varieties of garnet are generally recognized:
pyrope, $Mg_3Al_2Si_3O_{12}$;
almandine, $Fe^{2+}_3Al_2Si_3O_{12}$;
spessartite, $Mn_3Al_2Si_3O_{12}$;
grossularite, $Ca_3Al_2Si_3O_{12}$;
andradite, $Ca_3(Fe^{3+},Ti)_2Si_3O_{12}$;
and uvarovite, $Ca_3Cr_2Si_3O_{12}$.
Varieties of garnet are used as gemstones and abrasives.

gas A state of matter in which the matter concerned occupies the whole of its container irrespective of its quantity. In an *ideal gas, which obeys the *gas laws exactly, the molecules themselves would have a negligible volume and negligible forces between them, and collisions between molecules would be perfectly elastic. In practice, however, the behaviour of real gases deviates from the gas laws because their molecules occupy a finite volume, there are small forces between molecules, and in polyatomic gases collisions are to a certain extent inelastic (*see* equation of state).

gas chromatography A technique for separating or analysing mixtures of gases by *chromatography. The apparatus consists of a very long tube containing the stationary phase. This may be a solid, such as kieselguhr (*gas–solid chromatography*, or *GSC*), or a nonvolatile liquid, such as a hydrocarbon oil coated on a solid support (*gas–liquid chromatography*, or *GLC*). The sample is often a volatile liquid mixture, which is vaporized and swept through the column by a carrier gas (e.g. hydrogen). The components of the mixture pass through the column at different rates and are detected as they leave, either by measuring the thermal conductivity of the gas or by a flame detector.

Gas chromatography is usually used for analysis; components can be identified by the time they take to pass through the column. It is sometimes also used for separating mixtures.

gas constant (universal molar gas constant) Symbol R. The constant that appears in the *universal gas equation* (*see* gas laws). It has the value 8.314 34 $J\,K^{-1}\,mol^{-1}$.

gas equation *See* gas laws.

gas laws Laws relating the temperature, pressure, and volume of an *ideal gas. *Boyle's law states that the pressure (p) of a specimen is inversely proportional to the volume (V) at constant temperature (pV = constant). The modern equivalent of *Charles' law states that the volume is directly proportional to the thermodynamic temperature (T) at constant pressure (V/T = constant); originally this law stated the constant expansivity of a gas kept at constant pressure. The pressure law states that the pressure is directly proportional to the thermodynamic temperature for a specimen kept at constant volume. The three laws can be combined in the *universal gas equation*, $pV = nRT$, where n is the amount of gas in the specimen and R is the *gas constant. The gas laws were first established experimentally for real gases, although they are obeyed by real gases to only a limited extent; they are obeyed best at high temperatures and low pressures. *See also* equation of state.

gasoline *See* petroleum.

gas thermometer A device for measuring temperature in which the working fluid is a gas. It provides the most accurate method of measuring temperatures in the range 2.5 to 1337 K. Using a fixed mass of gas a *constant-volume thermometer* measures the pressure of a fixed volume of gas at relevant temperatures, usually by means of a mercury manometer and a barometer.

Gattermann reaction A variation of the *Sandmeyer reaction for preparing chloro- or bromoarenes by reaction of the diazonium compound. In the Gattermann reaction the aromatic amine is added to sodium nitrite and the halogen acid (10°C), then fresh copper powder (e.g. from Zn + $CuSO_4$) is added and the solution warmed. The diazonium salt then forms the haloarene, e.g.

$$C_6H_5N_2{}^+Cl^- \rightarrow C_6H_5Cl + N_2$$

The copper acts as a catalyst. The reaction is easier to perform than the Sandmeyer reaction and takes place at lower temperature, but generally gives lower yields. It was discovered in 1890 by the German chemist Ludwig Gattermann (1860–1920).

gauche conformation *See* conformation.

Gay Lussac's law 1. When gases combine chemically the volumes of the reactants and the volume of the product, if it is gaseous, bear simple relationships to each other when measured under the same conditions of temperature and pressure. The law was first stated in 1808 by J. L. Gay Lussac (1778–1850) and led to *Avogadro's law. **2.** *See* Charles' law.

gaylussite A mineral consisting of a hydrated mixed carbonate of sodium and calcium, $Na_2CO_3.CaCO_3.5H_2O$.

Geiger counter (Geiger–Müller counter) A device used to detect and measure ionizing radiation. It consists of a tube containing a low-pressure gas (usually argon or neon and methane) and a cylindrical hollow cathode through the centre of which runs a fine-wire anode. A potential difference of about 1000 volts is maintained between the electrodes. An ionizing particle or photon passing through a window into the tube will cause an ion to be produced and the high p.d. will accelerate it towards its appropriate electrode, causing an avalanche of further ionizations by collision. The consequent current pulses can be counted in electronic circuits or simply amplified to work a small loudspeaker in the instrument. It was first devised in 1908 by Hans Geiger (1882–1947). Geiger and W. Müller produced an improved design in 1928.

gel A lyophilic *colloid that has coagulated to a rigid or jelly-like solid. In a gel, the disperse medium has formed a loosely-held network of linked molecules through the dispersion medium. Examples of gels are silica gel and gelatin.

gelatin(e) A colourless or pale yellow water-soluble protein obtained by boiling collagen with water and evaporating the solution. It swells when water is added and dissolves in hot water to form a solution that sets to a gel on cooling. It is used in photographic emulsions and adhesives, and in jellies and other foodstuffs.

gel filtration A type of column *chromatography in which a mixture of liquids is passed down a column containing a gel.

Small molecules in the mixture can enter pores in the gel and move slowly down the column; large molecules, which cannot enter the pores, move more quickly. Thus, mixtures of molecules can be separated on the basis of their size. The technique is used particularly for separating proteins but it can also be applied to other polymers and to cell nuclei, viruses, etc.

gelignite A high explosive made from nitroglycerin, cellulose nitrate, sodium nitrate, and wood pulp.

gem Designating molecules in which two functional groups are attached to the same atom in a molecule. For example, chloral hydrate, $CCl_3CH(OH)_2$, is a gem diol in which both hydroxyl groups are on the same carbon atom.

geochemistry The scientific study of the chemical composition of the earth. It includes the study of the abundance of the earth's elements and their isotopes and the distribution of the elements in environments of the earth (lithosphere, atmosphere, biosphere, and hydrosphere).

geometrical isomerism *See* isomerism.

geraniol An alcohol, $C_9H_{15}CH_2OH$, present in a number of essential oils.

germanium Symbol Ge. A lustrous hard metalloid element belonging to group IV of the periodic table; a.n. 32; r.a.m. 72.59; r.d. 5.36; m.p. 937°C; b.p. 2830°C. It is found in zinc sulphide and in certain other sulphide ores, and is mainly obtained as a by-product of zinc smelting. It is also present in some coal (up to 1.6%). Small amounts are used in specialized alloys but the main use depends on its semiconductor properties. Chemically, it forms compounds in the +2 and +4 oxidation states, the germanium(IV) compounds being the more stable. The element also forms a large number of organometallic compounds. Predicted in 1871 by Mendeleev (eka-silicon), it was discovered by Winkler in 1886.

German silver (nickel silver) An alloy of copper, zinc, and nickel, often in the proportions 5:2:2. It resembles silver in appearance and is used in cheap jewellery and cutlery and as a base for silver-plated wire. *See also* electrum.

Gibbs free energy (Gibbs function) *See* free energy. It is named after the US chemist J. W. Gibbs (1839–1903).

gibbsite A mineral form of hydrated *aluminium hydroxide ($Al(OH)_3$). It is named after the US mineralogist George Gibbs (d. 1833).

giga- Symbol G. A prefix used in the metric system to denote one thousand million times. For example, 10^9 joules = 1 gigajoule (GJ).

glacial ethanoic (*or* acetic) acid *See* ethanoic acid.

glass Any noncrystalline solid; i.e. a solid in which the atoms are random and have no long-range ordered pattern. Glasses are often regarded as supercooled liquids. Characteristically they have no definite melting point, but soften over a range of temperatures.

The common glass used in windows, bottles, etc., is *soda glass*, which is made by heating a mixture of lime (calcium oxide), soda (sodium carbonate), and sand (silicon (IV) oxide). It is a form of calcium silicate. Borosilicate glasses (e.g. *Pyrex*) are made by incorporating some boron oxide, so that silicon atoms are replaced by boron atoms. They are tougher than soda glass and more resistant to temperature changes, hence their use in cooking utensils and laboratory apparatus. Glasses for special purposes (e.g. optical glass) have other elements added (e.g. barium, lead).

glass electrode A type of *half cell having a glass bulb containing an acidic solution of fixed pH, into which dips a platinum wire. The glass bulb is thin enough for hydrogen ions to diffuse through. If the bulb is placed in a solution containing hydrogen ions, the electrode potential depends on the hydrogen-ion concentration. Glass electrodes are used in pH measurement.

glass fibres Melted glass drawn into thin fibres some 0.005 mm–0.01 mm in diameter. The fibres may be spun into threads and woven into fabrics, which are then impregnated with resins to give a material that is both strong and corrosion resistant for use in car bodies and boat building.

glauberite A mineral consisting of a mixed sulphate of sodium and calcium, $Na_2SO_4.CaSO_4$.

Glauber's salt *Sodium sulphate decahydrate, $Na_2SO_4.10H_2O$, used as a laxative. It is named after J. R. Glauber (1604–68).

GLC (gas–liquid chromatography) *See* gas chromatography.

globulin Any of a group of globular proteins that are generally insoluble in water and present in blood, eggs, milk, and as a reserve protein in seeds. Blood serum globulins comprise four types: α_1-, α_2-, and β-globulins, which serve as carrier proteins; and γ-globulins, which include the immunoglobulins responsible for immune responses.

gluconic acid An optically active hydroxycarboxylic acid, $CH_2(OH)(CHOH)_4COOH$. It is the carboxylic acid corresponding to the aldose sugar glucose, and can be made by the action of certain moulds.

glucose (dextrose; grape sugar) A white crystalline sugar, $C_6H_{12}O_6$, occurring widely in nature. Like other *monosaccharides, glucose is optically active: most naturally occurring glucose is dextrorotatory. Glucose and its derivatives are crucially important in the energy metabolism of living organisms. Glucose is also a constituent of many polysaccharides, most notably starch and cellulose. These yield glucose when broken down, for example by enzymes during digestion.

glutamic acid *See* amino acid.

glutamine *See* amino acid.

glyceride (acylglycerol) A fatty-acid ester of glycerol. Esterification can occur at one, two, or all three hydroxyl groups of the glycerol molecule producing mono-, di-, and triglycerides respectively. *Triglycerides are the major constituent of fats and oils found in living organisms. Alternatively, one of the hydroxyl groups may be esterified with a phosphate group forming a phosphoglyceride (*see* phospholipid) or to a sugar forming a *glycolipid*.

glycerine *See* glycerol.

glycerol (glycerine; propane-1,2,3,-triol) A trihydric alcohol, $HOCH_2CH(OH)CH_2OH$. Glycerol is a colourless sweet-tasting viscous liquid, miscible with water but insoluble in ether. It is widely distributed in all living organisms as a constituent of the *glycerides, which yield glycerol when hydrolysed.

glycine *See* amino acid.

glycogen (animal starch) A *polysaccharide consisting of a highly branched polymer of glucose occurring in animal tissues, especially in liver and muscle cells. It is the major store of carbohydrate energy in animal cells and is present as granular clusters of minute particles.

glycol *See* ethane-1,2-diol.

glycolipid *See* glyceride.

goethite A yellow-brown mineral, FeO.OH, crystallizing in the orthorhombic system. It is formed as a result of the oxidation and hydration of iron minerals or as a direct precipitate from marine or fresh water (e.g. in swamps and bogs). Most *limonite is composed largely of cryptocrystalline goethite. Goethite is mined as an ore of iron.

gold Symbol Au. A soft yellow malleable metallic *transition element; a.n. 79; r.a.m. 196.967; r.d. 19.32; m.p. 1064.43°C; b.p. 2807±2°C. It is found as the free metal in gravel or in quartz veins, and is also present in some lead and copper sulphide ores. It also occurs combined with silver in the telluride sylvanite, $(Ag,Au)Te_2$. It is used in jewellery, dentistry, and electronic devices. Chemically, it is unreactive, being unaffected by oxygen. It reacts with chlorine at 200°C to form gold(III) chloride. It forms a number of complexes with gold in the +1 and +3 oxidation states.

Goldschmidt process A method of extracting metals by reducing the oxide with aluminium powder, e.g.

$$Cr_2O_3 + 2Al \rightarrow 2Cr + Al_2O_3$$

The reaction can also be used to produce molten iron (*see* thermite). It was discovered by the German chemist Hans Goldschmidt (1861–1923).

Gooch crucible A porcelain dish with a perforated base over which a layer of asbestos is placed, used for filtration in gravimetric analysis.

graft copolymer *See* polymer.

Graham's law The rates at which gases diffuse is inversely proportional to the square roots of their densities. This principle is made use of in the diffusion method of separating isotopes. The law was formulated in 1829 by Thomas Graham (1805–69).

gram Symbol g. One thousandth of a kilogram. The gram is the fundamental unit of mass in *c.g.s. units and was formerly used in such units as the *gram-atom*, *gram-molecule*, and *gram-equivalent*, which have now been replaced by the *mole.

graphite *See* carbon.

gravimetric analysis A type of quantitative analysis that depends on weighing. For instance, the amount of silver in a solution of silver salts could be measured by adding excess hydrochloric acid to precipitate silver chloride, filtering the precipitate, washing, drying, and weighing.

gray Symbol Gy. The derived SI unit of absorbed *dose of ionizing radiation (*see* radiation units). It is named after the British radiobiologist L. H. Gray (1905–65).

greenockite A mineral form of cadmium sulphide, CdS.

green vitriol *See* iron(II) sulphate.

Grignard reagents A class of organometallic compounds of magnesium, with the general formula RMgX, where R is an organic group and X a halogen atom (e.g. CH_3MgCl, C_2H_5MgBr, etc.). They actually have the structure $R_2Mg.MgCl_2$, and can be made by reacting a haloalkane with magnesium in ether; they are rarely isolated but are extensively used in organic synthesis, when they are made in one reaction mixture. Grignard reagents have a number of reactions that make them useful in organic synthesis. With methanal they give a primary alcohol

$$CH_3MgCl + HCHO \rightarrow CH_3CH_2OH$$

Other aldehydes give a secondary alcohol

$$CH_3CHO + CH_3MgCl \rightarrow (CH_3)_2CHOH$$

With alcohols, hydrocarbons are formed

$$CH_3MgCl + C_2H_5OH \rightarrow C_2H_5CH_3$$

Water also gives a hydrocarbon

$$CH_3MgCl + H_2O \rightarrow CH_4$$

The compounds are named after their discoverer, the French chemist F. A. V. Grignard (1871–1935).

ground state The lowest stable energy state of a system, such as a molecule, atom, or nucleus. *See* energy level.

group (in chemistry) *See* periodic table.

group 0 elements *See* noble gases.

group I elements A group of elements in the *periodic table, divided into two subgroups: group IA (the main group, *see* alkali metals) and group IB. Group IB consists of the *coinage metals, copper, silver, and gold, which are usually classified with the *transition elements.

group II elements A group of elements in the *periodic table, divided into two subgroups: group IIA (the main group, *see* alkaline-earth metals) and group IIB. Group IIB consists of the three metals zinc (Zn), cadmium (Cd), and mercury (Hg), which have two *s*-electrons outside filled *d*-subshells. Moreover, none of their compounds have unfilled *d*-levels, and the metals are regarded as nontransition elements. They are sometimes called the *zinc group*. Zinc and cadmium are relatively electropositive metals, forming compounds containing divalent ions Zn^{2+} or Cd^{2+}. Mercury is more unreactive and also unusual in forming mercury(I) compounds, which contain the ion $Hg_2{}^{2+}$.

group III elements A group of elements in the *periodic table, divided into subgroup IIIB (the main group) and subgroup IIIA. The IIIA subgroup consists of scandium (Sc), yttrium (Y), and lanthanum (La), which are generally classified with the *lanthanoids, and actinium (Ac), generally classified with the *actinoids. The term *group III elements* usually refers to the main-group elements: boron (B), aluminium (Al), gallium (Ga), indium (In), and thallium (Tl), which all have outer electronic configurations ns^2np^1 with no partly filled inner levels. They are the first members of the *p*-block. The group differs from the alkali metals and alkaline-earth metals in displaying a considerable variation in properties as the group is descended.

Boron has a small atomic radius and a relatively high ionization energy. In consequence its chemistry is largely covalent and it is generally classed as a metalloid. It forms a large number of volatile hydrides, some of which have the uncommon bonding characteristic of *electron-deficient compounds. It also forms a weakly acidic oxide. In some ways, boron resembles silicon (*see* diagonal relationship).

As the group is descended, atomic radii increase and ionization energies are all lower than for boron. There is an increase in polar interactions and the formation of distinct M^{3+} ions. This increase in metallic character is clearly illustrated by the increasing basic character of the hydroxides: boron hydroxide is acidic, aluminium and gallium hydroxides are amphoteric, indium hydroxide is basic, and thallium forms only the oxide. As the elements of group III have a vacant *p*-orbital they display many electron-acceptor properties. For example, many boron compounds form adducts with donors such as ammonia and organic amines (acting as Lewis acids). A large number of complexes of the type $[BF_4]^-$, $[AlCl_4]^-$, $[InCl_4]^-$, $[TlI_4]^-$ are known and the heavier members can expand their coordination numbers to six as in $[AlF_6]^{3-}$ and $[TlCl_6]^{3-}$. This acceptor property is also seen in bridged dimers of the type Al_2Cl_6. Another feature of group III is the increasing stability of the monovalent state down the group. The electron configuration ns^2np^1 suggests that only one electron could be lost or shared in forming compounds. In fact, for the lighter members of the group the energy required to promote an electron from the *s*-subshell to a vacant *p*-subshell is small. It is more than compensated for by the resulting energy gain in forming three bonds rather than one. This energy gain is less important for the heavier members of the group. Thus, aluminium forms compounds of the type AlCl in the gas phase at high temperatures. Gallium similarly forms such compounds and gallium(I) oxide (Ga_2O) can be isolated. Indium has a number of known indium(I) compounds (e.g. InCl, In_2O, $In^I[In^{III}Cl_6]$). Thallium has stable monovalent compounds. In aqueous solution, thallium(I) compounds are more stable than the corresponding thallium (III) compounds. *See* inert-pair effect.

group IV elements A group of elements in the *periodic table, divided into subgroup IVB (the main group) and subgroup IVA. The IVA subgroup consists of titanium (Ti), zirconium (Zr), and hafnium (Hf), which are generally classified with the *transition elements. The term *group IV elements* usually refers to the main-group elements: carbon (C), silicon (Si), germanium (Ge), tin (Sn), and lead (Pb), which all have outer electronic configurations ns^2np^2 with no partly filled inner levels.

The main valency of the elements is 4, and the members of the group show a variation from nonmetallic to metallic behaviour in moving down the group. Thus, carbon is a nonmetal and forms an acidic oxide (CO_2) and a neutral oxide. Carbon compounds are mostly covalent. One allotrope (diamond) is an insulator, although graphite is a fairly good conductor. Silicon and germanium are metalloids, having semiconductor properties. Tin is a metal, but does have a nonmetallic allotrope (grey tin). Lead is definitely a metal. Another feature of the group is the tendency to form divalent compounds as the size of the atom increases. Thus carbon has only the highly reactive carbenes. Silicon forms analogous silylenes. Germanium has an unstable hydroxide (Ge $(OH)_2$), a sulphide (GeS), and halides. The sulphide and halides disproportionate to germanium and the germanium(IV) compound. Tin has a number of tin(II) compounds, which are moderately reducing, being oxidized to the tin(IV) compound. Lead has a stable lead(II) state. *See* inert-pair effect.

In general, the reactivity of the elements increases down the group from carbon to lead. All react with oxygen on heating. The first four form the dioxide; lead forms the monoxide (i.e. lead(II) oxide, PbO). Similarly, all will react with chlorine to form the tetrachloride (in the case of the first four) or the dichloride (for lead). Carbon is the only one capable of reacting directly with hydrogen. The hydrides all exist from the stable methane (CH_4) to the unstable plumbane (PbH_4).

group V elements A group of elements in the *periodic table, divided into subgroup VB (the main group) and subgroup VA. The VA subgroup consists of vanadium (V), niobium (Nb), and tantalum (Ta), which are generally classified with the *transition elements. The term *group V elements* usually refers to the main-group elements: nitrogen (N), phosphorus (P), arsenic (As), antimony (Sb), and bismuth (Bi), which all have outer electronic configurations ns^2np^3 with no partly filled inner levels.

The lighter elements (N and P) are nonmetals; the heavier elements are metalloids. The lighter elements are electronegative in character and have fairly large ionization energies. Nitrogen has a valency of 3 and tends to form covalent compounds. The other elements have available *d*-sublevels and can promote an *s*-electron into one of these to form compounds with the V oxidation state. Thus, they have two oxides

P_2O_3, P_2O_5, Sb_2O_3, Sb_2O_5, etc. In the case of bismuth, the pentoxide Bi_2O_5 is difficult to prepare and unstable – an example of the increasing stability of the III oxidation state in going from phosphorus to bismuth. The oxides also show how there is increasing metallic (electropositive) character down the group. Nitrogen and phosphorus have oxides that are either neutral (N_2O, NO) or acidic. Bismuth trioxide (Bi_2O_3) is basic. Bismuth is the only member of the group that forms a well-characterized positive ion Bi^{3+}.

group VI elements A group of elements in the *periodic table, divided into group VIB (the main group) and group VIA. The VIA subgroup consists of chromium (Cr), molybdenum (Mo), and tungsten (W), which are generally classified with the *transition elements. The term *group VI elements* usually refers to the main-group elements: oxygen (O), sulphur (S), selenium (Se), tellurium (Te), and polonium (Po), which all have outer electronic configurations ns^2np^4 with no partly filled inner levels. They are also called the *chalcogens*.

The configurations are just two electrons short of the configuration of a noble gas and the elements are characteristically electronegative and almost entirely nonmetallic. Ionization energies are high, (O 1314 to Po 813 kJ mol^{-1}) and monatomic cations are not known. Polyatomic cations do exist, e.g. O_2^+, S_8^{2+}, Se_8^{2+}, Te_4^{2+}. Electronegativity decreases down the group but the nearest approach to metallic character is the occurrence of 'metallic' allotropes of selenium, tellurium, and polonium along with some metalloid properties, in particular, marked photoconductivity. The elements of group VI combine with a wide range of other elements and the bonding is largely covalent. The elements all form hydrides of the type XH_2. Apart from water, these materials are all toxic foul-smelling gases; they show decreasing thermal stability with increasing relative atomic mass of X. The hydrides dissolve in water to give very weak acids (acidity increases down the group). Oxygen forms the additional hydride H_2O_2 (hydrogen peroxide), but sulphur forms a range of sulphanes, such as H_2S_2, H_2S_4, H_2S_6.

Oxygen forms the fluorides O_2F_2 and OF_2, both powerful fluorinating agents; sulphur forms analogous fluorides along with some higher fluorides, S_2S_2, SF_2, SF_4, SF_6, S_2F_{10}. Selenium and tellurium form only the higher fluorides MF_4 and MF_6; this is in contrast to the formation of lower valence states by heavier elements observed in groups III, IV, and V. The chlorides are limited to M_2Cl_2 and MCl_4; the bromides are similar except that sulphur only forms S_2Br_2. All metallic elements form oxides and sulphides and many form selenides.

group VII elements A group of elements in the *periodic table divided into two subgroups: group VIIA and group VIIB. Group VIIB is the main group (*see* halogens) and group VIIA consists of the elements manganese (Mn), technetium (Tc), and rhenium (Re), which are usually classified with the *transition elements.

group VIII elements A group of nine elements in the *periodic table consisting of the *platinum metals – cobalt (Co), rhodium (Rh), iridium (Ir), nickel (Ni), palladium (Pd), and platinum (Pt) – together with iron (Fe), ruthenium (Ru), and osmium (Os). These metals are now classified with the *transition elements.

GSC (gas–solid chromatography) *See* gas chromatography.

guanidine A crystalline basic compound $HN{:}C(NH_2)_2$, related to urea.

guanine A *purine derivative. It is one of the major component bases of *nucleotides and the nucleic acids *DNA and *RNA.

gum Any of a variety of substances obtained from plants. Typically they are insoluble in organic solvents but form gelatinous or sticky solutions with water. Gum resins are mixtures of gums and natural resins. Gums are produced by the young xylem vessels of some plants (mainly trees) in response to wounding or pruning. The exudate hardens when it reaches the plant surface and thus provides a temporary protective seal while the cells below divide to form a permanent repair. Excessive gum formation is a symptom of some plant diseases. *See also* mucilage.

guncotton *See* cellulose nitrate.

gun metal A type of bronze usually containing 88–90% copper, 8–10% tin, and 2–4% zinc. Formerly used for cannons, it is still used for bearings and other parts that require high resistance to wear and corrosion.

gunpowder An explosive consisting of a mixture of potassium nitrate, sulphur, and charcoal.

gypsum A monoclinic mineral form of hydrated *calcium sulphate, $CaSO_4.2H_2O$. It occurs in five varieties: *rock gypsum*, which is often red stained and granular; *gypsite*, an impure earthy form occurring as a surface deposit; *alabaster*, a pure fine-grained translucent form; *satin spar*, which is fibrous and silky; and *selenite*, which occurs as transparent crystals in muds and clays. It is used in the building industry and in the manufacture of cement, rubber, paper, and plaster of Paris.

H

Haber process An industrial process for producing ammonia by reaction of nitrogen with hydrogen:

$$N_2 + 3H_2 \rightleftharpoons 2NH_3$$

The reaction is reversible and exothermic, so that a high yield of ammonia is favoured by low temperature (*see* Le Chatelier's principle). However, the rate of reaction would be too slow for equilibrium to be reached at normal temperatures, so an optimum temperature of about 450°C is used, with a catalyst of iron containing potassium and aluminium oxide promoters. The higher the pressure the greater the yield, although there are technical difficulties in using very high pressures. A pressure of about 250 atmospheres is commonly employed.

The process is of immense importance for the fixation of nitrogen for fertilizers. It was developed in 1908 by the German chemist Fritz Haber (1886–1934) and was developed for industrial use by Carl Bosch (1874–1940), hence the alternative name *Haber–Bosch process*. The nitrogen is obtained from liquid air. Formerly, the hydrogen was from *water gas and the water–gas shift reaction (the *Bosch process*) but now the raw material (called *synthesis gas*) is obtained by steam *reforming natural gas.

habit *See* crystal habit.

haem (heme) An iron-containing molecule that binds with proteins as a *cofactor or *prosthetic group to form the *haemoproteins*. These are *haemoglobin, *myoglobin, and the *cytochromes. Essentially, haem comprises a *porphyrin with its four nitrogen atoms holding the iron(II) atom as a chelate. This iron can reversibly bind oxygen (as in haemoglobin and myoglobin) or (as in the cytochromes) conduct electrons by conversion between the iron(II) and iron (III) series.

haematite A mineral form of iron(III) oxide, Fe_2O_3. It is the most important ore of iron and usually occurs in two main forms: as a massive red kidney-shaped ore (*kidney ore*) and as grey to black metallic crystals known as *specular iron ore*. Haematite is the major red colouring agent in rocks; the largest deposits are of sedimentary origin. In industry haematite is also used as a polishing agent (jeweller's rouge) and in paints.

haemoglobin One of a group of globular proteins occurring widely in animals as oxygen carriers in blood. Vertebrate haemoglobin comprises two pairs of polypeptide chains (forming the *globin* protein) with each chain folded to provide a binding site for a *haem group. Each of the four haem groups binds one oxygen molecule to form *oxyhaemoglobin*. Dissociation occurs in oxygen-depleted tissues: oxygen is released and haemoglobin is reformed. The haem groups also bind other inorganic molecules, including carbon monoxide. This binds more strongly than oxygen and competes with it (hence its toxicity). In vertebrates, haemoglobin is contained in the red blood cells (erythrocytes).

hafnium Symbol Hf. A silvery lustrous metallic *transition element; a.n. 72; r.a.m. 178.49; r.d. 13.3; m.p. 2230±20°C; b.p. 4602°C. The element is found with zirconium and is extracted by formation of the chloride and reduction by the Kroll process. It is used in tungsten alloys in filaments and electrodes and as a neutron absorber. The metal forms a passive oxide layer in air. Most of its compounds are hafnium(IV) complexes; less stable hafnium(III) complexes also exist. The element was first reported by Urbain in 1911, and its existence was finally established by D. Coster and G. C. de Hevesey in 1923.

half cell An electrode in contact with a solution of ions, forming part of a *cell. Various types of half cell exist, the simplest consisting of a metal electrode immersed in a solution of metal ions. Gas half cells have a gold or platinum plate in a solution with gas bubbled over the metal plate. The commonest is the *hydrogen half cell. Half cells can also be formed by a metal in contact with an insoluble salt or oxide and a solution. The *calomel half cell is an example of this. Half cells are commonly referred to as *electrodes*.

half-thickness The thickness of a specified material that reduces the intensity of a beam of radiation to half its original value.

half-width Half the width of a spectrum line (or in some cases the full width) measured at half its height.

halide A compound of a halogen with another element or group. The halides of typical metals are ionic (e.g. sodium fluoride, Na^+F^-). Metals can also form halides in which the bonding is largely covalent (e.g. aluminium chloride, $AlCl_3$). Organic compounds are also sometimes referred to as halides; e.g. the alkyl halides (*see* haloalkanes) and the *acyl halides. Halides are named *fluorides*, *chlorides*, *bromides*, or *iodides*.

halite (rock salt) Naturally occurring *sodium chloride (common salt, NaCl), crystallizing in the cubic system. It is chiefly colourless or white (sometimes blue) when pure but the presence of impurities may colour it grey, pink, red, or brown. Halite often occurs in association with anhydrite and gypsum.

Hall–Heroult cell An electrolytic cell used industrially for the extraction of aluminium from bauxite. The bauxite is first purified by dissolving it in sodium hydroxide and filtering off insoluble constituents. Aluminium hydroxide is then precipitated (by adding CO_2) and this is decomposed by heating to obtain pure Al_2O_3. In the

Hall–Heroult cell, the oxide is mixed with cryolite (to lower its melting point) and the molten mixture electrolysed using graphite anodes. The cathode is the lining of the cell, also of graphite. The electrolyte is kept in a molten state (about 850°C) by the current. Molten aluminium collects at the bottom of the cell and can be tapped off. Oxygen forms at the anode, and gradually oxidizes it away. The cell is named after the US chemist Charles Martin Hall (1863–1914), who discovered the process in 1886, and the French chemist Paul Heroult (1863–1914), who discovered it independently in the same year.

haloalkanes (alkyl halides) Organic compounds in which one or more hydrogen atoms of an alkane have been substituted by halogen atoms. Examples are chloromethane, CH_3Cl, dibromoethane, CH_2BrCH_2Br, etc. Haloalkanes can be formed by direct reaction between alkanes and halogens using ultraviolet radiation. They are usually made by reaction of an alcohol with a halogen carrier.

haloform reaction A reaction for producing haloforms from methyl ketones. An example is the production of chloroform from propanone using sodium chlorate(I) (or bleaching powder):

$$CH_3COCH_3 + 3NaOCl \rightarrow CH_3COCl_3 + 3NaOH$$

The substituted ketone then reacts to give chloroform (trichloromethane):

$$CH_3COCCl_3 + NaOH \rightarrow NaOCOCH_3 + CHCl_3$$

The reaction can also be used for making carboxylic acids, since $RCOCH_3$ gives the product NaOCOR. It is particularly useful for aromatic acids as the starting ketone can be made by a Friedel–Crafts acylation.

The reaction of methyl ketones with sodium iodate(I) gives iodoform (triiodomethane), which is a yellow solid with a characteristic smell. This reaction is used in the *iodoform test* to identify methyl ketones. It also gives a positive result with a secondary alcohol of the formula $RCH(OH)CH_3$ (which is first oxidized to a methylketone) or with ethanol (oxidized to ethanal, which also undergoes the reaction).

haloforms The four compounds with formula CHX_3, where X is a halogen atom. They are *chloroform* ($CHCl_3$), and, by analogy, *fluoroform* (CHF_3), *bromoform* ($CHBr_3$), and *iodoform* (CHI_3). The systematic names are trichloromethane, trifluoromethane, etc.

halogenating agent *See* halogenation.

halogenation A chemical reaction in which a halogen atom is introduced into a compound. Halogenations are described as *chlorination*, *fluorination*, *bromination*, etc., according to the halogen involved. Halogenation reactions may take place by direct reaction with the halogen. This occurs with alkanes, where the reaction involves free radicals and requires high temperature, ultraviolet radiation, or a chemical initiator; e.g.

$$C_2H_6 + Br_2 \rightarrow C_2H_5Br + HBr$$

The halogenation of aromatic compounds can be effected by electrophilic substitution using an aluminium chloride catalyst:

$$C_6H_6 + Cl_2 \rightarrow C_6H_5Cl + HCl$$

Halogenation can also be carried out using compounds, such as phosphorus halides (e.g. PCl_3) or sulphur dihalide oxides (e.g. $SOCl_2$), which react with –OH groups. Such compounds are called *halogenating agents*. Addition reactions are also referred to as halogenations; e.g.

$$C_2H_4 + Br_2 \rightarrow CH_2BrCH_2Br$$

halogens (group VII elements) A group of elements in the *periodic table (group VIIB): fluorine (F), chlorine (Cl), bromine (Br), iodine (I), and astatine (At). All have a characteristic electron configuration of noble gases but with outer ns^2np^5 electrons. The outer shell is thus one electron short of a noble-gas configuration. Consequently, the halogens are typical nonmetals; they have high electronegativities – high electron affinities and high ionization energies. They form compounds by gaining an electron to complete the stable configuration; i.e. they are good oxidizing agents. Alternatively, they share their outer electrons to form covalent compounds, with single bonds.

All are reactive elements with the reactivity decreasing down the group. The electron affinity decreases down the group and other properties also show a change from fluorine to astatine. Thus, the melting and boiling points increase; at 20°C, fluorine and chlorine are gases, bromine a liquid, and iodine and astatine are solids. All exist as diatomic molecules.

The name 'halogen' comes from the Greek 'salt-producer', and the elements react with metals to form ionic halide salts. They also combine with nonmetals, the activity decreasing down the group: fluorine reacts with all nonmetals except nitrogen and the noble gases helium, neon, and argon; iodine does not react with any noble gas, nor with carbon, nitrogen, oxygen, or sulphur. The elements fluorine to iodine all react with hydrogen to give the acid, with the activity being greatest for fluorine, which reacts explosively. Chlorine and hydrogen react slowly at room temperature in the dark (sunlight causes a free-radical chain reaction). Bromine and hydrogen react if heated in the presence of a catalyst. Iodine and hydrogen react only slowly and the reaction is not complete. There is a decrease in oxidizing ability down the group from fluorine to iodine. As a consequence, each halogen will displace any halogen below it from a solution of its salt, for example:

$$Cl_2 + 2Br^- \rightarrow Br_2 + 2Cl^-$$

The halogens also form a wide variety of organic compounds in which the halogen atom is linked to carbon. In general, the aryl compounds are more stable than the alkyl compounds and there is decreasing resistance to chemical attack down the group from the fluoride to the iodide.

Fluorine has only a valency of 1, although the other halogens can have higher oxidation states using their vacant *d*-electron levels. There is also evidence for increasing metallic behaviour down the group. Chlorine and bromine form compounds with oxygen in which the halogen atom is assigned a positive oxidation state. Only iodine, however, forms positive ions, as in $I^+NO_3^-$.

hardening of oils The process of converting unsaturated esters of *fatty acids into (more solid) saturated esters by hydrogenation using a nickel catalyst. It is used in the manufacture of margarine from vegetable oils.

hardness of water The presence in water of dissolved calcium or magnesium ions, which form a scum with soap and prevent the formation of a lather. The main cause of hard water is dissolved calcium hydrogencarbonate ($Ca(HCO_3)_2$), which is formed in limestone or chalk regions by the action of dissolved carbon dioxide on calcium carbonate. This type is known as *temporary hardness* because it is removed by boiling:

$$Ca(HCO_3)_2(aq) \rightarrow CaCO_3(s) + H_2O(l) + CO_2(g)$$

The precipitated calcium carbonate is the 'fur' (or 'scale') formed in kettles, boilers, pipes, etc. In some areas, hardness also results from dissolved calcium sulphate ($CaSO_4$), which cannot be removed by boiling (*permanent hardness*).

Hard water is a considerable problem in washing, reducing the efficiency of boilers, heating systems, etc., and in certain industrial processes. Various methods of *water softening* are used. In public supplies, the temporary hardness can be removed by adding lime (calcium hydroxide), which precipitates calcium carbonate

$$Ca(OH)_2(aq) + Ca(HCO_3)_2(aq) \rightarrow 2CaCO_3(s) + 2H_2O(l)$$

This is known as the *Clark process* (or as '*clarking*'). It does not remove permanent hardness. Both temporary and permanent hardness can be treated by precipitating calcium carbonate by added sodium carbonate – hence its use as a washing soda and in bath salts. Calcium (and other) ions can also be removed from water by ion-exchange using zeolites (e.g. *Permutit*). This method is used in small domestic water-softeners. Another technique is not to remove the Ca^{2+} ions but to complex them and prevent them reacting further. For domestic use polyphosphates (containing the ion $P_6O_{18}^{6-}$, e.g. *Calgon*) are added. Other sequestering agents are also used for industrial water. *See also* sequestration.

heat capacity (thermal capacity) The ratio of the heat supplied to an object or specimen to its consequent rise in temperature. The *specific heat capacity* is the ratio of the heat supplied to unit mass of a substance to its consequent rise in temperature. The *molar heat capacity* is the ratio of the heat supplied to unit amount of a substance to its consequent rise in temperature. In practice, heat capacity (C) is measured in joules per kelvin, specific heat capacity (c) in $J\,K^{-1}\,kg^{-1}$, and molar heat capacity (C_m) in $J\,K^{-1}\,mol^{-1}$. For a gas, the values of c and C_m are commonly given either at *constant volume*, when only its *internal energy is increased, or at *constant pressure*, which requires a greater input of heat as the gas is allowed to expand and do work against the surroundings. The symbols for the specific and molar heat capacities at constant volume are c_v and C_v, respectively; those for the specific and molar heat capacities at constant pressure are c_p and C_p.

heat of atomization The energy required to dissociate one mole of a given substance into atoms.

heat of combustion The energy liberated when one mole of a given substance is completely oxidized.

heat of formation The energy liberated or absorbed when one mole of a compound is formed from its constituent elements.

heat of neutralization The energy liberated in neutralizing one mole of an acid or base.

heat of reaction The energy liberated or absorbed as a result of the complete chemical reaction of molar amounts of the reactants.

heat of solution The energy liberated or absorbed when one mole of a given substance is completely dissolved in a large volume of solvent (strictly, to infinite dilution).

heavy hydrogen *See* deuterium.

heavy spar A mineral form of *barium sulphate, $BaSO_4$.

heavy water (deuterium oxide) Water in which hydrogen atoms, ^1H, are replaced by the heavier isotope deuterium, ^2H. It is a colourless liquid, which forms hexagonal crystals on freezing. Its physical properties differ from those of 'normal' water; r.d. 1.105; m.p. 3.8°C; b.p. 101.4°C. Deuterium oxide occurs to a small extent (about 0.003% by weight) in natural water, from which it can be separated by fractional distillation or by electrolysis. It is particularly useful in the nuclear industry because of its ability to reduce the energies of fast neutrons to thermal energies and because its quenching cross-section is lower than that of hydrogen and consequently it does not appreciably reduce the neutron flux. In the laboratory it is used for *labelling other molecules for studies of reaction mechanisms.

hecto- Symbol h. A prefix used in the metric system to denote 100 times. For example, 100 coulombs = 1 hectocoulomb (hC).

Heisenberg uncertainty principle *See* uncertainty principle.

helium Symbol He. A colourless odourless gaseous nonmetallic element belonging to group 0 of the periodic table (*see* noble gases); a.n. 2; r.a.m. 4.0026; d. 0.178 g dm^{-3}; m.p. −272.2°C (at 20 atm.); b.p. −268.93°C. The element has the lowest boiling point of all substances and can be solidified only under pressure. Natural helium is mostly helium−4, with a small amount of helium−3. There are also two short-lived radioactive isotopes: helium−5 and −6. It occurs in ores of uranium and thorium and in some natural-gas deposits. It has a variety of uses, including the provision of inert atmospheres for welding and semiconductor manufacture, as a refrigerant for superconductors, and as a diluent in breathing apparatus. It is also used in filling balloons. Chemically it is totally inert and has no known compounds. It was discovered in the solar spectrum in 1868 by Lockyer.

Helmholtz free energy *See* free energy.

hemiacetals *See* acetals.

hemihydrate A crystalline hydrate containing two molecules of compound per molecule of water (e.g. $2CaSO_4.H_2O$).

hemiketals *See* ketals.

henry Symbol H. The *SI unit of inductance equal to the inductance of a closed circuit in which an e.m.f. of one volt is produced when the electric current in the circuit varies uniformly at a rate of one ampere per second. It is named after Joseph Henry (1797−1878), a US physicist.

heptahydrate A crystalline hydrate that has seven molecules of water per molecule of compound.

heptane A liquid straight-chain alkane obtained from petroleum, C_7H_{16}; r.d. 0.684; m.p. 182°C; b.p. 371°C. In standardizing *octane numbers, heptane is given a value zero.

heptaoxodiphosphoric(V) acid *See* phosphoric(V) acid.

heptavalent (septivalent) Having a valency of seven.

hertz Symbol Hz. The *SI unit of frequency equal to one cycle per second. It is named after Heinrich Hertz (1857−94), a German physicist.

Hess's law If reactants can be converted into products by a series of reactions, the sum of the heats of these reactions (with due regard to their sign) is equal to the heat of reaction for direct conversion from reactants to products. More generally, the overall energy change in going from reactants to products does not depend on the route taken. The law can be used to obtain thermodynamic data that cannot be measured directly. For example, the heat of formation of ethane can be found by considering the reactions:

$$2C(s) + 3H_2(g) + 3\tfrac{1}{2}O_2(g) \rightarrow 2CO_2(g) + 3H_2O(l)$$

The heat of this reaction is $2\Delta H_C + 3\Delta H_H$, where ΔH_C and ΔH_H are the heats of combustion of carbon and hydrogen respectively, which can be measured. By Hess' law, this is equal to the sum of the energies for two stages:

$$2C(s) + 3H_2(g) \rightarrow C_2H_6(g)$$

(the heat of formation of ethane, ΔH_f) and

$$C_2H_6(g) + 3\tfrac{1}{2}O_2 \rightarrow 2CO_2(g) + 3H_2O(l)$$

(the heat of combustion of ethane, ΔH_E). As ΔH_E can be measured and as

$$\Delta H_f + \Delta H_E = 2\Delta H_c + 3\Delta H_H$$

ΔH_f can be found. Another example is the use of the *Born–Haber cycle to obtain lattice energies. The law was first put forward in 1840 by the Russian chemist Germain Henri Hess (1802–50). It is sometimes called the *law of constant heat summation* and is a consequence of the law of conservation of energy.

hetero atom An odd atom in the ring of a heterocyclic compound. For instance, nitrogen is the hetero atom in pyridine.

heterocyclic *See* cyclic.

heterogeneous catalysis *See* catalysis.

heterolytic fission The breaking of a bond in a compound in which the two fragments are oppositely charged ions. For example, $HCl \rightarrow H^+ + Cl^-$. *Compare* homolytic fission.

heteropolar bond *See* chemical bond.

heteropolymer *See* polymer.

Heusler alloys Ferromagnetic alloys containing no ferromagnetic elements. The original alloys contained copper, manganese, and tin and were first made by Conrad Heusler (19th-century mining engineer).

hexadecane (cetane) A colourless liquid straight-chain alkane hydrocarbon, $C_{16}H_{34}$, used in standardizing *cetane ratings of Diesel fuel.

hexadecanoate *See* palmitate.

hexadecanoic acid *See* palmitic acid.

hexagonal close packing *See* close packing.

hexagonal crystal *See* crystal system.

hexanedioate (adipate) A salt or ester of hexanedioic acid.

hexanedioic acid (adipic acid) A carboxylic acid, $(CH_2)_4(COOH)_2$; r.d. 1.366; m.p. 149°C; b.p. 265°C (100 mmHg). It is used in the manufacture of *nylon 6,6. *See also* polymerization.

hexanoate (caproate) A salt or ester of hexanoic acid.

hexanoic acid (caproic acid) A liquid fatty acid, $CH_3(CH_2)_4COOH$; r.d. 0.93; m.p. −3.4°C; b.p. 205°C. Glycerides of the acid occur naturally in cow and goat milk and in some vegetable oils.

hexose A *monosaccharide that has six carbon atoms in its molecules.

high-speed steel A steel that will remain hard at dull red heat and can therefore be used in cutting tools for high-speed lathes. It usually contains 12–22% tungsten, up to 5% chromium, and 0.4–0.7% carbon. It may also contain small amounts of vanadium, molybdenum, and other metals.

histidine *See* amino acid.

histochemistry The study of the distribution of the chemical constituents of tissues by means of their chemical reactions. It utilizes such techniques as staining, light and electron microscopy, autoradiography, and *chromatography.

holmium Symbol Ho. A soft silvery metallic element belonging to the *lanthanoids; a.n. 67; r.a.m. 164.93; r.d. 8.795 (20°C); m.p. 1472°C; b.p. 2700°C. It occurs in apatite, xenotime, and some other rare-earth minerals. There is one natural isotope, holmium-165; eighteen artificial isotopes have been produced. There are no uses for the element, which was discovered by P. T. Cleve and J. L. Soret in 1879.

homocyclic *See* cyclic.

homologous series A series of related chemical compounds that have the same functional group(s) but differ in formula by a fixed group of atoms. For instance, the simple carboxylic acids: methanoic (HCOOH), ethanoic (CH_3COOH), propanoic (C_2H_5COOH), etc., form a homologous series in which each member differs from the next by CH_2. Successive members of such a series are called *homologues*.

homolytic fission The breaking of a bond in a compound in which the fragments are uncharged free radicals. For example, $Cl_2 \rightarrow Cl\cdot + Cl\cdot$. *Compare* heterolytic fission.

homopolar bond *See* chemical bond.

homopolymer *See* polymer.

hormone A substance that is manufactured and secreted in very small quantities into the bloodstream by an endocrine gland or a specialized nerve cell and regulates the growth or functioning of a specific tissue or organ in a distant part of the body. For example, the hormone insulin controls the rate and manner in which glucose is used by the body.

hornblende Any of a group of common rock-forming minerals of the amphibole group with the generalized formula:

$$(Ca,Na)_2(Mg,Fe,Al)_5(Al,Si)_8O_{22}(OH,F)_2$$

Hornblendes consist mainly of calcium, iron, and magnesium silicate.

hybrid orbital *See* orbital.

hydracid *See* binary acid.

hydrate A substance formed by combination of a compound with water. *See* water of crystallization.

hydrated alumina *See* aluminium hydroxide.

hydrated aluminium hydroxide. *See* aluminium hydroxide.

hydration *See* solvation.

hydrazine A colourless liquid or white crystalline solid, N_2H_4; r.d. 1.01 (liquid); m.p. 1.4°C; b.p. 113.5°C. It is very soluble in water and soluble in ethanol. Hydrazine is prepared by the *Raschig synthesis* in which ammonia reacts with sodium(I) chlorate (sodium hypochlorite) to give NH_2Cl, which then undergoes further reaction with ammonia to give N_2H_4. Industrial production must be carefully controlled to avoid a side reaction leading to NH_4Cl. The compound is a weak base giving rise to two series of salts, those based on $N_2H_5^+$, which are stable in water (sometimes written in the form $N_2H_4.HCl$ rather than $N_2H_5^+Cl^-$), and a less stable and extensively hydrolysed series based on $N_2H_6^{2+}$. Hydrazine is a powerful reducing agent and reacts violently with many oxidizing agents, hence its use as a rocket propellant.

hydrazoic acid *See* hydrogen azide.

hydrazones Organic compounds containing the group $=C:NNH_2$, formed by condensation of substituted hydrazines with with aldehydes and ketones. *Phenylhydrazones* contain the group $=C:NNHC_6H_5$.

hydrobromic acid *See* hydrogen bromide.

hydrocarbons Chemical compounds that contain only carbon and hydrogen. A vast number of different hydrocarbon compounds exist, the main types being the *alkanes, *alkenes, *alkynes, and *arenes.

hydrochloric acid *See* hydrogen chloride.

hydrochloride *See* amine salts.

$$R-C(=O)-R' + H_2N-NH-R'' \xrightarrow{-H_2O} R-C(=N-NH-R'')-R'$$

ketone hyrdazine hydrazone

Formation of a hydrazone from a ketone. The same reaction occurs with an aldehyde (R′=H). If R″=C_6H_5, the product is phenylhydrazone

hydrocyanic acid *See* hydrogen cyanide.

hydrofluoric acid *See* hydrogen fluoride.

hydrogen Symbol H. A colourless odourless gaseous chemical element; a.n. 1; r.a.m. 1.008; d. 0.0899 g dm^{-3}; m.p. −259.14°C; b.p. −252.87°C. It is the lightest element and the most abundant in the universe. It is present in water and in all organic compounds. There are three isotopes: naturally occurring hydrogen consists of the two stable isotopes hydrogen-1 (99.985%) and *deuterium. The radioactive *tritium is made artificially. The gas is diatomic and has two forms: *orthohydrogen*, in which the nuclear spins are parallel, and *parahydrogen*, in which they are antiparallel. At normal temperatures the gas is 25% parahydrogen. In the liquid it is 99.8% parahydrogen. The main source of hydrogen is steam *reforming of natural gas. It can also be made by the Bosch process (*see* Haber process) and by electrolysis of water. The main use is in the Haber process for making ammonia. Hydrogen is also used in various other industrial processes, such as the reduction of oxide ores, the refining of petroleum, the production of hydrocarbons from coal, and the hydrogenation of vegetable oils. Considerable interest has also been shown in its potential use in a 'hydrogen fuel economy' in which primary energy sources not based on fossil fuels (e.g. nuclear, solar, or geothermal energy) are used to produce electricity, which is employed in electrolysing water. The hydrogen formed is stored as liquid hydrogen or as metal *hydrides. Chemically, hydrogen reacts with most elements. It was discovered by Henry Cavendish in 1776.

hydrogenation 1. A chemical reaction with hydrogen; in particular, an addition reaction in which hydrogen adds to an unsaturated compound. Nickel is a good catalyst for such reactions. **2.** The process of converting coal to oil by making the carbon in the coal combine with hydrogen to form hydrocarbons. *See* Fischer–Tropsch process; Bergius process.

hydrogen azide (hydrazoic acid; azoimide) A colourless liquid, HN_3; r.d. 1.09; m.p. −80°C; b.p. 37°C. It is highly toxic and a powerful reductant, which explodes in the presence of oxygen and other oxidizing agents. It may be prepared by the reaction of sodium amide and sodium nitrate at 175°C followed by distillation of a mixture of the resulting sodium azide and a dilute acid. *See also* azides.

hydrogen bond A type of electrostatic interaction between molecules occurring in molecules that have hydrogen atoms bound to electronegative atoms (F, N, O). It is a strong dipole–dipole attraction caused by the electron-withdrawing properties of the electronegative atom. Thus, in the water molecule the oxygen atom attracts the electrons in the O–H bonds. The hydrogen atom has no inner shells of electrons to shield the nucleus, and there is an electrostatic interaction between the hydrogen proton and a lone pair of electrons on an oxygen atom in a neighbouring molecule. Each oxygen atom has two lone pairs and can make hydrogen bonds to two different hydrogen atoms. The strengths of hydrogen bonds are about one tenth of the strengths of normal covalent bonds. Hydrogen bonding does, however, have significant effects on physical properties. Thus it accounts for the unusual properties of *water and for the relatively high boiling points of H_2O, HF, and NH_3 (compared with H_2S, HCl, and PH_3). It is also of great importance in living organisms. Hydrogen bonding occurs between bases in the chains of DNA. It also occurs between the C=O and N–H groups in proteins, and is responsible for maintaining the secondary structure.

hydrogen bromide A colourless gas, HBr; m.p. −86°C; b.p. −66.4°C. It can be made by direct combination of the elements using a platinum catalyst. It is a strong acid dissociating extensively in solution (*hydrobromic acid*).

hydrogencarbonate (bicarbonate) A salt of *carbonic acid in which one hydrogen atom has been replaced; it thus contains the hydrogencarbonate ion HCO_3^-.

hydrogen chloride A colourless fuming gas, HCl; m.p. −114°C; b.p. −85°C. It can be prepared in the laboratory by heating sodium chloride with concentrated sulphuric acid (hence the former name *spirits of salt*). Industrially it is made directly from the elements at high temperature and used in the manufacture of PVC and other chloro compounds. It is a strong acid and dissociates fully in solution (*hydrochloric acid*).

hydrogen cyanide (hydrocyanic acid; prussic acid) A colourless liquid or gas, HCN, with a characteristic odour of almonds; r.d. 0.699 (liquid at 22°C); m.p. −14°C; b.p. 26°C. It is an extremely poisonous substance formed by the action of acids on metal cyanides. Industrially, it is made by catalytic oxidation of ammonia and methane with air and is used in producing acrylate plastics. Hydrogen cyanide is a weak acid ($K_a = 2.1 \times 10^{-9}$ mol dm^{-3}). With organic carbonyl compounds it forms *cyanohydrins.

hydrogen electrode *See* hydrogen half cell.

hydrogen fluoride A colourless liquid, HF; r.d. 0.99; m.p. −83°C; b.p. 19.5°C. It can be made by the action of sulphuric acid on calcium fluoride. The compound is an extremely corrosive fluorinating agent, which attacks glass. It is unlike the other hydrogen halides in being a liquid (a result of *hydrogen-bond formation). It is also a weaker acid than the others because the small size of the fluorine atom means that the H–F bond is shorter and stronger. Solutions of hydrogen fluoride in water are known as *hydrofluoric acid.*

hydrogen half cell (hydrogen electrode) A type of *half cell in which a metal foil is immersed in a solution of hydrogen ions and hydrogen gas is bubbled over the foil. The standard hydrogen electrode, used in measuring standard *electrode potentials, uses a platinum foil with a 1.0 M solution of hydrogen ions, the gas at 1 atmosphere pressure, and a temperature of 25°C. It is written Pt(s)|H_2(g), H^+(aq), the effective reaction being $H_2 \rightarrow 2H^+ + 2e$.

hydrogen iodide A colourless gas, HI; m.p. −51°C; b.p. −36°C. It can be made by direct combination of the elements using a platinum catalyst. It is a strong acid dissociating extensively in solution (*hydroiodic acid*). It is also a reducing agent.

hydrogen ion *See* acids; pH.

hydrogen peroxide A colourless or pale blue viscous unstable liquid, H_2O_2; r.d. 1.44; m.p. −0.89°C; b.p. 151.4°C. As with water, there is considerable hydrogen bonding in the liquid, which has a high dielectric constant. It can be made in the laboratory by adding dilute acid to barium peroxide at 0°C. Large quantities are made commercially by electrolysis of $KHSO_4.H_2SO_4$ solutions. Another industrial process involves catalytic oxidation (using nickel, palladium, or platinum with an anthraquinone) of hydrogen and water in the presence of oxygen. Hydrogen peroxide readily decomposes in light or in the presence of metal ions to give water and oxygen. It is usually supplied in solutions designated by volume strength. For example, 20-volume hydrogen peroxide would yield 20 volumes of oxygen per volume of solution. Although the *peroxides are formally salts of H_2O_2, the compound is essentially neutral. Thus, the acidity constant of the ionization

$$H_2O_2 + H_2O \leftrightharpoons H_3O^+ + HO_2^-$$

is 1.5×10^{-12} mol dm^{-3}. It is a strong oxidizing agent, hence its use as a mild antiseptic and as a bleaching agent for cloth, hair, etc. It has also been used as an oxidant in rocket fuels.

hydrogen spectrum The atomic spectrum of hydrogen is characterized by lines corresponding to radiation quanta of sharply defined energy. A graph of the frequencies at which these lines occur against the ordinal number that characterizes their position in the series of lines, produces a smooth curve indicating that they obey a formal law. In 1885 J. J. Balmer (1825–98) discovered the law having the form:

$$1/\lambda = R(1/n_1{}^2 + 1/n_2{}^2)$$

This law gives the so-called *Balmer series* of lines in the visible spectrum in which n_1 = 2 and n_2 = 3,4,5..., λ is the wavelength associated with the lines, and R is the *Rydberg constant.

In the *Lyman series*, discovered by Theodore Lyman (1874–1954), n_1 = 1 and the lines fall in the ultraviolet. The Lyman series is the strongest feature of the solar spectrum as observed by rockets and satellites above the earth's atmosphere. In the *Paschen series*, discovered by F. Paschen (1865–1947), n_1 = 3 and the lines occur in the far infrared. The *Brackett series* also occurs in the far infrared, with n_1 = 4.

hydrogensulphate (bisulphate) A salt containing the ion HSO_4^- or an ester of the type $RHSO_4$, where R is an organic group. It was formerly called *hydrosulphate*.

hydrogen sulphide (sulphuretted hydrogen) A gas, H_2S, with an odour of rotten eggs; r.d. 1.54 (liquid); m.p. −85.5°C; b.p. −60.7°C. It is soluble in water and ethanol and may be prepared by the action of mineral acids on metal sulphides, typically hydrochloric acid on iron(II) sulphide (*see* Kipp's apparatus). Solutions in water (known as *hydrosulphuric acid*) contain the anions HS^- and minute traces of S^{2-} and are weakly acidic. Acid salts (those containing the HS^- ion) are known as *hydrogensulphides* (formerly *hydrosulphides*). In acid solution hydrogen sulphide is a mild reducing agent. Hydrogen sulphide has an important role in traditional qualitative chemical analysis, where it precipitates metals with insoluble sulphides (in acid solution: Cu, Pb, Hg, Cd, Bi, As, Sb, Sn). The formation of a black precipitate with alkaline solutions of lead salts may be used as a test for hydrogen sulphide but the characteristic smell is usually sufficient. Hydrogen sulphide is exceedingly poisonous (more toxic than hydrogen cyanide).

The compound burns in air with a blue flame to form sulphur(IV) oxide (SO_2); solutions of hydrogen sulphide exposed to the air undergo oxidation but in this case only to elemental sulphur. North Sea gas contains some hydrogen sulphide (from S-proteins in plants) as do volcanic emissions.

hydrogensulphite (bisulphite) A salt containing the ion $^-HSO_3$ or an ester of the type $RHSO_3$, where R is an organic group.

hydroiodic acid *See* hydrogen iodide.

hydrolysis A chemical reaction of a compound with water. For instance, salts of weak acids or bases hydrolyse in aqueous solution, as in

$$Na^{+}\,{}^{-}OOCCH_3 + H_2O \leftrightharpoons Na^+ + OH^- + CH_3COOH$$

The reverse reaction of *esterification is another example. *See also* solvolysis.

hydromagnesite A mineral form of basic *magnesium carbonate, $3MgCO_3.Mg(OH)_2.3H_2O$.

hydrophilic Having an affinity for water. *See* lyophilic.

hydrophobic Lacking affinity for water. *See* lyophobic.

hydroquinone *See* benzene-1,4-diol.

hydrosol A sol in which the continuous phase is water. *See* colloid.

hydrosulphate *See* hydrogensulphate.

hydrosulphide *See* hydrogen sulphide.

hydrosulphuric acid *See* hydrogen sulphide.

hydroxide A metallic compound containing the ion OH^- (*hydroxide ion*) or containing the group –OH (hydroxyl group) bound to a metal atom. Hydroxides of typical metals are basic; those of *metalloids are amphoteric.

hydroxycerussite *See* lead(II) carbonate hydroxide.

hydroxyl group The group –OH in a chemical compound.

2-hydroxypropanoic acid *See* lactic acid.

hygroscopic Describing a substance that can take up water from the atmosphere. *See also* deliquescence.

hyperfine structure *See* fine structure.

hypertonic solution A solution that has a higher osmotic pressure than some other solution. *Compare* hypotonic solution.

hypochlorite *See* chlorates.

hypochlorous acid *See* chloric(I) acid.

hypophosphorus acid *See* phosphinic acid.

hyposulphite *See* sulphinate.

hyposulphurous acid *See* sulphinic acid.

hypotonic solution A solution that has a lower osmotic pressure than some other solution. *Compare* hypertonic solution.

I

ice *See* water.

ice point The temperature at which there is equilibrium between ice and water at standard atmospheric pressure (i.e. the freezing or melting point under standard conditions). It was used as a fixed point (0°) on the Celsius scale, but the kelvin and the International Practical Temperature Scale are based on the *triple point of water.

ideal crystal A single crystal with a perfectly regular lattice that contains no impurities, imperfections, or other defects.

ideal gas (perfect gas) A hypothetical gas that obeys the *gas laws exactly. An ideal gas would consist of molecules that occupy negligible space and have negligible forces between them. All collisions made between molecules and the walls of the container or between molecules and other molecules would be perfectly elastic, because the molecules would have no means of storing energy except as translational kinetic energy.

ideal solution *See* Raoult's law.

ignition temperature The temperature to which a substance must be heated before it will burn in air.

imides Organic compounds containing the group –CO.NH.CO.– (the *imido group*).

imido group *See* imides.

imines Compounds containing the group –NH– (the *imino group*) joined to two other groups; i.e. secondary *amines.

imino group *See* imines.

Imperial units The British system of units based on the pound and the yard. The former f.p.s. system was used in engineering and was loosely based on Imperial units; for all scientific purposes *SI units are now used. Imperial units are also being replaced for general purposes by metric units.

implosion An inward collapse of a vessel, especially as a result of evacuation.

incandescence The emission of light by a substance as a result of raising it to a high temperature.

indene A colourless flammable hydrocarbon, C_9H_8; r.d. 1.01; m.p. –35°C; b.p. 182°C. Indene is an aromatic hydrocarbon with a five-membered ring fused to a benzene ring. It is present in coal tar and is used as a solvent and raw material for making other organic compounds.

indeterminacy *See* uncertainty principle.

indicator A substance used to show the presence of a chemical substance or ion by its colour. *Acid–base indicators* are compounds, such as phenolphthalein and methyl orange, that change colour on going from acidic to basic solutions. They are usually weak acids in which the un-ionized form HA has a different colour from the negative ion A^-. In solution the indicator dissociates slightly

$$HA \leftrightharpoons H^+ + A^-$$

In acid solution the concentration of H^+ is high, and the indicator is largely undissociated HA; in alkaline solutions the equilibrium is displaced to the right and A^- is formed. Useful acid–base indicators show a sharp colour change over a range of about 2 pH units. In titration, the point at which the reaction is complete is the *equivalence point* (i.e. the point at which equivalent quantities of acid and base are added). The *end point* is the point at which the indicator just changes colour. For accuracy, the two must be the same. During a titration the pH changes sharply close to the equivalence point, and the indicator used must change colour over the same range.

Other types of indicator can be used for other reactions. Starch, for example, is used in iodine titrations because of the deep blue complex it forms. *Oxidation–reduction indicators* are substances that show a reversible colour change between oxidized and reduced forms. *See also* absorption indicator.

indigo A blue vat dye, $C_{16}H_{10}N_2O_2$. It occurs as the glucoside *indican* in the leaves of plants of the genus *Indigofera*, from which it was formerly extracted. It is now made synthetically.

indium Symbol In. A soft silvery element belonging to group IIIA of the periodic table; a.n. 49; r.a.m. 114.82; r.d. 7.31 (20°C); m.p. 156.6°C; b.p. 2080±2°C. It occurs in zinc blende and some iron ores and is obtained from zinc flue dust in total quantities of about 40 tonnes per annum. Naturally occurring indium consists of 4.23% indium–113 (stable) and 95.77% indium–115 (half-life 6×10^{14} years). There are a further five short-lived radioisotopes. The uses of the metal are small – some special-purpose electroplates and some special fusible alloys. Several semiconductor compounds are used, such as InAs, InP, and InSb. With only three electrons in its valency shell, indium is an electron acceptor; it forms stable indium(I), indium(II), and indium(III) compounds. The element was discovered in 1863 by Reich and Richter.

inductive effect The effect of a group or atom of a compound in pulling electrons towards itself or in pushing them away. Inductive effects can be used to explain some aspects of organic reactions. For instance, electron-withdrawing groups, such as $-NO_2$, $-CN$, $-CHO$, $-COOH$, and the halogens substituted on a benzene ring, reduce the electron density on the ring and decrease its susceptibility to further (electrophilic) substitution. Electron-releasing groups, such as $-OH$, $-NH_2$, $-OCH_3$, and $-CH_3$, have the opposite effect.

inert gases *See* noble gases.

inert-pair effect An effect seen especially in groups III and IV of the periodic table, in which the heavier elements in the group tend to form compounds with a valency two lower than the expected group valency. It is used to account for the existence of thallium(I) compounds in group III and lead(II) in group IV. In forming compounds, elements in these groups promote an electron from a filled *s*-level state to an empty *p*-level. The energy required for this is more than compensated for by the extra energy gain in forming two more bonds. For the heavier elements, the bond strengths or lattice energies in the compounds are lower than those of the lighter elements. Consequently the energy compensation is less important and the lower valence states become favoured.

inhibition A reduction in the rate of a catalysed reaction by substances called *inhibitors*. In biochemical reactions, in which the catalysts are *enzymes, if the inhibitor molecules resemble the substrate molecules they may bind to the active site of the enzyme, so preventing normal enzymatic activity. Alternatively they may form a complex with the substrate–enzyme intermediate or irreversibly destroy the enzyme configuration and active-site properties. The toxic effects of many substances are produced in this way. Inhibition by reaction products (*feedback inhibition*) is important in the control of enzyme activity.

inner Describing a chemical compound formed by reaction of one part of a molecule with another part of the same molecule. Thus, a lactam is an inner amide; a lactone is an inner ester.

inner transition series *See* transition elements.

inorganic chemistry The branch of chemistry concerned with compounds of elements other than carbon. Certain simple carbon compounds, such as CO, CO_2, CS_2, and carbonates and cyanides, are usually treated in inorganic chemistry.

insulin A hormone, secreted by the islets of Langerhans in the pancreas, that promotes the uptake of glucose by body cells and thereby controls its concentration in the blood. Underproduction of insulin results in the accumulation of large amounts of glucose in the blood and its subsequent excretion in the urine. This condition, known as *diabetes mellitus*, can be treated successfully by insulin injections. Insulin is a protein, the whole structure of which is now known.

intermediate bond *See* chemical bond.

intermetallic compound A compound consisting of two or more metallic elements present in definite proportions in an alloy.

intermolecular forces Weak forces occurring between molecules. *See* van der Waals' forces; hydrogen bond.

internal conversion A process in which an excited atomic nucleus decays to the *ground state and the energy released is transferred by electromagnetic coupling to one of the bound electrons of that atom rather than being released as a photon. The coupling is usually with an electron in the K-, L-, or M-shell of the atom, and this *conversion electron* is ejected from the atom with a kinetic energy equal to the difference between the nuclear transition energy and the binding energy of the electron. The resulting ion is itself in an excited state and usually subsequently emits an Auger electron or an X-ray photon.

internal energy Symbol *U*. The total of the kinetic energies of the atoms and molecules of which a system consists and the potential energies associated with their mutual interactions. It does not include the kinetic and potential energies of the system as a whole nor their nuclear energies or other intra-atomic energies. The value of the absolute internal energy of a system in any particular state cannot be measured; the significant quantity is the change in internal energy, ΔU. For a closed system (i.e. one that is not being replenished from outside its boundaries) the change in internal energy is equal to the heat absorbed by the system (Q) from its surroundings, less the work done (W) by the system on its surroundings, i.e. $\Delta U = Q - W$. *See also* energy; heat; thermodynamics.

interstitial *See* defect.

interstitial compound A compound in which ions or atoms of a nonmetal occupy interstitial positions in a metal lattice. Such compounds often have metallic properties. Examples are found in the *carbides, *borides, and *silicides.

Invar A tradename for an alloy of iron (63.8%), nickel (36%), and carbon (0.2%) that has a very low expansivity over a a restricted temperature range. It is used in watches and other instruments to reduce their sensitivity to changes in temperature.

inversion A chemical reaction involving a change from one optically active configuration to the opposite configuration. The Walden inversion is an example. *See* nucleophilic substitution.

iodic acid Any of various oxoacids of iodine, such as iodic(V) acid and iodic(VII) acid. When used without an oxidation state specified, the term usually refers to iodic(V) acid (HIO_3).

iodic(V) acid A colourless or very pale yellow solid, HIO_3; r.d. 4.63; decomposes at 110°C. It is soluble in water but insoluble in pure ethanol and other organic solvents. The compound is obtained by oxidizing iodine with concentrated nitric acid, hydrogen peroxide, or ozone. It is a strong acid and a powerful oxidizing agent.

iodic(VII) acid (periodic acid) A hygroscopic white solid, H_5IO_6, which decomposes at 138°C and is very soluble in water, ethanol, and ethoxyethane. Iodic(VII) acid may be prepared by electrolytic oxidation of concentrated solutions of iodic(V) acid at low temperatures. It is a weak acid but a strong oxidizing agent.

iodide *See* halide.

iodine Symbol I. A dark violet nonmetallic element belonging to group VII of the periodic table (*see* halogens); a.n. 53; r.a.m. 126.9045; r.d. 4.94; m.p. 113.5°C; b.p. 183.45°C. The element is insoluble in water but soluble in ethanol and other organic solvents. When heated it gives a violet vapour that sublimes. Iodine is required as a trace element (*see* essential element) by living organisms; in animals it is concentrated in the thyroid gland as a constituent of thyroid hormones. The element is present in sea water and was formerly extracted from seaweed. It is now obtained from oil-well brines (displacement by chlorine). There is one stable isotope, iodine–127, and fourteen radioactive isotopes. It is used in medicine as a mild antiseptic (dissolved in ethanol as *tincture of iodine*), and in the manufacture of iodine compounds. Chemically, it is less reactive than the other halogens and the most electropositive (metallic) halogen. It was discovered in 1812 by Courtois.

iodine(V) oxide (iodine pentoxide) A white solid, I_2O_5; r.d. 4.79; decomposes at 310°C. It dissolves in water to give iodic(V) acid and also acts as an oxidizing agent.

iodine value A measure of the amount of unsaturation in a fat or vegetable oil (i.e. the number of double bonds). It is obtained by finding the percentage of iodine by weight absorbed by the sample in a given time under standard conditions.

iodoethane (ethyl iodide) A colourless liquid *haloalkane, C_2H_5I; r.d. 1.9; m.p. −108°C; b.p. 72°C. It is made by reacting ethanol with a mixture of iodine and red phosphorus.

iodoform *See* tri-iodomethane.

Iodoform test *See* haloform reaction.

iodomethane (methyl iodide) A colourless liquid haloalkane, CH_3I; r.d. 2.28; m.p. −66.45°C; b.p. 42.4°C. It can be made by reacting methanol with a mixture of iodine and red phosphorus.

ion An atom or group of atoms that has either lost one or more electrons, making it positively charged (an anion), or gained one or more electrons, making it negatively charged (a cation). *See also* ionization.

ion exchange The exchange of ions of the same charge between a solution (usually aqueous) and a solid in contact with it. The process occurs widely in nature, especially in the absorption and retention of water-soluble fertilizers by soils. For example, if a potassium salt is dissolved in water and applied to soil, potassium ions are absorbed by the soil and sodium and calcium ions are released from it.

The soil, in this case, is acting as an ion exchanger. Synthetic *ion-exchange resins* consist of various copolymers having a cross-linked three-dimensional structure to which ionic groups have been attached. An *anionic resin* has negative ions built into its structure and therefore exchanges positive ions. A *cationic resin* has positive ions built in and exchanges negative ions. Ion-exchange resins, which are used in sugar refining to remove salts, are synthetic organic polymers containing side groups that can be ionized. In anion exchange, the side groups are ionized basic groups, such as $-NH_3^+$ to which anions X^- are attached. The exchange reaction is one in which different anions in the solution displace the X^- from the solid. Similarly, cation exchange occurs with resins that have ionized acidic side groups such as $-COO^-$ or $-SO_2O^-$, with positive ions M^+ attached.

Ion exchange also occurs with inorganic polymers such as *zeolites, in which positive ions are held at sites in the silicate lattice. These are used for water-softening, in which Ca^{2+} ions in solution displace Na^+ions in the zeolite. The zeolite can be regenerated with sodium chloride solution. *Ion-exchange membranes* are used as separators in electrolytic cells to remove salts from sea water (*see also* desalination) and in producing deionized water.

ionic bond *See* chemical bond.

ionic crystal *See* crystal.

ionic product The product of the concentrations of ions present in a given solution taking the stoichiometry into account. For a sodium chloride solution the ionic product is $[Na^+][Cl^-]$; for a calcium chloride solution it is $[Ca^{2+}][Cl^-]^2$. In pure water, there is an equilibrium with a small amount of self-ionization:

$$H_2O \rightleftharpoons H^+ + OH^-$$

The equilibrium constant of this dissociation is given by

$$K_W = [H^+][OH^-]$$

since the concentration $[H_2O]$ can be taken as constant. K_W is referred to as the ionic product of water. It has the value 10^{-14} $mol^2\,dm^{-6}$ at 25°C. In pure water (i.e. no added acid or added alkali) $[H^+] = [OH^-] = 10^{-7}\ mol\,dm^{-3}$. *See also* solubility product; pH scale.

ionic radius A value assigned to the radius of an ion in a crystalline solid, based on the assumption that the ions are spherical with a definite size. X-ray diffraction can be used to measure the internuclear distance in crystalline solids. For example, in NaF the Na – F distance is 0.231 nm, and this is assumed to be the sum of the Na^+ and F^- radii. By making certain assumptions about the shielding effect that the inner electrons have on the outer electrons, it is possible to assign individual values to the ionic radii – Na^+ 0.096 nm; F^- 0.135 nm. In general, negative ions have larger ionic radii than positive ions. The larger the negative charge, the larger the ion; the larger the positive charge, the smaller the ion.

ionic strength Symbol *I*. A function expressing the effect of the charge of the ions in a solution, equal to the sum of the molality of each type of ion present multiplied by the square of its charge. $I = \Sigma m_i z_i^2$.

ionization The process of producing *ions. Certain molecules (*see* electrolytes) ionize in solution; for example, *acids ionize when dissolved in water (*see also* solvation):

$$HCl \rightarrow H^+ + Cl^-$$

Electron transfer also causes ionization in certain reactions; for example, sodium and chlorine react by the transfer of a valence electron from the sodium atom to the chlorine atom to form the ions that constitute a sodium chloride crystal:

$$Na + Cl \rightarrow Na^+Cl^-$$

Ions may also be formed when an atom or molecule loses one or more electrons as a result of energy gained in a collision with another particle or a quantum of radiation (*see* photoionization). This may occur as a result of the impact of *ionizing radiation or of *thermal ionization and the reaction takes the form

$$A \rightarrow A^+ + e$$

Alternatively, ions can be formed by electron capture, i.e.

$$A + e \rightarrow A^-$$

ionization gauge A vacuum gauge consisting of a three-electrode system inserted into the container in which the pressure is to be measured. Electrons from the cathode are attracted to the grid, which is positively biased. Some pass through the grid but do not reach the anode, as it is maintained at a negative potential. Some of these electrons do, however, collide with gas molecules, ionizing them and converting them to positive ions. These ions are attracted to the anode; the resulting anode current can be used as a measure of the number of gas molecules present. Pressure as low as 10^{-6} pascal can be measured in this way.

ionization potential (IP) Symbol *I*. The minimum energy required to remove an electron from a specified atom or molecule to such a distance that there is no electrostatic interaction between ion and electron. Originally defined as the minimum potential through which an electron would have to fall to ionize an atom, the ionization potential was measured in volts. It is now, however, defined as the energy to effect an ionization and is conveniently measured in electronvolts (although this is not an SI unit).

The energy to remove the least strongly bound electron is the *first ionization potential*. Second, third, and higher ionization potentials can also be measured, although there is some ambiguity in terminology. Thus, in chemistry the second ionization potential is often taken to be the minimum energy required to remove an electron from the singly charged ion; the second IP of lithium would be the energy for the process

$$Li^+ \rightarrow Li^{2+} + e$$

In physics, the second ionization potential is the energy required to remove an electron from the next to highest energy level in the neutral atom or molecule; e.g.

$$Li \rightarrow Li^{*+} + e,$$

where Li^{*+} is an excited singly charged ion produced by removing an electron from the K-shell.

ionizing radiation Radiation of sufficiently high energy to cause *ionization in the medium through which it passes. It may consist of a stream of high-energy particles (e.g. electrons, protons, alpha-particles) or short-wavelength electromagnetic radiation (ultraviolet, X-rays, gamma-rays). This type of radiation can cause extensive damage to the molecular structure of a substance either as a result of the direct transfer of energy to its atoms or molecules or as a result of the secondary electrons released by ionization. In biological tissue the effect of ionizing radiation can be very serious, usually as a consequence of the ejection of an electron from a water molecule and the oxidizing or reducing effects of the resulting highly reactive species:

$$H_2O \rightarrow e^- + H_2O^* + H_2O^+ \rightarrow$$
$$\cdot OH + H_3O^+ + \cdot H,$$

where the dot before a radical indicates an unpaired electron and an * denotes an excited species.

ion-microprobe analysis A technique for analysing the surface composition of solids. The sample is bombarded with a narrow beam (as small as 2 μm diameter) of high-energy ions. Ions ejected from the surface by sputtering are detected by mass spectrometry. The technique allows quantitative analysis of both chemical and isotopic composition for concentrations as low as a few parts per million.

ion pair A pair of oppositely charged ions produced as a result of a single ionization; e.g.

$$HCl \rightarrow H^+ + Cl^-.$$

Sometimes a positive ion and an electron are referred to as an ion pair, as in

$$A \rightarrow A^+ + e^-.$$

ion pump A type of *vacuum pump that can reduce the pressure in a container to about 1 nanopascal by passing a beam of electrons through the residual gas. The gas is ionized and the positive ions formed are attracted to a cathode within the container where they remain trapped. The pump is only useful at very low pressures, i.e. below about 1 micropascal. The pump has a limited capacity because the absorbed ions eventually saturate the surface of the cathode. A more effective pump can be made by simultaneously producing a film of metal by ion impact (sputtering), so that fresh surface is continuously produced. The device is then known as a *sputter-ion pump*.

IP *See* ionization potential.

iridium Symbol Ir. A silvery metallic *transition element (*see also* platinum metals); a.n. 77; r.a.m. 192.20; r.d. 22.42; m.p. 2410°C; b.p. 4130°C. It occurs with platinum and is mainly used in alloys with platinum and osmium. The element forms a range of iridium(III) and iridium(IV) complexes. It was discovered in 1804 by Tennant.

iron Symbol Fe. A silvery malleable and ductile metallic *transition element; a.n. 26; r.a.m. 55.847; r.d. 7.87; m.p. 1535°C; b.p. 2750°C. The main sources are the ores *haematite (Fe_2O_3), *magnetite (Fe_3O_4), limonite ($FeO(OH)_n H_2O$), ilmenite ($FeTiO_3$), siderite ($FeCO_3$), and pyrite (FeS_2). The metal is smelted in a *blast furnace to give impure *pig iron, which is further processed to give *cast iron, *wrought iron, and various types of *steel. The pure element has three crystal forms: *alpha-iron*, stable below 906°C with a body-centred-cubic structure; *gamma-iron*, stable between 906°C and 1403°C with a nonmagnetic face-centred-cubic structure; and *delta-iron*, which is the body-centred-cubic form above 1403°C. Alpha-iron is ferromagnetic up to its Curie point (768°C). The element has nine isotopes (mass numbers 52–60), and is the fourth most abundant in the earth's crust. It is required as a trace element (*see* essential element) by living organisms. Iron is quite reactive, being oxidized by moist air, displacing hydrogen from dilute acids, and combining with nonmetallic elements. It forms ionic salts and numerous complexes with the metal in the +2 or +3 oxidation states. Iron(VI) also exists in the ferrate ion FeO_4^{2-}, and the element also forms complexes in which its oxidation number is zero (e.g. $Fe(CO)_5$).

iron(II) chloride A green-yellow deliquescent compound, $FeCl_2$; hexagonal; r.d. 3.16; m.p. 670°C. It also exists in hydrated forms: $FeCl_2.2H_2O$ (green monoclinic; r.d. 2.36) and $FeCl_2.4H_2O$ (blue-green monoclinic deliquescent; r.d. 1.93). Anhydrous iron(II) chloride can be made by passing a stream of dry hydrogen chloride over the heated metal; the hydrated forms can be made using dilute hydrochloric acid or by recrystallizing with water. It is converted into iron(III) chloride by the action of chlorine.

iron(III) chloride A black-brown solid, $FeCl_3$; hexagonal; r.d. 2.9; m.p. 306°C; decomposes at 315°C. It also exists as the hexahydrate $FeCl_3.6H_2O$, a brown-yellow deliquescent crystalline substance (m.p. 37°C; b.p. 280–285°C). Iron(III) chloride is prepared by passing dry chlorine over iron wire or steel wool. The reaction proceeds with incandescence when started and iron (III) chloride sublimes as almost black iridescent scales. The compound is rapidly hydrolysed in moist air. In solution it is partly hydrolysed; hydrolysis can be suppressed by the addition of hydrochloric acid. The compound dissolves in many organic solvents, forming solutions of low electrical conductivity: in ethanol, ethoxyethane, and pyridine the molecular weight corresponds to $FeCl_3$ but is higher in other solvents corresponding to Fe_2Cl_6. The vapour is also dimerized. In many ways the compound resembles aluminium chloride, which it may replace in Friedel–Crafts reactions.

iron(II) oxide A black solid, FeO; cubic; r.d. 5.7; m.p. 369°C. It can be obtained by heating iron(II) oxalate; the carbon monoxide formed produces a reducing atmosphere thus preventing oxidation to iron(III) oxide. The compound has the sodium chloride structure, indicating its ionic nature, but the crystal lattice is deficient in iron(II) ions and it is nonstoichiometric. Iron(II) oxide dissolves readily in dilute acids.

iron(III) oxide A red-brown to black insoluble solid, Fe_2O_3; trigonal; r.d. 5.24; m.p. 1565°C. There is also a hydrated form, $Fe_2O_3xH_2O$, which is a red-brown powder; r.d. 2.44–3.60. (*See* rusting.)

Iron(III) oxide occurs naturally as *haematite and can be prepared by heating iron(III) hydroxide or iron(II) sulphate. It is readily reduced on heating with carbon or in a stream of carbon monoxide, hydrogen, or coal gas:

$$Fe_2O_3 + 3C \rightarrow 2Fe + 3CO$$

iron pyrites *See* pyrite.

iron(II) sulphate An off-white solid, $FeSO_4.H_2O$; monoclinic; r.d. 2.970. There is also a heptahydrate, $FeSO_4.7H_2O$; blue-green monoclinic; r.d. 1.898; m.p. 64°C. The heptahydrate is the best known iron(II) salt and is sometimes called *green vitriol* or *copperas*. It is obtained by the action of dilute sulphuric acid on iron in a reducing atmosphere. The anhydrous compound is very hygroscopic. It decomposes at red heat to give iron(III) oxide, sulphur trioxide, and sulphur dioxide. A solution of iron(II) sulphate is gradually oxidized on exposure to air, a basic iron(III) sulphate being deposited.

iron(III) sulphate A yellow hygroscopic compound, $Fe_2(SO_4)_3$; rhombic; r.d. 3.097; decomposes above 480°C. It is obtained by heating an aqueous acidified solution of iron(II) sulphate with hydrogen peroxide:

$$2FeSO_4 + H_2SO_4 + H_2O_2 \rightarrow Fe_2(SO_4)_3 + 2H_2O$$

On crystallizing, the hydrate $Fe_2(SO_4)_3.9H_2O$ is formed. The acid sulphate $Fe_2(SO_4)_3.H_2SO_4.8H_2O$ is deposited from solutions containing a sufficient excess of sulphuric acid.

irreversible process *See* reversible process.

irreversible reaction *See* chemical reaction.

isentropic process Any process that takes place without a change of *entropy. The quantity of heat transferred, δQ, in a reversible process is proportional to the change in entropy, δS, i.e. $\delta Q = T\delta S$, where T is the thermodynamic temperature. Therefore, a reversible *adiabatic process is isentropic, i.e. when $\delta Q = 0$, δS also equals 0.

iso- Prefix denoting that a compound is an *isomer, e.g. isopentane ($CH_3CH(CH_3)C_2H_5$, 2-methylbutane) is an isomer of pentane.

isobar A curve on a graph indicating readings taken at constant pressure.

isocyanate *See* cyanic acid.

isocyanic acid *See* cyanic acid.

isocyanide *See* isonitrile.

isocyanide test A test for primary amines by reaction with an alcoholic solution of potassium hydroxide and trichloromethane.

$$RNH_2 + 3KOH + CHCl_3 \rightarrow RNC + 3KCl + 3H_2O$$

The isocyanide RNC is recognized by its unpleasant smell. This reaction of primary amines is called the *carbylamine reaction*.

isoelectronic Describing compounds that have the same numbers of valence electrons. For example, nitrogen (N_2) and carbon monoxide (CO) are isoelectronic molecules.

isoleucine *See* amino acid.

isomerism The existence of chemical compounds (*isomers*) that have the same molecular formulae but different molecular structures or different arrangements of atoms in space. In *structural isomerism* the molecules have different molecular structures: i.e. they may be different types of compound or they may simply differ in the position of the functional group in the molecule. Structural isomers generally have different physical and chemical properties. In *stereoisomerism*, the isomers have the same formula and functional groups, but differ in the arrangement of groups in space. Optical isomerism is one form of this (*see* optical activity). Another type is *cis–trans isomerism* (formerly *geometrical isomerism*), in which the isomers have different positions of groups with respect to a double bond or central atom (see illustration).

ISOMERISM

1-chloropropane　　　　2-chloropropane

structural isomers in which the functional group has different positions

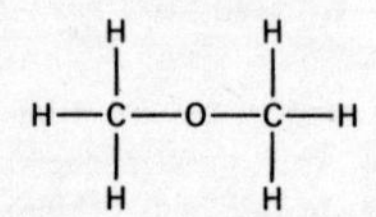

methoxymethane　　　　ethanol

structural isomers in which the functional groups are different

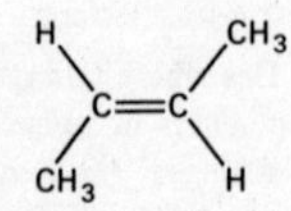

trans-but-2-ene　　　　*cis*-but-2-ene

cis–trans isomers in which the groups are distributed on a double bond

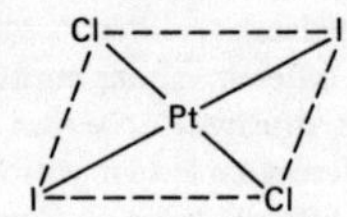

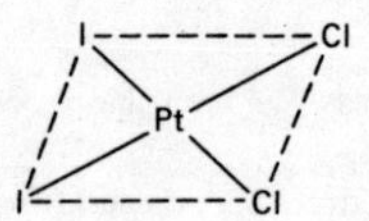

cis–trans isomers in a square-planar complex

keto form　　　　enol form

keto–enol tautomerism

Isomerism

isomers *See* isomerism.

isometric 1. (in crystallography) Denoting a system in which the axes are perpendicular to each other, as in cubic crystals. **2.** Denoting a line on a graph illustrating the way in which temperature and pressure are interrelated at constant volume.

isomorphism The existence of two or more substances (*isomorphs*) that have the same crystal structure, so that they are able to form *solid solutions.

isonitrile (isocyanide; carbylamine) An organic compound containing the group –NC, in which the bonding is to the nitrogen atom.

iso-octane *See* octane; octane number.

isoprene A colourless liquid diene, CH_2:C(CH_3)CH:CH_2. The systematic name is *2-methylbuta-1,3-diene*. It is the structural unit in *terpenes and natural *rubber, and is used in making synthetic rubbers.

isotactic polymer *See* polymer.

isothermal process Any process that takes place at constant temperature. In such a process heat is, if necessary, supplied or removed from the system at just the right rate to maintain constant temperature. *Compare* adiabatic process.

isotonic Describing solutions that have the same osmotic pressure.

isotope One of two or more atoms of the same element that have the same number of protons in their nucleus but different numbers of neutrons. Hydrogen (1 proton, no neutrons), deuterium (1 proton, 1 neutron), and tritium (1 proton, 2 neutrons) are isotopes of hydrogen. Most elements in nature consist of a mixture of isotopes. *See* isotope separation.

isotope separation The separation of the *isotopes of an element from each other on the basis of slight differences in their physical properties. For laboratory quantities the most suitable device is often the mass spectrometer. On a larger scale the methods used include gaseous diffusion (widely used for separating isotopes of uranium in the form of the gas uranium hexafluoride), distillation (formerly used to produce heavy water), electrolysis (requiring cheap electrical power), thermal diffusion (formerly used to separate uranium isotopes, but now considered uneconomic), centrifuging (a method in which there is renewed interest), and laser methods (involving the excitation of one isotope and its subsequent separation by electromagnetic means).

isotopic number (neutron excess) The difference between the number of neutrons in an isotope and the number of protons.

isotropic Denoting a medium whose physical properties are independent of direction. *Compare* anisotropic.

J

jade A hard semiprecious stone consisting either of jadeite or nephrite. *Jadeite*, the most valued of the two, is a sodium aluminium pyroxene, $NaAlSi_2O_6$. It is prized for its intense translucent green colour but white, green and white, brown, and orange varieties also occur. The only important source of jadeite is in the Mogaung region of upper Burma. *Nephrite* is one of the amphibole group of rock-forming minerals. It occurs in a variety of colours, including green, yellow, white, and black. Important sources include the Soviet Union, New Zealand, Alaska, China, and W USA.

jadeite *See* jade.

Jahn–Teller effect If a likely structure of a nonlinear molecule or ion would have degenerate orbitals (i.e. two molecular orbitals with the same energy levels) the actual structure of the molecule or ion is distorted so as to split the energy levels ('raise' the degeneracy). The effect is observed in inorganic complexes. For example, the ion $[Cu(H_2O)_6]^{2+}$ is octahedral and the six ligands might be expected to occupy equidistant positions at the corners of a regular octahedron. In fact, the octahedron is distorted, with four ligands in a square and two opposite ligands further away. If the 'original' structure has a centre of symmetry, the distorted structure must also have a centre of symmetry. The effect was predicted theoretically by H. A. Jahn and Edward Teller in 1937.

jasper An impure variety of *chalcedony. It is associated with iron ores and as a result contains iron oxide impurities that give the mineral its characteristic red or reddish-brown colour. Jasper is used as a gemstone.

jet A variety of *coal that can be cut and polished and is used for jewellery, ornaments, etc.

jeweller's rouge Red powdered haematite, iron(III) oxide, Fe_2O_3. It is a mild abrasive used in metal cleaners and polishes.

joule Symbol J. The *SI unit of work and energy equal to the work done when the point of application of a force of one newton moves, in the direction of the force, a distance of one metre. 1 joule = 10^7 ergs = 0.2388 calorie. It is named after James Prescott Joule (1818–89).

Joule's law The *internal energy of a given mass of gas is independent of its volume and pressure, being a function of temperature alone. This law applies only to *ideal gases (for which it provides a definition of thermodynamic temperature) as in a real gas intermolecular forces would cause changes in the internal energy should a change of volume occur. *See also* Joule–Thomson effect.

Joule–Thomson effect (Joule–Kelvin effect) The change in temperature that occurs when a gas expands through a porous plug into a region of lower pressure. For most real gases the temperature falls under these circumstances as the gas has to do internal work in overcoming the intermolecular forces to enable the expansion to take place. This is a deviation from *Joule's law. There is usually also a deviation from *Boyle's law, which can cause either a rise or a fall in temperature since any increase in the product of pressure and volume is a measure of external work done. At a given pressure, there is a particular temperature, called the *inversion temperature* of the gas, at which the rise in temperature from the Boyle's law deviation is balanced by the fall from the Joule's law deviation. There is then no temperature change. Above the inversion temperature the gas is heated by expansion, below it, it is cooled. The effect was discovered by James Joule working in collaboration with William Thomson (later Lord Kelvin; 1824–1907).

K

kainite A naturally occurring double salt of magnesium sulphate and potassium chloride, $MgSO_4.KCl.3H_2O$.

kalinite A mineral form of *aluminium potassium sulphate $(Al_2(SO_4)_3.K_2SO_4.24H_2O)$.

kaolin (china clay) A soft white clay that is composed chiefly of the mineral kaolinite (*see* clay minerals). It is formed during the weathering and hydrothermal alteration of other clays or feldspar. Kaolin is mined in the UK, France, Czechoslovakia, and USA. Besides its vital importance in the ceramics industry it is also used extensively as a filler in the manufacture of rubber, paper, paint, and textiles and as a constituent of medicines.

katharometer An instrument for comparing the thermal conductivities of two gases by comparing the rate of loss of heat from two heating coils surrounded by the gases. The instrument can be used to detect the presence of a small amount of an impurity in air and is also used as a detector in gas chromatography.

Kekulé structure A proposed structure of *benzene in which the molecule has a hexagonal ring of carbon atoms linked by alternating double and single bonds. It was suggested in 1865 by Friedrich August Kekulé (1829–69).

kelvin Symbol K. The *SI unit of thermodynamic *temperature equal to the fraction 1/273.16 of the thermodynamic temperature of the *triple point of water. The magnitude of the kelvin is equal to that of the degree celsius (centigrade), but a temperature expressed in degrees celsius is numerically equal to the temperature in kelvins less 273.15 (i.e. °C = K − 273.15). The *absolute zero of temperature has a temperature of 0 K (−273.15°C). The former name *degree kelvin* (symbol °K) became obsolete by international agreement in 1967. The unit is named after Lord Kelvin (1824–1907).

Kelvin effect *See* Thomson effect.

keratin Any of a group of fibrous *proteins occurring in hair, feathers, hooves, and horns. Keratins have coiled polypeptide chains that combine to form supercoils of several polypeptides linked by disulphide bonds between adjacent cysteine amino acids.

kerosine *See* petroleum.

ketals Organic compounds, similar to *acetals, formed by addition of an alcohol to a ketone. If one molecule of ketone (RR′CO) reacts with one molecule of alcohol R″OH, then a *hemiketal* is formed. The rings of ketose sugars are hemiketals. Further reaction produces a full ketal ($RR'C(OR'')_2$).

keto–enol tautomerism A form of tautomerism in which a compound containing a $-CH_2-CO-$ group (the *keto form* of the molecule) is in equilibrium with one containing the $-CH=C(OH)-$ group (the *enol*). It occurs by migration of a hydrogen atom between a carbon atom and the oxygen on an adjacent carbon. *See* isomerism.

keto form *See* keto–enol tautomerism.

ketohexose *See* monosaccharide.

ketone body Any of three compounds, acetoacetic acid (3-oxobutanoic acid, CH_3COCH_2COOH), *β*-hydroxybutyric acid (3-hydroxybutanoic acid, $CH_3CH(OH)CH_2COOH$), and acetone or (propanone, CH_3COCH_3), produced by the liver as a result of the metabolism of body fat deposits. Ketone bodies are normally used as energy sources by peripheral tissues.

ketones Organic compounds that contain the group −CO− linked to two hydrocarbon groups. The *ketone group* is a carbonyl group with two single bonds to other carbon atoms. In systematic chemical nomenclature, ketone names end with the suffix -*one*. Examples are propanone (acetone), CH_3COCH_3, and butanone (methyl ethyl ketone), $CH_3COC_2H_5$. Ketones can be made by oxidizing secondary alcohols to convert the C−OH group to C=O. Certain ketones form addition compounds with sodium hydrogensulphate(IV) (sodium hydrogensulphite). They also form addition compounds with hydrogen cyanide to give *cyanohydrins and with alcohols to give *ketals. They undergo condensation reactions to yield *oximes, *hydrazones, phenylhydrazones, and *semicarbazones. These are reactions that they share with aldehydes. Unlike aldehydes, they do not affect Fehling's solution or Tollen's reagent and do not easily oxidize. Strong oxidizing agents produce a mixture of carboxylic acids; butanone, for example, gives ethanoic and propanoic acids.

ketopentose *See* monosaccharide.

ketose *See* monosaccharide.

kieselguhr A soft fine-grained deposit consisting of the siliceous skeletal remains of diatoms, formed in lakes and ponds. Kieselguhr is used as an absorbent, filtering material, filler, and insulator.

kieserite A mineral form of *magnesium sulphate monohydrate, $MgSO_4.H_2O$.

kilo- Symbol k. A prefix used in the metric system to denote 1000 times. For example, 1000 volts = 1 kilovolt (kV).

kilogram Symbol kg. The *SI unit of mass defined as a mass equal to that of the international platinum–iridium prototype kept by the International Bureau of Weights and Measures at Sèvres, near Paris.

kimberlite A rare igneous rock that often contains diamonds. It occurs as narrow pipe intrusions but is often altered and fragmented. It consists of olivine and phlogopite mica, usually with calcite, serpentine, and other minerals. The chief occurrences of kimberlite are in South Africa, especially at Kimberley (after which the rock is named), and in the Yakutia area of Siberia.

kinematic viscosity Symbol ν. The ratio of the *viscosity of a liquid to its density. The SI unit is $m^2 s^{-1}$.

kinetic effect A chemical effect that depends on reaction rate rather than on thermodynamics. For example, diamond is thermodynamically less stable than graphite; its apparent stability depends on the vanishingly slow rate at which it is converted. *Overvoltage in electrolytic cells is another example of a kinetic effect. *Kinetic isotope effects* are changes in reaction rates produced by isotope substitution. For example, if the slow step in a chemical reaction is the breaking of a C–H bond, the rate for the deuterated compound would be slightly lower because of the lower vibrational frequency of the C–D bond. Such effects are used in investigating the mechanisms of chemical reactions.

kinetic energy *See* energy.

kinetic theory A theory, largely the work of Count Rumford (1753–1814), James Joule (1818–89), and James Clerk Maxwell (1831–79), that explains the physical properties of matter in terms of the motions of its constituent particles. In a gas, for example, the pressure is due to the incessant impacts of the gas molecules on the walls of the container. If it is assumed that the molecules occupy negligible space, exert negligible forces on each other except during collisions, are perfectly elastic, and make only brief collisions with each other, it can be shown that the pressure p exerted by one mole of gas containing n molecules each of mass m in a container of volume V, will be given by:

$$p = nm\bar{c}^2/3V,$$

where $\bar{c}^2$ is the mean square speed of the molecules. As according to the *gas laws for one mole of gas: $pV = RT$, where T is the thermodynamic temperature, and R is the molar *gas constant, it follows that:

$$RT = nm\bar{c}^2/3$$

Thus, the thermodynamic temperature of a gas is proportional to the mean square speed of its molecules. As the average kinetic *energy of translation of the molecules is $m\bar{c}^2/2$, the temperature is given by:

$$T = (m\bar{c}^2/2)(2n/3R)$$

The number of molecules in one mole of any gas is the *Avogadro constant, N_A; therefore in this equation $n = N_A$. The ratio R/N_A is a constant called the *Boltzmann constant (k). The average kinetic energy of translation of the molecules of one mole of any gas is therefore $3kT/2$. For monatomic gases this is proportional to the *internal energy (U) of the gas, i.e.

$$U = N_A 3kT/2$$

and as $k = R/N_A$

$$U = 3RT/2$$

For diatomic and polyatomic gases the rotational and vibrational energies also have to be taken into account (*see* degrees of freedom).

In liquids, according to the kinetic theory, the atoms and molecules still move around at random, the temperature being proportional to their average kinetic energy. However, they are sufficiently close to each other for the attractive forces between molecules to be important. A molecule that approaches the surface will experience a resultant force tending to keep it within the liquid. It is, therefore, only some of the fastest moving molecules that escape; as a result the average kinetic energy of those that fail to escape is reduced. In this way evaporation from the surface of a liquid causes its temperature to fall.

In a crystalline solid the atoms, ions, and molecules are able only to vibrate about the fixed positions of a *crystal lattice; the attractive forces are so strong at this range that no free movement is possible.

Kipp's apparatus A laboratory apparatus for making a gas by the reaction of a solid with a liquid (e.g. the reaction of hydrochloric acid with iron sulphide to give hydrogen sulphide). It consists of three interconnected glass globes arranged vertically, with the solid in the middle globe. The upper and lower globes are connected by a tube and contain the liquid. The middle globe has a tube with a tap for drawing off gas. When the tap is closed, pressure of gas forces the liquid down in the bottom reservoir and up into the top, and reaction does not occur. When the tap is opened, the release in pressure allows the liquid to rise into the middle globe, where it reacts with the solid. It is named after Petrus Kipp (1808–64).

Kjeldahl's method A method for measuring the percentage of nitrogen in an organic compound. The compound is boiled with concentrated sulphuric acid and copper(II) sulphate catalyst to convert any nitrogen to ammonium sulphate. Alkali is added and the mixture heated to distil off ammonia. This is passed into a standard acid solution, and the amount of ammonia can then be found by estimating the amount of unreacted acid by titration. The amount of nitrogen in the original specimen can then be calculated. The method was developed by the Danish chemist Johan Kjeldahl (1849–1900).

Kohlrausch's law If a salt is dissolved in water, the conductivity of the (dilute) solution is the sum of two values – one depending on the positive ions and the other on the negative ions. The law, which depends on the independent migration of ions, was deduced experimentally by the German chemist Friedrich Kohlrausch (1840–1910).

Kolbe's method A method of making alkanes by electrolysing a solution of a carboxylic acid salt. For a salt Na^+RCOO^-, the carboxylate ions lose electrons at the cathode to give radicals:

$$RCOO^- - e \rightarrow RCOO\cdot$$

These decompose to give alkyl radicals

$$RCOO\cdot \rightarrow R\cdot + CO_2$$

Two alkyl radicals couple to give an alkane

$$R\cdot + R\cdot \rightarrow RR$$

The method can only be used for hydrocarbons with an even number of carbon atoms, although mixtures of two salts can be electrolysed to give a mixture of three products. The method was discovered by the German chemist Herman Kolbe (1818–84), who electrolysed pentanoic acid (C_4H_9COOH) in 1849 and obtained a hydrocarbon, which he assumed was the substance 'butyl' C_4H_9 (actually octane, C_8H_{18}).

Kovar A tradename for an alloy of iron, cobalt, and nickel with an *expansivity similar to that of glass. It is therefore used in making glass-to-metal seals, especially in circumstances in which a temperature variation can be expected.

Krebs cycle (citric acid cycle; tricarboxylic acid cycle; TCA cycle) A cyclical series of biochemical reactions that is fundamental to the metabolism of aerobic organisms, i.e. animals, plants, and many microorganisms. The enzymes of the Krebs cycle are in close association with the components of the *electron transport chain. The two-carbon acetyl coenzyme A (acetyl CoA) reacts with the four-carbon oxaloacetate to form the six-carbon citric acid. In a series of seven reactions, this is reconverted to oxaloacetate and produces two molecules of carbon dioxide. Most importantly, the cycle generates one molecule of guanosine triphosphate (GTP – equivalent to 1 ATP) and reduces three molecules of the coenzyme NAD to NADH and one molecule of the coenzyme FAD to $FADH_2$. NADH and $FADH_2$ are then oxidized by the electron transport chain to generate three and two molecules of ATP respectively. This gives a net yield of 12 molecules of ATP per molecule of acetyl CoA.

Acetyl CoA can be derived from carbohydrates (via *glycolysis), fats, or certain amino acids. (Other amino acids may enter the cycle at different stages.) Thus the Krebs cycle is the central 'crossroads' in the complex system of metabolic pathways and is involved not only in degradation and energy production but also in the synthesis of biomolecules. It is named after its principal discoverer, Sir Hans Adolf Krebs (1900–81).

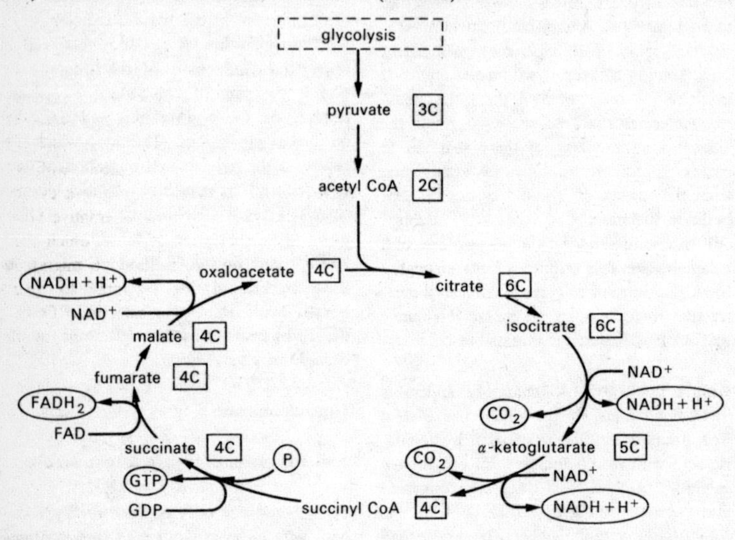

Krebs cycle

Kroll process A process for producing certain metals by reducing the chloride with magnesium metal, e.g.

$$TiCl_4 + 2Mg \rightarrow Ti + 2MgCl_2$$

krypton Symbol Kr. A colourless gaseous element belonging to group 0 (the *noble gases) of the periodic table; a.n. 36; r.a.m. 83.80; d. 3.73 $g\,m^{-3}$; m.p. −156.6°C; b.p. −152.3°C. Krypton occurs in air (0.0001% by volume) from which it can be extracted by fractional distillation of liquid air. Usually, the element is not isolated but is used with other inert gases in fluorescent lamps, etc. The element has five natural isotopes (mass numbers 78, 80, 82, 83, 84) and there are five radioactive isotopes (76, 77, 79, 81, 85). Krypton−85 (half-life 10.76 years) is produced in fission reactors and it has been suggested that an equilibrium amount will eventually occur in the atmosphere. The element is inert and forms no stable compounds (KrF_2 has been reported).

Kupfer nickel A naturally occurring form of nickel arsenide, NiAs; an important ore of nickel.

L

labelling The process of replacing a stable atom in a compound with a radioisotope of the same element to enable its path through a biological or mechanical system to be traced by the radiation it emits. In some cases a different stable isotope is used and the path is detected by means of a mass spectrometer. A compound containing either a radioactive or stable isotope is called a *labelled compound.* If a hydrogen atom in each molecule of the compound has been replaced by a tritium atom, the compound is called a *tritiated compound.* A radioactive labelled compound will behave chemically and physically in the same way as an otherwise identical stable compound, and its presence can easily be detected using a Geiger counter. This process of *radioactive tracing* is widely used in chemistry, biology, medicine, and engineering. For example, it can be used to follow the course of the reaction of a carboxylic acid with an alcohol to give an ester, e.g.

$$CH_3COOH + C_2H_5OH \rightarrow C_2H_5COOCH_3 + H_2O$$

To determine whether the noncarbonyl oxygen in the ester comes from the acid or the alcohol, the reaction is performed with the labelled compound $CH_3CO^{18}OH$, in which the oxygen in the hydroxyl group of the acid has been 'labelled' by using the ^{18}O isotope. It is then found that the water product is $H_2{}^{18}O$; i.e. the oxygen in the ester comes from the alcohol, not the acid.

labile Describing a chemical compound in which certain atoms or groups can easily be replaced by other atoms or groups. The term is applied to coordination complexes in which ligands can easily be replaced by other ligands in an equilibrium reaction.

lactams Organic compounds containing a ring of atoms in which the group −NH.CO.− forms part of the ring. Lactams can be formed by reaction of an $-NH_2$ group in one part of a molecule with a −COOH group in the other to give a cyclic amide. They can exist in an alternative tautomeric form, the *lactim* form, in which the hydrogen atom on the nitrogen has migrated to the oxygen of the carbonyl to give −N=C(OH)−. The pyrimidine base uracil is an example of a lactam.

lactate A salt or ester of lactic acid (i.e. a 2-hydroxypropanoate).

lactic acid (2-hydroxypropanoic acid) A clear odourless hygroscopic syrupy liquid, $CH_3CH(OH)COOH$, with a sour taste; r.d. 1.206; m.p. 18°C; b.p. 122°C. It is prepared by the hydrolysis of ethanal cyanohydrin or

amino acid $\xrightarrow{-H_2O}$ γ lactam (γ butyrolactam) ⇌ γ lactim

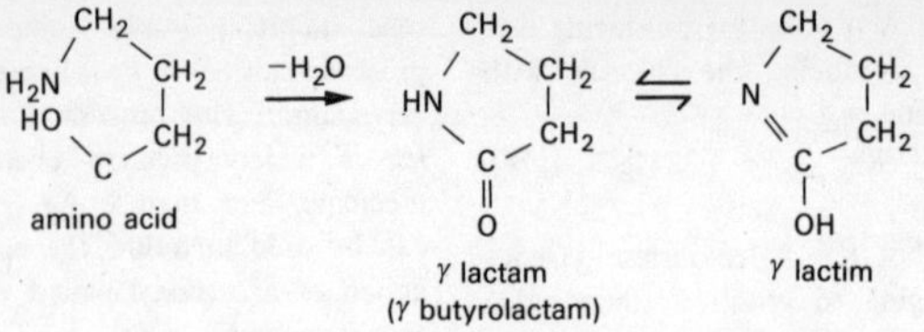

Lactam formation

the oxidation of propan-1,2-diol using dilute nitric acid. Lactic acid is manufactured by the fermentation of lactose (from milk) and used in the dyeing and tanning industries. It is an alpha hydroxy *carboxylic acid. *See also* optical activity.

Lactic acid is produced from pyruvic acid in active muscle tissue when oxygen is limited and subsequently removed for conversion to glucose by the liver. During strenuous exercise it may build up in the muscles, causing cramplike pains. It is also produced by fermentation in certain bacteria and is characteristic of sour milk.

lactims *See* lactams.

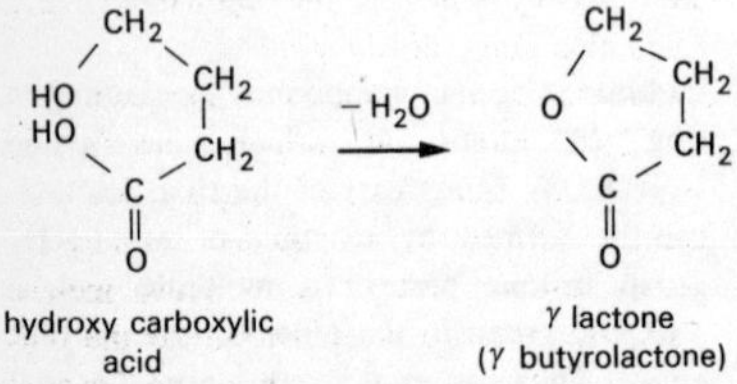

hydroxy carboxylic acid $\xrightarrow{-H_2O}$ γ lactone (γ butyrolactone)

Lactone formation

lactones Organic compounds containing a ring of atoms in which the group –CO.O– forms part of the ring. Lactones can be formed (or regarded as formed) by reaction of an –OH group in one part of a molecule with a –COOH group in the other to give a cyclic ester. This type of reaction occurs with γ-hydroxy carboxylic acids such as the compound $CH_2(OH)CH_2CH_2COOH$ (the hydroxyl group is on the third carbon from the carboxyl group). The resulting γ-lactone has a five-membered ring. Similarly, δ-lactones have six-membered rings. β-lactones, with a four-membered ring, are not produced directly from β-hydroxy acids, but can be synthesized by other means.

lactose (milk sugar) A sugar comprising one glucose molecule linked to a galactose molecule. Lactose is manufactured by the mammary gland and occurs only in milk. For example, cows' milk contains about 4.7% lactose. It is less sweet than sucrose (cane sugar).

Ladenburg benzene An (erroneous) structure for *benzene proposed by Albert Ladenburg (1842–1911), in which the six carbon atoms were arranged at the corners of a triangular prism and linked by single bonds to each other and to the six hydrogen atoms.

laevorotatory Designating a chemical compound that rotates the plane of plane-polarized light to the left (anticlockwise for someone facing the oncoming radiation). *See* optical activity.

laevulose *See* fructose.

LAH Lithium aluminium hydride; *see* lithium tetrahydroaluminate (III).

lake A pigment made by combining an organic dyestuff with an inorganic compound (usually an oxide, hydroxide, or salt). Absorption of the organic compound on the inorganic substrate yields a coloured complex, as in the combination of a dyestuff with a *mordant. Lakes are used in paints and printing inks.

Lamb shift A small energy difference between two levels ($^2S_{1/2}$ and $^2P_{1/2}$) in the *hydrogen spectrum. The shift results from the quantum interaction between the atomic electron and the electromagnetic radiation. It was first explained by Willis Eugene Lamb (1913–).

lamellar solids Solid substances in which the crystal structure has distinct layers (i.e. has a layer lattice). The *micas are an example of this type of compound. *Intercalation compounds* are lamellar compounds formed by interposition of atoms, ions, etc., between the layers of an existing element or compound. For example, graphite is a lamellar solid. With strong oxidizing agents (e.g. a mixture of concentrated sulphuric and nitric acids) it forms a nonstoichiometric 'graphitic oxide', which is an intercalation compound having oxygen atoms between the layers of carbon atoms.

lamp black A finely divided (microcrystalline) form of carbon made by burning organic compounds in insufficient oxygen. It is used as a black pigment and filler.

lanolin An emulsion of purified wool fat in water, containing cholesterol and certain terpene alcohols and esters. It is used in cosmetics.

lansfordite A mineral form of *magnesium carbonate pentahydrate, $MgCO_3.5H_2O$.

lanthanides *See* lanthanoids.

lanthanoid contraction *See* lanthanoids.

lanthanoids (lanthanides; lanthanons; rare-earth elements) A series of elements in the *periodic table, generally considered to range in proton number from cerium (58) to lutetium (71) inclusive. The lanthanoids all have two outer *s*-electrons (a $6s^2$ configuration), follow lanthanum, and are classified together because an increasing proton number corresponds to increase in number of $4f$ electrons. In fact, the $4f$ and $5d$ levels are close in energy and the filling is not smooth. The outer electron configurations are as follows:

57 lanthanum (La) $5d^16s^2$
58 cerium (Ce) $4f5d^16s^2$ (or $4f^26s^2$)
59 praseodymium (Pr) $4f^36s^2$
60 neodymium (Nd) $4f^46s^2$
61 promethium (Pm) $4f^56s^2$
62 samarium (Sm) $4f^66s^2$
63 europium (Eu) $4f^76s^2$
64 gadolinium (Gd) $4f^75d^16s^2$
65 terbium (Tb) $4f^96s^2$
66 dysprosium (Dy) $4f^{10}6s^2$
67 holmium (Ho) $4f^{11}6s^2$
68 erbium (Er) $4f^{12}6s^2$
69 thulium (Tm) $4f^{13}6s^2$
70 ytterbium (Yb) $4f^{14}6s^2$
71 lutetium (Lu) $4f^{14}5d^16s^2$

Note that lanthanum itself does not have a $4f$ electron but it is generally classified with the lanthanoids because of its chemical similarities, as are yttrium (Yt) and scandium (Sc). Scandium, yttrium, and lanthanum are *d*-block elements; the lanthanoids and *actinoids make up the *f*-block.

The lanthanoids are sometimes simply called the *rare earths*, although strictly the 'earths' are their oxides. Nor are they particularly rare: they occur widely, usually together. All are silvery very reactive metals. The *f*-electrons do not penetrate to the outer part of the atom and there is no *f*-orbital participation in bonding (unlike the *d*-orbitals of the main *transition elements) and the elements form few coordination compounds. The main compounds contain M^{3+} ions. Cerium also has the highly oxidizing Ce^{4+} state and europium and ytterbium have a M^{2+} state.

The $4f$ orbitals in the atoms are not very effective in shielding the outer electrons from the nuclear charge. In going across the series the increasing nuclear charge causes a contraction in the radius of the M^{3+} ion – from 0.1061 nm in lanthanum to 0.0848 nm in lutetium. This effect, the *lanthanoid con-*

traction, accounts for the similarity between the transition elements zirconium and hafnium.

lanthanons *See* lanthanoids.

lanthanum Symbol La. A silvery metallic element belonging to group IIIB of the periodic table and often considered to be one of the *lanthanoids; a.n. 57; r.a.m. 138.91; r.d. 6.146 (20°C); m.p. 918°C; b.p. 3464°C. Its principal ore is bastnasite, from which it is separated by an ion-exchange process. There are two natural isotopes, lanthanum–139 (stable) and lanthanum–138 (half-life 10^{10}–10^{15} years). The metal, being pyrophoric, is used in alloys for lighter flints and the oxide is used in some optical glasses. The largest use of lanthanum, however, is as a catalyst in cracking crude oil. Its chemistry resembles that of the lanthanoids. The element was discovered by C. G. Mosander in 1839.

lapis lazuli A blue rock that is widely used as a semiprecious stone and for ornamental purposes. It is composed chiefly of the deep blue mineral *lazurite* embedded in a matrix of white calcite and usually also contains small specks of pyrite. It occurs in only a few places in crystalline limestones as a contact metamorphic mineral. The chief source is Afghanistan; lapis lazuli also occurs near Lake Baikal in Siberia and in Chile. It was formerly used to make the artists' pigment ultramarine.

lattice The regular arrangement of atoms, ions, or molecules in a crystalline solid. *See* crystal lattice.

lattice energy A measure of the stability of a *crystal lattice, given by the energy that would be released per mole if atoms, ions, or molecules of the crystal were brought together from infinite distances apart to form the lattice. *See* Born–Haber cycle.

lattice vibrations The periodic vibrations of the atoms, ions, or molecules in a *crystal lattice about their mean positions. On heating, the amplitude of the vibrations increases until they are so energetic that the lattice breaks down. The temperature at which this happens is the melting point of the solid and the substance becomes a liquid. On cooling, the amplitude of the vibrations diminishes. At *absolute zero a residual vibration persists, associated with the *zero-point energy of the substance. The increase in the electrical resistance of a conductor is due to increased scattering of the free conduction electrons by the vibrating lattice particles.

laughing gas *See* dinitrogen oxide.

lauric acid *See* dodecanoic acid.

law of chemical equilibrium *See* equilibrium constant.

law of constant composition *See* chemical combination.

law of definite proportions *See* chemical combination.

law of mass action *See* mass action.

law of multiple proportions *See* chemical combination.

law of octaves (Newlands' law) An attempt at classifying elements made by J. A. R. Newlands (1837–98) in 1863. He arranged 56 elements in order of increasing atomic mass in groups of eight, pointing out that each element resembled the element eight places from it in the list. He drew an analogy with the notes of a musical scale. *Newlands' octaves* were groups of similar elements distinguished in this way: e.g. oxygen and sulphur; nitrogen and phosphorus; and fluorine, chlorine, bromine, and iodine. In some cases it was necessary to put two elements in the same position. The proposal was rejected at the time. *See* periodic table.

law of reciprocal proportions *See* chemical combination.

lawrencium Symbol Lr. A radioactive metallic transuranic element belonging to the *actinoids; a.n. 103; mass number of only known isotope 257 (half-life 8 seconds). The element was identified by A. Ghiorso and associates in 1961. The alternative name *unniltrium* has been proposed.

laws of chemical combination *See* chemical combination.

layer lattice A crystal structure in which the atoms are chemically bonded in plane layers, with relatively weak forces between atoms in adjacent layers. Graphite and micas are examples of substances having layer lattices (i.e. they are *lamellar solids).

lazurite *See* lapis lazuli.

L-dopa *See* dopa.

L–D process *See* basic-oxygen process.

leaching Extraction of soluble components of a solid mixture by percolating a solvent through it.

lead Symbol Pb. A heavy dull grey soft ductile metallic element belonging to *group IV of the periodic table; a.n. 82; r.a.m. 207.19; r.d. 11.35; m.p. 327.5°C; b.p. 1740°C. The main ore is the sulphide galena (PbS); other minor sources include anglesite ($PbSO_4$), cerussite ($PbCO_3$), and litharge (PbO). The metal is extracted by roasting the ore to give the oxide, followed by reduction with carbon. Silver is also recovered from the ores. Lead has a variety of uses including building construction, lead-plate accumulators, bullets, and shot, and is a constituent of such alloys as solder, pewter, bearing metals, type metals, and fusible alloys. Chemically, it forms compounds with the +2 and +4 oxidation states, the lead (II) state being the more stable.

lead(II) acetate *See* lead(II) ethanoate.

lead–acid accumulator An accumulator in which the electrodes are made of lead and the electrolyte consists of dilute sulphuric acid. The electrodes are usually cast from a lead alloy containing 7–12% of antimony (to give increased hardness and corrosion resistance) and a small amount of tin (for better casting properties). The electrodes are coated with a paste of lead(II) oxide (PbO) and finely divided lead; after insertion into the electrolyte a 'forming' current is passed through the cell to convert the PbO on the negative plate into a sponge of finely divided lead. On the positive plate the PbO is converted to lead(IV) oxide (PbO_2). The equation for the overall reaction during discharge is:

$$PbO_2 + 2H_2SO_4 + Pb \rightarrow 2PbSO_4 + 2H_2O$$

The reaction is reversed during charging. Each cell gives an e.m.f. of about 2 volts and in motor vehicles a 12-volt battery of six cells is usually used. The lead–acid battery produces 80–120 kJ per kilogram. *Compare* nickel–iron accumulator.

lead(II) carbonate A white solid, $PbCO_3$, insoluble in water; rhombic; r.d. 6.6. It occurs as the mineral *cerussite, which is isomorphous with aragonite and may be prepared in the laboratory by the addition of cold ammonium carbonate solution to a cold solution of a lead(II) salt (acetate or nitrate). It decomposes at 315°C to lead(II) oxide and carbon dioxide.

lead(II) carbonate hydroxide (white lead; basic lead carbonate) A powder, $2PbCO_3.Pb(OH)_2$, insoluble in water, slightly soluble in aqueous carbonate solutions; r.d. 6.14; decomposes at 400°C. Lead(II) carbonate hydroxide occurs as the mineral *hydroxycerussite* (of variable composition). It was previously manufactured from lead in processes using spent tanning bark or horse manure, which released carbon dioxide. It is currently made by electrolysis of mixed solutions (e.g. ammonium nitrate, nitric acid, sulphuric acid, and acetic acid) using lead anodes. For the highest grade product the lead must be exceptionally pure (known in the trade as 'corroding lead') as small amounts of metallic impurity impart grey or pink discolorations. The material was used widely in paints, both for art work and for commerce, but it has the disadvantage of reacting with hydrogen sulphide in industrial atmospheres and producing black lead sulphide. The poisonous nature of lead compounds has also contributed to the declining importance of this material.

lead-chamber process An obsolete method of making sulphuric acid by the catalytic oxidation of sulphur dioxide with air using a potassium nitrate catalyst in water. The process was carried out in lead containers (which was expensive) and only produced dilute acid. It was replaced in 1876 by the *contact process.

lead dioxide *See* lead(IV) oxide.

lead(II) ethanoate (lead(II) acetate) A white crystalline solid, $Pb(CH_3COO)_2$, soluble in water and slightly soluble in ethanol. It exists as the anhydrous compound (r.d. 3.25; m.p. 280°C), as a trihydrate, $Pb(CH_3COO)_2.3H_2O$ (monoclinic; r.d. 2.55; loses water at 75°C), and as a decahydrate, $Pb(CH_3COO)_2.10H_2O$ (rhombic; r.d. 1.69). The common form is the trihydrate. Its chief interest stems from the fact that it is soluble in water and it also forms a variety of complexes in solution. It was once known as *sugar of lead* because of its sweet taste.

lead(IV) ethanoate (lead tetra-acetate) A colourless solid, $Pb(CH_3COO)_4$, which decomposes in water and is soluble in pure ethanoic acid; monoclinic; r.d. 2.228; m.p. 175°C. It may be prepared by dissolving dilead(II) lead(IV) oxide in warm ethanoic acid. In solution it behaves essentially as a covalent compound (no measurable conductivity) in contrast to the lead(II) salt, which is a weak electrolyte.

lead(IV) hydride *See* plumbane.

lead monoxide *See* lead(II) oxide.

lead(II) oxide (lead monoxide) A solid yellow compound, PbO, which is insoluble in water; m.p. 886°C. It exists in two crystalline forms: *litharge* (tetrahedral; r.d. 9.53) and *massicot* (rhombic; r.d. 8.0). It can be prepared by heating the nitrate, and is manufactured by heating molten lead in air. If the temperature used is lower than the melting point of the oxide, the product is massicot; above this, litharge is formed. Variations in the temperature and in the rate of cooling give rise to crystal vacancies and red, orange, and brown forms of litharge can be produced. The oxide is amphoteric, dissolving in acids to give lead(II) salts and in alkalis to give *plumbates.

lead(IV) oxide (lead dioxide) A dark brown or black solid with a rutile lattice, PbO_2, which is insoluble in water and slightly soluble in concentrated sulphuric and nitric acids; r.d. 9.375; decomposes at 290°C. Lead(IV) oxide may be prepared by the oxidation of lead(II) oxide by heating with alkaline chlorates or nitrates, or by anodic oxidation of lead(II) solutions. It is an oxidizing agent and readily reverts to the lead(II) oxidation state, as illustrated by its conversion to Pb_3O_4 and PbO on heating. It reacts with hydrochloric acid to evolve chlorine. Lead(IV) oxide has been used in the manufacture of safety matches and was widely used until the mid-1970s as an adsorbent for sulphur dioxide in pollution monitoring.

lead(II) sulphate A white crystalline solid, $PbSO_4$, which is virtually insoluble in water and soluble in solutions of ammonium salts; r.d. 6.2; m.p. 1170°C. It occurs as the mineral *anglesite*; it may be prepared in the laboratory by adding any solution containing sulphate ions to solutions of lead(II) ethanoate. The material known as *basic lead (II) sulphate* may be made by shaking together lead(II) sulphate and lead(II) hydroxide in water. This material has been used in white paint in preference to lead(II) carbonate hydroxide, as it is not so susceptible to discoloration through reaction with hydrogen sulphide. The toxicity of lead compounds has led to a decline in the use of these compounds.

lead(II) sulphide A black crystalline solid, PbS, which is insoluble in water; r.d. 7.5; m.p. 1114°C. It occurs naturally as the metallic-looking mineral *galena (the principal ore of lead). It may be prepared in the laboratory by the reaction of hydrogen sulphide with soluble lead(II) salts. Lead(II) sulphide has been used as an electrical rectifier.

lead tetra-acetate *See* lead(IV) ethanoate.

lead tetraethyl(IV) (tetraethyl lead) A colourless liquid, $Pb(C_2H_5)_4$, insoluble in water, soluble in benzene, ethanol, ether, and petroleum; r.d. 1.659; m.p. −137°C; b.p. 200°C. It may be prepared by the reaction of hydrogen and ethene with lead but a more convenient laboratory and industrial method is the reaction of a sodium–lead alloy with chloroethane. A more recent industrial process is the electrolysis of ethylmagnesium chloride (the Grignard reagent) using a lead anode and slowly running additional chloroethane onto the cathode. Lead tetraethyl is used in fuel for internal-combustion engines (along with 1,2-dibromoethane) to increase the *octane number and reduce knocking. Mounting pressure from groups concerned about environmental pollution has led to a reduction of permitted levels of lead tetraethyl in many countries.

Leblanc process An obsolete process for manufacturing sodium carbonate. The raw materials were sodium chloride, sulphuric acid, coke, and limestone (calcium carbonate), and the process involved two stages. First the sodium chloride was heated with sulphuric acid to give sodium sulphate:

$$2NaCl(s) + H_2SO_4(l) \rightarrow Na_2SO_4(s) + 2HCl(g)$$

The sodium sulphate was then heated with coke and limestone:

$$Na_2SO_4 + 2C + CaCO_3 \rightarrow Na_2CO_3 + CaS + 2CO_2$$

Calcium sulphide was a by-product, the sodium carbonate being extracted by crystallization. The process, invented in 1783 by the French chemist Nicolas Leblanc (1742–1806), was the first for producing sodium carbonate synthetically (earlier methods were from wood ash and other vegetable sources). By the end of the 19th century it had been largely replaced by the *Solvay process.

lechatelierite A mineral form of *silicon (IV) oxide, SiO_2.

Le Chatelier's principle If a system is in equilibrium, any change imposed on the system tends to shift the equilibrium to nullify the effect of the applied change. The principle, which is a consequence of the law of conservation of energy, was first stated in 1888 by Henri Le Chatelier (1850–1936). It is applied to chemical equilibria. For example, in the gas reaction

$$2SO_2 + O_2 \rightleftharpoons 2SO_3$$

an increase in pressure on the reaction mixture displaces the equilibrium to the right, since this reduces the total number of molecules present and thus decreases the pressure. The standard enthalpy change for the forward reaction is negative (i.e. the reaction is exothermic). Thus, an increase in temperature displaces the equilibrium to the left since this tends to reduce the temperature. The *equilibrium constant thus falls with increasing temperature.

Leclanché cell A primary *voltaic cell consisting of a carbon rod (the anode) and a zinc rod (the cathode) dipping into an electrolyte of a 10–20% solution of ammonium chloride. *Polarization is prevented by using a mixture of manganese dioxide mixed with crushed carbon, held in contact with the anode by means of a porous bag or pot; this reacts with the hydrogen produced. This wet form of the cell, devised in 1867 by Georges Leclanché (1839–82), has an e.m.f. of about 1.5 volts. The *dry cell based on it is widely used in torches, radios, and calculators.

leucine *See* amino acid.

leuco form *See* dyes.

Lewis acid and base *See* acid.

***l*-form** *See* optical activity.

Liebig condenser A laboratory condenser having a straight glass tube surrounded by a coaxial glass jacket through which cooling water is passed. It is named after the German organic chemist Justus von Liebig (1803–73).

ligand An ion or molecule that donates a pair of electrons to a metal atom or ion in forming a coordination *complex. Molecules that function as ligands are acting as Lewis bases (*see* acid). For example, in the complex hexaquocopper(II) ion $[Cu(H_2O)_6]^{2+}$ six water molecules coordinate to a central Cu^{2+} ion. In the tetrachloroplatinate(II) ion $[PtCl_4]^{2-}$, four Cl^- ions are coordinated to a central Pt^{2+} ion. A feature of such ligands is that they have lone pairs of electrons, which they donate to empty metal orbitals. A certain class of ligands also have empty *p*- or *d*-orbitals in addition to their lone pair of electrons and can produce complexes in which the metal has low oxidation state. A double bond is formed between the metal and the ligand: a sigma bond by donation of the lone pair from ligand to metal, and a pi bond by *back donation* of electrons on the metal to empty *d*-orbitals on the ligand. Carbon monoxide is the most important such ligand, forming metal carbonyls (e.g. $Ni(CO)_4$).

The examples given above are examples of *monodentate* ligands (literally: 'having one tooth'), in which there is only one point on each ligand at which coordination can occur. Some ligands are *polydentate*; i.e. they have two or more possible coordination points. For instance, 1,2-diaminoethane, $H_2NC_2H_4NH_2$, is a *bidentate* ligand, having two coordination points. Certain polydentate ligands can form *chelates.

ligand-field theory A modification of *crystal-field theory in which the overlap of orbitals is taken into account. *See also* complex.

ligase Any of a class of enzymes that catalyse the formation of covalent bonds using the energy released by the cleavage of ATP. Ligases are important in the synthesis and repair of many biological molecules, including DNA.

lignin A complex organic polymer that is deposited within the cellulose of plant cell walls during secondary thickening. Lignification makes the walls woody and therefore rigid.

lignite *See* coal.

lime *See* calcium oxide.

limestone A sedimentary rock that is composed largely of carbonate minerals, especially carbonates of calcium and magnesium. *Calcite and *aragonite are the chief minerals; *dolomite is also present in the dolomitic limestones. There are many varieties of limestones but most are deposited in shallow water. *Organic limestones* (e.g. *chalk) are formed from the calcareous skeletons of organisms; *precipitated limestones* include oolite, which is composed of ooliths – spherical bodies formed by the precipitation of carbonate around a nucleus; and *clastic limestones* are derived from fragments of pre-existing calcareous rocks.

limewater A saturated solution of *calcium hydroxide in water. When carbon dioxide gas is bubbled through limewater, a 'milky' precipitate of calcium carbonate is formed:

$$Ca(OH)_2(aq) + CO_2(g) \rightarrow CaCO_3(s) + H_2O(l)$$

If the carbon dioxide continues to be bubbled through, the calcium carbonate eventually redissolves to form a clear solution of calcium hydrogencarbonate:

$$CaCO_3(s) + CO_2(g) + H_2O(g) \rightarrow Ca(HCO_3)_2(aq)$$

If cold limewater is used the original calcium carbonate precipitated has a calcite structure; hot limewater yields an aragonite structure.

limonite A generic term for a group of hydrous iron oxides, mostly amorphous. *Goethite and *haematite are important constituents, together with colloidal silica, clays, and manganese oxides. Limonite is formed by direct precipitation from marine or fresh water in shallow seas, lagoons, and bogs (thus it is often called *bog iron ore*) and by oxidation of iron-rich minerals. It is used as an ore of iron and as a pigment.

linear molecule A molecule in which the atoms are in a straight line, as in carbon dioxide, O=C=O.

line spectrum *See* spectrum.

Linnz–Donnewitz process *See* basic-oxygen process.

linoleic acid A liquid polyunsaturated *fatty acid with two double bonds, $CH_3(CH_2)_4CH{:}CHCH_2CH{:}CH(CH_2)_7COOH$. Linoleic acid is abundant in many plant fats and oils, e.g. linseed oil, groundnut oil, and soya-bean oil. It is an *essential fatty acid.

linolenic acid A liquid polyunsaturated *fatty acid with three double bonds in its structure:

$CH_3CH_2CH{:}CHCH_2CH{:}CHCH_2CH{:}CH(CH_2)_7COOH$. Linolenic acid occurs in certain plant oils, e.g. linseed and soya-bean oil, and in algae. It is one of the *essential fatty acids.

linseed oil A pale yellow oil pressed from flax seed. It contains a mixture of glycerides of fatty acids, including linoleic acid and linolenic acid. It is a *drying oil, used in oil paints, varnishes, linoleum, etc.

lipase An enzyme secreted by the pancreas and the glands of the small intestine of vertebrates that catalyses the breakdown of fats into fatty acids and glycerol.

lipid Any of a diverse group of organic compounds, occurring in living organisms, that are insoluble in water but soluble in organic solvents, such as chloroform, benzene, etc. Lipids are broadly classified into two categories: *complex lipids*, which are esters of long-chain fatty acids and include the *glycerides (which constitute the *fats and *oils of animals and plants), glycolipids, *phospholipids, and *waxes; and *simple lipids*, which do not contain fatty acids and include the *steroids and *terpenes.

Lipids have a variety of functions in living organisms. Fats and oils are a convenient and concentrated means of storing food energy in plants and animals. Phospholipids and *sterols, such as cholesterol, are major components of cell membranes. Waxes provide vital waterproofing for body surfaces. Terpenes include vitamins A, E, and K, and phytol (a component of chlorophyll) and occur in essential oils, such as menthol and camphor. Steroids include the adrenal hormones, sex hormones, and bile acids.

Lipids can combine with proteins to form *lipoproteins*, e.g. in cell membranes. In bacterial cell walls, lipids may associate with polysaccharides to form *lipopolysaccharides*.

lipoic acid A vitamin of the *vitamin B complex. It is one of the *coenzymes involved in the decarboxylation of pyruvate by the enzyme pyruvate dehydrogenase. Good sources of lipoic acid include liver and yeast.

lipolysis The breakdown of storage lipids in living organisms. Most long-term energy reserves are in the form of triglycerides in fats and oils. When these are needed, e.g. during starvation, lipase enzymes convert the triglycerides into glycerol and the component fatty acids. These are then transported to tissues and oxidized to provide energy.

lipoprotein *See* lipid.

liquation The separation of mixtures of solids by heating to a temperature at which lower-melting components liquefy.

liquefaction of gases The conversion of a gaseous substance into a liquid. This is usually achieved by one of four methods or by a combination of two of them:
(1) by vapour compression, provided that the substance is below its *critical temperature;
(2) by refrigeration at constant pressure, typically by cooling it with a colder fluid in a countercurrent heat exchanger;
(3) by making it perform work adiabatically against the atmosphere in a reversible cycle;
(4) by the *Joule–Thomson effect.

Large quantities of liquefied gases are now used commercially, especially *liquefied petroleum gas and liquefied natural gas.

liquefied petroleum gas (LPG) Various petroleum gases, principally propane and butane, stored as a liquid under pressure. It is used as an engine fuel and has the advantage of causing very little cylinder-head deposits.

Liquefied natural gas (LNG) is a similar product and consists mainly of methane. However, it cannot be liquefied simply by pressure as it has a low critical temperature of 190 K and must therefore be cooled to below this temperature before it will liquefy. Once liquefied it has to be stored in well-insulated containers. It provides a convenient form in which to ship natural gas in bulk from oil wells to users. It is also used as an engine fuel.

liquid A phase of matter between that of a crystalline solid and a *gas. In a liquid, the large-scale three-dimensional atomic (or ionic or molecular) regularity of the solid is absent but, on the other hand, so is the total disorganization of the gas. Although liquids have been studied for many years there is still no comprehensive theory of the liquid state. It is clear, however, from diffraction studies that there is a short-range structural regularity extending over several molecular diameters. These bundles of ordered atoms, molecules, or ions move about in relation to each other, enabling liquids to have almost fixed volumes, which adopt the shape of their containers.

liquid crystal A substance that flows like a liquid but has some order in its arrangement of molecules. *Nematic crystals* have long molecules all aligned in the same direction, but otherwise randomly arranged. *Cholesteric* and *smectic* liquid crystals also have aligned molecules, which are arranged in distinct layers. In cholesteric crystals, the axes of the molecules are parallel to the plane of the layers; in smectic crystals they are perpendicular.

litharge *See* lead(II) oxide.

lithia *See* lithium oxide.

lithium Symbol Li. A soft silvery metal, the first member of *group I of the periodic table; a.n. 3; r.a.m. 6.939; r.d. 0.534; m.p. 180.54°C; b.p. 1340°C. It is a rare element found in spodumene ($LiAlSi_2O_6$), petalite ($LiAlSi_4O_{10}$), the mica lepidolite, and certain brines. It is usually extracted by treatment with sulphuric acid to give the sulphate, which is converted to the chloride. This is mixed with a small amount of potassium chloride, melted, and electrolysed. The stable isotopes are lithium–6 and lithium–7. Lithium–5 and lithium–8 are short-lived radioisotopes. The metal is used to remove oxygen in metallurgy and as a constituent of some Al and Mg alloys. It is also used in batteries and is a potential tritium source for fusion research. Lithium salts are used in psychomedicine. The element reacts with oxygen and water; on heating it also reacts with nitrogen and hydrogen. Its chemistry differs somewhat from that of the other group I elements because of the small size of the Li^+ ion.

lithium aluminium hydride *See* lithium tetrahydroaluminate(III).

lithium carbonate A white solid, Li_2CO_3; r.d. 2.11; m.p. 735°C; decomposes above 1200°C. It is produced commercially by treating the ore with sulphuric acid at 250°C and leaching the product to give a solution of lithium sulphate. The carbonate is then obtained by precipitation with sodium carbonate solution. Lithium carbonate is used in the prevention and treatment of manic-depressive disorders. It is also used industrially in ceramic glazes.

lithium deuteride *See* lithium hydride.

lithium hydride A white solid, LiH; cubic; r.d. 0.82; m.p. 686°C; decomposes at about 850°C. It is produced by direct combination of the elements at temperatures above 500°C. The bonding in lithium hydride is believed to be largely ionic; i.e. Li^+H^- as supported by the fact that hydrogen is released from the anode on electrolysis of the molten salt. The compound reacts violently and exothermically with water to yield hydrogen and lithium hydroxide. It is used as a reducing agent to prepare other hydrides and the ^2H isotopic compound, *lithium deuteride*, is particularly valuable for deuterating a range of organic compounds. Lithium hydride has also been used as a shielding material for thermal neutrons.

lithium hydrogencarbonate A compound, $LiHCO_3$, formed by the reaction of carbon dioxide with aqueous lithium carbonate and known only in solution. It has found medicinal uses similar to those of lithium carbonate and is sometimes included in proprietary mineral waters.

lithium hydroxide A white crystalline solid, LiOH, soluble in water, slightly soluble in ethanol and insoluble in ether. It is known as the monohydrate (monoclinic; r.d. 1.51) and in the anhydrous form (tetragonal, r.d. 1.46; m.p. 450°C; decomposes at 924°C). The compound is made by reacting lime with lithium salts or lithium ores. Lithium hydroxide is basic but has a closer resemblance to group II hydroxides than to the other group I hydroxides (an example of the first member of a periodic group having atypical properties).

lithium oxide (lithia) A white crystalline compound, Li_2O; cubic; r.d. 2.01; m.p. 1700°C. It can be obtained from a number of lithium ores; the main uses are in lubricating greases, ceramics, glass and refractories, alkaline storage batteries, and as a flux in brazing and welding.

lithium sulphate A white or colourless crystalline material, Li_2SO_4, soluble in water and insoluble in ethanol. It forms a monohydrate (monoclinic; r.d. 1.88) and an anhydrous form, which exists in α- (monoclinic), β- (hexagonal) and γ- (cubic) forms; r.d. 2.23. The compound is prepared by the reaction of the hydroxide or carbonate with sulphuric acid. It is not isomorphous with other group I sulphates and does not form alums.

lithium tetrahydroaluminate(III) (lithium aluminium hydride; LAH) A white or light grey powder, $LiAlH_4$; r.d. 0.917; decomposes at 125°C. It is prepared by the reaction of excess lithium hydride with aluminium chloride. The compound is soluble in ethoxyethane, reacts violently with water to release hydrogen, and is widely used as a powerful reducing agent in organic chemistry. It should always be treated as a serious fire risk in storage.

litmus A water-soluble dye extracted from certain lichens. It turns red under acid conditions and blue under alkaline conditions, the colour change occurring over the pH range 4.5–8.3 (at 25°C). It is not suitable for titrations because of the wide range over which the colour changes, but is used as a rough *indicator of acidity or alkalinity, both in solution and as litmus paper (absorbent paper soaked in litmus solution).

litre Symbol l. A unit of volume in the metric system regarded as a special name for the cubic decimetre. It was formerly defined as the volume of 1 kilogram of pure water at 4°C at standard pressure, which is equivalent to 1.000 028 dm^3.

lixivation The separation of mixtures by dissolving soluble constituents in water.

localized bond A *chemical bond in which the electrons forming the bond remain between (or close to) the linked atoms. *Compare* delocalization.

lodestone *See* magnetite.

logarithmic scale 1. A scale of measurement in which an increase or decrease of one unit represents a tenfold increase or decrease in the quantity measured. Decibels and pH measurements are common examples of logarithmic scales of measurement. **2.** A scale on the axis of a graph in which an increase of one unit represents a tenfold increase in the variable quantity. If a curve $y = x^n$ is plotted on graph paper with logarithmic scales on both axes, the result is a straight line of slope n, i.e. $\log y = n \log x$, which enables n to be determined.

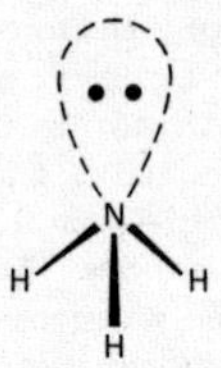

Lone pair of electrons in ammonia

lone pair A pair of electrons having opposite spin in an orbital of an atom. For instance, in ammonia the nitrogen atom has five electrons, three of which are used in forming single bonds with hydrogen atoms. The other two occupy a filled atomic orbital and constitute a lone pair. The orbital containing these electrons is equivalent to a single bond (sigma orbital) in spatial orientation, accounting for the pyramidal shape of the molecule. In the water molecule, there are two lone pairs on the oxygen atom. In considering the shapes of molecules, repulsions between bonds and lone pairs can be taken into account:

lone pair–lone pair → lone pair–bond → bond–bond.

long period *See* periodic table.

Loschmidt's constant (Loschmidt number) The number of particles per unit volume of an *ideal gas at STP. It has the value $2.687\,19 \times 10^{25}\ m^{-3}$ and was first worked out by Joseph Loschmidt (1821–95).

lowering of vapour pressure A reduction in the saturated vapour pressure of a pure liquid when a solute is introduced. If the solute is a solid of low vapour pressure, the decrease in vapour pressure of the liquid is proportional to the concentration of particles of solute; i.e. to the number of dissolved molecules or ions per unit volume. To a first approximation, it does not depend on the nature of the particles. *See* colligative property; Raoult's law.

L-series *See* absolute configuration.

lumen Symbol lm. The SI unit of *luminous flux equal to the flux emitted by a uniform point source of 1 candela in a solid angle of 1 steradian.

luminescence The emission of light by a substance for any reason other than a rise in its temperature. In general, atoms of substances emit *photons of electromagnetic energy when they return to the *ground state after having been in an excited state (*see* excitation). The causes of the excitation are various. If the exciting cause is a photon, the process is called *photoluminescence*; if it is an electron it is called *electroluminescence*. *Chemiluminescence* is luminescence resulting from a chemical reaction (such as the slow oxidation of phosphorus); *bioluminescence* is the luminescence produced by a living organism (such as a firefly). If the luminescence persists significantly after the exciting cause is removed it is called *phosphorescence*; if it does not it is called *fluorescence*. This distinction is arbitrary since there must always be some delay; in some definitions a persistence of more than 10 nanoseconds (10^{-8} s) is treated as phosphorescence.

lutetium Symbol Lu. A silvery metallic element belonging to the *lanthanoids; a.n. 71; r.a.m. 174.97; r.d. 9.842 (20°C); m.p. 1663°C; b.p. 3402°C. Lutetium is the least abundant of the elements and the little quantities that are available have been obtained by processing other metals. There are two natural isotopes, lutetium–175 (stable) and lutetium–176 (half-life 2.2×10^{10} years). There are no uses for the element, which was first identified by G. Urban in 1907.

lux Symbol lx. The SI unit of illuminance equal to the illumination produced by a luminous flux of 1 lumen distributed uniformly over an area of 1 square metre.

lye *See* potassium hydroxide.

Lyman series *See* hydrogen spectrum.

lyophilic Having an affinity for a solvent ('solvent-loving'; if the solvent is water the term *hydrophilic* is used). *See* colloid.

lyophobic Lacking any affinity for a solvent ('solvent-hating'; if the solvent is water the term *hydrophobic* is used). *See* colloid.

lysine *See* amino acid.

M

macromolecular crystal A crystalline solid in which the atoms are all linked together by covalent bonds. Carbon (in diamond), boron nitride, and silicon carbide are examples of substances that have macromolecular crystals. In effect, the crystal is a large molecule (hence the alternative description *giant-molecular*), which accounts for the hardness and high melting point of such materials.

macromolecule A very large molecule. Natural and synthetic polymers have macromolecules, as do such substances as haemoglobin. *See also* colloids.

Magnadur Tradename for a ceramic material used to make permanent magnets. It consists of sintered iron oxide and barium oxide.

Magnalium Tradename for an aluminium-based alloy of high reflectivity for light and ultraviolet radiation that contains 1–2% of copper and between 5% and 30% of magnesium. Strong and light, these alloys also sometimes contain other elements, such as tin, lead, and nickel.

magnesia *See* magnesium oxide.

magnesite A white, colourless, or grey mineral form of *magnesium carbonate, $MgCO_3$, crystallizing in the trigonal system. It is formed as a replacement mineral of magnesium-rich rocks when carbon dioxide is available. Magnesite is mined both as an ore for magnesium and as a source of magnesium carbonate. It occurs in Austria, USA, Greece, Norway, India, Australia, and South Africa.

magnesium Symbol Mg. A silvery metallic element belonging to group II of the periodic table (*see* alkaline-earth metals); a.n. 12; r.a.m. 24.321; r.d. 1.74; m.p. 651°C; b.p. 1107°C. The element is found in a number of minerals, including magnesite ($MgCO_3$), dolomite ($MgCO_3.CaCO_3$), and carnallite ($MgCl_2.KCl.6H_2O$). It is also present in sea water, and it is an *essential element for living organisms. Extraction is by electrolysis of the fused chloride. The element is used in a number of light alloys (e.g. for aircraft). Chemically, it is very reactive. In air it forms a protective oxide coating but when ignited it burns with an intense white flame. It also reacts with the halogens, sulphur, and nitrogen. Magnesium was first isolated by Bussy in 1828.

magnesium bicarbonate *See* magnesium hydrogencarbonate.

magnesium carbonate A white compound, $MgCO_3$, existing in anhydrous and hydrated forms. The anhydrous material (trigonal; r.d. 2.96) is found in the mineral *magnesite. There is also a trihydrate, $MgCO_3.3H_2O$ (rhombic; r.d. 1.85), which occurs naturally as *nesquehonite*, and a pentahydrate, $MgCO_3.5H_2O$ (monoclinic; r.d. 1.73), which occurs as *lansfordite*. Magnesium carbonate also occurs in the mixed salt *dolomite ($CaCO_3.MgCO_3$) and as *basic magnesium carbonate* in the two minerals *artinite* ($MgCO_3.Mg(OH)_2.3H_2O$) and *hydromagnesite* ($3MgCO_3.Mg(OH)_2.3H_2O$). The anhydrous salt can be formed by heating magnesium oxide in a stream of carbon dioxide:

$$MgO(s) + CO_2(g) \rightarrow MgCO_3(s)$$

Above 350°C, the reverse reaction predominates and the carbonate decomposes. Magnesium carbonate is used in making magnesium oxide and is a drying agent (e.g. in table salt). It is also used as a medical antacid and laxative (the basic carbonate is used) and is a component of certain inks and glasses.

magnesium chloride A white solid compound, $MgCl_2$. The anhydrous salt (hexagonal; r.d. 2.32; m.p. 714°C; b.p. 1412°C) can be prepared by the direct combination of dry chlorine with magnesium:

$$Mg(s) + Cl_2(g) \rightarrow MgCl_2(s)$$

The compound also occurs naturally as a constituent of carnallite ($KCl.MgCl_2$). It is a deliquescent compound that commonly forms the hexahydrate, $MgCl_2.6H_2O$ (monoclinic; r.d. 1.57). When heated, this hydrolyses to give magnesium oxide and hydrogen chloride gas. The fused chloride is electrolysed to produce magnesium and it is also used for fireproofing wood, in magnesia cements and artificial leather, and as a laxative.

magnesium hydrogencarbonate (magnesium bicarbonate) A compound, Mg $(HCO_3)_2$, that is stable only in solution. It is formed by the action of carbon dioxide on a suspension of magnesium carbonate in water:

$$MgCO_3(s) + CO_2(g) + H_2O(l) \rightarrow Mg(HCO_3)_2(aq)$$

On heating, this process is reversed. Magnesium hydrogencarbonate is one of the compounds responsible for temporary *hardness in water.

magnesium hydroxide A white solid compound, $Mg(OH)_2$; trigonal; r.d. 2.36; decomposes at 350°C. Magnesium hydroxide occurs naturally as the mineral *brucite* and can be prepared by reacting magnesium sulphate or chloride with sodium hydroxide solution. It is used in the refining of sugar and in the processing of uranium. Medicinally it is important as an antacid (*milk of magnesia*) and as a laxative.

magnesium oxide (magnesia) A white compound, MgO; cubic; r.d. 3.58; m.p. 2800°C. It occurs naturally as the mineral *periclase* and is prepared commercially by thermally decomposing the mineral *magnesite:

$$MgCO_3(s) \rightarrow MgO(s) + CO_2(g)$$

It has a wide range of uses, including reflective coatings on optical instruments and aircraft windscreens and in semiconductors. Its high melting point makes it useful as a refractory lining in metal and glass furnaces.

magnesium peroxide A white solid, MgO_2. It decomposes at 100°C to release oxygen and also releases oxygen on reaction with water:

$$2MgO_2(s) + 2H_2O \rightarrow 2Mg(OH)_2 + O_2$$

The compound is prepared by reacting sodium peroxide with magnesium sulphate solution and is used as a bleach for cotton and silk.

magnesium sulphate A white soluble compound, $MgSO_4$, existing as the anhydrous compound (rhombic; r.d. 2.66; decomposes at 1124°C) and in hydrated crystalline forms. The monohydrate $MgSO_4.H_2O$ (monoclinic; r.d. 2.45) occurs naturally as the mineral *kieserite*. The commonest hydrate is the heptahydrate, $MgSO_4.7H_2O$ (rhombic; r.d. 1.68), which is called *Epsom salt(s)*, and occurs naturally as the mineral *epsomite*. This is a white powder with a bitter saline taste, which loses $6H_2O$ at 150°C and $7H_2O$ at 200°C. It is used in sizing and fireproofing cotton and silk, in tanning leather, and in the manufacture of fertilizers, explosives, and matches. In medicine, it is used as a laxative. It is also used in veterinary medicine for treatment of local inflammations and infected wounds.

magnetic moment The ratio between the maximum torque (T_{max}) exerted on a magnet, current-carrying coil, or moving charge situated in a magnetic field and the strength of that field. It is thus a measure of the strength of a magnet or current-carrying coil. In the Sommerfeld approach this quantity (also called *electromagnetic moment* or *magnetic area moment*) is the ratio T_{max}/B. In the Kennelly approach the quantity (also called *magnetic dipole moment*) is T_{max}/H.

In the case of a magnet placed in a magnetic field of field strength H, the maximum torque T_{max} occurs when the axis of the magnet is perpendicular to the field. In the case of a coil of N turns and area A carrying a current I, the magnetic moment can be shown to be $m = T/B = NIA$ or $m = T/H = \mu NIA$. Magnetic moments are measured in $A\,m^2$.

An orbital electron has an orbital magnetic moment IA, where I is the equivalent current as the electron moves round its orbit. It is given by $I = q\omega/2\pi$, where q is the electronic charge and ω is its angular velocity. The orbital magnetic moment is therefore $IA = q\omega A/2\pi$, where A is the orbital area. If the electron is spinning there is also a spin magnetic moment (*see* spin); atomic nuclei also have magnetic moments (*see* nuclear moment).

magnetic quantum number *See* atom.

magnetism A group of phenomena associated with magnetic fields. Whenever an electric current flows a magnetic field is produced; as the orbital motion and the *spin of atomic electrons are equivalent to tiny current loops, individual atoms create magnetic fields around them, when their orbital electrons have a net *magnetic moment as a result of their angular momentum. The magnetic moment of an atom is the vector sum of the magnetic moments of the orbital motions and the spins of all the electrons in the atom. The macroscopic magnetic properties of a substance arise from the magnetic moments of its component atoms and molecules. Different materials have different characteristics in an applied magnetic field; there are four main types of magnetic behaviour:

(a) In *diamagnetism* the magnetization is in the opposite direction to that of the applied field, i.e. the susceptibility is negative. Although all substances are diamagnetic, it is a weak form of magnetism and may be masked by other, stronger, forms. It results from changes induced in the orbits of electrons in the atoms of a substance by the applied field, the direction of the change opposing the applied flux. There is thus a weak negative susceptibility (of the order of $-10^{-8}\ m^3\ mol^{-1}$) and a relative permeability of slightly less than one.

(b) In *paramagnetism* the atoms or molecules of the substance have net orbital or spin magnetic moments that are capable of being aligned in the direction of the applied field. They therefore have a positive (but small) susceptibility and a relative permeability slightly in excess of one. Paramagnetism occurs in all atoms and molecules with unpaired electrons; e.g. free atoms, free radicals, and compounds of transition metals containing ions with unfilled electron shells. It also occurs in metals as a result of the magnetic moments associated with the spins of the conducting electrons.

(c) In *ferromagnetic* substances, within a certain temperature range, there are net atomic magnetic moments, which line up in such a way that magnetization persists after the removal of the applied field. Below a certain temperature, called the *Curie point* (or Curie temperature) an increasing magnetic field applied to a ferromagnetic substance will cause increasing magnetization to a high value, called the *saturation magnetization*. This is because a ferromagnetic substance consists of small (1–0.1 mm across) magnetized regions called *domains*. The total magnetic moment of a sample of the substance is the vector sum of the magnetic moments of the component domains. Within each domain the individual atomic magnetic moments are spontaneously aligned by *exchange forces*, related to whether or not the atomic electron spins are parallel or antiparallel. However, in an unmagnetized piece of ferromagnetic material the magnetic moments of the domains themselves are not aligned; when an external field is applied those domains that are aligned with the field increase in size at the expense of the others. In a very strong field all the domains are lined up in the direction of the field and provide the high observed magnetization. Iron, nickel, cobalt, and their alloys are ferromagnetic. Above the Curie point, ferromagnetic materials become paramagnetic.

(d) Some metals, alloys, and transition-element salts exhibit another form of magnetism called *antiferromagnetism*. This occurs below a certain temperature, called the *Néel temperature*, when an ordered array of atomic magnetic moments spontaneously forms in which alternate moments have opposite directions. There is therefore no net resultant magnetic moment in the absence of an applied field. In manganese fluoride, for example, this antiparallel arrangement occurs below a Néel temperature of 72 K. Below this temperature the spontaneous ordering opposes the normal tendency of the magnetic moments to align with the applied field. Above the Néel temperature the substance is paramagnetic.

A special form of antiferromagnetism is *ferrimagnetism*, a type of magnetism exhibited by the *ferrites. In these materials the magnetic moments of adjacent ions are antiparallel and of unequal strength, or the number of magnetic moments in one direction is greater than those in the opposite direction. By suitable choice of rare-earth ions in the ferrite lattices it is possible to design ferrimagnetic substances with specific magnetizations for use in electronic components.

magnetite A black mineral form of iron oxide crystallizing in the cubic system. It is a mixed iron(II)-iron(III) oxide, Fe_3O_4, and is one of the major ores of iron. It is strongly magnetic and some varieties, known as *lodestone*, are natural magnets; these were used as compasses in the ancient world. Magnetite is widely distributed and occurs as an accessory mineral in almost all igneous and metamorphic rocks. The largest deposits of the mineral occur in N Sweden.

magnetochemistry The branch of physical chemistry concerned with measuring and investigating the magnetic properties of compounds. It is used particularly for studying transition-metal complexes, many of which are paramagnetic because they have unpaired electrons. Measurement of the magnetic susceptibility allows the magnetic moment of the metal atom to be calculated, and this gives information about the bonding in the complex.

magneton A unit for measuring magnetic moments of nuclear, atomic, or molecular magnets. The *Bohr magneton* μ_B has the value of the classical magnetic moment of an electron, given by

$$\mu_B = eh/4\pi m_e = 9.274 \times 10^{-24} \text{ A m}^2,$$

where e and m_e are the charge and mass of the electron and h is the Planck constant. The *nuclear magneton*, μ_N is obtained by replacing the mass of the electron by the mass of the proton and is therefore given by

$$\mu_N = \mu_B . m_e/m_p = 5.05 \times 10^{-27} \text{ A m}^2.$$

malachite A secondary mineral form of copper carbonate–hydroxide, $CuCO_3.Cu(OH)_2$. It is bright green and crystallizes in the monoclinic system but usually occurs as aggregates of fibres or in massive form. It is generally found with *azurite in association with the more important copper ores and is itself mined as an ore of copper (e.g. in Zaïre). It is also used as an ornamental stone and as a gemstone.

maleic acid *See* butenedioic acid.

malic acid (2-hydroxybutanedioic acid) A crystalline solid, HOOCCH(OH)CH_2COOH. L-malic acid occurs in living organisms as an intermediate metabolite in the *Krebs cycle and also (in certain plants) in photosynthesis. It is found especially in the juice of unripe fruits, e.g. green apples.

maltose (malt sugar) A sugar consisting of two linked glucose molecules that results from the action of the enzyme amylase on starch. Maltose occurs in barley seeds following germination and drying, which is the basis of the malting process used in the manufacture of beer and malt whisky.

malt sugar *See* maltose.

manganate(VI) A salt containing the ion $MnO_4{}^{2-}$. Manganate(VI) ions are dark green; they are produced by manganate(VII) ions in basic solution.

manganate(VII) (permanganate) A salt containing the ion $MnO_4{}^{-}$. Manganate(VII) ions are dark purple and strong oxidizing agents.

manganese Symbol Mn. A grey brittle metallic *transition element, a.n. 25; r.a.m. 54.94; r.d. 7.4; m.p. 1244°C; b.p. 2040°C. The main sources are pyrolusite (MnO_2) and rhodochrosite ($MnCO_3$). The metal can be extracted by reduction of the oxide using magnesium (*Kroll process) or aluminium (*Goldschmidt process). Often the ore is mixed with iron ore and reduced in an electric furnace to produce ferromanganese for use in alloy steels. The element is quite electropositive, reacting with water and dilute acids to give hydrogen. It combines with oxygen, nitrogen, and other metals when heated. Salts of manganese contain the element in the +2 and +3 oxidation states. Manganese(II) salts are the more stable. It also forms compounds in higher oxidation states, such as manganese(IV) oxide and manganate(VI) and manganate(VII) salts. The element was discovered in 1774 by Scheele.

manganese(IV) oxide (manganese dioxide) A black oxide made by heating manganese(II) nitrate. The compound also occurs naturally as pyrolusite. It is a strong oxidizing agent, used as a depolarizing agent in voltaic cells.

manganic compounds Compounds of manganese in its +3 oxidation state; e.g. manganic oxide is manganese(III) oxide, Mn_2O_3.

manganin A copper alloy containing 13–18% of manganese and 1–4% of nickel. It has a high electrical resistance, which is relatively insensitive to temperature changes. It is therefore suitable for use in resistance wire.

manganous compounds Compounds of manganese in its +2 oxidation state; e.g. manganous oxide is manganese(II) oxide, MnO.

mannitol A polyhydric alcohol, $CH_2OH(CHOH)_4CH_2OH$, found in some plants, used as a sweetener in certain foodstuffs.

mannose A *monosaccharide, $C_6H_{12}O_6$, stereoisomeric with glucose, that occurs naturally only in polymerized forms called *mannans*. These are found in plants, fungi, and bacteria, serving as food energy stores.

manometer A device for measuring pressure differences, usually by the difference in height of two liquid columns. The simplest type is the U-tube manometer, which consists of a glass tube bent into the shape of a U. If a pressure to be measured is fed to one side of the U-tube and the other is open to the atmosphere, the difference in level of the liquid in the two limbs gives a measure of the unknown pressure.

marble A metamorphic rock composed of recrystallized *calcite or *dolomite. Pure marbles are white but such impurities as silica or clay minerals result in variations of colour. Marble is extensively used for building purposes and ornamental use; the pure white marble from Carrara in Italy is especially prized by sculptors. The term is applied commercially to any limestone or dolomite that can be cut and polished.

Markovnikoff's rule When an acid HA adds to an alkene, a mixture of products can be formed if the alkene is not symmetrical. For instance, the reaction between $C_2H_5CH{:}CH_2$ and HCl can give $C_2H_5CH_2CH_2Cl$ or $C_2H_5CHClCH_3$. In general, a mixture of products occurs in which one predominates over the other. In

1870, W. Markovnikoff proposed the rule that the main product would be the one in which the hydrogen atom adds to the carbon having the larger number of hydrogen atoms (the latter product above). This occurs when the mechanism is *electrophilic addition, in which the first step is addition of H^+. The electron-releasing effect of the alkyl group (C_2H_5) distorts the electron-distribution in the double bond, making the carbon atom furthest from the alkyl group negative. This is the atom attacked by H^+ giving the carbonium ion $C_2H_5C^+HCH_3$, which further reacts with the negative ion Cl^-.

In some circumstances *anti-Markovnikoff* behaviour occurs, in which the opposite effect is found. This happens when the mechanism involves free radicals and is common in addition of hydrogen bromide when peroxides are present.

marsh gas Methane formed by rotting vegetation in marshes.

Marsh's test A chemical test for arsenic in which hydrochloric acid and zinc are added to the sample, arsine being produced by the nascent hydrogen generated. Gas from the sample is led through a heated glass tube and, if arsine is present, it decomposes to give a brown deposit of arsenic metal. The arsenic is distinguished from antimony (which gives a similar result) by the fact that antimony does not dissolve in sodium chlorate(I) (hypochlorite). The test was devised in 1836 by the British chemist James Marsh.

martensite A solid solution of carbon in alpha-iron (*see* iron) formed when *steel is cooled too rapidly for pearlite to form from austenite. It is responsible for the hardness of quenched steel.

mascagnite A mineral form of *ammonium sulphate, $(NH_4)_2SO_4$.

mass A measure of a body's *inertia, i.e. its resistance to acceleration. According to Newton's laws of motion, if two unequal masses, m_1 and m_2, are allowed to collide, in the absence of any other forces both will experience the same force of collision. If the two bodies acquire accelerations a_1 and a_2 as a result of the collision, then $m_1a_1 = m_2a_2$. This equation enables two masses to be compared. If one of the masses is regarded as a standard of mass, the mass of all other masses can be measured in terms of this standard. The body used for this purpose is a 1-kg cylinder of platinum–iridium alloy, called the international standard of mass.

mass action The law of mass action states that the rate at which a chemical reaction takes place at a given temperature is proportional to the product of the *active masses* of the reactants. The active mass of a reactant is taken to be its molar concentration. For example, for a reaction

$$A + B \rightarrow C$$

the rate is given by

$$R = k[A][B]$$

where k is the *rate constant. The principle was introduced by C. M. Guldberg and P. Waage in 1863. It is strictly correct only for ideal gases. In real cases *activities can be used.

mass concentration *See* concentration.

massicot *See* lead(II) oxide.

mass number *See* nucleon number.

mass spectrum *See* spectrum.

masurium A former name for *technetium.

Maxwell–Boltzmann distribution A law describing the distribution of speeds among the molecules of a gas. In a system consisting of N molecules that are independent of each other except that they exchange energy on collision, it is clearly impossible to say what velocity any particular molecule will have. However, statistical statements regarding certain functions of the molecules were worked out by James Clerk Maxwell (1831–79) and Ludwig Boltzmann (1844–1906). One form of their law states that $n = N\exp(-E/RT)$, where n is the number of molecules with energy in excess of E, T is the thermodynamic temperature, and R is the *gas constant.

Mcleod gauge A vacuum pressure gauge in which a relatively large volume of a low-pressure gas is compressed to a small volume in a glass apparatus. The volume is reduced to an extent that causes the pressure to rise sufficiently to support a column of fluid high enough to read. This simple device, which relies on *Boyle's law, is suitable for measuring pressures in the range 10^3 to 10^{-3} pascal.

mean free path The average distance travelled between collisions by the molecules in a gas, the electrons in a metallic crystal, the neutrons in a moderator, etc. According to the *kinetic theory the mean free path between elastic collisions of gas molecules of diameter d (assuming they are rigid spheres) is $1/\sqrt{2}n\pi d^2$, where n is the number of molecules per unit volume in the gas. As n is proportional to the pressure of the gas, the mean free path is inversely proportional to the pressure.

mean free time The average time that elapses between the collisions of the molecules in a gas, the electrons in a crystal, the neutrons in a moderator, etc. *See* mean free path.

mega- Symbol M. A prefix used in the metric system to denote one million times. For example, 10^6 volts = 1 megavolt (MV).

melamine A white crystalline compound, $C_3N_6H_6$. Melamine is a cyclic compound having a six-membered ring of alternating C and N atoms, with three NH_2 groups. It can be copolymerized with methanal to give thermosetting *melamine resins*, which are used particularly for laminated coatings.

melting point (m.p.) The temperature at which a solid changes into a liquid. A pure substance under standard conditions of pressure (usually 1 atmosphere) has a single reproducible melting point. If heat is gradually and uniformly supplied to a solid the consequent rise in temperature stops at the melting point until the fusion process is complete.

Mendeleev's law *See* periodic law.

mendelevium Symbol Md. A radioactive metallic transuranic element belonging to the *actinoids; a.n. 101; mass number of the only known isotope 256 (half-life 1.3 hours). It was first identified by A. Ghiorso, G. T. Seaborg, and associates in 1955. The alternative name *unnilunium* has been proposed.

Mendius reaction A reaction in which an organic nitrile is reduced by nascent hydrogen (e.g. from sodium in ethanol) to a primary amine:

$$RCN + 2H_2 \rightarrow RCH_2NH_2$$

menthol A white crystalline terpene alcohol, $C_{10}H_{19}OH$; r.d. 0.89; m.p. 42°C; b.p. 212°C. It has a minty taste and is found in certain essential oils (e.g. peppermint) and used as a flavouring.

mercaptans *See* thiols.

mercapto group *See* thiols.

mercuric compounds Compounds of mercury in its +2 oxidation state; e.g. mercuric chloride is mercury(II) chloride, $HgCl_2$.

mercurous compounds Compounds of mercury in its +1 oxidation state; e.g. mercury(I) chloride is mercurous chloride, HgCl.

mercury Symbol Hg. A heavy silvery liquid metallic element belonging to the *zinc group; a.n. 80; r.a.m. 200.59; r.d. 13.55; m.p. −38.87°C; b.p. 356.58°C. The main ore is the sulphide cinnabar (HgS), which can be decomposed to the elements. Mercury is used in thermometers, barometers, and other scientific apparatus, and in dental amalgams. The element is less reactive than zinc and cadmium and will not displace hydrogen from acids. It is also unusual in forming mercury(I) compounds containing the Hg_2^{2+} ion, as well as mercury(II) compounds containing Hg^{2+} ions. It also forms a number of complexes and organomercury compounds (e.g. the *Grignard reagents).

mercury cell A primary *voltaic cell consisting of a zinc anode and a cathode of mercury(II) oxide (HgO) mixed with graphite. The electrolyte is potassium hydroxide (KOH) saturated with zinc oxide, the overall reaction being:

$$Zn + HgO \rightarrow ZnO + Hg$$

The e.m.f. is 1.35 volts and the cell will deliver about 0.3 ampere-hour per cm^3.

mercury(I) chloride A white salt, Hg_2Cl_2; r.d. 7.0; m.p. 302°C; b.p. 384°C. It is made by heating mercury(II) chloride with mercury and is used in calomel cells (so called because the salt was formerly called *calomel*) and as a fungicide.

mercury(II) chloride A white salt, $HgCl_2$; r.d. 5.4; m.p. 276°C; b.p. 303°C. It is made by reacting mercury with chlorine and used in making other mercury compounds.

mercury(II) oxide A yellow or red oxide of mercury, HgO. The red form is made by heating mercury in oxygen at 350°C; the yellow form, which differs from the red in particle size, is precipitated when sodium hydroxide solution is added to a solution of mercury(II) nitrate. Both forms decompose to the elements at high temperature. The black precipitate formed when sodium hydroxide is added to mercury(I) nitrate solution is sometimes referred to as mercury(I) oxide (Hg_2O) but is probably a mixture of HgO and free mercury.

mercury(II) sulphide A red or black compound, HgS, occurring naturally as the minerals cinnabar (red) and metacinnabar (black). It can be obtained as a black precipitate by bubbling hydrogen sulphide through a solution of mercury(II) nitrate. The red form is obtained by sublimation. The compound is also called *vermilion* (used as a pigment).

meso form *See* optical activity.

mesomerism A former name for *resonance in molecules.

meta- 1. Prefix designating a benzene compound in which two substituents are in the 1,3 positions on the benzene ring. The abbreviation *m-* is used; for example, *m*-xylene is 1,3-dimethylbenzene. *Compare* ortho-; para-. **2.** Prefix designating a lower oxo acid, e.g. metaphosphoric acid. *Compare* ortho-.

metabolism The sum of the chemical reactions that occur within living organisms. The various compounds that take part in or are formed by these reactions are called *metabolites*. In animals many metabolites are obtained by the digestion of food, whereas in plants only the basic starting materials (carbon dioxide, water, and minerals) are externally derived. The synthesis (*anabolism) and breakdown (*catabolism) of most compounds occurs by a number of reaction steps, the reaction sequence being termed a *metabolic pathway*. Some pathways (e.g.

*glycolysis) are linear; others (e.g. the *Krebs cycle) are cyclic.

metaboric acid *See* boric acid.

metal Any of a class of chemical elements that are typically lustrous solids that are good conductors of heat and electricity. Not all metals have all these properties (e.g. mercury is a liquid). In chemistry, metals fall into two distinct types. Those of the *s*- and *p*-blocks (e.g. sodium and aluminium) are generally soft silvery reactive elements. They tend to form positive ions and so are described as electropositive. This is contrasted with typical nonmetallic behaviour of forming negative ions. The *transition elements (e.g. iron and copper) are harder substances and generally less reactive. They form coordination complexes. All metals have oxides that are basic.

metaldehyde A solid compound, $C_4O_4H_4(CH_3)_4$, formed by polymerization of ethanal (acetaldehyde) in dilute acid solutions below 0°C. The compound, a tetramer of ethanal, is used in slug pellets and as a fuel for portable stoves.

metal fatigue A cumulative effect causing a metal to fail after repeated applications of stress, none of which exceeds the ultimate tensile strength. The *fatigue strength* (or *fatigue limit*) is the stress that will cause failure after a specified number (usually 10^7) of cycles. The number of cycles required to produce failure decreases as the level of stress or strain increases. Other factors, such as corrosion, also reduce the fatigue life.

metallic bond A chemical bond of the type holding together the atoms in a solid metal or alloy. In such solids, the atoms are considered to be ionized, with the positive ions occupying lattice positions. The valence electrons are able to move freely (or almost freely) through the lattice, forming an 'electron gas'. The bonding force is electrostatic attraction between the positive metal ions and the electrons. The existence of free electrons accounts for the good electrical and thermal conductivities of metals. *See also* energy bands.

metallic crystal A crystalline solid in which the atoms are held together by *metallic bonds. Metallic crystals are found in some *interstitial compounds as well as in metals and alloys.

metallized dye *See* dyes.

metallocene A type of organometallic complex in which one or more aromatic rings (e.g. $C_5H_5^-$ or C_6H_6) coordinate to a metal ion or atom by the pi electrons of the ring. *Ferrocene was the first such compound to be discovered.

metallography The microscopic study of the structure of metals and their alloys. Both optical microscopes and electron microscopes are used in this work.

metalloid (semimetal) Any of a class of chemical elements intermediate in properties between metals and nonmetals. The classification is not clear cut, but typical metalloids are boron, silicon, germanium, arsenic, and tellurium. They are electrical semiconductors and their oxides are amphoteric.

metallurgy The branch of applied science concerned with the production of metals from their ores, the purification of metals, the manufacture of alloys, and the use and performance of metals in engineering practice. *Process metallurgy* is concerned with the extraction and production of metals, while *physical metallurgy* concerns the mechanical behaviour of metals.

metamict state The amorphous state of a substance that has lost its crystalline structure as a result of the radioactivity of uranium or thorium. *Metamict minerals* are minerals whose structure has been disrupted by this process. The metamictization is caused by alpha-particles and the recoil nuclei from radioactive disintegration.

metaphosphoric acid *See* phosphoric(VI) acid.

metaplumbate *See* plumbate.

metastable state A condition of a system in which it has a precarious stability that can easily be disturbed. It is unlike a state of stable equilibrium in that a minor disturbance will cause a system in a metastable state to fall to a lower energy level. A book lying on a table is in a state of stable equilibrium; a thin book standing on edge is in metastable equilibrium. Supercooled water is also in a metastable state. It is liquid below 0°C; a grain of dust or ice introduced into it will cause it to freeze. An excited state of an atom or nucleus that has an appreciable lifetime is also metastable.

metastannate *See* stannate.

metathesis *See* double decomposition.

methacrylate A salt or ester of methacrylic acid (2-methylpropenoic acid).

methacrylate resins *Acrylic resins obtained by polymerizing 2-methylpropenoic acid or its esters.

methacrylic acid *See* 2-methylpropenoic acid.

methanal (formaldehyde) A colourless gas, HCHO; r.d. 0.815 (at −20°C); m.p. −92°C; b.p. −19°C. It is the simplest *aldehyde, made by the catalytic oxidation of methanol (500°C; silver catalyst) by air. It forms two polymers: *methanal trimer and polymethanal. *See also* formalin.

methanal trimer A cyclic trimer of methanal, $C_3O_3H_6$, obtained by distillation of an acidic solution of methanal. It has a six-membered ring of alternating $-O-$ and $-CH_2-$ groups.

methane A colourless odourless gas, CH_4; m.p. −182.5°C; b.p. −161°C. Methane is the simplest hydrocarbon, being the first member of the *alkane series. It is the main constituent of natural gas (~99%) and as such is an important raw material for producing other organic compounds. It can be converted into methanol by catalytic oxidation.

methanide *See* carbide.

methanoate (formate) A salt or ester of methanoic acid.

methanoic acid (formic acid) A colourless pungent liquid, HCOOH; r.d. 1.2; m.p. 8°C; b.p. 101°C. It can be made by the action of concentrated sulphuric acid on the sodium salt (sodium methanoate), and occurs naturally in ants and stinging nettles. Methanoic acid is the simplest of the *carboxylic acids.

methanol (methyl alcohol) A colourless liquid, CH_3OH; r.d. 0.79; m.p. −98°C; b.p. 64°C. It is made by catalytic oxidation of methane (from natural gas) using air. Methanol is used as a solvent (*see* methylated spirits) and as a raw material for making methanal (mainly for urea–formaldehyde resins). It was formerly made by the dry distillation of wood (hence the name *wood alcohol*).

methionine *See* amino acid.

methoxy group The organic group CH_3O-.

methyl acetate *See* methyl ethanoate.

methyl alcohol *See* methanol.

methylamine A colourless flammable gas, CH_3NH_2; m.p. −92°C; b.p. −7°C. It can be made by a catalytic reaction between methanol and ammonia and is used in the manufacture of other organic chemicals.

methylated spirits A mixture consisting mainly of ethanol with added methanal (~9.5%), pyridine (~0.5%), and blue dye. The additives are included to make the ethanol undrinkable so that it can be sold without excise duty for use as a solvent and a fuel (for small spirit stoves).

methylation A chemical reaction in which a methyl group (CH_3-) is introduced in a molecule. A particular example is the replacement of a hydrogen atom by a methyl group, as in a *Friedel–Crafts reaction.

methylbenzene (toluene) A colourless liquid, $CH_3C_6H_5$; r.d. 0.9; m.p. −94°C; b.p. 111°C. Methylbenzene is derived from benzene by replacement of a hydrogen atom by a methyl group. It can be obtained from coal tar or made from methylcyclohexane (extracted from crude oil) by catalytic dehydrogenation. Its main uses are as a solvent and as a raw material for producing TNT.

methyl bromide *See* bromomethane.

2-methylbuta-1,3-diene *See* isoprene.

methyl chloride *See* chloromethane.

methylene The highly reactive *carbene, $:CH_2$. The divalent CH_2 group in a compound is the *methylene group*.

methyl ethanoate (methyl acetate) A colourless volatile fragrant liquid, CH_3COOCH_3; r.d. 0.92; m.p. −98°C; b.p. 54°C. A typical *ester, it can be made from methanol and methanoic acid and is used mainly as a solvent.

methyl ethyl ketone *See* butanone.

methyl group *or* **radical** The organic group CH_3-.

methyl methacrylate An ester of methacrylic acid (2-methylpropenoic acid), $CH_2{:}C(CH_3)COOCH_3$, used in making *methacrylate resins.

methyl orange An organic dye used as an acid–base *indicator. It changes from red below pH 3.1 to yellow above pH 4.4 (at 25°C) and is used for titrations involving weak bases.

methylphenols (cresols) Organic compounds having a methyl group and a hydroxyl group bound directly to a benzene ring. There are three isomeric methylphenols with the formula $CH_3C_6H_4OH$, differing in the relative positions of the methyl and hydroxyl groups. A mixture of the three can be obtained by distilling coal tar and is used as a germicide and antiseptic.

2-methylpropenoic acid (methacrylic acid) A white crystalline unsaturated carboxylic acid, $CH_2{:}C(CH_3)COOH$, used in making *methacrylate resins.

methyl red An organic dye similar in structure and use to methyl orange. It changes from red below pH 4.4 to yellow above pH 6.0 (at 25°C).

metre Symbol m. The SI unit of length, being the length of the path travelled by light in vacuum during a time interval of $1/(2.99\,792\,458 \times 10^8)$ second. This definition, adopted by the General Conference on Weights and Measures in October, 1983, replaced the 1967 definition based on the krypton lamp, i.e. 1 650 763.73 wavelengths in a vacuum of the radiation corresponding to the transition between the levels $2p^{10}$ and $5d^5$ of the nuclide krypton–86. This definition (in 1958) replaced the older definition of a metre based on a platinum–iridium bar of standard length. When the *metric system was introduced in 1791 in France, the metre was intended to be one ten-millionth of the earth's meridian quadrant passing through Paris. However, the original geodetic surveys proved the impracticability of such a standard and the original platinum metre bar, the *mètre des archives*, was constructed in 1793.

metric system A decimal system of units originally devised by a committee of the French Academy, which included J. L. Lagrange and P. S. Laplace, in 1791. It was based on the *metre, the gram defined in terms of the mass of a cubic centimetre of water, and the second. This centimetre-gram-second system (*see* c.g.s. units) later gave way for scientific work to the metre-kilogram-second system (*see* m.k.s. units) on which *SI units are based.

mica Any of a group of silicate minerals with a layered structure. Micas are composed of linked SiO_4 tetrahedra with cations and hydroxyl groupings between the layers. The general formula is $X_2Y_{4-6}Z_8O_{20}(OH,F)_4$, where X = K,Na,Ca; Y = Al,Mg,Fe,Li; and Z = Si,Al. The three main mica minerals are

*muscovite, $K_2Al_4(Si_6Al_2O_{20})(OH,F)_4$

*biotite, $K_2(Mg,Fe^{2+}$

lepidolite, $K_2(Li,Al)_{5-6}(Si_{6-7}Al_{2-1}O_{20})(OH,F)_4$.

Micas have perfect basal cleavage and the thin cleavage flakes are flexible and elastic. Flakes of mica are used as electrical insulators and as the dielectric in capacitors.

micelle An aggregate of molecules in a *colloid. For example, when soap or other *detergents dissolve in water they do so as micelles – small clusters of molecules in which the nonpolar hydrocarbon groups are in the centre and the hydrophilic polar groups are on the outside solvated by the water molecules.

micro- Symbol μ. A prefix used in the metric system to denote one millionth. For example, 10^{-6} metre = 1 micrometre (μm).

microbalance A sensitive *balance capable of weighing masses of the order 10^{-6} to 10^{-9} kg.

microscope A device for forming a magnified image of a small object. The *simple microscope* consists of a biconvex magnifying glass or an equivalent system of lenses, either hand-held or in a simple frame. The *compound microscope* uses two lenses or systems of lenses, the second magnifying the real image formed by the first. The lenses are usually mounted at the opposite ends of a tube that has mechanical controls to move it in relation to the object. An optical condenser and mirror, often with a separate light source, provide illumination of the object. The widely used *binocular microscope* consists of two separate instruments fastened together so that one eye looks through one while the other eye looks through the other. This gives stereoscopic vision. *See also* electron microscope.

migration 1. The movement of a group, atom, or double bond from one part of a molecule to another. **2.** The movement of ions under the influence of an electric field.

milk of magnesia *See* magnesium hydroxide.

milk sugar *See* lactose.

milli- Symbol m. A prefix used in the metric system to denote one thousandth. For example, 0.001 volt = 1 millivolt (mV).

mineral A naturally occurring substance that has a characteristic chemical composition and, in general, a crystalline structure. The term is also often applied generally to organic substances that are obtained by mining (e.g. coal, petroleum, and natural gas) but strictly speaking these are not minerals, being complex mixtures without definite chemical formulas. Rocks are composed of mixtures of minerals. Minerals may be identified by the properties of their crystal system, hardness (measured on the Mohs' scale), relative density, lustre, colour, cleavage, and fracture. Many names of minerals end in *-ite*.

mineral acid A common inorganic acid, such as hydrochloric acid, sulphuric acid, or nitric acid.

mirabilite A mineral form of *sodium sulphate, $Na_2SO_4.10H_2O$.

misch metal An alloy of cerium (50%), lanthanum (25%), neodymium (18%), praseodymium (5%), and other rare earths. It is used alloyed with iron (up to 30%) in lighter flints, and in small quantities to improve the malleability of iron. It is also added to copper alloys to make them harder, to aluminium alloys to make them stronger, to magnesium alloys to reduce creep, and to nickel alloys to reduce oxidation.

Mitscherlich's law (law of isomorphism) Substances that have the same crystal structure have similar chemical formulae. The law can be used to determine the formula of an unknown compound if it is isomorphous with a compound of known formula.

mixture A system of two or more distinct chemical substances. *Homogeneous mixtures* are those in which the atoms or molecules are interspersed, as in a mixture of gases or in a solution. *Heterogeneous mixtures* have distinguishable phases, e.g. a mixture of iron filings and sulphur. In a mixture there is no redistribution of valence electrons, and the components retain their individual chemical properties. Unlike compounds, mixtures can be separated by physical means (distillation, crystallization, etc.).

m.k.s. units A *metric system of units devised by A. Giorgi (and sometimes known as *Giorgi units*) in 1901. It is based on the metre, kilogram, and second and grew from the earlier *c.g.s. units. The electrical unit chosen to augment these three basic units was the ampere and the permeability of space (magnetic constant) was taken as 10^{-7} Hm^{-1}. To simplify electromagnetic calculations the magnetic constant was later changed to $4\pi \times 10^{-7}$ Hm^{-1} to give the *rationalized MKSA system*. This system, with some modifications, formed the basis of *SI units, now used in all scientific work.

mmHg A unit of pressure equal to that exerted under standard gravity by a height of one millimetre of mercury, or 133.322 pascals.

molal concentration *See* concentration.

molality *See* concentration.

molar Denoting that an extensive physical property is being expressed per *amount of substance, usually per mole. For example, the molar heat capacity of a compound is the heat capacity of that compound per unit amount of substance; in SI units it would be expressed in $J\ K^{-1}\ mol^{-1}$.

molar conductivity Symbol Λ. The conductivity of that volume of an electrolyte that contains one mole of solution between electrodes placed one metre apart.

molar heat capacity *See* heat capacity.

molarity *See* concentration.

molar volume (molecular volume) The volume occupied by a substance per unit amount of substance.

mole Symbol mol. The SI unit of *amount of substance. It is equal to the amount of substance that contains as many elementary units as there are atoms in 0.012 kg of carbon-12. The elementary units may be atoms, molecules, ions, radicals, electrons, etc., and must be specified. 1 mole of a compound has a mass equal to its *relative molecular mass expressed in grams.

molecular beam A beam of atoms, ions, or molecules at low pressure, in which all the particles are travelling in the same direction and there are few collisions between them. They are formed by allowing a gas or vapour to pass through an aperture into an enclosure, which acts as a collimator by containing several additional apertures and vacuum pumps to remove any particles that do not pass through the apertures. Molecular beams are used in studies of surfaces and chemical reactions and in spectroscopy.

molecular distillation Distillation in high vacuum (about 0.1 pascal) with the condensing surface so close to the surface of the evaporating liquid that the molecules of the liquid travel to the condensing surface without collisions. This technique enables very much lower temperatures to be used than are used with distillation at atmospheric pressure and therefore heat-sensitive substances can be distilled. Oxidation of the distillate is also eliminated as there is no oxygen present.

molecular formula *See* formula.

molecularity The number of molecules involved in forming the activated complex in a step of a chemical reaction. Reactions are said to be *unimolecular*, *bimolecular*, or *trimolecular* according to whether 1, 2, or 3 molecules are involved.

molecular orbital *See* orbital.

molecular sieve Porous crystalline substances, especially aluminosilicates (*see* zeolite), that can be dehydrated with little change in crystal structure. As they form regularly spaced cavities, they provide a high surface area for the adsorption of smaller molecules. The general formula of these substances is $M_nO.Al_2O_3.xSiO_2.yH_2O$, where M is a metal ion and n is twice the reciprocal of its valency. Molecular sieves are used as drying agents and in the separation and purification of fluids. They can also be loaded with chemical substances, which remain separated from any reaction that is taking place around them, until they are released by heating or by displacement with a more strongly adsorbed substance. They can thus be used as cation exchange mediums and as catalysts and catalyst supports.

molecular volume *See* molar volume.

molecular weight *See* relative molecular mass.

molecule One of the fundamental units forming a chemical compound; the smallest part of a chemical compound that can take part in a chemical reaction. In most covalent compounds, molecules consist of groups of atoms held together by covalent or coordinate bonds. Covalent substances that form *macromolecular crystals have no discrete molecules (in a sense, the whole crystal is a molecule). Similarly, ionic compounds do not have single molecules, being collections of oppositely charged ions.

mole fraction Symbol X. A measure of the amount of a component in a mixture. The mole fraction of component A is given by $X_A = n_A/N$, where n_A is the amount of substance of A (for a given entity) and N is the total amount of substance of the mixture (for the same entity).

Molisch's test *See* alpha-naphthol test.

molybdenum Symbol Mo. A silvery hard metallic *transition element; a.n. 42; r.a.m. 95.94; r.d. 10.22; m.p. 2610°C; b.p. 5560°C. It is found in molybdenite (MoS_2), the metal being extracted by roasting to give the oxide, followed by reduction with hydrogen. The element is used in alloy steels. Molybdenum(IV) sulphide (MoS_2) is used as a lubricant. Chemically, it is unreactive, being unaffected by most acids. It oxidizes at high temperatures and can be dissolved in molten alkali to give a range of molybdates and polymolybdates. Molybdenum was discovered in 1778 by Scheele.

Mond process A method of obtaining pure nickel by heating the impure metal in a stream of carbon monoxide at 50–60°C. Volatile nickel carbonyl ($Ni(CO)_4$) is formed, and this can be decomposed at higher temperatures (180°C) to give pure nickel. The method was invented by the German–British chemist Ludwig Mond (1839–1909).

Monel metal An alloy of nickel (60–70%), copper (25–35%), and small quantities of iron, manganese, silicon, and carbon. It is used to make acid-resisting equipment in the chemical industry.

monoclinic *See* crystal system.

monohydrate A crystalline compound having one molecule of water per molecule of compound.

monomer A molecule (or compound) that joins with others in forming a dimer, trimer, or polymer.

monosaccharide (simple sugar) A carbohydrate that cannot be split into smaller units by the action of dilute acids. Monosaccharides are classified according to the number of carbon atoms they possess: *trioses* have three carbon atoms; *tetroses*, four; *pentoses*, five; *hexoses*, six; etc. Each of these is further divided into *aldoses* and *ketoses*, depending on whether the molecule contains an aldehyde group (–CHO) or a ketone group (–CO–). For example glucose, having six carbon atoms and an aldehyde group, is an *aldohexose* whereas fructose is a *ketohexose*. These aldehyde and ketone groups confer reducing properties on monosaccharides: they can be oxidized to yield sugar acids. They also react with phosphoric acid to produce phosphate esters (e.g. in *ATP), which are important in cell metabolism. Monosaccharides can exist as either straight-chain or ring-shaped molecules. They also exhibit *optical activity, giving rise to both dextrorotatory and laevorotatory forms.

monotropy *See* allotropy.

monovalent (univalent) Having a valency of one.

mordant A substance used in certain dyeing processes. Mordants are often inorganic oxides or salts, which are absorbed on the fabric. The dyestuff then forms a coloured complex with the mordant, the colour depending on the mordant used as well as the dyestuff. *See also* lake.

morphine An alkaloid present in opium. It is an analgesic and narcotic, used medically for the relief of severe pain.

mosaic gold *See* tin(IV) sulphide.

Moseley's law The frequencies of the lines in the *X-ray spectra of the elements are related to the atomic numbers of the elements. If the square roots of the frequencies of corresponding lines of a set of elements are plotted against the atomic numbers a straight line is obtained. The law was discovered by H. G. Moseley (1887–1915).

moss agate *See* agate.

multiple proportions *See* chemical combination.

multiplet 1. A spectral line formed by more than two (*see* doublet) closely spaced lines. **2.** A group of elementary particles that are identical in all respects except that of electric charge.

Mumetal The original trade name for a ferromagnetic alloy, containing 78% nickel, 17% iron, and 5% copper, that had a high permeability and a low coercive force. More modern versions also contain chromium and molybdenum. These alloys are used in some transformer cores and for shielding various devices from external magnetic fields.

Muntz metal A form of *brass containing 60% copper, 39% zinc, and small amounts of lead and iron. Stronger than alpha-brass, it is used for hot forgings, brazing rods, and large nuts and bolts. It is named after G. F. Muntz (1794–1857).

muscovite (white mica; potash mica) A mineral form of potassium aluminosilicate, $K_2Al_4(Si_6Al_2)O_{20}(OH,F)_4$; one of the most important members of the *mica group of minerals. It is chemically complex and has a sheetlike crystal structure. It is usually silvery-grey in colour, sometimes tinted with green, brown, or pink. Muscovite is a common constituent of certain granites and pegmatites. It is also common in metamorphic and sedimentary rocks. It is widely used in industry, for example in the manufacture of electrical equipment and as a filler in roofing materials, wallpapers, and paint.

mustard gas A highly poisonous gas, $(ClCH_2CH_2)_2S$; dichlorodiethyl sulphide. It is made from ethene and disulphur dichloride (S_2Cl_2), and used as a war gas.

mutarotation Change of optical activity with time as a result of spontaneous chemical reaction.

myoglobin A globular protein occurring widely in muscle tissue as an oxygen carrier. It comprises a single polypeptide chain and a *haem group, which reversibly binds a molecule of oxygen. This is only relinquished at relatively low external oxygen concentrations, e.g. during strenuous exercise when muscle oxygen demand outpaces supply from the blood. Myoglobin thus acts as an emergency oxygen store.

N

NAD (nicotinamide adenine dinucleotide) A *coenzyme, derived from the B vitamin *nicotinic acid, that participates in many biological dehydrogenation reactions. NAD is characteristically loosely bound to the enzymes concerned. It normally carries a positive charge and can accept one hydrogen atom and two electrons to become the reduced form, *NADH*. NADH is generated during the oxidation of food; it then gives up its two electrons (and single proton) to the electron transport chain, thereby reverting to NAD^+ and generating three molecules of ATP per molecule of NADH.

NADP (*nicotinamide adenine dinucleotide phosphate*) differs from NAD only in possessing an additional phosphate group. It functions in the same way as NAD although anabolic reactions (*see* anabolism) generally use NADPH (reduced NADP) as a hydrogen donor rather than NADH. Enzymes tend to be specific for either NAD or NADP as coenzyme.

nano- Symbol n. A prefix used in the metric system to denote 10^{-9}. For example, 10^{-9} second = 1 nanosecond (ns).

naphthalene A white volatile solid, $C_{10}H_8$; r.d. 1.14; m.p. 802°C; b.p. 218°C. Naphthalene is an aromatic hydrocarbon with an odour of mothballs and is obtained from crude oil. It is a raw material for making certain synthetic resins.

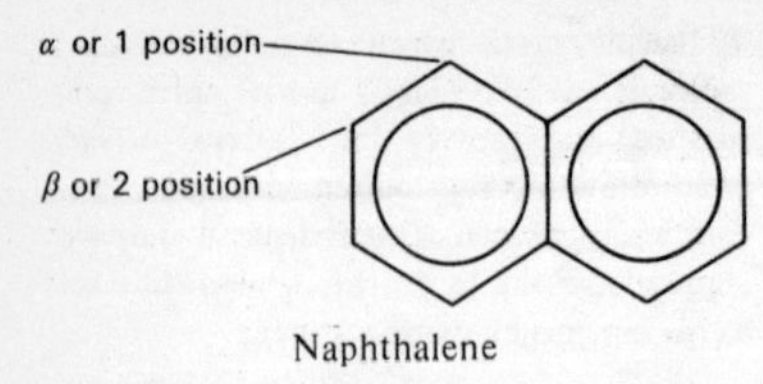

Naphthalene

naphthols Two phenols derived from naphthalene with the formula $C_{10}H_7OH$, differing in the position of the –OH group. The most important is naphthalen-2-ol (β-naphthol), with the –OH in the 2-position. It is a white solid (r.d. 1.2; m.p. 122°C; b.p. 285°C) used in rubber as an antioxidant. Naphthalen-2-ol will couple with diazonium salts at the 1-position to form red *azo compounds, a reaction used in testing for primary amines (by making the diazonium salt and adding naphthalen-2-ol).

naphthyl group The group $C_{10}H_7$– obtained by removing a hydrogen atom from naphthalene. There are two forms depending on whether the hydrogen is removed from the 1- or 2-position.

nascent hydrogen A reactive form of hydrogen generated *in situ* in the reaction mixture (e.g. by the action of acid on zinc). Nascent hydrogen can reduce elements and compounds that do not readily react with 'normal' hydrogen. It was once thought that the hydrogen was present as atoms, but this is not the case. Probably hydrogen molecules are formed in an excited state and react before they revert to the ground state.

natron A mineral form of hydrated sodium carbonate, $Na_2CO_3.H_2O$.

Natta process *See* Ziegler process.

natural gas A naturally occurring mixture of gaseous hydrocarbons that is found in porous sedimentary rocks in the earth's crust, usually in association with *petroleum deposits. It consists chiefly of methane (about 85%), ethane (up to about 10%), propane (about 3%), and butane. Carbon dioxide, nitrogen, oxygen, hydrogen sulphide, and sometimes helium may also be present. Natural gas, like petroleum, originates in the decomposition of organic matter. It is widely used as a fuel and also to produce carbon black and some organic chemicals. Natural gas occurs on every continent, the major reserves occurring in the USSR, USA, Algeria, Canada, and the Middle East.

nematic crystal *See* liquid crystal.

neodymium Symbol Nd. A soft silvery metallic element belonging to the *lanthanoids; a.n. 60; r.a.m. 144.24; r.d. 7.004 (20°); m.p. 1016°C; b.p. 3068°C. It occurs in bastnasite and monazite, from which it is recovered by an ion-exchange process. There are seven naturally occurring isotopes, all of which are stable, except neodymium-144, which is slightly radioactive (half-life 10^{10}–10^{15} years). Seven artificial radioisotopes have been produced. The metal is used to colour glass violet-purple and to make it dichroic. It is also used in mischmetal (18% neodymium). It was discovered by C. A. von Welsbach in 1885.

neon Symbol Ne. A colourless gaseous element belonging to group I of the periodic table (the *noble gases); a.n. 10; r.a.m. 20.79; d. 0.9 g dm^{-3}; m.p. –248.67°C; b.p. –246.05°C. Neon occurs in air (0.0018% by volume) and is obtained by fractional distillation of liquid air. It is used in discharge tubes and neon lamps, in which it has a characteristic red glow. It forms hardly any compounds (neon fluorides have been reported). The element was discovered in 1898 by Sir William Ramsey and M. W. Travers.

neoprene A synthetic rubber made by polymerizing the compound 2-chlorobuta-1,2-diene. Neoprene is often used in place of natural rubber in applications requiring resistance to chemical attack.

nephrite *See* jade.

neptunium Symbol Np. A radioactive metallic transuranic element belonging to the *actinoids; a.n. 93; r.a.m. 237.0482. The most stable isotope, neptunium–237, has a half-life of 2.2×10^6 years and is produced in small quantities as a by-product by nuclear reactors. Other isotopes have mass numbers 229–236 and 238–241. The only other relatively long-lived isotope is neptunium–236 (half-life 5×10^3 years). The element was first produced by McMillan and Abelson in 1940.

neptunium series *See* radioactive series.

Nernst heat theorem A statement of the third law of *thermodynamics in a restricted form: if a chemical change takes place between pure crystalline solids at *absolute zero there is no change of entropy.

nesquehonite A mineral form of *magnesium carbonate trihydrate, $MgCO_3.3H_2O$.

Nessler's reagent A solution of mercury (II) iodide (HgI_2) in potassium iodide and potassium hydroxide. It is used in testing for ammonia, with which it forms a brown coloration or precipitate.

neutral Describing a compound or solution that is neither acidic nor basic. A neutral solution is one that contains equal numbers of both protonated and deprotonated forms of the solvent.

neutralization The process in which an acid reacts with a base to form a salt and water.

neutron A neutral hadron that is stable in the atomic nucleus but decays into a proton, an electron, and an antineutrino with a mean life of 12 minutes outside the nucleus. Its rest mass is slightly greater than that of the proton, being $1.674\,92 \times 10^{-27}$ kg. Neutrons occur in all atomic nuclei except normal hydrogen. The neutron was first reported in 1932 by James Chadwick (1891–1974).

neutron number Symbol N. The number of neutrons in an atomic nucleus of a particular nuclide. It is equal to the difference between the *nucleon number and the *atomic number.

Newlands' law *See* law of octaves.

newton Symbol N. The *SI unit of force, being the force required to give a mass of one kilogram an acceleration of 1 $m\,s^{-2}$. It is named after Sir Isaac Newton (1642–1727).

niacin *See* nicotinic acid.

Nichrome Tradename for a group of nickel–chromium alloys used for wire in heating elements as they possess good resistance to oxidation and have a high resistivity. Typical is Nichrome V containing 80% nickel and 19.5% chromium, the balance consisting of manganese, silicon, and carbon.

nickel Symbol Ni. A malleable ductile silvery metallic *transition element; a.n. 28; r.a.m. 58.70; r.d. 8.9; m.p. 1450°C; b.p. 2840°C. It is found in the minerals pentlandite (NiS), pyrrhoite ((Fe,Ni)S), and garnierite ($(Ni,Mg)_6(OH)_6Si_4O_{11}.H_2O$). Nickel is also present in certain iron meteorites (up to 20%). The metal is extracted by roasting the ore to give the oxide, followed by reduction with carbon monoxide and purification by the *Mond process. Alternatively electolysis is used. Nickel metal is used in special steels, in Invar, and, being ferromagnetic, in magnetic alloys, such as *Mumetal. It is also an effective catalyst, particularly for hydrogenation reactions (*see also* Raney nickel). The main compounds are formed with nickel in the +2 oxidation state; the +3 state also exists (e.g. the black oxide, Ni_2O_3). Nickel was discovered by A. F. Cronstedt in 1751.

nickel carbonyl A colourless volatile liquid, $Ni(CO)_4$; m.p. −25°C; b.p. 43°C. It is formed by direct combination of nickel metal with carbon monoxide at 50–60°C. The reaction is reversed at higher temperatures, and the reactions are the basis of the *Mond process for purifying nickel. The nickel in the compound has an oxidation state of zero, and the compound is a typical example of a complex with pi-bonding *ligands, in which filled *d*-orbitals on the nickel overlap with empty *p*-orbitals on the carbon.

nickel–iron accumulator (Edison cell; NIFE cell) A *secondary cell devised by Thomas Edison (1847–1931) having a positive plate of nickel oxide and a negative plate of iron both immersed in an electrolyte of potassium hydroxide. The reaction on discharge is

$$2NiOOH.H_2O + Fe \rightarrow 2Ni(OH)_2 + Fe(OH)_2,$$

the reverse occurring during charging. Each cell gives an e.m.f. of about 1.2 volts and produces about 100 kJ per kilogram during each discharge. *Compare* lead–acid accumulator.

nickel(II) oxide A green powder, NiO; r.d. 6.6. It can be made by heating nickel (II) nitrate or carbonate with air excluded.

nickel(III) oxide (nickel peroxide; nickel sesquioxide) A black or grey powder, Ni_2O_3; r.d. 4.8. It is made by heating nickel(II) oxide in air and used in *nickel–iron accumulators.

nickel silver *See* German silver.

nicotinamide adenine dinucleotide *See* NAD.

nicotine A colourless poisonous *alkaloid present in tobacco. It is used as an insecticide.

nicotinic acid (niacin) A vitamin of the *vitamin B complex. It can be manufactured by plants and animals from the amino acid tryptophan. The amide derivative, nicotinamide, is a component of the coenzymes *NAD and NADP. These take part in many metabolic reactions as hydrogen acceptors. Deficiency of nicotinic acid causes the disease pellagra in humans. Apart from tryptophan-rich protein, good sources are liver and groundnut and sunflower meals.

NIFE cell *See* nickel–iron accumulator.

niobium Symbol Nb. A soft ductile grey-blue metallic transition element; a.n. 41; r.a.m. 92.91; r.d. 8.57; m.p. 2468°C; b.p. 4742°C. It occurs in several minerals, including niobite ($Fe(NbO_3)_2$), and is extracted by several methods including reduction of the complex fluoride K_2NbF_7 using sodium. It is used in special steels and in welded joints (to increase strength). Niobium–zirconium alloys are used in superconductors. Chemically, the element combines with the halogens and oxidizes in air at 200°C. It forms a number of compounds and complexes with the metal in oxidation states 2, 3, or 5. The element was discovered by Charles Hatchett in 1801 and first isolated by Blomstrand in 1864. Formerly, it was called *columbium*.

nitrate A salt or ester of nitric acid.

nitration A type of chemical reaction in which a nitro group ($-NO_2$) is added to or substituted in a molecule. Nitration can be carried out by a mixture of concentrated nitric and sulphuric acids. An example is electrophilic substitution of benzene (and benzene compounds), where the electrophile is the nitryl ion NO_2^+.

nitre (saltpetre) Commercial *potassium nitrate; the name was formerly applied to natural crustlike efflorescences, occurring in some arid regions.

nitre cake *See* sodium hydrogensulphate.

nitric acid A colourless corrosive poisonous liquid, HNO_3; r.d. 1.50; m.p. −42°C; b.p. 83°C. Nitric acid may be prepared in the laboratory by the distillation of a mixture of an alkali-metal nitrate and concentrated sulphuric acid. The industrial production is by the oxidation of ammonia to nitrogen monoxide, the oxidation of this to nitrogen dioxide, and the reaction of nitrogen dioxide with water to form nitric acid and nitrogen monoxide (which is recycled). The first reaction (NH_3 to NO) is catalysed by platinum or platinum/rhodium in the form of fine wire gauze. The oxidation of NO and the absorption of NO_2 to form the product are noncatalytic and proceed with high yields but both reactions are second-order and slow. Increases in pressure reduce the selectivity of the reaction and therefore rather large gas absorption towers are required. In practice the absorbing acid is refrigerated to around 2°C and a commercial 'concentrated nitric acid' at about 67% is produced.

Nitric acid is a strong acid (highly dissociated in aqueous solution) and dilute solutions behave much like other mineral acids. Concentrated nitric acid is a strong oxidizing agent. Most metals dissolve to form nitrates but with the evolution of nitrogen oxides. Concentrated nitric acid also reacts with several nonmetals to give the oxo acid or oxide. Nitric acid is generally stored in dark brown bottles because of the photolytic decomposition to dinitrogen tetroxide. *See also* nitration.

nitric oxide *See* nitrogen monoxide.

nitrides Compounds of nitrogen with a more electropositive element. Boron nitride is a covalent compound having macromolecular crystals. Certain electropositive elements, such as lithium, magnesium, and calcium, react directly with nitrogen to form ionic nitrides containing the N^{3-} ion. Transition elements form a range of interstitial nitrides (e.g. Mn_4N, W_2N), which can be produced by heating the metal in ammonia.

nitrification A chemical process in which nitrogen (mostly in the form of ammonia) in plant and animal wastes and dead remains is oxidized at first to nitrites and then to nitrates. These reactions are effected mainly by the bacteria *Nitrosomonas* and *Nitrobacter* respectively. Unlike ammonia, nitrates are readily taken up by plant roots; nitrification is therefore a crucial part of the *nitrogen cycle. Nitrogen-containing compounds are often applied to soils deficient in this element, as fertilizer. *Compare* denitrification.

nitriles (cyanides) Organic compounds containing the group −CN bound to an organic group. Nitriles are made by reaction between potassium cyanide and haloalkanes in alcoholic solution, e.g.

$$KCN + CH_3Cl \rightarrow CH_3CN + KCl$$

An alternative method is dehydration of amides

$$CH_3CONH_2 - H_2O \rightarrow CH_3CN$$

They can be hydrolysed to amides and carboxylic acids and can be reduced to amines.

nitrite A salt or ester of nitrous acid; the nitrite ion, NO_2^-, has a bond angle of 115°.

nitrobenzene A yellow oily liquid, $C_6H_5NO_2$; r.d. 1.2; m.p. 6°C; b.p. 211°C. It is made by the *nitration of benzene using a mixture of nitric and sulphuric acids.

nitrocellulose *See* cellulose nitrate.

nitro compounds Organic compounds containing the group $-NO_2$ (the *nitro group*) bound to a carbon atom in a benzene ring. Nitro compounds are made by *nitration reactions. They can be reduced to aromatic amines (e.g. nitrobenzene can be reduced to phenylamine).

nitrogen Symbol N. A colourless gaseous element belonging to *group V of the periodic table; a.n. 7; r.a.m. 14.0067; d. 1.2506 g dm^{-3}; m.p. −209.86°C; b.p. −195.8°C. It occurs in air (about 78% by volume) and is an essential constituent of proteins and nucleic acids in living organisms (*see* nitrogen cycle). Nitrogen is obtained for industrial purposes by fractional distillation of liquid air. Pure nitrogen can be obtained in the laboratory by heating a metal azide. There are two natural isotopes: nitrogen−14 and nitrogen−15 (about 3%). The element is used in the *Haber process for making ammonia and is also used to provide an inert atmosphere in welding and metallurgy. The gas is diatomic and relatively inert − it reacts with hydrogen at high temperatures and with oxygen in electric discharges. It also forms *nitrides with certain metals. Nitrogen was discovered in 1772 by D. Rutherford.

nitrogen cycle One of the major cycles of chemical elements in the environment. Nitrates in the soil are taken up by plant roots and may then pass along food chains into animals. Decomposing bacteria convert nitrogen-containing compounds (especially ammonia) in plant and animal wastes and dead remains back into nitrates, which are released into the soil and can again be taken up by plants (*see* nitrification). Though nitrogen is essential to all forms of life, the huge amount present in the atmosphere is not directly available to most organisms (*compare* carbon cycle). It can, however, be assimilated by some specialized bacteria and algae (*see* nitrogen fixation) and is thus made available to other organisms indirectly. Lightning flashes also make some nitrogen available to plants by causing the combination of atmospheric nitrogen and oxygen to form oxides of nitrogen, which enter the soil and form nitrates. Some nitrogen is returned from the soil to the atmosphere by denitrifying bacteria (*see* denitrification).

nitrogen dioxide *See* dinitrogen tetroxide.

nitrogen fixation A chemical process in which atmospheric nitrogen is assimilated into organic compounds in living organisms and hence into the *nitrogen cycle. The ability to fix nitrogen is limited to certain bacteria (e.g. *Azotobacter*) and blue-green algae (e.g. *Anabaena*). *Rhizobium* bacteria are able to fix nitrogen in association with cells in the roots of leguminous plants such as peas and beans, in which they form characteristic root nodules; cultivation of legumes is therefore one way of increasing soil nitrogen. Various chemical processes are used to fix atmospheric nitrogen in the manufacture of fertilizers. These include the *Birkeland−Eyde process, the cyanamide process (*see* calcium dicarbide), and the *Haber process.

nitrogen monoxide (nitric oxide) A colourless gas, NO; m.p. −90.8°C; b.p. −88.5°C. It is soluble in water, ethanol, and ether. In the liquid state nitrogen monoxide is blue in colour (r.d. 1.26). It is formed in many reactions involving the reduction of nitric acid, but more convenient reactions for the preparation of reasonably pure NO are reactions of sodium nitrite, sulphuric acid, and either sodium iodide or iron(II) sulphate. Nitrogen monoxide reacts readily with oxygen to give nitrogen dioxide and with the halogens to give the nitrosyl halides XNO (X = F,Cl,Br). It is oxidized to nitric acid by strong oxidizing agents and reduced to dinitrogen oxide by reducing agents. The molecule has one unpaired electron, which accounts for its paramagnetism

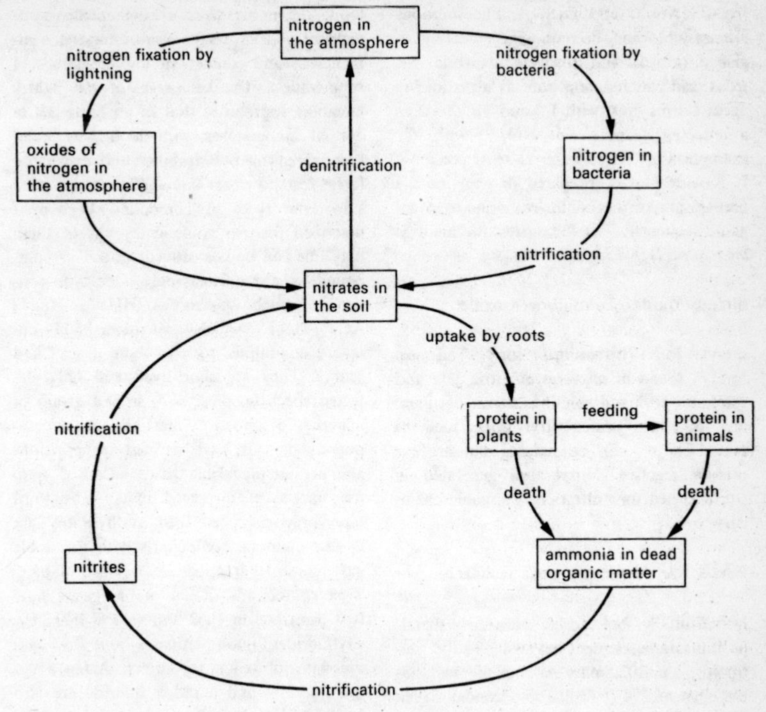

The nitrogen cycle

and for the blue colour in the liquid state. This electron is relatively easily removed to give the *nitrosyl ion* NO^+, which is the ion present in such compounds as $NOClO_4$, $NOBF_4$, $NOFeCl_4$, $(NO)_2PtCl_6$ and a ligand in complexes such as $Co(CO)_3NO$.

nitrogenous base A basic compound containing nitrogen. The term is used especially of organic ring compounds, such as adenine, guanine, cytosine, and thymine, which are constituents of nucleic acids. *See* amine salts.

nitroglycerine An explosive made by reacting 1,2,3-trihydroxypropane (glycerol) with a mixture of concentrated sulphuric and nitric acids. Despite its name and method of preparation, it is not a nitro compound, but an ester of nitric acid, $CH_2(NO_3)CH(NO_3)CH_2(NO_3)$. It is used in dynamites.

nitro group *See* nitro compounds.

nitronium ion *See* nitryl ion.

nitrosyl ion The ion NO^+. *See* nitrogen monoxide.

nitrous acid A weak acid, HNO_2, known only in solution and in the gas phase. It is prepared by the action of acids upon nitrites, preferably using a combination that removes the salt as an insoluble precipitate (e.g. $Ba(NO_2)_2$ and H_2SO_4). The solutions are unstable and decompose on heating to give nitric acid and nitrogen monoxide. Nitrous acid can function both as an oxidizing agent (forms NO) with I^- and Fe^{2+}, or as a reducing agent (forms NO_3^-) with, for example, Cu^{2+}; the latter is most common. It is widely used (prepared *in situ*) for the preparation of diazonium compounds in organic chemistry. The full systematic name is *dioxonitric(III) acid.*

nitrous oxide *See* dinitrogen oxide.

nitryl ion (nitronium ion) The ion NO_2^+, found in mixtures of nitric acid and sulphuric acid and solutions of nitrogen oxides in nitric acid. Nitryl salts, such as $NO_2^+ClO_4^-$, can be isolated but are extremely reactive. Nitryl ions generated *in situ* are used for *nitration in organic chemistry.

NMR *See* nuclear magnetic resonance.

nobelium Symbol No. A radioactive metallic transuranic element belonging to the *actinoids; a.n. 102; mass number of most stable element 254 (half-life 55 seconds). Seven isotopes are known. The element was first identified with certainty by A. Ghiorso and G. T. Seaborg in 1966. The alternative name *unnilbium* has been proposed.

noble gases (inert gases; rare gases; group 0 elements) A group of monatomic gaseous elements forming group 0 (sometimes called group VIII) of the *periodic table: helium (He), neon (Ne), argon (Ar), krypton (Kr), xenon (Xe), and radon (Rn). The electron configuration of helium is $1s^2$. The configurations of the others terminate in ns^2np^6 and all inner shells are fully occupied. The elements thus represent the termination of a period and have closed-shell configuration and associated high ionization energies (He 2370 to Rn 1040 kJ mol^{-1}) and lack of chemical reactivity. Being monatomic the noble gases are spherically symmetrical and have very weak interatomic interactions and consequent low enthalpies of vaporization. The behaviour of the lighter members approaches that of an ideal gas at normal temperatures; with the heavier members increasing polarizability and dispersion forces lead to easier liquefaction under pressure. Four types of 'compound' have been described for the noble gases but of these only one can be correctly described as compounds in the normal sense. One type consists of such species as HHe^+, He_2^+, Ar_2^+, $HeLi^+$, which form under highly energetic conditions, such as those in arcs and sparks. They are short-lived and only detected spectroscopically. A second group of materials described as inert-gas–metal compounds do not have defined compositions and are simply noble gases adsorbed onto the surface of dispersed metal. The third type, previously described as 'hydrates' are in fact clathrate compounds with the noble gas molecule trapped in a water lattice. True compounds of the noble gases were first described in 1962 and several fluorides, oxyfluorides, fluoroplatinates, and fluoroantimonates of xenon are known. A few krypton fluorides and a radon fluoride are also known although the short half-life of radon and its intense alpha activity restrict the availability of information. Apart from argon, the noble gases are present in the atmosphere at only trace levels. Helium may be found along with natural gas (up to 7%), arising from the radioactive decay of heavier elements (via alpha particles).

nonbenzenoid aromatics Aromatic compounds that have rings other than benzene rings. Examples are the cyclopentadienyl anion, $C_5H_5^-$, and the tropyllium cation, $C_7H_7^+$.

nonmetal Any of a class of chemical elements that are typically poor conductors of heat and electricity and that do not form positive ions. Nonmetals are electronegative elements, such as carbon, nitrogen, oxygen, phosphorus, sulphur, and the halogens. They form compounds that contain negative ions or covalent bonds. Their oxides are either neutral or acidic.

nonpolar compound A compound that has covalent molecules with no permanent dipole moment. Examples of nonpolar compounds are methane and benzene.

nonpolar solvent *See* solvent.

nonrelativistic quantum theory *See* quantum theory.

nonstoichiometric compound (Berthollide compound) A chemical compound in which the elements do not combine in simple ratios. For example, rutile (titanium(IV) oxide) is often deficient in oxygen, having such a formula as $TiO_{1.8}$.

noradrenaline (norepinephrine) A hormone produced by the adrenal glands and also secreted from nerve endings in the sympathetic nervous system as a chemical transmitter of nerve impulses. Many of its general actions are similar to those of *adrenaline, but it is more concerned with maintaining normal body activity than with preparing the body for emergencies.

Nordhausen sulphuric acid *See* disulphuric(VI) acid.

norepinephrine *See* noradrenaline.

normal Having a concentration of one gram equivalent per dm^3.

N.T.P. *See* s.t.p.

nuclear magnetic resonance (NMR) The absorption of electromagnetic radiation at a suitable precise frequency by a nucleus with a nonzero magnetic moment in an external magnetic field. The phenomenon occurs if the nucleus has nonzero *spin, in which case it behaves as a small magnet. In an external magnetic field, the nucleus's magnetic moment vector precesses about the field direction but only certain orientations are allowed by quantum rules. Thus, for hydrogen (spin of ½) there are two possible states in the presence of a field, each with a slightly different energy. Nuclear magnetic resonance is the absorption of radiation at a photon energy equal to the difference between these levels, causing a transition from a lower to a higher energy state. For practical purposes, the difference in energy levels is small and the radiation is in the radiofrequency region of the electromagnetic spectrum. It depends on the field strength.

NMR can be used for the accurate determination of nuclear moments. It can also be used in a sensitive form of magnetometer to measure magnetic fields. In medicine, NMR tomography is being developed, in which images of tissue are produced by magnetic-resonance techniques.

The main application of NMR is as a technique for chemical analysis and structure determination, known as *NMR spectroscopy*. It depends on the fact that the electrons in a molecule shield the nucleus to some extent from the field, causing different atoms to absorb at slightly different frequencies (or at slightly different fields for a fixed frequency). Such effects are known as *chemical shifts*. In an NMR spectrometer, the sample is subjected to a strong field, which can be varied in a controlled way over a small region. It is irradiated with radiation at a fixed frequency, and a detector monitors the field at the sample. As the field changes, absorption corresponding to transitions occurs at certain values, and this causes oscillations in the field, which induce a signal in the detector. The most common nucleus studied is ^1H. For instance, an NMR spectrum of ethanol (CH_3CH_2OH) has three peaks in the ratio 3:2:1, corresponding to the three different hydrogen-atom environments. The peaks also have a fine structure caused by interaction between spins in the molecule. Other nuclei can also be used for NMR spectroscopy (e.g. ^{13}C, ^{14}N, ^{19}F) although these generally have lower magnetic moment and natural abundance than hydrogen. *See also* electron spin resonance.

nucleon A *proton or a *neutron.

nucleon number (mass number) Symbol A. The number of *nucleons in an atomic nucleus of a particular nuclide.

nucleophile An ion or molecule that can donate electrons. Nucleophiles are often oxidizing agents and Lewis bases. They are either negative ions (e.g. Cl^-) or molecules that have electron pairs (e.g. NH_3). In organic reactions they tend to attack positively charged parts of a molecule. *Compare* electrophile.

nucleophilic addition A type of addition reaction in which the first step is attachment of a *nucleophile to a positive (electron-deficient) part of the molecule. *Aldehydes and *ketones undergo reactions of this type because of polarization of the carbonyl group (carbon positive).

nucleophilic substitution A type of substitution reaction in which a *nucleophile displaces another group or atom from a compound. For example, in

$$CR_3Cl + OH^- \rightarrow CR_3OH + Cl^-$$

the nucleophile is the OH^- ion. There are two possible mechanisms of nucleophilic substitution. In *S_N1 reactions,* a positive carbonium ion is first formed:

$$CR_3Cl \rightarrow CR_3^+ + Cl^-$$

This then reacts with the nucleophile

$$CR_3^+ + OH^- \rightarrow CR_3OH$$

The CR_3^+ ion is planar and the OH^- ion can attack from either side. Consequently, if the original molecule is optically active (the three R groups are different) then a racemic mixture of products results.

The alternative mechanism, the *S_N2 reaction,* is a concerted reaction in which the nucleophile approaches from the side of the R groups as the other group (Cl in the example) leaves. In this case the configuration of the molecule is inverted. If the original molecule is optically active, the product has the opposite activity, an effect known as *Walden inversion.*

nucleoside An organic compound consisting of a nitrogen-containing *purine or *pyrimidine base linked to a sugar (ribose or deoxyribose). An example is *adenosine. *Compare* nucleotide.

nucleotide An organic compound consisting of a nitrogen-containing *purine or *pyrimidine base linked to a sugar (ribose or deoxyribose) and a phosphate group. *DNA and *RNA are made up of long chains of nucleotides (i.e. *polynucleotides*). *Compare* nucleoside.

nucleus The central core of an atom that contains most of its mass. It is positively charged and consists of one or more nucleons (protons or neutrons). The positive charge of the nucleus is determined by the number of protons it contains (*see* atomic number) and in the neutral atom this is balanced by an equal number of electrons, which move around the nucleus. The simplest nucleus is the hydrogen nucleus, consisting of one proton only. All other nuclei also contain one or more neutrons. The neutrons contribute to the atomic mass (*see* nucleon number) but not to the nuclear charge. The most massive nucleus that occurs in nature is uranium–238, containing 92 protons and 146 neutrons. The symbol used for this *nuclide is $^{238}_{92}U$, the upper figure being the nucleon number and the lower figure the atomic number. In all nuclei the nucleon number (A) is equal to the sum of the atomic number (Z) and the neutron number (N), i.e. $A = Z + N$.

nuclide 1. An atomic nucleus as characterized by its *atomic number and its *neutron number. An *isotope refers to a series of different atoms that have the same atomic number but different neutron numbers (e.g. uranium–238 and uranium–235 are isotopes of uranium), whereas a nuclide refers only to a particular nuclear species (e.g. the nuclides uranium–235 and plutonium–239 are fissile). **2.** The atom to which a specified nucleus belongs.

nylon Any of various synthetic polyamide fibres having a protein-like structure formed by the condensation between an amino group of one molecule and a carboxylic acid group of another. There are three main nylon fibres, nylon 6, nylon 6,6, and nylon 6,10. Nylon 6, for example Enkalon and Celon, is formed by the self-condensation of 6-aminohexanoic acid. Nylon 6,6, for example Bri nylon, is made by polycondensation of hexanedioic acid (adipic acid) and 1,6-diaminohexane (hexamethylenediamine) having an average formula weight between 12 000 and 15 000. Nylon 6,10 is prepared by polymerizing decanedioic acid and 1,6-diaminohexane.

O

occlusion 1. The trapping of small pockets of liquid in a crystal during crystallization. **2.** The absorption of a gas by a solid such that atoms or molecules of the gas occupy interstitial positions in the solid lattice. Palladium, for example, can occlude hydrogen.

ochre A yellow or red mineral form of iron (III) oxide, Fe_2O_3, used as a pigment.

octadecanoate *See* stearate.

octadecanoic acid *See* stearic acid.

octadecenoic acid A straight-chain unsaturated fatty acid with the formula $C_{17}H_{33}COOH$. *Cis-octadec-9-enoic acid* (*see* oleic acid) has the formula $CH_3(CH_2)_7CH{:}CH(CH_2)_7COOH$. The glycerides of this acid are found in many natural fats and oils.

octahydrate A crystalline hydrate having eight molecules of water per molecule of compound.

octane A straight-chain liquid *alkane, C_8H_{18}; r.d. 0.7; m.p. 216°C; b.p. 399°C. It is present in petroleum. The compound is isomeric with 2,2,4-trimethylpentane, $(CH_3)_3CCH_2CH(CH_3)_2$, *iso-octane*). *See* octane number.

octane number A number that provides a measure of the ability of a fuel to resist knocking when it is burnt in a spark-ignition engine. It is the percentage by volume of iso-octane (C_8H_{18}; 2,2,4-trimethylpentane) in a blend with normal heptane (C_7H_{16}) that matches the knocking behaviour of the fuel being tested in a single cylinder four-stroke engine of standard design. *Compare* cetane number.

octanoic acid (caprylic acid) A colourless liquid straight-chain saturated *carboxylic acid, $CH_3(CH_2)_6COOH$; b.p. 237°C.

octavalent Having a valency of eight.

octave *See* law of octaves.

octet A stable group of eight electrons in the outer shell of an atom (as in an atom of an inert gas).

ohm Symbol Ω. The derived *SI unit of electrical resistance, being the resistance between two points on a conductor when a constant potential difference of one volt, applied between these points, produces a current of one ampere in the conductor. The former *international ohm* (sometimes called the 'mercury ohm') was defined in terms of the resistance of a column of mercury. The unit is named after Georg Ohm (1787–1854).

oil Any of various viscous liquids that are generally immiscible with water. Natural plant and animal oils are either volatile mixtures of terpenes and simple esters (e.g. *essential oils) or are *glycerides of fatty acids. Mineral oils are mixtures of hydrocarbons (e.g. *petroleum).

oil of vitriol *See* sulphuric acid.

oil sand (tar sand; bituminous sand) A sandstone or porous carbonate rock that is impregnated with hydrocarbons. The largest deposit of oil sand occurs in Alberta, Canada (the Athabasca tar sands); there are also deposits in the Orinoco Basin of Venezuela, the USSR, USA, Madagascar, Albania, Trinidad, and Romania.

oil shale A fine-grained carbonaceous sedimentary rock from which oil can be extracted. The rock contains organic matter – *kerogen* – which decomposes to yield oil when heated. Deposits of oil shale occur on every continent, the largest known reserves occurring in Colorado, Utah, and Wyoming in the USA. Commercial production of oil from oil shale is generally considered to be uneconomic unless the price of petroleum rises above the recovery costs for oil from oil shale. However, threats of declining conventional oil resources have resulted in considerable interest and developments in recovery techniques.

oleate A salt or ester of *oleic acid.

olefines *See* alkenes.

oleic acid An unsaturated *fatty acid with one double bond, $CH_3(CH_2)_7CH{:}CH(CH_2)_7COOH$; r.d. 0.9; m.p. 13°C. Oleic acid is one of the most abundant constituent fatty acids of animal and plant fats, occurring in butterfat, lard, tallow, groundnut oil, soyabean oil, etc. Its systematic chemical name is *cis-octadec-9-enoic acid.*

oleum *See* disulphuric(VI) acid.

oligopeptide *See* peptide.

olivine An important group of rock-forming silicate minerals crystallizing in the orthorhombic system. Olivine conforms to the general formula $(Mg,Fe)_2SiO_4$ and comprises a complete series from pure magnesium silicate (forsterite, Mg_2SiO_4) to pure iron silicate (fayalite, Fe_2SiO_4). It is green, brown-green, or yellow-green in colour.

onium ion An ion formed by adding a proton to a neutral molecule, e.g. the hydroxonium ion (H_3O^+) or the ammonium ion (NH_4^+).

opal A hydrous amorphous form of silica. Many varieties of opal occur, some being prized as gemstones. Common opal is usually milk white but the presence of impurities may colour it yellow, green, or red. Precious opals, which are used as gemstones, display the property of *opalescence* – a characteristic internal play of colours resulting from the interference of light rays within the stone. Black opal has a black background against which the colours are displayed. The chief sources of precious opals are Australia and Mexico. Geyserite is a variety deposited by geysers or hot springs. Another variety, diatomite, is made up of the skeletons of diatoms.

open chain *See* chain.

open-hearth process A traditional method for manufacturing steel by heating together scrap, pig iron, hot metal, etc., in a refractory-lined shallow open furnace heated by burning producer gas in air.

optical activity The ability of certain substances to rotate the plane of plane-polarized light as it passes through a crystal, liquid, or solution. It occurs when the molecules of the substance are asymmetric, so that they can exist in two different structural forms each being a mirror image of the other. The two forms are *optical isomers* or *enantiomers*. The existence of such forms is also known as *enantiomorphism* (the mirror images being *enantiomorphs*). One form will rotate the light in one direction and the other will rotate it by an equal amount in the other. The two possible forms are described as *dextrorotatory or *laevorotatory according to the direction of rotation, and prefixes are used to designate the isomer, as in *d*-tartaric and *l*-tartaric acids. An equimolar mixture of the two forms is not optically active. It is called a *racemic mixture* (or *racemate*) and designated by *dl*-. In addition, certain molecules can have a *meso form* in which one part of the molecule is a mirror image of the other. Such molecules are not optically active.

Molecules that show optical activity have no plane of symmetry. The commonest case of this is in organic compounds in which a carbon atom is linked to four different groups. An atom of this type is said to be a *chiral centre*. Asymmetric molecules showing optical activity can also occur in inorganic compounds. For example, an octahedral complex in which the central ion coordinates to eight different ligands would be optically active. Many naturally occurring compounds show optical isomerism and usually only one isomer occurs naturally. For instance, glucose is found in the dextrorotatory form. The other isomer, *l*-glucose, can be synthesized in the laboratory, but cannot be synthesized by living organisms. *See also* absolute configuration.

optical glass Glass used in the manufacture of lenses, prisms, and other optical parts. It must be homogeneous and free from bubbles and strain. Optical *crown glass* may contain potassium or barium in place of the sodium of ordinary crown glass and has a refractive index in the range 1.51 to 1.54. *Flint glass* contains lead oxide and has a refractive index between 1.58 and 1.72. Higher refractive indexes are obtained by adding lanthanoid oxides to glasses; these are now known as lanthanum crowns and flints.

optical isomers *See* optical activity.

optical rotary dispersion (ORD) The effect in which the amount of rotation of plane-polarized light by an optically active compound depends on the wavelength. A graph of rotation against wavelength has a characteristic shape showing peaks or troughs.

optical rotation Rotation of plane-polarized light. *See* optical activity.

orbit The path of an electron as it travels round the nucleus of an atom. *See* orbital.

orbital A region in which an electron may be found in an atom or molecule. In the original *Bohr theory of the atom the electrons were assumed to move around the nucleus in circular orbits, but further advances in quantum mechanics led to the view that it is not possible to give a definite path for an electron. According to *wave mechanics, the electron has a certain probability of being in a given element of space. Thus for a hydrogen atom the electron can be anywhere from close to the nucleus to out in space but the maximum probability occurs in a spherical shell around the nucleus with a radius equal to the Bohr radius of the atom. The probabilities of finding an electron in different regions can be obtained by solving the Schrödinger wave equation to give the wave function ψ, and the probability is then proportional to $|\psi|^2$. Thus the idea of electrons in fixed orbits has been replaced by that of a probability distribution around

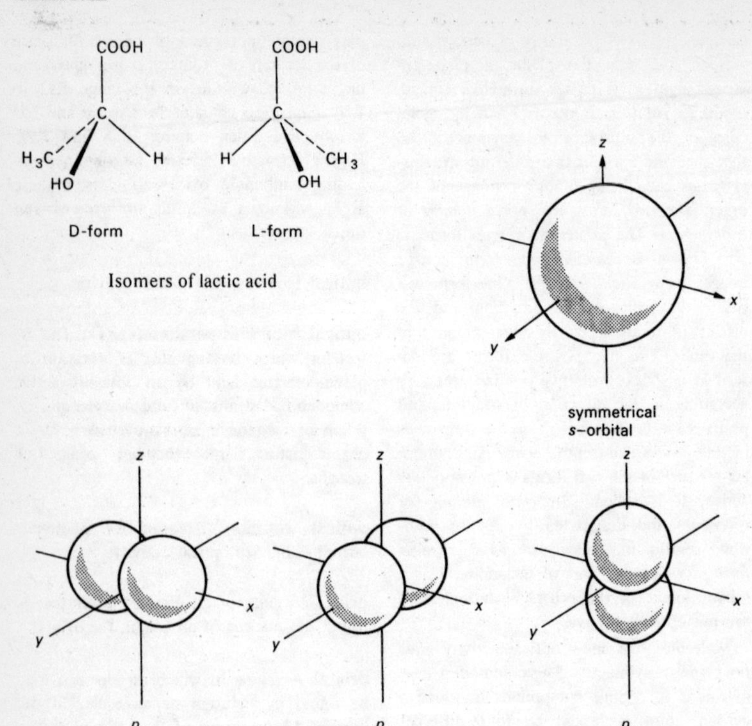

three equivalent *p*-orbitals, each having 2 lobes

Atomic orbitals

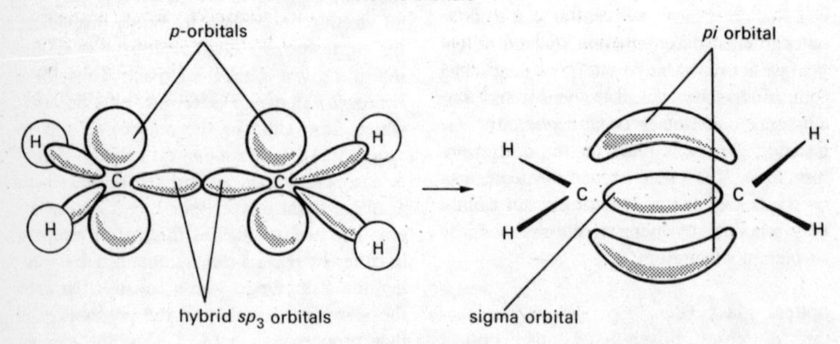

Molecular orbitals: formation of the double bond in ethene

Orbitals

the nucleus – an *atomic orbital*. Alternatively, the orbital can be thought of as an electric charge distribution (averaged over time). In representing orbitals it is convenient to take a surface enclosing the space in which the electron is likely to be found with a high probability.

The possible atomic orbitals correspond to subshells of the atom. Thus there is one *s*-orbital for each shell (orbital quantum number $l = 0$). This is spherical. There are three *p*-orbitals (corresponding to the three values of l) and five *d*-orbitals. The shapes of orbitals depend on the value of l. For instance, *p*-orbitals each have two lobes; most *d*-orbitals have four lobes.

In molecules, the valence electrons move under the influence of two nuclei (in a bond involving two atoms) and there are corresponding *molecular orbitals* for electrons. It is convenient in considering these to regard them as formed by overlap of atomic orbitals. In a hydrogen molecule the *s*-orbitals on the two atoms overlap and form a molecular orbital between the two nuclei. This is an example of a *sigma orbital*. In a double bond, as in ethene, one bond is produced by overlap along the line of axes to form a sigma orbital. The other is produced by sideways overlap of the lobes of the *p*-orbitals (see illustration). The resulting molecular orbital has two parts, one on each side of the sigma orbital – this is a *pi orbital*. In fact, the combination of two atomic orbitals produces two molecular orbitals with different energies. The one of lower energy is the *bonding orbital*, holding the atoms together; the other is the *antibonding orbital*, which would tend to push the atoms apart. In the case of valence electrons, only the lower (bonding) orbital is filled.

In considering the formation of molecular orbitals it is often useful to think in terms of *hybrid* atomic orbitals. For instance, carbon has in its outer shell one *s*-orbital and three *p*-orbitals. In forming methane (or other tetrahedral molecules) these can be regarded as combining to give four equivalent sp^3 hybrid orbitals, each with a lobe directed to a corner of a tetrahedron. It is these that overlap with the *s*-orbitals on the hydrogen atoms. In ethene, two *p*-orbitals combine with the *s*-orbital to give three sp^2 hybrids with lobes in a plane pointing to the corners of an equilateral triangle. These form the sigma orbitals in the C–H and C–C bonds. The remaining *p*-orbitals (one on each carbon) form the pi orbital. In ethyne, sp^2 hybridization occurs to give two hybrid orbitals on each atom with lobes pointing along the axis. The two remaining *p*-orbitals on each carbon form two pi orbitals. Hybrid atomic orbitals can also involve *d*-orbitals. For instance, square-planar complexes use sp^2d hybrids; octahedral complexes use sp^3d^2.

orbital quantum number *See* atom.

order In the expression for the rate of a chemical reaction, the sum of the powers of the concentrations is the overall order of the reaction. For instance, in a reaction

$$A + B \rightarrow C$$

the rate equation may have the form

$$R = k[A][B]^2$$

This reaction would be described as *first order* in A and *second order* in B. The overall order is three. The order of a reaction depends on the mechanism and it is possible for the rate to be independent of concentration (*zero order*) or for the order to be a fraction. *See also* molecularity, pseudo order.

ore A naturally occurring mineral from which a metal can be extracted, usually on a commercial basis. The metal may be present in the ore as the native metal, but more commonly it occurs in a combined form as an oxide, sulphide, sulphate, silicate, etc.

ore dressing *See* beneficiation.

organic chemistry The branch of chemistry concerned with compounds of carbon.

organo- Prefix used before the name of an element to indicate compounds of the elements containing organic groups (with the element bound to carbon atoms). For example, lead(IV) tetraethyl is an organolead compound.

organometallic compound A compound in which a metal atom or ion is bound to an organic group. Organometallic compounds may have single metal–carbon bonds, as in the aluminium alkyls (e.g. $Al(CH_3)_3$). In some cases, the bonding is to the pi electrons of a double bond, as in complexes formed between platinum and ethene, or to the pi electrons of a ring, as in ferrocene.

ornithine (orn) An *amino acid, $H_2N(CH_2)_3CH(NH_2)COOH$, that is not a constituent of proteins but is important in living organisms as an intermediate in the reactions of the *urea cycle and in arginine synthesis.

ornithine cycle *See* urea cycle.

orpiment A natural yellow mineral form of arsenic(III) sulphide, As_2S_3. The name is also used for the synthetic compound, which is used as a pigment.

ortho- 1. Prefix indicating that a benzene compound has two substituted groups in the 1,2 positions (i.e. on adjacent carbon atoms). For instance, orthodichlorobenzene is 1,2-dichlorobenzene. **2.** Prefix formerly used to indicate the most hydrated form of an acid. For example, phosphoric(V) acid, H_3PO_4, was called orthophosphoric acid to distinguish it from the lower metaphosphoric acid, HPO_3 (which is actually $(HPO_3)_n$).

orthoboric acid *See* boric acid.

orthoclase *See* feldspars.

orthohydrogen *See* hydrogen.

orthophosphoric acid *See* phosphoric(V) acid.

orthoplumbate *See* plumbate.

orthorhombic *See* crystal system.

orthosilicate *See* silicate.

orthostannate *See* stannate.

osmiridium A hard white naturally occurring alloy consisting principally of osmium (17–48%) and iridium (49%). It also contains small quantities of platinum, rhodium, and ruthenium. It is used for making small items subject to wear, e.g. electrical contacts or the tips of pen nibs.

osmium Symbol Os. A hard blue-white metallic *transition element; a.n. 76; r.a.m. 190.2; r.d. 22.57; m.p. 3045°C; b.p. 5027°C. It is found associated with platinum and is used in certain alloys with platinum and iridium (*see* osmiridium). Osmium forms a number of complexes in a range of oxidation states.

osmium(IV) oxide (osmium tetroxide) A yellow solid, OsO_4, made by heating osmium in air. It is used as an oxidizing agent in organic chemistry, as a catalyst, and as a fixative in electron microscopy.

osmometer *See* osmosis.

osmosis The passage of a solvent through a *semipermeable membrane* separating two solutions of different concentrations. A semipermeable membrane is one through which the molecules of a solvent can pass but the molecules of most solutes cannot. There is a thermodynamic tendency for solutions separated by such a membrane to become equal in concentration, the water (or other solvent) flowing from the weaker to the stronger solution. Osmosis will stop when the two solutions reach equal concentration, and can also be stopped by applying a pressure to the liquid on the stronger-solution side of the membrane. The pressure required to stop the flow from a pure solvent into a solution is a characteristic of the solution, and is called the *osmotic pressure* (symbol Π). Osmotic pressure depends only on the concentration of particles in the solution, not on their nature (i.e. it is a *colligative property). For a solution of n moles in volume V at thermodynamic temperature T, the osmotic pressure is given by $\Pi V = RT$, where R is the gas constant. Osmotic-pressure measurements are used in finding the relative molecular masses of compounds, particularly macromolecules. A device used to measure osmotic pressure is called an *osmometer*.

The distribution of water in living organisms is dependent to a large extent on osmosis, water entering the cells through their membranes. A cell membrane is not truly semipermeable as it allows the passage of certain solute molecules; it is described as *differentially permeable*.

osmotic pressure *See* osmosis.

Ostwald's dilution law An expression for the degree of dissociation of a weak electrolyte. For example, if a weak acid dissociates in water

$$HA \rightleftharpoons H^+ + A^-$$

the dissociation constant K_a is given by

$$K_a = \alpha^2 n/(1 - \alpha)V$$

where α is the degree of dissociation, n the initial amount of substance (before dissociation), and V the volume. If α is small compared with 1, then $\alpha^2 = KV/n$; i.e. the degree of dissociation is proportional to the square root of the dilution. The law was first put forward by W. Ostwald (1853–1932) to account for electrical conductivities of electrolyte solutions.

overpotential A potential that must be applied in an electrolytic cell in addition to the theoretical potential required to liberate a given substance at an electrode. The value depends on the electrode material and on the current density. It is a kinetic effect occurring because of the significant activation energy for electron transfer at the electrodes, and is particularly important for the liberation of such gases as hydrogen and oxygen. For example, in the electrolysis of a solution of zinc ions, hydrogen (E = 0.00 V) would be expected to be liberated at the cathode in preference to zinc (E = −0.76 V). In fact, the high overpotential of hydrogen on zinc (about 1 V under suitable conditions) means that zinc can be deposited instead.

oxalate A salt or ester of *oxalic acid.

oxalic acid (ethanedioic acid) A crystalline solid, $(COOH)_2$, that is slightly soluble in water. Oxalic acid is strongly acidic and very poisonous. It occurs in certain plants, e.g. sorrel and the leaf blades of rhubarb.

oxidant *See* oxidizing agent.

oxidation *See* oxidation–reduction.

oxidation number (oxidation state) *See* oxidation–reduction.

oxidation–reduction (redox) Originally, *oxidation* was simply regarded as a chemical reaction with oxygen. The reverse process – loss of oxygen – was called *reduction*. Reaction with hydrogen also came to be regarded as reduction. Later, a more general idea of oxidation and reduction was developed in which oxidation was loss of electrons and reduction was gain of electrons. This wider definition covered the original one. For example, in the reaction

$$4Na(s) + O_2(g) \rightarrow 2Na_2O(s)$$

the sodium atoms lose electrons to give Na^+ ions and are oxidized. At the same time, the oxygen atoms gain electrons and are reduced. These definitions of oxidation and reduction also apply to reactions that do not involve oxygen. For instance in

$$2Na(s) + Cl_2(g) \rightarrow 2NaCl(s)$$

the sodium is oxidized and the chlorine reduced. Oxidation and reduction also occurs at the electrodes in *cells.

This definition of oxidation and reduction applies only to reactions in which electron transfer occurs – i.e. to reactions involving ions. It can be extended to reactions between covalent compounds by using the concept of *oxidation number* (or *state*). This is a measure of the electron control that an atom has in a compound compared to the atom in the pure element. An oxidation number consists of two parts:

(1) Its sign, which indicates whether the control has increased (negative) or decreased (positive).

(2) Its value, which gives the number of electrons over which control has changed.

The change of electron control may be complete (in ionic compounds) or partial (in covalent compounds). For example, in SO_2 the sulphur has an oxidation number +4, having gained partial control over 4 electrons compared to sulphur atoms in pure sulphur. The oxygen has an oxidation number −2, each oxygen having lost partial control over 2 electrons compared to oxygen atoms in gaseous oxygen. Oxidation is a reaction involving an increase in oxidation number and reduction involves a decrease. Thus in

$$2H_2 + O_2 \rightarrow 2H_2O$$

the hydrogen in water is +1 and the oxygen −2. The hydrogen is oxidized and the oxygen is reduced.

The oxidation number is used in naming inorganic compounds. Thus in H_2SO_4, sulphuric(VI) acid, the sulphur has an oxidation number of +6. Compounds that tend to undergo reduction readily are *oxidizing agents; those that undergo oxidation are *reducing agents.

oxides Binary compounds formed between elements and oxygen. Oxides of nonmetals are covalent compounds having simple molecules (e.g. CO, CO_2, SO_2) or giant molecular lattices (e.g. SiO_2). They are typically acidic or neutral. Oxides of metals are ionic, containing the O^{2-} ion. They are generally basic or *amphoteric. Various other types of ionic oxide exist (*see* ozonides; peroxides; superoxides).

oxidizing acid An acid that can act as a strong oxidizing agent as well as an acid. Nitric acid is a common example. It is able to attack metals, such as copper, that are below hydrogen in the electromotive series, by oxidizing the metal:

$$2HNO_3 + Cu \rightarrow CuO + H_2O + 2NO_2$$

This is followed by reaction between the acid and the oxide:

$$2HNO_3 + CuO \rightarrow Cu(NO_3)_2 + H_2O$$

oxidizing agent (oxidant) A substance that brings about oxidation in other substances. It achieves this by being itself reduced. Oxidizing agents contain atoms with high oxidation numbers; that is the atoms have suffered electron loss. In oxidizing other substances these atoms gain electrons.

oximes Compounds containing the group C:NOH, formed by reaction of an aldehyde or ketone with hydroxylamine (H_2NOH). Ethanal (CH_3CHO), for example, forms the oxime CH_3CH:NOH.

oxo- Prefix indicating the presence of oxygen in a chemical compound.

oxo acid An *acid in which the acidic hydrogen atom(s) are bound to oxygen atoms. Sulphuric acid is an example: the two acidic hydrogens are on the −OH groups bound to the sulphur. *Compare* binary acid.

oxonium ion An ion of the type R_3O^+, in which R indicates hydrogen or an organic

$$\mathrm{R{-}C({=}O){-}R'} \quad + \quad \mathrm{H_2N{-}O{-}H} \xrightarrow{-\mathrm{H_2O}} \mathrm{R{-}C(R'){=}N{-}O{-}H}$$

ketone hydroxylamine oxime

Formation of an oxime from a ketone
The same reaction occurs with an aldehyde (R′=H)

group. The hydroxonium ion, H_3O^+, is formed when *acids dissociate in water.

oxo process An industrial process for making aldehydes by reaction between alkanes, carbon monoxide, and hydrogen (cobalt catalyst using high pressure and temperature).

oxyacetylene burner A welding or cutting torch that burns a mixture of oxygen and acetylene (ethyne) in a specially designed jet. The flame temperature of about 3300°C enables all ferrous metals to be welded. For cutting, the point at which the steel is to be cut is preheated with the oxyacetylene flame and a powerful jet of oxygen is then directed onto the steel. The oxygen reacts with the hot steel to form iron oxide and the heat of this reaction melts more iron, which is blown away by the force of the jet.

oxygen Symbol O. A colourless odourless gaseous element belonging to *group VI of the periodic table; a.n. 8; r.a.m. 15.9994; d. 1.429 g dm^{-3}; m.p. −214.4°C; b.p. −183°C. It is the most abundant element in the earth's crust (49.2% by weight) and is present in the atmosphere (28% by volume). Atmospheric oxygen is of vital importance for all organisms that carry out aerobic respiration. For industrial purposes it is obtained by fractional distillation of liquid air. It is used in metallurgical processes, in high-temperature flames (e.g. for welding), and in breathing apparatus. The common form is diatomic (*dioxygen*, O_2); there is also a reactive allotrope *ozone (O_3). Chemically, oxygen reacts with most other elements forming *oxides. The element was discovered by Priestley in 1774.

oxyhaemoglobin *See* haemoglobin.

ozone (trioxygen) A colourless gas, O_3, soluble in cold water and in alkalis; m.p. −192.7°C; b.p. −111.9°C. Liquid ozone is dark blue in colour and is diamagnetic (dioxygen, O_2, is paramagnetic). The gas is made by passing oxygen through a silent electric discharge and is usually used in mixtures with oxygen. It is produced in the stratosphere by the action of high-energy ultraviolet radiation on oxygen and its presence there acts as a screen for ultraviolet radiation (*see* ozone layer). It is a powerful oxidizing agent and is used to form ozonides by reaction with alkenes and subsequently by hydrolysis to carbonyl compounds.

ozone layer (ozonosphere) A layer of the *earth's atmosphere in which most of the atmosphere's ozone is concentrated. It occurs 15–50 km above the earth's surface and is virtually synonymous with the stratosphere. In this layer most of the sun's ultraviolet radiation is absorbed by the ozone molecules, causing a rise in the temperature of the stratosphere and preventing vertical mixing so that the stratosphere forms a stable layer. By absorbing most of the solar ultraviolet radiation the ozone layer protects living organisms on earth. The fact that the ozone layer is thinnest at the equator is believed to account for the high equatorial incidence of skin cancer as a result of exposure to unabsorbed solar ultraviolet radiation.

ozonides 1. A group of compounds formed by reaction of ozone with alkali metal hydroxides and formally containing the ion

O_3^-. **2.** Unstable compounds formed by the addition of ozone to the C=C double bond in alkenes. *See* ozonolysis.

ozonolysis A reaction of alkenes with ozone to form an ozonide. It was once used to investigate the structure of alkenes by hydrolysing the ozonide to give aldehydes or ketones. For instance

$$R_2C{:}CHR' \rightarrow R_2CO + R'CHO$$

These could be identified, and the structure of the original alkene determined.

P

palladium Symbol Pd. A soft white ductile *transition element (*see also* platinum metals); a.n. 46; r.a.m. 106.4; r.d. 12.26; m.p. 1551±1°C; b.p. 3140±1°C. It occurs in some copper and nickel ores and is used in jewellery and as a catalyst for hydrogenation reactions. Chemically, it does not react with oxygen at normal temperatures. It dissolves slowly in hydrochloric acid. Palladium is capable of occluding 900 times its own volume of hydrogen. It forms few simple salts, most compounds being complexes of palladium(II) with some palladium(IV). It was discovered by Woolaston in 1803.

palmitate (hexadecanoate) A salt or ester of palmitic acid.

palmitic acid (hexadecanoic acid) A saturated fatty acid, $CH_3(CH_2)_{14}COOH$; r.d. 0.85; m.p. 63°C; b.p. 351°C. Glycerides of palmitic acid occur widely in plant and animal oils and fats.

pantothenic acid A vitamin of the *vitamin B complex. It is a constituent of coenzyme A, which performs a crucial role in the oxidation of fats, carbohydrates, and certain amino acids. Deficiency rarely occurs because the vitamin occurs in many foods, especially cereal grains, peas, egg yolk, liver, and yeast.

papain A protein-digesting enzyme occurring in the fruit of the West Indian papaya tree (*Carica papaya*). It is used as a digestant and in the manufacture of meat tenderizers.

paper chromatography A technique for analysing mixtures by *chromatography, in which the stationary phase is absorbent paper. A spot of the mixture to be investigated is placed near one edge of the paper and the sheet is suspended vertically in a solvent, which rises through the paper by capillary action carrying the components with it. The components move at different rates, partly because they absorb to different extents on the cellulose and partly because of partition between the solvent and the moisture in the paper. The paper is removed and dried, and the different components form a line of spots along the paper. Colourless substances are detected by using ultraviolet radiation or by spraying with a substance that reacts to give a coloured spot (e.g. ninhydrin gives a blue coloration with amino acids). The components can be identified by the distance they move in a given time.

para- 1. Prefix designating a benzene compound in which two substituents are in the 1,4 positions, i.e. directly opposite each other, on the benzene ring. The abbreviation *p*- is used; for example, *p*-xylene is 1,4-dimethylbenzene. *Compare* ortho-; meta-. **2.** Prefix denoting the form of diatomic molecules in which the nuclei have opposite spins, e.g. parahydrogen. *Compare* ortho-.

paraffin *See* petroleum.

paraffins *See* alkanes.

paraffin wax *See* petroleum.

paraformaldehyde *See* methanal.

parahydrogen *See* hydrogen.

paraldehyde *See* ethanal.

partial pressure *See* Dalton's law.

partition If a substance is in contact with two different phases then, in general, it will have a different affinity for each phase. Part of the substance will be absorbed or dissolved by one and part by the other, the relative amounts depending on the relative affinities. The substance is said to be *partitioned* between the two phases. For example, if two immiscible liquids are taken and a third compound is shaken up with them, then an equilibrium is reached in which the concentration in one solvent differs from that in the other. The ratio of the concentrations is the *partition coefficient* of the system. The *partition law* states that this ratio is a constant for given liquids.

partition coefficient *See* partition.

pascal The *SI unit of pressure equal to one newton per square metre.

Paschen series *See* hydrogen spectrum.

passive Describing a solid that has reacted with another substance to form a protective layer, so that further reaction stops. The solid is said to have been 'rendered passive'. For example, aluminium reacts spontaneously with oxygen in air to form a thin layer of *aluminium oxide, which prevents further oxidation. Similarly, pure iron forms a protective oxide layer with concentrated nitric acid and is not dissolved further.

***p*-block elements** The block of elements in the periodic table consisting of the main groups III (B to Tl), IV (C to Pb), V (N to Bi), VI (O to Po), VII (F to At) and 0 (He to Rn). The outer electronic configurations of these elements all have the form ns^2np^x where $x = 1$ to 6. Members at the top and on the right of the *p*-block are nonmetals (C, N, P, O, F, S, Cl, Br, I, At). Those on the left and at the bottom are metals (Al, Ga, In, Tl, Sn, Pb, Sb, Bi, Po). Between the two, from the top left to bottom right, lie an ill-defined group of metalloid elements (B, Si, Ge, As, Te).

peacock ore *See* bornite.

pearl ash *See* potassium carbonate.

pearlite *See* steel.

penicillin An antibiotic derived from the mould *Penicillium notatum*; specifically it is known as *penicillin G* and belongs to a class of similar substances called penicillins. They produce their effects by disrupting synthesis of the bacterial cell wall, and are used to treat a variety of infections caused by bacteria.

pentahydrate A crystalline hydrate containing five molecules of water per molecule of compound.

pentane A straight-chain alkane hydrocarbon, C_5H_{12}; r.d. 0.63; m.p. −129.7°C; b.p. 36.1°C. It is obtained by distillation of petroleum.

pentanoic acid (valeric acid) A colourless liquid *carboxylic acid, $CH_3(CH_2)_3COOH$; r.d. 0.9; m.p. −34°C; b.p. 185°C. It is used in the perfume industry.

pentavalent (quinquevalent) Having a valency of five.

pentlandite A mineral consisting of a mixed iron–nickel sulphide, $(Fe,Ni)_9S_8$, crystallizing in the cubic system; the chief ore of nickel. It is yellowish-bronze in colour with a metallic lustre. The chief occurrence of the mineral is at Sudbury in Ontario, Canada.

pentose A sugar that has five carbon atoms per molecule. *See* monosaccharide.

pentyl group *or* **radical** The organic group $CH_3CH_2CH_2CH_2CH_2-$, derived from pentane.

pepsin An enzyme, secreted by cells lining the interior of the vertebrate stomach, that catalyses the breakdown of proteins.

peptide Any of a group of organic compounds comprising two or more amino acids linked by *peptide bonds*. These bonds are formed by the reaction between adjacent carboxyl (–COOH) and amino ($-NH_2$) groups with the elimination of water (see illustration). *Dipeptides* contain two amino acids, *tripeptides* three, and so on. *Polypeptides contain more than ten and usually 100–300. Naturally occurring *oligopeptides* (of less than ten amino acids) include the tripeptide glutathione and the pituitary hormones vasopressin and oxytocin, which are octapeptides. Peptides also result from protein breakdown, e.g. during digestion.

per- Prefix indicating that a chemical compound contains an excess of an element, e.g. a peroxide.

perchlorate *See* chlorates.

perchloric acid *See* chloric(VII) acid.

perdisulphuric acid *See* peroxosulphuric (VI) acid.

perfect solution *See* Raoult's law.

period *See* periodic table.

periodic acid *See* iodic(VII) acid.

periodic law The principle that the physical and chemical properties of elements are a periodic function of their proton number. The concept was first proposed in 1869 by the Russian chemist Dimitri Mendeleev (1834–1907), using relative atomic mass rather than proton number, as a culmination of efforts to rationalize chemical properties by J. W. Döbereiner (1817), J. A. R. Newlands (1863), and Lothar Meyer (1864). One of the major successes of the periodic law was its ability to predict chemical and physical properties of undiscovered elements and unknown compounds that were later confirmed experimentally. *See* periodic table.

periodic table A table of elements arranged in order of increasing proton number to show the similarities of chemical elements with related electronic configurations. (The original form was proposed by Dimitri Mendeleev in 1869 using relative atomic masses.) In the modern *short form* (see Appendix) the *lanthanoids and *actinoids are not shown. The elements fall into vertical columns, known as *groups*. Going down a group, the atoms of the elements all have the same outer shell structure, but an increasing number of inner shells. Traditionally, the alkali metals are shown on the left of the table and the groups are numbered IA to VIIA, IB to VIIB, and 0 (for the noble gases). It is now more common to classify all the elements in the middle of the table as *transition elements and to regard the nontransition elements as *main-group* elements, numbered from I to VII, with the noble gases in group 0. Horizontal rows in

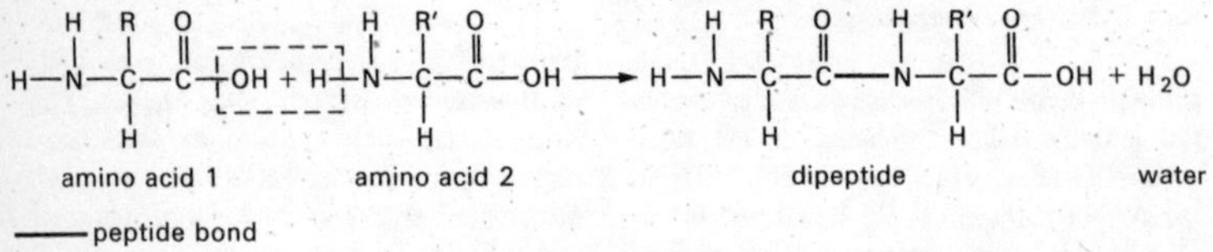

```
   H  R  O              H  R' O              H  R  O   H  R' O
   |  |  ||             |  |  ||             |  |  ||  |  |  ||
H—N—C—C—OH  +  H—N—C—C—OH  ——►  H—N—C—C——N—C—C—OH  +  H2O
      |                    |                    |         |
      H                    H                    H         H
 amino acid 1        amino acid 2               dipeptide              water
```

—— peptide bond

Formation of a peptide bond

the table are *periods*. The first three are called *short periods*; the next four (which include transition elements) are *long periods*. Within a period, the atoms of all the elements have the same number of shells, but with a steadily increasing number of electrons in the outer shell. The periodic table can also be divided into four *blocks* depending on the type of shell being filled: the *s-block, the *p-block, the *d-block, and the *f-block.

There are certain general features of chemical behaviour shown in the periodic table. In moving down a group, there is an increase in metallic character because of the increased size of the atom. In going across a period, there is a change from metallic (electropositive) behaviour to nonmetallic (electronegative) because of the increasing number of electrons in the outer shell. Consequently, metallic elements tend to be those on the left and towards the bottom of the table; nonmetallic elements are towards the top and the right.

There is also a significant difference between the elements of the second short period (lithium to fluorine) and the other elements in their respective groups. This is because the atoms in the second period are smaller and their valence electrons are shielded by a small $1s^2$ inner shell. Atoms in the other periods have inner *s*- and *p*-electrons shielding the outer electrons from the nucleus. Moreover, those in the second period only have *s*- and *p*-orbitals available for bonding. Heavier atoms can also promote electrons to vacant *d*-orbitals in their outer shell and use these for bonding. *See also* diagonal relationship; inert-pair effect.

Permalloys A group of alloys of high magnetic permeability consisting of iron and nickel (usually 40–80%) often with small amounts of other elements (e.g. 3–5% molybdenum, copper, chromium, or tungsten). They are used in thin foils in electronic transformers, for magnetic shielding, and in computer memories.

permanent gas A gas, such as oxygen or nitrogen, that was formerly thought to be impossible to liquefy. A permanent gas is now regarded as one that cannot be liquefied by pressure alone at normal temperatures (i.e. a gas that has a critical temperature below room temperature).

permanent hardness *See* hardness of water.

permanganate *See* manganate(VII).

permonosulphuric(VI) acid *See* peroxosulphuric(VI) acid.

Permutit Tradename for a *zeolite used for water softening.

peroxides 1. A group of inorganic compounds that contain the $O_2{}^{2-}$ ion. They are notionally derived from hydrogen peroxide, H_2O_2, but these ions do not exist in aqueous solution due to extremely rapid hydrolysis to OH^-.

peroxodisulphuric acid *See* peroxosulphuric(VI) acid.

peroxomonosulphuric(VI) acid *See* peroxosulphuric(VI) acid.

peroxosulphuric(VI) acid The term commonly refers to *peroxomonosulphuric(VI) acid*, H_2SO_5, which is also called *permonosulphuric(VI) acid* and *Caro's acid*. It is a crystalline compound made by the action of hydrogen peroxide on concentrated sulphuric acid. It decomposes in water and the crystals decompose, with melting, above 45°C. The compound *peroxodisulphuric acid*, $H_2S_2O_8$, also exists (formerly called *perdisulphuric acid*). It is made by the high-current electrolysis of sulphate solutions. It decomposes at 65°C (with melting) and is hydrolysed in water to give the mono acid and sulphuric acid. Both peroxo acids are very powerful oxidizing agents. *See also* sulphuric acid (for structural formulas).

Perspex Tradename for a form of *polymethylmethacrylate.

peta- Symbol P. A prefix used in the metric system to denote one thousand million million times. For example, 10^{15} metres = 1 petametre (Pm).

petrochemicals Organic chemicals obtained from petroleum or natural gas.

petroleum A naturally occurring oil that consists chiefly of hydrocarbons with some other elements, such as sulphur, oxygen, and nitrogen. In its unrefined form petroleum is known as *crude oil* (sometimes *rock oil*). Petroleum is believed to have been formed from the remains of living organisms that were deposited, together with rock particles and biochemical and chemical precipitates, in shallow depressions, chiefly in marine conditions. Under burial and compaction the organic matter went through a series of processes before being transformed into petroleum, which migrated from the source rock to become trapped in large underground reservoirs beneath a layer of impermeable rock. The petroleum often floats above a layer of water and is held under pressure beneath a layer of *natural gas.

Petroleum reservoirs are discovered through geological exploration: commercially important oil reserves are detected by exploratory narrow-bore drilling. The major known reserves of petroleum are in Saudi Arabia, USSR, China, Kuwait, Iran, Iraq, Mexico, USA, United Arab Emirates, Libya, and Venezuela. The oil is actually obtained by the sinking of an oil well. Before it can be used it is separated by fractional distillation in oil refineries. The main fractions obtained are:

(1) *Refinery gas* A mixture of methane, ethane, butane, and propane used as a fuel and for making other organic chemicals.

(2) *Gasoline* A mixture of hydrocarbons containing 5 to 8 carbon atoms, boiling in the range 40–180°C. It is used for motor fuels and for making other chemicals.

(3) *Kerosine* (or *paraffin oil*) A mixture of hydrocarbons having 11 or 12 carbon atoms, boiling in the range 160–250°C. Kerosine is a fuel for jet aircraft and for oil-fired domestic heating. It is also cracked to produce smaller hydrocarbons for use in motor fuels.

(4) *Diesel oil* (or *gas oil*) A mixture of hydrocarbons having 13 to 25 carbon atoms, boiling in the range 220–350°C. It is a fuel for diesel engines.

The residue is a mixture of higher hydrocarbons. The liquid components are obtained by vacuum distillation and used in lubricating oils. The solid components (*paraffin wax*) are obtained by solvent extraction. The final residue is a black tar containing free carbon (*asphalt* or *bitumen*).

petroleum ether A colourless volatile flammable mixture of hydrocarbons (not an ether), mainly pentane and hexane. It boils in the range 30–70°C and is used as a solvent.

pewter An alloy of lead and tin. It usually contains 63% tin; pewter tankards and food containers should have less than 35% of lead so that the lead remains in solid solution with the tin in the presence of weak acids in the food and drink. Copper is sometimes added to increase ductility and antimony is added if a hard alloy is required.

pH *See* pH scale.

phase A homogeneous part of a heterogeneous system that is separated from other parts by a distinguishable boundary. A mixture of ice and water is a two-phase system. A solution of salt in water is a single-phase system.

phase diagram A graph showing the relationship between solid, liquid, and gaseous *phases over a range of conditions (e.g. temperature and pressure). *See* steel (illustration).

phase rule For any system at equilibrium, the relationship $P + F = C + 2$ holds, where P is the number of distinct phases, C the number of components, and F the number of degrees of freedom of the system. The relationship derived by Josiah Willard Gibbs in 1876, is often called the *Gibbs phase rule.*

phase space *See* statistical mechanics.

phenol (carbolic acid) A white crystalline solid, C_6H_5OH; r.d. 1.1; m.p. 42°C; b.p. 182°C. It is made by the *cumene process or by the *Raschig process and is used to make a variety of other organic chemicals. *See also* phenols.

phenolphthalein A dye used as an acid-base *indicator. It is colourless below pH 8 and red above pH 9.6. It is used in titrations involving weak acids and strong bases. It is also used as a laxative.

phenols Organic compounds that contain a hydroxyl group (–OH) bound directly to a carbon atom in a benzene ring. Unlike normal alcohols, phenols are acidic because of the influence of the aromatic ring. Thus, phenol itself (C_6H_5OH) ionizes in water:

$$C_6H_5OH \rightarrow C_6H_5O^- + H^+$$

Phenols are made by fusing a sulphonic acid salt with sodium hydroxide to form the sodium salt of the phenol. The free phenol is liberated by adding sulphuric acid.

phenylalanine *See* amino acid.

phenylamine (aniline; aminobenzene) A colourless oily liquid aromatic *amine, $C_6H_5NH_2$, with an 'earthy' smell; r.d. 1.027; m.p. –6.2°C; b.p. 185°C. The compound turns brown on exposure to sunlight. It is basic, forming the *phenylammonium* (or *anilinium*) *ion*, $C_6H_5NH_3^+$, with strong acids. It is manufactured by the reduction of nitrobenzene or by the addition of ammonia to chlorobenzene using a copper(II) salt catalyst at 200°C and 55 atm. The compound is used extensively in the rubber industry and in the manufacture of drugs and dyes.

phenylammonium ion The ion $C_6H_5NH_3^+$, derived from *phenylamine.

phenylethene (styrene) A liquid hydrocarbon, $C_6H_5CH{:}CH_2$; r.d. 0.9; m.p. –31°C; b.p. 145°C. It can be made by dehydrogenating ethylbenzene and is used in making polystyrene.

phenyl group The organic group C_6H_5-, present in benzene.

phenylhydrazones *See* hydrazones.

phenylmethanol (benzyl alcohol) A liquid aromatic alcohol, $C_6H_5CH_2OH$; r.d. 1.04; m.p. –15.3°C; b.p. 205.4°C. It is used mainly as a solvent.

3-phenylpropenoic acid *See* cinnamic acid.

Phillips process A process for making high-density polyethene by polymerizing ethene at high pressure (30 atmospheres) and 150°C. The catalyst is chromium(III) oxide supported on silica and alumina.

phlogiston theory A former theory of combustion in which all flammable objects were supposed to contain a substance called *phlogiston*, which was released when the object burned. The existence of this hypothetical substance was proposed in 1669 by Johann Becher, who called it 'combustible earth' (*terra pinguis*: literally 'fat earth'). For example, according to Becher, the conversion of wood to ashes by burning was explained on the assumption that the original wood consisted of ash and *terra pinguis*, which was released on burning. In the early 18th century Georg Stahl renamed the substance *phlogiston* (from the Greek for 'burned') and extended the theory to include the calcination (and corrosion) of metals. Thus, metals were thought to be composed of *calx* (a powdery residue) and phlogiston; when a metal was heated, phlogiston was set free and the calx remained. The process could be reversed by heating the metal over charcoal (a substance believed to be rich in phlogiston, because combustion almost totally consumed it). The calx would absorb the phlogiston released by the burning charcoal and become metallic again.

The theory was finally demolished by Antoine Lavoisier, who showed by careful experiments with reactions in closed containers that there was no *absolute* gain in mass – the gain in mass of the substance was matched by a corresponding loss in mass of the air used in combustion. After experiments with Priestley's dephlogisticated air, Lavoisier realized that this gas, which he named oxygen, was taken up to form a calx (now called an oxide). The role of oxygen in the new theory was almost exactly the opposite of phlogiston's role in the old. In combustion and corrosion phlogiston was released; in the modern theory, oxygen is taken up to form an oxide.

phosgene *See* carbonyl chloride.

phosphagen A compound found in animal tissues that provides a reserve of chemical energy in the form of high-energy phosphate bonds. The most common phosphagens are creatine phosphate, occurring in vertebrate muscle and nerves, and arginine phosphate, found in most invertebrates. During tissue activity (e.g. muscle contraction) phosphagens give up their phosphate groups, thereby generating *ATP from ADP. The phosphagens are then reformed when ATP is available.

phosphates Salts based formally on phosphorus(V) oxoacids and in particular salts of *phosphoric(V) acid, H_3PO_4. A large number of polymeric phosphates also exist, containing P–O–P bridges. These are formed by heating the free acid and its salts under a variety of conditions; as well as linear polyphosphates, cyclic polyphosphates and or cross-linked polyphosphates ultraphosphates are known.

phosphatide *See* phospholipid.

phosphide A binary compound of phosphorus with a more electropositive element. Phosphides show a wide range of properties. Alkali and alkaline earth metals form ionic phosphides, such as Na_3P and Ca_3P_2, which are readily hydrolysed by water. The other transition-metal phosphides are inert metallic-looking solids with high melting points and electrical conductivities.

phosphine A colourless highly toxic gas, PH_3; m.p. –133°C; b.p. –87.7°C; slightly soluble in water. Phosphine may be prepared by reacting water or dilute acids with calcium phosphide or by reaction between yellow phosphorus and concentrated alkali. Solutions of phosphine are neutral but phosphine does react with some acids to give phosphonium salts containing PH_4^+ ions, analogous to the ammonium ions. Phosphine prepared in the laboratory is usually contaminated with diphosphine and is spontaneously flammable but the pure compound is not so. Phosphine can function as a ligand in binding to transition-metal ions. Dilute gas mixtures of very pure phosphine and the rare gases are used for doping semiconductors.

phosphinic acid (hypophosphorus acid) A white crystallline solid, H_3PO_2; r.d. 1.493; m.p. 26.5°C; decomposes above

130°C. It is soluble in water, ethanol, and ethoxyethane. Salts of phosphinic acid may be prepared by boiling white phosphorus with the hydroxides of group I or group II metals. The free acid is made by the oxidation of phosphine with iodine. It is a weak monobasic acid in which it is the –O–H group that is ionized to give the ion $H_2PO_2^-$. The acid and its salts are readily oxidized to the orthophosphate and consequently are good reducing agents.

phosphite *See* phosphonic acid.

phospholipid (phosphatide) One of a group of lipids having both a phosphate group and one or more fatty acids. *Glycerophospholipids* are based on *glycerol; the three hydroxyl groups are esterified with two fatty acids and a phosphate group, which may itself be bound to one of a variety of simple organic groups. *Sphingophospholipids* are based on the alcohol sphingosine and contain only one fatty acid linked to an amino group. With their hydrophilic polar phosphate groups and long hydrophobic hydrocarbon 'tails', phospholipids readily form membrane-like structures in water (*see* micelle). They are a major component of cell membranes.

phosphonate *See* phosphonic acid.

phosphonic acid (phosphorous acid; orthophosphorous acid) A colourless to pale-yellow deliquescent crystalline solid, H_3PO_3; r.d. 1.65; m.p. 73.6°C; decomposes at 200°C; very soluble in water and soluble in alcohol. Phosphonic acid may be crystallized from the solution obtained by adding ice-cold water to phosphorus(III) oxide or phosphorus trichloride. The structure of this material is unusual in that it contains one direct P–H bond and is more correctly written $(HO)_2HPO$. The acid is dibasic, giving rise to the ions $H_2PO_3^-$ and HPO_3^{2-} (*phosphonates*; formerly *phosphites*), and has moderate reducing properties. On heating it gives phosphine and phosphoric(V) acid.

phosphonium ion The ion PH_4^+, or the corresponding organic derivatives of the type R_3PH^+, RPH_3^+. The phosphonium ion PH_4^+ is formally analogous to the ammonium ion NH_4^+ but PH_3 has a much lower proton affinity than NH_3 and reaction of PH_3 with acids is necessary for the production of phosphonium salts.

phosphor A substance that is capable of *luminescence (including phosphorescence). Phosphors that release their energy after a short delay of between 10^{-10} and 10^{-4} second are sometimes called *scintillators*.

phosphor bronze An alloy of copper containing 4% to 10% of tin and 0.05% to 1% of phosphorus as a deoxidizing agent. It is used particularly for marine purposes and where it is exposed to heavy wear, as in gear wheels. *See also* bronze.

phosphorescence *See* luminescence.

phosphoric(V) acid (orthophosphoric acid) A white rhombic solid, H_3PO_4; r.d. 1.834; m.p. 42.35°C; loses water at 213°C; very soluble in water and soluble in ethanol. Phosphoric(V) acid is very deliquescent and is generally supplied as a concentrated aqueous solution. It is the most commercially important derivative of phosphorus, accounting for over 90% of the phosphate rock mined. It is manufactured by two methods; the *wet process*, in which the product contains some of the impurities originally present in the rock and applications are largely in the fertilizer industry, and the *thermal process*, which produces a much purer product suitable for the foodstuffs and detergent industries. In the wet process the phosphate rock, $Ca_3(PO_4)_2$, is treated with sulphuric acid and the calcium sulphate removed either as gypsum or the hemihydrate. In the thermal process, molten phosphorus is sprayed and burned in a mixture of air and steam. Phosphoric(V) acid is a weak tribasic acid, which is best visualized as $(HO)_3PO$. Its full systematic name is *tetraoxo-phosphoric(V) acid.* It gives rise to three series of salts containing *phosphate(V)* ions based on the anions $[(HO)_2PO_2]^-$, $[(HO)PO_3]^{2-}$, and PO_4^{3-}. These salts are acidic, neutral, and alkaline in character respectively and phosphate ions often feature in buffer systems. There is also a wide range of higher acids and acid anions in which there is some P–O–P chain formation. The simplest of these is *pyrophosphoric acid* (technically *heptaoxodiphosphoric(V) acid*), $H_4P_2O_7$, produced by heating phosphoric (V) acid (solid) and phosphorus(III) chloride oxide. *Metaphosphoric acid* is a glassy polymeric solid $(HPO_2)_x$.

phosphorous acid *See* phosphonic acid.

phosphorus Symbol P. A nonmetallic element belonging to *group V of the periodic table; a.n. 15; r.a.m. 30.9738; r.d. 1.82 (white), 2.20 (red); m.p. 44.1°C (α-white); b.p. 280°C (α-white). It occurs in various phosphate rocks, from which it is extracted by heating with carbon (coke) and silicon (IV) oxide in an electric furnace (1500°C). Calcium silicate and carbon monoxide are also produced. Phosphorus has a number of allotropic forms. The α-white form consists of P_4 tetrahedra (there is also a β-white form stable below −77°C). If α-white phosphorus is dissolved in lead and heated at 500°C a violet form is obtained. Red phosphorus, which is a combination of violet and white phosphorus, is obtained by heating α-white phosphorus at 250°C with air excluded. There is also a black allotrope, which has a graphite-like structure, made by heating white phosphorus at 300°C with a mercury catalyst. The element is highly reactive. If forms metal *phosphides and covalently bonded phosphorus(III) and phosphorus(V) compounds. Phosphorus is an *essential element for living organisms. It was discovered by Brandt in 1669.

phosphorus(III) bromide (phosphorus tribromide) A colourless fuming liquid, PBr_3; r.d. 2.85; m.p. −40°C; b.p. 173°C. It is prepared by passing bromine vapour over phosphorus but avoiding an excess, which would lead to the phosphorus(V) bromide. Like the other phosphorus(III) halides, PBr_3 is pyramidal in the gas phase. In the liquid phase the P–Br bonds are labile; for example, PBr_3 will react with PCl_3 to give a mixture of products in which the halogen atoms have been redistributed. Phosphorus (III) bromide is rapidly hydrolysed by water to give phosphonic acid and hydrogen bromide. It reacts readily with many organic hydroxyl groups and is used as a reagent for introducing bromine atoms into organic molecules.

phosphorus(V) bromide (phosphorus pentabromide) A yellow readily sublimable solid, PBr_5, which decomposes below 100°C and is soluble in benzene and carbon tetrachloride (tetrachloromethane). It may be prepared by the reaction of phosphorus(III) bromide with bromine or the direct reaction of phosphorus with excess bromine. It is very readily hydrolysed to give hydrogen bromide and phosphoric(V) acid. An interesting feature of this material is that in the solid state it has the structure $[PBr_4]^+Br^-$. It is used in organic chemistry as a brominating agent.

phosphorus(III) chloride (phosphorus trichloride) A colourless fuming liquid, PCl_3; r.d. 1.57; m.p. −112°C; b.p. 75.5°C. It is soluble in ether and in carbon tetrachloride but reacts with water and with ethanol. It may be prepared by passing chlorine over excess phosphorus (excess chlorine contaminates the product with phosphorus (V) chloride). The molecule is pyramidal in the gas phase and possesses weak electron-pair donor properties. It is hydrolysed violently by water to phosphonic acid and hydrogen chloride. Phosphorus(III) chloride is an important starting point for the synthesis of a variety of inorganic and organic derivatives of phosphorus.

phosphorus(V) chloride (phosphorus pentachloride) A yellow-white rhombic solid, PCl_5, which fumes in air; r.d. 3.6; m.p. 148°C (under pressure); sublimes at 160–162°C. It is decomposed by water to give hydrogen chloride and phosphoric(V) acid. It is soluble in organic solvents. The compound may be prepared by the reaction of chlorine with phosphorus(III) chloride. Phosphorus(V) chloride is structurally interesting in that in the gas phase it has the expected trigonal bipyramidal form but in the solid phase it consists of the ions $[PCl_4]^+[PCl_6]^-$. The same ions are detected when phosphorus(V) chloride is dissolved in polar solvents. It is used in organic chemisty as a chlorinating agent.

phosphorus(III) chloride oxide (phosphorus oxychloride; phosphoryl chloride) A colourless fuming liquid, $POCl_3$; r.d. 1.67; m.p. 2°C; b.p. 105.3°C. It may be prepared by the reaction of phosphorus(III) chloride with oxygen or by the reaction of phosphorus(V) oxide with phosphorus(V) chloride. Its reactions are very similar to those of phosphorus(III) chloride. Hydrolysis with water gives phosphoric(V) acid. Phosphorus(III) chloride oxide has a distorted tetrahedral shape and can act as a donor towards metal ions, thus giving rise to a series of complexes.

phosphorus(III) oxide (phosphorus trioxide) A white or colourless waxy solid, P_4O_6; r.d. 2.13; m.p. 23.8°C; b.p. 173.8°C. It is soluble in ether, chloroform, and benzene but reacts with cold water to give phosphonic acid, H_3PO_3, and with hot water to give phosphine and phosphoric(V) acid. The compound is formed when phosphorus is burned in an oxygen-deficient atmosphere (about 50% yield). As it is difficult to separate from white phosphorus by distillation, the mixture is irradiated with ultraviolet radiation to convert excess white phosphorus into the red form, after which the oxide can be separated by dissolution in organic solvents. Although called a trioxide for historical reasons, phosphorus(III) oxide consists of P_4O_6 molecules of tetrahedral symmetry in which each phosphorus atom is linked to the three others by an oxygen bridge. The chemistry is very complex. Above 210°C it decomposes into red phosphorus and polymeric oxides. It reacts with chlorine and bromine to give oxo-halides and with alkalis to give phosphonates (*see* phosphonic acid).

phosphorus(V) oxide (phosphorus pentoxide; phosphoric anhydride) A white powdery and extremely deliquescent solid, P_4O_{10}; r.d. 2.39; m.p. 580°C under pressure; sublimes at 360°C. It reacts violently with water to give phosphoric(V) acid. It is prepared by burning elemental phosphorus in a plentiful supply of oxygen, then purified by sublimation. The hexagonal crystalline form consists of P_4O_{10} molecular units; these have the phosphorus atoms arranged tetrahedrally, each P atom linked to three others by oxygen bridges and having in addition one terminal oxygen atom. The compound is used as a drying agent and as a dehydrating agent; for example, amides are converted into nitrites and sulphuric acid is converted to sulphur trioxide.

phosphorus oxychloride *See* phosphorus (III) chloride oxide.

phosphorus pentabromide *See* phosphorus (V) bromide.

phosphorus pentachloride *See* phosphorus (V) chloride.

phosphorus tribromide *See* phosphorus (III) bromide.

phosphorus trichloride *See* phosphorus (III) chloride.

phosphorus trioxide *See* phosphorus(III) oxide.

phosphoryl chloride *See* phosphorus(III) chloride oxide.

photochemical reaction A chemical reaction caused by light or ultraviolet radiation. The incident photons are absorbed by reactant molecules to give excited molecules or free radicals, which undergo further reaction.

photochemistry The branch of chemistry concerned with *photochemical reactions.

photochromism A change of colour occurring in certain substances when exposed to light. Photochromic materials are used in sunglasses that darken in bright sunlight.

photoconductive effect *See* photoelectric effect.

photoelectric effect The liberation of electrons from a metal surface exposed to electromagnetic radiation. The number of electrons emitted depends on the intensity of the radiation. The kinetic energy of the electrons emitted depends on the frequency of the radiation. The effect is a quantum process in which the radiation is regarded as a stream of *photons, each having an energy hf, where h is the Planck constant and f is the frequency of the radiation. A photon can only eject an electron if the photon energy exceeds the *work function, ϕ, of the solid, i.e. if $hf_0 = \phi$ an electron will be ejected; f_0 is the minimum frequency (or *threshold frequency*) at which ejection will occur. For many metals the photoelectric effect occurs at ultraviolet frequencies or above, but for some materials (having low work functions) it occurs with light. The maximum kinetic energy, E_m, of the photoelectron is given by *Einstein's equation: $E_m = hf - \phi$.

Apart from the liberation of electrons from metals other phenomena are also referred to as photoelectric effects. These are the *photoconductive effect* and the *photovoltaic effect*. In the photoconductive effect, an increase in the electrical conductivity of a semiconductor is caused by radiation as a result of the excitation of additional free charge carriers by the incident photons. *Photoconductive cells*, using such photosensitive materials as cadmium sulphide, are widely used as radiation detectors and light switches (e.g. to switch on street lighting).

In the photovoltaic effect, an e.m.f. is produced between two layers of different materials as a result of irradiation. The effect is made use of in *photovoltaic cells*, most of which consist of $p-n$ semiconductor junctions. When photons are absorbed near a $p-n$ junction new free charge carriers are produced (as in photoconductivity); however, in the photovoltaic effect the electric field in the junction region causes the new charge carriers to move, creating a flow of current in an external circuit without the need for a battery.

photoelectron spectroscopy A technique for determining the *ionization potentials of molecules. The sample is a gas or vapour irradiated with a narrow beam of ultraviolet radiation (usually from a helium source at 58.4 nm, 21.21 eV photon energy). The photoelectrons produced in accordance with *Einstein's equation are passed through a slit into a vacuum region, where they are deflected by magnetic or electrostatic fields to give an energy spectrum. The photoelectron spectrum obtained has peaks corresponding to the ionization potentials of the molecule (and hence the orbital energies). The technique also gives information on the vibrational energy levels of the ions formed. *ESCA* (electron spectroscopy for chemical analysis) is a similar analytical technique in which a beam of X-rays is used. In this case, the electrons ejected are from the inner shells of the atoms. Peaks in the electron spectrum for a particular element show characteristic chemical shifts, which depend on the presence of other atoms in the molecule.

photoionization The *ionization of an atom or molecule as a result of irradiation by electromagnetic radiation. For a photoionization to occur the incident photon of the radiation must have an energy in excess of the *ionization potential of the species being irradiated. The ejected photoelectron will have an energy, E, given by $E = hf - I$, where h is the Planck constant, f is the frequency of the incident radiation, and I is the ionization potential of the irradiated species.

photoluminescence *See* luminescence.

photolysis A chemical reaction produced by exposure to light or ultraviolet radiation. Photolytic reactions often involve free radicals, the first step being homolytic fission of a chemical bond. *See* flash photolysis.

photon A particle with zero rest mass consisting of a *quantum of electromagnetic radiation. The photon may also be regarded as a unit of energy equal to hf, where h is the *Planck constant and f is the frequency of the radiation in hertz. Photons travel at the speed of light. They are required to explain the photoelectric effect and other phenomena that require light to have particle character.

photosensitive substance **1.** Any substance that when exposed to electromagnetic radiation produces a photoconductive, photoelectric, or photovoltaic effect. **2.** Any substance, such as the emulsion of a photographic film, in which electromagnetic radiation produces a chemical change.

photosynthesis The chemical process by which green plants synthesize organic compounds from carbon dioxide and water in the presence of sunlight. It occurs in the chloroplasts (most of which are in the leaves) and there are two principal series of reactions. In the *light reactions*, which require the presence of light, energy from sunlight is absorbed by *photosynthetic pigments (chiefly the green pigment *chlorophyll) and converted into chemical energy. In the ensuing *dark reactions*, which can take place either in light or darkness, this chemical energy is used in the production of simple organic compounds from carbon dioxide and water. Further chemical reactions convert these compounds into chemicals useful to the plant. Photosynthesis can be summarized by the equation:

$$CO_2 + 2H_2O \rightarrow [CH_2O] + H_2O + O_2$$

Since virtually all other forms of life are directly or indirectly dependent on plants for food, photosynthesis is the basis for all life on earth. Furthermore virtually all the atmospheric oxygen has originated from oxygen released during photosynthesis.

photosynthetic pigments The plant pigments responsible for the capture of light energy during the light reactions of *photosynthesis. The green pigment *chlorophyll is the principal light receptor, absorbing blue and red light. However the *carotenoids and various other pigments also absorb light energy and pass this on to the chlorophyll molecules.

pH scale A logarithmic scale for expressing the acidity or alkalinity of a solution. To a first approximation, the pH of a solution can be defined as $-\log_{10}c$, where c is the concentration of hydrogen ions in moles per cubic decimetre. A neutral solution at 25°C has a hydrogen-ion concentration of 10^{-7} mol dm^{-3}, so the pH is 7. A pH below 7 indicates an acid solution; one above 7 indicates an alkaline solution. More accurately, the pH depends not on the concentration of hydrogen ions but on their *activity, which cannot be measured experimentally. For practical purposes, the pH scale is defined by using a hydrogen electrode in the solution of interest as one half of a cell, with a reference electrode (e.g. a calomel electrode) as the other half cell. The pH is then given by $(E - E_R)F/2.303RT$, where E is the e.m.f. of the cell and E_R the standard electrode potential of the reference electrode. In practice, a glass electrode is more convenient than a hydrogen electrode.

pH stands for 'potential of hydrogen'. The scale was introduced by S. P. Sørensen in 1909.

phthalic acid A colourless crystalline dicarboxylic acid, $C_6H_4(COOH)_2$; r.d. 1.6; m.p. 207°C. The two –COOH groups are substituted on adjacent carbon atoms of the ring, the technical name being *benzene-1,2-dicarboxylic acid*. The acid is made from *phthalic anhydride* (benzene-1,2-dicarboxylic anhydride, $C_8H_4O_3$), which is made by the catalytic oxidation of naphthalene. The anhydride is used in making plasticizers and polyester resins.

phthalic anhydride *See* phthalic acid.

physical chemistry The branch of chemistry concerned with the effect of chemical structure on physical properties. It includes chemical thermodynamics and electrochemistry.

physics The study of the laws that determine the structure of the universe with reference to the matter and energy of which it consists. It is concerned not with chemical changes that occur but with the forces that exist between objects and the interrelationship between matter and energy. Traditionally, the study was divided into separate fields: heat, light, sound, electricity and magnetism, and mechanics. Since the turn of the century, however, quantum mechanics and relativistic physics have become increasingly important; the growth of modern physics has been accompanied by the studies of atomic physics, nuclear physics, and particle physics. The physics of astronomical bodies and their interactions is known as *astrophysics*, the physics of the earth is known as *geophysics*, and the study of the physical aspects of biology is called *biophysics*.

physisorption *See* adsorption.

pi bond *See* orbital.

pico- Symbol p. A prefix used in the metric system to denote 10^{-12}. For example, 10^{-12} farad = 1 picofarad (pF).

picrate A salt or ester of picric acid.

picric acid (2,4,6-trinitrophenol) A yellow highly explosive nitro compound, $C_6H_2(NO_2)_3$; r.d. 1.8; m.p. 122°C.

pi electron An electron in a pi orbital. *See* orbital.

pig iron The impure form of iron produced by a blast furnace, which is cast into pigs (blocks) for converting at a later date into cast iron, steel, etc. The composition depends on the ores used, the smelting procedure, and the use to which the pigs will later be put.

pi orbital *See* orbital.

pipette A graduated tube used for transferring measured volumes of liquid.

pirsonnite A mineral consisting of a hydrated mixed carbonate of sodium and calcium, $Na_2CO_3.CaCO_3.2H_2O$.

pitch A black or dark-brown residue resulting from the distillation of coal tar, wood tar, or petroleum (bitumen). The term is also sometimes used for the naturally occurring petroleum residue (asphalt). Pitch is used as a binding agent (e.g. in road tars), for waterproofing (e.g. in roofing felts), and as a fuel.

pitchblende *See* uraninite.

pK value A measure of the strength of an acid on a logarithmic scale. The pK value is given by $\log_{10}(1/K_a)$, where K_a is the acid dissociation constant. pK values are often used to compare the strengths of different acids.

Planck constant Symbol *h*. The fundamental constant equal to the ratio of the energy of a quantum of energy to its frequency. It has the value $6.626\,196 \times 10^{-34}$ J s. It is named after Max Planck (1858–1947).

plane-polarized light *See* polarization of light.

plaster of Paris The hemihydrate of *calcium sulphate, $2CaSO_4.H_2O$, prepared by heating the mineral gypsum. When ground to a fine powder and mixed with water, plaster of Paris sets hard, forming interlocking crystals of gypsum. The setting results in an increase in volume and so the plaster fits tightly into a mould. It is used in pottery making, as a cast for setting broken bones, and as a constituent of the plaster used in the building industry.

plasticizer A substance added to a synthetic resin to make it flexible. *See* plastics.

plastics Materials that can be shaped by applying heat or pressure. Most plastics are made from polymeric synthetic *resins, although a few are based on natural substances (e.g. cellulose derivatives or shellac). They fall into two main classes. *Thermoplastic materials* can be repeatedly softened by heating and hardened again on cooling. *Thermosetting materials* are initially soft, but change irreversibly to a hard rigid form on heating. Plastics contain the synthetic resin mixed with such additives as pigments, plasticizers (to improve flexibility), antioxidants and other stabilizers, and fillers.

platinum Symbol Pt. A silvery white metallic *transition element (*see also* platinum metals); a.n. 78; r.a.m. 195.09; r.d. 21.37; m.p. 1772°C; b.p. ~3800°C. It occurs in some nickel and copper ores and is also found native in some deposits. The main source is the anode sludge obtained in copper–nickel refining. The element is used in jewellery, laboratory apparatus (e.g. thermocouples, electrodes, etc.), electrical contacts, and in certain alloys (e.g. with iridium or rhodium). It is also a hydrogenation catalyst. The element does not oxidize nor dissolve in hydrochloric acid. Most of its compounds are platinium(II) or platinium(IV) complexes.

platinum black Black finely divided platinum metal produced by vacuum evaporation and used as an absorbent and a catalyst.

platinum metals The three members of the second and third transition series immediately proceeding silver and gold: ruthenium (Ru), rhodium (Rh), and palladium (Pd); and osmium (Os), iridium (Ir), and platinum (Pt). These elements, together with iron, cobalt, and nickel, were formerly classed as group VIII of the periodic table. The platinum-group metals are relatively hard and resistant to corrosion and are used in jewellery and in some industrial applications (e.g. electrical contacts). They have certain chemical similarities that justify classifying them together. All are resistant to chemical attack. In solution they form a vast range of complex ions. They also form coordination compounds with carbon monoxide and other pi-bonding ligands. A number of complexes can be made in which a hydrogen atom is linked directly to the metal. The metals and their organic compounds have considerable catalytic activity. *See also* transition elements.

pleochroic Denoting a crystal that appears to be of different colours, depending on the direction from which it is viewed. It is caused by polarization of light as it passes through an anisotropic medium.

plumbago *See* carbon.

plumbane (lead(IV) hydride) An extremely unstable gas, PbH_4, said to be formed by the action of acids on magnesium–lead alloys. It was first reported in 1924, although doubts have since been expressed about the existence of the compound. It demonstrates the declining stability of the hydrides in group IV. More stable organic derivatives are known; e.g. trimethyl plumbane, $(CH_3)_3PbH$.

plumbate A compound formed by reaction of lead oxides (or hydroxides) with alkali. The oxides of lead are amphoteric (weakly acidic) and react to give plumbate ions. With the lead(IV) oxide, reaction with molten alkali gives the plumbate(IV) ion

$$PbO_2 + 2OH^- \rightarrow PbO_3^{2-} + H_2O$$

In fact, various ions are present in which the lead is bound to hydroxide groups, the principal one being the hexahydroxoplumbate(IV) ion $Pb(OH)_6^{2-}$. This is the negative ion present in crystalline 'trihydrates' of the type $K_2PbO_3.3H_2O$. Lead(II) oxide gives the trihydroxoplumbate(II) ion in alkaline solutions

$$PbO(s) + OH^-(aq) + H_2O(l) \rightarrow Pb(OH)_3^{2-}(aq)$$

Plumbate(IV) compounds were formerly referred to as *orthoplumbates* (PbO_4^{4-}) or *metaplumbates* (PbO_3^{2-}). Plumbate(II) compounds were called *plumbites*.

plumbic compounds Compounds of lead in its higher (+4) oxidation state; e.g. plumbic oxide is lead(IV) oxide, PbO_2.

plumbite *See* plumbate.

plumbous compounds Compounds of lead in its lower (+2) oxidation state; e.g. plumbous oxide is lead(II) oxide, PbO.

plutonium Symbol Pu. A dense silvery radioactive metallic transuranic element belonging to the *actinoids; a.n. 94; mass number of most stable isotope 244 (half-life 7.6×10^7 years); r.d. 19.84; m.p. 641°C; b.p. 3232°C. Thirteen isotopes are known, by far the most important being plutonium–239 (half-life 2.44×10^4 years), which undergoes nuclear fission with slow neutrons and is therefore a vital power source for nuclear weapons and some nuclear reactors. About 20 tonnes of plutonium are produced annually by the world's nuclear reactors. The element was first produced by Seaborg, McMillan, Kennedy, and Wahl in 1940.

poise A *c.g.s. unit of viscosity equal to the tangential force in dynes per square centimetre required to maintain a difference in velocity of one centimetre per second between two parallel planes of a fluid separated by one centimetre. 1 poise is equal to 10^{-1} N s m^{-2}.

poison 1. Any substance that is injurious to the health of a living organism. **2.** A substance that prevents the activity of a catalyst. **3.** A substance that absorbs neutrons in a nuclear reactor and therefore slows down the reaction. It may be added intentionally for this purpose or may be formed as a fission product and need to be periodically removed.

polar compound A compound that is either ionic (e.g. sodium chloride) or that has molecules with a large permanent dipole moment (e.g. water).

polariscope (polarimeter) A device used to study optically active substances (*see* optical activity). The simplest type of instrument consists of a light source, collimator, polarizer, and analyser. The specimen is placed between polarizer and analyser, so that any rotation of the plane of polarization of the light can be assessed by turning the analyser.

polarization **1.** The process of confining the vibrations of the vector constituting a transverse wave to one direction. In unpolarized radiation the vector oscillates in all directions perpendicular to the direction of propagation. *See* polarization of light. **2.** The formation of products of the chemical reaction in a *voltaic cell in the vicinity of the electrodes resulting in increased resistance to current flow and, frequently, to a reduction in the e.m.f. of the cell. *See also* depolarization. **3.** The partial separation of electric charges in an insulator subjected to an electric field. **4.** The separation of charge in a polar *chemical bond.

polarization of light The process of confining the vibrations of the electric vector of light waves to one direction. In unpolarized light the electric field vibrates in all directions perpendicular to the direction of propagation. After reflection or transmission through certain substances (*see* Polaroid) the electric field is confined to one direction and the radiation is said to be *plane-polarized light*. The plane of plane-polarized light can be rotated when it passes through certain substances (*see* optical activity).

In *circularly polarized light*, the tip of the electric vector describes a circular helix about the direction of propagation with a frequency equal to the frequency of the light. The magnitude of the vector remains constant. In *elliptically polarized light*, the vector also rotates about the direction of propagation but the amplitude changes; a projection of the vector on a plane at right angles to the direction of propagation describes an ellipse. Circularly and elliptically polarized light are produced using a retardation plate.

polar molecule A molecule that has a dipole moment; i.e. one in which there is some separation of charge in the *chemical bonds, so that one part of the molecule has a positive charge and the other a negative charge.

Polaroid A doubly refracting material that plane-polarizes unpolarized light passed through it. It consists of a plastic sheet strained in a manner that makes it birefringent by aligning its molecules. Sunglasses incorporating a Polaroid material absorb light that is vibrating horizontally – produced by reflection from horizontal surfaces – and thus reduce glare.

polar solvent *See* solvent.

pollution An undesirable change in the physical, chemical, or biological characteristics of the natural environment, brought about by man's activities. It may be harmful to human or nonhuman life. Pollution may affect the soil, rivers, seas, or the atmosphere. There are two main classes of pollutants: those that are *biodegradable* (e.g. sewage), i.e. can be rendered harmless by natural processes and need therefore cause no permanent harm if adequately dispersed or treated; and those that are *nonbiodegradable* (e.g. heavy metals (such as lead) and *DDT), which eventually accumulate in the environment and may be concentrated in food chains. Other forms of pollution in the environment include noise (e.g. from jet aircraft, traffic, and industrial processes) and thermal pollution (e.g. the release of excessive waste heat into lakes or rivers causing harm to wildlife). Recent pollution problems include the disposal of radioactive waste; *acid rain* resulting from industrial emissions of sulphates; increasing levels of human waste; and high levels of carbon dioxide in the atmosphere. Attempts to contain or prevent pollution include strict regulations concerning factory emissions, the use of smokeless fuels, and the banning of certain pesticides.

polonium Symbol Po. A rare radioactive metallic element of group VIA of the periodic table; a.n. 84; r.a.m. 210; r.d. 9.32; m.p. 254°C; b.p. 962°C. The element occurs in uranium ores to an extent of about 100 micrograms per 1000 kilograms. It has 27 isotopes, more than any other element. The longest-lived isotope is polonium–209 (half-life 103 years). Polonium has attracted attention as a possible heat source for spacecraft as the energy released as it decays is $1.4 \times 10^5\ J\,kg^{-1}\,s^{-1}$. It was discovered by Marie Curie in 1898 in a sample of pitchblende.

poly- Prefix indicating a polymer, e.g. polyethene. Sometimes brackets are used in polymer names to indicate the repeated unit, e.g. poly(ethene).

polyamide A type of condensation polymer produced by the interaction of an amino group of one molecule and a carboxylic acid group of another molecule to give a protein-like structure. The polyamide chains are linked together by hydrogen bonding.

polychloroethene (PVC; polyvinyl chloride) A tough white solid material, which softens with the application of a plasticizer, manufactured from chloroethene by heating in an inert solvent using benzoyl peroxide as an initiator, or by the free-radical mechanism initiated by heating chloroethene under water with potassium persulphate or hydrogen peroxide. The polymer is used in a variety of ways, being easy to colour and resistant to fire, chemicals, and weather.

polycyclic Denoting a compound that has two or more rings in its molecules. Polycyclic compounds may contain single rings (as in phenylbenzene, $C_6H_5.C_6H_5$) or fused rings (as in naphthalene, $C_{10}H_8$).

polydioxoboric(III) acid *See* boric acid.

polyester A condensation polymer formed by the interaction of polyhydric alcohols and polybasic acids. Linear polyesters are saturated thermoplastics and linked by dipole–dipole attraction as the carbonyl groups are polarized. They are extensively used as fibres (e.g. *Terylene*). Unsaturated polyesters readily copolymerize to give thermosetting products. They are used in the manufacture of glass-fibre products.

polyethene (polyethylene; polythene) A flexible waxy translucent polyalkene thermoplastic made in a variety of ways producing a polymer of varying characteristics. In the ICI process, ethene containing a trace of oxygen is subjected to a pressure in excess of 1500 atmospheres and a temperature of 200°C. Low-density polyethene (r.d. 0.92) has a formula weight between 50 000 and 300 000, softening at a temperature around 110°C, while the high-density polythene (r.d. 0.945–0.96) has a formula weight up to 3 000 000, softening around 130°C. The low-density polymer is less crystalline, being more atactic. Polyethene is used as an insulator; it is acid resistant and is easily moulded and blown. *See* Phillips process; Ziegler process.

polyethylene *See* polyethene.

polyhydric alcohol An *alcohol that has several hydroxyl groups per molecule.

polymer A substance having large molecules consisting of repeated units (the monomers). There are a number of natural polymers, such as polysaccharides. Synthetic polymers are extensively used in *plastics. Polymers do not have a definite formula since they consist of chains of different lengths. The various types of polymer are shown in the illustration.

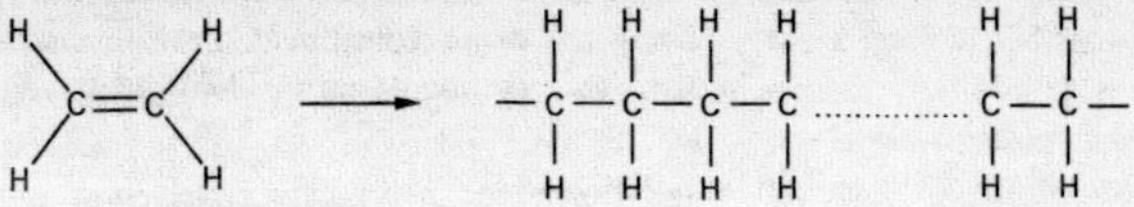

Addition polymerization of ethene to form polyethene a homopolymer

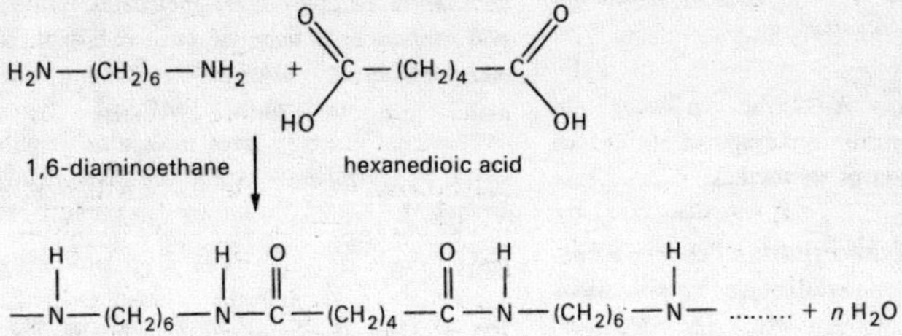

Condensation polymerization to form nylon: a heteropolymer

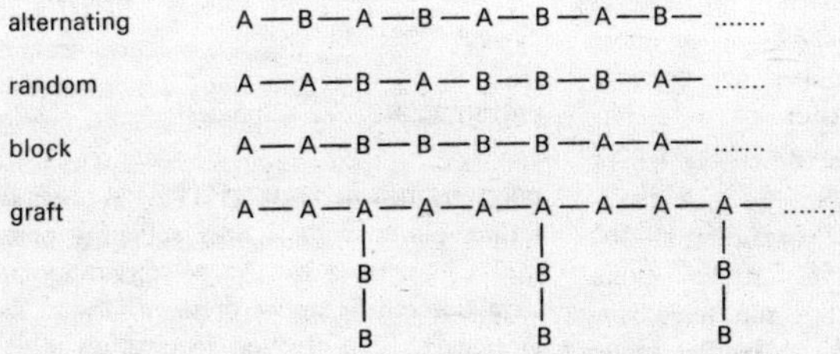

Types of copolymer depending on the arrangement of the monomers A and B

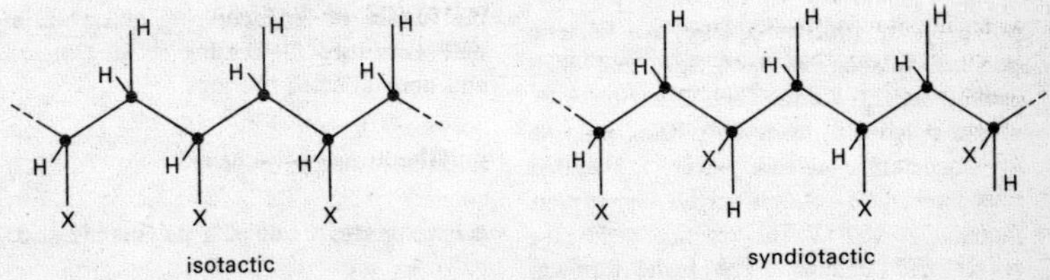

Types of stereospecific polymer

Polymers

polymerization A chemical reaction in which molecules join together to form a polymer. If the reaction is an addition reaction, the process is *addition polymerization*; condensation reactions cause *condensation polymerization*, in which a small molecule is eliminated during the reaction. Polymers consisting of a single monomer are *homopolymers*; those formed from two different monomers are *copolymers*.

polymethanal A solid polymer of methanal, formed by evaporation of an aqueous solution of methanal.

polymethylmethacrylate A clear thermoplastic acrylic material made by polymerizing methyl methacrylate. The technical name is *poly(methyl 2-methylpropenoate)*. It is used in such materials as *Perspex*.

polymorphism The existence of chemical substances in two (*dimorphism*) or more physical forms. *See* allotropy.

polypeptide A *peptide comprising ten or more amino acids. Polypeptides that constitute proteins usually contain 100–300 amino acids. Shorter ones include certain antibiotics, e.g. gramicidin, and some hormones, e.g. ACTH, which has 39 amino acids. The properties of a polypeptide are determined by the type and sequence of its constituent amino acids.

polypropene (polypropylene) An isotactic polymer existing in both low and high formula-weight forms. The lower-formula-weight polymer is made by passing propene at moderate pressure over a heated phosphoric acid catalyst spread on an inert material at 200°C. The reaction yields the trimer and tetramer. The higher-formula-weight polymer is produced by passing propene into an inert solvent, heptane, which contains a trialkyl aluminium and a titanium compound. The product is a mixture of isotactic and atactic polypropene, the former being the major constituent. Polypropene is used as a thermoplastic moulding material.

polypropylene *See* polypropene.

polysaccharide Any of a group of carbohydrates comprising long chains of monosaccharide (simple-sugar) molecules. *Homopolysaccharides* consist of only one type of monosaccharide; *heteropolysaccharides* contain two or more different types. Polysaccharides may have molecular weights of up to several million and are often highly branched. Some important examples are starch, glycogen, and cellulose.

polystyrene A clear glasslike material manufactured by free-radical polymerization of phenylethene (styrene) using benzoyl peroxide as an initiator. It is used as both a thermal and electrical insulator and for packing and decorative purposes.

polysulphides *See* sulphides.

polytetrafluoroethane (PTFE) A thermosetting plastic with a high softening point (327°C) prepared by the polymerization of tetrafluoroethane under pressure (45–50 atmospheres). The reaction requires an initiator, ammonium peroxosulphate. The polymer has a low coefficient of friction and its 'anti-stick' properties are probably due to its helical structure with the fluorine atoms on the surface of an inner ring of carbon atoms. It is used for coating cooking utensils and nonlubricated bearings.

polythene *See* polyethene.

polythionate A salt of a polythionic acid.

polythionic acids Oxo acids of sulphur in which the general formula is $HO.SO_2.S_n.SO_2.OH$, where $n = 0–4$. *See also* sulphuric acid.

polyurethane A polymer containing the urethane group –NH.CO.O–, prepared by reacting di-isocyanates with appropriate diols or triols. A wide range of polyurethanes can be made, and they are used in adhesives, durable paints and varnishes, plastics, and rubbers. Addition of water to the polyurethane plastics turns them into foams.

polyvinyl chloride *See* polychloroethene.

porphyrin Any of a group of related organic compounds characterized by the possession of a cyclic group of four linked nitrogen-containing rings (a *pyrrole* nucleus). Porphyrins differ in the nature of their side-chain groups. They include the chlorophylls and the haem groups of haemoglobin, myoglobin, and the cytochromes.

potash Any of a number of potassium compounds, such as the carbonate or the hydroxide.

potash alum *See* aluminium potassium sulphate; alums.

potash mica *See* muscovite.

potassium Symbol K. A soft silvery metallic element belonging to group I of the periodic table (*see* alkali metals); a.n. 19; r.a.m. 39.098; r.d. 0.87; m.p. 63.7°C; b.p. 774°C. The element occurs in seawater and in a number of minerals, such as sylvite (KCl), carnallite ($KCl.MgCl_2.6H_2O$), and kainite ($MgSO_4.KCl.3H_2O$). It is obtained by electrolysis. The metal has few uses but potassium salts are used for a wide range of applications. Potassium is an essential element for living organisms. Chemically, it is highly reactive, resembling sodium in its behaviour and compounds. It also forms an orange-coloured superoxide, KO_2, which contains the O_2^- ion. Potassium was discovered by Sir Humphry Davy in 1807.

potassium–argon dating A *dating technique for certain rocks that depends on the decay of the radioisotope potassium–40 to argon–40, a process with a half-life of about 1.27×10^{10} years. It assumes that all the argon–40 formed in the potassium-bearing mineral accumulates within it and that all the argon present is formed by the decay of potassium–40. The mass of argon–40 and potassium–40 in the sample is estimated and the sample is then dated from the equation:

$$^{40}Ar = 0.1102\ ^{40}K(e^{\lambda t} - 1),$$

where λ is the decay constant and t is the time in years since the mineral cooled to about 300°C, when the ^{40}Ar became trapped in the crystal lattice. The method is effective for micas, feldspar, and some other minerals.

potassium bicarbonate *See* potassium hydrogencarbonate.

potassium bichromate *See* potassium dichromate.

potassium bromide A white or colourless crystalline solid, KBr, slightly hygroscopic and soluble in water and very slightly soluble in ethanol; cubic; r.d. 2.73; m.p. 734°C; b.p. 1435°C. Potassium bromide may be prepared by the action of bromine on hot potassium hydroxide solution or by the action of iron(III) bromide or hydrogen bromide on potassium carbonate solution. It is used widely in the photographic industry and is also used as a sedative. Because of its range of transparency to infrared radiation, KBr is used both as a matrix for solid samples and as a prism material in infrared spectroscopy.

potassium carbonate (pearl ash; potash) A translucent (granular) or white (powder) deliquescent solid known in the anhydrous and hydrated forms. K_2CO_3 (monoclinic; r.d. 2.4; m.p. 891°C) decomposes without boiling. $2K_2CO_3.3H_2O$ (monoclinic; r.d.

2.04) dehydrates to $K_2CO_3.H_2O$ above 100°C and to K_2CO_3 above 130°C. It is prepared by the Engel–Precht process in which potassium chloride and magnesium oxide react with carbon dioxide to give the compound *Engel's salt*, $MgCO_3.KHCO_3.4H_2O$. This is decomposed in solution to give the hydrogencarbonate, which can then be calcined to K_2CO_3. Potassium carbonate is soluble in water (insoluble in alcohol) with significant hydrolysis to produce basic solutions. Industrial uses include glasses and glazes, the manufacture of soft soaps, and in dyeing and wool finishing. It is used in the laboratory as a drying agent.

potassium chlorate A colourless crystalline compound, $KClO_3$, which is soluble in water and moderately soluble in ethanol; monoclinic; r.d. 2.33; m.p. 360°C; decomposes above 400°C giving off oxygen. The industrial route to potassium chlorate involves the fractional crystallization of a solution of potassium chloride and sodium chlorate but it may also be prepared by electrolysis of hot concentrated solutions of potassium chloride. It is a powerful oxidizing agent finding applications in weedkillers and disinfectants and, because of its ability to produce oxygen, it is used in explosives, pyrotechnics, and matches.

potassium chloride A white crystalline solid, KCl, which is soluble in water and very slightly soluble in ethanol; cubic; r.d. 1.98; m.p. 772°C; sublimes at 1500°C. Potassium chloride occurs naturally as the mineral *sylvite* (KCl) and as *carnallite* ($KCl.MgCl_2.6H_2O$); it is produced industrially by fractional crystallization of these deposits or of solutions from lake brines. It has the interesting property of being more soluble than sodium chloride in hot water but less soluble in cold. It is used as a fertilizer, in photography, and as a source of other potassium salts, such as the chlorate and the hydroxide. It has low toxicity.

potassium chromate A bright yellow crystalline solid, K_2CrO_4, soluble in water and insoluble in alcohol; rhombic; r.d. 2.73; m.p. 971°C; decomposes without boiling. It is produced industrially by roasting powdered chromite ore with potash and limestone and leaching the resulting cinder with hot potassium sulphate solution. Potassium chromate is used in leather finishing, as a textile mordant, and in enamels and pigments. In the laboratory it is used as an analytical reagent and as an indicator. Like other chromium(III) compounds it is toxic when ingested or inhaled.

potassium chromium sulphate (chrome alum) A violet or ruby-red crystalline solid, $K_2SO_4.Cr_2(SO_4)_3.24H_2O$, soluble in water and insoluble in ethanol; cubic or octahedral; r.d. 1.826; m.p. 89°C; loses $10H_2O$ at 100°C, $12H_2O$ at 400°C. Six water molecules surround each of the chromium(III) ions and the remaining ones are hydrogen bonded to the sulphate ions. Like all alums, the compound may be prepared by mixing equimolar quantities of the constituent sulphates. *See* alums.

potassium cyanide (cyanide) A white crystalline or granular deliquescent solid, KCN, soluble in water and in ethanol and having a faint characteristic odour of almonds (due to hydrolysis forming hydrogen cyanide at the surface); cubic; r.d. 1.52; m.p. 634°C. It is prepared industrially by the absorption of hydrogen cyanide in potassium hydroxide. The compound is used in the extraction of silver and gold, in some metal-finishing processes and electroplating, as an insecticide and fumigant (source of HCN), and in the preparation of cyanogen derivatives. In the laboratory it is used in analysis, as a reducing agent, and as a stabilizing *ligand for low oxidation states. The salt itself is highly toxic and aqueous solutions of potassium cyanide are strongly hydrolysed to give rise to the slow release of equally toxic hydrogen cyanide gas.

potassium dichromate (potassium bichromate) An orange-red crystalline solid, $K_2Cr_2O_7$, soluble in water and insoluble in alcohol; monoclinic or triclinic; r.d. 2.67; monoclinic changes to triclinic at 241.6°C; m.p. 396°C; decomposes above 500°C. It is prepared by acidification of crude potassium chromate solution (the addition of a base to solutions of potassium dichromate reverses this process). The compound is used industrially as an oxidizing agent in the chemical industry and in dyestuffs manufacture, in electroplating, pyrotechnics, glass manufacture, glues, tanning, photography and lithography, and in ceramic products. Laboratory uses include application as an analytical reagent and as an oxidizng agent. Potassium dichromate is toxic and considered a fire risk on account of its oxidizing properties.

potassium dioxide *See* potassium superoxide.

potassium hydride A white or greyish white crystalline solid, KH; r.d. 1.43–1.47. It is prepared by passing hydrogen over heated potassium and marketed as a light grey powder dispersed in oil. The solid decomposes on heating and in contact with moisture and is an excellent reducing agent. Potassium hydride is a fire hazard because it produces hydrogen on reaction with water.

potassium hydrogencarbonate (potassium bicarbonate) A white crystalline solid, $KHCO_3$, soluble in water and insoluble in ethanol; r.d. 2.17; decomposes about 120°C. It occurs naturally as *calcinite* and is prepared by passing carbon dioxide into saturated potassium carbonate solution. It is used in baking, soft-drinks manufacture, and in CO_2 fire extinguishers. Because of its buffering capacity, it is added to some detergents and also used as a laboratory reagent.

potassium hydrogentartrate (cream of tartar) A white crystalline acid salt, $HOOC(CHOH)_2COOK$. It is obtained from deposits on wine vats (argol) and used in baking powders.

potassium hydroxide (caustic potash; lye) A white deliquescent solid, KOH, often sold as pellets, flakes, or sticks, soluble in water and in ethanol and very slightly soluble in ether; rhombic; r.d. 2.02; m.p. 405°C; b.p. 1320°C. It is prepared industrially by the electrolysis of concentrated potassium chloride solution but it can also be made by heating potassium carbonate or sulphate with slaked lime, $Ca(OH)_2$. It closely resembles sodium hydroxide but is more soluble and is therefore preferred as an absorber for carbon dioxide and sulphur dioxide. It is also used in the manufacture of soft soap, other potassium salts, and in Ni–Fe and alkaline storage cells. Potassium hydroxide is extremely corrosive to body tissues and especially damaging to the eyes.

potassium iodate A white crystalline solid, KIO_3, soluble in water and insoluble in ethanol; monoclinic; r.d. 3.9; m.p. 561°C. It may be prepared by the reaction of iodine with hot concentrated potassium hydroxide or by careful electrolysis of potassium iodide solution. It is an oxidizing agent and is used as an analytical reagent. Some potassium iodate is used as a food additive.

potassium iodide A white crystalline solid, KI, with a strong bitter taste, soluble in water, ethanol, and acetone; cubic; r.d. 3.12; m.p. 686°C; b.p. 1330°C. It may be prepared by the reaction of iodine with hot potassium hydroxide solution followed by separation from the iodate (which is also formed) by fractional crystallization. In solution it has the interesting property of dissolving iodine to form the triiodide ion I_3^-, which is brown. Potassium iodide is widely used as an analytical reagent, in photography, and also as an additive to table salt to prevent goitre and other disorders due to iodine deficiency.

potassium manganate(VII) (potassium permanganate) A compound, $KMnO_4$, forming purple crystals with a metallic sheen, soluble in water (intense purple solution), acetone, and methanol, but decomposed by ethanol; r.d. 2.70; decomposition begins slightly above 100°C and is complete at 240°C. The compound is prepared by fusing manganese(IV) oxide with potassium hydroxide to form the manganate and electrolysing the manganate solution using iron electrodes at about 60°C. An alternative route employs production of sodium manganate by a similar fusion process, oxidation with chlorine and sulphuric acid, then treatment with potassium chloride to crystallize the required product.

Potassium manganate(VII) is widely used as an oxidizing agent and as a disinfectant in a variety of applications, and as an analytical reagent.

potassium monoxide A grey crystalline solid, K_2O; cubic; r.d. 2.32; decomposition occurs at 350°C. It may be prepared by the oxidation of potassium metal with potassium nitrate. It reacts with ethanol to form potassium ethoxide (KOC_2H_5), and with liquid ammonia to form potassium hydroxide and potassamide (KNH_2).

potassium nitrate (saltpetre) A colourless rhombohedral or trigonal solid, KNO_3, soluble in water, insoluble in alcohol; r.d. 2.109; transition to trigonal form at 129°C; m.p. 334°C; decomposes at 400°C. It occurs naturally as *nitre* and may be prepared by the reaction of sodium nitrate with potassium chloride followed by fractional crystallization. It is a powerful oxidizing agent (releases oxygen on heating) and is used in gunpowder and fertilizers.

potassium nitrite A white or slightly yellow deliquescent solid, KNO_2, soluble in water and insoluble in ethanol; r.d. 1.91; m.p. 300°C; may explode at 600°C. Potassium nitrite reacts with cold dilute mineral acids to give nitrous acid and is also able to behave as a reducing agent (if oxidized to the nitrate) or as an oxidizing agent (if reduced to nitrogen). It is used in organic synthesis because of its part in diazotization, and in detecting the presence of the amino groups in organic compounds.

potassium permanganate *See* potassium manganate(VII).

potassium sulphate A white crystalline powder, K_2SO_4, soluble in water and insoluble in ethanol; rhombic or hexagonal; r.d. 2.66; m.p. 1072°C. It occurs naturally as *schönite* (Strassfurt deposits) and in lake brines, from which it is separated by fractional crystallization. It has also been produced by the Hargreaves process, which involves the oxidation of potassium chloride with sulphuric acid. In the laboratory it may be obtained by the reaction of either potassium hydroxide or potassium carbonate with sulphuric acid. Potassium sulphate is used in cements, in glass manufacture, as a food additive, and as a fertilizer (source of K^+) for chloride-sensitive plants, such as tobacco and citrus.

potassium sulphide A yellow-red or brown-red deliquescent solid, K_2S, which is soluble in water and in ethanol but insoluble in diethyl ether; cubic; r.d. 1.80; m.p. 840°C. It is made industrially by reducing potassium sulphate with carbon at high temperatures in the absence of air. In the laboratory it may be prepared by the reaction of hydrogen sulphide with potassium hydroxide. The pentahydrate is obtained on crystallization. Solutions are strongly alkaline due to hydrolysis. It is used as an analytical reagent and as a depilatory. Potassium sulphide is generally regarded as a hazardous chemical with a fire risk; dusts of K_2S have been known to explode.

potassium sulphite A white crystalline solid, K_2SO_3, soluble in water and very sparingly soluble in ethanol; r.d. 1.51; decomposes on heating. It is a reducing agent and is used as such in photography and in the food and brewing industries, where it prevents oxidation.

potassium superoxide (potassium dioxide) A yellow paramagnetic solid, KO_2, produced by burning potassium in an excess of oxygen; it is very soluble (by reaction) in water, soluble in ethanol, and slightly soluble in diethyl ether; m.p. 490°C. When treated with cold water or dilute mineral acids, hydrogen peroxide is obtained. The compound is a powerful oxidizing agent and on strong heating releases oxygen with the formation of the monoxide, K_2O.

potential barrier A region containing a maximum of potential that prevents a particle on one side of it from passing to the other side. According to classical theory a particle must possess energy in excess of the height of the potential barrier to pass it. However, in quantum theory there is a finite probability that a particle with less energy will pass through the barrier (*see* tunnel effect). A potential barrier surrounds the atomic nucleus and is important in nuclear physics; a similar but much lower barrier exists at the interface between semiconductors and metals and between differently doped semiconductors. These barriers are important in the design of electronic devices.

potentiometric titration A titration in which the end point is found by measuring the potential on an electrode immersed in the reaction mixture.

powder metallurgy A process in which powdered metals or alloys are pressed into a variety of shapes at high temperatures. The process started with the pressing of powdered tungsten into incandescent lamp filaments in the first decade of this century and is now widely used for making self-lubricating bearings and cemented tungsten carbide cutting tools.

The powders are produced by atomization of molten metals, chemical decomposition of a compound of the metal, or crushing and grinding of the metal or alloy. The parts are pressed into moulds at pressures ranging from 140×10^6 Pa to 830×10^6 Pa after which they are heated in a controlled atmosphere to bond the particles together (*see* sintering).

praseodymium Symbol Pr. A soft silvery metallic element belonging to the *lanthanoids; a.n. 59; r.a.m. 140.91; r.d. 6.769 (20°C); m.p. 934°C; b.p. 3512°C. It occurs in bastnasite and monazite, from which it is recovered by an ion-exchange process. The only naturally occurring isotope is praseodymium-141, which is not radioactive; however, fourteen radioisotopes have been produced. It is used in mischmetal, a rare-earth alloy containing 5% praseodymium, for use in lighter flints. Another rare-earth mixture containing 30% praseodymium is used as a catalyst in cracking crude oil. The element was discovered by C. A. von Welsbach in 1885.

precipitate A suspension of small solid particles produced in a liquid by chemical reaction.

precipitation 1. All liquid and solid forms of water that are deposited from the atmosphere; it includes rain, drizzle, snow, hail, dew, and hoar frost. **2.** The formation of a precipitate.

precursor A compound that leads to another compound in a series of chemical reactions.

pressure The force acting on unit area of a surface or the ratio of force to area. It is measured in *pascals in SI units. *Absolute* *pressure* is pressure measured on a gauge that reads zero at zero pressure rather than at atmospheric pressure. *Gauge pressure* is measured on a gauge that reads zero at atmospheric pressure.

pressure gauge Any device used to measure *pressure. Three basic types are in use: the liquid-column gauge (e.g. the mercury barometer and the manometer), the expanding-element gauge (e.g. the Bourdon gauge and the aneroid barometer), and the electrical transducer. In the last category the strain gauge is an example. Capacitor pressure gauges also come into this category. In these devices, the pressure to be measured displaces one plate of a capacitor and thus alters its capacitance.

primary alcohol *See* alcohols.

primary amine *See* amines.

primary cell A *voltaic cell in which the chemical reaction producing the e.m.f. is not satisfactorily reversible and the cell cannot therefore be recharged by the application of a charging current. *See* Daniell cell; Leclanché cell; Weston cell; mercury cell. *Compare* secondary cell.

producer gas (air gas) A mixture of carbon monoxide and nitrogen made by passing air over very hot carbon. Usually some steam is added to the air and the mixture contains hydrogen. The gas is used as a fuel in some industrial processes.

product *See* chemical reaction.

progesterone A hormone, produced primarily by the corpus luteum of the ovary but also by the placenta, that prepares the inner lining of the uterus for implantation of a fertilized egg cell. If implantation fails, the corpus luteum degenerates and progesterone production ceases accordingly. If implantation occurs, the corpus luteum continues to secrete progesterone, under the influence of luteinizing hormone and prolactin, for several months of pregnancy, by which time the placenta has taken over this function. During pregnancy, progesterone maintains the constitution of the uterus and prevents further release of eggs from the ovary. Small amounts of progesterone are produced by the testes. *See also* progestogen.

progestogen One of a group of naturally occurring or synthetic hormones that maintain the normal course of pregnancy. The best known is *progesterone. In high doses progestogens inhibit secretion of luteinizing hormone, thereby preventing ovulation, and alter the consistency of mucus in the vagina so that conception tends not to occur. They are therefore used as major constituents of oral contraceptives.

prolactin (lactogenic hormone; luteotrophic hormone; luteotrophin) A hormone produced by the anterior pituitary gland. In mammals it stimulates the mammary glands to produce milk and the corpus luteum of the ovary to secrete the hormone *progesterone.

proline *See* amino acid.

promethium Symbol Pm. A soft silvery metallic element belonging to the *lanthanoids; a.n. 61; r.a.m. 145; r.d. 7.26 (20°C); m.p. 1042°C; b.p. 3000°C. The only naturally occurring isotope, promethium-147, has a half-life of only 2.52 years. Eighteen other radioisotopes have been produced, but they have very short half-lives. The only known source of the element is nuclear-waste material. Promethium-147 is of interest as a beta-decay power source but the promethium-146 and -148, which emit penetrating gamma radiation, must first be removed. It was discovered by J. A. Marinsky, L. E. Glendenin, and C. D. Coryell in 1947.

promoter A substance added to a catalyst to increase its activity.

proof A measure of the amount of alcohol (ethanol) in drinks. *Proof spirit* contains 42.28% ethanol by weight (about 57% by volume). Degrees of proof express the percentage of proof spirit present, so 70° proof spirit contains 0.7 × 57% alcohol.

propanal (propionaldehyde) A colourless liquid *aldehyde, C_2H_5CHO; m.p. −185.2°C; b.p. 48.8°C.

propane A colourless gaseous hydrocarbon, C_3H_8; m.p. −190°C; b.p. −42°C. It is the third member of the *alkane series and is obtained from petroleum. Its main use is as bottled gas for fuel.

propanoic acid (propionic acid) A colourless liquid *carboxylic acid, CH_3CH_2COOH; r.d. 0.99; m.p. −20.8°C; b.p. 141°C. It is used to make calcium propanate – an additive in bread.

propanol Either of two *alcohols with the formula C_3H_7OH. Propan-1-ol is $CH_3CH_2CH_2OH$ and propan-2-ol is $CH_3CH(OH)CH_3$. Both are colourless volatile liquids. Propan-2-ol is used in making propanone (acetone).

propanone (acetone) A colourless flammable volatile compound, CH_3COCH_3; r.d. 0.79; m.p. −95.4°C, b.p. 56.2°C. The simplest *ketone, propanone is miscible with water. It is made by oxidation of propan-2-ol (*see* propanol) or is obtained as a by-product in the manufacture of phenol from cumene; it is used as a solvent and as a raw material for making plastics.

propenal (acrolein) A colourless pungent liquid unsaturated aldehyde, CH_2:CHCHO; r.d. 0.84; m.p. −87°C; b.p. 53°C. It is made from propene and can be polymerized to give acrylate resins.

propene (propylene) A colourless gaseous hydrocarbon, CH_3CH:CH_2; m.p. −81°C; b.p. 48.8°C. It is an *alkene obtained from petroleum by cracking alkanes. Its main use is in the manufacture of polypropene.

propenoate (acrylate) A salt or ester of *propenoic acid.

propenoic acid (acrylic acid) An unsaturated liquid *carboxylic acid, CH_2:CHCOOH; m.p. 7°C; b.p. 141°C. It readily polymerizes and it is used in the manufacture of *acrylic resins.

propenonitrile (acrylonitrile; vinyl cyanide) A colourless liquid, H_2C:CHCN; r.d. 0.81; m.p. −83.5°C. It is an unsaturated nitrile, made from propene and used to make acrylic resins.

propionaldehyde *See* propanal.

propylene *See* propene.

propyl group The organic group $CH_3CH_2CH_2$–.

prostaglandin Any of a group of organic compounds derived from *essential fatty acids and causing a range of physiological effects in animals. Prostaglandins have been detected in most body tissues. They act at very low concentrations to cause the contraction of smooth muscle; natural and synthetic prostaglandins are used to induce abortion or labour in humans and domestic animals. Two prostaglandin derivatives have antagonistic effects on blood circulation: *thromboxane A_2* causes blood clotting while *prostacyclin* causes blood vessels to dilate. Inflammation in allergic reactions and other diseases is also thought to involve prostaglandins.

prosthetic group A tightly bound nonpeptide inorganic or organic component of a protein. Prosthetic groups may be lipids, carbohydrates, metal ions, phosphate groups, etc. Some *coenzymes are more correctly regarded as prosthetic groups.

protactinium Symbol Pa. A radioactive metallic element belonging to the *actinoids; a.n. 91; r.a.m. 231.036; r.d. 15.37 (calculated); m.p. <1600°C (estimated). The most stable isotope, protactinium–231, has a half-life of 3.43×10^4 years; ten other radioisotopes are known. Protactinium–231 occurs in all uranium ores as it is a member of the uranium series. Protactinium has no practical applications; it was discovered by Lise Meitner and Otto Hahn in 1917.

protamine Any of a group of proteins of relatively low molecular weight found in association with the chromosomal *DNA of vertebrate male germ cells. They contain a single polypeptide chain comprising about 67% arginine. Protamines are thought to protect and support the chromosomes.

protease (peptidase; proteinase; proteolytic enzyme) Any enzyme that catalyses the splitting of proteins into smaller *peptide fractions and amino acids, a process known as *proteolysis*. Examples are *pepsin and *trypsin. Several proteases, acting sequentially, are normally required for the complete digestion of a protein to its constituent amino acids.

protein Any of a large group of organic compounds found in all living organisms. Proteins comprise carbon, hydrogen, oxygen, and nitrogen and most also contain sulphur; molecular weights range from 6000 to several million. Protein molecules consist of one or several long chains (*polypeptides) of *amino acids linked in a characteristic sequence. This sequence is called the *primary structure* of the protein. These polypeptides may undergo coiling or pleating, the nature and extent of which is described as the *secondary structure*. The three-dimensional shape of the coiled or pleated polypeptides is called the *tertiary structure*. *Quaternary structure* specifies the structural relationship of the component polypeptides.

Proteins may be broadly classified into globular proteins and fibrous proteins. Globular proteins have compact rounded molecules and are usually water-soluble. Of prime importance are the *enzymes, proteins that catalyse biochemical reactions. Other globular proteins include the antibodies, which combine with foreign substances in the body; the carrier proteins, such as haemoglobin; the storage proteins (e.g. casein in milk and albumin in egg white), and certain hormones (e.g. insulin). Fibrous proteins are generally insoluble in water and consist of long coiled strands or flat sheets, which confer strength and elasticity. In this category are keratin and collagen. Actin and myosin are the principal fibrous proteins of muscle, the interaction of which brings about muscle contraction.

When heated over 50°C or subjected to strong acids or alkalis, proteins lose their specific tertiary structure and may form insoluble coagulates (e.g. egg white). This usually inactivates their biological properties.

protein synthesis The process by which living cells manufacture proteins from their constituent amino acids, in accordance with the genetic information carried in the DNA of the chromosomes. This information is encoded in messenger *RNA, which is transcribed from DNA in the nucleus of the cell: the sequence of amino acids in a particular protein is determined by the sequence of nucleotides in messenger RNA. At the ribosomes the information carried by messenger RNA is translated into the sequence of amino acids of the protein in the process of translation.

proteolysis The enzymic splitting of proteins. *See* protease.

proteolytic enzyme *See* protease.

proton An elementary particle that is stable, bears a positive charge equal in magnitude to that of the *electron, and has a mass of $1.672\,614 \times 10^{-27}$ kg, which is 1836.12 times that of the electron. The proton is a hydrogen ion and occurs in all atomic nuclei.

protonic acid An *acid that forms positive hydrogen ions (or, strictly, oxonium ions) in aqueous solution. The term is used to distinguish 'traditional' acids from Lewis acids or from Lowry–Br o/nsted acids in non-aqueous solvents.

proton number *See* atomic number.

prussic acid *See* hydrogen cyanide.

pseudoaromatic A compound that has a ring of atoms containing alternating double and single bonds, yet does not have the characteristic properties of *aromatic compounds. Such compounds do not obey the Huckel rule. Cyclooctatetraene (C_8H_8), for instance, has a ring of eight carbon atoms with conjugated double bonds, but the ring is not planar and the compound acts like an alkene, undergoing addition reactions.

pseudohalogens A group of compounds, including cyanogen $(CN)_2$ and thiocyanogen $(SCN)_2$, that have some resemblance to the halogens. Thus, they form hydrogen acids (HCN and HSCN) and ionic salts containing such ions as CN^- and SCN^-.

pseudo order An order of a chemical reaction that appears to be less than the true order because of the experimental conditions used. Pseudo orders occur when one reactant is present in large excess. For example, a reaction of substance A undergoing hydrolysis may appear to be proportional only to [A] because the amount of water present is so large.

PTFE *See* polytetrafluoroethane.

ptyalin An enzyme that digests carbohydrates (*see* amylase). It is present in mammalian saliva and is responsible for the initial stages of starch digestion.

pumice A porous volcanic rock that is light and full of cavities due to expanding gases that were liberated from solution in the lava while it solidified. Pumice is often light enough to float on water. It is usually acid (siliceous) in composition, and is used as an abrasive and for polishing.

pump A device that imparts energy to a fluid in order to move it from one place or level to another or to raise its pressure (*compare* vacuum pump). *Centrifugal pumps* and turbines have rotating impellers, which increase the velocity of the fluid, part of the energy so acquired by the fluid then being converted to pressure energy. Displacement pumps act directly on the fluid, forcing it to flow against a pressure. They include piston, plunger, gear, screw, and cam pumps.

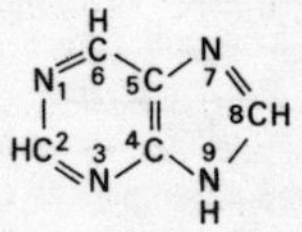

Purine

purine An organic nitrogenous base, sparingly soluble in water, that gives rise to a group of biologically important derivatives, notably *adenine and *guanine, which occur in nucleotides and nucleic acids (DNA and RNA).

PVC *See* polychloroethene.

pyridine A colourless liquid with a strong unpleasant smell, C_5H_5N; r.d. 0.9; m.p. −42°C; b.p. 115°C. Pyridine is an aromatic heterocyclic compound present in coal tar. It is used in making other organic chemicals.

pyridoxine *See* vitamin B complex.

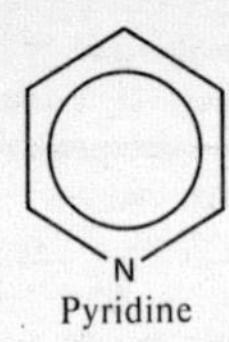

Pyridine

Pyrimidine

pyrimidine An organic nitrogenous base, sparingly soluble in water, that gives rise to a group of biologically important derivatives, notably *uracil, *thymine, and *cytosine, which occur in *nucleotides and nucleic acids (DNA and RNA).

pyrite (iron pyrites) A mineral form of iron(II) sulphide, FeS_2. Superficially it resembles gold in appearance, hence it is also known as *fool's gold*, but it is harder and more brittle than gold (which may be cut with a knife). Pyrite crystallizes in the cubic system, is brass yellow in colour, has a metallic lustre, and a hardness of 6–6.5 on the Mohs' scale. It is the most common and widespread of the sulphide minerals and is used as a source of sulphur for the production of sulphuric acid. Sources include the Rio Tinto mines in Spain.

pyro- Prefix denoting an oxo acid that could be obtained from a lower acid by dehydration of two molecules. For example, pyrosulphuric acid is $H_2S_2O_7$ (i.e. $2H_2SO_4$ minus H_2O).

pyroboric acid *See* boric acid.

pyroelectricity The property of certain crystals, such as tourmaline, of acquiring opposite electrical charges on opposite faces when heated. In tourmaline a rise in temperature of 1 K at room temperature produces a polarization of some 10^{-5} C m^{-2}.

pyrolysis Chemical decomposition occurring as a result of high temperature.

pyrometric cones *See* Seger cones.

pyrometry The measurement of high temperatures (beyond the range of thermometers) using a *pyrometer*. Modern *narrow-band* or *spectral* pyrometers use infrared-sensitive photoelectric cells behind filters that exclude visible light. In the *optical pyrometer* (or disappearing filament pyrometer) the image of the incandescent source is focused in the plane of a tungsten filament that is heated electrically. A variable resistor is used to adjust the current through the filament until it blends into the image of the source, when viewed through a red filter and an eyepiece. The temperature is then read from a calibrated ammeter or a calibrated dial on the variable resistor. In the *total-radiation pyrometer* radiation emitted by the source is focused by a concave mirror onto a blackened foil to which a thermopile is attached. From the e.m.f. produced by the thermopile the temperature of the source can be calculated.

pyrophoric Igniting spontaneously in air. *Pyrophoric alloys* are alloys that give sparks when struck. *See* misch metal.

pyrophosphoric acid *See* phosphoric(V) acid.

pyrosilicate *See* silicate.

pyrosulphuric acid *See* disulphuric(VI) acid.

pyroxenes A group of ferromagnesian rock-forming silicate minerals. They are common in basic igneous rocks but may also be developed by metamorphic processes in gneisses, schists, and marbles. Pyroxenes have a complex crystal chemistry; they are composed of continuous chains of silicon and oxygen atoms linked by a variety of other elements. They are related to the *amphiboles, from which they differ in cleavage angles. The general formula is $X_{1-p}Y_{1+p}Z_2O_6$, where X = Ca,Na; Y = Mg,Fe^{2+},Mn,Li,Al,Fe^{3+},Ti; and Z = Si,Al. Orthorhombic pyroxenes (*orthopyroxenes*), $(Mg,Fe)_2Si_2O_6$, vary in composition between the end-members enstatite ($Mg_2Si_2O_6$) and orthoferrosilite ($Fe_2Si_2O_6$). Monoclinic pyroxenes (*clinopyroxenes*), the larger group, include diopside, $CaMgSi_2O_6$; hedenbergite, $CaFe^{2+}Si_2O_6$; johannsenite, $CaMnSi_2O_6$; augite, $(Ca,Mg,Fe,Ti,Al)_2(Si,Al)_2O_6$; aegirine, $NaFe^{3+}Si_2O_6$; jadeite (*see* jade); and pigeonite $(Mg,Fe^{2+},Ca)(Mg,Fe^{2+})Si_2O_6$.

pyruvic acid (2-oxopropanoic acid) A colourless liquid organic acid, $CH_3COCOOH$; m.p. 13°C. It is an important intermediate compound in metabolism, being produced during glycolysis.

Q

quadrivalent Having a valency of four.

qualitative analysis *See* analysis.

quantitative analysis *See* analysis.

quantum The minimum amount by which certain properties, such as energy or angular momentum, of a system can change. Such properties do not, therefore, vary continuously, but in integral multiples of the relevant quantum. This concept forms the basis of the *quantum theory. In waves and fields the quantum can be regarded as an excitation, giving a particle-like interpretation to the wave or field. Thus, the quantum of the electromagnetic field is the *photon and the graviton is the quantum of the gravitational field. *See* quantum mechanics.

quantum mechanics A system of mechanics that was developed from *quantum theory and is used to explain the properties of atoms and molecules. Using the energy *quantum as a starting point it incorporates Heisenberg's *uncertainty principle and the de Broglie wavelength to establish the wave–particle duality on which *Schrödinger's equation is based. This form of quantum mechanics is called *wave mechanics*. An alternative but equivalent formalism, *matrix mechanics*, is based on mathematical operators.

quantum number *See* atom; spin.

quantum state The state of a quantized system as described by its quantum numbers. For instance, the state of a hydrogen *atom is described by the four quantum numbers n, l, m, m_s. In the ground state they have values 1, 0, 1, and ½ respectively.

quantum statistics A statistical description of a system of particles that obeys the rules of *quantum mechanics rather than classical mechanics. In quantum statistics, energy states are considered to be quantized; if the particles are treated as indistinguishable, *Bose–Einstein statistics* apply and the particles are called *bosons*. All known bosons have an angular momentum nh, where n is zero or an integer and h is the Planck constant. For identical bosons the *wave function is always symmetric. For the particular case in which not more than one particle may appear in any of the cells into which the particles are distributed, *Fermi–Dirac statistics* apply and the particles are called *fermions*. All known fermions have a total angular momentum $(n + \frac{1}{2})h$ and any wave function that involves identical fermions is always antisymmetric.

quantum theory The theory devised by Max Planck (1858–1947) in 1900 to account for the emission of the black-body radiation from hot bodies. According to this theory energy is emitted in quanta (*see* quantum), each of which has an energy equal to $h\nu$, where h is the *Planck constant and ν is the frequency of the radiation. This theory led to the modern theory of the interaction between matter and radiation known as *quantum mechanics, which generalizes and replaces classical mechanics and Maxwell's electromagnetic theory. In *nonrelativistic quantum theory* particles are assumed to be neither created nor destroyed, to move slowly relative to the speed of light, and to have a mass that does not change with velocity. These assumptions apply to atomic and molecular phenomena and to some aspects of nuclear physics. *Relativistic quantum theory* applies to particles that have zero rest mass or travel at or near the speed of light.

quartz The most abundant and common mineral, consisting of crystalline silica (silicon dioxide, SiO_2), crystallizing in the trigonal system. It has a hardness of 7 on the Mohs' scale. Well-formed crystals of quartz are six-sided prisms terminating in six-sided pyramids. Quartz is ordinarily colourless and transparent, in which form it is known as *rock crystal.* Coloured varieties, a number of which are used as gemstones, include *amethyst, citrine quartz (yellow), rose quartz (pink), milk quartz (white), smoky quartz (grey-brown), *chalcedony, *agate, and *jasper. Quartz occurs in many rocks, especially igneous rocks such as granite and quartzite (of which it is the chief constituent), metamorphic rocks such as gneisses and schists, and sedimentary rocks such as sandstone and limestone. The mineral has the property of being piezoelectric and hence is used to make oscillators for clocks, radios, and radar instruments. It is also used in optical instruments and in glass, glaze, and abrasives.

quaternary ammonium compounds *See* amine salts.

quenching 1. (in metallurgy) The rapid cooling of a metal by immersing it in a bath of liquid in order to improve its properties. Steels are quenched to make them harder but some nonferrous metals are quenched for other reasons (copper, for example, is made softer by quenching). **2.** (in physics) The process of inhibiting a continuous discharge in a *Geiger counter so that the incidence of further ionizing radiation can cause a new discharge. This is achieved by introducing a quenching vapour, such as ether or a halogen gas, into the tube.

quicklime *See* calcium oxide.

quinhydrone electrode A *half-cell consisting of a platinum electrode in an equimolar solution of quinone (cyclohexadiene-1,4-dione) and hydroquinone (benzene-1,4-diol). It depends on the oxidation–reduction reaction

$$C_6H_4(OH)_2 \rightleftharpoons C_6H_4O_2 + 2H^+ + 2e$$

quinol *See* benzene-1,4-diol.

quinone 1. *See* cyclohexadiene-1,4-dione. **2.** Any similar compound containing C=O groups in an unsaturated ring.

R

racemate *See* racemic mixture.

racemic mixture (racemate) A mixture of equal quantities of the *d*- and *l*-forms of an optically active compound. Racemic mixtures are denoted by the prefix *dl* (e.g. *dl*-lactic acid). A racemic mixture shows no *optical activity.

racemization A chemical reaction in which an optically active compound is converted into a *racemic mixture.

rad *See* radiation units.

radiation **1.** Energy travelling in the form of electromagnetic waves or photons. **2.** A stream of particles, especially alpha- or beta-particles from a radioactive source or neutrons from a nuclear reactor.

radiation units Units of measurement used to express the activity of a radionuclide and the dose of ionizing radiation. The units *curie, roentgen, rad,* and *rem* are not coherent with SI units but their temporary use with SI units has been approved while the derived SI units *becquerel, gray,* and *sievert* become familiar.

The becquerel (Bq), the SI unit of activity, is the activity of a radionuclide decaying at a rate, on average, of one spontaneous nuclear transition per second. Thus 1 Bq = 1 s^{-1}. The former unit, the curie (Ci), is equal to 3.7×10^{10} Bq. The curie was originally chosen to approximate the activity of 1 gram of radium-226.

The gray (Gy), the SI unit of absorbed dose, is the absorbed dose when the energy per unit mass imparted to matter by ionizing radiation is 1 joule per kilogram. The former unit, the rad (rd), is equal to 10^{-2} Gy.

The sievert (Sv), the SI unit of dose equivalent, is the dose equivalent when the absorbed dose of ionizing radiation multiplied by the stipulated dimensionless factors is 1 J kg^{-1}. As different types of radiation cause different effects in biological tissue a weighted absorbed dose, called the *dose equivalent*, is used in which the absorbed dose is modified by multiplying it by dimensionless factors stipulated by the International Commission on Radiological Protection. The former unit of dose equivalent, the rem (originally an acronym for *r*oentgen *e*quivalent *m*an), is equal to 10^{-2} Sv.

In SI units, exposure to ionizing radiation is expressed in coulombs per kilogram, the quantity of X- or gamma-radiation that produces ion pairs carrying 1 coulomb of charge of either sign in 1 kilogram of pure dry air. The former unit, the roentgen (R), is equal to 2.58×10^{-4} C kg^{-1}.

radical A group of atoms, either in a compound or existing alone. *See* free radical; functional group.

radioactive age The age of an archaeological or geological specimen as determined by a process that depends on a radioactive decay. *See* carbon dating; fission-track dating; potassium–argon dating; rubidium–strontium dating; uranium–lead dating.

radioactive series A series of radioactive nuclides in which each member of the series is formed by the decay of the nuclide before it. The series ends with a stable nuclide. Three radioactive series occur naturally, those headed by thorium–232 (*thorium series*), uranium–235 (*actinium series*), and uranium–238 (*uranium series*). All three series end with an isotope of lead. The *neptunium series* starts with the artificial isotope plutonium–241, which decays to neptunium–237, and ends with thallium–81.

radioactive tracing *See* labelling.

radioactivity The spontaneous disintegration of certain atomic nuclei accompanied by the emission of alpha particles (helium nuclei), beta particles (electrons), or gamma radiation (short-wave electromagnetic waves).

Natural radioactivity is the result of the spontaneous disintegration of naturally occurring radioisotopes; these radioisotopes can be arranged in three *radioactive series. The rate of disintegration is uninfluenced by chemical changes or any normal changes in their environment. However, radioactivity can be induced in many nuclides by bombarding them with neutrons or other particles. *See also* ionizing radiation; radiation units.

radiocarbon dating *See* carbon dating.

radiochemistry The branch of chemistry concerned with radioactive compounds and with ionization. It includes the study of compounds of radioactive elements and the preparation and use of compounds containing radioactive atoms. *See* labelling; radiolysis.

radioisotope (radioactive isotope) An isotope of an element that is radioactive. *See* labelling.

radiolysis The use of ionizing radiation to produce chemical reactions. The radiation used includes alpha particles, electrons, neutrons, X-rays, and gamma rays from radioactive materials or from accelerators. Energy transfer produces ions and excited species, which undergo further reaction. A particular feature of radiolysis is the formation of short-lived solvated electrons in water and other polar solvents.

radiometric dating (radioactive dating) *See* dating techniques; radioactive age.

radionuclide (radioactive nuclide) A *nuclide that is radioactive.

radium Symbol Ra. A radioactive metallic element belonging to *group II of the periodic table; a.n. 88; r.a.m. 226.0254; r.d. ~5; m.p. 700°C; b.p. 1140°C. It occurs in uranium ores (e.g. pitchblende). The most stable isotope is radium–226 (half-life 1602 years), which decays to radon. It is used as a radioactive source in research and, to some extent, in radiotherapy. The element was isolated from pitchblende in 1898 by Marie and Pierre Curie.

radon Symbol Rn. A colourless radioactive gaseous element belonging to group 0 of the periodic table (the *noble gases); a.n. 86; r.a.m. 222; d. 9.73 g dm^{-3}; m.p. −71°C; b.p. −61.8°C. At least 20 isotopes are known, the most stable being radon–222 (half-life 3.8 days). It is formed by decay of radium–226 and undergoes alpha decay. It is used in radiotherapy. As a noble gas, radon is practically inert, although radon fluoride can be made. It was first isolated by Ramsey and Gray in 1908.

raffinate A liquid purified by solvent extraction.

r.a.m. *See* relative atomic mass.

Raney nickel A spongy form of nickel made by the action of sodium hydroxide on a nickel–aluminium alloy. The sodium hydroxide dissolves the aluminium leaving a highly active form of nickel with a large surface area. The material is a black pyrophoric powder saturated with hydrogen. It is an extremely efficient catalyst, especially for hydrogenation reactions at room temperature. It was discovered in 1927 by the American chemist M. Raney.

ranksite A mineral consisting of a mixed sodium carbonate, sodium sulphate, and potassium chloride, $2Na_2CO_3.9Na_2SO_4.KCl$.

Raoult's law The partial vapour pressure of a solvent is proportional to its mole fraction. If p is the vapour pressure of the solvent (with a substance dissolved in it) and X the mole fraction of solvent (number of moles of solvent divided by total number of moles) then $p = p_0X$, where p_0 is the vapour pressure of the pure solvent. A solution that obeys Raoult's law is said to be an *ideal solution*. In general the law holds only for dilute solutions, although some mixtures of liquids obey it over a whole range of concentrations. Such solutions are *perfect solutions* and occur when the intermolecular forces between molecules of the pure substances are similar to the forces between molecules of one and molecules of the other. Deviations in Raoult's law for mixtures of liquids cause the formation of *azeotropes. The law was discovered by the French chemist François Raoult (1830–1901).

rare-earth elements *See* lanthanoids.

rarefaction A reduction in the pressure of a fluid and therefore of its density.

rare gases *See* noble gases.

Raschig process An industrial process for making chlorobenzene (and phenol) by a gas-phase reaction between benzene vapour, hydrogen chloride, and oxygen (air) at 230°C

$$2C_6H_6 + 2HCl + O_2 \rightarrow 2H_2O + 2C_6H_5Cl$$

The catalyst is copper(II) chloride. The chlorobenzene is mainly used for making phenol by the reaction

$$C_6H_5Cl + H_2O \rightarrow HCl + C_6H_5OH$$

This reaction proceeds at 430°C with a silicon catalyst.

Raschig synthesis *See* hydrazine.

rate constant (velocity constant) Symbol k. The constant in an expression for the rate of a chemical reaction in terms of concentrations (or activities). For instance, in a simple unimolecular reaction

$$A \rightarrow B$$

the rate is proportional to the concentration of A, i.e.

$$\text{rate} = k[A]$$

where k is the rate constant, which depends on the temperature. The equation is the *rate equation* of the reaction, and its form depends on the reaction mechanism.

rate-determining step The slowest step in a chemical reaction that involves a number of steps. In such reactions, there is often a single step that is appreciably slower than the other steps, and the rate of this determines the overall rate of the reaction.

rationalized units A system of units in which the defining equations have been made to conform to the geometry of the system in a logical way. Thus equations that involve circular symmetry contain the factor 2π, while those involving spherical symmetry contain the factor 4π. *SI units are rationalized; c.g.s. units are unrationalized.

rayon A textile made from cellulose. There are two types, both made from wood pulp. In the viscose process, the pulp is dissolved in carbon disulphide and sodium hydroxide to give a thick brown liquid containing cellulose xanthate. The liquid is then forced through fine nozzles into acid, where the xanthate is decomposed and a cellulose filament is produced. The product is *viscose rayon*. In the acetate process cellulose acetate is made and dissolved in a solvent. The solution is forced through nozzles into air, where the solvent quickly evaporates leaving a filament of *acetate rayon*.

reactant *See* chemical reaction.

reaction *See* chemical reaction.

reactive dye *See* dyes.

reagent A substance reacting with another substance. Laboratory reagents are compounds, such as sulphuric acid, hydrochloric acid, sodium hydroxide, etc., used in chemical analysis or experiments.

realgar A red mineral form of arsenic(II) sulphide, As_2S_2.

real gas A gas that does not have the properties assigned to an *ideal gas. Its molecules have a finite size and there are forces between them (*see* equation of state).

rearrangement A type of chemical reaction in which the atoms in a molecule rearrange to form a new molecule.

reciprocal proportions *See* chemical combination.

recombination process The process in which a neutral atom or molecule is formed by the combination of a positive ion and a negative ion or electron; i.e. a process of the type:

$$A^+ + B^- \rightarrow AB.$$

or

$$A^+ + e^- \rightarrow A$$

In recombination, the neutral species formed is usually in an excited state, from which it can decay with emission of light or other electromagnetic radiation.

rectification The process of purifying a liquid by *distillation. *See* fractional distillation.

rectified spirit A constant-boiling mixture of *ethanol (95.6°) and water; it is obtained by distillation.

red lead *See* dilead(II) lead(IV) oxide.

redox *See* oxidation–reduction.

reducing agent (reductant) A substance that brings about reduction in other substances. It achieves this by being itself oxidized. Reducing agents contain atoms with low oxidation numbers; that is the atoms have gained electrons. In reducing other substances, these atoms lose electrons.

reductant *See* reducing agent.

reduction *See* oxidation–reduction.

refinery gas *See* petroleum.

refining The process of purifying substances or extracting substances from mixtures.

refluxing A laboratory technique in which a liquid is boiled in a container attached to a condenser (*reflux condenser*), so that the liquid continuously flows back into the container. It is used for carrying out reactions over long periods in organic synthesis.

reforming The conversion of straight-chain alkanes into branched-chain alkanes by *cracking or by catalytic reaction. It is used in petroleum refining to produce hydrocarbons suitable for use in gasoline. Benzene is also manufactured from alkane hydrocarbons by catalytic reforming. *Steam reforming* is a process used to convert methane (from natural gas) into a mixture of carbon monoxide and hydrogen, which is used to synthesize organic chemicals. The reaction

$$CH_4 + H_2O \rightarrow CO + 3H_2$$

occurs at about 900°C using a nickel catalyst.

Regnault's method A technique for measuring gas density by evacuating and weighing a glass bulb of known volume, admitting gas at known pressure, and reweighing. The determination must be carried out at constant known temperature and the result corrected to standard temperature and pressure. The method is named after the French chemist Henri Victor Regnault (1810–78).

relative atomic mass (atomic weight; r.a.m.) Symbol A_r. The ratio of the average mass per atom of the naturally occurring form of an element to 1/12 of the mass of a carbon-12 atom.

relative density (r.d.) The ratio of the *density of a substance to the density of some reference substance. For liquids or solids it is the ratio of the density (usually at 20°C) to the density of water (at its maximum density). This quantity was formerly called *specific gravity*. Sometimes relative densities of gases are used; for example, relative to dry air, both gases being at s.t.p.

relative molecular mass (molecular weight) Symbol M_r. The ratio of the average mass per molecule of the naturally occurring form of an element or compound to 1/12 of the mass of a carbon-12 atom. It is equal to the sum of the relative atomic masses of all the atoms that comprise a molecule.

rem *See* radiation units.

rennin An enzyme secreted by cells lining the stomach in mammals that is responsible for clotting milk. It acts on a soluble milk protein (*caseinogen*), which it converts to the insoluble form casein. This ensures that milk remains in the stomach long enough to be acted on by protein-digesting enzymes.

resin A synthetic or naturally occurring *polymer. Synthetic resins are used in making *plastics. Natural resins are acidic chemicals secreted by many trees (especially conifers) into ducts or canals. They are found either as brittle glassy substances or dissolved in essential oils. Their functions are probably similar to those of gums and mucilages.

resolution The process of separating a racemic mixture into its optically active constituents. In some cases the crystals of the two forms have a different appearance, and the separation can be done by hand. In general, however, physical methods (distillation, crystallization, etc.) cannot be used because the optical isomers have identical physical properties. The most common technique is to react the mixture with a compound that is itself optically active, and then separate the two. For instance, a racemic mixture of *l*-A and *d*-A reacted with *l*-B, gives two compounds AB that are not optical isomers (they are diastereoisomers) and can be separated and reconverted into the pure *l*-A and *d*-A. Biological techniques using bacteria that convert one form but not the other can also be used.

resonance The representation of the structure of a molecule by two or more conventional formulae. For example, the formula of methanal can be represented by a covalent structure $H_2C{=}O$, in which there is a double bond in the carbonyl group. It is known that in such compounds the oxygen has some negative charge and the carbon some positive charge. The true bonding in the molecule is somewhere between $H_2C{=}O$ and the ionic compound $H_2C^+O^-$. It is said to be a *resonance hybrid* of the two, indicated by

$$H_2C{=}O \leftrightarrow H_2C^+O^-$$

The two possible structures are called *canonical forms*, and they need not contribute equally to the actual form. Note that the double-headed arrow does not imply that the two forms are in equilibrium.

retinol *See* vitamin A.

retort 1. A laboratory apparatus consisting of a glass bulb with a long neck. **2.** A vessel used for reaction or distillation in industrial chemical processes.

reverberatory furnace A metallurgical furnace in which the charge to be heated is kept separate from the fuel. It consists of a shallow hearth on which the charge is heated by flames that pass over it and by radiation reflected onto it from a low roof.

reverse osmosis A method of obtaining pure water from water containing a salt, as in *desalination. Pure water and the salt water are separated by a semipermeable membrane and the pressure of the salt water is raised above the osmotic pressure, causing water from the brine to pass through the membrane into the pure water.

This process requires a pressure of some 25 atmospheres, which makes it difficult to apply on a large scale.

reversible process Any process in which the variables that define the state of the system can be made to change in such a way that they pass through the same values in the reverse order when the process is reversed. It is also a condition of a reversible process that any exchanges of energy, work, or matter with the surroundings should be reversed in direction and order when the process is reversed. Any process that does not comply with these conditions when it is reversed is said to be an *irreversible process*. All natural processes are irreversible, although some processes can be made to approach closely to a reversible process.

R_F value (in chromatography) The distance travelled by the solvent front divided by the distance travelled by a given component. For a given system at a known temperature, it is a characteristic of the component and can be used to identify components.

rhe A unit of fluidity equal to the reciprocal of the *poise.

rhenium Symbol Re. A silvery-white metallic *transition element; a.n. 75; r.a.m. 186.2; r.d. 21.0; m.p. 3180°C; b.p. 5620°C. The element is obtained as a by-product in refining molybdenum, and is used in certain alloys (e.g. rhenium–molybdenum alloys are superconducting). The element forms a number of complexes with oxidation states in the range 1–7.

rheology The study of the deformation and flow of matter.

rheopexy The process by which certain thixotropic substances set more rapidly when they are stirred, shaken, or tapped. Gypsum in water is such a *rheopectic substance*.

rhodium Symbol Rh. A silvery-white metallic *transition element; a.n. 45; r.a.m. 102.9; r.d. 12.4; m.p. 1966°C; b.p. 3727°C. It occurs with platinum and is used in certain platinum alloys (e.g. for thermocouples) and in plating jewellery and optical reflectors. Chemically, it is not attacked by acids (dissolves only slowly in aqua regia) and reacts with nonmetals (e.g. oxygen and chlorine) at red heat. Its main oxidation state is +3 although it also forms complexes in the +4 state. The element was discovered in 1803 by W. H. Wollaston.

riboflavin *See* vitamin B complex.

ribonucleic acid *See* RNA.

ribose A *monosaccharide, $C_5H_{10}O_5$, rarely occurring free in nature but important as a component of *RNA (ribonucleic acid). Its derivative *deoxyribose*, $C_5H_{10}O_4$, is equally important as a constituent of *DNA (deoxyribonucleic acid), which carries the genetic code in chromosomes.

ring A closed chain of atoms in a molecule. In compounds, such as naphthalene, in which two rings share a common side, the rings are *fused rings*. *Ring closures* are chemical reactions in which one part of a chain reacts with another to form a ring, as in the formation of *lactams and *lactones.

RNA (ribonucleic acid) A complex organic compound (a nucleic acid) in living cells that is concerned with *protein synthesis. In some viruses, RNA is also the hereditary material. Most RNA is synthesized in the nucleus and then distributed to various parts of the cytoplasm. An RNA molecule consists of a long chain of *nucleotides in which the sugar is *ribose and the bases are adenine, cytosine, guanine, and uracil (*compare* DNA).

Rochelle salt Potassium sodium tartrate tetrahydrate, $KNaC_4H_4O_6.4H_2O$. A colour-

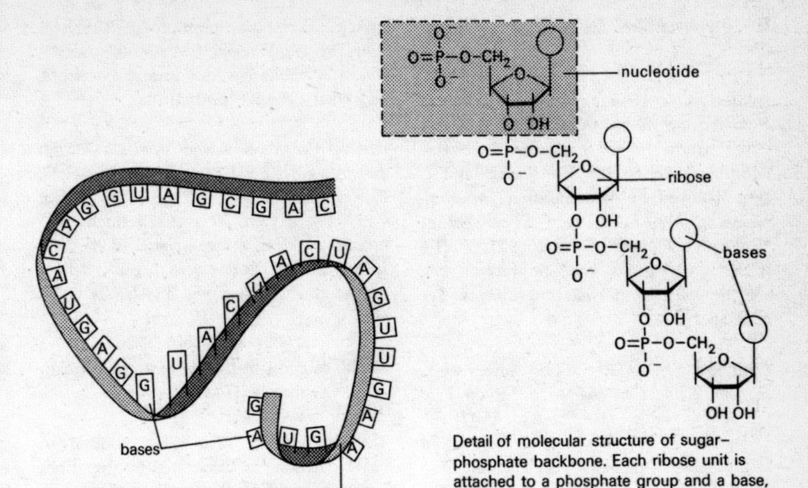

Single-stranded structure of RNA

Detail of molecular structure of sugar–phosphate backbone. Each ribose unit is attached to a phosphate group and a base, forming a nucleotide

adenine (A)

guanine (G)

cytosine (C)

uracil (U)

The four bases of RNA

Molecular structure of RNA

less crystalline salt used for its piezoelectric properties.

rock An aggregate of mineral particles that makes up part of the earth's crust. It may be consolidated or unconsolidated (e.g. sand, gravel, mud, shells, coral, and clay).

rock crystal *See* quartz.

rock salt *See* halite.

roentgen The former unit of dose equivalent (*see* radiation units). It is named after the discoverer of X-rays, W. K. Roentgen (1845–1923).

Rose's metal An alloy of low melting point (about 100°C) consisting of 50% bismuth, 25–28% lead, and 22–25% tin.

R–S convention *See* absolute configuration.

rubber A polymeric substance obtained from the sap of the tree *Hevea brasiliensis*. Crude natural rubber is obtained by coagulating and drying the sap (latex), and is then modified by *vulcanization and compounding with fillers. It is a polymer of *isoprene containing the unit $-CH_2C(CH_3)$ $:CHCH_2-$. Various synthetic rubbers can also be made by polymerizing alkenes. *See* neoprene.

rubidium Symbol Rb. A soft silvery-white metallic element belonging to *group I of the periodic table; a.n. 37; r.a.m. 85.47; r.d. 1.53; m.p. 38.4°C; b.p. 688°C. It is found in a number of minerals (e.g. lepidolite) and in certain brines. The metal is obtained by electrolysis of molten rubidium chloride. The naturally occurring isotope rubidium-87 is radioactive (*see* rubidium–strontium dating). The metal is highly reactive, with properties similar to those of other group I elements, igniting spontaneously in air. It was discovered spectroscopically by R. W. Bunsen and G. R. Kirchhoff in 1861.

rubidium–strontium dating A method of dating geological specimens based on the decay of the radioisotope rubidium-87 into the stable isotope strontium-87. Natural rubidium contains 27.85% of rubidium-87, which has a half-life of 4.7×10^{11} years. The ratio $^{87}Rb/^{87}Sr$ in a specimen gives an estimate of its age (up to several thousand million years).

ruby The transparent red variety of the mineral *corundum, the colour being due to the presence of traces of chromium. It is a valuable gemstone, more precious than diamonds. The finest rubies are obtained from Mogok in Burma, where they occur in metamorphic limestones; Sri Lanka and Thailand are the only other important sources. Rubies have been produced synthetically by the Verneuil flame-fusion process. Industrial rubies are used in lasers, watches, and other precision instruments.

rusting Corrosion of iron (or steel) to form a hydrated iron(III) oxide $Fe_2O_3.xH_2O$. Rusting occurs only in the presence of both water and oxygen. It is an electrochemical process in which different parts of the iron surface act as electrodes in a cell reaction. At the anode, iron atoms dissolve as Fe^{2+} ions:

$$Fe(s) \rightarrow Fe^{2+}(aq) + 2e$$

At the cathode, hydroxide ions are formed:

$$O_2(aq) + 2H_2O(l) + 4e \rightarrow 4OH^-(aq)$$

The $Fe(OH)_2$ in solution is oxidized to Fe_2O_3. Rusting is accelerated by impurities in the iron and by the presence of acids or other electrolytes in the water.

ruthenium Symbol Ru. A hard white metallic *transition element; a.n. 44; r.a.m. 101.07; r.d. 12.41; m.p. 2310°C; b.p. 3900°C. It is found associated with platinum and is used as a catalyst and in certain platinum alloys. Chemically, it dissolves in fused alkalis but is not attacked by acids. It reacts with oxygen and halogens at high temperatures. It also forms complexes with a range of oxidation states. The element was isolated by K. K. Klaus in 1844.

rutile A mineral form of titanium(IV) oxide, TiO_2.

Rydberg spectrum An absorption spectrum of a gas in the ultraviolet region, consisting of a series of lines that become closer together towards shorter wavelengths, merging into a continuous absorption region. The absorption lines correspond to electronic transitions to successively higher energy levels. The onset of the continuum corresponds to photoionization of the atom or molecule, and can thus be used to determine the ionization potential.

S

saccharide *See* sugar.

saccharose *See* sucrose.

Sachse reaction A reaction of methane at high temperature to produce ethyne:

$$2CH_4 \rightarrow C_2H_2 + 3H_2$$

The reaction occurs at about 1500°C, the high temperature being obtained by burning part of the methane in air.

sacrificial protection (cathodic protection) The protection of iron or steel against corrosion (*see* rusting) by using a more reactive metal. A common form is galvanizing (*see* galvanized iron), in which the iron surface is coated with a layer of zinc. Even if the zinc layer is scratched, the iron does not rust because zinc ions are formed in solution in preference to iron ions. Pieces of magnesium alloy are similarly used in protecting pipelines, etc.

sal ammoniac *See* ammonium chloride.

salicylic acid (1-hydroxybenzoic acid) A naturally occurring carboxylic acid found in certain plants, HOC_6H_4COOH; r.d. 1.4; m.p. 157°C. It is used in making aspirin and in the foodstuffs and dyestuffs industries.

saline Describing a chemical compound that is a salt, or a solution containing a salt. *See also* physiological saline.

salinometer An instrument for measuring the salinity of a solution. There are two main types: one is a type of *hydrometer to measure density; the other is an apparatus for measuring the electrical conductivity of the solution.

sal soda Anhydrous *sodium carbonate, Na_2CO_3.

salt A compound formed by reaction of an acid with a base, in which the hydrogen of the acid has been replaced by metal or other positive ions. Typically, salts are crystalline ionic compounds such as Na^+Cl^- and $NH_4{}^+NO_3{}^-$. Covalent metal compounds, such as $TiCl_4$, are also often regarded as salts.

salt bridge An electrical connection made between two half cells. It usually consists of a glass U-tube filled with agar jelly containing a salt, such as potassium chloride. A strip of filter paper soaked in the salt solution can also be used.

salt cake Industrial *sodium sulphate.

saltpetre *See* nitre.

samarium Symbol Sm. A soft silvery metallic element belonging to the *lanthanoids; a.n. 62; r.a.m. 150.35; r.d. 7.52 (20°C); m.p. 1073°C; b.p. 1791°C. It occurs in monazite and bastnatite. There are seven naturally occurring isotopes, all of which are stable except samarium-147, which is weakly radioactive (half-life 2.5×10^{11} years). The metal is used in special alloys for making nuclear-reactor parts as it is a neutron absorber. Samarium oxide (Sm_2O_3) is used in small quantities in special optical glasses. The largest use of the element is in the ferromagnetic alloy $SmCo_5$, which produces permanent magnets five times stronger than any other material. The element was discovered by François Lecoq de Boisbaudran in 1879.

sand Particles of rock with diameters in the range 0.06–2.00 mm. Most sands are composed chiefly of particles of quartz, which are derived from the weathering of quartz-bearing rocks.

Sandmeyer reaction A reaction of diazonium salts used to prepare chloro- or bromo-substituted aromatic compounds. The method is to diazotize an aromatic amide at low temperature and add an equimolar solution of the halogen acid and copper(I) halide. A complex of the diazonium salt and copper halide forms, which decomposes when the temperature is raised. The copper halide acts as a catalyst in the reaction of the halide ions from the acid, for example

$$C_6H_5N_2{}^+(aq) + Cl^-(aq) + CuCl(aq) \rightarrow C_6H_5Cl(l) + N_2(g) + CuCl(aq)$$

The reaction was discovered in 1884 by the German chemist Traugott Sandmeyer (1854–1922). *See also* Gattermann reaction.

sandwich compound A transition-metal complex in which a metal atom or ion is 'sandwiched' between two rings of atoms. *Ferrocene was the first such compound to be prepared. *See also* metallocene.

saponification The reaction of esters with alkalis to give alcohols and salts of carboxylic acids:

$$RCOOR' + OH^- \rightarrow RCOO^- + R'OH$$

See esterification; soap.

sapphire Any of the gem varieties of *corundum except ruby, especially the blue variety, but other colours of sapphire include yellow, brown, green, pink, orange, and purple. Sapphires are obtained from igneous and metamorphic rocks and from alluvial deposits. The chief sources are Sri Lanka, Kashmir, Burma, Thailand, East Africa, the USA, and Australia. Sapphires are used as gemstones and record-player styluses. They are synthesized by the Verneuil flame-fusion process.

saturated 1. (of a compound) Consisting of molecules that have only single bonds (i.e. no double or triple bonds). Saturated compounds can undergo substitution reactions but not addition reactions. *Compare* unsaturated. **2.** (of a solution) Containing the maximum equilibrium amount of solute at a given temperature. In a saturated solution the dissolved substance is in equilibrium with undissolved substance; i.e. the rate at which solute particles leave the solution is exactly balanced by the rate at which they dissolve. A solution containing less than the equilibrium amount is said to be *unsaturated*. One containing more than the equilibrium amount is *supersaturated*. Supersaturated solutions can be made by slowly cooling a saturated solution. Such solutions are metastable; if a small crystal seed is added the excess solute crystallizes out of solution. **3.** (of a vapour) *See* vapour pressure.

saturation *See* supersaturation.

***s*-block elements** The elements of the first two main groups of the *periodic table; i.e. groups IA (Li, Na, K, Rb, Cs, Fr) and IIA (Be, Mg, Ca, Sr, Ba, Ra). The outer electronic configurations of these elements all have inert-gas structures plus outer ns^1 (IA) or ns^2 (IIA) electrons. The term thus excludes elements with incomplete inner *d*-levels (transition metals) or with incomplete inner *f*-levels (lanthanoids and actinoids) even though these often have outer ns^2 or occasionally ns^1 configurations. Typically, the *s*-block elements are reactive metals forming stable ionic compounds containing M^+ or M^{2+} ions. *See* alkali metals; alkaline-earth metals.

scandium Symbol Sc. A rare soft silvery metallic element belonging to group IIIB of the periodic table; a.n. 21; r.a.m. 44.956; r.d. 2.985 (alpha form), 3.19 (beta form); m.p. 1540°C; b.p. 2850°C. Scandium often occurs in *lanthanoid ores, from which it can be separated on account of the greater solubility of its thiocyanate in ether. The only natural isotope, which is not radioactive, is scandium-45, and there are nine radioactive isotopes, all with relatively short half-lives. Because of the metal's high reactivity and high cost no substantial uses have been found for either the metal or its compounds. Predicted in 1869 by Mendeleev, and then called *ekaboron*, the oxide (called *scandia*) was isolated by Nilson in 1879.

scanning electron microscope *See* electron microscope.

scheelite A mineral form of calcium tungstate, $CaWO_4$, used as an ore of tungsten. It occurs in contact metamorphosed deposits and vein deposits as colourless or white tetragonal crystals.

Schiff's base A compound formed by a condensation reaction between an aromatic amine and an aldehyde or ketone, for example

$$RNH_2 + R'CHO \rightarrow RN{:}CHR' + H_2O$$

The compounds are often crystalline and are used in organic chemistry for characterizing aromatic amines (by preparing the Schiff's base and measuring the melting point). They are named after the German chemist Hugo Schiff (1834–1915).

Schiff's reagent A reagent used for testing for aldehydes and ketones; it consists of a solution of fuchsin dye that has been decolorized by sulphur dioxide. Aliphatic aldehydes restore the pink immediately, whereas aromatic ketones have no effect on the reagent. Aromatic aldehydes and aliphatic ketones restore the colour slowly.

schönite A mineral form of potassium sulphate, K_2SO_4.

Schrödinger equation An equation used in wave mechanics (*see* quantum mechanics) for the wave function of a particle. The time-independent Schrödinger equation is

$$\Delta^2\psi + 8\pi^2 m(E - U)\psi/h^2 = 0$$

where ψ is the wave function; Δ^2 the Laplace operator, h the Planck constant, m the particle's mass, E its total energy, and U its potential energy.

secondary alcohol *See* alcohols.

secondary amine *See* amines.

secondary cell A *voltaic cell in which the chemical reaction producing the e.m.f. is reversible and the cell can therefore be charged by passing a current through it. *See* accumulator. *Compare* primary cell.

second-order reaction *See* order.

sedimentation The settling of the solid particles through a liquid either to produce a concentrated slurry from a dilute suspension or to clarify a liquid containing solid particles. Usually this relies on the force of gravity, but if the particles are too small or the difference in density between the solid and liquid phases is too small, a *centrifuge may be used. In the simplest case the rate of sedimentation is determined by Stokes's law, but in practice the predicted rate is rarely reached. Measurement of the rate of sedimentation in an *ultracentrifuge can be used to estimate the size of macromolecules.

seed A crystal used to induce other crystals to form from a gas, liquid, or solution.

Seger cones (pyrometric cones) A series of cones used to indicate the temperature inside a furnace or kiln. The cones are made from different mixtures of clay, limestone, feldspars, etc., and each one softens at a different temperature. The drooping of the vertex is an indication that the known softening temperature has been reached and thus the furnace temperature can be estimated.

selenides Binary compounds of selenium with other more electropositive elements. Selenides of nonmetals are covalent (e.g. H_2Se). Most metal selenides can be prepared by direct combination of the elements. Some are well-defined ionic compounds (containing Se^{2-}), while others are nonstoichiometric interstitial compounds (e.g. Pd_4Se, $PdSe_2$).

selenium Symbol Se. A metalloid element belonging to group VI of the periodic table; a.n. 34; r.a.m. 78.96; r.d. 4.79 (grey); m.p. 217°C (grey); b.p. 684.9°C. There are a number of allotropic forms, including grey, red, and black selenium. It occurs in sulphide ores of other metals and is obtained as a by-product (e.g. from the anode sludge in electrolytic refining). The element is a semiconductor; the grey allotrope is light-sensitive and is used in photocells, xerography, and similar applications. Chemically, it resembles sulphur, and forms compounds with selenium in the +2, +4, and +6 oxidation states. Selenium was discovered in 1817 by J. J. Berzelius.

selenium cell Either of two types of *photoelectric cell; one type relies on the photoconductive effect, the other on the photovoltaic effect (*see* photoelectric effect). In the photoconductive selenium cell an external e.m.f. must be applied; as the selenium changes its resistance on exposure to light, the current produced is a measure of the light energy falling on the selenium. In the photovoltaic selenium cell, the e.m.f. is generated within the cell. In this type of cell, a thin film of vitreous or metallic selenium is applied to a metal surface, a transparent film of another metal, usually gold or platinum, being placed over the selenium. Both types of cell are used as light meters in photography.

Seliwanoff's test A biochemical test to identify the presence of ketonic sugars, such as fructose, in solution. It was devised by the Russian chemist F. F. Seliwanoff. A few drops of the reagent, consisting of resorcinol crystals dissolved in equal amounts of water and hydrochloric acid, are heated with the test solution and the formation of a red precipitate indicates a positive result.

semicarbazones Organic compounds containing the unsaturated group $=C:N.NH.CO.NH_2$. They are formed when aldehydes or ketones react with semicarbazide ($H_2N.NH.CO.NH_2$). Semicarbazones are crystalline compounds with relatively high melting points. They are used to identify aldehydes and ketones in quantitative analysis: the semicarbazone derivative is made and identified by its melting point. Semicarbazones are also used in separating ketones from reaction mixtures: the derivative is crystallized out and hydrolysed to give the ketone.

semimetal *See* metalloid.

semipermeable membrane A membrane that is permeable to molecules of the solvent but not the solute in *osmosis. Semipermeable membranes can be made by supporting a film of material (e.g. cellulose) on a wire gauze or porous pot.

semipolar bond *See* chemical bond.

septivalent (heptavalent) Having a valency of seven.

sequestration The process of forming coordination complexes of an ion in solution. Sequestration often involves the formation of chelate complexes, and is used to prevent the chemical effect of an ion without removing it from the solution (e.g. the sequestration of Ca^{2+} ions in water softening). It is also used as a way of supplying ions in a protected form, e.g. the use of sequestered iron solutions for plants in regions having alkaline soil.

serine *See* amino acid.

serpentine Any of a group of hydrous magnesium silicate minerals with the general composition $Mg_3Si_2O_5(OH)_4$. Serpentine is monoclinic and occurs in two main forms: *chrysotile*, which is fibrous and the chief source of *asbestos; and *antigorite*, which occurs as platy masses. It is generally green or white with a mottled appearance, sometimes resembling a snakeskin – hence the name. It is formed through the metamorphic alteration of ultrabasic rocks rich in olivine, pyroxene, and amphibole. *Serpentinite* is a rock consisting mainly of serpentine; it is used as an ornamental stone.

sesqui- Prefix indicating a ratio of 2:3 in a chemical compound. For example, a sesquioxide has the formula M_2O_3.

sexivalent (hexavalent) Having a valency of six.

sherardizing The process of coating iron or steel with a zinc corrosion-resistant layer by heating the iron or steel in contact with zinc dust to a temperature slightly below the melting point of zinc. At a temperature of about 371°C the two metals amalgamate to form internal layers of zinc–iron alloys and an external layer of pure zinc. The process was invented by Sherard Cowper-Coles (d. 1935).

short period *See* periodic table.

sial The rocks that form the earth's continental crust. These are granite rock types rich in *si*lica (SiO_2) and *al*uminium (Al), hence the name. *Compare* sima.

side chain *See* chain.

side reaction A chemical reaction that occurs at the same time as a main reaction but to a lesser extent, thus leading to other products mixed with the main products.

siderite A brown or grey-green mineral form of iron(II) carbonate, $FeCO_3$, often with magnesium and manganese substituting for the iron. It occurs in sedimentary deposits or in hydrothermal veins and is an important iron ore. It is found in England, Greenland, Spain, N Africa, and the USA.

siemens Symbol S. The SI unit of electrical conductance equal to the conductance of a circuit or element that has a resistance of 1 ohm. $1\ S = 10^{-1}\ \Omega$. The unit was formerly called the mho or reciprocal ohm. It is named after Sir William Siemens (1823–83).

sievert The SI unit of dose equivalent (*see* radiation units).

sigma electron An electron in a sigma orbital. *See* orbital.

silane (silicane) 1. A colourless gas, SiH_4, which is insoluble in water; d. 1.44 g dm^{-3}; r.d. 0.68 (liquid); m.p. −185°C; b.p. −112°C. Silane is produced by reduction of silicon tetrachloride using lithium tetrahydridoaluminate(III). It is also formed by the reaction of magnesium silicide (Mg_2Si) with acids, although other silicon hydrides are also produced at the same time. Silane itself is stable in the absence of air but is spontaneously flammable, even at low temperatures. It is used for controlled silicon doping of semiconductors in the electronics industry. It is also a reducing agent and has been used for the removal of corrosion in inaccessible plants (e.g. pipes in nuclear reactors). **2.** (*or* **silicon hydride)** Any of a class of compounds of silicon and hydrogen. They have the general formula Si_nH_{2n+2}. The first three in the series are silane itself (SiH_4), *disilane* (Si_2H_6), and *trisilane* (Si_3H_8). The compounds are analogous to the alkanes but are much less stable and only the lower members of the series can be prepared in any quantity (up to Si_6H_{14}). No silicon hydrides containing double or triple bonds exist (i.e. there are no analogues of the alkenes and alkynes).

silica *See* silicon(IV) oxide.

silica gel A rigid gel made by coagulating a sol of sodium silicate and heating to drive off water. It is used as a support for catalysts and also as a drying agent because it readily absorbs moisture from the air. The gel itself is colourless but, when used in desiccators, etc., a blue cobalt salt is added. As moisture is taken up, the salt turns pink, indicating that the gel needs to be regenerated (by heating).

silicane *See* silane.

silicate Any of a group of substances containing negative ions composed of silicon and oxygen. The silicates are a very extensive group and natural silicates form the major component of most rocks (*see* silicate minerals). The basic structural unit is the tetrahedral SiO_4 group. This may occur as a simple discrete $SiO_4{}^{4-}$ anion as in the *orthosilicates*, e.g. *phenacite* (Be_2SiO_4) and *willemite* (Zn_2SiO_4). Many larger silicate species are also found. These are composed of SiO_4 tetrahedra linked by sharing oxygen atoms as in the *pyrosilicates*, $Si_2O_7{}^{6-}$, e.g. $Sc_2Si_2O_7$. The linking can extend to such forms as benitoite, $BaTiSi_3O_9$, or alternatively infinite chain anions, which are single strand (*pyroxenes) or double strand (*amphiboles). Spodumene, $LiAl(SiO_3)_2$, is a pyroxene and the asbestos minerals are amphiboles. Infinite two-dimensional sheets are also possible, as in the various *micas and the linking can extend to full three-dimensional framework structures, often with substituted trivalent atoms in the lattice. The *zeolites are examples of this.

SiO_4^{4-} as in Be_2SiO_4 (phenacite)

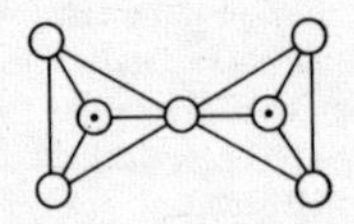

$Si_2O_5^{2-}$ as in $Sc_2Si_2O_7$ (thortveitite)

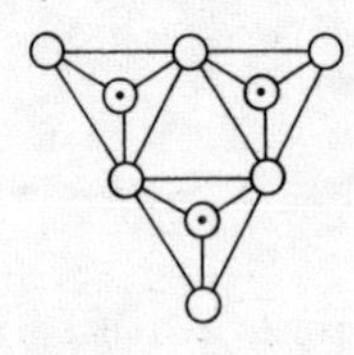

$Si_3O_9^{6-}$ as in $BaTiSi_3O_9$ (bentonite)

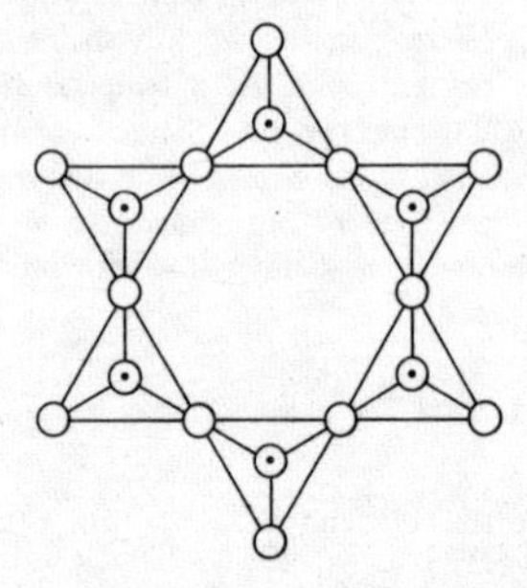

$Si_6O_{18}^{12-}$ as in $Be_3Al_2Si_6O_{18}$ (beryl)

Structure of some discrete silicon ions

silicate minerals A group of rock-forming minerals that make up the bulk of the earth's outer crust (about 90%) and constitute one-third of all minerals. All silicate minerals are based on a fundamental structural unit – the SiO_4 tetrahedron (*see* silicate). They consist of a metal (e.g. calcium, magnesium, aluminium) combined with silicon and oxygen. The silicate minerals are classified on a structural basis according to how the tetrahedra are linked together. The six groups are: nesosilicates (e.g. olivine and *garnet); sorosilicates (e.g. hemimorphite); cyclosilicates (e.g. axinite, *beryl, and *tourmaline); inosilicates (e.g. *amphiboles and *pyroxenes); phyllosilicates (e.g. *micas, *clay minerals, and *talc); and tektosilicates (e.g. *feldspars and *feldspathoids). Many

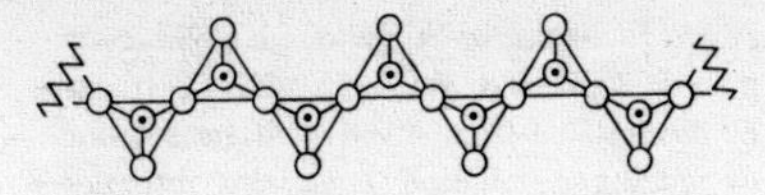

single chain : pyroxenes

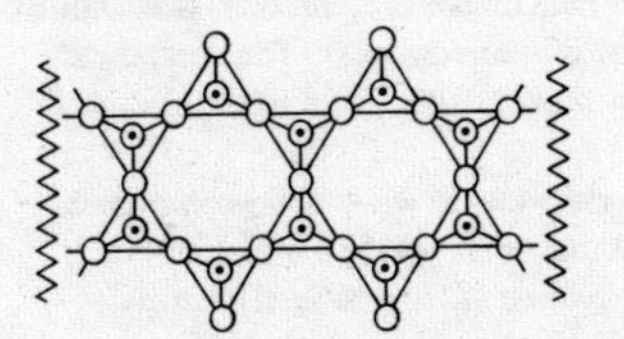

double chain : amphiboles

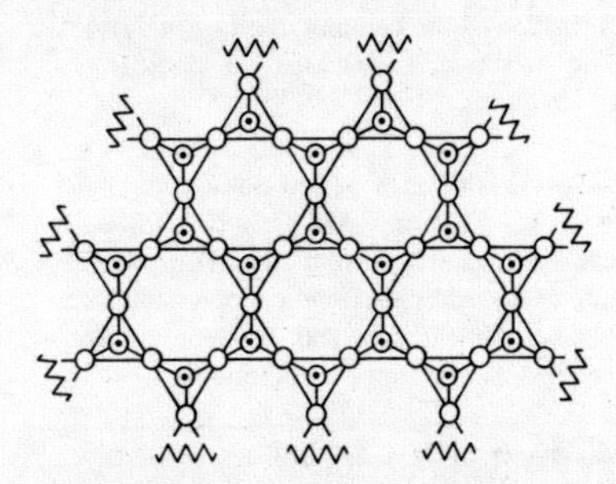

sheet : micas

Structure of some polymeric silicate ions

silicate minerals are of economic importance.

silicide A compound of silicon with a more electropositive element. The silicides are structurally similar to the interstitial carbides but the range encountered is more diverse. They react with mineral acids to form a range of *silanes.

silicon Symbol Si. A metalloid element belonging to *group IV of the periodic table; a.n. 14; r.a.m. 28.086; r.d. 2.33; m.p. 1410°C; b.p. 2355°C. Silicon is the second most abundant element in the earth's crust (25.7% by weight) occurring in various forms of silicon(IV) oxide (e.g. *quartz) and in *silicate minerals. The element is extracted by reducing the oxide with carbon in an electric furnace and is used extensively for its semiconductor properties. It has a diamond-like crystal structure; an amorphous form also exists. Chemically, silicon is less reactive than carbon. The element combines with oxygen at red heat and is also dissolved by molten alkali. There is a large number of organosilicon compounds (e.g. *siloxanes) although silicon does not form the range of silicon–hydrogen compounds and derivatives that carbon does (*see* silanes). The element was identified by Lavoisier in 1787 and first isolated in 1823 by Berzelius.

silicon carbide (carborundum) A black solid compound, SiC, insoluble in water and soluble in molten alkali; r.d. 3.217; m.p. *c.* 2700°C. Silicon carbide is made by heating silicon(IV) oxide with carbon in an electric furnace (depending on the grade required sand and coke may be used). It is extremely hard and is widely used as an abrasive. The solid exists in both zinc blende and wurtzite structures.

silicon dioxide *See* silicon(IV) oxide.

silicones Polymeric compounds containing chains of silicon atoms alternating with oxygen atoms, with the silicon atoms linked to organic groups. A variety of silicone materials exist, including oils, waxes, and rubbers. They tend to be more resistant to temperature and chemical attack than their carbon analogues.

silicon hydride *See* silane.

silicon(IV) oxide (silicon dioxide; silica) A colourless or white vitreous solid, SiO_2, insoluble in water and soluble (by reaction) in hydrofluoric acid and in strong alkali; m.p. 1610°C; b.p. 2230°C. The following forms occur naturally: *cristobalite* (cubic or tetragonal crystals; r.d. 2.32); *tridymite* (rhombic; r.d. 2.26); *quartz (hexagonal; r.d.

2.63–2.66); *lechatelierite* (r.d. 2.19). Quartz has two modifications: α-quartz below 575°C and β-quartz above 575°C; above 870°C β-quartz is slowly transformed to tridymite and above 1470°C this is slowly converted to cristobalite. Various forms of silicon(IV) oxide occur widely in the earth's crust; yellow sand for example is quartz with iron(III) oxide impurities and flint is essentially amorphous silica. The gemstones amethyst, opal, and rock crystal are also forms of quartz.

Silica is an important commercial material in the form of *silica brick*, a highly refractive furnace lining, which is also resistant to abrasion and to corrosion. Silicon(IV) oxide is also the basis of both clear and opaque silica glass, which is used on account of its transparency to ultraviolet radiation and its resistance to both thermal and mechanical shock. A certain proportion of silicon(IV) oxide is also used in ordinary glass and in some glazes and enamels. It also finds many applications as a drying agent in the form of *silica gel.

siloxanes A group of compounds containing silicon atoms bound to oxygen atoms, with organic groups linked to the silicon atoms, e.g. $R_3SiOSiR_3$, where R is an organic group. *Silicones are polymers of siloxanes.

silver Symbol Ag. A white lustrous soft metallic *transition element; a.n. 47; r.a.m. 107.87; r.d. 10.5; m.p. 961.93°C; b.p. 2212°C. It occurs as the element and as the minerals argentite (Ag_2S) and horn silver (AgCl). It is also present in ores of lead and copper, and is extracted as a by-product of smelting and refining these metals. The element is used in jewellery, tableware, etc., and silver compounds are used in photography. Chemically, silver is less reactive than copper. A dark silver sulphide forms when silver tarnishes in air because of the presence of sulphur compounds. Silver(I) ionic salts exist (e.g. $AgNO_3$, AgCl) and there are a number of silver(II) complexes.

silver(I) bromide A yellowish solid compound, AgBr; r.d. 6.5; m.p. 432°C. It can be precipitated from silver(I) nitrate solution by adding a solution containing bromide ions. It dissolves in concentrated ammonia solutions (but, unlike the chloride, does not dissolve in dilute ammonia). The compound is used in photographic emulsions.

silver(I) chloride A white solid compound, AgCl; r.d. 5.6; m.p. 445°C; b.p. 1550°C. It can be precipitated from silver(I) nitrate solution by adding a solution of chloride ions. It dissolves in ammonia solution (due to formation of the complex ion $[Ag(NH_3)_2]^+$). The compound is used in photographic emulsions.

silver(I) iodide A yellow solid compound, AgI; r.d. 6.01; m.p. 556°C; b.p. 1506°C. It can be precipitated from silver(I) nitrate solutions by adding a solution of iodide ions. Unlike the chloride and bromide, it does not dissolve in ammonia solutions.

silver-mirror test *See* Tollen's reagent.

silver(I) nitrate A colourless solid, $AgNO_3$; r.d. 4.3; m.p. 212°C. It is an important silver salt because it is water-soluble. It is used in photography. In the laboratory, it is used as a test for chloride, bromide, and iodide ions and in volumetric analysis of chlorides using an *absorption indicator.

silver(I) oxide A brown slightly water-soluble amorphous powder, Ag_2O; r.d. 7.14. It can be made by adding sodium hydroxide solution to silver(I) nitrate solution. Silver(I) oxide is strongly basic and is also an oxidizing agent. It is used in certain reactions in preparative organic chemistry; for example, moist silver(I) oxide converts haloalkanes into alcohols; dry silver oxide converts haloalkanes into ethers. The compound decomposes to the elements at 160°C and can be reduced by hydrogen to silver. With ozone it gives the oxide AgO (which is diamagnetic and probably $Ag^{I}Ag^{III}O_2$).

sima The rocks that form the earth's oceanic crust and underlie the upper crust. These are basaltic rock types rich in *si*lica (SiO_2) and *ma*gnesium (Mg), hence the name. The sima is denser and more plastic than the *sial that forms the continental crust.

single bond *See* chemical bond.

sintered glass Porous glass made by sintering powdered glass, used for filtration of precipitates in gravimetric analysis.

sintering The process of heating and compacting a powdered material at a temperature below its melting point in order to weld the particles together into a single rigid shape. Materials commonly sintered include metals and alloys, glass, and ceramic oxides. Sintered magnetic materials, cooled in a magnetic field, make especially retentive permanent magnets.

SI units Système International d'Unités: the international system of units now recommended for all scientific purposes. A coherent and rationalized system of units derived from the *m.k.s. units, SI units have now replaced *c.g.s. units and *Imperial units. The system has seven *base units* and two *supplementary units* (see Appendix), all other units being derived from these nine units. There are 18 derived units with special names. Each unit has an agreed symbol (a capital letter or an initial capital letter if it is named after a scientist, otherwise the symbol consists of one or two lower-case letters). Decimal multiples of the units are indicated by a set of prefixes; whenever possible a prefix representing 10 raised to a power that is a multiple of three should be used.

slag Material produced during the *smelting or refining of metals by reaction of the flux with impurities (e.g. calcium silicate formed by reaction of calcium oxide flux with silicon dioxide impurities). The liquid slag can be separated from the liquid metal because it floats on the surface. *See also* basic slag.

slaked lime *See* calcium hydroxide.

slurry A paste consisting of a suspension of a solid in a liquid.

smectic *See* liquid crystal.

smelting The process of separating a metal from its ore by heating the ore to a high temperature in a suitable furnace in the presence of a reducing agent, such as carbon, and a fluxing agent, such as limestone. Iron ore is smelted in this way so that the metal melts and, being denser than the molten *slag, sinks below the slag, enabling it to be removed from the furnace separately.

smoke A fine suspension of solid particles in a gas.

S_N1 reaction *See* nucleophilic substitution.

S_N2 reaction *See* nucleophilic substitution.

SNG Substitute (or synthetic) natural gas; a mixture of gaseous hydrocarbons produced from coal, petroleum, etc., and suitable for use as a fuel. Before the discovery of natural gas *coal gas was widely used as a domestic and industrial fuel. This gave way to natural gas in the early part of this century in the US and other countries where natural gas was plentiful. The replacement of coal gas occurred somewhat later in the UK and other parts of Europe. More recently, interest has developed in ways of manufacturing hydrocarbon gas fuels. The main sources are coal and the naphtha fraction of petroleum. In the case of coal three methods have been used: (1) pyrolysis – i.e. more efficient forms of destructive distillation, often with further hydrogenation of the hydrocarbon products; (2) heating the coal with hydrogen and catalysts to give hydrocarbons – a process known as *hydroliquefaction* (*see also* Bergius process); (3) producing carbon monoxide and hydrogen and obtaining hydrocarbons by the *Fischer–Tropsch process. SNG from naptha is made by steam *reforming.

soap A substance made by boiling animal fats with sodium hydroxide. The reaction involves the hydrolysis of *glyceride esters of fatty acids to glycerol and sodium salts of the acids present (mainly the stearate, oleate, and palmitate), giving a soft semisolid with *detergent action. Potassium hydroxide gives a more liquid product (*soft soap*). By extension, other metal salts of long-chain fatty acids are also called soaps. *See also* saponification.

soda Any of a number of sodium compounds, such as caustic soda (NaOH) or, especially, washing soda ($Na_2CO_3.10H_2O$).

soda ash Anhydrous *sodium carbonate, Na_2CO_3.

soda lime A mixed hydroxide of sodium and calcium made by slaking lime with caustic soda solution (to give NaOH + Ca $(OH)_2$) and recovering greyish white granules by evaporation. The material is produced largely for industrial adsorption of carbon dioxide and water, but also finds some applications in pollution and effluent control. It is also used as a laboratory drying agent.

sodamide *See* sodium amide.

sodium Symbol Na. A soft silvery reactive element belonging to group I of the periodic table (*see* alkali metals); a.n. 11; r.a.m. 22.9898; r.d. 0.97; m.p. 97.5°C; b.p. 892°C. Sodium occurs as the chloride in sea water and in the mineral halite. It is extracted by electrolysis in a *Downs cell. The metal is used as a reducing agent in certain reactions and liquid sodium is also a coolant in nuclear reactors. Chemically, it is highly reactive, oxidizing in air and reacting violently with water (it is kept under oil). It dissolves in liquid ammonia to form blue solutions containing solvated electrons. Sodium is a major element required by living organisms. The element was first isolated by Humphry Davy in 1807.

sodium acetate *See* sodium ethanoate.

sodium aluminate A white solid, $NaAlO_2$ or $Na_2Al_2O_4$, which is insoluble in ethanol and soluble in water giving strongly alkaline solutions; m.p. 1800°C. It is manufactured by heating bauxite with sodium carbonate and extracting the residue with water, or it may be prepared in the laboratory by adding excess aluminium to hot concentrated sodium hydroxide. In solution the ion $Al(OH)_4^-$ predominates. Sodium aluminate is used as a mordant, in the production of zeolites, in effluent treatment, in glass manufacture, and in cleansing compounds.

sodium amide (sodamide) A white crystalline powder, $NaNH_2$, which decomposes in water and in warm ethanol, and has an odour of ammonia; m.p. 210°C; b.p. 400°C. It is produced by passing dry ammonia over metallic sodium at 350°C. It reacts with red-hot carbon to give sodium cyanide and with nitrogen(I) oxide to give sodium azide.

sodium azide A white or colourless crystalline solid, NaN_3, soluble in water and slightly soluble in alcohol; hexagonal; r.d. 1.846; decomposes on heating. It is made by the action of nitrogen(I) oxide on hot sodamide ($NaNH_2$) and is used as an organic reagent and in the manufacture of detonators.

sodium benzenecarboxylate (sodium benzoate) A colourless crystalline or white amorphous powder, C_6H_5COONa, soluble in water and slightly soluble in ethanol. It is made by the reaction of sodium hydroxide with benzoic acid and is used in the dyestuffs industry and as a food preservative. It was formerly used as an antiseptic.

sodium benzoate *See* sodium benzenecarboxylate.

sodium bicarbonate *See* sodium hydrogencarbonate.

sodium bisulphate *See* sodium hydrogensulphate.

sodium bisulphite *See* sodium hydrogensulphite.

sodium bromide A white crystalline solid, NaBr, known chiefly as the dihydrate (monoclinic; r.d. 2.17), and as the anhydrous salt (cubic; r.d. 3.20; m.p. 747°C; b.p. 1390°C). The dihydrate loses water at about 52°C and is very slightly soluble in alcohol. Sodium bromide is prepared by the reaction of bromine on hot sodium hydroxide solution or of hydrogen bromide on sodium carbonate solution. It is a sedative and is also used in photographic processing and in analytical chemistry.

sodium carbonate Anhydrous sodium carbonate (*soda ash*, *sal soda*) is a white powder, which cakes and aggregates on exposure to air due to the formation of hydrates. The monohydrate, $Na_2CO_3.H_2O$, is a white crystalline material, which is soluble in water and insoluble in alcohol; r.d. 1.55; loses water at 109°C; m.p. 851°C. The decahydrate, $Na_2CO_3.10H_2O$, (*washing soda*) is a translucent efflorescent crystalline solid; r.d. 1.44; loses water at 32–34°C to give the monohydrate; m.p. 851°C. Sodium carbonate may be manufactured by the *Solvay process or by suitable crystallization procedures from any one of a number of natural deposits, such as:

trona $5(Na_2CO_3.NaHCO_3.2H_2O)$,
natron $(Na_2CO_3.10H_2O)$,
ranksite $(2Na_2CO_3.9Na_2SO_4.KCl)$,
pirsonnite $(Na_2CO_3.CaCO_3.2H_2O)$, and
gaylussite $(Na_2CO_3.CaCO_3.5H_2O)$.

The method of extraction is very sensitive to the relative energy costs and transport costs in the region involved. Sodium carbonate is used in photography, in cleaning, in pH control of water, in textile treatment, glasses and glazes, and as a food additive and volumetric reagent. *See also* sodium sesquicarbonate.

sodium chlorate(V) A white crystalline solid, $NaClO_3$; cubic; r.d. 2.49; m.p. 250°C. It decomposes above its melting point to give oxygen and sodium chloride. The compound is soluble in water and in ethanol and is prepared by the reaction of chlorine on hot concentrated sodium hydroxide. Sodium chlorate is a powerful oxidizing agent and is used in the manufacture of matches and soft explosives, in calico printing, and as a garden weedkiller.

sodium chloride (common salt) A colourless crystalline solid, NaCl, soluble in water and very slightly soluble in ethanol; cubic; r.d. 2.17; m.p. 801°C; b.p. 1413°C. It occurs as the mineral *halite (rock salt) and in natural brines and sea water. It has the interesting property of a solubility in water that changes very little with temperature. It is used industrially as the starting point for a range of sodium-based products (e.g. Solvay process for Na_2CO_3, Castner–Kellner process for NaOH), and is known universally as a preservative and seasoner of foods. Sodium chloride has a key role in biological systems in maintaining electrolyte balances.

sodium cyanide A white or colourless crystalline solid, NaCN, deliquescent, soluble in water and in liquid ammonia, and slightly soluble in ethanol; cubic; m.p. 564°C; b.p. 1496°C. Sodium cyanide is now made by absorbing hydrogen cyanide in sodium hydroxide or sodium carbonate solution. The compound is extremely poisonous because it reacts with the iron in haemoglobin in the blood, so preventing oxygen reaching the tissues of the body. It is used in the extraction of precious metals and in electroplating industries. Aqueous solutions are alkaline due to salt hydrolysis.

sodium dichromate A red crystalline solid, $Na_2Cr_2O_7.2H_2O$, soluble in water and insoluble in ethanol. It is usually known as the dihydrate (r.d. 2.52), which starts to lose water above 100°C; the compound decomposes above 400°C. It is made by melting chrome iron ore with lime and soda ash and acidification of the chromate thus formed. Sodium dichromate is cheaper than the corresponding potassium compound but has the disadvantage of being hygroscopic. It is used as a mordant in dyeing, as an oxidizing agent in organic chemistry, and in analytical chemistry.

sodium dihydrogenorthophosphate *See* sodium dihydrogenphosphate(V).

sodium dihydrogenphosphate(V) (sodium dihydrogenorthophosphate) A colourless crystalline solid, NaH_2PO_4, soluble in water and insoluble in alcohol, known as the monohydrate (r.d. 2.04) and the dihydrate (r.d. 1.91). The dihydrate loses one water molecule at 60°C and the second molecule of water at 100°C, followed by decomposition at 204°C. The compound may be prepared by treating sodium carbonate with an equimolar quantity of phosphoric acid or by neutralizing phosphoric acid with sodium hydroxide. It is used in the preparation of sodium phosphate (Na_3PO_4), in baking powders, as a food additive, and as a constituent of buffering systems. Both sodium dihydrogenphosphate and trisodium phosphate enriched in ^{32}P have been used to study phosphate participation in metabolic processes.

sodium dioxide *See* sodium superoxide.

sodium ethanoate (sodium acetate) A colourless crystalline compound, CH_3COONa, which is known as the anhydrous salt (r.d. 1.52; m.p. 324°C) or the trihydrate (r.d. 1.45; loses water at 58°C). Both forms are soluble in water and in ethoxyethane, and slightly soluble in ethanol. The compound may be prepared by the reaction of ethanoic acid (acetic acid) with sodium carbonate or with sodium hydroxide. Because it is a salt of a strong base and a weak acid, sodium ethanoate is used in buffers for pH control in many laboratory applications, in foodstuffs, and in electroplating. It is also used in dyeing, soaps, pharmaceuticals, and in photography.

sodium fluoride A crystalline compound, NaF, soluble in water and very slightly soluble in ethanol; cubic; r.d. 2.56; m.p. 993°C; b.p. 1695°C. It occurs naturally as villiaumite and may be prepared by the reaction of sodium hydroxide or of sodium carbonate with hydrogen fluoride. The reaction of sodium fluoride with concentrated sulphuric acid may be used as a source of hydrogen fluoride. The compound is used in ceramic enamels and as a preservative agent for fermentation. It is highly toxic but in very dilute solution (less than 1 part per million) it is used in the fluoridation of water for the prevention of tooth decay on account of its ability to replace OH groups with F groups in the material of dental enamel.

sodium formate *See* sodium methanoate.

sodium hexafluoraluminate A colourless monoclinic solid, Na_3AlF_6, very slightly soluble in water; r.d. 2.9; m.p. 1000°C. It changes to a cubic form at 580°C. The compound occurs naturally as the mineral *cryolite but a considerable amount is manufactured by the reaction of aluminium fluoride wth alumina and sodium hydroxide or directly with sodium aluminate. Its most important use is in the manufacture of aluminium in the *Hall–Heroult cell. It is also used in the manufacture of enamels, opaque glasses, and ceramic glazes.

sodium hydride A white crystalline solid, NaH; cubic; r.d. 0.92; decomposes above 300°C (slow); completely decomposed at 800°C. Sodium hydride is prepared by the reaction of pure dry hydrogen with sodium at 350°C. Electrolysis of sodium hydride in molten LiCl/KCl leads to the evolution of hydrogen; this is taken as evidence for the ionic nature of NaH and the presence of the hydride ion (H^-). It reacts violently with water to give sodium hydroxide and hydrogen, with halogens to give the halide and appropriate hydrogen halide, and ignites spontaneously with oxygen at 230°C. It is a powerful reducing agent with several laboratory applications.

sodium hydrogencarbonate (bicarbonate of soda; sodium bicarbonate) A white crystalline solid, $NaHCO_3$, soluble in water and slightly soluble in ethanol; monoclinic; r.d. 2.159; loses carbon dioxide above 270°C. It is manufactured in the *Solvay process and may be prepared in the laboratory by passing carbon dioxide through sodium carbonate or sodium hydroxide solution. Sodium hydrogencarbonate reacts with acids to give carbon dioxide and, as it does not have strongly corrosive or strongly basic properties itself, it is employed in bulk for the treatment of acid spillage and in medicinal applications as an antacid. Sodium hydrogencarbonate is also used in baking powders (and is known as *baking soda*), dry-powder fire extinguishers, and in the textiles, tanning, paper, and ceramics industries. The hydrogencarbonate ion has an important biological role as an intermediate between atmospheric CO_2/H_2CO_3 and the carbonate ion CO_3^{2-}. For water-living organisms this is the most important and in some cases the only source of carbon.

sodium hydrogensulphate (sodium bisulphate) A colourless solid, $NaHSO_4$, known in anhydrous and monohydrate forms. The anhydrous solid is triclinic (r.d. 2.435; m.p. >315°C). The monohydrate is monoclinic and deliquescent (r.d. 2.103; m.p. 59°C). Both forms are soluble in water and slightly soluble in alcohol. Sodium hydrogensulphate was originally made by the reaction between sodium nitrate and sulphuric acid, hence its old name of *nitre cake*. It may be manufactured by the reaction of sodium hydroxide with sulphuric acid, or by heating equimolar proportions of sodium chloride and concentrated sulphuric acid. Solutions of sodium hydrogensulphate are acidic. On heating the compound decomposes (via $Na_2S_2O_7$) to give sulphur trioxide. It is used in paper making, glass making, and textile finishing.

sodium hydrogensulphite (sodium bisulphite) A white solid, $NaHSO_3$, which is very soluble in water (yellow in solution) and slightly soluble in ethanol; monoclinic; r.d. 1.48. It decomposes on heating to give sodium sulphate, sulphur dioxide, and sulphur. It is formed by saturating a solution of sodium carbonate with sulphur dioxide. The compound is used in the brewing industry and in the sterilization of wine casks. It is a general antiseptic and bleaching agent. *See also* aldehydes.

sodium hydroxide (caustic soda) A white transluscent deliquescent solid, NaOH, soluble in water and ethanol but insoluble in ether; r.d. 2.13; m.p. 318°C; b.p. 1390°C. Hydrates containing 7, 5, 4, 3.5, 3, 2, and 1 molecule of water are known.

Sodium hydroxide was formerly made by the treatment of sodium carbonate with lime but its main source today is from the electrolysis of brine using mercury cells or any of a variety of diaphragm cells. The principal product demanded from these cells is chlorine (for use in plastics) and sodium hydroxide is almost reduced to the status of a by-product. It is strongly alkaline and finds many applications in the chemical industry, particularly in the production of soaps and paper. It is also used to adsorb acidic gases, such as carbon dioxide and sulphur dioxide, and is used in the treatment of effluent for the removal of heavy metals (as hydroxides) and of acidity. Sodium hydroxide solutions are extremely corrosive to body tissue and are particularly hazardous to the eyes.

sodium iodide A white crystalline solid, NaI, very soluble in water and soluble in both ethanol and ethanoic acid. It is known in both the anhydrous form (cubic; r.d. 3.67; m.p. 661°C; b.p. 1304°C) and as the dihydrate (monoclinic; r.d. 2.45). It is prepared by the reaction of hydrogen iodide with sodium carbonate or sodium hydroxide in solution. Like potassium iodide, sodium iodide in aqueous solution dissolves iodine to form a brown solution containing the I_3^- ion. It finds applications in photography and is also used in medicine as an expectorant and in the administration of radioactive iodine for studies of thyroid function and for treatment of diseases of the thyroid.

sodium methanoate (sodium formate) A colourless deliquescent solid, HCOONa, soluble in water and slightly soluble in ethanol; monoclinic; r.d. 1.92; m.p. 253°C; decomposes on further heating. The monohydrate is also known. The compound may be produced by the reaction of carbon monoxide with solid sodium hydroxide at 200°C and 10 atmospheres pressure; in the laboratory it can be conveniently prepared by the reaction of methanoic acid and sodium hydroxide. Its uses are in the production of oxalic acid (ethanedioic acid) and methanoic acid and in the laboratory it is a convenient source of carbon monoxide.

sodium monoxide A whitish-grey deliquescent solid, Na_2O; r.d. 2.27; sublimes at 1275°C. It is manufactured by oxidation of the metal in a limited supply of oxygen and purified by sublimation. Reaction with water produces sodium hydroxide. Its commercial applications are similar to those of sodium hydroxide.

sodium nitrate (Chile saltpetre) A white solid, $NaNO_3$, soluble in water and in ethanol; trigonal; r.d. 2.261; m.p. 306°C; decomposes at 380°C. A rhombohedral form is also known. It is obtained from deposits of caliche or may be prepared by the reaction of nitric acid with sodium hydroxide or sodium carbonate. It was previously used for the manufacture of nitric acid by heating with concentrated sulphuric acid. Its main use is in nitrate fertilizers.

sodium nitrite A yellow hygroscopic crystalline compound, $NaNO_2$, soluble in water, slightly soluble in ether and in ethanol; rhombohedral; r.d. 2.17; m.p. 271°C; decomposes above 320°C. It is formed by the thermal decomposition of sodium nitrate and is used in the preparation of nitrous acid (reaction with cold dilute hydrochloric acid). Sodium nitrite is used in organic *diazotization and as a corrosion inhibitor.

sodium orthophosphate *See* trisodium phosphate(V).

sodium peroxide A whitish solid (yellow when hot), Na_2O_2, soluble in ice-water and decomposed in warm water or alcohol; r.d. 2.80; decomposes at 460°C. A crystalline octahydrate (hexagonal) is obtained by crystallization from ice-water. The compound is formed by the combustion of sodium metal in excess oxygen. At normal temperatures it reacts with water to give sodium hydroxide and hydrogen peroxide. It is a powerful oxidizing agent reacting with iodine vapour to give the iodate and periodate, with carbon at 300°C to give the carbonate, and with nitrogen(II) oxide to give the nitrate. It is used as a bleaching agent in wool and yarn processing, in the refining of oils and fats, and in the production of wood pulp.

sodium sesquicarbonate A white crystalline hydrated double salt, $Na_2CO_3.NaHCO_3.2H_2O$, soluble in water but less alkaline than sodium carbonate; r.d. 2.12; decomposes on heating. It may be prepared by crystallizing equimolar quantities of the constituent materials; it also occurs naturally as *trona* and in Searles Lake brines. It is widely used as a detergent and soap builder and, because of its mild alkaline properties, as a water-softening agent and bath-salt base. *See also* sodium carbonate.

sodium sulphate A white crystalline compound, Na_2SO_4, usually known as the anhydrous compound (orthorhombic; r.d. 2.67; m.p. 888°C) or the decahydrate (monoclinic; r.d. 1.46; loses water at 100°C). The decahydrate is known as *Glauber's salt*. A metastable heptahydrate ($Na_2SO_4.7H_2O$) also exists. All forms are soluble in water, dissolving to give a neutral solution. The compound occurs naturally as
mirabilite ($Na_2SO_4.10H_2O$),
threnardite (Na_2SO_4), and
glauberite ($Na_2SO_4.CaSO_4$).

Sodium sulphate may be produced industrially by the reaction of magnesium sulphate with sodium chloride in solution followed by crystallization, or by the reaction of concentrated sulphuric acid with solid sodium chloride. The latter method was used in the *Leblanc process for the production of alkali and has given the name *salt cake* to impure industrial sodium sulphate. Sodium sulphate is used in the manufacture of glass and soft glazes and in dyeing to promote an even finish. It also finds medicinal application as a purgative and in commercial aperient salts.

sodium sulphide A yellow-red solid, Na_2S, formed by the reduction of sodium sulphate with carbon (coke) at elevated temperatures. It is a corrosive and readily oxidized material of variable composition and usually contains polysulphides of the type Na_2S_2, Na_2S_3, and Na_2S_4, which cause the variety of colours. It is known in an anhydrous form (r.d. 1.85; m.p. 1180°C) and as a nonahydrate, $Na_2S.9H_2O$ (r.d. 1.43; decomposes at 920°C). Other hydrates of sodium sulphide have been reported. The compound is deliquescent, soluble in water with extensive hydrolysis, and slightly soluble in alcohol. It is used in wood pulping, dyestuffs manufacture, and metallurgy on account of its reducing properties. It has also been used for the production of sodium thiosulphate (for the photographic industry) and as a depilatory agent in leather preparation. It is a strong skin irritant.

sodium sulphite A white solid, Na_2SO_3, existing in an anhydrous form (r.d. 2.63) and as a heptahydrate (r.d. 1.59). Sodium sulphite is soluble in water and because it is readily oxidized it is widely used as a convenient reducing agent. It is prepared by reacting sulphur dioxide with either sodium carbonate or sodium hydroxide. Dilute mineral acids reverse this process and release sulphur dioxide. Sodium sulphite is used as a bleaching agent in textiles and in paper manufacture. Its use as an antioxidant in some canned foodstuffs gives rise to a slightly sulphurous smell immediately on opening, but its use is prohibited in meats or foods that contain vitamin B_1. Sodium sulphite solutions are occasionally used as biological preservatives.

sodium superoxide (sodium dioxide) A whitish-yellow solid, NaO_2, formed by the reaction of sodium peroxide with excess oxygen at elevated temperatures and pressures. It reacts with water to form hydrogen peroxide and oxygen.

sodium thiosulphate (hypo) A colourless efflorescent solid, $Na_2S_2O_3$, soluble in water but insoluble in ethanol, commonly encountered as the pentahydrate (monoclinic; r.d. 1.73; m.p. 42°C), which loses water at 100°C to give the anhydrous form (r.d. 1.66). It is prepared by the reaction of sulphur dioxide with a suspension of sulphur in boiling sodium hydroxide solution. Aqueous solutions of sodium thiosulphate are readily oxidized in the presence of air to sodium tetrathionate and sodium sulphate. The reaction with dilute acids gives sulphur and sulphur dioxide. It is used in the photographic industry and in analytical chemistry.

soft soap *See* soap.

soft water *See* hardness of water.

sol A *colloid in which small solid particles are dispersed in a liquid continuous phase.

solder An alloy used to join metal surfaces. A *soft solder* melts at a temperature in the range 200–300°C and consists of a tin–lead alloy. The tin content varies between 80% for the lower end of the melting range and 31% for the higher end. *Hard solders* contain substantial quantities of silver in the alloy. *Brazing solders* are usually alloys of copper and zinc, which melt at over 800°C.

solid A state of matter in which there is a three-dimensional regularity of structure, resulting from the proximity of the component atoms, ions, or molecules and the strength of the forces between them. True solids are crystalline (*see also* amorphous). If a crystalline solid is heated, the kinetic energy of the components increases. At a specific temperature, called the *melting point*, the forces between the components become unable to contain them within the crystal structure. At this temperature, the lattice breaks down and the solid becomes a liquid.

solid solution A crystalline material that is a mixture of two or more components, with ions, atoms, or molecules of one component replacing some of the ions, atoms, or molecules of the other component in its normal crystal lattice. Solid solutions are found in certain alloys. For example, gold and copper form solid solutions in which some of the copper atoms in the lattice are replaced by gold atoms. In general, the gold atoms are distributed at random, and a range of gold–copper compositions is possible. At a certain composition, the gold and copper atoms can each form regular individual lattices (referred to as *superlattices*). Mixed crystals of double salts (such as alums) are also examples of solid solutions. Compounds can form solid solutions if they are isomorphous (*see* isomorphism).

solubility The quantity of solute that dissolves in a given quantity of solvent to form a saturated solution. Solubility is measured in kilograms per metre cubed, moles per kilogram of solvent, etc. The solubility of a substance in a given solvent depends on the temperature. Generally, for a solid in a liquid, solubility increases with temperature; for a gas, solubility decreases. *See also* concentration.

solubility product Symbol K_s. The product of the concentrations of ions in a saturated solution. For instance, if a compound A_xB_y is in equilibrium with its solution

$$A_xB_y(s) \rightleftharpoons xA^+(aq) + yB^-(aq)$$

the equilibrium constant is

$$K_c = [A^+]^x[B^-]^y/[A_xB_y]$$

Since the concentration of the undissolved solid can be put equal to 1, the solubility product is given by

$$K_s = [A^+]^x[B^-]^y$$

The expression is only true for sparingly soluble salts. If the product of ionic concentrations in a solution exceeds the solubility product, then precipitation occurs.

solute The substance dissolved in a solvent in forming a *solution.

solution A homogeneous mixture of a liquid (the *solvent) with a gas or solid (the *solute*). In a solution, the molecules of the solute are discrete and mixed with the molecules of solvent. There is usually some interaction between the solvent and solute molecules (*see* solvation). Two liquids that can mix on the molecular level are said to be *miscible*. In this case, the solvent is the major component and the solute the minor component. *See also* solid solution.

solvation The interaction of ions of a solute with the molecules of solvent. For instance, when sodium chloride is dissolved in water the sodium ions attract polar water molecules, with the negative oxygen atoms pointing towards the positive Na^+ ion. Solvation of transition-metal ions can also occur by formation of coordinate bonds, as in the hexaquocopper(II) ion $[Cu(H_2O)_6]^{2+}$. Solvation is the process that causes ionic solids to dissolve, because the energy released compensates for the energy necessary to break down the crystal lattice. It occurs only with polar solvents. Solvation in which the solvent is water is called *hydration*.

Solvay process (ammonia–soda process) An industrial method of making sodium carbonate from calcium carbonate and sodium chloride. The calcium carbonate is first heated to give calcium oxide and carbon dioxide, which is bubbled into a solution of sodium chloride in ammonia. Sodium hydrogencarbonate is precipitated:

$$H_2O + CO_2(g) + NaCl(aq) + NH_3(aq) \rightarrow NaHCO_3(s) + NH_4Cl(aq)$$

The sodium hydrogencarbonate is heated to give sodium carbonate and carbon dioxide. The ammonium chloride is heated with calcium oxide (from the first stage) to regenerate the ammonia. The process was patented in 1861 by the Belgian chemist Ernest Solvay (1838–1922).

solvent A liquid that dissolves another substance or substances to form a *solution. *Polar solvents* are compounds such as water and liquid ammonia, which have dipole moments and consequently high dielectric constants. These solvents are capable of dissolving ionic compounds or covalent compounds that ionize (*see* solvation). *Nonpolar solvents* are compounds such as ethoxyethane and benzene, which do not have permanent dipole moments. These do not dissolve ionic compounds but will dissolve nonpolar covalent compounds.

solvent extraction The process of separating one constituent from a mixture by dissolving it in a solvent in which it is soluble but in which the other constituents of the mixture are not. The process is usually carried out in the liquid phase, in which case it is also known as *liquid–liquid extraction.* In liquid–liquid extraction, the solution containing the desired constituent must be immiscible with the rest of the mixture. The process is widely used in extracting oil from oil-bearing materials.

solvolysis A reaction between a compound and its solvent. *See* hydrolysis.

sorption *Absorption of a gas by a solid.

sorption pump A type of vacuum pump in which gas is removed from a system by absorption on a solid (e.g. activated charcoal or a zeolite) at low temperature.

species A chemical entity, such as a particular atom, ion, or molecule.

specific **1.** Denoting that an extensive physical quantity so described is expressed per unit mass. For example, the *specific latent heat* of a body is its latent heat per unit mass. When the extensive physical quantity is denoted by a capital letter (e.g. L for latent heat), the specific quantity is denoted by the corresponding lower-case letter (e.g. *l* for specific latent heat). **2.** In some older physical quantities the adjective 'specific' was added for other reasons (e.g. specific gravity, specific resistance). These names are now no longer used.

specific activity *See* activity.

specific gravity *See* relative density; specific.

specific heat capacity *See* heat capacity.

spectrograph *See* spectroscope.

spectrometer Any of various instruments for producing a spectrum and measuring the wavelengths, energies, etc., involved. A simple type, for visible radiation, is a spectroscope equipped with a calibrated scale allowing wavelengths to be read off or calculated. In the X-ray to infrared region of the electromagnetic spectrum, the spectrum is produced by dispersing the radiation with a prism or diffraction grating (or crystal, in the case of hard X-rays). Some form of photoelectric detector is used, and the spectrum can be obtained as a graphical plot, which shows how the intensity of the radiation varies with wavelength. Such instruments are also called *spectrophotometers.* Spectrometers also exist for investigating the gamma-ray region and the microwave and radio-wave regions of the spectrum (*see* electron spin resonance; nuclear magnetic resonance). Instruments for obtaining spectra of particle beams are also called spectrometers (*see* spectrum; photoelectron spectroscopy).

spectrophotometer *See* spectrometer.

spectroscope An optical instrument that produces a *spectrum for visual observation. The first such instrument was made by R. W. Bunsen; in its simplest form it consists of a hollow tube with a slit at one end by which the light enters and a collimating lens at the other end to produce a parallel beam, a prism to disperse the light, and a telescope for viewing the spectrum (see illustration). In the *spectrograph,* the spectroscope is provided with a camera to record the spectrum. For a broad range of spectroscopic work, from the ultraviolet to the infrared, a diffraction grating is used instead of a prism. *See also* spectrometer.

spectroscopy The study of methods of producing and analysing *spectra using *spectroscopes, *spectrometers, spectrographs, and spectrophotometers. The interpretations of the spectra so produced can be used for chemical analysis, examining atomic and molecular energy levels and molecular structures, and for determining the composition and motions of celestial bodies.

spectrum 1. A distribution of entities or properties arrayed in order of increasing or decreasing magnitude. For example, a beam of ions passed through a mass spectrograph, in which they are deflected according to their charge-to-mass ratios, will have a range of masses called a *mass spectrum.* A *sound spectrum* is the distribution of energy over a range of frequencies of a particular source. **2.** A range of electromagnetic energies arrayed in order of increasing or decreasing wavelength or frequency (*see* electromagnetic spectrum). The *emission spectrum* of a body or substance is the characteristic range of radiations it emits when it is heated, bombarded by electron or ions, or absorbs photons. The *absorption spectrum* of a substance is produced by examining, through the substance and through a spectroscope, a continuous spectrum of radiation. The energies removed from the continuous spectrum by the absorbing medium show up as black lines or bands; with a substance capable of emitting a spectrum these are in exactly the same positions in the spectrum as the emission lines and bands would occur in the emission spectrum.

Emission and absorption spectra may show a *continuous spectrum*, a *line spectrum*, or a *band spectrum*. A continuous spectrum contains an unbroken sequence of frequencies over a relatively wide range; it is produced by incandescent solids, liquids, and compressed gases. Line spectra are discontinuous lines produced by excited atoms and ions as they fall back to a lower energy level. Band spectra (closely grouped bands of lines) are characteristic of molecular gases or chemical compounds. *See also* spectroscopy.

speculum An alloy of copper and tin formerly used in reflecting telescopes to make the main mirror as it could be cast, ground, and polished to make a highly reflective surface. It has now been largely replaced by silvered glass for this purpose.

sphalerite (zinc blende) A mineral form of zinc sulphide, ZnS, crystallizing in the cubic system; the principal ore of zinc. It is usually yellow-brown to brownish-black in colour and occurs, often with galena, in metasomatic deposits and also in hydrothermal veins and replacement deposits. Sphalerite is mined on every continent, the chief sources including the USA, Canada, Mexico, the USSR, Australia, Peru, and Poland.

spiegel (spiegeleisen) A form of *pig iron containing 15–30% of manganese and 4–5% of carbon. It is added to steel in a Bessemer converter as a deoxidizing agent and to raise the manganese content of steel.

spin (intrinsic angular momentum) Symbol *s*. The part of the total angular momentum of a particle, atom, nucleus, etc., that can continue to exist even when the particle is apparently at rest, i.e. when its translational motion is zero and therefore its orbital angular momentum is zero. A molecule, atom, or nucleus in a specified energy level, or a particular elementary particle, has a particular spin, just as it has a particular charge or mass. According to *quantum theory, this is quantized and is restricted to multiples of $h/2\pi$, where h is the *Planck constant. Spin is characterized by a quantum number s. For example, for an electron $s = \pm\frac{1}{2}$, implying a spin of $+h/4\pi$ when it is spinning in one direction and $-h/4\pi$ when it is spinning in the other. Because of their spin, charged particles also have their own intrinsic magnetic moments and in a magnetic field the spin of the particles lines up at an angle to the direction of the field, precessing around this direction.

spinel A group of oxide minerals with the general formula $F^{2+}R_2{}^{3+}O_4$, where F^{2+} = Mg, Fe, Zn, Mn, or Ni and R^{3+} = Al, Fe, or Cr, crystallizing in the cubic system. The spinels are divided into three series: spinel ($MgAl_2O_4$), *magnetite, and *chromite. They occur in high-temperature igneous or metamorphic rocks.

spirits of salt Hydrogen chloride, so-called because it can be made by adding sulphuric acid to common salt (sodium chloride).

spontaneous combustion Combustion in which a substance produces sufficient heat within itself, usually by a slow oxidation process, for ignition to take place without the need for an external high-temperature energy source.

sputtering The process by which some of the atoms of an electrode (usually a cathode) are ejected as a result of bombardment by heavy positive ions. Although the process is generally unwanted, it can be used to produce a clean surface or to deposit a uniform film of a metal on an object in an evacuated enclosure.

square-planar Describing a *coordination compound in which four ligands positioned at the corners of a square coordinate to a metal ion at the centre of the square.

stabilization energy The amount by which the energy of a delocalized chemical structure is less than the theoretical energy of a structure with localized bonds. It is obtained by subtracting the experimental heat of formation of the compound (in $kJ\ mol^{-1}$) from that calculated on the basis of a classical structure with localized bonds.

stable equilibrium *See* equilibrium.

staggered conformation *See* conformation.

stainless steel A form of *steel containing at least 11–12% of chromium, a low percentage of carbon, and often some other elements, notably nickel and molybdenum. Stainless steel does not rust or stain and therefore has a wide variety of uses in industrial, chemical, and domestic environments. A particularly successful alloy is the steel known as 18–8, which contains 18% Cr, 8% Ni, and 0.08% C.

stalactites and stalagmites Accretions of calcium carbonate in limestone caves. Stalactites are tapering cones or pendants that hang down from the roofs of caves; stalagmites are upward projections from the cave floor and tend to be broader at their bases than stalactites. Both are formed from drips of water containing calcium carbonate in solution and may take thousands of years to grow.

standard cell A *voltaic cell, such as a *Clarke cell, or *Weston cell, used as a standard of e.m.f.

standard electrode An electrode (a half cell) used in measuring electrode potential. *See* hydrogen half cell.

standard electrode potential *See* electrode potential.

standard solution A solution of known concentration for use in volumetric analysis.

standard state A state of a system used as a reference value in thermodynamic measurements. Standard states involve a reference value of pressure (usually 1 atmosphere, 101.325 kPa) or concentration (usually 1 M). Thermodynamic functions are designated as 'standard' when they refer to changes in which reactants and products are all in their standard and their normal physical state. For example, the standard molar enthalpy of formation of water at 298 K is the enthalpy change for the reaction

$$H_2(g) + \tfrac{1}{2}O_2(g) \rightarrow H_2O(l)$$

$\Delta H_{298} = -285.83\ kJ\ mol^{-1}$. Note the superscript is used to denote standard state and the temperature should be indicated.

standard temperature and pressure *See* s.t.p.

stannane *See* tin(IV) hydride.

stannate A compound formed by reaction of tin oxides (or hydroxides) with alkali. Tin oxides are amphoteric (weakly acidic) and react to give stannate ions. Tin(IV) oxide with molten alkali gives the stannate(IV) ion

$$SnO_2 + 2OH^- \rightarrow SnO_3^{2-} + H_2O$$

In fact, there are various ions present in which the tin is bound to hydroxide groups, the main one being the hexahydroxostannate (IV) ion, $Sn(OH)_6^{2-}$. This is the negative ion present in crystalline 'trihydrates' of the type $K_2Sn_2O_3.3H_2O$.

Tin(II) oxide gives the trihydroxostannate (II) ion in alkaline solutions

$$SnO(s) + OH^-(aq) + H_2O(l) \rightarrow Sn(OH)_3^-(aq)$$

Stannate(IV) compounds were formerly referred to as *orthostannates* (SnO_4^{4-}) or *metastannates* (SnO_3^{2-}). Stannate(II) compounds were called *stannites*.

stannic compounds Compounds of tin in its higher (+4) oxidation state; e.g. stannic chloride is tin(IV) chloride.

stannite *See* stannate.

stannous compounds Compounds of tin in its lower (+2) oxidation state; e.g. stannous chloride is tin(II) chloride.

starch A *polysaccharide consisting of various proportions of two glucose polymers, *amylose and *amylopectin. It occurs widely in plants, especially in roots, tubers, seeds, and fruits, as a carbohydrate energy store. Starch is therefore a major energy source for animals. When digested it ultimately yields glucose. Starch granules are insoluble in cold water but disrupt if heated to form a gelatinous solution. This gives an intense blue colour with iodine solutions and starch is used as an *indicator in certain titrations.

state of matter One of the three physical states in which matter can exist, i.e. *solid, *liquid, or *gas. Plasma is sometimes regarded as the fourth state of matter.

stationary phase *See* chromatography.

stationary state A state of a system when it has an energy level permitted by *quantum mechanics. A transition from one stationary state to another can only occur by the emission or absorption of an appropriate quanta of energy in the form of photons.

statistical mechanics The branch of physics in which statistical methods are applied to the microscopic constituents of a system in order to predict its macroscopic properties. The earliest application of this method was Boltzmann's attempt to explain the thermodynamic properties of gases on the basis of the statistical properties of large assemblies of molecules.

In classical statistical mechanics, each particle is regarded as occupying a point in *phase space*, i.e. to have an exact position and momentum at any particular instant. The probability that this point will occupy any small volume of the phase space is taken to be proportional to the volume. The Maxwell–Boltzmann law gives the most probable distribution of the particles in phase space.

With the advent of quantum theory, the exactness of these premises was disturbed (by the Heisenberg uncertainty principle). In the *quantum statistics that evolved as a result, the phase space is divided into cells, each having a volume h^f, where h is the Planck constant and f is the number of degrees of freedom of the particles. This new concept led to Bose–Einstein statistics, and for particles obeying the Pauli exclusion principle, to Fermi–Dirac statistics.

steam distillation A method of distilling liquids that are immiscible with water by bubbling steam through them. It depends on the fact that the vapour pressure (and hence the boiling point) of a mixture of two immiscible liquids is lower than the vapour pressure of either pure liquid.

steam point The temperature at which the maximum vapour pressure of water is equal to the standard atmospheric pressure (101 325 Pa). On the Celsius scale it has the value 100°C.

stearate (octadecanoate) A salt or ester of stearic acid.

stearic acid (octadecanoic acid) A solid saturated *fatty acid, $CH_3(CH_2)_{16}COOH$; r.d. 0.94; m.p. 69.6°C; b.p. 376°C (with decomposition). It occurs widely (as *glycerides) in animal and vegetable fats.

steel Any of a number of alloys consisting predominantly of iron with varying proportions of carbon (up to 1.7%) and, in some cases, small quantities of other elements (*alloy steels*), such as manganese, silicon, chromium, molybdenum, and nickel. Steels con-

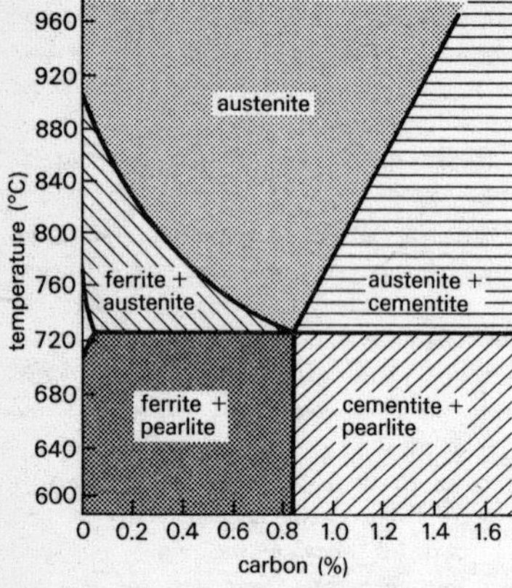

Phase diagram for steel

taining over 11–12% of chromium are known as *stainless steels.

Carbon steels exist in three stable crystalline phases: *ferrite* has a body-centred cubic crystal, *austenite* has a face-centred cubic crystal, and *cementite* has an orthorhombic crystal. *Pearlite* is a mixture of ferrite and cementite arranged in parallel plates. The phase diagram shows how the phases form at different temperatures and compositions.

Steels are manufactured by the *basic-oxygen process (L–D process), which has largely replaced the *Bessemer process and the *open-hearth process, or in electrical furnaces.

step A single stage in a chemical reaction. For example, the addition of hydrogen chloride to ethene involves three steps:

$$HCl \rightarrow H^+ + Cl^-$$
$$H^+ + C_2H_4 \rightarrow CH_3CH_2^+$$
$$CH_3CH_2^+ + Cl^- \rightarrow CH_3CH_2Cl$$

steradian Symbol sr. The supplementary *SI unit of solid angle equal to the solid angle that encloses a surface on a sphere equal to the square of the radius of the sphere.

stere A unit of volume equal to 1 m^3. It is not now used for scientific purposes.

stereochemistry The branch of chemistry concerned with the structure of molecules and the way the arrangement of atoms and groups affects the chemical properties.

stereoisomerism *See* isomerism.

stereoregular Describing a *polymer that has a regular pattern of side groups along its chain.

stereospecific Describing chemical reactions that give products with a particular arrangement of atoms in space. An example of a stereospecific reaction is the *Ziegler process for making polyethene.

steric effect An effect in which the rate or path of a chemical reaction depends on the size or arrangement of groups in a molecule.

steric hindrance An effect in which a chemical reaction is slowed down or prevented because large groups on a reactant molecule hinder the approach of another reactant molecule.

steroid Any of a group of lipids derived from a saturated compound called cyclopentanoperhydrophenanthrene, which has a nucleus of four rings. Some of the most important steroid derivatives are the steroid alcohols, or sterols. Other steroids include the bile acids, which aid digestion of fats in the intestine; the sex hormones (androgens and oestrogens); and the corticosteroid hormones, produced by the adrenal cortex. *Vitamin D is also based on the steroid structure.

sterol Any of a group of *steroid-based alcohols having a hydrocarbon side-chain of 8–10 carbon atoms. Sterols exist either as free sterols or as esters of fatty acids. Animal sterols (*zoosterols*) include cholesterol and lanosterol. The major plant sterol (*phytosterol*) is beta-sitosterol, while fungal sterols (*mycosterols*) include ergosterol.

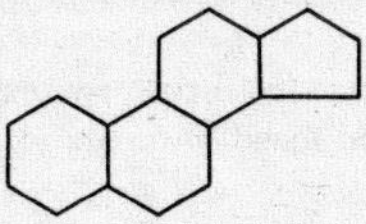

steroid nucleus

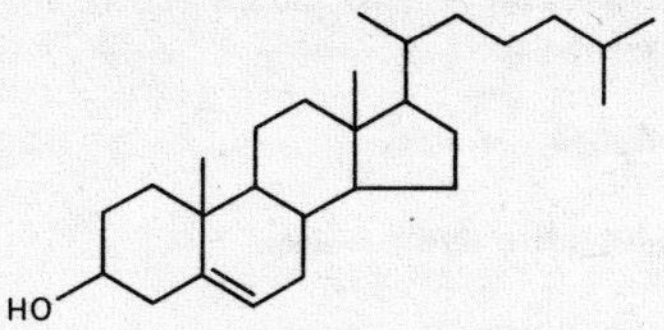

HO

cholesterol (a sterol)

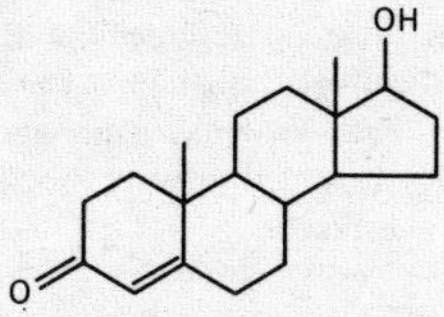

OH

O

testosterone (an androgen)

Steroid structure

stoichiometric Describing chemical reactions in which the reactants combine in simple whole-number ratios.

stoichiometric coefficient *See* chemical equation.

stoichiometric compound A compound in which atoms are combined in exact whole-number ratios. *Compare* nonstoichiometric compound.

stoichiometric mixture A mixture of substances that can react to give products with no excess reactant.

stoichiometric sum *See* chemical equation.

stoichiometry The relative proportions in which elements form compounds or in which substances react.

s.t.p. Standard temperature and pressure, formerly known as *N.T.P.* (normal temperature and pressure). The standard conditions used as a basis for calculations involving quantities that vary with temperature and pressure. These conditions are used when comparing the properties of gases. They are 273.15 K (or 0°C) and 101 325 Pa (or 760.0 mmHg).

straight chain *See* chain.

strong acid An *acid that is completely dissociated in aqueous solution.

strontia *See* strontium oxide.

strontianite A mineral form of *strontium carbonate, $SrCO_3$.

strontium Symbol Sr. A soft yellowish metallic element belonging to group II of the periodic table (*see* alkaline-earth elements); a.n. 38; r.a.m. 87.62; r.d. 2.54; m.p. 800°C; b.p. 1300°C. The element is found in the minerals strontianite ($SrCO_3$) and celestine ($SrSO_4$). It can be obtained by roasting the ore to give the oxide, followed by reduction with aluminium (i.e. the *Goldschmidt process). The element, which is highly reactive, is used in certain alloys and as a vacuum getter. The isotope strontium-90 is present in radioactive fallout (half-life 28 years), and can be metabolized with calcium so that it collects in bone. Strontium was discovered by Klaproth and Hope in 1798 and isolated by Humphry Davy in 1808.

strontium bicarbonate *See* strontium hydrogencarbonate.

strontium carbonate A white solid, $SrCO_3$; orthorhombic; r.d. 3.7; decomposes at 1770°C. It occurs naturally as the mineral *strontianite* and is prepared industrially by boiling celestine (strontium sulphate) with ammonium carbonate. It can also be prepared by passing carbon dioxide over strontium oxide or hydroxide or by passing the gas through a solution of strontium salt. It is a phosphor, used to coat the glass of cathode-ray screens, and is also used in the refining of sugar, as a slagging agent in certain metal furnaces, and to provide a red flame in fireworks.

strontium chloride A white compound, $SrCl_2$. The anhydrous salt (cubic; r.d. 3.05; m.p. 872°C; b.p. 1250°C) can be prepared by passing chlorine over heated strontium. It is deliquescent and readily forms the hexahydrate, $SrCl_2.6H_2O$ (r.d. 2.67). This can be made by neutralizing hydrochloric acid with strontium carbonate, oxide, or hydroxide. Strontium chloride is used for military flares.

strontium hydrogencarbonate (strontium bicarbonate) A compound, $Sr(HCO_3)_2$, which is stable only in solution. It is formed by the action of carbon dioxide on a suspension of strontium carbonate in water. On heating, this process is reversed.

strontium oxide (strontia) A white compound, SrO; r.d. 4.7; m.p. 2430°C, b.p. 3000°C. It can be prepared by the decomposition of heated strontium carbonate, hydroxide, or nitrate, and is used in the manufacture of other strontium salts, in pigments, soaps and greases, and as a drying agent.

strontium sulphate A white solid, $SrSO_4$; r.d. 3.96; m.p. 1605°C. It can be made by dissolving strontium oxide, hydroxide, or carbonate in sulphuric acid. It is used as a pigment in paints and ceramic glazes and to provide a red colour in fireworks.

structural formula *See* formula.

strychnine A colourless poisonous crystalline alkaloid found in certain plants.

styrene *See* phenylethene.

sublimate A solid formed by sublimation.

sublimation A direct change of state from solid to gas.

subshell *See* atom.

substantive dye *See* dyes.

substantivity The affinity of a dye for its substrate.

substituent 1. An atom or group that replaces another in a substitution reaction. **2.** An atom or group regarded as having replaced a hydrogen atom in a chemical derivative. For example, dibromobenzene ($C_6H_4Br_2$) is a derivative of benzene with bromine substituents.

substitution reaction (displacement reaction) A reaction in which one atom or molecule is replaced by another atom or molecule. *See* electrophilic substitution; nucleophilic substitution.

substrate The substance upon which an *enzyme acts in biochemical reactions.

succinic acid (butanedioic acid) A crystalline solid, $HOOC(CH_2)_2COOH$, that is soluble in water. It occurs in living organisms as an intermediate in metabolism.

sucrose (cane sugar; beet sugar; saccharose) A sugar comprising one molecule of glucose linked to a fructose molecule. It occurs widely in plants and is particularly abundant in sugar cane and sugar beet (15–20%), from which it is extracted and refined for table sugar. If heated to 200°C, sucrose becomes caramel.

sugar (saccharide) Any of a group of water-soluble *carbohydrates of relatively low molecular weight and typically having a sweet taste. The simple sugars are called *monosaccharides. More complex sugars comprise between two and ten monosaccharides linked together: *disaccharides contain two, trisaccharides three, and so on. The name is often used to refer specifically to *sucrose (table sugar).

sugar of lead *See* lead(II) ethanoate.

sulpha drugs *See* sulphonamides.

sulphanes Compounds of hydrogen and sulphur containing chains of sulphur atoms. They have the general formula H_2S_n. The simplest is hydrogen sulphide, H_2S; other members of the series are H_2S_2, H_2S_3, H_2S_4, etc. *See* sulphides.

sulphate A salt or ester of sulphuric(VI) acid. Organic sulphates have the formula R_2SO_4, where R is an organic group. Sulphate salts contain the ion SO_4^{2-}.

sulphides 1. Inorganic compounds of sulphur with more electropositive elements. Compounds of sulphur with nonmetals are covalent compounds, e.g. hydrogen sulphide (H_2S). Metals form ionic sulphides containing the S^{2-}ion; these are salts of hydrogen sulphide. *Polysulphides* can also be produced containing the polymeric ion S_x^{2-}. **2.** (*or* **thio ethers)** Organic compounds that contain the group –S– linked to two hydrocarbon groups. Organic sulphides are named from the linking groups, e.g. dimethyl sulphide (CH_3SCH_3), ethyl methyl sulphide ($C_2H_5SCH_3$). They are analogues of ethers in which the oxygen is replaced by sulphur (hence the alternative name) but are generally more reactive than ethers. Thus they react with halogen compounds to form *sulphonium compounds and can be oxidized to *sulphoxides.

sulphinate (dithionite; hyposulphite) A salt containing the negative ion $S_2O_4^{2-}$, usually formed by the reduction of sulphites with excess SO_2. Solutions are not very stable and decompose to give thiosulphate and hydrogensulphite ions. The structure is $^-O_2S\text{-}SO_2^-$.

sulphininic acid (dithionous acid; hyposulphurous acid) An unstable acid, $H_2S_2O_4$, known in the form of its salts (sulphinates). *See also* sulphuric acid.

sulphite A salt or ester derived from sulphurous acid. The salts contain the trioxosulphate(IV) ion SO_3^{2-}. Sulphites generally have reducing properties.

sulphonamides Organic compounds containing the group $-SO_2.NH_2$. The sulphonamides are amides of sulphonic acids. Many have antibacterial action and are also known as *sulpha drugs*, including
sulphadiazine, $NH_2C_6H_4SO_2NHC_4H_3N_2$,
sulphathiazole, $NH_2C_6H_4SO_2NHC_5H_2NS$,
and several others. They act by preventing bacteria from reproducing and are used to treat a variety of bacterial infections, especially of the gut and urinary system.

sulphonate A salt or ester of a sulphonic acid.

sulphonation A type of chemical reaction in which a $-SO_3H$ group is substituted on a benzene ring to form a *sulphonic acid. The reaction is carried out by refluxing with concentrated sulphuric(VI) acid for a long period. It can also occur with cold disulphuric(VI) acid ($H_2S_2O_7$). Sulphonation is an example of electrophilic substitution in which the electrophile is a sulphur trioxide molecule, SO_3.

sulphonic acids Organic compounds containing the $-SO_2.OH$ group. Sulphonic acids are formed by reaction of aromatic hydrocarbons with concentrated sulphuric acid. They are strong acids, ionizing completely in solution to form the sulphonate ion, $-SO_2.O^-$.

sulphonium compounds Compounds containing the ion R_3S^+ (sulphonium ion), where R is any organic group. Sulphonium compounds can be formed by reaction of organic sulphides with halogen compounds. For example, diethyl sulphide, $C_2H_5SC_2H_5$, reacts with chloromethane, CH_3Cl, to give diethylmethylsulphonium chloride, $(C_2H_5)_2.CH_3.S^+Cl^-$.

sulphoxides Organic compounds containing the group $=S=O$ (*sulphoxide group*) linked to two other groups, e.g. dimethyl sulphoxide, $(CH_3)_2SO$.

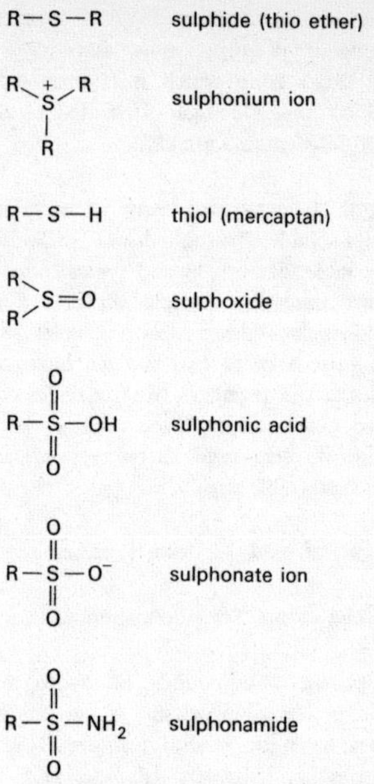

R—S—R	sulphide (thio ether)
R_3S^+	sulphonium ion
R—S—H	thiol (mercaptan)
$R_2S{=}O$	sulphoxide
$R{-}S({=}O)_2{-}OH$	sulphonic acid
$R{-}S({=}O)_2{-}O^-$	sulphonate ion
$R{-}S({=}O)_2{-}NH_2$	sulphonamide

Examples of organic sulphur compounds

sulphur Symbol S. A yellow nonmetallic element belonging to *group VI of the periodic table; a.n. 16; r.a.m. 32.06; r.d. 2.07 (rhombic); m.p. 112.8°C; b.p. 444.674°C. The element occurs in many sulphide and sulphate minerals and native sulphur is also found in Sicily and the USA (obtained by the *Frasch process). It is an essential element for living organisms.

Sulphur has various allotropic forms. Below 95.6°C the stable crystal form is rhombic; above this temperature the element transforms into a triclinic form. These crystalline forms both contain cyclic S_8 molecules. At temperatures just above its melting point, molten sulphur is a yellow liquid containing S_8 rings (as in the solid form). At about 160°C, the sulphur atoms form chains and the liquid becomes more viscous and dark brown. If the molten sulphur is cooled quickly from this temperature (e.g. by pouring into cold water) a reddish-brown solid known as *plastic sulphur* is obtained. Above 200°C the viscosity decreases. Sulphur vapour contains a mixture of S_2, S_4, S_6, and S_8 molecules. *Flowers of sulphur* is a yellow powder obtained by subliming the vapour. It is used as a plant fungicide. The element is also used to produce sulphuric acid and other sulphur compounds.

sulphur dichloride *See* disulphur dichloride.

sulphur dichloride dioxide (sulphuryl chloride) A colourless liquid, SO_2Cl_2; r.d. 1.67; m.p. −54.1°C; b.p. 69°C. It decomposes in water but is soluble in benzene. The compound is formed by the action of chlorine on sulphur dioxide in the presence of an iron(III) chloride catalyst or sunlight. It is used as a chlorinating agent and a source of the related fluoride, SO_2F_2.

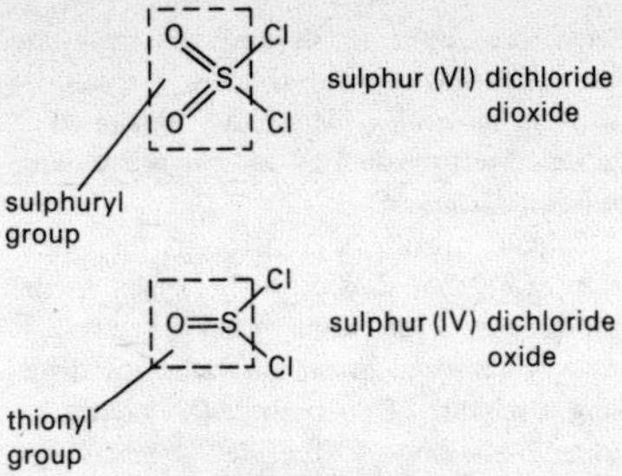

Oxychlorides of sulphur

sulphur dichloride oxide (thionyl chloride) A colourless fuming liquid, $SOCl_2$; m.p. −99.5°C; b.p. 75.7°C. It hydrolyses rapidly in water but is soluble in benzene. It may be prepared by the direct action of sulphur on chlorine monoxide or, more commonly, by the reaction of phosphorus (V) chloride with sulphur dioxide. It is used as a chlorinating agent in synthetic organic chemistry (replacing −OH groups with Cl).

sulphur dioxide (sulphur(IV) oxide) A colourless liquid or pungent gas, SO_2, formed by sulphur burning in air; r.d. 1.43 (liquid); m.p. −72.7°C; b.p. −10°C. It can be made by heating iron sulphide (pyrites) in air. The compound is a reducing agent and is used in bleaching and as a fumigant and food preservative. Large quantities are also used in the *contact process for manufacturing sulphuric acid. It dissolves in water to give a mixture of sulphuric and sulphurous acids.

sulphuretted hydrogen *See* hydrogen sulphide.

sulphuric acid (oil of vitriol) A colourless oily liquid, H_2SO_4; r.d. 1.84; m.p. 10.36°C; b.p. 338°C. The pure acid is rarely used; it is commonly available as a 96–98% solution (m.p. 3.0°C). The compound also forms a range of hydrates: $H_2SO_4.H_2O$ (m.p. 8.62°C); $H_2SO_4.2H_2O$ (m.p. −38/39°C); $H_2SO_4.6H_2O$ (m.p. −54°C); $H_2SO_4.8H_2O$ (m.p. −62°C). Its full systematic name is *tetraoxosulphuric(VI) acid.*

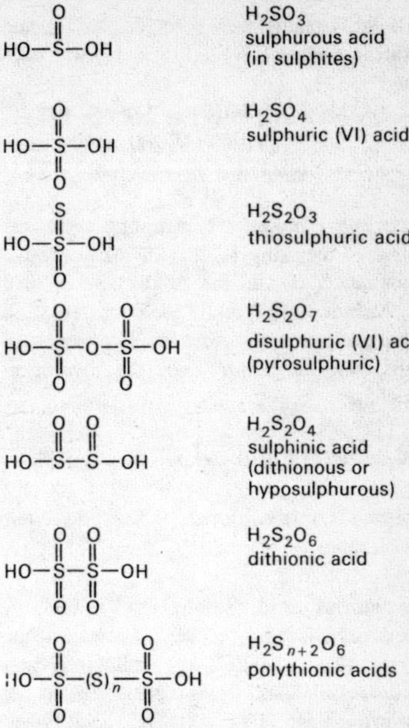

Structures of some oxo acids of sulphur

Until the 1930s, sulphuric acid was manufactured by the *lead-chamber process, but this has now been replaced by the *contact process (catalytic oxidation of sulphur dioxide). More sulphuric acid is made in the UK than any other chemical product; production levels (UK) are commonly 12 000 to 13 000 tonnes per day. It is extensively used in industry, the main applications being fertilizers (32%), chemicals (16%), paints and pigments (15%), detergents (11%), and fibres (9%).

In concentrated sulphuric acid there is extensive hydrogen bonding and several competing equilibria, to give species such as H_3O^+, HSO_4^-, $H_3SO_4^+$, and $H_2S_2O_7$. Apart from being a powerful protonating agent (it protonates chlorides and nitrates producing hydrogen chloride and nitric acid), the compound is a moderately strong oxidizing agent. Thus, it will dissolve copper:

$$Cu(s) + H_2SO_4(l) \rightarrow CuO(s) + H_2O(l) + SO_2(g)$$

$$CuO(s) + H_2SO_4(l) \rightarrow CuSO_4(aq) + H_2O(l)$$

It is also a powerful dehydrating agent, capable of removing H_2O from many organic compounds (as in the production of acid *anhydrides). In dilute solution it is a strong dibasic acid forming two series of salts, the sulphates and the hydrogen-sulphates.

sulphuric(IV) acid *See* sulphurous acid.

sulphur monochloride *See* disulphur dichloride.

sulphurous acid (sulphuric(IV) acid) A weak dibasic acid, H_2SO_3, known in the form of its salts: the sulphites and hydrogensulphites. It is considered to be formed (along with sulphuric acid) when sulphur dioxide is dissolved in water. It is probable, however, that the molecule H_2SO_3 is not present and that the solution contains hydrated SO_2. It is a reducing agent. The systematic name is *trioxosulphuric(IV) acid. See also* sulphuric acid.

sulphur(IV) oxide *See* sulphur dioxide.

sulphur(VI) oxide *See* sulphur trioxide.

sulphur trioxide (sulphur(VI) oxide) A colourless fuming solid, SO_3, which has three crystalline modifications. In decreasing order of stability these are: α, r.d. 1.97; m.p. 16.83°C; b.p. 44.8°C; β, m.p. 16.24°C; sublimes at 50°C; r.d. 2.29; γ, m.p. 16.8°C; b.p. 44.8°C. All are polymeric, with linked SO_4 tetrahedra: the γ-form has an icelike structure and is obtained by rapid quenching of the vapour; the β-form has infinite helical chains; and the α-form has infinite chains with some cross-linking of the SO_4 tetrahedra. Even in the vapour, there are polymeric species, and not discrete sulphur trioxide molecules (hence the compound is more correctly called by its systematic name *sulphur(VI) oxide*).

Sulphur trioxide is prepared by the oxidation of sulphur dioxide with oxygen in the presence of a vanadium(V) oxide catalyst. It may be prepared in the laboratory by distilling a mixture of concentrated sulphuric acid and phosphorus(V) oxide. It reacts violently with water to give sulphuric(VI) acid and is an important intermediate in the preparation of sulphuric acid and oleum.

sulphuryl chloride *See* sulphur dichloride dioxide.

sulphuryl group The group $=SO_2$, as in *sulphur dichloride oxide.

sulphydryl group *See* thiols.

superlattice *See* solid solution.

supernatant liquid The clear liquid remaining when a precipitate has settled.

superoxides A group of inorganic compounds that contain the O_2^- ion. They are formed in significant quantities only for sodium, potassium, rubidium, and caesium. They are very powerful oxidizing agents and react vigorously with water to give oxygen gas and OH^- ions. The superoxide ion has an unpaired electron and is paramagnetic and coloured (orange).

superphosphate A commercial phosphate mixture consisting mainly of monocalcium phosphate. Single-superphosphate is made by treating phosphate rock with sulphuric acid; the product contains 16–20% 'available' P_2O_5:

$$Ca_{10}(PO_4)_6F_2 + 7H_2SO_4 = 3Ca(H_2PO_4)_2 + 7CaSO_4 + 2HF.$$

Triple-superphosphate is made by using phosphoric(V) acid in place of sulphuric acid; the product contains 45–50% 'available' P_2O_5:

$$Ca_{10}(PO_4)_6F_2 + 14H_3PO_4 = 10Ca(H_2PO_4)_2 + 2HF$$

superplasticity The ability of some metals and alloys to stretch uniformly by several thousand percent at high temperatures, unlike normal alloys, which fail after being stretched 100% or less. Since 1962, when this property was discovered in an alloy of zinc and aluminium (22%), many alloys and ceramics have been shown to possess this property. For superplasticity to occur, the metal grain must be small and rounded and the alloy must have a slow rate of deformation.

supersaturated solution *See* saturated.

supersaturation 1. The state of the atmosphere in which the relative humidity is over 100%. This occurs in pure air where no condensation nuclei are available. Supersaturation is usually prevented in the atmosphere by the abundance of condensation nuclei (e.g. dust, sea salt, and smoke particles). **2.** The state of any vapour whose pressure exceeds that at which condensation usually occurs (at the prevailing temperature).

supplementary units *See* SI units.

surface tension Symbol γ. The property of a liquid that makes it behave as if its surface is enclosed in an elastic skin. The property results from intermolecular forces: a molecule in the interior of a liquid experiences a force of attraction from other molecules equally from all sides, whereas a molecule at the surface is only attracted by molecules below it in the liquid. The surface tension is defined as the force acting over the surface per unit length of surface perpendicular to the force. It is measured in newtons per metre. It can equally be defined as the energy required to increase the surface area by one square metre, i.e. it can be measured in joules per metre squared (which is equivalent to $N\,m^{-1}$).

The property of surface tension is responsible for the formation of liquid drops, soap bubbles, and meniscuses, as well as the rise of liquids in a capillary tube (*capillarity*), the absorption of liquids by porous substances, and the ability of liquids to wet a surface.

surfactant (surface active agent) A substance, such as a *detergent, added to a liquid to increase its spreading or wetting properties by reducing its *surface tension.

suspension A mixture in which small solid or liquid particles are suspended in a liquid or gas.

sylvite (sylvine) A mineral form of *potassium chloride, KCl.

syndiotactic *See* polymer.

synthesis The formation of chemical compounds from more simple compounds.

synthesis gas *See* Haber process.

synthetic Describing a substance that has been made artificially; i.e. one that does not come from a natural source.

Système International d'Unités *See* SI units.

T

tactic polymer *See* polymer.

talc A white or pale-green mineral form of magnesium silicate, $Mg_3Si_4O_{10}(OH)_2$, crystallizing in the triclinic system. It forms as a secondary mineral by alteration of magnesium-rich olivines, pyroxenes, and amphiboles of ultrabasic rocks. It is soaplike to touch and very soft, having a hardness of 1 on the Mohs' scale. Massive fine-grained talc is known as *soapstone* or *steatite*. Talc in powdered form is used as a lubricant, as a filler in paper, paints, and rubber, and in cosmetics, ceramics, and French chalk. It occurs chiefly in the USA, USSR, France, and Japan.

tannic acid A yellowish complex organic compound present in certain plants. It is used in dyeing as a mordant.

tannin One of a group of complex organic chemicals commonly found in leaves, unripe fruits, and the bark of trees. Their function is uncertain though the unpleasant taste may discourage grazing animals. Some tannins have commercial uses, notably in the production of leather and ink.

tantalum Symbol Ta. A heavy blue-grey metallic *transition element; a.n. 73; r.a.m. 180.948; r.d. 16.63; m.p. 2996°C; b.p. 5427°C. It is found with niobium in the ore columbite–tantalite $(Fe,Mn)(Ta,Nb)_2O_6$. It is extracted by dissolving in hydrofluoric acid, separating the tantalum and niobium fluorides to give K_2TaF_7, and reduction of this with sodium. The element contains the stable isotope tantalum–181 and the long-lived radioactive isotope tantalum–180 (0.012%; half-life $>10^7$ years). There are several other short-lived isotopes. The element is used in certain alloys and in electronic components. Tantalum parts are also used in surgery because of the unreactive nature of the metal (e.g. in pins to join bones). Chemically, the metal forms a passive oxide layer in air. It forms complexes in the +2, +3, +4, and +5 oxidation states. Tantalum was identified in 1802 by Ekeberg and first isolated in 1820 by Berzelius.

tar Any of various black semisolid mixtures of hydrocarbons and free carbon, produced by destructive distillation of *coal or by *petroleum refining.

tar sand *See* oil sand.

tartaric acid A crystalline naturally occurring carboxylic acid, $(CHOH)_2(COOH)_2$; r.d. 1.8; m.p. 170°C. It can be obtained from tartar (potassium hydrogen tartrate) deposits from wine vats, and is used in baking powders and as a foodstuffs additive. The compound is optically active (*see* optical activity). The systematic name is *2,3-dihydroxybutanedioic acid.*

tartrate A salt or ester of *tartaric acid.

tautomerism A type of *isomerism in which the two isomers (*tautomers*) are in equilibrium. *See* keto–enol tautomerism.

TCA cycle *See* Krebs cycle.

technetium Symbol Tc. A radioactive metallic *transition element; a.n. 43; m.p. 2171°C; b.p. 4876°C. The element can be detected in certain stars and is present in the fission products of uranium. It was first made by Perrier and Segré by bombarding molybdenum with deuterons to give technetium–97. The most stable isotope is technetium–99 (half-life 2.12×10^5 years); this is used to some extent in labelling for medical diagnosis. There are sixteen known isotopes. Chemically, the metal has properties intermediate between manganese and rhenium.

Teflon Tradename for a form of *polytetrafluoroethene.

tellurides Binary compounds of tellurium with other more electropositive elements. Compounds of tellurium with nonmetals are covalent (e.g. H_2Te). Metal tellurides can be made by direct combination of the elements and are ionic (containing Te^{2-}) or nonstoichiometric interstitial compounds (e.g. Pd_4Te, $PdTe_2$).

tellurium Symbol Te. A silvery metalloid element belonging to *group VI of the periodic table; a.n. 52; r.a.m. 127.60; r.d. 6.24 (crystalline); m.p. 451°C; b.p. 1390±3°C. It occurs mainly as *tellurides in ores of gold, silver, copper, and nickel and it is obtained as a by-product in copper refining. There are eight natural isotopes and nine radioactive isotopes. The element is used in semiconductors and small amounts are added to certain steels. Tellurium is also added in small quantities to lead. Its chemistry is similar to that of sulphur. It was discovered by Franz Müller in 1782.

temperature The property of a body or region of space that determines whether or not there will be a net flow of heat into it or out of it from a neighbouring body or region and in which direction (if any) the heat will flow. If there is no heat flow the bodies or regions are said to be in *thermodynamic equilibrium* and at the same temperature. If there is a flow of heat, the direction of the flow is from the body or region of higher temperature. Broadly, there are two methods of quantifying this property. The empirical method is to take two or more reproducible temperature-dependent events and assign *fixed points* on a scale of values to these events. For example, the Celsius temperature scale uses the freezing point and boiling point of water as the two fixed points, assigns the values 0 and 100 to them, respectively, and divides the scale between them into 100 degrees. This method is serviceable for many practical purposes (*see* temperature scales), but lacking a theoretical basis it is awkward to use in many scientific contexts. In the 19th century, Lord Kelvin proposed a thermodynamic method to specify temperature, based on the measurement of the quantity of heat flowing between bodies at different temperatures. This concept relies on an absolute scale of temperature with an *absolute zero of temperature, at which no body can give up heat. He also used Sadi Carnot's concept of an ideal frictionless perfectly efficient heat engine (*see* Carnot cycle). This Carnot engine takes in a quantity of heat q_1 at a temperature T, and exhausts heat q_2 at T_2, so that $T_1/T_2 = q_1/q_2$. If T_2 has a value fixed by definition, a Carnot engine can be run between this fixed temperature and any unknown temperature T_1, enabling T_1 to be calculated by measuring the values of q_1 and q_2. This concept remains the basis for defining *thermodynamic temperature*, quite independently of the nature of the working substance. The unit in which thermodynamic temperature is expressed is the *kelvin. In practice thermodynamic temperatures cannot be measured directly; they are usually inferred from measurements with a gas thermometer containing a nearly ideal gas. This is possible because another aspect of thermodynamic temperature is its relationship to the *internal energy of a given amount of substance. This can be shown most simply in the case of an ideal monatomic gas, in which the internal energy per mole (U) is equal to the total kinetic energy of translation of the atoms in one mole of the gas (a monatomic gas has no rotational or vibrational energy). According to *kinetic theory, the thermodynamic temperature of such a gas is given by $T = 2U/3R$, where R is the universal *gas constant.

temperature scales A number of empirical scales of *temperature have been in use: the *Celsius scale is widely used for many purposes and in certain countries the *Fahrenheit scale is still used. These scales both rely on the use of *fixed points*, such as the freezing point and the boiling point of water, and the division of the *fundamental interval* between these two points into units of temperature (100 degrees in the case of the Celsius scale and 180 degrees in the Fahrenheit scale).

However, for scientific purposes the scale in use is the *International Practical Temperature Scale (1968)*, which is designed to conform as closely as possible to thermodynam-

	T/K	t/°C
triple point of equilibrium hydrogen	13.81	−259.34
temperature of equilibrium hydrogen when its vapour pressure is 25/76 standard atmosphere	17.042	−256.108
b.p. of equilibrium hydrogen	20.28	−252.87
b.p. of neon	27.102	−246.048
triple point of oxygen	54.361	−218.789
b.p. of oxygen	90.188	−182.962
triple point of water	273.16	0.01
b.p. of water	373.15	100
f.p. of zinc	692.73	419.58
f.p. of silver	1235.08	961.93
f.p. of gold	1337.58	1064.43

Temperature scales

ic temperature and is expressed in the unit of thermodynamic temperature, the *kelvin. The eleven fixed points of the scale are given in the table, with the instruments specified for interpolating between them. Above the freezing point of gold, a radiation pyrometer is used, based on Planck's law of radiation. The scale is expected to be refined in the late 1980s.

tempering The process of increasing the toughness of an alloy, such as steel, by heating it to a predetermined temperature, maintaining it at this temperature for a predetermined time, and cooling it to room temperature at a predetermined rate. In steel, the purpose of the process is to heat the alloy to a temperature that will enable the excess carbide to precipitate out of the supersaturated solid solution of *martensite and then to cool the saturated solution fast enough to prevent further precipitation or grain growth. For this reason steel is quenched rapidly by dipping into cold water.

temporary hardness *See* hardness of water.

tera- Symbol T. A prefix used in the metric system to denote one million million times. For example, 10^{12} volts = 1 teravolt (TV).

terbium Symbol Tb. A silvery metallic element belonging to the *lanthanoids; a.n. 65; r.a.m. 158.92; r.d. 8.23 (20°C); m.p. 1365°C; b.p. 3230°C. It occurs in apatite and xenotime, from which it is obtained by an ion-exchange process. There is only one natural isotope, terbium−159, which is stable. Seventeen artificial isotopes have been identified. There are virtually no uses for this element although its potentialities are still being explored. It was discovered by C. G. Mosander in 1843.

ternary compound A chemical compound containing three different elements.

terpenes A group of unsaturated hydrocarbons present in plants (*see* essential oil). Terpenes consist of isoprene units, CH_2:C(CH_3)CH:CH_2. Monoterpenes have two units, $C_{10}H_{16}$, sesquiterpenes three units, $C_{15}H_{24}$, diterpenes four units, $C_{20}H_{32}$, etc.

tertiary alcohol *See* alcohols.

tertiary amine *See* amines.

tervalent (trivalent) Having a valency of three.

Terylene Tradename for a type of *polyester used in synthetic fibres.

tesla Symbol T. The SI unit of magnetic flux density equal to one weber of magnetic flux per square metre, i.e. 1 T = 1 Wb m^{-2}. It is named after Nikola Tesla (1870–1943), Croatian-born US electrical engineer.

tetrachloroethene A colourless nonflammable volatile liquid, CCl_2:CCl_2; r.d. 1.6; m.p. −22°C; b.p. 121°C. It is used as a solvent.

tetrachloromethane (carbon tetrachloride) A colourless volatile liquid with a characteristic odour, virtually insoluble in water but miscible with many organic liquids, such as ethanol and benzene; r.d. 1.586; m.p. −23°C; b.p. 76.8°C. It is made by the chlorination of methane (previously by chlorination of carbon disulphide). The compound is a good solvent for waxes, lacquers, and rubbers and the main industrial use is as a solvent, but safer substances (e.g. 1,1,1-trichloroethane) are increasingly being used. Moist carbon tetrachloride is partly decomposed to phosgene and hydrogen chloride and this provides a further restriction on its use.

tetraethyl lead *See* lead tetraethyl(IV).

tetragonal *See* crystal system.

tetrahedral angle The angle between the bonds in a *tetrahedral compound (approximately 109° for a regular tetrahedron).

tetrahedral compound A compound in which four atoms or groups situated at the corners of a tetrahedron are linked (by covalent or coordinate bonds) to an atom at the centre of the tetrahedron. *See also* complex.

tetrahydrate A crystalline hydrate containing four molecules of water per molecule of compound.

tetrahydroxomonoxodiboric(III) acid *See* boric acid.

tetraoxophosphoric(V) acid *See* phosphoric(V) acid.

tetraoxosulphuric(IV) acid *See* sulphuric acid.

tetravalent (quadrivalent) Having a valency of four.

thallium Symbol Tl. A greyish metallic element belonging to *group IIIA of the periodic table; a.n. 81; r.a.m. 204.39; r.d. 11.85 (20°C); m.p. 303.3°C; b.p. 1460°C. It occurs in zinc blende and some iron ores and is recovered in small quantities from lead and zinc concentrates. The naturally occurring isotopes are thallium–203 and thallium–205; eleven radioisotopes have been identified. It has few uses – experimental alloys for special purposes and some minor uses in electronics. The sulphate has been used as a rodenticide. Thallium(I) compounds resemble those of the alkali metals. Thallium(III) compounds are easily reduced to the thallium(I) state and are therefore strong oxidizing agents. The element was discovered by Sir William Crookes in 1861.

thermal capacity *See* heat capacity.

thermal equilibrium *See* equilibrium.

thermite A stoichiometric powdered mixture of iron(III) oxide and aluminium for the reaction:

$$2Al + Fe_2O_3 \rightarrow Al_2O_3 + 2Fe$$

The reaction is highly exothermic and the increase in temperature is sufficient to melt the iron produced. It has been used for localized welding of steel objects (e.g. railway lines) in the *Thermit process*. Thermite is also used in incendiary bombs.

thermochemistry The branch of physical chemistry concerned with heats of chemical reaction, heats of formation of chemical compounds, etc.

thermodynamics The study of the laws that govern the conversion of energy from one form to another, the direction in which heat will flow, and the availability of energy to do work. It is based on the concept that in an isolated system anywhere in the universe there is a measurable quantity of energy called the *internal energy (U) of the system. This is the total kinetic and potential energy of the atoms and molecules of the system of all kinds that can be transferred directly as heat; it therefore excludes chemical and nuclear energy. The value of U can only be changed if the system ceases to be isolated. In these circumstances U can change by the transfer of mass to or from the system, the transfer of heat (Q) to or from the system, or by the work (W) being done on or by the system. For an adiabatic ($Q = 0$) system of constant mass, $\Delta U = W$. By convention, W is taken to be positive if work is done on the system and negative if work is done by the system. For nonadiabatic systems of constant mass, $\Delta U = Q + W$. This statement, which is equivalent to the law of conservation of energy, is known as the *first law of thermodynamics*.

All natural processes conform to this law, but not all processes conforming to it can occur in nature. Most natural processes are irreversible, i.e. they will only proceed in one direction (*see* reversible process). The direction that a natural process can take is the subject of the *second law of thermodynamics*, which can be stated in a variety of ways. R. Clausius (1822–88) stated the law in two ways: "heat cannot be transferred from one body to a second body at a higher temperature without producing some other effect" and "the entropy of a closed system increases with time". These statements introduce the thermodynamic concepts of *temperature (T) and *entropy (S), both of which are parameters determining the direction in which an irreversible process can go. The temperature of a body or system determines whether heat will flow into it or out of it; its entropy is a measure of the unavailability of its energy to do work. Thus T and S determine the relationship between Q and W in the statement of the first law. This is usually presented by stating the second law in the form $\Delta U = T\Delta S - W$.

The second law is concerned with changes in entropy (ΔS). The *third law of thermodynamics* provides an absolute scale of values for entropy by stating that for changes involving only perfect crystalline solids at *absolute zero, the change of the total entropy is zero. This law enables absolute values to be stated for entropies.

One other law is used in thermodynamics. Because it is fundamental to, and assumed by, the other laws of thermodynamics it is usually known as the *zeroth law of thermodynamics*. This states that if two bodies are each in thermal equilibrium with a third body, then all three bodies are in thermal equilibrium with each other. *See also* enthalpy; free energy.

thermoluminescence *Luminescence produced in a solid when its temperature is raised. It arises when free electrons and holes, trapped in a solid as a result of exposure to ionizing radiation, unite and emit photons of light. The process is made use of in *thermoluminescent dating*, which assumes that the number of electrons and holes trapped in a sample of pottery is related to the length of time that has elapsed since the pottery was fired. By comparing the luminescence produced by heating a piece of pottery of unknown age with the luminescence produced by heating similar materials of known age, a fairly accurate estimate of the age of an object can be made.

thermoluminescent dating *See* thermoluminescence.

thermometer An instrument used for measuring the *temperature of a substance. A number of techniques and forms are used in thermometers depending on such factors as the degree of accuracy required and the range of temperatures to be measured, but they all measure temperature by making use of some property of a substance that varies with temperature. For example, *liquid-in-glass thermometers* depend on the expansion of a liquid, usually mercury or alcohol coloured with dye. These consist of a liquid-filled glass bulb attached to a partially filled capillary tube. In the *bimetallic thermometer* the unequal expansion of two dissimilar metals that have been bonded together into a narrow strip and coiled is used to move a pointer round a dial. The *gas thermometer*, which is more accurate than the liquid-in-glass thermometer, measures the variation in the pressure of a gas kept at constant volume. The *resistance thermometer* is based on the change in resistance of conductors or semiconductors with temperature change. Platinum, nickel, and copper are the metals most commonly used in resistance thermometers.

thermoplastic *See* plastics.

thermosetting *See* plastics.

thermostat A device that controls the heating or cooling of a substance in order to maintain it at a constant temperature. It consists of a temperature-sensing instrument connected to a switching device. When the temperature reaches a predetermined level the sensor switches the heating or cooling source on or off according to a predetermined program. The sensing thermometer is often a bimetallic strip that triggers a simple electrical switch. Thermostats are used for space-heating controls, in water heaters and refrigerators, and to maintain the environment of a scientific experiment at a constant temperature.

thiamin(e) *See* vitamin B complex.

thin-layer chromatography A technique for the analysis of liquid mixtures using *chromatography. The stationary phase is a thin layer of an absorbing solid (e.g. alumina) prepared by spreading a slurry of the solid on a plate (usually glass) and drying it in an oven. A spot of the mixture to be analysed is placed near one edge and the plate is stood upright in a solvent. The solvent rises through the layer by capillary action carrying the components up the plate at different rates (depending on the extent to which they are absorbed by the solid). After a given time, the plate is dried and the location of spots noted. It is possible to identify constituents of the mixture by the distance moved in a given time. The technique needs careful control of the thickness of the layer and of the temperature.

thiocyanate A salt or ester of cyanic acid.

thiocyanic acid An unstable gas, HSCN.

thio ethers *See* sulphides.

thiol group *See* thiols.

thiols (mercaptans; thio alcohols) Organic compounds that contain the group –SH (called the *thiol group*, *mercapto group*, or *sulphydryl group*). Thiols are analogues of alcohols in which the oxygen atom is replaced by a sulphur atom. They are named according to the parent hydrocarbon; e.g. ethane thiol (C_2H_5SH). A characteristic property is their strong disagreeable odour. For example the odour of garlic is produced by ethane thiol. Unlike alcohols they are acidic, reacting with alkalis and certain metals to form saltlike compounds. The older name, mercaptan, comes from their ability to react with ('seize') mercury.

thionyl chloride *See* sulphur dichloride oxide.

thionyl group The group =SO, as in *sulphur dichloride oxide.

thiosulphate A salt containing the ion $S_2O_3^{2-}$ formally derived from thiosulphuric acid. Thiosulphates readily decompose in acid solution to give elemental sulphur and hydrogensulphite (HSO_3^-) ions.

thiosulphuric acid An unstable acid, $H_2S_2O_3$, formed by the reaction of sulphur trioxide with hydrogen sulphide. *See also* sulphuric acid.

thiourea A white crystalline solid, $(NH_2)_2CS$; r.d. 1.4; m.p. 180°C. It is used as a fixer in photography.

thoria *See* thorium.

thorium Symbol Th. A grey radioactive metallic element belonging to the *actinoids; a.n. 90; r.a.m. 232.038; r.d. 11.5–11.9 (17°C); m.p. 1740–1760°C; b.p. 4780–4800°C. It occurs in monazite sand in Brazil, India, and USA. The isotopes of thorium have mass numbers from 223 to 234 inclusive; the most stable isotope, thorium-232, has a half-life of 1.39×10^{10} years. It has an oxidation state of (+4) and its chemistry resembles that of the other actinoids. It can be used as a nuclear fuel for breeder reactors as thorium-232 captures slow neutrons to breed uranium-233. Thorium dioxide (*thoria*, ThO_2) is used on gas mantles and in special refractories. The element was discovered by J. J. Berzelius in 1829.

thorium series *See* radioactive series.

threnardite A mineral form of *sodium sulphate, Na_2SO_4.

threonine *See* amino acid.

thulium Symbol Tm. A soft grey metallic element belonging to the *lanthanoids; a.n. 69; r.a.m. 168.934; r.d. 9.321 (20°C); m.p. 1545°C; b.p. 1950°C. It occurs in apatite and xenotime. There is one natural isotope, thulium-169, and seventeen artificial isotopes have been produced. There are no uses for the element, which was discovered by P. T. Cleve in 1879.

thymine A *pyrimidine derivative and one of the major component bases of *nucleotides and the nucleic acid *DNA.

tin Symbol Sn. A silvery malleable metallic element belonging to *group IV of the periodic table; a.n. 50; r.a.m. 118.69; r.d. 7.29; m.p. 231.97°C; b.p. 2270°C. It is found as tin(IV) oxide in ores, such as cassiterite, and is extracted by reduction with carbon. The metal (called *white tin*) has a powdery nonmetallic allotrope *grey tin*, into which it changes below 18°C. The formation of this allotrope is called *tin plague*; it can be reversed by heating to 100°C. The natural element has 21 isotopes (the largest number of any element); five radioactive isotopes are also known. The metal is used as a thin protective coating for steel plate and is a constituent of a number of alloys (e.g. phosphor bronze, gun metal, solder, Babbitt metal, and pewter). Chemically it is reactive. It combines directly with chlorine and oxygen and displaces hydrogen from dilute acids. It also dissolves in alkalis to form *stannates. There are two series of compounds with tin in the +2 and +4 oxidation states.

tin(II) chloride A white solid, $SnCl_2$, soluble in water and ethanol. It exists in the anhydrous form (rhombic; r.d. 3.95; m.p. 246°C; b.p. 652°C) and as a dihydrate,

$SnCl_2.2H_2O$ (monoclinic; r.d. 2.71; m.p. 37.7°C). The compound is made by dissolving metallic tin in hydrochloric acid and is partially hydrolysed in solution.

$$Sn^{2+} + H_2O \rightleftharpoons SnOH^+ + H^+$$

Excess acid must be present to prevent the precipitation of basic salts. In the presence of additional chloride ions the pyramidal ion $[SnCl_3]^-$ is formed; in the gas phase the $SnCl_2$ molecule is bent. It is a reducing agent in acid solutions and oxidizes slowly in air:

$$Sn^{2+} \rightarrow Sn^{4+} + 2e$$

tin(IV) chloride A colourless fuming liquid, $SnCl_4$, hydrolysed in cold water, decomposed by hot water, and soluble in ethers; r.d. 2.226; m.p. −33°C; b.p. 114°C. Tin(IV) chloride is a covalent compound, which may be prepared directly from the elements. It dissolves sulphur, phosphorus, bromine, and iodine, and there is evidence for the presence of species such as $SnCl_2I_2$. In hydrochloric acid and in chloride solutions the coordination is extended from four to six by the formation of the $SnCl_6{}^{2-}$ ion.

tincture A solution with alcohol as the solvent (e.g. tincture of iodine).

tin(IV) hydride (stannane) A highly reactive and volatile gas (b.p. −53°), SnH_4, which decomposes on moderate heating (150°C). It is prepared by the reduction of tin chlorides using lithium tetrahydridoaluminate(III) and is used in the synthesis of some organo-tin compounds. The compound has reducing properties.

tin(IV) oxide (tin dioxide) A white solid, SnO_2, insoluble in water; tetrahedral; r.d. 6.95; m.p. 1630°C; sublimes at about 1850°C. Tin(IV) oxide is trimorphic: the common form, which occurs naturally as the ore *cassiterite, has a rutile lattice but hexagonal and rhombic forms are also known. There are also two so-called dihydrates, $SnO_2.2H_2O$, known as α- and β-stannic acid. These are essentially tin hydroxides. Tin (IV) oxide is amphoteric, dissolving in molten alkalis to form *stannates; in the presence of sulphur, thiostannates are produced.

tin plague *See* tin.

tin(II) sulphide A grey-black cubic or monoclinic solid, SnS, virtually insoluble in water; r.d. 5.22; m.p. 882°C; b.p. 1230°C. It has a layer structure similar to that of black phosphorus. Its heat of formation is low and it can be made by heating the elements together. Above 265°C it slowly decomposes (disproportionates) to tin(IV) sulphide and tin metal. The compound reacts with hydrochloric acid to give tin(II) chloride and hydrogen sulphide.

tin(IV) sulphide (mosaic gold) A bronze or golden yellow crystalline compound, SnS_2, insoluble in water and in ethanol; hexagonal; r.d. 4.5; decomposes at 600°C. It is prepared by the reaction of hydrogen sulphide with a soluble tin(IV) salt or by the action of heat on thiostannic acid, H_2SnS_3. The golden-yellow form used for producing a gilded effect on wood is prepared by heating tin, sulphur, and ammonium chloride.

titania *See* titanium(IV) oxide.

titanium Symbol Ti. A white metallic *transition element; a.n. 22; r.a.m. 47.9; r.d. 4.507; m.p. 1660±10°C, b.p. 3280°C. The main sources are rutile (TiO_2) and, to a lesser extent, ilmenite ($FeTiO_3$). The element also occurs in numerous other minerals. It is obtained by heating the oxide with carbon and chlorine to give $TiCl_4$, which is reduced by the *Kroll process. The main use is in a large number of strong light corrosion-resistant alloys for aircraft, ships, chemical plant, etc. The element forms a passive oxide coating in air. At higher temperatures it reacts with oxygen, nitrogen, chlorine, and other nonmetals. It dissolves in dilute acids. The main compounds are titanium(IV) salts and complexes; titanium(II) and titanium(III) compounds are also known. The element was first discovered by Gregor in 1789.

titanium dioxide *See* titanium(IV) oxide.

titanium(IV) oxide (titania; titanium dioxide) A white oxide, TiO_2, occurring naturally in various forms, particularly the mineral rutile. It is used as a white pigment and as a filler for plastics, rubber, etc.

titration A method of volumetric analysis in which a volume of one reagent is added to a known volume of another reagent slowly from a burette until an end point is reached (*see* indicator). The volume added before the end point is reached is noted. If one of the solutions has a known concentration, that of the other can be calculated.

TNT *See* trinitrotoluene.

tocopherol *See* vitamin E.

Tollen's reagent A reagent used in testing for aldehydes. It is made by adding sodium hydroxide to silver nitrate to give silver(I) oxide, which is dissolved in aqueous ammonia (giving the complex ion $[Ag(NH_3)_2]^+$). The sample is warmed with the reagent in a test tube. Aldehydes reduce the complex Ag^+ ion to metallic silver, forming a bright silver mirror on the inside of the tube (hence the name *silver-mirror test*). Ketones give a negative result.

toluene *See* methylbenzene.

topaz A variably coloured aluminium silicate mineral, $Al_2(SiO_4)(OH,F)_2$, that forms orthorhombic crystals. It occurs chiefly in acid igneous rocks, such as granites and pegmatites. Topaz is valued as a gemstone because of its transparency, variety of colours (the wine-yellow variety being most highly prized), and great hardness (8 on the Mohs' scale). When heated, yellow or brownish topaz often becomes a rose-pink colour. The main sources of topaz are Brazil, the USSR, and the USA.

torr A unit of pressure, used in high-vacuum technology, defined as 1 mmHg. 1 torr is equal to 133.322 pascals. The unit is named after Evangelista Torricelli (1609–47).

total-radiation pyrometer *See* pyrometry.

tourmaline A group of minerals composed of complex cyclosilicates containing boron with the general formula $NaR_3^{2+}Al_6B_3Si_6O_{27}(H,F)_4$, where R = Fe^{2+}, Mg, or (Al + Li). The crystals are trigonal, elongated, and variably coloured, the two ends of the crystals often having different colours. Tourmaline is used as a gemstone and because of its double refraction and piezoelectric properties is also used in polarizers and some pressure gauges.

trace element *See* essential element.

tracing (radioactive tracing) *See* labelling.

transamination A biochemical reaction in amino acid metabolism in which an amine group is transferred from an amino acid to a keto acid to form a new amino acid and keto acid. The coenzyme required for this reaction is pyridoxal phosphate.

trans effect An effect in the substitution of inorganic square-planar complexes, in which certain ligands in the original complex are able to direct the incoming ligand into the trans position. The order of ligands in decreasing trans-directing power is:

$CN^- > NO_2 > I^- > Br^- > Cl^- > NH_3 > H_2O$.

transition elements A set of elements in the *periodic table in which filling of electrons in an inner *d*- or *f*-level occurs. With increasing proton number, electrons fill atomic levels up to argon, which has the electron configuration $1s^2 2s^2 2p^6 3s^2 3p^6$. In this shell, there are 5 *d*-orbitals, which can each contain 2 electrons. However, at this point the subshell of lowest energy is not the 3*d* but the 4*s*. The next two elements, potassium and calcium, have the configurations $[Ar]4s^1$ and $[Ar]4s^2$ respectively. For the next element, scandium, the 3*d* level is of lower energy than the 4*p* level, and scandium has the configuration $[Ar]3d^1 4s^2$. This filling of the inner *d*-level continues up to zinc $[Ar]3d^{10}4s^2$, giving the first transition series. There is a further series of this type in the next period of the table: between yttrium ($[Kr]4d5s^2$) and cadmium ($[Kr]4d^{10}5s^2$). This is the second transition series. In the next period of the table the situation is rather more complicated. Lanthanum has the configuration $[Xe]5d^1 6s^2$. The level of lowest energy then becomes the 4*f* level and the next element, cerium, has the configuration $[Xe]4f^1 5d^1 6s^2$. There are 7 of these *f*-orbitals, each of which can contain 2 electrons, and filling of the *f*-levels continues up to lutetium ($[Xe]4f^{14}5d^1 6s^2$). Then the filling of the 5*d* levels continues from hafnium to mercury. The series of 14 elements from cerium to lutetium is a 'series within a series', sometimes called an *inner transition series*. This one is the *lanthanoid series. In the next period there is a similar inner transition series, the *actinoid series, from thorium to lawrencium. Then filling of the *d*-level continues from element 104 onwards.

In fact, the classification of chemical elements is valuable only in so far as it illustrates chemical behaviour, and it is conventional to use the term 'transition elements' in a more restricted sense. The elements in the inner transition series from cerium (58) to lutetium (71) are called the lanthanoids; those in the series from thorium (90) to lawrencium (103) are the actinoids. These two series together make up the *f*-block in the periodic table. It is also common to include scandium, yttrium, and lanthanum with the lanthanoids (because of chemical similarity) and to include actinium with the actinoids. Of the remaining transition elements, it is usual to speak of three *main transition series*: from titanium to copper; from zirconium to silver; and from hafnium to gold. All these elements have similar chemical properties that result from the presence of unfilled *d*-orbitals in the element or (in the case of copper, silver, and gold) in the ions. The elements from 104 to 109 and the undiscovered elements 110 and 111 make up a fourth transition series. The elements zinc, cadmium, and mercury have filled *d*-orbitals both in the elements and in compounds, and are usually regarded as nontransition elements (*see* group II elements).

The elements of the three main transition series are all typical metals (in the nonchemical sense), i.e. most are strong hard materials that are good conductors of heat and electricity and have high melting and boiling points. Chemically, their behaviour depends on the existence of unfilled *d*-orbitals. They exhibit variable valency, have coloured compounds, and form *coordination compounds. Many of their compounds are paramagnetic as a result of the presence of unpaired electrons. Many of them are good catalysts. They are less reactive than the *s*-and *p*-block metals.

transition point (transition temperature) **1.** The temperature at which one crystalline form of a substance changes to another form. **2.** The temperature at which a substance changes phase. **3.** The temperature at which a substance becomes superconducting. **4.** The temperature at which some other change, such as a change of magnetic properties, takes place.

transition state (activated complex) The association of atoms of highest energy formed during a chemical reaction. The transition state can be regarded as a short-lived intermediate that breaks down to give the products. For example, in a S_N2 substitution reaction, one atom or group approaches the molecule as the other leaves. The transition state is an intermediate state in which both attacking and leaving groups are partly bound to the molecule, e.g.

$$B + RA \rightarrow B\text{---}R\text{---}A \rightarrow BR + A$$

In the theory of reaction rates, the reactants are assumed to be in equilibrium with this activated complex, which decomposes to the products.

transmission electron microscope *See* electron microscope.

transport number Symbol *t*. The fraction of the total charge carried by a particular type of ion in the conduction of electricity through electrolytes.

transuranic elements Elements with an atomic number greater than 92, i.e. elements above uranium in the *periodic table.

triacylglycerol *See* triglyceride.

triazine *See* azine.

triboluminescence *Luminescence caused by friction; for example, some crystalline substances emit light when they are crushed as a result of static electric charges generated by the friction.

tribromomethane (bromoform) A colourless liquid *haloform, $CHBr_3$; r.d. 2.9; m.p. 8°C; b.p. 150°C.

tricarbon dioxide (carbon suboxide) A colourless gas, C_3O_2, with an unpleasant odour; r.d. 1.114 (liquid at 0°C); m.p. −111.3°C; b.p. 7°C. It is the acid anhydride of malonic acid, from which it can be prepared by dehydration using phosphorus (V) oxide. The molecule is linear (O:C:C:C:O).

tricarboxylic acid cycle *See* Krebs cycle.

trichloroethanal (chloral) A liquid aldehyde, CCl_3CHO; r.d. 1.51; m.p. −57.5°C; b.p. 97.8°C. It is made by chlorinating ethanal and used in making DDT. *See also* 2,2,2-trichloroethanediol.

2,2,2-trichloroethanediol (chloral hydrate) A colourless crystalline solid, $CCl_3CH(OH)_2$; r.d. 1.91; m.p. 57°C; b.p. 96.3°C. It is made by the hydrolysis of trichloroethanal and is unusual in having two −OH groups on the same carbon atom. Gem diols of this type are usually unstable; in this case the compound is stabilized by the presence of the three Cl atoms. It is used as a sedative.

trichloromethane (chloroform) A colourless volatile sweet-smelling liquid *haloform, $CHCl_3$; r.d. 1.48; m.p. −63.5°C; b.p. 61°C. It can be made by chlorination of methane (followed by separation of the mixture of products) or by the haloform reaction. It is an effective anaesthetic but can cause liver damage and it has now been replaced by other halogenated hydrocarbons. Chloroform is used as a solvent and raw material for making other compounds.

triclinic *See* crystal system.

tridymite A mineral form of *silicon(IV) oxide, SiO_2.

triglyceride (triacylglycerol) An ester of glycerol (propane-1,2,3-triol) in which all three hydroxyl groups are esterified with a fatty acid. Triglycerides are the major constituent of fats and oils and provide a concentrated food energy store in living organisms as well as cooking fats and oils, margarines, soaps, etc. Their physical and chemical properties depend on the nature of their constituent fatty acids. In *simple triglycerides* all three fatty acids are identical; in *mixed triglycerides* two or three different fatty acids are present.

trihydric alcohol *See* triol.

tri-iodomethane (iodoform) A yellow volatile solid sweet-smelling *haloform, CHI_3; r.d. 4.1; m.p. 115°C. It is made by the haloform reaction.

tri-iron tetroxide (ferrosoferric oxide) A black magnetic oxide, Fe_3O_4; r.d. 5.2. It is formed when iron is heated in steam and also occurs naturally as the mineral *magnetite. The oxide dissolves in acids to give a mixture of iron(II) and iron(III) salts.

trimethylaluminium (aluminium trimethyl) A colourless liquid, $Al(CH_3)_3$, which ignites in air and reacts with water to give aluminium hydroxide and methane, usually with extreme vigour; r.d. 0.752; m.p. 0°C; b.p. 130°C. Like other aluminium alkyls it may be prepared by reacting a Grignard reagent with aluminium trichloride. Aluminium alkyls are used in the *Ziegler process for the manufacture of high-density polyethene (polythene).

2,4,6-trinitrophenol *See* picric acid.

trinitrotoluene (TNT) A yellow highly explosive crystalline solid, $CH_3C_6H_2(NO_2)_3$; r.d. 1.6; m.p. 81°C. It is made by nitrating toluene (methylbenzene), the systematic name being 1-methyl-2,4,6-trinitrobenzene.

triol (trihydric alcohol) An *alcohol containing three hydroxyl groups per molecule.

trioxoboric(III) acid *See* boric acid.

trioxosulphuric(IV) acid *See* sulphurous acid.

trioxygen *See* ozone.

triple bond *See* chemical bond.

triple point The temperature and pressure at which the vapour, liquid, and solid phases of a substance are in equilibrium. For water the triple point occurs at 273.16 K and 611.2 Pa. This value forms the basis of the definition of the *kelvin and the thermodynamic *temperature scale.

trisilane *See* silane.

trisodium phosphate(V) (sodium orthophosphate) A colourless crystalline compound, Na_3PO_4, soluble in water and insoluble in ethanol. It is known both as the decahydrate (octagonal; r.d. 2.54) and the dodecahydrate (trigonal; r.d. 1.62) The dodecahydrate loses water at about 76°C and the decahydrate melts at 100°C. Trisodium phosphate may be prepared by boiling sodium carbonate with the stoichiometric amount of phosphoric acid and subsequently adding sodium hydroxide to the disodium salt thus formed. It is useful as an additive for high-pressure boiler feed water (for removal of calcium and magnesium as phosphates), in emulsifiers, as a water-softening agent, and as a component in detergents and cleaning agents. Sodium phosphate labelled with the radioactive isotope ^{32}P is used in the study of the role of phosphate in biological processes and is also used (intravenously) in the treatment of polycythaemia.

tritiated compound *See* labelling.

tritium Symbol T. An isotope of hydrogen with mass number 3; i.e. the nucleus contains 2 neutrons and 1 proton. It is radioactive (half-life 12.3 years), undergoing beta decay to helium-3. Tritium is used in *labelling.

trivalent (tervalent) Having a valency of three.

trona A mineral form of sodium sesquicarbonate, consisting of a mixed hydrated sodium carbonate and sodium hydrogencarbonate, $Na_2CO_3.NaHCO_3.2H_2O$.

tropyllium ion The positive ion $C_7H_7^+$, having a ring of seven carbon atoms. The ion is symmetrical and has characteristic properties of *aromatic compounds.

trypsin An enzyme that digests proteins (*see* protease). It is secreted in an inactive form (*trypsinogen*) by the pancreas into the duodenum. There, trypsinogen is acted on by an enzyme (*enterokinase*) produced in the duodenum to yield trypsin. The active enzyme plays an important role in the digestion of proteins in the anterior portion of the small intestine.

trypsinogen *See* trypsin.

tryptophan *See* amino acid.

tungsten Symbol W. A white or grey metallic *transition element (formerly called *wolfram*); a.n. 74; r.a.m. 183.85; r.d. 19.3; m.p. 3410°C; b.p. 5660°C. It is found in a number of ores, including the oxides wolframite, $(Fe,Mn)WO_4$, and scheelite, $CaWO_4$. The ore is heated with concentrated sodium hydroxide solution to form a soluble *tungstate*. The oxide WO_3 is precipitated from this by adding acid, and is reduced to the metal using hydrogen. It is used in various alloys, especially high-speed steels (for cutting tools) and in lamp filaments. Tungsten forms a protective oxide in air and can be oxidized at high temperature. It does not dissolve in dilute acids. It forms compounds in which the oxidation state ranges from +2 to +6. The metal was first isolated by J. J. and F. d'Elhuyer in 1783.

tungsten carbide A black powder, WC, made by heating powdered tungsten metal with lamp black at 1600°C. It is extremely hard (9.5 on Mohs' scale) and is used in dies and cutting tools. A ditungsten carbide, W_2C, also exists.

tunnel effect An effect in which electrons are able to tunnel through a narrow potential barrier that would constitute a forbidden region if the electrons were treated as classical particles. That there is a finite probability of an electron tunnelling from one classically allowed region to another arises as a consequence of *quantum mechanics. The effect is made use of in the tunnel diode.

turpentine An oily liquid extracted from pine resin. It contains pinene, $C_{10}H_{16}$, and other terpenes and is mainly used as a solvent.

turquoise A mineral consisting of a hydrated phosphate of aluminium and copper, $CuAl_6(PO_4)_4(OH)_8.4H_2O)$, that is prized as a semiprecious stone. It crystallizes in the triclinic system and is generally blue in colour, the 'robin's egg' blue variety being the most sought after. It usually occurs in veinlets and as masses and is formed by the action of surface waters on aluminium-rich rocks. The finest specimens are obtained from Iran.

tyrosine *See* amino acid.

U

ultracentrifuge A high-speed centrifuge used to measure the rate of sedimentation of colloidal particles or to separate macromolecules, such as proteins or nucleic acids, from solutions. Ultracentrifuges are electrically driven and are capable of speeds up to 60 000 rpm.

ultramicroscope A form of microscope that uses the Tyndall effect to reveal the presence of particles that cannot be seen with a normal optical microscope. Colloidal particles, smoke particles, etc., are suspended in a liquid or gas in a cell with a black background and illuminated by an intense cone of light that enters the cell from the side and has its apex in the field of view. The particles then produce diffraction-ring systems, appearing as bright specks on the dark background.

ultraviolet radiation (UV) Electromagnetic radiation having wavelengths between that of violet light and long X-rays, i.e. between 400 nanometres and 4 nm. In the range 400–300 nm the radiation is known as the *near ultraviolet*. In the range 300–200 nm it is known as the *far ultraviolet*. Below 200 nm it is known as the *extreme ultraviolet* or the *vacuum ultraviolet*, as absorption by the oxygen in the air makes the use of evacuated apparatus essential. The sun is a strong emitter of UV radiation but only the near UV reaches the surface of the earth as the ozone in the atmosphere absorbs all wavelengths below 290 nm.

Most UV radiation for practical use is produced by various types of mercury-vapour lamps. Ordinary glass absorbs UV radiation and therefore lenses and prisms for use in the UV are made from quartz.

uncertainty principle (Heisenberg uncertainty principle; principle of indeterminism) The principle that it is not possible to know with unlimited accuracy both the position and momentum of a particle. This principle, discovered in 1927 by Werner Heisenberg (1901–76), is usually stated in the form: $\Delta x \Delta p_x > h/2\pi$, where Δx is the uncertainty in the x-coordinate of the particle, Δp_x is the uncertainty in the x-component of the particle's momentum, and h is the *Planck constant. An explanation of the uncertainty is that in order to locate a particle exactly, an observer must be able to bounce off it a photon of radiation; this act of location itself alters the position of the particle in an unpredictable way. To locate the position accurately, photons of short wavelength would have to be used. These would have associated large momenta and cause a large effect on the position. On the other hand, using long-wavelength photons would have less effect on the particle's position, but would be less accurate because of the longer wavelength. The principle has had a profound effect on scientific thought as it appears to upset the classical relationship between cause and effect at the atomic level.

uniaxial crystal A double-refracting crystal (*see* double refraction) having only one *optic axis.

unimolecular reaction A chemical reaction or step involving only one molecule. An example is the decomposition of dinitrogen tetroxide:

$$N_2O_4 \rightarrow 2NO_2$$

Molecules colliding with other molecules acquire sufficient activation energy to react, and the activated complex only involves the atoms of a single molecule.

unit A specified measure of a physical quantity, such as length, mass, time, etc., specified multiples of which are used to express magnitudes of that physical quantity. For scientific purposes previous systems of units have now been replaced by *SI units.

unit cell The group of particles (atoms, ions, or molecules) in a crystal that is repeated in three dimensions in the *crystal lattice. *See also* crystal system.

univalent (monovalent) Having a valency of one.

universal constants *See* fundamental constants.

universal indicator A mixture of acid–base *indicators that changes colour (e.g. red-yellow-orange-green-blue) over a range of pH.

unsaturated 1. (of a compound) Having double or triple bonds in its molecules. Unsaturated compounds can undergo addition reactions as well as substitution. *Compare* saturated. **2.** (of a solution) *See* saturated.

unstable equilibrium *See* equilibrium.

uracil A *pyrimidine derivative and one of the major component bases of *nucleotides and the nucleic acid *RNA.

uraninite A mineral form of uranium(IV) oxide, containing minute amounts of radium, thorium, polonium, lead, and helium. When uraninite occurs in a massive form with a pitchy lustre it is known as *pitchblende*, the chief ore of uranium. Uraninite occurs in Saxony (East Germany), Romania, Norway, the UK (Cornwall), E Africa (Zaïre), USA, and Canada (Great Bear Lake).

uranium Symbol U. A white radioactive metallic element belonging to the *actinoids; a.n. 92; r.a.m. 238.03; r.d. 19.05 (20°C); m.p. 1132±1°C; b.p. 3818°C. It occurs as *uraninite, from which the metal is extracted by an ion-exchange process. Three isotopes are found in nature: uranium–238 (99.28%), uranium–235 (0.71%), and uranium–234 (0.006%). As uranium–235 undergoes nuclear fission with slow neutrons it is the fuel used in nuclear reactors and nuclear weapons; uranium has therefore assumed enormous technical and political importance since their invention. It was discovered by M. H. Klaproth in 1789.

uranium(VI) fluoride (uranium hexafluoride) A volatile white solid, UF_6; r.d. 5.1; m.p. 64.5°C. It is used in the separation of uranium isotopes by gas diffusion.

uranium hexafluoride *See* uranium(VI) fluoride.

uranium–lead dating A group of methods of *dating certain rocks that depends on the decay of the radioisotopes uranium–238 to lead–206 (half-life 4.5×10^9 years) or the decay of uranium–235 to lead–207 (half-life 7.1×10^8 years). One form of uranium–lead dating depends on measuring the ratio of the amount of helium trapped in the rock to the amount of uranium present (since the decay $^{238}U \rightarrow {}^{206}Pb$ releases eight alpha-particles). Another method of calculating the age of the rocks is to measure the ratio of radiogenic lead (^{206}Pb, ^{207}Pb, and ^{208}Pb) present to nonradiogenic lead (^{204}Pb). These methods give reliable results for ages of the order 10^7–10^9 years.

uranium(IV) oxide A black solid, UO_2; r.d. 10.9; m.p. 3000°C. It occurs naturally as *uraninite and is used in nuclear reactors.

uranium series *See* radioactive series.

urea (carbamide) A white crystalline solid, $CO(NH_2)_2$; r.d. 1.3; m.p. 133°C. It is soluble in water but insoluble in certain organic solvents. Urea is the major end product of nitrogen excretion in mammals, being synthesized by the *urea cycle. Urea is synthesized industrially from ammonia and carbon dioxide for use in *urea–formaldehyde resins and pharmaceuticals, as a source of nonprotein nitrogen for ruminant livestock, and as a nitrogen fertilizer.

urea cycle (ornithine cycle) The series of biochemical reactions that converts ammonia to *urea during the excretion of metabolic nitrogen. Urea formation occurs in mammals and, to a lesser extent, in some other animals. The liver converts ammonia to the much less toxic urea, which is excreted in solution in urine.

urea–formaldehyde resins Synthetic resins made by copolymerizing urea with formaldehyde (methanal). They are used as adhesives or thermosetting plastics.

urethane resins (polyurethanes) Synthetic resins containing the repeating group –NH–CO–O–. There are numerous types made by copolymerizing isocyanate esters with polyhydric alcohols. They have a variety of uses in plastics, paints, and solid foams.

uric acid The end product of purine breakdown in most primates, birds, terrestrial reptiles, and insects and also (except in primates) the major form in which metabolic nitrogen is excreted. Being fairly insoluble, uric acid can be expelled in solid form, which conserves valuable water in arid environments. The accumulation of uric acid in the synovial fluid of joints causes gout.

UV *See* ultraviolet radiation.

vacancy *See* defect.

vacuum A space in which there is a low pressure of gas, i.e. relatively few atoms or molecules. A *perfect vacuum* would contain no atoms or molecules, but this is unobtainable as all the materials that surround such a space have a finite *vapour pressure. In a *soft* (or *low*) *vacuum* the pressure is reduced to about 10^{-2} pascal, whereas a *hard* (or *high*) vacuum has a pressure of 10^{-2}–10^{-7} pascal. Below 10^{-7} pascal is known as an *ultrahigh vacuum*. *See also* vacuum pump.

vacuum distillation Distillation under reduced pressure. The depression in the boiling point of the substance distilled means that the temperature is lower, which may prevent the substance from decomposing.

vacuum pump A pump used to reduce the gas pressure in a container. The normal laboratory rotary oil-seal pump can maintain a pressure of 10^{-1} Pa. For pressures down to 10^{-7} Pa a *diffusion pump is required. *Ion pumps can achieve a pressure of 10^{-9} Pa and a *cryogenic pump combined with a diffusion pump can reach 10^{-13} Pa.

valence *See* valency.

valence band *See* energy bands.

valence electron An electron in one of the outer shells of an atom that takes part in forming chemical bonds.

valency (valence) The combining power of an atom or radical, equal to the number of hydrogen atoms that the atom could combine with or displace in a chemical compound (hydrogen has a valency of 1). It is equal to the ionic charge in ionic compounds; for example, in Na_2S, sodium has a valency of 1 (Na^+) and sulphur a valency of 2 (S^{2-}). In covalent compounds it is equal to the number of bonds formed; in CO_2 both carbon and oxygen have a valency of 2.

valine *See* amino acid.

vanadium Symbol V. A silvery-white metallic *transition element; a.n. 23; r.a.m. 50.94; r.d. 6.1; m.p. 1890°C; b.p. 3380°C. It occurs in a number of complex ores, including vanadinite ($Pb_5Cl(VO_4)_3$) and carnotite ($K_2(ClO_2)_2(VO_4)_2$). The pure metal can be obtained by reducing the oxide with calcium. The element is used in a large number of alloy steels. Chemically, it reacts with nonmetals at high temperatures but is not affected by hydrochloric acid or alkalis. It forms a range of complexes with oxidation states from +2 to +5. Vanadium was discovered in 1801 by del Rio, who allowed himself to be persuaded that what he had discovered was an impure form of chromium. The element was rediscovered and named by Sefström in 1880.

vanadium(V) oxide (vanadium pentoxide) A crystalline compound, V_2O_5, used extensively as a catalyst in industrial gas-phase oxidation processes.

vanadium pentoxide *See* vanadium(V) oxide.

van der Waals' equation *See* equation of state.

van der Waals' force An attractive force between atoms or molecules, named after J. D. van der Waals (1837–1923). The force accounts for the term a/V^2 in the van der Waals equation (*see* equation of state). These forces are much weaker than those arising from valence bonds and are inversely proportional to the seventh power of the distance between the atoms or molecules. They are the forces responsible for nonideal behaviour of gases and for the lattice energy of molecular crystals. There are three factors causing such forces: (1) dipole–dipole interaction, i.e. electrostatic attractions between two molecules with permanent dipole moments; (2) dipole-induced dipole interactions, in which the dipole of one molecule polarizes a neighbouring molecule; (3) dispersion forces arising because of small instantaneous dipoles in atoms.

van't Hoff factor Symbol i. A factor appearing in equations for *colligative properties, equal to the ratio of the number of actual particles present to the number of undissociated particles. It was first suggested by Jacobus van't Hoff (1852–1911).

van't Hoff's isochore An equation for the variation of equilibrium constant with temperature

$$(\mathrm{d}\log_e K)/\mathrm{d}T = \Delta H/RT^2$$

where T is the thermodynamic temperature and ΔH the enthalpy of the reaction.

vapour density The density of a gas or vapour relative to hydrogen, oxygen, or air. Taking hydrogen as the reference substance, the vapour density is the ratio of the mass of a particular volume of a gas to the mass of an equal volume of hydrogen under identical conditions of pressure and temperature. Taking the density of hydrogen as 1, this ratio is equal to half the relative molecular mass of the gas.

vapour pressure The pressure exerted by a vapour. All solids and liquids give off vapours, consisting of atoms or molecules of the substances that have evaporated from the condensed forms. These atoms or molecules exert a vapour pressure. If the substance is in an enclosed space, the vapour pressure will reach a maximum value that depends only on the nature of the substance and the temperature. This maximum value occurs when there is a dynamic equilibrium between the atoms or molecules escaping from the liquid or solid and those that strike the surface of the liquid or solid and return to it. The vapour is then said to be a *saturated vapour* and the pressure it exerts is the *saturated vapour pressure.*

verdigris A green patina of basic copper salts formed on copper. The composition of verdigris varies depending on the atmospheric conditions, but includes the basic carbonate $CuCO_3.Cu(OH)_2$, the basic sulphate $CuSO_4.Cu(OH)_2.H_2O$, and sometimes the basic chloride $CuCl_2.Cu(OH)_2$.

vermiculite *See* clay minerals.

vicinal (vic) Designating a molecule in which two atoms or groups are linked to adjacent atoms. For example, 1,2-dichloroethane (CH_2ClCH_2Cl) is a vic dihalide.

Victor Meyer's method A method of measuring vapour density, devised by Victor Meyer (1848–97). A weighed sample in a small tube is dropped into a heated bulb with a long neck. The sample vaporizes and displaces air, which is collected over water and the volume measured. The vapour density can then be calculated.

villiaumite A mineral form of sodium fluoride, NaF.

vinyl acetate *See* ethenyl ethanoate.

vinyl chloride *See* chloroethene.

vinyl group The organic group CH_2:CH–.

virial equation A gas law that attempts to account for the behaviour of real gases, as opposed to an ideal gas. It takes the form

$$pV = RT + Bp + Cp^2 + Dp^3 + \dots,$$

where *B*, *C*, and *D* are known as *virial coefficients.*

viscosity A measure of the resistance to flow that a fluid offers when it is subjected to shear stress. For a Newtonian fluid, the force, *F*, needed to maintain a velocity gradient, dv/dx, between adjacent planes of a fluid of area *A* is given by: $F = \eta A(dv/dx)$, where η is a constant, called the coefficient of viscosity. In SI units it has the unit pascal second (in the c.g.s. system it was measured in *poise). Non-Newtonian fluids, such as clays, do not conform to this simple model. *See also* kinematic viscosity.

vitamin One of a number of organic compounds required by living organisms in relatively small amounts to maintain normal health. There are some 14 generally recognized major vitamins: the water-soluble *vitamin B complex (containing 9) and *vitamin C and the fat-soluble *vitamin A, *vitamin D, *vitamin E, and *vitamin K. Most B vitamins and vitamin C occur in plants, animals, and microorganisms; they function typically as *coenzymes. Vitamins A, D, E, and K occur only in animals, especially vertebrates, and perform a variety of metabolic roles. Animals are unable to manufacture many vitamins themselves and must have adequate amounts in the diet. Foods may contain vitamin precursors (called *provitamins*) that are chemically changed to the actual vitamin on entering the body. Many vitamins are destroyed by light and heat, e.g. during cooking.

vitamin A (retinol) A fat-soluble vitamin that cannot be synthesized by mammals and other vertebrates and must be provided in the diet. Green plants contain precursors of the vitamin, notably carotenes, that are converted to vitamin A in the intestinal wall and liver. The aldehyde derivative of vitamin A, *retinal*, is a constituent of the visual pigment rhodopsin. Deficiency affects the eyes, causing night blindness, xerophthalmia, and eventually total blindness. The role of vitamin A in other aspects of metabolism is less clear.

vitamin B complex A group of water-soluble vitamins that characteristically serve as components of *coenzymes. Plants and many microorganisms can manufacture B vitamins but dietary sources are essential for most animals. Heat and light tend to destroy B vitamins.

Vitamin B_1 (*thiamin(e)*) is a precursor of the coenzyme thiamine pyrophosphate, which functions in carbohydrate metabolism. Deficiency leads to beriberi in humans and to polyneuritis in birds. Good sources include brewer's yeast, wheatgerm, beans, peas, and green vegetables.

Vitamin B_2 (*riboflavin*) occurs in green vegetables, yeast, liver, and milk. It is a constituent of the coenzymes *FAD and

FMN, which have an important role in the metabolism of all major nutrients as well as in the oxidative phosphorylation reactions of the electron transport chain. Deficiency of B_2 causes inflammation of the tongue and lips and mouth sores.

Vitamin B_6 (pyridoxine) is widely distributed in cereal grains, yeast, liver, milk, etc. It is a constituent of a coenzyme (pyridoxal phosphate) involved in amino acid metabolism. Deficiency causes retarded growth, dermatitis, convulsions, and other symptoms.

Vitamin B_{12} (cyanocobalamin) is manufactured only by microorganisms and natural sources are entirely of animal origin. Liver is especially rich in it. One form of B_{12} functions as a coenzyme in a number of reactions, including the oxidation of fatty acids and the synthesis of DNA. It also works in conjunction with *folic acid (another B vitamin) in the synthesis of the amino acid methionine and it is required for normal production of red blood cells. Vitamin B_{12} can only be absorbed from the gut in the presence of a glycoprotein called intrinsic factor; lack of this factor or deficiency of B_{12} results in pernicious anaemia.

Other vitamins in the B complex include *nicotinic acid, *pantothenic acid, *folic acid, *biotin, and *lipoic acid. *See also* choline.

vitamin C (ascorbic acid) A colourless crystalline water-soluble vitamin found especially in citrus fruits and green vegetables. Most organisms synthesize it from glucose but man and other primates and various other species must obtain it from their diet. It is required for the maintenance of healthy connective tissue; deficiency leads to scurvy. Vitamin C is readily destroyed by heat and light.

vitamin D A fat-soluble vitamin occurring in the form of two steroid derivatives: *vitamin D_2* (*ergocalciferol*, or *calciferol*), found in yeast; and *vitamin D_3* (*cholecalciferol*), which occurs in animals. Vitamin D_2 is formed from a steroid by the action of ultraviolet light and D_3 is produced by the action of sunlight on a cholesterol derivative in the skin. Fish-liver oils are the major dietary source. The active form of vitamin D is manufactured in response to the secretion of parathyroid hormone, which occurs when blood calcium levels are low. It causes increased uptake of calcium from the gut, which increases the supply of calcium for bone synthesis. Vitamin D deficiency causes rickets in growing animals and osteomalacia in mature animals. Both conditions are characterized by weak deformed bones.

vitamin E (tocopherol) A fat-soluble vitamin consisting of several closely related compounds, deficiency of which leads to a range of disorders in different species, including muscular dystrophy, liver damage, and infertility. Good sources are cereal grains and green vegetables. Vitamin E prevents the oxidation of unsaturated fatty acids in cell membranes, so maintaining their structure.

vitamin K A fat-soluble vitamin consisting of several related compounds that act as coenzymes in the synthesis of several proteins necessary for blood clotting. Deficiency of vitamin K, which leads to extensive bleeding, is rare because a form of the vitamin is manufactured by intestinal bacteria. Green vegetables and egg yolk are good sources.

volt Symbol V. The SI unit of electric potential, potential difference, or e.m.f. defined as the difference of potential between two points on a conductor carrying a constant current of one ampere when the power dissipated between the points is one watt. It is named after Alessandro Volta (1745–1827).

voltaic cell (galvanic cell) A device that produces an e.m.f. as a result of chemical reactions that take place within it. These reactions occur at the surfaces of two electrodes, each of which dips into an electrolyte. The first voltaic cell, devised by Alessandro Volta (1745–1827), had electrodes of two different metals dipping into brine. *See* primary cell; secondary cell.

voltameter (coulometer) 1. An electrolytic cell formerly used to measure quantity of electric charge. The increase in mass (m) of the cathode of the cell as a result of the deposition on it of a metal from a solution of its salt enables the charge (Q) to be determined from the relationship $Q = m/z$, where z is the electrochemical equivalent of the metal. **2.** Any other type of electrolytic cell used for measurement.

volume Symbol V. The space occupied by a body or mass of fluid.

volumetric analysis A method of quantitative analysis using measurement of volumes. For gases, the main technique is in reacting or absorbing gases in graduated containers over mercury, and measuring the volume changes. For liquids, it involves *titrations.

vulcanite (ebonite) A hard black insulating material made by the vulcanization of rubber with a high proportion of sulphur (up to 30%).

vulcanization A process for hardening rubber by heating it with sulphur or sulphur compounds.

Wacker process A process for the manufacture of ethanal by the air oxidation of ethene. A mixture of air and ethene is bubbled through a solution containing palladium(II) chloride and copper(II) chloride. The Pd^{2+} ions form a complex with the ethene in which the ion is bound to the pi electrons in the C=C bond. This decreases the electron density in the bond, making it susceptible to nucleophilic attack by water molecules. The complex formed breaks down to ethanal and palladium metal. The Cu^{2+} ions oxidize the palladium back to Pd^{2+}, being reduced to Cu^{+} ions in the process. The air present oxidizes Cu^{+} back to Cu^{2+}. Thus the copper(II) and palladium(II) ions effectively act as catalysts in the process, which is now the main source of ethanal and, by further oxidation, ethanoic acid. It can also be applied to other alkenes.

warfarin 3-(alpha-acetonylbenzyl)-4-hydroxycoumarin: a synthetic anticoagulant used both therapeutically in clinical medicine and, in lethal doses, as a rodenticide.

washing soda *Sodium carbonate decahydrate, $Na_2CO_3.10H_2O$.

water A colourless liquid, H_2O; r.d. 1.000 (4°C); m.p. 0.000°C; b.p. 100.000°C. In the gas phase water consists of single H_2O molecules in which the H-O-H angle is 105°. The structure of liquid water is still controversial; hydrogen bonding of the type H_2O ... H-O-H imposes a high degree of structure and current models supported by X-ray scattering studies have short-range ordered regions, which are constantly disintegrating and re-forming. This ordering of the liquid state is sufficient to make the density of water at about 0°C higher than that of the relatively open-structured ice; the maximum density occurs at 3.98°C. This accounts for the well-known phenomenon of ice floating on water and the contraction of water below ice, a fact of enormous biological significance for all aquatic organisms.

Ice has nine distinct structural modifications of which ordinary ice, or ice I, has an open structure built of puckered six-membered rings in which each H_2O unit is tetrahedrally surrounded by four other H_2O units.

Because of its angular shape the water molecule has a permanent dipole moment and in addition it is strongly hydrogen bonded and has a high dielectric constant. These properties combine to make water a powerful solvent for both polar and ionic compounds. Species in solution are frequently strongly hydrated and in fact ions frequently written as, for example, Cu^{2+} are essentially $[Cu(H_2O)_6]^{2+}$. Crystalline *hydrates are also common for inorganic substances; polar organic compounds, particularly those with O-H and N-H bonds, also form hydrates.

Pure liquid water is very weakly dissociated into H_3O^+ and OH^- ions by self ionization:

$$H_2O \rightleftharpoons H^+ + OH^-$$

(*see* ionic product) and consequently any species that increases the concentration of the positive species, H_3O^+, is acidic and species increasing the concentration of the negative species, OH^-, are basic (*see* acid). The phenomena of ion transport in water and the division of materials into *hydrophilic* (water loving) and *hydrophobic* (water hating) substances are central features of almost all biological chemistry. A further property of water that is of fundamental importance to the whole planet is its strong absorption in the infrared range of the spectrum and its transparency to visible and near ultraviolet radiation. This allows solar radiation to reach the earth during hours of daylight but restricts rapid heat loss at night. Thus atmospheric water prevents violent diurnal oscillations in the earth's ambient temperature.

water gas A mixture of carbon monoxide and hydrogen produced by passing steam over hot carbon (coke):

$$H_2O(g) + C(s) \rightarrow CO(g) + H_2(g)$$

The reaction is strongly endothermic but the reaction can be used in conjunction with that for *producer gas for making fuel gas. The main use of water gas before World War II was in producing hydrogen for the *Haber process. Here the above reaction was combined with the *water-gas shift reaction* to increase the amount of hydrogen:

$$CO + H_2O \rightleftharpoons CO_2 + H_2$$

Most hydrogen for the Haber process is now made from natural gas by steam *reforming.

water glass A viscous colloidal solution of sodium silicates in water, used to make silica gel and as a size and preservative.

water of crystallization Water present in crystalline compounds in definite proportions. Many crystalline salts form hydrates containing 1, 2, 3, or more molecules of water per molecule of compound, and the water may be held in the crystal in various ways. Thus, the water molecules may simply occupy lattice positions in the crystal, or they may form bonds with the anions or the cations present. In copper sulphate pentahydrate ($CuSO_4.5H_2O$), for instance, each copper ion is coordinated to four water molecules through the lone pairs on the oxygen to form the *complex $[Cu(H_2O)_4]^{2+}$. Each sulphate ion has one water molecule held by hydrogen bonding. The difference between the two types of bonding is demonstrated by the fact that the pentahydrate converts to the monohydrate at 100°C and only becomes anhydrous above 250°C. *Water of constitution* is an obsolete term for water combined in a compound (as in a metal hydroxide $M(OH)_2$ regarded as a hydrated oxide $MO.H_2O$).

water softening *See* hardness of water.

watt Symbol W. The SI unit of power, defined as a power of one joule per second. In electrical contexts it is equal to the rate of energy transformation by an electric current of one ampere flowing through a conductor the ends of which are maintained at a potential difference of one volt. The unit is named after James Watt (1736–1819).

wave function A function $\psi(x,y,z)$ appearing in *Schrödinger's equation in wave mechanics. The wave function is a mathematical expression involving the coordinates of a particle in space. If the Schrödinger equation can be solved for a particle in a given system (e.g. an electron in an atom) then, depending on the boundary conditions, the solution is a set of allowed wave functions (*eigenfunctions*) of the particle, each corresponding to an allowed energy level (*eigenvalue*). The physical significance of the wave function is that the square of its absolute value, $|\psi|^2$, at a point is proportional to the probability of finding the particle in a small element of volume, dxdydz, at that point. For an electron in an atom, this gives rise to the idea of atomic and molecular *orbitals.

wave mechanics *See* quantum mechanics.

wax Any of various solid or semisolid substances. There are two main types. Mineral waxes are mixtures of hydrocarbons with high molecular weights. Paraffin wax, obtained from *petroleum, is an example. Waxes secreted by plants or animals are mainly esters of fatty acids and usually have a protective function.

weak acid An *acid that is only partially dissociated in aqueous solution.

weber Symbol Wb. The SI unit of magnetic flux equal to the flux that, linking a circuit of one turn, produces in it an e.m.f. of one volt as it is reduced to zero at a uniform rate in one second. It is named after Wilhelm Weber (1804–91).

Weston cell (cadmium cell) A type of primary *voltaic cell, which is used as a standard; it produces a constant e.m.f. of 1.0186 volts at 20°C. The cell is usually made in an H-shaped glass vessel with a mercury anode covered with a paste of cadmium sulphate and mercury(I) sulphate in one leg and a cadmium amalgam cathode covered with cadmium sulphate in the other leg. The electrolyte, which connects the two electrodes by means of the bar of the H, is a saturated solution of cadmium sulphate. In some cells sulphuric acid is added to prevent the hydrolysis of mercury sulphate.

white arsenic *See* arsenic(III) oxide.

white mica *See* muscovite.

white spirit A liquid mixture of hydrocarbons obtained from petroleum, used as a solvent for paint ('turpentine substitute').

Williamson's synthesis Either of two methods of producing ethers, both named after the British chemist Alexander Williamson (1824–1904).

1. The dehydration of alcohols using concentrated sulphuric acid. The overall reaction can be written

$$2ROH \rightarrow H_2O + ROR$$

The method is used for making ethoxyethane ($C_2H_5OC_2H_5$) from ethanol by heating at 140°C with excess of alcohol (excess acid at 170°C gives ethene). Although the steps in the reaction are all reversible, the ether is distilled off so the reaction can proceed to completion. This is *Williamson's continuous process*. In general, there are two possible mechanisms for this synthesis. In the first (favoured by primary alcohols), an alkylhydrogen sulphate is formed

$$ROH + H_2SO_4 \rightleftharpoons ROSO_3H + H_2O$$

This reacts with another alcohol molecule to give an oxonium ion

$$ROH + ROSO_3H \rightarrow ROHR^+$$

This loses a proton to give ROR.

The second mechanism (favoured by tertiary alcohols) is formation of a carbonium ion

$$ROH + H^+ \rightarrow H_2O + R^+$$

This is attacked by the lone pair on the other alcohol molecule

$$R^+ + ROH \rightarrow ROHR^+$$

and the oxonium ion formed again gives the product by loss of a proton.

The method can be used for making symmetric ethers (i.e. having both R groups the same). It can successfully be used for mixed ethers only when one alcohol is primary and the other tertiary (otherwise a mixture of the three possible products results).

2. A method of preparing ethers by reacting a haloalkane with an alkoxide. The reaction, discovered in 1850, is a nucleophilic substitution in which the negative alkoxide ion displaces a halide ion; for example:

$$RI + {}^{-}OR' \rightarrow ROR' + I^{-}$$

A mixture of the reagents is refluxed in ethanol. The method is particularly useful for preparing mixed ethers, although a possible side reaction under some conditions is an elimination to give an alcohol and an alkene.

witherite A mineral form of *barium carbonate, $BaCO_3$.

Wöhler's synthesis A synthesis of urea performed by the German chemist Friedrich Wöhler (1800–82) in 1828. He discovered that urea ($CO(NH_2)_2$) was formed when a solution of ammonium isocyanate (NH_4NCO) was evaporated. At the time it was believed that organic substances such as urea could be made only by living organisms, and its production from an inorganic compound was a notable discovery. It is sometimes (erroneously) cited as ending the belief in vitalism.

wolfram *See* tungsten.

wolframite (iron manganese tungsten) A mineral consisting of a mixed iron–manganese tungstate, $(FeMn)WO_4$, crystallizing in the monoclinic system; the principal ore of tungsten. It commonly occurs as blackish or brownish tabular crystal groups. It is found chiefly in quartz veins associated with granitic rocks. China is the major producer of wolframite.

wood alcohol *See* methanol.

Wood's metal A low-melting (71°C) alloy of bismuth (50%), lead (25%), tin (12.5%), and cadmium (12.5%). It is used for fusible links in automatic sprinkler systems. The melting point can be changed by varying the composition.

Woodward–Hoffmann rules Rules governing the formation of products during certain types of organic concerted reactions. The theory of such reactions was put forward in 1969 by the American chemists Robert Burns Woodward (1917–79) and Roald Hoffmann (1937–), and is concerned with the way that orbitals of the reactants change continuously into orbitals of the products during reaction and with conservation of orbital symmetry during this process. It is sometimes known as *frontier-orbital theory*.

work function A quantity that determines the extent to which thermionic or photoelectric emission will occur according to the Richardson equation or Einstein's photoelectric equation. It is sometimes expressed as a potential difference (symbol ϕ) in volts and sometimes as the work to be done by the exciting electron (symbol W) in electronvolts or joules. The former has been called the *work function potential* and the latter the *work function energy*.

work hardening An increase in the hardness of metals as a result of working them cold. It causes a permanent distortion of the crystal structure and is particularly apparent with iron, copper, aluminium, etc., whereas with lead and zinc it does not occur as these metals are capable of recrystallizing at room temperature.

wrought iron A highly refined form of iron containing 1–3% of slag (mostly iron silicate), which is evenly distributed throughout the material in threads and fibres so that the product has a fibrous structure quite dissimilar to that of crystalline cast iron. Wrought iron rusts less readily than other forms of metallic iron and it welds and works more easily. It is used for chains, hooks, tubes, etc.

Wurtz reaction A reaction to prepare alkanes by reacting a haloalkane with sodium:

$$2RX + 2Na \rightarrow 2NaX + RR$$

The haloalkane is refluxed with sodium in dry ether. The method is named after the French chemist Charles-Adolphe Wurtz (1817–84). The analogous reaction using a haloalkane and a haloarene, for example:

$$C_6H_5Cl + CH_3Cl + 2Na \rightarrow 2NaCl + C_6H_5CH_3$$

is called the *Fittig reaction* after the German chemist Rudolph Fittig (1835–1910).

X

xanthates Salts or esters containing the group –SCS(OR), where R is an organic group. Cellulose xanthate is an intermediate in the manufacture of *rayon by the viscose process.

xenon Symbol Xe. A colourless odourless gas belonging to group 0 of the periodic table (*see* noble gases); a.n. 54; r.a.m. 131.30; d. 5.887 $g\,dm^{-3}$; m.p. –111.9°C; b.p. –107.1°C. It is present in the atmosphere (0.00087%) from which it is extracted by distillation of liquid air. There are nine natural isotopes with mass numbers 124, 126, 128–132, 134, and 136. Seven radioactive isotopes are also known. The element is used in fluorescent lamps and bubble chambers. Several compounds of xenon are now known, including $XePtF_6$, XeF_2, XeF_4, $XeSiF_6$, XeO_2F_2, and XeO_3. The element was discovered in 1898 by Ramsey and Travers.

X-ray crystallography The use of *X-ray diffraction to determine the structure of crystals or molecules. The technique involves directing a beam of X-rays at a crystalline sample and recording the diffracted X-rays on a photographic plate. The diffraction pattern consists of a pattern of spots on the plate, and the crystal structure can be worked out from the positions and intensities of the diffraction spots. X-rays are diffracted by the electrons in the molecules and if molecular crystals of a compound are used, the electron density distribution in the molecule can be determined.

X-ray diffraction The diffraction of X-rays by a crystal. The wavelengths of X-rays are comparable in size to the distances between atoms in most crystals, and the repeated pattern of the crystal lattice acts like a diffraction grating for X-rays. Thus, a crystal of suitable type can be used to disperse X-rays in a spectrometer. X-ray diffraction is also the basis of X-ray crystallography.

X-ray fluorescence The emission of *X-rays from excited atoms produced by the impact of high-energy electrons, other particles, or a primary beam of other X-rays. The wavelengths of the fluorescent X-rays can be measured by an X-ray spectrometer as a means of chemical analysis. X-ray fluorescence is used in such techniques as *electron-probe microanalysis.

X-rays Electromagnetic radiation of shorter wavelength than ultraviolet radiation and longer wavelength than gamma radiation. The range of wavelengths is 10^{-11} m to 10^{-9} m. Atoms of all the elements emit a characteristic *X-ray spectrum* when they are bombarded by electrons. The X-ray photons are emitted when the incident electrons knock an inner orbital electron out of an atom. When this happens an outer electron falls into the inner shell to replace it, losing potential energy (ΔE) in doing so. The wavelength λ of the emitted photon will then be given by $\lambda = ch/\Delta E$, where c is the speed of light and h is the Planck constant.

X-rays can pass through many forms of matter and they are therefore used medically and industrially to examine internal structures. X-rays are produced for these purposes by an X-ray tube.

X-ray spectrum *See* X-rays.

xylenes *See* dimethylbenzenes.

Y

yeasts A group of unicellular fungi many of which belong to the Ascomycetes. Certain species of the genus *Saccharomyces* are used in the baking and brewing industries.

ytterbium Symbol Yb. A silvery metallic element belonging to the *lanthanoids; a.n. 70; r.a.m. 173.04; r.d. 6.966 (20°C); m.p. 819°C; b.p. 1196°C. It occurs in gadolinite, monazite, and xenotime. There are seven natural isotopes and ten artificial isotopes are known. There are no uses for the element, which was discovered by J. D. G. Marignac in 1878.

yttrium Symbol Y. A silvery-grey metallic element belonging to group IIIB of the periodic table; a.n. 39; r.a.m. 88.905; r.d. 4.469 (20°C); m.p. 1522°C; b.p. 3338°C. It occurs in uranium ores and in *lanthanoid ores, from which it can be extracted by an ion exchange process. The natural isotope is yttrium–89, and there are 14 known artificial isotopes. The metal is used in superconducting alloys and in alloys for strong permanent magnets (in both cases, with cobalt). The oxide (Y_2O_3) is used in colour-television phosphors, neodymium-doped lasers, and microwave components. Chemically it resembles the lanthanoids, forming ionic compounds containing Y^{3+} ions. The metal is stable in air below 400°C. It was discovered in 1828 by Friedrich Wöhler.

Z

Zeeman effect The splitting of the lines in a spectrum when the source of the spectrum is exposed to a magnetic field. It was discovered in 1896 by P. Zeeman (1865–1943). In the *normal Zeeman effect* a single line is split into three if the field is perpendicular to the light path or two lines if the field is parallel to the light path. This effect can be explained by classical electromagnetic principles in terms of the speeding up and slowing down of orbital electrons in the source as a result of the applied field. The *anomolous Zeeman effect* is a complicated splitting of the lines into several closely spaced lines, so called because it does not agree with classical predictions. This effect is explained by quantum mechanics in terms of electron spin.

zeolite A natural or synthetic hydrated aluminosilicate with an open three-dimensional crystal structure, in which water molecules are held in cavities in the lattice. The water can be driven off by heating and the zeolite can then absorb other molecules of suitable size. Zeolites are used for separating mixtures by selective absorption – for this reason they are often called *molecular sieves*. They are also used in sorption pumps for vacuum systems and certain types (e.g. *Permutit*) are used in ion-exchange (e.g. water-softening).

zero order *See* order.

zeroth law of thermodynamics *See* thermodynamics.

Ziegler process An industrial process for the manufacture of high-density polyethene using catalysts of titanium(IV) chloride ($TiCl_4$) and aluminium alkyls (e.g. triethylaluminium, $Al(C_2H_5)_3$). The process was introduced in 1953 by the German chemist Karl Ziegler (1898–1973). It allowed the manufacture of polythene at lower temperatures (about 60°C) and pressures (about 1 atm.) than used in the original process. Moreover, the polyethene produced had more straight-chain molecules, giving the product more rigidity and a higher melting point than the earlier low-density polyethene. The reaction involves the formation of a titanium alkyl in which the titanium can coordinate directly to the pi bond in ethene.

In 1954 the process was developed further by the Italian chemist Giulio Natta (1903–79), who extended the use of Ziegler's catalysts (and similar catalysts) to other alkenes. In particular he showed how to produce stereospecific polymers of propene.

Ziesel reaction A method of determining the number of methoxy ($-OCH_3$) groups in an organic compound. The compound is heated wih excess hydriodic acid, forming an alcohol and iodomethane

$$R-O-CH_3 + HI \rightarrow ROH + CH_3I$$

The iodomethane is distilled off and led into an alcoholic solution of silver nitrate, where it precipitates silver iodide. This is filtered and weighed, and the number of iodine atoms and hence methoxy groups can be calculated. The method was developed by S. Ziesel in 1886.

zinc Symbol Zn. A blue-white metallic element; a.n. 30; r.a.m. 65.38; r.d. 7.1; m.p. 419.57°C; b.p. 907°C. It occurs in sphalerite (or zinc blende, ZnS), which is found associated with the lead sulphide, and in smithsonite ($ZnCO_3$). Ores are roasted to give the oxide and this is reduced with carbon (coke) at high temperature, the zinc vapour being condensed. Alternatively, the oxide is dissolved in sulphuric acid and the zinc obtained by electrolysis. There are five stable isotopes (mass numbers 64, 66, 67, 68, and 70) and six radioactive isotopes are known. The metal is used in galvanizing and in a number of alloys (brass, bronze, etc.). Chemically it is a reactive metal, combining with oxygen and other nonmetals and reacting with dilute acids to release hydrogen. It also dissolves in alkalis to give *zincates. Most of its compounds contain the Zn^{2+} ion.

zincate A salt formed in solution by dissolving zinc or zinc oxide in alkali. The formula is often written $ZnO_2{}^{2-}$ although in aqueous solution the ions present are probably complex ions in which the Zn^{2-} is coordinated to OH^- ions. $ZnO_2{}^{2-}$ ions may exist in molten sodium zincate, but most solid 'zincates' are mixed oxides.

zinc blende A mineral form of *zinc sulphide, ZnS, the principal ore of zinc (*see* sphalerite). The *zinc-blende structure* is the crystal structure of this compound (and of other compounds). It has zinc atoms surrounded by four sulphur atoms at the corners of a tetrahedron. Each sulphur is similarly surrounded by four zinc atoms. The crystals belong to the cubic system.

zinc chloride A white crystalline compound, $ZnCl_2$. The anhydrous salt, which is deliquescent, can be made by the action of hydrogen chloride gas on hot zinc, r.d. 2.9; m.p. 290°C; b.p. 732°C. It has a relatively low melting point and sublimes easily, indicating that it is a molecular compound rather than ionic. Various hydrates also exist. Zinc chloride is used as a catalyst, dehydrating agent, and flux for hard solder. It was once known as *butter of zinc*.

zinc group The group of elements in the periodic table consisting of zinc (Zn), cadmium (Cd), and mercury (Hg). *See* group II elements.

zincite A mineral form of *zinc oxide, ZnO.

zinc oxide A powder, white when cold and yellow when hot, ZnO; r.d. 5.5; m.p. 1975°C. It occurs naturally as a reddish orange ore *zincite*, and can also be made by oxidizing hot zinc in air. It is amphoteric, forming *zincates with bases. It is used as a pigment (*Chinese white*) and a mild antiseptic in zinc ointments. An archaic name is *philosopher's wool*.

zinc sulphate A white crystalline water-soluble compound made by heating zinc sulphide ore in air and dissolving out and recrystallizing the sulphate. The common form is the heptahydrate, $ZnSO_4.7H_2O$; r.d. 1.9. This loses water above 30°C to give the hexahydrate and more water is lost above 100°C to form the monohydrate. The anhydrous salt forms at 450°C and this decomposes above 500°C. The compound, which was formerly called *white vitriol*, is used as a mordant and as a styptic (to check bleeding).

zinc sulphide A yellow-white water-soluble solid, ZnS. It occurs naturally as *sphalerite (*see also* zinc blende) and wurtzite. The compound sublimes at 1180°C. It is used as a pigment and phosphor.

zirconia *See* zirconium.

zirconium Symbol Zr. A grey-white metallic *transition element; a.n. 40; r.a.m. 91.22; r.d. 6.44; m.p. 1853°C; b.p. 4376°C. It is found in zircon ($ZrSiO_4$; the main source) and in baddeleyite (ZnO_2). Extraction is by chlorination to give $ZrCl_4$ which is purified by solvent extraction and reduced with magnesium (Kroll process). There are five natural isotopes (mass numbers 90, 91, 92, 94, and 96) and six radioactive isotopes are known. The element is used in nuclear reactors (it is an effective neutron absorber) and in certain alloys. The metal forms a passive layer of oxide in air and burns at 500°C. Most of its compounds are complexes of zirconium(IV). *Zirconium(IV) oxide* (*zirconia*) is used as an electrolyte in fuel cells. The element was identified in 1789 by Klaproth and was first isolated by Berzelius in 1824.

zirconium(IV) oxide *See* zirconium.

zone refining A technique used to reduce the level of impurities in certain metals, alloys, semiconductors, and other materials. It is based on the observation that the solubility of an impurity may be different in the liquid and solid phases of a material. To take advantage of this observation, a narrow molten zone is moved along the length of a specimen of the material, with the result that the impurities are segregated at one end of the bar and the pure material at the other. In general, if the impurities lower the melting point of the material they are moved in the same direction as the molten zone moves, and vice versa.

zwitterion (ampholyte ion) An ion that has a positive and negative charge on the same group of atoms. Zwitterions can be formed from compounds that contain both acid groups and basic groups in their molecules. For example, aminoethanoic acid (the amino acid glycine) has the formula $H_2N.CH_2.COOH$. However, under neutral conditions, it exists in the different form of the zwitterion $^+H_3N.CH_2.COO^-$, which can be regarded as having been produced by an internal neutralization reaction (transfer of a proton from the carboxyl group to the amino group). Aminoethanoic acid, as a consequence, has some properties characteristic of ionic compounds; e.g. a high melting point and solubility in water. In acid solutions, the positive ion $^+H_3NCH_2COOH$ is formed. In basic solutions, the negative ion $H_2NCH_2COO^-$ predominates. The name comes from the German *zwei*, two.

Appendix 1 SI units

Table 1.1 Base and supplementary SI units

Physical quantity	*Name*	*Symbol*
length	metre	m
mass	kilogram	kg
time	second	s
electric current	ampere	A
thermodynamic temperature	kelvin	K
luminous intensity	candela	cd
amount of substance	mole	mol
*plane angle	radian	rad
*solid angle	steradian	sr

*supplementary units

Table 1.2 Derived SI units with special names

Physical quantity	*Name of SI unit*	*Symbol of SI unit*
frequency	hertz	Hz
energy	joule	J
force	newton	N
power	watt	W
pressure	pascal	Pa
electric charge	coulomb	C
electric potential difference	volt	V
electric resistance	ohm	Ω
electric conductance	siemens	S
electric capacitance	farad	F
magnetic flux	weber	Wb
inductance	henry	H
magnetic flux density (magnetic induction)	tesla	T
luminous flux	lumen	lm
illuminance	lux	lx
absorbed dose	gray	Gy
activity	becquerel	Bq
dose equivalent	sievert	Sv

Table 1.3 Decimal multiples and submultiples to be used with SI units

Submultiple	*Prefix*	*Symbol*	*Multiple*	*Prefix*	*Symbol*
10^{-1}	deci	d	10	deca	da
10^{-2}	centi	c	10^{2}	hecto	h
10^{-3}	milli	m	10^{3}	kilo	k
10^{-6}	micro	μ	10^{6}	mega	M
10^{-9}	nano	n	10^{9}	giga	G
10^{-12}	pico	p	10^{12}	tera	T
10^{-15}	femto	f	10^{15}	peta	P
10^{-18}	atto	a	10^{18}	exa	E

Table 1.4 Conversion of units to SI units

From	*To*	*Multiply by*
in	m	2.54×10^{-2}
ft	m	0.3048
sq. in	m^2	6.4516×10^{-4}
sq. ft	m^2	9.2903×10^{-2}
cu. in	m^3	1.63871×10^{-5}
cu. ft	m^3	2.83168×10^{-2}
l(itre)	m^3	10^{-3}
gal(lon)	m^3	$4.546\ 09 \times 10^{-3}$
gal(lon)	l(itre)	4.546 09
miles/hr	$m\ s^{-1}$	0.477 04
km/hr	$m\ s^{-1}$	0.277 78
lb	kg	0.453 592
$g\ cm^{-3}$	$kg\ m^{-3}$	10^{3}
lb/in^3	$kg\ m^{-3}$	$2.767\ 99 \times 10^{4}$
dyne	N	10^{-5}
kgf	N	9.806 65
poundal	N	0.138 255
lbf	N	4.448 22
mmHg	Pa	133.322
atmosphere	Pa	$1.013\ 25 \times 10^{5}$
hp	W	745.7
erg	J	10^{-7}
eV	J	$1.602\ 10 \times 10^{-19}$
kW h	J	3.6×10^{6}
cal	J	4.1868

Appendix 2 Fundamental constants

Constant	*Symbol*	*Value in SI units*
acceleration of free fall	g	$9.806\ 65\ \text{m s}^{-2}$
Avogadro constant	L, N_A	$6.022\ 52 \times 10^{23}\ \text{mol}^{-1}$
Boltzmann constant	$k = R/N_A$	$1.380\ 622 \times 10^{-23}\ \text{J K}^{-1}$
electric constant	ϵ_0	$8.854 \times 10^{-12}\ \text{F m}^{-1}$
electronic charge	e	$1.602\ 192 \times 10^{-19}\ \text{C}$
electronic rest mass	m_e	$9.109\ 558 \times 10^{-31}\ \text{kg}$
Faraday constant	F	$9.648\ 670 \times 10^{4}\ \text{C mol}^{-1}$
gas constant	R	$8.314\ 34\ \text{J K}^{-1}\ \text{mol}^{-1}$
gravitational constant	G	$6.664 \times 10^{-11}\ \text{N m}^{2}\ \text{kg}^{-2}$
Loschmidt's constant	N_L	$2.687\ 19 \times 10^{25}\ \text{m}^{-3}$
magnetic constant	μ_0	$4\pi \times 10^{-7}\ \text{H m}^{-1}$
neutron rest mass	m_n	$1.674\ 92 \times 10^{-27}\ \text{kg}$
Planck constant	h	$6.626\ 196 \times 10^{-34}\ \text{J s}$
proton rest mass	m_p	$1.672\ 614 \times 10^{-27}\ \text{kg}$
speed of light	c	$2.997\ 924\ 58 \times 10^{8}\ \text{m s}^{-1}$
Stefan–Boltzmann constant	σ	$5.6697 \times 10^{-8}\ \text{W m}^{-2}\ \text{K}^{-4}$

Appendix 3 The solar system

Planet	*Equatorial diameter (km)*	*Mean distance from sun (10^6 km)*	*Sidereal period*
Mercury	4878	57.91	87.969 days
Venus	12 100	108	224.7 days
Earth	12 756	149.60	365.256 days
Mars	6762	227.94	686.980 days
Jupiter	142 700	778	11.86 years
Saturn	120 800	1430	29.46 years
Uranus	51 800	2869.6	84.01 years
Neptune	49 400	4496	164.8 years
Pluto	3500	5900	248.4 years

The Periodic Table

IA	IIA	IIIA	IVA	VA	VIA	VIIA	VIII			IB	IIB	IIIB	IVB	VB	VIB	VIIB	0
				1 H													2 He
3 Li	4 Be											5 B	6 C	7 N	8 O	9 F	10 Ne
11 Na	12 Mg	←				Transition elements					→	13 Al	14 Si	15 P	16 S	17 Cl	18 Ar
19 K	20 Ca	21 Sc	22 Ti	23 V	24 Cr	25 Mn	26 Fe	27 Co	28 Ni	29 Cu	30 Zn	31 Ga	32 Ge	33 As	34 Se	35 Br	36 Kr
37 Rb	38 Sr	39 Y	40 Zr	41 Nb	42 Mo	43 Tc	44 Ru	45 Rh	46 Pd	47 Ag	48 Cd	49 In	50 Sn	51 Sb	52 Te	53 I	54 Xe
55 Cs	56 Ba	57* La	72 Hf	73 Ta	74 W	75 Re	76 Os	77 Ir	78 Pt	79 Au	80 Hg	81 Ti	82 Pb	83 Bi	84 Po	85 At	86 Rn
87 Fr	88 Ra	89† Ac															

s-block (IA–IIA); d-block (IVA–IIB); p-block (IIIB–0)

*Lanthanides	57 La	58 Ce	59 Pr	60 Nd	61 Pm	62 Sm	63 Eu	64 Gd	65 Tb	66 Dy	67 Ho	68 Er	69 Tm	70 Yb	71 Lu
†Actinides	89 Ac	90 Th	91 Pa	92 U	93 Np	94 Pu	95 Am	96 Cm	97 Bk	98 Cf	99 Es	100 Fm	101 Md	102 No	103 Lr

f-block (58–71, 90–103)

First Edition	1894
Second Edition	1894
Third Edition	1896
Fourth Edition	1899
Fifth Edition	1902
Sixth Edition	1905
Seventh Edition	1910
Eighth Edition	1920
Ninth Edition	1922
Tenth Edition	1924
Eleventh Edition	1931
Twelfth Edition	1945
Thirteenth Edition	1957
Second Impression . . *August*	1961
Fourteenth Edition	1963
Second Impression . . *December*	1965
Fifteenth Edition	1967
Sixteenth Edition	1971
Seventeenth Edition	1975

First to Tenth Editions prepared by the Author, Sir M. D. CHALMERS, K.C.B., C.S.I. (Draftsman of the Act).

CHALMERS'

SALE OF GOODS ACT, 1893

CHALMERS'

SALE OF GOODS ACT, 1893

INCLUDING

THE FACTORS ACTS, 1889 & 1890

SEVENTEENTH EDITION

BY

MICHAEL MARK, M.A., B.C.L.

OF THE INNER TEMPLE, BARRISTER

with assistance from

JONATHAN MANCE, B.A.

OF THE MIDDLE TEMPLE, BARRISTER

LONDON

BUTTERWORTHS

1975

ENGLAND: BUTTERWORTH & CO. (PUBLISHERS) LTD.
LONDON: 88 KINGSWAY, WC2B 6AB

AUSTRALIA: BUTTERWORTHS PTY. LTD.
SYDNEY: 586 PACIFIC HIGHWAY, CHATSWOOD, NSW 2067
MELBOURNE: 343 LITTLE COLLINS STREET, 3000
BRISBANE: 240 QUEEN STREET, 4000

CANADA: BUTTERWORTH & CO. (CANADA) LTD.
TORONTO: 2265 MIDLAND AVENUE, SCARBOROUGH M1P 4S1

NEW ZEALAND: BUTTERWORTHS OF NEW ZEALAND LTD.
WELLINGTON: 26–28 WARING TAYLOR STREET, 1

SOUTH AFRICA: BUTTERWORTH & CO. (SOUTH AFRICA) (PTY.) LTD.
DURBAN: 152–154 GALE STREET

ISBN—Casebound: 0 406 56444 2
Limp: 0 406 56445 0

PREFACE

ALTHOUGH it is only four years since the last edition was published, new legislation, in particular the Supply of Goods (Implied Terms) Act, 1973, has made necessary this new edition.

In addition to dealing with the amendments to the Sale of Goods Act, 1893 introduced by the Supply of Goods (Implied Terms) Act, 1973, I have taken the opportunity to expand or rewrite the commentary to a large number of other sections of the Sale of Goods Act, and to some of the more important provisions of the Factors Act, 1889 and other legislation affecting the sale of goods. In particular, I have rewritten or expanded the commentary to ss. 6, 12, 14, 16, 20, 48, 49, 50, 52, 53, 54 and 55 of the Sale of Goods Act, to ss. 2, 8 and 9 of the Factors Act, to s. 3 of the Law Reform (Miscellaneous Provisions) Act, 1933 (dealing with the award of interest) and to ss. 2 and 3 of the Misrepresentation Act, 1967.

While I have sought to preserve the brief, clear exposition of principles which was a feature of earlier editions, I have departed from earlier practice by considering some of the more important problems which have not yet been finally resolved by the courts. I have also severely pruned the references to the Civil Law and to Continental Codes which, as appears from Sir Mackenzie Chalmers' Introduction to the 1894 edition, reproduced *post*, were once of considerable value, but which have become largely otiose.

The passages on international sales of goods have also been expanded and a new passage on confirming houses added. I am once again indebted to Mr. Jonathan Mance of the Middle Temple for revising this part of the book. His views have also been of considerable help to me in rewriting much of the commentary to the Sale of Goods Act.

A major innovation is the new passage on statutory controls. Previous editions have concentrated exclusively on the civil rights and liabilities of buyer and seller and have made only passing references to obligations imposed by statute under threat of criminal proceedings. Frequently, however, it is no longer possible for a buyer or a seller to be fully aware of his rights and obligations without considering the effect of such statutes.

The object of Section C of this edition is to enable businessmen and consumers and their legal advisers to see in outline the range of controls imposed upon sellers and the consequences which might flow from a breach of those controls. I have also sought to examine in greater detail some of the provisions, in particular those of the Trade Descriptions Act, 1968, which affect almost all sales in the course of a business, and those common to many statutes which deal with the question who may be liable if a statutory offence is committed and what defences are available to a person charged with an offence.

Finally, I have included in Appendix I a commentary on s. 35 of the Powers of Criminal Courts Act, 1973 which enables a court before whom an offender is convicted of an offence to award compensation to any person who has suffered loss as a result of that offence or any other offence which the court takes into consideration in determining sentence. This provision may well be of considerable assistance to consumers whose loss is clear but is too small to be worth pursuing in the civil courts.

I have sought to state the law on the basis of reports available to me as at August 1st, 1975.

I should like to thank the staff of Butterworths for their help and co-operation throughout in producing this new edition.

MICHAEL MARK

Lincoln's Inn,
August 1975

INTRODUCTION TO FIRST EDITION (1894)

It is difficult to know whether to call this little book a first edition or a second edition. It is a first edition of the Sale of Goods Act, 1893, but it is a reproduction of my book on the Sale of Goods, published in 1890, which was in substance a commentary on the Sale of Goods Bill. The clauses of the Bill, with a few verbal alterations, formed the large type propositions of the book. But though the language of the propositions remains the same, its effect is now very different. Those propositions were only law in so far as they were correct and logical inductions from the decided cases. Now the position is reversed. The propositions have become sections of the Act, and the decided cases are only law in so far as they are correct and logical deductions from the language of the Act. Each case, therefore, must be tested with reference to the Act itself. But it may be none the less useful to the reader to call his attention to the decisions which formed the basis of the various sections, and which were intended to be reproduced in the Act. In so far as the law is unaltered, they are still in point as illustrations.

The history of the Act is as follows: The Bill was originally drafted by me in 1888. I then settled it in consultation with Lord Herschell, who kindly consented to take charge of it. In 1889, Lord Herschell introduced it in the House of Lords, not to press it on, but to get criticisms on it. In 1890 there was no opportunity of proceeding with it, but in 1891 the Bill was again introduced in the Lords, and referred to a Select Committee. It had in the meantime been criticised by Lord Bramwell, Mr. Walter Ker, and other friends, and the Bar Committee had submitted a valuable memorandum on it. In the Lords it was carefully considered by a Select Committee, consisting of Lords Herschell, Halsbury, Bramwell, and Watson. A question arose as to its extension to Scotland, so the Bill stood over till 1892. It was then again introduced in the Lords, and extended to Scotland, on the advice of Lord Watson, who had consulted various Scotch legal authorities. Professor Richard Brown and Mr. Spens of Glasgow took an infinity of pains to suggest the necessary amendments. In 1893 the Bill was again passed through the Lords in the form in which it was settled in 1892. It was then considered by a Select Committee of the House of Commons and further amended. The Committee consisted of Sir Charles Russell, A.-G., Sir R. Webster, Q.C., Mr. Asher, Q.C. (the Scotch Solicitor-General), Mr. Shiress Will, Q.C., Mr. Bousfield, Q.C., Mr. Ambrose, Q.C., and Mr. Mather. Some of the amendments introduced by the Commons were modified on its return to the Lords, and it was finally settled in its present form.

The Bill, in its original form, was drafted on the same lines as the Bills of Exchange Bill. On Lord Herschell's advice it endeavoured to reproduce as exactly as possible the existing law, leaving any amendments that might seem desirable to be introduced in Committee on the authority of the Legislature. So far as England is concerned, the conscious changes effected in the law have been very slight. They are pointed out in the notes to the various sections. As regards Scotland, in some cases the Scottish rule has been saved or enacted for Scotland, in others it has been modified, while in others the English rule has been adopted. These points are noted under the sections as they arise. Scottish law differs from English law mainly by adhering to the Roman law in matters where English law has developed a rule of its own. The Mercantile Law Commission of 1855 reported on this question, and recommended that on certain points the Scotch rule should be adopted in England, while on other points the English rule should be adopted in Scotland. The recommendations of the Commission were partially embodied in the English and Scotch Mercantile Law Amendment Acts of 1856. The result was curious. Either by accident or design certain rules were enacted for England which resembled, but did not reproduce, the Scotch law, while other rules were enacted for Scotland which resembled, but did not reproduce, the English law. The present Act has carried the process of assimilation somewhat further. It is perhaps to be regretted that the process has not been completed; but future legislation may accomplish that. It is always easier to amend an Act than to alter common law. Legislation, too, is cheaper than litigation. Moreover, in mercantile matters, the certainty of the rule is often of more importance than the substance of the rule (*a*). If the parties know beforehand what their legal position is, they can provide for their particular wants by express stipulation. Sale is a consensual contract, and the Act does not seek to prevent the parties from making any bargain they please.

(*a*) Cf. *Lockyer* v. *Offley* (1786), 1 T. R. 252, at p. 259.

Its object is to lay down clear rules for the case where the parties have either formed no intention, or failed to express it.

As regards this edition, I have not attempted to expound or criticise the mass of cases which illustrate or are modified by the Act. Such a work could hardly be undertaken with any prospect of success until the Act has been for some time in operation. I have only sought to indicate the sources of the various provisions in the Act, and to elucidate the general principles of the law of sale by citations from eminent judges. Our common law is rich in the exposition of principles, and these expositions lose none of their value now that the law is codified. A rule can never be appreciated apart from the reasons on which it is founded.

I have compared the main propositions of the English law with the corresponding provisions of the Code Napoleon, which is the model on which most of the Continental Codes have been framed (*b*). On the one hand, the scope and effect of a principle are often best brought out by contrast; on the other hand, where any rule of municipal law is found to be generally adopted in other countries, there is a strong presumption that the rule is founded on broad grounds of expediency, and that its application should not be narrowed. The Roman lawyers were justified in attaching a peculiar value to those rules of law which were *juris gentium*. I have also made frequent reference to Pothier's *Traité du Contrat de Vente*. Although published more than a century ago—for Pothier died in 1772—it is still, probably, the best reasoned treatise on the Law of Sale that has seen the light of day. "The authority of Pothier," says BEST, C.J., "is as high as can be had next to the decision of a court of justice in this country" (*c*). This statement must obviously be taken with the qualification that it only holds good when Pothier is discussing some principle of general application; for the law he was particularly dealing with was French law, as modified by the custom of Orleans, before the Code Napoleon.

(*b*) Since the fourth edition of this book was published, Germany has enacted a new civil code for her empire. The German Civil Code of 1900 is a notable addition to legal theory and practice. Its main provisions relating to sale are collected in a sub-heading of the chapter dealing with "Obligatory Relations." But the scheme of the German Code is so widely different from the lines on which the Sale of Goods Act is framed, that any attempt at comparison would be futile, unless explained by a detailed discussion, which would be unsuitable to a small manual such as this.

(*c*) *Cox* v. *Troy* (1822), 5 B & Ald. 481; *cf. M'Lean* v. *Clydesdale Bank* (1883), 9 App. Cas., at p. 105, *per* LORD BLACKBURN.

The references to the Civil Law need little comment. It is the foundation of the Scottish law, and it is an inexhaustible store of legal principles. There is hardly a judgment of importance on the law of sale in which reference is not made to the Civil Law. "The Roman law," says TINDAL, C.J., "forms no rule binding in itself on the subjects of these realms; but in deciding a case upon principle, where no direct authority can be cited from our books, it affords no small evidence of the soundness of the conclusion at which we have arrived, if it prove to be supported by that law—the fruit of the researches of the most learned men, the collective wisdom of ages, and the groundwork of the municipal law of most of the countries of Europe" (*d*). My task of reference in this edition has been much facilitated by Dr. Moyle's excellent monograph on the *Contract of Sale in the Civil Law*, published in 1892.

To facilitate reference to contemporaneous reports, the date of each case cited has been given.

M. D. CHALMERS.

BIRMINGHAM COUNTY COURT,
1894.

(*d*) *Acton* v. *Blundell* (1843), 12 M. & W., at p. 324; *cf. Keighley* v. *Durant*, [1901] A. C., at p. 244.

CONTENTS

PAGE
PREFACE TO SEVENTEENTH EDITION . . . v
INTRODUCTION TO FIRST EDITION (1894) . . . vii
TABLE OF STATUTES . . . xv
TABLE OF CASES . . . xxi

SECTION A
RELATIONSHIP OF SALE AND CONTRACT 1

SECTION B
INTERNATIONAL CONTRACTS OF SALE 33

SECTION C
STATUTORY CONTROLS OVER THE SALE OF GOODS
I STATUTORY PRICE CONTROLS . . . 51
II CONSUMER PROTECTION . . . 54

SECTION D
THE SALE OF GOODS ACT, 1893 73

PART I
FORMATION OF THE CONTRACT
Contract of Sale

SECT.
1. Sale and agreement to sell . . . 74
2. Capacity to buy and sell . . . 85

Formalities of the Contract

3. Contract of sale, how made . . . 93
[4. Contract of sale for £10 and upwards—*Repealed*] . . . 94

Subject Matter of Contract

5. Existing or future goods . . . 94
6. Goods which have perished . . . 97
7. Goods perishing before sale but after agreement to sell 99

The Price

8. Ascertainment of price . . . 100
9. Agreement to sell at valuation . . . 103

Conditions and Warranties

10. Stipulations as to time . . . 105
11. When condition to be treated as warranty . . . 108
12. Implied undertaking as to title, &c. . . . 111
13. Sale by description . . . 117
14. Implied undertakings as to quality or fitness . . . 122
15. Sale by sample . . . 137

PART II

EFFECTS OF THE CONTRACT

SECT. *Transfer of Property as between Seller and Buyer* PAGE

16. Goods must be ascertained 140
17. Property passes when intended to pass 143
18. Rules for ascertaining intention 146
19. Reservation of right of disposal 154
20. Risk *primâ facie* passes with property 157

Transfer of Title

21. Sale by person not the owner 161
22. Market overt 169
23. Sale under voidable title 171
24. [Revesting of property in stolen goods on conviction of offender—*Repealed*] 175
25. Seller or buyer in possession after sale 175
26. Effect of writs of execution 177

PART III

PERFORMANCE OF THE CONTRACT

27. Duties of seller and buyer 179
28. Payment and delivery are concurrent conditions . . 180
29. Rules as to delivery 182
30. Delivery of wrong quantity 186
31. Instalment deliveries 188
32. Delivery to carrier 191
33. Risk where goods are delivered at distant place . . 194
34. Buyer's right of examining the goods 195
35. Acceptance 197
36. Buyer not bound to return rejected goods 203
37. Liability of buyer for neglecting or refusing delivery of goods 204

PART IV

RIGHTS OF UNPAID SELLER AGAINST THE GOODS

38. Unpaid seller defined 205
39. Unpaid seller's rights 206
40. Attachment by seller in Scotland 208

Unpaid Seller's Lien

41. Seller's lien 208
42. Part delivery 210
43. Termination of lien 211

Stoppage in Transitu

44. Right of stoppage in transitu 213
45. Duration of transit 214
46. How stoppage in transitu is effected 219

SECT. *Re-sale by Buyer or Seller* PAGE

47. Effect of sub-sale or pledge by buyer 220
48. Sale not generally rescinded by lien or stoppage in transitu. 223

PART V

ACTIONS FOR BREACH OF THE CONTRACT

Remedies of the Seller

49. Action for price 226
50. Damages for non-acceptance 231

Remedies of the Buyer

51. Damages for non-delivery 235
52. Specific performance 243
53. Remedy for breach of warranty 245
54. Interest and special damages 254

PART VI

SUPPLEMENTARY

55. Exclusion of implied terms and conditions 259
55A. Conflict of laws 265
56. Reasonable time a question of fact 266
57. Rights, &c., enforceable by action 266
58. Auction sales 266
59. Payment into Court in Scotland when breach of warranty alleged 270
[60. Repeals—*Repealed.*] 271
61. Savings 271
62. Interpretation of terms 273
[63. Commencement—*Repealed.*] 288
64. Short title 288

THE FACTORS ACT, 1889 289

Preliminary 290
Dispositions by Mercantile Agents 295
Dispositions by Sellers and Buyers of Goods 304
Supplemental 311

THE FACTORS (SCOTLAND) ACT, 1890 . 315

APPENDIX I.—STATUTES

Bills of Lading Act, 1855 316
Bills of Sale Act, 1878, s. 4 319
Finance Act, 1901, s. 10 322
Law Reform (Miscellaneous Provisions) Act, 1934, s. 3 . . 323
Law Reform (Frustrated Contracts) Act, 1943 326
Exchange Control Act, 1947, s. 33 (1) 330
Misrepresentation Act, 1967 332
Theft Act, 1968 339
The Auctions (Bidding Agreements) Act, 1969, ss. 3, 4 . . 343
Finance Act, 1972, s. 42 344
Powers of Criminal Courts Act, 1973, s. 35 345

TABLE OF STATUTES

References in this Table to "*Stats*" are to Halsbury's Statutes of England (Third Edition) showing the volume and page at which the annotated text of the Act will be found. Page references printed in bold type indicate where the Act is set out in part or in full.

PAGE

Consumer Credit Act 1974 (44 *Stats*. 746)—*contd*.
11–13 63
16 . . . . . 3, 63
(5) 64
20 63
21–42, 55 64
67–73, 76–79 . . . 64
94–104, 107–110 . . 64
137, 138 63,65
s. 139 63
(2) 65
140 63
145, 147–150, 158, 159 . 64
161 (1), (2), (4) . . . 71
162, 163 72
164 71
165 72
166 70
167 69
168 67
169 65
170 55
171 (7) 65
173 261
181 . . . . . 63, 64
187 63
189 (1) 63, 64
(2) . . . 64, 274
192 312
(1) 63
(2) 137
(3) 136
(3) 4, 94, 103, 111, 124, 127, 136, 176, 180, 308
194 (4) 82
Sch. 1 69
3 . . . 4, 63, 103, 180
para. 44 . . . 137
4 . 111, 124, 127, 136, 176, 308, 312
5 . 82, 94, 103, 124, 180

Consumer Protection Act 1961 (30 *Stats*. 47)—
s. 1 (a) 55
2 55
(1), (2) 55
3 55
(2) 66
(2B) 67, 68
(3) 65

Consumer Protection Act 1971 (41 *Stats*. 1275)—
s. 1 66
3 (2) 66
(3) 65

PAGE

Corporate Bodies' Contracts Act 1960 (7 *Stats*. 14) . . 94
s. 1 (3) 94

Counter-Inflation Act 1973 (43 *Stats*. 1577) . . . 2, 51
s. 2, 5 51
6 51
(4), (5) 51
15, 17 51
Sch. 3, para. 3. . . . 51
4 51

County Courts Act 1888—
s. 156 163

County Courts Act 1959 (7 *Stats*. 302)—
s. 1 (2) 329
97 50
135 163
140 (3) 179

Criminal Appeal Act 1968 (8 *Stats*. 687)—
s. 30 . . 340, 341, 347
(4) 341
42 . . 340, 341, 347

Criminal Justice Act 1972 (42 *Stats*. 99)—
s. 1 346
6 (2) 340, 341
(3)–(5) 341

Criminal Law Act 1967 (8 *Stats*. 552)—
Sch. 3, Pt. I 171
Pt. III . . 169, 171

Currency and Bank Notes Act 1954 (2 *Stats*. 770)—
s. 1 228

Customs and Excise Act 1952 (9 *Stats*. 55)—
s. 1 (1) 323

Decimal Currency Act 1969 (40 *Stats*. 316)—
Sch. 2 80

European Communities Act 1972 (42 *Stats*. 78)—
s. 2 52

Exchange Control Act 1947 (22 *Stats*. 900) . . . **330–332**
s. 33 42
(1) 81, 109

Factors Act 1823—
s. 14 313

PAGE

Factors Act 1825 . . 293, 313
s. 1 304
2 296
3 302
4 . . . . 296, 301

Factors Act 1842 . . 164, 313
s. 1 299
2 . . . . 303, 308
3 302
4 292, 293, 296, 303, 308

Factors Act 1877 . 221, 290, 313
s. 2 300
5 311

Factors Act 1889 (1 *Stats*. 94) 162, 183, 206, 278, **289–314**, 315, 319
s. 1 (2) 177, 278
(5) 83
2 173
5 79
8 . 115, 177, 241, 243
9 . 34, 82, 115, 173, 177, 241, 243
10 . . . . 222, 241

Factors (Scotland) Act 1890 . 162, 278, 282, 314, **315**
s. 1 222
(2) 303

Fair Trading Act 1973 (43 *Stats*. 1618) . . . 54, 65, 70
s. 1–3 61
13, 14, 17 62
22 . . . 55, 62, 71
23 . . 62, 65, 69, 71
24 66
25 67
26 . . . . 54, 55
27, 28 71
29, 30, 32 72
34 (1)–(3) 62
35 62
37 63
41 62
130 71
131 70
132 65
137 280

Farm and Garden Chemicals Act 1967 (1 *Stats*. 577) . 54

Finance Act 1901 (9 *Stats*. 33)—
s. 1 345
10 . . . . **2**, **322**

Finance Act 1902 (9 *Stats*. 35)—
s. 7 322

PAGE

Finance Act 1908 (19 *Stats*. 340)—
s. 4 323

Finance Act 1972 (42 *Stats*. 159)—
s. 42 . . . 2, 323, **344–345**

Finance (No. 2) Act 1964 (9 *Stats*. 380)—
s. 3 323
(8) 323

Food and Drugs Act 1955 (14 *Stats*. 15; 21 *Stats*. 478) 54, 56, 65, 70
s. 8 (4) 69
9 72
(4) 72
87, 88 70
91 71
100, 105 72
106, 107, 109 . . . 70
113 (1) 67, 68
(3) 67
115 (1) 283
127 70
211 72

Food and Drugs (Adulteration) Act 1928 54

Hire-Purchase Act 1964 (30 *Stats*. 55)—
s. 27 . 162, 296, 308, 309
28 . . 162, 308, 309
29 . . . . 162, 308
(2) 296

Hire-Purchase Act 1965 (30 *Stats*. 1) . 4, 26, 82, 94, 103, 107, 111, 180, 225
s. 5 . . . . 4, 94, 103
6 . . . . 4, 94, 103
7–10 . . . . . 4, 94
53 . . . . 162, 312
54 . . . . 176, 308
58 (3), (4) 124

Hire-Purchase Act (Northern Ireland) 1966—
s. 65 (3) 124

Hire-Purchase (Scotland) Act 1965—
s. 54 (3), (5) 124

Hotel Proprietors Act 1956 (17 *Stats*. 816) 162

Infants Relief Act 1874 (17 *Stats*. 412) 89
s. 1 87

Innkeepers Act 1878 (17 *Stats*. 815) 162

PAGE

Interpretation Act 1889 (32 *Stats.* 434) 73
s. 1 74
19 . . . 74, 294
20 . . . . 74, 93
Landlord and Tenant Act 1954 (18 *Stats.* 553, 726) . . . 75
Larceny Act 1916—
s. 45 341
Law of Property Act 1925 (27 *Stats.* 341)—
s. 61 105
72 75
136 28
188 168
Law Reform (Enforcement of Contracts) Act 1954 (7 *Stats.* 14)—
s. 1 93
Law Reform (Frustrated Contracts) Act 1943 (7 *Stats.* 9) . 90, 109, **326–330**
s. 1 . . . . 20, 100
2 21
(2), (5) 21
Law Reform (Miscellaneous Provisions) Act 1934 (25 *Stats.* 752)—
s. 3 . 26, 28, 46, 231, 255, **323–326**
Laws in Wales Act 1542 (6 *Stats.* 460)—
s. 47 170
Limitation Act 1939 (19 *Stats.* 60)—
s. 26 333
Magistrates' Court Act 1952 (21 *Stats.* 181)—
s. 29 346
104 71
Marine Insurance Act 1906 (17 *Stats.* 836)—
s. 31 103
55 194
Married Women (Restraint upon Anticipation) Act 1949 (17 *Stats.* 131) . . . 89
Matrimonial Proceedings and Property Act 1970 (40 *Stats.* 759)—
s. 41 (1) 92
Matrimonial Proceedings (Magistrates' Courts) Act 1960 (17 *Stats.* 241) . . 92

PAGE

Mental Health Act 1959 (25 *Stats.* 42) 86
Mercantile Law Amendment (Scotland) Act 1856—
s. 1 . . . . 145, 178
2 208
3 208
6 127
Merchandise Marks Act 1926 . 54
Merchant Shipping Act 1894 (31 *Stats.* 57) 283
s. 24, 26, 65 93
Merchant Shipping Act 1965 (31 *Stats.* 678)—
Sch. 2 93
Misrepresentation Act 1967 (22 *Stats.* 675) . 198, **332–338**
s. 1 6
2 (1), (2) 7
3 . . . . 7, 264
4 (1) 109
(2) . . . 196, 197
Partnership Act 1890 (24 *Stats.* 500)—
s. 5, 6 168
Plant Varieties and Seeds Act 1964 (1 *Stats.* 630)—
s. 17 54
(5) 55
Post Office Act 1969 (25 *Stats.* 470)—
s. 64 218
Powers of Criminal Courts Act 1973 (43 *Stats.* 288)—
s. 35 . . 27, 55, 69, **345**
38 346
Prices Act 1974 . . . 2, 51
Prices Act 1975 . . 51, 61
Profiteering Prevention Act 1948 (Australia) . . . 104
Profiteering Prevention Act 1959 (Australia) . . . 104
Purchase Tax Act 1963 (26 *Stats.* 633)—
s. 35 345
Registration of Business Names Act 1916 (37 *Stats.* 867) . 91
s. 8 91
Resale Prices Act 1964 (37 *Stats.* 158)—
s. 1, 5 52
11 (1) 52

PAGE

Restrictive Trade Practices Act 1956 (37 *Stats.* 77) . . .52, 65
s. 25 (1), (2), (4) . . 52
26 (1) 52

Sale of Horses Act 1555 . 171

Sale of Horses Act 1588 . 171

Sale of Land by Auction Act 1867 (2 *Stats.* 499) . . 268
s. 5 269

Sheriffs Act 1887 (30 *Stats.* 577)—
s. 10, 29 178

Stamp Duty Act 1891—
Sch. 1 77

Statute Law Revision Act 1908 . . 271, 288, 313, 314

Statute of Frauds (1677) (7 *Stats.* 6)—
s. 4 . . . . 49, 198
15, 16 178
17 293

Statute of Frauds (Amendment) Act 1828 (7 *Stats.* 7)—
s. 7 95

Supply of Goods (Implied Terms) Act 1973 (43 *Stats.* 1337 263
s. 1 . . . . 112, 113
2 117
3 . . . . 124, 127
4 . . . . 10, 261
5 (1) 265
(2) 50
6 272
7 (1) 274, 275
(2) 129, 286
14 111
(1), (2) 111
15 111
(1) 111
16 (1) 80

Supreme Court of Judicature Act 1873—
s. 25 (6) 28

Supreme Court of Judicature (Consolidation) Act 1925 (7 *Stats.* 573)—
s. 102 50
225 273

Theft Act 1968 (8 *Stats.* 782) 165, 174, 297, 339–342, 345
s. 13 280
28 . . . . 175, 347
31 (2) 175
34 (2) 280
Sch. 3, Pt. III . 175, 342

PAGE

Trade Descriptions Act 1968 (37 *Stats.* 948) . 5, 7, 54, 65, 66, 68, 69, 99, 103, 121, 130, 346
s. 1 . . 5, 56, 60, 61,66
(1) 58, 68
(2), (5) 60
2 . . . 56, 57, 61
(3)–(5) . . . 56
3 61
(2), (3) . . . 57
4 (1)–(3) . . . 58
5 (1)–(3) . . . 59
6 . . . 59, 60, 61
7 59
8, 9 60
11 (1)–(3) 60
12 66
18 69
19 71
(1), (2), (4) . . 71
20, 21 65
22 (2) 71
23 . . . . 66, 71
24 . . . 66, 67, 68
(3) 58
26 71
(3), (5) 71
27 71
28 72
29, 33 72
35 55
39 (1) 59

Trade Descriptions Act 1972 (42 *Stats.* 2249) . 54, 65
s. 1 60

Trading Stamps Act 1964 (37 *Stats.* 170)—
s. 3, 4 80
10 280

Trading with the Enemy Act 1939 (38 *Stats.* 359) . . 89
s. 7 90

Uniform Law on International Sales Act 1967 (30 *Stats.* 128)—
s. 1 (4) 50
(5) 276

Unsolicited Goods and Services Act 1971 (41 *Stats.* 1277) 54, 65
s. 1 61
(2) 61
2 (1), (2) 61
5 65

Weights and Measures Act 1963 (39 *Stats.* 720) . . 54, 65, 71
s. 9 55
21 56

Weights and Measures Act 1963
(39 *Stats.* 720)—*contd.*

	PAGE
22	56
(1)	56
23	56
24	56
(1), (2)	56
26 (1)	67
27 (1)	67, 68
(4)	67
32	71
41	70
48, 49	72
50	65
51	70
52	69

TABLE OF CASES

In the following Table references are given to the English and Empire Digest where a digest of the case may be found.

PAGE

A.B.D. (Metals and Waste), Ltd. *v.* Anglo-Chemical and Ore Co., Ltd., [1955] 2 Lloyd's Rep. 456 233

A/S Tankexpress *v.* Compagnie Financiere Belge Des Petroles S.A., [1948] 2 All E. R. 939; [1949] A. C. 76; [1949] L. J. R. 170; 93 Sol. Jo. 26, H. L.; 41 Digest (Repl.) 221, *482* 357

Abernethie *v.* Kleiman, Ltd., [1969] 2 All E. R. 790; [1970] 1 Q. B. 10; [1969] 2 W. L. R. 1364; 113 Sol. Jo. 307; 20 P. & C. R. 56, C. A.; 31 (2) Digest (Reissue) 951, *7747* 274

Acebal *v.* Levy (1834), 10 Bing. 376; 4 M. & Scott, 217; 3 L. J. C. P. 98; 39 Digest (Repl.) 466, *176* 101

Acmé Wood Co. *v.* Sutherland Innes & Co. (1904), 9 Com. Cas. 170; 39 Digest (Repl.) 684, *1790* 359

Acraman *v.* Morrice (1849), 8 C. B. 449; 19 L. J. C. P. 57; 14 L. T. O. S. 292; 14 Jur. 69; 39 Digest (Repl.) 619, *1320* 147

Adair & Co., Ltd. *v.* Birnbaum, [1938] 4 All E. R. 775; [1939] 2 K. B. 149; 108 L. J. K. B. 452; 160 L. T. 244; 55 T. L. R. 234; 83 Sol. Jo. 31; 44 Com. Cas. 47, C. A.; 39 Digest (Repl.) 504, *491* . . 356

Adams *v.* Sprung (1961), The Guardian, 8th November, C. A. . . . 203

Agius *v.* Great Western Colliery Co., [1899] 1 Q. B. 413; 68 L. J. Q. B. 312; 80 L. T. 140; 47 W. R. 403, C. A.; 39 Digest (Repl.) 588, *1090* 253

Agricultores Federados Argentinos Sociedad Cooperative *v.* Ampro S. A. Commerciale Industrielle et Financiere, [1965] 2 Lloyd's Rep. 157 34

Agroexport State Enterprise for Foreign Trade *v.* Compagnie Européene de Céréals, [1974] 1 Lloyd's Rep. 499 42

Ahmad *v.* Ali, [1947] A. C. 414, P. C. 253

Aird *v.* Pullan 1904, 7 F. (Ct. of Sess.) 258; 39 Digest (Repl.) 582, **592* 109

Aitken & Co. *v.* Boullen, 1908, S. C. 490; 39 Digest (Repl.) 694, **1036* 138, 188

Ajayi (Emmanuel Ayodeji) (trading under name and style of Colony Carrier Co.) *v.* R. T. Briscoe (Nigeria), Ltd., [1964] 3 All E. R. 556; [1964] 1 W. L. R. 1326; 108 Sol. Jo. 857, P. C.; Digest Cont. Vol. B p. 252, *1213a* 23

Ajello *v.* Worsley, [1898] 1 Ch. 274; [1895–99] All E. R. Rep. 1222; 67 L. J. Ch. 172; 77 L. T. 783; 14 T. L. R. 168; 39 Digest (Repl.) 642, *1506* 52, 94, 279

Akerib *v.* Booth, [1961] 1 All E. R. 380; [1961] 1 W. L. R. 367; 105 Sol. Jo. 231, C. A.; 31 (1) Digest (Reissue) 316, *2541* 363

Alan (W. J.) & Co., Ltd. *v.* El Naso Export and Import Co., [1972] 2 All E. R. 127; [1972] 2 Q. B. 189; [1972] 2 W. L. R. 800; 116 Sol. Jo. 139; [1972] 1 Lloyd's Rep. 313, C. A.; 12 Digest (Reissue) 337, *2425* 23, 44

Albazero, The, Owners of *Albacruz v.* Owners of *Albazero*, [1974] 2 All E. R. 906 . . . . 37, 156, 157

Albeko Schumaschinen A.-G. *v.* Kamborian Shoe Machine Co., [1961] 111 L. Jo. 519 48

Alcock *v.* Smith, [1892] 1 Ch. 238; 61 L. J. Ch. 161; 66 L. T. 126; 8 T. L. R. 222; 36 Sol. Jo. 199, C. A.; 11 Digest (Reissue) 413, *480* 49

Aldridge *v.* Johnson (1857), 26 L. J. Q. B. 296; 7 E. & B. 885; 3 Jur. N. S. 913; 5 W. R. 703; 39 Digest (Repl.) 607, *1205* . . 79, 143, 148

PAGE
Alewyn *v.* Pryor (1826), Ry. & M. 406; 39 Digest (Repl.) 492, *401* . 356
Alexander *v.* Gardner (1835), 1 Bing. N. C. 671; 1 Scott 281; 4 L. J. C. P. 223; 39 Digest (Repl.) 603, *1182* 192, 226, 357
Alexander *v.* Rayson, [1936] 1 K. B. 169; 105 L. J. K. B. 148; 154 L. T. 205; 52 T. L. R. 131; 80 Sol. Jo. 15, C. A.; 31 (1) Digest (Reissue) 97, *778* 18
Alexander *v.* Vanderzee (1872), L. R. 7 C. P. 530; 20 W. R. 871; 39 Digest (Repl.) 519, *597* 355
Alexander *v.* Webber, [1922] 1 K. B. 642; 91 L. J. K. B. 320; 126 L. T. 512; 38 T. L. R. 42; 39 Digest (Repl.) 826, *2885*. 27
Alexandria Cotton and Trading Co. (Sudan), Ltd. *v.* Cotton Co. of Ethiopia, Ltd., [1963] 1 Lloyd's Rep. 576 . . . . . .43, 106
Algoma Truck and Tractor Sales *v.* Bert's Auto Supply Ltd., [1968] 2 O. R. 153; Digest Cont. Vol. C p. 855, **409a* 368
Alison & Co. *v.* Wallsend Slipway & Engineering Co. (1927), 43 T. L. R. 323, C. A.; 39 Digest (Repl.) 574, *1000* 10, 253
All Trades Distributors, Ltd. *v.* Agencies Kaufman, Ltd. (1969), 113 Sol. Jo. 995, C. A.. 231
Allard & Co. (Rubber), Ltd. *v.* R. J. Hawkins & Co., [1958] 1 Lloyd's Rep. 184 366
Allen *v.* Hopkins (1844), 13 M. & W. 94; 13 L. J. Ex. 316; 3 L. T. O. S. 204; 39 Digest (Repl.) 526, *639*. 31
Allen *v.* Pink (1838), 4 M. & W. 140; 1 H. & N. 207; 7 L. J. Ex. 206; 17 Digest (Reissue) 382, *1468* 5
Allen *v.* Stein (1790), M. 4949. 208
Alves *v.* Hodgson (1797), 7 Term Rep. 241; 2 Esp. 527; 11 Digest (Reissue) 469, *805* 49
Amco *v.* Wade, [1968] Qd. R. 445 281
American Commerce Co. *v.* Boehm, Ltd. (1919), 35 T. L. R. 224; 39 Digest (Repl.) 502, *478* 323, 360
Amos and Wood, Ltd. *v.* Kaprow, [1948] W. N. 71; 64 T. L. R. 110; 92 Sol. Jo. 153, C. A.; 39 Digest (Repl.) 496, *432* 106
Anchor Line (Henderson, Brothers), Ltd., *Re.*, [1936] 2 All E. R. 941; [1937] Ch. 1; 105 L. J. Ch. 330; 155 L. T. 100; 80 Sol. Jo. 572, C. A.; 39 Digest (Repl.) 448, *35* 144, 146, 224
Ancona *v.* Rogers (1876), 1 Ex. D. 285; [1874–80] All E. R. Rep. 369; 46 L. J. Ex. 121; 35 L. T. 115, C. A.; 7 Digest (Repl.) 116, *691* 184, 321
Anderson *v.* Croall & Sons, Ltd., 1904, 11 S. L. T. 163; 6 F. (Ct. of Sess.) 153; 3 Digest (Repl.) 48, **190*. 268
Anderson *v.* Morice (1874), L. R. 10 C. P. 58; 44 L. J. C. P. 10; 31 L. T. 605; 23 W. R. 180; 2 Asp. M. L. C. 424; reversed (1875), L. R. 10 C. P. 609; 44 L. J. C. P. 341; 33 L. T. 355; affirmed (1876), 1 App. Cas. 713; 46 L. J. C. P. 11; 35 L. T. 566, H. L.; 39 Digest (Repl.) 621, *1335* 152,161, 363
Anderson *v.* Ryan, [1967] I. R. 34; Digest Cont. Vol. C p. 859, **1597a* 116
Andrews *v.* Belfield (1857), 2 C. B. N. S. 799; 39 Digest (Repl.) 451, *55* 365
Andrews *v.* Hopkinson, [1956] 3 All E. R. 422; [1957] 1 Q. B. 229; [1956] 3 W. L. R. 732; 100 Sol. Jo. 768; 26 Digest (Repl.) 666, *36* 122
Andrews Bros., Ltd. *v.* Singer & Co., Ltd., [1934] 1 K. B. 17; [1933] All E. R. Rep. 479; 103 L. J. K. B. 90; 150 L. T. 172; 50 T. L. R. 33, C. A.; 39 Digest (Repl.) 572, *987* 119, 361, 367
Anglesey (Marquis), *Re.*, Willmot *v.* Gardner, [1901] 2 Ch. 548; 70 L. J. Ch. 810; 85 L. T. 179; 49 W. R. 708; 45 Sol. Jo. 738, C. A.; 12 Digest (Reissue) 144, *835*. 15, 324
Anglia Television *v.* Reed, [1971] 3 All E. R. 690; [1972] 1 Q. B. 60; [1971] 3 W. L. R. 528; 115 Sol. Jo. 723, C. A.; 17 Digest (Reissue) 111, *165* 251, 252
Anglo-African Shipping Co., Ltd. *v.* J. Mortimer, Ltd., [1962] 1 Lloyd's Rep. 81; on appeal, [1962] 1 Lloyd's Rep. 610, C. A. 34, 46

PAGE

Anglo-Cyprian Trade Agencies, Ltd. *v.* Paphos Wine Industries, Ltd., [1951] 1 All E. R. 873; 95 Sol. Jo. 336; 51 Digest (Repl.) 917, *4644* 256

Anglo-Egyptian Navigation Co. *v.* Rennie (1875), L. R. 10 C. P. 271; 44 L. J. C. P. 130; 32 L. T. 467; 23 W. R. 626; on appeal, L. R. 10 C. P. 571; 39 Digest (Repl.) 444, *5* 78

Anglo-Russian Merchant Traders & John Batt & Co. (London), *Re*, [1917] 2 K. B. 679; 86 L. J. K. B. 1360; 116 L. T. 805; 61 Sol. Jo. 591, C. A.; 12 Digest (Reissue) 500, *3496* 42

Angus *v.* McLachlan (1883), 23 Ch. D. 330; [1881–5] All E. R. Rep. 845; 52 L. J. Ch. 587; 48 L. T. 863; 32 Digest (Repl.) 275, *211* . . 213

Annie Johnson, The, [1918] P. 154; 87 L. J. P. 127; 118 L. T. 721; 14 Asp. M. L. C. 301; 39 Digest (Repl.) 648, *1543* 154

Anthony *v.* Halstead (1877), 37 L. T. 433; 2 Digest (Repl.) 355, *386* . 368

Ant. Jurgens Margarinefabreiken *v.* Louis Dreyfus & Co., [1914] 3 K. B. 40; 83 L. J. K. B. 1344; 111 L. T. 248; 19 Com. Cas. 333; 39 Digest (Repl.) 747, *2248* 221

Appleby *v.* Myers (1867), L. R. 2 C. P. 651; [1861–73] All E. R. 452; 36 L. J. C. P. 331; 16 L. T. 669; 12 Digest (Reissue) 764, *5461* . 95, 100

Appleby *v.* Sleep, [1968] 2 All E. R. 265; [1968] 1 W. L. R. 948; 112 Sol. Jo. 380; 66 L. G. R. 555; Digest Cont. Vol. C p. 386, *188a* 2

Arab Bank, Ltd. *v.* Barclays Bank (D. C. & O.), [1954] 2 All E. R. 226; [1954] A. C. 495; [1954] 2 W. L. R. 1022; 98 Sol. Jo. 350, H. L.; 2 Digest (Repl.) 223, *343* 90

Archbolds (Freightage), Ltd. *v.* S. Spanglett, Ltd., [1961] 1 All E. R. 417; [1961] 1 Q. B. 374; [1961] 2 W. L. R. 170; 105 Sol. Jo. 149, C. A.; 12 Digest (Reissue) 331, *2388* 18

Arcos *v.* E. & A. Ronaasen & Son, [1933] A. C. 470; [1933] All E. R. Rep. 646; 102 L. J. K. B. 346; 149 L. T. 98; 49 T. L. R. 231; 77 Sol. Jo. 99; 38 Com. Cas. 166, H. L.; 39 Digest (Repl.) 718, *2041* 122, 366

Arcos, Ltd. and Russo-Norwegian Onega Wood Co. *v.* Aronson (1930) 36 Ll. L. Rep. 108 100

Ardath Tobacco Co. *v.* Ocker (1931), 47 T. L. R. 177; 33 Digest (Repl.) 491, *482* 170

Argentina, The (1867), L. R. 1 A. & E. 370; 16 L. T. 743; 1 Digest (Repl.) 186, *723* 319

Armaghdown Motors, Ltd. *v.* Gray Motors, Ltd., [1963] N. Z. L. R. 5: 119, 367

Armory *v.* Delamirie (1721), 1 Str. 505; 3 Digest (Repl.) 67, *83* . . 164

Armstrong *v.* Jackson, [1917] 2 K. B. 822; [1916–17] All E. R. Rep. 1117; 86 L. J. K. B. 1375; 177 L. T. 479; 33 T. L. R. 444; 1 Digest (Repl.) 532, *1610* 75

Arnhold Karberg *v.* Blythe Green & Co. *See* Karberg (Arnhold) *v.* Blythe Green & Co.

Arnold *v.* Cheque Bank (1876), 1 C. P. D. 578; 45 L. J. C. P. 562; 34 L. T. 729; 40 J. P. 711; 24 W. R. 759; 3 Digest (Repl.) 260, *738*. 31

Arnold Meyer (N. V.) *v.* Aune. *See* Meyer (N. V. Arnold Otto) *v.* Aune

Aron & Co. *v.* Comptoir Wegimont, [1921] 3 K. B. 435; 90 L. J. K. B. 1233; 26 Com. Cas. 303; 39 Digest (Repl.) 521, *603* . 107, 108, 355

Aronson *v.* Mologa Holzindustrie A.-G. Leningrad (1927), 32 Com. Cas. 276, C. A.; 39 Digest (Repl.) 826, *2879* 242

Aruna Mills, Ltd. *v.* Dhanrajmal Gobindram (or Dhrajmal Gobindram), [1968] 1 All E. R. 113; [1968] 1 Q. B. 655; [1968] 2 W. L. R. 101; 111 Sol. Jo. 924; [1968] 1 Lloyd's Rep. 304; 17 Digest (Reissue) 142, *363* 239, 256, 258

Aryeh *v.* Lawrence Kostoris & Son, Ltd., [1967] 1 Lloyd's Rep. 63, C. A 237, 248

Ashford Shire Council *v.* Dependable Motors Pty., Ltd., [1961] All E. R. 96; [1961] A. C. 336; [1960] 3 W. L. R. 999; 104 Sol. Jo. 1055, P. C.; 39 Digest (Repl.) 547, *805* 123

PAGE

Ashforth *v.* Redford (1873), L. R. 9 C. P. 20; 43 L. J. C. P. 57; 39 Digest (Repl.) 734, *2132* 359
Ashington Piggeries, Ltd. *v.* Christopher Hill, Ltd. *See* Hill (Christopher) Ltd. *v.* Ashington Piggeries, Ltd.
Ashmore *v.* Cox, [1899] 1 Q. B. 436; 68 L. J. Q. B. 72; 4 Com. Cas. 48; 15 T. L. R. 55; 39 Digest (Repl.) 520, *602* 235, 355
Ashmore Benson Pease & Co., Ltd. *v.* A. V. Dawson, Ltd., [1973] 2 All E. R. 856; [1973] 1 W. L. R. 828; 117 Sol. Jo. 203; [1973] 2 Lloyd's Rep. 21, C. A. 18
Ashworth *v.* Wells (1898), 78 L. T. 136; 14 T. L. R. 227, C. A.; 39 Digest (Repl.) 591, *1117* 247, 249, 251
Asiatic Petroleum Co. *v.* Lennard's Carrying Co., [1914] 1 K. B. 419; 85 L. J. K. B. 1075; 114 L. T. 645; 32 T. L. R. 367; 18 Com. Cas. 23, 328, C. A.; 41 Digest (Repl.) 295, *1073*. 279
Assunzione, The, [1954] 1 All E. R. 278; [1954] P. 150; [1954] 2 W. L. R. 234; 98 Sol. Jo. 107, C. A.; 11 Digest (Reissue) 461, *768* 47
Astley Industrial Trust *v.* Grimley, [1963] 2 All E. R. 33; [1963] 1 W. L. R. 584; 107 Sol. Jo. 474, C. A.; Digest Cont. Vol. A p. 645, *36ba* 122, 367
Astley Industrial Trust, Ltd. *v.* Miller (Oakes, third party), [1968] 2 All E. R. 36; Digest Cont. Vol. C p. 1036, *588a* . 242, 296, 298
Astra Trust, Ltd. *v.* Adams and Williams, [1969] 1 Lloyd's Rep. 81; Digest Cont. Vol. C p. 887, *3877a* 365
Atkinson *v.* Bell (1828), 8 B. & C. 277; 2 M. & Ry. 292; 39 Digest (Repl.) 612, *1250* 151
Attenborough *v.* St. Katherine's Dock Co. (1878), 3 C. P. D. 450; 47 L. J. C. P. 763; 38 L. T. 404, C. A.; 39 Digest (Repl.) 653, *1567* . 173
Attenborough *v.* Solomon, [1913] A. C. 76; [1911–13] All E. R. Rep. 155; 82 L. J. Ch. 178; 107 L. T. 833; 29 T. L. R. 79, H. L.; 23 Digest (Repl.) 394, *4649* 283
Attenborough & Inland Revenue Comrs., *Re* (1855), 11 Exch. 461; 37 Digest (Repl.) 7, *34* 82
Attorney-General *v.* De Keyser's Royal Hotel, [1920] A. C. 508; [1920] All E. R. Rep. 80; 89 L. J. Ch. 417; 122 L. T. 691; 36 T. L. R. 600, H. L.; 17 Digest (Reissue) 485, *100* 31
Attorney-General *v.* Pritchard (1928), 97 L. J. K. B. 561; 44 T. L. R. 490; 26 Digest (Repl.) 674, *72* 81, 180
Attwood *v.* Emery (1856), 1 C. B. N. S. 110; 26 L. J. C. P. 73; 5 W. R. 19; *sub nom.* Atwood *v.* Emery, 28 L. T. O. S. 85; 39 Digest (Repl.) 680, *1747* 358
Avery *v.* Bowden (1855), 5 E. & B. 714; (1856), 6 E. & B. 953; 26 L. J. Q. B. 3; 28 L. T. O. S. 145; 3 Jur. N. S. 238; 5 W. R. 45; 12 Digest (Reissue) 418, *3055* 238
Azémar *v.* Casella (1867), L. R. 2 C. P. 677; 36 L. J. C. P. 263; 16 L. T. 571, Ex. Ch.; 39 Digest (Repl.) 534, *695* . . . 117, 367

B. and P. Wholesale Distributors *v.* Marko, [1953] C. L. Y. 3266 195, 197
Badham *v.* Lambs, Ltd., [1945] 2 All E. R. 295; [1946] K. B. 45; 115 L. J. K. B. 180; 173 L. T. 139; 61 T. L. R. 569; 89 Sol. Jo. 381; 45 Digest (Repl.) 77, *242* 54
Badische Anilin und Soda Fabrik *v.* Basle Chemical Works, [1898] A. C. 200; 67 L. J. Ch. 14; 77 L. T. 573; 14 T. L. R. 82; 39 Digest (Repl.) 609, *1223* 149, 191, 218
Badische Anilin und Soda Fabrik *v.* Hickson, [1906] A. C. 419; 75 L. J. Ch. 621; 95 L. T. 68; 22 T. L. R. 641; 39 Digest (Repl.) 605, *1189* 141, 144
Bagueley *v.* Hawley (1867), L. R. 2 C. P. 625; 36 L. J. C. P. 328; 17 L. T. 116; 39 Digest (Repl.) 525, *638*. 95

PAGE

Baily *v.* De Crespigny (1869), L. R. 4 Q. B. 180; 38 L. J. Q. B. 98; 19 L. T. 681; 12 Digest (Reissue) 463, *3336* 109
Baindail (otherwise Lawson) *v.* Baindail, [1946] 1 All E. R. 342; [1946] P. 122; 115 L. J. P. 65; 174 L. T. 320; 62 T. L. R. 263; 90 Sol. Jo. 151, C. A.; 11 Digest (Reissue) 505, *985* 91
Baird *v.* Wells (1890), 44 Ch. D. 661; [1886–90] All E. R. Rep. 666; 59 L. J. Ch. 673; 63 L. T. 312; 8 (2) Digest (Reissue) 623, *51* . . 80
Baker *v.* Firminger (1859), 28 L. J. Ex. 130; 39 Digest 540, *1512* . 182
Baker *v.* Keen (1819), 2 Stark. 501; 28 (2) Digest (Reissue) 748, *829* . 92
Baldey *v.* Parker (1823), 2 B. & C. 37; 3 D. & R. 220; 1 L. J. O. S. K. B. 229; 39 Digest (Repl.) 455, *80* 212, 213
Baldry *v.* Marshall, [1925] 1 K. B. 260; [1924] All E. R. Rep. 155; 94 L. J. K. B. 192; 132 L. T. 326, C. A.; 39 Digest (Repl.) 551, *827* 10, 110, 133, 360
Ballantine & Co. *v.* Cramp and Bosman (1923), 129 L. T. 502; [1923] All E. R. Rep. 579; 39 Digest (Repl.) 531, *685* 190, 365
Ballett *v.* Mingay, [1943] 1 All E. R. 143; [1943] K. B. 281; 112 L. J. K. B. 193; 168 L. T. 34, C. A.; 28 (2) Digest (Reissue) 707, *425* 88
Bank Line *v.* Capel & Co., [1919] A. C. 435; [1918–19] All E. R. Rep. 504; 88 L. J. K. B. 211; 120 L. T. 129, H. L.; 12 Digest (Reissue) 490, *3461* 20
Bank of England *v.* Vagliano Brothers, [1891] A. C. 107; [1891–4] All E. R. Rep. 93; 60 L. J. Q. B. 145; 64 L. T. 353; 7 T. L. R. 333; 39 W. R. 657; H. L.; 3 Digest (Repl.) 179, *302* 73
Bankart *v.* Bowers, [1866] L. R. 1, C. P. 484; 39 Digest (Repl.) 662, *1627* 181
Banner, *Ex parte,* [1876] 2 Ch. D. 278; 47 L. J. Bk. 73; 34 L. T. 199; 39 Digest (Repl.) 635, *1457* 156, 226
Bannerman *v.* White (1861), 10 C. B. N. S. 844; 31 L. J. C. P. 28; 4 L. T. 740; 8 Jur. N. S. 282; 9 W. R. 784; 39 Digest (Repl.) 554, *842* 5, 109, 245, 353
Banque Belge Pour L'Etranger *v.* Hambrouck, [1921] 1 K. B. 321, 326; 90 L. J. K. B. 322; 26 Com. Cas. 72; 37 T. L. R. 76, C. A.; 3 Digest (Repl.) 255, *713* 101, 164, 281
Barber *v.* Meyerstein (1870), L. R. 4 H. L. 317; 39 L. J. C. P. 187; 22 L. T. 808; 18 W. R. 1041, 3 Mar. L. C. 449, H. L.; 41 Digest (Repl.) 267, *821* 319
Barclays Bank, Ltd. *v.* Customs and Excise Commissioners, [1963] 1 Lloyd's Rep. 81 319
Barclays Bank *v.* Okenarhe, [1966] 2 Lloyd's Rep. 87. 172
Barclays Bank, Ltd. *v.* Quistclose Investments, Ltd., [1968] 3 All E. R. 651; [1970] A. C. 567; [1968] 3 W. L. R. 1097; 112 Sol. Jo. 903, H. L.; Digest Cont. Vol. C p. 35, *401a* 258
Barker *v.* Furlong, [1891] 2 Ch. 172; 60 L. J. Ch. 368; 64 L. T. 411; 39 W. R. 621; 7 T. L. R. 406; 3 Digest (Repl.) 49, *342*. . . . 268
Barker *v.* Inland Truck Sales (1970), 11 D. L. R. (3d) 469 . . 13, 127, 201
Barker (Junior) & Co., Ltd. *v.* Agius, Ltd. (1927), 43 T. L. R. 751; 33 Com. Cas. 120; 28 Ll. L. R. 282; 39 Digest (Repl.) 576, *1011* . 110, 186, 187, 200, 360, 366
Barnard *v.* Faber, [1893] 1 Q. B. 340; 62 L. J. Q. B. 159; 68 L. T. 179; 41 W. R. 193; 9 T. L. R. 160; 4 R. 201, C. A.; 29 Digest (Repl.) 61, *160* 351
Barnes & Co. *v.* Toye (1884), 13 Q. B. D. 410; 53 L. J. Q. B. 567; 48 J. P. 664; 51 L. T. 292; 33 W. R. 15; 28 (2) Digest (Reissue) 694, *262* 85
Barnett, *Ex parte, Re* Tamplin. *See* Tamplin, *Re, Ex parte* Barnett
Barnett *v.* Javeri & Co., [1916] 2 K. B. 390; 85 L. J. K. B. 1703; 15 L. T. 217; 13 Asp. M. L. C. 424; 22 Com. Cas. 5; 39 Digest (Repl.) 494, *412* 357
Barr *v.* Gibson (1838), 3 M. & W. 390; 7 L. J. Ex. 124; 39 Digest (Repl.) 530, *667* 98, 121, 126

PAGE

Barr's Contract, *Re*, *Re* Moorwell Holdings *v.* Barr, [1956] 2 All E. R. 853; [1956] 1 Ch. 551; [1956] 1 W. L. R. 918; 100 Sol. Jo. 550; 40 Digest (Repl.) 120, *944* 105

Barrow, *Ex parte*, *Re* Worsdell (1877), 6 Ch. D. 783; 46 L. J. Bcy. 71; 36 L. T. 325; 25 W. R. 466; 3 Asp. M. L. C. 387; 39 Digest (Repl.) 755, *2339* 215

Barrow *v.* Arnaud (1846), 8 Q. B. 595, 604; 10 Jur. 319; 39 Digest (Repl.) 816, *2794* 232, 237, 248

Barrow, Lane and Ballard *v.* Phillip Phillips & Co., [1929] 1 K. B. 574; [1928] All E. R. Rep. 74; 98 L. J. K. B. 193; 140 L. T. 670; 34 Com. Cas. 119; 45 T. L. R. 133; 39 Digest (Repl.) 495, *420* . 97, 98, 99

Barry *v.* Van den Hurk, [1920] 2 K. B. 709; 89 L. J. K. B. 899; 123 L. T. 719; 36 T. L. R. 663; 17 Digest (Reissue) 204, *756* . . 258

Bartholomew *v.* Markwick (1864), 15 C. B. N. S. 711; 33 L. J. C. P. 145; 3 New Rep. 386; 9 L. T. 651; 10 Jur. N. S. 615; 12 W. R. 314; 39 Digest (Repl.) 733, *2123* 230

Bartlett *v.* Holmes (1853), 13 C. B. 630; 22 L. J. C. P. 182; 1 C. L. R. 159; 21 L. T. O. S. 104; 17 Jur. 858; 1 W. R. 334; 39 Digest (Repl.) 682, *1775* 180, 357

Bartlett *v.* Sidney Marcus, Ltd., [1965] 2 All E. R. 753; [1965] 1 W. L. R. 1013; 109 Sol. Jo. 451, C. A.; Digest Cont. Vol. B p. 630, *781a* . 133

Barton Thompson & Co. *v.* Vigers Brothers (1906), 110 L. T. 667, n.; 19 Com. Cas. 175; 39 Digest (Repl.) 638, *1471*. 155

Batchelor *v.* Vyse (1834), 4 Moo. & S. 552; 4 Digest (Repl.) 79, *681* . 162

Bazeley *v.* Forder (1868), L. R. 3 Q. B. 559; 9 B. & S. 599; 37 L. J. Q. B. 237; 18 L. T. 756; 32 J. P. 550; 27 (1) Digest (Reissue) 232, *1696* 92

Beale *v.* Taylor, [1967] 3 All E. R. 253; [1967] 1 W. L. R. 1193; 111 Sol. Jo. 668, C. A.; Digest Cont. Vol. C p. 851, *685a* . . 119, 367

Beaver Specialty, Ltd. *v.* Donald H. Bain, Ltd. (1974) 39 D. L. R. (3d) 574 194

Beck & Co. *v.* Szymanowski & Co., [1924] A. C. 43; [1923] All E. R. Rep. 244; 93 L. J. K. B. 25; 130 L. T. 387; 29 Com. Cas. 50, H. L.; 39 Digest (Repl.) 576, *1010* 362

Beckett *v.* Kingston Brothers (Butchers), Ltd., [1970] 1 All E. R. 715; [1970] 1 Q. B. 606; [1970] 2 W. L. R. 558; 134 J. P. 270; 114 Sol. Jo. 9; 68 L. G. R. 244; [1970] R. P. C. 135; Digest Cont. Vol. C p. 1025, *1142b*. 67

Beckett *v.* Tower Assets Co., [1891] 1 Q. B. 1; 60 L. J. Q. B. 56; 7 T. L. R. 21; on appeal, [1891] 1 Q. B. 638, C. A.; 39 Digest (Repl.) 445, *15* 82

Beebee & Co. *v.* Turners Successors (1931), 48 T. L. R. 61; 21 Digest (Repl.) 619, *1060* 179

Beer *v.* Walker (1877), 46 L. J. Q. B. 677; [1874–80] All E. R. Rep. 1139; 37 L. T. 278; 39 Digest (Repl.) 541, *768* . . . 130, 195, 283

Behn *v.* Burness (1863), 3 B. & S. 751; 32 L. J. Q. B. 204; 8 L. T. 207; 9 Jur. N. S. 620; 12 Digest (Reissue) 522, *3612* . 108, 109, 245, 354

Behnke *v.* Bede Shipbuilding Co., Ltd., [1927] 1 K. B. 649; [1927] All E. R. Rep. 689; 96 L. J. K. B. 649; 136 L. T. 666; 32 Com. Cas. 134; 43 T. L. R. 170; 39 Digest (Repl.) 449, *38* 244

Behrend & Co. *v.* Produce Broker's Co., Ltd., [1920] 3 K. B. 530; [1920] All E. R. Rep. 125; 90 L. J. K. B. 143; 124 L. T. 281; 25 Com. Cas. 286; 36 T. L. R. 775; 39 Digest (Repl.) 722, *2053* . 186, 187, 203

Beldessi *v.* Island Equipment, Ltd. (1974), 41 D. L. R. (3d) 147 . . 201

Belgian Grain and Produce Co. *v.* Cox & Co. (France), Ltd. (1919), 1 Ll. L. Rep. 256, C. A. 46

Bell *v.* Lever Brothers, Ltd., [1932] A. C. 161; [1931] All E. R. Rep. 1; 101 L. J. K. B. 129; 146 L. T. 258; 48 T. L. R. 133; 76 Sol. Jo. 50; 37 Com. Cas. 98, H. L.; 35 Digest (Repl.) 23, *140* . . 75, 97

PAGE
Belsize Motor Supply Co. *v.* Cox, [1914] 1 K. B. 244; [1911–13] All E. R. Rep. 1084; 83 L. J. K. B. 261; 110 L. T. 151; 26 Digest (Repl.) 660, *16* 82, 177, 309
Belvoir Finance Co., Ltd. *v.* Harold G. Cole, Ltd., [1969] 2 All E. R. 904; [1969] 1 W. L. R. 1877; Digest Cont. Vol. C p. 417, *40b* 296, 309
Belvoir Finance Co., Ltd. *v.* Stapleton, [1970] 3 All E. R. 664; [1971] 1 Q. B. 210; [1970] 3 W. L. R. 530; 114 Sol. Jo. 719, C. A.; Digest Cont. Vol. C p. 419, *81e* 18, 19
Benabu & Co. *v.* Produce Brokers' Co., Ltd. (1921), 37 T. L. R. 609; on appeal, 37 T. L. R. 851; 26 Com. Cas. 335, C. A.; 2 Digest (Repl.) 589, *1201* 366
Bennett *v.* Griffin Finance, [1967] 1 All E. R. 515; [1967] 2 Q. B. 46; [1967] 2 W. L. R. 561; 111 Sol. Jo. 150, C. A.; Digest Cont. Vol. C p. 418, *54a* 76, 115
Bennett, Ltd. *v.* Kreeger (1925), 41 T. L. R. 609; 17 Digest (Reissue) 164, *470* 247, 253
Bentley *v.* Vilmont. *See* Vilmont *v.* Bentley.
Bentley Brothers *v.* Metcalfe & Co., [1906] 2 K. B. 548; 75 L. J. K. B. 891; 95 L. T. 596; 22 T. L. R. 676, C. A.; 12 Digest (Reissue) 772, *5506* 280
Bentley (Dick) Productions, Ltd. *v.* Harold Smith (Motors), Ltd., [1965] 2 All E. R. 65; [1965] 1 W. L. R. 623; 109 Sol. Jo. 329, C. A.; Digest Cont. Vol. B p. 629, *559a* 5, 6
Bentsen *v.* Taylor, [1893] 2 Q. B. 274; 63 L. J. Q. B. 15; 69 L. T. 487; 42 W. R. 8; 9 T. L. R. 552; 4 R. 510, C. A.; 12 Digest (Reissue) 431, *3114* 23, 108
Bentworth Finance, Ltd. *v.* Lubert, [1967] 2 All E. R. 810; [1968] 1 Q. B. 680; [1967] 3 W. L. R. 378; 111 Sol. Jo. 272, C. A.; Digest Cont. Vol. C p. 417, *33a* 294
Berg *v.* Sadler and Moore, [1937] 1 All E. R. 637; [1937] 2 K. B. 158; 106 L. J. K. B. 593; 156 L. T. 334; 53 T. L. R. 430; 81 Sol. Jo. 158, 42 Com. Cas. 228, C. A.; 39 Digest (Repl.) 802, *2702* . . 236
Berg & Sons *v.* Landauer (1925), 42 T. L. R. 142; 39 Digest 461, *885* 107, 355
Bergheim *v.* Blaenavon Co. (1875), L. R. 10 Q. B. 319; 44 L. J. Q. B. 92; 32 L. T. 451; 23 W. R. 618; 39 Digest (Repl.) 813, *2772*. 235
Berk *v.* International Explosives Co. (1901), 7 Com. Cas. 20; 39 Digest (Repl.) 691, *1849* 364
Berndtson *v.* Strang (1867), L. R. 4 Eq. 481; on appeal (1868) L. R. 3 Ch. App. 588; 37 L. J. Ch. 665; 19 L. T. 40; 16 W. R. 1025; 3 Mar. L. C. 154, 39 Digest (Repl.) 707, *2444* . . . 211, 216
Besseler Waechter Glover & Co. *v.* South Derwent Coal Co., Ltd., [1938] 1 K. B. 408; [1937] 4 All E. R. 552; 107 L. J. K. B. 365; 158 L. T. 12; 54 T. L. R. 140; 43 Com. Cas. 86; 12 Digest (Reissue) 439, *3171* 22
Bethell *v.* Clark (1887), 19 Q. B. D. 553; affirmed (1888), 20 Q. B. D. 615; 57 L. J. Q. B. 302; 59 L. T. 808; 36 W. R. 611; 3 Asp. M. L. C. 346; *sub nom. Re* Bethell & Co. & Clark, 4 T. L. R. 401, C. A.; 39 Digest (Repl.) 754, *2328* 214, 215, 217
Betterbee *v.* Davis (1811), 3 Camp. 70; 12 Digest (Reissue) 398, *2912*. 228
Bevan *v.* Bevan, [1955] 2 All E. R. 206; [1955] 2 Q. B. 227; [1955] 2 W. L. R. 948; 99 Sol. Jo. 306; 2 Digest (Repl.) 266, *603* . . . 90
Beverley *v.* Lincoln Gas Co. (1837), 6 Ad. & El. 829; 2 Nev. & P. K. B. 283; Will. Woll. & Dav. 519; L. J. Q. B. 113; 39 Digest (Repl.) 627, *1383* 147
Bevington *v.* Dale (1902), 7 Com. Cas. 112; 17 Digest (Reissue) 61, *660* 158, 159, 368
Bexwell *v.* Christie (1776), Cowp. 395; 3 Digest (Repl.) 15, *113* . . . 267

PAGE

Bianchi *v.* Nash (1836), 1 M. & W. 545; 1 Tyr. & G. 916; 5 L. J. Ex. 262; 39 Digest (Repl.) 623, *1353* 75, 77

Biddell Brothers *v.* Clemens (E) Horst Co., [1911] 1 K. B. 214; reversed [1911] 1 K. B. 934; [1911–13] All E. R. Rep. 93; 80 L. J. K. B. 584; 1104 L. T. 577; 27 T. L. R. 331; 55 Sol. Jo. 383, 12 Asp. M. L. C. 1; 16 Com. Cas. 197, C. A.; affirmed *sub nom.* Clemens (E) Horst Co. *v.* Biddell Brothers, [1912] A. C. 18; [1911–13] All E. R. Rep. 93; 81 L. J. K. B. 42; 105 L. T. 563; 28 T. L. R. 42; 56 Sol. Jo. 50; 12 Asp. M. L. C. 90; 17 Com. Cas. 55, H. L.; 39 Digest (Repl.) 704, *1945* . . . 36, 38, 39, 138, 180, 181, 185, 195, 277, 358

Biddlecombe *v.* Bond (1835), 4 Ad. & El. 322; 5 L. J. K. B. 47; 17 Digest (Reissue) 313, *795* 287

Bigge *v.* Parkinson (1862), 7 H. & N. 955; 31 L. J. Ex. 301; 8 Jur. N. S. 1014; 10 W. R. 349; *sub nom.* Smith *v.* Parkinson, 7 L. T. 92; 39 Digest (Repl.) 574, *992* 10

Biggerstaff *v.* Rowatt's Wharf, Ltd., [1896] 2 Ch. 93; 65 L. J. Ch. 536; 74 L. T. 473, C. A.; 12 Digest (Reissue) 277, *2016*. 257

Biggin & Co., Ltd. *v.* Permanite, Ltd., [1951] 2 All E. R. 191; [1951] 2 K. B. 314; [1951] 2 T. L. R. 159; 95 Sol. Jo. 414, C. A.; 39 Digest (Repl.) 827, *2894* 249, 250, 253

Biggs *v.* Evans, [1894] 1 Q. B. 88; 69 L. T. 723; 58 J. P. 84; 10 T. L. R. 59; 1 Digest (Repl.) 386, *511* 292, 296, 297

Bill *v.* Bament (1841), 9 M. & W. 36; 11 L. J. Ex. 81; 39 Digest (Repl.) 463, *146* 213

Birch *v.* Paramount Estates, Ltd. (1956), 16 Estates Gazette 396. . 5

Bird *v.* Brown (1850), 4 Exch. 786; 19 L. J. Ex. 154; 14 L. T. O. S. 399; 14 Jur. 132; 39 Digest (Repl.) 766, *2442* 216

Bishop *v.* Shillito (1819), 2 B. & Ald. 329, n.; 39 Digest (Repl.) 634, *1444* 75, 278

Bishopsgate Motor Finance Corporation, Ltd. *v.* Transport Brakes, Ltd., [1949] 1 All E. R. 37; [1949] 1 K. B. 322; [1949] L. J. R. 741; 65 T. L. R. 66; 93 Sol. Jo. 71, C. A.; 33 Digest (Repl.) 489, *441* 169, 171

Bissell *v.* Beard (1873), 28 L. T. 740; 39 Digest (Repl.) 523, *621* . . 105

Black *v.* Smallwood, [1966] A. L. R. 744 174

Blackburn Bobbin Co., Ltd. *v.* Allen & Sons Ltd., [1918] 1 K. B. 540; affirmed, [1918] 2 K. B. 467; 87 L. J. K. B. 1085; 119 L. T. 215; 23 Com. Cas. 471; 34 T. L. R. 508, C. A.; 12 Digest (Reissue) 502, *3502* 19, 21

Blairmore Sailing Ship Co. *v.* Macredie, [1898] A. C. 593; 67 L. J. P. C. 96; 79 L. T. 217; 14 T. L. R. 513; 8 Asp. M. L. 429; 3 Com. Cas. 241; 24 R. 893, H. L.; 12 Digest (Reissue) 373, *2701* . . 279

Blandford, *Ex parte* (1893), 10 Morr. 231 321

Bloxam *v.* Morley (1825), 4 B. & C. 951 209

Bloxam *v.* Sanders (1825), 4 B. & C. 941; 7 D. & R. 407; 39 Digest (Repl.) 777, *2530* 181, 204, 207, 209

Blundell-Leigh *v.* Attenborough, [1921] 1 K. B. 382; reversed, [1921] 3 K. B. 235; [1921] All E. R. Rep. 525; 90 L. J. K. B. 1005; 125 L. T. 386; 37 T. L. R. 567, C. A.; 37 Digest (Repl.) 6, *25* . 83, 115, 277

Blyth Shipbuilding and Dry Dock Co., *Re*, [1926] Ch. 494; [1926] All E. R. Rep. 373; 95 L. J. Ch. 350; 134 L. T. 643, C. A.; 7 Digest (Repl.) 429, *358* 144, 151, 154

Blythswood Motors, Ltd. *v.* Raeside, 1966 S. L. T. 13. 232

Boks & Co. *v.* Rayner & Co. (1921), 37 T. L. R. 519, affirmed (1921), 37 T. L. R. 800, C. A.; 39 Digest (Repl.) 716, *2023* 35

Bollinger (J.) *v.* Costa Brava Wine Co., Ltd., [1959] 3 All E. R. 800; [1960] Ch. 262; [1959] 3 W. L. R. 966; 103 Sol. Jo. 1028; [1960] R. P. C. 16; 46 Digest (Repl.) 227, *1485* 54

PAGE

Bolt and Nut Co. (Tipton), Ltd. *v*. Rowland Nicholls & Co., Ltd., [1964] 1 All E. R. 137; [1964] 2 Q. B. 10; [1964] 2 W. L. R. 98; 107 Sol. Jo. 909, C. A.; 3rd Digest Supp.. 230

Bolton *v*. Lancs. & Yorks. Railway (1866), L. R. 1 C. P. 431; 35 L. J. C. P. 147; 13 T. L. 764; 12 Jur. N. S. 317; 14 W. R. 430; 39 Digest (Repl.) 765, *2423* 207, 211, 214, 215, 216

Bonython *v*. Commonwealth of Australia, [1951] A. C. 201; 66 (Pt. 2) T. L. R. 969; 94 Sol. Jo. 821, P. C.; 35 Digest (Repl.) 189, *33* . 47

Bonzi *v*. Stewart (1842), 4 Man. & G. 295; 11 L. J. C. P. 228; 5 Scott 1; 1 Digest (Repl.) 394, *556* 303

Booth Steamship Co. *v*. Cargo Fleet Iron Co., [1916] 2 K. B. 570; [1916–17] All E. R. Rep. 938; 85 L. J. K. B. 1577; 115 L. T. 199; 32 T. L. R. 535; 22 Com. Cas. 9, C. A.; 39 Digest (Repl.) 774, *2494* 219, 220, 272

Borrowman *v*. Drayton (1876), 2 Ex. D. 15; 46 L. J. Ex. 273; 35 L. T. 727, C. A.; 39 Digest (Repl.) 692, *1855* 117, 363

Borrowman *v*. Free (1878), 4 Q. B. D. 500; 48 L. J. Q. B. 65; 40 L. T. 25, C. A.; 39 Digest (Repl.) 721, *2049* 38

Borthwick (Thomas), (Glasgow), Ltd. *v*. Bunge & Co., Ltd., [1969] 1 Lloyd's Rep. 17 355

Borthwick (Thomas) (Glasgow), Ltd. *v*. Faure Fairclough, Ltd., [1968] 1 Lloyd's Rep. 16 . . . . 17, 19

Boshali *v*. Allied Commercial Exporters, Ltd., [1961] Nigeria L. R. 917, P. C. 120

Bostock *v*. Nicholson, [1904] 1 K. B. 725; 73 L. J. K. B. 524; 91 L. T. 626; 9 Com. Cas. 200; 20 T. L. R. 342, C. A.; 39 Digest (Repl.) 589, *1094* 125, 247, 251, 254

Boswell *v*. Kilborn (1862), 15 Moore, P. C. C. 309; 8 Jur. N. S. 443; 6 L. T. 79; 10 W. R. 517; 39 Digest (Repl.) 600, *1162*. . 142, 147

Boulton *v*. Jones (1857), 2 H. & N. 564; 27 L. J. Ex. 117; 30 L. T. O. S. 188; 3 Jur. N. S. 1156; 39 Digest (Repl.) 786, *2602* . . . 173

Bourne *v*. Seymour (1855), 16 C. B. 337; 24 L. J. C. P. 202; 25 L. T. O. S. 162; 1 Jur. N. S. 1001; 3 W. R. 511; 39 Digest (Repl.) 566, *951* . 364

Bowes *v*. Shand (1877), 2 App. Cas. 455; 46 L. J. Q. B. 561; 36 L. T. 857; 25 W. R. 730; 5 Asp. M. L. C. 461, H. L.; 39 Digest (Repl.) 519, *598* 107, 121, 355

Bowmaker (Commercial), Ltd. *v*. Day, [1965] 2 All E. R. 856, n.; [1965] 1 W. L. R. 1396; 109 Sol. Jo. 853; Digest Cont. Vol. B p. 334, *77ha* 117, 247

Boyd *v*. Siffken (1809), 2 Camp. 326; 39 Digest (Repl.) 492, *397* . . 356

Boyd (David T.) & Co., Ltd. *v*. Louca, [1973] 1 Lloyd's Rep. 209 34, 355

Bradford & Sons *v*. Price Brothers (1923), 92 L. J. K. B. 871; 129 L. T. 408; 39 T. L. R. 272; 1 Digest (Repl.) 422, *794* 230

Brading *v*. F. McNeill & Co., Ltd., [1946] 1 Ch. 145; 115 L. J. Ch. 109; 174 L. T. 140; 62 T. L. R. 191; 90 Sol. Jo. 128; 17 Digest (Reissue) 105, *126* 241

Bradley *v*. Ramsay & Co. (1912), 106 L. T. 771; 28 T. L. R. 388, C. A.; 39 Digest (Repl.) 628, *1406* 31

Bradshaw *v*. Booth's Marine, Ltd., [1973] 2 O. R. 646. 280

Brady *v*. Oastler (1864), 3 H. & C. 112; 33 L. J. Ex. 300; 11 L. T. 681; 11 Jur. N. S. 22; 39 Digest (Repl.) 817, *2798* 237

Bragg *v*. Villanova (1923), 40 T. L. R. 154; 39 Digest (Repl.) 717, *2024* 35, 195

Braithwaite *v*. Foreign Hardwood Co., [1905] 2 K. B. 543; 74 L. J. K. B. 688; 92 L. T. 637; 21 T. L. R. 413; 10 Asp. M. L. C. 52; 10 Com. Cas. 189, C. A.; 39 Digest (Repl.) 508, *523*. 23, 189, 204

Brandon *v*. Leckie (1972), 29 D. L. R. (3d) 633. 310

Brandt *v*. Lawrence (1876), 1 Q. B. D. 344; 46 L. J. Q. B. 327; 24 W. R. 749, C. A.; 39 Digest (Repl.) 660, *1601* 355

PAGE
Brandt *v.* Liverpool, Brazil and River Plate Steam Navigation Co., [1924] 1 K. B. 575; [1923] All E. R. Rep. 656; 93 L. J. K. B. 646; 130 L. T. 392; 16 Asp. M. L. C. 262; 29 Com. Cas. 57, C. A.; 41 Digest (Repl.) 251, *698* 317
Brandt & Co. *v.* Morris & Co., [1917] 2 K. B. 784; [1916–17] All E. R. Rep. 925; 87 L. J. K. B. 101; 117 L. T. 196; 23 Com. Cas. 137; 39 Digest (Repl.) 807, *2739* 41
Brauer & Co. (Great Britain), Ltd. *v.* James Clark (Brush Materials), Ltd., [1952] 2 All E. R. 497; [1952] 2 T. L. R. 349; 96 Sol. Jo. 548, C. A.; 39 Digest (Repl.) 809, *2748* 20, 42, 358
Breckwoldt *v.* Hanna (1963), 5 W. I. R. 356; Digest Cont. Vol. A p. 1305, **1107a* 202
Brice *v.* Bannister (1878), 3 Q. B. D. 569; 47 L. J. Q. B. 722; 38 L. T. 739; 26 W. R. 670, C. A.; 7 Digest (Repl.) 436, *385* 29
Bridge *v.* Campbell Discount Co., Ltd., [1962] 1 All E. R. 385; [1962] A. C. 600; [1962] 2 W. L. R. 439; 106 Sol. Jo. 94, H. L.; Digest Cont. Vol. A p. 648, *39a* 239
Bridge *v.* Wain (1816), 1 Stark. 504; 39 Digest (Repl.) 591, *1111*. . 240
Brinsmead *v.* Harrison (1871), L. R. 6 C. P. 584; [1861–73] All E. R. Rep. 465; 40 L. J. C. P. 281; 24 L. T. 798; 19 W. R. 956; on appeal, [1872] L. R. 7 C. P. 547; 41 L. J. C. P. 190; 27 L. T. 99; 20 W. R. 784; 46 Digest (Repl.) 525, *688* 30, 31
Bristol and West of England Bank *v.* Midland Railway, [1891] 2 Q. B. 653; 61 L. J. Q. B. 115; 65 L. T. 234; 40 W. R. 148; 7 Asp. M. L. C. 69, C. A.; 46 Digest (Repl.) 492, *405* 317, 319
Bistol Tramways Co. *v.* Fiat Motors, Ltd., [1910] 2 K. B. 831; [1908–10] All E. R. Rep. 113; 79 L. J. K. B. 1107; 103 L. T. 443; 26 T. L. R. 629, C. A.; 39 Digest (Repl.) 511, *541* . 4, 73, 125, 131, 133
British-American Continental Bank, *Re*, [1923] 1 Ch. 276; [1923] All E. R. Rep. 52; 92 L. J. Ch. 241; 128 L. T. 727, C. A.; 17 Digest (Reissue) 205, *757* 258
British and Beningtons, Ltd. *v.* North West Cachar Tea Co., [1923] A. C. 48; [1922] All E. R. Rep. 224; 92 L. J. K. B. 62; 128 L. T. 422; 20 Com. Cas. 265; 13 Ll. L. R. 67; H. L.; 39 Digest (Repl.) 664, *1643* 22, 38, 189, 204
British and Foreign Mar. Ins. Co. Ltd. *v.* Gaunt, [1921] 2 A. C. 41; [1921] All E. R. Rep. 447; 90 L. J. K. B. 801; 125 L. T. 491; 26 Com. Cas. 247, H. L.; 29 Digest (Repl.) 117, *620* 40
British Bank for Foreign Trade, Ltd. *v.* Novimex, Ltd., [1949] 1 All E. R. 155; [1949] 1 K. B. 623; [1949] L. J. R. 658; 93 Sol. Jo. 146, C. A.; 12 Digest (Reissue) 777, *5533* 101
British Crane Hire Corporation *v.* Ipswich Plant Hire, Ltd., [1974] 1 All E. R. 1059; [1974] 2 W. L. R. 856; 118 Sol. Jo. 387, C. A. . 15, 16
British Fermentation Products, Ltd. *v.* British Italian Trading Co., Ltd., [1942] 2 All E. R. 256; [1942] 2 K. B. 145; 111 L. J. K. B. 589; 167 L. T. 91; 106 J. P. 178; 58 T. L. R. 255; 86 Sol. Jo. 195; 40 L. G. R. 179; 25 Digest (Repl.) 112, *329*. 68
British Imex Industries, Ltd. *v.* Midland Bank, Ltd., [1958] 1 All E. R. 264; [1958] 1 Q. B. 542; [1958] 2 W. L. R. 103; 102 Sol. Jo. 69; [1957] 2 Lloyd's Rep. 591; 39 Digest (Repl.) 706, *1958*. . . . 44
British Motor Body Co. *v.* Shaw, 1914 S. C. 922; 40 Digest (Repl.) 452, **317* 240
British Oil and Cake Co. *v.* Burstall (1923), 39 T. L. R. 406; 67 Sol. Jo. 577; 39 Digest (Repl.) 828, *2896* 250
British Road Services, Ltd. *v.* Arthur V. Crutchley & Co., Ltd., [1968] 1 All E. R. 811; [1968] 1 Lloyd's Rep. 271, C. A.; Digest Cont. Vol. C p. 31, *183a* 160

PAGE

British Russian Gazette and Trade Outlook, Ltd. *v.* Associated Newspapers, Ltd., [1933] 2 K. B. 616; [1933] All E. R. Rep. 320; 102 L. J. K. B. 775; 149 L. T. 545, C. A.; 1 Digest (Repl.) 759, *2965* . 24
British Waggon Co. *v.* Lea (1880), 5 Q. B. D. 149; [1874–80] All E. R. Rep. *1135*; 49 L. J. Q. B. 321; 42 L. T. 437; 44 J. P. 440; 28 W. R. 349, C. A.; 12 Digest (Reissue) 725, *5247* 30
British Westinghouse Electric Co. *v.* Underground Railway, [1912] A. C. 673; [1911–13] All E. R. Rep. 63; 81 L. J. K. B. 1132; 107 L. T. 325; 56 Sol. Jo. 734, H. L.; 39 Digest (Repl.) 592, *1119* 233, 255, 256
Brogden *v.* Marriott (1836), 3 Bing. N. C. 88; 2 Scott, 712; 5 L. J. C. P. 302; 39 Digest (Repl.) 500, *458*. 103
Brogden *v.* Metropolitan Ry. Co. (1877), 2 App. Cas. 666, H. L.; 39 Digest (Repl.) 453, *66* 93
Brooke Tool Manufacturing Co. *v.* Hydraulic Gears Co. (1920), 89 L. J. K. B. 263; 122 L. T. 126; 12 Digest (Reissue) 516, *3576* . .21, 107
Broome *v.* Pardess Co-operative Society of Orange Growers, Est. (1900), Ltd., [1940] 1 All E. R. 603; 109 L. J. K. B. 571; 162 L. T. 399; 56 T. L. R. 430; 84 Sol. Jo. 333; 45 Com. Cas. 151; 12 Digest (Reissue) 753, *5404* 195
Brown *v.* Byrne (1854), 3 E. & B. 703; 23 L. J. Q. B. 313; 17 Digest (Reissue) 48, *552* 4, 100
Brown *v.* Muller (1872), L. R. 7 Ex. 319; 41 L. J. Ex. 214; 27 L. T. 272; 21 W. R. 384; 39 Digest (Repl.) 818, *2811* . . . 231, 235
Brown (B. S.) & Sons, Ltd. *v.* Craiks, Ltd., [1970] 1 All E. R. 823; [1970] 1 W. L. R. 752; 114 Sol. Jo. 282; 1970 S. L. T. 41; Digest Cont. Vol. C p. 855, *859a* 128, 129
Browne *v.* Hare (1853), 3 H. & N. 484; on appeal (1859), 4 H. & N. 822; 29 L. J. Ex. 6; 7 W. R. 619; *sub nom.* Hare *v.* Browne, 33 L. T. O. S. 334; 5 Jur. N. S. 711; 39 Digest (Repl.) 630 . . . 35, 155
Bruner *v.* Moore, [1904] 1 Ch. 305; 73 L. J. Ch. 377; 89 L. T. 738; 20 T. L. R. 125; 48 Sol. Jo. 131; 17 Digest (Reissue) 317, *842* . 22, 105
Bryans *v.* Nix (1839), 4 M. & W. 775; 1 H. & H. 480; 8 L. J. Ex. 137; 39 Digest (Repl.) 608, *1219* 192
Buckley *v.* Lever Brothers, Ltd., [1953] 4 D. L. R. 16; [1953] O. R. 704; [1953] O. W. N. 658; 39 Digest (Repl.) 543, **349* 79, 132
Buckman *v.* Levi (1813), 3 Camp. 414; 39 Digest (Repl.) 682, *1770* 192, 193
Budberg *v.* Jerwood and Ward (1934), 51 T. L. R. 99; 78 Sol. Jo. 878; 1 Digest (Repl.) 387, *518*. 292
Budd *v.* Fairmaner (1831), 8 Bing. 48; 5 C. & P. 78; 1 Moo. & S. 74; 1 L. J. C. P. 16; 2 Digest (Repl.) 348, *325* 368
Buddle *v.* Green (1857), 27 L. J. Ex. 33; 39 Digest (Repl.) 191, *413* 95, 179, 180, 183
Bull *v.* Robison (1854), 10 Exch. 342; 24 L. J. Ex. 165; 39 Digest (Repl.) 581, *1842* 195
Bunney *v.* Poyntz (1833), 4 B. & Ad. 568; 1 Nev. & M. K. B. 229; 2 L. J. K. B. 55; 39 Digest (Repl.) 745, *2228* . . . 206, 209
Bunting *v.* Tory (1948), 64 T. L. R. 353; 39 Digest (Repl.) 832, *2928* . 251
Burdick *v.* Sewell (1884), 13 Q. B. 159, C. A.; *sub nom.* Sewell *v.* Burdick; 10 App. Cas. 74; [1881–5] All E. R. Rep. 223; 54 L. J. Q. B. 156; 52 L. T. 445; 33 W. R. 461; 1 T. L. R. 128; 5 Asp. M. L. C. 376, H. L.; 41 Digest (Repl.) 240, *621* . . 83, 222, 283, 316, 317
Burnand *v.* Haggis (1863), 14 C. B. N. S. 45; 2 New Rep. 126; 32 L. J. C. P. 189; 8 L. T. 320; 9 Jur. N. S. 1325; 11 W. R. 644; 2 Digest (Repl.) 342, *300*. 88
Burrough's Business Machines, Ltd. *v.* Feed Rite Mills (1962), Ltd. (1974), 42 D. L. R. (3d) 303 201
Bussey *v.* Barnett (1842), 9 M. & W. 312; 11 L. J. Ex. 211; 39 Digest (Repl.) 729, *2099* 181

PAGE

Butterworth *v.* Kingsway Motors, Ltd., [1954] 2 All E. R. 694; [1954] 1 W. L. R. 1286; 98 Sol. Jo. 717; 39 Digest (Repl.) 581, *1038* 115, 175

Cahn *v.* Pockett's Bristol Steam Channel Co., [1899] 1 Q. B. 643; 68 L. J. Q. B. 515; 80 L. T. 269; 15 T. L. R. 247; 47 W. R. 422; 43 Sol. Jo. 331; 8 Asp. M. L. C. 517; 4 Com. Cas. 168, C. A.; 39 Digest (Repl.) 636, *1462* 155, 156, 174, 176, 220, 309

Cain *v.* Moon, [1896] 2 Q. B. 283; 65 L. J. Q. B. 587; 74 L. T. 728; 25 Digest (Repl.) 594, *307* 78, 277

Calcutta Co *v.* De Mattos (1863), 32 L. J. Q. B. 322 . 9, 81, 180, 191, 229, 357, 369

Callot *v.* Nash (1923), 39 T. L. R. 291; 27 Digest (Reissue) 219, *1530* . 92

Cammel *v.* Sewell (1860), 5 H. & N. 728; 29 L. J. Ex. 350, Ex. Ch.; 11 Digest (Reissue) 610, *1547* 49

Cammel Laird & Co. Ltd. *v.* Manganese Bronze and Brass Co., Ltd., [1934] A. C. 402; [1934] All E. R. Rep. 1; 103 L. J. K. B. 289; 151 L. T. 142; 50 L. T. R. 350; 39 Com. Cas. 194, H. L.; 39 Digest (Repl.) 544, *784* 78, 129, 135, 365

Campbell *v.* Mersey Docks (1863), 14 C. B. N. S. 412; 2 New Rep. 32; 8 L. T. 245; 11 W. R. 596; 39 Digest (Repl.) 602, *1175*. . . . 148

Campbell Mostyn (Provisions), Ltd. *v.* Barnett Trading Co., [1954] C. L. Y. 2985, C. A. 232, 234, 249

Canada Atlantic Grain Export Co. *v.* Eilers (1929), 35 Com. Cas. 90; 35 Ll. L. R. 206; 39 Digest (Repl.) 574, *1001* 129, 361

Cantière San Rocco *v.* Clyde Shipbuilding and Engineering Co., [1924] A. C. 226; 93 L. J. P. C. 86; 130 L. T. 610; 1923 Sc. (H. L.); 12 Digest (Reissue) 503, **1835* 90, 327

Cantière Meccanico Brindisino *v.* Constant (1912), 17 Com. Cas. 182 . 39

Cap Palos, The, [1921] p. 458; [1921] All E. R. Rep. 249; 91 L. J. P. 11; 126 L. T. 82; 37 T. L. R. 921; 15 Asp. M. L. C. 403, C. A.; 42 Digest (Repl.) 783, *5531* 12

Car and Universal Finance Co., Ltd. *v.* Caldwell, [1964] 1 All E. R. 290; [1965] 1 Q. B. at p. 535, [1964] 2 W. L. R. at p. 606; 108 Sol. Jo. 15, C. A.; Digest Cont. Vol. B p. 633, *1577* 8, 171, 172

Carapanayoti *v.* Comptoir Commercial André and Cie S.A., [1972] 1 Lloyd's Rep. 139; Digest Supp. 356

Carnforth Co., *Ex parte* (1876), 4 Ch. D. 108; 46 L. J. Ch. 115; 35 L. T. 776, C. A.; 39 Digest (Repl.) 698, *1899* 287

Carter, *Ex parte*, *Re* Hamilton Young & Co., [1905] 2 K. B. 772; 74 L. J. K. B. 905; 93 L. T. 591; 54 W. R. 260; 21 T. L. R. 757; 12 Mans. 365, C. A.; 7 Digest (Repl.) 30, *149* 321

Carter *v.* Crick (1859), 4 H. &. N. 412; 28 L. J. Ex. 238; 33 L. T. O. S. 166; 7 W. R. 507; 39 Digest (Repl.) 510, *533* 365

Carver & Co., *Re* (1911), 17 Com. Cas. 59; 39 Digest (Repl.) 722, *2052* 183, 241

Case (J. I.) Threshing and Machine Co., *Re* (1919), 49 D. L. R. 30. . 361

Cassaboglou *v.* Gibbs (1883), 11 Q. B. D. 797; 52 L. J. Q. B. 538; 48 L. T. 850; 32 W. R. 138, C. A.; 39 Digest (Repl.) 766, *2436* 206, 207

Cassidy (Peter) Seed Co., Ltd. *v.* Osuustukkukauppa I. L., [1957] 1 All E. R. 484; [1957] 1 W. L. R. 273; 101 Sol. Jo. 149; [1957] 1 Lloyd's Rep. 25; 39 Digest (Repl.) 660, *1599* 41, 42, 358

Castle *v.* Playford (1872), L. R. 7 Ex. 98; 41 L. J. Ex. 44; 26 L. T. 315; 20 W. R. 440; 39 Digest (Repl.) 496, *426* 103, 152, 157

Castle *v.* Sworder (1860), 29 L. J. Ex. 235; reversed (1861), 6 H. & N. 828; 30 L. J. Ex. 310; 4 L. T. 865; 8 Jur. N. S. 233; 9 W. R. 697; 39 Digest (Repl.) 468, *192* 195, 196

PAGE

Caswell *v.* Coare (1809), 1 Taunt. 566; 2 Camp. 82; 2 Digest (Repl.) 360, *418* 202, 203
Celia S.S. v. S.S. Volturro, [1921] 2 A. C. 544; [1921] All E. R. Rep. 110; 90 L. J. P. 385; 126 L. T. 1; 27 Com. Cas. 46; 37 T. L. R. 969, H. L.; 17 Digest (Reissue) 204, *753* 239, 258
Cellulose Acetate Silk Co., Ltd. *v.* Widnes Foundry (1925), Ltd., [1933] A. C. 20; 101 L. J. K. B. 694; 147 L. T. 401; 48 T. L. R. 595; 39 Com. Cas. 61, H. L.; 17 Digest (Reissue) 178, *553* 239
Central Farmers, Ltd. *v.* Smith, 1974 S. L. T. (Sh. Ct.) 87 201
Central Newbury Car Auctions *v.* Unity Finance, [1956] 3 All E. R. 905; [1957] 1 Q. B. 371; [1956] 3 W. L. R. 927; 100 Sol. Jo. 927, C. A.; 21 Digest (Repl.) 484, *1713* 166, 293
Chai Sau Yin *v.* Liew Kwee San, [1962] A. C. 304; [1962] 2 W. L. R. 765; 106 Sol. Jo. 217, P. C.; 17 Digest (Reissue) 539, **201* . . 18
Chalmers, *Ex parte, Re* Edwards (1873), 8 Ch. App. 289; 42 L. J. Bcy. 37; 28 L. T. 325; 21 W. R. 349; 39 Digest (Repl.) 697, *1894* 205, 208, 210, 211, 214
Champanhac & Co., Ltd. *v.* Waller & Co., Ltd., [1948] 2 All E. R. 724; 39 Digest (Repl.) 558, *877* . . . 10, 118, 120, 138, 139, 360
Chanter *v.* Hopkins (1838), 4 M. & W. 399; 8 L. J. Ex. 14; 39 Digest (Repl.) 550, *818* 4, 121, 353
Chanter *v.* Lesse (1839), 5 M. & W. 698; 9 L. J. Ex. 327 330
Chao (Trading as Zung Fu Co.) *v.* British Traders and Shippers, Ltd., (No. 2), [1954] 3 All E. R. 165; *sub nom.* Kwei Tek Chao *v.* British Traders and Shippers, Ltd., [1954] 2 Q. B. 459; [1954] 3 W. L. R. 496; 98 Sol. Jo. 592; 39 Digest (Repl.) 811, *2755*. 240
Chao (Trading as Zung Fu Co.) *v.* British Traders and Shippers, Ltd. (N. V. Handelsmaatschappij J. Smits Imports–Export, Third Party), [1954] 1 All E. R. 779; [1954] 1 Lloyd's Rep. 16; *sub nom.* Kwei Tek Chao *v.* British Traders and Shippers, Ltd., [1954] 2 Q. B. 459; 98 Sol. Jo. 163; [1954] 2 W. L. R. 365; 39 Digest (Repl.) 720, *2047* 7, 38, 45, 106, 197, 201, 247, 357
Chapelton *v.* Barry Urban District Council, [1940] 1 All E. R. 356; [1940] 1 K. B. 532; 109 L. J. K. B. 213; 162 L. T. 169; 104 J. P. 165; 56 T. L. R. 331; 84 Sol. Jo. 185; 38 L. G. R. 149, C. A.; 3 Digest (Repl.) 96, *240* 16
Chaplin *v.* Leslie Frewin (Publishers), [1965] 3 All E. R. 764; [1966] Ch. at p. 79; [1966] 2 W. L. R. at p. 47; 109 Sol. Jo. 871, C. A.; Digest Cont. Vol. B p. 146, *313a* 87
Charles Duval *v.* Gans. *See* Duval (Charles) *v.* Gans.
Chaplin *v.* Rogers (1800), 1 East, 192; 39 Digest (Repl.) 462, *139*. . 184
Chapman *v.* Morton (1843), 11 M. & W. 534; 12 L. J. Ex. 292; 1 L. T. O. S. 148; 39 Digest (Repl.) 714, *2009* . . . 197, 201
Chapman *v.* Speller (1850), 14 Q. B. 621; 19 L. J. Q. B. 239; 15 L. T. O. S. 158; 14 Jur. 625; 39 Digest (Repl.) 525, *636* 95
Chapman *v.* Withers (1888), 20 Q. B. D. 824; 57 L. J. Q. B. 457; 2 Digest (Repl.) 359, *416* 362
Chappell & Co., Ltd. *v.* Nestlé Co., Ltd., [1959] 2 All E. R. 701; [1960] A. C. 87; [1959] 3 W. L. R. 168; 103 Sol. Jo. 561, H. L.; Digest Cont. Vol. A p. 312, *569a*. 79
Chapple *v.* Cooper (1844), 13 M. & W. 252; 13 L. J. Ex. 286; 28 (2) Digest (Reissue) 695, *304* 88, 89
Charron *v.* Montreal Trust (1958), 15 D. L. R. (2d.) 240; [1958] O. R. 597; [1958] O. W. N. 357; 11 Digest (Reissue) 469, **457* . . . 91
Charter *v.* Sullivan, [1957] 1 All E. R. 809; [1957] 2 Q. B. 117; [1957] 2 W. L. R. 528; 101 Sol. Jo. 265, C. A.; 39 Digest (Repl.) 795, *2665* 231, 232, 233, 234, 235, 237

PAGE

Charterhouse Credit Co. *v.* Tolly, [1963] 2 All E. R. 432; [1963] 2 Q. B. 683; [1963] 2 W. L. R. 1168; 107 Sol. Jo. 234, C. A.; Digest Cont. Vol. A p. 649, *43b* 122
Cheetham & Co. *v.* Thornham Spinning Co., [1964] 2 Lloyd's Rep. 17 144, 146
Chelmsford Auctions, Ltd. *v.* Poole, [1973] 1 All E. R. 810; [1973] Q. B. 542; [1973] 2 W. L. R. 219; 117 Sol. Jo. 122; 225 Estates Gazette 1369, C. A.; Digest Supp. 268
Chesham *v.* Beresford Hotel (1913), 29 T. L. R. 584; 29 Digest (Repl.) 27, *312* 162
Chess (Oscar), Ltd. *v.* Williams, [1957] 1 All E. R. 325; [1957] 1 W. L. R. 370; 101 Sol. Jo. 186, C. A.; 39 Digest (Repl.) 514, *559* . . 5, 108
Chesterman *v.* Lamb (1834), 2 Ad. & El. 129; 4 Nw. & M. 195; 2 Digest (Repl.) 360, *424* 202, 203
Chettiar *v.* Chettiar, [1962] 1 All E. R. 494; [1962] A. C. 294; [1962] 2 W. L. R. 548; 106 Sol. Jo. 110, P. C.; 12 Digest (Reissue) 341, *2461* 19
Chinery *v.* Viall (1860), 5 H. & N. 288; 2 L. T. 466; 29 L. J. Ex. 180; 39 Digest (Repl.) 823, *2864* 181, 213, 242
Christopher Hill, Ltd. *v.* Ashington Piggeries, Ltd. *See* Hill (Christopher), Ltd. *v.* Ashington Piggeries, Ltd.
Churchill (V. L.) & Co., Ltd., and Lonberg *Re*, [1941] 3 All E. R. 137; 165 L. T. 274, C. A.; 2 Digest (Repl.) 248, *494* 90
Churchill and Sim *v.* Goddard, [1936] 1 All E. R. 675; [1937] 1 K. B. 92; 105 L. J. K. B. 571; 154 L. T. 586; 52 T. L. R. 356; 80 Sol. Jo. 285; 41 Com. Cas. 309, C. A.; 1 Digest (Repl.) 718, *2653* . . . 155
City Bank *v.* Barrow (1880), 5 App. Cas. 664; 43 L. T. 393, H. L.; 1 Digest (Repl.) 384, *494* 49, 164, 291
City Fur Manufacturing Co., Ltd. *v.* Fureenbond (Brokers), London, Ltd. (1937), 1 All E. R. 799; 81 Sol. Jo. 218; 39 Digest (Repl.) 656, *1585* 176, 306
City Motors (1933), Proprietary Ltd. *v.* Southern Aeriel Proprietary, Ltd. (1961), 106 C. L. R. 477 143
Claddagh Steamship Co. *v.* Stevens & Co., 1919, S. C. 132 (H. L.); 1 S. L. T. 31; 56 Sc. L. R. 619; 39 Digest (Repl.) 791, **1397* . 190
Clark *v.* Bulmer (1843), 11 M. & W. 243; 1 Dow. & L. 367; 12 L. J. Ex. 463; 39 Digest (Repl.) 444, *4* 78
Clark *v.* Cox, McEuen & Co., [1921] 1 K. B. 139; 89 L. J. K. B. 596; 122 L. T. 647; 15 Asp. M. L. C. 5; 25 Com. Cas. 94, C. A.; 39 Digest (Repl.) 566, *946* 20, 357
Clarke, *Re* (1887), 36 Ch. D. 348 245
Clarke *v.* Army and Navy C. S. Ltd., [1903] 1 K. B. 155; 72 L. J. K. B. 153; 88 L. T. 1; 19 T. L. R. 80, C. A.; 39 Digest (Repl.) 563, *923* . 271
Clarke *v.* Hutchins (1811), 14 East, 475; 39 Digest (Repl.) 702, 1938 . 192
Clarke *v.* Reilly (1962), 96 I. L. T. R. 96 79
Clarke *v.* Spence (1836), 4 Ad. & El. 448; 1 Har. & W. 760; 6 Nev. & M. K. B. 399; 5 L. J. K. B. 161; 39 Digest (Repl.) 615, *1282* . . 151
Clarke *v.* Westrope (1856), 18 C. B. 765; 25 L. J. C. P. 287; 20 J. P. 728; 39 Digest (Repl.) 505, *499* 104
Clay *v.* Yates (1856), 1 H. & N. 73; 25 L. J. Ex. 237; 27 L. T. O. S. 126; 2 Jur. N. S. 908; 4 W. R. 557; 39 Digest (Repl.) 450, *51* . . 78
Clayton *v.* Le Roy, [1911] 2 K. B. 1031; 81 L. J. K. B. 49; 104 L. T. 419; 75 J. P. 229; 27 T. L. R. 206; reversed, [1911] 2 K. B. 1046; [1911–13] All E. R. Rep. 284; 81 L. J. K. B. 58; 105 L. T. 430; 75 J. P. 521; 27 T. L. R. 479, C. A.; 37 Digest (Repl.) 11, *67* . 170
Clemens (E.) Horst & Co. *v.* Biddell Brothers. *See* Biddell Brothers *v.* Clemens Horst & Co.
Clever *v.* Kirkman (1875), 33 L. T. 672; 24 W. R. 159; 17 Digest (Reissue) 383, *1481*. 4
Clifford *v.* Watts (1870), L. R. 5 C. P. 577; 40 L. J. C. P. 36; 22 L. T. 717; 18 W. R. 925; 12 Digest (Reissue) 472, *3393*. 97, 98

PAGE

Clode *v.* Barnes, [1974] 1 All E. R. 1166; [1974] 1 W. L. R. 544; 138 J. P. 371; 118 Sol. Jo. 257 58
Clough *v.* London and North Western Railway (1871), L. R. 7 Exch. 26; [1861–73] All E. R. Rep. 647; 41 L. J. Ex. 17; 25 L. T. 708; 20 W. R. 189; 39 Digest (Repl.) 652, *1566* 171
Clyde Cycle Co. *v.* Hargreaves (1898), 78 L. T. 296; 14 T. L. R. 338; 28 (2) Digest (Reissue) 696, *310*. 89
Cobbold *v.* Caston (1824), 1 Bing. 399; 8 Moore, C. P. 456; 2 L. J. O. S. C. P. 38; 39 Digest (Repl.) 447, *21* 77
Cochrane *v.* Moore (1890), 25 Q. B. D. 57; [1886–90] All E. R. Rep. 731; 59 L. J. Q. B. 377; 63 L. T. 153; 54 J. P. 804, C. A.; 25 Digest (Repl.) 555, *50* 78, 85, 146
Cockerell *v.* Aucompte (1857), 2 C. B. N. S. 440; 26 L. J. C. P. 194; 3 Jur. N. S. 844; 5 W. R. 633; 39 Digest (Repl.) 688, *1827* . . 364
Coddington *v.* Paleologo (1867), L. R. 2 Exch. 193; 37 L. J. Ex. 73; 15 L. T. 581; 15 W. R. 961; 39 Digest (Repl.) 658, *1592* . 10, 191, 358
Coggs *v.* Bernard (1703), 2 Ld. Raym. 909; 1 Com. 133; 1 Smith L. C. (13th Edn.), 175, 201; 12 Digest (Reissue) 274, *1983* . . . 83
Cohen *v.* Mitchell (1890), 25 Q. B. D. 262; 59 L. J. Q. B. 409, C. A.; 5 Digest (Repl.) 790, *6699* 166
Cohen *v.* Roche, [1927] 1 K. B. 169; 95 L. J. K. B. 945; 136 L. T. 219; 42 T. L. R. 674; 70 Sol. Jo. 942; 39 Digest (Repl.) 482, *306*. 31, 244, 269
Cohen (George) & Sons & Co., Ltd. *v.* Jamieson and Paterson, 1963 S. C. 289; Digest Cont. Vol. A p. 1304, **615a* 270
Coldman *v.* Hill, [1919] 1 K. B. 443; [1918–19] All E. R. Rep. 434; 88 L. J. K. B. 491; 120 L. T. 412; 35 T. L. R. 146; 63 Sol. Jo. 166, C. A.; 22 Digest (Reissue) 197, *1657*. 160
Cole *v.* North-Western Bank (1875), L. R. 10 C. P. 354; [1874–80] All E. R. Rep. 486; 44 L. J. C. P. 233; 32 L. T. 433; 1 Digest (Repl.) 385, *501*. . . . 164, 289, 290, 291, 296, 301, 303, 304, 308, 313
Collett *v.* Co-operative Wholesale Society, Ltd., [1970] 1 All E. R. 274; [1970] 1 W. L. R. 250; 134 J. P. 227; 114 Sol. Jo. 9; 68 L. G. R. 158; Digest Cont. Vol. C p 1057, *146b* 56
Colley *v.* Overseas Exporters, [1921] 3 K. B. 302; [1921] All E. R. Rep. 596; 90 L. J. K. B. 1301; 126 L. T. 58; 26 Com. Cas. 325; 37 T. L. R. 797; 39 Digest (Repl.) 671, *1698* 84, 227, 229
Collins, *Ex parte*, Lees, *Re* (1875), 10 Ch. App. 367; 44 L. J. Bcy. 78; 32 L. T. 108; 7 Digest (Repl.) 83, *474*. 348
Collins Trading Co., Pty., Ltd. *v.* Maher, [1969] V. R. 20; Digest Cont. Vol. C p. 851, **39a* 78
Collyer *v.* Isaacs (1881), 19 Ch. D. 342; [1881–5] All E. R. Rep. 828; 51 L. J. Ch. 14; 45 L. T. 567, C. A.; 20 Digest (Repl.) 272, *178* . 96
Colonial Bank *v.* Whinney (1885), 30 Ch. D. 261; on appeal (1886), 11 App. Cas. 426; [1886–90] All E. R. Rep. 468; 56 L. J. Ch. 53; 55 L. T. 362, H. L.; 8 (2) Digest (Reissue) 489, *1* . . 161, 164, 281
Colonial Ins. Co. *v.* Adelaide Mar. Ins. Co. (1886), 12 App. Cas. 128; 56 L. J. P. C. 19; 56 L. T. 173; 35 W. R. 636; 3 T. L. R. 252; 6 Asp. M. L. C. 94; 39 Digest (Repl.) 608, *1215* . . . 161, 186, 190, 363
Columbia Phonograph Co. *v.* Regent Fittings Co. (1913), 30 R. P. C. 484; 36 Digest (Repl.) 1023, *3694* 53
Commercial Fibres *v.* Zaibaida, [1975] 1 Lloyd's Rep. 27 199
Commonwealth Trust *v.* Akotey, [1926] A. C. 72; [1925] All E. R. Rep. 270; 94 L. J. P. C. 167; 134 L. T. 33; 41 T. L. R. 641, P. C.; 39 Digest (Repl.) 772, *2487* 165
Compagnie d'Armement Maritime S. A. *v.* Compagnie Tunisienne de Navigations S. A., [1970] 3 All E. R. 71; [1970] 3 W. L. R. 389; 114 Sol. Jo. 618, H. L.; 11 Digest (Reissue) 457, *759* 48

PAGE

Comptoir Commercial Anversois *v.* Power, Son & Co., Re, [1920] 1 K. B. 868; [1918–19] All E. R. Rep. 661; 89 L. J. K. B. 849; 122 L. T. 567; 36 T. L. R. 101, C. A.; 39 Digest (Repl.) 570, *975* . . 20, 357

Comptoir D'Achat et de Vente du Boerenbond Belge S/A *v.* Luis de Ridder, Limitada (The Julia), [1949] 1 All E. R. 269; [1949] A. C. 293; [1949] L. J. R. 513; 65 T. L. R. 126; 93 Sol. Jo. 101, H. L.; 39 Digest (Repl.) 640, *1488* . . . 33, 37, 41, 158, 159

Congreve *v.* Evetts (1854), 10 Exch. 298; 23 L. J. Ex. 273; 18 Jur. 655; 7 Digest (Repl.) 125, *727* 96, 278

Consolidated Co. *v.* Curtis & Son, (1892) 1 Q. B. 495; 61 L. J. Q. B. 325; 23 Digest (Repl.) 50, *353* 161, 268

Continental Contractors, Ltd. *v.* Medway Oil and Storage Co. (1925), 23 Ll. L. Rep. 55, C. A. 38, 189

Contract and Trading Co. (Southern), Ltd. *v.* Barbey, [1959] 3 All E. R. 846; [1960] A. C. 244; [1959] 2 W. L. R. 15; 103 Sol. Jo. 1027, H. L.; 17 Digest (Reissue) 509, *204* 332

Conway Brothers and Savage *v.* Mulhern & Co., Ltd. (1901), 17 T. L. R. 730; 39 Digest (Repl.) 502, *475* 323

Cooden Engineering *v.* Stanford, [1952] 2 All E. R. 915; [1953] 1 Q. B. 86; [1952] T. L. R. 822; 96 Sol. Jo. 802, C. A.; 17 Digest (Reissue) 184, *584* 239

Cooke, *Ex parte, Re* Strachan (1876), 4 Ch. D. 123; 46 L. J. Bcy. 52; 35 L. T. 649; 41 J. P. 180; 25 W. R. 171, C. A.; 5 Digest (Repl.) 776, *6613* 312

Cooke & Sons *v.* Eshelby (1887), 12 App. Cas. 271; [1886–90] All E. R. Rep. 791; 56 L. J. Q. B. 505; 56 L. T. 673, H. L.; 1 Digest (Repl.) 665, *2328* 313

Cooper, *Ex parte* (1879), 11 Ch. D. 68; 48 L. J. Bcy. 49; 40 L. T. 105; 27 W. R. 518; 4 Asp. M. L. C. 63, C. A.; 39 Digest (Repl.) 763, *2403* 210, 211, 216

Cooper, *Re, Ex parte* Trustee of Property of Bankrupt *v.* Registrar and High Bailiff of Peterborough and Huntingdon County Courts, [1958] 2 All E. R. 97; [1958] Ch. 922; [1958] 3 W. L. R. 468; 102 Sol. Jo. 635; 5 Digest (Repl.) 892, *7419* 178

Cooper *v.* Bill (1865), 3 H. &. C. 722; 12 L. T. 466; 34 L. J. Ex. 161; 39 Digest (Repl.) 746, *2239* 212

Cooper *v.* Cooper (1888), 13 App. Cas. 88; 59 L. T. 1, H. L.; 22 Digest (Reissue) 677, *7239* 91

Cooper *v.* Micklefield Coal and Lime Co. (1912), 107 L. T. 457; 12 Digest (Reissue) 724, *5244* 29

Cooper *v.* Shepherd (1846), 3 C. B. 266; 4 Dow. & L. 218; 15 L. J. C. P. 237; 7 L. T. O. S. 282; 10 Jur. 758; 46 Digest (Repl.) 525, *686* . 30

Cooper *v.* Shuttleworth (1856), 25 L. J. Ex. 114; 39 Digest (Repl.) 502, *506* 104

Cooper *v.* Stubbs, [1925] 2 K. B. 753; [1925] All E. R. Rep. 643; 94 L. J. K. B. 903; 133 L. T. 582; 41 T. L. R. 614; 69 Sol. Jo. 743; 10 T. C. 29, C. A.; 28 (1) Digest (Reissue) 25, *95* 103

Cooper *v.* Willomatt (1845), 1 C. B. 672; [1843–60] All E. R. Rep. 556; 14 L. J. C. P. 219; 5 L. T. O. S. 173; 9 Jur. 598; 3 Digest (Repl.) 177, *372* 161, 243

Coote *v.* Jecks (1872), L. R. 13 Eq. 597; 41 L. J. Ch. 599; 7 Digest (Repl.) 31, *152* 321

Cordova Land Co. *v.* Victor Brothers Incorporated, [1966] 1 W. L. R. 793; 110 Sol. Jo. 290 . . . . 130, 195

Cork Distilleries Co. *v.* G. S. Railway (1874), L. R. 7 H. L. 269; 8 (1) Digest (Reissue) 36, *204* 370

Corn Products Co. *v.* Fry, [1917] W. N. 224; 39 Digest (Repl.) 502, *477* 323

Cornish *v.* Abington (1859), 4 H. & N. 549; 28 L. J. Ex. 262; 21 Digest (Repl.) 369, *1105* 165

PAGE
Cornwal *v*. Wilson (1750), 1 Ves. Sen. 509; 1 Digest (Repl.) 469, *1153* 201, 202
Cort *v*. Ambergate Ry. Co. (1851), 17 Q. B. 127; 20 L. J. Q. B. 460; 17 L. T. O. S. 179; 15 Jur. 877; 39 Digest (Repl.) 659, *1597* 23, 231
Cory *v*. Thames Iron Works Co. (1868), L. R. 3 Q. B. 181; [1861–73] All E. R. Rep. 597; 37 L. J. Q. B. 68; 17 L. T. 495; 16 W. R. 456; 39 Digest (Repl.) 815, *2781* 241
Cory Bros. & Co., Ltd. *v*. Universe Petroleum Co., Ltd. (1933), 46 Ll. L. R. 309; 39 Digest (Repl.) 490, *386*. 365
Cottee *v*. Douglas Seaton (Used Cars), Ltd., [1972] 3 All E. R. 750; [1972] 1 W. L. R. 1408; 137 J. P. 1; 116 Sol. Jo. 821; [1972] R. T. R. 509; Digest Supp. 58, 66
Cotton, *Re*, *Ex parte* Cooke (1913), 108 L. T. 310; 57 Sol. Jo. 343, C. A.; 5 Digest (Repl.) 775, *6612*. 312
Couchman *v*. Hill, [1947] 1 All E. R. 103; [1947] K. B. 554; [1948] L. J. R. 295; 176 L. T. 278; 63 T. L. T. 81, C. A.; 2 Digest (Repl.) 348, *333* 5, 110, 354, 360
County of Durham Electrical Power Co. *v*. Inland Revenue, [1909] 2. K. B. 604; 78 L. J. K. B. 1158; 101 L. T. 51; 73 J. P. 425; 25 T. L. R. 672; 8 L. G. R. 1088, C. A.; 39 Digest (Repl.) 318, *641* . 280
County of Lancaster S.S. *v*. Sharp (1889), 24 Q. B. D. 158; 59 L. J. Q. B. 22; 61 L. T. 692; 6 Asp. M. L. C. 448; 41 Digest (Repl.) 466, *2434*. 359
Coupe *v*. Guyett, [1973] 2 All E. R. 1058; [1973] 1 W. L. R. 669; 137 J. P. 694; 117 Sol. Jo. 415; Digest Supp. 66
Couston *v*. Chapman (1872), L. R. 2 Sc. & Div. 250 H. L.; 39 Digest (Repl.) 561, *898* 109, 140, 203, 245, 267
Couturier *v*. Hastie (1856), 5 H. L. Cas. 673; 25 L. J. Ex. 253; 28 L. T. O. S. 240; 2 Jur. N. S. 1241; 39 Digest (Repl.) 494, *417* 37, 97, 98
Covas *v*. Bingham (1853), 2. E. & B. 836; 2 C. L. R. 212; 23 L. J. Q. B. 26; 18 J. P. 569; 18 Jur. 596; 39 Digest (Repl.) 491, *393* . 257, 364
Coventry *v*. Gladstone (1868), L. R. 6 Eq. 44; 37 L. J. Ch. 492; 16 W. R. 837; 39 Digest (Repl.) 753, *2323* 215, 222
Coventry *v*. Great Eastern Ry. Co. (1883), 11 Q. B. D. 776; 52 L. J. Q. B. 694; 49 L. T. 641, C. A.; 21 Digest (Repl.) 485, *1720* . . 166
Cowas-Jee *v*. Thompson (1845), 3 Moo. Ind. App. 422; 5 Moo. P. C. C. 165; 39 Digest (Repl.) 684, *1792* 293
Cowern *v*. Nield, [1912] 2 K. B. 419; [1911–13] All E. R. Rep. 425; 81 L. J. K. B. 865; 106 L. T. 884; 28 T. L. R. 423; 39 Digest (Repl.) 826, *2878* 88
Cox *v*. Hoare (1906), 95 L. T. 121, affirmed (1907), 96 L. T. 719, C. A.; 39 Digest (Repl.) 479, *273* 273
Cox *v*. Prentice (1815), 3 M. & S. 344; 1 Digest (Repl.) 770, *3031* . 97, 98, 102
Cox *v*. Todd (1825), 7 Dow. and Ry. K. B. 131; 4 L. J. O. S. K. B. 34; 39 Digest (Repl.) 675, *1720* 358
Crane *v*. London Dock Co. (1864), 5 B. & S. 313; [1861–73] All E. R. Rep. 439; 4 New. Rep. 94; 33 L. J. Q. B. 224; 10 L. T. 372; 28 J. P. 565; 10 Jur. N. S. 984; 12 W. R. 745; 33 Digest (Repl.) 490 *451* 169, 170
Cranston *v*. Marshall (1850), 5 Exch. 395; 19 L. J. Ex. 340; 8 (1) Digest (Reissue) 163, *958* 351
Craven *v*. Ryder (1816), 6 Taunt. 433; [1814–23] All E. R. Rep. 432; 2 Marsh 127; Holt, N. P. 100; 39 Digest (Repl.) 769, *2457* . . 221
Cremer *v*. General Carriers, S.A., [1974] 1 All E. R. 1; [1974] 1 W. L. R. 341; 117 Sol. Jo. 873; [1973] 2 Lloyd's Rep. 366. 325
Crichton *v*. Love, 1908, S. C. 818; 45 Sc. L. R. 600; 16 S. L. T. 7; 39 Digest (Repl.) 548, **396* 126
Croft *v*. Lumley (1858), 6 H. L. Cas. 672; [1843–60] All E. R. Rep. 162; 27 L. J. Q. B. 321; 31 L. T. O. S. 382; 22 J. P. 639; 4 Jur. N. S. 903; 6 W. R. 523; 31 (2) Digest (Reissue) 684, *5614* 191

PAGE

Croom and Arthur *v.* Stewart & Co. (1905), 7 F. (Ct. of Sess.) 563; 42 Sc. L. R. 437; 12 S. L. T. 799; 21 Digest (Repl.) 442, **1087* . . 201

Cross *v.* Eglin (1831), 2 B. &. Ad. 106; 9 L. J. O. S. K. B. 145; 39 Digest (Repl.) 690, *1837* 363

Cross *v.* Williams (1862), 7 H. & N. 675; 31 L. J. Ex. 145; 6 L. T. 675; 1 Digest (Repl.) 757, *2954* 93

Crows Transport, Ltd. *v.* Phoenix Assurance Co., Ltd., [1965] 1 All E. R. 596; [1965] 1 W. L. R. 383; 109 Sol. Jo. 70; [1965] 1 Lloyd's Rep. 139, C. A.; Digest Cont. Vol. B p. 457, *3578a* 217

Crowther *v.* Solent Motor Co., [1975] 1 All E. R. 139; [1975] 1 W. L. R. 30, C. A. 134

Crozier & Co. *v.* Auerbach, [1908] 2 K. B. 161; 77 L. J. K. B. 373; 99 L. T. 225; 24 T. L. R. 409; 52 Sol. Jo. 335, C. A.; 39 Digest (Repl.) 580, *1036* 38

Cullinane *v.* British " Rema " Manufacturing Co., Ltd., [1953] 2 All E. R. 1257; [1954] 1 Q. B. 292; [1953] 3 W. L. R. 923; 97 Sol. Jo. 811, C. A.; 39 Digest (Repl.) 592, *1122* 252

Cuming *v.* Brown (1808), 9 East 506; 39 Digest (Repl.) 771, *2475* . . 311

Cummings *v.* London Bullion Co., Ltd., [1951] W. N. 102; 95 Sol. Jo. 157; reversed, [1952] 1 All E. R. 383; [1952] 1 K. B. 327; [1952] 1 T. L. R. 385; 96 Sol. Jo. 148, C. A.; 17 Digest (Reissue) 509, *203* 259, 332

Cunard Steamship Co., Ltd. *v.* Buerger, [1927] A. C. 1; [1926] All E. R. Rep. 103; 96 L. J. K. B. 18; 135 L. T. 494; 42 T. L. R. 653; 17 Asp. M. L. C. 92; 32 Com. Cas. 183, H. L.; 41 Digest (Repl.) 458, *2383* 12

Cundy *v.* Lindsay (1878), 3 App. Cas. 459; [1874–80] All E. R. Rep. 1149; 47 L. J. Q. B. 481; 38 L. T. 573; 14 Cox, C. C. 93; 33 Digest (Repl.) 489, *436* 162, 171, 172, 173

Cunliffe *v.* Harrison (1851), 6 Exch. 903; 20 L. J. Ex. 325; 17 L. T. O. S. 189; 39 Digest (Repl.) 466, *178* 186, 187

Cunningham & Co., Ltd., *Re,* Attenborough's Case (1885), 28 Ch. D. 682; 54 L. J. Ch. 448; 7 Digest (Repl.) 32, *159* 293

Cunningham, Ltd. *v.* Robert Munro & Co., Ltd. (1922), 28 Com. Cas. 42 D. C.; 39 Digest (Repl.) 649, *1550* 34, 35, 158

Curtis *v.* Maloney, [1950] 2 All E. R. 982; [1951] 1 K. B. 736; 66 (pt. 2) T. L. R. 869; 94 Sol. Jo. 761, C. A.; 21 Digest (Repl.) 609, *992* 114, 162

Cusack *v.* Robinson (1861), 1 B. & S. 299; 30 L. J. Q. B. 261; 4 L. T. 506; 7 Jur. N. S. 542; 9 W. R. 735; 39 Digest (Repl.) 457, *93* 210, 212, 213

Cuthbert *v.* Cumming (1855), 11 Exch. 405; 24 L. J. Ex. 310, Ex. Ch.; 17 Digest (Reissue) 67, *711* 363

Cutter *v.* Powell (1795), 6 Term Rep. 320; 2 Smith's L. C. (13th Edn.) 1; 12 Digest (Reissue) 146, *844* 182

Cutter *v.* Powell (1795), 2 Smith's L. C. (7th Edn.) 30 351

Dagenham (Thames) Dock Co., *Re, Ex parte* Hulse (1873), 8 Ch. App. 1022; 43 L. J. Ch. 261; 38 J. P. 180; 21 W. R. 898; 40 Digest (Repl.) 243, *2043* 27

Daimler Co., Ltd. *v.* Continental Tyre & Rubber Co. (Gt. Britain), Ltd., [1916] 2 A. C. 387; [1916–17] All E. R. Rep. 191; 85 L. J. K. B. 1333; 114 L. T. 1049; 22 Com. Cas. 32, H. L.; 2 Digest (Repl.) 219, *315* 89

D'Almeida Araujo (J.) Lda. *v.* Sir Frederick Becker & Co., Ltd., [1953] 2 All E. R. 288; [1953] 2 Q. B. 329; [1953] 3 W. L. R. 57; 97 Sol. Jo. 404; [1953] 2 Lloyd's Rep. 30; 11 Digest (Reissue) 464, *782* . 49

Daniel *v.* Rogers, [1918] 2 K. B. 228; [1918–19] All E. R. Rep. 696; 87 L. J. K. B. 1149; 119 L. T. 212; 62 Sol. Jo. 583, C. A.; 43 Digest (Repl.) 211, *1399* 91

PAGE

Daniels and Daniels *v.* White & Sons & Tarbard, [1938] 4 All E. R. 258; 160 L. T. 128; 82 Sol. Jo. 912; 36 Digest (Repl.) 87, *467* . 119, 133
Darbishire *v.* Warran, [1963] 3 All E. R. 310; [1963] 1 W. L. R. 1067; 107 Sol. Jo. 631; [1963] 2 Lloyd's Rep. 187, C. A.; 17 Digest (Reissue) 128, *269* 117
Darlington (Peter) Partners, Ltd. *v.* Gosho Co., Ltd., [1964] 1 Lloyd's Rep. 149 366
Daulatram Remeshwarlall *v.* European Grain and Shipping, Ltd., [1971] 1 Lloyd's Rep. 368 9
Davey *v.* Paine Brothers (Motors), Ltd., [1954] N. Z. L. R. 1122 167, 174
David T. Boyd & Co., Ltd. *v.* Louca. *See* Boyd (David T.) & Co., Ltd. *v.* Louca
David Taylor & Son, Ltd. *v.* Barnett. *See* Taylor (David) & Son, Ltd. *v.* Barnett
Davies *v.* British Geon, Ltd., [1956] 3 All E. R. 389; [1957] 1 Q. B. 1; [1956] 3 W. L. R. 679; 100 Sol. Jo. 650, C. A.; 50 Digest (Repl.) 381 275
Davies *v.* Customs and Excise Commissioners, [1975] 1 All E. R. 309; [1975] 1 W. L. R. 204; 119 Sol. Jo. 100; [1975] S. T. C. 28 . . 79
Davis *v.* Hedges (1871), L. R. 6 Q. B. 687; 40 L. J. Q. B. 276; 25 L. T. 155; 20 W. R. 60; 39 Digest (Repl.) 586, *1073* 246
Davis *v.* Reilly, [1898] 1 Q. B. 1; 66 L. J. Q. B. 844; 77 L. T. 399; 12 Digest (Reissue) 589, *4144* 230
Davis Contractors *v.* Fareham Urban District Council, [1956] 2 All E. R. 145; [1956] A. C. 696; [1956] 3 W. L. R. 37; 100 Sol. Jo. 378; 54 L. G. R. 289, H. L.; 12 Digest (Reissue) 507, *3518* . . . 21
Davis & Co. *v.* Afa-Minerva (E.M.I.), Ltd., [1974] 2 Lloyd's Rep. 27 . 334
Dawood (Ebrahim), Ltd. *v.* Heath, [1961] 2 Lloyd's Rep. 512; Digest Cont. Vol. B p. 635, *1869a* 186, 188, 257, 366
Dawson *v.* Collis (1851), 10 C. B. 523; 2 L. M. & P. 14; 20 L. J. C. P. 116; 39 Digest (Repl.) 561, *896* 245
Dawson (Clapham) Ltd. *v.* Dutfield, [1936] 2 All E. R. 232; 39 Digest (Repl.) 784, *2590* 79
Debenham *v.* Mellon (1880), 6 App. Cas. 24; 50 L. J. Q. B. 155; 43 L. T. 673; 27 (1) Digest (Reissue) 216, *1495* 91
Debtor, a, *Re*, [1908] 1 K. B. 344; 77 L. J. K. B. 409; 98 L. T. 652; 52 Sol. Jo. 174; 15 Mans. 1, C. A.; 12 Digest (Reissue) 240, *1610* . 230
Decro-Wall International S. A. *v.* Practitioners in Marketing, Ltd., [1971] 2 All E. R. 216; [1971] 1 W. L. R. 361; 115 Sol. Jo. 171, C. A. 190, 191
De Gorter *v.* Attenborough (1904), 21 T. L. R. 19; 1 Digest (Repl.) 392, *540* 297
Delaney *v.* Wallis (1884), 13 L. R. Ir. 31; 15 Cox, C. C. 525; 33 Digest (Repl.) 492, **74* 169, 242
De Lassalle *v.* Guildford, [1901] 2 K. B. 215; [1901] All E. R. Rep. 495; 70 L. J. K. B. 533; 84 L. T. 549; 17 T. L. R. 384; C. A., 39 Digest (Repl.) 513, *555* 4, 5
Demby Hamilton & Co., Ltd. *v.* Barden, [1949] 1 All E. R. 435; 39 Digest (Repl.) 647, *1536* 157, 158, 159
Dempster (R. and J.), Ltd. *v.* Motherwell Bridge and Engineering Co., 1964 S. L. T. 353 . . . . .78, 102
Denbigh Cowan & Co. and Atcherley & Co., *Re* (1921), 90 L. J. K. B. 836; [1921] All E. R. Rep. 245; 125 L. T. 388, C. A.; 39 Digest (Repl.) 710, *1982* 39, 40, 360
Dennant *v.* Skinner and Collom, [1948] 2 All E. R. 29; [1948] 2 K. B. 164; [1948] L. J. R. 1576; 39 Digest (Repl.) 650, *1557* . 143, 144
Denny *v.* Skelton (1916), 115 L. T. 305; 86 L. J. K. B. 280; 13 Asp. M. L. C. 437; 39 Digest (Repl.) 604, *1185* 278
Derfflinger No. 2, The (1918), 87 L. J. P. C. 195; 118 L. T. 521; 14 Asp. M. L. C. 267; P. C.; 37 Digest (Repl.) 470, *202* . . . 155

PAGE

Derry *v.* Peek (1889), 14 App. Cas. 337; [1886–90] All E. R. Rep. 1; 58 L. J. Ch. 864; 61 L. T. 265, H. L.; 35 Digest (Repl.) 45, *395* . . 7, 286

Devaux *v.* Connolly (1849), 8 C. B. 640; 19 L. J. C. P. 71; 39 Digest (Repl.) 824, *2869* 257

Dexters, Ltd. *v.* Hill Crest Oil Co. (Bradford), Ltd., [1926] 1 K. B. 348; [1925] All E. R. Rep. 273; 95 L. J. K. B. 386; 134 L. T. 494; 42 T. L. R. 212; 31 Com. Cas. 161, C. A.; 2 Digest (Repl.) 587, *1175* . 253

Diamond *v.* B. C. Thoroughbred Breeders Society and Boyd (1966), 52 D. L. R. (2d.) 146 . . . . 200, 367

Diamond Alkali Corporation *v.* Bourgeois, [1921] 3 K. B. 443; [1921] All E. R. Rep. 283; 26 Com. Cas. 310; 91 L. J. K. B. 147; 126 L. T. 379; 15 Asp. M. L. C. 455; 39 Digest (Repl.) 707, *1960*. . 39

Dick Bentley Productions, Ltd. *v.* Harold Smith (Motors), Ltd. *See* Bentley (Dick) Productions, Ltd. *v.* Harold Smith (Motors), Ltd.

Dickenson *v.* Naul (1833), 4 B. & Ad. 638; 1 N. & M. 721; 3 Digest (Repl.) 41, *299* 31

Dies *v.* British and International Mining and Finance Corporation, Ltd., [1939] 1 K. B. 724; 108 L. J. K. B. 398; 160 L. T. 563; 39 Digest (Repl.) 826, *2881* 27

Diestal *v.* Stevenson & Co., [1906] 2 K. B. 345; 75 L. J. K. B. 797; 96 L. T. 10; 22 L. T. R. 673; 12 Com. Cas. 1; 39 Digest (Repl.) 813, *2774* 233, 235

Di Ferdinando *v.* Simon, Smits & Co., [1920] 3 K. B. 409, 414; [1920] All E. R. Rep. 347; 89 L. J. K. B. 1039; 124 L. T. 117; 26 Com. Cas. 37; 36 T. L. R. 797, C. A.; 17 Digest (Reissue) 204, *754* 239, 258

Dimmock *v.* Hallett (1866), 2 Ch. App. 21; 37 L. J. Ch. 146; 15 L. T. 374; 31 J. P. 163; 12 Jur. N. S. 953; 15 Digest (Repl.) 53, *339* . 5

Discount Records, Ltd. *v.* Barclays Bank, Ltd. (1974), Times, 18th July 45

Dixon, *Ex parte, Re* Henley (1876), 4 Ch. D. 133; [1874–80] All E. R. Rep. 1004; 46 L. J. Bcy. 20; 35 L. T. 644; 25 W. R. 105; C. A.; 1 Digest (Repl.) 667, *2344* 314

Dixon *v.* Baldwen (1804), 5 East 175; 39 Digest (Repl.) 754, *2333* . 215

Dixon *v.* Fletcher (1837), 3 M. & W. 146; M. & H. 342; 39 Digest (Repl.) 688, *1815* 186

Dixon *v.* London Small Arms Co. (1876), 1 App. Cas. 632; 46 L. J. Q. B. 617; 35 L. T. 559, H. L.; 36 Digest (Repl.) 843, *1926* . . . 76

Dixon *v.* Yates (1833), 5 B. & Ad. 313; [1824–34] All E. R. Rep. 744; 2 N. & M. 177; 2 L. J. K. B. 198; 39 Digest (Repl.) 745, *2226*. 146, 210, 211, 221

Dixon (Peter) & Sons *v.* Henderson & Co., [1919] 2 K. B. 778; 87 L. J. K. B. 683; 118 L. T. 328; 23 Com. Cas. 294, C. A.; 12 Digest (Reissue) 496, *3484* 20

Dixon Kerly, Ltd. *v.* Robinson, [1965] 2 Lloyd's Rep. 404 . . . 135

Doak *v.* Bedford, [1964] 1 All E. R. 311; [1964] 2 Q. B. 587; [1964] 2 W. L. R. 545; 128 J. P. 230; 108 Sol. Jo. 76; 62 L. G. R. 249; 30 Digest (Reissue) 91, *682* 148

Dobel *v.* Stevens (1825), 3 B. & C. 623; 5 Dow. & Ry. K. B. 490; 3 L. J. O. S. K. B. 89; 40 Digest (Repl.) 50, *313* 4

Dobell & Co., Ltd. *v.* Barber and Garratt, [1931] 1 K. B. 219; 100 L. J. K. B. 65; 144 L. T. 266, C. A.; 2 Digest (Repl.) 159, *1164* 250, 251

Doble *v.* David Greig, Ltd., [1972] 2 All E. R. 195; [1972] 1 W. L. R. 703; 136 J. P. 469; 116 Sol. Jo. 217; 70 L. G. R. 411 . . . 57, 59, 61

Docker *v.* Hyams, [1969] 3 All E. R. 808; [1969] 1 W. L. R. 1060; 113 Sol. Jo. 381; [1969] 1 Lloyd's Rep. 487, C. A.; Digest Cont. Vol. C p. 851, *58a* 361, 365

Dodsley *v.* Varley (1840), 12 Ad. & El. 632; 4 Per. & Dav. 448; 5 Jur. 316; 39 Digest (Repl.) 463, *151* 212, 213

Dodson *v.* Wentworth (1842), 4 Man. & G. 1080; 5 Scott, N. R. 821; 12 L. J. C. P. 59; 6 Jur. 1066; 39 Digest (Repl.) 761, *2381* . . 215

PAGE

Doe *v.* Bowater, Ltd., [1916] W. N. 185 . . . . . . 254, 364
Doe *v.* Donston (1818), 1 B. & Ald. 230; 21 Digest (Repl.) 605, *977* . 162
Dominion Coal Co. *v.* Dominion Iron & Steel Co., [1909] A. C. 293; 78 L. J. P. C. 115; 100 L. T. 245; 25 T. L. R. 309; P. C., 39 Digest (Repl.) 535, *708* 189, 245, 367
Dona Mari, The, [1973] 2 Lloyd's Rep. 366 161
Donald *v.* Suckling (1866), L. R. 1 Q. B. 585; 7 B. & S. 783; 35 L. J. Q. B. 232; 14 L. T. 772; 17 Digest (Reissue) 107, *143* . . 83, 283
Donnelly *v.* Rowlands, [1971] 1 All E. R. 9; [1971] 1 W. L. R. 1600; 135 J. P. 100; 69 L. G. R. 42; *sub nom.* Donelly *v.* Rowlands, 114 Sol. Jo. 863; Digest Supp. 57
Doolubdass Pettamberdass *v.* Ramloll Thackoorseydass (1850), 5 Moo. Ind. App. 109, P. C.; 3 Digest (Repl.) 23, *166* 269
Dorab Ally *v.* Abdool Azeez (1878), L. R. 5 Ind. App. 116 . . . 114
Doyle *v.* Olby (Ironmongers), Ltd., [1969] 2 All E. R. 119; [1969] 2 Q. B. 158; [1969] 2 W. L. R. 673; 113 Sol. Jo. 128, C. A.; Digest Cont. Vol. C p. 706, *598a* 7, 334
Drake, *Ex parte*, *Re* Ware (1877), 5 Ch. D. 866; 46 L. J. Bcy. 105; 36 L. T. 667; 25 W. R. 641, C. A.; 46 Digest (Repl.) 525, *683* . 30, 31
Drummond *v.* Van Ingen (1887), 12 App. Cas. 284; 56 L. J. Q. B. 563; 57 L. T. 1; 36 W. R. 20; 3 T. L. R. 541, H. L.; 39 Digest (Repl.) 559, *885* 10, 138, 139
Drury *v.* Victor Buckland, Ltd., [1941] 1 All E. R. 269; 85 Sol. Jo. 117, C. A.; 39 Digest (Repl.) 539, *748*. 122
Dublin City Distillery Co. *v.* Doherty, [1914] A. C. 823; 83 L. J. P. C. 265; 111 L. T. 81; 58 Sol. Jo. 413, H. L.; 39 Digest (Repl.) 467, *183* 277
Du Jardin *v.* Beadman Brothers, Ltd., [1952] 2 All E. R. 160; [1952] 2 Q. B. 712; [1952] 1 T. L. R. 1601; 96 Sol. Jo. 414; 1 Digest (Repl.) 392, *538* 174, 295
Dumenil (Peter) & Co., Ltd. *v.* James Ruddin, Ltd., [1953] 2 All E. R. 294; [1953] 1 W. L. R. 815; 97 Sol. Jo. 437, C. A.; 12 Digest (Reissue) 414, *3047* 190
Duncan *v.* Topham (1849), 8 C. B. 225; 18 L. J. C. P. 310; 13 L. T. O. S. 304; 39 Digest (Repl.) 679, *1743* 358
Duncan Fox & Co. *v.* Schrempft, [1915] 3 K. B. 355; 84 L. J. K. B. 2206; 113 L. T. 600; 20 Cam. Cas. 337; 31 T. L. R. 491, C. A.; 12 Digest (Reissue) 492, *3469* 39
Dunkirk Colliery *v.* Lever (1878), 9 Ch. D. 20; 39 L. T. 239; 17 Digest (Reissue) 126, *259* 233
Dunlop *v.* Grote (1845), 2 C. & K. 153; 39 Digest (Repl.) 786, *2596* . 227
Dunlop *v.* Higgins (1848), 1 H. L. Cas. 381; 12 Jur. 295; 17 Digest (Reissue) 64, *685* 237
Dunlop *v.* Lambert (1839), 6 Cl. & Fin. 600; Mall. & Rob. 663; 39 Digest (Repl.) 611, *1244* 191, 368
Dunlop Pneumatic Tyre Co., Ltd. *v.* New Garage and Motor Co., Ltd., [1915] A. C. 79; [1914–15] All E. R. Rep. 333; 83 L. J. K. B. 1574; 111 L. T. 862; 30 T. L. R. 625; 17 Digest (Reissue) 178, *551* . . 239
Dunlop Pneumatic Tyre Co. *v.* Selfridge & Co., [1915] A. C. 847; [1914–15] All E. R. Rep. 333; 84 L. J. K. B. 1680; 113 L. T. 386; 31 T. L. R. 399; 59 Sol. Jo. 439, H. L.; 39 Digest (Repl.) 645, *1523* 53
Dupont *v.* British South Africa Co. (1901), 18 T. L. R. 24; 39 Digest (Repl.) 634, *1445* 40
Dutton *v.* Solomonson (1803), 3 B. &. P. 582; 39 Digest (Repl.) 701, *1925* 191
Duval (Charles) *v.* Gans, [1904] 2 K. B. 685; 73 L. J. K. B. 907; 91 L. T. 308; 20 T. L. R. 705, C. A.; 12 Digest (Reissue) 568, *3963* . 228
Dyal Singh *v.* Kenyan Insurance, Ltd., [1954] 1 All E. R. 847; [1954] A. C. 287; [1954] 2 W. L. R. 607; 98 Sol. Jo. 231, P. C.; 44 Digest (Repl.) 248, *716* 162

PAGE

Eastern Distributors, Ltd. *v.* Goldring, [1957] 2 All E. R. 525; [1957] 2 Q. B. 600; [1957] 3 W. L. R. 237; 101 Sol. Jo. 533, C. A.; 39 Digest (Repl.) 656, *1582* . . . 166, 167, 176, 306
Eastgate, *Re, Ex parte* Ward, [1905] 1 K. B. 465; [1904–07] All E. R. Rep. 890; 74 L. J. K. B. 324; 92 L. T. 207; 21 T. L. R. 198; 53 W. R. 432; 12 Mans. 11; 39 Digest (Repl.) 829, *2902* . . 172, 226
Eaton's (Mrs.) Car Sales *v.* Thomesen, [1973] N. Z. L. R. 686 . . 253
Ebbw Vale Steel Co. *v.* Blaina Iron Co. (1901), 6 Com. Cas. 33, C. A.; 39 Digest (Repl.) 518, *593* 189
Eberle's Hotels & Restaurant Co. *v.* Jonas (1887), 18 Q. B. D. 459; 56 L. J. Q. B. 278; 35 W. R. 467; 3 T. L. R. 421, C. A.; 43 Digest (Repl.) 522, *667* 30
Ebrahim Dawood, Ltd. *v.* Heath. *See* Dawood (Ebrahim), Ltd. *v.* Heath.
Edelstein *v.* Schuler, [1902] 2 K. B. 144; [1900–03] All E. R. Rep. 884; 71 L. J. K. B. 572; 87 L. T. 204; 18 T. L. R. 597; 7 Com. Cas. 172; 1 Digest (Repl.) 788, *3160* 271
Edwards, *Re, Ex parte* Chalmers (1873), 8 Ch. App. (289); 42 L. J. Bcy. 37; 28 L. T. 325; 21 W. R. 349; 39 Digest (Repl.) 697, *1894* 205, 208, 210, 211, 214
Edwards *v.* Skyways, Ltd., [1964] 1 All E. R. 494; [1964] 1 W. L. R. 349; 108 Sol. Jo. 279; 12 Digest (Reissue) 24, *19* 1
Edwards, Ltd. *v.* Vaughan (1910), 26 T. L. R. 545, C. A.; 39 Digest (Repl.) 624, *1358* 149, 152, 309
Eichengruen *v.* Mond, [1940] 3 All E. R. 148; [1940] Ch. 785; 109 L. J. Ch. 468; 163 L. T. 219; 56 T. L. R. 845; 84 Sol. Jo. 426, C. A.; 2 Digest (Repl.) 243, *452* 90
Elbinger Actien Gesellschaft *v.* Armstrong (1874), L. R. 9 Q. B. 473; 43 L. J. Q. B. 211; 30 L. T. 871; 38 J. P. 774; 23 W. R. 127; 39 Digest (Repl.) 810, *2752* 239, 240, 241
Electric Construction Co. *v.* Hurry and Young (1897), 24 R. (Ct. of Sess.) 312; 34 Sc. L. R. 295; 4 S. L. T. 287; 39 Digest (Repl.) 578, **565* 201
Elkington & Co., Ltd. *v.* Amery, [1936] 2 All E. R. 86; 80 Sol. Jo. 465, C. A.; 28 (2) Digest (Reissue) 697, *330* 89
Ellen *v.* Topp (1851), 6 Exch. 424; 20 L. J. Ex. 241; 15 Jur. 451; 12 Digest (Reissue) 525, *3634* 108
Elliman Sons & Co., *v.* Carrington & Son Ltd., [1901] 2 Ch. 275; 70 L. J. Ch. 577; 84 L. T. 858; 49 W. R. 532; 45 Sol. Jo. 536; 39 Digest (Repl.) 644, *1517* 52
Elliot *v.* Grey, [1959] 3 All E. R. 733; [1960] 1 Q. B. 367; [1959] 3 W. L. R. 956; 124 J. P. 58; 103 Sol. Jo. 921; 57 L. G. R. 357; 45 Digest (Repl.) 65, *187* 74
Ellis *v.* Hunt (1789), 3 Term Rep. 464; 39 Digest (Repl.) 754, *2331* . 184
Ellis *v.* Thompson (1838), 3 M. & W. 445; 1 Horn & H. 131; 7 L. J. Ex. 185; 39 Digest (Repl.) 675, *1726* 183
Ellis *v.* Stenning, [1932] 2 Ch. 81; [1932] All E. R. Rep. 597; 101 L. J. Ch. 401; 147 L. T. 449; 76 Sol. Jo. 232; 21 Digest (Repl.) 457, *1576* 31
Elmdove, Ltd. *v.* Keech (Trading as Electronic Services) (1969), 113 Sol. Jo. 871, C. A. 107
Elmore *v.* Stone (1809), 1 Taunt. 458; 39 Digest (Repl.) 467, *184* 184, 277
Elorio, The, 1916 S. C. 882 245
Elphick *v.* Barnes (1880), 5 C. P. D. 321; 49 L. J. C. P. 698; 39 Digest (Repl.) 496, *422* 99, 147, 152, 157, 158
Elson *v.* Prices Tailors, Ltd., [1963] 1 All E. R. 231; [1963] 1 W. L. R. 287; 107 Sol. Jo. 195; 40 Tax Cas. 671; 41 A. T. C. 345; 28 (1) Digest (Reissue) 53, *213* 27
Elwin *v.* O'Reagan and Maxwell, [1971] N. Z. L. R. 1124 . . . 310
Emanuel *v.* Dane (1812), 3 Camp. 299; 39 Digest (Repl.) 581, *1038* . 80

PAGE

Emanuel *v.* Sammut, [1959] 2 Lloyd's Rep. 629 36
Embiricos *v.* Anglo-Austrian Bank, [1905] 1 K. B. 677; 74 L. J. K. B. 326; 92 L. T. 305; 10 Com. Cas. 99, C. A.; 3 Digest (Repl.) 257, *725* 49
Emmanuel Ayodeji Ajayi (Trading under name and style of Colony Carrier Co.) *v.* R. T. Briscoe (Nigeria), Ltd. *See* Ajayi (Emmanuel Ayodeji) (Trading under name and style of Colony Carrier Co.) *v.* R. T. Briscoe (Nigeria), Ltd.
Emmerson *v.* Heelis (1809), 2 Taunt. 38; 39 Digest (Repl.) 302, *472* . 267
Enrico Furst & Co. *v.* W. E. Fischer, Ltd. *See* Furst (Enrico) & Co. *v.* W. E. Fischer, Ltd.
Erie Natural Gas Co. *v.* Carroll, [1911] A. C. 105; 80 L. J. P. C. 59; 103 L. T. 678; 17 Digest (Reissue) 104, *119* 237, 238
Ertel Bieber & Co. *v.* Rio Tinto Co., [1918] A. C. 260; [1918–19] All E. R. Rep. 127; 87 L. J. K. B. 531; 118 L. T. 181; 23 Com. Cas. 243; 34 T. L. R. 208, H. L.; 2 Digest (Repl.) 434, *92* . 20, 329
Esmail *v.* Rosenthal & Sons, Ltd., [1964] 109 Sol. Jo. 839; [1964] 2 Lloyd's Rep. 447, C. A.; affirmed *sub nom.* Rosenthal (J.) & Sons, Ltd. *v.* Esmail (Trading as M. H. Esmail & Sons) [1965] 2 All E. R. 860; [1965] 1 W. L. R. 1117; 109 Sol. Jo. 553; [1965] 2 Lloyd's Rep. 171, H. L.; Digest Cont. Vol. B p. 632, *1011a* 362
Esposito *v.* Bowden (1857), 7 E. & B. 763; [1843–60] All E. R. Rep. 39; 27 L. J. Q. B. 17; 3 Jur. N. S. 1209; 2 Digest (Repl.) 257, *564* . 20
Esso *v.* Customs and Excise Commissioners (1975), Times, 31st January, C. A. 78, 130
Esso Petroleum Co., Ltd. *v.* Harper's Garage (Stourport), Ltd., [1967] 1 All E. R. 699; [1968] A. C. 269; [1967] 2 W. L. R. 871; 111 Sol. Jo. 174, H. L.; Digest Cont. Vol. C p. 985, *132a* . . . 52, 271
European Grain and Shipping, Ltd. *v.* J. H. Rayner & Co., Ltd., [1970] 2 Lloyd's Rep. 239 358
Evans *v.* British Doughnut Co., Ltd., [1944] 1 All E. R. 158; [1944] K. B. 102; 113 L. J. K. B. 177; 170 L. T. 248; 108 J. P. 59; 60 T. L. R. 156; 42 L. G. R. 270; 46 Digest (Repl.) 171, *1138* . . 57

F. and B. Transport, Ltd. *v.* White Truck Sales Manitoba, Ltd. (1965), 51 W. W. R. 124; Digest Cont. Vol. B p. 544, **179b* . . 362, 367
F.M.C. (Meat), Ltd. *v.* Fairfield Cold Stores, Ltd., [1971] 2 Lloyd's Rep. 221 325
Fairclough, Dodd and Jones *v.* J. H. Vantol, Ltd. *See* Vantol, Ltd. *v.* Fairclough, Dodd and Jones.
Fairmaner *v.* Budd (1831), 7 Bing. 574; 12 Digest (Reissue) 617, *4348*. 80
Falk, *Ex parte*. *See* Kiell, *Re Ex parte* Falk.
Farina *v.* Home (1846), 16 M. & W. 119; 16 L. J. Ex. 73; 8 L. T. O. S. 277; 39 Digest (Repl.) 459, *111*. 183, 277
Farmeloe *v.* Bain (1876), 1 C. P. D. 445; 45 L. J. C. P. 264; 34 L. T. 324; 39 Digest (Repl.) 775, *2513* 221
Farmers' and Settlers' Co-operative Society, Ltd., The, *Re*, City Bank of Sydney *v.* Barden (1908), 9 S. R. (N. S. W.) 41; 1 Digest (Repl.) 396 **416* 299
Farnworth Finance Facilities, Ltd. *v.* Attryde, [1970] 2 All E. R. 774; [1970] 1 W. L. R. 1053; 114 Sol. Jo. 354, C. A.; Digest Cont. Vol. C p. 417, *35a* 11, 12
Farquharson Brothers & Co. *v.* King & Co., [1902] A. C. 325; [1900–03] All E. R. Rep. 120; 71 L. J. K. B. 667; 86 L. T. 810, 18 T. L. R. 665, H. L.; 21 Digest (Repl.) 367, *1092* . . 163, 164, 165, 166
Fawcett *v.* Smethurst (1914), 84 L. J. K. B. 473; 112 L. T. 309; 31 T. L. R. 85; 59 Sol. Jo. 220; 28 (2) Digest (Reissue) 705, *410*. . 88
Federspiel *v.* Twigg, [1957] 1 Lloyd's Rep. 240 . . . 143, 148, 153, 154

PAGE

Feise *v.* Wray (1802), 3 East. 93; 39 Digest (Repl.) 741, *2193* . . 205
Feliciana, The (1915), 59 Sol. Jo. 546; 39 Digest (Repl.) 750, *2287* . 287
Felston Tile Co., Ltd. *v.* Winget, Ltd., [1936] 3 All E. R. 473, C. A. ; 26 Digest (Repl.) 660, *13* 82
Fenton *v.* Scotty's Car Sales, Ltd., [1968] N. Z. L. R. 929; Digest Cont. Vol. C p. 924, **794Aa* 18
Fenwick *v.* Macdonald Fraser & Co., 1904, 6 F. (Ct. of Sess.) 850; 3 Digest (Repl.) 46, **183* 267
Ferens *v.* O'Brien (1883), 11 Q. B. D. 21; 52 L. J. M. C. 70; 15 Cox, C. C. 332; 15 Digest (Repl.) 1085, *10,735* 280
Ferrier, *Re, Ex parte* Trustee *v.* Donald, [1944] 295; 114 L. J. Ch. 15; 60 T. L. R. 295; 88 Sol. Jo. 171; 39 Digest (Repl.) 624, *1364* 147, 149, 179
Fessard *v.* Mugnier (1865), 18 C. B. N. S. 286; 34 L. J. C. P. 126; 12 Digest (Reissue) 566, *3948* 228
Fibrosa Société Anonyme *v.* Fairbairn Lawson Combe Barbour, Ltd., [1941] 2 All E. R. 300; [1942] 1 K. B. 12; 110 L. J. K. B. 666; 165 L. T. 73; 57 T. L. R. 547; 46 Com. Cas. 229; 70 Ll. L. Rep. 30; reversed, [1942] 2 All E. R. 122; [1943] A. C. 32; 111 L. J. K. B. 433; 167 L. T. 101; 58 T. L. R. 308; 86 Sol. Jo. 232; 73 Ll. L Rep. 45; 12 Digest (Reissue) 495, *3479* 22, 328
Field *v.* Lelean (1861), 6 H. & N. 617; L. J. Ex. 168; 4 L. T. 121; 7 Jur. N. S. 918; 9 W. R. 387; 44 Digest (Repl.) 429, *355* . 4, 182, 209, 357
Findlay Bannatyne & Co. *v.* Donaldson (1842), 5 Bell, S. C. App. 105, H. L. 270
Finlay & Co., Ltd. *v.* Kwik Hoo Tong Handel Maatschappij, [1929] 1 K. B. 400; [1928] All E. R. Rep. 110; 98 L. J. K. B. 251; 140 L. T. 389; 34 Com. Cas. 143; 45 T. L. R. 149, C. A.; 17 Digest (Reissue) 127, *265* 38, 253, 255
Fischel *v.* Scott (1854), 15 C. B. 69; 2 C. L. R. 1774; 39 Digest (Repl.) 493, *408* 356
Fisher, Ltd. *v.* Eastwoods, Ltd., [1936] 1 All E. R. 421; 12 Digest (Reissue) 435, *3133* 22
Fisher, Reeves & Co. *v.* Armour & Co., [1920] 3 K. B. 614; 90 L. J. K. B. 172; 124 L. T. 122; 36 T. L. R. 800; 64 Sol. Jo. 698; 15 Asp. M. L. C. 91; 26 Com. Cas. 46, C. A.; 39 Digest (Repl.) 535, *709* 41, 197, 201, 359
Fixby Engineering Co., Ltd. *v.* Auchlocham Sand and Gravel Co., Ltd., 1974 S. L. T. (Sh. Ct.) 58. 191
Fletcher *v.* Budgen, [1974] 2 All E. R. 1243; [1974] 1 W. L. R. 1056; 138 J. P. 582; 118 Sol. Jo. 498; 59 Cr. App. Rep. 234 58
Fletcher *v.* Sledmore (1973), L. G. R. 179. 58
Fletcher *v.* Tayleur (1885), 17 C. B. 21; 25 L. J. C. P. 65; 26 L. T. O. S. 60; 39 Digest (Repl.) 815, *2779*. 241
Flory *v.* Denny (1852), 7 Exch. 581; 21 L. J. Ex. 223; 19 L. T. O. S. 158; 7 Digest (Repl.) 5, *9*. 83
Foaminol Laboratories *v.* British Artid Plastics, [1941] 2 All E. R. 393; 17 Digest (Reissue) 156, *432* 252, 254
Foley *v.* Classique Coaches, Ltd., [1934] 2 K. B. 1; [1934] All E. R. Rep. 88; 103 L. J. K. B. 550; 151 L. T. 242, C. A.; 40 Digest (Repl.) 359, *2880* 102
Foley Motors, Ltd. *v.* McGhee, [1970] N. Z. L. R. 649. 337
Folkes *v.* King, [1923] 1 K. B. 282; [1922] All E. R. Rep. 658; 128 L. T. 405; 39 T. L. R. 77; 28 Com. Cas. 110, C. A.; 1 Digest (Repl.) 390, *531* 172, 174, 295
Forbes *v.* Smith (1863), 2 New Rep. 19; 11 W. R. 574; 39 Digest (Repl.) 784, *2582* 227
Ford *v.* Yates (1841), 2 Man. & G. 549; 2 Scott N. R. 645; 10 L. J. C. P. 117; 39 Digest (Repl.) 660, *1603* 182

PAGE

Ford Motor Co. of Canada, Ltd. *v.* Haley, [1967] S. C. R. 437; 60 W. W. R. 497; Digest Cont. Vol. C p. 856, **586a* 248

Foreman and Ellams, Ltd. *v.* Blackburn, [1928] 2 K. B. 60; [1928] All E. R. Rep. 512; 97 L. J. K. B. 355; 139 L. T. 68; 33 Com. Cas. 358; 39 Digest (Repl.) 706, *1957* 355

Forget *v.* Ostigny, [1895] A. C. 318; 64 L. J. P. C. 62; 72 L. T. 399, P. C.; 16 Digest (Repl.) 193, *804* 103

Forrestt *v.* Aramayo (1900), 83 L. T. 335; 9 Asp. M. L. C. 134, C. A.; 39 Digest (Repl.) 671, *1697* 182, 205

Forsyth *v.* Jervis (1816), 1 Stark. 437; 39 Digest (Repl.) 789, *2619* . 79

Foster *v.* Driscoll, [1929] 1 K. B. 470; [1928] All E. R. Rep. 130; 98 L. J. K. B. 282; 140 L. T. 479; 45 T. L. R. 185, C. A.; 6 Digest (Repl.) 65, *588* 48

Fothergill *v.* Rowland (1873), L. R. 17 Eq. 132; 43 L. J. Ch. 252; 29 L. T. 414; 38 J. P. 244; 22 W. R. 42; 28 (2) Digest (Reissue) 1047, *686* 245

Fragano *v.* Long (1825), 4 B. & C. 219; [1824–34] All E. R. Rep. 171; 6 Dow. & Ry. K. B. 283; 3 L. J. O. S. K. B. 177; 39 Digest (Repl.) 609, *1227* 40

France *v.* Gaudet (1871), L. R. 6 Q. B. 199; 40 L. J. Q. B. 121; 19 W. R. 622; 39 Digest (Repl.) 824, *2865* 242

Francis, *Ex parte* (1887), 4 Morr. 146; 39 Digest (Repl.) 762, *2389* . 215

Francis *v.* Trans-Canada Trailer Sales, Ltd. (1969), 6 D. L. R. (3d) 705; Digest Cont. Vol. C p. 860, **1969fb* 361

Franklin *v.* Neate (1844), 13 M. & W. 481; 14 L. J. Ex. 59; 4 L. T. O. S. 214; 39 Digest (Repl.) 447, *26*. 283

Frebold *v.* Circle Products, Ltd. (1970), 114 Sol. Jo. 262; [1970] 1 Lloyd's Rep. 499, C. A.; Digest Cont. Vol. C p. 857, *1794a* 33, 34, 35, 36

Freedhoff *v.* Pomalift Industries, Ltd. (1971), 13 D. L. R. (3d) 523 . 254

Freeman *v.* Appleyard (1862), 1 New Rep. 30; 32 L. J. Ex. 175; 1 Digest (Repl.) 389, *523* 281

Freeman *v.* Consolidated Motors, Ltd. (1968), 69 D. L. R. (2d) 581 . 6, 201

Freeman *v.* Cooke (1848), 2 Exch. 654; [1843–60] All E. R. Rep. 185; 18 L. J. Ex. 114; 12 Digest (Reissue) 85, *437* 165

Freeth *v.* Burr (1874), L. R. 9 C. P. 208; [1874–80] All E. R. Rep. 751; 43 L. J. C. P. 91; 29 L. T. 773; 22 W. R. 370; 39 Digest (Repl.) 696, *1887* 190

French *v.* Gething, [1921] 3 K. B. 280; affirmed, [1922] 1 K. B. 236; [1921] All E. R. Rep. 415; 91 L. J. K. B. 276; 126 L. T. 301; 38 T. L. R. 77; 66 Sol. Jo. 140; [1922] B. & C. R. 30, C. A.; 27 (1) Digest (Reissue) 188, *1254* 278

Frost *v.* Aylesbury Dairy Co., [1905] 1 K. B. 608; [1904–07] All E. R. Rep. 132; 74 L. J. K. B. 386; 92 L. T. 527; 53 W. R. 354; 21 T. L. R. 300; 39 Digest (Repl.) 546, *792* 125

Fuentes *v.* Montis (1868), L. R. 3 C. P. 268; 37 L. J. C. P. 137; 18 L. T. 21; 16 W. R. 900; affirmed, L. R. 4 C. P. 93; 38 L. J. C. P. 95; 19 L. T. 364; 17 W. R. 208; 1 Digest (Repl.) 391, *537* 289, 300, 313

Fuld (No. 3), *In the Estate of,* Hartley *v.* Fuld, [1965] 3 All E. R. 776; [1968] P. 675; [1966] 2 W. L. R. 717; 110 Sol. Jo. 133; 11 Digest (Reissue) 445, *697* 49

Fuller *v.* Abrahams (1821), 6 Moore 316; 3 Br. & B. 116; 3 Digest (Repl.) 23, *171* 269

Funduck and Horncastle, *Re* (1974), 39 D. L. R. (3d) 94 . . . 307

Furniss *v.* Scholes, [1974] R. T. R. 133 57

First (Enrico) & Co. *v.* W. E. Fischer, Ltd., [1960] 2 Lloyd's Rep. 340 43, 44, 107

PAGE

Gabarron *v.* Kreeft (1875), L. R. 10 Ex. 274; 44 L. J. Ex. 274; 33 L. T. 365; 24 W. R. 146; 3 Asp. M. L. C. 36; 39 Digest (Repl.) 599, *1154* 35

Gabbiano, The, [1940] P. 166; 109 L. J. P. 74; 165 L. T. 5; 56 T. L. R. 774; 84 Sol. Jo. 394; 45 Com. Cas. 235; 19 Asp. M. L. C. 371; 37 Digest (Repl.) 470, *203* 33

Gabriel Wade and English, Ltd. *v.* Arcos Ltd. (1929), 34 Ll. L. Rep. 306 187

Galbraith and Grant *v.* Block, [1922] 2 K. B. 155; [1922] All E. R. Rep. 443; 91 L. J. K. B. 649; 127 L. T. 521; 38 T. L. R. 669; 66 Sol. Jo. 596; 39 Digest (Repl.) 666, *1664* 184, 185, 192

Gale *v.* New, [1937] 4 All E. R. 645; 54 T. L. R. 213; 82 Sol. Jo. 14, C. A.; 39 Digest (Repl.) 616, *1295* 141

Gallagher *v.* Shilcock, [1949] 1 All E. R. 921; [1949] 2 K. B. 765; [1949] L. J. R. 1721; 65 T. L. R. 496; 93 Sol. Jo. 302, 39 Digest (Repl.) 776, *2522* 27, 224

Gallaher (or Gallacher), Ltd. *v.* Supersafe Supermarkets, Ltd., [1964] L. R. 5 R. P. 89; 108 Sol. Jo. 878; Digest Cont. Vol. B p. 703, *154* 53

Gapp *v.* Bond (1887), 19 Q. B. D. 200; 56 L. J. Q. B. 438; 57 L. T. 437, C. A.; 7 Digest (Repl.) 29, *144*. 321

Gardano and Giamperi *v.* Greek Petroleum, George Mamidakis & Co., [1961] 3 All E. R. 919; [1962] 1 W. L. R. 40; 106 Sol. Jo. 76; [1961] 2 Lloyd's Rep. 259; 41 Digest (Repl.) 271, *876*. . . . 317

Gardiner *v.* Gray (1815), 4 Camp. 144; 39 Digest (Repl.) 556, *861*. . 139

Garnac Grain Co. Inc. *v.* H. M. F. Faure and Fairclough, Ltd., [1967] 2 All E. R. 353; [1968] A. C. 1130; [1967] 3 W. L. R. 143; 113 Sol. Jo. 434; [1967] 1 Lloyd's Rep. 405, H. L.; Digest Cont. Vol. C p. 11, *2725a*. 238

Gattorno *v.* Adams (1862), 12 C. B. N. S. 560; 39 Digest (Repl.) 519, *595* 355

Gavin's Trustee *v.* Fraser, 1920 S. C. 674; 39 Digest (Repl.) 449, **32* . 272

Geddling *v.* Marsh, [1920] 1 K. B. 668; [1920] All E. R. Rep. 631; 89 L. J. K. B. 526; 122 L. T. 775; 36 T. L. R. 337; 39 Digest (Repl.) 542, *780* 125, 130

General and Finance Facilities, Ltd. *v.* Hughes (1966), 110 Sol. Jo. 847 166

General Trading Co., *Re* (1911), 16 Com. Cas. 95; 39 Digest (Repl.) 517, *586* 106

Genn *v.* Winkel (1912), 107 L. T. 434; [1911–13] All E. R. Rep. 910; 28 T. L. R. 483; 56 Sol. Jo. 612; 17 Com. Cas. 323, C. A.; 39 Digest (Repl.) 629, *1411* 149

George Cohen & Sons & Co., Ltd. *v.* Jamieson and Patterson. *See* Cohen (George) & Sons & Co., Ltd. *v.* Jamieson and Patterson.

George Reid Inc. *v.* Nelson Machinery Co. *See* Reid (George) Inc. *v.* Nelson Machinery Co.

George Whitechurch, Ltd. *v.* Cavanagh. *See* Whitechurch (George), Ltd. *v.* Cavanagh.

Gian Singh & Co., Ltd. *v.* Banque de l'Indochine. *See* Singh (Gian) & Co., Ltd. *v.* Banque de l'Indochine.

Gibbes, *Ex parte* (1875), 1 Ch. D. 101; 45 L. J. Bk. 10; 33 L. T. 479; 39 Digest (Repl.) 757, *2349* 215

Gibraltar Packers, Ltd. *v.* Basic Economy and Development Corporation, [1966] 1 Lloyd's Rep. 615 100

Gibson *v.* Carruthers (1841), 8 M. & W. 321; [1835–42] All E. R. Rep. 565; 11 L. J. Ex. 138; 39 Digest (Repl.) 775, *2514* 214

Gieve, *Re*, [1899] 1 Q. B. 794; 68 L. J. Q. B. 509; 80 L. T. 438; 47 W. R. 441; 15 T. L. R. 251; 43 Sol. Jo. 334; 6 Mans, 136, C. A.; 44 Digest (Repl.) 440, *446* 103

Gillard *v.* Brittan (1841), 1 Dowl. N. S. 424; 8 M. & W. 575; 11 L. J. Ex. 133; 39 Digest (Repl.) 780, *2552* 242

Gillett *v.* Hill (1834), 2 Cr. & M. 530; 4 Tyr. 290; 3 L. J. Ex. 145; 39 Digest (Repl.) 600, *1161* 142, 166

PAGE

Ginzberg *v.* Barrow Haematite Steel Co., Ltd. and McKellor, [1966] 1 Lloyd's Rep. 343 . . . . 37, 143

Glasgow and South Western Rail. Co. *v.* Boyd and Forrest, [1915] A. C. 526; 84 L. J. P. C. 157, P. C.; 35 Digest (Repl.) 83, *765* . . 6

Glass's Fruit Markets, Ltd. *v.* A. Southwell & Son (Fruit), Ltd., [1969] 2 Lloyd's Rep. 398 365

Gloucestershire County Council *v.* Richardson, [1968] 3 All E. R. 1181; [1969] 1 A. C. 480; [1968] 3 W. L. R. 645; 112 Sol. Jo. 759; 67 L. G. R. 15, H. L.; Digest Cont. Vol. C p. 60, *254b* . . . 77, 122

Glyn Mills & Co. *v.* East and West India Docks (1882), 7 App. Cas. 591; [1881–85] All E. R. Rep. 674; 52 L. J. Q. B. 146; 47 L. T. 309; 31 W. R. 201; 4 Asp. M. L. C. 580, H. L.; 41 Digest (Repl.) 266, *812* 319

Glynn *v.* Margestson & Co., [1893] A. C. 351; [1891–4] All E. R. Rep. 693; 62 L. K. J. Q. B. 466; 69 L. T. 1; 9 T. L. R. 437; 7 Asp. M. L. C. 366; 1 R. 193, H. L.; 41 Digest (Repl.) 163, *82* 12

Godley *v.* Perry, [1960] 1 All E. R. 36; [1960] 1 W. L. R. 9; 104 Sol. Jo. 16; 39 Digest (Repl.) 529, *665* 119, 140

Godts *v.* Rose (1885), 17 C. B. 229; 25 L. J. C. P. 61; 1 Jur. N. S. 1173; 26 L. T. O. S. 240; 4 W. R. 129; 39 Digest (Repl.) 602, *1174* 148, 183, 278, 357

Golding Davis & Co., *Ex parte*, *Re* Knight (1880), 13 Ch. D. 628; 42 L. T. 270; 28 W. R. 481, C. A.; 39 Digest (Repl.) 769, *2459* 221, 223

Goldshede *v.* Cottrell (1836), 2 M. & W. 20; 6 L. J. Ex. 26; 6 Digest (Repl.) 329, *2404* 230

Gompertz *v.* Denton (1832), 1 Cr. & M. 207; 1 Dowl. 623; 3 Tyr. 233; 2 L. J. Ex. 82; 39 Digest (Repl.) 581, *1043* 245

Goodbody, *Re* (1899), 82 L. T. 484; 9 Asp. M. L. C. 69; 5 Com. Cas. 59, C. A.; 39 Digest (Repl.) 710, *1977* 356

Goodlock *v.* Cousins, [1897] 1 Q. B. 558; 66 L. J. Q. B. 558; 76 L. T. 313, C. A.; 21 Digest (Repl.) 609, *990*. 163

Goodyear Tyre and Rubber (Great Britain), Ltd. *v.* Lancashire Batteries, Ltd., [1958] 2 All E. R. 324; [1958] 1 W. L. R. 655; L. R. R. P. 22; 102 Sol. Jo. 453; reversed, [1958] 3 All E. R. 7; L. R. 1 R. P. at p. 29; [1958] 1 W. L. R. 857; 102 Sol. Jo. 581, C. A.; 45 Digest (Repl.) 402, *150* 53

Gordon *v.* Whitehouse (1856), 18 C. B. 747; 25 L. J. C. P. 300; 39 Digest (Repl.) 503, *485* 230

Gore *v.* Gibson (1845), 13 M. & W. 623; 14 L. J. Ex. 151; 4 L. T. O. S. 319; 9 Jur. 140; 39 Digest (Repl.) 453, *64* 87

Gorrissen *v.* Perrin (1857), 2 C. B. N. S. 681; 27 L. J. C. P. 29; 29 L. T. O. S. 227; 3 Jur. N. S. 867; 5 W. R. 709; 39 Digest (Repl.) 493, *407* 356, 363

Gosling *v.* Anderson (1972), Times, 8th February 334

Goss *v.* Lord Nugent (1833), 5 B. & Ad. 58; [1824–34] All E. R. Rep. 305; 2 N. & M. 28; 2 L. J. K. B. 127; 12 Digest (Reissue) 438, *3166* 4, 22

Graff *v.* Evans (1882), 8 Q. B. D. 373; [1881–85] All E. R. Rep. 211; 51 L. J. M. 25; 46 L. T. 347; 46 J. P. 262; 30 W. R. 380 D. C.; 8 (2) Digest (Reissue) 637, *137*. 80

Grafton *v.* Armitage (1845), 2 C. B. 336; 15 L. J. P. C. 20; 6 L. T. O. S. 152; 9 Jur. 1039; Digest Cont. Vol. B p. 31, *274Aa* 78

Grant *v.* Australian Knitting Mills, Ltd., [1936] A. C. 85; 105 L. J. P. C. 6; 154. L T. 18; 52 T. L. R. 38; 79 Sol. Jo. 815; P. C.; 36 Digest (Repl.) 86, *461* 119, 130, 134

Grant *v.* Norway (1851), 10 C. B. 665; 20 L. J. C. P. 93; 16 L. T. O. S. 504; 15 Jur. 296; 41 Digest (Repl.) 539, *3167* 318

Grantham *v.* Hawley (1615), Hob. Rep. 132; 2 Roll. 48, pl. 20; 39 Digest (Repl.) 610, *1229* 96

Gratitudine, The (1801) 3 Ch. Rob. 240; 1 Digest (Repl.) 142, *294* . 162

PAGE

Graves *v.* Legg (1854), 9 Exch. 709; [1843–60] All E. R. Rep. 399; 23 L. J. Ex. 228; 2 C. L. R. 1266; 23 L. T. O. S. 254; 39 Digest (Repl.) 576, *1012* 108, 109, 355
Graves *v.* Weld (1833), 5 B. & Ad. 105; 2 N. & M. 725; 2 L. J. K. B. 176; 2 Digest (Repl.) 40, *195* 281
Greaves *v.* Ashlin (1813), 3 Camp. 426; 39 Digest (Repl.) 677, *1733* . 204
Grébert-Borgnis *v.* Nugent (1885), 15 Q. B. D. 85; 54 L. J. Q. B. 511; 1 T. L. R. 434, C. A.; 39 Digest (Repl.) 811, *2757* . 235, 239, 240, 253
Green *v.* Arcos Ltd. (1931), 47 T. L. R. 336, C. A.; 39 Digest (Repl.) 723, *2062* 188, 362
Green *v.* Baverstock (1863), 14 C. B. N. S. 204; 32 L. J. C. P. 181; 3 Digest (Repl.) 14, *101* 267, 359
Greenwood *v.* Bennett, [1973] 1 Q. B. 195; [1972] 3 All E. R. 586; [1972] 3 W. L. R. 691; 116 Sol. Jo. 762, C. A.; Digest Supp. 117, 162
Gregg *v.* Wells (1839), 10 Ad. & El. 90; 2 Per. & Dav. 296; L. J. Q. B. 193; 21 Digest (Repl.) 457, *1575*. 165
Grenfell *v.* E. B. Meyrowitz, Ltd., [1936] 2 All E. R. 1313, C. A.; 39 Digest (Repl.) 535, *711*. 121
Grice *v.* Richardson (1877), 3 App. Cas. 319; 47 L. J. P. C. 48; 37 L. T. 677; 36 W. R. 358, P. C.; 39 Digest (Repl.) 745, *2224*. . . 209
Griffith *v.* Brymer (1903), 19 T. L. R. 434; 47 Sol. Jo. 493; 12 Digest (Reissue) 457, *3310* 97
Griffiths, *Re, Ex parte* Board of Trade (1890), 60 L. J. Q. B. 235 . . 274
Griffiths *v.* Perry (1859), 28 L. J. Q. B. 204; 1 E. & E. 680; 5 Jur. N. S. 107; 39 Digest (Repl.) 775, *2515* . 205, 208, 209, 210, 213, 214
Griffiths *v.* Peter Conway, Ltd., [1939] 1 All E. R. 685, C. A.; 39 Digest (Repl.) 552, *830* 135
Grigg *v.* National Guardian Assurance Co., [1891] 3 Ch. 206; 61 L. J. Ch. 11; 64 L. T. 787; 7 Digest (Repl.) 21, *94* 321
Grimoldby *v.* Wells (1875), L. R. 10 C. P. 391; 44 L. J. P. C. 203; 32 L. T. 490; 39 Digest (Repl.) 561, *899* 140, 203
Grissell *v.* Bristowe (1868), L. R. 3 C. P. 112; 37 L. J. C. P. 89; 17 L. T. 564; 16 W. R. 428; reversed, L. R. 4 C. P. 36; 38 L. J. C. P. 10; 19 L. T. 390, Ex. Ch.; 1 Digest (Repl.) 660, *2308* 81
Grist *v.* Bailey, [1966] 2 All E. R. 875; [1967] Ch. 532; [1966] 3 W. L. R. 618; 110 Sol. Jo. 791; Digest Cont. Vol. B p. 545, *120a*. . . . 99
Groom *v.* Barber, [1915] 1 K. B. 316; [1914–15] All E. R. Rep. 194; 84 L. J. K. B. 318; 112 L. T. 301; 20 Com. Cas. 71; 31 T. L. R. 66; 39 Digest (Repl.) 704, *1946* 37, 40, 359
Guaranty Trust Co. of New York *v.* Hannay & Co., [1918] 2 K. B. 623; [1914–15] All E. R. Rep. 24; 87 L. J. K. B. 1223; 119 L. T. 321; 23 Com. Cas. 399; 34 T. L. R. 427, C. A.; 6 Digest (Repl.) 211, *1458* 319
Guardhouse *v.* Blackburn (1866), L. R. 1 P. & D. 109; [1861–73] All E. R. Rep. 680; 35 L. J. P. & M. 116; 14 L. T. 69; 12 Jur. N. S. 278; 14 W. R. 463; 17 Digest (Reissue) 385, *1493* 4
Gulamali Abdul Hussain *v.* Mohammed Yousuf, [1953] 1 Mad. L. J. R. 504 38
Gunn *v.* Bolckow, Vaughan & Co. (1875), 10 Ch. App. 491; 44 L. J. Ch. 732; 32 L. T. 781; 39 Digest (Repl.) 641, *1500* . . 206, 209, 230, 293
Gurney *v.* Behrend (1854), 3 E. & B. 622; [1843–60] All E. R. Rep. 520; 23 L. J. Q. B. 265; 23 L. T. O. 89; 18 Jur. 856; 39 Digest (Repl.) 772, *2479* 319
Gurr *v.* Cuthbert (1843), 12 L. J. Ex. 309; 39 Digest (Repl.) 823, *2857* 213
Gwillim *v.* Daniell (1835), 2 C. M. & R. 61; 1 Gale 143; 4 L. J. Ex. 174; 39 Digest (Repl.) 691, *1845* 364

PAGE

Hadley *v.* Baxendale (1854), 9 Exch. 341; [1843–60] All E. R. Rep. 461; 23 L. J. Ex. 179; 18 Jur. 358; 2 C. L. R. 517; 17 Digest (Reissue) 101, *109* 236

Haegerstrand *v.* Annie Thomas Steamship Co. (1905), 10 Com. Cas. 67; 39 Digest (Repl.) 451, *57* 368

Hale *v.* Rawson (1858), 4 C. B. N. S. 85; 27 L. J. C. P. 189; 31 L. T. O. S. 59; 4 Jur. N. S. 363; 6 W. R. 339; 39 Digest (Repl.) 491, *389* 19, 94, 356

Halfway Garage (Nottingham) *v.* Lepley (1964), Guardian, February 8th C. A. 299, 307

Hall, *Re, Ex parte* Close (1884), 14 Q. B. D. 386; 54 L. J. Q. B. 43; 7 Digest (Repl.) 30, *148* 293

Hall *v.* Farmer, [1970] 1 All E. R. 729; [1970] 1 W. L. R. 366; 114 Sol. Jo. 106; 68 L. G. R. 165; Digest Cont. Vol. C p. 1057, *131a* . . 69

Hall, Ltd. *v.* Pim Junior & Co., Ltd. (1928), 139 L. T. 50; [1928] All E. R. Rep. 763; 33 Com. Cas. 324; 30 Ll. L. Rep. 159, H. L.; 39 Digest (Repl.) 817, *2803* 253

Hallas *v.* Robinson (1885), 15 Q. B. D. 288; 54 L. J. Q. B. 364; 45 L. T. 403, C. A.; 7 Digest (Repl.) 152, *841* 96

Hallet's Estates, *Re*, Knatchball *v.* Hallett (1879), 13 Ch. D. 696; [1874–80] All E. R. Rep. 793; 49 L. J. Ch. 415; 42 L. T. 421; 28 W. R. 732, C. A.; 1 Digest (Repl.) 655, *2278* 312

Halliday *v.* Holgate (1868), L. R. 3 Ex. Ch. 299; 37 L. J. Ex. 174; 18 L. T. 656; 5 Digest (Repl.) 950, *7742* 83, 283

Hamilton *v.* Vaughan-Sherrin Electrical Engineering Co., [1894] 3 Ch. 589; 63 L. J. Ch. 795; 71 L. T. 325; 43 W. R. 126; 10 T. L. R. 642; 38 Sol. Jo. 663; 8 R. 750; 9 Digest (Repl.) 285, *1783* . . 87

Hamlyn *v.* Talisker Distillery, [1894] A. C. 202; [1891–94] All E. R. Rep. 849; 71 L. T. 1; 58 J. P. 540; 10 T. L. R. 479; 6 R. 188, H. L.; 2 Digest (Repl.) 448, *173* 48

Hammer and Barrow *v.* Coca-cola, [1962] N. Z. L. R. 723 . 193, 197, 198, 199, 200

Hammerton Cars *v.* Redbridge London Borough, [1974] 2 All E. R. 216; [1974] 1 W. L. R. 484; 118 Sol. Jo. 240 347

Hammond *v.* Bussey (1887), 20 Q. B. D. 79; 57 L. J. Q. B. 58, C. A.; 39 Digest (Repl.) 587, *1088* 235, 246, 253

Handel (N. V.) My. J. Smits Import-Export *v.* English Exporters (London), Ltd., [1955] 1 Lloyd's Rep. 317, C. A.; 11 Digest (Reissue) 468, *800* 33

Hannah *v.* Peel, [1945] 2 All E. R. 288; [1945] K. B. 509; 114 L. J. K. B. 533; 61 T. L. R. 502; 89 Sol. Jo. 307; 3 Digest (Repl.) 69, *87* 104

Hans a Nord, The, [1974] 2 Lloyd's Rep. 216 . . . . . 9, 191, 202

Hanson *v.* Armitage (1822), 5 B. & Ald. 557; 1 C. &. P. 273, n.; 1 Dow. & Ry. K. B. 128; 39 Digest (Repl.) 468, *195* 193

Hanson *v.* Meyer (1805), 6 East, 614; 2 Smith K. B. 670; 39 Digest (Repl.) 620, *1330* 147

Hansson *v.* Hamel and Horley (1921), 91 L. J. K. B. 65; 26 Com. Cas. 236, C. A.; affirmed, [1922] 2 A. C. 36; [1922] All E. R. Rep. 237; 91 L. J. K. B. 433; 38 T. L. R. 466; 127 L. T. 74; 66 Sol. Jo. 421; 15 Asp. M. L. C. 546; 27 Com. Cas. 321, H. L.; 39 Digest (Repl.) 706, *1956* 39, 317, 355

Harbutt's Plasticine, Ltd. *v.* Wayne Tank and Pump Co., Ltd., [1970] 1 All E. R. 225; [1970] 1 Q. B. 447; [1970] 2 W. L. R. 198; 114 Sol. Jo. 29; [1970] 1 Lloyd's Rep. 15, C. A.; 17 Digest (Reissue) 110, *164* 9, 11, 12, 13, 28, 128, 238, 325

Hardie and Lane, Ltd. *v.* Chilton, [1928] 2 K. B. 306; [1928] All E. R. Rep. 36; 97 L. J. K. B. 539; 139 L. T. 275; 44 T. L. R. 470; 45 Digest (Repl.) 316, *265* 52

PAGE

Hardman *v.* Booth (1863), 1 H. & C. 803; 1 New Rep. 240; 32 L. J. Ex. 105; 7 L. T. 638; 9 Jur. N. S. 81; 11 W. R. 239; 39 Digest (Repl.) 653, *1570* 173

Hardwick, *Re, Ex parte* Hubbard (1886), 17 Q. B. D. 690; 55 L. J. Q. B. 490; 7 Digest (Repl.) 22, *96* 82, 83

Hardwick Game Farm *v.* Suffolk Agricultural and Poultry Producers Association, Ltd., [1964] 2 Lloyd's Rep. 227; on appeal, [1966] 1 All E. R. 309; [1966] 1 W. L. R. 287; 110 Sol. Jo. 11; [1966] 1 Lloyd's Rep. 197. C. A.; affirmed Henry Kendall & Sons *v.* William Lillico & Sons, Ltd., [1968] 2 All E. R. 444; [1969] 2 A. C. 31; [1968] 3 W. L. R. 110; 112 Sol. Jo. 562; [1968] 1 Lloyd's Rep. 547, H. L.; Digest Cont. Vol. B p. 17, *1166b* . 12, 16, 125, 128, 129, 130, 133, 134, 135, 136, 251, 361, 362, 366

Hardy & Co. *v.* Hillerns and Fowler, [1923] 2 K. B. 490; [1923] All E. R. Rep. 275; 92 L. J. K. B. 930; 129 L. T. 674; 39 T. L. R. 547; 29 Com. Cas. 30, C. A.; 39 Digest (Repl.) 562, *905* . 196, 197, 198, 200, 202

Hare *v.* Nicoll, [1966] 1 All E. R. 285; [1966] 2 Q. B. 1303; [1966] 2 W. L. R. 441; 110 Sol. Jo. 11, C. A.; Digest Cont. Vol. B p. 665, *300a* 105, 106

Hargreave *v.* Spink, [1892] 1 Q. B. 25; 61 L. J. Q. B. 318; 65 L. T. 650; 40 W. R. 254; 8 T. L. R. 18; 33 Digest (Repl.) 491, *481* . 170

Harland and Wolff *v.* Burstall (1901), 84 L. T. 324; 17 T. L. R. 338; 9 Asp. M. L. C. 184; 6 Com. Cas. 113; 39 Digest (Repl.) 684, *1800* 186, 364

Harling *v.* Eddy, [1951] 2 All E. R. 212; [1951] 2 K. B. 739; [1951] 2 T. L. R. 245; 95 Sol. Jo. 501, C. A.; 2 Digest (Repl.) 349, *334* 5, 108, 110, 354, 360

Harlow and Jones *v.* Panex (International), Ltd., [1967] 2 Lloyd's Rep. 509 . . . . 23, 34, 233, 234, 255

Harnor *v.* Groves (1855), 15 C. B. 667; 24 L. J. C. P. 53; 39 Digest (Repl.) 580, *1037* 197, 198

Harper & Co. *v.* Mackechnie & Co., [1925] 2 K. B. 423; 95 L. J. K. B. 162; 134 L. T. 90; 31 Com. Cas. 21; 39 Digest (Repl.) 713, *1999* . 39

Harris *v.* Nickerson (1873), L. R. 8 Q. B. 286; 42 L. J. Q. B. 171; 28 L. T. 414; 3 Digest (Repl.) 51, *362* 268

Harris & Sons *v.* Plymouth Varnish and Colour Co., Ltd. (1933), 49 T. L. R. 521; 38 Com. Cas. 316; 39 Digest (Repl.) 529, *663* . . 130

Harrison *v.* Knowles and Foster, [1917] 2 K. B. 606; 86 L. J. K. B. 1490; 117 L. T. 363; 33 T. L. R. 467; 61 Sol. Jo. 695; 22 Com. Cas. 293; affirmed, [1918] 1 K. B. 608; [1918–19] All E. R. Rep. 306; 87 L. J. K. B. 680; 23 Com. Cas. 282; 34 T. L. R. 235; C. A.; 39 Digest (Repl.) 514, *557* 108, 354, 361, 364

Harrison *v.* Luke (1845), 14 M. & W. 139; 14 L. J. Ex. 248; 39 Digest (Repl.) 784, *2591* 79

Harrison and Jones, Ltd. *v.* Bunten and Lancaster, Ltd., [1953] 1 All E. R. 903; [1953] 1 Q. B. 646; [1953] 2 W. L. R. 840; 97 Sol. Jo. 281; 35 Digest (Repl.) 113, *126* 120

Harrison and Micks, *Re,* [1917] 1 K. B. 755; 86 L. J. K. B. 573; 116 L. T. 606; 22 Com. Cas. 273; 39 Digest (Repl.) 689, *1831* . 363, 364

Hart *v.* Herwig (1873), 8 Ch. App. 860; 42 L. J. Ch. 457; 29 L. T. 47; 21 W. R. 663; 2 Asp. M. L. C. 63; 2 Digest (Repl.) 173, *34* . . 244

Hart *v.* Mills (1846), 15 M. &. W. 85; 15 L. J. Ex. 200; 39 Digest (Repl.) 688, *1816* 186

Hartley *v.* Hymans, [1920] 3 K. B. 475; [1920] All E. R. Rep. 328; 90 L. J. K. B. 14; 124 L. T. 31; 36 T. L. R. 805; 25 Com. Cas. 365; 39 Digest (Repl.) 673, *1713* 22, 24, 107

Harvey, *Re, Ex parte* Harvey & Co. (1890), 7 Morr. 138; 5 Digest (Repl.) 967, *7828* 76

PAGE

Harvey *v.* Facey, [1893] A. C. 552; 62 L. J. P. C. 127; 69 L. T. 504; 42 W. R. 129; 9 T. L. R. 612; 1 R. 428, P. C.; 40 Digest (Repl.) 11, *1* 76

Hastings *v.* Pearson, [1893] 1 Q. B. 62; 62 L. J. Q. B. 75; 67 L. T. 553; 57 J. L. 70; 41 W. R. 127; 9 T. L. R. 18; 36 Sol. Jo. 866; 5 R. 26, C. A.; 1 Digest (Repl.) 387, *515* 291

Hatfield *v.* Phillips (1842), 9 M. & W. 647; affirmed, (1845), 14 M. & W. 665; 12 Cl. & Fin. 343; 10 Jur. 189; 41 Digest (Repl.) 260, *765* . 301

Hathesing *v.* Laing (1873), L. R. 17 Eq. 92; 43 L. J. Ch. 233; 29 L. T. 734; 6 Digest (Repl.) 427, *3004*. 293

Havering London Borough *v.* Stevenson, [1970] 3 All E. R. 609; [1970] 1 W. L. R. 1375, D. C.; Digest Cont. Vol. C p 1024, *1116f* : 128, 274

Hawes *v.* Watson (1824), 2 B. & C. 540; 4 Dow. & Ry. K. B. 22; Ry & M. 6; 2 L. J. O. S. K. B. 83; 39 Digest (Repl.) 759, *2369* . . 212

Hayman *v.* Flewker (1863), 32 L. J. C. P. 132; 1 Digest (Repl.) 384, *496* 292

Hayman *v.* M'Lintock, 1907 S. C. 936; 44 Sc. L. R. 691; 39 Digest (Repl.) 599, **741* 141, 142, 143, 183, 222, 283

Hayward Brothers *v.* Daniel (1904), 91 L. T. 319; 12 Digest (Reissue) 473, *3396* 97

Head *v.* Tattersall (1871), L. R. 7 Exch. 7; 41 L. J. Ex. 4; 25 L. T. 631; 39 Digest (Repl.) 625, *1366* . 81, 152, 157, 158, 160, 203, 362

Head (Philip) & Sons, Ltd. *v.* Showfronts, Ltd. (1970), 113 Sol. Jo. 978; [1970] 1 Lloyd's Rep. 140; Digest Cont. Vol. C p. 856, *1188a* 78, 146, 148, 150

Healing (Sales) Pty., Ltd. *v.* Inglis Electrix Pty., Ltd. (1968), 42 A. L. J. R. 280 . . . . .116, 242, 246

Healy *v.* Howlett & Sons, [1917] 1 K. B. 337; 86 L. J. K. B. 252; 116 L. T. 591; 39 Digest (Repl.) 601, *1166* 142, 150, 195

Heap *v.* Motorists' Advisory Agency, [1923] 1 K. B. 577; [1922] All E. R. Rep. 251; 92 L. J. K. B. 553; 129 L. T. 146; 39 T. L. R. 150; 39 Digest (Repl.) 650, *1556* 172, 295, 299, 307

Heaven and Kesterton, Ltd. *v.* Establissements Francois Albiac et cie, [1956] 2 Lloyd's Rep. 316 255

Hedley Byrne & Co. *v.* Heller & Partners, [1963] 2 All E. R. 575; [1964] A. C. 465; [1963] 3 W. L. R. 101; 107 Sol. Jo. 454; [1963] 1 Lloyd's Rep. 485, H. L.; Digest Cont. Vol. A p. 51, *1117a* . . . 333, 335

Heilbutt *v.* Hickson (1872), L. R. 7 C. P. 438; 41 L. J. C. P. 228; 27 L. T. 336; 20 W. R. 1005; 39 Digest (Repl.) 561, *902*: 74, 109, 138, 140, 141, 195, 197, 201, 203, 245, 246, 352

Heilbutt Symons & Co. *v.* Buckleton, [1913] A. C. 30; [1911–13] All E. R. Rep. 83; 82 L. J. K. B. 245; 107 L. T. 769, H. L.; 39 Digest (Repl.) 513, *556* 5, 108

Helbert Wagg & Co., Ltd., *Re, Re* Prudential Assurance Co., Ltd. *See Re* Wagg (Helbert) & Co., Ltd., *Re* Prudential Assurance Co.

Helby *v.* Matthews, [1895] A. C. 471; [1895–99] All E. R. Rep. 821; 64 L. J. Q. B. 465; 72 L. T. 841, H. L.; 26 Digest (Repl.) 660, *14*: 82, 162, 309

Helfand *v.* Royal Canadian Art Pottery (1970), 11 D. L. R. (3d) 404 . 368

Hellings *v.* Russell (1875), 33 L. T. 380; 1 Digest (Repl.) 384, *495*. . 291

Helps *v.* Winterbottom (1831), 2 B. & Ad. 431; 9 L. J. O. S. K. B. 258; 39 Digest (Repl.) 735, *2137* 227

Henderson & Co. *v.* Williams, [1895] 1 Q. B. 521; 74 L. J. Q. B. 308; 72 L. T. 98, C. A.; 21 Digest (Repl.) 449, *1525*. 165

Henley, *Re, Ex parte* Dixon (1876), 4 Ch. D. 133; [1874–80] All E. R. Rep. 1004; 46 L. J. Bcy. 20; 35 L. T. 644; 25 W. R. 105, C. A.; 1 Digest (Repl.) 667, *2344* 314

Henry *v.* Hammond, [1913] 2 K. B. 515; 82 L. J. K. B. 575; 108 L. T. 729; 29 T. L. R. 340; 57 Sol. Jo. 358; 12 Asp. M. L. C. 332; 1 Digest (Repl.) 530, *1605* 312

Henry, Ltd. *v.* Wilhelm G. Clasen, [1973] 1 Lloyd's Rep. 159, C. A. . 359

PAGE

Henry Kendall & Sons *v.* William Lillico & Sons, Ltd. *See* Kendall (Henry) & Sons *v.* William Lillico & Sons, Ltd.

Heron *v.* Dilworth Equipment, Ltd. (1963), 36 D. L. R. (2d) 462 . . 361

Heseltine *v.* Siggers (1848), 1 Exch. 856; 18 L. J. Ex. 166; 39 Digest (Repl.) 494, *416* 95

Heskell *v.* Continental Express, Ltd., [1950] 1 All E. R. 1033; 83 Ll. L. Rep. 438; 93 Sol. Jo. 339; 17 Digest (Reissue) 155, *429*. . 106

Hewlings *v.* Graham (1901), 70 L. J. Ch. 568; 84 L. T. 497; 28 (2) Digest (Reissue) 697, *323*. 89

Heyward's Case (1595), 2 Coke 35 a.; Poph. 95; 2 And. 202; 17 Digest (Reissue) 411, *1752* 153

Heyworth *v.* Hutchinson (1867), L. R. 2 Q. B. 447; 36 L. J. Q. B. 270; 39 Digest (Repl.) 507, *512* 139, 354, 367

Hibbert *v.* McKiernan, [1948] 1 All E. R. 860; [1948] 2 K. B. 142; [1948] L. J. R. 1521; 112 J. P. 287; 64 T. L. R. 256; 92 Sol. Jo. 259; 46 L. G. R. 238; 15 Digest (Repl.) 1086, *10,752* 165

Hibblewhite *v.* M'Morine (1839), 5 M. & W. 462; 8 L. J. Ex. 271; 13 Jur. 509; 39 Digest (Repl.) 490, *381* 95

Hickman *v.* Haynes (1875), L. R. 10 C. P. 598; 44 L. J. C. P. 358; 32 L. T. 873; 23 W. R. 872; 39 Digest (Repl.) 779, *2687* . . 22, 232, 239

Hill *v.* Perrott (1810), 3 Taunt. 274; 12 Digest (Reissue) 143, *827*. . 31

Hill (Christopher), Ltd. *v.* Ashington Piggeries, Ltd., [1969] 3 All E. R. 1496; [1969] 2 Lloyd's Rep. 425, C. A.; reversed, Ashington Piggeries, Ltd. *v.* Christopher Hill, Ltd., [1971] 1 All E. R. 847, H. L.; Digest Cont. Vol. C p. 852, *726a* . 81, 118, 120, 132, 134, 135, 136, 140, 360, 366, 367

Hill (D. J.) & Co. Pty., Ltd. *v.* Walter H. Wright Pty., Ltd., [1971] V. R. 749 16

Hill & Sons *v.* Showell & Sons (1918), 87 L. J. K. B. 1106; 119 L. T. 651; 62 Sol. Jo. 715, H. L.; 39 Digest (Repl.) 813, *2768* . 234, 255

Hillas & Co., Ltd. *v.* Arcos, Ltd. (1932), 147 L. T. 503; [1932] All E. R. Rep. 494; 147 L. T. 503; 38 Com. Cas. 23, H. L.; 39 Digest (Repl.) 448, *34* 16, 101

Hills and Grant, Ltd. *v.* Hodson, [1934] Ch. 53; 103 L. J. Ch. 17; 150 L. T. 16; 3 Digest (Repl.) 12, *89*. 269

Hinchcliffe *v.* Barwick (1880), 5 Ex. D. 177; 49 L. J. Ex. 495; 42 L. T. 492, C. A.; 2 Digest (Repl.) 345, *310* 362, 367

Hinde *v.* Liddell (1875), L. R. 10 Q. B. 265; 44 L. J. Q. B. 105; 32 L. T. 449; 23 W. R. 650; 39 Digest (Repl.) 810, *2753* . . 235, 238

Hiort *v.* London and North-Western Railway (1879), 4 Ex. D. 188; 48 L. J. Ex. 545; 40 L. T. 674, C. A.; 17 Digest (Reissue) 87, *29*. . 242

Hire Purchase Furnishing Co. *v.* Richens (1887), 20 Q. B. D. 387; 58 L. T. 460; 36 W. R. 365; 4 T. L. R. 184, C. A.; 12 Digest (Reissue) 290, *2097* 19

Hoare *v.* Rennie (1859), 5 H. & N. 19; 29 L. J. Ex. 73; 1 L. T. 104; 8 W. R. 80; 39 Digest (Repl.) 699, *1908* 190

Hocker *v.* Waller (1924), 29 Com. Cas. 296; 1 Digest (Repl.) 581, *1859*. 75

Hodgens *v.* Hodgens (1837), 4 Cl. & Fin. 323; 11 Bli. N. S. 62, H. L.; 28 (2) Digest (Reissue) 746, **372* 92

Hodgson *v.* Loy (1797), 7 Term Rep. 440; 39 Digest (Repl.) 743, *2201* 205

Hodgson & Co., *Re*, [1920] W. N. 198; 39 Digest (Repl.) 497, *436*. . 258

Holden *v.* Bostock (1902), 50 W. R. 323; 18 T. L. R. 317; 46 Sol. Jo. 265, C. A.; 39 Digest (Repl.) 589, *1105* 247, 251

Holland Colombo Trading Society, Ltd. *v.* Segu Mohamed Alawdeen, [1954] 2 Lloyd's Rep. 45, P. C.. 39

Holliday *v.* Morgan (1858), 1 E. & E. 1; 28 L. J. Q. B. 9; 2 Digest (Repl.) 354, *379* 367

PAGE

Hollier *v.* Rambler Motors, [1972] 1 All E. R. 399; [1972] 2 Q. B. 71; [1972] 2 W. L. R. 401; 116 Sol. Jo. 158; [1972] R. T. R. 190, C. A.; Digest Supp. 16

Hollins *v.* Fowler (1875), L. R. 7 H. L. 757; [1874–80] All E. R. Rep. 118; 44 L. J. Q. B. 169; 33 L. T. 73; 1 Digest (Repl.) 787, *3155* . 163, 165, 242

Holmes *v.* Wilson (1839), 10 Ad. & El. 503; 46 Digest (Repl.) 353, *26* . 30

Holroyd *v.* Marshall (1862), 10 H. L. Cas. 191; [1861–73] All E. R. Rep. 414; 33 L. J. Ch. 193; 7 L. T. 172; 9 Jur. N. S. 213; 11 W. R. 171 H. L.; 39 Digest (Repl.) 821, *2834* 96

Honck *v.* Muller (1881), 7 Q. B. D. 92; 50 L. J. Q. B. 529; 45 L. T. 202, C. A.; 39 Digest (Repl.) 700, *1917* . . . 9, 10, 180, 190, 330

Hong Guan & Co., Ltd. *v.* R. Jumabhay & Sons, Ltd., [1960] 2 All E. R. 100; [1960] A. C. 684; [1960] 2 W. L. R. 754; 104 Sol. Jo. 367; [1960] 1 Lloyd's Rep. 405, P. C.; 12 Digest (Reissue) 502, *3505* 357

Hong Kong Fir Co., Ltd. *v.* Kawasaki Kisen Kaisha, Ltd., [1962] 1 All E. R. 474; [1962] 2 Q. B. at p. 41; [1962] 2 W. L. R. 474; 106 Sol. Jo. 35; [1961] 2 Lloyd's Rep, 478, C. A.; 41 Digest (Repl.) 363, *1553* 9, 20, 110, 348

Hookway *v.* Alfred Isaacs & Sons, [1954] 1 Lloyd's Rep. 491 . 140, 367

Hooper *v.* Gumm (1867), 2 Ch. App. 282; 36 L. J. Ch. 605; 16 L. T. 107; 15 W. R. 464; 2 Mar. L. C. 481; 39 Digest (Repl.) 650, *1552* . 282

Hope *v.* Hayley (1856), 5 E. & B. 830; 25 L. J. Q. B. 155; 7 Digest (Repl.) 125, *728* 96

Hopkins *v.* Tanqueray (1854), 15 C. B. 130; [1843–60] All E. R. Rep. 96; 23 L. J. C. P. 162; 23 L. T. O. S. 144; 18 Jur. 608; 2 W. R. 475; 39 Digest (Repl.) 510, *532* 5, 353

Horn *v.* Minister of Food, [1948] 2 All E. R. 1036; 65 T. L. R. 106; 39 Digest (Repl.) 554, *849* 98, 158, 159

Houghland *v.* Low (R. R.) (Luxury Coaches), Ltd., [1962] 2 All E. R. 159; [1962] 1 Q. B. 694; [1962] 2 W. L. R. 1015; 106 Sol. Jo. 243, C. A.; Digest Cont. Vol. A p. 46, *71a* 160, 287

Houlditch *v.* Desanges (1818), 2 Stark. 337; 39 Digest (Repl.) 747, *2251* 212

Houndsditch Warehouse Co., Ltd. *v.* Waltex, Ltd., [1944] 2 All E. R. 518; [1944] K. B. 579; 113 L. J. K. B. 547; 171 L. T. 275; 60 T. L. R. 517; 17 Digest (Reissue) 127, *267* 255

Household Machines, Ltd. *v.* Cosmos Exporters, Ltd., [1946] 2 All E. R. 622; [1947] K. B. 217; [1947] L. J. R. 578; 176 L. T. 49; 62 T. L. R. 757; 39 Digest (Repl.) 811, *2758* 240

Howard *v.* Castle (1796), 6 Term Rep. 642; 3 Digest (Repl.) 13, *95* . 267

Howard *v.* Harris (1884), 1 Cab. & El. 253; 3 Digest (Repl.) 60, *37* . 160

Howcroft and Watkins *v.* Perkins (1900), 16 T. L. R. 217; 39 Digest (Repl.) 534, *699* 360

Howe *v.* Smith (1884), 27 Ch. D. 89; [1881–5] All E. R. Rep. 201; 53 L. J. Ch. 1055; 50 L. T. 573; 48 J. P. 773; 32 W. L. R. 802, C.A.; 40 Digest (Repl.) 241, *2027* 27

Howell *v.* Coupland (1876) 1 Q. B. D. 258; [1874–80] All E. R. Rep. 878; 46 L. J. Q. B. 147; 33 L. T. 832; 40 J. P. 276; 24 W. R. 470; C. A.; 39 Digest (Repl.) 489, *379* 95, 100

Howell *v.* Evans (1926), 134 L. T. 570; 42 T. L. R. 310; 39 Digest (Repl.) 739, *2173* 189

Howes *v.* Ball (1827), 7 B. & C. 481; [1824–34] All E. R. 382; 1 M. & Ry. 288; 6 L. J. O. S. K. B. 106; 39 Digest (Repl.) 779, *2549*. 83, 283

Hubbard, *Ex parte, Re* Hardwick (1886), 17 Q. B. D. 690; 55 L. J. Q. B. 490, C. A.; 7 Digest (Repl.) 22, *96* 82, 83

Hughes, *Ex parte* (1892), 67 L. T. 598; 7 Asp. M. L. C. 249; 9 Morr. 294; 39 Digest (Repl.) 751, *2312* 215

PAGE

Hull Rope Works Co. *v.* Adams (1896), 65 L. J. Q. B. 114; 73 L. T. 446; 26 Digest (Repl.) 659, *6* 309
Humble *v.* Mitchell (1839), 11 Ad. & El. 205; 2 Ry. & Can. Cas. 70; 3 Per & Dav. 141; 9 L. J. Q. B. 29; 3 Jur. 1188; 39 Digest (Repl.) 450, *46* 281
Hunt *v.* De Blaquiere (1829), 5 Bing. 550; 3 Moo. & P. 108; 7 L. J. O. S., C. P. 198; 27 (1) Digest (Reissue) 230, *1674* . . . 92
Hunt and Winterbotham (West of England), Ltd. *v.* British Road Services (Parcels), Ltd., [1962] 1 All E. R. 111; [1962] 1 Q. B. 617; [1962] 2 W. L. R. 172; 105 Sol. Jo. 1124; [1961] 2 Lloyd's Rep. 422, C. A.; Digest Cont. Vol. A p. 1222, *57a* 14
Hutchison *v.* Bowker (1839), 5 M. & W. 535; 9 L. J. Ex. 24; 12 Digest (Reissue) 77, *411* 365
Hutton *v.* Lippert (1883), 8 App. Cas. 309; 52 L. J. P. C. 54; 49 L. T. 64, P. C.; 39 Digest (Repl.) 331, **432* 76
Hvalfangerselskabet Polaris Aktieselskap *v.* Unilever Co., Ltd. (1933), 77 Sol. Jo. 388; 39 Com. Cas. 1, H. L.; 39 Digest (Repl.) 490, *384* 364
Hydraulic Engineering Co. *v.* McHaffie (1878), 4 Q. B. D. 670, C. A.; 39 Digest (Repl.) 680, *1748* 240, 252, 253, 358
Hyslop *v.* Shirlaw 1905, 7 F. (Ct. of Sess.) 875; 42 Sc. L. R. 668; 13 S. L. T. 209; 39 Digest (Repl.) 565, **488* 366

Ian Stach, Ltd. *v.* Baker Bosly Ltd. *See* Stach (Ian), Ltd. *v.* Baker Bosly, Ltd.
Imperial Bank *v.* London and St. Katherine Docks Co. (1877), 5 Ch. D. 195; 46 L. J. Ch. 335; 36 L. T. 233; 39 Digest (Repl.) 742, *2199* 206
Imperial Loan Co. *v.* Stone [1892] 1 Q. B. 599; [1891–94] All E. R. Rep. 412; 61 L. J. Q. B. 449; 66 L. T. 556; 56 J. P. 436; 8 T. L. R. 408, C. A.; 33 Digest (Repl.) 590, *58* 86
Imperial Tobacco Co. (of G. B. and Ireland), Ltd. *v.* Parslay, [1936] 2 All E. R. 515; 52 T. L. R. 585; 80 Sol. Jo. 464; C. A., 17 (Reissue) 183, *582* 239
Industrial Acceptance Corporation *v.* Whiteshall Finance Corporation (1966), 57 D. L. R. (2d) 670 302
Ingham *v.* Emes, [1955] 2 All E. R. 740; [1955] 2 Q. B. 366; [1955] 3 W L. R. 245; 99 Sol. Jo. 490, C. A.; Digest Cont. Vol. B p. 33, *282m* 77, 122
Inglis *v.* Robertson, [1898] A. C. 616; 67 L. J. P. C. 108; 79 L. T. 224; 14 T. L. R. 517, H. L.; 44 Digest (Repl.) 194, *65* . . . 74, 290, 301
Inglis *v.* Stock. *See* Stock *v.* Inglis.
Ingram *v.* Little, [1960] 3 All E. R. 332; [1961] 1 Q. B. 31; [1960] 3 W. L. R. 504; 104 Sol. Jo. 704, C. A.; 35 Digest (Repl.) 104, *70*: 172, 173, 174
International Fibre Syndicate *v.* Dawson 1901, 3 F., (Ct. of Sess.), 32; H. L.; 12 Digest (Reissue) 726, *5256* 29
International Sponge Co. *v.* Andrew Watts & Sons, [1911] A. C. 279; 81 L. J. P. C. 12; 16 Com. Cas. 224; 27 T. L. R. 364; 1 Digest (Repl.) 420, *790* 228, 230
Ireland *v.* Livingston (1872), L. R. 5 H. L. 395; [1861–73] All E. R. Rep. 585; 41 L. J. Q. B. 201; 27 L. T. 79; 1 Asp. M. L. C. 389, H. L.; 39 Digest (Repl.) 688, *1810* 36, 205, 363
Ironmonger & Co. *v.* Dyne (1928), 44 T. L. R. 497; 3 Digest (Repl.) 347, *1140* 103
Isaacs, *Re, Ex parte* Miles (1885), 15 Q. B. D. 39; 54 L. J. Q. B. 566, C. A.; 39 Digest (Repl.) 755, *2335* 215, 217
Isaacs *v.* Hardy (1884), Cab. & El. 287; 29 Digest (Repl.) 445, *10* . 78
Isherwood *v.* Whitmore (1843), 11 M. & W. 347; 12 L. J. Ex. 318; 1 L. T. O. S. 81; 7 Jur. 535; 39 Digest (Repl.) 716, *2019* . . 196

PAGE

Jacks & Co. *v.* Palmers Shipbuilding & Iron Co. (1928), 98 L. J. K. B. 366; 140 L.T. 473; 34 Com. Cas. 107, C. A.; 12 Digest (Reissue) 422, *3073* 359
Jackson *v.* Allaway (1844), 6 Man. & G. 942; 7 Scott N. R. 875; 13 L. J. C. P. 84; 2 L. T. O. S. 328; 8 Jur. 63; 39 Digest (Repl.) 663, *1631* 182
Jackson *v.* Rotax Motor & Cycle Co., [1910] 2 K. B. 937; 80 L. J. K. B. 38; 103 L. T. 411, C. A.; 39 Digest (Repl.) 555, *856*: 11. 108, 125, 138, 188, 191, 358
Jackson *v.* Watson & Sons, [1909] 2 K. B. 193; 78 L. J. K. B. 587; 100 L. T. 799; 25 T. L. R. 454; 53 Sol. Jo. 447; C. A., 39 Digest (Repl.) 589, *1103* 125, 247
Jacobs *v.* Booth's Distillery (1901), 85 L. T. 262; 50 W. R. 49, H. L. 271
Jacobs *v.* Credit Lyonnais (1884), 12 Q. B. D. 589; [1881–5] All E. R. Rep. 151; 53 L. J. Q. B. 156; 50 L. T. 194; 32 W. R. 761, C. A.; 12 Digest (Reissue) 491, *3466* 48
Jacobs *v.* Morris, [1902] 1 Ch. 816; 71 L. J. Ch. 363; 86 L. T. 275; 50 W. R. 371; 18 T. L. R. 384; 46 Sol. Jo. 315, C. A.; 1 Digest (Repl.) 350, *298* 5, 361
Jager *v.* Tolme and Runge and London Produce Clearing House, Ltd., [1916] 1 K. B. 939; 85 L. J. K. B. 1116; 114 L. T. 647; 32 T. L. R. 291, C. A.; 12 Digest (Reissue) 492, *3471* 359
Jamal *v.* Moolla Dawood, Sons & Co., [1916] 1 A. C. 175; 85 L. J. P. C. 29; 114 L. T. I.; 32 T. L. R. 79; 60 Sol. Jo. 139, P. C.; 17 Digest (Reissue) 127, *261* 249
James *v.* Griffin (1837), 2 M. & W. 623; 6 L. J. Ex. 241; 39 Digest (Repl.) 765, *2422* 216, 218
James Miller & Partners, Ltd. *v.* Whitworth Street Estates, Ltd. *See* Miller (James) & Partners, Ltd. *v.* Whitworth Street Estates, Ltd.
Janesich *v.* Attenborough & Son (1910), 102 L. T. 605; 26 T. L. R. 278; 1 Digest (Repl.) 393, *543* 297, 298
Janson *v.* Driefontein Consolidated Mines, Ltd., [1902] A. C. 484; [1900–3] All E. R. Rep. 426; 71 L. J. K. B. 857; 87 L. T. 372; 51 W. R. 142; 18 T. L. R. 796; 7 Com. Cas. 268, H. L.; 12 Digest (Reissue) 296, *2132* 89, 90
Jarvis *v.* Swan Tours, Ltd., [1973] 1 All E. R. 71; [1973] Q. B. 233; [1972] 3 W. L. R. 954; 116 Sol. Jo. 822, C. A.; Digest Supp. . 334
Jebara *v.* Ottoman Bank, [1927] 2 K. B. 254; 96 L. J. K. B. 581; 137 L. T. 101; 43 T. L. R. 369; 32 Com Cas. 228; C. A., on appeal *sub nom.* Ottoman Bank *v.* Jebara, [1928] A. C. 269; [1928] All E. R. Rep. 243; 97 L. J. K. B. 502; 139 L. T. 194; 44 T. L. R. 525; 72 Sol. Jo. 516; 33 Com. Cas. 260, H. L.; 1 Digest (Repl.) 271, *620* 20, 168, 224, 271
Jefford *v.* Gee, [1970] 1 All E. R. 1202; [1970] 2 Q. B. 130; [1970] 2 W. L. R. 702; 114 Sol. Jo. 206; [1970] 1 Lloyd's Rep. 107, C. A.; Digest Cont. Vol. C p. 709, *182a* 326
Jenkyns *v.* Brown (1849), 14 Q. B. 496; 19 L. J. Q. B. 286; 14 L. T. O. S. 395; 14 Jur. 505; 39 Digest (Repl.) 630, *1420* 283
Jenkyns *v.* Usborne (1844), 7 M. & Gr. 678; 8 Scott N. R. 505; 13 L. J. C. P. 196; 3 L. T. O. S. 300; 8 Jur. 1139, 39 Digest (Repl.) 763, *2401* 206, 308
Jenner *v.* Smith (1869), L. R. 4 C. P. 270; 39 Digest 489, *1078* . . 148
Jennings *v.* Rundall (1799), 8 Term Rep. 335; 28 (1) Digest (Reissue) 704, *405* 88
Jerome *v.* Bentley & Co., [1952] 2 All E. R. 114; [1952] 2 T. L. R. 58; 96 Sol. Jo. 463; 1 Digest (Repl.) 655, *2276*. 165, 166
Jetley *v.* Hill (1884), Cab. & El. 239; 27 (1) Digest (Reissue) 213, *1456* 92
Jewan *v.* Whitworth (1866), L. R. 2 Eq. 692; 36 L. J. Ch. 127; 1 Digest (Repl.) 395, *561* 302

PAGE

Jewelowski *v.* Propp, [1944] 1 All E. R. 483; [1944] 1 K. B. 510; 113 L. J. K. B. 335; 171 L. T. 234; 60 T. L. R. 559; 17 Digest (Reissue) 132, *300* 255
Jewry *v.* Busk (1814), 5 Taunt. 302; 12 Digest (Reissue) 518, *3595* . 101
Joachimson *v.* Swiss Bank, [1921] 3 K. B. 110; [1921] All E. R. Rep. 92; 90 L. J. K. B. 973; 125 L. T. 338; 37 T. L. R. 534; 26 Com. Cas. 196, C. A.; 3 Digest (Repl.) 193, *376* 228
Joblin *v.* Watkins and Roseveare (Motors), Ltd., [1949] 1 All E. R. 47; 64 T. L. R. 464; 1 Digest (Repl.) 389, *526* . . . 82, 293, 297
John *v.* Matthews, [1970] 2 All E. R. 643; [1970] 2 Q. B. 443; [1970] 2 W. L. R. 1246; 134 J. P. 526; 114 Sol. Jo. 472; Digest Cont. Vol. C p. 1024, *1116c* 60
John Mackintosh & Sons *v.* Baker's Bargain Stores (Seaford), Ltd. *See* Mackintosh (John) & Sons *v.* Baker's Bargain Stores (Seaford), Ltd.
Johnson *v.* Crédit Lyonnais (1877), 3 C. P. D. 32; 48 L. J. Q. B. 241; 37 L. T. 657, C. A.; 1 Digest (Repl.) 385, *500*: . . 304, 305, 313
Johnson *v.* Lancashire Railway (1878), 3 C. P. D. 499; 39 L. T. 448; 39 Digest (Repl.) 824, *2866* 242
Johnson *v.* Macdonald (1842), 9 M. & W. 600; 12 L. J. Ex. 99; 5 Jur. 264; 39 Digest (Repl.) 493, *404* 85, 356
Johnson *v.* Raylton (1881), 7 Q. B. D. 438; 50 L. J. Q. B. 753; 45 L. T. 374; 30 W. R. 350, C. A.; 39 Digest (Repl.) 569, *970* 29, 365
Johnson *v.* Stear (1863), 15 C. B. N. S. 330; 33 L. J. C. P. 130; 9 L. T. 538; 10 Jur. N. S. 99; 12 W. R. 347; 37 Digest (Repl.) 11, *96* 242
Johnson *v.* Taylor Brothers, [1920] A. C. 144; 87 L. J. K. B. 227; 122 L. T. 130; 64 Sol. Jo. 82; 36 T. L. R. 62; 25 Com. Cas. 69; 39 Digest (Repl.) 799, *2691* 37, 39, 228
Johnstone *v.* Marks (1887), 19 Q. B. D. 509; 57 L. J. Q. B. 6; 35 W. R. 806; 3 T. L. R. 810, D. C.; 28 (2) Digest (Reissue) 694, *263* . 85
Jolly *v.* Rees (1863), 15 C. B. N. S. 628; 3 New Rep. 473; 33 L. J. C. P. 177; 12 W. R. 473; 27 (1) Digest (Reissue) 218, *1522* . . . 92
Jonassohn *v.* Young (1863), 4 B. & S. 296; 2 New Rep. 390; 32 L. J. Q. B. 385; 10 Jur. N. S. 43; 11 W. R. 962; 39 Digest (Repl.) 790, *2629* 190
Jones *v.* Bright (1829), 5 Bing. 533; 3 Moo. & P. 155; Dan & Ll. 304; 7 L. J. O. S. C. P. 213; 39 Digest (Repl.) 553, *832*. 352
Jones *v.* Gibbons (1853), 8 Exch. 920; 22 L. J. Ex. 247; 1 W. R. 348; 39 Digest (Repl.) 679, *1745* 185, 235, 358
Jones *v.* Gordon (1877), 2 App. Cas. 616; 47 L. J. Bk. 1; 37 L. T. 477; 6 Digest (Repl.) 131, *974* 286, 287
Jones *v.* Just (1868), L. R. 3 Q. B. 197; 9 B. & S. 141; 37 L. J. Q. B. 89; 18 L. T. 208; 16 W. R. 643; 39 Digest (Repl.) 531, *675*: 126, 246, 249
Jones *v.* Tankerville, [1909] 2 Ch. 440; 78 L. J. Ch. 674; 101 L. T. 202; 25 T. L. R. 714; 39 Digest (Repl.) 822, *2849* . . . 151
Jones *v.* Tarleton (1842), 9 M. & W. 675 213
Jones Brothers (Holloway), Ltd. *v.* Woodhouse, [1923] 2 K. B. 117; 92 L. J. K. B. 638; 129 L. T. 317; 21 Digest (Repl.) 593, *814* . . 162
Jordeson & Co. *v.* London Hardwood Co. (1913), 110 L. T. 666; 19 Com. Cas. 161; 39 Digest (Repl.) 638, *1472* 155
Joseph *v.* Lyons (1884), 15 Q. B. D. 280; 54 L. J. Q. B. 1; 51 L. T. 740, C. A.; 7 Digest (Repl.) 152, *840* 96
Joseph Travers & Sons, Ltd. *v.* Longel, Ltd. *See* Travers (Joseph) & Sons, Ltd. *v.* Longel, Ltd.
Josling *v.* Irvine (1861), 6 H. & N. 512; 30 L. J. Ex. 78; 4 L. T. 251; 39 Digest (Repl.) 817, *2797* 235
Josling *v.* Kingsford (1863), 13 C. B. N. S. 447; 32 L. J. C. P. 94; 7 L. T. 790; 9 Jur. N. S. 947; 11 W. R. 377; 39 Digest (Repl.) 531, *674* 117
Joyce *v.* Swann (1864), 17 C. B. N. S. 84; 39 Digest (Repl.) 501, *463* 155

PAGE

Kaltenbach *v.* Lewis (1883), 24 Ch. D. 54, C. A.; affirmed (1885), 10 App. Cas. 617; 55 L. J. Ch. 58; 53 L. T. 787, H. L.; 1 Digest (Repl.) 666, *2341* 299, 302, 313
Kaltenbach *v.* Mackenzie (1878), 3 C. P. D. 467; 46 L. J. C. P. 9; 39 L. T. 215, C. A.; 29 Digest (Repl.) 327, *2478* 162
Karberg (Arnhold) *v.* Blythe Green & Co., [1916] 1 K. B. 495; 85 L. J. K. B. 665; 114 L. T. 152; 32 T. L. R. 186; 60 Sol. Jo. 156; 13 Asp. M. L. C. 235; 21 Com. Cas. 174, C. A.; 39 Digest (Repl.) 807, *1959* . 20, 37, 39
Karsales (Harrow) *v.* Wallis, [1956] 2 All E. R. 866; [1956] 1 W. L. R. 936; 100 Sol. Jo. 548, C. A.; 26 Digest (Repl.) 666, *35* . 12, 119
Kasler & Cohen *v.* Slavonski, [1928] 1 K. B. 78; 96 L. J. K. B. 850; 137 L. T. 641; 17 Digest (Reissue) 166, *483* 253
Kaufman *v.* Gerson, [1904] 1 K. B. 591; [1904–7] All E. R. Rep. 896; 73 L. J. K. B. 320; 90 L. T. 608; 52 W. R. 420; 20 T. L. R. 277; 48 Sol. Jo. 296, C. A.; 12 Digest (Reissue) 119, *648* 48
Kayford, Ltd., *Re* (1974), 118 Sol. Jo. 752 . . . . . . 142, 258
Keane *v.* Mount Vernon Colliery Co., [1933] A. C. 309; 102 L. J. P. C. 97; 149 L. T. 73; 49 T. L. R. 306; 77 Sol. Jo. 157; 26 B. W. C. C. 245, H. L.; 34 Digest (Repl.) 621, *4263* 85, 89
Keighley *v.* Bryan (1894), 70 L. T. 155; 7 Asp. M. L. C. 418, C. A.; 39 Digest (Repl.) 709, *1976* 365
Keighley, Maxsted & Co. *v.* Durant, [1901] A. C. 240; [1900–03] All E. R. Rep. 40; 70 L. J. K. B. 662; 87 L. T. 777; 17 T. L. R. 527, H. L.; 1 Digest (Repl.) 457, *1074* 271
Keith *v.* Burrows (1876), 1 C. P. D. 722; 45 L. J. Q. B. 876; 35 L. T. 508; 25 W. R. 43; 3 Asp. M. L. C. 280; on appeal (1877), 2 App. Cas. 636, H. L.; 39 Digest (Repl.) 48, *576* 82
Kemp *v.* Baerselman, [1906] 2 K. B. 604; 75 L. J. K. B. 873; 50 Sol. Jo. 615 C. A.; 12 Digest (Reissue) 724, *5243* 29
Kemp *v.* Falk. *See* Kiell, *Re, Ex parte* Falk.
Kemp *v.* Ismay (1909), 100 L. T. 996; 14 Com. Cas. 202; 39 Digest (Repl.) 754, *2329* 215
Kemp *v.* Pryor (1802), 7 Ves. Jun. 237 202
Kemp *v.* Tolland, [1956] 2 Lloyd's Rep. 681 325
Kempler *v.* Bravingtons, Ltd. (1925), 133 L. T. 680; 41 T. L. R. 519; 69 Sol. Jo. 639, C. A.; 39 Digest (Repl.) 625, *1369* . . . 149, 368
Kendall *v.* Marshall, Stevens & Co. (1883), 11 Q. B. D. 356; 52 L. J. Q. B. 313; 48 L. T. 951; 31 W. R. 597, C. A.; 39 Digest (Repl.) 756, *2340* 215
Kendall (Henry) & Sons *v.* William Lillico & Sons, Ltd. *See* Hardwick Game Farm *v.* Suffolk Agricultural and Poultry Producers Association, Ltd.
Kennedy *v.* Panama Mail Co. (1867), L. R. 2 Q. B. 580; 8 B. & S. 571; 36 L. J. Q. B. 260; 17 L. T. 62; 12 Digest (Reissue) 285, *2068* . 4, 353
Kenyon, Son and Craven, Ltd. *v.* Baxter Hoare & Co., Ltd., [1971] 2 All E. R. 708; [1971] 1 W. L. R. 519; [1971] 1 Lloyd's Rep. 232; Digest Supp. 11, 12, 13
Khan *v.* Duché (1905), 10 Com. Cas. 87; 39 Digest (Repl.) 717, *2026*: 195, 358
Kiddell *v.* Burnard (1842), 9 M. & W. 668; 1 Car. & M. 291; 11 L. J. Ex. 268; 2 Digest (Repl.) 353, *361* 367
Kidston *v.* Monceau Iron Works Co. (1902), 88 L. T. 556; 18 T. L. R. 320; 7 Com. Cas. 82; 39 Digest (Repl.) 518, *585* 358
Kiell, *Re, Ex parte* Falk (1880), 14 Ch. D. 446; 42 L. T. 780; 28 W. R. 785; 4 Asp. M. L. C. 280, C. A.; on appeal *sub nom.* Kemp *v.* Falk (1882), 7 App. Cas. 573; 52 L. J. Ch. 167; 47 L. T. 454; 31 W. R. 125; 5 Asp. M. L. C. 1, H. L.; 39 Digest (Repl.) 771, *2473*: 207, 210, 211, 215, 216, 217, 218, 219, 220, 221, 222, 223, 293

PAGE

Kier & Co., Ltd. *v.* Whitehead Iron and Steel Co., Ltd., [1938] 1 All E. R. 591; 150 L. T. 228; 54 T. L. R. 452; 82 Sol. Jo. 235; 60 Ll. L. Rep. 177; 39 Digest (Repl.) 692, *1853* . . . 364

Kilpin *v.* Ratley, [1892] 1 Q. B. 582; 66 L. T. 797; 56 J. P. 565; 40 W. R. 749; 8 T. L. R. 290; 25 Digest (Repl.) 556, *62* . . . 78

King's Norton Metal Co., Ltd. *v.* Edridge Merrett & Co., Ltd. (1897), 14 T. L. R. 98, C. A.; 39 Digest (Repl.) 654, *1572* 174

Kingdom *v.* Cox (1848), 5 C. B. 522; 17 L. J. P. C. 155; 12 Jur. 336; 39 Digest (Repl.) 678, *1737* 187

Kingsbury *v.* Collins & Elmes (1827), 4 Bing. 202; 2 Digest (Repl.) 41, *222* 281

Kingsley *v.* Sterling Industrial Securities, Ltd., [1966] 2 All E. R. 414; [1967] 2 Q. B. 747; [1966] 2 W. L. R. at p. 1277; 110 Sol. Jo. 267; Digest Cont. Vol. B p. 333, *21a* 76

Kinnear *v.* Brodie, 1901 3 F. (Ct. of Sess.) 540; 38 Sc. L. R. 336; 8 S. L. T. 475; 2 Digest (Repl.) 359, **506* 245

Kirkham *v.* Attenborough, [1897] 1 Q. B. 201; [1895–99] All E. R. Rep. 450; 66 L. J. Q. B. 149; 75 L. T. 543; 45 W. R. 213; 13 T. L. R. 131; 41 Sol. Jo. 141, C. A.; 39 Digest (Repl.) 624, *1359* . 149, 368

Kirshenboim *v.* Salmon and Gluckstein, [1898] 2 Q. B. 19; 67 L. J. Q. B. 601; 78 L. T. 658; 62 J. P. 439; 46 W. R. 573; 14 T. L. R. 395; 42 Sol. Jo. 538; 19 Cox C. C. 127; 46 Digest (Repl.) 170, *1134* . . 57

Kitson *v.* Hardwick (1872), L. R. 7 C. P. 473; 26 L. T. 846; 4 Digest (Repl.) 257, *2327* 75

Kitto *v.* Bilbie, Hobson & Co. (1895), 72 L. T. 266; 11 T. L. R. 214; 2 Mans. 122; 15 R. 188; 5 Digest (Repl.) 1171, *9457* 307

Kleinert *v.* Abosso Gold Mining Co. (1913), 58 Sol. Jo. 45, P. C.; 39 Digest (Repl.) 800, *2696* 23

Kleinjan Holst N.V. Rotterdam *v.* Bremer Handelsgesellschaft m.b.h. Hambourg, [1972] 2 Lloyd's Rep. 11; Digest Supp. 38

Knight, *Re*, *Ex parte* Golding Davis & Co., Ltd. (1880), 13 Ch. D. 628; 42 L. T. 270; 28 W. R. 481, C. A.; 39 Digest (Repl.) 769, *2459*: 221, 223

Knights *v.* Wiffen (1870), L. R. 5 Q. B. 660; 40 L. J. K. B. 51; 23 L. T. 610; 19 W. R. 244; 39 Digest (Repl.) 769, *2463* . 165, 213, 221

Koninklijke Bunge *v.* Compagnie Continentale D'Importation, [1973] 2 Lloyd's Rep. 44 . . . . 356, 358

Koppel *v.* Koppel (Wide Claimant), [1966] 2 All E. R. 187; [1966] 1 W. L. R. 802; 110 Sol. Jo. 229; 198 Estates Gazette 25, C. A.; Digest Cont. Vol. B p. 632, *1307a* 321

Korvine's Trust, *Re*, [1921] 1 Ch. 343; 90 L. J. Ch. 192; 124 L. T. 500; 65 Sol. Jo. 205; 11 Digest (Reissue) 417, *502* 49

Koufos *v.* Czarnikow, Ltd., Heron II, [1967] 3 All E. R. 686; [1969] 1 A. C. 350; [1967] 3 W. L. R. 1491; 111 Sol. Jo. 848; [1967] 2 Lloyd's Rep. 457, H. L.; Digest Cont. Vol. C p. 882, *1754a* . 241

Kreuger *v.* Blanck (1870), L. R. 5 Exch. 179; 39 L. J. Ex. 160; 23 L. T. 128; 18 W. R. 813; 3 Mar. L. C. 470; 39 Digest (Repl.) 692, *1854* 363

Kronprincessin Cecilie, The (1917), 33 T. L. R. 292, P. C.; 39 Digest (Repl.) 630, *1418* 76

Kruger (Cruger) *v.* Wilcox (Wilcocks) (1755), Tudor's Merc. Cas. (3rd Edn.) 370; Ambl. 252; 1 Digest (Repl.) 638, *2155* . . . 205

Krupp *v.* Orconera Iron Ore Co. (1919), 88 L. J. Ch. 304; 120 L. J. 386; 35 T. L. R. 234; 2 Digest (Repl.) 272, *625* 20

Kuenigl *v.* Donnersmarck, [1955] 1 All E.R. 46; [1955] 1 Q. B. 515; [1955] 2 W. L. R. 82; 99 Sol. Jo. 60; 2 Digest (Repl.) 221, *326* 90

Kuhlirz *v.* Lambert Bros. Ltd. (1913), 18 Com. Cas. 217; 108 L. T. 565; 1 Digest (Repl.) 544, *1689* 169

Kum *v.* Wah Tat Bank, [1971] 1 Lloyd's Rep. 439; Digest Supp.: 17, 192, 293

PAGE

Kursell *v.* Timber Operators and Contractors, Ltd., [1927] 1 K. B. 298; 95 L. J. K. B. 569; 135 L. T. 223; 42 T. L. R. 435, C. A.; 2 Digest (Repl.) 127, *952* 20, 142, 151, 288

Kwei Tek Chao *v.* British Traders & Shippers, Ltd. etc. *See* Chao (Trading as Zung Fu Co.) *v.* British Traders & Shippers, Ltd. etc.

Kwik Hoo Tong Handel Maatschappij (N. V.) *v.* James Finlay & Co., Ltd., [1927] A. C. 604; [1928] All E. R. Rep. 110; 96 L. J. K. B. 902; 137 L. T. 458; 33 Com. Cas. 55, H. L.; 50 Digest (Repl.) 343, *694* 48

Kymer *v.* Suwercropp (1807), 1 Camp. 109; 39 Digest (Repl.) 783, *2568* 226

Lacis *v.* Cashmarts, [1969] 2 Q. B. 400; [1969] 2 W. L. R. 329; 112 Sol. Jo. 1005; Digest Cont. Vol. C p. 262, *10,386b* . . . 145, 146

Ladenburg & Co. *v.* Goodwin, Ferreira & Co., [1912] 3 K. B. 275; 81 L. J. K. B. 1174; 107 L. T. 587; 18 Com. Cas. 16; 28 T. L. R. 541; 10 Digest (Repl.) 813, *5280* 83

Laidler *v.* Burlinson (1837), 2 M. & W. 602; Murp. & H. 109; 6 L. J. Ex. 160; 39 Digest (Repl.) 614, *1271* 151

Laing (Sir J.) *v.* Barclay & Co., [1908] A. C. 35; 77 L. J. P. C. 33; 97 L. T. 816; Asp. M. L. C. 583, H. L.; 39 Digest (Repl.) 612, *1256* 144

Lake *v.* Simmons, [1926] 2 K. B. 51; 95 L. J. K. B. 586; 135 L. T. 129; 42 T. L. R. 425; 70 Sol. Jo. 584; 31 Com. Cas. 271; [1926] W. C. & I. Rep. 252; reversed, [1927] A. C. 487; [1927] All E. R. Rep. 49; 96 L. J. K. B. 621; 137 L. T. 233; 43 T. L. R. 417; 33 Com. Cas. 16; 71 Sol. Jo. 369; 35 Digest (Repl.) 103, *69* . 172, 174, 296

Lakelse Dairy Produce, Ltd. *v.* General Dairy Machinery, Ltd. (1970), 10 D. L. R. (3d) 277; Digest Cont. Vol. C p. 854, **364b* . . 254

Lamb *v.* Attenborough (1862), 31 L. J. Q. B. 41; 1 B. & S. 831; 8 Jur. N. S. 280; 1 Digest (Repl.) 386, *510* 291

Lamb *v.* Wright & Co., [1924] 1 K. B. 857; [1924] All E. R. Rep. 220; 93 L. J. K. B. 366; 130 L. T. 703; 40 T. L. R. 290; 5 Digest (Repl.) 844, *7105* 162

Lamb (W. T.) & Sons *v.* Goring Brick Co., Ltd., [1932] 1 K. B. 710; [1931] All E. R. Rep. 314; 101 L. J. K. B. 214; 146 L. T. 318; 48 T. L. R. 160; 37 Com. Cas. 73, C. A.; 39 Digest (Repl.) 643, *1513* 76

Lambert *v.* G. & C. Finance Corporation (1963), 107 Sol. Jo. 666; 39 Digest (Repl.) 597, *1142* 146, 298

Lambton, *Ex parte*, *Re* Lindsay (1875), 10 Ch. App. 405; 44 L. J. Bk. 81; 32 L. T. 380; 5 Digest (Repl.) 770, *6582* . . . 205, 209

Lamond *v.* Davall (1847), 9 Q. B. 1030; 16 L. J. Q. B. 136; 11 Jur. 266; 39 Digest (Repl.) 451, *54* 81, 203, 223, 233

Lancashire Waggon Co. *v.* Fitzhugh (1861), 30 L. J. Ex. 231; 6 H. & N. 502; 46 Digest (Repl.) 473, *196* 242

Lancashire Waggon Co., Ltd. *v.* Nuttall (1879), 42 L. T. 465; 44 J. P. 536, C. A.; 3 Digest (Repl.) 667, *37* 191

Lancaster *v.* Turner & Co., [1924] 2 K. B. 222; [1924] All E. R. Rep. 189; 93 L. J. K. B. 1024; 131 L. T. 525; 29 Com. Cas. 207, C. A.; 39 Digest (Repl.) 504, *488* 9, 356

Landauer *v.* Asser, [1905] 2 K. B. 184; 74 L. J. K. B. 659; 93 L. T. 20; 21 T. L. R. 429; 10 Com. Cas. 265; 29 Digest (Repl.) 113, *582* 40, 359

Landauer & Co. *v.* Craven and Speeding Brothers, [1912] 2 K. B. 94; [1911–13] All E. R. Rep. 338; 81 L. J. K. B. 650; 106 L. T. 298; 56 Sol. Jo. 274; 17 Com. Cas. 193; 39 Digest (Repl.) 705, *1955* . 36, 37

La Neuville *v.* Nourse (1813), 3 Camp. 351; 39 Digest (Repl.) 554, *844* 80

PAGE

Lang *v.* Crude Rubber Washing Co. (1911), [1939] 2 K. B. 173 n.; 108 L. J. K. B. 460 n.; 160 L. T. 249 n.; 44 Com. Cas. 62 n.; 39 Digest (Repl.) 504, *490* 356

Lang *v.* Smyth (1831), 7 Bing. 284; 5 M. & P. 78; 9 L. J. O. S. C. P. 91; 1 Digest (Repl.) 399, *595* 281

Langen and Wind, Ltd. *v.* Bell, [1972] 1 All E. R. 296; [1972] Ch. 685; [1972] 2 W. L. R. 170; 115 Sol. Jo. 966; Digest Supp. . . 244

Langley, Beldon and Gaunt, Ltd. *v.* Morley [1965] 1 Lloyd's Rep. 297 . 219

Langton *v.* Higgins (1859), 4 H. & N. 402; 28 L. J. Ex. 252; 33 L. T. O. S. 166; 7 W. R. 489; 39 Digest (Repl.) 610, *1231* . . 96, 148, 242

Lanificio Di Manerbio, S. A. *v.* Gold (1932), 76 Sol. Jo. 289 . . 323

Lascelles *v.* Rathbun (1919), 35 T. L. R. 347; 63 Sol. Jo. 410, C. A.; 8 (2) Digest (Reissue) 629, *90* 93

Latch *v.* Wedlake (1840), 11 Ad. & El. 959; 3 Per. & Dav. 499; 9 L. J. Q. B. 201; 17 Digest (Reissue) 385, *1495* 4

Laurie & Morewood *v.* Dudin & Sons, [1926] 1 K. B. 223; [1925] All E. R. Rep. 414; 95 L. J. K. B. 191; 134 L. T. 309; 31 Com. Cas. 96; 42 T. L. R. 149; 39 Digest (Repl.) 639, *1482* . . 73, 141, 166

Law and Bonar *v.* American Tobacco Co., [1916] 2 K. B. 605; 85 L. J. K. B. 1714; 115 L. T. 612; 21 Com. Cas. 250; 39 Digest (Repl.) 713, *1998* 33, 40, 194

Leaf *v.* International Galleries, [1950] 1 All E. R. 693; [1950] 2 K. B. 86; 66 (pt. 1) T. L. R. 1031, C. A.; 39 Digest (Repl.) 831, *2919* . 6, 7, 197, 200, 271, 366

Leask *v.* Scott (1877), 2 Q. B. D. 376; [1874–80] All E. R. Rep. 722; 46 L. J. Q. B. 576; 36 L. T. 784; 25 W. R. 654; 3 Asp. M. L. C. 469, C. A.; 39 Digest (Repl.) 772, *2481* 222

Leather Cloth Co. *v.* Hieronimus (1875), L. R. 10 Q. B. 140; 44 L. J. Q. B. 54; 23 L. T. 307; 39 Digest (Repl.) 474, *241* 24

Leavey (J.) & Co., Ltd. *v.* George Hirst & Co., Ltd., [1943] 2 All E. R. 581; [1944] 1 K. B. 24; 113 L. J. K. B. 229; 169 L. T. 353; 60 T. L. R. 72; 88 Sol. Jo. 7, C. A.; 39 Digest (Repl.) 810, *2754*: 239, 240

Lebeaupin *v.* R. Crispin & Co., [1920] 2 K. B. 714; [1920] All E .R. Rep. 353; 89 L. J. K. B. 1024; 124 L. T. 124; 36 T. L. R. 739; 64 Sol. Jo. 652; 25 Com. Cas. 335; 12 Digest (Reissue) 473, *3398* . 258, 357

Lee *v.* Bayes (1856), 18 C. B. 599; 25 L. J. C. P. 249; 27 L. T. O. S. 157; 20 J. P. 694; 2 Jur. N. S. 1093; 33 Digest (Repl.) 489, 437 . 161, 171

Lee *v.* Butler, [1893] 2 Q. B. 318; [1891–94] All E. R. Rep. 1200; 62 L. J. Q. B. 591; 69 L. T. 370, C. A.; 26 Digest (Repl.) 659, *5*: 74, 81, 309

Lee *v.* Griffin (1861), 1 B. & S. 272; 30 L. J. Q. B. 252; 4 L. T. 546; 7 Jur. N. S. 1302; 9 W. R. 702; 39 Digest (Repl.) 444, *9* . . . 78

Leeming *v.* Snaith (1851), 16 Q. B. 275; 20 L. J. Q. B. 164; 16 L. T. O. S. 362; 15 Jur. 988; 39 Digest (Repl.) 690, *1838* 364

Lees, *Re*, Collins, *Ex parte* (1875), 10 Ch. App. 367; 44 L. J. Bk. 78; 32 L. T. 108; 7 Digest (Repl.) 83, *474* 348

Leigh *v.* Paterson (1818), 8 Taunt. 540; 2 Moore, C. P. 588; 39 Digest (Repl.) 818, *2804* 235, 238

Leith's Estate, *Re* (1866), L. R. 1 P. C. 296; 4 Moo. P. C. C. N. S. 158; 36 L. J. P. C. 17; 12 Jur. N. S. 967; 15 W. R. 534; 39 Digest (Repl.) 98, *1128* 213

Leroux *v.* Brown (1852), 12 C. B. 801; 22 L. J. P. C. 1; 16 Jur. 1021; 1 W. R. 22; 12 Digest (Reissue) 203, *1285* 49

Leslie *v.* Sheill, [1914] 3 K. B. 607; [1914–15] All E. R. Rep. 511; 83 L. J. K. B. 1145; 111 L. T. 106; 30 T. L. R. 460; 58 Sol. Jo. 453, C. A.; 12 Digest (Reissue) 696, *5019* 88

Lester *v.* Balfour Williamson Merchant Shippers, [1953] 1 All E. R. 1146; [1953] 2 Q. B. 168; [1953] 2 W. L. R. 1068; 117 J. P. 308; 97 Sol. Jo. 353; 51 L. G. R. 357; [1953] 2 Lloyd's Rep. 13; 47 Digest (Repl.) 797, *148* 69

PAGE

Lesters Leather and Skin Co., Ltd. *v.* Home and Overseas Brokers, Ltd. [1948] W. N. 437; 64 T. L. R. 569; 92 Sol. Jo. 646, C. A.; 39 Digest (Repl.) 819, *2820* 255
L'Estrange *v.* Graucob, Ltd., [1934] 2 K. B. 394; [1934] All E. R. Rep. 16; 103 L. J. K. B. 730; 152 L. T. 164; 12 Digest (Reissue) 78, *415*: 120, 361
Levey *v.* Goldberg, [1922] 1 K. B. 688; 91 L. J. K. B. 551; 127 L. T. 298; 28 Com. Cas. 244; 38 T. L. R. 446; 12 Digest (Reissue) 443, *3197* 22, 182
Levy *v.* Green (1859), 8 E. & B. 575; on appeal, 1 E. & E. 969; 28 L. J. Q. B. 319; 33 L. T. O. S. 241; 5 Jur. N. S. 1245; 7 W. R. 486; 39 Digest (Repl.) 693, *1865* 186
Levy *v.* Langridge (1838), 4 M. & W. 337; 1 H. & N. 325; 7 L. J. Ex. 387; 39 Digest (Repl.) 564, *927* 7
Lewis *v.* Averay, [1971] 3 All E. R. 907; [1972] 1 Q. B. 198; [1971] 3 W. L. R. 603; 115 Sol. Jo. 755, C. A.; Digest Supp. . 172, 173, 174
Lewis *v.* Clifton (1854), 14 C. B. 245; 23 L. J. C. P. 68; 22 L. T. O. S. 259; 18 Jur. 291; 2 W. R. 230; 39 Digest (Repl.) 806, *2729* . . 235
Lewis *v.* Davison (1839), 4 M. & W. 654; 1 Horn. & H. 425; 8 L. J. Ex. 78; 3 J. P. 167; 3 Jur. 387; 12 Digest (Reissue) 290, *2096* . . 19
Libau Wood Co. *v.* H. Smith & Sons, Ltd. (1930), 37 Ll. L. Rep. 296 . . . . 37, 38, 197, 200
Lickbarrow *v.* Mason (1793), 6 East, 22, n.; 4 Bro. Parl. Cas. 57; 2 Term Rep. 63; Smith L. C. (13th Edn.) 703, H. L.; 39 Digest (Repl.) 750, *2279* 214, 222
Lindley *v.* Lacy (1864) 17 C. B. N. S. 578; 5 New Rep. 51; 34 L. J. C. P. 17; 11 L. T. 273; 10 Jur. N. S. 1103; 13 W. R. 80; 17 Digest (Reissue) 383, *1477* 4
Lindsay *Re*, Lambton, *Ex parte* (1875), 10 Ch. App. 405; 32 L. T. 380; 23 W. R. 602; 2 Asp. M. L. C. 525; *sub nom. Re* Lindsay, *Ex parte* Greener, 44 L. J. Bcy. 81; 5 Digest (Repl.) 770, *6582*: 205, 209
Lindsay & Co. *v.* European Grain and Shipping Agency (1963), 107 Sol. Jo. 435; [1963] 1 Lloyd's Rep. 437, C. A.; Digest Cont. Vol. B 631, *888a* 361
Lipton *v.* Powell, [1921] 2 K. B. 51; 90 L. J. K. B. 366; 125 L. T. 89; 37 T. L. R. 289; 12 Digest (Reissue) 369, *2679* . . . 17, 18
Litt *v.* Cowley (1816), 7 Taunt. 169; 2 Marsh, 457; Holt N. P. 338; 39 Digest (Repl.) 774, *2495* 219, 220
Lloyd del Pacifico *v.* Board of Trade (1930), 46 T. L. R. 476; 35 Com. Cas. 325; 35 Ll. L. Rep. 217, C. A.; 2 Digest (Repl.) 720, *2326* 282, 360
Lloyds and Scottish Finance, Ltd. *v.* Modern Cars and Caravans (Kingston), Ltd., [1964] 2 All E. R. 732; [1966] 1 Q. B. 764; [1964] 3 W. L. R. 859; 108 Sol. Jo. 859; Digest Cont. Vol. B p. 630, *653a* 115, 179, 251, 253
Lloyds and Scottish Finance, Ltd. *v.* Williamson, [1965] 1 All E. R. 641; [1965] 1 W. L. R. 404; 109 Sol. Jo. 10, C. A.; Digest Cont. Vol. B p. 9, *864a* 166, 313
Lloyds Bank, Ltd. *v.* Bank of America, National Trust and Savings Association, [1938] 2 All E. R. 63; [1938] 2 K. B. 147; 107 L. J. K. B. 538; 158 L. T. 301; 54 T. L. R. 599; 82 Sol. Jo. 312; 43 Com. Cas. 209, C. A.; 1 Digest (Repl.) 387, *519* 296
Lockett *v.* Nicklin (1848), 2 Exch. 93; 19 L. J. Ex. 403; 39 Digest (Repl.) 487, *356* 93, 182
Loder *v.* Kekulé (1857), 3 C. B. N. S. 128; 27 L. J. C. P. 27; 30 L. T. O. S. 64; 4 Jur. N. S. 93; 5 W. R. 884; 39 Digest (Repl.) 591, *1113* 246, 249
Loders and Nucoline, Ltd. *v.* Bank of New Zealand (1929), 45 T. L. R. 203; 33 Ll. L. Rep. 70; 39 Digest (Repl.) 827, *2893* 39

PAGE

Loftus *v.* Roberts (1902), 18 T. L. R. 532, C. A.; 12 Digest (Reissue) 65, *338* 101
Lombank Ltd. *v.* Excell, [1963] 3 All E. R. 486; [1964] 1 Q. B. 415; [1963] 3 W. L. R. 700, C. A.; 17 Digest (Reissue) 185, *587* . 239
London and N. W. Railway *v.* Bartlett (1861), 31 L. J. Ex. 92; [1861–73] All E. R. Rep. 843; 7 H. & N. 400; 5 L. T. 399; 2 J. P. 167; 39 Digest (Repl.) 670, *1691* 215
London and Yorkshire Bank *v.* White (1895), 11 T. L. R. 570; 7 Digest (Repl.) 23, *106* 83
London Armoury Co., Ltd. *v.* Ever Ready Co. (Great Britain), Ltd., [1941] 1 All E. R. 364; [1941] 1 K. B. 742; 110 L. J. K. B. 531; 165 L. T. 221; 57 T. L. R. 343; 46 Digest (Repl.) 170, *1132* . . 54
London Jewellers, Ltd. *v.* Attenborough, [1934] 2 K. B. 206; [1934] All E. R. Rep. 270; 103 L. J. K. B. 429; 151 L. T. 124; 50 T. L. R. 436; 78 Sol. Jo. 413; 39 Com. Cas. 290, C. A.; 30 Digest (Reissue) 257, *625* 149
London Plywood and Timber Co., Ltd. *v.* Nasic Oak Extract Factory and Steam Sawmills Co., Ltd., [1939] 2 K. B. 343; 108 L. J. K. B. 587; 55 T. L. R. 826; 83 Sol. Jo. 607; 39 Digest (Repl.) 694, *1868* 188
Long *v.* Lloyd, [1958] 2 All E. R. 402; [1958] 1 W. L. R. 753; 102 Sol. Jo. 488, C. A.; 39 Digest (Repl.) 829, *2915* . . 6, 200, 201
Longbottom & Co. *v.* Bass, Walker & Co., [1922] W. N. 245, C. A.; 39 Digest (Repl.) 697, *1892* 182, 206, 211
Lord *v.* Price (1874), L. R. 9. Ex. 54; 43 L. J. Ex. 49; 30 L. T. 271; 22 W. R. 318; 39 Digest (Repl.) 743, *2213* . . . 223, 224
Lorymer *v.* Smith (1882), 1 B. & C. 1; 2 Dow. & Ry. K. B. 23; 1 L. J. O. S. K. B. 7; 39 Digest (Repl.) 559, *881* . . 137, 195
Lowe *v.* Lombank, Ltd., [1960] 1 All E. R. 611; [1960] 1 W. L. R. 196; 104 Sol. Jo. 210, C. A.; Digest Cont. Vol. A p. 644, *36a* . 262
Lowther *v.* Harris, [1927] 1 K. B. 393; [1926] All E. R. Rep. 352; 96 L. J. K. B. 170; 136 L. T. 377; 43 T. L. R. 24; 1 Digest (Repl.) 386, *512* 292, 296
Lucas *v.* Bristow (1858), E. B. & E. 907; 27 L. J. Q. B. 364; 5 Jur. N. S. 68; 17 Digest (Reissue) 65, *697* 365
Lucy *v.* Mouflet (1860), 29 L. J. Ex. 110; 5 H. & N. 229; 39 Digest (Repl.) 562, *913* 201, 202, 203
Lumley *v.* Ravenscroft, [1895] 1 Q. B. 683; 64 L. J. Q. B. 441; 72 L. T. 382; 59 J. P. 277; 43 W. R. 584; 39 Sol. Jo. 345; 14 R. 347, C. A.; 28 (2) Digest (Reissue) 688, *210* 87
Lunn *v.* Thornton (1845), 1 C. B. 379; 14 L. J. C. P. 161; 4 L. T. O. S. 417; 9 Jur. 350; 39 (Digest) (Repl). 494, *415* 95
Luxor (Eastbourne), Ltd. *v.* Cooper, [1941] 1 All E. R. 33; [1941] A. C. 108; 110 L. J. K. B. 131; 164 L. T. 313; 57 T. L. R. 213; 85 Sol. Jo. 105; 46 Com. Cas. 120, H. L.; 1 Digest (Repl.) 587, *1881* . 14, 15
Lyons *v.* Hoffnung (1890), 15 App. Cas. 391; [1886–90] All E. R. Rep. 1012; 59 L. J. P. C. 79; 63 L. T. 293; 39 W. R. 390; 6 Asp. M. L. C. 551; P. C.; 39 Digest (Repl.) 758, *2361* . . . 157, 215, 217
Lyons & Co. *v.* May and Baker, [1923] 1 K. B. 685; 92 L. J. K. B. 675; 129 L. J. 413; 39 Digest (Repl.) 743, *2209* 206
Lysney *v.* Selby (1705), 2 Ld. Raym. 1118; 40 Digest (Repl.) 50, *312* 352

Maas *v.* Pepper, [1905] A. C. 102; 74 L. J. K. B. 452; 92 L. T. 371; 21 T. L. R. 304; 7 Digest (Repl.) 19, *85* 76, 272
Macauley and Cullen *v.* Horgan, [1925] 2 Ir. R. 1; 39 Digest (Repl.) 816, **1523* 239
McBain *v.* Wallace & Co. (1881), 6 App. Cas. 588; 45 L. T. 261; 30 W. R. 65, H. L.; 39 Digest (Repl.) 665, *1653* . . 84, 145, 272
McCarthy Milling Co., Ltd. *v.* Elder Packing Co., Ltd. (1973), 33 D. L. R. (No. 3d) 52 102

PAGE

McCollin *v.* Gilpin (1881), 6 Q. B. D. 516; 44 L. T. 914; 45 J. P. 828; 29 W. R. 408, C. A.; 17 Digest (Reissue) 355, *1221* 365

McConnel *v.* Murphy (1873), L. R. 5 P. C. 203; 28 L. T. 713; 39 Digest (Repl.) 689, *1829* 188, 364

McCutcheon (*or* M'Cutcheon) *v.* David MacBrayne, [1964] 1 All E. R. 430; [1964] 1 W. L. R. 125; 108 Sol. Jo. 93; [1964] 1 Lloyd's Rep. 16; 1964 S. C. (H. L.) 28, H. L.; Digest Cont. Vol. B 71, *254b* . . 15, 16

Macdonald *v.* Longbottom (1860), 29 L. J. Q. B. 256; [1843–60] All E. R. Rep. 1050; 1 E. & E. 987; 6 Jur. N. S. 724; 2 L. T. 606; 39 Digest (Repl.) 490, *382* 365

McDougall *v.* Aeromarine of Emsworth, Ltd., [1958] 3 All E. R. 431; [1958] 1 W. L. R. 1126; 102 Sol. Jo. 860; [1958] 2 Lloyd's Rep. 345; 39 Digest (Repl.) 615, *1281* 107

McDowall *v.* Snowball 1904, 7 F. (Ct. of Sess.) 35; 39 Digest 615, (Repl.) 752, **1237* 359

McEntire *v.* Crossley Brothers, [1895] A. C. 457; [1895–99] All E. R. Rep. 829; 64 L. J. C. P. 129; 72 L. T. 731; 2 Mans. 334; 11 R. 207, H. L.; 39 Digest (Repl.) 595, *1128* 85, 224, 230, 309

McEwan *v.* Smith (1849), 2 H. L. Cas. 309; 13 Jur. 265; 39 Digest (Repl.) 639, *1477* . . . 185, 213, 221, 308

Mack and Edwards (Sales), Ltd. *v.* McPhail Brothers (1968), 112 Sol. Jo. 211, C. A. 101, 102

Mackay *v.* Dick (1881), 6 App. Cas. 251; 29 W. R. 541, H. L.; 39 Digest (Repl.) 785, *2593* 23

Mackender *v.* Feldia A.-G., [1966] 3 All E. R. 847; [1967] 2 Q. B. 590; [1967] 2 W. L. R. 119; 110 Sol. Jo. 811; C. A.; 50 Digest (Repl.) 341, *689* 48

M'Kenzie *v.* Hancock (1826), Ry. & M. 436; 2 Digest (Repl.) 360, *423* 202

Mackintosh *v.* Mitcheson (1849), 4 Exch. 175; 18 L. J. Ex. 385; 13 L. T. O. S. 344; 42 Digest (Repl.) 735, *4884* 93

Mackintosh (*or* Macintosh) (John) & Sons, Ltd. *v.* Baker's Bargain Stores (Seaford), Ltd., [1965] 3 All E. R. 412; L. R. 5 R. P. 305; [1965] 1 W. L. R. 1182; 109 Sol. Jo. 678; Digest Cont. Vol. B p. 704, *154a* 53

MacLean *v.* People's Gas Supply Co., Ltd., [1940] 4 D. L. R. 433 . 130

MacLean (R. G.), Ltd. *v.* Canadian Vickers, Ltd. (1969), 5 D. L. R. (3d) 100; (1971), 15 D. L. R. (3d) 15; Digest Cont. Vol. C p. 856, **581a* 13, 361

Macleod *v.* Kerr, 1965 S. L. T. 358 . 8, 173

Macklin *v.* Newbury Sanitary Laundry (1919), 63 Sol. Jo. 337, 39 Digest (Repl.) 796, *2671* 234

McLay *v.* Perry (1881), 44 L. T. 152; 39 Digest (Repl.) 689, *1830* . 364

Maclean *v.* Dunn (1828), 4 Bing. 722; 1 Moo. & P. 761; 6 L. J. O. S. C. P. 184; 1 M. & P. 761; 39 Digest (Repl.) 778, *2537* . . . 223, 234

McManus *v.* Fortescue, [1907] 2 K. B. 1; [1904–07] All E. R. Rep. 707; 76 L. J. K. B. 393; 96 L. T. 444; 23 T. L. R. 292, C. A.; 3 Digest (Repl.) 48, *334* 268

McMaster & Co. *v.* Cox., McEuen & Co., 1920, S. C. 566; 57 S. L. R. 504; 1920, 2 S. L. T. 122; reversed, 1921, S. C. (H. L.) 24; 58 S. L. R. 70; 1921; 1 S. L. T. 84, H. L.; 39 Digest (Repl.) 794, *2660* 20, 41

MacNab *v.* Alexanders of Greenock, 1971 S. L. T. 121 68

Macnee *v.* Gorst (1867), L. R. 4 Eq. 315; 15 W. R. 1197; 1 Digest (Repl.) 395, *562* 302

Macpherson Train & Co., Ltd. *v.* Howard Ross & Co., Ltd., [1955] 2 All E. R. 445; [1955] 1 W. L. R. 640; 99 Sol. Jo. 385; [1955] 1 Lloyd's Rep. 518; 39 Digest (Repl.) 521, *606*: 106, 108, 119, 356, 366

McRae *v.* Commonwealth Disposals Commission (1951), 84 C. L. R. 377 98

PAGE

Maddison *v.* Alderson (1883), 8 App. Cas. 467; [1881–85) All E. R. Rep. 742; 52 L. J. Q. B. 737; 49 L. T. 303, H. L.; 12 Digest (Reissue) 203, *1289* 7

Madeleine Vionnet et Cie *v.* Wills, [1939] 4 All E. R. 136; [1940] 1 K. B. 72; 109 L. J. K. B. 22; 16 L. T. 311; 56 T. L. R. 15; 83 Sol. Jo. 849, C. A.; 35 Digest (Repl.) 197, *72* 258

Madell *v.* Thomas & Co., [1891] 1 Q. B. 230; 60 L. J. Q. B. 227; 64 L. T. 9, C. A.; 7 Digest (Repl.) 55, *288* 76

Maerkle *v.* British and Continental Fur Co., Ltd., [1954] 3 All E. R. 50 [1954] 1 W. L. R. 1242; 98 Sol. Jo. 588, C. A.; 2 Digest (Repl.) 270, *616* 90

Magennis *v.* Fallon (1828), 2 Mol. 561; 2 L. R. Ir. 167; 40 Digest (Repl.) 72, **376* 5

Mahmoud and Ispahani, *Re*, [1921] 2 K. B. 716; [1921] All E. R. Rep. 217; 90 L. J. K. B. 821; 125 L. T. 161; 37 T. L. R. 489; 26 Com. Cas. 215, C. A.; 39 Digest (Repl.) 794, *2659* 18

Maine Spinning Co. *v.* Sutcliffe & Co. (1917), 87 L. J. K. B. 382; 118 L. T. 351; 34 T. L. R. 154; 23 Com. Cas. 216; [1916–17] All E. R. Rep. 537; 39 Digest (Repl.) 670, *1690* 22, 35

Mainprice *v.* Westley (1865), B. & S. 420; 34 L. J. Q. B. 229; 13 L. T. 560; 3 Digest (Repl.) 47, *332* 267

Malas (Trading as Hamzeh Malas & Sons) *v.* British Imex Industries, Ltd., [1958] 1 All E. R. 262; [1958] 2 Q. B. 127; [1958] 2 W. L. R. 100; 102 Sol. Jo. 68; [1957] 2 Lloyd's Rep. 549, C. A.; 3 Digest (Repl.) 285, *857* 45

Male *v.* Roberts (1790), 3 Esp. 163; 6 R. R. 823 91

Malik Co. *v.* Central European Trading Agency, Ltd., [1974] 2 Lloyd's Rep. 279 42

Malmberg *v.* Evans & Co. (1924), 29 Com. Cas. 235; affirmed, 30 Com. Cas. 107; C. A., 39 Digest (Repl.) 709, *1969* 40

Manbre Saccharine Co. *v.* Corn Products Co., [1919] 1 K. B. 198; [1918–19] All E. R. Rep. 980; 88 L. J. K. B. 402; 120 L. T. 113; 24 Com. Cas. 89; 35 T. L. R. 94; 39 Digest (Repl.) 705, *1951*: 40, 118, 366

Manchester Liners, Ltd. *v.* Rea, Ltd., [1922] 2 A. C. 74; [1922] All E. R. Rep. 605; 91 L. J. K. B. 504; 127 L. T. 405; 38 T. L. R. 526, 66 Sol. Jo. 421; 27 Com. Cas. 274, H. L.; 39 Digest (Repl.) 547, *804* 126, 135, 284

Manchester, Sheffield and Lincolnshire Ry. & Co. *v.* North Central Wagon Co. *See* North Central Wagon Co. *v.* Manchester, Sheffield and Lincolnshire Rail. Co.

Manchester Ship Canal Co. *v.* Horlock, [1914] 2 Ch. 199; 83 L. J. Ch. 87; 111 L. T. 260; 30 T. L. R. 500, C. A.; 12 Digest (Reissue) 756, *5422* 93, 282

Manders *v.* Williams (1849), 4 Exch. 339; [1843–60] All E. R. Rep. 545; 18 L. J. Ex. 437; 39 Digest (Repl.) 625, *1368* 77, 162

Mann *v.* D'Arcy, [1968] 2 All E. R. 172; [1968] 1 W. L. R. 893; 112 Sol. Jo. 310; Digest Cont. Vol. C p. 763, *400a* 168

Maple Flock Co. Ltd. *v.* Universal Furniture Products (Wembley) Ltd., [1934] 1 K. B. 148; [1933] All E. R. Rep. 15; 103 L. J. K. B. 513; 150 L. T. 69; 50 T. L. R. 58; 39 Com. Cas. 89, C. A.; 39 Digest (Repl.) 698, *1904* 191

Marcel, (Furriers), Ltd. *v.* Tapper, [1953] 1 All E. R. 15; [1953] 1 W. L. R. 49; 97 Sol. Jo. 10; 39 Digest (Repl.) 445, *12*. . . . 78

Maredelanto Compania Maviera S. A. *v.* Bergbau-Handel G.m.b.H.; [1970] 3 All E. R. 125; [1971] 1 Q. B. at p. 183; [1970] 3 W. L. R. 601; 114 Sol. Jo. 548, C. A.; Digest Cont. Vol. C p. 881, *1316a*: 238, 355

PAGE

Margarine Union G.m.b.H. *v.* Cambay Prince S.S. Co., Ltd., [1967] 3 All E. R. 775; [1969] 1 Q. B. 219; [1967] 3 W. L. R. 1569; 111 Sol. Jo. 943; [1967] 2 Lloyd's Rep. 315; Digest Cont. Vol. C p. 857, *1539a* 161, 317, 319

Margaronis Navigation Agency, Ltd. *v.* Henry W. Peabody & Co. of London, Ltd., [1964] 3 All E. R. 333; [1965] 2 Q. B. 430; [1964] 3 W. L. R. 873; 108 Sol. Jo. 562; [1964] 2 Lloyd's Rep. 153, C. A.; 41 Digest (Repl.) 202 . . . . 187, 363

Market Overt, The Case of (1596), 5 Co. Rep. 83 b; 33 Digest (Repl.) 491, *467* 169

Marlborough Hill, The *v.* Cowan & Sons, Ltd., [1921] 1 A. C. 444; 90 L. J. P. C. 87; 124 L. T. 675; 26 Com. Cas. 121; 37 T. L. R. 190; P. C.; 1 Digest (Repl.) 174, *586* 316

Marleau *v.* People's Gas Supply Co., Ltd., [1940] 4 D. L. R. 433 . 280

Marles *v.* Phillip Trant & Sons, Ltd. (No. 2), [1953] 1 All E. R. 651; [1954] 1 Q. B. 29; [1953] 2 W. L. R. 564; 97 Sol. Jo. 189, C. A.; 2 Digest (Repl.) 157, *1152* 18

Marsh *v.* Commissioner of Police, [1944] 2 All E. R. 392; [1945] K. B. 43; 171 L. T. 331; 61 T. L. R. 15; 88 Sol. Jo. 391; 42 L. G. R. 305, C. A.; 29 Digest (Repl.) 25, *284* 162

Marshall *v.* Nicoll, 1919, S. C. 129; 56 Sc. L. R. 615, H. L.; 39 Digest (Repl.) 819, *2824* 235, 240

Marshall (W. E.) & Co. *v.* Lewis and Peat (Rubber), Ltd., [1963] 1 Lloyd's Rep. 562 276

Marson *v.* Short (1835), 2 Bing. N. C. 118; 2 Scott, 243; 1 Hodges 260; 4 L. J. C. P. 270; 39 Digest (Repl.) 316, 622 74

Marten *v.* Whale, [1917] 2 K. B. 480; 86 L. J. K. B. 1305; 117 L. T. 137; 33 T. L. R. 330, C. A.; 39 Digest (Repl.) 451, *58* . . 74, 77, 82, 177, 309

Martin *v.* Hogan (1917), 24 C. L. R. 231; 39 Digest (Repl.) 711, **1064* 228

Martin *v.* Reid (1862), 31 L. J. C. P. 126; 11 C. B. N. S. 730; 5 L. T. 727; 37 Digest (Repl.) 4, *10* 162

Martindale *v.* Smith (1841), 1 Q. B. 389; 10 L. J. Q. B. 155; 1 G. & D. 1; 5 Jur. 932; 39 Digest (Repl.) 518, *589* 105, 205

Martineau *v.* Kitching (1872), L. R. 7 Q. B. 436; 41 L. J. Q. B. 227; 26 L. T. 336; 39 Digest (Repl.) 498, *446* . . 103, 152, 157, 158, 159

Marum, *Ex parte*, *Re* Wilson and Wilson (1915), 84 L. J. K. B. 1893; 113 L. T. 1116; 2 Digest (Repl.) 243, *458* 90

Marvin *v.* Wallis (1856), 6 E. & B. 726; 25 L. J. Q. B. 369; 27 L. T. O. S. 182; 2 Jur. N. S. 689; 39 Digest (Repl.) 468, *191* . . 277

Maschinenfabrik Seydelmann K. G. *v.* Presswood Brothers, Ltd., [1966] 1 O. R. 316, C. A. 19

Mash and Murrell, Ltd. *v.* Joseph I. Emanuel, Ltd., [1961] 1 All E. R. 485; [1961] 1 W. L. R. 862; 105 Sol. Jo. 468; [1961] 1 Lloyd's Rep. 46; reversed, [1962] 1 All E. R. 77; [1962] 1 W. L. R. 16; 105 Sol. Jo. 1007; [1961] 2 Lloyd's Rep. 326, C. A.; 39 Digest (Repl.) 545, *788* 35, 130, 195, 283

Mason *v.* Burningham, [1949] 2 All E. R. 134; [1949] 2 K. B. 545; [1949] L. J. R. 1430; 65 T. L. R. 466; 93 Sol. Jo. 496; C. A., 13 Digest (Repl.) 475, *992* 117

Mason *v.* Clarke, [1955] 1 All E. R. 914; [1956] A. C. 778; [1955] 2 W. L. R. 853; 99 Sol. Jo. 274, H. L.; 25 Digest (Repl.) 374, *48* 18

Matsoukis *v.* Priestman [1915] K. B. 681; [1914–15] All E. R. Rep. 1077; 84 L. J. C.P. 967; 113 L. T. 48; 20 Com. Cas. 252; 13 Asp. M. L. C. 68; 39 Digest (Repl.) 523, *622* 20

Matthews *v.* Baxter (1873), L. R. 8 Exch. 132; 42 L. J. Ex. 73; 21 W. R. 389; *sub nom.* Mathews *v.* Baxter, 28 L. T. 169; 12 Digest (Reissue) 188, *1124* 87

PAGE

May and Butcher *v.* R., [1934] 2 K. B. 17, n.; [1929] All E. R. Rep. 679; 103 L. J. K. B. 556, n.; 151 L. T. 246, n.; H. L., 39 Digest (Repl.) 448, *33* 101, 102

May and Hassell, Ltd. *v.* Exporters of Moscow (1940), 45 Com. Cas. 128; 39 Digest (Repl.) 521, *605* 356

Mear *v.* Baker (1954), 118 J. P. 483; 46 Digest (Repl.) 168, *1118* . 57, 61

Mechan & Son *v.* Bow McLachlan & Co., 1910, S. C. 758; 47 Sc. L. R. 650; 1 S. L. T. 406; 39 Digest (Repl.) 464, **75* 197, 198

Mechans, Ltd. *v.* Highland Marine Charters, Ltd., 1964 S. C. 48; 1964 S. L. T. 27; Digest Cont. Vol. B p. 635, **1121a* . 109, 199, 363

Medway Oil and Storage Co., Ltd., *v.* Silica Gel. Corporation (1928), 33 Com. Cas. 195, H. L.; 39 Digest (Repl.) 548, *807* . 119, 134, 135

Melachrino *v.* Nickoll and Knight, [1920] 1 K. B. 693; [1918–19] All E. R. Rep. 857; 89 L. J. K. B. 906; 122 L. T. 565; 25 Com. Cas. 103; 36 T. L. R. 143; 17 Digest (Reissue) 130, *279* 255

Melrose *v.* Hastie (1851), 13 Dunl. Sess. Cas. 880; 39 Digest (Repl.) 776, **1281* 208

Mendelssohn *v.* Normand, Ltd., [1969] 2 All E. R. 1215; [1970] 1 Q. B. 177; [1969] 3 W. L. R. 139; 113 Sol. Jo. 263, C. A.; Digest Cont. Vol. C p. 32, *196a* 5, 16, 361, 363

Mercantile Bank of India, Ltd. *v.* Central Bank of India, Ltd., [1938] 1 All E. R. 52; [1938] A. C. 287, 107 L. J. P. C. 25; 158 L. T. 269; 54 T. L. R. 208; 81 Sol. Jo. 1020, P. C.; 3 Digest (Repl.) 311, *975* 165

Mercantile Credit Co., Ltd. *v.* Hamblin, [1964] 3 All E. R. 592; [1965] 2 Q. B. 242; [1964] 3 W. L. R. 798; 108 Sol. Jo. 674, C. A.; Digest Cont. Vol. B p. 633, *1552a* 163, 166, 167

Mercantile Credit, Ltd. *v.* Upton & Son Pty., Ltd. (1974), 48 A. L. J. R. 301 305

Merchant Banking Co. *v.* Phoenix Bessemer Steel Co. (1877), 5 Ch. D. 205; 46 L. J. Ch. 418; 36 L. T. 395; 25 W. R. 457; 39 Digest (Repl.) 711, *1987* 207, 211, 221

Merchant Shipping Co. *v.* Armitage (1873), L. R. 9 Q. B. 99; 43 L. J. Q. B. 44; 29 L. T. 809; 2 Asp. M. L. C. 185; 35 Digest (Repl.) 209, *161* 229

Meredith *v.* Meigh (1853), 2 E. & B. 364; 22 L. J. Q. B. 401; 17 Jur. 649; 39 Digest (Repl.) 468, *197* 193

Mersey Steel and Iron Co. *v.* Naylor (1882), 9 Q. B. D. 648; 51 L. J. Q. B. 576; 47 L. T. 369; 31 W. R. 80; C. A., affirmed (1884), 9 App. Cas. 434; [1881–85] All E. R. Rep. 365; 53 L. J. Q. B. 497; 51 L. T. 637; 39 Digest (Repl.) 518, *592* 189, 191, 204

Mess *v.* Duffus (1901), 6 Com. Cas. 165; 39 Digest (Repl.) 664, *1642* 210

Messers, Ltd. *v.* Morrison's Export Co., Ltd., [1939] 1 All E. R. 92; 55 T. L. R. 245; 83 Sol. Jo. 75; 39 Digest (Repl.) 536, *726* . 356, 366

Metropolitan Water Board *v.* Dick Kerr & Co., [1918] A. C. 119; [1916–17] All E. R. Rep. 122; 37 L. J. K. B. 370; 117 L. T. 766; 23 Com. Cas. 148; 34 T. L. R. 113; H. L.; 12 Digest (Reissue) 506, *3513* 20

Meyer *v.* Everth (1814), 4 Camp. 22; 39 Digest (Repl.) 556, *862* . . 139

Meyer (N. V. Arnold Otto) *v.* Aune, [1939] 3 All E. R. 168; 55 T. L. R. 876; 83 Sol. Jo. 626; [1939] 64 Ll. L. Rep. 121; 39 Digest (Repl.] 707, *1962* 39

Meyer, Ltd. *v.* Kivisto (1929), 142 L. T. 480; 46 T. L. R. 162; 74 Sol. Jo. 58; 35 Ll. L. Rep. 265; 39 Digest (Repl.) 514, *558*. . . . 362

Meyer, Ltd. *v.* Osakeyhtio Carelia Timber Co. (1930), 37 Ll. L. Rep. 212; 36 Com. Cas. 17, C. A.; 39 Digest (Repl.) 528, *662* . 355, 362

Meyer, Ltd. *v.* Travaru A/BH. Cornelius (1930), 46 T. L. R. 553; 74 Sol. Jo. 466; 37 Ll. L. Rep. 204; 39 Digest (Repl.) 536, *723*: 356, 362, 366

PAGE

Michael *v.* Hart & Co., [1902] 1 K. B. 482; 71 L. J. K. B. 265; 86 L. T. 474; 50 W. R. 308; 18 T. L. R. 254, C. A.; *sub nom.* Hart & Co. *v.* Michael (1903) 89 L. T. 422, H. L.; 44 Digest (Repl.) 433, *385* . 238

Michelin Tyre Co. *v.* Macfarlane (Glasgow), Ltd. 1917, 55 Sc. L. R. 35, 1917 2 S. L. T. 205, H. L. 76

Microbeads AG *v.* Vinhurst Road Markings, Ltd. (1974), Times, 13th November, C. A. 113, 115, 116

Mildred *v.* Maspons (1883), 8 App. Cas. 874; 53 L. J. Q. B. 33, H. L.; 11 Digest (Reissue) 480, *872* 304

Miles, *Ex parte, Re* Isaacs (1885), 15 Q. B. D. 39; 54 L. J. Q. B. 566, C. A.; 39 Digest (Repl.) 755, *2335* 215, 217

Miles *v.* Gorton (1834), 2 Cr. & M. 504; 3 L. J. Ex. 155; 4 Tyr. 295; 39 Digest (Repl.) 743, *2207* 209, 210

Milgate *v.* Kebble (1841), 3 M. & Gr. 100; 3 Scott M. R. 358; 10 L. J. C. P. 277; 39 Digest (Repl.) 777, *2533* . . 184, 223, 243, 283

Millar & Co. *v.* Taylor & Co., [1916] 1 K. B. 402; 85 L. J. K. B. 346; 114 L. T. 216; 32 T. L. R. 161, C. A.; 12 Digest (Reissue) 491, *3467* 20

Millar's Machinery Co., Ltd., *v.* Way & Son (1935), 40 Com. Cas. 204, C. A.; 39 Digest (Repl.) 801, *2697* 240

Millars Karri & Jarrah Co. *v.* Weddell & Co. (1908); 100 L. T. 128; 11 Asp. M. L. C. 184; 14 Com. Cas. 25; 39 Digest (Repl.) 700, *1911.* 190

Miller *v.* Borner, [1900] 1 Q. B. 691; 69 L. J. Q. B. 49; 82 L. T. 258; 48 W. R. 588; 9 Asp. M. L. C. 31; 5 Com. Cas. 175; 41 Digest (Repl.) 203, *332* 363

Miller (James) & Partners, Ltd. *v.* Whitworth Street Estates, Ltd., [1970] 1 All E. R. 796; [1970] A. C. 583; [1970] 2 W. L. R. 728; 114 Sol. Jo. 225; [1970] 1 Lloyd's Rep. 269; 214 Estates Gazette 1111, H. L.; Digest Cont. Vol. C p. 29, *1949a* 48, 49

Millet *v.* Van Heek & Co., [1921] 2 K. B. 369; [1921] All E. R. Rep. 519; 90 L. J. K. B. 671; 125 L. T. 51; 37 T. L. R. 411; 65 Sol. Jo. 356, C. A.; 39 Digest (Repl.) 674, *1718* 236, 238, 358

Mills *v.* Stokman (1967), 116 C. L. R. 61 281

Minister of Supply and Development *v.* Servicemen's Co-operative Joinery Manufacturers, Ltd. (1951), 82 C. L. R. 621 230

Minster Trust, Ltd., *v.* Traps Tractors, Ltd., [1954] 3 All E. R. 136; [1954] 1 W. L. R. 963, 98 Sol. Jo. 456; 39 Digest (Repl.) 509, *524* 249

Mirabita *v.* Imperial Ottoman Bank (1878), 3 Ex. D. 164; 47 L. J. Q. B. 418; 38 L. T. 597; 3 Asp. M. L. C. 591, C. A.; 39 Digest (Repl.) 637, *1467* 36, 148, 154, 155, 156

Miramichi, The [1915] P. 71; 84 L. J. P. 105; 122 L. T. 349; 31 T. L. R. 72; 59 Sol. Jo. 107; 13 Asp. M. L. C. 21; 1 P. Cas. 137; 37 Digest (Repl.) 469, *194* 155

Mischeff *v.* Springett, [1942] 2 All E. R. 349; [1942] 2 K. B. 331; 111 L. J. K. B. 690; 17 Digest (Reissue) 527, *273* 75

Miss Gray, Ltd. *v.* Cathcart (Earl), [1922] 38 T. L. R. 562; 27 (1) Digest (Reissue) 217, *1502* 91

Mitchell *v.* Jones (1905), 24 N. Z. L. R. 932; 39 Digest (Repl.) 657, **897* 306

Mitchell Cotts & Co. (Middle East), Ltd. *v.* Hairco, Ltd., [1943] 2 All E. R. 552; 169 L. T. 349; 60 T. L. R. 31; 87 Sol. Jo. 447, C. A.; 39 Digest (Repl.) 790, *2635* 41

Mitchell-Henry *v.* Norwich Union, [1918] 2 K. B. 67; [1918–19] All E. R. Rep. 1142; 87 L. J. K. B. 695; 119 L. T. 111; 34 T. L. R. 359; 12 Digest (Reissue) 589, *4153* 228

Modern Light Cars, Ltd. *v.* Seals, [1934] 1 K. B. 32; [1933] All E. R. Rep. 539; 102 L. J. K. B. 680; 149 L. T. 285; 49 T. L. R. 503; 77 Sol. Jo. 420; 26 Digest (Repl.) 658, *1* 309

Modiano Brothers & Son *v.* Pearson & Co., Ltd. (1929), 34 Ll. L. Rep. 52 10, 360

PAGE

Mody *v.* Gregson (1868), L. R. 4 Ex. 49; 38 L. J. Ex. 12; 19 L. T. 458; 17 W. R. 176; 39 Digest (Repl.) 553, *835* . . . 10, 138, 139

Mogileff, The, [1921] P. 236; 90 L. J. P. 329; 37 T. L. R. 549; 65 Sol. Jo. 581; 1 Digest (Repl.) 152, *384* 93

Mollett *v.* Robinson (1870), L. R. 5 C. P. 646; 39 L. J. C. P. 290; 23 L. T. 185; 18 W. R. 1160; affirmed (1872), L. R. 7 C. P. 84; 41 L. J. C. P. 65; reversed *sub nom.* Robinson *v.* Mollett (1875), L. R. 7 H. L. 802; 44 L. J. C. P. 362; 33 L. T. 544; 17 Digest (Reissue) 25, *273* 17

Molling & Co. *v.* Dean & Son, Ltd. (1901), 18 T. L. R. 217; 39 Digest (Repl.) 716, *2021* 251, 252

Monarch SS Co., Ltd. *v.* AB. Karlshamns Oljefabriker, [1949] 1 All E. R. 1; [1949] A. C. 196; [1949] L. J. R. 772; 65 T. L. R. 217; 93 Sol. Jo. 117; 82 Lloyd L. R. 137; [1949] S. C. (H. L.) 1; [1949] S. L. T. 51; 41 Digest (Repl.) 362, *1549* 252

Mondel *v.* Steel (1841), 8 M. & W. 858; [1835–42] All E. R. Rep. 511; 1 Dowl. N. S. 1; 10 L. J. Ex. 426; 39 Digest (Repl.) 583, *1054*: 245, 246

Monforts *v.* Marsden (1895), 12 R. P. C. 266; 39 Digest (Repl.) 527, *647* 113, 114, 116

Monk *v.* Whittenbury (1831) 2 B. & Ad. 484; 1 M. & Rob. 81; 1 Digest (Repl.) 383, *491* 296

Moody *v.* Pall Mall Deposit and Forwarding Co. (1917), 33 T. L. R. 306; 1 Digest (Repl.) 386, *508* 300

Moorcock, The (1889), 14 P. D. 64; [1886–90] All E. R. Rep. 530; 58 L. J. P. 73, C. A.; 12 Digest (Reissue) 751, *5395* 102

Moore *v.* Campbell (1854), 10 Exch. 323; 23 L. J. Ex. 310; 2 C. L. R. 1084; 39 Digest (Repl.) 485, *335* 4, 188, 364

Moore *v.* Ray, [1950] 2 All E. R. 561; [1951] K. B. 58; 114 J. P. 486; 66 (pt. 2) T. L. R. 721; 94 Sol. Jo. 596; 49 L. G. R. 187; 25 Digest (Repl.) 113, *333* 68

Moore & Co., *v.* Landauer & Co. [1921] 1 K. B. 73; affirmed, [1921] 2 K. B. 519; [1921] All E. R. Rep. 466; 26 Com. Cas. 267; 90 L. J. K. B. 731; 125 L. T. 372; 37 T. L. R. 452, C. A.; 39 Digest (Repl.) 528, *660* 22, 118, 186, 187, 366

Moore Nettlefold & Co. *v.* Singer Manufacturing Co., [1904] 1 K. B. 821; [1904–07] All E. R. Rep. 810; 73 L. J. K. B. 457; 90 L. T. 469; 20 T. L. R. 366, C. A.; 13 Digest (Repl.) 476, *1004* . . 75

Moralice (London), Ltd. *v.* E. D. & F. Man, [1954] 2 Lloyd's Rep. 526 44

Mordaunt Brothers *v.* British Oil and Cake Mills, [1910] 2 K. B. 502; 79 L. J. K. B. 967; 103 L. T. 217; 54 Sol. Jo. 654; 15 Com. Cas. 285; 39 Digest (Repl.) 769, *2462* 221

Morel Brothers & Co., Ltd. *v.* Earl of Westmorland, [1904] A. C. 11; [1900–03] All E. R. Rep. 397; 73 L. J. K. B. 93; 89 L. T. 702; 20 T. L. R. 38, H. L. 27 (1) Digest (Reissue) 211, *1434* . . . 92

Morgan *v.* Bain (1874), L. R. 10 C. P. 15; 44 L. J. C. P. 47; 31 L. T. 616; 23 W. R. 239; 39 Digest (Repl.) 697, *1895* 210

Morgan *v.* Gath (1865), 3 H. & C. 748; 34 L. J. Ex. 165; 13 L. T. 96; 11 Jur. N. S. 654; 13 W. R. 756; 39 Digest (Repl.) 715, *2016*: 186, 293

Morgan *v.* Griffith (1871), L. R. 6 Ex. Ch. 70; 40 L. J. Ex. 46; 23 L. T. 783; 19 W. R. 957; 12 Digest (Reissue) 160, *930* . . 4

Morgan *v.* Russell & Sons, [1909] 1 K. B. 357; 78 L. J. K. B. 187; 100 L. T. 118; 25 T. L. R. 120; 53 Sol. Jo. 136; 39 Digest (Repl.) 443, *2* 76, 281

Morgan *v.* Thorne (1841), 7 M. & W. 400; 10 L. J. Ex. 125; 5 Jur. 294; 28 (2) Digest (Reissue) 663, *1* 88

Morison *v.* Gray (1824), 2 Bing. 260; 9 Moore C. P. 484; 3 L. J. O. S. C. P. 261; 39 Digest (Repl.) 742, *2197* 205

Morison *v.* Lockhart, 1912, S. C. 1017; 39 Digest (Repl.) 617, **782* . 288

PAGE

Morley *v.* Attenborough (1849), 3 Exch. 500; 18 L. J. Ex. 148; 12 L. T. O. S. 532; 13 J. P. 427; 13 Jur. 282; 39 Digest (Repl.) 526, *640* 113, 126
Morris *v.* Baron & Co., [1918] A. C. 1; 87 L. J. K. B. 145; 118 L. T. 34, H. L.; 39 Digest (Repl.) 487, *354* 4, 22
Morris *v.* Levison (1876), 1 C. P. D. 155; 45 L. J. Q. B. 409; 34 L. T. 576; 24 W. R. 517; 3 Asp. M. L. C. 171; 41 Digest (Repl.) 489, *2622* 364
Morris *v.* C. W. Martin & Sons, Ltd., [1965] 2 All E. R. 725; [1966] 1 Q. B. 716; [1965] 3 W. L. R. 276; 109 Sol. Jo. 451; [1965] 2 Lloyd's Rep. 63, C. A.; Digest Cont. Vol. B p. 30, *151a* 160
Morris Motors, Ltd. *v.* Lilley (Trading as G. & L. Motors), [1959] 3 All E. R. 737; [1959] 1 W. L. R. 1184; 103 Sol. Jo. 1003; 39 Digest (Repl.) 577, *1017* 367
Morris Motors Ltd. *v.* Phelan, [1960] 2 All E. R. 208 n.; [1960] 1 W. L. R. 352, 566, n.; 104 Sol. Jo. 310; [1960] R. P. C. 209; 39 Digest (Repl.) 577, *1018* 367
Morrisson *v.* Robertson, 1908 S. C. 332; 45 Scl L. R. 264; 39 Digest (Repl.) 655, **893* 174
Morritt, *Re* (1866), 18 Q. B. D. 222; 56 L. J. Q. B. 139; 56 L. T. 42; 35 W. R. 277; 3 T. L. R. 266, C. A.; 35 Digest (Repl.) 287, *79* . . 83
Mortimer *v.* Bell (1865), L. R. 1 Ch. App. 10; 35 L. J. Ch. 25; 3 Digest (Repl.) 14, *103* 267
Mortimore *v.* Wright (1846), 6 M. & W. 482; 9 L. J. Ex. 158; 4 Jur. 465; 28 (2) Digest (Reissue) 748, *832*. 92
Morton *v.* Lamb (1797), 7 Term Rep. 125; 39 Digest (Repl.) 662, *1625* 181
Moss *v.* Hancock, [1899] 2 Q. B. 111; 68 L. J. Q. B. 657; 80 L. T. 693; 63 J. P. 517; 47 W. R. 698; 15 T. L. R. 353; 43 Sol. Jo. 479; 19 Cox. C. C. 324, D. C.; 39 Digest (Repl.) 449, *36* 281
Moss *v.* Sweet (1851), 16 Q. B. 493; 20 L. J. Q. B. 167; 16 L. T. O. S. 341; 15 Jur. 536; 39 Digest (Repl.) 627, *1386* 147
Motor Credits (Hire Finance), Ltd. *v.* Pacific Motors Pty., Ltd. (1963) 109 C. L. R. 87, on appeal *sub nom.* Pacific Motor Auctions Pty., Ltd. *v.* Motor Credits (Hire Finance), Ltd., [1965] 2 All E. R. 105; [1965] A. C. 867; [1965] 2 W. L. R. 881; 109 Sol. Jo. 210; [1965] A. L. R. 1084, P. C.; Digest Cont. Vol. B p. 634, *1582a* 163, 166, 167, 305, 306
Mouat *v.* Betts Motors, Ltd., [1958] 3 All E. R. 402; [1959] A. C. 71; [1958] 3 W. L. R. 598; 102 Sol. Jo. 810; [1958] 2 Lloyd's Rep. 321, P. C.; 17 Digest (Reissue) 536, **150* 235, 236, 237
Mount (D. F.), Ltd. *v.* Jay and Jay (Provisions) Co., Ltd., [1959] 3 All E. R. 307; [1960] 1 Q. B. 159; [1959] 3 W. L. R. 537; 103 Sol. Jo. 636; 39 Digest (Repl.) 607, *1201* 175, 221, 307
Mowbray, Robinson & Co. *v.* Rosser (1922), 91 L. J. K. B. 524; 126 L. T. 748; 38 T. L. R. 413; 66 Sol. Jo. 315; C. A.; 41 Digest (Repl.) 532, *3078* 355
Moyce *v.* Newington (1878), 4 Q. B. D. 32; 48 L. J. Q. B. 125; 39 L. T. 535; 43 J. P. 191; 27 W. R. 319; 14 Cox. C. C. 182; 39 Digest (Repl.) 829, *2904* 171
Muhammed Tasa el Sheikh Ahmad *v.* Ali. *See* Ahmad *v.* Ali.
Muirhead *v.* Dickson 1905, 7 F. (Ct. of Sess.) 686 81
Muller Brothers *v.* G. M. Power Plant Co. (1963), Guardian, 9th May . 355
Muller, Maclean & Co. *v.* Leslie & Anderson, [1921] W. N. 235; 39 Digest (Repl.) 786, *2598* 229
Mulliner *v.* Florence (1878), 3 Q. B. D. 484; 47 L. J. Q. B. 700; 38 L. T. 167; 42 J. P. 293; 26 W. R. 385, C. A.; 32 Digest (Repl.) 273, 196 . . . . 213, 283

PAGE

Munro *v.* Willmott, [1948] 2 All E. R. 983; [1949] 1 K. B. 295; [1949] L. J. R. 471; 64 T. L. R. 627; 92 Sol. Jo. 662; 1 Digest (Repl.) 328, *137* 168
Munro & Co. *v.* Bennet & Son, 1911 S. C. 337; 39 Digest (Repl.) 724, **1129* 201
Munro & Co., Ltd. *v.* Meyer, [1930] 2 K. B. 312; [1930] All E. R. Rep. 241; 99 L. J. K. B. 703; 143 L. T. 565; 35 Com. Cas. 232; 39 Digest (Repl.) 533, *692* 118, 191, 360, 361
Murdoch & Co. *v.* Greig 1889, 16 R. (Ct. of Sess.) 396; 26 Digest (Repl.) 672, **36*. 80
Murdoch *v.* Wood, [1921] W. N. 299 311
Murgatroyd *v.* Wright, [1907] 2 K. B. 333; 76 L. J. K. B. 747; 97 L. T. 108; 23 T. L. R. 517; 14 Mons. 201, D. C.; 13 Digest (Repl.) 451, *746* 179
Murray *v.* Stair (Earl) (1823), 2 B. & C. 82; 3 Dow. & Ry. K. B. 278; 7 Digest (Repl.) 228, *665* 4
Myers (G. H.) & Co. *v.* Brent Cross Service Co., [1934] 1 K. B. 46; [1933] All E. R. Rep. 9; 103 L. J. K. B. 123; 150 L. T. 96, D. C. 3 Digest (Repl.) 102, *285* 122

N. V. Handel My. J. Smits Import-Export *v.* English Exporters (London), Ltd., [1955] 1 Lloyd's Rep. 317, C. A.; 11 Digest (Reissue) 468, *800* 33
N. V. Kwik Hoo Tong Handel Maatschappij *v.* James Finlay & Co., Ltd., [1927] A. C. 604; [1928] All E. R. Rep. 110; 96 L. J. K. B. 902; 137 L. T. 458; 33 Com. Cas. 55, H. L.; 50 Digest (Repl.) 343, *694* . 48
Naish *v.* Gore, [1971] 3 All E. R. 737; 136 J. P. 1; [1972] R. T. R. 102; Digest Supp. 68
Nanka Bruce *v.* Commonwealth Trust, [1926] A. C. 77; 94 L. J. P. C. 169; 134 L. T. 35; 39 Digest (Repl.) 654, *1574* 149, 152
Napier *v.* Dexters, Ltd. (1926), 26 Ll. L. Rep. 62 . 34, 148, 154, 227, 228
Nash *v.* Inman, [1908] 2 K. B. 1; [1908–10] All E. R. Rep. 317; 77 L. J. K. B. 626; 98 L. T. 658; 24 L. T. R. 401, C. A.; 28 (1) Digest (Reissue) 133, *395* 86, 89
National Mercantile Bank *v.* Rymill (1881), 44 L. T. 767, C. A.; 3 Digest (Repl.) 50, *351* 268
National Phonograph Co. of Australia *v.* Manck, [1911] A. C. 336; 80 L. J. P. C. 105; 104 L. T. 5; 27 T. L. R. 239; 28 R. P. C. 229; 39 Digest (Repl.) 642, *1502* 53
Navulshaw *v.* Brownrigg (1852), 2 De. G. M. & G. 441; 21 L. J. Ch. 908; 20 L. T. O. S. 25; 16 Jur. 979; 1 Digest (Repl.) 393, *545* 299
Neill *v.* Whitworth (1866), L. R. 1 C. P. 684; 35 L. J. C. P. 304; 12 Jur. N. S. 671; 14 L. T. 670; 14 W. R. 844; 39 Digest (Repl.) 493, *411*: 356, 359
Nelson *v.* William Chalmers & Co. 1913, S. C. 441; 50 Sc. L. R. 363; 1 S. L. T. 190; 39 Digest (Repl.) 615, **778* 109
Nevill, *Re, Ex parte* White (1871), 6 Ch. App. 397; 40 L. J. Bcy. 73; 24 L. T. 45; on appeal *sub nom.* Towle & Co. *v.* White (1873), 29 L. T. 78; 21 W. R. 465; H. L.; 39 Digest (Repl.) 625, *1370* . . 76
New *v.* Swain (1828), 1 Dan. & Ll. 193; 39 Digest (Repl.) 744, *2223* . 209
New Zealand Shipping Co. *v.* Société des Ateliers et Chantiers de France [1919] A. C. 1; [1918–19] All E. R. Rep. 552; 87 L. J. K. B. 746; 118 L. T. 731; 34 T. L. R. 460; 62 Sol. Jo. 519; 14 Asp. M. L. C. 291, H. L.; 39 Digest (Repl.) 808, *2744* 20, 23, 81
Newborne *v.* Sensolid (Great Britain), Ltd., [1953] 1 All E. R. 708; [1954 1 Q. B. 45; [1953] 2 W. L. R. 596; 97 Sol. Jo. 209, C. A.; 1 Digest (Repl.) 718, *2651* 174

PAGE

Newbridge Rhondda Brewery Co. *v.* Evans (1902), 86 L. T. 453; 18 T. L. R. 396, D. C.; 39 Digest (Repl.) 502, *476* 323

Newman *v.* Bourne and Hollingsworth (1915), 31 T. L. R. 209; 3 Digest (Repl.) 69, *95* 160

Newman Industries *v.* Indo-British Industries, [1956] 2 Lloyd's Rep. 219; reversed, [1957] 1 Lloyd's Rep. 211, C. A.; Digest Cont. Vol. B 629, *439a* 44

Newman *v.* Jones (1886), 17 Q. B. D. 132; 35 L. J. M. C. 113; 55 L. T. 327; 8 (2) Digest (Reissue) 638, *143* 80

Newman *v.* Lipman, [1950] 2 All E. R. 832; [1951] 1 K. B. 333; 114 J. P. 561; [1951] 1 T. L. R. 190; 94 Sol. Jo. 673; 49 L. G. R. 457, D. C.; 26 Digest (Repl.) 482, *1690* 78

Newtons of Wembley, Ltd. *v.* Williams, [1964] 3 All E. R. 532; [1965] 1 Q. B. 560; [1964] 3 W. L. R. 888; 108 Sol. Jo. 619, C. A.; Digest Cont. Vol. B p. 634, *1588a* 8, 172, 300, 310

Niblett *v.* Confectioners' Materials Co., [1921] 3 K. B. 387; [1921] All E. R. Rep. 459; 90 L. J. K. B. 984; 125 L. T. 552; 37 T. L. R. 653, C. A.; 39 Digest (Repl.) 527, *648* . 113, 114, 115, 116, 130, 283

Nichol *v.* Godts (1854), 10 Exch. 191; 2 C. L. R. 1468; 23 L. J. Ex. 314; 23 L. T. O. S. 162; 39 Digest (Repl.) 560, *892* . . 117, 120, 366

Nicholls *v.* Tavistock Urban Council, [1923] 2 Ch. 18; 92 L. J. Ch. 233; 128 L. T. 565; 87 J. P. 98; 67 Sol. Jo. 298; 21 L. G. R. 194; 33 Digest (Repl.) 455, *62* 269

Nicholson *v.* Bradfield Union (1866), L. R. 1 Q. B. 620; 7 B. & S. 774; 35 L. J. Q. B. 176; 14 L. T. 830; 30 J. P. 549; 12 Jur. N. S. 686; 14 W. R. 731; 39 Digest (Repl.) 727, *2085* 186

Nicholson *v.* Harper, [1895] 2 Ch. 415; [1895–99] All E. R. Rep. 882; 64 L. J. Ch. 672; 73 L. T. 19; 59 J. P. 727; 43 W. R. 550; 11 T. L. R. 435; 39 Sol. Jo. 524; 13 R. 567; 39 Digest (Repl.) 535, *1461* . 306

Nicholson and Venn *v.* Smith-Marriott (1947), 177 L. T. 189; 39 Digest (Repl.) 529, *664* 110, 360, 366

Nickoll *v.* Ashton, [1900] 2 Q. B. 298; 69 L. J. Q. B. 640; 82 L. T. 761; 16 T. L. R. 370; 5 Com. Cas. 252; 9 Asp. M. L. C. 94; on appeal, [1901] 2 K. B. 126; [1900–03] All E. R. Rep. 928; 70 L. J. K. B. 600; 84 L. T. 804; 49 W. R. 513; 17 T. L. 467; 6 Com. Cas. 150; 9 Asp. M. C. 209, C. A.; 39 Digest (Repl.) 819, *2817* . 238, 255, 355

Nicol *v.* Cooper (1896), 1 Com. Cas. 410 168

Nicolene, Ltd. *v.* Simmonds, [1953] 1 All E. R. 822; [1953] 1 Q. B. 543; 97 Sol. Jo. 247; 1 Lloyd's L. R. 189, C. A.; 12 Digest (Reissue) 93, *483* 101

Ningchow, The, [1916] P. 221; 115 L. T. 554; 31 T. L. R. 470; 13 Asp. M. L. C. 509; 37 Digest (Repl.) 472, *223* 283

Noordam, The, (No. 2), [1920] A. C. 904; 89 L. J. P. 234; 123 L. T. 477; 36 T. L. R. 581; 15 Asp. M. L. C.; 37 Digest (Repl.) 494, *477* . 280

Norman *v.* Bennett, [1974] 3 All E. R. 351; [1974] 1 W. L. R. 1229; 138 J. P. 746; 118 Sol. Jo. 697 59

Norman *v.* Phillips (1845), 14 M. & W. 277; 14 L. J. Ex. 306; 9 Jur. 832; 39 Digest (Repl.) 459, *114* 193

North Central Wagon Co. *v.* Manchester, Sheffield and Lincolnshire Rail. Co. (1887), 35 Ch. D. 191; 56 L. J. Ch. 609; 56 L. T. 755; 35 W. R. 443; 3 T. L. R. 206, C. A.; affirmed *sub nom.* Manchester Sheffield and Lincolnshire Rail. Co. *v.* North Central Wagon Co. (1888), 13 App. Cas. 554, 58 L. J. Ch. 219; 59 L. T. 730; 37 W. R. 305; 4 T. L. R. 728, H. L.; 7 Digest (Repl.) 6, *12* 82, 83, 321

North-Western Bank, *Ex parte*, *Re* Slee (1872), L. R. 15 Eq. 69; 42 L. J. Bk. 6; 27 L. T. 461; 7 Digest (Repl.) 30, *146* 83

North-Western Rubber Co., *Re*, [1908] 2 K. B. 907; 78 L. J. K. B. 51; 99 L. T. 680; 17 Digest (Repl.) 44, *523* 367

PAGE

North-Western Salt Co. *v.* Electrolytic Alkali Co., [1913] 3 K. B. 422, C. A.; reversed, [1914] A. C. 461; [1914–15] All E. R. Rep. 752; 83 L. J. K. B. 530; 110 L. T. 852; 30 T. L. R. 313; 58 Sol. Jo. 338, H. L.; 39 Digest (Repl.) 643, *1512* 17
Northern Steel and Hardware Co. *v.* Batt & Co. (1917), 33 T. L. R. 516, C. A.; 39 Digest (Repl.) 712, *1995* 194
Nudgee Bakery Pty., Ltd.'s Agreement, *Re*, [1971] Qd. R. 24 . . . 104
Nugent *v.* Smith (1876), 1 C. P. D. 423; 45 L. J. C. P. 697; 34 L. T. 827, C. A.; 3 Digest (Repl.) 55, *2* 21
Nyberg *v.* Handelaar, [1892] 2 Q. B. 202; [1891–94] All E. R. Rep. 1109; 61 L. J. Q. B. 309; 67 L. T. 361, C. A.; 1 Digest (Repl.) 654, *2275*: 168, 283

Oakes *v.* Turquand and Harding. *See* Overend, Gurney & Co., *Re*, Oakes *v.* Turquand and Harding, Peek *v.* Turquand and Harding.
Ocean Tramp Tankers Corporation *v.* V/O Soufracht, [1964] 1 All E.R. 161; [1964] 2 Q. B. 226; [1964] 2 W. L. R. 114; 107 Sol. Jo. 931; [1963] 2 Lloyd's Rep. 381, C.A.; 12 Digest (Reissue) 499, *3491* 21
O'Connor *v.* Donnelly, [1944] Ir. Jur. Rep. 1 121
O'Dea *v.* Crowhurst (1899), 68 L. J. Q. B. 655; 80 L. T. 491; 63 J. P. 424; 15 T. L. R. 320; 19 Cox, C. C. 260; 33 Digest (Repl.) 494, *507* 284
Odessa, The, [1916] 1 A. C. 145; 85 L. J. P. C. 49; 114 L. T. 10; 32 T. L. R. 103; 60 Sol. Jo. 292; 13 Asp. M. L. C. 215, P. C.; 37 Digest (Repl.) 13, *88* 83, 283
Ogg *v.* Shuter (1875), 1 C. P. D. 47; 45 L. J. Q. B. 44; 33 L. T. 492; 24 W. R. 100; 3 Asp. M. L. C. 77, C. A.; 39 Digest (Repl.) 631, *1422* 35, 154, 155
Ogle *v.* Earl Vane (1868), L. R. 3 Q. B. 272; 9 B. & S. 182; 37 L. J. Q. B. 77; 16 W. L. R. 463; 39 Digest (Repl.) 819, *2815* . . . 22, 239
Old Colony Trust Co. *v.* Lawyers' Title and Trust Co. (1924), 297 F. 152 45
Oldfield *v.* Lowe (1829), 9 B. & C. 73; 7 L. J. O. S. K. B. 142; 39 Digest (Repl.) 613, *1258* 151
Olearia Tirrena, S. p. A. *v.* N. V. Algemeene Oliehandel, *The Osterbek*, [1972] 2 Lloyd's Rep. 341, Digest Supp. 34
Oleificio Zucchi S. p. A. *v.* Northern Sales, Ltd., [1965] 2 Lloyd's Rep. 496; 28 M. L. R. 189 . . . . 130, 195
Ollet *v.* Jordan, [1918] 2 K. B. 41; [1918–19] All E. R. Rep. 1069; 97 L. J. K. B. 934; 119 L. T. 50; 25 Digest (Repl.) 119, *395* . . . 195
Oppenheimer *v.* Attenborough, [1908] 1 K. B. 221; [1904–07] All E. R. Rep. 1016; 77 L. J. K. B. 209; 98 L. T. 94; 24 T. L. R. 115; 13 Com. Cas. 125, C. A.; 1 Digest (Repl.) 392, *541* . . 291, 295, 297, 298
Oppenheimer *v.* Frazer and Wyatt, [1907] 2 K. B. 50; [1904–07] All E. R. Rep. 143; 76 L. J. K. B. 806; 97 L. T. 3; 12 Com. Cas. 289; 23 T. L. R. 410, C. A.; 1 Digest (Repl.) 390, *530* . . 168, 174, 295
Orchard *v.* Simpson (1857), 2 C. B. N. S. 299; 39 Digest (Repl.) 499, *450* 359
Orient Co., Ltd. *v.* Brekke, [1913] 1 K. B. 531; 82 L. J. K. B. 427; 108 L. T. 507; 18 Com. Cas. 101; 39 Digest (Repl.) 710, *1981* . 39, 180
Ornstein *v.* Hickerson (1941) 40 F. Supp. 305 41
Orteric, The, [1920] A. C. 724, 733; 89 L. J. P. 209; 123 L. T. 448; 15 Asp. M. L. C. 10, P. C.; 37 Digest (Repl.) 469, *192* . . 76, 155, 272
Oscar Chess, Ltd. *v.* Williams. *See* Chess (Oscar), Ltd. *v.* Williams.
Osterbek, The. *See* Olearia Tirrena, S. p. A. *v.* N. V. Algemeene Oliehandel, The *Osterbek*.
Ottoman Bank *v.* Jebara. *See* Jebara *v.* Ottoman Bank.
Ottoman Bank of Nicosia *v.* Chakarian, [1937] 4 All E. R. 570; [1938] A. C. 260, 107 L. J. P. C. 15; 158 L. T. 1; 54 T. L. R. 122, P. C.; 35 Digest (Repl.) 198, *79* 258
Overbrooke Estates, Ltd. *v.* Glencombe Properties, Ltd., [1974] 3 All E. R. 511; [1974] 1 W. L. R. 1335; 118 Sol. Jo. 775 . . 5, 336, 361

PAGE

Overend, Gurney & Co., *Re*, Oakes *v.* Turquand and Harding, Peek *v.* Turquand and Harding (1867), L. R. 2 H. L. 325; 36 L. J. Ch. 949; 16 L. T. 808; 15 W. R. 1201, H. L.; 10 Digest (Repl.) 1046, *7244* . 6
Oxendale *v.* Wetherell (1829), 4 Man. & Ry. 429; 9 B. & C. 386; 7 L. J. O. S. K. B. 264; 39 Digest (Repl.) 694, *1871* . . 186, 330

P. and O. Steam Navigation Co. *v.* Shand (1865), 3 Moo. P. C. C. N. S. 272; 6 New Rep. 387; 12 L. T. 808; 11 Jur. N. S. 771; 13 W. R. 1049; 2 Mar. L. C. 244, P. C.; 11 Digest (Reissue) 472, *822* . . 48
Pacific Motor Auctions Proprietary, Ltd. *v.* Motor Credits (Hire Finance), Ltd. *See* Motor Credits (Hire Finance), Ltd. *v.* Pacific Motor Pty., Ltd.
Paddington Burial Board *v.* Inland Revenue Commissioners (1884), 13 Q. B. D. 9; 53 L. J. Q. B. 224; 50 L. T. 211; 48 J. P. 311; 32 W. R. 551; 2 Tax Cas. 46; 7 Digest (Repl.) 573, *240* 274
Page *v.* Cowasjee (1866), L. R. 1 P. C. 127; 3 Moo. P. C. C. N. S. 499; 12 Jur. N. S. 361; 14 L. T. 176; 39 Digest (Repl.) 780, *2553*: 223, 225, 242
Pagnan and Fratelli *v.* Corbisa Industrial Agropacuaria Limitada, [1971] 1 All E. R. 165; [1970] 1 W. L. R. 1306; 114 Sol. Jo. 568, C. A.; Digest Cont. Vol. C p. 858, *2768a* 256
Panchaud Frères S. A. *v.* Etablissements General Grain Co., [1970] 1 Lloyd's Rep. 53, C. A.; Digest Cont. Vol. C p. 856, *1046a* . . 23, 38
Panchaud Frères S. A. *v.* R. Pagnan and Fratelli, [1974] 1 Lloyd's Rep. 394 325
Panola O. H. G. *v.* Circle Products, Ltd. *See* Frebold *v.* Circle Products, Ltd.
Panoutsos *v.* Raymond Hadley Corporation of New York [1917] 2 K. B. 473; [1916–17] All E. R. Rep. 448; 86 L. J. K. B. 1325; 117 L. T. 330; 33 T. L. R. 436; 22 Com. Cas. 308, C. A.; 39 Digest (Repl.) 576, *1009* 23, 107
Pantanassa, The, Norsk Bjergningskompagni A/S *v.* Owners of S. S. Pantanassa (Her Cargo and Freight), [1970] 1 All E. R. 848; [1970] P. 187; [1970] 2 W. L. R. 981; 114 Sol. Jo. 372; 1 Lloyd's Rep. 153; Digest Cont. Vol. C p. 885, *3392a* 40, 357
Parchim, The, [1918] A. C. 157; 87 L. J. P. 18; 117 L. T. 738; 34 T. L. R. 53; 14 Asp. M. L. C. 196, P. C.; 39 Digest (Repl.) 634, *1454* . 33, 35, 36, 49, 85, 144, 155, 157, 159
Parfitt *v.* Jepson (1877), 46 L. J. Q. B. 529; 36 L. T. 251; 3 Digest (Repl.) 13, *90* 268
Parker *v.* Palmer (1821), 4 B. & Ald. 387; 39 Digest (Repl.) 558, *874* 137, 139, 197
Parker *v.* Schuller (1901), 17 T. L. R. 299, C. A. 38
Parkinson *v.* Lee (1802), 2 East 314; 39 Digest (Repl.) 559, *883* . . 140
Parsons *v.* New Zealand Shipping Co., [1901] 1 Q. B. 548; 70 L. J. K. B. 404; 84 L. T. 218; 49 W. R. 355; 17 T. L. R. 274; 9 Asp. M. L. C. 170; 6 Com. Cas. 41, C. A.; 41 Digest (Repl.) 254, *722* . . . 318
Parsons *v.* Sexton (1874), 4 C. B. 899; 16 L. J. C. P. 181; 9 L. T. O. S. 410; 11 Jur. 849; 39 Digest (Repl) 584, *1055* 245
Partabmull Rameshwar *v.* Sethia K. C. (1944), Ltd., [1951] 2 All E. R. 352 n.; 95 Sol. Jo. 528; [1951] 2 Lloyd's Rep. 89; 12 Digest (Reissue) 753, *5406* 42
Pasley *v.* Freeman (1789), 3 Term Rep. 51; 39 Digest (Repl.) 512, *547* 352
Paton *v.* Payne 1897, 34 Sc. L. R. 112, H. L.; 39 Digest (Repl.) 522, *614* 107
Paton's Trustees *v.* Finlayson, 1923 S. C. 872; 39 Digest (Repl.) 744, **1217* 212, 281
Patrick *v.* Russo-British Grain Export Co., [1927] 2 K. B. 535; [1927] All E. R. Rep. 692; 97 L. J. K. B. 60; 137 L. T. 815; 33 Com. Cas. 60; 45 T. L. R. 724; 33 Com. Cas. 60; 39 Digest (Repl.) 812, *2765* 240

PAGE

Patten *v*. Thomas Motors Pty., Ltd. (1965) 66 S. R. (N. S. W.) 458 . 115
Patterson Motors *v*. Riley [1966] 56 D. L. R. (2d) 278 . . . 166
Paul *v*. Dod (1846), 2 C. B. 800; 15 L. J. C. P. 177; 10 Jur. 335; 39 Digest (Repl.) 734, *2133* 230
Paul *v*. Pim & Co., [1922] 2 K. B. 360; 91 L. J. K. B. 556; 27 Com. Cas. 98; 38 T. L. R. 95; 66 Sol. Jo. 93; 27 Com. Cas. 98; 39 Digest (Repl.) 693, *1866* 363
Pavia & Co., S. *P*. A. *v*. Thurmann-Neilson, [1952] 1 All E. R. 492; [1952] 2 Q. B. 84; [1952] W. N. 116; [1952] 1 T. L. R. 586; 96 Sol. Jo. 193; [1952] 1 Lloyd's Rep. 153, C. A.; 39 Digest (Repl.) 734, *2129* 43, 106
Payne *v*. Cave (1789), 3 Term Rep. 148; 3 Digest (Repl.) 20, *148* . 267
Payne *v*. Elsden (1900), 17 T. L. R. 161; 39 Digest (Repl.) 527, *650*: 112, 114
Payne *v*. Leconfield (1882), 51 L. J. Q. B. 646; 30 W. R. 814; 3 Digest (Repl.) 6, *32* 268
Payne *v*. Lillico & Sons (1920), 36 T. L. R. 569; 3 Ll. L. R. 110; 39 Digest (Repl.) 688, *1825* 187, 365
Paynter *v*. James (1867), L. R. 2 C. P. 348; 15 L. T. 660; 15 W. R. 493; 2 Marr. L. C. 450; affirmed (1868), 18 L. T. 449; 16 W. R. 768; 41 Digest (Repl.) 505, *2770* 181, 357
Payzu, Ltd. *v*. Saunders, [1919] 2 K. B. 581; [1918–19] All E. R. Rep. 961; 89 L. J. K. B. 17; 121 L. T. 563; 35 T. L. R. 657, C. A.; 39 Digest (Repl.) 699, *1907* 189, 255, 360
Pearce *v*. Brain, [1929] 2 K. B. 310; [1929] All E. R. Rep. 627; 98 L. J. K. B. 559; 141 L. T. 264; 45 T. L. R. 501; 73 Sol. Jo. 402; 93 J. P. Jo. 380; 28 (2) Digest (Reissue) 679, *168* 87
Pearce *v*. Brooks (1866), L. R. 1 Exch. 213; 4 H. & C. 358; 35 L. J. Ex. 134; 14 L. T. 288; 30 J. P. 295; 12 Jur. N. S. 342; 14 W. R. 614; 12 Digest (Reissue) 320, *2314* 18
Pearl Mill Co. *v*. Ivy Tannery Co., [1919] 1 K. B. 78; [1918–19] All E. R. Rep. 702; 88 L. J. K. B. 134; 120 L. T. 28; 24 Com. Cas. 169; 39 Digest (Repl.) 805, *2728* 22, 185, 358
Pearson, *Ex parte*, *Re* Wiltshire Iron Co. (1868), 3 Ch. App. 443; 37 L. J. Ch. 554; 18 L. T. 423; 16 W. R. 682; 10 Digest (Repl.) 904, *6142* 192
Pearson *v*. Dawson (1858), E. B. & E. 448; 27 L. J. Q. B. 248; 4 Jur. N. S. 1015; 39 Digest (Repl.) 769, *2461* 221
Pearson *v*. Rose and Young. Ltd., [1950] 2 All E. R. 1027; [1951] 1 K. B. 275; 66 (pt. 2) T. L. R. 886; 94 Sol. Jo. 778, C. A.; 1 Digest (Repl.) 387, *520* 174, 295, 297, 298
Pearson & Son, Ltd. *v*. Dublin Corporation, [1907] A. C. 351 [1904–7] All E. R. Rep. 255; 78 L. J. P. C. 91; 100 L. T. 499; 25 T. L. R. 360; 53 Sol. Jo. 319, H. L.; 38 Digest (Repl.) 408, *666* . 333, 335
Peirce *v*. London Horse and Carriage Repository, [1922] W. N. 170; 39 Digest (Repl.) 654, *1576* 172
Pelhams (Materials), Ltd. *v*. Mercantile Commodities Syndicate, [1953] 2 Lloyd's Rep. 281 196
Penarth Dock Engineering Co., Ltd. *v*. Pounds, [1963] 1 Lloyd's Rep. 359 204
Pennell *v*. Deffell (1853), 4 De. G. M. & G. 372; 1 Eq. Rep. 579; 23 L. J. Ch. 115; 22 L. T. O. S. 126; 18 Jur. 273; 1 W. R. 499; 5 Digest (Repl.) 777, *6622* 312
Percival Ltd. *v*. L. C. C. Asylums Committee (1918), 87 L. J. K. B. 677; 82 J. P. 157; 16 L. G. R. 367; 39 Digest (Repl.) 792, *2645* . 364
Perkins *v*. Bell. [1893] 1 Q. B. 193; [1891–94] All E. R. Rep. 62; L. J. Q. B. 91; 67 L. T. 792; 41 W. R. 195; 9 T. L. R. 147; 37 Sol. Jo. 130; C. A.; 39 Digest (Repl.) 561, *904* 140
Perlak Petroleum Maatschappij *v*. Deen, [1924] 1 K. B. 111; 93 L. J. K. B. 158; 130 L. T. 234, C. A.; 18 Digest (Repl.) 208, *1807* 49

PAGE

Peruvian Guano Co. *v.* Dreyfus Bros. & Co., [1892] A. C. 166; 61 L. J. Ch. 749; 66 L. T. 536; 8 T. L. R. 327; 7 Asp. M. L. C. 225, H. L.; 17 Digest (Reissue) 150, *407* 31
Peter Cassidy Seed Co., Ltd. *v.* Onunstuklenkanppa I. L. *See* Cassidy (Peter) Seed Co., Ltd. *v.* Onunstuklenkanppa I. L.
Peter Darlington Partners, Ltd. *v.* Gosho Co. *See* Darlington (Peter) Partners, Ltd. *v.* Gosho Co., Ltd.
Peter Dixon & Sons *v.* Henderson & Co., [1919] 2 K. B. 778; 87 L. J. K. B. 683; 118 L. T. 328; 23 Com. Cas. 294, C. A.; 12 Digest (Reissue) 496, *3484* 20
Peter Dumenil & Co., Ltd. *v.* James Ruddin, Ltd. *See* Dumenil (Peter) & Co., Ltd. *v.* Ruddin (James), Ltd.
Peters *v.* Fleming (1840), 6 M. & W. 42; 9 L. J. Ex. 81; 28 (2) Digest (Reissue) 692, *248* 85
Peto *v.* Blades (1814), 5 Taunt. 657; 39 Digest (Repl.) 525, *635* . . 114
Petrofina (Great Britain), Ltd. *v.* Martin, [1966] 1 All E. R. 126; [1966] Ch. 146; [1966] 2 W. L. R. 318; 109 Sol. Jo. 1009, C. A.; Digest Cont. Vol. B p. 701, *130a* 52, 271
Pettitt *v.* Mitchell (1842), 4 M. & Gr. 819; 5 Scott (N. R.) 721; 12 L. J. C. P. 9; 39 Digest (Repl.) 559, *878* 195
Peyrae v. Wilkinson, [1924] 2 K. B. 166; [1923] All E. R. Rep. 180; 93 L. J. K. B. 121; 130 L. T. 511; 35 Digest (Repl.) 197, *71* . 258
Pfizer Corporation *v.* Ministry of Health, [1965] 1 All E. R. 450; [1965] A. C. 512; [1965] 2 W. L. R. 387; 109 Sol. Jo. 149; [1965] R. P. C. at p. 284, H. L.; Digest Cont. Vol. B p. 588, *1932a* 2, 280
Pharmaceutical Society of Great Britain *v.* Boots Cash Chemists (Southern), Ltd., [1953] 1 All E. R. 482, [1953] 1 Q. B. 401; [1953] 2 W. L. R. 427; 117 J. P. 132; 97 Sol. Jo. 149, C. A.; 33 Digest (Repl.) 561, *263* 76
Pharmaceutical Society *v.* White, [1901] 1 K. B. 601; 70 L. J. K. B. 386; 84 L. T. 188; 65 J. P. 340; 49 W. R. 407; 17 T. L. R. 262; 45 Sol. Jo. 293, C. A.; 33 Digest (Repl) 560, *253* . . . 284
Phelps *v.* Comber (1885), 29 Ch. D. 813; 54 L. J. Ch. 1017; 52 L. T. 873; 33 W. R. 829; 5 Asp. M. L. C. 428, C. A.; 39 Digest (Repl.) 768, *2454* 214, 219, 223
Phillip Head & Sons, Ltd. *v.* Showfronts, Ltd. *See* Head (Philip) & Sons, Ltd. *v.* Showfronts, Ltd.
Phillips *v.* Brooks, [1919] 2 K. B. 243; [1918–19] All E. R. Rep. 246; 24 Com. Cas. 263; 88 L. J. K. B. 953; 121 L. T. 249; 35 T. L. R. 470; 24 Com. Cas. 263; 39 Digest (Repl.) 653, *1568* 172
Phillips *v.* Huth (1840), 6 M. & W. 572; 10 L. J. Ex. 65; 1 Digest (Repl.) 395, *557* 301
Phillpots *v.* Evans (1839), 5 M. & W. 475; 9 L. J. Ex. 33; 39 Digest (Repl.) 798, *2683* 232
Phoenix Distributors *v.* Clarke, [1966] 2 Lloyd's Rep. 285 130
Pickard *v.* Sears (1837), 6 Ad. & El. 469; 2 Nev. & P. K. B. 488; Will. Woll. & Dav. 678; 35 Digest (Repl.) 537, *2169* 165
Pickford *v.* Grand Junction Railway (1841), 8 M. & W. 372; 9 Dowl. 766; 8 (1) Digest (Reissue) 17, *80* 181
Pignataro *v.* Gilroy, [1919] 1 K. B. 459; 23 Com. Cas. 174; 88 L. J. K. B. 726; 120 L. T. 480; 35 T. L. R. 191; 63 Sol. Jo. 265; 24 Com. Cas. 174; 39 Digest (Repl.) 607, *1203* 150
Pigot *v.* Cubley (1864), 33 L. J. C. P. 134; 15 C. B. N. S. 701; 3 New Rep. 607; 9 L. T. 804; 10 Jur. N. S. 318; 12 W. R. 467; 37 Digest (Repl.) 14, *114* 162
Pigott (J. H.) & Son *v.* Docks and Inland Waterways Executive, [1953] 1 All E. R. 22; [1953] 1 Q. B. 338; [1953] 1 W. L. R. 94; 97 Sol. Jo. 30; [1952] 2 Lloyd's Rep. 499; 42 Digest (Repl.) 1115, *9320* 356

PAGE

Pilkington *v.* Wood, [1953] 2 All E. R. 810; [1953] Ch. 770; [1953] 3 W. L. R. 522; 97 Sol. Jo. 572; 43 Digest (Repl.) 120, *1091* . . . 115
Pinnock Brothers *v.* Lewis and Peat, [1923] 1 K. B. 690; 28 Com. Cas. 210; 92 L. J. K. B. 695; 129 L. T. 320; 39 T. L. R. 212; 39 T. L. R. 212; 67 Sol. Jo. 501; 28 Com. Cas. 210; 39 Digest (Repl.) 594, *1124* 118, 247, 251, 361, 366
Plaimar, Ltd. *v.* Waters Trading Co., Ltd. (1945), 73 C. L. R. 304 . 40
Plasticmoda S. P. A. *v.* Davidsons, Ltd., [1952] 1 Lloyd's Rep. 527: 23, 106, 107
Plasycoed Collieries Co. *v.* Partridge, Jones & Co., [1912] 2 K. B. 345; 81 L. J. K. B. 723; 106 L. T. 426; 3 Digest (Repl.) 113, *341* . 75
Playford *v.* Mercer (1870), 22 L. T. 41; 3 Mar, L. C. 335; 39 Digest (Repl.) 685, *1796* 359
Plischke and Sohne G. m. b. H. *v.* Allison Bros., Ltd., [1936] 2 All E. R. 1009; 39 Digest (Repl.) 761, *2383* 218
Pocahontas Fuel Co. *v.* Ambatielos (1922) 27 Com. Cas. 148; 42 Digest (Repl.) 735, *4895* 15, 93
Podar Trading Co., Ltd., Bombay *v.* Francois Tagher, Barcelona, [1949] 2 All E. R. 62; [1949] 2 K. B. 277; [1949] L. J. R. 1470; 65 T. L. R. 433; 93 Sol. Jo. 406, D. C.; 2 Digest (Repl.) 599, *1278* . 358
Polar Refrigeration Service, Ltd. *v.* Moldenhauer (1967), 61 D. L. R. (2d) 462 . . . . 147, 201
Polenghi Brothers *v.* Dried Milk Co. (1904), 92 L. T. 64; 53 W. R. 318; 21 T. L. R. 118; 49 Sol. Jo. 120; 10 Com. Cas. 42; 39 Digest (Repl.) 562, *907* 40, 138, 195, 229, 230, 357
Pollock *v.* Macrae, 1922 S. C. H. L. 192; 39 Digest (Repl.) 802, **1467* 10 109, 362
Polsky *v.* S. and A. Services, [1951] 1 All E. R. 185; affirmed, [1951] 1 All E. R. 1062; 95 Sol. Jo. 414, C. A.; 7 Digest (Repl.) 18, *79* 76
Pontifex *v.* Midland Railway (1877), 3 Q. B. D. 23; 47 L. J. Q. B. 28; 37 L. T. 403; 26 W. R. 209; 39 Digest (Repl.) 774, *2506* 220
Poole *v.* Smith's Car Sales (Balham), Ltd., [1962] 2 All E. R. 482; [1962] 1 W. L. R. 744; 106 Sol. Jo. 284, C. A.; 39 Digest (Repl.) 624, *1363* 147, 152, 158
Pordage *v.* Cole (1669) 1 Wms. Saund. 319; 1 Lev. 274; 2 Keb 542; T. Raym. 183; 1 Sid. 423; 12 Digest (Reissue) 523, *3624* . . . 182, 227
Port Line Ltd. *v.* Ben Line Steamers, Ltd., [1958] 1 All E. R. 787; [1958] 2 Q. B. 146; [1958] 2 W. L. R. 551; [1958] 1 Lloyd's Rep. 290; 7 Digest (Repl.) 18, *79* 20
Portalis *v.* Tetley (1867), L. R. 5 Eq. 140; 37 L. J. Ch. 139; 17 L. T. 344; 1 Digest (Repl.) 388, *521* 299
Porter *v.* Freudenberg, Kreglinger *v.* S. Samuel & Rosenfeld, *Re* Merlin's Patents, [1915] 1 K. B. 857; [1914–15] All E. R. Rep. 918; 84 L. J. K. B. 1001; 112 L. T. 313; 20 Com. Cas. 189, 31 T. L. R. 162, C. A.; 2 Digest (Repl.) 213, *270* 89, 90
Porter *v.* Latec Proprietary, Ltd. (1964), 38 A. L. J. R. 184 . . . 173
Porter Spiers *v.* Cameron, [1950] C. L. Y. (Sc.) 5324 270
Postmaster-General *v.* W. H. Jones & Co. (London), Ltd., [1957] N. Z. L. R. 829; 39 Digest (Repl.) 749, **1227* 218
Potts & Co., Ltd. *v.* Brown, Macfarlane & Co., Ltd. (1925), 30 Com. Cas. 64, H. L.; 39 Digest (Repl.) 679, *1741* 22, 197
Poulton *v.* Anglo-American Oil Co. (1911), 27 T. L. R. 216; C. A.; 39 Digest (Repl.) 747, *2240* 209
Poulton *v.* Lattimore (1829), 9 B. & C. 259; 4 M. & Ry. 208; 7 L. J. O. S. K. B. 225; 39 Digest (Repl.) 583, *1047* 245
Pound (A. V.) & Co., Ltd. *v.* (M. W.) Hardy & Co. Incorporated, [1956] 1 All E. R. 639; [1956] A. C. 588; [1956] 2 W. L. R. 683; 100 Sol. Jo. 208; [1956] 1 Lloyd's Rep. 255, H. L.; 39 Digest (Repl.) 659, *1598* 41, 42

PAGE

Powell *v.* Horton (1836), 2 Bing. N. C. 668; 2 Hodges, 12; 3 Scott, 110; 5 L. J. C. P. 204; 39 Digest (Repl.) 534, *702* 365
Power *v.* Barham (1836), 4 Ad. & E.1 473; 1 Har. & W. 683; 6 Nev. & M. K. B. 62; 5 L. J. K. B. 88; 39 Digest (Repl.) 565, *938* . . 366
Prager *v.* Blatspiel, [1924] 1 K. B. 566; [1924] All E. R. Rep. 524; 93 L. J. K. B. 410; 130 L. T. 672; 40 T. L. R. 287; 68 Sol. Jo. 460; 39 Digest (Repl.) 777, *2526* 168, 224, 271
President of India *v.* Metcalfe Shipping Co., Ltd., [1969] 1 All E. R. 861; [1969] 2 Q. B. 123; [1969] 2 W. L. R. 125; 113 Sol. Jo. 69; [1969] 1 Lloyd's Rep. 32; affirmed, [1969] 3 All E. R. 1549; [1970] 1 Q. B. 289; [1969] 3 W. L. R. 1120; 113 Sol. Jo. 792; [1969] 2 Lloyd's Rep. 476, C. A.; Digest Cont. Vol. C p. 25, *156c* . 33, 144, 146, 161
Price and Pearce *v.* Bank of Scotland, 1912 S. C. (H. L.) 19; Digest (Repl.) 773, *2488* 287
Priest *v.* Last, [1903] 2 K. B. 148; 72 L. J. K. B. 657, C. A. . . . 125
Prinz Adalbert, The, [1917] A. C. 586; 86 L. J. P. C. 165; 116 L. T. 802; 33 T. L. R. 490; 61 Sol. Jo. 610; 14 Asp. M. L. C. 81, P. C.; 41 Digest (Repl.) 263, *791* 76, 155, 277, 318
Pritchett Co. *v.* Currie, [1916] 2 Ch. 515; [1916–17] All E. R. Rep. 705; 85 L. J. Ch. 753; 115 L. T. 325, C. A.; 39 Digest (Repl.) 604, *1184* 288
Produce Brokers Co. *v.* Olympia Oil and Cake Co., (1914), 84 L. J. K. B. 1153, C. A.; on appeal, [1916] 1 A. C. 314; [1914–15] All E. R. Rep. 133; 85 L. J. K. B. 160; 114 L. T. 94; 21 Com. Cas. 320; 32 T. L. R. 115, H. L.; 39 Digest (Repl.) 605, *1190* . . 98, 367
Produce Brokers Co. *v.* Olympia Oil and Cake Co., [1917] 1 K. B. 320; [1916–17] All E. R. Rep. 753; *sub nom.* Olympia Oil and Cake Co. *v.* Produce Brokers Co., 86 L. J. K. B. 421; 116 L. T. 1; 33 T. L. R. 95, C. A.; 39 Digest (Repl.) 604, *1186* 4, 17, 98
Produce Brokers Co. *v.* Weis & Co. [1918], 87 L. J. K. B. 472; 118 L. T. 111; 12 Digest (Reissue) 500, *3493* 19, 39
Pym *v.* Campbell (1856) 6 E. & B. 370; 25 L. J. Q. B. 277; 2 Jur. N. S. 641; 4 W. R. 528; *sub nom.* Pim *v.* Campbell, 27 L. T. O. S. 122; 17 Digest (Reissue) 257, *224* 4
Pyrene *v.* Scindia Navigation Co., Ltd., [1954] 2 All E. R. 158; [1954] 2 Q. B. 402; [1954] 2 W. L. R. 1005; 98 Sol. Jo. 354; [1954] 1 Lloyd's Rep. 321; 41 Digest (Repl.) 237, *593* 34

Qualter Hall & Co., Ltd. *v,* Board of Trade, [1961] 3 All E. R. 389; [1962] Ch. 273; [1961] 3 W. L. R. 825; 105 Sol. Jo. 884, C. A.; 44 Digest (Repl.) 242, *660* 74

R. *v.* Bicester Justices, *Ex parte* Unigate, Ltd., [1975] 1 All E. R. 449; [1975] 1 W. L. R. 207; 139 J. P. 216; 118 Sol. Jo. 829 . . . 69
R. *v.* Bonner, [1970] 2 All E. R. 97; [1970] 1 W. L. R. 838; 134 J. P. 429; 114 Sol. Jo. 188, C. A.; Digest Cont. Vol. C p. 264, *10,596a* . 168
R. *v.* Bradburn (1973), 57 Cr. App. Rep. 948, C. A. 346
R. *v.* Chruch (1970) 114 Sol. Jo. 907, C. A. 341
R. *v.* Churchill (No. 2), [1966] 2 All E. R. 215; [1967] 1 Q. B. 190; [1966] 2 W. L. R. 1116; 110 Sol. Jo. 526, C. C. A.; Digest Cont. Vol. C p. 186, *893a* 69
R. *v.* Daly, [1974] 1 All E. R. 290; [1974] 1 W. L. R. 133; 138 J. P. 245; 118 Sol. Jo. 66; 58 Cr. App. Rep. 333, C. A. 346
R. *v.* Denyer, [1926] 2 K. B. 258; 95 L. J. K. B. 699; 134 L. T. 637; 42 T. L. R. 452, C. C. A.; 15 Digest (Repl.) 1124, *11,217* . . 52
R. *v.* Eaton, [1966] Crim. L. R. 333 . . . . 149, 292

PAGE

R. *v.* Ferguson, [1970] 2 All E. R. 820; [1970] 1 W. L. R. 1246, 134 J. P. 608; 114 Sol. Jo. 472, C. A.; Digest Cont. Vol. C p. 233, *5965b* 341
R. *v.* Grundy, [1974] 1 All E. R. 292; [1974] 1 W. L. R. 139; 138 J. P. 242; 118 Sol. Jo. 34, C. A. 347
R. *v.* Haesler, [1973] R. T. R. 486, C. A. 69
R. *v.* Jones, [1898] 1 Q. B. 119; 67 L. J. Q. B. 41; 77 L. T. 503; 46 W. R. 191; 14 T. L. R. 79; 42 Sol. Jo. 82; 19 Cox, C. C. 87; 39 Digest (Repl.) 661, *1606* 182
R. *v.* Justelius, [1973] 1 N. S. W. L. R. 471 147
R. *v.* Kneeshaw, [1974] 1 All E. R. 896; [1975] Q. B. 57; [1974] 2 W. L. R. 432; 138 J. P. 291; 118 Sol. Jo. 218; 58 Cr. App. Rep. 439, C. A.: 346, 347
R. *v.* Oddy, [1974] 2 All E. R. 666; [1974] 1 W. L. R. 1212; 138 J. P. 515; 118 Sol. Jo. 329; 59 Cr. App. Rep. 66, C. A. 346, 347
R. *v.* Parker, [1970] 2 All E. R. 458; [1970] 1 W. L. R. 1003; 134 J. P. 497; 114 Sol. Jo. 396, C. A.; Digest Cont. Vol. C p. 227, *5385a* . 341
R. *v.* Reiss, [1957] Crim. L. R. 404 . . . . 57, 61
R. *v.* Salem Mohammed Monsoor Ali (1972), 56 Cr. App. Rep. 301, C. A. . 347
R. *v.* Thomson Holidays, Ltd., [1974] 1 All E. R. 823; [1974] Q. B. 592; [1974] 2 W. L. R. 371; 138 J. P. 284; 118 Sol. Jo. 96; 58 Cr. App. Rep. 429, C. A. 346, 347
Raahe O/Y Osakeytio *v.* Goddard (1935), 154 L. T. 124; [1935] All E. R. Rep. 277; 80 Sol. Jo. 93; 39 Digest (Repl.) 694, *1867* . . 188
Raffoul *v.* Essanda, Ltd. (1970), s. 72 S. R. (N. S. W.) 633 . . 167, 298
Rafuse Motors *v.* Mardo Constructions (1963), 41 D. L. R. (2d) 340; 48 M. P. R. 296; Digest Cont. Vol. A 1304, **640b* 201
Rainbow *v.* Howkins, [1904] 2 K. B. 322; 73 L. J. K. B. 641; 91 L. T. 149; 20 T. L. R. 508; 3 Digest (Repl.) 47, *333* 268
Ralli Brothers *v.* Compania Naviera Sota y Aznar., [1920] 1 K. B. 614; affirmed, [1920] 2 K. B. 287; [1920] All E. R. Rep. 427; 89 L. J. K. B. 999; 123 L. T. 375; 15 Asp. M. L. C. 33; 36 T. L. R. 456; 64 Sol. Jo. 462, 25 Com. Cas. 227, C. A.; 11 Digest (Reissue) 476, *843* 48
Ramsay *v.* Margrett, [1894] 2 Q. B. 18; 63 L. J. Q. B. 513; 70 L. T. 788, C. A.; 27 (1) Digest (Reissue) 121, *812* 83
Ramsden *v.* Gray (1849), 7 C. B. 961 235
Randall *v.* Newson (1877), 2 Q. B. D. 102; 46 L. J. Q. B. 259; 36 L. T. 164, C. A.; 25 W. R. 313; 39 Digest (Repl.) 539, *745* . 126, 140, 246
Randall *v.* Raper (1858), 27 L. J. Q. B. 266; E. B. & E. 84; 31 L. T. O. S. 81; 4 Jur. N. S. 662; 6 W. R. 445; 39 Digest (Repl.) 587, *1085* 246
Rapalli *v.* K. L. Take, Ltd., [1958] 2 Lloyd's Rep. 469, C. A.; Digest Cont. Vol. A p. 1304, *777a* 121, 133, 187, 366
Raphael *v.* Bank of England (1855), 17 C. B. 161; 25 L. J. C. P. 33; 3 Digest (Repl.) 138, *69* 299
Rappaport *v.* London Plywood and Timber Co. Ltd., [1940] 1 All E. R. 576; 56 T. L. R. 378; 84 Sol. Jo. 344; 39 Digest (Repl.) 809, *2747* 241
Rasnoimport (V/O) *v.* Guthrie & Co., Ltd., [1966] 1 Lloyd's Rep. 1. 318
Ratcliffe *v.* Evans, [1892] 2 Q. B. 524; 61 L. J. Q. B. 535; 66 L. T. 794; 17 Digest (Reissue) 84, *21* 256
Rawlings *v.* General Trading Co., [1921] 1 K. B. 635; 90 L. J. Q. B. 404; 124 L. T. 562; 37 T. L. R. 252, C. A.; 3 Digest (Repl.) 22, *163* 18, 269
Rawson *v.* Johnson (1801), 1 East, 203; 39 Digest (Repl.) 663, *1630* 181
Rayner & Co., Ltd. *v.* Hambros Bank, Ltd., [1942] 2 All E. R. 694; [1943] 1 K. B. 37; 112 L. J. K. B. 27; 167 L. T. 380; 59 T. L. R. 21, C. A.; 3 Digest (Repl.) 285, *856* 44
Read *v.* Hutchinson (1813), 3 Camp. 352; 6 Digest (Repl.) 302, *2205* . 79

PAGE

Reddall *v.* Union Castle Steamship Co. (1914), 84 L. J. K. B. 360; 112 L. T. 910; 13 Asp. M. L. C. 51; 20 Com. Cas. 86; 39 Digest (Repl.) 757, *2352* 218
Redgrave *v.* Hurd (1881), 20 Ch. D. 1; 51 L. J. Ch. 113; 45 L. T. 485; 30 W. R. 251, C. A.; 40 Digest (Repl.) 396, *3173* 131
Redhead *v.* Westwood (1888), 59 L. T. 293; 4 T. L. R. 671; 7 Digest (Repl.) 17, *76* 83
Reed *v.* Mestaer (1802), 2 Comyn on Contract, 229 230
Rees *v.* Munday, [1974] 3 All E. R. 506; [1974] 1 W. L. R. 1284; 138 J. P. 767; 118 Sol. Jo. 697 . . . . 59, 99
Reeve *v.* Whitmore (1863), 4 De G. J. & S. 1; 33 L. J. Ch. 63; 7 Digest (Repl.) 123, *716* 96
Reeves *v.* Armour. *See* Fisher, Reeves *v.* Armour & Co.
Reeves *v.* Capper (1838), 5 Bing. N. C. 136; [1835–42] All E. R. Rep. 164; 6 Scott, 877; 2 Jur. 1067; 21 Digest (Repl.) 594, *820* . . . 83
Regazzoni *v.* K. C. Sethia (1944), Ltd., [1957] 3 All E. R. 286; [1958] A. C. 301; [1957] 3 W. L. R. 752; 101 Sol. Jo. 848; [1957] 2 Lloyd's Rep. 289, H. L.; 11 Digest (Reissue) 484, *892* 48
Reid *v.* Commissioner of Police of the Metropolis, [1973] 2 All E. R. 97; [1973] Q. B. 551; [1973] 2 W. L. R. 576; *sub nom.* Reid *v.* Metropolitan Police Commissioner, 117 Sol. Jo. 244, C. A.; Digest Supp.: 73, 170
Reid *v.* Macbeth & Gray, [1904] A. C. 223; 73 L. J. P. C. 57; 90 L. T. 422; 20 T. L. R. 316; 6 F. (Ct. of Sess.) (H. L.) 25; 41 Sc. L. R. 369; 11 S. L. T. 783; 39 Digest (Repl.) 598, **723* . . . 143, 144
Reid (George) Inc. *v.* Nelson Machinery Co., [1973] 2 W. W. R. 597 . 367
Rein *v.* Stein, [1892] 1 Q. B. 753; 61 L. J. Q. B. 401; 66 L. T. 469, C. A.; 12 Digest (Reissue) 568, *3959* 228
Rendell *v.* Turnbull (1908), 27 N. Z. L. R. 1067; 39 Digest (Repl.) 495, **168* 98
Renton (G. H.) & Co., Ltd. *v.* Palmyra Trading Corporation of Panama, [1956] 3 All E. R. 957; [1957] A. C. 149; [1957] 2 W. L. R. 45; 101 Sol. Jo. 43; [1956] 2 Lloyd's Rep. 379, H. L.; 41 Digest (Repl.) 419, *2045* 14
Restrictive Trades Practices (Registrar) *v.* W. H. Smith & Son, Ltd., [1969] 3 All E. R. 1065; [1969] 1 W. L. R. 1460; 113 Sol. Jo. 686, C. A., Digest Cont. Vol. C p. 988, *155a*. 65
Reuter *v.* Sala (1879), 4 C. P. D. 239; 48 L. J. Q. B. 492; 40 L. T. 476; 27 W. R. 631; 4 Asp. M. L. C. 121, C. A.; 39 Digest (Repl.) 520, *600* 107, 188, 191, 355, 364
Rhodes, *Re*, Rhodes (1890), 44 Ch. D. 94; [1886–90] All E. R. Rep. 871; 59 L. J. Ch. 298, 62 L. T. 342; 38 W. R. 385, C. A.; 33 Digest (Repl.) 594, *105* 86
Rice *v.* Reed, [1900] 1 Q. B. 54; [1895–99] All E. R. Rep. 262; 69 L. J. Q. B. 33; 81 L. T. 410, C. A.; 12 Digest (Reissue) 633, *4519* . . . 31
Richards *v.* Delbridge (1874) L. R. 18 Eq. 11; 43 L. J. Ch. 459; 22 W. R. 584; 25 Digest (Repl.) 588, *276* 79
Riches *v.* Westminster Bank, Ltd., [1943] 2 All E. R. 725, C. A.; 35 Digest (Repl.) 211, *180* 326
Richmond London Borough *v.* Motor Sales (Hounslow), Ltd. (1971), 135 J. P. 236; 69 L. G. R. 266; *sub nom.* Richmond upon Thames London Borough Council *v.* Motor Sales (Hounslow), Ltd., 115 Sol. Jo. 156; [1971] R. T. R. 116; Digest Supp. 68
Rickards (Charles), Ltd. *v.* Oppenheim, [1950] 1 All E. R. 420; [1950] 1 K. B. 616; 66 (pt. 1) T. L. R. 435; 94 Sol. Jo. 161, C. A.; 12 Digest (Reissue) 383, *2777* 23, 107, 239
Rigge *v.* Burbidge (1846), 15 M. & W. 598; 4 Dow. & L. 1; 15 L. J. Ex. 309; 39 Digest (Repl.) 585, *1072* 246
Ripon City, The, [1897] P. 226; 66 L. J. P. 110; 77 L. T. 98; 1 Digest (Repl.) 743, *2842* 81, 93
Robbins of Putney *v.* Meek, [1971] R. T. R. 345; Digest Supp. . 254, 255

PAGE

Roberts, *Re*, Evans *v.* Roberts (1887), 36 Ch. D. 196; 56 L. J. Ch. 952; 57 L. T. 79; 51 J. P. 757; 35 W. R. 684; 3 T. L. R. 678; 3 Digest (Repl.) 22, *1878* 321

Roberts *v.* Gray, [1913] 1 K. B. 520; [1911–13] All E. R. Rep. 870; 82 L. J. K. B. 362; 108 L. T. 232; 29 T. L. R. 149, C. A.; 28 (2) Digest (Reissue) 682, *181* 88

Roberts and Cooper, Ltd. *v.* Salvesen & Co., 1918 S. C. 794; 17 Digest (Reissue) 190, **449* 27

Robertson *v.* Dicicco (1972), 116 Sol. Jo. 744; 70 L. G. R. 589; [1972] R. T. R. 431; Digest Supp. 61

Robin and Rambler Coaches *v.* Turner, [1947] 2 All E. R. 284; 39 Digest (Repl.) 652, *1563* 171

Robinson *v.* Graves [1935] 1 K. B. 579; [1935] All E. R. Rep. 935; 104 L. J. K. B. 441; 153 L. T. 26; 51 T. L. R. 334; 79 Sol. Jo. 180; 40 Com. Cas. 217; C. A.; 39 Digest (Repl.) 445, *11* 78

Robinson *v.* Mollett. *See* Mollett *v.* Robinson.

Robophone Facilities, Ltd. *v.* Blank, [1966] 3 All E. R. 128; [1966] 1 W. L. R. 1428; 110 Sol. Jo. 544, C. A.; Digest Cont. Vol. B p. 43, *366a* 239

Rodger *v.* Comptoir d'Escompte (1869), L. R. 2 P. C. 393; 5 Moore P. C. N. S. 538; 38 L. J. P. C. 30; 21 L. T. 33; 17 W. R. 468; 3 Mar. L. C. 271; 39 Digest (Repl.) 758, *2363* 222

Rodocanachi *v.* Milburn (1886), 18 Q. B. D. 67; 56 L. J. Q. B. 202; 56 L. T. 594; 35 W. R. 241; 3 T. L. R. 115; 6 Asp. M. L. C. 100, C. A.; 39 Digest (Repl.) 811, *2760* 241

Rodriguez *v.* Speyer Brothers, [1919] A. C. 59; [1918–19] All E. R. Rep. 884; 88 L. J. K. B. 147; 119 L. T. 409; 34 T. L. R. 628; 62 Sol. Jo. 765, H. L.; 39 Digest (Repl.) 63, 723 90

Rohde *v.* Thwaites (1827), 6 B. & C. 388; 9 Dow & Ry. K. B. 293; 5 L. J. O. S. K. B. 163; 39 Digest (Repl.) 601, *1170* 75, 148

Rolls *v.* Miller (1884), 27 Ch. D. 71; [1881–5] All E. R. Rep. 915; 53 L. J. Ch. 682; 50 L. T. 597; 32 W. R. 806, C. A.; 31 (1) Digest (Reissue) 395, *3146* 274

Roots *v.* Lord Dormer (1832), 4 B. & Ad. 77; 1 Nev. & M. K. B. 667; 39 Digest (Repl.) 302, *473* 267

Roper *v.* Johnson (1873), L. R. 8 C. P. 167; 42 L. J. C. P. 65; 28 L. T. 296; 39 Digest (Repl.) 818, *2809* 231, 235, 238

Rose and Frank *v.* Crompton (J. R.) Brothers, Ltd., [1925] A. C. 445; [1924] All E. R. Rep. 245; 94 L. J. K. B. 120; 132 L. T. 641; 30 Com. Cas. 163, C. A.; 39 Digest (Repl.) 781, *2556* 1

Rose (Frederick E.) (London), Ltd., *v.* Wm. H. Pim Junr. & Co., Ltd. [1953] 2 All E. R. 739; [1953] 2 Q. B. 450; [1953] 3 W. L. R. 497; 97 Sol. Jo. 556, C. A.; 35 Digest (Repl.) 147, *369* 244

Rosenthal (J.) & Sons, Ltd. *v.* Esmail (Trading as H. M. H. Esmail & Sons), [1965] 2 All E. R. 860; [1965] 1 W. L. R. 1117; 109 Sol. Jo. 553; [1965] 2 Lloyd's Rep. 171, H. L.; Digest Cont. Vol. B p. 632, *1011a* 108, 110, 200

Rosevear China Clay Co., *Ex parte* (1879), 11 Ch. D. 560; on appeal 11 Ch. D. at p. 565; 48 L. J. Bcy. 100; 40 L. T. 730; 27 W. R. 591; 4 Asp. M. L. C. 144, C. A.; 39 Digest (Repl.) 685, *1793* . 215, 216

Ross Brothers *v.* Shaw & Co., [1917] 2 I. R. 367; 39 Digest (Repl.) 677, **974* 358

Roth & Co. *v.* Taysen Townsend & Co. and Grant & Co. (1895), 73 L. T. 628; 12 T. L. R. 100; 8 Asp. M. L. C. 120; 1 Com. Cas. 240; on appeal (1896), 12 T. L. R. 211; 1 Com. Cas. 306, C. A.; 39 Digest (Repl.) 798, *2684* 233, 238, 255

Rourke *v.* Short (1856), 5 E. & B. 904; 25 L. J. Q. B. 196; 26 L. T. O. S. 235; 2 Jur. N. S. 352; 39 Digest (Repl.) 500, *459* 76, 103

Routledge *v.* McKay, [1954] 1 All E. R. 855; [1954] 1 W. L. R. 615; 98 Sol. Jo. 247; 47 R. & I. T. 224 C. A.; 39 Digest (Repl.) 316, *627* . 5

PAGE

Rowland *v.* Divall, [1923] 2 K. B. 500; [1923] All E. R. Rep. 270; 92 L. J. K. B. 1041; 129 L. T. 757; 67 Sol. Jo. 703, C. A.; 39 Digest (Repl.) 526, *646* 113, 114, 115, 257
Ruben (E. & S.), Ltd. *v.* Faire Brothers & Co., Ltd., [1949] 1 All E. R. 215; [1949] 1 K. B. 254; [1949] L. J. R. 800; 93 Sol. Jo. 103; 39 Digest (Repl.) 725, *2075* 118, 137, 196, 197, 199, 366
Rugg *v.* Minett (1809), 11 East, 210; 39 Digest (Repl.) 495, *421* . 147, 157
Rusholme Bolton and Roberts Hadfield, Ltd. *v.* S. G. Read & Co. (London), Ltd., [1955] 1 All E. R. 180; [1955] 1 W. L. R. 146; 99 Sol. Jo. 132; 1 Digest (Repl.) 721, *2673* 43, 46, 47
Ryan *v.* Ridley (1902), 19 T. L. R. 45; 8 Com. Cas. 105; 39 Digest (Repl.) 778, *2535* 234, 357
Ryan *v.* Sams (1848), 12 Q. B. D. 460; 17 L. J. Q. B. 271; 12 Jur. 745; 27 (1) Digest (Reissue) 241, *1759* 92
Ryder *v.* Wombwell (1868), L. R. 4. Ex. 32; 38 L. J. Ex. 8; 19 L. T. 491, Ex. Ch.; 28 (2) Digest (Reissue) 691, *228* 85

Sachs *v.* Miklos, [1948] 1 All E. R. 67; [1948] 2 K. B. 23; [1948] L. J. R. 1012; 64 J. R. 181, C. A.; 1 Digest (Repl.) 328, *136* 168
Sadler Brothers Co. *v.* Merideth, [1963] 2 Lloyd's 293 217
Saffaron *v.* Société Minière Cafrika (1958), 100 C. L. R. 231 . . . 44
Sainsbury, Ltd. *v.* Street, [1972] 3 All E. R. 1127; [1972] 1 W. L. R. 834; 116 Sol. Jo. 483 . . . . 95, 97, 99, 100
St. John Shipping Corporation *v.* Joseph Rank, Ltd., [1956] 3 All E. R. 683; [1957] 1 Q. B. 267; [1956] 3 W. L. R. 870; 100 Sol. Jo. 841; [1956] 2 Lloyd's Rep. 413, 41 Digest (Repl.) 337, *1327* . . . 18
St. Margaret's Trusts *v.* Castle, [1964] C. L.Y. 1685, C. A. . . 76, 82, 292
Sajan Singh *v.* Sardana Ali, [1960] 1 All E. R. 269; [1960] A. C. 167; [1960] 2 W. L. R. 180; 104 Sol. Jo. 84, P. C.; 12 Digest (Reissue) 350, *2524* 19
Saks *v.* Tilley (1915), 32 T. L. R. 148, C. A.; 39 Digest (Repl.) 596, *1135* 144, 155
Sale Continuation, Ltd. *v.* Austin Taylor & Co., Ltd., [1967] 2 All E. R. 1092; [1968] 2 Q. B. 849; [1967] 3 W. L. R. 1427; 111 Sol. Jo. 472; [1967] 2 Lloyd's Rep. 403; Digest Cont. Vol. C p. 38, *854a* . 46
Salter *v.* Woollams (1841), 2 M. & G. 650; Drinkwater, 146; 3 Scott, N. R. 59; 10 L. J. C. P.; 145; 39 Digest (Repl.) 683, *1781* . 180, 184
Samuels *v.* Davis, [1943] 2 All E. R. 3; [1943] K. B. 526; 112 L. J. K. B. 561; 168 L. T. 296, C. A.; 39 Digest (Repl.) 541, *763* . . 77, 122
Sanday & Co. *v.* Keighley, Maxted & Co. (1922), 91 L. J. K. B. 624, 127 L. T. 327; 38 T. L. R. 561; 66 Sol. Jo. 437; 27 Com. Cas. 296; 15 Asp. M. L. C. 596; 39 Digest (Repl.) 680, *1750* . . . 106, 355
Sanday & Co., Ltd. *v.* Shipton, Anderson & Co., Ltd. (1926), 25 Ll. L. Rep. 508 24
Sandeman & Sons *v.* Tyzack & Branfoot S. S. Co., Ltd., [1913] A. C. 680; [1911–13] All E. R. Rep. 1013; 83 L. J. P. C. 23; 109 L. T. 580; 29 T. L. R. 694; 57 Sol. Jo. 752; 12 Asp. M. L. C. 437, H. L.; 41 Digest (Repl.) 440, *2212* 143
Sanders *v.* Jameson (1848), 2 Car. & Kir. 557; 39 Digest (Repl.) 559, *879* 197
Sanders *v.* Maclean (1883), 11 Q. B. D. 327; 52 L. J. Q. B. 481; 49 L. T. 462; 31 W. R. 698; 5 Asp. M. L. C. 160, C. A.; 39 Digest (Repl.) 518, *591* 36, 185, 277, 289, 318, 319, 357
Sandford *v.* Dairy Supplies, Ltd., [1941] N. Z. L. R. 141 230
Sargant & Sons *v.* Paterson & Co. (1923), 129 L. T. 471; 39 T. L. R. 378; 39 Digest (Repl.) 809, *2746* 19
Saunt *v.* Belcher and Gibbons (1920), 90 L. J. K. B. 541; [1920] All E. R. Rep. 142; 26 Com. Cas. 115; 39 Digest (Repl.) 716, *2022* . . 197

PAGE

Scaliaris *v.* Ofverberg & Co. (1921), 37 T. L. R. 307, C. A.; 39 Digest (Repl.) 536, *724* 35, 197
Scammell (G.) and Nephew, Ltd. *v.* Ouston, [1941] 1 All E. R. 14; [1941] A. C. 251; 110 L. J. K. B. 197; 164 L. T. 379; 57 T. L. R. 280; 85 Sol. Jo. 224; 46 Com. Cas. 190, H. L.; 39 Digest (Repl.) 448, *31* . 101
Scarf *v.* Jardine (1882), 7 App. Cas. 345; [1881–85] All E. R. Rep. 651; 51 L. J. Q. B. 612; 47 L. T. 258; 30 W. R. 893, H. L.; 21 Digest (Repl.) 299, *633* 8
Schaffenius *v.* Goldberg, [1916] 1 K. B. 284; [1914–15] All E. R. Rep. 387; 85 L. J. K. B. 374; 113 L. T. 949; 32 T. L. R. 133; 60 Sol. Jo. 105, C. A.; 2 Digest (Repl.) 242, *445* 90
Schawel *v.* Reade, [1913] 2 I. R. 64, H. L. 368
Schering, Ltd. *v.* Stockholms Enskilda Bank Aktiebolag, [1946] 1 All E. R. 36; [1946] A. C. 219; 115 L. J. Ch. 58; 174 L. T. 49; 62 T. L. R. 122, H. L.; 12 Digest (Reissue) 482, *3429* 90
Schneider *v.* Foster (1857), 2 H. & N. 4; 39 Digest (Repl.) 736, *2150* . 227
Scholfield *v.* Emerson Brantingham Implements (1918) 43 D. L. R. 509 (Sup. Ct. of Canada) 201
Schorsch Meier G.m.b.H. *v.* Hennin, [1975] 1 All E. R. 152; [1974] 3 W. L. R. 823; 118 Sol. Jo. 881; [1975] 1 C. M. L. R. 20, C. A. . . 258
Schotsmans *v.* Lancashire and Yorkshire Railway Co. (1867), 2 Ch. App. 332; 36 L. J. Ch. 361; 16 L. T. 189; 51 W. R. 537; 2 Mar. L. C. 485; 39 Digest (Repl.) 773, *2493* . 157, 207, 212, 215, 216, 217, 218, 220
Schuler A.G. *v.* Wickman Machine Tool Sales, Ltd., [1973] 2 All E. R. 39; [1974] A. C. 235; [1973] 2 W. L. R. 683; 117 Sol. Jo. 340, H. L.; 17 Digest (Repl.) 378, *1428* 9
Scott *v.* Brown & Co., [1892] 2 Q. B. 724; [1891–94] All E. R. Rep. 654; 61 L. J. Q. B. 738; 67 L. T. 782; 57 J. P. 213; 41 W. R. 116; 8 T. L. R. 755; 36 Sol. Jo. 698; 4 R. 42, C. A.; 12 Digest (Reissue) 292, *2108* . 269
Scott *v.* England (1844), 2 D. & L. 520; 14 L. J. Q. B. 43; 9 Jur. 87; 39 Digest (Repl.) 786, *2600* 226
Scott *v.* Hanson (1829) 1 Russ & M. 128; 40 Digest (Repl.) 52, *337* . 5
Scott (Donald) & Co. *v.* Barclays Bank, [1923] 2 K. B. 1; 28 Com. Cas. 253; 92 L. J. K. B. 772; 129 L. T. 108; 39 T. L. R. 198; 67 Sol. Jo. 456, C. A.; 39 Digest (Repl.) 708, *1964* 39
Scriven Brothers *v.* Schmoll Fils & Co. Inc. (1924), 40 T. L. R. 676; 39 Digest (Repl.) 801, *2701* 278
Scrivener *v.* Great Northern Railway (1871), 19 W. R. 388; 39 Digest (Repl.) 747, *2252* 212
Seath *v.* Moore (1886), 11 App. Cas. 350; 55 L. J. P. C. 54; 54 L. T. 690; 5 Asp. M. L. C. 586, P. C.; 39 Digest (Repl.) 618, *1312*: 145, 146, 151
Seaton *v.* Benedict (1828), 5 Bing. 28; 2 Moo. & P. 66; 6 L. J. O. S. C. P. 208; 27 (1) Digest (Reissue) 208, *1407* 92
Seton, Laing & Co. *v.* Lafone (1887), 19 Q. B. D. 68; 56 L. J. Q. B. 415; 57 L. T. 457; 35 W. R. 749; 3 T. L. R. 624, C. A.; 3 Digest (Repl.) 109, *314* 165
Sewell *v.* Burdick *See* Burdick *v.* Sewell.
Seymour, *Re*, [1913] 1 Ch. 475; 82 L. J. Ch. 233; 108 L. T. 392, C. A.; 17 Digest (Reissue) 243, *69* 78
Shackleton, *Re*, *Ex parte* Whittaker (1875), 10 Ch. App. 446; 44 L. J. Bcy. 91; 32 L. T. 443; 23 W. R. 555; 5 Digest (Repl.) 740, *6403* . . 225
Shankland *v.* Robinson & Co., 1920 S. C. (H. L.) 103; 57 Sc. L. R. 400, H. L.; 39 Digest (Repl.) 790, *2628* 267, 287
Sharpe & Co. *v.* Nosawa & Co., [1917] 2 K. B. 814; 87 L. J. K. B. 33; 118 L. T. 91; 22 Com. Cas. 286; 39 Digest (Repl.) 704, *1947* . 36, 39, 235, 239
Shaw *v.* Holland (1846), 15 M. & W. 136; 15 L. J. Ex. 87; 10 Jur. 100; 9 Digest (Repl.) 371, *2383* 235

PAGE

Sheldon *v.* Cox (1824), 3 B. & C. 420; 5 Dow. & Ry. K. B. 277; 39 Digest (Repl.) 496, *428* 79
Shell Mex, Ltd. *v.* Elton Cop. Dyeing Co. (1928), 34 Com. Cas. 39; 39 Digest (Repl.) 605, *1188* 229, 244
Shepherd *v.* Harrison (1871), L. R. 5 H. L. 116; 40 L. J. Q. B. 148; 24 L. T. 857; 20 W. R. 1; 1 Asp. M. L. C. 66, H. L.; 39 Digest (Repl.) 637, *1465* 155
Shepherd *v.* Kain (1821), 5 B. & Ald. 240; 39 Digest (Repl.) 532, *686* 117, 360
Sherratt *v.* Geralds the American Jewellers (1970), 114 Sol. Jo. 147, 68 L. G. R. 256; Digest Cont. Vol. C p. 1023, *1116b* 68
Shipton *v.* Casson (1826), 5 B. & C. 378; 8 Dow. & Ry. K. B. 130; 4 L. J. O. S. K. B. 199; 39 Digest (Repl.) 686, *1802* 186
Shipton Anderson & Co. *v.* Weil Bros., [1912] 1 K. B. 574; 17 Com. Cas. 153; 81 L. J. K. B. 910; 106 L. T. 272; 28 T. L. R. 269; 39 Digest (Repl.) 690, *1840* 187
Shipton, Anderson & Co., and Harrison Brothers & Co., *Re,* [1915] 3 K. B. 676; 113 L. T. 1009; 84 L. J. K. B. 2137; 31 T. L. R. 598; 26 Com. Cas. 138, D. C.; 12 Digest (Reissue) 500, *3492* . . 20
Shirlaw *v.* Southern Foundries (1926), Ltd., [1939] 2 All E. R. 113; [1939] 2 K. B. 206; 108 L. J. K. B. 747; 160 L. T. 353; 55 T. L. R. 611; 83 Sol. Jo. 357, C. A.; affirmed *sub nom.* Southern Foundries *v.* Shirlaw [1940] 2 All E. R. 445; [1940] A. C. 701; 109 L. J. K. B. 461; 164 L. T. 251; 56 T. L. R. 637; 84 Sol. Jo. 464, H. L.; 9 Digest (Repl.) 557, *3689* 15, 104
Short *v.* Simpson (1866), L. R. 1 C. P. 248; Har. & Ruth. 181; 35 L. J. C. P. 147: 13 L. T. 674; 12 Jur. N. S. 258; 14 W. R. 307; 41 Digest (Repl.) 272 *891* 319
Siffken *v.* Wray (1805), 6 East 371; 2 Smith K. B. 480; 39 Digest (Repl.) 742, *2198* 206, 207
Simon *v.* Pawson and Leafs, Ltd. (1932), 148 L. T. 154; 38 Com. Cas. 151, C. A.; 17 Digest (Reissue) 154, *421* 254
Simond *v.* Braddon (1857), 2 C. B. 251, 324; 26 L. J. C. P. 198; 3 Jur. N. S. 719; 5 W. R. 594; 39 Digest (Repl.) 533, *694* 356
Simonin *v.* Mallac (1860), 2 Sw. & Ir. 67; 29 L. J. P. M. & A. 97; 2 L. T. 327; 6 Jur. N. S. 561; 11 Digest (Reissue) 539, *1173* 91
Simpson *v.* Crippin (1872), L. R. 8 Q. B. 14; 42 L. J. Q. B. 28; 27 L. T. 546; 8 W. R. 80; 39 Digest (Repl.) 700, *1914* 190
Simpson *v.* Nicholls (1838), 3 M. & W. 240; 6 Dowl. 355; 1 Horn & H. 12; 7 L. J. Ex. 117; 2 J. P. 135; 2 Jur. 82; 45 Digest (Repl.) 245, *130* 87
Sims *v.* Marryat (1851), 17 Q. B. 281; 20 L. J. Q. B. 454; 39 Digest (Repl.) 538, *741* 113
Sims & Co. *v.* Midland Railway (1912), [1913] 1 K. B. 103; 82 L. J. K. B. 67; 107 L. T. 700; 29 T. L. R. 81; 18 Com. Cas. 44; 8 (1) Digest (Reissue) 29, *168* 162, 168
Sinason-Teicher Inter-American Grain Corpn. *v.* Oilcakes and Oilseeds Trading Co., Ltd., [1954] 3 All E. R. 468; [1954] 1 W. L. R. 1394; 98 Sol. Jo. 804; [1954] 2 Lloyd's Rep. 327, C. A.; 39 Digest (Repl.) 732, *2118* 43, 106
Singh *v.* Ali. *See* Sajan Singh *v.* Sardara Ali.
Singh (Gian) & Co., Ltd. *v.* Banque de l'Indochine, [1974] 2 All E. R. 754; [1974] 1 W. L. R. 1234, 118 Sol. Jo. 644, P. C. 44
Sky Petroleum, Ltd. *v.* V.I.P. Petroleum, Ltd., [1974] 1 All E. R. 954; [1974] 1 W. L. R. 576; 118 Sol. Jo. 311 245
Slade *v.* Sinclair and Wilcox, Ltd. (1930), 74 Sol. Jo. 122 253
Slater *v.* Hoyle and Smith, [1920] 2 K. B. 11; [1918–19] All E. R. Rep. 654; 89 L. J. K. B. 401; 122 L. T. 611; 36 T. L. R. 132; 25 Com. Cas. 140, C. A.; 39 Digest (Repl.) 592, *1120* . . 241, 247, 248, 249

PAGE

Slavonski *v.* La Pelleterie de Roubain Société Anonyme (1927), 137 L. T. 645; 17 Digest (Reissue) 166, *483* 253
Slee, *Re, Ex parte* North Western Bank (1872), L. R. 15 Eq. 69; 42 L. J. Bcy. 6; 27 L. T. 461; 1 Digest (Repl.) 390, *528* 83
Smeaton Hanscomb & Co. *v.* Sassoon I. Setty Son & Co., [1953] 2 All E. R. 1471; [1953] 1 W. L. R. 1468; 97 Sol. Jo. 862; [1953] 2 Lloyd's Rep. 580; 39 Digest (Repl.) 582, *1046* 362, 366
Smedleys, Ltd. *v.* Breed, [1974] 2 All E. R. 21; [1974] A. C. 839; [1974] 2 W. L. R. 575; 138 J. P. 439; 118 Sol. Jo. 363, H. L. 70
Smeed *v.* Foord (1859), 28 L. J. Q. B. 178; 1 E. & E. 602; 5 Jur. N. S. 291; 17 Digest (Reissue) 154, *417* 235, 241
Smith *v.* Baker (1873), L. R. 8. C. P. 350; 42 L. J. C. P. 155; 28 L. T. 637; 37 J. P. 567; 45 Digest (Repl.) 317, *273* 31
Smith *v.* Blandy (1825), Ry. & M. 257; 39 Digest (Repl.) 496, *430* . 359
Smith *v.* Chance (1819), 2 B. & Ald. 753; [1814–23] All E. R. Rep. 770; 39 Digest (Repl.) 659, *1596* 184
Smith *v.* Green (1875), 1 C. P. D. 92; 45 L. J. C. P. 28; 33 L. T. 572; 40 J. P. 103; 24 W. R. 142; 2 Digest (Repl.) 405, *732* 246
Smith *v.* Hamilton, [1950] 2 All E. R. 928; [1951] Ch. 174; 66 (pt. 2) T. L. R. 937; 94 Sol. Jo. 724; 40 Digest (Repl.) 242, *2042* . . . 105
Smith *v.* Jeffryes (1846), 15 M. & W. 561; 15 L. J. Ex. 325; 7 L. T. O. S. 231; 17 Digest (Reissue) 358, *1248* 365
Smith *v.* Johnson (1899), 15 T. L. R. 179; 39 Digest (Repl.) 589, *1104* . 251
Smith *v.* Mercer (1867), L. R. 3 Ex. 51; 37 L. J. Ex. 24; 17 L. T. 317; 39 Digest (Repl.) 736, *2153* 359
Smith *v.* Myers (1871), L. R. 7 Q. B. 139; 41 L. J. Q. B. 91; 26 L. T. 103; 20 W. R. 186; 1 Asp. M. L. C. 222; 39 Digest (Repl.) 493, *410* . 356
Smith *v.* Peters (1875), L. R. 20 Eq. 511; 44 L. J. Ch. 613; 23 W. R. 783; 39 Digest (Repl.) 506, *507* 105
Smyth (Ross T.) & Co., Ltd. *v.* Bailey, Son & Co., [1940] 3 All E. R. 60, 164 L. T. 102; 45 Com. Cas. 292; *sub nom.* Bailey, Son & Co. *v.* Smyth (Ross T.) & Co., 56 T. L. R. 825; 84 Sol. Jo. 572; 67 Ll. L. Rep. 147, H. L.; 39 Digest (Repl.) 611, *1239* . . . 36, 37, 148, 154
Smyth (Ross T.) & Co., Ltd. (Liverpool) *v.* Lindsay (W. N.) Ltd., (Leith), [1953] 2 All E. R. 1064; [1953] 1 W. L. R. 1280; 97 Sol. Jo. 744; [1953] 2 Lloyd's Rep. 378; 12 Digest (Reissue) 465, *3346* . 20, 42
Snee *v.* Prescot (1743), 1 Atk. 245; 39 Digest (Repl.) 765, *2426* . . 219
Snell *v.* Unity Finance, Ltd., [1963] 3 All E. R. 50; [1964] 2 Q. B. 203; [1963] 3 W. L. R. 559; 107 Sol. Jo. 533, C. A.; Digest Cont. Vol. A p. 325, *929a* 17
Snook *v.* London and West Riding Investments, Ltd., [1967] 1 All E. R. 518; [1967] 2 Q. B. 786; [1967] 2 W. L. R. 1020; 111 Sol. Jo. 71, C. A.; Digest Cont. Vol. C p. 416, *21b* 76, 167
Snowball, *Ex parte, Re,* Douglas (1872), 7 Ch. App. 534; 41 L. J. Bk. 49; 26 L. T. 894; 5 Digest (Repl.) 988, *7967* 299
Sobell Industries, Ltd. *v.* Cory Brothers Co., Ltd., [1955] 2 Lloyd's Rep. 82; Digest Cont. Vol. B p. 11, *2676a* 43, 46
Société Anonyme Scholefield (1902), 7 Com. Cas. 114, C. A.; 39 Digest (Repl.) 690, *1834* 364
Société D'Advances Commerciales (London), Ltd. *v.* Besse (A.) & Co. (London), Ltd., [1952] 1 T. L. R. 644; [1952] 1 Lloyd's Rep. 242; 39 Digest (Repl.) 571, *987* 42
Société des Hotels Le Touquet *v.* Cummings, [1922] 1 K. B. 451; [1921] All E. R. Rep. 408; 91 L. J. K. B. 288; 126 L. T. 513; 38 T. L. R. 221; 66 Sol. Jo. 269, C. A.; 35 Digest (Repl.) 195, *57* . . . 259
Société Metallurgique d'Anbrives and Villerupt *v.* British Bank for Foreign Trade (1922), 11 Ll. L. Rep. 168 . . . . 45, 46
Social Services Tax Act, *Re, Re* Central Heat Distribution, Ltd. (1970), 74 W. W. R. 246 280

PAGE

Solle *v.* Butcher, [1949] 2 All E. R. 1107; [1950] 1 K. B. 671; 66 (pt. 1) T. L. R. 448, C. A.; 31 (2) Digest (Reissue) 1042, *8232* . . 97, 99

Somes *v.* British Empire Shipping Co. (1860), 30 L. J. Q. B. 229; 8 H. L. Cas. 338; 2 L. T. 547; 6 Jur. N. S. 761; 32 Digest (Repl.) 295, *397* . 209

Soper *v.* Arnold (1889) 14 App. Cas. 429; 59 L. J. Ch. 214; 61 L. T. 702; 40 Digest (Repl.) 242, *2028* 27

Soproma S. p. A. *v.* Marine and Animal By-Products Corporation, [1966] 1 Lloyd's Rep. 367 . . . . 39, 44, 107

Sorfareren, The (1915), 85 L. J. P. 121; 114 L. T. 46; 32 T. L. R. 108; 13 Asp. M. L. C. 223; 1 Br. & Col. Pr. Cas. 589; on appeal (1917), 117 L. T. 259, P. C.; 39 Digest (Repl.) 632, *1429* 35

Sorrell *v.* Smith, [1925] A. C. 700; [1925] All E. R. 1; 94 L. J. Ch. 347; 133 L. T. 370; 41 T. L. R. 529; 69 Sol. Jo. 641, H. L.; 45 Digest (Repl.) 301, *182* 52

Sorrentino Fratelli *v.* Buerger, [1915] 1 K. B. 307; [1914] W. N. 427; 84 L. J. K. B. 725; 112 L. T. 294; 20 Com. Cas. 132; affirmed, [1915] 3 K. B. 367; 84 L. J. K. B. 1937; 113 L. T. 840; 13 Asp. M. L. C. 164; 21 Com. Cas. 33, C. A.; 41 Digest (Repl.) 348, *1420* 29

Sottomayor *v.* De Barros (1877), 3 P. D. 1; [1874–80] All E. R. Rep. 94; 47 L. J. P. 23; 37 L. T. 415; 26 W. R. 455, C. A.; 11 Digest (Reissue) 511, *1018* 91

Sottomayor *v.* De Barros (1879), 5 P. D. 94; [1874–80] All E. R. Rep. 97; 49 L. J. P. 1; 41 L. T. 281; 27 W. R. 917; 11 Digest (Reissue) 512, *1019* 91

South Australian Insurance Co. *v.* Randell (1869), L. R. 3 P. C. 101; 6 Moore, P. C. N. S. 341; 22 L. T. 843; 39 Digest (Repl.) 447, *28* . 76, 79

South-West Suburban Water Co. *v.* St. Marylebone Guardians, [1904] 2 K. B. 174; 73 L. J. K. B. 347; 68 J. P. 257; 52 W. R. 378; 20 T. L. R. 299; 48 Sol. Jo. 397; 2 L. G. R. 567; 47 Digest (Repl.) 600, *149* 274

Southern Foundries *v.* Shirlaw. *See* Shirlaw *v.* Southern Foundries.

Southern Industrial Trust, Ltd. *v.* Brooke House Motors, Ltd. (1968) 112 Sol. Jo. 798, C. A. 18

Southwell *v.* Lewis (1880), 45 J. P. 206; 25 Digest (Repl.) 138, *553* . 274

Sovfracht (V/O) *v.* Van Udens Scheepvaart En Agentuur Maatschappij (N. V. Gebr.), [1943] 1 All E. R. 76; [1943] A. C. 203; 112 L. J. K. B. 32; 168 L. T. 323; 59 T. L. R. 101; 86 Sol. Jo. 376; [1942] W. N. 223, H. L.; 2 Digest (Repl.) 220, *322* 89, 90

Spalding *v.* Ruding (1843), 6 Beav. 376; 12 L. J. Ch. 503; 1 L. T. O. S. 384; 7 Jur. 733; on appeal (1846), 15 L. J. Ch. 374; 39 Digest (Repl.) 770, *2470* 222

Spartali *v.* Benecke (1850), 10 C. B. 212; 19 L. J. C. P. 293; 15 L. T. O. S. 183; 39 Digest (Repl.) 661, *1609* 182, 209, 357

Speight *v.* Gaunt (1883), 9 App. Cas. 1; 53 L. J. Ch. 419; 50 L. T. 330; 48 J. P. 84; 32 W. R. 435, H. L. 47 Digest (Repl.) 387, *3462* . . 289

Spence *v.* Union Marine Insurance Co. (1868), L. R. 3 C. P. 427; 37 L. J. C. P. 169; 18 L. T. 632; 29 Digest (Repl.) 299, *2264* . . 143

Spencer *v.* North Country Finance Co., [1963] C. L. Y. 212 . . . 166

Spencer Trading Co., Ltd. *v.* Devon, [1947] 1 All E. R. 284; 39 Digest (Repl.) 550, *812* 132

Spicer *v.* Cooper (1841), 1 Q. B. 424; 1 Gal. & Dav. 52; 10 L. J. Q. B. 241; 5 Jur. 1036; 39 Digest (Repl.) 481, *289* 363

Spiro *v.* Lintern, [1973] 3 All E. R. 319; [1973] 1 W. L. R. 1002; 117 Sol. Jo. 584; 227 Estates Gazette 2045, C. A.; Digest Supp. . 23, 167

Springer *v.* Great Western Ry. Co., [1921] 1 K. B. 257; [1920] All E. R. Rep. 361; 89 L. J. K. B. 1010; 124 L. T. 79, C. A.; 8 (1) Digest (Reissue) 39, *218* 164, 168

Square *v.* Model Farm Dairies (Bournemouth), Ltd., [1939] 1 All E. R. 259; [1939] 2 K. B. 365; 108 L. J. K. B. 198; 160 L. T. 165; 55 T. L. R. 384; 83 Sol. Jo. 152, C. A.; 25 Digest (Repl.) 141, *571* . 54

PAGE

Stach (Ian), Ltd. *v.* Baker Bosley, Ltd., [1958] 1 All E. R. 542; [1958] 2 Q. B. 130; [1958] 2 W. L. R. 419; 102 Sol. Jo. 177; [1958] 1 Lloyd's Rep. 127; 39 Digest (Repl.) 733, *2125* . . . 34, 35, 43, 106

Stadium Finance, Ltd. *v.* Robbins, [1962] 2 All E. R. 633; [1962] 2 Q. B. 664; [1962] 3 W. L. R. 453; 106 Sol. Jo. 369, C. A.; Digest Cont. Vol. A p. 7, *509a* 295, 298

Stamp *v.* United Dominions Trust (Commercial), Ltd., [1967] 1 All E. R. 251; [1967] 1 Q. B. 418; [1967] 2 W. L. R. 541; 131 J. P. 177; 110 Sol. Jo. 906; Digest Cont. Vol. C p. 233, *5926a* 341

Stanton *v.* Richardson (1872), L. R. 7 C. P. 421; L. J. C. P. 180; 27 L. T. 513; 21 W. R. 71; 1 Asp. M. L. C. 449; on appeal, (1874) L. R. 9 C. P. 390; affirmed, (1875), 45 L. J. C. P. 78; 33 L. T. 193; 24 W. R. 324; 3 Asp. M. L. C. 23, H. L.; 41 Digest (Repl.) 155, *16* . . 352

Stapleton, *Ex parte* (1879), 10 Ch. D. 586; [1874–80] All E. R. Rep. 1182; 14 L. T. 14, C. A.; 39 Digest (Repl.) 778, *2543* . . . 205, 210, 223, 232, 234

Stapleton *v.* Haymen (1864), 2 H. & C. 918; 3 New Rep. 481; 33 L. J. Ex. 170; 9 L. T. 655; 10 Jur. N. S. 497; 12 W. R. 317; 1 Mar. L. C. 416; 42 Digest (Repl.) 638, *3916* 282

Startup *v.* Cortazzi (1835), 2 Cr. M. & R. 165; 5 Tyr. 697; 4 L. J. Ex. 218; 39 Digest (Repl.) 813, *2769* 237

Startup *v.* Macdonald (1843) 6 Man. & G. 593; 7 Scott, N. R. 269; 12 L. J. Ex. 477 Ex. Ch.; 12 Digest (Reissue) 396, *2886* . 185, 228, 358

Staunton *v.* Wood (1851), 16 Q. B. 638; 16 L. T. O. S. 486; 15 Jur. 1123; 39 Digest (Repl.) 522, *612* 357, 358

Steam Herring Fleet, Ltd. *v.* Richards & Co., Ltd. (1901), 17 T. L. R. 731; 39 Digest (Repl.) 816, *2783* 252

Steedman *v.* Drinkle, [1916] 1 A. C. 275; [1914–15] All E. R. Rep. 298; 85 L. J. P. 79; 114 L. T. 248; 32 T. L. R. 231; P. C., 40 Digest (Repl.) 142, *1088* 27

Steedman *v.* Frigidaire Corpn., [1933] 1 D. L. R. 161 6

Steel Wing & Co., Ltd., *Re*, [1921] 1 Ch. 349; [1920] All E. R. Rep. 292; 90 L. J. Ch. 116; 124 L. T. 664; 65 Sol. Jo. 240; [1920] B. & C. R. 160; 8 (2) Digest (Reissue) 502, *86* 29

Steels and Busks *v.* Bleecker Bik & Co., [1956] 1 Lloyd's Rep. 228 . 117 121, 135, 140, 367

Stein *v.* Hambro's Bank of Northern Commerce (1921), 9 Ll. L. Rep. 507 46

Stein, Forbes & Co. *v.* County Tailoring Co. (1916), 86 L. J. K. B. 448; 115 L. T. 215; 13 Asp. M. L. C. 422; 39 Digest (Repl.) 794, *2658* . 226, 229, 230, 358

Steinberg *v.* Scala (Leeds) Ltd., [1923] 2 Ch. 452; [1923] All E. R. Rep. 239; 92 L. J. K. B. 944; 129 L. T. 624; 39 T. L. R. 542; 67 Sol. Jo. 656, C. A.; 12 Digest (Reissue) 281, *2034* 87

Steinberger *v.* Atkinson & Co. (1914), 31 T. L. R. 110; 39 Digest (Repl.) 744, *2220* 189, 211

Stephenson *v.* Thompson, [1924] 2 K. B. 240; [1924] All E. R. Rep. 736; 93 L. J. K. B. 771; 131 L. T. 279; 88 J. P. 142; 40 T. L. R. 513; 68 Sol. Jo. 536; 22 L. G. R. 359; [1924] B. & C. R. 170; C. A.; 30 Digest (Reissue) 268, *751* 320

Stephens *v.* Wilkinson (1831) 2 B. & Ad. 320; 9 L. J. O. S. K. B. 231; 39 Digest (Repl.) 780, *2551* 242

Sterns *v.* Vickers, [1923] 1 K. B. 78; [1922] All E. R. Rep. 126; 92 L. J. K. B. 331; 128 L. T. 402, C. A.; 39 Digest (Repl.) 647, *1535* 158, 159

Stewart (Robert) & Sons, Ltd. *v.* Carapamayoti & Co., Ltd., [1962] 1 All E. R. 418; [1962] 1 W. L. R. 34; 106 Sol. Jo. 16; [1961] 2 Lloyd's Rep. 387; 17 Digest (Reissue) 178, *554* 239

Stewart *v.* Kennedy (1890), 15 App. Cas. 75; 17 R. (H.L.) 1; 27 Sc. L. R. 386; 39 Digest (Repl.) 822, **1570* 245

PAGE

Stirling *v.* Maitland (1864), 5 B. & S. 841; [1861–73] All E. R. Rep. 358; 5 New Rep. 46; 34 L. J. Q. B. 1; 11 L. T. 337; 29 J. P. 115; 11 Jur. N. S. 218; 13 W. R. 76; 12 Digest (Reissue) 761, *5448* . . . 104
Stock *v.* Inglis (1884), 12 Q. B. D. 564; 52 L. J. Q. B. 356; 51 L. T. 449; affirmed *sub nom.* Inglis *v.* Stock, (1885) 10 App. Cas. 263; [1881–85] All E. R. Rep. 668; 54 L. J. Q. B. 582; 52 L. T. 821; 33 W. R. 877; 5 Asp. M. L. C. 422, H. L.; 39 Digest (Repl.) 649, *1548* . 34, 142, 157, 161
Stock *v.* Urey, [1954] N. I. 71; 39 Digest (Repl.) 527, **288* . . . 117
Stocks *v.* Wilson, [1913] 2 K. B. 235; 82 L. J. K. B. 598; 108 L. T. 834; 29 T. L. R. 352; 28 (2) Digest (Reissue) 690, *225* . . . 87, 88, 89
Stockloser *v.* Johnson, [1954] 1 All E. R. 630; [1954] 1 Q. B. 476; [1954] 2 W. L. R. 439; 98 Sol. Jo. 178, C. A.; 20 Digest (Repl.) 548, *2565* 27
Stoneham *Re*, Stoneham *v.* Stoneham, [1919] 1 Ch. 149; [1918–19] All E. R. Rep. 1051; 88 L. J. Ch. 77; 120 L. T. 341; 25 Digest (Repl.) 556, *63* 78
Stoneleigh Finance, Ltd. *v.* Phillips, [1965] 1 All E. R. 513; [1965] 2 Q. B. 537; [1965] 2 W. L. R. 508; 109 Sol. Jo. 68, C. A.; Digest Cont. Vol. B p. 106, *5282a* 76, 163, 166, 171, 297
Stoveld *v.* Hughes (1811), 14 East, 308; 39 Digest (Repl.) 769, *2460* . 221
Strachan, *Re, Ex parte* Cooke (1876), 4 Ch. D. 123; 46 L. J. Bcy. 52; 35 L. T. 649; 41 J. P. 180; 25 W. R. 171, C. A.; 5 Digest (Repl.) 776, *6613* 312
Strass *v.* Spillers and Baker, Ltd., [1911] 2 K. B. 759; 80 L. J. K. B. 1218; 104 L. T. 284; 16 Com. Cas. 166; 29 Digest (Repl.) 113, *583*: 40, 359
Street *v.* Blay (1831), 2 B. & Ad. 456; [1824–34] All E. R. Rep. 329; 39 Digest (Repl.) 581, *1040* 245, 352
Stucley *v.* Baily (1862), 1 H. & C. 405; 31 L. J. Ex. 483; 10 W. R. 720; 39 Digest (Repl.) 513, *550* 353
Suisse Atlantique Société D'Armement Maritime S. A. *v.* N. V. Rotterdamsche Kolen Centrale, [1966] 2 All E. R. 61; [1967] 1 A. C. 361; [1966] 2 W. L. R. 944; 110 Sol. Jo. 367; [1966] 1 Lloyd's Rep. 529, H. L.; Digest Cont. Vol. B p. 652, *2413a* . 11, 12, 13, 14, 127, 128, 265
Sullivan *v.* Constable (1932) 48 T. L. R. 369 C. A.; 39 Digest (Repl.) 577, *1014* 108, 110
Sumner, Permain & Co. *v.* Webb & Co. [1922] 1 K. B. 55; [1921] All E. R. Rep. 680; 91 L. J. K. B. 228; 126 L. T. 294; 38 T. L. R. 45; 66 Sol. Jo. (W.R.) 27 Com. Cas. 105; 39 Digest (Repl.) 556, *858* . . 130
Sutherland *v.* Allhusen (1866) 14 L. T. 666; 2 Mar. L. C. 349; 39 Digest (Repl.) 566, *949* 34
Sutro (L.) & Co., Heilbut Symons & Co., *Re*, [1917] 2 K. B. 348; 86 L. J. K. B. 1226; 116 L. T. 545; 33 T. L. R. 359; 14 Asp. M. L. C. 34; 23 Com. Cas. 21, C. A.; 39 Digest (Repl.) 669, *1685* . . 17, 35
Sutton London Borough *v.* Perry Sanger & Co., Ltd. (1971), 135 J. P. Jo. 239 68
Swan and Edgar, Ltd. *v.* Mathieson (1910), 103 L. T. 832; [1908–10] All E. R. Rep. 523; 27 T. L. R. 153; 27 (1) Digest (Reissue) 228, *1656* 92
Swain *v.* Shepherd (1832), 1 Mood. & Rob. 223; 39 Digest (Repl.) 624, *1361* 147
Sweet *v.* Pym (1800), 1 East 4; 39 Digest (Repl.) 745, *2233* . . 207
Sweeting *v.* Turner (1871), L. R. 7 Q. B. 310; 41 L. J. Q. B. 58; 25 L. T. 796; 36 J. P. 597; 20 W. R. 185; 39 Digest (Repl.) 642, *1507* . 157, 161
Syers *v.* Jonas (1848), 2 Exch. 111; 39 Digest (Repl.) 557, *863* . 4, 139, 245
Sykes (F. & G.) (Wessex), Ltd. *v.* Fine Fare Ltd., [1966] 2 Lloyd's Rep. 205; on appeal; [1967] 1 Lloyd's Rep. 53, C. A. 102
Syndic in Bankruptcy of S. N. Khoury *v.* Khayat, [1943] 2 All E. R. 406; [1943] A. C. 507; P. C.; 6 Digest (Repl.) 24, *167* . . . 258
Sztejn *v.* J. Henry Schroder Banking Corpn. (1941), 31 N. Y. S. (2d) 631 45

PAGE

T.C. Industrial Plant Pty., Ltd. *v.* Roberts (Queensland) Pty., Ltd. (1963), 37 A. L. J. R. 289 252
Tailby *v.* Official Receiver (1888), 13 App. Cas.; 523; 58 L. J. Q. B. 75; 60 L. T. 162; H. L.; 7 Digest (Repl.) 126, *733* 96
Tamplin, *Re, Ex parte* Barnett (1890), 59 L. J. Q. B. 194; 62 L. T. 264; 38 W. R. 351; 6 T. L. R. 206; 7 Morr. 70, D. C.; 36 Digest (Repl.) 507, *724* 168
Tamvaco *v.* Lucas (1859), 1 E. & E. 581, 592; *sub nom.* Tanvaco *v.* Lucas 28 L. J. Q. B. 150, 301; 1 L. T. 161; 5 Jur N. S. 731; 1258; 7 W. R. 568; 39 Digest (Repl.) 688, *1823* 186, 357
Tancred *v.* Steel Co. of Scotland (1890), 15 App. Cas. 125, 62 L. T. 738, H. L.; 39 Digest (Repl.) 690, *1835* 364
Tansley *v.* Turner (1835), 2 Bing. N. C. 151; 1 Hodg. 267; 2 Scott, 238,; 4 L. J. P. C. 272; 39 Digest (Repl.) 620, *1327* 152
Tarleton Engineering *v.* Nattrass, [1973] 3 All E. R. 699; [1973] 1 W. L. R. 1261; 137 J. P. 837; 117 Sol. Jo. 745; Digest Supp. . . 58, 66
Tarling *v.* Baxter (1827), 6 B. & C. 360; 5 L. J. O. S. K. B. 164; 39 Digest (Repl.) 617, *1298* 146, 148, 157
Taylor *v.* Bullen (1850), 5 Exch. 779; 20 L. J. Ex. 21; 16 L. T. O. S. 154; 39 Digest (Repl.) 532, *690* 360
Taylor *v.* Caldwell (1863), 3 B. & S. 826; 2 New Rep. 198; 32 L. J. Q. B. 164; 8 L. T. 356; 27 J. P. 710; 11 W. R. 726; 12 Digest (Reissue) 460, *3329* 21, 95, 100
Taylor *v.* Combined Buyers, Ltd, [1924] N. Z. L. R. 627 367
Taylor *v.* Great Eastern Railway, [1901] 1 K. B. 774; 70 L. J. K. B. 499; 84 L. T. 770; 49 W. R. 431; 17 T. L. R. 394; 45 Sol. Jo. 381; 6 Com. Cas. 121; 39 Digest (Repl.) 488, *360* 220
Taylor *v.* Kymer (1832), 3 B. & Ad. 320; 1 L. J. K. B. 114; 1 Digest (Repl.) 639, *2165* 303
Taylor *v.* Oakes, Roncoroni & Co. (1922), 38 T. L. R. 349; affirmed; 127 L. T. 267; 38 T. L. R. 517; 66 Sol. Jo. 556; 27 Com. Cas. 261, C. A.; 39 Digest (Repl.) 727, *2083* 38, 189
Taylor *v.* Smith, [1974] Crim. L. R. 200 . . . . 58, 61
Taylor *v.* Thompson, [1930] W. N. 16; 26 Digest (Repl.) 660, *11* . 81
Taylor & Co. *v.* Landauer & Co., [1940] 4 All E. R. 335; 164 L. T. 299; 57 T. L. R. 47; 85 Sol. Jo. 119; 12 Digest (Reissue) 501, *3500* . . 42
Taylor & Sons, Ltd. *v.* Bank of Athens (1922), 91 L. J. K. B. 776; 128 L. T. 795; 27 Com. Cas. 142; 39 Digest (Repl.) 590, *1109* . 247, 255
Taylor (David) & Son, Ltd. *v.* Barnett [1953] 1 All E. R. 843; [1953] 1 W. L. R. 562; 97 Sol. Jo. 226, C. A.; 39 Digest (Repl.) 497, *439* 17
Tebbits Bros. *v.* Smith (1917), 33 T. L. R. 508, C. A.; 39 Digest (Repl.) 691, *1843* 364
Teheran-Europe Co., Ltd. *v.* S. T. Belton (Tractors), Ltd., [1968] 2 All E. R. 886; [1968] 2 Q. B. 545; [1968] 3 W. L. R. 205; 112 Sol. Jo. 501; [1968] 2 Lloyd's Rep. 37, C. A.; Digest Cont. Vol. C p. 854, *804a* 134, 136, 367
Telegrafo, The (1871), L. R. 3 P. C. 673; Moore P. C. N. S. 43; 40 L. J. Adm. 18; 24 L. T. 748; 1 Digest (Repl.) 183, *694* 162
Tennants (Lancashire) Ltd., *v.* Wilson & Co., [1917] A. C. 495; 86 L. J. K. B. 1191; 116 L. T. 780; 23 Com. Cas. 41; 12 Digest (Reissue) 496, *3483* 20
Tesco Stores, Ltd. *v.* Roberts, [1974] 3 All E. R. 74; [1974] 1 W. L. R. 1253; 138 J. P. 689; 118 Sol. Jo. 627 283
Tesco Supermarkets, Ltd. *v.* Nattrass, [1971] 2 All E. R. 127, [1972] A. C. 153; [1971] 2 W. L. R. 1166; 135 J. P. 289; 115 Sol. Jo. 285; 69 L. G. R. 403, H. L.; Digest Supp. 65, 67, 68, 69
Thelman Frēres *v.* Texas Flour Mill Co. (1900), 5 Com. Cas. 321, C. A. . 355
Thol *v.* Henderson (1881), 8 Q. B. D. 457; 46 L. T. 483; 46 J. P. 422; 39 Digest (Repl.) 827, *2891* 236, 237

PAGE

Thomas *v.* Alper, [1953] C. L. Y. 3277 184
Thomas Borthwick (Glasgow), Ltd. *v.* Bunge & Co., Ltd. *See* Borthwick (Thomas) (Glasgow), Ltd. *v.* Bunge & Co., Ltd.
Thomas Borthwick (Glasgow), Ltd. *v.* Faure Fairclough, Ltd. *See* Borthwick (Thomas) (Glasgow), Ltd. *v.* Faure Fairclough, Ltd.
Thomas Gabriel & Sons *v.* Churchill and Sim. *See* Gabriel (Thomas) & Sons *v.* Churchill and Sim.
Thomas Young & Sons, Ltd. *v.* Hobson & Partners. *See* Young (Thomas) & Sons, Ltd. *v.* Hobson & Partners.
Thompson *v.* Ball (1948), 92 Sol. Jo. 272; 25 Digest (Repl.) 84, *117* . 2
Thompson *v.* Dominy (1845), 14 M. & W. 403; 14 L. J. Ex. 320; 8 (2) Digest (Reissue) 517, *209* 317
Thompson and Shacknell, Ltd. *v.* Veale (1896), 74 L. T. 130, C. A.; 26 Digest (Repl.) 659, *7* 309
Thompson (W. L.), Ltd. *v.* Robinson (R.) (Gunmakers), Ltd., [1955] 1 All E. R. 154; [1955] Ch. 177; [1955] 2 W. L. R. 185; 99 Sol. Jo. 76; 39 Digest (Repl.) 795, *2664* 232, 233, 234, 235
Thoreson *v.* Capital Credit Corporation Ltd. and Active Bailiff Service, Ltd. (1962), 37 D. L. R. (2d.) 317; 41 W. L. R. 1; reversed (1964), 46 W. W. R. 210; Digest Cont. Vol. A p. 6, **68a* . 292, 299, 312
Thornett *v.* Haines (1846), 15 M. & W. 367; 15 L. J. Ex. 230; 3 Digest (Repl.) 13, *98* 267, 359
Thornett and Fehr *v.* Beers & Sons, [1919] 1 K. B. 486; 88 L. J. K. B. 684; 120 L. T. 570; 24 Com. Cas. 133; 39 Digest (Repl.) 555, *852*: 131, 195
Thornett and Fehr & Yuills, *Re*, [1921] 1 K. B. 219; 90 L. J. K. B. 361; 124 L. T. 218; 37 T. L. R. 31; 26 Com. Cas. 59; 39 Digest (Repl.) 806, *2732* 97, 188, 365
Thornley *v.* Tuckwell (Butchers), Ltd., [1964] Crim. L. R. 127: 57, 61, 148, 366
Thornton *v.* Shoe Lane Parking, Ltd.; [1971] 1 All E. R. 686; [1971] 2 Q. B. 163; [1971] 2 W. L. R. 585; 115 Sol. Jo. 75; [1971] 1 Lloyd's Rep. 289; [1971] R. T. R. 79, C. A.; 12 Digest (Reissue) 78, *418* . 13
Three Rivers Trading Co., Ltd. *v.* Gwinear and District Farmers Ltd. (1967), 11 Sol. Jo. 831, C. A.; 17 Digest (Reissue) 58, *639* . . . 363
Thurnell *v.* Balbirnie (1837), 2 M. & W. 786; Murp. & H. 235; 6 L. J. Ex. 255; 1 Jur. 358; 39 Digest (Repl.) 505, *492* 104
Tiffin *v.* Pitcher, [1969] C. L. Y. 3234 147
Tigress, The (1863), 32 L. J. P. M. & A. 97; Brown & Lush 38; 1 New Rep. 449; 9 Jur. N. S. 361; 8 L. T. 117; 11 W. R. 538; 1 Mar. L. C. 323; 39 Digest (Repl.) 774, *2502* 219, 220
Tilley *v.* Bowman Ltd., [1910] 1 K. B. 745; [1908–10] All E. R. Rep. 952; 79 L. J. K. B. 547; 102 L. T. 318; 39 Digest (Repl.) 830, *2916*: 172, 226
Tingey & Co., Ltd. *v.* Chambers, Ltd., [1967] N. Z. L. R. 785 . . . 298
Titmus *v.* Littlewood, [1916] 1 K. B. 732; [1916–17] All E. R. Rep. 720; 85 L. J. K. B. 738; 114 L. T. 614; 32 T. L. R. 278; 30 Digest (Reissue) 91, *678* 284
Todd *v.* Armsur (1882), 9 R. (Ct. of Sess.) 901; 33 Digest (Repl.) 492, **72*: 49, 169
Toepfer *v.* Continental Grain Co., [1973] 1 Lloyd's Rep. 289 . . . 366
Tolhurst *v.* Associated Portland Cement Manufacturers [1902] 2 K. B. 660; 71 L. J. K. B. 949; 51 W. R. 81; 87 L. T. 465; 18 T. L. R. 827; affirmed, [1903] A. C. 414; [1900–03] All E. R. Rep. 386; 72 L. J. K. B. 834; 89 L. T. 196; 19 T. L. R. 677; 12 Digest (Reissue) 724, *5242* 28, 29
Tooke *v.* Hollingworth (1793), 5 Term. Rep. 215; affirmed *sub nom.* Hollingworth *v.* Tooke (1795), 2 Hy. Bl. 501; 5 Digest (Repl.) 758, *6512* 226
Torkington *v.* Magee, [1902] 2 K. B. 427; [1900–03] All E. R. Rep. 991; 71 L. J. K. B. 712; 87 L. T. 304; 18 T. L. R. 703; reversed, [1903] 1 K. B. 644; 72 L. J. K. B. 336; 88 L. T. 443; 19 T. L. R. 331, C. A.; 1 Digest (Repl.) 95, *705* 281

PAGE

Torni, The, [1932] p. 27; 48 T. L. R. 195; affirmed, [1932] P. 78, 90; [1932] All E. R. Rep. at p. 384; 101 L. J. P. 44; 147 L. T. 208; 48 T. L. R. 471; 18 Asp. M. L. C. 315, C. A.; 22 Digest (Reissue) 683, *7292* 48
Toulmin *v.* Hedley (1845), 2 Car. & Kir. 157; 39 Digest (Repl.) 716, *2018* 195
Towle & Co. *v.* White (1873), 29 L. T. 78; 21 W. R. 465, H. L.; 1 Digest (Repl.) 313, *29* 76
Townley *v.* Crump (1835), 4 Ad. & El. 58; 1 Har. & W. 564; 5 Nev. & M. 606; 5 L. J. K. B. 14; 39 Digest (Repl.) 761, *2384* 209
Trading Society Kwik Hoo *v.* Sugar Commission (1923), 129 L. T. 500 356
Trans Trust S. P. R. L. *v.* Danubian Trading Co., Ltd., [1952] 1 All E. R. 970; [1952] 2 Q. B. 297; [1952] 1 T. L. R. 1066; 96 Sol. Jo. 312; [1952] 1 Lloyd's Rep. 348, C. A.; 39 Digest (Repl.) 781, *2557* . 253
Travers (Joseph) & Sons, Ltd. *v.* Longel, Ltd. (1947), 64 T. L. R. 150; 39 Digest (Repl.) 560, *887* 118, 120
Tredegar Iron Co. *v.* Hawthorn (1902), 18 T. L. R. 716, C. A.; 39 Digest (Repl.) 798, *2685* 238
Tremoille *v.* Christie (1893), 69 L. T. 338; 37 Sol. Jo. 650; 1 Digest (Repl.) 386, *513* 291
Triefus & Co. *v.* Post Office, [1957] 2 All E. R. 387; [1957] 2 Q. B. 352; [1957] 3 W. L. R. 1; 101 Sol. Jo. 463, C. A.; 37 Digest (Repl.) 216, *4* 218
Trollope and Colls, Ltd. *v.* North West Metropolitan Regional Hospital Board, [1973] 2 All E. R. 260; [1973] 1 W. L. R. 601; 117 Sol. Jo. 355, H. L.; Digest Supp. 15
Trollope and Colls, Ltd. and Holland and Hannen and Cubitts, Ltd. (Trading as Nuclear Civil Constructors) (a firm) *v.* Atomic Power Constructions, Ltd., [1962] 3 All E. R. 1035; [1963] 1 W. L. R. 333; 107 Sol. Jo. 254; Digest Cont. Vol. A 72, *22a* 5
Tsakiroglou & Co., Ltd. *v.* Noblee and Thorl. G.m.b.H., [1961] 2 All E. R. 179; [1962] A. C. 93; [1961] 2 W. L. R. 633; 105 Sol. Jo. 346; [1961] 1 Lloyd's Rep. 329 H. L.; 39 Digest (Repl.) 569, *967* . . 19, 21, 192
Tucker *v.* Humphrey (1828), 4 Bing. 516; 6 L. J. O. S. C. P. 92; 39 Digest (Repl.) 760, *2376* 205
Tucker *v.* Linger (1883), 8 App. Cas. 508; 52 L. J. Ch. 941; 49 L. T. 373; 48 J. P. 4; 32 W. R. 40, H. L.; 17 Digest (Reissue) 53, *594* . . 17
Tunstall *v.* Steigmann, [1962] 2 All E. R. 417; [1962] 2 Q. B. 593; [1962] 2 W. L. R. 1045; 106 Sol. Jo. 282; 182 Estates Gazette 459, C. A.; 31 (2) Digest (Reissue) 967, *7801* 75
Turley *v.* Bates (1863) 2 H. & C. 200; 3 New Rep. 478, 10 L. T. 35; 10 Jur. N. S. 368; 12 W. R. 438; *sub nom.* Furley *v.* Bates, 33 L. J. Ex. 43; 39 Digest (Repl.) 620, *1329* 146, 152
Turner *v.* Liverpool Docks (1851), 6 Exch. 543; 20 L. J. Ex. 393; 17 L. T. O. S. 212; 39 Digest (Repl.) 631, *1426* 35, 154
Turner *v.* Sampson (1911), 27 T. L. R. 200; 1 Digest (Repl.) 432, *877* . 297
Tyers *v.* Rosedale Ferryhill Iron Co. (1875), L. R. 10 Ex. 195; 44 L. J. Ex. 130; 33 L. T. 56; 23 W. R. 871; 39 Digest (Repl.) 698, *1903*: 23, 239
Tzortzis *v.* Monark Line A/B, [1968] 1 All E. R. 949; [1968] 1 W. L. R. 406; 112 Sol. Jo. 108; [1968] 1 Lloyd's Rep. 337, C. A.; 11 Digest (Reissue) 461, *775* 48

Underwood *v.* Burgh Castle Brick and Cement Syndicate, [1922] 1 K. B. 343; [1921] All E. R. Rep. 515; 91 L. J. K. B. 355; 126 L. T. 401; 38 T. L. R. 44, C. A.; 39 Digest (Repl.) 618, *1314* . . . 149, 288
Union Bank *v.* Munster (1887), 37 Ch. D. 51; 57 L. J. Ch. 124; 57 L. T. 877; 52 J. P. 453; 36 W. R. 72; 4 T. L. R. 29; 3 Digest (Repl.) 393, *467* 267
Union Credit Bank, Ltd., and Davies *v.* Mersey Docks and Harbour Board, [1899] 2 Q. B. 205; 68 L. J. Q. B. 842; 81 L. T. 44; 4 Com. Cas. 227; 1 Digest (Repl.) 367, *396* 293

PAGE

United Australia, Ltd. *v.* Barclays Bank, [1940] 4 All E. R. 20; [1941] A. C. 1; 109 L. J. K. B. 919; 57 T. L. R. 13, H. L.; 3 Digest (Repl.) 224, *542* 31

United Baltic Corporation *v.* Burgett and Newsam (1921), 8 Ll. L. Rep. 190, C. A. 39

United Dominions Trust (Commercial), Ltd. *v.* Eagle Aircraft Services, Ltd., [1968] 1 All E. R. 104; [1968] 1 W. L. R. 74; 111 Sol. Jo. 849, C. A.; Digest Cont. Vol. C p. 406, *18d* 105, 107

United Fresh Meat Co. *v.* Charterhouse Cold Storage, [1974] 2 Lloyd's Rep. 286 325

United Railways of Havana and Regla Warehouses, Ltd., *Re*, [1960] 2 All E. R. 332; [1961] A. C. 1007; [1960] 2 W. L. R. 969; 104 Sol. Jo. 466, H. L.; Digest Cont. Vol. A 231, *862a* 47

United States of America *v.* Dollfus Mieg. et Compagnie S.A., [1952] 1 All E. R. 572; [1952] A. C. 582; [1952] 1 T. L. R. 541; 96 Sol. Jo. 180, H. L.; 1 Digest (Repl.) 58, *433* 30

United States of America *v.* Motor Trucks, Ltd., [1924] A. C. 196; 93 L. J. P. C. 46; 130 L. T. 129; 39 T. L. R. 723, P. C., 35 Digest (Repl.) 137, *307* 244

United States Steel Products Co. *v.* Great Western Ry. Co., [1916] 1 A. C. 189; [1914–15] All E. R. Rep. 1049; 85 L. J. K. B. 1; 113 L. T. 886; 21 Com. Cas. 105; 31 T. L. R. 561; 39 Digest (Repl.) 749, *2272* 219

Unity Finance *v.* Hammond (1965), 109 Sol. Jo. 70, C. A. 2

Universal Cargo Carriers Corporation *v.* Citati, [1957] 2 All E. R. 70; [1957] 2 Q. B. 401; [1957] 2 W. L. R. 713; 101 Sol. Jo. 320; [1957] 1 Lloyd's Rep. 174; affirmed [1957] 3 All E. R. 234 [1957] 1 W. L. R. 979; 101 Sol. Jo. 762; [1957] 2 Lloyd's Rep. 191, C. A.; Digest Cont. Vol. A p. 290, *2979a* 104, 189

Universal Steam Navigation Co. *v.* McKelvie, [1923] A. C. 492; 92 L. J. K. B. 647; 129 L. T. 395; 39 T. L. R. 480; 1 Digest (Repl.) 731, *2751* 271

Universal Stock Exchange *v.* Strachan, [1896] A. C. 166; [1895–99] All E. R. Rep. 751; 65 L. J. Q. B. 429; 74 L. T. 468; 60 J. P. 468; 44 W. R. 497, H. L.; 44 Digest (Repl.) 440, *442* 103

Urquhart, Lindsay & Co. *v.* Eastern Bank, [1922] 1 K. B. 318; [1921] All E. R. Rep. 340; 91 L. J. K. B. 274; 126 L. T. 534; 27 Com. Cas. 124; 39 Digest (Repl.) 499, *456* 45, 46

Valentini *v.* Canali (1889), 24 Q. B. D. 166; [1886–90] All E. R. Rep. 883; 59 L. J. Q. B. 74; 61 L. T. 731; 54 J. P. 295; 38 W. R. 331; 6 T. L. R. 75; 28 (2) Digest (Reissue) 689, *218* 87

Valpy *v.* Gibson (1847), 4 C. B. 837; 16 L. J. C. P. 241; 9 L. T. O. S. 434; 11 Jur. 826; 39 Digest (Repl.) 748, *2265* 101, 212, 215

Valpy *v.* Oakeley (1851), 16 Q. B. 941; 20 L. J. Q. B. 380; 17 L. T. O. S. 124; 16 Jur. 38; 39 Digest (Repl.) 817, *2795* . . . 210, 213, 237

Vamvakas *v.* Custodian of Enemy Property [1952] 1 All E. R. 629; [1952] 2 Q. B. 183; [1952] 1 T. L. R. 638; 96 Sol. Jo. 213; 2 Digest (Repl.) 265, *601* 90

Van Casteel *v.* Booker (1848), 2 Exch. 691; 18 L. J. Ex. 9; 12 L. T. O. S. 65; 39 Digest (Repl.) 629, *1415* 205

Van den Hurk *v.* Martens & Co. [1920] 1 K. B. 850; 89 L. J. K. B. 545; 123 L. T. 110; 25 Com. Cas. 170; 39 Digest (Repl.) 592, *1121* . 195, 197, 249

Van der Sterren *v.* Cybernetics (Holdings), Pty., Ltd. (1970), 44 A. L. J. R. 157 . . . . 12, 13, 362

PAGE

Vantol, Ltd. *v.* Fairclough Dodd and Jones, [1955] 3 All E. R. 750; [1955] 1 W. L. R. 1302; 99 Sol. Jo. 888; [1955] 1 Lloyd's Rep. 546; C. A.; reversed *sub nom.* Fairclough Dodd and Jones *v.* J. H. Vantol, Ltd., [1956] 3 All E. R. 921; [1957] 1 W. L. R. 136; 101 Sol. Jo. 86; [1956] 2 Lloyd's Rep. 437, H. L.; 39 Digest (Repl.) 568, *962* 36, 356

Varley *v.* Whipp, [1900] 1 Q. B. 513; 69 L. J. Q. B. 513; 69 L. J. Q. B. 333; 48 W. R. 363; 44 Sol. Jo. 263; 39 Digest (Repl.) 528, *654*. 117, 119, 121, 146, 367

Verschures Creameries *v.* Hull and Netherlands S. S. Co., [1921] 2 K. B. 608; [1921] All E. R. Rep. 215; 91 L. J. K. B. 39; 125 L. T. 165, C. A.; 45 Digest (Repl.) 316, *269* 31

Vesta, The, [1921] 1 A. C. 774; 90 L. J. P. 250; 125 L. T. 261; 37 T. L. R. 505; 15 Asp. M. L. C. 194, P. C.; 37 Digest (Repl.) 472, *224* 81, 155

Vic Mill, Ltd., *Re*, [1913] 1 Ch. 465; 82 L. J. Ch. 251; 108 L. T. 144; 57 Sol. Jo. 404, C. A.; 39 Digest (Repl.) 797, *2676* 231, 232

Vickers *v.* Hertz (1871), L. R. 2 Sc. & Div. 113; 1 Digest (Repl.) 383, *492* 314

Vickers *v.* Vickers (1867), L. R. 4 Eq. 529; 36 L. J. Ch. 946; 39 Digest (Repl.) 506, *505* 104, 105

Victoria Laundry (Windsor), Ltd. *v.* Newman Industries, Ltd., [1949] 1 All E. R. 997; [1949] 2 K. B. 528; 65 T. L. R. 274; 93 Sol. Jo. 371, C. A.; 39 Digest (Repl.) 816, *2790* . . 236, 237, 241, 256

Victory Motors, Ltd. *v.* Bayda (1973), 3 W. W. R. 747 232

Vigers *v.* Sanderson, [1901] 1 K. B. 608; 70 L. J. K. B. 383; 84 L. T. 464; 17 T. L. R. 316; 6 Com. Cas. 99; 39 Digest (Repl.) 575, *1004* 367

Villars, *Ex parte* (1874), 9 Ch. App. 432; 43 L. J. Bcy. 76; 30 L. T. 348; C. A. 21 Digest (Repl.) 602, *927* 114

Vilmont *v.* Bentley (1886), 18 Q. B. D. 322; 56 L. J. Q. B. 128; 56 L. T. 318; 51 J. P. 436; 35 W. R. 238; 3 T. L. R. 185, C. A.; affirmed *sub nom.* Bentley *v.* Vilmont (1887), 12 App. Cas. 471; 57 L. J. Q. B. 18; 57 L. T. 854; H. L.; 37 Digest (Repl.) 174, *59* 169

Vincentelli *v.* Rowlatt (1911), 105 L. T. 411; 12 Asp. M. L. C. 34; 16 Com. Cas. 310; 39 Digest (Repl.) 713, *2003* 40

Vita Food Products, Incorporated *v.* Unus Shipping Co., Ltd., [1939] 1 All E. R. 513; [1939] A. C. 277; 108 L. J. P. C. 40; 160 L. T. 579; 55 T. L. R. 402; 83 Sol. Jo. 295; 44 Com. Cas. 123; 19 Asp. M. L. C. 257, P. C.; 11 Digest (Reissue) 456, *755* 48

Wackerbarth *v.* Masson (1812), 3 Camp. 270; 39 Digest (Repl.) 681, *1764* 359

Waddington & Sons *v.* Neale & Sons (1907), 96 L. T. 786; 23 T. L. R. 464; 1 Digest (Repl.) 392, *542* 297

Wade (Gabriel) and English, Ltd. *v.* Arcos, Ltd., *See* Gabriel Wade and English, Ltd. *v.* Arcos, Ltd.

Wagg (Helbert) & Co., Ltd., *Re*, *Re* Prudential Assurance Co., Ltd., [1956] 1 All E. R. 129; [1956] Ch. 323; [1956] 2 W. L. R. 183; 100 Sol. Jo. 53; 11 Digest (Reissue) 345, *25* 48

Wah Tat Banking, Ltd. and Overseas Chinese Banking Corporation, Ltd. *v.* Chan Chen Kum and Hua Siang Steamship Co., [1967] 2 Lloyd's Rep. 437; Digest Cont. Vol. C p. 880, **79a* 165

Wait, *Re*, [1927] 1 Ch. 606; [1926] All E. R. Rep. 433; 96 L. J. Ch. 179; 136 L. T. 552; 43 T. L. R. 150; 71 Sol. Jo. 105; 17 Asp. M. L. C. 222; 32 Com. Cas. 134, C. A.; 39 Digest (Repl.) 821, *2836* . 73, 96, 100, 142, 143, 225, 243, 245

Wait *v.* Baker (1848), 2 Exch. 1; 17 L. J. Ex. 307; 39 Digest (Repl.) 601, *1173* 35, 151, 153, 154, 191

PAGE

Waite *v.* Redpath Dorman Long, Ltd., [1971] 1 All E. R. 513; [1971] 1 Q. B. 294; [1970] 3 W. L. R. 1034; 114 Sol. Jo. 886; Digest Supp. . 326

Walkers, Winser and Ham and Shaw, Son & Co., *Re,* [1904] 2 K. B. 152; 73 L. J. K. B. 325; 90 L. T. 454; 53 W. R. 79; 9 Com. Cas. 174; 20 T. L. R. 274; 48 Sol. Jo. 277; 39 Digest (Repl.) 563, *919* . 139, 366

Waller *v.* Drakeford (1853), 1 E. & B. 749; 22 L. J. Q. B. 274; 17 J. P. 663; 17 Jur. 853; *sub nom.* Drakeford *v.* Waller, 1 Saund. & M. 114; 21 L. T. O. S. 87; 39 Digest (Repl.) 786, *2601* 166

Wallersteiner *v.* Moir (No. 2) (1975), 119 Sol. Jo. 97 . . . 325, 326

Wallis *v.* Russell, [1902] 2 I. R. 585; 36 I. L. T. 67, C. A.; 39 Digest (Repl.) 546, **374* 73

Wallis, Son and Wells *v.* Pratt, [1910] 2 K. B. 1003; 79 L. J. K. B. 1013; 103 L. T. 118; 26 T. L. R. 572, C. A.; reversed, [1911] A. C. 394; [1911–13] All E. R. Rep. 989; 80 L. J. K. B. 1058; 105 L. T. 146; 27 T. L. R. 431; 55 Sol. Jo. 496, H. L.; 12 Digest (Reissue) 522, *3613*: 6, 10, 73, 108, 109, 110, 118, 197, 198, 353, 354, 360, 366

Walton *v.* Mascall (1844) 13 M. & W. 452; 2 Dow. & L. 410; 14 L. J. Ex. 54; 1 Digest (Repl.) 66, *482* 228

Ward, *Ex parte. See* Eastgate, *Re, Ex parte* Ward.

Ward *v.* Hobbs (1877), 3 Q. B. D. 150; 47 L. J. Q. B. 90; 37 L. T. 654; 26 W. R. 151, C. A.; affirmed (1878), 4 App. Cas. 13; 48 L. J. Q. B. 281; 40 L. T. 73; 43 J. P. 252; 27 W. R. 114, H. L.; 2 Digest (Repl.) 348, *332* 121, 287, 360

Ward (R. V.), Ltd. *v.* Bignall [1967] 2 All E. R. 449; [1967] 1 Q. B. 534; [1967] 2 W. L. R. 1050; 111 Sol. Jo. 190, C. A.; Digest Cont. Vol. C p. 858, *2522a* 27, 107, 150, 224, 225

Wardar's (Import and Export) Co., Ltd. *v.* W. Norwood & Sons, Ltd. [1968] 2 All E. R. 602; [1968] 2 Q. B. 663; [1968] 2 W. L. R. 1440; 112 Sol. Jo. 310; [1968] 2 Lloyd's Rep. 1, C. A.; Digest Cont. Vol. C p. 856, *1216a* 142, 153, 154

Warde *v.* Stuart (1856), 1 C. B. N. S. 88, 5 W. R. 6; 1 Digest (Repl.) 615, *2018* 365

Ware, *Re, Ex parte* Drake (1877), 5 Ch. D. 866; 46 L. J. Bcy. 105; 36 L. T. 667; 25 W. R. 641; C. A.; 46 Digest (Repl.) 525, *683* . 30, 31

Warlow *v.* Harrison (1859), 1 E. & E. 309; [1843–60] All E. R. Rep. 620; 29 L. J. Q. B. 19; 3 Digest (Repl.) 47, *331* 267

Warman *v.* Southern Counties Car Finance Corporation, Ltd., [1949] 1 All E. R. 711; [1949] 2 K. B. 576; [1949] L. J. R. 1182; 93 Sol. Jo. 319; 26 Digest (Repl.) 666, *33* 114

Warming's Used Cars, Ltd. *v.* Tucker, [1956] S. A. S. R. 249 . 112, 114

Watson, *Ex parte* (1877), 5 Ch. D. 35; 46 L. J. Bcy 97; 36 L. T. 75, 25 W. R. 409, 3 Asp. M. L. C. 396, C. A.; 39 Digest (Repl.) 753, *2327*: 215, 219

Watson, *Ex parte, Re* Companies Acts (1888), 21 Q. B. D. 301; 57 L. J. Q. B. 609; 21 Digest (Repl.) 378, *1141* 286

Watson, *Re, Ex parte* Official Receiver in Bankruptcy (1890), 25 Q. B. D. 27; 59 L. J. Q. B. 394; 63 L. T. 209; 38 W. R. 567; 6 T. L. R. 332; 7. Morr. 155, C. A.; 7 Digest (Repl.) 6, *14* 76

Watson *v.* Denton (1835), 7 C. & P. 85; 17 Digest (Reissue) 139, *340* . 202

Watson *v.* Park Royal, [1961] 2 All E. R. 346; [1961] 1. W. L. R. 727; 105 Sol. Jo. 301; 46 Digest (Repl.) 212, *1408* 91

Watts *v.* Friend (1830), 10 B. & C. 446; L. & Welsb. 193; 5 Man. & Ry. K. B. 439; 8 L. J. O. S. K. B. 181; 39 Digest (Repl.) 454, *73* . . 94

Waugh *v.* Morris (1873), L. R. 8. Q. B. 202; [1861–73] All E. R. Rep. 941; 42 L. J. Q. B. 57; 28 L. T. 265; 21 W. R. 438; 1 Asp. M. L. C. 573; 12 Digest (Reissue) 289, *2089* 19

Wayne's Merthyr Steam Coal & Iron Co. *v.* Morewood & Co. (1877), 46 L. J. Q. B. 746; 39 Digest (Repl.) 735, *2138* . . 230, 231, 359

Webster *v.* Higgin, [1948] 2 All E. R. 127; 92 Sol. Jo. 454, C. A.; 26 Digest (Repl.) 665, *30* 10, 361

PAGE

Weiner *v.* Gill, [1905] 2 K. B. 172; 74 L. J. F. B. 845; 92 L. T. 843; 53 W. R. 553; 21 T. L. R. 478; 10 Com. Cas. 213; affirmed [1906] 2 K. B. 574; [1904–07] All E. R. Rep. 773; 75 L. J. K. B. 916; 95 L. T. 438; 50 Sol. Jo. 632; 22 T. L. R. 699, 11 Com. Cas. 240, C. A.; 39 Digest (Repl.) 627, *1377* 149, 368

Weiner *v.* Harris, [1910] 1 K. B. 285; [1908–10] All E. R. Rep. 405; 79 L. J. K. B. 342; 101 L. T. 647; 26 T. L. R. 96; 54 Sol. Jo. 81; 15 Com. Cas. 39, C. A.; 39 Digest (Repl.) 626, *1372* . . 76, 290, 291, 368

Weir *v.* Bell (1878), 3 Ex. D. 238; 47 L. J. Q. B. 704; 38 L. T. 929; 26 W. R. 746, C. A.; 35 Digest (Repl.) 29, *199* 286

Weis & Co. *v.* Produce Brokers (1921), 6 Ll. L. Rep. 164; on appeal 7 Ll. L. Rep. 211, C. A.; 41 Digest (Repl.) 237, *586* 39

Weis & Co., Ltd. & Credit Colonial et Commercial, Antwerp, *Re*, [1916] 1 K. B. 346; 85 L. J. K. B. 533; 114 L. T. 168; 13 Asp. M. L. C. 242 21 Com. Cas. 186; 39 Digest (Repl.) 671, *1700* 19

Wertheim *v.* Chicoutimi Pulp Co., [1911] A. C. 301; [1908–10] All E. R. Rep. 707; 80 L. J. P. C. 91; 104 L. T. 226; 16 Com. Cas. 297, P. C.; 39 Digest (Repl.) 814, *2778* 239, 241, 254

West (H.W.) Ltd. *v.* McBlain, [1950] N. I. 144 115

West Ham Union *v.* Pearson (1890), 62 L. T. 638; 54 L. P. 645 . . 87

Wester Moffat Colliery Co. *v.* Jeffrey & Co., 1911 S. C. 346 . . . 165

Western Credit Pty., Ltd. *v.* Dragon Motors Pty., Ltd., [1973] W. A. R. 184 177

Western Tractors *v.* Dyck (1969), 7 D. L. R. (3d) 535; Digest Cont. Vol. C p. 854, **364a* 13, 361

Westzinthus, *Re*, (1833), 5 B. & Ad. 817; 2 Nev. & M. K. B. 644; 3 L. J. K. B. 56; 39 Digest (Repl.) 770, *2469* 222

Wheatley's Trustee *v.* Wheatley, Ltd. (1901), 85 L. T. 491; 4 Digest (Repl.) 74, *629* 76

Whistler *v.* Forster (1863), 32 L. J. C. P. 161; 14 C. B. N. S. 248; 8 L. T. 317; 6 Digest (Repl.) 50, *422* 164

White, *Ex parte*, *Re* Nevill (1870), L. R. 6 Ch. App. 397; 40 L. J. Bcy. 73; 24 L. T. 45; on appeal *sub nom.* Towle & Co. *v.* White (1873) 29 L. T. 78; 21 W. R. 465, H. L.; 39 Digest (Repl.) 625, *1370* . . 76

White *v.* John Warwick & Co., [1953] 2 All E. R. 1021; [1953] 1 W. L. R. 1285; 97 Sol. Jo. 740, C. A.; 3 Digest (Repl.) 96, *241* . . . 363

White *v.* Williams, [1912] A. C. 814; 82 L. J. P. C. 11; 107 L. T. 99; 28 T. L. R. 521; 12 Asp. M. L. C. 208; 17 Com. Cas. 309, P. C.; 39 Digest (Repl.) 684, *1786* 359

White and Carter (Councils), Ltd. *v.* MacGregor, [1961] 3 All E. R. 1178; [1962] A. C. 413; [1962] 2 W. L. R. 17; 105 Sol. Jo. 1104, H. L.; 12 Digest (Reissue) 434, *3126* 256

White Hudson & Co. *v.* Asian Organization, Ltd., [1965] 1 All E. R. 1040; [1964] 1 W. L. R. 1466, P. C. 57

White Sea Timber Trust, Ltd. *v.* W. W. North, Ltd. (1932) 148 L. T. 263; [1932] All E. R. Rep. 136; 49 T. L. R. 142; 77 Sol. Jo. 30; 18 Asp. M. L. C. 367; 39 Digest (Repl.) 719, *2042* . . . 362, 366

Whitehead *v.* Anderson (1842), 9 M. & W. 518; 11 L. J. Ex. 157; 39 Digest (Repl.) 759, *2364* 215, 219

Whitehorn Brothers *v.* Davison, [1911] 1 K. B. 463; [1908] All E. R. Rep. 885; 80 L. J. K. B. 425; 104 L. T. 234, C. A.; 39 Digest (Repl.) 654, *1575* 115, 171, 172, 173, 174, 175, 299

Whitehouse *v.* Frost (1810), 12 East 614; 39 Digest (Repl.) 666, *1658* . 141

Whiteley *v.* Hilt, [1918] 2 K. B. 808; 87 L. J. K. B. 1058; 119 L. T. 632; 34 T. L. R. 592; 62 Sol. Jo. 717, C. A.; 39 Digest (Repl.) 820, *2826* . 31, 177, 242, 244

Wickberg *v.* Shatsky (1969), 4 D. L. R. (3d) 540 174

Wickens Motors (Gloucester) *v.* Hall, [1972] 3 All E. R. 759; [1972] 1 W. L. R. 1418; 116 Sol. Jo. 744; [1972] R. T. R. 519; Digest Supp. . 58

PAGE

Wickham Holdings, Ltd. *v.* Brooke House Motors, Ltd., [1967] 1 All E. R. 117; [1967] 1 W. L. R. 295, C. A.; Digest Cont. Vol. C p. 419, *81d* 177, 242
Widenmeyer *v.* Burn Stewart & Co., Ltd., 1966 S. L. T. 215 . 80, 148, 157
Wiehe *v.* Dennis Brothers (1913), 29 T. L. R. 250; 3 Digest (Repl.) 66, *75* 158, 160
Wigglesworth *v.* Dallison (1779), 1 Doug. K. B. 201; 1 Smith L. C. 597, 23; 17 Digest (Reissue) 40, *449* 4
Wigram *v.* Buckley, [1894] 3 Ch. 483; 63 L. J. Ch. 689; 71 L. T. 287; 43 W. R. 147; 10 T. L. R. 654; 38 Sol. Jo. 680; 7 R. 469, C. A.; 39 Digest (Repl.) 43, *506* 179
Wilensko Slaski Towarzystwo Drewno *v.* Fenwick & Co. (West Hartlepool), Ltd. [1938] 2 All E. R. 429; 54 T. L. R. 1019; 44 Com. Cas. 1; 61 Ll. L. Rep. 249; 39 Digest (Repl.) 723, *2065* 188
Wilkinson *v.* Barclay, [1946] 2 All E. R. 337 n., C. A.; 17 Digest (Reissue) 501, *173* 110
Wilkinson *v.* King (1809), 2 Camp. 335; 33 Digest (Repl.) 491, *476*: 170, 243
Wilks *v.* Atkinson (1815), 1 Marsh 412; 6 Taunt. 11; 39 Digest (Repl.) 315, *615* 181, 182
Willets *v.* Chaplin & Co. (1923), 39 T. L. R. 222; 3 Digest (Repl.) 116, *364* 162
Williams *v.* Agius, [1914] A. C. 510; [1914–15] All E. R. Rep. 97; 83 L. J. K. B. 715; 110 L. T. 865; 90 Com. Cas. 200; 30 T. L. R. 351; 58 Sol. Jo. 377; 19 Com. Cas. 200; H. L., 39 Digest (Repl.) 812, *2761* 236, 237, 241
Williams *v.* Atlantic Assurance Co., Ltd. [1933] 1 K. B. 81; [1932] All E. R. Rep. 32; 102 L. J. K. B. 241; 148 L. T. 313; 37 Com. Cas. 304; 18 Asp. M. L. C. 334, C. A.; 8 (2) Digest (Reissue) 502, *87* . 29
Williams *v.* Reynolds (1865), 6 B. & S. 495; 34 L. J. Q. B. 221; 6 New Rep. 293; 12 L. T. 729; 11 Jur. N. S. 973; 13 W. R. 940; 39 Digest (Repl.) 817, *2799* 236, 237
Wills & Sons *v.* Cunningham & Co., [1924] 2 K. B. 220; [1923] All E. R. Rep. 299; 93 L. J. K. B. 1008; 131 L. T. 400; 40 T. L. R. 108; 39 Digest (Repl.) 809, *2745* 20, 357
Wilson *v.* Dunville (1879), 4 L. R. Ir. 249; 39 Digest (Repl.) 591, **660* 246
Wilson *v.* Rickett, Cockerell & Co., Ltd., [1954] 1 All E. R. 868; [1954] 1 Q. B. 598; [1954] 2 W. L. R. 629; Sol. Jo. 233, C. A.; 39 Digest (Repl.) 556, *857* 130
Wilson and Wilson, *Re, Ex parte* Marum (1915), 84 L. J. K. B. 1893; 113 L. T. 116; [1915] H. B. R. 189; 2 Digest (Repl.) 243, *458* . . 90
Wilson *v.* Wright, [1937] 4 All E. R. 371; 82 Sol. Jo. 14; 59 Ll. L. Rep. 86, C. A.; 39 Digest (Repl.) 523, *618* 108, 355
Wilson Holgate & Co. *v.* Belgian Grain Co., [1920] 2 K. B. 1; 89 L. J. K. B. 300; 122 L. T. 534; 35 T. L. R. 530; 14 Asp. M. L. C. 566; 25 Com. Cas. 1; 39 Digest (Repl.) 708, *1965* 40
Wiltshire Iron Co., *Re, Ex parte* Pearson (1868), 3 Ch. App. 443; 37 L. J. Ch. 554; 18 L. T. 423; 16 W. R. 682; 10 Digest (Repl.) 904, *6142* 192
Wimble *v.* Rosenberg & Sons, [1913] 3 K. B. 743; 82 L. J. K. B. 1251; 109 L. T. 294; 29 T. L. R. 752; 57 Sol. Jo. 784; 12 Asp. M. L. C. 373; 18 Com. Cas. 65, 302, C. A.; 39 Digest (Repl.) 703, *1943* . . 35, 73, 193, 194
Windschuegl *v.* A. Pickering & Co., Ltd. (1950), 84 Ll. L. Rep. 89 . . 42
Winnipeg Fish Co. *v.* Whitman Fish Co. (1909) 41 S. C. R. 453 . 202
Withers *v.* Reynolds (1831), 2 B. & Ad. 882; 1 L. J. K. B. 30; 39 Digest (Repl.) 517, *588* 190
Wolf & Sons *v.* Carr Parker & Co. (1915), 31 T. L. R. 407, C. A.; 2 Digest (Repl.) 241, *439* 90
Wong *v.* Beaumont Property Trust, Ltd., [1964] 2 All E. R. 119; [1965] 1 Q. B. 173; [1964] 2 W. L. R. 1325; 108 Sol. Jo. 237, C. A.; Digest Cont. Vol. B p. 230, *241a* 19

PAGE
Wood *v.* Baxter (1883), 49 L. T. 45; 39 Digest (Repl.) 672, *1705* . 180
Wood *v.* Rowcliffe (1846), 6 Hare, 183; 1 Digest (Repl.) 385, *502* . 291
Wood *v.* Smith (1829), 5 Man. & Ry. K. B. 124; 4 Car. & P. 45; 8 L. J. O. S. K. B. 50; 39 Digest (Repl.) 515, *562* . . . 363
Wood *v.* Tassell (1844), 6 Q. B. 234; 39 Digest (Repl.) 682, *1767* . 184
Wood Components of London *v.* James Webster, Ltd., [1959] 2 Lloyd's Rep. 200 365
Woodhouse A. C. Israel Cocoa, Ltd. S. A. *v.* Nigerian Produce Marketing Co., Ltd., [1971] 1 All E. R. 665; [1971] 1 W. L. R. 272; 15 Sol. Jo. 56; (1970), Times, November 25th, C. A. 23
Woodland *v.* Fuller (1840), 11 A. & E. 859; [1835–42] All E. R. Rep. 343; 3 Per. & Dav. 570; 9 L. J. Q. B. 181; 4 Jur, 743; 21 Digest (Repl.) 527, *272* 179
Woodley *v.* Coventry (1863), 32 L. J. Ex. 185; 2 H. & C. 164; 8 L. T. 249; 3 Digest (Repl.) 110, *324* 221
Woods *v.* Russell (1822), 5 B. & Ald. 942; [1814–23] All E. R. Rep. 274; 1 D. & R. 587; 39 Digest (Repl.) 613, *1262* 151
Woodward, *Re, Ex parte* Huggins (1886), 54 L. T. 683; 3 Morr. 75; 2 Digest (Repl.) 340, *294* 77
Woodward & Co. *v.* Wolfe, [1936] 3 All E. R. 529; 155 L. T. 619; 53 T. L. R. 87; 80 Sol. Jo. 976; 17 Digest (Reissue) 59, *647* . . 103
Woolfe *v.* Horne (1877), 2 Q. B. D. 355; 46 L. J. Q. B. 534; 36 L. T. 705; 41 J. P. 501; 25 W. R. 728; C. A.; 39 Digest (Repl.) 524, *627*: 179, 358
Worcester Works Finance, Ltd. *v.* Cooden Engineering Co., Ltd., [1971] 3 All E. R. 708; [1972] 1 Q. B. 210; [1971] 3 W. L. R. 661; 115 Sol. Jo. 605, C. A.; 30 Digest (Reissue) 268, *752* . 297, 305, 306, 307
Workman Clark & Co. *v.* Lloyd Brazileño, [1908] 1 K. B. 968, 979, C. A.; 77 L. J. K. B. 953; 99 L. T. 477; 24 T. L. R. 458; 11 Asp. M. L. C. 126, C. A.; 39 Digest (Repl.) 782, *2561* 189, 229
World Beauty, The, Owners of *Andros Springs v.* Owners of *World Beauty*, [1969] 3 All E. R. 158; [1970] P. 144; [1969] 3 W. L. R. 110; 113 Sol. Jo. 363; [1969] 1 Lloyd's Rep. 350, C. A.; Digest Cont. Vol. C p. 890, *7304a* 251
Worsdell, *Re, Ex parte* Barrow (1877), 6 Ch. D. 783; 46 L. J. Bcy. 71; 36 L. T. 325; 25 W. R. 466; 3 Asp. M. L. C. 387; 39 Digest (Repl.) 755, *2339* 215
Wren *v.* Holt, [1903] 1 K. B. 610; 72 L. J. K. B. 340; 88 L. T. 282; 67 J. P. 191; 51 W. R. 435; 19 T. L. R. 292, C. A.; 39 Digest (Repl.) 555, *851* 125, 271
Wrightson *v.* McArthur and Hutchisons, [1921] 1 K. B. 807; [1921] All E. R. Rep. 261; 90 L. J. K. B. 842; 125 L. T. 383; 37 T. L. R. 575; 7 Digest (Repl.) 22, *101* 184
Wylde *v.* Legge (1901), 84 L. T. 121, D. C.; 3 Digest (Repl.) 659, 8 . 309
Wyllie *v.* Povah (1907), 23 T. L. R. 687; 12 Com. Cas. 317; 39 Digest (Repl.) 675, *1723* 81, 357

Xenos *v.* Wickham (1866), L. R. 2 H. L. 296; 36 L. J. C. P. 113; 16 L. T. 800; 16 W. R. 38; 2 Mar. L. C. 537, H. L.; 17 Digest (Reissue) 251, *153* 151

Yangtsze Ins. Association *v.* Lukmanjee, [1918] A. C. 585; 23 Com. Cas. 299; 87 L. J. C. P. 111; 34 T. L. R. 320; 29 Digest (Repl.) 109, *548*: 40, 358
Yates *v.* Pym (1816), 6 Taunt. 446; [1914–23] All E. R. Rep. 784; *sub nom.* Yeats *v.* Pim, 2 Marsh 141; 39 Digest (Repl.) 573, *988* . 365
Yeoman Credit, Ltd. *v.* Apps, [1961] 2 All E. R. 281; [1962] 2 Q. B. 508; 105 Sol. Jo. 567, C. A.; Digest Cont. Vol. A 648, *43a* . . 122, 362
Yin *v.* San. *See* Chai Sau Yin *v.* Liew Kwee San.

PAGE

York Glass Co., Ltd. *v.* Jubb (1926), 134 L. T. 36; [1925] All E. R. Rep. 285; 42 T. L. R. 1 C. A.; 33 Digest (Repl.) 590, *59* 86
Yorkshire Railway Wagon Co. *v.* Maclure (1882), 21 Ch. D. 309; 51 L. J. Ch. 857; 47 L. T. 290; 30 W. R. 761, C. A.; 26 Digest (Repl.) 17, *55* 76
Young *v.* Matthews (1866), L. R. 2 C. P. 127; 36 L. J. C. P. 61; 15 L. T. 182; 39 Digest (Repl.) 618, *1313* 146, 147
Young (Thomas) & Sons, Ltd. *v.* Hobson & Partners (1949), 65 T. L. R. 365, C. A.; 39 Digest (Repl.) 703, *1944* 192, 193, 369
Young and Marten, Ltd. *v.* McManus Childs, Ltd., [1968] 2 All E. R. 1169; [1969] 1 A. C. 454; [1968] 3 W. L. R. 630; 112 Sol. Jo. 744; 67 L. G. R. 1, H. L.; Digest Cont. Vol. C p. 33, *285a* . . 77, 122, 134
Youngmann *v.* Briesemann (1893), 67 L. T. 642, 644, C. A. . . . 213
Yuill *v.* Robson, [1908] 1 K. B. 270; 77 L. J. K. B. 259; 98 L. T. 364; 13 Com. Cas. 166; 24 T. L. R. 180, C. A.; 29 Digest (Repl.) 10, *474* . 40

Zagury *v.* Furnell (1809), 2 Camp. 239; 39 Digest (Repl.) 621, *1338* . 147
Zien *v.* Field (1964), 41 D. L. R. (2d.) 394; 43 W. W. R. 577; reversed *sub nom.* Field *v.* Zien, [1963] S. C. R. 632; 46 W. W. R. 150; 12 Digest (Reissue) 522, **1889* . . . .

SECTION A

RELATIONSHIP OF SALE AND CONTRACT

Relation of sale to contract generally

The contract of sale is governed in part by principles peculiar to itself, and in part by principles common to contracts generally. The Sale of Goods Act, 1893, except incidentally, deals only with the first-mentioned principles. The principles of law which govern contracts generally are outside its scope, but are expressly saved in their application to contracts of sale by s. 61, *post,* unless they are inconsistent with the express provisions of the Act. Further, s. 55, *post,* permits parties to contract out of most of the provisions of the Act, expressly or by necessary implication. Since 1973, there have been stringent restrictions on the capacity of the parties to contract out of the provisions of ss. 12–15 of the Act—see s. 55 (3)–(11), *post.* Apart from these restrictions, however, the Act in no way prevents the parties from making what bargain they please (*a*), and by giving legal effect in particular to commercial usage, the development of the law merchant is not impeded. Nor does the Act touch or alter the ordinary law relating to construction and interpretation of written agreements or the admission of parol evidence in relation thereto. Finally, the Act does not touch the tortious liability, in conversion or negligence, of the seller of goods.

If the law of contract were codified, the Sale of Goods Act would form a single chapter in the code.

Intention to create legal relations

An agreement to sell may be made, by express and appropriate terms, binding in honour only, and not in law, but in business dealings a heavy onus of proof lies on the person who alleges that there was no intention to enter into legal relations (*b*). Moreover, specific orders given and accepted under such a general agreement may be enforceable (*c*).

Consideration

Section 1 of the Sale of Goods Act defines a contract of sale as a contract whereby the seller transfers or agrees to transfer the property in goods to the buyer for a money consideration called the price. " Property " and " goods " are defined by s. 62 (1) of the Sale of Goods Act. The goods may be either existing goods owned or

---

(*a*) Restrictions are imposed by other legislation: see for example the statutes relating to price control and resale price maintenance, p. 51, *post,* and the consumer protection legislation, p. 54, *post.*

(*b*) *Edwards* v. *Skyways, Ltd.*, [1964] 1 All E. R. 494; [1964] 1 W. L. R. 349.

(*c*) *Rose and Frank* v. *Crompton Brothers,* [1925] A. C. 445.

possessed by the seller, or goods to be manufactured or acquired by the seller after the making of the contract—see s. 5 of the Sale of Goods Act.

The price is described as " a money consideration ", but may by agreement be paid partly with other goods (which are regarded for this purpose as the equivalent of money): see further, p. 79, *post.*

The distinction between contracts for the sale of goods and various similar contracts are examined, in the comment to s. 1 of the Sale of Goods Act, *post.*

The rules for determining the price where none is agreed by the parties, or where they agree to leave it to the valuation of a third party, are considered in ss. 8 and 9 of the Sale of Goods Act.

Statutory price control.—Recently, statutory restrictions have been imposed on the prices which may be charged on certain goods sold in the course of a business. Currently, these restrictions are principally to be found in the Counter-Inflation Act, 1973 (*d*). Where goods are unlawfully sold at a price in excess of the maximum permitted under that Act, the contract is not invalid, but the person paying the price may recover the excess unless he is himself liable to punishment in connection with the offence (*e*).

Alteration of customs or excise duties and of value added tax.—By s. 10 of the Finance Act, 1901 (*post,* Appendix I), unless otherwise agreed, where after the making of the contract and before the delivery of the goods, a new or increased customs or excise duty is imposed, the amount may be added to the price; and, conversely, where the duty is repealed or lowered, the amount may be deducted from the price. In case of disagreement as to the amount it may be fixed by the Commissioners of Customs and Excise. A similar provision is made by s. 42 of the Finance Act, 1972 (*post,* Appendix I) in respect of variations in the rate of value added tax between the date of the contract and the date on which the goods are supplied.

Offer and acceptance

The rules of offer and acceptance apply to contracts of sale in the same way as to contracts generally. The contract of sale must be founded on mutual consent (*f*). The offer and acceptance may

(*d*) See also Prices Act, 1974, and p. 51, *post.*

(*e*) S. I. 1973, No. 660, para. 3.

(*f*) Thus the supply of drugs to a patient by the Ministry of Health or by a chemist on payment of a two shilling charge in accordance with a statutory duty is not a sale of goods: *Pfizer Corporation* v. *Ministry of Health,* [1965] 1 All E. R. 450; [1965] A. C. 512; *Appleby* v. *Sleep,* [1968] 2 All E. R. 265; [1968] 1 W. L. R. 948, D. C. See also *Thompson* v. *Ball* (1948), 92 Sol. Jo. 272, D. C. (licensee reluctantly served whisky to food and drugs sampling officer after first refusing to do so because he feared he might otherwise commit the offence of obstructing the officer: no sale). The parties must be agreed as to the terms of the contract including the terms upon which the price is to be paid: *cf. Unity Finance* v. *Hammond* (1965), 109 Sol. Jo. 70 (a hire purchase agreement where the parties were not *ad idem* as to the size of the weekly payments).

be express, or may be implied in whole or in part from the conduct of the parties.

Frequently, goods are bought and sold without any agreement being made as to the price of the goods. In such a case, s. 8 (2) of the Sale of Goods Act provides that the buyer must pay a reasonable price. The circumstances in which s. 8 (2) applies, or in which apart from s. 8 (2) the court will imply a term that the goods should be sold at a reasonable price are discussed in the comment to s. 8 (2), *post*.

Mistake

Section 61 (2) of the Sale of Goods Act preserves the rules of the common law relating to the effect of mistake on a contract for the sale of goods, save insofar as they are inconsistent with the express provisions of the Act (*g*).

The law of mistake is of little practical importance in relation to the sale of goods, but questions occasionally arise where a fraudulent buyer has disguised his identity, in relation to s. 23 of the Sale of Goods Act, for if by reason of the fraud the contract is void for mistake, the fraudulent buyer acquires no title to the goods, and therefore can give no title even to one who buys from him in good faith. See further the comment to s. 23, *post*.

Questions may also arise whether a contract for the sale of specific goods is void or may be avoided because at the time it was made the goods did not exist and had never existed, or perhaps were so defective that the seller could not perform his obligations under the contract by delivering them. See further the comment to s. 6, *post*.

Capacity

Section 2 of the Sale of Goods Act specifically provides that capacity to buy and sell is regulated by the general law concerning capacity to contract, and to transfer and acquire property, subject to a proviso relating to the sale of necessaries. The general law concerning capacity to contract and to transfer and acquire property is discussed in the comment to s. 2 of the Sale of Goods Act, *post*.

Formalities

There are no special formalities required for most contracts for the sale of goods, except where the sale is made on credit terms to an individual and the credit is less than £5000. In such cases, except insofar as the agreement is exempted by regulations made under s. 16 of the Consumer Credit Act, 1974, the seller must comply with the

---

(*g*) The saving may relate to s. 6 of the Sale of Goods Act, but see the comment to that section, *post*.

requirements of that Act (*h*). See further s. 3 of the Sale of Goods Act, *post.*

If, however, the parties have put a contract of sale into writing, the ordinary rules of evidence apply. The general rule is that parol evidence is inadmissible to contradict, add to, or vary the terms of the written instrument; but such evidence is admissible to explain it, and, in explaining it, to annex incidents thereto (*i*).

The general rule is subject to certain exceptions. Thus (1) Parol evidence is always admissible to avoid a contract, as for instance by showing that it was induced by fraud (*j*), or founded on such mistake as to prevent what appears to be a contract from ever having been a contract at all (*k*); (2) Parol evidence is admissible to show that a written agreement which purports to be unconditional was in fact executed with the intention that it should only take effect as a contract on the performance of a condition precedent (*l*); (3) A written agreement may in general (*m*) be modified by a subsequent oral agreement; (4) A written agreement can always be rescinded or abrogated by a subsequent oral agreement (*n*); (5) Parol evidence may be admitted to show that a written agreement is subject to a collateral oral warranty, provided that this is not inconsistent with the writing (*o*).

(*h*) Under Consumer Credit Act, 1974, Sched. 3, the relevant provisions will come into force on dates to be appointed. As yet none of the provisions have been brought into force. The Act also repeals the Hire-Purchase Act, 1965 from a date to be appointed under s. 192 (4) of that Act. As yet no date has been appointed, and credit sale and conditional sale agreements coming within the Hire-Purchase Act, 1965 must comply with the formalities required by ss. 5–10 of that Act.

(*i*) *Phipson on Evidence* (11th ed.), Chs. 43–44; *Cross on Evidence* (4th ed.), pp. 532 *et seq.*; *Bristol Tramways Co.* v. *Fiat Motors, Ltd.*, [1910] 2 K. B. 831, at p. 838, *per* FARWELL, L. J. As to incidents annexed by usage, see *Syers* v. *Jonas* (1848), 2 Exch. 111 (usage to sell by sample); *Brown* v. *Byrne* (1854), 3 E. & B. 703 (usage to deduct discount); *Field* v. *Lelean* (1861), 6 H. & N. 617, Ex. Ch. (usage not to deliver till time of payment arrives); *Produce Brokers Co.* v. *Olympia Oil and Cake Co.* [1917], 1 K. B. 320, C. A. (appropriation by original seller passed on the sub-buyer after loss); but see as to the limitations of usage, p. 17, *post.* See further, notes to *Wigglesworth* v. *Dallison*, 1 Smith L. C. (13th ed.) 597.

(*j*) *Chanter* v. *Hopkins* (1838), 4 M. & W. 399, at p. 406; *Kennedy* v. *Panama Co.* (1867), L. R. 2 Q. B. 580; *Dobel* v. *Stevens* (1825), 3. L. J. (O. S.) K. B. 89.

(*k*) P. 3, *ante.*

(*l*) *Moore* v. *Campbell* (1854), 10 Exch. 323; *Murray* v. *Earl of Stair* (1823), 2 B. & C. 82; *Latch* v. *Wedlake* (1840), 11 Ad. & El. 959; *Clever* v. *Kirkman* (1875), 33 L. T. 672; *Lindley* v. *Lacey* (1864), 17 C. B., N. S. 578; *Guardhouse* v. *Blackburn* (1866), L. R. 1 P. & D. 109; *Pym* v. *Campbell* (1856), 25 L. J. Q. B. 277.

(*m*) But not if the agreement is one required by law to be in writing: *Goss* v. *Lord Nugent* (1833), 2. L. J. K. B. 127.

(*n*) Even if it is one required by law to be in writing: *Morris* v. *Baron & Co.*, [1918] A. C. 1, H. L.

(*o*) *Morgan* v. *Griffith* (1851), L. R. 6 Ex. 70; *De Lassalle* v. *Guildford*, [1901] 2 K. B. 215.

Moreover, the general rule only applies where the writing is intended to constitute the contract itself, and not merely a note or memorandum of it (*p*).

An oral express term will override a written exemption clause (*q*), if made by the seller or by an agent with actual or ostensible authority to agree it (*r*).

A contract of sale, like other contracts, may be made retrospective in effect, so as to activate earlier dealings between the parties (*s*).

Representations

In the course of the formation of a contract of sale, representations may be made by either party which may have the following consequences:—

(1) The representation may be a mere expression of opinion, or a mere commendation by the seller of his wares—what is often called a " puff ". Such a representation will give rise to no civil liability (*a*). An expression of opinion or a puff may however amount to or be treated as a trade description for the purposes of the Trade Descriptions Act, 1968, so that if the opinion is wrong or the puff is unjustified, the representants may be guilty of the offence of applying a false trade description under s. 1 of that Act (*b*).

(2) The representation may or may not become incorporated into the contract. Whether or not it does will depend upon the intention of the parties (*c*). If made in the course of negotiations for

---

(*p*) *Allen* v. *Pink* (1838) 4 M. & W. 140.

(*q*) *Couchman* v. *Hill*, [1947] 1 All E. R. 103, C. A.; [1947] K. B. 554, applied and approved in *Harling* v. *Eddy*, [1951] 2 All E. R. 212, C. A.; [1951], 2 K. B. 739.

(*r*) An agent will not have such ostensible authority if the buyer is put on notice by a written document that he has no such authority before the contract is made or the terms agreed: *Jacobs* v. *Morris*, [1902] 1 Ch. 816, C. A.; *Overbrooke Estates, Ltd.* v. *Glencombe Properties, Ltd.*, [1974] 3 All E. R. 511; [1974] 1 W. L. R. 1335, explaining *Mendelssohn* v. *Normand, Ltd.*, [1969] 2 All E. R. 1215; [1970] 1 Q. B. 177, C. A.

(*s*) *Trollope and Colls, Ltd.* v. *Atomic Power Constructions*, [1962] 3 All E. R. 1035, 1040; [1963] 1 W. L. R. 333, 339-340.

(*a*) *Scott* v. *Hanson* (1829), 1 Russ. & M. 128; *Dimmock* v. *Hallett* (1866), 36 L. J. Ch. 146; *Magennis* v. *Fallon* (1828), 2 Mol. 561 at p. 588; Spencer Bower & Turner: *Actionable Misrepresentation* (3rd ed.), paras. 29–35, 49–51.

(*b*) See further p. 56, *post*.

(*c*) *Hopkins* v. *Tanqueray* (1854), 15 C. B. 130; *Bannerman* v. *White* (1861), 10 C. B. N. S. 844; *Routledge* v. *McKay*, [1954] 1 All E. R. 855; [1954] 1 W. L. R. 615, C. A.; *De Lassalle* v. *Guildford*, [1901] 2 K. B. 215, C. A.; criticised as incomplete in *Heilbutt* v. *Buckleton*, [1913] A. C. 30; *Couchman* v. *Hill*, [1947] 1 All E. R. 103; [1947] K. B. 554, C. A.; *Birch* v. *Paramount Estates Ltd.*, cited in *Oscar Chess* v. *Williams*, [1957] 1 All E. R. 325; [1957] 1 W. L. R. 370, C. A.; *Dick Bentley Productions, Ltd.* v. *Harold Smith* (*Motors*), *Ltd.*, [1965] 2 All E. R. 65; [1965] 1 W. L. R. 623, C. A. A representation which is not incorporated in the contract is sometimes (and confusingly) called a " mere " representation.

a contract for the purpose of inducing the other party to act on it, and it actually induces him to act on it, it is *prima facie* a term of the contract (*d*).

(a) If it is a condition, the other party will be entitled (*e*) to rescind if it is broken ;

(b) If it is only a warranty, the other party will in the event of a breach be entitled only to damages for the breach of warranty.

(c) If it is neither a condition nor a warranty, but an " intermediate " term, the rights of the other party upon its breach will depend upon all the circumstances (*f*).

(d) The right, if any, of the other party to rescind for misrepresentation is not lost just because the misrepresentation has become a term of the contract (*g*).

(3) Whether or not incorporated into the contract, the representation may be made before or at the time of contracting about some existing fact or past event, and be one of the causes inducing the making of the contract (*h*). In that event,

(a) if the representation prove to be false in a material respect (regardless of whether or not made fraudulently), equity will *prima facie* (*i*) allow the party who has been misled to rescind

---

(*d*) It is for the maker of the representation to rebut this inference by showing that it really was innocent, in that he was in fact innocent of fault in making it, and that it would not be reasonable in the circumstances for him to be bound by it: *Dick Bentley Productions* v. *Harold Smith, supra, per* Lord DENNING, M.R. As to the classification of terms, see p. 8, *post*.

(*e*) Subject of course to s. 11 (1) (a) and s. 11 (1) (c) of the Sale of Goods Act, *post*: see generally *Wallis* v. *Pratt & Haynes*, [1911] A. C. 394.

(*f*) See p. 9, *post*.

(*g*) Misrepresentation Act, 1967, s. 1, *post*, Appendix I.

(*h*) As to inducement and materiality see Spencer Bower and Turner: *Actionable Misrepresentation*, Ch. VI.

(*i*) But subject to the limitations imposed on the equitable remedy : thus there will be no rescission if the plaintiff has delayed too long (*Leaf* v. *International Galleries*, [1950] 1 All E. R. 693 ; [1950] 2 K. B. 86, C. A.) ; or has otherwise affirmed the contract (*Long* v. *Lloyd*, [1958] 2 All E. R. 402 ; [1958] 1 W. L. R. 753; and see the cases on affirmation for the purposes of s. 18, rule 4 and s. 35 of the Sale of Goods Act, *post*); or where *restitutio in integrum* is no longer possible (*Glasgow South Western Rail. Co.* v. *Boyd and Forrest*, [1915] A. C. 526; *cf. Freeman* v. *Consolidated Motors, Ltd.* (1968), 69 D. L. R. (2d) 581 (trade-in car sold by motor dealer before buyer sought to rescind contract of which trade-in formed part)); or where third party rights would be prejudiced (*Oakes* v. *Turquand* (1867), L. R. 2 H. L. 325); as to the onus of proof, see *Steedman* v. *Frigidaire Corpn.*, [1933] 1 D. L. R. 161, P. C.; *Zien* v. *Field* (1964), 41 D. L. R. (2d) 394. Prior to 1967 it was doubtful whether a party could rescind for innocent (as opposed to fraudulent) misrepresentation once the contract had been executed. It is now settled by s. 1 of the Misrepresentation Act, 1967, *post*, Appendix I, that the right to rescind will not be lost just because the contract has been performed. See generally Spencer Bower & Turner, *op. cit.*, chaps. XIII–XIV.

the contract (*k*); but where the misrepresentation was not fraudulent, the court or an arbitrator has a discretion to award damages in lieu of rescission if of the opinion that it would be equitable to do so, having regard to the nature of the misrepresentation and the loss that would be caused by it if the contract were upheld, as well as to the loss that rescission would cause to the other party (*l*);

(b) if the representation was made fraudulently, the party deceived will be entitled to damages for the tort of deceit (*m*), and also to rescind the contract (*n*);

(c) if the representation was not made fraudulently, but was made in such circumstances that had it been fraudulent, the person making it would have been liable to damages, then that person will be so liable unless he proves that he had reasonable grounds to believe and did believe up to the time the contract was made that the facts represented were true (*o*);

(d) the representation may create an estoppel against its maker(*p*);

(e) whether or not the representation was made fraudulently the representor may be guilty of an offence under the Trade Descriptions Act, 1968 if the misrepresentation was made in the course of a business (*q*).

(*k*) In theory, therefore, a party faced with a breach of condition has two distinct and independent rights to rescind: one for breach of condition and the other for material misrepresentation. In practice, however, it is difficult to imagine a case where the latter remedy would avail him if the former did not: as DENNING, L.J., pointed out in *Leaf* v. *International Galleries*, [1950] 1 All E. R. 693, at p. 695; [1950] 2 K. B. 86, at pp. 90–91, "it is to be remembered that an innocent misrepresentation is much less potent than a breach of condition . . . if a claim to reject for breach of condition is barred, it seems to me *a fortiori* that a claim to rescission on the ground of innocent misrepresentation is also barred." It is possible that the effect of s. 3 of the Misrepresentation Act, 1967, *post*, Appendix I, is that the right to rescind for breach of condition may be excluded, while the right to rescind for misrepresentation is not. See further the comment to that section, *post*.

(*l*) Misrepresentation Act, 1967, s. 2 (2), *post*, Appendix I.

(*m*) *Levy* v. *Langridge* (1838), 4 M. & W. 337, Ex. Ch.; *Derry* v. *Peek* (1889), 14 App. Cas. 337, and see the notes to s. 62 (2) of the Sale of Goods Act, *post*. Damages for deceit should compensate the plaintiff for all the actual damage directly flowing from the fraudulent inducement and are not limited to such as may reasonably be supposed to have been in the contemplation of the parties: *Doyle* v. *Olby*, [1969] 2 All E. R. 119; [1969] 2 Q. B. 158, C. A.

(*n*) But the goods must be rejected within a reasonable time after the fraud is discovered: *Kwei Tek Chao* v. *British Traders*, [1954] 1 All E. R. 779; [1954] 2 Q. B. 459.

(*o*) Misrepresentation Act, 1967, s. 2 (1), *post*, Appendix I.

(*p*) *Maddison* v. *Alderson* (1883), 8 App. Cas. 467, at p. 473; *cf.* the comment to s. 21 of the Sale of Goods Act, *post*.

(*q*) See p. 56, *post*.

Although the decision to rescind should normally be communicated to the other contracting party (*r*), this may not be essential at any rate where that other party is a rogue who has deliberately absconded. In such a case it is sufficient if the rescinding party takes all possible steps to recover the goods (*s*). It may be that this is also the position in other cases where the communication of the intention to rescind cannot readily be effected (*t*).

Terms of the contract

Conditions and warranties contrasted.—The Sale of Goods Act draws throughout a distinction between the terms " condition " and " warranty," and defines the circumstances under which a condition may be treated as a warranty.

This distinction had often been insisted on, but seldom observed by judges and text-writers before the Act. As used in the Act, " condition " is the equivalent of the old term " dependent covenant," while "warranty " is equivalent to the old term " independent covenant".

A warranty is defined by s. 62 (1) of the Sale of Goods Act (as regards England and Ireland, but not Scotland) as an agreement with reference to goods which are the subject of a contract of sale, but collateral to the main purpose of such contract, the breach of which gives rise to a claim for damages, but not to a right to reject the goods and treat the contract as repudiated.

No express definition of " condition " is given by the Act, but it is defined by inference as a term in the contract the breach of which gives rise to a right to repudiate the contract. Whether a given stipulation is a condition or a warranty is a question of construction and intention, and not of terminology: see s. 11 (1) of the Sale of Goods Act, *post.*

Relation to general law of contract.—Sections 10 to 15 of the Sale of Goods Act deal with conditions and warranties peculiar to the law of sale. But the Act must be regarded as a single chapter in the general law of contract, and it therefore does not attempt to deal with the law of representations, conditions, and warranties to the extent that they are governed by considerations common to the whole field of contract. In so far as sale is regulated by the general law of contract, the rules which apply are saved by s. 61 (2), *post,* except insofar as they are inconsistent with the express provisions of the Act.

(*r*) *Scarf* v. *Jardine* (1882), 7 App. Cas. 345, at p. 361; *Kerr on Fraud and Mistake* (7th ed.), p. 530; Spencer Bower & Turner, *op. cit.*, para. 239.

(*s*) *Car and Universal Finance Co., Ltd.* v. *Caldwell,* [1964] 1 All E. R. 290; [1965] 1 Q. B. 525, C. A.; *Newtons of Wembley, Ltd.* v. *Williams,* [1964] 3 All E. R. 532; [1965] 1 Q. B. 560, C. A. But *cf. MacLeod* v. *Kerr,* 1965 S. L. T. 358, where the Court of Session took a different view; see also 29 M. L. R. 442.

(*t*) *Car and Universal Finance Co., Ltd.* v. *Caldwell, supra.*

In other classes of contract, many terms may be neither conditions nor warranties as defined above, but "intermediate terms", and the question whether a breach by one party of such an "intermediate" term entitles the other party to treat the contract as repudiated turns upon the seriousness of the breach in all the circumstances (*u*). In such other contracts, the nature of the term is determined from the contract and the surrounding circumstances.

In contracts for the sale of goods, where terms are implied by ss. 12–15 of the Act, they must be either conditions or warranties although similar terms in other contracts may be of the intermediate kind (*a*). Indeed, the whole Act, and particularly s. 11 (1) (*b*), appears to have been drafted on the assumption that a term has either to be a condition or a warranty, but there is nothing in the Act so inconsistent with the existence of such intermediate terms as to prevent the ordinary rules of common law from applying in respect of them except in the case of terms implied by ss. 12–15 of the Act (*b*).

Express agreement.—Sale is a consensual contract, and the parties may alter at will the obligations which the law implies from the general nature of the contract. BLACKBURN, J., discussing the correlative obligations of payment and delivery, says

> " There is no rule of law to prevent the parties from making any bargain they please " (*c*) ;

and BRETT, L.J., says,

> " Merchants are not bound to make their contracts according to any rule of law " (*d*).

Interpretation and construction.—There is no canon of construction peculiar to contracts of sale.

> " The rule of construction applicable in general to all written contracts is, that they are to be construed according to the real intention of the parties, to be collected from the language they have used ;

(*u*) See the judgment of DIPLOCK, L.J., in *Hong Kong Fir Co. Ltd.* v. *Kawasaki*, [1962] 1 All E. R. 474, at pp. 485–9 ; [1962] 2 Q. B. 26, at pp. 65–73. *Schuler A G* v. *Wickman Machine Tool Sales, Ltd.*, [1973] 2 All E. R. 39; [1974] A. C. 235, H. L.; *Daulatram Rameshwarlall* v. *European Grain & Shipping, Ltd.*, [1971] 1 Lloyd's Rep. 368.

(*a*) Cf. *Harbutt's Plasticine, Ltd.* v. *Wayne Tank and Pump Co., Ltd.*, [1970] 1 All E. R. 225; [1970] 1 Q. B. 447, C. A.

(*b*) *Cehave N.V.* v. *Bremer Handelgesellschaft m.b.h.* (1975), *Times*, 22nd July, C.A.; *Daulatram Rameshwarlall* v. *European & Shipping, Ltd.*, *supra.*

(*c*) *Calcutta Co.* v. *De Mattos* (1863), 32 L. J. Q. B. 322, at p. 328; see the passage cited at length, *post*, Appendix II, Note C.

(*d*) *Honck* v. *Muller* (1881), 7 Q. B. D. 92, at p. 103, C. A. For an extreme application of this principle, see *Lancaster* v. *Turner & Co.*, [1924] 2 K. B. 222 (effect of " invoicing back clause " in case of seller's default).

that effect is to be given, if possible, to every word used, and that every word is to be interpreted according to its natural and ordinary meaning, unless such construction would be contrary to the manifest intention of the parties, or would necessarily lead to some contradiction or absurdity. But this rule, though applicable to contracts in general, must be received with some qualification, when the contract or a portion of the contract in question consists of an incomplete sentence, ambiguous in its terms, and upon which a literal construction of every word would either be impracticable or would leave the contract indeterminable and uncertain. And such is the case with the contract in question, which I think is to be construed according to what we can collect to have been the substantial intention of the parties, applying our common sense, and such knowledge as we may possess, to the language in which they have expressed themselves " (*e*).

Section 55 (2) of the Sale of Goods Act, *post* (*f*), expressly provides that an express condition or warranty does not negative a condition or warranty implied by the Sale of Goods Act unless inconsistent with it. But this probably does no more than restate the position at common law (*g*).

Exemption clauses.—Subject to the principles set out below in relation to fundamental breach, at common law a party to a contract may by a suitably worded exemption clause (*h*) exclude or limit his liability for breach of any term of the contract. Formerly, this was also the case in a contract for the sale of goods, and before 1973 s. 55 of the Act, *post*, provided that any implied condition or warranty may be negatived or varied by express agreement, course of dealing or usage. That section has now been amended by s. 4 of the Supply of Goods (Implied Terms) Act, 1973 so as severely to restrict attempts by a seller to limit or exclude his liability in respect of the conditions and warranties implied by ss. 12–15 of the Act (*i*). The restrictions apply to all contracts for the sale of goods except contracts for the international sale of goods, which are excluded by s. 61 (6) of the Act, *post* (*k*).

(*e*) *Per* KELLY, C.B. in *Coddington* v. *Paleologo* (1867), L. R. 2 Exch. 193, at p. 200 ; *cf. Honck* v. *Muller* (1881), 7 Q. B. D. 92, at p. 103, *per* Lord ESHER ; *Champanhac & Co., Ltd.* v. *Waller & Co., Ltd.*, [1948] 2 All E. R. 724.

(*f*) Formerly s. 14 (4). The present s. 55 (2) was introduced by s. 4 of the Supply of Goods (Implied Terms) Act, 1973.

(*g*) *Mody* v. *Gregson* (1868), L. R. 4 Exch. 49, at p. 53, Ex. Ch. ; approved, *Drummond* v. *Van Ingen* (1887), 12 App. Cas. 284, at p. 294, *per* Lord HERSCHELL. *Cf. Bigge* v. *Parkinson* (1862), 7 H. & N. 955.

(*h*) In several cases the need for clear language in provisions designed to limit or exclude the ordinary rights of the buyer has been emphasised: see, *e.g. Wallis* v. *Pratt* [1911] A. C. 394; *Pollock* v. *McRae*, 1922 S. C. 192, H. L.; *Baldry* v. *Marshall*, [1925] 1 K. B. 260; *Alison & Co.* v. *Wallsend Slipway and Engineering Co.* (1927), 43 T. L. R. 323, C. A.; *Modiano Bros. & Son* v. *Pearson & Co., Ltd.* (1929), 34 Ll. L. Rep. 52; *Webster* v. *Higgin*, [1948] 2 All E. R. 127.

(*i*) See further the comment to s. 55, *post*.

(*k*) For the definition of a contract for the international sale of goods, see s. 62 (1), *post*.

Fundamental breach and exemption clauses.—A fundamental breach of contract is a breach or a number of breaches rendering the purported performance of a contract quite different from that which the contract contemplated (*l*). In determining whether a breach of contract is fundamental, it is necessary to look not only at the breach, but at the events resulting from that breach. A breach which may be easily and cheaply remedied but which could entail serious consequences if not remedied will probably not be fundamental if remedied before any risk of those consequences arises (*m*). But once the other party to the contract has been put in hazard by the breach, then this hazard will be taken into account, and the greater the hazard, the more likely it is that the breach will be held to be fundamental (*n*). Insofar as fundamental breaches may be divided into breaches of fundamental terms and other fundamental breaches (*o*) a breach by the seller of any of the conditions implied by ss. 12–15 of the Sale of Goods Act will presumably be a breach of fundamental term.

But it would seem that not every breach of a fundamental term will amount to a fundamental breach. In *Harbutt's Plasticine, Ltd.* v. *Wayne Tank and Pump Co., Ltd.* (*p*), the defendants designed and erected in the plaintiffs' factory a system for conveying and storing molten wax. This was wholly unsuitable for the purpose for which it was intended because it incorporated plastic pipelines which could not withstand the heat of molten wax. As a result the plaintiffs' factory burnt down. The Court of Appeal held that there had been a fundamental breach of contract, but it would seem that if the defect had been discovered before the accident, the defendants would have been entitled to remedy the defect at their own expense by substituting steel pipelines, although the contract provided for plastic ones, and although the system was unfit for the purpose for which it was intended with plastic pipelines. The breach would not then have been a fundamental one (*q*).

(*l*) *Suisse Atlantique Société D'Armement Maritime S.A.* v. *N.V. Rotterdamsche Kolen Centrale*, [1966] 2 All E. R. 61; [1967] 1 A. C. 361, H. L. *Cf.* the doctrine of *vis major* or frustration, *post*, p. 19.

(*m*) *Harbutt's Plasticine, Ltd.* v. *Wayne Tank and Pump Co., Ltd.*, [1970] 1 All E. R. 225; [1970] 1 Q. B. 447, C. A.

(*n*) *Ibid.*, and see *Farnworth Finance Facilities, Ltd.* v. *Attryde*, [1970] 2 All E. R. 774; [1970] 1 W. L. R. 1053, C. A.; *Kenyon, Son & Craven, Ltd.* v. *Baxter Hoare & Co., Ltd.*, [1971] 1 W. L. R. 519, at p. 530.

(*o*) See *ibid.*, *per* Lord UPJOHN, at [1966] 2 All E. R. 85; [1967] 1 A. C. 421; and *cf. per* Viscount DILHORNE, at [1966] 2 All E. R. 68; [1967] 1 A. C. 393.

(*p*) [1970] 1 All E. R. 225; [1970] 1 Q. B. 447, C. A.

(*q*) [1970] 1 All E. R. at pp. 235, 239; [1970] 1 Q. B. at pp. 466, 471; *cf. Farnworth Finance Facilities, Ltd.* v. *Attryde*, *supra*. See further, p. 9, *ante*. Had the *Harbutt's Plasticine* case been one for the sale of goods, however, the seller would probably not have been entitled to remedy the defect for there would have been a breach of the condition implied by s. 14 (2) of the Sale of Goods Act which would have entitled the buyer to rescind at least once the goods had been tendered: *cf. Jackson* v. *Rotax Motors and Cycle Co.*, [1910] 2 K. B. 937, C. A.

In recent years the view has been put forward that no exemption clause could relieve a party to a contract from liability for a fundamental breach (*r*). But it has now been settled by the House of Lords in *Suisse Atlantique Société D'Armement Maritime S.A.* v. *N.V. Rotterdamsche Kolen Centrale* (*s*) that this view was wrong. To cover a case of fundamental breach, an exemption clause could, it seems, take one of three forms: it might limit liability in the event of such a breach, for example, by limiting the quantum of damages payable or by limiting the time within which a particular remedy would be available to an injured party (*t*); or it might exclude liability; or it might make permissible what would otherwise be a fundamental breach (*u*). Yet, apparently, even where such a clause exists a breach by the protected party which amounts to a repudiation of the contract may be treated by the other party as putting an end to the contract (*a*). Should he elect to affirm the contract, however (*b*), he will continue to be bound by its terms—including the exemption clause—and damages will be limited accordingly (*c*).

(*r*) *E.g., Karsales (Harrow), Ltd.* v. *Wallis*, [1956] 2 All E. R. 866; [1956] 1 W. L. R. 936.

(*s*) [1966] 2 All E. R. 61; [1967] 1 A. C. 361; and see 29 M. L. R. 546.

(*t*) *Suisse Atlantique Société Maritime S.A.* v. *N.V. Rotterdamsche Kolen Centrale*, [1966] 2 All E. R. at p. 92; [1967] 1 A. C. at p. 431; *H. & E. Van der Sterren* v. *Cibernetics (Holdings) Pty., Ltd.* (1970), 44 A. L. J. R. 157 (H.C. of Aus.).

(*u*) *Cf.* the division adopted by DIPLOCK, L.J., in *Hardwick Game Farm* v. *Suffolk Agricultural and Poultry Producers Association, Ltd.*, [1966] 1 All E. R. 309, 346; [1966] 1 W. L. R. 287, 341. The third case is not strictly speaking an exemption clause, but permits the performance of the contract in a way which would otherwise be held not to have been contemplated by the parties. For an unsuccessful attempt to construe a clause in this way, see *Glynn* v. *Margetson & Co.*, [1893] A. C. 351.

(*a*) *Suisse Atlantique Société D'Armement Maritime S.A.* v. *N.V. Rotterdamsche Kolen Centrale*, [1966] 2 All E. R. 61; [1967] 1 A. C. 361, H. L. *Harbutt's Plasticine, Ltd.* v. *Wayne Tank and Pump Co., Ltd.*, [1970] 1 All E. R. 225; [1970] 1 Q. B. 447, C. A.; *Farnworth Finance Facilities, Ltd.* v. *Attryde*, [1920] 2 All E. R. 774; [1970] 1 W. L. R. 1053, C. A. But in considering what amounts to a repudiation it would seem necessary to look at the terms of the exemption clause—see *per* Lord WILBERFORCE at [1966] 2 All E. R. 94; [1967] 1 A. C. 431; *Kenyon, Son & Craven, Ltd.* v. *Baxter Hoare & Co., Ltd.*, [1971] 1 W. L. R. 519, 531; and whilst the wilfulness of a breach is an important factor in deciding whether it is repudiatory (see at [1966] 2 All E. R. 68, 70, 94; [1967] 1 A. C. 394, 397, 434), it would seem to be possible for a party to exempt himself even from wilful breach by a clearly worded exemption clause (see at [1966] 2 All E. R. 79, 92; [1967] 1 A. C. 410, 434); and see also *The Cap Palos*, [1921] P. 458, 471, 472, *per* ATKIN, L. J.; *Cunard Steamship Co.* v. *Buerger*, [1927] A. C. 1, 13, *per* Lord PARMOOR. See further *Chitty on Contracts*, 23rd ed., para. 738; 86 L. Q. R. 513; 1970 C. L. J. 221.

(*b*) *E.g.* by failing to accept the repudiation within a reasonable time.

(*c*) *Suisse Atlantique Société D'Armement Maritime S.A.* v. *N.V. Rotterdamsche Kolen Centrale*, [1966] 2 All E. R. 69, 71; [1967] 1 A. C. 395, 398. *Harbutt's Plasticine, Ltd.* v. *Wayne Tank & Pump Co., Ltd.*, [1970] 1 All E. R. 225; [1970] 1 Q. B. 447, C. A.

But the courts will only interpret exemption clauses as extending to cases of fundamental breach if they are worded in clear and unambiguous terms (*d*). Indeed, even when they are so worded, they must still be construed in conjunction with the contract as a whole (*e*), and their terms may have to be read down if they cannot be applied literally without creating an absurdity or defeating the main object of the contract (*f*).

> " There is a strong, though rebuttable, presumption that in inserting a clause of exclusion or limitation in their contract the parties are not contemplating breaches of fundamental terms and such clauses do not apply to relieve a party from the consequences of such a breach even where the contract continues in force. This result has been achieved by a robust use of a well-known canon of construction, that wide words which taken in isolation would bear one meaning must be so construed as to give business efficacy to the contract and the presumed intention of the parties, upon the footing that both parties are intending to carry out the contract fundamentally " (*g*).

The principles involved in determining whether an exemption clause ought to be read down would appear to be those involved in determining whether any term ought to be implied into a contract (*h*). Thus the language which the parties have used in an exemption clause is not so to be modified unless it is necessary to give effect to what the parties must be understood to have intended (*i*).

In construing a contract with a view to ascertaining what the parties intended, the court may be more inclined to read down a clause in standard printed conditions imposed upon an ignorant buyer, than a clause in clear language agreed between two parties negotiating on an equal footing (*k*).

(*d*) *Suisse Atlantique Société D'Armement Maritime S.A.* v. *N.V. Rotterdamsche Kolen Centrale, supra.*

(*e*) *Ibid.*, [1966] 2 All E. R. at pp. 67, 81; [1967] 1 A. C. at pp. 392, 413–414; applied in *Western Tractors* v. *Dyck* (1969), 7 D. L. R. (3d) 535; *Barker* v. *Inland Truck Sales* (1970), 11 D. L. R. (3d) 469.

(*f*) *Ibid.*, [1966] 2 All E. R. at p. 71; [1967] 1 A. C. at p. 398, per Lord REID; *Harbutt's Plasticine, Ltd.* v. *Wayne Tank and Pump Co., Ltd.*, [1970] 1 All E. R. 225, at p. 235; [1970] 1 Q. B. 447, at p. 467, per Lord DENNING, M.R.; *H. & E. Van der Sterren* v. *Cibernetics (Holdings) Pty., Ltd.* (1970), 44 A. L. J. R. 157 (H. C. of Aus.).

(*g*) *Ibid.*, [1966] 2 All E. R. 89; [1967] 1 A. C. 427, *per* Lord UPJOHN.

(*h*) *Post*, p. 15; and see *R. G. Maclean, Ltd.* v. *Canada Vickers, Ltd.* (1971), 15 D. L. R. (3d) 15.

(*i*) *H. & E. Van der Sterren* v. *Cibernetics (Holdings) Pty., Ltd.* (1970), 44 A. L. J. R. 157 (Sale of goods; Clause limiting time within which buyer could make claim in relation to the goods sold; see p. 308, *post.*

(*k*) *Ibid.*, and see *Thornton* v. *Shoe Lane Parking, Ltd.*, [1971] 1 All E. R. 686; [1971] 2 Q. B. 163, C. A.; but *cf. Kenyon Son & Craven, Ltd.* v. *Baxter Hoare & Co., Ltd.*, [1971] 2 All E. R. 708, at p. 720; [1971] 1 W. L. R. 519, at p. 533.

These principles of construction do not apply to an agreed damage clause inserted for the benefit of both parties (*l*).

In applying these principles, it would seem that while a suitably drafted clause may expressly provide for the delivery of beans instead of peas, or may limit the damage in the event of peas not being delivered, it is still not possible to go so far as to provide that the seller is not obliged to deliver the contract goods at all, or that he shall not be liable for failure to do so.

> " One may safely say that the parties cannot, in a contract, have contemplated that the clause should have so wide an ambit as in effect to deprive one party's stipulations of all force; to do so would be to reduce the contract to a mere declaration of intent " (*m*).

Implied terms.—A distinction must be drawn between the implication of terms

> " based on an intention imputed to the parties from their actual circumstances ",

as discussed below, and that of a term

> " which does not depend on the actual intention of the parties but on a rule of law, such as the terms, warranties or conditions which, if not expressly excluded, the law imports, as for instance under the Sale of Goods Act and the Marine Insurance Act " (*n*).

There are many cases in which the Courts must pronounce upon rights and obligations of parties to a contract which have not been expressly provided for by the parties themselves and are not implied by operation of law. In such cases, the Court is often asked to imply a term into the contract, and it will generally do so, though always with great caution, if it is clear from the nature of the transaction that the parties must have intended such a term to be a part of the contract, and if in all the circumstances it is a reasonable term (though this last consideration alone is not enough) ; but the Court will always be careful not to make for the parties a contract they have not made themselves.

> " The court does not make a contract for the parties. The court will not even improve the contract which the parties have made for themselves, however desirable the improvement might be. The court's function is to interpret and apply the contract which the

---

(*l*) *Suisse Atlantique Société D'Armement Maritime S.A.* v. *N.V. Rotterdamsche Kolen Centrale*, [1966] 2 All E. R. 61; [1967] 1 A. C. 361; and *cf. G. H. Renton & Co., Ltd.* v. *Palmyra Trading Corporation of Panama*, [1956] 3 All E. R. 957; [1957] A. C. 149.

(*m*) *Ibid.*, [1966] 2 All E. R. 92; [1967] 1 A. C. 432, *per* Lord WILBERFORCE. As to the onus of proof where a fundamental breach is alleged, see *Hunt and Winterbotham* (*West of England*), *Ltd.* v. *British Road Services* (*Parcels*), *Ltd.*, [1962] 1 All E. R. 111; [1962] 1 Q. B. 617, C. A., and the cases there cited; *Chitty on Contracts*, 23rd ed., Vol. I, para. 748.

(*n*) *Luxor* v. *Cooper*, [1941] 1 All E. R. 33, at p. 52 ; [1941] A. C. 108, at p. 137, *per* Lord WRIGHT.

parties have made for themselves . . . An unexpressed term can be implied if and only if the court finds that the parties must have intended that term to form part of their contract: it is not enough for the court to find that such a term would have been adopted by the parties as reasonable men if it had been suggested to them: it must have been a term that went without saying, a term necessary to give business efficacy to the contract, a term which, although tacit, formed part of the contract which the parties made for themselves " (*o*).

But the court is not concerned to decide what the parties actually intended. A term will not be implied merely because both parties independently intended such a term to be incorporated in the contract, if in fact it was not referred to by either party and could not be described as "going without saying".

" The judicial task is not to discover the actual intention of each party; it is to decide what each was reasonably entitled to conclude from the attitude of the other " (*p*).

Course of dealing.—The phrase " course of dealing " means:

" that past business between the parties raises an implication as to the terms to be implied into a fresh contract, where no express provision is made on the point at issue " (*q*).

Thus where a tradesman regularly charged a customer interest on debts due for over three years and the customer never objected, the customer was taken to have agreed to pay interest on all such debts with that tradesman (*r*).

The problem whether a term or terms should be implied into a contract as a result of a course of dealing, where no term could have been implied if the contract had been an isolated transaction, has most frequently arisen in recent years where one of the parties has written " Conditions of Contract " but the parties conclude a contract without any express reference to them, for example because they are contained in a document which is given to the other party only after the contract is concluded.

In such cases, the terms will be incorporated only if the parties by their conduct must be taken to have agreed that they should form

(*o*) *Trollope and Colls, Ltd.* v. *North West Metropolitan Regional Hospital Board,* [1973] 2 All E. R. 260, at p. 267; [1973] 1 W. L. R. 601, at p. 609, per Lord PEARSON.

(*p*) *McCutcheon* v. *David MacBrayne,* [1964] 1 All E. R. 430, at p. 433; [1964] 1 W. L. R. 125, at p. 128, *per* Lord REID; *British Crane Hire Corporation* v. *Ipswich Plant Hire Ltd.,* [1974] 1 All E. R. 1059; [1975] Q. B. 303, C. A. For the " officious bystander " text see *Shirlaw* v. *Southern Foundries,* [1939] 2 All E. R. 113, at p. 124; [1939] 2 K. B. 206, at p. 227; affirmed *sub nom. Southern Foundries* v. *Shirlaw,* [1940] 2 All E. R. 445; [1940] A. C. 701; *Luxor* v. *Cooper,* [1941] 1 All E. R. 33, at p. 52; [1941] A. C. 108, at p. 137, *per* Lord WRIGHT.

(*q*) *Pocahontas Fuel Co.* v. *Ambatielos* (1922), 27 Com. Cas. 148, at p. 152, *per* MCCARDIE, J.

(*r*) *Re Marquis of Anglesey, Willmot* v. *Gardner,* [1901] 2 Ch. 548, C. A.

part of the contract (*s*). It is irrelevant that the other party has not read the terms on the documents previously supplied to him, provided that he knew that that the documents contained terms which the party supplying the documents intended should be part of the contract between them, or at least that the party supplying the documents had taken reasonable steps to draw them to his attention (*t*).

But the terms will not normally be incorporated if the contract in question is materially different from the earlier contracts (*u*), nor if the earlier contracts were few and infrequent (*a*), nor if the document containing the terms is not one which the other party would expect to contain terms, and in fact he did not realise that it contained them (*b*), particularly if the party seeking to impose the terms ought to have realised that they would not or might not have come to the attention of the representative of the other party responsible for entering into the contracts (*c*).

Even where no course of dealing can be established, however, the standard terms of one party may be incorporated into a contract if from the conduct of the parties a common understanding can be shown to have existed between them that the contract should be on the usual conditions of one of them. Such an understanding may exist between parties of equal bargaining power in the same trade where the other party knows, and it is common knowledge in the trade, that such conditions are always imposed even though they are sometimes sent to the other party only after the agreement is made (*d*).

(*s*) *Hardwick Game Farm* v. *Suffolk Agricultural Association Ltd.*, [1966] 1 All E. R. 309, at pp. 322, 344–345; [1966] 1 W. L. R. 287, at pp. 308, 338–340; *sub nom. Henry Kendall & Sons* v. *William Lillico & Sons, Ltd.*, [1968] 2 All E. R. 444, at pp. 474–475, 481, 496; [1969] 2 A. C. 31, at pp. 104–105, 112–113, 130, H. L. See also as to the effect of terms contained in an invoice or confirmation note, 33 M. L. R. 518; Spencer Bower & Turner, *Estoppel by Representation* (2nd ed.), pp. 95, *et seq.*; and see the judgment of Lord WRIGHT in *Hillas & Co.* v. *Arcos Ltd.* (1932), 38 Com. Cas. 23.

(*t*) *Henry Kendall & Sons* v. *William Lillico & Sons, Ltd., supra*; *Mendelssohn* v. *Normand*, [1969] 2 All E. R. 1215; [1970] 1 Q. B. 177, C. A.

(*u*) *McCutcheon* v. *David MacBrayne*, [1964] 1 All E. R. 430; [1964] 1 W. L. R. 125, H. L. But it may be possible to infer a common understanding that the purchase was to be on the current conditions of one of the parties provided these were reasonable: *Farmway, Ltd.* v. *Bird*, unreported as yet, 3rd October 1974, C. A.

(*a*) *Hollier* v. *Rambler Motors*, [1972] 1 All E. R. 399; [1972] 2 Q. B. 71, C. A. (3 or 4 occasions over 5 years).

(*b*) *Cf. Chapelton* v. *Barry Urban District Council*, [1940] 1 All E. R. 356; [1940] 1 K. B. 532, C. A.

(*c*) See *D. J. Hill & Co. Pty., Ltd.* v. *Walter H. Wright Pty., Ltd.*, [1971] V. R. 749; and compare *British Crane Hire Corporation, Ltd.* v. *Ipswich Plant Hire, Ltd.*, [1974] 1 All E. R. 1059; [1975] Q. B. 303, C. A. (two previous transactions, neither of which were known to the Manager placing the order: *semble* no sufficient course of dealing).

(*d*) *British Crane Hire Corporation, Ltd.* v. *Ipswich Plant Hire, Ltd., supra.* The equality of bargaining power is presumably relevant to the question whether the other party can be taken to have assented to the terms being incorporated into the contract, but it would not seem to be essential.

Usage.—Usage, says Lord BLACKBURN, when it is properly proved—

" is to be considered as part of the agreement ; and if the agreement be in writing, though the custom is not written it is to be treated exactly as if that unwritten customary clause had been written out at length " (*e*).

A usage may affect transactions either in a particular trade or a particular place and binds all those who participate in such transactions whatever the nature of their callings and in whatever capacity they do business (*f*).

To have effect, the usage must—

(1) be legal (*e*) ;
(2) be reasonable (*e*) ;
(3) be sufficiently certain that it would not be void for uncertainty if an express term of the contract (*f*);
(4) be generally accepted by those who habitually do business in the trade or market concerned and be so generally known that an outsider, who makes reasonable enquiries could not fail to be aware of it (*g*);
(5) not have been expressly or impliedly excluded by the agreement of the parties, as, for instance, where it is inconsistent with a written term of the contract (*h*) ;
(6) not be—

" so inconsistent with the nature of the contract to which it is sought to be applied as that it would change its nature altogether, or as to change its intrinsic character " (*i*).

Illegality

The common law rules relating to the effect of illegality upon a contract are saved by s. 61 (2) of the Sale of Goods Act, *post*.

When illegality is apparent on the face of the transaction the Court must take cognisance of it, though the defence of illegality is not set up (*k*); and similarly the Court must take cognisance when the illegality becomes apparent from the evidence (*l*). But when

---

(*e*) *Tucker* v. *Linger* (1883), 8 App. Cas. 508, at p. 511.

(*f*) *Kum* v. *Wah Tat Bank*, [1971] 1 Lloyd's Rep. 439, P. C.

(*g*) *Kum* v. *Wah Tat Bank*, *supra;* and see *Robinson* v. *Mollet* (1875), L. R. 7 H. L. 802.

(*h*) *Produce Brokers Co.* v. *Olympia Oil & Cake Co.*, [1916] 1 A. C. 314 ; *Re Sutro & Co.*, [1917] 2 K. B. 348, C. A.; *Thomas Borthwick* (*Glasgow*), *Ltd.*, v. *Faure Fairclough, Ltd.*, [1968] 1 Lloyd's Rep. 16; *Kum* v. *Wah Tat Bank*, *supra*.

(*i*) *Per* BRETT, J., in *Robinson* v. *Mollet* (*supra*), at p. 819. See also *per* WILLES, J. in the same case at (1870), L. R. 5 C. P. 646, *sub nom. Mollet* v. *Robinson*.

(*k*) *Re North West Salt Co.*, [1913] 3 K. B. 422, C. A.; reversed, [1914] A. C. 461.

(*l*) *Snell* v. *Unity Finance, Ltd.*, [1963] 3 All E. R. 50; [1964] 2 Q. B. 203, C. A.; *Lipton* v. *Powell*, [1921] 2 K. B. 51, at p. 58; as to the duty of an arbitrator, see *David Taylor & Son, Ltd.* v. *Barnett*, [1953] 1 All E. R. 843; [1953] 1 W. L. R. 562, C. A.

the illegality is not apparent, it must be pleaded (*m*). Speaking generally, a contract is illegal when it is prohibited by statute or is contrary to public policy, as where it is for an illegal or immoral purpose which is known to both parties (*n*), or is intended by both parties to be performed in an illegal manner (*o*). If only one party has in mind the illegal or immoral purpose, or the intention to perform the contract in an illegal manner, and the other party does not know of his illicit designs, the contract is not illegal (*p*), and may be enforced by the innocent party, but not by the guilty party (*q*). Where, on the other hand, the contract itself (as opposed to its purpose or mode of performance) is prohibited by statute, and the prohibition is made in the public interest, neither party can enforce the contract, even though one party was unaware of the facts constituting the illegality (*r*).

(*m*) *Lipton* v. *Powell*, [1921] 2 K. B. 51.

(*n*) It is sufficient if the facts which give rise to the illegality are known to both parties even if one or both of them fails to realise that those facts constitute an offence: *Belvoir Finance Co.* v. *Stapleton*, [1970] 3 All E. R. 664; [1971] 1 Q. B. 210, C. A.; *cf. Southern Industrial Trust, Ltd.* v. *Brooke House Motors, Ltd.* (1968), 112 Sol. Jo. 798, C. A. (H.P. transaction: illegality of H.P. agreement through fraud of seller and customer; finance company entitled to enforce sale agreement with seller, being unaware of the fraud or illegality); *Mason* v. *Clarke*, [1955] 1 All E. R. 914; [1955] A. C. 778, H. L. (receipt required for unlawful purpose of which party giving receipt was unaware; innocent party not debarred from enforcing contract). As to good morals the rule is *ex turpi causâ non oritur actio*; *cf. Pearce* v. *Brooks* (1866), L. R. 1 Exch. 213 (supply of goods, not necessaries, to known prostitute). See the judgment of ATKIN, L.J., in *Rawlings* v. *General Trading Co.*, [1921] 1 K. B. 635, C. A., as to restraint of trade.

(*o*) See *Archbolds (Freightage), Ltd.* v. *Spanglett, Ltd.*, [1961] 1 All E. R. 417; [1961] 1 Q. B. 374, C. A.; *Mason* v. *Clarke, supra*. It is sufficient to debar the innocent party from enforcing the contract if, after the contract has been entered into, he becomes aware that the other party intends to perform the contract in a manner which the innocent party knows to be unlawful, and participates in the unlawful performance, for example by standing by and sanctioning it when by objecting he might prevent it: *Ashmore Benson Pease & Co., Ltd.* v. *A. V. Dawson, Ltd.*, [1973] 2 All E. R. 856; [1973] 1 W. L. R. 828, C. A.

(*p*) *Marles* v. *Philip Trant & Sons, Ltd.*, [1953] 1 All E. R. 651; [1954] 1 Q. B. 29, C. A.

(*q*) *Alexander* v. *Rayson*, [1936] 1 K. B. 169; *St. John Shipping Corporation* v. *Joseph Rank, Ltd.*, [1956] 3 All E. R. 683, at p. 687; [1957] 1 Q. B. 267, at p. 283. But if the guilty party did not have the unlawful intention to perform the contract illegally when the contract was made, he can enforce the contract, provided that in so doing he does not have to disclose that he committed an illegal act: *St. John Shipping Corpn.* v. *Joseph Rank, Ltd., supra*.

(*r*) *Re Mahmoud and Ispahani*, [1921] 2 K. B. 716, C. A.; *Archbolds (Freightage), Ltd.* v. *S. Spanglett, Ltd.*, [1961] 1 All E. R. 417; [1961] 1 Q. B. 374, C. A.; *Yin* v. *Sam*, [1962] A. C. 304, P. C. See also *St. John Shipping Corpn.* v. *Joseph Rank, Ltd., supra*; *Fenton* v. *Scotty's Car Sales*, [1968] N. Z. L. R. 929.

Normally it is for the person alleging illegality to prove it (*s*). Where a contract is capable of two constructions and the effect of one is to render it illegal, the other construction will be given to it *ut res magis valeat quam pereat* (*t*). So too, where a contract can be performed either legally or illegally the presumption is that it will be performed legally (*u*). Where the property in goods is transferred in pursuance of an illegal transaction, then, at least if the contract has been fully executed, it remains in the transferee notwithstanding the illegality (*a*), even if the transferee has never taken possession of the goods (*b*). As to illegality by subsequent legislation or outbreak of war, see p. 20, *post*.

Many statutory offences in connection with the sale of goods have been created in recent years. In general, however, the statutes creating the offences have provided that a contract is not to be void or unenforceable by reason only of such offences (*c*).

Impossibility

There are cases common to the whole field of contract where performance is excused on the ground of impossibility. As a general rule, if a man makes a contract he must fulfil its conditions or pay damages. It is no excuse that he cannot get the goods he has contracted to deliver (*d*), or that he can only obtain them at a prohibitive price, or that they can only be shipped by a much longer sea route than the parties contemplated at the time of contracting (*e*), or that they cannot be shipped because the agreed shipping line failed to provide a vessel to load in the agreed month of shipment (*f*). If he wishes

---

(*s*) *Hire Purchase Furnishing Co.* v. *Richens* (1887), 20 Q. B. D. 387.

(*t*) *Cf. Wong* v. *Beaumont Property Trust, Ltd.*, [1964] 2 All E. R. 119; [1965] 1 Q. B. 173, C. A. (construction of lease).

(*u*) *Lewis* v. *Davidson* (1839), 4 M. & W. 654, at p. 657; *Waugh* v. *Morris* (1873), L. R. 8 Q. B. 202, at p. 208; applied in *Maschinenfabrik Seydelmann K. G.* v. *Presswood Brothers, Ltd.*, [1966] 1 O. R. 316.

(*a*) *Singh* v. *Ali*, [1960] 1 All E. R. 269; [1960] A. C. 167, P. C.; *Chettiar* v. *Chettiar*, [1962] 1 All E. R. 494; [1962] A. C. 294, P. C.

(*b*) *Belvoir Finance Co.* v. *Stapleton*, [1970] 3 All E. R. 664; [1971] 1 Q. B. 210, C. A.

(*c*) See further p. 55, *post*. As to sales on Sunday, see 17 Halsbury's Laws (3rd ed.) tit. " Sunday Trading ".

(*d*) *Hale* v. *Rawson* (1858), 4 C. B., N. S., 85; *Re Weis & Co.*, [1916] 1 K. B. 346; *Blackburn Bobbin Co.* v. *Allen & Sons, Ltd.*, [1918] 2 K. B. 467, C. A.; *Produce Brokers Co.* v. *Weis & Co.* (1918), 87 L. J. K. B. 472; *Sargant & Sons* v. *Paterson & Co.* (1923), 129 L. T. 471 (intended port of shipment closed by enemy).

(*e*) *Tsakiroglou & Co. Ltd.* v. *Noblee & Thorl G.m.b.H.*, [1961] 2 All E. R. 179; [1962] A. C. 93, H. L.

(*f*) *Thomas Borthwick (Glasgow), Ltd.* v. *Faure Fairclough, Ltd.*, [1968] 1 Lloyd's Rep. 16.

to be safe he must protect himself by express stipulation (*g*). But there are cases of *vis major*, independent of any particular contract, which excuse performance. For example, if the goods which a man has contracted to sell are requisitioned by the Government before delivery, he is absolved from his obligation to deliver them (*h*).

Supervening illegality.—Again, if a contract, legal in its inception, is prohibited by or under an Act of Parliament before its fulfilment, the obligation is dissolved (*i*). A declaration of war by this country operates as an Act of Parliament prohibiting all intercourse with the enemy (*k*). If a contract made before war involves intercourse with the enemy for its due performance, the contract is dissolved, and a clause purporting to suspend its operation during war is null and void (*l*). As to who is an enemy, see p. 89, *post*.

The provisions of s. 1 of the Law Reform (Frustrated Contracts) Act, 1943, apply to these cases, because the impossibility does not arise from the fact that the goods have perished.

(*g*) *New Zealand Shipping Co.* v. *Société des Ateliers de France*, [1919] A. C 1 ; *per* Lord LOREBURN in *Tennants* (*Lancashire*) *Ltd.* v. *Wilson & Co.*, [1917] A. C. 495, at p. 510. The following cases illustrate how particular stipulations of this kind have been construed by the courts: *Matsoukis* v. *Priestman & Co.*, [1915] 1 K. B. 681 (provision for payment of £10 in respect of any day in which performance was late, "force majeure" excepted); *Peter Dixon & Sons* v. *Henderson & Co.*, [1919] 2 K. B. 778, C. A. (suspension of delivery if "hindered" by war, etc.); *Clark* v. *Cox McEuen & Co.* (*1919*), [1921] 1 K. B. 139, C. A. (contract to be cancelled if vessel lost before declaration of shipment); *Re Comptoir Commercial Anversois*, [1920] 1 K. B. 868 C. A. (contract to be at an end in the event of "hostilities preventing shipment"); *Wills & Sons* v. *Cunningham & Co.*, [1924] 2 K. B. 220 (scope of clause "unforeseen circumstances excepted"); *Brauer* v. *Clark*, [1952] 2 All E. R. 497, C. A. (sale subject to export licence); *Smyth* v. *Lindsay*, [1953] 2 All E. R. 1064 (contract to be cancelled if fulfilment rendered impossible by "prohibition of export"); and see note "Licences", p. 41, *post*, and further examples at p. 358, *post*.

(*h*) *Re Shipton, Anderson & Co.*, [1915] 3 K. B. 676, and see *Bank Line* v. *Capel & Co.*, [1919] A. C. 435 (charter party frustrated by requisitioning of ship); but *cf. Port Line, Ltd.* v. *Ben Line Steamers, Ltd.*, [1958] 1 All E. R. 787 ; [1958] 2 Q. B. 146 (a time charter for thirty months of which a little over seventeen months have expired is not frustrated by a requisitioning which would be expected to, and does in fact, last three to four months); and see also *Hong Kong Fir Shipping Co. Ltd.*, v. *Kawasaki Kisen Kaisha, Ltd.*, [1962] 1 All E. R. 474; [1962] 2 Q. B. 26, C. A. As to a temporary embargo on export, see *Millar & Co.* v. *Taylor & Co.*, [1916] 1 K. B. 402, C. A.

(*i*) *Metropolitan Water Board* v. *Dick Kerr & Co.*, [1918] A. C. 119 (building contract); *MacMaster* v. *Cox & Co.*, 1920 S.C. 566 ; reversed, 1921 S. C. 24 (H. L.) (sale) ; *Kursell* v. *Timber Operators Ltd.*, [1927] 1 K. B. 298.

(*k*) *Esposito* v. *Bowden* (1857), 7 E. & B. 763, at p. 781, Ex. Ch.

(*l*) *Ertel Bieber & Co.* v. *Rio Tinto Co.*, [1918] A. C. 260 ; *Ottoman Bank* v. *Jebara*, [1928] A. C. 269 (completion did not involve intercourse with enemy), As to c.i.f. contracts, see *Karberg* v. *Blythe Green & Co.*, [1916] 1 K. B. 495. C. A.; *Krupp* v. *Orconera Iron Ore Co.* (1919), 88 L. J. Ch. 304.

Continental lawyers recognise *force majeure* as excusing the performance of a contract on the principle of act of state. English judges formerly rested the defence of *vis major* on the fiction of an implied condition forming the basis of the contract between the parties (*m*). This fiction has now largely been abandoned. The current test is whether a situation has arisen which renders performance of the contract a thing radically different from that which the parties undertook (*n*). Though *vis major* may excuse performance of a contract, it cannot be relied on to justify the variation of a contract (*o*).

Law Reform (*Frustrated Contracts*) *Act, 1943.*—The Act does not deal in any way with the general law of frustration as a cause of the discharge of contracts. What it does is to make certain provisions about what shall happen once a contract governed by English law has become impossible of performance or been otherwise frustrated, and the parties thereto have for that reason been discharged from the further performance of the contract. The Act applies only to contracts governed by English law, and it applies to contracts to which the Crown is a party in like manner as to contracts between subjects, s. 2 (2), *post,* Appendix I. There are certain contracts to which the Act does not apply at all, and special provision is made for contracts capable of being severed, and for contracts containing express provisions about frustration; see s. 2, *post,* Appendix I.

The Act applies to contracts of sale of goods in the same way as it applies to other contracts governed by English law, but subject to the exceptions contained in s. 2 (5) (c), *post,* Appendix I, which provides that the Act shall not apply to contracts to which s. 7 of the Sale of Goods Act applies or to any other contract for the sale, or sale and delivery, of specific goods, where the contract is frustrated through the goods perishing.

No doubt the reason for thus excluding the application of the Act in the two cases there specified is that the case of impossibility by the perishing of the goods is peculiar to the law of sale, and the special rules applicable to the perishing of specific goods have been long established and are well understood by the commercial community, so that there is no point in disturbing them. It is now clear that s. 7 of the Sale of Goods Act does not apply to specifically described goods.

Rescission or variation

A contract of sale may be rescinded by a subsequent agreement,

---

(*m*) *Taylor* v. *Caldwell* (1863), 3 B. & S. 826, and see the authorities collected in *Blackburn Bobbin Co.* v. *Allen & Sons,* [1918] 1 K. B. 540; affirmed [1918], 2 K. B. 467, C. A. As to *vis major* and " act of God," see *Nugent* v. *Smith* (1876), 1 C. P. D. 423, at p. 429, C. A.

(*n*) *Davis Contractors* v. *Fareham Urban District Council,* [1956] 2 All E. R. 145; [1956] A. C. 696; *Tsakiroglou & Co., Ltd.* v. *Noblee & Thorl G.m.b.H.,* [1961] 2 All E. R. 179; [1962] A. C. 93; *Ocean Tramp Tankers Corporation* v. *V/O Sovfracht,* [1964] 1 All E. R. 161; [1964] 2 Q. B. 226, C. A.

(*o*) *Brooke Tool Manufacturing Co.* v. *Hydraulic Gears Co.* (1920), 89 L. J. K. B. 263 (delivery delayed by war conditions; buyers refused to accept delayed delivery. Sellers have no action for the price).

which may be oral (*p*) or in writing, and which may rescind the former contract either expressly or by necessary implication. The latter will be the case if the second contract is inconsistent with the first to an extent which goes to its very root (*q*). Rescission may also be implied from the parties' course of conduct (*r*) or from the fact that neither party has insisted upon performance for a long time after the contract was made (*s*). Variation of the original contract must be carefully distinguished from rescission:

> "In the first case there are no such executory clauses in the second arrangement as would enable you to sue upon that alone if the first did not exist; in the second you could sue on the second arrangement alone, and the first contract is got rid of either by express words to that effect, or because, the second dealing with the same subject-matter as the first but in a different way, it is impossible that the two should be both performed" (*t*).

Waiver.—A distinction must also be drawn between variation of a contract and the mere waiver by one party of the performance of a stipulation in the contract. If a stipulation operates for the benefit of both parties it can only be altered by mutual consent (*u*). But a party may always waive unilaterally a stipulation which is for his own benefit alone (*a*). Thus if a seller at the request of the buyer postpones delivery, or a buyer at the request of the seller does not insist upon delivery at the stipulated time, there is no variation of the contract though it may affect the measure of damages (*b*). Where

---

(*p*) *Goss* v. *Lord Nugent* (1833), 5 B. & Ad. 58; *Morris* v. *Baron & Co.* [1918] A. C. 1; *British and Beningtons, Ltd.* v. *North West Cachar Tea Co.*, [1923] A. C. 48.

(*q*) *Morris* v. *Baron & Co.*, [1918] A. C. 1, at p. 27; *British and Beningtons Ltd.* v. *N. W. Cachar Tea Co.*, [1923] A. C. 48.

(*r*) *Bruner* v. *Moore*, [1904] 1 Ch. 305, at p. 312; *Hartley* v. *Hymans*, [1920] 3 K. B. 475.

(*s*) *Pearl Mill Co.* v. *Ivy Tannery Co.*, [1919] 1 K. B. 78; *Fisher* v. *Eastwoods*, [1936] 1 All E. R. 421.

(*t*) *Morris* v. *Baron & Co.*, [1918] A. C. 1, at p. 26; *Goss* v. *Lord Nugent* (1833), 5 B. & Ad. 58, at pp. 65, 66. It will be found that all the cases in which evidence of the new oral arrangement was rejected were cases of attempted variation.

(*u*) *Maine Spinning Co.* v. *Sutcliffe & Co.* (1917), 87 L. J. K. B. 382 ("delivery d.o.b. Liverpool"); *Fibrosa S.A.* v. *Fairbairn Lawson Combe Barbour, Ltd.*, [1941] 2 All E. R. 300; [1942] 1 K. B. 12, C. A. ("delivery c.i.f. Gdynia").

(*a*) See, e.g., *Hartley* v. *Hymans*, [1920] 3 K. B. 475. As to waiver of a condition before examination of the goods, see *per* ATKIN, L.J., in *Moore & Co.* v. *Landauer & Co.* (1921), 26 Com. Cas. 267, at p. 276, C. A.

(*b*) *Hickman* v. *Haynes* (1875), L. R. 10 C. P. 598 (forbearance of seller); *Potts & Co.* v. *Brown McFarlane & Co.* (1925), 30 Com. Cas. 64, H. L. (demand of inspection extending time of delivery); *Levey* v. *Goldberg*, [1922] 1 K. B. 688 (postponement of time of delivery at buyer's request); *Besseler Waechter Glover & Co.* v. *South Derwent Coal Co., Ltd.*, [1938] 1 K. B. 408 (postponement of delivery dates); *cf. Ogle* v. *Vane* (1868), L. R. 3 Q. B. 272, Ex. Ch. (forbearance of buyer to buy in against the seller on the latter failing to deliver at the contract time). See also p. 238, *post*.

the fulfilment of a condition by one party is prevented by the other, the condition is waived (*c*); and the wrongful repudiation of a contract by one party may operate as a waiver of conditions precedent to be performed by the other (*d*).

It seems that where a party does waive a stipulation, he cannot afterwards treat the non-performance of that stipulation as giving him the right to rescind the contract (*e*), and where there is what may be called a continuing waiver he must give the other party reasonable notice of his intention to insist in the future on the performance of the stipulation, in order that the other party may have an opportunity of complying with it (*f*).

A party will be treated as having waived a stipulation if his conduct is reasonably interpreted by the other party as amounting to a waiver, provided that it clearly and unambiguously bears that interpretation, even if the first party did not intend to waive the stipulation (*g*).

Where a party is already in breach of a stipulation in a contract, the other party cannot, strictly speaking, waive performance of that stipulation. He may, however, waive one or more of his remedies arising out of the breach. Thus, if a buyer is entitled to reject goods on a certain ground, but so conducts himself as to lead the other to believe that he is not relying on that ground, he cannot afterwards reject the goods on that ground, at any rate if it would be unfair or unjust to allow him to do so (*h*).

(*c*) *Mackay* v. *Dick* (1881), 6 App. Cas. 251; followed, *Kleinert* v. *Abosso Mining Co.* (1913), 58 Sol. Jo. 45, P. C. *Cf. New Zealand Shipping Co.* v. *Société des Ateliers*, [1919] A. C. 1, at p. 6, *per* Lord FINLAY.

(*d*) *Cort* v. *Ambergate Rail. Co.* (1851), 17 Q. B. 127; *Braithwaite* v. *Foreign Hardwood Co.*, [1905] 2 K. B. 543, C. A. See s. 31, *post*, and notes.

(*e*) *Harlow and Jones* v. *Panex (International), Ltd.*, [1967] 2 Lloyd's Rep. 509. See also *Bentsen* v. *Taylor*, [1893] 2 Q. B. 274, 283.

(*f*) *Tyers* v. *Rosedale Co.* (1875), L. R. 10 Exch. 195, Ex. Ch.: *Panoutsos* v. *Raymond Hadley Corporation of New York*, [1917] 2 K. B. 473, applied in *Plasticmoda S.P.A.* v. *Davidsons, Ltd.*, [1952] 1 Lloyd's Rep. 527. See also *Rickards* v. *Oppenheim*, [1950] 1 All E. R. 420, C. A.; [1950] 1 K. B. 616. As to quasi-estoppel, see now *Ajayi* v. *R. T. Briscoe (Nigeria), Ltd.*, [1964] 3 All E. R. 556; [1964] 1 W. L. R. 1326, P. C., where the earlier decisions are reviewed; *W. J. Alan & Co., Ltd.* v. *El Nasr Export and Import Co.*, [1972] 2 All E. R. 127; [1972] 2 Q. B. 189, C. A.

(*g*) *Woodhouse A.C. Israel Cocoa, Ltd. S.A.* v. *Nigerian Produce Marketing Co.*, [1972] 2 All E. R. 271; [1972] A. C. 741, H. L.; *cf. Spiro* v. *Lintern*, [1973] 3 All E. R. 319; [1973] 1 W. L. R. 1002, C. A.

(*h*) *Panchaud Freres S.A.* v. *Etablissements General Grain Co.*, [1970] 1 Lloyd's Rep. 53, C. A., where Lord DENNING, M.R., at p. 57, treated this rule as an example of estoppel by conduct. WINN, L.J., on the other hand considered, at p. 59, that it was not an example of estoppel or quasi-estoppel, but may rather be "an inchoate doctrine stemming from the manifest convenience of consistency in pragmatic affairs negativing any liberty to blow hot and cold in commercial conduct." See also *Bentsen* v. *Taylor*, [1893] 2 Q. B. 274, 283; *Toepfer* v. *Cremer* (1975), *Times*, 19th March, C. A.; *The Vladimir Ilich*, [1975] 1 Lloyd's Rep. 322.

An accord and satisfaction after breach is often referred to as waiver. Thus a new contract may be made after breach, and such new contract in itself may be an accord and satisfaction, though often it is merely an accord, and there is no satisfaction until it is performed. In this last case it cannot be sued on, and does not operate to discharge the original contract. Where, however, the new contract is itself the accord and satisfaction it can be sued on, and also provides a complete answer to an action on the original contract. An assent to accept a substituted mode of performance has also been held not to constitute a variation of the contract (*i*).

Performance of the contract

The law relating to the performance of a contract for the sale of goods is codified in Part III of the Sale of Goods Act, 1893. The rules there laid down apply in the absence of contrary agreement between the parties.

Duties of the seller.—The basic duty of the seller is to deliver the goods to the buyer in accordance with the terms of the contract (*k*). Delivery is defined by s. 62 (1) of the Sale of Goods Act as the voluntary transfer of possession from one person to another (*l*). The rules as to delivery are set out in ss. 28–32 of the Sale of Goods Act, *post.*

It is also of the essence of a contract of sale that the seller should transfer the property in the goods to the buyer. The time at which the property is transferred depends upon the intention of the parties, provided that where the contract is for the sale of unascertained goods, no property in the goods can be transferred to the buyer until the goods are ascertained (*m*). Rules for ascertaining the intention of the parties are set out in ss. 17 and 18 of the Sale of Goods Act, *post.*

Where the seller is not the absolute owner of the goods at the date of the contract of sale, he may still be able to perform his obligations under the contract if—

(1) He is authorised by the owner of the goods to sell them on his behalf;

(2) he possesses some special common law or statutory power of

---

(*i*) *Leather Cloth Co.* v. *Hieronimus* (1875), L. R. 10 Q. B. 140 (change of route); doubted in *Hartley* v. *Hymans*, [1920] 3 K. B. 475, at p. 493; *Sanday & Co., Ltd.* v. *Shipton Anderson & Co., Ltd.* (1926), 25 Ll. L. Rep. 508; see also the judgment of SCRUTTON, L.J., in *British Russian Gazette* v. *Associated Newspapers*, [1933] 2 K. B. 616, 643, C. A.

(*k*) Sale of Goods Act, 1893, s. 27, *post.*

(*l*) See further as to the different forms which delivery may take, post, p. 276.

(*m*) Sale of Goods Act, 1893, s. 16, *post.* For the extent to which a buyer can acquire an equitable interest in goods before the property in them passes to him, see the comment to s. 16.

sale or is selling under the order of a court of competent jurisdiction (*n*);

(3) he acquires the right to sell the goods before the time when the property is to pass (*o*).

In each of the above cases the seller is acting perfectly lawfully in selling the goods, even though in the first two cases they do not belong to him.

(4) He is selling goods which he has no right to sell, but the buyer obtains a good title by reason either of the law of estoppel or of one of the statutory provisions enabling an apparent owner to dispose of goods as if he were the true owner (*p*).

In such a case, the true owner will have no recourse against the buyer (who will generally therefore have no cause for complaint on that ground against the seller (*q*)), but the true owner will normally be able to recover damages from the seller who will have converted his property (*r*).

Duties of the buyer.—It is the duty of the buyer to accept the goods and to pay for them in accordance with the terms of the contract (*s*). In the absence of agreement to the contrary, payment and delivery are concurrent conditions (*t*).

As to payment generally, see the comment to s. 49 of the Sale of Goods Act, *post*. As to payment by banker's irrevocable letter of credit, see *post*, p. 42.

Rules as to acceptance are set out in ss. 34–37 of the Sale of Goods Act, *post*.

Remedies for breach of contract

Rights of unpaid seller against the goods.—The unpaid seller retains rights over the goods so long as he retains possession of them, even if the property in them has passed to the buyer. Where the property in the goods has passed to the buyer, the unpaid seller in possession has a lien over the goods for the price, and where the property has not passed to the buyer, he has a right of withholding delivery similar to and co-extensive with his right of lien. These rights will be lost if they are once waived (*a*).

(*n*) For examples of such powers of sale, see note (*d*) at p. 162, *post*. The validity of such sales is not affected by anything in the Sale of Goods Act, 1893: see s. 21 (2) (b) of that Act.

(*o*) Sale of Goods Act, 1893, s. 12 (1).

(*p*) Such statutory provisions are unaffected by the Sale of Goods Act, 1893: see s. 21 (2) (a)) of that Act.

(*q*) But he may be able to recover any costs incurred in resisting the true owner's claim—see the comment to s. 12 of the Sale of Goods Act, *post*.

(*r*) For the true owner's right to waive the tort and claim the price from the seller in an action for money had and received, see *post*, p. 31.

(*s*) Sale of Goods Act, 1893, s. 27, *post*.

(*t*) *Ibid.*, s. 28, *post*.

(*a*) *Ibid.*, ss. 39–43, *post*.

The unpaid seller has also the right, if the seller becomes insolvent, to stop the goods in transit after he has parted with possession of them (*b*), but circumstances in which this right is of value are rare.

Where the unpaid seller exercises his right of lien or stoppage in transit, he acquires a limited right to re-sell the goods in the circumstances set out in s. 48 (3) of the Sale of Goods Act, 1893, *post*. He may, of course, under the terms of the contract of sale, reserve for himself a more extensive right to re-sell the goods if the buyer makes default, and s. 48 (4) of the Sale of Goods Act, *post*, provides for such a case (*c*). He may also provide under the terms of the contract of sale that although the buyer obtains possession of the goods, property in them shall not pass to him until the price is paid or some other condition is satisfied. In this case, if the buyer repudiates his obligations under the contract the seller may terminate it and claim back the goods (*d*).

Other rights of the seller.—In the circumstances set out in s. 49 of the Sale of Goods Act, *post*, and the comment thereto the seller may sue the buyer for the price of the goods. Except in those cases, he is entitled only to damages for non-acceptance of the goods sold, which are calculated in accordance with the principles set out in s. 50 of the Sale of Goods Act, *post*. If the buyer wrongfully fails to take delivery of the goods or delays in so doing, the seller may also be entitled under s. 37 of the Sale of Goods Act, *post*, to a reasonable charge for the care and custody of the goods.

The seller's right to special damages and to interest on the price or on damages awarded are preserved by s. 54 of the Sale of Goods Act, *post*. The principles on which such damages or interest are awarded are those which apply in contract generally. As to special damages, see further the comment to s. 54 of the Sale of Goods Act, *post*. As to interest, see further s. 3 of the Law Reform (Miscellaneous Provisions) Act, 1934 and the comment thereon, *post*, Appendix I.

As to mitigation of damages, see p. 255, *post*.

As to the rate of exchange where the price or charges have to be calculated in a foreign currency, see p. 258, *post*.

Deposit.—Sometimes part of the price is prepaid by way of security when the contract is entered into. The money so prepaid is called a deposit. The return of the deposit in case the sale goes off is usually a matter of agreement, but in the absence of a different

---

(*b*) *Ibid.*, ss. 44–46, *post*.

(*c*) As to re-sale by buyer and seller generally, where a right of lien or stoppage in transit has been exercised, see *ibid.*, ss. 47–48, *post*.

(*d*) P. 224, *post*. But his rights in this respect may be restricted if the agreement is within the Consumer Credit Act, 1974, p. 63, *post*, or, pending its repeal under the provisions of the Consumer Credit Act, is within the Hire-Purchase Act, 1965.

agreement the deposit is forfeited if the sale goes off through the buyer's fault (*e*). "The deposit," says Lord MACNAGHTEN,

> "serves two purposes: if the purchase is carried out it goes against the purchase-money; but its primary purpose is this: it is a guarantee that the purchaser means business" (*f*).

In deciding whether a payment in advance is a deposit or a part payment which can be recovered if the sale goes off, regard must be had to all the surrounding circumstances (*g*).

Remedies of the buyer.—If the seller fails to deliver any goods, or delivers goods which the buyer rejects as not being in conformity with the contract, the buyer may have available one or more of the following remedies:

(1) He may sue for damages for non-delivery—see s. 51 of the Sale of Goods Act, 1893, *post*.

(2) If the price has been paid by him, he may recover it in an action for money had and received for a consideration which has wholly failed, and in special circumstances a seller might be held to be a trustee of the purchase price for the buyer until he delivers the goods. See further p. 257, *post*.

(3) Where the buyer cannot be adequately compensated in damages, he may sometimes sue for specific performance of the contract. This right is set out in s. 52 of the Sale of Goods Act, *post*, in the case of specific or ascertained goods. The Sale of Goods Act does not deal with specific performance of a contract for the sale of unascertained goods, but in special cases the remedy is available on general equitable principles (*h*).

(4) If the goods are specific or ascertained, and under the terms of the contract the buyer is entitled to immediate possession of them, he may claim them in an action for conversion or detinue —see p. 241, *post*.

(5) In a clear case, if the seller is prosecuted and convicted of an offence committed by him in connection with the contract, the buyer may be able to obtain compensation under s. 35 of the Powers of

(*e*) *Howe* v. *Smith* (1884), 27 Ch. D. 89, C. A. See the history of the law of earnest and deposit traced by FRY, L.J., at p. 94. *Cf. Roberts and Cooper, Ltd.* v. *Salvesen & Co.*, 1918 S. C. 794; *Alexander* v. *Webber*, [1922] 1 K. B. 642 (secret agreement for commission between seller and buyer's servant; buyer entitled to recover his deposit because contract avoided by seller's fraud); *Dies* v. *British and Finance Corporation, Ltd.*, [1939] 1 K. B. 724; *cf. R. V. Ward, Ltd.* v. *Bignall*, [1967] 2 All E. R. 449; [1967] 1 Q. B. 534, C. A., overruling *Gallagher* v. *Shilcock*, [1949] 1 All E. R. 921; [1949] 2 K. B. 765. But note that a deposit may be recoverable where it would be unconscionable in equity for it to be forfeited: *Stockloser* v. *Johnson*, [1954] 1 All E. R. 630; [1954] 1 Q. B. 476, C. A.; *Steedman* v. *Drinkle*, [1916] 1 A. C. 275, P. C.; *Re Dagenham* (*Thames*) *Dock Co., ex p. Hulse* (1873), 8 Ch. App. 1022.

(*f*) *Soper* v. *Arnold* (1889), 14 App. Cas. 429, at p. 435.

(*g*) *Elson* v. *Prices Tailors, Ltd.*, [1963] 1 All E. R. 231; [1963] 1 W. L. R. 287.

(*h*) See further the comment to s. 52, *post*.

Criminal Courts Act, 1973 (*i*) for any loss or damage resulting from that offence. He may also be able to recover compensation if the seller is convicted of another offence and asks for an offence committed in connection with the contract to be taken into account by the Court in determining sentence.

Where the seller is in breach of some other condition of the contract, the buyer may treat the contract as repudiated by such breach, at least in the great majority of cases (*k*), and may then recover damages for breach of contract. He may also recover the price, if he has paid it, in an action for money had and received for a consideration which has wholly failed—see p. 257, *post*.

If the buyer accepts the goods delivered although they are not in conformity with the contract, or otherwise elects, or is bound (*l*), to treat a breach of condition as a breach of warranty, then he may sue for damages for breach of warranty (*m*), or he may set off the damages in diminution or extinction of the price, and claim in respect of any further damage he may have suffered by reason of the breach (*n*).

As to the measure of damages for breach of warranty, see s. 53 of the Sale of Goods Act, *post*. As to special damages and mitigation of damages, see s. 54 of the Sale of Goods Act, *post*. As to interest, see s. 3 of the Law Reform (Miscellaneous Provisions) Act, 1934, Appendix I, *post*.

Assignment of contract of sale

There is no provision in the Sale of Goods Act which deals with the assignment of contracts of sale. Such contracts can therefore be assigned subject to the conditions applicable in the case of other contracts. If the assignment is to be a legal assignment the provisions of s. 136 of the Law of Property Act, 1925, must be complied with. But this section, which replaces s. 25 (6) of the Judicature Act, 1873,

> " does not enlarge the class of choses in action, the assignability of which was previously recognised either at law or in equity " (*o*).

The simplest cases arise where the goods have been delivered and the right to receive the price is assigned, or where, the price having been paid, the right to receive the goods is assigned. In both these cases the assignor has performed his part of the contract, and the obligation of the other party to the contract is varied to this extent only, that he must pay the price or deliver the goods to the assignee

(*i*) Appendix I, *post*.

(*k*) *Cf. Harbutt's Plasticine, Ltd.* v. *Wayne Tank and Pump Co., Ltd.*, [1970] 1 All E. R. 225; [1970] 1 Q. B. 447, C. A.; and see p. 9, *ante*.

(*l*) See s. 11 of the Sale of Goods Act, *post*.

(*m*) *Ibid.*, s. 53 (1) (b).

(*n*) *Ibid.*, s.53 (1) (a).

(*o*) *Tolhurst* v. *Associated Portland Cement Manufacturers, Ltd.*, [1902] 2 K. B. 660, at p. 676.

instead of to the person with whom he contracted. The rule is the same where the price is not due at the date of the assignment; the right to receive it when due can be assigned (*p*). Apparently an assignment of part of a debt or other legal chose in action is not effectual to transfer the legal right to that part which is assigned, but operates in equity only, with the result that the assignee could not sue in his own name (*q*).

But the contract may provide for a series of deliveries, or the quantity, quality, or the time of the delivery may depend on the wishes or needs of the parties. In these cases, where neither party has performed his part of the contract, there can be no assignment where the duties which were owed to the assignor would be made more onerous by being transferred to the assignee, or where the rights to be assigned are essentially personal. Thus where A agreed to supply B with all the eggs he should require for manufacturing purposes for a year, and B purported to assign his right to a limited company, it was held that the contract was personal, for A's liability was measured by the personal requirements of B (*r*).

Similarly where deliveries of coal were to be made by a company over a long period, and credit was to be given, the company were held to be entitled to refuse to supply an assignee on the ground that the giving of credit was a personal matter in that particular case (*s*). The obligation to pay, like obligations to do work or deliver goods, cannot always be performed vicariously (*t*).

But where A contracted to supply for fifty years all the chalk that a company required at a certain works, and that company assigned the benefit of the contract to a larger company, it was held that the latter could enforce the contract because the amount of the deliveries was to be ascertained by reference to the needs of certain stated works, and was thus not personal to the original contracting company (*u*). The first company did not get rid of its liability to pay should the second company fail to do so, nor is the original contractor bound to look to the assignee in any way for the discharge of duties owed to him (*v*).

(*p*) *Brice* v. *Bannister* (1878), 3 Q. B. D. 569.

(*q*) *Re Steel Wing Co. Ltd.*, [1921] 1 Ch. 349; *Williams* v. *Atlantic Assurance Co. Ltd.*, [1933] 1 K. B. 81.

(*r*) *Kemp* v. *Baerselman*, [1906] 2 K. B. 604; see also *International Fibre Syndicate* v. *Dawson* (1901), 3 F. (Ct. of Sess.) 32.

(*s*) *Cooper* v. *Micklefield Coal & Lime Co.* (1912), 107 L. T. 457.

(*t*) *Ibid.*, at p. 458. See also *Johnson* v. *Raylton* (1881), 7 Q. B. D. 438, C. A., where there was held to be an implied term that a manufacturer contracting to supply specific goods must supply goods of his own manufacture.

(*u*) *Tolhurst* v. *Associated Portland Cement Manufacturers*, [1903] A. C. 414; see also the judgment of COLLINS, M.R., in the same case in the Court of Appeal, [1902] 2 K. B. 660, at pp. 668, 669.

(*v*) See the useful summary by ATKIN, J., in *Sorrentino Fratelli* v. *Buerger*, [1915] 1 K. B. 307, at pp. 312, 313.

In any case it is only the benefit of a contract which can be assigned. The assignor cannot get rid of his obligations. The most he can do is to perform them through his assignee, he himself remaining liable (*w*).

Assignment must be carefully distinguished from novation which is the substitution, with the consent of all parties, of a new contract for a previously existing one. The parties to the new contract may be different from those to the old one.

Quasi-Contracts of Sale

The Sale of Goods Act deals only with contracts of sale, properly so called. But there are certain quasi-contract of sale which require to be noted. By a quasi-contract of sale is meant a transaction to which, independently of the will of the parties, the law annexes consequences similar to those which result from a sale (*a*). For example:—

(1) *Satisfied judgment in trover, trespass, and detinue.*—In trover and in trespass to goods, when the substance of the action is conversion, judgment recovered by the plaintiff for damages estimated on the footing of the full value of the goods and satisfied, either voluntarily by the defendant or by execution, operates as a sale of the goods by the plaintiff to the defendant as from the time when the judgment is satisfied (*b*).

In detinue also the defendant had at common law the power to defeat the plaintiff's claim to the return of the goods and to satisfy the judgment by paying their value (*c*), so that, in the words of JESSEL, M.R.,

> " the theory of the judgment in an action of detinue is that it is a kind of involuntary sale of the plaintiff's goods to the defendant. The plaintiff wants to get his goods back and the Court gives him the next best thing, that is the value of the goods " (*d*).

This defect was remedied by s. 78 of the Common Law Procedure Act, 1854 now replaced by R. S. C., O. 45, rr. 4, 13 (3); but the

---

(*w*) *British Waggon Co.* v. *Lea* (1880), 5 Q. B. D. 149, C. A.; especially at p. 154.

(*a*) As to quasi-contracts, see *Chitty on Contracts* (23rd ed.), chap. 29. As to implied contracts where there is consent in fact, though not expressed, see ss. 3 and 8 of the Sale of Goods Act, *post*.

(*b*) *U.S.A.* v. *Dollfus Mieg*, [1952] 1 All E. R. 572; [1952] A. C. 582 at p. 622 *per* Lord TUCKER, (trover); *Cooper* v. *Shepherd* (1846), 3 C. B. 226 (trover); *Brinsmead* v. *Harrison* (1871), L. R. 6 C. P. 584 (trover); and *cf.* the note to *Holmes* v. *Wilson* (1829), 10 Ad. & El. 503, at p. 511. It may of course be that there is someone who claims through the defendant and thus has a better title than he. If, for example, A takes goods from B and sells them to C, it is clear that satisfaction made by A to B will operate as a sale to C.

(*c*) See *Eberle's Hotels Co.* v. *Jonas* (1887), 18 Q. B. D. 459, at p. 468.

(*d*) *Ex p. Drake* (1877), 5 Ch. D. 866, at p. 871, C. A.

Court still has a discretion and may refuse to compel the defendant to return the goods (*e*).

An unsatisfied judgment does not transfer the property in any of these cases (*f*).

(2) *Waiver of tort.*—It has been held that where a plaintiff has been induced, by the fraud of a third person, to sell goods to an insolvent buyer, and such third person has afterwards obtained the goods himself, the plaintiff may waive the tort, and treat the transaction as a sale to such third person (*g*).

Again, there are other cases where, when one person has wrongfully obtained possession of the goods of another, as when they have been sold to him by a third person who had no right to sell them, the owner of the goods may waive the tort, and treat the transaction as a sale by himself to the person who has got the goods (*h*). So, too, as against the person who has thus wrongfully sold the goods and received the price, " the owner may waive the tort and recover the proceeds in an action for money had and received " (*i*). " If the servant of a company," says Lord SUMNER,

> " acting ultra vires of the company converts a stranger's chattel, and, having sold it, pays the proceeds into the company's account as its servant, I suppose an action for conversion would lie against the servant, and for money had and received against the company " (*k*).

(*e*) *Peruvian Guano Co.* v. *Dreyfus Bros. & Co.*, [1892] A. C. 166, at p. 176 ; *Whiteley* v. *Hilt*, [1918] 2 K. B. 808, at pp. 819, 824, C. A. ; *Cohen* v. *Roche*, [1927] 1 K. B. 169, at pp. 179–181.

(*f*) *Ex p. Drake, supra* (detinue) ; *Ellis* v. *Stenning*, [1932] 2 Ch. 81 (trover). In *Bradley* v. *Ramsay & Co.* (1912), 106 L. T. 771, C. A., where goods were sent on approval to the defendant and the plaintiff claimed their return in an action for detinue, judgment was entered by consent for the plaintiff ; but on the facts (distinguishing *Brinsmead* v. *Harrison* (*supra*)) it was held to be in substance a judgment for the price of the goods, and consequently, though it was not satisfied, the property in the goods was transferred.

(*g*) *Hill* v. *Perrott* (1810), 3 Taunt. 274.

(*h*) *Cf. Dickenson* v. *Naul* (1833), 4 B. & Ad. 638 ; *Allen* v. *Hopkins* (1844), 13 M. & W. 94.

(*i*) *Arnold* v. *Cheque Bank* (1876), 1 C. P. D. 578, at p. 585. This will be the position where " the plaintiff in truth treats the wrongdoer as having acted as his agent, overlooks the wrong, and by consent of both parties is content to receive the proceeds," *per* Lord ATKIN in *United Australia, Ltd.* v. *Barclays Bank*, [1940] 4 All E. R. 20, at p. 36 ; [1941] A. C. 1 at p. 28 ; overruling the dictum of BOVILL, C.J., in *Smith* v. *Baker*, [1873] L. R. 8 C. P. 350 at p. 355 ; and the dicta of A. L. SMITH, L.J., in *Rice* v. *Reed*, [1900] 1 Q. B. 54 at pp. 65–66, C. A. ; and distinguishing *Verschures Creameries Ltd.* v. *Hull and Netherlands Steamship Co. Ltd.*, [1921] 2 K. B. 608. Where, however, there is no true election between two inconsistent rights, the institution of proceedings against one tortfeasor for money had and received will not bar a subsequent action against another for a separate tort, at any rate so long as there is no satisfied judgment in the first action : *United Australia, Ltd.* v. *Barclays Bank* (*supra*).

(*k*) *Att.-General* v. *De Keyser's Hotel*, [1920] A. C. 508, at p. 556.

(3) *Sale by estoppel.*—So, too, there may be a sale by estoppel. Suppose a defendant sells specific goods to one person, and the documents of title to the goods to another person, he would be liable to both, though a doubt might arise as to which person would be entitled to the goods themselves. Likewise, a person holding himself out as the buyer may be liable as such, and a person who stands by and lets his goods be sold is bound by the sale (*l*).

For further instances of quasi-contract, see the notes to s. 2 of the Sale of Goods Act, *post*.

(*l*) As to estoppel and the passing of property, see the comment to s. 21, of the Sale of Goods Act, *post*.

SECTION B

INTERNATIONAL CONTRACTS OF SALE

A. Customary international contracts

The most common forms of international contract for the sale of goods are f.o.b. (free on board), c.i.f. (at a price to cover cost, insurance and freight) and ex-ship. Others are f.a.s. (free alongside), f.o.b. stowed (free on board and stowed), c. and f. (where the seller is not obliged to arrange insurance), delivered (where the seller undertakes that the goods will be delivered at, for example, the buyer's place of business) and ex-works. The ordinary incidents of f.o.b., c.i.f. and ex-ship contracts are described below. These incidents *prima facie* follow if a contract is described as f.o.b., c.i.f. or ex-ship (*a*). But upon a true construction of the remainder of the contract they may be, and commonly are, varied, sometimes to the point where the abbreviation f.o.b., c.i.f. or ex-ship merely indicates the mode of calculating the price. Thus hybrid f.o.b./c.i.f. contracts (*b*) or c.i.f./ex-ship contracts (*c*) are common.

A contract may, on its true construction be capable of performance at one or the other party's option either as f.o.b. contract or as a c.i.f. contract (*d*).

Save for delivered and ex-works contracts, the forms of contract mentioned above contemplate the shipment of goods by sea. Nonetheless f.o.b., c.i.f. and the other abbreviations are sometimes used in relation to contracts for sale where, for example, the goods are to be sent by road or air or combined transport from the seller to the buyer. The abbreviations will in such a case retain importance as indications as to the division of costs, as to responsibility for insurance (*dd*), as to risk, and, usually, as to the time when property is to pass. But many of the incidents of a contract for carriage by sea cannot be relevant in such a context. For example, it is unlikely that there will be a document of title equivalent to a bill of lading. Thus,

---

(*a*) *N.V. Handel My. J. Smits Import-Export* v. *English Exporters* (*London*) *Ltd.*, [1957] 1 Lloyd's Rep. 517 (f.o.b.); *Frebold* v. *Circle Products, Ltd.*, [1970] 1 Lloyd's Rep. 499, C. A. (f.o.b.); *Comptoir d'Achat et de Vente du Boerenbond Belge S/A* v. *Luis de Ridder*, [1949] 1 All E. R. 269, at p. 274; [1949] A. C. 293, at p. 311 (c.i.f.). In *Law and Bonar* v. *British American Tobacco*, [1916] 2 K. B. 605, a printed clause inconsistent with what were otherwise c.i.f. terms was disregarded.

(*b*) See *The Parchim*, [1918] A. C. 157, P. C.; *President of India* v. *Metcalfe Shipping Co., Ltd.*, [1969] 1 All E. R. 861; [1969] 2 Q. B. 123; on appeal, [1969] 3 All E. R. 1549; [1970] 1 Q. B. 289, C.A.

(*c*) See *The Gabbiano*, [1940] P. 166.

(*d*) See *Comptoir d'Achat* v. *Luis de Ridder, supra.*

(*dd*) *Lindon Tricotagefabrik* v. *White & Meacham*, [1975] 1 Lloyd's Rep. 384 (Sale c.i.f. Customer's warehouse; goods stolen before delivery: sellers not entitled to price because they failed to tender insurance documents).

the seller will not have the protection of a document which he can hold until the price is paid, so preventing the buyer from obtaining the goods in the meantime. A seller will not therefore be fully protected in such a case unless he agrees with both the buyer and the carrier that actual delivery by the carrier to the buyer is to be withheld pending payment of the price (*e*) or unless he obtains collateral protection (*f*).

(1) **F.o.b. contracts.**—In the absence of provision to the contrary (*g*), the buyer must nominate a ship and the port of shipment (*h*), and notify the seller when it will be likely to arrive (*i*). This is a condition precedent to the seller's duty to bring goods to the port (*j*). The arrival and readiness to load of the ship by the date agreed (if any) is *prima facie* of the essence of an f.o.b. contract (*k*). On the ship's arrival, the seller must deliver the goods on board at his own expense (*l*). Thereafter the goods are at the buyer's risk, and he is responsible for the freight and any subsequent charges (*m*). " The seller," says HAMILTON, L.J.,

---

(*e*) In *Frebold* v. *Circle Products*, [1970] 1 Lloyd's Rep. 499, C. A., such a provision was treated as reserving the seller's right of possession only, and not the property in the goods, after shipment.

(*f*) *E.g.* the right to draw on a letter of credit under which the Bank agrees to accept an air or road consignment note, or the intervention of a confirming house or of a guarantor. If the seller reserves title but not possession under the contract, this will not assist him if the buyer resells or otherwise disposes of the goods to somebody who takes them in good faith and without notice of the seller's interest, at least where English law is the proper law of the second disposition—see s. 9 of the Factors Act, 1889, *post*.

(*g*) Provision to the contrary is very common: see *Ian Stach* v. *Baker Bosley*, [1958] 1 All E. R. 542, at p. 544; [1958] 2 Q. B. 130, at p. 132. The classic f.o.b. contract has developed into a " flexible instrument ": *Pyrene* v. *Scindia Navigation Co., Ltd.*, [1954] 2 All E. R. 158; [1954] 2 Q. B. 402.

(*h*) An f.o.b. contract which stipulated a range of ports by the words " f.o.b. . . . good Danish port " was thus not void for uncertainty: *David T. Boyd & Co., Ltd.* v. *Louca*, [1973] 1 Lloyd's Rep. 209.

(*i*) *J. J. Cunningham, Ltd.* v. *Munro & Co., Ltd.* (1922), 28 Com. Cas. 42, at p. 45; *Harlow and Jones, Ltd.* v. *Panex* (*International*), *Ltd.*, [1967] 2 Lloyd's Rep. 509, esp. at pp. 526–528. The buyer may change his mind and later nominate a different vessel: *Agricultores Federados Argentinos Sociedad Cooperative* v. *Ampro S.A. Commerciale Industrielle et Financière*, [1965] 2 Lloyd's Rep. 157. Where shipment is to be within a certain period, the buyer must notify the seller of the likely arrival date of the nominated ship in such time as will enable the seller to bring forward and ship the goods within that period: *F. E. Napier* v. *Dexters, Ltd.* (1926), 26 Ll. L. Rep. 184, C. A.

(*j*) *Sutherland* v. *Allhusen* (1866), 14 L. T. 666; *Anglo-African Shipping Co., Ltd.* v. *J. Mortimer, Ltd.*, [1962] 1 Lloyd's Rep. 81 at p. 92; and see on appeal, [1962] 1 Lloyd's Rep. 610, 615, 622–623, C. A.

(*k*) *The Osterbek*, [1972] 2 Lloyd's Rep. 341, at p. 351.

(*l*) Strictly across the ship's rail: *Pyrene* v. *Scindia Navigation Co., Ltd.*, [1954] 2 All E. R. 158, at p. 161; [1954] 2 Q. B. 402, at p. 414.

(*m*) *Stock* v. *Inglis* (1884), 12 Q. B. D. 564, at p. 573; affirmed sub nom. *Inglis* v. *Stock* (1885), 10 App. Cas. 263, at p. 271; see also 19 Mod. L. R. 417, and, for a discussion on the various types of f.o.b. contract, 3 Business L. R. 256.

" puts the goods safely on board, pays the charge for doing so, and for the buyer's protection, but not under a mandate to send, gives up possession of them to the ship 'only upon the terms of a reasonable and ordinary bill of lading or other contract of carriage (*n*). There his contractual liability as seller ceases, and delivery to the buyer is complete so far as he is concerned ' " (*o*).

The buyer under an f.o.b. contract cannot claim delivery of the goods before shipment (*p*). There is no rule that the goods must be examined before shipment. The proper place for examination depends on the circumstances of the particular contract (*q*). Upon shipment the goods must be in such a condition that they can endure the normal journey and can be merchantable, and fit for any relevant purpose made known to the seller, on arrival and for a reasonable time thereafter to enable them to be disposed of or used (*r*). It has been stated that, if the seller ships goods in circumstances in which he retains all rights over and title to goods, he does not fulfil his obligations under an f.o.b. contract (*s*). There are, however, many cases where a seller under an f.o.b. contract has been held to retain rights over and title to the goods after shipment to secure the contract price, for example by taking the bill of lading to his own order (*t*). It is a question of fact in each case whether the seller, by so acting, retains not only the possession of the goods (*u*), but also the title to them (*v*), but as the price is sufficiently secured by retaining possession of the goods, special circumstances would seem to be

---

(*n*) See s. 32 (2) of the Sale of Goods Act, *post*.

(*o*) *Wimble* v. *Rosenberg & Sons*, [1913] 3 K. B. 743, at p. 757. *Cf. Re L. Sutro, Heilbut Symons & Co.*, [1917] 2 K. B. 348, C. A. (printed contract alleged usage as to alternative route); *Cunningham, Ltd.* v. *Robert Munro & Co., Ltd.* (1922), 28 Com. Cas. 42, at p. 45 (general statement of principles of law applicable to f.o.b. contract); see also *Ian Stach* v. *Baker Bosley, Ltd.*, [1958] 1 All E. R. 542; [1958] 2 Q. B. 130.

(*p*) *Maine Spinning Co.* v. *Sutcliffe & Co.* (1917), 87 L. J. K. B. 382.

(*q*) *Boks & Co.* v. *Rayner & Co.* (1921), 37 T. L. R. 519, affirmed by C. A., *ibid.*, 800; *cf. Scaliaris* v. *Ofverberg & Co.* (1921), 37 T. L. R. 307, C.A.; *Bragg* v. *Villanova* (1923), 40 T. L. R. 154 (goods shipped f.o.b. Tarragona, but not inspected there).

(*r*) *Mash and Murell* v. *Joseph I. Emmanuel Ltd.*, [1961] 1 All E. R. 485; [1961] 1 W. L. R. 862; reversed on facts, [1962] 1 All E. R. 77, n; [1962] 1 W. L. R. 16, C. A. This implied term applies also to c.i.f. and other contracts where the goods are to be subject to transit before use; see, *e.g.*, *Commercial Fibres* v. *Zabaida*, [1975] 1 Lloyd's Rep. 27. See further, s. 33 of the Sale of Goods Act, *post*.

(*s*) *Wait* v. *Baker* (1848), 2 Exch. 1, at p. 8; *Browne* v. *Hare* (1853), 3 H. & N. 484; (1859), 4 H. & N. 822. See also *Gabarron* v. *Kreeft* (1875), L. R. 10 Ex. 274.

(*t*) *The Parchim*, [1918] A. C. 157, P. C.; *Frebold* v. *Circle Products, Ltd.*, [1970] 1 Lloyd's Rep. 499, C. A.

(*u*) As in *The Sorfareren* (1915), 114 L. T. 46, and in the cases cited in the preceding note.

(*v*) As in *Wait* v. *Baker*, supra; *Turner* v. *Liverpool Docks Trustees* (1851), 6 Exch. 543; *Ogg* v. *Shuter* (1875), 1 C. P. D. 47.

required to rebut the *prima facie* inference that the parties intend the title to the goods to pass on shipment (*w*).

Where payment is by irrevocable letter of credit, property passes at the latest when payment is made against presentation of the documents by the seller to the bank.

(2) C.i.f. contracts.—Under an ordinary c.i.f. contract the seller has

> " firstly to ship at the port of shipment goods of the description contained in the contract (*x*) ; secondly, to procure a contract of affreightment under which the goods will be delivered at the destination contemplated by the contract (*a*); thirdly, to arrange for an insurance upon the terms current in the trade which will be available to the buyer ; fourthly, to make out an invoice as described by BLACKBURN, J., in *Ireland* v. *Livingston* (*b*), or in some similar form ; and finally to tender these documents to the buyer (*c*), so that he may know what freight he has to pay, and obtain delivery of the goods if they arrive, or recover for their loss if they are lost on the voyage " (*d*).

(*w*) *Frebold* v. *Circle Products, Ltd.*, [1970] 1 Lloyd's Rep. 499, C. A. This inference necessarily weakens the contrary inference to be drawn in a bill of lading case under s. 19 (2) of the Sale of Goods Act, *post*, if the bill is made out to the seller's order, but in principle s. 19 (2) would seem to cover f.o.b. contracts; see *The Parchim*, supra, at pp. 170–171. It has been said that the retention of possession alone would be inadequate to enable the seller to raise finance—see per Lord WRIGHT in *Smyth* v. *Bailey*, [1940] 3 All E. R. 60, at p. 68. But the right to the physical possession of the goods can be transferred for this purpose, as in *Mirabita* v. *Imperial Ottoman Bank* (1878), 3 Ex. D. 164, and *The Parchim*, *supra*. For conflicting views as to the extent to which a seller may retain rights over goods after placing them on board ship, see Carver, *British Shipping Laws*, 12th ed., Vol. 3, para. 1066; Sassoon, *British Shipping Laws*, Vol. 5, para. 391. As to the reservation of a right of disposal of goods generally after they have been appropriated to a contract, see s. 19 of the Sale of Goods Act, 1893, *post*.

(*x*) Within the time stated in the contract, if any, or within a reasonable time: *Landauer & Co.* v. *Craven and Speeding Brothers*, [1912] 2 K. B. 94. Alternatively, the seller may acquire the goods afloat after their shipment from the port of shipment: *Ross T. Smyth & Co.* v. *Bailey Son & Co.*, [1940] 3 All E. R. 60, at pp. 67–68; see also *Vantol* v. *Fairclough Dodd and Jones*, [1955] 1 Lloyd's Rep. 546, 552; [1956] 2 Lloyd's Rep. 437, 447, H. L.; *Lewis Emanuel* v. *Sammut*, [1959] 2 Lloyd's Rep. 629.

(*a*) But the seller does not promise that the goods will be delivered at their destination. The risk of loss or damage is on the buyer as from the shipment. The seller does however give the implied undertaking as to their condition on shipment discussed at note (r) on p. 35.

(*b*) (1872), L. R. 5 H. L. 395, at p. 406.

(*c*) Tender must be as soon as possible after shipment: *Sharpe & Co.* v. *Nosawa & Co.*, [1917] 2 K. B. 814 (reviewing earlier cases). *Quaere* what the position is when the seller acquires goods afloat. There is no absolute duty to tender the documents prior to the vessel's arrival, or before charges are incurred on the goods at the port of destination: *Sanders* v. *Maclean* (1883), 11 Q. B. D. 327.

(*d*) *Biddell Brothers* v. *E. Clemens Horst Co.*, [1911] 1 K. B. 214, at p. 220, *per* HAMILTON, J., and see *ibid.*, [1911] 1 K. B. 934, at p. 952, *per* KENNEDY, L.J., whose dissenting judgment was approved, [1912] A. C. 18 ; *Johnson* v.

It is sometimes said that a c.i.f. contract is a contract for the sale of documents rather than a sale of goods; but the cases show that it is a contract for the sale of insured goods, lost or not lost, to be implemented by the transfer of proper documents.

In the normal c.i.f. contract the property in the goods only passes upon transfer of the documents (*e*), except in special cases where no right of disposal is reserved and property passes on shipment (*f*). Further, in a c.i.f. sale of goods " in bulk " (*g*) property cannot pass until the goods sold have become ascertained usually on or after discharge (*h*).

The risk in the goods passes from shipment in the case of specific goods (*i*). With a sale of unascertained or future goods, there must be a measure of appropriation of goods to the contract before the risk can pass (*k*). Once such appropriation occurs, the risk passes retrospectively as from shipment, but it would seem that if the goods have already been lost before appropriation, the appropriation will not be effective (*l*).

The transfer of proper documents has four elements: first, all the documents called for in the contract must be tendered (*m*); second, the documents must be of the type discussed below and be valid on tender (*n*); third, they must in the case of the bill of lading show, and in the case of the policy and invoice relate to, goods of the contractual description (*m*); fourth, the statements in the documents must be accurate. As to the third and fourth heads,

---

Taylor Bros., [1920] A. C. 144, at p. 156 (suggesting six requisites). See also *Landauer* v. *Craven*, [1912] 2 K. B. 94; *Groom* v. *Barber*, [1915] 1 K. B. 316; *Karberg* v. *Blythe Green & Co.*, [1916] 1 K. B. 495, at p. 513, C. A.

(*e*) " In this course of business, the general property in the goods remains in the seller until he transfers the bills of lading ": *Smyth* v. *Bailey*, [1940] 3 All E. R. 60, H. L., *per* Lord WRIGHT, at p. 68. Where payment is cash against documents, property will not pass until payment is made, despite endorsement and delivery of the bill of lading: *Ginzberg* v. *Barrow Haematite Steel Co., Ltd.*, [1966] 1 Lloyd's Rep. 343, at p. 348, and s. 19 (3) of the Sale of Goods Act.

(*f*) See *The Albazero*, [1974] 2 All E. R. 906; affirmed on appeal by Court of Appeal (as yet unreported).

(*g*) *I.e.*, the sale of an unascertained part of a larger cargo on board a vessel.

(*h*) See the comment to s. 16 of the Sale of Goods Act, *post*.

(*i*) Or retrospectively as from shipment in the case of specific goods acquired afloat: see *Comptoir D'Achat et de Vente du Boerenland Belge S/A* v. *Luis de Ridder*, (1949) 1 All E. R. 269, at p. 274; [1949] A. C. 293, at p. 309, *per* Lord PORTER.

(*k*) " Appropriation " is here used in the sense of the seller binding himself to deliver particular goods or documents relating to them or to make delivery of part of a particular bulk shipment or documents relating to it; see *Benjamin's Sale of Goods*, paras. 1503, 1549.

(*l*) See *Couturier* v. *Hastie* (1856), 5 H. L. Cas. 673, H. L., where the contract would seem to have been a c.i.f. contract. See also Benjamin, *op. cit.*, paras. 1549–1550, 1565.

(*m*) *Libau Wood Co.* v. *H. Smith & Sons* (1930), 37 Ll. L. Rep. 296.

(*n*) *Arnhold Karberg & Co.* v. *Blythe Green & Co.*, [1916] 1 K. B. 495, C. A.

it is not enough to tender documents which purport to relate to the shipment of the contractual goods. If they do so only because they are inaccurate, their tender is a breach of contract by the seller. The seller is also in breach for not shipping the contractual goods. The two breaches are separate (*o*). The fact that the goods shipped do not comply with the contract does not necessarily make the documents inaccurate. Many contracts call for documents giving only the barest description, and this description may be perfectly accurate. But it would seem that a buyer can reject even accurate documents, if he can show that the goods to which they relate do not comply with the contract, since the seller owes a duty to ship or acquire afloat goods of the contract description, and is not entitled to tender documents relating to any but such goods (*p*).

A buyer who initially rejects documents alleging wrongly that they are inaccurate or inadequate may subsequently justify his rejection by showing non-compliance of the goods with the contract (*q*).

Provided he can do so within a reasonable time of shipment, a seller whose tender of documents is justifiably rejected may obtain and tender fresh documents (*r*). If necessary, and provided there has been no appropriation of particular goods to the contract (*s*), these may relate to fresh goods (*t*).

If the goods shipped do not correspond with the contract description, there is a breach of the contract when the goods are shipped (*u*). In case of non-delivery, the seller is in breach for failing to ship, or acquire afloat, the contractual goods, and the

---

(*o*) *Finlay & Co., Ltd.* v. *Kwik Hoo Tong Handel Matschappij*, [1929] 1 K. B. 400; *Chao* v. *British Traders and Shippers*, [1954] 1 All E. R. 779; [1954] 2 Q. B. 459; but see *Panchaud Frères S.A.* v. *Établissement General Grain Co.*, [1970] 1 Lloyd's Rep. 53, C. A. As to the measure of damages recoverable when a buyer pays against inaccurate documents believing them to be accurate, see *Chao* v. *British Traders and Shippers, supra*.

(*p*) *Cf. Taylor* v. *Oakes, Roncoroni & Co.* (1922), 38 T. L. R. 349, at p. 351; affirmed on appeal, 38 T. L. R. 517. The buyer cannot claim to wait until the goods arrive in order to inspect them before deciding whether or not to accept the documents: *E. Clemens Horst* v. *Biddell Brothers*, [1912] A. C. 18, H. L. For a contrary view to that in the text, see *Gulamali Abdul Hussain* v. *Mohammed Yousuf*, [1953] 1 Mad. L. J. R. 504, at p. 507.

(*q*) *Taylor* v. *Oakes, Roncoroni & Co., supra*; *British and Beningtons, Ltd.* v. *North West Cachar Tea Co.*, [1923] A. C. 48, H. L.; *Continental Contractors* v. *Medway* (1925), 23 Ll. L. Rep. 55, 124.

(*r*) *Libau Wood Co.* v. *H. Smith & Sons, Ltd.* (1930), 37 Ll. L. Rep. 296. See also *The Vladimir Ilich*, [1975] 1 Lloyd's Rep. 322 (impossible appropriation by seller).

(*s*) See *Kleinjan Holst N.V. Rotterdam* v. *Bremer Handelsgesellschaft m.b.h. Hambourg*, [1972] 2 Lloyd's Rep. 11.

(*t*) *Borrowman* v. *Free* (1878), 4 Q. B. D. 500.

(*u*) *Parker* v. *Schuller* (1901), 17 T. L. R. 299, C. A.; *Crozier & Co.* v. *Auerbach*, [1908] 2 K. B. 161, C. A.; *Libau Wood Co.* v. *H. Smith & Sons, Ltd.* (1930), 37 Ll. L. Rep. 296.

contract is broken at the time when (v) and the place where (w) the documents ought to have been tendered.

Damages for non-delivery fall *prima facie* to be calculated by the market value of similar goods at the latter time and place (a).

Apart from contrary provision, the contract of affreightment (almost invariably evidenced by a bill of lading) must cover the whole transit and be transferable, so that the buyer may be in a position to claim delivery and to sue for any breach (b). A " received for shipment " bill of lading is not good tender. The bill must evidence shipment " on board " (c), except in trades where a received for shipment bill of lading is shown to be usual (d).

The buyer may refuse to accept an enemy bill of lading though the contract was entered into before the outbreak of war (e).

The goods must be covered by an effective policy of insurance (f), and even if the goods arrive safely this condition is not dispensed with (g). On the other hand, there is a valid tender under the contract if the proper documents be tendered after knowledge that the goods have been lost, but the goods must be covered by a policy which can be assigned to the buyer. An open cover taken out by

---

(v) *Sharpe & Co.* v. *Nosawa & Co.*, [1917] 2 K. B. 814; distinguished on the special facts, *Produce Brokers Co.* v. *Weis & Co.* (1918), 87 L. J. K. B. 472.

(w) *Biddell Bros.* v. *E. Clemens Horst Co.*, [1911] 1 K. B. 934, 961, C. A. And the non-delivery of the documents may be treated as the breach for the purpose of R. S. C., O. 11, r. 1 (g), which was altered to its present form to meet the decision in *Johnson* v. *Taylor Bros.*, [1920] A. C. 144.

(a) *Sharpe & Co.* v. *Nosawa & Co.*, [1917] 2 K. B. 814; and see the comment to s. 51 of the Sale of Goods Act, 1893, *post*.

(b) *Hansson* v. *Hamel & Horley* (1921), 26 Com. Cas. 236, C. A.; affirmed [1922] 2 A. C. 36; *N.V. Arnold Meyer* v. *Aune* (1939), 64 Ll. L. Rep. 121. As to the presence of transhipment terms, see *Holland Colombo Trading Society, Ltd.* v. *Segu Mohamed Alawdeen*, [1954] 2 Lloyd's Rep. 45, at p. 53; *Soproma S.p.A.* v. *Marine and Animal By-Products Corpn.*, [1966] 1 Lloyd's Rep. 367, at p. 388-389.

(c) *Diamond Alkali Corporation* v. *Bourgeois*, [1921] 3 K. B. 443.

(d) *Weis* v. *Produce Brokers* (1921), 7 Ll. L. Rep. 211, C. A.; *United Baltic Corpn.* v. *Burgett & Newsam* (1921), 8 Ll. L. Rep. 190, 191, C. A.

(e) *Karberg* v. *Blythe Green & Co.*, [1916] 1 K. B. 495, C. A.; *cf. Duncan Fox & Co.* v. *Schrempft*, [1915] 3 K. B. 355, C. A.

(f) *Cantiere Meccanico Brindisino* v. *Constant* (1912), 17 Com. Cas. 182, 192; *Loders & Nucoline, Ltd.* v. *Bank of New Zealand* (1929), 33 Ll. L. Rep. 70 (term in contract providing for " insurance at contract price " is not satisfied by separate insurances on freight and goods); *Diamond Alkali Corporation* v. *Bourgeois*, [1921] 3 K. B. 443 (certificate of insurance insufficient); *Scott (Donald) & Co.* v. *Barclays Bank*, [1923] 2 K. B. 1, C. A. (certificate of insurance not an " approved insurance policy "). Even if a certificate of insurance is accepted, the seller impliedly warrants that the assertions in it are true, and that he will produce, or procure the production of, the policy referred to in the certificate: *Harper & Co.* v. *Mackechnie & Co.*, [1925] 2 K. B. 423.

(g) *Orient Co., Ltd.* v. *Brekke*, [1913] 1 K. B. 531. *Cf. Re Denbigh Cowan & Co. & Atcherley & Co.* (1921), 90 L. J. K. B. 836, C. A.

the seller protecting all goods shipped by him is not enough (*h*). A contract by the seller to insure cattle sent abroad against "all risks" is not satisfied by taking out a Lloyds' "all risks" policy if the policy contains the "free of capture and seizure" clause (*i*). If the seller has effected a proper insurance under a c.i.f. policy, he may retain for his own benefit "increased value" policies which he has subsequently taken out (*k*). If an English buyer purchases goods from a foreigner abroad on c.i.f. terms, he cannot demand an English policy, but he may reject the goods if the policy tendered does not specify the risks (*m*).

(3) "Ex-ship" contracts.—"In the case of a sale 'ex-ship'," says Lord SUMNER,

> "the seller has to cause delivery to be made to the buyer from a ship which has arrived at the port of delivery, and has reached a place therein which is usual for delivery of goods of the kind in question. The seller has therefore to pay the freight or otherwise to release the shipowner's lien, and to furnish the buyer with an effectual direction to the ship to deliver. Till this is done the buyer is not bound to pay for the goods. Till this is done he may have an insurable interest in profits, but none that can correctly be described as an interest 'upon goods,' nor any interest which the seller, as seller, is bound to insure for him" (*n*).

The inclusion of terms based on the assumption that the goods will arrive does not necessarily make what is otherwise a c.i.f. contract into an ex-ship contract (*o*). The contract will, however, be an ex-ship contract if it is quite clear that arrival is a condition of payment (*p*). This will generally, if not always, be the case

---

(*h*) *Manbre Saccharine Co.* v. *Corn Products Co.*, [1919] 1 K. B. 198. *Cf. Wilson Holgate & Co.* v. *Belgian Grain Co.*, [1920] 2 K. B. 1, as to an alleged usage.

(*i*) *Yuill* v. *Robson*, [1908] 1 K. B. 270, C. A. (cattle sent from Buenos Aires and seized by Natal Government on account of disease). But distinguish *Vincentelli* v. *Rowlett* (1911), 105 L. T. 411 ("all risks" policy not covering improper stowage for which ship is liable). As to scope and limits of ordinary "all risks" policy, see *British and Foreign Marine Ins. Co.* v. *Gaunt*, [1921] 2 A. C. 41, at p. 57. As to war risks, see *Groom* v. *Barber*, [1915] 1 K. B. 316, when the contract ran "war risk for buyer's account"; *Law and Bonar* v. *American Tobacco Co.*, [1916] 2 K. B. 605 (c.i.f. contract made before war).

(*k*) *Strass* v. *Spillers & Bakers*, [1911] 2 K. B. 759; but distinguish *Landauer* v. *Asser*, [1905] 2 K. B. 184.

(*m*) *Malmberg* v. *Evans & Co.* (1924), 29 Com. Cas. 235; affirmed 30 Com. Cas. 107, C. A., on the particular facts without deciding the general point.

(*n*) *Yangtsze Ins. Association* v. *Lukmanjee*, [1918] A. C. 585, at p. 589, P. C.

(*o*) *Re Denbigh Cowan & Co. & Atcherley & Co.* (1921), 90 L. J. K. B. 836; *Plaimar, Ltd.* v. *Waters Trading Co., Ltd.* (1945), 72 C. L. R. 304 (High Ct. of Aus.); *Fragano* v. *Long* (1825), 4 B. & C. 219, considered in *The Pantanassa*, [1970] 1 All E. R. 848; [1970] P. 187.

(*p*) *Dupont* v. *British S.A. Co.* (1901), 18 T. L. R. 24; *Polenghi* v. *Dried Milk Co.* (1904), 10 Com. Cas. 42; *The Pantanassa*, [1970] 1 All E. R. 848, where under a contract nominally c.i.f. a buyer is, in accordance with a usual practice, merely given a delivery order addressed by the seller to his agents at the port of arrival (*q*).

(4) " Ex-store " contracts.—A contract to deliver " ex-store " does not cover a delivery " ex-lighter " or " ex-quay "; but the exact scope of the contract is somewhat indefinite (*r*).

B. Licences

The import or export of goods under international contracts of sale frequently requires, in modern times, the permission of a governmental authority in the form of an import or export licence. Where this is the case, the parties will usually provide in the contract which of them is to apply for the necessary licences and what is to happen if the application is refused.

If the contract is altogether silent about licences, or is expressed to be " subject to licences " without providing who is to obtain them, a term is usually implied (*s*) making this the duty of one party or the other (*t*). The tendency is to cast the duty upon the party best qualified, by knowledge of the necessary facts or otherwise, to obtain the licence (*u*). Thus, in a sale of turpentine f.a.s. Lisbon, where the seller knew that the buyer wished to export to East Germany, and where the only person who could apply for the necessary export licence was the seller's supplier, whose identity the seller deliberately withheld from the buyer, it was held to be the seller's duty to procure the licence (*v*). The buyer was not bound to change the contemplated destination to one for which an export licence could be obtained, at least where he was only requested to do so towards the end of the contractual shipment period. On the other hand, where jute was sold f.o.b. Dundee, and the seller had no knowledge of its destination, which might have been to another U.K. port or to a foreign port, the duty to apply for an export licence was held to be on the buyer (*a*).

Once it is determined, from the words of the contract or by implication, who is to apply for the licences, there is a separate question, again depending on the circumstances of the particular

---

(*q*) *See Comptoir d'Achat* v. *Luis de Ridder*, [1949] 1 All E. R. 269; [1949] A. C. 293, H. L.

(*r*) *Fisher Reeves & Co.* v. *Armour & Co.*, [1920] 3 K. B. 614, C. A.

(*s*) See the note on implied terms, p. 14, *ante*.

(*t*) *Cassidy* v. *Osuustukkukauppa*, [1957] 1 All E. R. 484.

(*u*) *Brandt* v. *Morris*, [1917] 2 K. B. 784, as explained in *Pound* v. *Hardy*, [1956] 1 All E. R. 639 ; [1956] A. C. 588.

(*v*) *Pound v. Hardy, supra.*

(*a*) *McMaster & Co.* v. *Cox, McEuen & Co.* 1921 S. C. (H. L.) 24 (where the export licence was required by legislation enacted after the contract was made). See also *Mitchell Cotts & Co.* v. *Hairco, Ltd.*, [1943] 2 All E. R. 552 (property having passed to buyer upon tender of shipping documents, buyer was importer of goods, and responsible for obtaining import licence).

case, whether the duty is an absolute one (*b*) or, more usually, whether it is only to use all reasonable diligence to obtain the necessary licences (*c*). Performance of the contract in the latter case is only excused if the duty has been performed but no licence has been obtained (*d*) or if no steps which could have been taken would have led to a licence being obtained (*e*). The other party may have a duty to co-operate with the party applying for a licence, for example by supplying necessary information, at least if requested to do so (*f*).

If an absolute prohibition of export supervenes upon a contract which provides for cancellation in such an event, and the seller receives notice during the contractual period for shipment that such a prohibition will be imposed, he is not discharged from performance unless he does everything possible to ship before the prohibition comes into force (*g*).

C. Exchange control

Section 33 of the Exchange Control Act, 1947, *post*, Appendix I, provides that where Treasury consent is required for the performance of any term of a contract, then, except insofar as it may be shown to be inconsistent with the intention of the parties, it is an implied condition in the contract that the term should not be performed except insofar as that consent is given or not required See further the comment to that section, *post*.

D. Payment by Banker's Irrevocable Letters of Credit

Where seller and buyer carry on business in different countries, the contract of sale frequently provides for the seller to be paid

---

(*b*) As in *Partabmull* v. *Sethia*, [1951] 2 All E. R. 352 *n*., H. L., and in *Cassidy* v. *Osuustukkukauppa*, [1957] 1 All E. R. 484.

(*c*) *Re Anglo-Russian Merchant Traders and Batt*, [1917] 2 K. B. 679, at p. 686, C. A.; *Brauer* v. *Clark*, [1952] 2 All E. R. 497, C. A.; *Pound* v. *Hardy* (*supra*). Where the contract is silent, the duty imposed is almost invariably only to use all reasonable diligence; and *semble* that where the contract is "subject to licence," the duty cannot be absolute—see the dictum of DEVLIN, J., in *Cassidy* v. *Osuustukkukauppa* (*supra*), at p. 488.

(*d*) *Taylor* v. *Landauer*, [1940] 4 All E. R. 335; *Société D'Avances Commerciales* v. *Besse*, [1952] 1 Lloyd's Rep. 242; *Brauer* v. *Clark* (*supra*).

(*e*) In either case, the onus of proof is upon the party upon whom the duty lay to show that he ought to be excused; this is a particularly difficult task in the latter case: *Windschluegl* v. *A. Pickering & Co.* (1950), 84 Ll. L. R. 89; at p. 95; *Société d'Avances Commerciales* v. *Besse, supra*. In *Agroexport State Enterprise for Foreign Trade* v. *Compagnie Européene de Céréals*, [1974] 1 Lloyd's Rep. 499, a government decision just before the shipment date to refuse further export licences was no excuse for a seller who had already had several months in which to obtain a licence. In *Malik Co.* v. *Central European Trading Agency, Ltd.*, [1974] 2 Lloyd's Rep. 279, the seller had failed to apply for several months, and although he did everything possible thereafter, he failed to discharge the onus upon him.

(*f*) *Pound* v. *Hardy*, [1956] 1 All E. R. 639; [1956] A. C. 588, at pp. 608, 611.

(*g*) *Smyth* v. *Lindsay*, [1953] 2 All E. R. 1064, distinguishing *Re Anglo-Russian Merchant Traders* v. *Batt* (*supra*), where the prohibition came into effect without prior notice.

through a bank. The most common provision is for payment by means of an irrevocable letter of credit, which has the considerable advantage to the seller that, provided he complies with its terms, he is assured of payment.

An irrevocable letter of credit is an undertaking issued by the buyer's bank by which the bank promises the seller to pay the price of goods upon presentation of specified documents, either in cash or by negotiating or accepting drafts presented by the seller with documents. The specified documents will almost invariably include the bill or bills of lading relating to the goods sold, and in the case of a c.i.f. contract the requisite policy of insurance, and may include such other documents as certificates of quality and origin.

The buyer's bank may instruct another bank in the seller's country to advise the seller of the opening of the letter of credit (*h*). This second bank may " confirm " the credit to the seller, thus adding its own personal liability to that of the issuing bank. Confirmation by a bank in the seller's own country is frequently a term of the contract (*i*).

Where the contract of sale stipulates for payment by an irrevocable letter of credit, it is the duty of the buyer to open such a credit prior to the earliest date for shipment, whether the contract is c.i.f. (*k*) or f.o.b. (*l*). It has been said that the credit must be opened a reasonable time before the first date for shipment (*m*), but this has been questioned (*n*). Failure to open the credit in time is a repudiation of the contract of sale (*o*).

The bank must comply strictly with the instructions given to it by the buyer as to the terms of the credit to be opened, and must take proper care to ensure that the documents when presented

---

(*h*) There may indeed be a chain of banks.

(*i*) See, *e.g.*, *Enrico Furst & Co.* v. *W. E. Fischer, Ltd.*, [1960] 2 Lloyd's Rep. 340. For the position of a " confirming house", see *Rusholme and Bolton and Hadfield* v. *Read, S. G.*, [1955] 1 All E. R. 180; *Sobell Industries* v. *Cony Bros.*, [1955] 2 Lloyd's Rep. 82.

(*k*) *Pavia & Co. S. P. A.* v. *Thurmann-Neilson,* [1952] 1 All E. R. 492; [1952] 2 Q. B. 84, C. A. See *Sinason-Teicher* v. *Oilcakes and Oilseeds Trading Co.*, [1954] 3 All E. R. 468, C. A. for an analogous case of a bank guarantee of payment upon presentation of documents.

(*l*) *Ian Stach* v. *Baker Bosley*, [1958] 1 All E. R. 542; [1958] 2 Q. B. 130. In the case of f.o.b. contracts there is no difference in this respect between "classic" f.o.b. contracts, where the seller need take no step towards shipment until he has had instructions from the buyer, and contracts where it is for the seller to arrange shipping space.

(*m*) *Sinason-Teicher Inter American Grain Corporation* v. *Oilcakes and Oilseeds Trading Co.*, [1954] 3 All E. R. 468, at p. 472, *per* DENNING L.J.

(*n*) *Ian Stach* v. *Baker Bosley, Ltd.*, [1958] 1 All E. R. 542, [1958] 2 Q. B. 130, at pp. 141, 144. See also *Alexandria Cotton and Trading Co. (Sudan), Ltd.* v. *Cotton Co. of Ethiopa, Ltd.*, [1963] 1 Lloyd's Rep. 576, where ROSKILL, J., said that he did not regard the rule as finally settled by the authorities.

(*o*) *Ian Stach* v. *Baker Bosley, supra.*

comply strictly with those terms (*p*). Otherwise the bank may sacrifice its right to indemnity from the buyer. But if the documents on reasonable examination appear on their face to comply, the bank is entitled to its indemnity even if the documents are in fact forgeries (*q*).

Bills of lading tendered by the seller to the bank must be clean, i.e. they must be free of express clauses indicating defects in the goods' condition upon shipment (*r*). " Received for shipment " bills of lading and bills incorporating the terms of a charterparty are not acceptable for tender under letters of credit (*s*).

The extent to which the seller may look to the buyer direct for the price will depend upon the construction of the sale contract (*t*). The *prima facie* position appears to be as follows:

(1) Payment by letter of credit is, where agreed, the primary mode of payment to which the seller must if possible look (*t*).

(2) If the seller is unable to obtain payment under the letter of credit, he may look to the buyer direct for payment unless either there is a clear stipulation to the contrary, or the seller was himself responsible for failing to obtain payment under the letter of credit (*u*).

(3) The buyer has a duty to the seller under the contract to provide a letter of credit in accordance with its terms and to take every reasonable step, at least, to ensure that the Bank will pay on presentation under the letter of credit. If the buyer fails to perform this duty, and as a result the Bank fails to pay,

---

(*p*) The " *de minimis* " rule does not apply: see *Rayner & Co., Ltd.* v. *Hambros Bank, Ltd.*, [1943] 1 K. B. 37; *Moralice* (*London*), *Ltd.* v. *E. D. & F. Man*, [1954] 2 Lloyd's Rep. 526.

(*q*) *Gian Singh & Co., Ltd.* v. *Banque de l'Indochine*, [1974] 2 All E. R. 754; [1974] 1 W. L. R. 1234, P. C.

(*r*) As to " clean " bills, see *British Imex Industries, Ltd.* v. *Midland Bank, Ltd.*, [1958] 1 All E. R. 264; [1958] 1 Q. B. 542.

(*s*) *Enrico Furst & Co.* v. *W. E. Fischer, Ltd.*, [1960] 2 Lloyd's Rep. 340, at pp. 345–346. The credit itself will usually expressly insist upon " on board " bills. See also Arts. 17 and 18 of the Uniform Customs and Practice for Documentary Credits (1962 Revision).

(*t*) *W. J. Alan & Co.* v. *El Nasr Export & Import Co.*, [1972] 2 All E. R. 127; [1972] 2 Q. B. 189, C. A., *Saffaron* v. *Société Minière Cafrika* (1958) 100 C. L. R. 231 (H. C. of Aus.), *Soproma S.p.A.* v. *Marine Animal By-Products Assocn.*, [1966] 1 Lloyd's Rep. 367.

(*u*) *Maran Road Saw Mill* v. *Austin Taylor Co., Ltd.*, [1975] 1 Lloyd's Rep. 156, at p. 159; *Newman Industries* v. *Indo-British Industries*, [1956] 2 Lloyd's Rep. 219; revd. on other grounds: [1957] 1 Lloyd's Rep. 211, C. A., where the bank's refusal to pay was arguably justified under the letter of credit, but SELLERS, J. held that even if it was justified, the seller could sue the buyer direct, because the buyer should have rectified the position with the bank so as to accord with the sale contract. In *Saffaron* v. *Société Minière Cafrika* (1958), 100 C. L. R. 231, the High Court of Australia suggested that the only circumstance in which a seller might forfeit the price would be if it was his fault that payment under the letter of credit had not been forthcoming. But even in such a case as that, there seems no good reason why the buyer should not pay the price unless he has been prejudiced by the seller's default.

then the seller would seem to be entitled to claim damages from the buyer equivalent to the price, even if the contract provides that he should not be entitled to look to the buyer direct for the price.

The opening of an irrevocable letter of credit gives rise to a contract between the issuing bank (and, if there is one, the confirming bank) on the one hand, and the seller on the other (*a*). This contract is governed by the terms of the letter of credit and is entirely separate from the contract of sale and from the buyer's contract with the issuing bank. Under this contract with the seller the bank must pay against documents (or honour or accept drafts as appropriate) provided the documents are on their face good. The bank cannot withhold payment because of the nature or condition of the goods. If the buyer has any complaint on this score, he must bring an action for damages against the seller (*b*). In general, he cannot prevent the seller from presenting the necessary documents to the bank for payment, nor can he prevent the bank from paying the seller (*b*).

In the United States, it has been held that the only exception to the bank's duty to pay exists if the documents contain statements fraudulently inserted by the seller or persons for whom he is responsible, or if the seller or other person presenting the documents knows that they contain inaccuracies in some respect material to the letter of credit (*c*). English dicta also support a right to reject in case of fraud (*d*). In principle, the bank would also seem to be justified in refusing payment if the contract of sale is illegal (*e*).

If the bank wrongfully fails to pay under the letter of credit,

---

(*a*) *Urquhart, Lindsay & Co., Ltd.* v. *Eastern Bank, Ltd.*, [1922] 1 K. B. 318; *Hamzeh Malas & Sons* v. *British Imex Industries, Ltd.*, [1958] 1 All E. R. 262; [1958] 2 Q. B. 127, C. A. Consideration can clearly be found in the seller's acting on the credit by bringing the goods forward for shipment.

(*b*) *Hamzeh Malas & Sons* v. *British Imex Industries, Ltd.*, *supra*.

(*c*) *Sztejn* v. *J. Henry Schroder Banking Corpn.* (1941), 31 N. Y. S. (2d) 631 (reviewing earlier decisions). Quaere what the position is if the documents have no validity because forged: see *Chao* v. *British Traders and Shippers, Ltd.*, [1954] 1 All E. R. 779; [1954] 2 Q. B. 459; *cf. Old Colony Trust Co.* v. *Lawyers' Title and Trust Co.* (1924), 297 F. 152, which appears to suggest that the bank would then always be able to reject them, even when presented by *bona fide* holders for value. The Uniform Commercial Code—Letters of Credit (1962 version), s. 5–114, assimilates forged documents with valid but fraudulent documents in those states which apply it.

(*d*) *Société Metallurgique d'Anbrives and Villerupt* v. *British Bank for Foreign Trade* (1922), 11 Ll. L. Rep. 168, at p. 170; *Hamzeh Malas & Sons* v. *British Imex Industries, Ltd.*, [1958] 1 All E. R. 262, at p. 263; [1958] 2 Q. B. 127, at p. 130, C. A. The position was left open in *Discount Records, Ltd.* v. *Barclays Bank, Ltd.*, [1975] 1 All E. R. 1071; [1975] 1 W. L. R. 315. See also Gutteridge & Megrah, *The Law of Bankers' Commercial Credits* (4th ed.), pp. 133–134; *Chitty on Contracts* (23rd ed.), Vol. 2, para. 450; 7 Malaya L. R. 24 (Ellinger); and contrast Davis, *The Law Relating to Commercial Letters of Credit* (3rd ed.), p. 156; *Paget's Law of Banking* (8th ed.), p. 660–661.

(*e*) See p. 17, *ante*.

the seller may recover from the bank the amount which should have been paid, usually with interest (*f*), upon retendering the documents (*g*). If instead the seller re-sells the goods and claims damages from the bank, then the seller would seem to be under a duty to the bank to take reasonable steps to mitigate his loss when re-selling (*h*). If the credit covers several instalments, failure to pay in respect of one shipment may amount to repudiation of the whole credit, in which case the seller may accept the repudiation and claim damages (*i*).

It is a term of the contract between the buyer and the bank (to be implied if not expressed) that the bank will not only open the credit, but will pay the seller in accordance with its terms, so that if the bank evinces an intention not to pay the seller, the buyer is relieved from his obligation to the bank to put it in funds to meet the credit (*k*).

E. Confirming Houses

Another means by which a buyer may provide his seller with security of payment is by use of a confirming house, usually carrying on its business in the seller's country. The buyer, who may or may not have been in direct contact with the seller, instructs the confirming house, in return for a commission, either simply to place the order with the seller in its own name or to add its confirmation to the buyer's order form and send the same to the seller.

In both cases the confirming house is, as regards the seller, a principal to the purchase (*l*) but, as regards the buyer, it acts as an agent with a corresponding right to indemnity (*m*). Its position is

(*f*) Now under the Law Reform (Miscellaneous Provisions) Act, 1934, s. 3, *post*, Appendix I. See also the cases cited at n. (*g*), *post*. There is also a right at law to interest on an overdue banker's draft, as with other negotiable instruments.

(*g*) As in *Belgian Grain and Produce Co.* v. *Cox & Co. (France), Ltd.* (1919), 1 Ll. L. Rep. 256, C. A.; *Stein* v. *Hambro's Bank of Northern Commerce* (1921), 9 Ll. L. Rep. 507. It was suggested in *Société Metallutgique d'Anbrives and Villerupt* v. *British Bank of Foreign Trade* (1922), 11 Ll. L. Rep. 168, that if the buyer would be able to reject the goods on arrival for non-compliance with the contract, damages against the bank for failure to pay against documents might be nominal, on the ground of circuity of action. *Sed quaere*.

(*h*) The seller's right to resell and claim damages would seem to be governed by principles analogous to those set out in s. 48 (3) of the Sale of Goods Act, 1893, *post*.

(*i*) *Urquhart, Lindsay & Co.* v. *Eastern Bank, Ltd.*, [1922] 1 K. B. 318, where it was said that the buyer owed no duty to mitigate his damages by letting the documents go against part payment.

(*k*) *Sale Continuation, Ltd.* v. *Austin Taylor & Co., Ltd.*, [1967] 2 All E. R. 1092; [1968] 2 Q. B. 849.

(*l*) *Rusholme Bolton and Roberts Hadfield, Ltd.* v. *S. G. Read & Co.*, [1955] 1 All E. R. 180; [1955] 1 W. L. R. 146; *Sobell Industries, Ltd.* v. *Cory Brothers, Co., Ltd.*, [1955] 2 Lloyd's Rep. 82; *cf. Maran Road Saw Mill* v. *Austin Taylor & Co., Ltd.*, [1975] 1 Lloyd's Rep. 156.

(*m*) *Anglo-African Shipping, Co., Ltd.* v. *J. Mortimer, Ltd.*, [1962] 1 Lloyd's Rep. 610, C. A., pointing out that confirming houses often act in additional capacities on behalf of the buyer, *e.g.* as forwarding agents.

thus distinct from that of a *del credere* agent (whose liability to the seller only arises in the event of the buyer's insolvency). It is, however, submitted that since in both cases the confirming house acts on the buyer's instructions for a commission, the buyer is also a party (whether disclosed or undisclosed) to the contract of purchase, and could (for example) be sued by the seller in the event of the confirming house's default (*n*).

A different situation exists where a confirming house intervenes in such a way as to give rise to a back-to-back sales first by the supplier to the confirming house, and then by the confirming house to the ultimate buyer. This is only distinguished from an ordinary sale and resale in that the confirming houses reward consists of a commission on the suppliers' sale price.

F. Conflict of laws

A distinction must be drawn here between (1) questions arising out of the law of contract, and (2) questions arising out of the law of property. The former concern such matters as capacity, formalities, essential validity, construction, discharge, and illegality; the latter cover such things as the passing of property and risk, lien, and stoppage *in transitu*.

(1) Questions arising out of the law of contract are generally determined by the " proper law " of the contract. The contract may or may not provide what the proper law shall be. Where it does not so provide, the proper law is " the system of law by reference to which the contract was made or that with which the transaction has the closest and most real connection " (*o*). In determining which system of law this is, it is necessary to consider the terms of the contract, the situation of the parties, and generally all the surrounding facts (*p*). These will include where the contract was made, where it is to be performed, the nature and situation of the subject matter, the language in which it is written, the system of

---

(*n*) *Rusholme* v. *S. G. Read, supra*, is sometimes regarded as an authority where the confirming house although instructed to purchase by the buyer was not authorised to create a contract between supplier and buyer—see *Benjamin's Sale of Goods*, para. 2123. This is not easy to follow in principle, and, it is submitted, PEARCE, J. certainly did not intend to decide that no contract existed at all between the supplier and Australian buyers, merely that there was no such contract made by the indent which such buyers had sent direct to the supplier prior to the confirming house's confirmation. PEARCE, J. expressly described the confirming house as the buyer's agent at pp. 183 and 150 and both agent and principal (whether disclosed or undisclosed) may be liable on a purchase—see *Bowstead on Agency* (13th ed.), Article 118.

(*o*) *Bonython* v. *Commonwealth of Australia*, [1951] A. C. 201, 219, P. C.; *re United Railways of Havana and Regla Warehouses*, [1960] 2 All E. R. 332; [1961] A. C. 1007; Cheshire's *Private International Law* (9th ed.), p. 201; Dicey and Morris *Conflict of Laws* (9th ed.) p. 721.

(*p*) *The Assunzione*, [1954] 1 All E. R. 278, 289; [1954] P. 150, 175.

law to which any legal terms used belong (*q*), the currency in which any payments are to be made, and the selection of arbitrators from some particular country (*r*).

The expressed intention of the parties as to the proper law of their contract will be given effect, save where their choice is illegal or contrary to public policy (*s*).

It is common for parties to choose English arbitration or jurisdiction in relation to contracts which have little or no other connection with England. Such choice is a strong indication that the parties also intend English law as the proper law (*t*), but it is not conclusive and may therefore be over ridden if other indications point clearly to some different system of law.

For questions as to *capacity*, see the note " conflict of laws " to s. 2, *post*.

A contract is valid as to *form* so long as it complies with the requisite formalities either of its proper law or of the *lex loci contractus* (*u*).

Questions of essential validity, construction and discharge are always governed by the proper law of the contract (*v*).

A contract which is *illegal* under its proper law is always void; and if it is valid under its proper law it will still not be enforced in England if it is contrary to English public policy (*a*) or, at least where English law is the proper law (*b*), in so far as performance would be illegal by the law of the place of performance (*c*) .

(*q*) *James Miller & Partners, Ltd.* v. *Whitworth Street Estates, Ltd.*, [1970] 1 All E. R. 796; [1970] A. C. 583, H. L.

(*r*) *Hamlyn* v. *Talisker Distillery*, [1894] A. C. 202; *Compagnie d'Armement Maritime S.A.* v. *Compagnie Tunisienne de Navigations S.A.*, [1970] 3 All E. R. 71; [1970] 3 W. L. R. 389, H. L.

(*s*) See, *e.g.*, Dicey and Morris, *loc cit.*, rule 146, sub-rule 1; Cheshire, *loc cit.*, pp. 205 *et seq.*; *N.V. Kwik Hoo Tong* v. *James Finlay & Co., Ltd.*, [1927] A. C. 604; *Vita Food Products, Incorporated* v. *Unus Shipping Co., Ltd.*, [1939] 1 All E. R. 513; [1939] A. C. 277, P.C. (refusing to follow the earlier decision of the Court of Appeal in *The Torni*, [1932] P. 78); *re Helbert Wagg & Co.*, [1956] 1 All E. R. 129, 136; [1956] Ch. 323, 341; *Mackender* v. *Feldia A.G.*, [1966] 3 All E. R. 847, C. A.; 56 L. Q. R. 320. See also s. 55 A of the Sale of Goods Act, *post*, which overrides the normal effect of choosing a foreign law.

(*t*) *Compagnie d'Armement Maritime S.A.* v. *Compagnie Tunisienne de Navigation S.A.*, [1970] 3 All E. R. 71; [1971] A. C. 572, H. L., explaining and disapproving statements in *Tzortzis* v. *Monark Line A/B*, [1968] 1 All E. R. 949; [1968] 1 W. L. R. 406, C. A.

(*u*) Dicey and Morris, *loc cit.*, rule 150; *Albeko Schumaschinen A.-G.* v. *Kamborian Shoe Machine Co.* (1961), 111 L. Jo. 519; 10 I. C. L. Q. 908.

(*v*) *P. & O.* v. *Shand* (1865), 3 Moo. P. C. C. N. S. 272 (essential validity); *Jacobs* v. *Crédit Lyonnais* (1884), 12 Q. B. D. 589, C. A. (discharge).

(*a*) *Kaufman* v. *Gerson*, [1904] 1 K. B. 591, C. A.

(*b*) Cheshire, *loc. cit.*, p. 233.

(*c*) *Ralli Bros.* v. *Compania Naviera &c.*, [1920] 1 K. B. 614; [1920] 2 K. B. 287, C. A.; *cf. Foster* v. *Driscoll*, [1929] 1 K. B. 470, C. A. (partnership agreement to smuggle intoxicating liquor into United States); *Regazzoni* v. *Sethia*, [1957] 3 All E. R. 286; [1958] A. C. 301; (c.i.f. contract illegal by law of country of export).

All rules of *procedure* and *evidence* are governed by the *lex fori*, the law of the court in which the action is brought. This includes the admissibility of evidence, the means of proving particular facts, the time within which an action must be brought, rights of set-off and counterclaim, remedies for breach, joinder of parties, the effect of recovering judgment, and execution. Hence in an action in England all English rules of procedure and evidence apply (*d*) and all such foreign rules are ignored. The test seems to be that if the foreign rule bars the remedy only, the English court ignores it as procedural; but if it makes the contract itself void, it will be recognised as a rule of substance (*e*).

(2) Passing to questions arising out of the law of property, subject to certain qualifications, the English rule has generally been taken to be that if personal property be disposed of in a manner binding according to the law of the country where it is (the *lex situs*), that disposition is binding in England (*f*). Thus where the master of a ship wrongfully sold the cargo by auction in Norway, but in such circumstances as to give a good title in Norway, the sale was held valid, although the cargo subsequently came to England (*g*); the title to a stolen horse bought in market overt in Ireland would be recognised in Scotland (*h*); and the validity of a pledge is determined by the law of the country where the goods were pledged (*i*). *Locus regit actum* is a rule of wide application.

G. Foreign law

In the absence of evidence to the contrary, foreign law is presumed to be the same as English law. If the foreign law differs, and the difference is relied on, it must be proved as a question of fact (*k*). But the effect of the evidence as to the foreign law is a

(*d*) See, *e.g.*, *Leroux* v. *Brown* (1852), 12 C. B. 801 (decided on a repealed part of s. 4 of the Statute of Frauds, 1677); *D'Almeida Araujo Lda.* v. *Becker & Co., Ltd.*, [1953] 2 All E. R. 288; [1953] 2 Q. B. 329; *In the Estate of Fuld (No. 3)*, *Hartley* v. *Fuld*, [1965] 3 All E. R. 776; [1966] 2 W. L. R. 717; and see *James Miller & Partners, Ltd.* v. *Whitworth Street Estates, Ltd.*, [1970] 1 All E. R. 796; [1970] A. C. 583, H. L. (English contract—arbitration in Scotland by Scottish arbiter governed as to procedure by Scottish law).

(*e*) *Alves* v. *Hodgson* (1797), 7 Term Rep. 241.

(*f*) *Embiricos* v. *Anglo-Austrian Bank*, [1905] 1 K. B. 677, at p. 683, C. A.; *Re Korvine's Trust*, [1921] 1 Ch. 343, 348 (*donatio mortis causâ*); Dicey and Morris, *Conflict of Laws* (9th ed.) Rules 81–82; Cheshire's *Private International Law* (8th ed.), p. 528; 22 I. C. L. Q. 213. British ships constitute a statutory exception to this rule.

(*g*) *Cammell* v. *Sewell* (1860), 5 H. & N. 728, Ex. Ch.; *cf. Alcock* v. *Smith*, [1892] 1 Ch. 238, C. A. (bill of exchange taken in execution abroad).

(*h*) *Todd* v. *Armsur*, 1882 9 R. (Ct. of Sess.) 901.

(*i*) *City Bank* v. *Barrow* (1880), 5 App. Cas. 664, at p. 677.

(*k*) *The Parchim*, [1918] A. C. 157, at pp. 160, 161, *per* Lord PARKER. As to expert evidence to prove the foreign law, see *Phipson on Evidence* (11th ed.), pp. 1290–1294; *Cross on Evidence* (4th ed.), pp. 556–559. *Cf. Perlak Petroleum Co.* v. *Deen* [1924], 1 K. B. 111, 115 (evidence of foreign law by non-expert inadmissible).

matter for the judge, and not for the jury, if any (*l*). In referring to foreign laws, it should be noted that some continental codes distinguish traders from non-traders, and modify or supplement the general law of the country as regards commercial transactions.

H. Uniform Law on International Sales Act, 1967

This Act is designed to incorporate into the law of the United Kingdom the Uniform law on the International Sale of Goods, which governs the obligations of the seller and buyer arising from a contract of sale to which it applies, and the Uniform Law for the Formation of Contracts for the International Sale of Goods, which is concerned with the formation of such contracts. The laws were the subject of Conventions drawn up at The Hague in 1964.

The two Uniform Laws relate to international sales as defined in Article 1 of the First Law. Save that the definition in Article 1 is limited to contracts between parties whose places of business are in different Contracting States, it is identical to that contained in s. 62 (1) of the Sale of Goods Act, 1893, *post*. By Article 4 of the First Law parties can also choose this law as the proper law of their contract whether or not the parties places of business or habitual residences are in different states and whether or not such states are parties to the Conventions, but only to the extent that it does not affect the application of ss. 12 to 15, 55 and 55A of the Sale of Goods Act, 1893 (*m*).

An Order in Council has been made (*n*) applying the Laws from 18th August 1972, but this would not seem to have had any great effect to date in that:

(1) The Order further declares pursuant to s. 1 (3) of the Act that the Laws will apply to a contract affected by this Act only if chosen by the parties to that contract; and

(2) Only six States have been declared to be contracting parties under the Order; Belgium, Israel, Italy, the Netherlands, San Marino, and the United Kingdom.

(*l*) Supreme Court of Judicature (Consolidation) Act, 1925, s. 102; (25 Halsbury's Statutes (3rd Edn.) 744) as to the High Court; County Courts Act, 1959, s. 97 (5 Halsbury's Statutes (3rd Edn.) 362), as to County Courts; the Administration of Justice Act, 1920, s. 15 (6 Halsbury's Statutes (3rd Edn.) 358), as to other Courts.

(*m*) Uniform Law on International Sales Act, 1967, s. 1 (4), as amended by s. 5 (2) of the Supply of Goods (Implied Terms) Act, 1973.

(*n*) S. I. 1972, No. 973.

SECTION C

STATUTORY CONTROLS OVER THE SALE OF GOODS

I STATUTORY PRICE CONTROLS

A. Counter-Inflation legislation

In recent years, successive governments have sought to reduce inflation by controlling the prices which may be charged for goods sold in the course of a business. At present control is principally exercised under the Counter Inflation Act, 1973, which applies for three years from 1st April, 1973 unless previously terminated by Order (*a*).

By s. 5 of that Act the Minister may make orders to ensure that the Price Commission receives notice of increases in any prices in time to consider whether the increases conform with the relevant provisions of a Price Code prepared by the Treasury under s. 2 of the Act (*b*). The notice may not exceed eight weeks, but the Order may provide that until the end of the period of notice it is a contravention of the Order to implement the increase.

By s. 6 of the Act, the Price Commission may restrict prices for the sale of goods in the course of a business by order or by notice given to the person or to each of the persons selling the goods. Persons affected must, however, be given the opportunity to make written representations to the Price Commission before any order is made or notice given (*c*).

The Minister and the Price Commission have very wide powers to obtain information (*d*).

Contravention of an order is punishable by fine (*e*). A contract is not invalid if the price is in excess of that permitted, but the person paying the price may recover the excess unless he himself is liable to punishment in connection with the offence (*f*).

The enforcement of the Act is entrusted to the local weights and measures authorities (*g*).

(*a*) Control over some fresh food and certain other goods is also exercised by the Secretary of State under the Prices Acts, 1974 and 1975, which will apply until, at latest, 31st March, 1976.

(*b*) Since the Act became law, a number of Orders have been made, amended and replaced. There have also been several versions of the Price Code.

(*c*) S. 6 (4), (5).

(*d*) S. 15.

(*e*) £400 on summary conviction; unlimited where the conviction is on indictment: see s. 17.

(*f*) Sch. 3, para. 3; S. I. 1973 No. 660.

(*g*) Sch. 4.

B. Underselling, &c.

At common law, unless he has otherwise bound himself by contract, a man may sell at any price he pleases, and to undersell trade rivals he may offer goods at a loss (*h*); moreover, so long as it is not in unreasonable restraint of trade (*i*) he can attach a condition to such a sale that the goods shall not be resold by the buyer below some specified price (*k*).

This position has now been radically altered by the combination of the Resale Prices Act, 1964 and Article 85 of the Treaty of Rome as enacted by s. 2 of the European Communities Act, 1972.

Under the 1964 Act, where a condition as to resale price maintenance is attached to a sale by a supplier to a dealer (*l*), it is now by statute unlawful and void (*m*) except in the case of:

(i) goods to be resold outside the United Kingdom (*n*);

(ii) goods exempted by the Restrictive Practices Court from the provisions of the Resale Prices Act, 1964, in exercise of its powers under s. 5 of that Act (*o*).

Even in those cases, however, re-sale price maintenance agreements will frequently be unenforceable by virtue of Article 85. Article 85 (1) applies to all agreements, decisions and concerted practices between undertakings which affect undertakings between member states and have as their object or effect the prevention, restriction or distortion of competition within the Common Market, including *inter alia* those fixing purchase or selling prices or any other trading conditions. Such agreements, decisions and practices must be notified to the Commission (*p*).

Agreements within category (i) mentioned above will thus commonly be caught by Article 85; so, too, agreements within category (ii) will commonly be caught insofar as they fix the price or conditions of sale in the United Kingdom from elsewhere in the Common Market. By concession (*q*) agreements to which only two undertakings are parties and which only restrict resale prices or conditions are not subject to notification, but they are still unenforceable if they materially affect trade between member states. The Commission

---

(*h*) *Ajello* v. *Worsley*, [1898] 1 Ch. 274, 280.

(*i*) *Petrofina (Great Britain), Ltd.* v. *Martin*, [1966] 1 All E. R. 126; [1966] Ch. 146, C. A.; *Esso Petroleum Co., Ltd.* v. *Harper's Garage (Stourport), Ltd.*, [1967] 1 All E. R. 699; [1968] A. C. 269, H. L.

(*k*) *Elliman* v. *Carrington*, [1901] 2 Ch. 275. As to enforcement of these conditions at common law, see *Sorrell* v. *Smith*, [1925] A. C. 700; *Hardie & Lane, Ltd.* v. *Chilton*, [1928] 2 K. B. 306, disapproving *R.* v. *Denyer*, [1926] 2 K. B. 258, C. C. A.

(*l*) See the definitions of "supplier" and "dealer" in s. 11 (1) of the Resale Prices Act, 1964; 37 Halsbury's Statutes (3rd Edn.) 167.

(*m*) Resales Prices Act, 1964, s. 1.

(*n*) *Ibid.*, s. 1.

(*o*) *I.e.*, books and medicaments.

(*p*) Common Market Regulations, Regulation 17 (1962), art. 4 (1).

(*q*) *Ibid.*, art. 4 (2).

has, however, stated that it considers minor agreements outside Article 85. For this purpose, minor agreements are agreements where (a) the products involved represent less than 5 per cent of the volume of such products in the area of the Market affected and (b) the aggregate turnover of the parties (if commercial undertakings) does not exceed 20,000,000 units of account (one unit being equal to a little over 1 U.S. dollar) (*r*). Infringements of Article 85 are punishable by fine. The commission may by decision (which would no doubt be enforced by injunction) require termination of an infringement (*s*).

Where resale price maintenance is still lawful any conditions as to the price at which goods sold by a supplier (*t*) may be resold can be enforced against any person not party to the original sale who subsequently acquires the goods with notice of the condition as if he had been a party to the sale (*u*). " Notice " here seems to mean no more than sufficient, rather than actual, notice, so that it is enough for the supplier to prove that he took all reasonable steps to bring the condition to the notice of the class of persons of which the defendant is one (*a*).

The remedy is by injunction or " any other relief which may be granted" (*b*). The court acts in accordance with acknowledged general principles in determining whether to grant an injunction (*c*). The remedy is not available against persons who acquire the goods otherwise than for the purpose of re-sale in the course of business, or their sub-purchasers (*d*), nor is it available against persons selling goods pursuant to an order of any court, or their sub-purchasers (*e*). " Condition as to price " includes conditions as to amount of discount or trading-in allowance for other goods taken in exchange on such re-sale (*f*).

(*r*) Notice of 27 May, 1970, J. O. 1970 C84/1.

(*s*) Regulation 17 (1962) arts. 3, 15 and 16.

(*t*) No definition of " supplier " is given in the Restrictive Trade Practices Act, 1956, which introduced this remedy.

(*u*) Restrictive Trade Practices Act, 1956, s. 25 (1); 37 Halsbury's Statutes (3rd Edn.) 106. At common law such a condition was not binding on subsequent buyers, even with notice (*Dunlop Pneumatic Tyre Co.* v. *Selfridge & Co.*, [1915] A. C. 847) except it seems in the case of patented articles known to be sold under a restrictive licence: (*National Phonograph Co. of Australia* v. *Menck*, [1911] A. C. 336, 347, P. C.; *Columbia Phonograph Co.* v. *Regent Fittings Co.* (1913), 30 R. P. C. 484).

(*a*) *Goodyear Tyre and Rubber Co.* (*Great Britain*) v. *Lancashire Batteries*, [1958] 3 All E. R. 7; [1958] 1 W. L. R. 857, C. A.

(*b*) Restrictive Trade Practices Act, 1956, s. 25 (4); 37 Halsbury's Statutes (3rd Edn.) 106.

(*c*) *Gallaher, Ltd.* v. *Supersafe Supermarkets* (1964), L. R. 5 R. P. 89; *cf.* 1956 Public Law 320, *n.* 15.

(*d*) Restrictive Trade Practices Act, 1956, s. 25 (2); 37 Halsbury's Statutes (3rd Edn.) 106.

(*e*) *John Mackintosh & Sons* v. *Baker's Bargain Stores* (*Seaford*), *Ltd.* [1965] 3 All E. R. 412; [1965] 1 W. L. R. 1182 (sale by liquidator of company being wound up by the court).

(*f*) *Ibid.*, s. 26 (1).

II CONSUMER PROTECTION

There is now a wide range of statutes for the protection of the consumer, governing a wide range of matters. Many deal with sales of specific types of goods (*g*). Others are of general application.

Those of general application or which are potentially of general application include the Consumer Protection Acts, 1961 and 1971; the Weights and Measures Act, 1963; the Trade Descriptions Acts, 1968 and 1972; the Unsolicited Goods and Services Act, 1971; the Fair Trading Act, 1973 and the Consumer Credit Act, 1974.

As most of these statutes have broadly similar provisions as to who may be liable and what defences may be available to a person charged with an offence, and as to their general enforcement, these aspects are discussed separately in relation to all the statutes.

The rights of the buyer where the seller has committed a statutory offence

The position of a buyer where the seller has committed a statutory offence in the course of selling goods to him would seem to be as follows:

(1) The buyer will be entitled to the same remedies for misrepresentation and for breach of contract, if any, as he would have had if the statutory offence had not existed.

(2) In some cases a warranty on the part of the seller is implied into the contract that he has complied with the provisions of the statute in question (*h*). In such cases, the buyer may be entitled to claim damages from the seller for breach of that warranty.

(3) Where the seller owes a statutory duty to the buyer to comply with a statutory requirement, and is in breach of that duty, the buyer (*i*) may claim damages from the seller in respect of its breach. But it would seem that in relation to consumer protection statutes, the seller will only owe such a duty to the buyer if the statute expressly so provides (*k*).

(*g*) See, *e.g.*, Food and Drugs Act, 1955; Plant Variety and Seeds Act, 1964; Farm and Garden Chemicals Act, 1967; Agriculture Act, 1970.

(*h*) See, *e.g.*, Plant Varieties and Seeds Act, 1964, s. 17; Agriculture Act, 1970, s. 72.

(*i*) The statutory duty may be owed to others as well as the buyer, as under the Consumer Protection Acts, 1961 and 1971, discussed p. 55, *post*, in which case those others may also claim damages for its breach.

(*k*) *Square* v. *Model Farm Dairies* (*Bournemouth*), *Ltd.*, [1939] 1 All E. R. 259; [1939] 2 K. B. 365, C. A. (Food and Drugs (Adulteration) Act, 1928); *London Armoury Co., Ltd.* v. *Ever Ready Co.* (*Great Britain*), *Ltd.*, [1941] 1 All E. R. 364; [1941] 1 K. B. 742 (Merchandise Marks Act, 1926); *Badhams* v. *Lambs, Ltd.*, [1945] 2 All E. R. 295; [1946] K. B. 45 (road traffic legislation regarding sale of car in unroadworthy condition); *J. Bollinger* v. *Costa Brava Wine Co., Ltd.*, [1959] 3 All E. R. 800, at p. 813; [1960] Ch. 262, at p. 287 (Merchandise Marks Acts); *H. P. Bulmer, Ltd.* v. *J. Bollinger*, (1975), *Times*, 20th May (Trade Descriptions Act, 1968). Section 26 (a) of the Fair Trading Act, 1973 expressly provides that contravention of an order made by the

(4) The buyer may in some cases be able to claim that the contract is void or unenforceable by the seller because it was illegal or because he intended to perform it in an illegal manner (*l*). But several of the statutes expressly provide that a contract shall not be unenforceable or void just because a seller commits an offence under that statute (*m*).

(5) If the seller is prosecuted for the offence, or if he is convicted of some other offence, and the offence from which the buyer has suffered loss is taken into account in determining sentence, then the buyer may be awarded compensation by the court which sentences the seller (*n*).

Consumer Protection Acts, 1961 and 1971

These Acts empower the Secretary of State to make such regulations in respect of prescribed classes of goods and component parts of goods "as are in his opinion expedient to prevent or reduce risk of death or personal injury" (*o*). As the scope of potential consumer protection extends only to risk of physical injury, their effect is more limited than the title of the Acts would suggest. The Secretary of State has no power under them to protect consumers' economic interests.

To date, such regulations have been made in respect of gas and electric fires, oil heaters, nightdresses, stands for carry cots, domestic electric appliances having a mains lead with three wires, cooking utensils, pencils and graphic instruments and toys.

Section 2 (1) of the 1961 Act prohibits the sale or possession for the purpose of selling of any goods which do not comply with the requirements of regulations made under s. 1 of the Act. Section 2 (2) deals in similar terms with component parts. The duty to comply with s. 2 is owed to every person who may be affected by a failure to comply with the requirements in question (*p*), and a person who commits an offence under s. 2 is not only liable to prosecution, but may be sued for breach of statutory duty by anybody affected by the contravention.

The Weights and Measures Act, 1963

Part I of this Act establishes units and standards of measurement. Part II provides that only these units may be " used for trade " (*q*).

Secretary of State under s. 22 of that Act, concerning a consumer trade practice, should not entitle any person affected to bring any civil claim. See further, p. 62, *post*.

(*l*) See generally, p. 17, *ante*.

(*m*) See, *e.g.*, Plant Variety and Seeds Act, 1964, s. 17 (5); Trade Descriptions Act, 1968, s. 35; Agriculture Act, 1970, ss. 68 (8), 71 (4); Fair Trading Act, 1973, s. 26; Consumer Credit Act, 1974, s. 170.

(*n*) Powers of Criminal Courts Act, 1973, s. 35, Appendix I, *post*.

(*o*) Consumer Protection Act, 1961, s. 1 (a).

(*p*) *Ibid.*, s. 3.

(*q*) " Used for trade " is defined very widely by s. 9, Weights and Measures Act, 1963.

Part III deals with public weighing and measuring equipment. Part IV contains important provisions for consumer protection.

Section 21 empowers the Department of Trade and Industry to make orders and regulations for certain transactions in goods: for example, to sell by weight or measure, net or gross, only in prescribed quantities, to require marking of quantities and, sometimes, to require certain written documents. Where under the Act goods may only be sold by quantity as expressed in a particular manner, and they are sold, exposed or offered for sale in some other manner, an offence is committed (*r*). It is also an offence to cause other people so to act (*s*). Section 23 provides that in certain circumstances, the quantity of goods sold must be stated in writing.

Section 24 prohibits short weight or measure. Section 24 (1) makes it an offence to deliver or cause to be delivered a lesser quantity than that purported to be sold or than corresponds with the price charged. Section 24 (2) makes it an offence for any person, in connection with the sale or purchase of goods, to make any misrepresentation (*t*) either by word of mouth or otherwise as to the quantity of the goods or to do any other act calculated to mislead (*u*) a person buying or selling the goods as to the quantity thereof.

Trade Descriptions Acts, 1968 and 1972

By s. 1 of the Trade Descriptions Act, 1968, any person who, in the course of a trade or business (*a*), either applies a false trade description to any goods or supplies or offers to supply any goods to which a false trade description is applied is, subject to the provisions of that Act, guilty of an offence.

Trade description.—A trade description is an indication, direct or indirect, and by whatever means given of any of the matters set out in s. 2 of the 1968 Act (*b*). These are quantity (*c*), size or gauge; method of manufacture, production, processing or re-conditioning; composition; fitness for purpose, strength, performance, behaviour or accuracy; any other physical characteristics; testing by any

---

(*r*) Weights and Measures Act, 1963, s. 22.

(*s*) *Ibid.*, s. 22 (1) (b).

(*t*) Such misrepresentation need not be made pursuant to a contract or by a party to a contract: *Collett* v. *Co-op Wholesale Society, Ltd.*, [1970] 1 All E. R. 274; [1970] 1 W. L. R. 250, D.C.

(*u*) An act is calculated to mislead if it is " likely to deceive or mislead ": *Collett* v. *Co-op Wholesale Society, Ltd.*, *supra.*

(*a*) See the note to the definition of " business " in s. 62 (1) of the Sale of Goods Act, 1893, *post*.

(*b*) Descriptions or marks applied in pursuance or certain Acts concerned with agriculture and horticulture and certain descriptions dealt with under the Food and Drugs Act, 1955 are expressly excluded by s. 2 (4)–(5) of the 1968 Act as amended.

(*c*) Defined by s. 2 (3) to include length, width, height, area, volume, capacity, weight and number.

person and results of such testing; approval by any person or conformity with a type approved by any person; place or date of manufacture, production, processing, or reconditioning; person by whom manufactured, produced, processed or reconditioned; or other history, including previous ownership or use.

" Indication " is much wider than " representation ", and covers conduct and signs of many different kinds (*d*).

False trade description.—A false trade description is a trade description which is false to a material degree or which, though not false, is misleading, that is to say, likely to be taken for such an indication of any of the matters specified in s. 2 of the 1968 Act as would be false to a material degree (*e*). The expression also includes anything which, though not a trade description, is likely to be taken for an indication of any of those matters, and, as such an indication, would be false to a material degree (*f*).

A trade description may be false even though there is no intent to mislead a customer (*g*) and even though the goods are of equal (*h*) or superior (*i*) quality to that indicated by the description. The test is whether the description is false, not whether the bargain is fair (*k*).

In deciding whether a trade description is false, the Courts will take a broad view of all the circumstances (*l*) and consider whether in those circumstances it is likely to mislead. An offence may be committed if a trade description is capable of being misunderstood, even though many more customers are not misled by it (*m*). In general, it seems that no evidence need be adduced that anybody was misled, but where it is not clear on the face of it that the description is false or misleading, then the prosecution must prove that members of the public were in fact misled (*m*).

(*d*) *Doble* v. *David Greig, Ltd.*, [1972] 2 All E. R. 195; [1972] 1 W. L. R. 703, D. C. But it must be an indication of one of the matters specified in s. 2: *Cadbury, Ltd.* v. *Halliday*, [1975] 2 All E. R. 226; [1975] 1 W. L. R. 649, D. C. ("extra value " not of itself a trade description).

(*e*) Trade Descriptions Act, 1968, s. 3 (2).

(*f*) *Ibid.*, s. 3 (3).

(*g*) *Mear* v. *Baker* (1954), 118 J. P. 483, D. C.; *R.* v. *Reiss*, [1957] Crim. L. R. 404.

(*h*) *Kirshenboim* v. *Salmon and Gluckstein*, [1898] 2 Q. B. 19, D. C.

(*i*) *Thornley* v. *Tuckwell (Butchers), Ltd.*, [1964] Crim. L. R. 127, D. C.

(*k*) *Furniss* v. *Scholes*, [1974] R. T. R. 133, D. C.

(*l*) See *e.g. Evans* v. *British Doughnut Co., Ltd.*, [1944] 1 All E. R. 158; [1944] K. B. 102, D. C.; *White Hudson & Co.* v. *Asian Organisation, Ltd.*, [1965] 1 All E. R. 1040; [1964] 1 W. L. R. 1466, P. C. (get up of sweets more important than printed words); *Donnelly* v. *Rowlands*, [1971] 1 All E. R. 9; [1971] 1 W. L. R. 1600, D. C. (dairyman used milk bottles embossed with name of owners of bottles and foil cap with his own name and address. Held: in context, and in absence of evidence that anybody had been misled, name of owner on bottles was not false trade description).

(*m*) *Doble* v. *David Greig, Ltd.*, [1972] 2 All E. R. 195; [1972] 1 W. L. R. 703, D. C.

Applying a trade description to goods.—A person applies a trade description to goods if he affixes or annexes it to or in any manner marks it on or incorporates it with the goods themselves or anything in or with which the goods are supplied; or if he places the goods in, on or with anything which the trade description has been affixed or annexed to, marked on or incorporated with, or if he places any such thing with the goods; or if he uses the trade description in any manner likely to be taken as referring to the goods (*n*). Also, where goods are supplied in pursuance of a request in which a trade description is used and the circumstances are such as to make it reasonable to infer that the goods are supplied as goods corresponding to that trade description, the person supplying the goods is deemed to have applied that trade description to the goods (*o*).

A person may apply a false trade description without realising that it is false (*p*). Indeed, it has been held that if one partner applies a false trade description, another partner who has no knowledge of the transaction in which it is applied may also be guilty of applying it (*q*).

A buyer as well as a seller may apply a trade description to goods, and if he does so in the course of a business, and it is false, then he commits the offence of applying a false trade description to goods (*r*).

But the courts have implied some limitations to the offence of applying a false trade description which are not to be found in the very wide wording of s. 1 (1) (a) of the 1968 Act. On a literal interpretation of that section, the context in which the trade description is applied would seem to be irrelevant. In *Wickens Motors (Gloucester)* v. *Hall* (*s*), however, it was held that no offence under s. 1 (1) (a) was committed if the false trade description was not applied in connection with the sale or supply of goods (*t*).

At least where the trade description is already on the goods or packaging when they come into the hands of the seller, he will not

---

(*n*) Trade Descriptions Act, 1968, s. 4 (1). An oral statement may amount to the use of a trade description: s. 4 (2).

(*o*) *Ibid.*, s. 4 (3).

(*p*) *Taylor* v. *Smith*, [1974] Crim. L. R. 200, D. C.; *Tarleton Engineering* v. *Nattrass*, [1973] 3 All E. R. 699; [1973] 1 W. L. R. 1261, D. C.

(*q*) *Clode* v. *Barnes*, [1974] 1 All E. R. 1166; [1974] 1 W. L. R. 544, D. C. This would seem to be because the knowledge of one partner is imputed to all the partners. But *cf. Cottee* v. *Douglas Seaton (Used Cars), Ltd.*, [1972] 3 All E. R. 750; [1972] 1 W. L. R. 1408, D. C. where a car had been damaged and the damage concealed by a previous owner. The seller was not aware of the damage when he sold the car, and it was held by the Divisional Court that he had not applied the false trade description. See also the defence under s. 24 (3) of the 1968 Act.

(*r*) *Fletcher* v. *Budgen*, [1974] 2 All E. R. 1243; [1974] 1 W. L. R. 1056, D. C.

(*s*) [1972] 3 All E. R. 759; [1972] 1 W. L. R. 1418, D. C.

(*t*) Or, presumably, the provision of services. But it is sufficient to constitute an offence if the person applying the false trade description has an interest in the fate of the goods: *Fletcher* v. *Sledmore* (1973), L. G. R. 179, where A sold a car to B, and then recommended the car to C who was considering buying it from B.

be held to have applied that description if he has disclaimed any responsibility as to its accuracy before the goods are supplied (*u*) in a form as bold, precise and compelling as the trade description itself (*a*). It would seem that the trade description is not applied if both parties know that it cannot be relied on and act on that basis (*b*).

Supplies or offers to supply any goods.—Goods are supplied when the seller notifies the buyer that they are available for delivery, or in the absence of such notification when they are actually delivered (*c*).

A person exposing goods for supply or having goods in his possession for supply is deemed to offer to supply them (*d*).

Trade descriptions used in advertisements.—Where in an advertisement (*e*) a trade description is used in relation to any class of goods, it is to be taken as referring to all goods of the class whether or not in existence when the advertisement is published:

(a) in determining whether the offence has been committed of applying a false trade description to any goods; and

(b) where goods of the class are supplied or offered to be supplied by a person publishing or displaying the advertisement, also in determining whether the offence has been committed of supplying or offering to supply any goods to which a false trade description is applied (*f*).

In determining for this purpose what goods come within any particular class, regard is to be had not only to the form and content of the advertisement, but also to the time, place, manner, and frequency of its publication, and all other matters making it likely or unlikely that a person to whom the goods are supplied would think of the goods as belonging to the class in relation to which the trade description is used in the advertisement (*g*).

Additional powers of Department of Trade and Industry.—The Department of Trade and Industry has power in certain circumstances by order:

(a) to assign definite meanings to expressions used in relation to goods (*h*);

(*u*) Or, presumably, contracted to be supplied.

(*a*) *Norman* v. *Bennett*, [1974] 3 All E. R. 351; [1974] 1 W. L. R. 1229, D. C.; *cf. Doble* v. *David Greig*, [1972] 2 All E. R. 195; [1972] 1 W. L. R. 703, D. C. (notice displayed at cash desk at supermarket held ineffective). See also *Zawadski* v. *Sleigh*, [1975] R. T. R. 113, D. C. (Disclaimer by auctioneer whether on own behalf or on behalf of seller, may be sufficient if bold enough to contradict false trade description.)

(*b*) *Norman* v. *Bennett*, *supra*.

(*c*) *Rees* v. *Munday*, [1974] 3 All E. R. 506; [1974] 1 W. L. R. 1284, D. C.

(*d*) Trade Descriptions Act, 1968, s. 6.

(*e*) Defined by s. 39 (1) of the Trade Descriptions Act, 1968 to include a catalogue, a price list and a circular.

(*f*) Trade Descriptions Act, 1968, s. 5 (1)–(2).

(*g*) *Ibid.*, s. 5 (3).

(*h*) *Ibid.*, s. 7.

(b) to impose requirements for securing that goods are marked with or accompanied by specified information or instructions and to regulate or prohibit the supply of goods with respect to which the requirements are not complied with (*i*);

(*c*) to require advertisements of goods to contain or refer to specified information concerning the goods (*k*).

False or misleading indications as to price of goods.—It is an offence for any person offering to supply goods (*l*) to give, by whatever means, any false indication (*m*) to the effect that the price at which goods are offered is equal to or less than either a recommended price or the price at which the goods or goods of the same description were previously offered by him, or that the price is less than such a price by a specified amount (*n*). It is also an offence for such a person to give any indication likely to be taken as an indication that the goods are being offered at a price less than that at which they are in fact being offered (*o*). These offences will only be committed, however, if the offer to supply goods is made in the course of a trade or business (*p*).

The Trade Descriptions Act, 1972.—This Act requires that the origin of certain imported goods should be conspicuously indicated in certain cases where those goods have applied to them a United Kingdom name or mark, or a name or mark which is likely to be taken for a United Kingdom name or mark (whether or not such a United Kingdom name or mark actually exists) (*q*).

(*i*) Trade Descriptions Act, 1968 s. 8. (*k*) *Ibid.*, s. 9.

(*l*) " Offer to supply " is defined by s. 6 of the 1968 Act: see *ante*, p. 59. In this context the expression also includes advertising goods as available for supply.

(*m*) See p. 57, *ante*. Anything likely to be taken as such an indication is sufficient for this purpose: see s. 11 (3) (c) of the 1968 Act. Advertising a price without any reference to value added tax may, if that tax is not included in the advertised price, be an offence: *Richards* v. *Westminster Motors, Ltd.*, (1975), *Times*, 17th June, D. C.

(*n*) Trade Descriptions Act, 1968, s. 11 (1). Unless the contrary is made clear, an indication that goods were previously offered at a particular price or at a higher price is to be treated as an indication that they were so offered by the person giving the indication and that that offer was made for at least 28 consecutive days within the preceding 6 months: s. 11 (3) (a). Also, unless the contrary is made clear, an indication as to a recommended price is to be treated as an indication that the price is recommended by the manufacturer or producer and that it is recommended generally for supply by retail in the area in which the goods are offered: s. 11 (3) (b).

(*o*) Trade Descriptions Act, 1968, s. 11 (2). See further, p. 57, *ante*.

(*p*) *John* v. *Matthews*, [1970] 2 All E. R. 643; [1970] 2 Q. B. 443, D. C.

(*q*) Section 1. The requirement does not apply if the name or mark is neither visible in the state in which the goods are supplied or offered nor likely to become visible on such inspection as may reasonably be expected to be made of the goods by a person to whom they are to be supplied (s. 1 (2)). The Secretary of State has power under s. 1 (5) to relax the requirements of s. 1 if satisfied that the interests of persons in the United Kingdom to whom such goods may be supplied would not be materially impaired by his so doing and that it is desirable for him to do so.

Offences under the Trade Descriptions Acts and civil liability for misrepresentation.—The offences created by the Trade Descriptions Acts are thus very wide. A seller may commit such an offence even though he has done nothing which would render him liable to the buyer for misrepresentation. In particular:

(1) A mere puff may be a false trade description (*r*).
(2) No representation need be made. Any indication is sufficient (*s*).
(3) There need be no intention that the trade description should be relied on by the buyer (*t*).
(4) The seller may be guilty of an offence even if the buyer suffers no loss (*u*).
(5) A potential seller may be guilty of an offence even if the goods are not sold (*v*), and nobody is influenced by it (*a*).

The Unsolicited Goods and Services Acts, 1971 and 1975

These Acts seeks to protect people to whom goods are sent or delivered without being asked for. Section 1 of the 1971 Act provides that in certain circumstances the receiver of unsolicited goods may use, deal with or dispose of them as if they were an unconditional gift to him. Any rights of the sender are thereby extinguished. The particular circumstances are defined by s. 1 (2).

It is a criminal offence for a person to make a demand for payment (*b*) for what he knows are unsolicited goods, if he has no reasonable cause to believe there is a right to payment (*c*). It is also a criminal offence in such a case to threaten to bring any legal proceedings with a view to obtaining payment for unsolicited goods, or to threaten to invoke or invoke or cause to be invoked any other collection procedure to obtain payment (*d*).

The Fair Trading Act, 1973

One of the objects of this Act is to protect the interests of consumers generally without having to introduce fresh legislation every time a new abuse appears.

Section 1 of the Act provides for the appointment of a Director-General of Fair Trading. His functions are set out in s. 2 of the Act,

(*r*) See, *e.g.*, *Robertson* v. *Dicicco*, [1972] R. T. R. 431 ("beautiful car"). But *cf.* Treitel (1968), 31 M. L. R. 662. A statement of opinion will also be a false trade description if it is likely to be taken for an indication of any of the matters specified in s. 2 of the 1968 Act: see s. 3 of that Act.

(*s*) *Doble* v. *David Greig, Ltd.*, [1972] 2 All E. R. 195; [1972] 1 W. L. R. 703, D. C.

(*t*) *Mear* v. *Baker* (1954), 118 J. P. 483, D. C.; *R.* v. *Reiss*, [1957] Crim. L. R. 404.

(*u*) *Thornley* v. *Tuckwell* (*Butchers*), *Ltd.*, [1964] Crim. L. R. 127, D. C.

(*v*) As where they are only offered for sale: see ss. 1 and 6 of the 1968 Act.

(*a*) *Taylor* v. *Smith*, [1974] R. T. R. 190, D. C.

(*b*) Orders made under the 1971 Act, as amended by the 1975 Act, set out regulations to be complied with, if an invoice or similar document is not to be treated as a demand for payment: 1975 S. I. Nos. 731, 732.

(*c*) Section 2 (1) of the 1971 Act.

(*d*) Section 2 (2) of the 1971 Act.

and include the keeping under general review of commercial activities which relate to the supply of goods and services and which may adversely affect the economic interests of consumers. He is also under a duty to receive and collate evidence relating to such activities which may adversely affect other interests of consumers.

The Act also creates the Consumer Protection Advisory Committee (*e*). The Secretary of State or any other Minister, or the Director-General of Fair Trading may refer to the Advisory Committee the question whether a "consumer trade practice" (*f*) affects the economic interests of consumers (*g*). If the Director-General refers such a question to the Advisory Committee, he may in certain circumstances include proposals for recommending to the Secretary of State that he should exercise his powers with respect to that consumer trade practice (*h*).

In such a case, the Secretary of State may, after receiving the report of the Advisory Committee, by an order made by statutory instrument give effect to the proposals as set out in the original reference or as modified by the Advisory Committee (*i*). Contravention of such an order is a criminal offence (*k*).

Additional functions of the Director-General for the protection of consumers are contained in Part III of the Act. Where a person carries on a business and in the course of that business persists in a course of conduct which is detrimental to consumers' interests (*l*) and which is unfair to consumers, the Director-General must use his best endeavours to obtain from such person a written assurance that he will discontinue that course of conduct (*m*). Two types of conduct are regarded as unfair for this purpose. First, if the course of conduct consists of contraventions of a duty enforceable by criminal proceedings, whether or not such duty relates to consumers, and whether or not a conviction in respect thereof has been secured (*n*). Second, an unfair course of conduct also consists of any breach of contract or other duty enforceable by civil proceedings, whether or not civil proceedings have in fact been brought (*o*).

If such an assurance is not forthcoming, or is given and is then not observed, the Director-General has power to bring proceedings against that person before the Restrictive Practices Court (*p*). If

---

(*e*) Fair Trading Act, 1973, s. 3.
(*f*) As defined in *ibid.*, s. 13.
(*g*) *Ibid.*, s. 14.
(*h*) *Ibid.*, s. 17.
(*i*) *Ibid.*, s. 22.
(*k*) *Ibid.*, s. 23.
(*l*) Whether those interests are economic interests or interests in respect of health, safety or other matters; s. 34 (1) (a).
(*m*) *Ibid.*, s. 34 (1).
(*n*) *Ibid.*, s. 34 (2).
(*o*) *Ibid.*, 34 (3).
(*p*) *Ibid.*, s. 35. In certain circumstances the proceedings may be brought in another court: see s. 41.

that person does not give a proper undertaking to the Court, and if it appears that he is likely to persist in his course of conduct, the Court may make an order under s. 37 of the Act directing him to refrain from continuing that course of conduct.

The rest of the Act is primarily concerned with monopoly situations and other uncompetitive practices, apart from Part XI which empowers the Secretary of State to make regulations concerning pyramid selling, and provides for the enforcement of those regulations.

The Consumer Credit Act, 1974

This Act, the various parts of which are to come into force on dates to be fixed (*q*), establishes for the protection of consumers a new system of credit control administered by the Director-General of Fair Trading. It replaces the previous legislation relating to moneylenders pawnbrokers and hire purchase traders and their transactions.

With certain limited exceptions (*r*), the Act is concerned only with consumer credit agreements. A consumer credit agreement is an agreement between an individual (*s*), the debtor, and any other person, the creditor, by which the creditor provides the debtor with credit not exceeding £5000 (*t*). Some provisions of the Act apply to all consumer credit agreements; others apply only to regulated agreements, that is consumer credit agreements which are not exempted under s. 16 of the Act. Regulated agreements are further subdivided by ss. 11–13 of the Act to take account of the purpose for which credit was given. The most important category for the purposes of the sale of goods is the debtor-creditor-supplier agreement which is a regulated consumer credit agreement to finance a transaction between the debtor and the creditor, or which is made under pre-existing arrangements between the creditor and a supplier (*u*) to finance a transaction between the debtor and the supplier.

(*q*) Consumer Credit Act, 1974, s. 192 (1), Sch. 3. None of the provisions for licensing businesses and regulating agreements are yet in force.

(*r*) See in particular the provisions as to extortionate credit bargains in ss. 137–140 of the Act.

(*s*) By s. 189 (1) of the Act, " individual " includes a partnership or other unincorporated body of persons not consisting entirely of bodies corporate. Companies and other bodies corporate cannot therefore be debtors under the Act except as members of such partnerships or other unincorporated bodies.

(*t*) Or such other sum as may be fixed by order under s. 181 of the Act. An item forming part of the total charge for the credit, as determined by regulations under s. 20 of the Act is not to be treated as credit even though time is allowed for its payment: s. 9 (4) of the Act.

(*u*) Or in some cases in contemplation of future arrangements: s. 12 of the Act. " Supplier " means any other person than the debtor and includes a person to whom the rights and duties of a supplier have passed by assignment or operation of law or (in relation to a prospective agreement) the prospective supplier: s. 189 (1) of the Act. " Arrangements between a creditor and a supplier " are defined by s. 187 of the Act.

Credit includes a cash loan and any other form of financial accommodation (*a*).

In relation to regulated agreements, the Act regulates the activities both of those who provide credit in the course of a business (*b*), and of those engaged in an ancillary credit business. An ancillary credit business is any business so far as it comprises or relates to credit brokerage, debt-adjusting, debt-counselling, debt-collecting or the operation of a credit reference agency. The various terms are defined in s. 145 of the Act, but the principal activity in relation to the sale of goods is credit brokerage, which is the effecting of introductions of individuals desiring to obtain credit to persons who in the course of their business enter into agreements providing individuals with credit not exceeding £5000 (*c*). Thus sellers who assist buyers to acquire their goods by introducing them to a finance house are credit brokers, and their dealings as credit brokers are regulated by the Act.

The detailed provisions of the Act are beyond the scope of this book. It provides for the licensing of most consumer credit businesses (*d*) and ancillary credit businesses (*e*) and regulates or empowers the Secretary of State to regulate virtually everything done in the course of such businesses. It also confers upon debtors rights which they would not otherwise have to be given information (*f*), to withdraw from agreements (*g*), to complete payments ahead of time and to terminate agreements (*h*). It also gives to a debtor who has obtained credit under a debtor-creditor-supplier agreement the right to hold the creditor liable for any misrepresentation or breach of contract by the supplier where the creditor is acting in the course of a business and the claim relates to items with a cash price of £30 to £10,000 each.

Extortionate credit agreements.—By s. 137 of the Act, the court has

---

(*a*) Section 9 (1).

(*b*) By s. 189 (1), "business" includes profession or trade, and by s. 189 (2), a person is not to be treated as carrying on a particular type of business merely because occasionally he enters into transactions belonging to a business of that type. *Cf.* the definition of business in s. 62 (1) of the Sale of Goods Act, 1893, *post*.

(*c*) Or such other sum as may be fixed by order under s. 181 of the Act. There is an exception where the agreement under which the credit is provided is exempted by an order of the Secretary of State under s. 16 (5) (a) of the Act on the ground that the number of payments to be made by the debtor does not exceed the number specified for that purpose in the order.

(*d*) A consumer credit business is defined by s. 189 (1) of the Act as "any business so far as it comprises or relates to the provision of credit under regulated consumer credit agreements".

(*e*) Sections 21–42, 147–150.

(*f*) Sections 55, 76–79, 103, 107–110, 157–159.

(*g*) Sections 67–73. (*h*) Sections 94–104.

power to re-open any agreement by which any person provides an individual with credit of any amount, and which is extortionate, so as to do justice between the parties.

A credit bargain is extortionate if it requires the debtor or a relative of his to make payments (whether unconditionally or on certain contingencies) which are grossly exorbitant or if it otherwise grossly contravenes ordinary principles of fair dealing. All relevant considerations are to be taken into account, and some are specified in s. 138 of the Act.

By s. 171 (7) of the Act, if it is alleged by a debtor or surety that a credit bargain is extortionate, it is for the creditor to prove the contrary.

If a credit bargain is re-opened, the court has wide powers to do whatever is necessary, including altering the terms of the credit agreement and setting aside the whole or any part of any obligation imposed on the debtor or any surety, for the purpose of relieving the debtor or surety from payment of any sum in excess of that fairly due and reasonable: s. 139 (2) of the Act.

Persons liable to prosecution

Offences by corporations.—Under most consumer protection statutes where an offence has been committed by a body corporate with the consent or connivance of, or is attributable to any neglect on the part of, any director, manager, secretary or other similar officer (*i*) of the body corporate, that person as well as the body corporate commits an offence (*k*). A similar provision is contained in s. 20 of the Trade Descriptions Act, 1968, but under that section the offence must have been committed with the consent ***and*** connivance of the officer charged.

Accomplices.—Section 21 of the Trade Descriptions Act, 1968 makes

---

(*i*) In a case under the Restrictive Trade Practices Act, 1956, " manager " and " officer " were held not to include the branch manager of a local retail outlet: *Registrar of Restrictive Trades Practices* v. *W. H. Smith & Son, Ltd.*, [1969] 3 All E. R. 1065; [1969] 1 W. L. R. 1460, C. A. *Cf. Tesco Supermarkets, Ltd.* v. *Nattrass*, [1971] 2 All E. R. 127; [1972] A. C. 153, H. L.

(*k*) See s. 107, Food and Drugs Act, 1955; s. 3 (3), Consumer Protection Acts, 1961 and 1971; s. 50, Weights and Measures Act, 1963; s. 5, Unsolicited Goods and Services Act, 1971; s. 132, Fair Trading Act, 1973; and s. 169, Consumer Credit Act, 1974.

The Food and Drugs Act, 1955, Consumer Protection Acts, 1961 to 1971, Weights and Measures Act, 1963 and Trade Descriptions Act, 1968 all provide that " director " in relation to any statutory body corporate under national ownership or in relation to a body corporate whose affairs are managed by its members means a member of that body corporate. The Unsolicited Goods and Services Act, 1971, the Fair Trading Act, 1973 and the Consumer Credit Act, 1974 have similar provisions, but only where the affairs of a body corporate are managed by its members.

it an offence in some circumstances for a person in the United Kingdom to assist in or induce the commission in any other country of certain acts in respect of goods which would constitute an offence under s. 1 or s. 12 of that Act if the acts were committed in the United Kingdom.

In addition, the usual common law rules as to accomplices will apply to offences under consumer protection statutes.

Act or default of another person.—Where the commission by any person of an offence under the Trade Descriptions Acts, 1968 and 1972, or under s. 23 of the Fair Trading Act, 1973, or under s. 3 of the Consumer Protection Act, 1961 (*l*) is due to the act or default of some other person, that other person is also guilty of the offence, whether or not proceedings are taken against the principal offender (*m*). For this purpose, the meaning of "other person" does not appear to be restricted in any way, and means any individual or corporate body. Thus, where the offence is committed by a corporate body, proceedings may be brought against any person, including an employee of the body, whatever his position in it.

There must be a sufficient connection between the act or default and the offence committed. What is a sufficient connection will no doubt have to be determined on the facts of each case, but in one case where A sold a car to B on which the mileage recorded was incorrect, and B resold it to C with the same mileage recorded, it was held that the false trade description applied by B was not due to the act or default of A (*n*).

The other person cannot be convicted under any of the foregoing provisions if the person alleged to have committed the principal offence is shown to have committed no offence (*o*). But it has been held that, in the case of an offence under the Trade Descriptions Act, 1968, there is still an offence capable of being due to the act or default of another person if the principal offender's only defence is under s. 24 of that Act (*p*). The position would seem to be the same under the Fair Trading Act, 1973 where the only defence is under s. 25 of that Act, which is identical to s. 24 of the Trade Descriptions Act. Under s. 3 (2B) of the Consumer Protection Act, 1961 (as amended) the defence is wider and less technical, and it is not clear whether the position would be the same (*q*).

(*l*) As amended by s. 1 of the Consumer Protection Act, 1971.

(*m*) Section 23, Trade Descriptions Act, 1968; s. 24, Fair Trading Act, 1973; s. 3 (2A), Consumer Protection Act, 1961.

(*n*) *Tarleton Engineering* v. *Nattrass*, [1973] 3 All E. R. 699; [1973] 1 W. L. R. 1261, D. C. See also *Cadbury, Ltd.* v. *Halliday*, [1975] 2 All E. R. 226; [1975] 1 W. L. R. 649, D. C. (manufacturer and retailer).

(*o*) *Cottee* v. *Douglas Seaton (Used Cars), Ltd.*, [1972] 3 All E. R. 750; [1972] 1 W. L. R. 1408, D. C.

(*p*) *Coupe* v. *Guyett*, [1973] 2 All E. R. 1058; [1973] 1 W. L. R. 669, D. C.

(*q*) See these defences considered, *post*.

Section 27 (4) of the Weights and Measures Act, 1963 and s. 113 (3) of the Food and Drugs Act, 1955 are similar in effect though different in phraseology and approach.

Defences

The offences created by the consumer protection legislation are in general offences of strict liability. A person may be guilty of an offence even though he has no intention to commit it. Where defences are created by the legislation, it is for the person charged to prove the facts which will constitute a defence.

Broadly similar defences are provided by s. 113 (1) of the Food and Drugs Act, 1955, s. 3 (2B) of the Consumer Protection Act, 1961 as amended by the Consumer Protection Act, 1971, ss. 26 (1) and 27 (1) of the Weights and Measures Act, 1963, s. 24 of the Trade Descriptions Act, 1968, s. 25 of the Fair Trading Act, 1973 and s. 168 of the Consumer Credit Act, 1974.

The due diligence defence.—Under s. 3 (2B) of the Consumer Protection Act, 1961, it is a defence for a person charged with an offence under that Act to prove that he took all reasonable precautions and exercised all due diligence to avoid the commission of the offence. These two requirements frequently appear in the defences available under the other Acts, but are there subject to further qualifications. While they must be taken together, the first, "reasonable precautions" refers to the setting up of a proper system of work, and the second, "due diligence" refers to the proper supervision of that system (*r*).

To establish that he took reasonable precautions and exercised due diligence, the person charged must show, and apparently need only show, that he acted without negligence (*s*). But the standard of care required would seem to be a high one.

A person running a business must devise a system which could rationally be said to be so designed that the commission of the offence could be avoided (*t*). A paper scheme, perfunctorily enforced would be inadequate (*u*). If, however, a proper scheme is devised and it is properly supervised, a failure in the scheme due to the default of an employee will not prevent the owner of the business from establishing the defence (*a*).

(*r*) *Tesco Supermarkets, Ltd.* v. *Nattrass*, [1971] 2 All E. R. 127; [1972] A. C. 153, H. L.

(*s*) *Ibid.*, at pp. 155 and 199, *per* Lord Diplock.

(*t*) *Ibid.*, at pp. 140 and 180, *per* Lord Morris.

(*u*) *Ibid.*, at pp. 135 and 174, *per* Lord Reid.

(*a*) *Ibid.*, where the appellants were able to rely on the defence provided by s. 24 of the Trade Descriptions Act, 1968 as the offence was committed due to the default of one of their shop managers. See also *Beckett* v. *Kingston Brothers (Butchers), Ltd.*, [1970] 1 All E. R. 715; [1970] 1 Q. B. 606, D. C., where the

It is a question of fact in all cases whether a person has taken reasonable precautions or exercised due diligence.

Act or default of another person.—Under s. 3 (2B) of the Consumer Protection Act, 1961, it is not necessary to show how the offence came to be committed, although it may be desirable to do so to help to show that all reasonable precautions were taken.

Under other statutory provisions the defence is only available if other requirements are satisfied.

Section 113 (1) of the Food and Drugs Act, 1955 and s. 27 (1) of the Weights and Measures Act, 1963 require not only that the person charged must prove that he has used all due diligence, but also that he must prove that the commission of the offence was due to the act or default (*b*) of another person. It is not sufficient to show that the offence *may* have been due to the other person's act or default (*c*).

Where a retailer seeks to rely for his defence on one of these sections to show that the contravention was due to the act or default of a middleman, the middleman may in his turn rely for his defence on the same section to show that it was due to the act or default of the manufacturer. The section, once used, is not exhausted (*d*).

The other person may be a servant or agent or other person who is

---

managing director of the company wrote a special letter to his area managers, requiring all turkeys which had been mistakenly labelled "Norfolk King Turkey" to be remarked "Danish", and this was held sufficient to exonerate the company when one of the turkeys was subsequently sold under the wrong label. But in *Sherratt* v. *Geralds the American Jewellers* (1970), 114 Sol. Jo. 147, jewellers obtained watches, marked "waterproof" from their suppliers. When charged under s. 1 (1) (b) of the Trade Descriptions Act, 1968 with supplying goods to which a false trade description was applied, they sought to rely upon s. 24 of that Act. It was held that they were not entitled to do so, because it was not enough for them to rely, however reasonably, on the reputation or experience of their wholesale suppliers. Where the person charged is only a dealer but not an expert, he is not excused from taking steps which to experts might be reasonable: *Sutton London Borough* v. *Perry Sanger & Co., Ltd.* (1971), 135 J. P. Jo. 239. Where a car is sold and the mileage indicated on the odometer amounts to a false trade description under the Trade Descriptions Act, 1968, the person charged can no longer say that he has taken all reasonable precautions to avoid the commission of an offence unless he has issued a disclaimer in a sufficiently clear form: *Simmons* v. *Potter*, [1975] Crim. L. R. 354, D. C., distinguishing *Naish* v. *Gore*, [1971] 3 All E. R. 737. For disclaimers, see *supra*, p. 59.

(*b*) In *Tesco Supermarkets, Ltd.* v. *Nattrass*, *supra*, Lord DIPLOCK, at pp. 153 and 196–197 defined a default in this context as a failure to act which constitutes a breach of a legal duty to act.

(*c*) *Moore* v. *Ray*, [1950] 2 All E. R. 561; [1951] K. B. 58.

(*d*) *British Fermentation Products, Ltd.* v. *British Italian Trading Co., Ltd.*, [1942] 2 All E. R. 256; [1942] 2 K. B. 145.

under the control of the person charged (*e*). Thus in the case of a company, an employee who is not within the " brain area " of the company is another person for this purpose, unless perhaps he is responsible for devising or supervising the system devised to avoid the commission of the offence (*f*).

Under both sections, the person charged must give the prosecution three days notice of his intention to bring the other person before the court, and must then bring him before the court. If it is then shown that the contravention was due to the act or default of that other person, he may be convicted (*g*) of the offence.

Penalties

Penalties for offences under the different statutes vary, Under s. 106 of the Food and Drugs Act, 1955, except where otherwise provided (*h*), the maximum penalty is a fine of £100 and three months imprisonment, plus £5 for each day in the case of a continuing offence. Under s. 52 of the Weights and Measures Act, 1963 penalties range from a maximum of £20 on summary conviction to £100 (£250 on a second conviction) and three months imprisonment on indictment.

Section 18 of the Trade Descriptions Act, 1968 and s. 23 of the Fair Trading Act, 1973 provide a maximum fine of £400 on summary conviction and an unlimited fine (*i*) plus imprisonment for up to two years on conviction on indictment. Section 167 of and Schedule 1 to the Consumer Credit Act, 1974 provides for fines of £50—£400 on summary conviction, and for an unlimited fine and imprisonment of from one to two years on conviction on indictment.

An offender may also be ordered as part of his sentence to compensate any person who has suffered loss as a result of the offence (*k*).

It has been said (*l*) that a sentence of imprisonment, whether suspended or not, is not normally justified for a contravention of a provision of the Trade Descriptions Act, 1968 unless the offender is guilty of dishonesty, and this will presumably also apply under other consumer protection legislation.

(*e*) *Hall* v. *Farmer*, [1970] 1 All E. R. 729, *per* Lord PARKER, C. J.

(*f*) *Tesco Supermarkets, Ltd.* v. *Nattrass*, [1971] 2 All E. R. 127; [1972] A. C. 153, H. L. But a person who sells only as an agent or broker for a disclosed principal cannot by selling be guilty of an act or default to which the commission of an offence by the principal is due: *Lester* v. *Balfour Williamson Merchant Shippers*, [1953] 1 All E. R. 1146; [1953] 2 Q. B. 168, D. C.

(*g*) Unless he has a defence: *e.g.*, that the information preferred against him was outside the time limit as in *R.* v. *Bicester Justices, ex parte Unigate, Ltd.*, [1975] 1 All E. R. 449; [1975] 1 W. L. R. 207, D. C.

(*h*) *E.g.* s. 8 (4) of that Act.

(*i*) See *R.* v. *Churchill* (*No. 2*) [1966] 2 All E. R. 215; [1967] 1 Q. B. 190.

(*k*) See s. 35, Powers of Criminal Courts Act, 1973, *post*, Appendix I.

(*l*) *R.* v. *Haesler*, [1973] R. T. R. 486, C. A.

Notification of conviction or judgment to Director-General of Fair Trading

Where a person is convicted of an offence in any criminal proceedings or has a judgment given against him in civil proceedings, and it appears to the court that having regard to the functions of the Director-General under Part III of the Fair Trading Act, 1973 or under the Consumer Credit Act, 1974 that the conviction or judgment should be brought to his attention, and that it may not be brought to his attention unless arrangements are made by the court, then the court may make such arrangements although the proceedings have been finally disposed of (*m*).

Enforcement of consumer protection legislation

The Food and Drugs Act, 1955.—Section 87 of the Food and Drugs Act, 1955 imposes a duty to prosecute in respect of contraventions of the provisions of that Act in some cases on the Minister of Agriculture, and in other cases on local councils, but enforcement of most provisions is in the hands of the food and drug authorities (*n*). The Act does not prohibit private prosecutions, but contemplates that prosecutions will ordinarily be instituted by food and drug authorities (*o*). There is no duty for these authorities to prosecute automatically in every case. They are required to use their discretion and consider whether the public interest will be served by a prosecution. Indeed, it has been said that where a prosecution does not serve the general interests of consumers, justices may think fit to grant an absolute discharge even though the Act has been contravened (*p*).

The Weights and Measures Act, 1963.—Private prosecutions may not be brought under this Act. The act is administered by local weights and measures authorities through their appointed inspectors (*q*), and it is provided by s. 51 of the Act that proceedings for any

---

(*m*) Fair Trading Act, 1973, s. 131; Consumer Credit Act, 1974, s. 166. A Home Office Circular (No. 10/1975) has been sent to Clerks to Magistrates' Courts setting out the information required. It is reproduced in N. L. J. Vol. 125, p. 131. Arrangements have also been made between the office of Fair Trading and local authorities for the outcome of prosecutions brought by local weights and measure authorities to be notified to the Director-General.

(*n*) As defined in s. 88 of the 1955 Act.

(*o*) *Smedleys, Ltd.* v. *Breed,* [1974] 2 All E. R. 21, 33; [1974] A. C. 839 *per* Viscount DILHORNE.

(*p*) *Ibid.*, at p. 34 and 856, *per* Viscount DILHORNE. *Cf.* s. 109 (1) of the 1955 Act which provides that certain government departments may initiate proceedings " where the general interests of consumers are affected ", and s. 127 of that Act which gives the Minister of Agriculture power to authorise a prosecution if he is of the opinion that the failure of a food and drugs authority to prosecute affects the general interests of consumers.

(*q*) Weights and Measures Act, 1963, s. 41.

offence under the Act shall not be instituted except by or on behalf of a local weights and measures authority or the chief officer of police for a police area.

The Trade Descriptions Acts, 1968 to 1972.—Private prosecutions may be brought under these Acts, but as with the Food and Drugs Act, 1955, it is clearly contemplated that private prosecutions will not normally be brought. The duty of enforcement is imposed on the local weights and measures authorities (*r*). If any such authority fails to discharge its functions properly, complaint may be made to the Department of Trade and Industry, which may cause a local enquiry to be held (*s*).

Where a local weights and measures authority intends to institute proceedings under these Acts, or under s. 23 of the Fair Trading Act, 1973, or under the Consumer Credit Act, 1974, it must give the Director-General of Fair Trading notice of such intended proceedings together with a summary of the facts on which the charge is to be founded (*t*).

Where a charge is being considered under the Trade Descriptions Acts in respect of facts which also constitute an offence under Part IV of the Weights and Measures Act, 1963, then except in the case of proceedings under s. 23 of the 1968 Act special provisions contained in the Weights and Measures Act must be complied with, and the time limit for prosecutions is three months instead of the normal twelve months (*u*) under s. 19 of the 1968 Act. There are also special restrictions on the admissibility of evidence where the act or omission alleged constitutes an offence under the food and drugs laws (*a*).

Powers of the relevant authorities

Under each Act, the relevant authorities have wide powers to make

---

(*r*) Trade Descriptions Act, 1968, s. 26. But s. 26 (5) provides that in Scotland the procurator fiscal shall be responsible for the initiation of proceedings. Similar provision as to enforcement is contained in s. 27 of the Fair Trading Act, 1973 in respect of an order made under s. 22 of that Act, and in s. 161 (1) of the Consumer Credit Act, 1974 in respect of that Act and regulations made under it. In addition to the local weights and measures authorities, the Director-General of Fair Trading also has a duty to enforce the provisions of the 1973 and 1974 Acts and orders and regulations made under them.

(*s*) Trade Descriptions Act, 1968, s. 26 (3). The Secretary of State has a similar power under s. 161 (4) of the Consumer Credit Act, 1974.

(*t*) Fair Trading Act, 1973, s. 130; Consumer Credit Act, 1974, s. 161 (2).

(*u*) The normal time limit in summary proceedings: see s. 19 (2) of the 1968 Act. The time limit in summary proceedings is only six months in some cases: see s. 19 (4) of the 1968 Act and s. 104 of the Magistrates Court Act, 1952. Where proceedings are brought by way of indictment, the time limit for commencing them is three years from the date of the offence or one year from its discovery by the prosecutor: s. 19 (1) of the 1968 Act.

(*a*) Trade Descriptions Act, 1968, s. 22 (2).

test purchases (*b*), and to enter premises and inspect, and in some cases seize, goods and documents (*c*). Where goods were seized, and during their detention they are lost or damaged or they deteriorate, the owner is entitled to be compensated for the loss suffered unless he is convicted of an offence (*d*).

It is an offence to obstruct the work of authorised officers (*e*).

(*b*) Weights and Measures Act, 1963, s. 32; Trade Descriptions Act, 1968, s. 27; Fair Trading Act, 1973, s. 28; Consumer Credit Act, 1974, s. 164; and *cf.* Food and Drugs Act, 1955, s. 91 (powers of sampling).

(*c*) Food and Drugs Act, 1955, ss. 9, 100; Weights and Measures Act, 1963, s. 48; Trade Descriptions Act, 1968, s. 28; Fair Trading Act, 1973, s. 29; Consumer Credit Act, 1974, s. 162.

(*d*) See Food and Drugs Act, 1955, ss. 9 (4), 211; Trade Descriptions Act, 1968, s. 33; Fair Trading Act, 1973, s. 32; Consumer Credit Act, 1974, s. 163.

(*e*) Food and Drugs Act, 1955, s. 105; Weights and Measures Act, 1963, s. 49; Trade Descriptions Act, 1968, s. 29; Fair Trading Act, 1973, s. 30; Consumer Credit Act, s. 165.

SECTION D

THE SALE OF GOODS ACT, 1893

(56 & 57 VICT. C. 71)

An Act for codifying the Law relating to the Sale of Goods (a). [20th February, 1894.]

Canon of construction.—The canon for construing a codifying Act was discussed by the House of Lords in a case on the Bills of Exchange Act, 1882 (*b*). " I think," says Lord HERSCHELL,

> " the proper course is in the first instance to examine the language of the statute, and to ask what is its natural meaning, uninfluenced by any considerations derived from the previous state of the law; and not to start with inquiring how the law previously stood, and then, assuming that it was probably intended to leave it unaltered, to see if the words of the enactment will bear an interpretation in conformity with this view."

But, of course, as he proceeds to point out, if any provision be of doubtful import, resort to the previous state of the law would be perfectly legitimate (*c*).

On the present Act, Lord ALVERSTONE, C.J., has said:

> " I think it is very important to bear in mind that the rights of people in regard to these matters depend now upon statute. To a large extent the old law, I will not say has been swept away, but it has become unnecessary to refer to it (*d*)."

In practice, however, the courts have sometimes applied the pre-1893 law even where the Act appears to be inconsistent with such a course (*e*).

The provisions of the Act must be read with and subject to the provisions of the Interpretation Act, 1889, which apply to all

---

(*a*) Short title " The Sale of Goods Act, 1893," see s. 64, *post*.

(*b*) 3 Halsbury's Statutes (3rd Edn.) 188.

(*c*) *Bank of England* v. *Vagliano*, [1891] A. C. 107, at p. 145. *Cf. Bristol Tramways Co.* v. *Fiat Motors Co., Ltd.*, [1910] 2 K. B. 831, at p. 836, C. A.; *Wimble* v. *Rosenberg & Sons*, [1913] 3 K. B., at p. 757, C. A.; *In re Wait*, [1927] 1 Ch. 606, at pp. 630–631; *Laurie & Morewood* v. *Dudin & Sons*, [1926] 1 K. B. 223, at p. 234. But note the criticism of PALLES, C.B., in *Wallis* v. *Russell*, [1902] 2 I. R. 585, at p. 590, and see the savings of common law in s. 61, *post*.

(*d*) *Wallis* v. *Pratt & Haynes*, [1911] A. C. 394, at p. 398.

(*e*) See *Reid* v. *Metropolitan Police Commissioner*, [1973] 2 All E. R. 97; [1973] 2 Q. B. 551, C. A. applying the rules as to market overt set out in *Co. Inst.*, Vol. II, p. 713, which appear wholly inconsistent with s. 22 of the Act, *post*.

later Acts of Parliament, unless expressly excluded. By s. 1 of that Act, unless the contrary intention appears, singular words include the plural, and *vice versâ* ; by s. 19 " person " includes any body of persons, whether corporate or unincorporate ; and by s. 20 references to " writing " include printing, lithography, photography, and other modes of representing or reproducing words in a visible form.

The Act is divided into Parts, and those Parts are again subdivided by various headings. Regard must be had to these divisions in construing the Act (*f*).

Scotland.—Scottish technical terms, which were inserted in the Bill when it was applied to Scotland, are printed in square brackets.

PART I

FORMATION OF THE CONTRACT

Contract of Sale

1. Sale and agreement to sell.—(1) A contract of sale of goods is a contract whereby the seller transfers or agrees to transfer the property in goods to the buyer for a money consideration, called the price (*g*). There may be a contract of sale between one part owner and another (*h*).

(2) A contract of sale may be absolute or conditional.

(3) Where under a contract of sale the property in the goods is transferred from the seller to the buyer the contract is called a sale ; but where the transfer of the property in the goods is to take place at a future time or subject to some condition thereafter to be fulfilled the contract is called an agreement to sell (*i*).

(*f*) *Inglis* v. *Robertson*, [1898] A. C. 616, at p. 630 ; and see *Qualter Hall & Co., Ltd.* v. *Board of Trade*, [1961] 3 All E. R. 389; [1962] Ch. 273, C. A.; *Elliott* v. *Grey*, [1959] 3 All E. R. 733; [1960] 1 Q. B. 367, D. C.

(*g*) See note " Hire purchase, credit-sale and conditional sale agreements", p. 81, *post*.

(*h*) As to sale by a co-owner to a third party, and the remedies of co-owners *inter se*, see *post*, notes to s. 21. As to sale of undivided moiety of specific chattel, creating co-ownership, see *Marson* v. *Short* (1835), 2 Bing. N. C. 118.

(*i*) *Cf. Heilbutt* v. *Hickson* (1872), L. R. 7 C. P. 438, at p. 449 ; *Lee* v. *Butler*, [1893] 2 Q. B. 318, C. A. ; *Marten* v. *Whale*, [1917] 2 K. B. 480, at p. 484, C. A. ; 34 Halsbury's Laws (3rd Edn.) 19.

(4) An agreement to sell becomes a sale when the time elapses or the conditions are fulfilled subject to which the property in the goods is to be transferred (*k*).

Definitions.—For " buyer," " contract of sale," " goods," " property," " seller " and " sale," see s. 62 (1), *post*.

Cross-references.—*Cf.* ss. 8 and 9, *post* (price), and s. 49, *post* (action for the price).

COMMENT

This section appears to be purely declaratory.

Sub-section (1).—*Seller and buyer must be different persons.*—The essence of sale is the transfer of the property in a thing from one person to another for a price. Hence it has been said that if a man purchase his own goods there is no sale, whether or not he knows that they are his (*l*). But one co-owner may sell to another, a partner may sell to his firm, and the firm may sell to a partner (*m*), and there are clearly certain quasi-exceptions to the rule; for instance, when a man's goods are sold under an execution or distress he may himself become the purchaser. So, too, a bankrupt may buy back his own goods from the trustee,

> " though the trustee, the auctioneer, or any one bearing a fiduciary character . . . is precluded from becoming a purchaser by the general policy of the law which prohibits an agent from selling to himself " (*n*).

The true position seems to be this: buyer and seller must be different persons, but where one person has by law the right to sell another person's goods, that other person may purchase his own goods. As to bids at auction on behalf of seller, see s. 58, *post*.

Cognate or ambiguous contracts.—Whether a given contract be a contract of sale or some other kind of contract is a question of substance and not of form. It is the duty of the court to look behind

---

(*k*) *Bishop* v. *Shillito* (1819), 2 Barn. & Ald. 329 *n*. (special condition); *Rohde* v. *Thwaites* (1827), 6 B. & C. 388, at p. 393 (appropriation of goods to contract); *Bianchi* v. *Nash* (1836), 1 M. & W. 545 (special condition). *Cf. Mischeff* v. *Springett*, [1942] 2 All E. R. 349; [1942] 2 K. B. 331.

(*l*) 2 *Black. Com.* (1st ed.), p. 450; *Bell* v. *Lever Bros. Ltd.*, [1932] A. C. 161, *per* Lord ATKIN at p. 218; *Moore* v. *Singer Manufacturing Co.*, [1904] 1 K. B. 820, C. A. (distrainor buying at auction goods distrained); *Plasycoed Collieries Co.* v. *Partridge, Jones & Co.*, [1912] 2 K. B. 345 (distrainor taking goods at appraised value and using them); *Tunstall* v. *Steigmann*, [1962] 2 All E. R. 417; [1962] 2 Q. B. 593, C. A. (where A sells his business to a limited company wholly owned and controlled by him, the business is no longer " carried on by " A for the purposes of the Landlord and Tenant Act, 1954).

(*m*) For power to assign personal property to self and others, see Law of Property Act, 1925, s. 72; 27 Halsbury's Statutes (3rd Edn.) 451.

(*n*) *Kitson* v. *Hardwick* (1872), L. R. 7 C. P. 473, at p. 478; *cf. Armstrong* v. *Jackson*, [1917] 2 K. B. 822 (broker); *Hocker* v. *Waller* (1924), 29 Com. Cas. 296, 298 (agent for sale on commission).

the form of a transaction and to ascertain its real substance (*o*), and where the true transaction is impeachable by or unenforceable against one of the parties to the litigation the court will treat as a nullity a document designed by the parties to it to mask the true nature of the transaction (*p*). Thus it depends on the real meaning and nature of the transaction whether it is to be construed as a contract of sale or an invitation to enter into a contract of sale (*q*); as a contract of sale or a mere guarantee for the price (*r*); as a contract of sale or a bailment on trust (*s*); as a contract of "sale or return," or a contract of *del credere* agency (*t*); as a contract of sale or a contract for sale on commission (*a*); as a contract of sale or a contract of loan on security or mortgage (*b*); as a contract of sale or a mere wager (*c*); as a contract of sale or a contract for work and materials; as a contract of sale or a contract of hiring, including the so-called hire-purchase agreement which is a contract of hiring coupled with an option to purchase (*d*); as a contract of sale or a contract to do work as an agent (*e*); as a contract of sale or a licence to get mineral products from land (*f*); as a contract of sale or a pledge to bankers (*g*); as a contract of sale or a contact

(*o*) *Stoneleigh Finance, Ltd.* v. *Phillips*, [1965] 1 All E. R. 513; [1965] 2 Q. B. 537, C. A.; *St. Margaret's Trusts* v. *Castle*, [1964] C. L. Y. 1685, C. A.

(*p*) But the design must be that of *both* parties: *Yorkshire Railway Wagon Co.* v. *Maclure* (1882), 21 Ch. D. 309, 314, *per* JESSEL, M.R.; *Stoneleigh Finance* v. *Phillips, supra; Kingsley* v. *Sterling Industrial Securities, Ltd.*, [1966] 2 All E. R. 414; [1966] 2 W. L. R. 1265, C. A.; *Bennett* v. *Griffiu Finance*, [1967] 1 All E. R. 515; [1967] 2 Q. B. 46, C. A.; *Snook* v. *London and West Riding Investments, Ltd.*, [1967] 1 All E. R. 518; [1967] 2 Q. B. 786, C. A.; *cf. Re Watson, Ex p. Official Receiver in Bankruptcy* (1890), 25 Q. B. D. 27, C. A.; and see the rule in the case of a bill of sale in 3 Halsbury's Laws of England (3rd Edn.), 259.

(*q*) *Harvey* v. *Facey*, [1893] A. C. 552; *Pharmaceutical Society of Great Britain* v. *Boots Cash Chemists (Southern), Ltd.*, [1953] 1 All E. R. 482, C. A.; [1953] 1 Q. B. 401.

(*r*) *Hutton* v. *Lippert* (1883), 8 App. Cas. 309, P. C.

(*s*) *South Australian Ins. Co.* v. *Randell* (1869), L. R. 3 P. C. 101.

(*t*) *Re Nevill, Ex. p. White* (1871), 6 Ch. App. 397; affirmed *sub nom. Towle & Co.* v. *White* (1873), 21 W. R. 465, H. L., *Weiner* v. *Harris*, [1910] 1 K. B. 285, C. A. See the English and Scottish cases reviewed in *Michelin Tyre Co.* v. *Macfarlane*, 1917 55 Sc. L. R. 35, H. L.

(*a*) *The Prinz Adalbert*, [1917] A. C. 586, P. C. For a hybrid contract of sale and agency, see *The Kronprincessin Cecilie* (1917), 33 T. L. R. 292, P. C.

(*b*) *Ex p. Harvey & Co.* (1890), 7 Morrell, 138; *Re Watson* (1890), 25 Q. B. D. 27, C. A.; *Madell* v. *Thomas*, [1891] 1 Q. B. 230, C. A.; *Wheatley's Trustee* v. *Wheatley Ltd.* (1901), 85 L. T. 491; *Maas* v. *Pepper*, [1905] A. C. 102 (bill of sale cases); *Polsky* v. *S. & A. Services*, [1951] 1 All E. R. 185; affirmed [1951] 1 All E. R. 1062, C. A. ("hire purchase" documents executed where true transaction was loan on security of motor car), and cases there cited.

(*c*) *Rourke* v. *Short* (1856), 5 E. & B. 904 (alternative price as a wager).

(*d*) See p. 81, *post*.

(*e*) *Dixon* v. *London Small Arms Co.* (1876), 1 App. Cas. 632, at pp. 645, 648, 654; *Lamb (W. T.) & Sons* v. *Goring Brick Co. Ltd.*, [1932] 1 K. B. 710.

(*f*) *Morgan* v. *Russell & Sons*, [1909] 1 K. B. 357.

(*g*) *The Orteric*, [1920] A. C. 724, P. C. (bill of exchange, with bill of lading attached, "sold" to bank).

of agistment (*h*); as a contract of sale or a contract to procure and carry goods (*i*); or as a contract of sale or an option in one party to force a sale (*k*). By s. 61 (4), *post*, it is expressly provided that

> "the provisions of this Act relating to contracts of sale do not apply to any transaction in the form of a contract of sale which is intended to operate by way of mortgage, pledge, charge, or other security."

The question whether a given contract be a contract of sale or some other allied form of contract, though often difficult to determine, is of practical and not merely of theoretical importance.

Work and materials.—The principal practical differences between such a contract and one of sale are as follows: first, if in writing, it must be properly stamped, whereas contracts of sale of goods are exempt from stamp duty by the First Schedule to the Stamp Act, 1891. Secondly, terms are implied into the former only in accordance with the general law of contract and not under this Act (*l*). Thirdly, terms implied into a contract for work and materials may be excluded or limited by a suitably worded exemption clause, whereas terms implied into a contract for the sale of goods by ss. 12–15 of this Act may not be excluded in many cases, and in other cases may be excluded only to a limited extent (*m*). Finally, the right to treat the contract as at an end as a result of a breach by the other party may in some cases depend upon the nature of the contract (*n*).

Opinions have differed much about the test for distinguishing between these two contracts, but the general rule deducible from the cases seems to be that if the main object of the contract is the transfer from A to B, for a price, of the property in a thing in which B had no previous property, then the contract is a contract of sale; but if the real substance of the contract is the performance of work by A for B, it is a contract for work and materials notwithstanding that the performance of the work necessitates the use of certain materials and that the property in those materials passes from A to B under the contract (*o*). Thus a contract to make two sets of false

---

(*h*) *Re Woodward, Ex p. Huggins* (1886), 54 L. T. 683.

(*i*) *Cobbold* v. *Caston* (1824), 1 Bing. 399.

(*k*) *Manders* v. *Williams* (1849), 4 Exch. 339; *Bianchi* v. *Nash* (1836), M. & W. 545; *Marten* v. *Whale*, [1917] 2 K. B. 480.

(*l*) For the nature of the conditions and warranties implied into a contract for work and materials and the circumstances in which such conditions and warranties will be implied, see *Young & Marten, Ltd.* v. *McManus Childs, Ltd.*, [1968] 2 All E. R. 1169; [1969] 1 A. C. 454, H. L.; *Gloucestershire County Council* v. *Richardson*, [1968] 2 All E. R. 1181; [1969] 1 A. C. 480, H. L.; *Samuels* v. *Davis*, [1943] 2 All E. R. 3; [1943] K. B. 526, C. A.; *Ingham* v. *Emes*, [1955] 2 All E. R. 740; [1955] 2 Q. B. 366, C. A.

(*m*) See s. 55, *post*.

(*n*) See p. 9, *ante*.

(*o*) The difficulty is an old one and was much debated by the Roman lawyers. See *Inst. III*, 24, 4, and De Zulueta, *The Roman Law of Sale*, pp. 15, 16. See also 1 L. Q. R. 8.

teeth to fit the person ordering them (*p*), a contract for a mink jacket to be made from skins selected by the customer (*q*) and a contract to provide and lay fitted carpets (*r*) have been held to be contracts of sale;' but a contract to print 500 copies of a treatise, the printer supplying the paper (*s*) and a contract for the manufacture and supply of oil storage tanks (*t*) have both been held to be contracts for work and materials. A contract to paint a picture may be one of sale or of work and materials, depending upon the facts of each case. (*u*).

This general rule would seem to be subject to an exception where the goods supplied are to be affixed to land before the property in them passes. Such a contract is not one by which the seller transfers or agrees to transfer the property in goods, for the goods lose their character of " goods " as defined by s. 62, *post*, upon being attached to the land. The contract cannot, therefore, be for the sale of goods (*v*).

Gift.—Where goods are transferred by one person to another without any price or other consideration being given in return, the transaction is called a gift (*w*). Where a gift of goods is not effected by deed (*x*), it is incomplete and ineffectual until delivery to the donee of the thing intended to be given. The intention to transfer the property is of no avail (*y*). So again, a gift must be distinguished

(*p*) *Lee* v. *Griffin* (1861), 30 L. J. Q. B. 252.

(*q*) *Marcel (Furriers), Ltd.* v. *Tapper*, [1953] 1 All E. R. 15; [1953] 1 W. L. R. 49.

(*r*) *Philip Head & Sons, Ltd.* v. *Showfronts, Ltd.*, [1970] 1 Lloyd's Rep. 140.

(*s*) *Clay* v. *Yates* (1856), 1 H. & N. 73.

(*t*) *R. and J. Dempster, Ltd.* v. *Motherwell Bridge and Engineering Co.*, 1964 S. L. T. 353; see also *Grafton* v. *Armitage* (1845), 2 C. B. 336 (contract for A to devise method of curving metal tubing; use by A of his own materials); *Anglo-Egyptian Navigation Co.* v. *Rennie* (1875), L. R. 10 C. P. 271 (contract to make and supply machinery for ship and to alter the engines of the ship according to specification); contrast *Cammell Laird & Co., Ltd.* v. *Manganese Bronze and Brass Co., Ltd.*, [1934] A. C. 402 (contract to make ships' propellers: sale of goods).

(*u*) *Isaacs* v. *Hardy* (1884), Cab. & El. 287; *Robinson* v. *Graves*, [1935] 1 K. B. 579, C. A.; and see *Newman* v. *Lipman*, [1950] 2 All E. R. 832; [1951] 1 K. B. 333, D. C., where the position of a photographer was argued but not decided.

(*v*) *Collins Trading Co. Pty.* v. *Maher*, [1969] V. R. 20 (oil heater to be installed in house); *cf. Clark* v. *Bulmer* (1843), 11 M. & W. 243 (steam engine to be completed and affixed to land). See also *Benjamin's Sale of Goods*, para. 36.

(*w*) As to " free gifts " given with goods sold, see *Esso* v. *Customs and Excise Commissioners*, [1975] 1 W. L. R. 406, C. A.

(*x*) *Re Seymour* [1913], 1 Ch. 475, C. A.

(*y*) *Cochrane* v. *Moore* (1890), 25 Q. B. D. 57, C. A. (attempted gift of undivided fourth part of a racehorse) ; see also 6 L. Q. R. 446 ; *Kilpin* v. *Ratley*, [1892] 1 Q. B. 582 ; *Cain* v. *Moon*, [1896] 2 Q. B. 283, 289 ; *Stoneham* v. *Stoneham*, [1919] 1 Ch. 149 (all cases on constructive delivery). As to the suggested rights of a donee, in the case of an incomplete gift, against persons other than the donor, see Pollock's *Essays in the Law*, chap. v. (gifts without delivery).

from a declaration of trust. In the case of a gift there must be a transfer of possession, in the case of a declaration of trust the trustee must retain, and intend to retain, the control. An inchoate gift falls between the two (*a*).

Exchange of goods or barter.—Where the consideration for the transfer of the property in goods from one person to another consists of the delivery of other goods, the contract is not a contract of sale, but is a contract of exchange or barter (*b*). But if the consideration for such a transfer consists partly of the delivery of goods (the goods being regarded as the equivalent of a certain sum of money) and partly of the payment of money, it is a contract of sale (*c*). Thus if the buyer in such a case fails to deliver the goods, the seller's proper remedy is not an action for detinue but an action for that part of the price which the goods represent (*d*). Similarly if the contract may be performed either by the delivery of goods or the payment of money, it seems that the contract is a contract of sale (*e*). But where the goods were supplied on the terms that they were to be paid for by the bill of a third party, without recourse to the buyer in the bill was dishonoured ,the contract was treated as one of exchange, rather than of sale (*f*). A contract for the supply of goods in exchange for vouchers redeemable by a third party has, however, been held to be a contract for the sale of goods (*g*).

When a statute refers in terms to contracts of sale (as, for instance, the Sale of Goods Act or the Stamp Act), it seems clear that it would have no application to contracts of exchange. Section 5 of the Factors Act, 1889, *post*, for its special purpose, draws a distinction between sales and exchanges. Apart from statute, however, it seems that the rules of law relating to sales apply in general

---

(*a*) *Cf. Richards* v. *Delbridge* (1874), L. R. 18 Eq. 11.

(*b*) *Harrison* v. *Luke* (1845), 14 M. & W. 139.

(*c*) *Forsyth* v. *Jervis* (1816), 1 Stark. 437; *Sheldon* v. *Cox* (1824), 3 B. & C. 420; *Aldridge* v. *Johnson* (1857), 7 E. & B. 885. It is not clear whether such an agreement is a sale where the goods are not regarded as the equivalent of a sum of money: see *Chappell & Co., Ltd.* v. *Nestlé Co., Ltd.*, [1959] 2 All E. R. 701, at p. 709; [1960] A. C. 87, at p. 109, *per* Lord Reid, but compare *Buckley* v. *Lever Brothers, Ltd.*, [1953] 4 D. L. R. 16 (special offer of clothes pegs for cash plus wrappers); *Clarke* v. *Reilly* (1962), 96 I. L. T. R. 96 (exchange of new car for old car plus cash).

(*d*) *Forsyth* v. *Jervis* (*supra*); *Dawson* (*Clapham*) *Ltd.* v. *Dutfield* [1936], 2 All E. R. 232.

(*e*) *South Australian Insurance Co.* v. *Randell* (1869), L. R. 3 P. C. 101.

(*f*) *Read* v. *Hutchinson* (1813), 3 Camp. 352.

(*g*) *Davies* v. *Customs and Excise Commissioners*, [1975] 1 All E. R. 309; [1975] 1 W. L. R. 204, D. C., where *Read* v. *Hutchinson*, *supra*, was not cited to the Court. The decision would seem to be justifiable only on the ground that if the vouchers were not met, the seller could claim the price from the person acquiring the goods—a possibility which it is unlikely that the parties contemplated.

to contracts of barter or exchange; but the question has been by no means fully worked out (*h*).

Trading stamps.—Section 3 of the Trading Stamps Act, 1964 (*i*) provides for the redemption of trading stamps for cash. If they are nevertheless redeemed for goods, s. 4 of the 1964 Act (*k*) implies warranties on the part of the promoter of the trading stamps scheme similar to the provisions of the new ss. 12 and 14 (2) of the Sale of Goods Act, 1893. These warranties are implied notwithstanding any terms to the contrary on which the redemption is made.

Liquor sold in clubs.—Where intoxicating liquor, which is *bonâ fide* the property of a club, is supplied to a member of the club, who pays for it, the transaction, though resembling a sale, is not a sale within the meaning of the Licensing Acts. In substance the member is consuming his own property, and the mode of payment is a matter of internal arrangement regulated by the rules of the club (*l*).

Sub-section (2).—As the contract of sale is consensual, it follows that it may be either absolute or conditional, as the parties may please. The conditions inserted by the parties may be either conditions precedent or conditions subsequent. In the apt phraseology of the Continental and Civil lawyers, a contract of sale may be either a sale pure and simple, transferring the property absolutely to the buyer, or it may be subject to a "suspensive" or "resolutive" condition (*m*). The division of conditions into those which are suspensive and those which are resolutive is convenient, because those terms mark clearly the distinction between an agreement for sale which is to become an actual sale on the fulfilment of a particular condition, and an actual sale passing the property to the buyer, but subject to defeasance on the happening of some specified event.

(*h*) See *Fairmaner* v. *Budd* (1831), 7 Bing. 574; *Emanuel* v. *Dane* (1812), 3 Camp. 299; *La Neuville* v. *Nourse* (1813), 3 Camp. 350. The Bill originally contained a clause applying its provisions *mutatis mutandis* to exchanges, but it was cut out by the Commons Select Committee. In Scotland the rule as to passing of risk in a contract of exchange has been held to be the same as it was under Scottish common law in a contract of sale: *Widenmeyer* v. *Burn Stewart & Co., Ltd.*, 1967 S. L. T. 129 (Ct.of Sess.) where the rules as to passing of property are also discussed.

(*i*) As amended by the Decimal Currency Act, 1969, Sch. 2; 1969 Annual Volume, Halsbury's Statutes (2nd Edn.) 171.

(*k*) As substituted by s. 16 (1) of the Supply of Goods (Implied Terms) Act, 1973.

(*l*) *Graff* v. *Evans* (1882), 8 Q. B. D. 373, 378; *cf. Newman* v. *Jones* (1886), 17 Q. B. D. 132 (unauthorised sale by steward to person not a member of club); as to status of member of proprietary club, see *Baird* v. *Wells* (1890), 44 Ch. D. 661; and see further Paterson's *Licensing Acts* (82nd ed.), pp. 887, *et seq.*

(*m*) French Civil Code, art. 1584, and see further arts. 1181–1184, defining these conditions; *cf.* De Zulueta, *The Roman Law of Sale*, pp. 28–30; *Murdoch* v. *Greig*, 1889 16 R. (Ct. of Sess.) 396.

When goods are sold by weight or measure, the weighing and measuring are suspensive conditions, and if goods be sent on approval, the approval of the buyer constitutes a suspensive condition (see s. 18, rules (3) and (4), *post*). But if goods be sold by auction with a condition that they may be re-sold if not paid for within twenty-four hours, the condition is resolutive (*n*). So too a term which allows the buyer to return goods if they do not correspond with a warranty is resolutive (*o*). As a contract of sale may be conditional, the parties may attach such consequences as they see fit to the non-fulfilment of the condition. They may stipulate that in the event of non-fulfilment both parties shall be discharged from their obligations, or that the seller shall not be bound to deliver, or that the buyer shall not be bound to pay the price (*p*). Conditions may also be implied by other statutes; see, for example, s. 33 (1) of the Exchange Control Act, 1947 (*q*).

In *Christopher Hill, Ltd.* v. *Ashington Piggeries, Ltd.* (*r*), the plaintiffs had agreed in May 1960 to supply food for the defendants' mink made up in accordance with the defendants' formula, orders to be placed by the defendants from time to time as they required the food. Subsequently the defendants placed a number of orders for the food. The Court of Appeal held that having regard to ss. 1 and 5 of the Sale of Goods Act, the agreement of May 1960 was a conditional contract of sale, the condition being the placing of orders by the defendants as and when they required the food. The subsequent orders were not themselves separate contracts of sale.

Hire-purchase, credit-sale and conditional sale agreements.—A hire-purchase agreement, whereby goods are let on hire with a proviso that they shall become the property of the hirer when a certain amount has been paid, may be an agreement for sale (*a*). If so, it will of course be a conditional agreement to buy for the purposes of the Sale of Goods Act, 1893. A conditional agreement

---

(*n*) See *Lamond* v. *Davall* (1847), 9 Q. B. 1030; and *cf.* s. 48 (4), *post.*

(*o*) *E.g., Head* v. *Tattersall* (1871), L. R. 7 Exch. 7; for a peculiar restrictive condition containing a term for novation, see *Grissell* v. *Bristowe* (1868), L. R. 3 C. P. 112. *Cf. The Ripon City*, [1897] P. 226 (conditional sale of ship); *Wyllie* v. *Povah* (1907), 12 Com. Cas. 317, 321 (goods sold while at sea, in case of non-arrival the contract to be void).

(*p*) *Calcutta Co.* v. *De Mattos* (1863), 32 L. J. Q. B. 322, at p. 328; 34 Halsbury's Laws (3rd Edn.) 10. *Cf. New Zealand Shipping Co.* v. *Société des Ateliers de France*, [1919] A. C. 1 (stipulation in a contract that it should be void in a certain event); *The Vesta*, [1921] 1 A. C. 774, at p. 784 (sale with a condition enabling buyer to call upon seller to re-purchase).

(*q*) *Post*, Appendix I.

(*r*) [1969] 3 All E. R. 1496, C. A.; reversed on other grounds: [1971] 1 All E. R. 847, H. L.

(*a*) *Lee* v. *Butler*, [1893] 2 Q. B. 318, C. A.; *Muirhead* v. *Dickson*, 1905 7 F. (Ct. of Sess.) 686; *A.-G.* v. *Pritchard* (1928), 97 L. J. K. B. 561; *Taylor* v. *Thompson*, [1930] W. N. 16.

to buy must be distinguished from an option to buy (*b*). By s. 62 of the Sale of Goods Act, 1893, " buyer " means a person who buys or agrees to buy goods. There must therefore be a consent to buy as well as a consent to sell. It follows that a hirer with an option to buy is not a " person having agreed to buy goods " within s. 9 of the Factors Act, 1889 (*c*). The hire-purchase transaction is essentially different, both in practice and in its legal incidents, from the ordinary sale of goods and it is therefore considered as being outside the scope of this book. Credit-sale agreements, however, although usually entered into in practice in circumstances similar to those of hire-purchase agreements, and although governed in some respects by the Hire-Purchase Act, 1965 (*d*) are for the most part governed by the general law relating to the sale of goods.

Conditional sale agreements are hybrid in character. They are contracts of sale and are thus generally subject to the Sale of Goods Act, but by virtue of the provisions of the Hire-Purchase Act, 1965 (*d*) certain sections of the Sale of Goods Act do not apply or are limited in their application. These agreements are also akin to hire-purchase agreements and several provisions relating to hire-purchase agreements are extended to them in the Hire-Purchase Act, 1965 (*d*).

To the extent that they are governed by the law relating to sale of goods, credit-sale and conditional sale agreements are dealt with in this book.

Mortgage of Goods.—A mortgage is sometimes called a conditional sale, but, as CAVE, J., has pointed out, a sale with a condition for re-sale to the original seller need have nothing to do with a mortgage (*e*). In a mortgage there is no sale of the goods and there is a debt always due from the mortgagor to the mortgagee, until it has been satisfied by payment or foreclosure (*e*). A mortgage of goods may be defined as a transfer of the general property in goods from mortgagor to mortgagee in order to secure a debt (*f*). This Act has no application to any mortgage, even though clad with the form of a contract of sale; see s. 61 (4), *post*. If a mortgage of goods

(*b*) See *Marten* v. *Whale*, [1917] 2 K. B. 480, C. A.

(*c*) *Helby* v. *Matthews*, [1895] A. C. 471; nor within s. 25 (2) of the Sale of Goods Act, 1893: *Belsize Motor Supply Co.* v. *Cox*, [1914] 1 K. B. 244; *cf. Felston Tile Co., Ltd.* v. *Winget, Ltd.*, [1936] 3 All E. R. 473, C. A.; *St. Margaret's Trust* v. *Castle*, [1964] C. L. Y. 1685, C. A.

(*d*) The Hire-Purchase Act, 1965 has been repealed by Sch. 5 of the Consumer Credit Act, 1974 as from a date to be appointed pursuant to s. 192 (4) of that Act. As yet no date has been appointed and the 1965 Act remains in force, For the effect of the Consumer Credit Act, 1974 on contracts for the sale of goods where credit is given by the seller, see p. 63, *ante*.

(*e*) *Beckett* v. *Tower Assets Co.*, [1891] 1 Q. B. 1, at p. 25; *cf. Manchester, Sheffield and Lincolnshire Rail Co.* v. *North Central Wagon Co.* (1888), 13 App. Cas. 554; and *Joblin* v. *Watkins and Roseveare Ltd.*, [1949] 1 All E. R. 47.

(*f*) *Keith* v. *Burrows* (1876), 1 C. P. D. 722, at p. 731, *per* LINDLEY, J.; *Ex p. Hubbard* (1886), 17 Q. B. D. 690 at p. 698, *per* BOWEN, J. A memorandum of a pledge of goods containing a power of sale is not a mortgage for stamp duty purposes: *Re Attenborough* (1855), 11 Exch. 461.

be in writing, it usually comes within the Bills of Sale Acts, 1878 and 1882 (*g*). But it seems that goods might be mortgaged by parol (*h*). If that be so, such a mortgage apparently comes under neither this Act nor the Bills of Sale Acts; for the latter Acts strike only at documents, and not at the transactions themselves (*i*).

Letters of hypothecation.—A letter of hypothecation of imported goods need not be registered as a bill of sale (*k*). The term "hypothecation" is somewhat loosely used. It usually denotes an equitable charge, without possession but not amounting to a mortgage (*l*).

Pledge.—A pledge may be defined as a delivery or bailment of goods by one person to another in order to secure payment of a debt (*m*). If the pledgor makes default in paying the debt the pledgee may sell the goods after notice to the pledgor (*n*). A pledge differs from a mortgage—(1) because the mortgagor may retain possession, while a pledge must be delivered to the pledgee; (2) because the mortgagee obtains the general property in the goods, while the pledgee only obtains a special property necessary to secure his rights (*o*). Note a wider definition of pledge given by s. 1 (5) of the Factors Act, 1889, *post*, for the purposes of that Act.

(*g*) 3 Halsbury's Statutes (3rd Edn.) 261.

(*h*) *Flory* v. *Denny* (1852), 7 Exch. 581 (no deed is necessary), and *cf. Reeves* v. *Capper* (1838), 5 Bing. N. C. 136; *Redhead* v. *Westwood* (1888), 59 L. T. 293; *Ramsay* v. *Margrett*, [1894] 2 Q. B. 18; *London and Yorkshire Bank* v. *White* (1895), 11 T. L. R. 570. There appears to be no statute requiring a mortgage of goods to be in writing.

(*i*) *North Central Wagon Co.* v. *Manchester, Sheffield and Lincolnshire Rail Co.* (1887), 35 Ch. D. 191, at p. 206, C. A.

(*k*) Bills of Sale Act, 1890; 3 Halsbury's Statutes (3rd Edn.) 270; as amended by Bills of Sale Act, 1891 (3 Halsbury's Statutes (3rd Edn.) 271. As to registration of charges created by limited companies, see ss. 95–106 of the Companies Act, 1948; 5 Halsbury's Statutes (3rd Edn.) 189 *et seq.*, and *Ladenburg & Co.* v. *Goodwin*, [1912] 3 K. B. 275 (hypothecation of proceeds of exported goods — charge on book debts); *Reed on Bills of Sale Acts* (14th ed.), pp. 78, 206.

(*l*) *Hart's Law of Banking* (4th ed.), p. 909; *Fisher and Lightwood's Law of Mortgages* (7th ed.), p. 112; *cf. Ex p. North Western Bank* (1872), L. R. 15 Eq. 69, at p. 73.

(*m*) *Story on Bailments* (9th ed.), pp. 7, 286; *cf. Blundell Leigh* v. *Attenborough*, [1921] 1 K. B. 382, 389 (nature of pledge and constructive delivery), reversed on inference of fact, but approved as to law, [1921] 3 K. B. 235, 239, C. A.

(*n*) *Story on Bailments* (9th ed.), p. 308, and see generally the notes to *Coggs* v. *Bernard*, 1 Smith L. C. (13th ed.), pp. 175, 201. The power of a pledgee to sell without the intervention of any court is peculiar to English law; see *The Odessa*, [1916] 1 A. C. 145, at p. 158, P. C. As to Scotland, see Green's *Encyclopædia of Scots Law*, tit. "Pledge."

(*o*) *Halliday* v. *Holgate* (1868), L. R. 3 Exch. 299, see *per* WILLES, J.; *Ex p. Hubbard* (1886), 17 Q. B. D. 690, at p. 698; *Re Morritt* (1866), 18 Q. B. D. 222, at p. 232; see a mere lien distinguished from a pledge, *Donald* v. *Suckling* (1866), L. R. 1 Q. B. at p. 612; *cf. Howes* v. *Ball* (1827), 7 B. & C. 481 (hypothecation); and see pledge and sale contrasted, *Burdick* v. *Sewell* (1884), 13 Q. B. D. 159, at p. 175, C. A., and 10 App. Cas. 74, at p. 93.

Sub-section (3).—The definition of " sale " has been criticised as failing to distinguish a contract, by which a *jus in personam* is created, from a conveyance, by which a *jus in rem* is transferred. But the language of English law is inveterate, an agreement to sell being known as an executory contract of sale, while a sale is known as an executed contract of sale.

The term " contract of sale " thus includes both actual sales and agreements for sale. It is important to distinguish clearly between the two classes of contract. An agreement to sell, or, as it is often called, an executory contract of sale, is a contract pure and simple ; whereas a sale, or, as it is called for distinction, an executed contract of sale, is a contract plus a conveyance. By an agreement to sell a *jus in personam* is created, by a sale a *jus in rem* also is transferred. Where goods have been sold (even though not delivered), and the buyer makes default, the seller may sue for the contract price on the count of " goods bargained and sold ", but where an agreement to buy is broken, the seller's normal remedy is an action for unliquidated damages (*p*). If an agreement to sell be broken by the seller, the buyer has only a personal remedy against the seller. The goods are still the property of the seller, and he can dispose of them as he likes; they may be taken in execution for his debts, and, if he becomes bankrupt, they pass to his trustee. But if there has been a sale, and the seller breaks his engagement to deliver the goods, the buyer has not only a personal remedy against the seller, but also remedies in respect of the goods themselves, such as the actions for conversion and detinue. In many cases, too he can follow the goods into the hands of third parties. Again, if there be an agreement for sale, and the goods are destroyed, the loss as a rule falls on the seller, while if there has been a sale, the loss as a rule falls upon the buyer, though the goods have never come into his possession. See s. 20, *post.*

Sub-section (4).—By s. 62, *post,* the term " sale " includes a bargain and sale, as well as a sale and delivery (*q*). According to the Civil Law which, with some statutory modifications, prevailed in Scotland before the Act, the property in the goods sold did not pass to the buyer until delivery (*r*), and then only if the price was paid or security found or credit given. But English law has rejected the objective test of delivery, and has adopted the rule that the property in the goods may be transferred by the contract itself if the parties

---

(*p*) For exceptions see the comment to s. 49, *post.*

(*q*) As to the old distinction between the action for goods bargained and sold and the action for goods sold and delivered, see Bullen and Leake, *Prec. of Plead.* (3rd ed.), pp. 38–9 ; see also *Colley* v. *Overseas Exporters,* [1921] 3 K. B. 302, 309–310, *per* McCardie, J.

(*r*) See *McBain* v. *Wallace* (1881), 6 App. Cas. 588, at pp. 605, 608, and Moyle's *Sale in the Civil Law,* p. 110.

so intend (*s*). The parties may make whatever bargain they please, and the law will give effect to it. When the parties express their intention clearly no difficulty arises. The contract may pass the property at once, or at a future time, or contingently on the performance of some condition (*t*). But in many cases the parties either form no intention on the point, or fail to express it. To meet such cases the Courts worked out a series of more or less artificial rules for determining when the property is to be deemed to pass, according to the imputed intention of the parties. These rules are now reproduced in s. 18 of the Act, *post.*

2. Capacity to buy and sell.—Capacity to buy and sell is regulated by the general law concerning capacity to contract, and to transfer and acquire property (*u*):

Provided that where necessaries are sold and delivered to an infant [or minor] or to a person who by reason of mental incapacity or drunkenness is incompetent to contract, he must pay a reasonable price therefor (*v*).

Necessaries in this section mean goods suitable to the condition in life of such infant [or minor] or other person (*a*), and to his actual requirements at the time of the sale and delivery (*b*).

ILLUSTRATION

Clothes to the value of £145, including eleven fancy waistcoats, are supplied by a tailor to an infant undergraduate at Cambridge. The tailor must show that the goods were suitable to the condition

---

(*s*) See *Blackburn on Sale* (1st ed.), pp. 187–197, who finds traces of the rule as far back as the time of Edward IV. The history of the question is treated exhaustively in the judgment in *Cochrane* v. *Moore* (1890), 25 Q. B. D. 57, C. A. (gift of a horse).

(*t*) *Johnson* v. *Macdonald* (1842), 9 M. & W. 600 (sale subject to arrival of ships with goods on board); *McEntire* v. *Crossley* (1895), A. C. 457, at p. 463 (hire-purchase). See s. 17, *post*, and *cf. The Parchim,* [1918] A. C. 157, *per* Lord PARKER at pp. 160, 161.

(*u*) English law, unlike some continental codes, draws no general distinction between traders and non-traders.

(*v*) *Ryder* v. *Wombwell* (1868), L. R. 4 Exch. 32, at p. 38, Ex. Ch. Lord BLANESBURGH in *Keane* v. *Mount Vernon Colliery Co.*, [1933] A. C. 309 at p. 328 said that the whole jurisprudence on this subject will be found in the judgment of WILLES, J., in this case (jewellery).

(*a*) *Ibid.*; *Peters* v. *Fleming* (1840), 6 M. & W. 42, at p. 46, *per* PARKE, B., and p. 48, *per* ALDERSON, B.

(*b*) *Barnes* v. *Toye* (1884), 13 Q. B. D. 410; *Johnstone* v. *Marks* (1887), 19 Q. B. D. 509, C. A.

in life of the infant and also that they were suitable to his actual requirements at the time of the sale and delivery (*c*).

Definitions.—For " delivery," " property," and " sale," see s. 62 (1), *post.*

Cross-reference.—*Cf.* s. 8, *post* (reasonable price).

COMMENT

Capacity to contract must be distinguished from authority to contract. Capacity means power to bind oneself; authority means power to bind another. Capacity is part of the law of status; authority is part of the law of principal and agent. Capacity is usually a question of law; authority is usually a question of fact. This section deals only with the question of capacity to buy and sell. As regards authority to buy or sell on behalf of another there appears to be nothing peculiar to the contract of sale, except the provisions of the Factors Acts, *post.* On this subject, therefore, the reader is referred to general works on the law of Agency and Partnership, though certain special cases of authority are, for convenience, considered below. Section 61 (2), *post,* contains an express saving for the law of principal and agent.

Capacity.—The section is probably declaratory. As COTTON, L.J., has pointed out, when necessaries are supplied to a person who is incompetent to contract, the obligation to pay for them is quasi-contractual and does not result from the agreement between the parties (*d*). He cannot bind himself to pay for them, but it is for his benefit that he should have them, and the law therefore will see that a fair price is paid therefor.

Mental patient.—As a rule, a contract made with a person who, at the time of contracting, is mentally disordered (*e*) to an extent such that his free and full consent is vitiated, cannot be enforced against him if the other contracting party knew the state of his mind; but such a person will still be under a quasi-contractual liability to pay for necessaries supplied to him (*f*).

(*c*) *Nash* v. *Inman,* [1908] 2 K. B. 1, at p. 5, C. A.

(*d*) *Re Rhodes* (1890), 44 Ch. D. 94, at pp. 105–107, C. A. See also *per* FLETCHER MOULTON, L.J., in *Nash* v. *Inman,* [1908] 2 K. B. 1, at p. 8, C. A.

(*e*) The expression comes from the Mental Health Act, 1959; 25 Halsbury's Statutes (3rd Edn.) 42 and now presumably takes the place of the expressions " lunatic " and " person of unsound mind " in former use.

(*f*) *Re Rhodes* (1890), 44 Ch. D. 94, C. A. (necessaries); *Imperial Loan Co.* v. *Stone,* [1892] 1 Q. B. 599, C. A. (promissory note signed by defendant as surety); *York Glass Co., Ltd.* v. *Jubb* (1926), 134 L. T. 36, C. A. (contract for purchase of land). As to contracts by mentally disordered persons generally, see 17 L. Q. R. 147.

Drunken man.—A contract made by a man is voidable at his option if, when he made it, he was so drunk as not to know what he was doing, and this fact was known to the other contracting party (*g*). But as POLLOCK, C.B., says, a drunkard is liable " when sober, for necessaries supplied to him when drunk " (*h*).

Infant or minor.—The term " minor " is the Scottish equivalent of the English term " infant." By s. 1 of the Infants Relief Act, 1874 (*i*)

> " all contracts . . . for goods supplied, or to be supplied, other than contracts for necessaries, and all accounts stated with infants shall be absolutely void."

These words are apparently to be interpreted as meaning that all contracts imposing upon the infant the obligation to pay for goods other than necessaries shall be unenforceable against him whether the goods have been supplied to him in pursuance of the contract or not. " Absolutely void," it is conceived, means irrevocably voidable at the option of the infant and does not mean that the adult party to the contract is not bound, which would alter the common law rule to the prejudice of the infant. But the infant cannot claim specific performance (*k*). Moreover, if an infant buys and pays for goods which are not necessaries he cannot recover the money he has paid (*l*), unless there has been a total failure of consideration (*m*). And if the infant has not paid for the goods which have been delivered to him he cannot be sued for their conversion, which would be possible if the property had not passed to him (*n*). It might, however, be held that in such cases the infant acquired the property not by virtue of the contract, but because there had been a delivery with the intention of passing the property (*o*) which would be operative even if the contract were void. Only a definite judical decision can settle the question. The language

---

(*g*) *Gore* v. *Gibson* (1845), 13 M. & W. 623; *Matthews* v. *Baxter* (1873), L. R. 8 Exch. 132.

(*h*) *Gore* v. *Gibson* (1845), 13 M. &. W. 623, at p. 625; see also *West Ham Union* v. *Pearson* (1890), 62 L. T. 638 (supply of necessaries to man suffering from delirium tremens).

(*i*) 17 Halsbury's Statutes (3rd Edn.) 412.

(*k*) *Lumley* v. *Ravenscroft*, [1895] 1 Q. B. 683, C. A.; 73 L. Q. R. 200; 74 L. Q. R. 99.

(*l*) *Hamilton* v. *Vaughan*, [1894] 3 Ch. 589, at p. 594. *Cf. Pearce* v. *Brain*, [1929] 2 K. B. 310 (a case of barter).

(*m*) *Valentini* v. *Canali* (1889), 24 Q. B. D. 166; *cf. Steinberg* v. *Scala* (*Leeds*) *Ltd.*, [1923] 2 Ch. 452 (contract for sale of shares voidable at the infant's option). And see *Chaplin* v. *Leslie Frewin* (*Publishers*), [1965] 3 All E. R. 764, 770; [1966] Ch. 71, 90, *per* Lord DENNING, M.R.

(*n*) *Stocks* v. *Wilson*, [1913] 2 K. B. 235.

(*o*) *Simpson* v. *Nicholls* (1838), 3 M. &. W. 240, at p. 244; and see the view expressed by LUSH, J., in *Stocks* v. *Wilson*, [1913] 2 K. B. 235.

of that Act is consistent with the view that an infant might be liable on an executory contract to supply him with necessaries, and in a case where an infant contracted to go on a billiard tour which involved the supply of necessaries to him, HAMILTON, L.J., said he could not see

> "why a contract which is in itself binding, because it is a contract for necessaries not qualified by unreasonable terms, can cease to be binding merely because it is still executory" (*p*).

But an infant has never yet been held liable for breach of contract to accept necessaries, nor for necessaries bargained and sold but not delivered, and it will be observed that by the language of s. 2 it is necessary that the goods should be sold and delivered. The definition of necessaries also assumes that the goods have been delivered. As the law makes the contract for the infant, and for his benefit, he is only liable to pay a reasonable price, and not any price he may have been led to agree to.

An infant may marry, so he may make himself liable for necessaries supplied to his family (*q*). Theoretically, at any rate, English law, unlike Scottish law, "knows of no distinction between infants of tender and of mature years" (*r*).

Even where an infant has obtained goods in pursuance of a contract induced by his fraud, the seller cannot recover the price of the goods in any form of action, though perhaps on equitable principles the infant may be compelled to return the goods, or, if he has sold them, to account for the proceeds of the sale (*s*). Conversely, where an infant has contracted to sell goods and has received the price but not delivered the goods, or has delivered goods of an inferior quality, he cannot be sued either for damages or for the return of the price as money had and received (*t*), and, it is submitted, the effect of this rule cannot be avoided by framing the action in tort (*u*).

(*p*) *Roberts* v. *Gray*, [1913] 1 K. B. 520, at p. 530, C. A.

(*q*) *Chapple* v. *Cooper* (1844), 13 M. & W. 252, at p. 259, *per* ALDERSON, B. (infant widow liable for husband's funeral expenses).

(*r*) *Morgan* v. *Thorne* (1841), 7 M. & W. 400, at p. 408, *per* PARKE, B. But obviously necessaries for a child of seven, and for a youth of seventeen, must stand on a different footing. As to Scotland, see Green's *Encyclopædia of Scots Law*, tits. "Minor" and "Sale."

(*s*) *Stocks* v. *Wilson*, [1913] 2 K. B. 235 ; *Leslie* v. *Sheill*, [1914] 3 K. B. 607. C. A. For the general rule that an infant's contract, unenforceable at law, cannot be enforced by the device of suing him in tort, see *Jennings* v. *Rundall* (1799), 8 Term Rep. 335 ; *Fawcett* v. *Smethurst* (1914), 84 L. J. K. B. 473 ; but *cf*. *Burnand* v. *Haggis* (1863), 14 C. B., N. S., 45 ; *Ballett* v. *Mingay*, [1943] 1 All E. R. 143 ; [1943] K. B. 281.

(*t*) *Cowern* v. *Nield*, [1912] 2 K. B. 419.

(*u*) *Leslie* v. *Sheill*, [1914] 3 K. B. 607, C. A. The contrary was suggested in *Cowern* v. *Nield* (*supra*), but this seems inconsistent with the views expressed in *Leslie* v. *Sheill*. See especially *per* KENNEDY, L.J., at p. 621.

Meaning of necessaries.—A somewhat artificial definition of necessaries has been gradually evolved by the cases, and is now embodied in the concluding paragraph of the section. It is for the judge to explain to the jury the legal meaning of necessaries, and for the jury to say whether in any particular case the goods supplied are necessaries. But it is a question of law for the judge to determine in the first instance whether there is any evidence to go to the jury (*v*), and most of the reported cases are cases deciding whether there was any evidence, having regard to all the circumstances, which was fit to be considered, and are not to be taken as deciding that a particular article is or is not a necessary (*a*). The burden of proving that the goods supplied were necessaries lies on the party seeking to enforce the contract (*b*).

Since the Infants Relief Act, 1874, a racing bicycle has been allowed as a necessary for an infant getting 21 shillings a week (*c*), but cartridges, and jewellery for a girl he was courting, were held not to be necessaries for an infant with an income of £7 a week (*d*). As was said long ago, " articles of mere luxury are always excluded, though luxurious articles of utility are sometimes allowed " (*e*).

Married women.—The Married Women (Restraint upon Anticipation) Act, 1949 (*f*), abolished the last remaining incapacities of married women.

Enemy.—At common law an alien enemy is a person who carries on business or voluntarily resides in hostile territory, which includes territory permanently occupied by the enemy (*g*). Locality, and not nationality, furnishes the test (*h*). But by statute enemy status may be imposed on enemy nationals and other persons having hostile associations outside enemy territory; see, *e.g.*, The Trading with the Enemy Act, 1939 (*i*).

(*v*) *Nash* v. *Inman*, [1908] 2 K. B. 1, C. A.; *cf. Stocks* v. *Wilson*, [1913] 2 K. B. 235, at p. 241.

(*a*) See, *e.g.*, *Elkington & Co. Ltd.* v. *Amery*, [1936] 2 All E. R. 86.

(*b*) *Nash* v. *Inman*, *supra*.

(*c*) *Clyde Cycle Co.* v. *Hargreaves* (1898), 78 L. T. 296.

(*d*) *Hewlings* v. *Graham* (1901), 70 L. J. Ch. 568.

(*e*) *Chapple* v. *Cooper* (1844), 13 M. & W. 252, at p. 258; cited with approval in *Keane* v. *Mount Vernon Colliery Co.*, [1933] A. C. 309, at pp. 326, 327.

(*f*) 17 Halsbury's Statutes (3rd Edn.) 131.

(*g*) *Sovfracht* v. *Van Udens Scheepvaart*, [1943] 1 All E. R. 76; [1943] A. C. 203.

(*h*) *Janson* v. *Driefontein Consolidated Mines*, [1902] A. C. 484, at p. 505; *Porter* v. *Freudenberg*, [1915] 1 K. B. 857, C. A. As to a British registered company under enemy control, which gives it enemy status, see *The Daimler Co.* v. *Continental Tyre Co.*, [1916] 2 A. C. 387, and see the Trading with the Enemy Act, 1939, as to enemy-controlled corporations.

(*i*) It is outside the scope of this book to deal with the general subject of trading with the enemy, as to which see generally Howard, *Trading with the Enemy* (1943).

An enemy may be sued, and, as he may be sued, he may defend and appeal if judgment is given against him when defendant, but as a general rule (except by licence from the Crown) an enemy cannot sue in a British court or be the *actor* in any legal proceeding (*k*).

If an enemy has a right of action which accrued before war, the right is suspended during war, but revives when peace is declared (*l*). For example, if goods are sold and delivered by a German firm to an English firm before war, no action for the price can be maintained during war (*m*). As soon as war is declared all commercial intercourse with the enemy is prohibited, and if and in so far as any contract is executory, and its due fulfilment would involve intercourse with the enemy, the contract is dissolved. See p. 20, *ante*, and the Law Reform (Frustrated Contracts) Act, 1943, in Appendix I, *post*.

(*k*) *Porter* v. *Freudenberg*, [1915] 1 K. B. 857, C. A.; *Ex p. Marum* (1915), 84 L. J. K. B. 1893. But an alien enemy may be joined *pro formâ* as a plaintiff in a partnership action: *Rodriguez* v. *Speyer Brothers*, [1919] A. C. 59; and an interned enemy may sue: *Schaffenius* v. *Goldberg*, [1916] 1 K. B. 284, C. A. For cases arising out of the last war, see *Sovfracht* v. *Van Udens Scheepvaart*, [1943] 1 All E. R. 76; [1943] A. C. 203 (affirming the general rules as to an enemy plaintiff's incapacity to sue as laid down in *Porter* v. *Freudenberg*, *supra*); *Eichengruen* v. *Mond*, [1940] 3 All E. R. 148, C. A.; [1940] Ch. 785 (power to strike out enemy plaintiff's statement of claim delivered before outbreak of war if frivolous and vexatious); *re Churchill & Co. Ltd. and Lonberg*, [1941] 3 All E. R. 137 (as to service on enemy defendant); *Schering, Ltd.* v. *Stockholms Enskilda Bank A.B.*, [1946] 1 All E. R. 36; [1946] A. C. 219 (war does not abrogate or discharge a debt incurred before the declaration of war and so obligation to pay and right to recover are only suspended). See also *Arab Bank, Ltd.* v. *Barclays Bank* (*D.C. & O.*), [1954] 2 All E. R. 226; [1954] A. C. 495; *Vamvakas* v. *Custodian of Enemy Property*, [1952] 1 All E. R. 629; [1952] 2 Q. B. 183 (difference between meaning of "enemy" under the Statute and at common law discussed). *Maerkle* v. *British & Continental Fur Co., Ltd.*, [1954] 3 All E. R. 50 (statement of claim struck out as disclosing no reasonable cause of action since the plaintiffs were "enemies" at time of transaction and by Trading with the Enemy (Custodian) Order, 1951, the property concerned vested in the custodian of enemy property); *Kuenigl* v. *Donnersmarck*, [1955] 1 All E. R. 46; [1955] 1 Q. B. 515 (an English company may be prejudiced if under enemy control in that it may have enemy character conferred upon it, but it remains English and still subject to English law). *Bevan* v. *Bevan*, [1955] 2 All E. R. 206; [1955] 2 Q. B. 227 ("enemy" plaintiff succeeded in action for maintenance due under separation agreement of 1932, since the agreement continued to subsist at common law—public policy not requiring its termination on the outbreak of war—and so defendant was subject to s. 7 of the 1939 Act and should have paid the sum due to the Custodian).

(*l*) *Janson* v. *Driefontein Consolidated Mines*, [1902] A. C. 484, at p. 499; *Porter* v. *Freudenberg*, [1915] 1 K. B. 857, at p. 873, C. A. Under the old form of pleading, the plea of "alien enemy" was usually raised by a dilatory plea and perhaps could not be raised by a plea in bar.

(*m*) *Wolf & Sons* v. *Carr Parker & Co.* (1915), 31 T. L. R. 407, C. A. Note a Scottish case, which goes further than English law by allowing a former alien enemy to recover a prepaid instalment of the price where the contract was dissolved by war; *Cantiere San Rocco* v. *Clyde Shipbuilding Co.*, [1924] A. C. 226.

Registration of business names.—Speaking generally, a person or firm may trade in any name that person or firm likes to adopt. But under the Registration of Business Names Act, 1916, an individual who carries on business in any name other than his own, and a firm whose firm name does not disclose the names of all the partners, must register the name in which the business is carried on and give the particulars required by the Act. By s. 8 of the Act, if default is made in complying with the provisions of the Act, the defaulter cannot maintain any action to enforce any contract made or entered into in relation to the business unless the court, for cause shown, grants him relief (*n*).

Conflict of laws.—There is only one English case dealing with this aspect of capacity: *Male* v. *Roberts* (*o*), decided over 160 years ago, in which Lord ELDON is reported to have referred the question to " the laws of that country where the contract arises." In that case, the *lex loci contractus* was the same as the proper law, and both the decision and the report are unsatisfactory in many respects. There are dicta in other English cases some of which favour the *lex loci contractus* as determining capacity (*p*), and some the *lex domicilii* (*q*).

Of the various views put forward in these cases, in decisions in other common law jurisdictions (*r*), and by modern text-book writers, the best seems to be that questions of capacity to enter into mercantile or commercial contracts should fall to be decided by the proper law of the contract (*s*).

Agency.—*Power of wife to bind husband.*—

Where husband and wife are living together, there is a presumption of fact that the wife has the husband's authority to pledge his credit for necessaries (*t*). This presumption may be rebutted by the husband in the following ways at least:

(a) by proof that the tradesman concerned had been expressly warned not to supply the wife with goods on credit;

(*n*) As to the scope of this enactment, see *Daniel* v. *Rogers*, [1918] 2 K. B. 228 C. A.; *Watson* v. *Park Royal*, [1961] 2 All E. R. 346; [1961] 1 W. L. R. 727. Note also s. 201 of the Companies Act, 1948; 5 Halsbury's Statutes (3rd Edn.) 271 (particulars with respect to directors).

(*o*) (1800), 3 Esp. 163.

(*p*) *E.g.*, *Sottomayor* v. *de Barros* (1877), 3 P. D. 1, 5; *Cooper* v. *Cooper* (1888), 13 App. Cas. 88, 99, 100.

(*q*) *E.g.*, *Simonin* v. *Mallac* (1860), 2 Sw. & Tr. 67; *Sottomayor* v. *de Barros* (1879), 5 P. D. 94; *Baindail* (*otherwise Lawson*) v. *Baindail*, [1946] 1 All E. R. 342; [1946] P. 122, C. A.

(*r*) See these reviewed in Dicey and Morris, *Conflict of Laws* (9th ed.) p. 768.

(*s*) *Charron* v. *Montreal Trust* (1958), 15 D. L. R. (2d.) 240.

(*t*) *Debenham* v. *Mellon* (1880), 6 App. Cas.24, especially at pp. 32, 36 and 37; *Miss Gray Ltd.* v. *Earl of Cathcart* (1922), 38 T. L. R. 562 (all the exceptions reviewed).

(b) by proof that the wife was already adequately supplied with necessaries (*u*);
(c) by proof that the wife had an adequate allowance for the purchase of necessaries (*v*);
(d) by proof that the wife had been expressly forbidden by the husband to pledge his credit (*w*), but if the husband has on previous occasions held his wife out as having authority to bind him, he will be estopped from relying on an express prohibition unless he has given express notice of it to the tradesman concerned (*x*).

But if the wife has separate estate and credit is given solely to her, the husband is not liable (*a*).

Formerly, a wife could also pledge her husband's credit for necessaries in certain circumstances as an agent of necessity (*b*). This power has now been abolished by s. 41 (1) of the Matrimonial Proceedings and Property Act, 1970 (*c*).

Divorced woman.—A woman who is divorced, or who is judicially separated from her husband, is on the same footing as a single woman. The same applies where a court of summary jurisdiction makes an order under the Matrimonial Proceedings (Magistrates' Courts) Act, 1960 (*d*) that a married woman shall no longer be bound to cohabit with her husband.

Mistress.—When a woman is living with a man as his wife, she has, while so living with him, the like authority to pledge his credit for necessaries as if she were married to him (*e*).

Parent and child.—A father is not liable for necessaries supplied to his infant child without his authority, nor is the mother, though she have separate estate (*f*). But authority or ratification is inferred on slight evidence (*g*).

(*u*) *Seaton* v. *Benedict* (1828), 5 Bing. 28.
(*v*) *Morel* v. *Earl of Westmorland*, [1904] A. C. 11.
(*w*) *Jolly* v. *Rees* (1863), 15 C. B., N. S., 628.
(*x*) *Jetley* v. *Hill* (1884), Cab. & El. 239; see also *Hunt* v. *De Blaquiere* (1829), 5 Bing. 550, at p. 560; *Swan & Edgar* v. *Mathieson* (1910), 103 L. T. 832.
(*a*) *Callot* v. *Nash* (1923), 39 T. L. R., 291. The mere fact that the wife has separate estate does not negative the husband's liability.
(*b*) See p. 27 of the 15th edition of the present work.
(*c*) 1970 vol., Halsbury's Statutes (3rd Edn.) 638.
(*d*) 17 Halsbury's Statutes (3rd Edn.) 241.
(*e*) *Ryan* v. *Sams* (1848), 12 Q. B. D. 460.
(*f*) *Mortimore* v. *Wright* (1846), 6 M. & W. 482, 486 (father); *Hodgens* v. *Hodgens* (1837), 4 Cl. & F. 323, 375, H. L. (mother with separate estate).
(*g*) *Baker* v. *Keen* (1819), 2 Stark. 501; *Bazeley* v. *Forder* (1868), L. R. 3 Q. B. 559.

Master of ship.—The master of a ship has implied authority to bind the owner for the price of necessaries supplied for the ship. (*h*).

Clubs and messes.—A member of a club is not liable for goods supplied to the club unless he has expressly or impliedly authorised the supply, and in determining that question the rules of the club are relevant (*i*). A commanding officer, as such, is not liable for goods ordered by the Mess Committee (*k*).

Formalities of the Contract

3. Contract of sale, how made.—Subject to the provisions of this Act and of any statute in that behalf (*l*), a contract of sale may be made in writing (either with or without seal), or by word of mouth, or partly in writing and partly by word of mouth (*m*), or may be implied from the conduct of the parties (*n*).

Provided that nothing in this section shall affect the law relating to corporations.

Definitions.—For " contract of sale," see s. 1 (1), *ante*, and s. 62 (1), *post*. For " writing," see s. 20 of the Interpretation Act, 1889.

COMMENT

A written offer to sell goods may be orally accepted, and *vice versâ*. Where a man goes into a restaurant, orders a dinner and eats it, obviously there is a sale though no mention be made of buying, or selling, or price. If, however, the parties have put a contract of sale into writing, the ordinary rules of evidence apply. See as to these rules, p. 4, *ante*.

(*h*) *Mackintosh* v. *Mitcheson* (1849), 4 Exch. 175. As to the nature of the master's contract when he draws a bill for the price, see *The Ripon City*, [1897] P. 226, at p. 231 ; *Pocahontas Fuel Co.* v. *Ambatielos* (1922), 27 Com. Cas. 148 (master's authority to draw, course of dealing). The price of necessaries supplied to a ship may be also recovered by action *in rem* against the ship: *The Mogileff*, [1921] P. 236.

(*i*) *Leake on Contracts* (8th ed.), p. 395 ; *Wertheimer on Clubs* (6th ed.), p. 63 *et seq.*; 6 Halsbury's Laws (4th Edn.) paras. 268 *et seq.*

(*k*) *Lascelles* v. *Rathbun* (1919), 35 T. L. R. 347, C. A. ; *cf. Cross* v. *Williams* (1862), 31 L. J. Ex. 145 (goods supplied to Volunteer Corps where there was authority).

(*l*) See the Merchant Shipping Act, 1894, ss. 24, 26, 65; as amended by the Merchant Shipping Act, 1965, Sch. 2; transfer of British ships and shares therein by bill of sale only. *Cf. Manchester Ship Canal Co.* v. *Horlock*, [1914] 2 Ch. 199, C. A. (ship sunk in fairway, sold by harbour authority).

(*m*) *Lockett* v. *Nicklin* (1848), 2 Exch. 93.

(*n*) *Brogden* v. *Metropolitan Rail. Co.* (1877), 2 App. Cas. 666.

Corporations.—Limited companies and, in England and Wales, all other corporations now contract exactly like private persons (*o*). But a contract which is required to be made in a special form, and which a private individual might vary or discharge without formalities, can only be varied or discharged by a limited company or other corporation in the manner in which it can be made (*p*).

[4.—Contract of sale for ten pounds and upwards.—*Repealed.*]

COMMENT

This section was repealed for all purposes by s. 1 of the Law Reform (Enforcement of Contracts) Act, 1954.

The formalities required for the making of credit-sale and conditional sale agreements are now set out in ss. 5–10 of the Hire-Purchase Act, 1965 (*q*).

Subject Matter of Contract

5. Existing or future goods.—(1) The goods which form the subject of a contract of sale may be either existing goods, owned or possessed by the seller, or goods to be manufactured or acquired by the seller after the making of the contract of sale, in this Act called " future goods " (*a*).

(2) There may be a contract for the sale of goods, the acquisition of which by the seller depends upon a contingency which may or may not happen (*b*).

(3) Where by a contract of sale the seller purports

(*o*) Companies Act, 1948, s. 32; Corporate Bodies' Contracts Act, 1960.

(*p*) Companies Act, 1948, s. 33 (3); Corporate Bodies' Contracts Act, 1960, s. 1 (3).

(*q*) 30 Halsbury's Statutes (3rd Edn.) 61. The Hire-Purchase Act, 1965 has been repealed by Schedule 5 to the Consumer Credit Act, 1974 as from a date to be appointed under s. 192 (4) of that Act. As yet no date has been appointed. The Consumer Credit Act provides for special formalities in the case of certain agreements where one party provides credit of any kind to another. These provisions are to come into force on a date to be appointed. As yet no date has been appointed. See further p. 63, *ante*.

(*a*) *Ajello* v. *Worsley*, [1898] 1 Ch. 274. (I have no piano by Ajello in stock and the maker has refused to supply me with one, yet I may contract to sell you such an article since I might be able to obtain one from another dealer).

(*b*) *Watts* v. *Friend* (1830), 10 B. & C. 446 (sale of crop not yet sown); *Hale* v. *Rawson* (1858), 4 C. B. N. S., 85 (sale of goods contingent on safe arrival of ship carrying them); see p. 19, *ante*, and see also comment to s. 1 (2), *ante*, and s. 7, *post*.

to effect a present sale of future goods, the contract operates as an agreement to sell the goods (*c*).

Definitions.—For " contract of sale," " goods," and " seller," see s. 62 (1), *post*. The definition of " future goods " is also repeated in that subsection.

History.—*Subsection* (1).—The Roman lawyers doubted whether an agreement to sell " future goods " constituted a contract of sale, but it is long since any such question has been raised in English law (*d*). The term " future goods " is not a very happy one, but the alternative " after-to-be-acquired goods " was impossible.

Subsection (2).—There is very little English authority on the point. " No doubt," says MARTIN, B., " a man may buy the chance of obtaining goods," but he then goes on to say that in the case he was dealing with the plaintiff bought the goods themselves (*e*). Perhaps the doubtful case of *Bagueley* v. *Hawley* (*f*) may be explained on the ground that the plaintiff there bought another man's bargain at an auction for what it was worth, and not the goods themselves (*g*). In *Howell* v. *Coupland* (*h*), there was a contract to supply 200 tons of potatoes to be grown on a particular farm and the crop failed. It was held that neither party was liable under the contract, as performance was impossible.

It was the opinion of SIR MCKENZIE CHALMERS that this rule was preserved by s. 7, *post*, but it has now been held (*i*) that it is either embodied in this sub-section or else is preserved as one of the common law principles retained by s. 61 (2), *post* (*j*).

Subsection (3).—There was one case in which it was supposed at common law that future goods could be assigned. It was said that a man might sell future goods which had a " potential existence," and that then the legal property in them would pass to the buyer as soon as they came into actual existence. Goods were

(*c*) *Lunn* v. *Thornton* (1845), 1 C. B. 379 (trover for furniture) ; *cf. Heseltine* v. *Siggers* (1848), 1 Exch. 856 (" bought and sold " in relation to stock may mean *agreed* to be bought or sold).

(*d*) *Hibblewhite* v. *M'Morine* (1839), 5 M. & W. 462, at p. 466. Section 7 of the Statute of Frauds (Amendment) Act, 1828 (Lord Tenterden's Act; 9 Geo. 4, c. 14, *repealed*), at any rate, concluded the question in England.

(*e*) *Buddle* v. *Green* (1857), 27 L. J. Ex. 33, at p. 34 (goods in hands of wharfinger).

(*f*) (1867), L. R. 2 C. P. 625.

(*g*) *Cf. Chapman* v. *Speller* (1850), 14 Q. B. 621 (sale by sheriff and sub-sale).

(*h*) (1876) 1 Q. B. D. 258.

(*i*) *Sainsbury, Ltd.* v. *Street* [1972] 3 All E. R. 1127; [1972] 1 W. L. R. 834.

(*j*) For those principles, see *Taylor* v. *Caldwell* (1863), 3 B. & S. 826; *Appleby* v. *Myers* (1867), L. R. 2 C. P. 651.

supposed to have a potential existence if they would naturally grow out of anything already owned by the seller. For instance, it was said a man might sell the wool to be grown on sheep which he then had, but not the wool on sheep which he was going to buy (*k*). There seems to be no rational distinction between one class of future goods and another (*l*).

COMMENT

Assignment of " future goods ".—The conditions under which an ordinary agreement to sell becomes a sale are dealt with in s. 1, *ante*, and ss. 16–20, *post*. But sometimes a contract purports presently to assign goods to be acquired in the future (*m*). In such case the legal property in the goods does not pass to the buyer unless and until the seller does some act irrevocably appropriating them to the contract (*n*), or the buyer takes possession of them under a licence to seize, which is equivalent to a delivery by the seller (*o*). But if the goods assigned be sufficiently described to become specific or ascertained goods on acquisition by the seller, the equitable interest in them passes to the buyer as soon as they are so acquired (*p*):

> " A man cannot in equity, any more than at law, assign what has no existence. A man can contract to assign property which is to come into existence in the future, and when it has come into existence equity, treating as done that which ought to be done, fastens upon that property, and the contract to assign thus becomes a complete assignment " (*q*).

It is only the equitable interest which passes to the buyer by the contract ; hence his rights are liable to be defeated if, before he gets the legal property in the goods, the seller disposes of them to a second purchaser without notice, who thus first obtains the legal estate (*r*).

If there is a contract for the sale of future goods and only part of those goods come into existence, it is a question of construction of

---

(*k*) *Grantham* v. *Hawley* (1615), Hobart Rep. 132, 2 Roll. 48, pl. 20.

(*l*) It might possibly be contended that goods having a potential existence do not come within the definition of " future goods," inasmuch as they accrue to the seller, and are not " acquired by " him.

(*m*) See such a contract distinguished from an agreement to sell *plus* a licence to seize: *Reeve* v. *Whitmore* (1863), 4 De G. J. & S. 1.

(*n*) *Langton* v. *Higgins* (1859), 4 H. & N. 402 (sale of future crop).

(*o*) *Congreve* v. *Evetts* (1854), 10 Exch. 298 ; *Hope* v. *Hayley* (1856), 5 E. B. 830.

(*p*) *Re Wait*, [1927] 1 Ch. 606, C. A., explaining *Holroyd* v. *Marshall* (1862), 10 H. L. Cas. 191; *cf. Tailby* v. *Official Receiver* (1888), 13 App. Cas. 523. But see further the discussion of *re Wait* in the comment to s. 16, *post*.

(*q*) *Collyer* v. *Isaacs* (1881), 19 Ch. D. 342, at pp. 351, 354, C. A.

(*r*) *Joseph* v. *Lyons* (1884), 15 Q. B. D. 280, C. A. ; *Hallas* v. *Robinson* (1885), 15 Q. B. D. 288, C. A. (bill of sale cases).

the contract in all the circumstances whether the seller is bound to deliver and the buyer to accept those goods or whether the contract is discharged (*s*).

6. Goods which have perished.—Where there is a contract for the sale of specific goods, and the goods without the knowledge of the seller have perished at the time when the contract is made, the contract is void (*t*).

Definitions.—For " contract of sale," see s. 1 (1), *ante*, and s. 62 (1), *post*. For " goods," " seller " and " specific goods," see s. 62 (1), *post*.

Cross-reference.—*Cf.* the narrower s. 7, *post*, which applies only to agreements to sell.

COMMENT

The rule in this section has at different times been said to be based either on the ground of common mistake (*u*), or on the ground of impossibility of performance, or on an implied condition precedent that the goods exist (*a*). The rule is confined to the case of specific goods. Generic goods, that is to say, goods defined by description only, come within the maxim *genus numquam perit* (*b*).

The rules of the common law are expressly preserved by s. 61 (2), *post*, except insofar as they are inconsistent with the express provisions of the Act. The common law rule embodied in this section appears to be that if the parties contract that one of them shall do something which without any fault on his part and without the knowledge of either of them at the time is impossible, then the contract is void (*c*), but it will not be void if one of the parties warrants

(*s*) *Sainsbury* v. *Street*, [1972] 3 All E. R. 1127; [1972] 1 W. L. R. 834 (seller not excused from delivering lesser quantity if buyer willing to take it).

(*t*) *Couturier* v. *Hastie* (1856), 5 H. L. Cas. 673 (cargo of corn).

(*u*) *Bell* v. *Lever Brothers*, [1932] A. C. 161, 217; and see *Barrow Lane and Ballard, Ltd.* v. *Phillip Phillips and Co., Ltd.*, [1929] 1 K. B. 574, at p. 582 (total failure of consideration and mistake).

(*a*) *Bell* v. *Lever Brothers*, *supra*, at p. 224; *Solle* v. *Butcher*, [1949] 2 All E. R. 1107, at p. 1119; [1950] 1 K. B. 671, at p. 681.

(*b*) *Re Thornett and Fehr and Yuills*, [1921] 1 K. B. 219; *Hayward Brothers* v. *Daniel* (1904), 91 L. T. 319.

(*c*) See *e.g. Cox* v. *Prentice* (1815), 3 M. & S. 344; *Couturier* v. *Hastie, supra; Clifford* v. *Watts* (1870), L. R. 5 C. P. 577; *Griffith* v. *Brymer* (1903), 19 T. L. R. 434 (contract to hire room to watch coronation of Edward VII; coronation cancelled before contract made; contract void). The mistake must be fundamental (*Bell* v. *Lever Bros., Ltd.*, [1932] A. C. 161, H. L.). There is, however, no clear authority as to what mistakes in relation to the sale of goods will be fundamental, other than a mistake as to their existence. See further *Benjamin's*

that performance is possible (*d*). Generally, a seller will not be treated as warranting the existence of goods (*e*), but such a warranty has been implied by the High Court of Australia in a case where the goods never existed (*f*).

There is no authority as to the meaning of " perish " in relation to this section, but it has been held at first instance in relation to s. 7 that rotten potatoes which cannot be used are still potatoes so long as they are in a form which permits of their being called potatoes (*g*). The case has, however, been criticised (*h*). Probably, the precise meaning of " perish " is academic, for if goods have so deteriorated that at common law the contract is void, or in the case of s. 7 avoided, then it will be irrelevant whether they have perished within the meaning of this section or s. 7.

If the basis of the rule is as stated above, there seems no reason why the situation should differ, in the absence of some special agreement, where for some other reason than the fact that goods have perished, by reason of facts for which neither party is responsible and of which neither was aware, the seller cannot perform and does not purport to perform his obligations under the contract (*i*).

Sale of Goods, paras. 218–224. " Void " in this context has commonly been assumed to mean of no effect whatsoever. But it is not always used in this sense, and both at common law and when used in statutes has been construed as meaning " voidable " at the option of one or both parties; see, *e.g.*, *Cox* v. *Prentice*, *supra*, *per* BAYLEY, J.; and see generally *Stroud's Judicial Dictionary* (4th ed.)., p. 2951.

(*d*) *Clifford* v. *Watts* (1870), L. R. 5 C. P. 577, at pp. 585, 588. This case was one of those cited by Sir McKenzie Chalmers, who does not appear to have regarded this section as altering the common law: see the 1st edition of this work at p. 10, and the 2nd edition at p. 16. See also *Produce Brokers Co.* v. *Olympia Oil and Cake Co.*, [1917] 1 K. B. 320, 330, C. A. (custom that buyer must accept seller's appropriation although goods at time of appropriation were lost); *Produce Brokers Co.* v. *Olympia Oil and Cake Co.* (1914), 84 L. J. K. B. 1153, C. A., and [1916] 1 A. C. 314 (arbitrator may find custom). *Cf.* Treitel, *Law of Contract* (3rd ed.) p. 220, where it is suggested that s. 6 cannot be excluded.

(*e*) *Barrow Lane and Ballard, Ltd.* v. *Phillip Phillips and Co., Ltd.*, [1929] 1 K. B. 574, at p. 582.

(*f*) *McRae* v. *Commonwealth Disposals Commission* (1950), 84 C. L. R. 377.

(*g*) *Horn* v. *Minister of Food*, [1948] 2 All E. R. 1036.

(*h*) Atiyah, *Sale of Goods* (4th ed.) p. 41. *Cf. Rendell* v. *Turnbull* (1908), 27 N. Z. L. R. 1067, where there was a contract for the sale of 75 tons of table-potatoes, which, unknown to the seller, were at the time of the contract unfit for human consumption. The contract was held to be void because the potatoes had ceased to be " table-potatoes".

(*i*) It has been said that mistakes as to quality are not fundamental (see *Benjamin's Sale of Goods*, para. 220, and cases there cited). But if the defect is such that it would be a breach of contract to tender the goods in that condition, and the defect is irremediable, it is difficult to see why the rights of the parties should turn on the philosophical question whether the goods still permit of being called by their original description. In *Couturier* v. *Hastie*, *supra*, it was assumed on the basis of *Barr* v. *Gibson* (1838), 3 M. & W. 390, that the contract would not be void if the goods still existed in a defective condition, but *Barr* v.

Indeed, even if the contract is not void, it may be set aside in equity " if the parties were under a common misapprehension either as to facts or as to their relative and respective rights, provided that the misapprehension was fundamental and that the party seeking to set it aside was not himself at fault " (*k*).

But if the goods are defective, and the seller nevertheless supplies them (*l*), he warrants that they are of the contract quality, and will no longer be able to treat the contract as void.

There may be some cases where upon the true construction of the contract the buyer may have the option of taking incomplete or damaged goods (*m*).

7. Goods perishing before sale but after agreement to sell.—Where there is an agreement to sell specific goods, and subsequently the goods, without any fault on the part of the seller or buyer, perish (*n*) before the risk passes to the buyer, the agreement is thereby avoided (*o*).

Definitions.—For " contract of sale," see s. 1 (1), *ante*, and s. 62 (1), *post*. For " buyer," " fault," " goods," " seller " and " specific goods," see s. 62 (1), *post*.

Gibson was decided on the basis that there was no implied warranty of the good quality or condition of the chattel sold—see at p. 399. While there are authorities that a mistake as to quality will not in general affect the validity of a contract (see *Benjamin's Sale of Goods*, para. 220) in none of those authorities would the defect in quality have meant that the seller could not perform his obligations under the contract. Moreover, if the seller does not warrant that the goods exist when the contract is made (see n. (*e*), *supra*) it is a little difficult to see why he should be responsible for their quality at that stage.

(*k*) *Solle* v. *Butcher*, [1949] 2 All E. R. 1107, at p. 1120; [1950] 1 K. B. 671, at p. 693, *per* DENNING, L.J. This statement of principle was held by GOFF, J., in *Grist* v. *Bailey*, [1966] 2 All E. R. 875; [1967] Ch. 532, to have been the basis of the decision in *Solle* v. *Butcher* and to be binding on him. But this remedy is discretionary and subject to the usual limitations on the equitable remedy of rescission.

(*l*) It has been held in relation to the Trade Descriptions Act, 1968 that goods are supplied when they are delivered or in some cases when the buyer is notified that they are available for delivery; *Rees* v. *Munday*, [1974] 3 All E. R. 506; [1974] 1 W. L. R. 1284, D. C. In the context of s. 14, *post*, " supplied " would seem to involve at least the seller parting with possession of the goods.

(*m*) *Cf. Sainsbury* v. *Street*, [1972] 3 All E. R. 1127; [1972] 1 W. L. R. 834. But see *Barrow Lane and Ballard* v. *Phillip Phillips & Co.*, [1929] 1 K. B. 574, at p. 583 (sale of 700 bags of nuts, 109 of which perished before the contract; buyer could not be compelled to take 591 bags).

(*n*) See comment to s. 6, *ante*.

(*o*) *Elphick* v. *Barnes* (1880), 5 C. P. D. 321 (death of a horse delivered on sale or return ").

Cross-references.—*Cf.* s. 20, *post* (risk generally), and s. 33, *post* (risk where distant delivery).

COMMENT

" Specific goods ".—It was the opinion of Sir McKenzie Chalmers that this section applies to specifically described goods, whether in existence at the time the contract was made or not. He cited the opinion of MELLISH, L.J., in *Howell* v. *Coupland* (*p*), where there was a contract to supply 200 tons of potatoes to be grown on a particular farm and the crop failed, that

> " This is not like the case of a contract to deliver so many goods of a particular kind, where no specific goods are to be sold. Here there was an agreement to sell and buy 200 tons out of a crop to be grown on specific land, so that it is an agreement to sell what will be, and may be called specific things ; therefore neither party is liable if the performance becomes impossible."

This involved giving to the word " specific " in the section a meaning which differs from that given in the definition clause, and it has now been held in *Sainsbury, Ltd.* v. *Street* (*q*), that this section does not apply to such a contract but that the decision in *Howell* v. *Coupland* would be covered by s. 5 (2), *ante*, or by the common law principles retained by s. 61 (2), *post*, of the Act (*r*).

See as to frustration of contracts generally, p. 19, *ante*.

The present section deals with a case of impossibility peculiar to the law of sale, and it is to be noted that the provisions of s. 1 of the Law Reform (Frustrated Contracts) Act, 1943 (*post*, Appendix I) do not apply to any contract to which s. 7 of the Sale of Goods Act applies (*s*).

The Price

8. Ascertainment of price.—(1) The price in a contract of sale may be fixed by the contract, or may be left to be fixed in manner thereby agreed, or may be determined by the course of dealing between the parties (*t*).

(*p*) (1876), 1 Q. B. D. 258, at p. 262.

(*q*) [1972] 3 All E. R. 1127; [1972] 1 W. L. R. 834, adopting the opinion of ATKIN, L.J. in *re Wait*, [1927] 1 Ch. 606, at p. 630, C. A.

(*r*) For these principles, see *Taylor* v. *Caldwell* (1863), 3 B. & S. 826 ; *Appleby* v. *Myers* (1867), L. R. 2 C. P. 651.

(*s*) Or to any contract for the sale of specific goods which is frustrated by the fact that the goods have perished: s. 2 (5) (*c*).

(*t*) *Arcos Ltd.* v. *Aronson* (1930), 36 Ll. L. Rep. 108 (price to be decided by arbitration); *Gibraltar Packers, Ltd.* v. *Basic Economy and Development Corporation*, [1966] 1 Lloyd's Rep. 615. As to usage to deduct discount, see *Brown* v. *Byrne* (1854), 3 E. & B. 703.

(2) Where the price is not determined in accordance with the foregoing provisions the buyer must pay a reasonable price (*u*). What is a reasonable price is a question of fact dependent on the circumstances of each particular case (*a*).

Definitions.—For " contract of sale," see s. 1 (1), *ante*, and s. 62 (1), *post*.

Cross-references.—*Cf.* s. 9, *post* (sale at valuation), and s. 49, *post* (action for price).

History.—The doctrine of implied or reasonable price seems to be an original development of English law. The rule of Roman law was that the price, or the mode of fixing it, must be expressed in the contract itself (*b*). If the price was not fixed, the contract was classed as innominate.

According to Roman law, the price must be a serious price (*c*), but English courts have always declined to enquire into the adequacy of any *bonâ fide* consideration.

COMMENT

The " price " of a thing is its agreed or estimated value expressed in terms of the currency of the country.

> " The technical term which has been invariably adopted for the numerical expression of the values of commodities, in terms of the standard, is ' price ' " (*d*).

Where the alleged contract provided for " prices to be agreed upon from time to time," the House of Lords held that there was no contract at all, but a mere agreement to agree, and s. 8(2) could only be brought into play where the contract was totally silent about the price (*e*). The principle seems to be that there is no

---

(*u*) *Valpy* v. *Gibson* (1847), 4 C. B. 837, 864; *cf. Jewry* v. *Busk* (1814), 5 Taunt. 302; *Mack and Edwards* (*Sales*), *Ltd.* v. *McPhail Brothers* (1968), 112 Sol. Jo. 211, C. A.

(*a*) *Acebal* v. *Levy* (1834), 10 Bing. 376. Such price may or may not be the market price, according to circumstances; *per* TINDAL, C.J., at p. 383.

(*b*) *Inst. III.*, 23.

(*c*) De Zulueta, *The Roman Law of Sale*, pp. 19–20.

(*d*) Scott's *Money and Banking*, p. 34. As to legal tender, see p. 228, *post*. As to varying meanings of the term " Currency," see *Banque Belge v. Hambrouck*, [1921] 1 K. B. 321, at p. 326. As to foreign currencies, see *post*, notes to s. 54.

(*e*) *May and Butcher* v. *R.*, [1934] 2 K. B. 17 *n.*, H. L. (decided in 1929). *Cf. Hillas* v. *Arcos* (1932), 38 Com. Cas. 23, H. L.; *Scammell* v. *Ouston*, [1941] 1 All E. R. 14; [1941] A. C. 251; *British Bank for Foreign Trades* v. *Novimex*, [1949] 1 All E. R. 155, C. A.; [1949] 1 K. B. 623; *Loftus* v. *Roberts* (1902), 18 T. L. R. 532 (explaining the older cases), and the judgment of DENNING, L.J., in *Nicolene* v. *Simmonds*, [1953] 1 All E. R. 822; [1953] 1 Q. B. 543.

contract unless the price is determined by the contract or is determinable otherwise than by a fresh agreement between the parties (*f*).

The further the parties have gone on with their contract, the more ready are the courts to imply any reasonable term to give effect to their intentions (*g*). In one recent case, the parties had disagreed as to the price, but the seller had nonetheless delivered, and the buyer accepted and used the goods. The Court of Appeal held that had the " contract " been executory, there would have been no contract, but since the goods had been delivered, there was a necessary implication from the conduct of the parties that the buyer would pay a reasonable price (*h*).

In another case, the contract was for the sale of certain land at a price fixed by the contract, and also of all the petrol required from time to time at a petrol filling station " at a price to be agreed by the parties in writing from time to time ". The contract contained an arbitration clause providing for the submission to arbitration of any dispute or difference on " the subject matter or construction " of the agreement; it was held that a term must be implied in the agreement, under the doctrine of *The Moorcock* (*i*), that the petrol should be sold at a reasonable price and that if any dispute arose as to what was a reasonable price it must be determined by arbitration (*k*).

Presumably if the price was subsequently fixed by the parties, the Court would hold that it was a reasonable price. Marine insurance

---

(*f*) In Scotland the Court of Session has held that, in very special circumstances and where the contract was not one of sale, the price might not be a material element at all, and the contract might still be enforceable even though it provided for " the prices to be mutually settled at a later and appropriate date ": *R. and J. Dempster, Ltd.* v. *Motherwell Bridge and Engineering Co., Ltd.*, 1964 S. L. T. 353. The clause of the Sale of Goods Bill originally provided that the price might " be left to be fixed by subsequent arrangement "; but these words were struck out in Committee.

(*g*) *F. & G. Sykes (Wessex), Ltd.* v. *Fine Fare, Ltd.*, [1967] 1 Lloyd's Rep. 53, C. A.

(*h*) *Mack and Edwards (Sales), Ltd.* v. *McPhail Brothers* (1968), 112 Sol. Jo. 211, C. A. This case is perhaps an example of a quasi-contractual obligation to pay a reasonable price for goods supplied in the absence of a binding contract —*cf.* the obligation of a person lacking capacity to contract to whom necessaries are supplied, p. 86, *ante*. In *McCarthy Milling Co., Ltd.* v. *Elder Packing Co., Ltd.* (1973), 33 D. L. R. (No. 3d.) 52, a price less than the full market price was agreed on the common assumption, for which the buyer was responsible, that the seller would receive the difference by way of Government subsidy. The subsidy was paid, but it later turned out that the goods were ineligible and the seller had to repay the subsidy. The buyer was held liable as a matter of quasi-contract to pay the amount of the subsidy to the seller. See also *Cox* v. *Prentice* (1815), 3 M. & S. 344.

(*i*) (1889), 14 P. D. 64, C. A.; but see the note " Implied Terms ", p. 14, *ante*.

(*k*) *Foley* v. *Classique Coaches Ltd.*, [1934] 2 K. B. 1, C. A.; and see *F. & G. Sykes (Wessex), Ltd.* v. *Fine Fare Ltd.*, [1967] 1 Lloyd's Rep. 53, C. A.; contrast the arbitration clause in *May and Butcher* v. *R.*, [1934] 2 K. B. 17 n., H. L.

policies are often effected " at a premium to be arranged " (*l*). A similar question arises there.

Yet another case is put by BLACKBURN, J., who says:

> " Where the price is not ascertained, and it could not be ascertained with precision in consequence of the thing perishing, nevertheless the seller may recover the price, if the risk is clearly thrown on the purchaser, by ascertaining the amount as nearly as you can " (*m*).

Such a case, however, would today fall within the section, since it arises only where the contract lays down the manner in which the price is to be fixed.

An alternative price, if in the nature of a wager, avoids the contract for being a gaming one (*n*).

For the effect on the price of an alteration of customs or excise duty, or of the rate of value added tax, see p. 2, *ante*.

For the law relating to deposits, see p. 26, *ante*.

For the law relating to price control and price maintenance, see p. 51, *ante*.

For the offence under the Trade Descriptions Act, 1968 of giving a misleading indication as to the price of goods, see p. 60, *ante*.

Credit sale and conditional sale agreements.—See the special provisions about these in ss. 5 and 6 of the Hire-Purchase Act, 1965 and in the Consumer Credit Act, 1974 (*o*).

9. Agreement to sell at valuation.—(1) Where there is an agreement to sell goods on the terms that the price is to be fixed by the valuation of a third

---

(*l*) See Marine Insurance Act, 1906, s. 31; 17 Halsbury's Statutes (3rd Edn.) 852.

(*m*) *Martineau* v. *Kitching* (1872), L. R. 7 Q. B. 436, at pp. 455, 456 (sugar sold at so much per cwt. and destroyed after it had come at buyer's risk but before it could be weighed). *Cf. Castle* v. *Playford* (1872), L. R. 7 Exch. 98, at pp. 99, 100.

(*n*) *Rourke* v. *Short* (1856), 5 El. & Bl. 904; *cf. Brogden* v. *Marriott* (1836), 3 Bing. N. C. 88; *Ironmonger & Co.* v. *Dyne* (1928), 44 T. L. R. 497, at p. 499. As to speculative contracts in " futures," see *Forget* v. *Ostigny*, [1895] A. C. 318, at p. 323, P. C.; and *Re Gieve*, [1899] 1 Q. B. 794, C. A. (wagering contract *plus* contract for sale of stock); *Universal Stock Exchange* v. *Strachan*, [1896] A. C. 166; *Cooper* v. *Stubbs*, [1925] 2 K. B. 753, C. A. (both sides must bet); *Woodward & Co.* v. *Wolfe* [1936] 3 All E. R. 529 (" futures " brokers do not bet).

(*o*) The Hire-Purchase Act, 1965 has been repealed by Schedule 5 to the Consumer Credit Act, 1974 from a date to be appointed under s. 192 (4) of that Act. So far no date has been appointed, and the Hire-Purchase Act, 1965 continues in force. There are also special provisions in the Consumer Credit Act, 1974 dealing with the price of goods sold on credit under agreements which are regulated agreements within the meaning of that Act (as to which see p. 63, *ante*). These provisions are to come into force on dates to be appointed under Sch. 3 to that Act. So far no such orders have been made.

party, and such third party cannot or does not make such valuation, the agreement is avoided (*p*); provided that if the goods or any part thereof have been delivered to and appropriated by the buyer he must pay a reasonable price therefor (*q*).

(2) Where such third party is prevented from making the valuation by the fault of the seller or buyer, the party not in fault may maintain an action for damages against the party in fault.

ILLUSTRATION

In 1966 A agreed to purchase from B for 5 years all the flour required by A for his bakery business at the maximum price fixed for the time being pursuant to the Australian Profiteering Prevention Acts, 1948 to 1959. In 1967 flour ceased to be subject to these maximum prices. Held: the agreement is suspended and is not enforceable at the suit of either party. No agreement can be implied to pay a reasonable price or a price fixed by every other body (*r*).

Definitions.—For " contract of sale," see s. 1 (1), *ante*, and s. 62 (1), *post*. For " action," " buyer," "delivery," " fault," " goods " and " seller," see s. 62 (1), *post.*

Cross-reference.—*Cf.* s. 8, *ante* (reasonable price).

COMMENT

The rule stated in subsection (2) appears to be a particular instance of the general common law rule stated by Lord ATKIN in *Southern Foundries* v. *Shirlaw* (*s*):

" if a party enters into an arrangement which can only take effect by the continuance of an existing set of circumstances . . . I look on the law to be that . . . there is an implied engagement on his part that he shall do nothing of his own motion to put an end to that state of circumstances, under which alone the agreement can be operative."

In a case before the Act where one of the parties prevented his valuer from acting, PAGE WOOD, V.C., refused him specific performance, on

(*p*) *Thurnell* v. *Balbirnie* (1837), 2 M. & W. 786 (damages); *Vickers* v. *Vickers* (1867), L. R. 4 Eq. 529 (specific performance); *Fry on Specific Performance* (6th ed.), §§ 357 *et seq.* If the valuer fails to act, there is no right of action, since there is no implied warranty that he shall act: *Cooper* v. *Shuttleworth* (1856), 25 L. J. Ex. 114.

(*q*) *Clarke* v. *Westrope* (1856), 18 C. B. 765.

(*r*) *Re Nudgee Bakery Pty., Ltd.'s Agreement*, [1971] Qd. R. 24.

(*s*) [1940] 2 All E. R. 445, at p. 454; [1940] A. C. 701 at p. 717, quoting and approving the words of COCKBURN, C.J., in *Stirling* v. *Maitland* (1864), 5 B. & S. 841, at p. 852. *Cf. Universal Cargo Carriers* v. *Citati*, [1957] 2 All E. R. 70; [1957] 2 Q. B. 401.

the ground that there was no contract to enforce, and saying that the Court had adopted this principle from the civil law (*t*).

In some cases the party in fault may be restrained from preventing the valuer from acting (*u*).

Valuation as used in this section must be distinguished from arbitration. Arbitration presupposes a dispute or difference, and while a reference to valuers may be an arbitration this will be so only where a dispute exists which it is intended to determine by a process of judicial enquiry. Essentially a valuation is intended to prevent a dispute (*a*).

Conditions and Warranties

10. Stipulations as to time.—(1) Unless a different intention appears from the terms of the contract stipulations as to time of payment are not deemed to be of the essence of a contract of sale (*b*). Whether any other stipulation as to time is of the essence of the contract or not depends on the terms of the contract (*c*).

(2) In a contract of sale " month " means primâ facie calendar month (*d*).

ILLUSTRATIONS

(1) Sale of beans ; goods to be shipped and bill of lading to be dated in December and/or January, and bill of lading to be *primâ facie*

(*t*) *Vickers* v. *Vickers* (1867), L. R. 4 Eq. 529, at p. 535. See *Inst. III.*, 23, where it is said "*Sin autem ille qui nominatus est vel non potuerit vel noluerit pretium definire, tunc pro nihilo esse venditionem quasi nullo pretio statuto.*"

(*u*) See *Fry on Specific Performance* (6th ed.), §§ 1157, 1158, and *Smith* v. *Peters* (1875), L. R. 20 Eq. 511 (mandatory order).

(*a*) *Russell on Arbitration* (18th ed.), pp. 45–46, 48–50; Hogg's *Arbitration*, pp. 9–11, where the cases are collected.

(*b*) *Martindale* v. *Smith* (1841), 1 Q. B. 389; and see *United Dominions Trust* (*Commercial*), *Ltd.* v. *Eagle Aircraft Services, Ltd.*, [1968] 1 All E. R. 104; [1968] 1 W. L. R. 74, C. A. For the rule in connection with sales of land, see *Smith* v. *Hamilton,* [1950] 2 All E. R. 928; [1951] Ch. 174; *Re Barr's Contract, Moorwell Holdings* v. *Barr,* [1956] 2 All E. R. 853; [1956] Ch. 551. For the rule in connection with an option to purchase shares, see *Hare* v. *Nicoll,* [1966] 1 All E. R. 285; [1966] 2 Q. B. 130, C. A.

(*c*) See the cases cited in notes (*m*) and (*n*), p. 107, *post.*

(*d*) *Cf.* Bills of Exchange Act, 1882, s. 14 (4) ; 3 Halsbury's Statutes (3rd Edn.) 197. In ordinary legal documents " month " at common law primarily meant lunar month: *Bruner* v. *Moore,* [1904] 1 Ch. 305, but in a particular trade it might by usage have a special trade meaning : *Bissell* v. *Beard* (1873), 28 L. T. 740 (iron trade). Now, by s. 61 of the Law of Property Act, 1925 (27 Halsbury's Statutes (3rd Edn.) 437), it means calendar month in all deeds, contracts, wills, orders and other instruments made after January 1st, 1926, " unless the context otherwise requires."

evidence of date of shipment. The buyer may refuse to accept the goods if the bill of lading is dated 2nd February, even though the goods were shipped on 30th January (*e*).

(2) Contract for sale of grain to be shipped from River Plate on ship " expected to be ready to load late September." The ship was not ready to load till the middle of November, and the sellers had no reasonable ground for stating that she would be ready in September. The buyer may refuse to accept the goods (*f*).

Definitions.—For " contract of sale," see s. 1 (1), *ante,* and s. 62 (1), *post.*

Cross-references.—Section 28 (payment and delivery concurrent conditions); s. 31 (payment in contracts involving delivery by instalments); s. 48 (3) (seller may make time of payment of the essence). For a general note on conditions and warranties, see p. 8, ante. For their history, see Note A, *post,* appendix II. *Cf.* s. 55, *post* (" unless a different intention appears ").

COMMENT

Stipulations as to time of payment.—When the contract provides for the opening of a confirmed credit, *primâ facie* it must be opened at the latest by the beginning of the agreed shipment period, and where a shipment date is given it must be opened a reasonable time before that date (*g*).

If a discount is agreed upon for payment within a stated period, the full price becomes payable at the end of that period if the reduced price has not then been paid (*h*).

In determining whether time is of the essence, the court will take into account the nature of the property involved and the intention of the parties (*i*).

Notwithstanding this section, where an unpaid seller has a lien over perishable goods or has stopped them in transit, time of payment is of the essence, and if no time of payment is specified, the buyer must tender the price within a reasonable time—see s. 48 (3),

(*e*) *Re General Trading Co.* (1911), 16 Com. Cas. 95. *Cf. Kwei Tek Chao* v. *British Traders and Shippers,* [1954] 1 All E. R. 779; [1954] 2 Q. B. 459; see also *Heskell* v. *Continental Express,* [1950] 1 All E. R. 1033, and *Macpherson Train* v. *Ross,* [1955] 2 All E. R. 445; [1955] 1 W. L. R. 640.

(*f*) *Sanday & Co.* v. *Keighley, Maxted & Co.* (1922), 91 L. J. K. B. 624.

(*g*) *Plasticmoda* v. *Davidsons,* [1952] 1 Lloyd's Rep. 527 ; *Pavia* v. *Thurmann-Nielson,* [1952] 1 All E. R. 492; [1952] 2 Q. B. 84, C. A. (c.i.f. contract) ; *Sinason-Teicher* v. *Oilcakes and Oilseeds Trading Co.*, [1954] 3 All E. R. 468, C. A. ; *Ian Stach* v. *Baker Bosley,* [1958] 1 All E. R. 542 ; [1958] 2 Q. B. 130 (f.o.b. contract). But see *Alexandria Cotton and Trading Co.* (*Sudan*), *Ltd.* v. *Cotton Co. of Ethiopia, Ltd.*, [1963] 1 Lloyd's Rep. 576, where ROSKILL, J., said that he did not regard the rule as finally settled by the authorities. See further, p. 43, *ante.*

(*h*) *Amos and Wood* v. *Kaprow* (1948), 64 T. L. R. 110.

(*i*) *Cf. Hare* v. *Nicoll,* [1966] 1 All E. R. 285; [1966] 2 Q. B. 130, C. A.

post. Likewise, where the goods are not perishable, such an unpaid seller may make time of the essence by giving notice to the buyer of his intention to re-sell, in which case the buyer must also tender the price within a reasonable time (*k*).

See the special provisions as to credit-sale and conditional sale agreements in the Hire-Purchase Act, 1965 (*l*).

Other stipulations as to time.—As regards stipulations other than those relating to the time of payment, time is of the essence of the contract, in most mercantile transactions (*m*) and in some non-mercantile transactions (*n*). Thus, where there was a contract for the sale of twenty-five tons of pepper, " name of vessel or vessels, marks and particulars to be declared within sixty days of date of bill of lading," COTTON, L.J., says,

> " It was argued that the rules of Courts of Equity are now to be regarded in all Courts, and that equity enforced contracts though the time fixed therein for completion had passed. This was in the cases of contracts, such as purchases and sales of land, where, unless a contrary intention could be collected from the contract, the Court presumed that time was not an essential condition. To apply this to mercantile contracts would be dangerous and unreasonable. We must therefore hold that the time within which the pepper was to be declared was an essential condition of the contract " (*o*).

Where a stipulation as to time is of the essence, the party in whose favour it operates may waive it unilaterally. He can then make it of the essence again only by giving reasonable notice, in clear terms, of his intention to do so (*p*).

(*k*) *R. V. Ward* v. *Bignall*, [1967] 2 All E. R. 449; [1967] 1 Q. B. 534, C.A.; and see p. 224, *post*.

(*l*) 30 Halsbury's Statutes (3rd Edn.) 61.

(*m*) *Bowes* v. *Shand* (1877), 2 App. Cas. 455, at p. 463, *per* Lord CAIRNS; *Elmdove, Ltd.* v. *Keech* (1969), 113 Sol. Jo. 871, C. A. (time presumed to be of the essence unless circumstances show otherwise); *cf. Paton* v. *Payne*, 1897 35 Sc. L. R. 112, H. L. (time not essential); *Hartley* v. *Hymans*, [1920] 3 K. B. 475, at p. 484 (time essential); *Brooke Tool Manufacturing Co.* v. *Hydraulic Gears Co.* (1920), 89 L. J. K. B. 263 (effect of delay caused by Government action where delivery by a certain date expressly stipulated); *Aron & Co.* v. *Comptoir Wegimont*, [1921] 3 K. B. 435 (effect of delay caused by strike where there is express obligation to ship during October). And see *United Dominions Trust (Commercial), Ltd.* v. *Eagle Aircraft Services, Ltd.*, [1968] 1 All E. R. 104; [1968] 1 W. L. R. 74, C. A.

(*n*) *McDougall* v. *Aeromarine of Emsworth*, [1958] 3 All E. R. 431; [1958] 1 W. L. R. 1126.

(*o*) *Reuter* v. *Sala* (1879), 4 C. P. D. 239, at pp. 246, 249, C. A. See also *Berg & Sons* v. *Landauer & Co.* (1925), 42 T. L. R. 142.

(*p*) *Rickards* v. *Oppenheim*, [1950] 1 All E. R. 420; [1950] 1 K. B. 616, C. A.; see also *Panoutsos* v. *Raymond Hadley*, [1917] 2 K. B. 473, C. A.; *Plasticmoda* v. *Davidsons*, [1952] 1 Lloyd's Rep. 527; *Furst (Enrico) & Co.* v. *W. E. Fischer, Ltd.*, [1960] 2 Lloyd's Rep. 340; *Soproma S.p.A.* v. *Marine and Animal By-Products Corporation*, [1966] 1 Lloyd's Rep. 367. As to waiver generally, see p. 22, *ante*.

When time of delivery is an essential condition it is often referred to in the cases as part of the description of the goods. But the Act treats the two conditions as distinct (*q*); *cf.* s. 13, *post.*

11. When condition to be treated as warranty.—
(1) In England or Ireland—

(a) Where a contract of sale is subject to any condition to be fulfilled by the seller, the buyer may waive the condition, or may elect to treat the breach of such condition as a breach of warranty, and not as a ground for treating the contract as repudiated (*r*):

(b) Whether a stipulation in a contract of sale is a condition, the breach of which may give rise to a right to treat the contract as repudiated, or a warranty, the breach of which may give rise to a claim for damages but not to a right to reject the goods and treat the contract as repudiated, depends in each case on the construction of the contract (*s*). A stipulation may be a condition, though called a warranty in the contract (*t*):

(c) Where a contract of sale is not severable (*u*),

(*q*) See *Aron & Co.* v. *Comptoir Wegimont* (*supra*), at p. 440, *per* McCardie, J.; *Wilson* v. *Wright*, [1937] 4 All E. R. 371, at p. 373, *per* Mackinnon, L. J.; and *Macpherson* v. *Ross*, [1955] 2 All E. R. 445.

(*r*) *Ellen* v. *Topp* [1851], 6 Exch. 424, at p. 431; *Behn* v. *Burness* (1863), 3 B. & S. 751, Ex. Ch.; *Wallis* v. *Pratt* [1911], A. C. 394; (applied, *Harling* v. *Eddy*, [1951] 2 All E. R. 212, C. A.; [1951] 2 K. B. 739) and judgment of Fletcher Moulton, L. J., in the Court below: [1910] 2 K. B. 1003, 1012, C. A.; *Sullivan* v. *Constable* (1932), 48 T. L. R. 369, C. A. (election by buyer to treat breach of condition as warranty); and see s. 53, *post*. As to waiver generally, see p. 22, *ante*.

(*s*) *Graves* v. *Legg* (1854), 9 Exch. 709; *Behn* v. *Burness* (1863), 3 B. & S. 751, at p. 754, Ex. Ch. For suggested tests for distinguishing a condition from a warranty, see *Heilbutt, Symons & Co.* v. *Buckleton*, [1913] A. C. 30; *Bentsen* v. *Taylor*, [1893] 2 Q. B. 274, at p. 281, *per* Bowen, L.J.; *Harrison* v. *Knowles and Foster*, [1917] 2 K. B. 606, at p. 610; *Wilson* v. *Wright*, [1937] 4 All E. R. 371, at p. 373, *per* Mackinnon, L.J.; *Oscar Chess* v. *Williams*, [1957] 1 All E. R. 325; [1957] 1 W. L. R. 370, C. A.; and see other cases cited in Appendix II, Note A. See also p. 9, *ante*.

(*t*) *Wallis* v. *Pratt*, [1911] A. C. 394, at p. 397.

(*u*) As to severable contracts, see *Jackson* v. *Rotax Motor and Cycle Co.*, [1910] 2 K. B. 937, 947, C. A.; *J. Rosenthal & Sons, Ltd.* v. *Esmail*, [1965] 2 All E. R. 860; [1965] 1 W. L. R. 1117, H. L.; and s. 31, *post*.

and the buyer has accepted the goods, or part thereof (*v*), [*or where the contract is for specific goods, the property in which has passed to the buyer*] (*a*), the breach of any condition to be fulfilled by the seller can only be treated as a breach of warranty, and not as a ground for rejecting the goods and treating the contract as repudiated (*b*), unless there be a term of the contract, express or implied, to that effect (*c*).

(2) In Scotland, failure by the seller to perform any material part of a contract of sale is a breach of contract, which entitles the buyer either within a reasonable time after delivery to reject the goods and treat the contract as repudiated, or to retain the goods and treat the failure to perform such material part as a breach which may give rise to a claim for compensation or damages (*d*).

(3) Nothing in this section shall affect the case of any condition or warranty, fulfilment of which is excused by law by reason of impossibility or otherwise (*e*).

Definitions.—For " contract of sale," see s. 1 (1), *ante*, and s. 62 (1), *post*. For " buyer," " delivery," " goods," " property," " seller," " specific goods," and " warranty," see s. 62 (1), *post*.

(*v*) *Graves* v. *Legg* (1854), 9 Exch. 709, at p. 717 ; *Behn* v. *Burness* (1863), 3 B. & S. 751, Ex. Ch. ; *Heilbutt* v. *Hickson* (1872), L. R. 7 C. P. 438, at p. 450. For the meaning of " acceptance ", see ss. 34(1) and 35, and the illustrations and comment to them, *post*.

(*a*) Repealed by s. 4 (1) of the Misrepresentation Act, 1967, *post*, Appendix I.

(*b*) In *Wallis* v. *Pratt*, [1910] 2 K. B. 1003, C. A., FLETCHER MOULTON, L.J. (whose dissenting judgment was upheld in the House of Lords : [1911] A. C. 394) said at p. 1013 that there was no reason to suppose that the Act intended that these should be the only modes in which a buyer could effectively bar himself from taking advantage of the choice of remedies given in the case of a breach of a condition.

(*c*) *Bannerman* v. *White*, [1861] 10 C. B. N.S. 844 ; and see 19 Mod. L. R. 315.

(*d*) *Couston* v. *Chapman* (1872), L. R. 2 Sc. & Div. 250 ; *Aird* v. *Pullan*, 1904 7 F. (Ct. of Sess.) 258 ; *Nelson* v. *William Chalmers & Co.*, 1913 S. C. 441 ; *Pollock* v. *Macrae*, 1922 S. C. 192 (H. L.) ; *Mechans, Ltd.* v. *Highland Marine Charters, Ltd.*, 1964 S. C. 48 (Ct. of Sess.).

(*e*) See ss. 6 and 7, *ante* ; *Baily* v. *De Crespigny* (1869), L. R. 4 Q. B. 180, at p. 185. As to dissolution of contracts by war, see cases cited, p. 20, *ante*. ; as to impossibility, see p. 19, *ante*. See also the Law Reform (Frustrated Contracts) Act, 1943, and s. 33 (1) of the Exchange Control Act, 1947, *post*, Appendix I.

Cross-references.—For a general note on conditions and warranties, see *ante,* p. 8. For the history of conditions and warranties, see Note A, *post,* Appendix II. *Cf.* ss. 16 to 18, *post* (passing of property), ss. 34 and 35, *post* (acceptance), and s. 53, *post* (remedy for breach of warranty).

Comment

Sub-section 1 (a).—See as to waiver, p. 22, *ante.*

Sub-section 1 (b).—The Act reflects, in the context of the law of sale of goods, the division of contractual terms into conditions and warranties which was contended for by judges and textbook writers at the time of its enactment. The modern common law has moved away from such a rigid division in relation to contracts generally, and it has now been held that these rules of common law prescribed by s. 61 (2), *post,* except insofar as they are inconsistent with the express provisions of the Act, apply to provisions in contracts for the sale of goods other than the terms implied by ss. 12–15, *post* (*g*).

Sub-section 1 (c).—Where, under this subsection, the buyer has lost his right to reject and sues for damages, he is still suing for a breach of condition, not for breach of warranty, and will not be barred by a clause in the contract which provides against actions for breach of warranty (*h*). For this sub-section to apply, the contract must not be severable at the time when the buyer has to decide how to treat the breach of condition. If the contract is not severable then, it is irrelevant that it could have been severed at an earlier time (*i*).

For the meaning of " acceptance," see ss. 34 (1) and 35, *post.* For the rules as to the passing of property in a sale of specific goods. see s. 18, *post.*

(*g*) *Cehave N.V.* v. *Bremer Handelgesellschaft mbh* (1975), *Times,* 22nd July, C. A. See further p. 9, *ante.* For a note on the history of conditions and warranties, see Appendix II, Note A, *post.*

(*h*) *Wallis* v. *Pratt,* [1911] A. C. 394 ; *Baldry* v. *Marshall,* [1925] 1 K. B. 260 ; *Barker (Junior) & Co.* v. *Agius, Ltd.* (1927), 33 Com. Cas. 120 ; *Sullivan* v. *Constable* (1932), 48 T. L. R. 369, C. A. ; *Couchman* v. *Hill,* [1947] 1 All E. R. 103 C. A. ; [1947] K. B. 554 ; *Harling* v. *Eddy,* [1951] 2 All E. R. 212, C. A. ; [1951] 2 K. B. 739 ; *Nicholson and Venn* v. *Smith Marriott* (1947), 177 L. T. 189 ; but *cf. Wilkinson* v. *Barclay,* [1946] 2 All E. R. 337, *n.,* C. A.

(*i*) *J. Rosenthal & Sons, Ltd.* v. *Esmail,* [1965] 2 All E. R. 860; [1965] 1 W. L. R. 1117, H. L.

Conditional sale agreements.—Section 11(1)(c) does not apply to conditional sale agreements which are agreements for consumer sales (*k*).

Sub-section (2).—In Scotland, no distinction was drawn between conditions and warranties, and the right of rejection was much larger than in England. This right is preserved by the Act. On the other hand, the *actio quanti minoris* was much restricted in Scotland, and when the buyer could return the goods he was not allowed to keep them and sue for damages. How he has this right, subject to the conditions specified by s. 59, *post*.

12. Implied undertakings as to title, &c.—(1) In every contract of sale, other than one to which sub-section (2) of this section applies, there is—

(*a*) an implied condition on the part of the seller that, in the case of a sale, he has a right to sell the goods, and in the case of an agreement to sell, he will have a right to sell the goods at the time when the property is to pass; and

(*b*) an implied warranty that the goods are free, and will remain free until the time when the property is to pass, from any charge or encumbrance not disclosed or known to the buyer before the contract is made and that the buyer will enjoy quiet possession of the goods except so far as it may be disturbed by the owner or other person entitled to the benefit of any charge or encumbrance so disclosed or known.

(2) In a contract of sale, in the case of which there appears from the contract or is to be inferred from the circumstances of the contract an intention that the seller should transfer only such title as he or a third person may have, there is—

(*k*) Supply of Goods (Implied Terms) Act, 1973, s. 14 (1). "Consumer sale" has the same meaning here as under s. 55, *post:* see s. 15 (1) of the 1973 Act. In the case of conditional sale agreements, a breach of condition to be fulfilled by the seller is, in England, Wales and Northern Ireland, to be treated as a breach of warranty if (but only if) it would be so treated if it formed part of a corresponding hire-purchase agreement; see s. 14 (2) of the 1973 Act. The relevant provisions of the 1973 Act at present define conditional sale agreement and hire-purchase agreement by reference to the Hire-Purchase Act, 1965, but by virtue of s. 192 (4) of and Sch. 4 to the Consumer Credit Act, 1974, from a date to be appointed ss. 14 and 15 of the 1973 Act have been amended to bring them into line with the Consumer Credit Act, 1974.

(*a*) an implied warranty that all charges or encumbrances known to the seller and not known to the buyer have been disclosed to the buyer before the contract is made; and

(*b*) an implied warranty that neither—

(i) the seller; nor

(ii) in a case where the parties to the contract intend that the seller should transfer only such title as a third person may have, that person; nor

(iii) anyone claiming through or under the seller or that third person otherwise than under a charge or encumbrance disclosed or known to the buyer before the contract is made;

will disturb the buyer's quiet possession of the goods (*l*).

ILLUSTRATIONS

(1) An auctioneer sells by auction a piano seized under a distress warrant. The buyer knows that the piano is sold under a distress. Prior to 1973, it was held that if the warrant turned out to be invalid, the auctioneer was not liable because the circumstances were such as to show an intention to exclude the operation of s. 12 (*m*). It would seem, however, that this defence is no longer available to the auctioneer, for he intends to transfer such title as the party distrained upon might have. There is therefore an implied warranty, which by virtue of s. 55 (3), *post,* cannot be excluded, that the party distrained on will not disturb the buyer's quiet possession of the goods: see s. 12 (2) (b) (ii).

(2) A offers a motor car to B, who cannot buy it himself. C asks B to buy the car and then immediately re-sell it to him. It turns out that the car was stolen. The circumstances were such as to exclude the operation of the original s. 12 as between B and C (*n*), and this would still seem to be the case as the claim by the true owner would not involve any breach by B of warranties implied by the new s. 12 (2).

(3) Sale of 3,000 tins of preserved milk, some of which are labelled "Nissly brand." This is admitted to be an infringement of the N. Company's trademark. The buyer can either reject the consignment, or take off the labels and claim damages for the reduced sale

---

(*l*) This section was substituted for the original s. 12 by s. 1 of the Supply of Goods (Implied Terms) Act, 1973. It implements recommendations of the Law Commission in its report on Amendments to the Sale of Goods Act, 1893 (Law Com. No. 24).

(*m*) *Payne* v. *Elsden* (1900), 17 T. L. R. 161.

(*n*) *Warming's Used Cars, Ltd.* v. *Tucker,* [1956] S. A. S. R. 249.

value: " if a vendor can be stopped by process of law from selling, he has not the right to sell " (*o*).

(4) In 1967, P files a specification for a patent for apparatus for marking roads. The specification is published in November 1970 and the patent is granted and sealed in 1972. In January 1970, M sells to V similar apparatus, which infringes P's patent. P takes steps to prevent V from using the apparatus. M is liable to V for damages for breach of the warranty of quiet possession (*p*).

(5) Purchase of motor-car which is used for some months. It then turns out that the car had been stolen, and the buyer restores it to its owner. Although he has had some months' use of it he can recover the full price from the seller as money paid on a consideration which has wholly failed (*q*).

Definitions.—For " contract of sale," see s. 1 (1), *ante*, and s. 62 (1), *post*. For " buyer," " goods," " property," " seller " and " warranty," see s. 62 (1), *post*.

Cross-references.—For a general note on conditions and warranties, see p. 8, *ante*. For their history, see Note A, *post*, Appendix II. For their effect, see s. 11, *ante*. *Cf.* ss. 16, 17 and 18, *post* (passing of property); ss. 21 to 26 and the Factors Acts, *post* (transfer of title); and s. 55, *post* (no contracting out).

COMMENT

Formerly the rule was stated to be that on a sale of specific goods there was no implied warranty of title, and that, in the absence of fraud, the seller was

" not liable for a bad title unless there was an express warranty, or an equivalent to it by declaration or conduct " (*r*).

But as Lord Campbell said in 1851, " the exceptions have well-nigh eaten up the rule " (*s*) and the original s. 12 (1) may be regarded as declaratory.

The present section was introduced by s. 1 of the Supply of Goods (Implied Terms) Act, 1973. The original section read:

" In a contract of sale, unless the circumstances are such as to show a different intention there is—

(1) An implied condition on the part of the seller that, in the case of a sale, he has a right to sell the goods, and that, in the case of an

---

(*o*) *Niblett* v. *Confectioners' Materials Co.*, [1921] 3 K. B. 387, C. A.; overruling *Monforts* v. *Marsden* (1895), 12 R. P. C. 266 (machine made under invalid patent). Note that " right to sell " is therefore wider than " right to pass the property in."

(*p*) *Microbeads AG* v. *Vinhurst Road Markings, Ltd.*, [1975] 1 All E. R. 529; [1975] 1 W. L. R. 218, C. A.

(*q*) *Rowland* v. *Divall*, [1923] 2 K. B. 500, 505, C. A.

(*r*) *Per* PARKE, B., in *Morley* v. *Attenborough* (1849), 3 Exch. 500, at p. 512 (auction sale of forfeited pledges).

(*s*) *Sims* v. *Marryat* (1851), 17 Q. B. 281, at p. 291 (sale of copyright).

agreement to sell, he will have a right to sell the goods at the time when the property is to pass:

(2) An implied warranty that the buyer shall have and enjoy quiet possession of the goods:

(3) An implieid warranty that the goods shall be free from any charge or encumbrance in favour of any third party, not declared or known to the buyer before or at the time when the contract is made."

The new sub-s. (1) is similar to the former sub-ss. (1)–(3), but while the former provisions would be excluded if the circumstances were such as to show a different intention, the condition and warranty implied by the new sub-s. (1) cannot be excluded, except in the case of a contract for the international sale of goods, unless the contract is one to which sub-s. (2) applies (*t*). In practice, the cases in which, before 1973, implied undertakings as to title were negatived mainly arose out of sales by sheriffs or forced sales by public auction, where the circumstances were such as to indicate that the seller was only selling such right as he might have in the goods. A sheriff selling an execution debtor's goods, or an auctioneer who, to the knowledge of the buyer, was selling goods seized under a distress warrant gave no implied undertaking as to title (*u*). He was only responsible if he *knew* that he had no title to sell (*a*).

The implied condition as to title.—The right to sell is wider than the right to pass property. If a seller can be stopped by process of law from selling, he has not the right to sell (*b*).

A contract of sale is a contract to transfer the property in goods (*c*). If the seller fails to transfer the property in the goods, there is a total failure of consideration and the buyer is entitled on discovering the true facts to treat the contract as at an end and recover the full purchase even though he has enjoyed the use of the goods (*d*). In such a case the buyer owes nothing to the seller for the use of the goods (*e*).

(*t*) See ss. 55 (3), 61 (6), *post*. Contracts for the international sale of goods are defined in s. 62 (1), *post*.

(*u*) *Ex p. Villars* (1874), 9 Ch. App. 432, at p. 437 (sheriff); *Payne* v. *Elsden* (1900), 17 T. L. R. 161 (*ante*, illustration 1); and see *Warming's Used Cars, Ltd.* v. *Tucker*, [1956] S. A. S. R. 249 (*ante*, illustration 2); *Niblett* v. *Confectioners' Materials Co.*, [1921] 3 K. B. 387, 401; *cf.* s. 5, *ante*.

(*a*) *Peto* v. *Blades* (1814), 5 Taunt. 657; *cf. Dorab Ally* v. *Abdool Azeze* (1878), L. R. 5 Ind. App. 116. See also the provisions of s. 15 of the Bankruptcy and Deeds of Arrangement Act, 1913 (3 Halsbury's Statutes (3rd Edn.) 18), as explained by the Court of Appeal in *Curtis* v. *Maloney*, [1950] 2 All E. R. 982; [1951] 1 K. B. 736.

(*b*) *Niblett* v. *Confectioners Materials Co.*, [1921] 3 K. B. 387, C. A.; overruling *Monforts* v. *Marsden* (1895), 12 R. P. C. 266 (machine made under invalid patent).

(*c*) Section 1, *ante*.

(*d*) *Rowland* v. *Divall*, [1923] 2 K. B. 500, C. A.

(*e*) *Warman* v. *Southern Counties Car Finance Corporation, Ltd.*, [1949] 1 All E. R. 711; [1949] 2 K. B. 576.

This is so even if, after the buyer has rescinded the contract, the property in the goods is acquired by the seller (*f*). But it may be different if, before the buyer rescinds, the seller acquires the property in the goods. In such a case, there is probably no total failure of consideration, for the buyer retains, and is entitled to retain the goods against all the world, but as the seller was in breach of the implied condition that he had the right to sell, the buyer may claim damages, and, subject to s. 11 (1) (c), may treat the contract as repudiated (*g*).

It would also seem clear (*h*) that a seller who has not got any right to sell, but who is able to give a good title to the buyer by reason of, for example, s. 8 or s. 9 of the Factors Act, 1889, *post,* is in breach of the condition implied by s. 12 (1) (a). The buyer may suffer no damage from the breach, as he gets a good title to the goods. But if he is put to expense establishing his title, this may be recoverable as damages from the seller, and in any event he is under no duty to the seller to mitigate his damages in a doubtful case by engaging in complicated litigation to determine whether he has indeed acquired a good title (*i*).

The warranty of freedom from encumbrances and of quiet possession.—This warranty is similar to those formerly implied by s. 12 (2) and (3) before 1973. It protects a buyer where the seller has a right to sell but the goods are subject to the rights of a third party (*k*).

(*f*) *Butterworth* v. *Kingsway Motors,* [1954] 2 All E. R. 694; [1954] 1 W. L. R. 1286.

(*g*) When the seller acquires the property in the goods, the contract with the buyer is " fed " so that the buyer then acquires a good title to the goods: *Whitehorn Brothers* v. *Davison,* [1911] 1 K. B. 463 at pp. 475, 481, 486, C. A.; *Blundell-Leigh* v. *Attenborough,* [1921] 3 K. B. 235 at pp. 240, 242, C. A.; *Butterworth* v. *Kingsway Motors,* [1954] 1 W. L. R. 1286, at p. 1295; *Patten* v. *Thomas Motors Pty., Ltd.* (1965) 66 S. R. (N. S. W.) 458. In *Bennett* v. *Griffin Finance,* [1967] 2 Q. B. 46, at p. 50, WINN, L.J., expressly reserved the point, but it was not argued before him and no authorities appear to have been cited. For a different view to that in the text see *West* (*H. W.*) *Ltd.* v. *McBlain,* [1950] N. I. 144. The statement in *Rowland* v. *Divall,* [1923] 2 K. B. 500, at pp. 506–507, that s. 11 (1) (*c*) was inapplicable to sales without title was made with reference to a situation in which there was a total failure of consideration, and there seems no reason why it should not apply to a situation where there is no such failure.

(*h*) But *cf.* ATKIN, L.J., in *Niblett* v. *Confectioners Materials Ltd.*, [1921] 3 K. B. 387, at p. 401; *Benjamin's Sale of Goods,* para. 264.

(*i*) *Pilkington* v. *Wood,* [1953] 2 All E. R. 810; [1953] Ch. 770. The buyer may in theory be able to rescind even if his title is not open to question, for the seller will still have had no right to sell the goods, and will therefore be in breach of the condition implied by section 12 (1) (*a*). In practice, there are not likely to be many occasions when the buyer will wish to avail himself of this right.

(*k*) *Lloyds and Scottish Finance, Ltd.* v. *Modern Cars and Caravans* (*Kingston*) *Ltd.*, [1964] 2 All E. R. 732; [1966] 1 Q. B. 764; *Microbeads A.G.* v. *Vinhurst Road Markings, Ltd.*, [1975] 1 All E. R. 529; [1975] 1 W. L. R. 218, C. A. (patent published after date of sale), *ante,* illustration (4).

The warranty as to quiet possession appears to give the buyer a contractual remedy if the seller himself wrongfully seizes the goods (*l*). It would also seem to allow the buyer a longer period of limitation than the other provisions of this section since time does not begin to run until the buyer's possession is disturbed.

It has been suggested (*m*) that this warranty ought to be limited in the same way as the covenent by a vendor for quiet enjoyment of real property, but this suggestion has now been disapproved by the Court of Appeal (*n*).

Contracts to transfer only such title as the seller or a third party may have.—Although sellers are not able to exclude the provisions of this section, they may make it clear to the buyer that they are only selling such title as they or some third person may have. If they make this clear, the condition and warranty implied by s. 12 (1) do not apply and instead the more limited warranties set out in s. 12 (2) will be implied. As these relate only to freedom from encumbrances and quiet possession, it follows that the condition implied by s. 12 (1) (a) that the seller has a right to sell can in practice be wholly excluded notwithstanding s. 55 (3) of the Act, if the seller makes it clear that he is only selling such title as he may have.

The intention may be expressed or it may be inferred from the circumstances. Circumstances from which such an intention may be inferred will presumably be similar to those (discussed above) in which, prior to 1973, the implied undertaking as to title was negatived (*o*). But the consequences will not necessarily be the same as before 1973, as appears from illustrations (1) and (2) to this section, *ante*.

Damages.—Damages for breach of the conditions and warranties implied by this section include consequential losses directly and naturally resulting from the breach. These have been held to include the cost of overhauling a typewriter subsequently returned to its

(*l*) *Monforts* v. *Marsden* (1895), 12 R. P. C. 266, at p. 269; *Niblett* v. *Confectioners' Materials Co. Ltd.*, [1921] 3 K. B. 387, at p. 403, *per* ATKIN, L.J.; *Healing (Sales) Pty., Ltd.* v. *Inglis Electric Pty., Ltd.* (1968), 121 C. L. R. 584 (H. C. of Aus.); but *cf. Anderson* v. *Ryan*, [1967] I. R. 34, where it was held that a buyer seeking to establish a breach of the former sub-s. (2) had to show that he did not get a right to quiet possession by virtue of the sale.

(*m*) *Niblett* v. *Confectioners' Materials Co., Ltd.*, [1921] 3 K. B. 387, at p. 403, per ATKIN, L.J.

(*n*) *Microbeads A.G.* v. *Vinhurst Road Markings, Ltd.*, [1975] 1 All E. R. 529; [1975] 1 W. L. R. 218, C. A.

(*o*) Under s. 12 (1) (b), the warranty implied is that the goods *are free* and will remain free until the time when the property is to pass from encumbrances not disclosed or known to the buyer. This is wider than the former s. 12 (3), under which the warranty was merely that the goods shall be free from encumbrances which are not declared or known to the buyer. The practical effect is probably the same. The seller will now be in breach of warranty if he does not disclose all charges and encumbrances affecting the goods, but it is difficult to conceive of a situation in which the buyer will suffer any loss recoverable under s. 12 (1) (b) which would not have been recoverable under the former s. 12 (2) or (3).

true owner (*p*), costs of litigation reasonably undertaken (*q*) and the costs of paying customs duty on a car which was impounded by the Customs and Excise because it had been illegally imported (*r*). See further as to general principles on assessing damages, ss. 53 and 54, *post*.

13. Sale by description.—(1) Where there is a contract for the sale of goods by description, there is an implied condition that the goods shall correspond with the description (*s*); and if the sale be by sample, as well as by description, it is not sufficient that the bulk of the goods corresponds with the sample if the goods do not also correspond with the description (*t*).

(2) A sale of goods shall not be prevented from being a sale by description by reason only that, being exposed for sale or hire, they are selected by the buyer (*u*).

ILLUSTRATIONS

(1) A agrees to buy a second-hand reaping machine, which he has never seen, but which the seller assures him to have been new the previous year, and to have been used only to cut about fifty acres. This is a sale by description, and if the machine does not correspond with the description, A may reject it (*a*).

(*p*) *Mason* v. *Burningham*, [1949] 2 All E. R. 134; [1949] 2 K. B. 545, C. A. But, following the decision in *Greenwood* v. *Bennett*, [1972] 3 All E. R. 586; [1973] 1 Q. B. 195, C. A., the plaintiff might well have been entitled to insist on being reimbursed the cost of the repairs before handing over the typewriter to the true owner, and if in any future case a purchaser has unreasonably failed to avail himself of such a right against the true owner, he may be precluded from recovering from the seller what he should have recovered from the owner.

(*q*) *Bowmaker* (*Commercial*) v. *Day*, [1965] 2 All E. R. 856n; [1965] 1 W. L. R. 1390.

(*r*) *Stock* v. *Urey*, [1954] N. I. 71; but it is difficult to see why the purchaser in that case was held to be entitled to pay more than the value of the car to the customs. In general, unless a purchaser can show that he acted reasonably in paying more than the value of the goods to recover them, he ought not to recover the excess from the seller: *cf. Darbishire* v. *Warran*, [1963] 3 All E. R. 310; [1963] 1 W. L. R. 1067, C. A.

(*s*) *Shepherd* v. *Kain* (1821), 5 B. & Ald. 240; *Josling* v. *Kingsford* (1863), 13 C. B., N. S., 447; *Borrowman* v. *Drayton* (1876), 2 Ex. D. 15, C. A. For examples of terms forming part of the description, see Appendix II, Note B, *post*.

(*t*) *Nichol* v. *Godts* (1854), 10 Exch. 191 (foreign refined rape oil); *Azémar* v. *Casella* (1867), L. R. 2 C. P. 677, Ex. Ch. (long staple Salem cotton; *Steels and Busks* v. *Bleecker Bik & Co.*, [1956] 1 Lloyd's Rep. 228 (pale crepe rubber, quality as previously delivered). See further s. 15, *post*, as to sale by sample.

(*u*) Sub-section (2) is new and was introduced by s. 2 of the Supply of Goods (Implied Terms) Act, 1973, implementing a recommendation of the Law Commission in its report on Amendments to the Sale of Goods Act, 1893 (Law Com. No. 24).

(*a*) *Varley* v. *Whipp*, [1900] 1 Q. B. 513.

(2) Contract for sale of large overboots, described as " waders ". A sample is sent to the buyer before he places his order. The " waders" were not articles known to ordinary commercial trading and formed part of the anti-gas equipment which had been made for the Government during the Second World War. The slightest examination of them would have cast great doubt upon any inference that they were waterproof. The buyer did not rely on the description. The sale is not a sale by description (*b*).

(3) Contract for the sale of East African copra cake. The goods are accepted and resold but are subsequently found to be so adulterated with castor oil as to be poisonous to cattle. The goods do not correspond with their description in that they are a mixture of copra cake and castor oil and are useless for any purpose for which copra cake could be expected to be used. The sellers are not protected by a clause in the contract to the effect that the " goods are not warranted free from defect rendering same unmerchantable which would not be apparent on reasonable examination, any statute or rule of law notwithstanding " (*c*).

(4) A sells to B a food compounded in accordance with an agreed formula. This is a sale by description. The formula includes herring meal. A uses herring meal which (unknown to either party) is contaminated by dimethyl-nitrosamine (D.M.N.A.) produced by a chemical reaction following the use of sodium nitrite as a preservative. D.M.N.A. (unknown to either party) is toxic to mink, for which, as A knows, B requires the food. The food corresponds with its description. D.M.N.A. is not an extraneous substance added to herring meal. It is just herring meal gone wrong (*d*).

(5) A sells seed to B as " English sainfoin " on the terms that " seller gives no warranty express or implied as to growth, description, or any other matters." B re-sells to C. The seed is sown and turns out to be giant sainfoin. The seed does not correspond with the description, and B may recover damages from A as in a case of breach of warranty (*e*).

(6) Contract for purchase of 3,000 tins of canned fruit from Australia, to be packed in cases each containing 30 tins. When the goods are tendered in London, a substantial part is tendered in cases containing 24 tins. The buyer may reject the whole (*f*).

(*b*) *Travers* v. *Longel* (1947), 64 T. L. R. 150.

(*c*) *Pinnock Brothers* v. *Lewis & Peat*, [1923] 1 K. B. 690; as explained in *Ashington Piggeries, Ltd.* v. *Christopher Hill, Ltd.*, [1971] 1 All E. R. 847; [1972] A. C. 441, H. L.; see also *Robert A. Munro & Co.* v. *Meyer*, [1930] 2 K. B. 312, at pp. 326–328 (" goods to be taken with all faults and defects at valuation" does not exclude the section and applies only to goods answering the trade description), applied *Champanhac & Co., Ltd.* v. *Waller & Co., Ltd.*, [1948] 2 All E. R. 724; *E. & S. Ruben, Ltd.* v. *Faire Bros. & Co., Ltd.*, [1949] 1 All E. R. 215; [1949] 1 K. B. 254. For further examples of exemption clauses as applied to terms forming part of the description, see Appendix II, Note B, *post*. See also the passage on fundamental breach and exemption clauses, p. 11, *ante*,

(*d*) *Ashington Piggeries, Ltd.* v. *Christopher Hill, Ltd.*, [1971] 1 All E. R. 847; [1972] A. C. 441, H. L.

(*e*) *Wallis* v. *Pratt*, [1911] A. C. 394, approving the dissenting judgment of FLETCHER MOULTON, L.J. at [1910] 2 K. B. 1011.

(*f*) *Re Moore & Co.* v. *Landauer & Co.*, [1921] 2 K. B. 519, C. A.; *cf. Manbre Saccharine Co.* v. *Corn Products Co.*, [1919] 1 K. B. 198, at p. 207.

(7) Contract for purchase of a number of "new Singer cars" providing that "all conditions and warranties implied by statute, common law, or otherwise" be excluded. One of the cars delivered is a used car: the sellers are in breach because the description "new Singer car" is an express, not an implied, term (*g*).

(8) Contract for sale of Australian canned peaches "afloat per s.s. Morton Bay due London approx. June 8th." Ship at that time already known not to be due until June 19th; in fact she arrives June 21st. Clause is part of description and buyers are entitled to reject (*h*).

Definitions.—For "contract of sale," see s. 1 (1), *ante*, and s. 62 (1), *post*.

Cross-references.—For the classification of terms in a contract for the sale of goods, the doctrine of fundamental breach, and the construction of exemption clause, see p. 8 *et seq.*, *ante*. For their effect see s. 11, *ante*. *Cf.* s. 18, rule 5, *post* (passing of property in unascertained goods sold by description). For restrictions on contracting out of the provisions of this section, see ss. 55, 61 (6), *post*. For exclusion clauses generally and for examples of terms forming part of the description see Appendix II, Note B, *post*.

COMMENT

Sale by description.—Where goods are described by the contract, and the buyer contracts in reliance on that description, there is a sale by description even if the goods be specific. CHANNELL, J., has said:

> "The term 'sale of goods by description' must apply to all cases where the purchaser has not seen the goods, but is relying on the description alone" (*i*).

And it may apply even where he has seen and selected (*k*) the goods, if the deviation of the goods from the description is not apparent (*l*). The only sales not by description are sales of specific goods *as such*. Specific goods may be sold as such when they are sold without any

---

(*g*) *Andrews Bros. Ltd.* v. *Singer & Co. Ltd.*, [1934] 1 K. B. 17. It would seem that the case might equally well have been decided on the basis of breach of a fundamental term: see *per* PARKER, L.J. in *Karsales* (*Harrow*) v. *Wallis*, [1956] 2 All E. R. 866, at p. 871 (a hire purchase case); and 77 L.Q.R. 98.

(*h*) *Macpherson Train* v. *Ross*, [1955] 2 All E. R. 445.

(*i*) *Varley* v. *Whipp*, [1900] 1 Q. B. 513, at p. 516; *Daniels and Daniels* v. *White and Sons and Tarbard*, [1938] 4 All E. R. 258; *Armaghdown Motors, Ltd.* v. *Grays Motors, Ltd.*, [1963] N. Z. L. R. 5.

(*k*) The new sub-s. (2) probably did not change the Law: see, *e.g.*, *Grant* v. *Australian Knitting Mills*, [1936] A. C., 85, at p. 100, P. C., and the cases cited in the next note.

(*l*) *Beale* v. *Taylor*, [1967] 3 All E. R. 253; [1967] 1 W. L. R. 1193, C. A. *Cf.* *Medway Oil and Storage Co.* v. *Silica Oil Corporation* (1928), 33 Com. Cas. 195, H. L.; *Godley* v. *Perry*, [1960] 1 All E. R. 36; [1960] 1 W. L. R. 9.

description, express or implied; or where any statement made about them is not essential to their identity; or where, though the goods are described, the description is not relied upon, as where the buyer buys the goods such as they are (*m*). The fact that a sale is a sale by sample does not prevent it from being a sale by description also (*n*). Where a sale is not a sale by sample but the seller sends a sample of the goods to the buyer, the sample will be taken into account as evidence of the description given by the seller (*o*).

Section 55, *post*, provides that, except in the case of contracts for the international sale of goods (*p*), a term which seeks to exclude or restrict the condition implied by this section is to be void in the case of a consumer sale, as there defined, and in other cases shall be unenforceable to the extent that it is shown that it would not be fair or reasonable to allow reliance on it (*q*)

But, quite apart from s. 55, an exemption clause will not normally be construed as relieving a seller from liability for failing to supply goods corresponding with the contract description. Thus, where there is a sale by description and the goods tendered or delivered do not answer the description, the seller cannot take advantage of any term in the contract merely to protect him from liability for defects in the things sold (*r*). It is for this reason that it is misleading to say, as the Act does, that it is a condition of the contract that the goods should correspond with the description. If this obligation were merely a condition of the contract, it would be possible to exclude it by such an exemption clause (*s*). But the true position is that if a seller delivers goods which do not correspond with the contract description he had not merely broken a condition of the contract but has entirely failed to perform it. As Lord ABINGER said, as long ago as 1838,

> " If a man offers to buy peas of another, and he sends him beans, he does not perform his contract ; but that is not a warranty ; there is no warranty that he should sell him peas ; the contract is to

---

(*m*) *Benjamin on Sale* (7th Edn.), p. 641, cited with approval by SELLERS, J. in *Travers* v. *Longel* (1947), 64 T. L. R. 150. *See* also *Christopher Hill, Ltd.* v. *Ashington Piggeries, Ltd.*, [1969] 3 All E. R. 1496, at p. 1511, C. A.; reversed on other grounds: [1971] 1 All E. R. 847; [1972] A. C. 441, H. L.

(*n*) *Nichol* v. *Godts* (1854), 10 Exch. 191; *Harrison and Jones* v. *Bunten and Lancaster,* [1953] 1 All E. R. 903; [1955] 1 Q. B. 646.

(*o*) *Boshali* v. *Allied Commercial Exporters, Ltd.*, [1961] Nigeria L. R. 917, P. C.

(*p*) Excluded from s. 55 by s. 61 (6), *post.* Contracts for the international sale of goods are defined by s. 62 (1), *post.*

(*q*) See further the comment to that section, *post.*

(*r*) Similarly where there is a sale by sample no such exemption clause can absolve the seller from his obligation to deliver goods which correspond with the sample : *Champanhac & Co., Ltd.* v. *Waller & Co., Ltd.*, [1948] 2 All E. R. 724.

(*s*) *L'Estrange* v. *Graucob, Ltd.*, [1934] 2 K. B. 394, shows clearly that a seller can effectively protect himself from liability for breach of a condition if he uses appropriate words.

sell peas, and if he sends him anything else in their stead, it is a non-performance of it " (*t*).

This rule is one particular manifestation of the doctrine of " fundamental breach," considered at length, p. 11, *ante*.

It follows that the crucial question to decide in a particular case is often whether words used in the contract with reference to the goods sold form part of the description under which the goods are sold, or amount only to a condition or warranty against the breach of which the seller can protect himself. Thus if a seller contracts to sell, " round mahogany logs " with an exemption clause which protects him against breaches of condition, and then tenders logs which are mahogany but square, the first question will be whether the shape of the logs formed part of their contractual description, and this will be a question of construction of the particular contract. Once it is decided that contractual words do form part of the description, it remains necessary to decide whether the defect complained of is sufficient to prevent the goods from answering that description, and this will be a question of fact in each case (*a*). Where there is a recognised trade description, the proper test is whether the goods comply with the description by the standard generally applied and accepted in the trade (*b*).

Once there is a misdescription however small then subject to the *de minimis* rule the buyer is entitled to reject the goods if he acts in time (*c*).

Failure on the part of the seller to deliver goods which correspond with the contract description may also have the effect of preventing the property from passing (*d*).

Relation to s. 14.—Where the article tendered answers the description, the buyer must, apart from some warranty, express or implied, take the risk as to its quality and condition (*e*). Thus a person purchasing " safety glass " flying goggles gets what he contracts to get if the glass is of the kind generally understood to be meant by that term, and it is not a condition of the contract that the glass should be absolutely safe (*f*).

The implied conditions of merchantableness and fitness for a particular purpose were formerly treated as part of the description,

---

(*t*) *Chanter* v. *Hopkins* (1838), 4 M. & W. 399, at p. 404 ; Lord BLACKBURN used the same illustration in *Bowes* v. *Shand* (1877), 2 App. Cas. 455, at p. 480.

(*a*) See the examples cited at Appendix II, Note B, *post*.

(*b*) *Steels and Busks* v. *Bleecker Bik & Co.*, [1956] 1 Lloyd's Rep. 228.

(*c*) *Rapalli* v. *Take*, [1958] 2 Ll. R. 469, C. A.

(*d*) *Varley* v. *Whipp*, [1900] 1 Q. B. 513 (though the possible effect of s. 11 (1) (*c*) does not appear to have been argued in this case). *Cf.* the dictum of GAVAN DUFFY, J., in *O'Connor* v. *Donnelly*, [1944] Ir. Jur. Rep. 1, at p. 8.

(*e*) *Barr* v. *Gibson* (1838), 3 M. & W. 390 ; *cf. Ward* v. *Hobbs* (1878), 4 App. Cas. 13.

(*f*) *Grenfell* v. *E. B. Meyrowitz, Ltd.*, [1936] 2 All E. R. 1313, C. A. But if a misleading description is used in the course of a business, an offence may be committed under the Trade Descriptions Act, 1968; see p. 56, *ante*.

but the Act treats them as distinct (see s. 14, *post*), and the fact that goods are merchantable is not now a proper test to apply in determining whether goods satisfy a contract description (*g*).

14. Implied undertakings as to quality or fitness.—

(1) Except as provided by this section, and s. 15 of this Act and subject to the provisions of any other enactment (*h*), there is no implied warranty or condition as to the quality or fitness for any particular purpose of goods supplied under a contract of sale (*i*).

(2) Where the seller sells goods in the course of a business, there is an implied condition that the goods supplied under the contract are of merchantable quality, except that there is no such condition—

(*a*) as regards defects specifically drawn to the buyer's attention before the contract is made; or

(*b*) if the buyer examines the goods before the contract is made, as regards defects which that examination ought to reveal.

(3) *Where the seller sells goods in the course of a business and the buyer, expressly or by implication, makes known to the seller any particular purpose for which the goods are being bought, there is an*

(*g*) *Arcos* v. *E. A. Ronaasen & Son*, [1933] A. C. 470.

(*h*) See, for example, note (*h*) at p. 54, *ante*.

(*i*) As to the words " supplied under a contract of sale," see the comment to sub-s. (2), *post*. For conditions imported into contracts for work and labour, see *Myers* v. *Brent Cross Service Co.*, [1934] 1 K. B. 46, D. C.; *Samuels* v. *Davis*, [1943] 2 All E. R. 3, C. A.; [1943] 1 K. B. 526; *Ingham* v. *Emes*, [1955] 2 All E. R. 740; [1955] 2 Q. B. 366, C. A.; *Young and Marten, Ltd.* v. *McManus Childs, Ltd.*, [1968] 2 All E. R. 1169; [1969] 1 A. C. 454, H. L.; *Gloucestershire County Council* v. *Richardson*, [1968] 3 All E. R. 1181; [1969] 1 A. C. 480, H. L.; as to contracts of hire and hire purchase, see *Drury* v. *Buckland*, [1941] 1 All E. R. 269, C. A., the remarks of McNair, J., in *Andrews* v. *Hopkinson*, [1956] 3 All E. R. 422, at p. 426; [1957] 1 Q. B. 229, at p. 237; *Yeoman Credit, Ltd.* v. *Apps*, [1961] 2 All E. R. 281; [1962] 2 Q. B. 508, C. A.; *Astley Industrial Trust* v. *Grimley*, [1963] 2 All E. R. 33; [1963] 1 W. L. R. 584, C. A.; *Charterhouse Credit Co.* v. *Tolly*, [1963] 2 All E. R. 432, 440; [1963] 2 Q. B. 683, 707, C. A.

implied condition that the goods supplied under the contract are reasonably fit for that purpose, whether or not that is a purpose for which such goods are commonly supplied, except where the circumstances show that the buyer does not rely (*j*), *or that it is unreasonable for him to rely, on the seller's skill or judgment* (*k*).

[(3) Where the seller sells goods in the course of a business and the buyer, expressly or by implication, makes known—

(*a*) to the seller, or

(*b*) where the purchase price or part of it is payable by instalments and the goods were previously sold by a credit-broker to the seller, to that credit broker.

any particular purpose for which the goods are being bought, there is an implied condition that the goods supplied under the contract are reasonably fit for that purpose, whether or not that is a purpose for which such goods are commonly supplied, except where the circumstances show that the buyer does not rely (*j*), or that it is unreasonable for him to rely, on the skill or judgment of the seller or credit-broker.

In this sub-section " credit-broker " means a person acting in the course of a business of credit brokerage carried on by him, that is a business of effecting introductions of individuals desiring to obtain credit—

(i) to persons carrying on any business so far as it relates to the provision of credit, or

(ii) to other persons engaged in credit brokerage (*kk*).]

(4) An implied condition or warranty as to quality

(*j*) *Ashford S. C.* v. *Dependable Motors*, [1961] 1 All E. R. 96; [1961] A. C. 336, P. C. (reliance by corporation: one agent makes known the purpose; another decides to buy).

(*k*) Repealed from a date yet to be appointed. See footnote (*l*) p. 124.

(*kk*) To come into force on a date to be appointed. See footnote (*l*), *post*.

or fitness for a particular purpose may be annexed to a contract of sale by usage.

(5) The foregoing provisions of this section apply to a sale by a person who in the course of a business is acting as agent for another as they apply to a sale by a principal in the course of a business, except where that other is not selling in the course of a business and either the buyer knows that fact or reasonable steps are taken to bring it to the notice of the buyer before the contract is made.

(6) *In the application of sub-section (3) above to an agreement for the sale of goods under which the purchase price or part of it is payable by instalments any reference to the seller shall include a reference to the person by whom any antecedent negotiations are conducted; and section 58 (3) and (5) of the Hire-Purchase Act 1965, section 54 (3) and (5) of the Hire-Purchase (Scotland) Act 1965 and section 65 (3) and (5) of the Hire-Purchase Act (Northern Ireland) 1966 (meaning of antecedent negotiations and related expressions) shall apply in relation to this sub-section as they apply in relation to each of those Acts, but as if a reference to any such agreement were included in the references in sub-section (3) of each of those sections to the agreements there mentioned (l).*

ILLUSTRATIONS

Sub-section (2).—(1) Plaintiff goes into a beer-house which he knows is tied to Holden & Co., and asks for beer. Beer contaminated

(*l*) This section was substituted for the original s. 14 by s. 3 of the Supply of Goods (Implied Terms) Act, 1973. It implements recommendations of the Law Commission in its report on Amendments to the Sale of Goods Act, 1893 (Law Com. No. 24). The section has since been further altered by Schs. 4 and 5 to the Consumer Credit Act, 1974, which substitutes a new sub-s. (3) for the sub-ss. (3) and (6) introduced by the 1973 Act. This is necessary to take account of the repeal by the 1974 Act of the Hire-Purchase Acts and the introduction of a new system of consumer credit control. The substitution will take effect from a date to be appointed under s. 192 (4) of the 1974 Act. At present no date has been appointed. The sub-sections which are to be repealed are printed in italics.

with arsenic is supplied to him, and he is injured in consequence. This is a breach of condition, and the seller is liable in damages (*m*).

(2) B by a written contract buys from Fiat Motors, Ltd. a Fiat motor omnibus, which he has inspected, and orders the chassis of six more. He explains orally that they are required for heavy traffic on hilly roads. When the cars are delivered they break down and are unfit for the traffic required. The seller is liable in damages because (1) the cars are not fit for the particular purpose required, and (2) the cars must be " merchantable," which they are not (*n*).

(3) Contract for the sale of 600 motor horns as required. The buyer accepts the first lot delivered. He rejects the rest as they are nearly all dented and scratched owing to bad packing. The motor horns could be repaired at a small expense. The contract is severable, and the buyer may reject the subsequent deliveries as the horns are unmerchantable (*o*).

(4) Plaintiff buys groundnut meal from the defendant for use as an ingredient in poultry foods. The meal is poisonous and many of the birds to which the poultry foods are fed die. The meal could safely be used in smaller inclusions for feeding to certain classes of poultry, and to certain other livestock. Before 1973, the meal was held to be of merchantable quality (*p*). But this may no longer be the case as it would not seem to be " as fit for the purpose or purposes for which goods of that kind are commonly bought as it is reasonable to expect ", see s. 62 (1A), *post*.

Sub-section (3).

(5) Plaintiff asks defendant for a hot-water bottle, and inquires whether it will stand boiling water. Defendant sells him an American rubber bottle, saying it will stand hot but not boiling water. The bottle, which was got for plaintiff's wife, bursts and injures her. The jury find that the bottle was not fit for use as a hot-water bottle. There is a breach of condition, and the seller is liable in damages (*q*).

(6) Plaintiff buys milk from a milk dealer for family use. The milk account book supplied to plaintiff contains a statement of the precautions taken to keep the milk pure. The milk contains the germs of typhoid fever, and plaintiff's wife becomes infected and dies. This is a breach of condition, and the milk dealer is liable in damages (*r*).

(7) A tells B that he can supply him with bunkering coal to suit his steamers. B says he will give an order, but that the order must come through L, the coal merchant with whom he deals. This

---

(*m*) *Wren* v. *Holt*, [1903] 1 K. B. 610, C. A. As to measure of damages, see *Bostock* v. *Nicholson*, [1904] 1 K. B. 725, C. A., and s. 53, *post*.

(*n*) *Bristol Tramways Co.* v. *Fiat Motors, Ltd.*, [1910] 2 K. B. 831, C. A.

(*o*) *Jackson* v. *Rotax Motor and Cycle Co.*, [1910] 2 K. B. 937, C. A.

(*p*) *Henry Kendall & Sons* v. *William Lillico & Sons, Ltd.*, [1968] 2 All E. R. 444; [1969] 2 A. C. 31, H. L.

(*q*) *Priest* v. *Last*, [1903] 2 K. B. 148, C. A. *Cf. Geddling* v. *Marsh* [1920] 1 K. B. 668 (mineral water bottle).

(*r*) *Frost* v. *Aylesbury Dairy Co.*, [1905] 1 K.B. 608, C. A.; *Jackson* v. *Watson & Sons*, [1909] 2 K. B. 193, C. A. (tinned salmon : death of plaintiff's wife).

conversation is repeated to L, who gives the order. If coal unfit for bunkering is supplied and rejected, A cannot sue L for the price (*s*).

(8) Sale of 500 tons of coal for bunkering steamship *Manchester*. It is known that the only available supply is from coal now at sea as there is a railway strike on. The coal is delivered, and found to be quite unsuitable for bunkering the *Manchester*. There is an implied condition of suitability, and the seller is liable in damages (*t*).

Definitions.—For " contract of sale," see s. 1 (1), *ante*, and s. 62 (1), *post*. For " business," " buyer," " goods," " quality," " seller " and " warranty ," see s. 62 (1), *post*. For " merchantable quality," see s. 62 (1A), *post*.

Cross-references.—For a general discussion as to terms of contracts, fundamental breach, and exemption clauses, see p. 8, *ante*. For the history of conditions and warranties, see Note A, *post*. Appendix II. For restrictions on exemption clauses, see s. 55, *post*. For examples of exemption clauses and of special terms as to quality, see Note B, *post*, Appendix II. *Cf*. s. 34, *post* (buyer's right to examine).

History.—In a case of 1838 where a ship was bought while on a voyage, and had stranded, though she was not a total wreck, PARKE, B. said:

> " In the bargain and sale of an existing chattel, by which the property passes, the law does not, in the absence of fraud, imply any warranty of the good quality or condition of the chattel so sold " (*u*).

The old rule of *caveat emptor*—let the buyer take care—probably owed its origin to the fact that in early times nearly all sales of goods took place in market overt (*a*). Its policy was at one time defended on the ground that it tended to diminish litigation (*b*), but its scope was limited in the course of the 19th century, as the Courts became more willing to imply terms that the goods should be of merchantable quality when the buyer had no opportunity of examining them (*c*) and that they should be fit for the purpose for which they were intended if the buyer made known that purpose to the seller so as to show that he relied on his skill and judgment (*d*). Its scope was further limited by the wording of the original s. 14 (*e*), particularly as subsequently interpreted by the Courts (*f*).

(*s*) *Crichton* v. *Love*, 1908 S. C. 818.

(*t*) *Manchester Liners, Ltd.* v. *Rea, Ltd.*, [1922] 2 A. C. 74.

(*u*) *Barr* v. *Gibson* (1838), 3 M. & W. 390, at p. 399, a decision perhaps even then already against the trend of judicial opinion : see *Benjamin on Sale* (8th ed.), p. 612.

(*a*) *Morley* v. *Attenborough* (1849), 3 Exch. 500, at p. 511.

(*b*) Mercantile Law Commission, 1855, 2nd Report, p. 10.

(*c*) *Jones* v. *Just* (1868), L. R. 3 Q. B. 197, reviewing the earlier authorities.

(*d*) *Randall* v. *Newson* (1877); 2 Q. B D. 102, C. A., reviewing the earlier. authorities.

(*e*) See the 2nd edition of this work at p. 30.

(*f*) See the comment to this section in the 16th edition of this work.

Ultimately, in 1969, the Law Commission recommended extensive changes to this section in its Report on Amendments to the Sale of Goods Act, 1893 (*g*), and these recommendations were largely adopted and enacted in s. 3 of the Supply of Goods (Implied Terms) Act, 1973, which substituted the present section for the original one.

The new section has since been amended by the Consumer Credit Act, 1974 (*h*) as a consequence of the repeal by that Act of the Hire Purchase Acts and the introduction of a new system of consumer credit control.

Now, under the Act, the implied conditions of fitness for purpose and of merchantable quality may often apply in sales of specific goods.

Scope for the rule of *caveat emptor* still remains where the circumstances envisaged by the present section are not satisfied.

Scotland.—In Scotland formerly, it was held that the seller guaranteed the buyer against all latent defects. But by s. 6 of the Mercantile Law Amendment (Scotland) Act, 1856, it was provided that if the seller did not know the goods to be defective or of bad quality, the goods, with all faults, should be at the risk of the purchaser unless there was an express warranty or unless the goods were expressly sold for a particular specified purpose. This enactment was intended to assimilate Scottish to English law, but it laid down a narrower rule for the former country. Now a uniform rule is laid down for both countries.

COMMENT

Exemption Clauses.—Section 55 of the Act, *post*, provides that except in the case of contracts for the international sale of goods (*i*), a term which seeks to exclude or restrict the conditions implied by this section is to be void in the case of a consumer sale, as there defined, and in other sales is to be unenforceable to the extent that it is shown that it would not be fair or reasonable to allow reliance on it (*k*).

Fundamental breach.—Insofar as fundamental breaches may be divided into breaches of fundamental terms and other fundamental breaches (*l*), a breach of either of the two conditions implied by this section would seem to be a breach of a fundamental term (*m*). It is doubtful, however, that every breach of either condition would be a fundamental breach. It would seem that a breach which

(*g*) Law Com. No. 24.

(*h*) Sch. 4. The amendments will take effect from a date to be fixed pursuant to s. 192 (4) of that Act.

(*i*) Excluded from s. 55 by s. 61 (6), *post*. Contracts for the international sale of goods are defined by s. 62 (1), *post*.

(*k*) See further the comment to s. 55, *post*.

(*l*) *Suisse Atlantique Société d'Armements Maritime S.A.* v. *N.V. Rotterdamsche Kolen Centrale,* [1966] 2 All E. R. at pp. 68, 85; [1967] 1 A. C. at pp. 421–422, 393, H. L.

(*m*) *Barker* v. *Inland Truck Sales* (1970), 11 D. L. R. (3d) 469.

could entail serious consequences if not remedied, but which may be easily and cheaply remedied, will probably not be fundamental if so remedied at the seller's expense before any risk of those consequences arises (*n*).

Sub-section (2).—The original sub-section read " Where goods are bought by description from a seller who deals in goods of that description (whether he be the manufacturer or not), there is an implied condition that the goods shall be of merchantable quality; provided that if the buyer has examined the goods, there shall be no implied condition as regards defects which such examination ought to have revealed ".

The condition implied by the present sub-section differs from the condition of merchantable quality formerly implied in the following respects:

(1) The original provision only applied when the goods were bought by description. It is no longer necessary that the goods should be bought by description.

(2) The original provision applied only where the seller dealt in goods of that description. Now it is sufficient that the seller sells the goods in the course of a business (*o*). This would include a sale of goods which are totally different from any the seller has dealt in previously. Indeed, it is unnecessary that the seller should ever have sold any goods previously.

It is not clear whether a sale of all or part of the capital equipment of a business will necessarily be a sale in the course of a business. But it is not necessary that the seller should be in business as a vendor or dealer. It is sufficient if the sale was an integral part of the business in question (*p*).

(3) Before 1973, the requirement that the goods should be of merchantable quality was frequently the subject of litigation. A number of definitions were put forward, but following two decisions of the House of Lords, it appeared that as a general rule (*q*), goods would be of merchantable quality if in the form

(*n*) *Harbutt's Plasticine, Ltd.* v. *Wayne Tank and Pump Co., Ltd.*, [1970] 1 All E. R. 225; [1970] 1 Q. B. 447, C. A. (work and materials; material unfit for purpose could have been replaced if fault discovered in time). *Cf. Suisse Atlantique Société d'Armements Maritime S.A.* v. *N.V. Rotterdamsche Kolen Centrale*, [1967] 1 A. C. at p. 392. See further p. 11, *ante*.

(*o*) " Business " is defined by s. 62 (1), *post*, to include a profession and the activities of any government department local authority or statutory undertaking. For the meaning of " in the course of a business " see *Borough of Havering* v. *Stevenson*, [1970] 3 All E. R. 609; [1970] 1 W. L. R. 1375, D. C.

(*p*) *Borough of Havering* v. *Stevenson*, *supra*.

(*q*) It would be dangerous to treat this definition as universally accurate: see *Henry Kendall & Sons* v. *William Lillico & Sons, Ltd.*, [1968] 2 All E. R. at p. 451; [1969] 2 A. C. at p. 77, *per* Lord REID; *B. S. Brown & Son, Ltd.* v. *Craiks, Ltd.*, [1970] 1 All E. R. at p. 825; [1970] 1 W. L. R. at p. 754, *per* Lord REID.

in which they were tendered they would be used by a reasonable man for some purpose for which goods of the same quality and same general character and designation would normally be used, so as to be saleable under the description by which they were sold at a price not substantially less than the contract price (*r*).

" Merchantable quality " is now defined by s. 62 (1A), *post*, which was introduced by s. 7 (2) of the Supply of Goods (Implied Terms) Act, 1973. This provides that goods are of merchantable quality if they are as fit for the purpose or purposes for which goods of that kind are commonly bought as it is reasonable to expect having regard to any description applied to them, the price (if relevant) and all the other relevant circumstances.

The pre-1973 decisions on the meaning of " merchantable quality " no longer represent the law, but a number of dicta in the course of those decisions are still relevant in considering how to apply the new statutory definition. Thus, if goods are offered for sale at a very cheap price, this is an indication that they might well not be as fit for the purposes for which such goods are commonly bought as most goods of that description (*s*) though that inference may not be made if the seller offers some other explanation for their cheapness, or if it is reasonable to draw some other inference from the circumstances (*t*).

On the other hand it may no longer be sufficient that the goods can be used for one purpose for which goods of that kind can normally be used, if it is reasonable in all the circumstances to expect them to be fit for some other purpose as well (*u*). Moreover, the degree of fitness which it is reasonable to expect would normally seem to be such a

---

(*r*) *B. S. Brown, Ltd.* v. *Craiks, Ltd.*, [1970] 1 All E. R. 823; [1970] 1 W. L. R. 752, H. L. Their Lordships adopted the definition given by Lord WRIGHT in *Cammell Laird & Co.* v. *Manganese Bronze and Brass Co., Ltd.*, [1934] A. C. 402, at p. 430, as amended by Lord REID in *Henry Kendall & Sons* v. *William Lillico & Sons, Ltd.*, [1968] 2 All E. R. at p. 451; [1969] 2 A. C. at pp. 76–77, but modified it in two respects: by allowing that the price element may be material (thus meeting criticisms of Lords GUEST, PEARCE and WILBERFORCE in *Kendall* v. *Lillico*); and by holding that it was not necessary that the goods should be usable for some purpose for which they—the contract goods—were normally used, provided they were usable for some purpose for which goods of the same general character and designation would normally be used. See also *Cehave N.V.* v. *Bremer Handelgesellschaft mbh* (1975), *Times*, 22nd July, C. A.; 85 L. Q. R. 74.

(*s*) *Cf. B. S. Brown, Ltd.* v. *Craiks, Ltd.*, [1970] 1 All E. R. 823, at pp. 825, 828, 830–831; [1970] 1 W. L. R. 752, at pp. 754–755, 757–758, 760.

(*t*) As, for example, if they are part of a clearance sale, or are described as shop soiled and no indication is given that they are sub-standard.

(*u*) Accordingly, cases such as *Kendall* v. *Lillico*, *ante*, Illustration (4), and *Canada Atlantic Grain Export Co.* v. *Eilers* (1929), 35 Com. Cas. 90 would probably be decided differently today. But see *Cehave N.V.* v. *Bremer Handelgesellschaft*, *supra*, *per* Lord DENNING, M. R.

degree as would enable them to be safely used in the light of existing knowledge (*a*).

When goods have only one use in the ordinary course of things, and are not fit for that use, they are not merchantable (*b*).

Where a buyer orders an article under a trade name, he will normally be taken to have ordered it as manufactured at the date of the order (*c*), and its merchantable quality must therefore be determined on that basis.

If there is a condition of merchantable quality and the seller knows that the goods will undergo transit of a substantial duration, the condition requires that the goods be in such a state at the start of the transit that they will remain merchantable throughout normal transit to their destination, and for a reasonable time thereafter to allow for their disposal (*d*).

The condition of merchantable quality extends to all goods supplied under the contract. This includes not only the goods contracted to be sold, but also all goods supplied in purported compliance with the contract (*e*) including containers (*f*).

If a trader who buys goods may be injuncted from selling them because the labels infringe another's trade marks, then they are not of merchantable quality (*g*). But it was held before 1973 that goods were not unmerchantable merely because their condition rendered them unsaleable in a foreign country to which the seller knew that

---

(*a*) As to the relevance before 1973 of after acquired knowledge as to the way the goods could safely be used, see *Kendall* v. *Lillico* at [1968] 2 All E. R. 478; [1969] 2 A. C. 108–109, *per* Lord GUEST, but *cf.* the dissenting judgment of Lord PEARCE at [1968] 2 All E. R. 486–487; [1969] 2 A. C. 118–119. Lord REID and Lord MORRIS, although agreeing in the result with Lord GUEST, do not appear to support his reasoning: see at [1968] 2 All E. R. 453, 469; [1969] 2 A. C. 79, 98. See further 86 L. Q. R. 167.

(*b*) *Grant* v. *Australian Knitting Mills, Ltd.*, [1936] A. C. 85.

(*c*) *Harris & Sons* v. *Plymouth Varnish and Colour Co., Ltd.*, (1933), 49 T. L. R. 521.

(*d*) *Beer* v. *Walker* (1877), 46 L. J. Q. B. 677; *Mash & Murrell* v. *Emanuel*, [1961] 1 All E. R. 485; [1961] 1 W. L. R. 862, reversed on the ground that the transit was not a normal one in fact: [1962] 1 All E. R. 77; [1962] 1 W. L. R. 16, C. A.); *Hardwick Game Farm* v. *Suffolk Agricultural and Poultry Producers Association, Ltd.*, [1964] 2 Lloyd's Rep. 227, 270, *per* HAVERS, J., *Cordova Land Co.* v. *Victor Brothers Incorporated*, [1966] 1 W. L. R. 793. But see *Oleificio Zucchi S.p.A.* v. *Northern Sales, Ltd.*, [1965] 2 Lloyd's Rep. 496, 517, 518; 28 M. L. R. 189; and *cf.* s. 33, *post*.

(*e*) *Wilson* v. *Rickett Cockerell & Co. Ltd.*, [1954] 1 All E. R. 868; [1954] 1 Q. B. 598, C. A.

(*f*) *Geddling* v. *Marsh*, [1920] 1 K. B. 668; *MacLean* v. *People's Gas Supply Co., Ltd.*, [1940] 4 D. L. R. 433 (Sup. Ct. of Can.), *Niblett* v. *Confectioners Materials Co.*, [1921] 3 K. B. 387, C. A. As to free gifts, see *Esso* v. *Customs and Excise Commissioners*, [1975] 1 W. L. R. 406, C. A.

(*g*) *Niblett* v. *Confectioners Materials Co.*, [1921] 3 K. B. 387, C. A. On the same principle, goods which cannot be resold in their existing packaging without infringing the Trade Descriptions Act, 1968 are probably not of merchantable quality if supplied to a trader for resale.

the buyer intended to export them (*h*). Those decisions turned on the meaning of merchantable quality under the original section 14 (2). The new definition is different, but unless there are special circumstances, as, for example, that the seller not only knew that the goods were intended for export, but also knew the legal requirements of the country to which they were being exported, it would not be reasonable to expect the goods to comply with special requirements of the local foreign law. The goods would therefore be of merchantable quality, as they would be as fit for the purpose or purposes for which such goods are commonly bought as it is reasonable to expect. Similarly, under the new sub-section there is no condition of merchantable quality as regards defects specifically drawn to the buyer's attention before the contract is made. There was no equivalent provision before 1973, but it seems clear that this was also the position under the earlier sub-section.

Examination.—Under the original sub-section there was no implied condition if the buyer had examined the goods as regards any defect which ought to have been revealed by such examination. The new proviso (b) makes it clear that this exception now only applies if the examination preceeded the contract.

Generally, the implied condition will only be excluded in respect of defects which ought to have been revealed by the examination actually made (*i*). It may be, however, that the buyer's conduct will show that he was willing to take the risk of any defects not revealed by a partial inspection which would have been apparent upon a full one, and in such a case the seller may not be liable for such defects (*k*).

Sub-section (3).—This sub-section replaces the original sub-section (1) which read " Where the buyer, expressly or by implication, makes known to the seller the particular purpose for which the goods are required, so as to show that the buyer relies on the seller's skill or judgment, and the goods are of a description which it is in the course of the seller's business to supply (whether he be the manufacturer or not), there is an implied condition that the goods shall be reasonably fit for such purpose, provided that in the case of a contract for the sale of a specific article under its patent or other trade name there

(*h*) *Sumner Permain & Co.* v. *Webb & Co., Ltd.*, [1922] 1 K. B. 55, C. A.; *Phoenix Distributors* v. *Clarke*, [1966] 2 Lloyds' Rep. 285; affirmed [1967] 1 Lloyd's Rep. 518, C. A.

(*i*) *Bristol Tramways Co.* v. *Fiat Motors, Ltd.*, [1910] 2 K. B. 831, C. A.; and *cf.* the rule that a misrepresentation made in negotiating a contract is no less actionable because an opportunity to examine has been afforded to the other party and has been refused—see *e.g.*, *Redgrave* v. *Hurd* (1881), 20 Ch. D. 1, C. A.

(*k*) *Cf. Thornett and Fehr* v. *Beer & Son*, [1919] 1 K. B. 486, where a full examination was agreed to and an opportunity for such an examination was afforded to the buyer, but without the seller's knowledge the buyer failed to avail himself of it. In view of the clear wording of the new sub-section, it is doubtful whether the condition will now be excluded in such a case except where the buyer is estopped from asserting that he did not examine the goods.

is no implied condition as to its fitness for any particular purpose ".

The present condition differs from the former condition in the following respects:

(1) The new provision applies whenever a seller sells goods in the course of a business (*l*). Before 1973 the goods had to be of a description which it was in the course of the seller's business to supply, but this requirement was broadly interpreted. Accordingly, even under the old law, generally, where a seller deliberately dealt in goods for the purpose of his business, he made it part of his business to supply such goods for the purposes of the original s. 14 (1) of Sale of Goods Act (*m*).

It was sufficient if the goods were of a kind which the seller supplied in the course of his business even though he had never sold goods of that exact description (*n*). It was not clear whether a person who sold goods in the course of his business of a kind which he had not previously supplied came within the original s. 14 (1), but the position probably was that if by entering the contract he was making it part of his business to supply such goods, and *a fortiori* if he had already acquired the goods with a view to selling them, then the sub-section applied (*o*), though in such a case it may have been harder for the buyer to show that he relied on the seller's skill and judgment, and the court may more readily have held the implied condition to have been excluded by the course of dealing between the parties.

Where the buyer relied on the seller's skill and judgment only in relation to certain components of the goods, it would seem to have been sufficient if it was in the course of the seller's business to supply those components (*p*).

(2) The condition was previously implied where the buyer made known to the seller the particular purpose for which the goods were required so as to show that he relied on the seller's skill and judgment (*q*). The new provision puts the burden on the seller to show that the buyer did not rely on his skill or judgment or that in all the circumstances it was unreasonable for the buyer to rely on his skill or judgment. Previously it was sufficient to prove that he did rely on the seller's skill and judgment, whether or not such reliance was reasonable.

(*l*) See comment to s. 14 (2), *ante*, and s. 62 (1), *post*.

(*m*) *Buckley* v. *Lever Brothers, Ltd.*, [1953] 4 D. L. R. 16 (" special offer " of clothes pegs by detergent manufacturer).

(*n*) *Ashington Piggeries, Ltd.* v. *Christopher Hill, Ltd.*, [1971] 1 All E. R. 847, H. L.; *Spencer Trading Co., Ltd.* v. *Devon*, [1947] 1 All E. R. 284.

(*o*) *Ashington Piggeries, Ltd.* v. *Christopher Hill, Ltd.*, *supra* at p. 876, *per* Lord WILBERFORCE, and *cf.* at p. 885, *per* Lord DIPLOCK.

(*p*) *Ashington Piggeries, Ltd.* v. *Christopher Hill, Ltd.*, [1971] 1 All E. R. at pp. 855, 868.

(*q*) The new sub-section reads " *any* particular purpose".

(3) The original sub-s. (1) was subject to a proviso which excluded any implied condition as to fitness for a particular purpose where a specified article was sold under its patent or trade name. This proviso has now been repealed, but in determining whether the circumstances show that a buyer did not rely, or that it was unreasonable for him to rely, on the seller's skill or judgment, the fact that he asked for a specific make of goods may still be relevant (*r*). Even before 1973, moreover, the proviso would not apply unless the buyer ordered the goods under a patent or trade name in such a way as to show that he was not relying on the seller's skill or judgment (*s*).

Any particular purpose.—There is no magic in the word "particular". A communicated purpose, if stated with reasonably sufficient precision, will be a particular purpose (*t*). A purpose may be in wide terms or it may be circumscribed or narrow. The less circumscribed the purpose, the less circumscribed as a rule, will be the range of goods which are reasonably fit for such purpose (*u*).

Reasonable fitness.—The implied undertakings as to fitness for a particular purpose and merchantable quality must be construed reasonably. Very minor complaints will be disregarded (*a*).

Reasonable fitness is a question of fact. In deciding the question of fact the rarity of the unsuitability should be weighed against the gravity of the consequence (*b*).

All relevant factors must be taken into account. In the case of second-hand goods a buyer should realise that defects will appear sooner or later. Thus, where a second-hand Jaguar was sold for £950 on the basis that it might need a repair to its clutch costing £25, but in fact a more serious repair was necessary costing £45, the car was held still to be reasonably fit for the purpose for which it was bought (*c*). But when another Jaguar was sold as being of exceptional quality and hardly run-in, although the representations were not taken literally, they were taken into account by the Court of Appeal in holding that the car was not fit for the purpose intended because the

---

(*r*) *Cf. Daniels and Daniels* v. *White & Sons, and Tarbard*, [1938] 4 All E. R. 258, at p. 263 ("I want a bottle of R. White's lemonade").

(*s*) *Baldry* v. *Marshall*, [1925] 1 K. B. 260, C. A.; *Bristol Tramways Co.* v. *Fiat Motors, Ltd.*, [1910] 2 K. B. 831, C. A.

(*t*) *Henry Kendall & Sons* v. *William Lillico & Sons, Ltd.*, [1968] 2 All E. R. at pp. 465, 475, 482–483, 490; [1969] 2 A. C. at pp. 93, 105, 114, 123.

(*u*) *Henry Kendall & Sons* v. *William Lillico & Sons Ltd.*, *per* Lord PEARCE at [1969] 2 A. C. 114; [1968] 2 All E. R. 483.

(*a*) *Bristol Tramways Co.* v. *Fiat Motor Co., Ltd.*, [1910] 2 K. B. 831; at p. 841, C. A.; *Rapalli* v. *K. L. Take, Ltd.*, [1958] 2 Lloyd's Rep. 469, C. A.

(*b*) *Henry Kendall & Son* v. *William Lillico & Sons*, [1968] 2 All E. R. at p. 483; [1969] 2 A. C. at p. 115, *per* Lord PEARCE.

(*c*) *Bartlett* v. *Sidney Marcus, Ltd.*, [1965] 2 All E. R. 753; [1965] 1 W. L. R. 1013, C. A.

engine was in fact " clapped out " at the time of the sale and broke up completely after being driven a further 2300 miles (*d*).

Reliance.—" The buyer's reliance is a question of fact to be answered by examining all that was said and done with regard to the proposed transaction on either side from its first inception to the conclusion of the agreement to purchase " (*e*).

Before 1973, in order to succeed, the buyer had to show that in all the circumstances a reasonable man in the shoes of the seller would have realised that he was being relied on (*f*). In many cases it was sufficient if the buyer made known to the seller the particular purpose for which he required the goods (*g*). The onus is now on the seller to show that the buyer did not rely on him or that it was unreasonable for the buyer to rely on him.

" It is clear that the reliance must be brought home to the mind of the seller, expressly or by implication. The reliance will seldom be express: it will usually arise by implication from the circumstances: thus to take a case of a purchase from a retailer, the reliance will in general be inferred from the fact that a buyer goes to the shop in the confidence that the tradesman has selected his stock with skill and judgment: the retailer need know nothing about the process of manufacture: it is immaterial whether he be manufacturer or not: the main inducement to deal with a good retail shop is the expectation that the tradesman will have bought the right goods of a good make . . ." (*h*).

But it would seem that no such inference can be drawn where two equally knowledgeable merchants are dealing with each other, just because one makes known to the other the purpose for which he requires the goods (*i*). In such a case, something more must be shown to prove reliance, as in *Henry Kendall & Sons* v. *William*

(*d*) *Crowther* v. *Solent Motor Co.*, [1975] 1 All E. R. 139; [1975] 1 W. L. R. 30, C. A.

(*e*) *Medway Oil and Storage Co., Ltd.* v. *Silica Gel Corpn.*, (1928) 33 Com. Cas. 195, at p. 196, *per* Lord SUMNER.

(*f*) *Henry Kendall & Sons* v. *William Lillico & Sons, Ltd.*, [1968] 2 All E. R. at p. 455; [1969] 2 A. C. at p. 81, *per* Lord REID.

(*g*) *Ashington Piggeries, Ltd.* v. *Christopher Hill, Ltd.*, [1971] 1 All E. R. at pp. 861, 877; *Teheran-Europe Co., Ltd.* v. *S. T. Belton (Tractors), Ltd.*, [1968] 2 All E. R. 886; [1968] 2 Q. B. 545, C. A. As to the position where the buyer asks for the goods of a specified manufacturer, *cf. Young and Marten* v. *McManus Childs, Ltd.*, [1968] 3 All E. R. 1169; [1969] 1 A. C. 454, H. L.

(*h*) *Grant* v. *Australian Knitting Mills, Ltd.*, [1936] A. C. 85, at p. 99. See also *Teheran-Europe Co., Ltd.* v. *S. T. Belton, Ltd.*, [1968] 2 All E. R. at p. 894; [1968] 2 Q. B. at p. 560, *per* DIPLOCK, L.J.

(*i*) *Henry Kendall & Sons* v. *William Lillico & Sons, Ltd.*, [1968] 2 All E. R. at pp. 457, 491–492; [1969] 2 A. C. at pp. 84, 124–125.

Lillico & Sons, Ltd. (*k*), where the sellers had acquired goods from a new source and recommended them to the buyers.

Reliance is not excluded because the seller may not himself have seen the goods he is selling (*l*). It has been said, however, that reliance will more readily be inferred where the article is to be used in an unchanged state, or where the seller specialises in the manufacture of such articles for a clear purpose, than where the material is a raw material or a material manufactured in bulk and capable of being used, and in fact used, for a large variety of purposes in the manufacture of other articles (*m*).

Partial Reliance.—The reliance on the seller's skill and judgment need not be exclusive of all reliance on anything else (*n*), and s. 14 has been held to cover a case in which the materials from which certain propellors were to be made, and certain dimensions, were specified by the buyer, other details being left to the seller's judgment. It was said that it was not necessary that the buyer should rely exclusively on the skill and judgment of the seller for every detail in the production of the goods, and that it is enough if reliance is placed on his skill and judgment to some substantial extent and that the unfitness of which complaint is made arises from matters in regard to which reliance was placed on the seller (*o*).

The condition implied is that the goods are reasonably fit for the purpose made known to the seller to the extent to which the buyer relied upon the seller's skill and judgment. A seller will not be held liable when the unsuitability arises from some special state of affairs relating to the buyer, of which the seller had no reason to be aware or in respect of which no reliance was placed on the seller (*p*). Thus, a buyer who asks for lawn seed for his garden cannot rely on this sub-section merely because his own lawn has some peculiar characteristic known only to him, unless he has made known that characteristic to the seller in circumstances

(*k*) [1968] 2 All E. R. 444; [1969] 2 A. C. 31, H. L., where it was said to be sufficient that a reasonable seller would have realised that he was inviting the buyer to rely on his skill and judgment.

(*l*) *Henry Kendall & Sons* v. *William Lillico & Sons, Ltd.*, *supra*, especially *per* Lord PEARCE at [1969] 2 A. C. 116; [1968] 2 All E. R. 484.

(*m*) *Steels and Busks* v. *Bleecker Bik & Co.*, [1956] 1 Lloyd's Rep. 228, at p. 235.

(*n*) *Medway Oil and Storage Co., Ltd.* v. *Silica Gel Corp.* (1928), 33 Com. Cas. 195, H. L.; *Manchester Liners* v. *Rea, Ltd.*, [1922] 2 A. C. 74.

(*o*) *Cammell Laird & Co., Ltd.* v. *Manganese Bronze & Brass Co., Ltd.*, [1934] A. C. 402; *Ashington Piggeries, Ltd.* v. *Christopher Hill, Ltd.*, [1971] 1 All E. R. 847; [1972] A. C. 441, H. L.: compare *Dixon Kerly, Ltd.* v. *Robinson* [1965] 2 Lloyd's Rep. 404.

(*p*) *Ashington Piggeries, Ltd.* v. *Christopher Hill, Ltd.*, [1971] 1 All E. R. 847; [1972] A. C. 441, H. L.; *Griffiths* v. *Peter Conway, Ltd.*, [1939] 1 All E. R. 685 (purchase of Harris tweed coat by woman with abnormally sensitive skin who did not disclose this fact to the seller).

which would bring the sub-section into force (*q*). But it would appear to be sufficient if the seed was not suitable for at least one type of lawn upon which the seed could fairly and reasonably have been expected to have been used (even, perhaps, if the lawn upon which it was used was not of that type) or if the seller should reasonably have contemplated that it was not unlikely that the seed would be used upon a lawn having a characteristic such as that possessed by the buyer's lawn (*r*).

Unreasonable to rely.—It is not possible to consider all the circumstances in which it may be unreasonable for the buyer to rely on the seller's skill or judgment, but in general it will probably be unreasonable for one party to rely on another if in all the circumstances he ought to appreciate that a reasonable man in the shoes of the seller would not realise that he was being relied on (*s*).

It may also be unreasonable to rely on a seller's skill or judgment in respect of matters which are or ought to be known to the buyer to be outside the normal scope of such a seller's experience. This may be the case either because the particular purpose is not one of which the seller can be expected to have sufficient knowledge to enable him to consider whether the goods are fit for that purpose (*t*) or because the buyer knows that the seller does not normally sell goods of the type in question and has no special skill or judgment in relation to such goods and did not originally acquire them for the purpose of resale (*u*).

Credit-Brokers.—The provision concerning credit-brokers was introduced by the Consumer Credit Act, 1974 (*a*), and will take effect from a date to be appointed under s. 192 (4) of that Act. It is designed to cover situations where a prospective buyer wishes to obtain goods on credit, and the prospective seller wishes to obtain the purchase price at once. The prospective seller therefore sells the goods to a finance company or other person willing to provide credit, and the finance company or other person then re-sells the goods to the prospective buyer on credit terms.

As a result of the amendment, if the prospective seller carries on a

---

(*q*) *Teheran-Europe Co., Ltd.* v. *S. T. Belton, Ltd.*, [1968] 2 All E. R. at pp. 895–896; [1968] 2 Q. B. at p. 563, *per* SACHS, L. J.; and see *Ashington Piggeries, Ltd.* v. *Christopher Hill, Ltd.*, [1971] 1 All E. R. 847, at p. 873, *per* Lord WILBERFORCE.

(*r*) *Cf. Christopher Hill, Ltd.* v. *Ashington Piggeries, Ltd.*, [1969] 3 All E. R. at p. 1523, C. A.; reversed on the facts, [1971] 1 All E. R. 847; [1972] A. C. 441, H. L., where the onus of proof is discussed.

(*s*) *Cf. Henry Kendall & Sons* v. *William Lillico & Sons*, [1968] 2 All E. R. 444, at p. 455; [1969] 2 A. C. 31, at p. 81.

(*t*) *Cf.* the cases cited at n. (*p*), *supra*.

(*u*) As where they form part of the capital assets of a business.

(*a*) S. 192 (3); Sch. 4.

business of credit-brokerage (*b*), and if the prospective buyer makes known to him the purpose for which the prospective buyer requires the goods, then the condition of fitness for purpose will be implied into the contract of sale between the prospective buyer and the finance company, except where the circumstances show that the prospective buyer did not rely, or that it was unreasonable for him to rely on the skill or judgment of the credit-broker.

It will be a question of fact in each case whether the prospective seller is or is not carrying on a business of credit-brokerage, and the onus would seem to be on the prospective buyer to prove that he does carry on such a business (*c*). If the prospective seller does not carry on such a business, then the condition will not be implied into the contract between the finance company and the prospective buyer just because the prospective buyer makes known to the prospective seller the purpose for which the goods are being bought. This will be so even though the prospective seller is acting in the course of a business is selling the goods to the finance company.

Sale by Sample

15. Sale by Sample.—(1) A contract of sale is a contract for sale by sample where there is a term in the contract, express or implied, to that effect.

(2) In the case of a contract for sale by sample—

(a) There is an implied condition that the bulk shall correspond with the sample in quality (*d*):

(b) There is an implied condition that the buyer shall have a reasonable opportunity of comparing the bulk with the sample (*e*):

(c) There is an implied condition that the goods shall be free from any defect, rendering them

---

(*b*) Such a business must be licensed under Part III of the Consumer Credit Act, 1974 from a date to be appointed under s. 192 (2) and para. 44 of Schedule 3 to that Act.

(*c*) For the definition of "business," see s. 62 (1), *post*.

(*d*) *Parker* v. *Palmer*, [1821] 4 B. & Ald. 387, at p. 391. Even if only a simple process is required to make the bulk correspond with the sample, it does not correspond with the sample : *Ruben* v. *Faire*, [1949] 1 All E. R. 215 ; [1949] 1 K. B. 254.

(*e*) *Lorymer* v. *Smith* (1822), 1 B. & C. 1.

unmerchantable (*f*), which would not be apparent on reasonable examination of the sample (*g*).

ILLUSTRATIONS

(1) Sale by sample, goods to be paid for in cash on arrival in exchange for shipping document. An opportunity to inspect is not a condition precedent to payment, and payment does not prejudice the right of rejection if the bulk does not correspond with the sample (*h*).

(2) Sale by sample of a quantity of maroon twill. A part of the twill supplied is inferior to sample. The buyer may reject the whole, or may retain the whole, claiming damages for the portion inferior to sample, but he cannot keep the part equal to the sample and reject the other part (*i*).

(3) Sale of Government surplus balloons, goods to be " as sample taken away " and to be sold " with all faults and imperfections." The sellers are still bound to deliver a bulk which corresponds with the sample, but *semble* the exemption clause may protect them against imperfections in the sample which would not be apparent on reasonable examination (*k*).

Definitions.—For " contract of sale," see s. 1 (1), *ante*, and s. 62 (1), *post*. For " buyer," " goods " and " quality," see s. 62 (1), *post*. For " unmerchantable " and " merchantable quality ", see s. 62 (1A), *post*.

Cross-references.—For a general note on conditions and warranties, see p. 8, *ante*. For their history, see Note A, *post*, Appendix II. For their effect, see s. 11, *ante*. *Cf*. s. 13 *ante* (sale by description and sample); s. 34, *post* (buyer's right to examine); and s. 55, *post* (exclusion of implied terms and conditions).

COMMENT

Exemption Clauses.—Section 55 of the Act, *post*, provides that, except in the case of contracts for the international sale of goods (*l*), a term which seeks to exclude or restrict the conditions implied by

(*f*) As to the term " merchantable," see notes to sub-s. (2) of the last section.

(*g*) *Heilbutt* v. *Hickson* (1872), L. R. 7 C. P. 438, at p. 456; *Mody* v. *Gregson* (1868), L. R. 4 Exch. 49; *Drummond* v. *Van Ingen* (1887), 12 App. Cas. 284 and *cf*. proviso to s. 14 (2), *ante*.

(*h*) *Polenghi* v. *Dried Milk Co.* (1904), 92 L. T. 64; *cf. Heilbutt* v. *Hickson* (1872), L. R. 7 C. P. 438; *Biddell Brothers* v. *E. Clemens Horst Co.*, [1911] 1 K. B. 934, *per* KENNEDY, L.J.; reversed, [1912] A. C. 18 (c.i.f. contract).

(*i*) *Aitken* v. *Boullen*, 1908 S. C. 490; *aliter* if the contract be severable: *Jackson* v. *Rotax Motor and Cycle Co.*, [1910] 2 K. B. 937, C. A.; and see s. 30 (3), *post*, and the cases about delivery of wrong quantity or mixed goods.

(*k*) *Champanhac & Co., Ltd.* v. *Waller & Co., Ltd.*, [1948] 2 All E. R. 724.

(*l*) Excluded from s. 55 by s. 61 (6), *post*. Contracts for the international sale of goods are defined by s. 62 (1), *post*.

this section is to be void in the case of a consumer sale, as there defined, and in other sales is to be unenforceable to the extent that it is shown that it would not be fair or reasonable to allow reliance on it (*m*).

Sub-section (1).—Evidence of usage is admissible to show that a sale was by sample, though the written contract may be silent on the point (*n*). On the other hand, the exhibition of a sample during the making of the contract does not necessarily make it a contract for sale by sample; it must be shown to be a term of the contract that the sale is a sale by sample, and if the contract is reduced to writing this term must be included in the writing (*o*).

"The office of a sample," says Lord MACNAGHTEN,

> "is to present to the eye the real meaning and intention of the parties with regard to the subject-matter of the contract which, owing to the imperfection of language, it may be difficult or impossible to express in words. The sample speaks for itself. But it cannot be treated as saying more than such a sample would tell a merchant of the class to which the buyer belongs, using due care and diligence, and appealing to it in the ordinary way, and with the knowledge possessed by merchants of that class at the time" (*p*).

Sub-section (2) (a).—Text writers and the older cases speak of the term that the bulk shall agree with the sample as a warranty, collateral to the contract (*q*). The Act, following Benjamin (*r*), prefers to treat it as a condition and not a warranty. The parties may, however, agree that the term shall be treated as a warranty and not as a condition (*s*). Subject to the effect of such a special agreement, however, if the seller fails to deliver goods which correspond with the sample, he is not merely in breach of a term of the contract, but is entirely failing to perform it, in the same way as if the goods do not correspond with their description. Accordingly, even where a seller may rely on an exemption clause, he cannot take advantage of a term in the contract designed merely to protect him from liability for defects in the thing sold (*t*).

(*m*) See further the comment to s. 55, *post*.

(*n*) *Syers* v. *Jonas* (1848), 2 Exch. 111.

(*o*) *Meyer* v. *Everth* (1814), 4 Camp. 22 (sugar described in bought note); *Gardiner* v. *Gray* (1815), 4 Camp. 144 (waste silk sold under written contract).

(*p*) *Drummond* v. *Van Ingen* (1887), 12 App. Cas. 284, at p. 297; *cf. Mody* v. *Gregson* (1868), L. R. 4 Exch. 49, at p. 53, *per* WILLES, J.

(*q*) *E.g., Parker* v. *Palmer* (1821), 4 B. & Ald. 387, at p. 391, *per* ABBOTT, C.J.

(*r*) *Benjamin on Sale* (4th ed.), p. 936.

(*s*) *Heyworth* v. *Hutchinson* (1867), L. R. 2 Q. B. 447; *Re Walkers and Shaw*, [1904] 2 K. B. 152.

(*t*) *Champanhac & Co., Ltd.* v. *Waller & Co., Ltd.*, [1948] 2 All E. R. 724 (*ante*, illustration (3)). See further as to fundamental breach and exemption clauses, p. 11, *ante*, and for examples of terms relating to sales by sample, see Appendix II, Note B, p. 367, *post*.

" Quality " here is confined to such qualities as are apparent on an ordinary examination of the sample as usually carried out in the trade (*u*).

Sub-section (2) (b).—*Primâ facie* the place of delivery is the place for comparing the bulk with the sample (*v*). But this presumption may be rebutted and BRETT, J., has expressed the opinion that

> " such a contract always contains an implied term that the goods may under certain circumstances be returned ; that such term necessarily contains certain varying or alternative applications, and, amongst others, the following, that, if the time of inspection, as agreed upon, be subsequent to the time agreed for the delivery of the goods, or if the place of inspection, as agreed upon, be different from the place of delivery, the purchaser may, upon inspection at such time and place, if the goods be not equal to sample, return them *then and there* on the hands of the seller " (*w*).

Sub-section (2) (c).—A reasonable examination is such an examination as, in the case of a sale to a trader, is ordinarily carried out in that trade (*x*). Reasonable does not mean practicable (*a*).

Before the Act, there was a conflict of judicial opinion whether a merchant who was not a manufacturer was to be responsible for a latent defect which examination of the sample failed to disclose (*b*). The Act draws no distinction between a manufacturer and anybody else, and in *Godley* v. *Perry* (*c*) a wholesaler and an importer were held to have been in breach of this sub-section.

PART II

EFFECTS OF THE CONTRACT

Transfer of Property as between Seller and Buyer

16. Goods must be ascertained.—Where there is a contract for the sale of unascertained goods no

(*u*) *Hookway* v. *Alfred Isaacs & Sons*, [1954] 1 Lloyd's Rep. 491; *Steels and Busks* v. *Bleecker Bik & Co.*, [1956] 1 Lloyd's Rep. 228; and see *Ashington Piggeries, Ltd.* v. *Christopher Hill, Ltd.*, [1971] 1 All E. R. 847, at pp. 856, 892; [1972] A. C. 441, 470–471, 514, H. L.

(*v*) *Perkins* v. *Bell*, [1893] 1 Q. B. 193, C. A. (barley delivered at T. station).

(*w*) *Heilbutt* v. *Hickson* (1872), L. R. 7 C. P. 438, at p. 456 ; *cf. Grimoldby* v. *Wells* (1875), L. R. 10 C. P. 391, at p. 395, *per* BRETT, J. This was also the law in Scotland: *Couston* v. *Chapman* (1872), L. R. 2 Sc. & Div. 250, at p. 254, *per* Lord CHELMSFORD (wine sold by auction).

(*x*) *Hookway* v. *Alfred Isaacs & Sons*, [1954] 1 Lloyd's Rep. 491; *Steels and Busks* v. *Bleecker Bik & Co.*, [1956] 1 Lloyd's Rep. 228; and see *Ashington Piggeries Ltd.*, [1971] 1 All E. R. 847 at pp. 856, 892; [1972] A. C. 441, at pp. 470–471, 514, H. L.

(*a*) *Godley* v. *Perry*, [1960] 1 All E. R. 36; [1960] 1 W. L. R. 9.

(*b*) *Cf. Parkinson* v. *Lee* (1802), 2 East, 314 (no liability) with *Randall* v. *Newson* (1877), 2 Q. B. D. 102, at p. 106 (liability).

(*c*) [1960] 1 All E. R. 36; [1960] 1 W. L. R. 9.

property in the goods is transferred to the buyer unless and until the goods are ascertained.

ILLUSTRATION

A, having 200 sacks of flour in a warehouse, sells 50 to B, receives the price and gives B a delivery order. B presents the delivery order at the warehouse and gets a storage warrant in exchange. Nothing is done to appropriate any particular 50 sacks to the contract. No property in any of the sacks passes to B, and if A becomes bankrupt his trustee can claim the whole of the flour (*d*).

Definitions.—For "contract of sale," see s. 1 (1) *ante*, and s. 62 (1), *post*. For "buyer," "goods" and "property," see s. 62 (1), *post*; and compare also "future goods" and "specific goods" defined in that section.

COMMENT

This section is declaratory. "In the case of executory contracts," said BOVILL, C.J.,

"where the goods are not ascertained or may not exist at the time of the contract, from the nature of the transaction, no property in the goods can pass to the purchaser by virtue of the contract itself; but where certain goods have been selected and appropriated by the seller, and have been approved and assented to by the buyer, then the case stands as to the vesting of the property very much in the same position as upon a contract for the sale of goods which are ascertained at the time of the bargain" (*e*).

"A contract to sell unascertained goods," says Lord LOREBURN "is not a complete sale, but a promise to sell" (*f*).

Unascertained goods.—Unascertained goods, that is to say, goods defined by description only, must be distinguished from specific goods, that is to say, goods identified and agreed upon at the time when the contract is made (*g*). Suppose A agrees to sell to B "fifty Southdown sheep," no property in any sheep can pass to B till the sheep are appropriated to the contract. A fulfils his contract by delivering at the appointed time any fifty Southdown sheep. But if he agreed to sell "*the* fifty Southdown sheep now in my field" he could not keep his contract by delivering any others, and the property might pass at once if the parties so intended.

(*d*) *Hayman* v. *M'Lintock*, 1907 S. C. 936; *Laurie & Morewood* v. *Dudin & Sons*, [1926] 1 K. B. 223; finally disposing of *Whitehouse* v. *Frost* (1810), 12 East, 614.

(*e*) *Heilbutt* v. *Hickson* (1872), L. R. 7 C. P. 438, at p. 449, BYLES, J. concurring. For a case about the passing of property, where a person paid a yearly sum for the property in scrap which he agreed to clear from artillery ranges, see *Gale* v. *New*, [1937] 4 All E. R. 645, C. A.

(*f*) *Badische Anilin Fabrik* v. *Hickson*, [1906] A. C. 419, at p. 421.

(*g*) See the definition of specific goods in s. 62 (1), *post*. Note also that specific goods may be sold by description; see comment to s. 13, *ante*.

Part of specific whole.—Although the buyer may have an undivided interest in the whole for insurance purposes (*h*), he can acquire no property in the goods before the goods are ascertained. The sale of an unascertained portion of a larger ascertained quantity of goods passes no property to the buyer till that portion is identified and appropriated to the contract. " If," says BAYLEY, B.,

> " I agree to deliver a certain quantity of oil as 10 out of 18 tons no one can say which part of the whole quantity I have agreed to deliver until a selection is made. There is no individuality until it has been divided " (*i*).

This is so even if the seller has contracted to sell the remainder of the goods to other buyers (*k*). Nor will he acquire any equitable interest in the whole (*l*). But a seller can confer such an interest on the buyer by creating a trust of the whole. No special form is required to create such a trust (*m*), but the person creating the trust must clearly evince his intention to create it, and must identify both the subject and object of the trust (*n*). In *Re Wait* (*o*), ATKIN, L.J., left open the possibility that a direction by the seller to the person holding the goods to deliver part of them to the buyer might in some circumstances amount to an equitable assignment of them which the buyer could enforce against the bulk (*p*).

Specific or ascertained goods.—Where the goods are specific or have been ascertained, the property in the goods passes, by virtue of s. 17, *post*, at such time as the parties to the contract intend it to pass, and in this context, the rules for ascertaining their intention set out in s. 18, *post*, may be relevant.

It would seem to be at least arguable, however, that even where the property has not passed, the buyer in these circumstances can

---

(*h*) *Inglis* v. *Stock* (1885), 10 App. Cas. 263, H. L.

(*i*) *Gillett* v. *Hill* (1834), 2 Cr. & M. 530, at p. 535 ; *Boswell* v. *Kilborn* (1862), 15 Moo. P. C. C. 309 (hops not separated from larger bulk) ; *Hayman* v. *M'Lintock*, 1907 S. C. 936 (sale of a number of sacks of flour from a larger number housed in a third party's store) ; *Kursell* v. *Timber Operators, Ltd.*, [1927] 1 K. B. 298 (all trees in a forest conforming to certain measurements); *in Re Wait*, [1927] 1 Ch. 606 (500 tons of wheat out of parcel of 1000 tons); *Wardar's (Import and Export) Co., Ltd.* v. *Norwood & Sons, Ltd.*, [1968] 2 All E. R. 602; [1968] 2 Q. B. 663, C. A. (600 cartons of kidneys out of a consignment of 1500 cartons).

(*k*) *Healy* v. *Howlett & Sons*, [1917] 1 K. B. 337, D. C.

(*l*) *Ibid.* The same conclusion must result from the decision in *Re Wait*, [1927] 1 Ch. 606, C. A.

(*m*) *Cf. Re Kayford, Ltd.*, [1975] 1 All E. R. 604; [1975] 1 W. L. R. 279, where the purchase price of goods which had not been delivered was held on the special facts of that case to be held on trust by the seller for the buyer.

(*n*) See *Snell's Principles of Equity* (27th ed.), p. 111.

(*o*) [1927] 1 Ch. 606, C. A.

(*p*) *Ibid.*, at pp. 633–634.

acquire an equitable interest in the property, so that, for example, if the seller becomes insolvent after the buyer has paid all or part of the price, the buyer may assert his interest in the goods rather than simply prove for his debt in the liquidation (*q*).

The buyer might also in some cases acquire a right to the immediate possession of the goods sufficient to enable him to bring an action against the seller for conversion if the seller fails to deliver them (*r*).

Goods which are part of a specific whole may become ascertained if they are set aside or marked in some way so as to be allocated to a particular contract (*s*). They may also become ascertained if the remainder of the bulk is disposed of leaving only sufficient goods to fulfil the contract with the buyer (*t*).

Goods mixed.—If the goods sold have been ascertained, and the property has passed to the buyer, his rights cannot be affected by the fact that the seller has afterwards mixed the goods sold with other goods (*u*).

17. Property passes when intended to pass.—(1) Where there is a contract for the sale of specific or ascertained goods (*v*) the property in them is transferred to the buyer at such time as the parties to the contract intend it to be transferred (*a*).

(*q*) In *Re Wait* [1927] 1 Ch. 606, the Court of Appeal held 2–1 that a buyer did not acquire such an interest merely by contracting to buy part of a parcel of goods. But Lord HANWORTH, M.R., one of the majority clearly contemplated that the buyer might claim an equitable interest in the goods once they were ascertained. In this he agreed with SARGANT, L.J., who dissented on the point at issue in the action. ATKIN, L.J., at p. 639, thought differently, on the ground that the common law position was altered by the Sale of Goods Act, but it does not appear to have been the intention of Sir MACKENZIE CHALMERS to alter the common law position in this respect when drafting the Act: see the 2nd edition of this work at p. 16, substantially reproduced in the comment to s. 5, *ante*, under the heading 'Assignment of "future goods"'. (*r*) P. 241, *post*.

(*s*) See the cases on appropriation reviewed in *Federspiel* v. *Twigg*, [1957] 1 Lloyd's Rep. 240. See also *Re Wait, supra, per* ATKIN, L.J., at p. 630.

(*t*) *Wait & James* v. *Midland Bank* (1926), 31 Com. Cas. 172.

(*u*) *Aldridge* v. *Johnson* (1857), 7 E. & B. 885; *Hayman* v. *M'Lintock*, 1907 S. C. 936; but *cf. Spence* v. *Union Mar. Ins. Co.* (1868), L. R. 3 C. P. 427, as to goods of different owners mixed during carriage, where a tenancy in common was held to arise. As to the limits of this doctrine, see *Sandeman & Sons* v. *Tyzack and Branfoot Steamship Co., Ltd.*, [1913] A. C. 680.

(*v*) Similar principles apply to a contract for the sale of unascertained goods: *Ginzberg* v. *Barrow Haematite Steel Co., Ltd. and McKellor*, [1966] 1 Lloyd's Rep. 343. "'Ascertained' probably means identified in accordance with the agreement after the time a contract of sale is made": *per* ATKIN, L.J., in *Re Wait*, [1927] 1 Ch. 606, at p. 630.

(*a*) *Reid* v. *Macbeth & Gray*, [1904] A. C. 223. *Cf. Dennant* v. *Skinner*, [1948] 2 All E. R. 29; [1948] 2 K. B. 164; see also *City Motors* (1933) *Proprietary Ltd.* v. *Southern Aerial Proprietary Ltd.* (1961), 106 C. L. R. 477.

(2) For the purpose of ascertaining the intention of the parties regard shall be had to the terms of the contract (*b*), the conduct of the parties, and the circumstances of the case (*c*).

ILLUSTRATIONS

(1) Contract to build a ship to be classed A1 at Lloyd's, to be paid for after completion. The vessel as she is constructed, and the materials from time to time intended for her, whether in the building yard or elsewhere, to be the property of the purchasers. This is a contract for a complete ship. The buyer obtains no property in iron plates at the railway station marked for the ship, and creditors of the buyer cannot attach them (*d*).

(2) Contract to build a ship, instalments of price to be paid as work proceeds, and buyer to have the right to inspect and superintend construction, delivery to be considered complete after satisfactory official trial. This is a contract for a complete ship, and no property passes to the buyer before the official trial (*e*).

(3) A in England agrees with B in England to sell him a quantity of patented dye stuffs, delivery to be made at Basle in Switzerland. A procures the goods abroad, and has them forwarded to Basle, where his agent holds them at B's disposal. If B assents to this, the property in the goods thereupon passes to B (*f*).

(4) B offers to buy a parcel of diamonds from a foreign firm. The diamonds are sent by post with a bill drawn on the buyer and an invoice marked " settled by acceptance." If the bill of exchange is not accepted, the ownership of the diamonds remains in the seller (*g*).

(5) Sale of car by auction. Buyer offers to pay by cheque, and is allowed to remove the car upon signing a form which provides that the property will not pass until the cheque has been met. The property has already passed at the fall of the hammer, and the form is of no effect (*h*).

(*b*) For cases depending on the inference to be drawn from the terms of the contract, see *Re Anchor Line* (*Henderson Brothers*), *Ltd.*, [1936] 2 All E. R. 941; [1937] 1 Ch. 1, C. A.; *Cheetham & Co.* v. *Thornham Spinning Co.*, [1964] 2 Lloyd's Rep. 17 (retention of shipping documents pending payment); *President of India* v. *Metcalfe Shipping Co., Ltd.*, [1969] 1 All E. R. 861; [1969] 2 Q. B. 123; on appeal, [1969] 3 All E. R. 1549; [1970] 1 Q. B. 289, C. A. (term as to passing of risk).

(*c*) *The Parchim*, [1918] A. C. 157, at pp. 161, 162, P. C.

(*d*) *Reid* v. *Macbeth & Gray*, [1904] A. C. 223; see at p. 230 for the principle. No question was raised in this case about the property in the uncompleted ship. *Cf. in Re Blyth Shipbuilding Co.*, [1926] Ch. 494, at p. 505.

(*e*) *Sir J. Laing* v. *Barclay & Co.*, [1908] A. C. 35.

(*f*) *Badische Anilin Fabrik* v. *Hickson*, [1906] A. C. 419 (there is no " vending " in England within the meaning of the Patent Acts).

(*g*) *Saks* v. *Tilley* (1915), 32 T. L. R. 148, C. A.; *cf.* s. 19 (3), *post.*

(*h*) *Dennant* v. *Skinner & Collom*, [1948] 2 All E. R. 29; [1948] 2 K. B. 164.

(6) A selects goods in a supermarket and takes them to the cash desk. The price is £185, but the cash register records only £85, being incapable of recording over £100. A pays £85, knowing it is not the true price, and removes the goods. The manager of the supermarket is supervising at the cash desk. Although normally it is the intention of the parties in this type of case that the property shall not pass until the price is paid, the property here passes to A upon payment of £85, because the manager intended the property so to pass (*i*).

Definitions.—For " contract of sale," see s. 1 (1), *ante*, and s. 62 (1), *post*. For " buyer," " goods," " property " and " specific goods," see s. 62 (1), *post*; compare also the definition of " future goods " in that section.

History.—By the Civil Law, the property in goods did not pass by virtue of a contract of sale until delivery (*k*). But, as soon as the parties were agreed on the subject-matter and the price, and there were no suspensive conditions, there was an *emptio perfecta*, the result of which was that the risk passed to the buyer, and he acquired a *jus ad rem*, though not a *jus in re*. The Scottish common law followed this rule, but by the Mercantile Law Amendment (Scotland) Act, 1856 s. 1 (now repealed), when goods had been sold but not delivered, the seller's creditors could not attach them, and a sub-vendee was entitled to demand the goods subject to satisfying the seller's lien for the price. The effect was, that when in England the property in goods would pass to the buyer, the same results followed in Scotland, though those results were arrived at in a different manner (*l*). Now, under the Act, the same rule applies to both countries.

COMMENT

By English law the property may pass by the contract itself, if such be the intention of the parties. In other words, the contract may include a conveyance. " Where, by the contract itself," says PARKE, B.,

" the vendor appropriates to the vendee a specific chattel, and the latter thereby agrees to take that specific chattel and to pay the stipulated price, the parties are then in the same situation as they would be after a delivery of goods in pursuance of a general contract. The very appropriation of the chattel is equivalent to delivery by the vendor, and the assent of the vendee to take the specific chattel and to pay the price is equivalent to his accepting

---

(*i*) *Lacis* v. *Cashmarts*, [1969] 2 Q. B. 400, D. C.
(*k*) Moyle's *Justinian* (5th ed.) p. 210, citing *Cod.* 2, 3, 20.
(*l*) *McBain* v. *Wallace* (1881), 6 App. Cas. 588, at p. 618; *Seath* v. *Moore* (1886), 11 App. Cas. 350, at pp. 370, 380. See, too, *Blackburn on Sale* (1st ed.), pp. 187–197.

possession. The effect of the contract, therefore, is to vest the property in the bargainee " (*m*).

Whether this be a satisfactory explanation or not, the rule is undoubted, and is as old as the year books (*n*). The section is thus declaratory. CHANNELL, J.'s comment on it was:

" It is impossible to imagine a clause more vague than this, but I think it correctly represents the state of the authorities when the Act was passed " (*o*).

18. Rules for ascertaining intention.—Unless a different intention appears (*p*), the following are rules for ascertaining the intention of the parties as to the time at which the property in the goods is to pass to the buyer.

Rule 1.—Where there is an unconditional contract for the sale of specific goods, in a deliverable state (*q*), the property in the goods passes to the buyer when the contract is made, and it is immaterial whether the time of payment or the time of delivery, or both, be postponed (*r*).

Rule 2.—Where there is a contract for the sale of specific goods and the seller is bound to do something to the goods, for the purpose of putting them into a deliverable state, the property does

---

(*m*) *Dixon* v. *Yates* (1833), 5 B. & Ad. 313, at p. 340.

(*n*) For a discussion of its policy, see 2nd Report of Mercantile Law Commission, 1855, pp. 9, 42; *Blackburn on Sale* (1st ed.), pp. 187–197; and for its history, see *Cochrane* v. *Moore* (1890), 25 Q. B. D. 57, C.A. (attempted gift of a fourth share of a horse).

(*o*) *Varley* v. *Whipp*, [1900] 1 Q. B. 513, at p. 517.

(*p*) *Turley* v. *Bates* (1863), 2 H. & C. 200 (intention of parties that property should pass immediately although the goods had still to be weighed); *Young* v. *Matthews* (1866), L. R. 2 C. P. 127 (intention of parties that property should pass immediately although some of the goods were still unfinished); *Re Anchor Line* (*Henderson Brothers*), *Ltd.*, [1936] 2 All E. R. 941; [1937] 1 Ch. 1 (terms of contract showing a different intention); *Lambert* v. *G. & C. Finance Corporation* (1963), 107 Sol. Jo. 666 (retention of car log book showed absence of intention to pass property in car); *Cheetham & Co.* v. *Thornham Spinning Co.*, [1964] 2 Lloyd's Rep. 17 (retention of shipping documents pending payment); *Lacis* v. *Cashmarts*, [1969] 2 Q. B. 400, D.C. (Illustration (6) to s. 17, *ante*); *President of India* v. *Metcalfe Shipping Co., Ltd.*, [1969] 1 All E. R. 861; [1969] 2 Q. B. 123; on appeal, [1969] 3 All E. R. 1549; [1970] 1 Q. B. 289, C. A. (term as to passing of risk); and see generally pp. 4, 9 and 14, *ante*.

(*q*) *Philip Head* v. *Showfronts, Ltd.*, [1970] 1 Lloyd's Rep. 140, *post*, Illustration 10.

(*r*) *Tarling* v. *Baxter* (1827), 6 B. & C. 360. See the rule stated and contrasted with the Civil Law and Scottish common law in *Seath* v. *Moore* (1886), 11 App. Cas. 350, at p. 370.

not pass until such thing be done (*s*), and the buyer has notice thereof.

Rule 3.—Where there is a contract for the sale of specific goods in a deliverable state, but the seller is bound to weigh, measure, test, or do some other act or thing with reference to the goods for the purpose of ascertaining the price, the property does not pass until such act or thing be done (*t*), and the buyer has notice thereof.

Rule 4.—When goods are delivered to the buyer on approval or " on sale or return " or other similar terms (*u*) the property therein passes to the buyer :

(a) When he signifies his approval or acceptance to the seller or does any other act adopting the transaction (*a*) :

(b) If he does not signify his approval or acceptance to the seller but retains the goods without giving notice of rejection, then, if a time has been fixed for the return of the goods, on the expiration of such time, and, if no time has been fixed, on the expiration of a reasonable time (*b*). What is a reasonable time is a question of fact.

(*s*) *Rugg* v. *Minett* (1809), 11 East, 210 ; *Acraman* v. *Morrice* (1849), 8 C. B. 449 ; *Boswell* v. *Kilborn* (1862), 15 Moo. P. C. C. 309. *Young* v. *Matthews* (1866), L. R. 2 C. P. 127, at p. 129.

(*t*) *Hanson* v. *Meyer* (1805), 6 East, 614 ; *Zagury* v. *Furnell* (1809), 2 Camp. 239. Goods of this kind were known in Scottish law as " fungibles. " *Cf.* Bell's *Principles of the Law of Scotland* (9th ed.), §137.

(*u*) *Polar Refrigeration Service, Ltd.* v. *Moldenhauer* (1967), 61 D. L. R. (2d) 463 (air-conditioning equipment: fitness for stated purpose a condition precedent); *Tiffin* v. *Pitcher*, [1969] C. L. Y. 3234 (sale of van conditional on replacement of engine and subsequent fitness for intended purpose); *R.* v. *Justelius*, [1973] 1 N. S. W. L. R. 471 (Advance subscription order for books, recipient to return them within 7 days or pay for them; company's practice to treat recipient as debtor if goods not returned. Held: " on similar terms" to sale or return, so that property in the books passed to recipient after 7 days).

(*a*) *Swain* v. *Shepherd* (1832), 1 M. & Rob. 233 ; Bell's *Inquiries into the Contract of Sale of Goods and Merchandise*, p. 111; *cf.* s. 35, *post.*

(*b*) *Moss* v. *Sweet* (1851), 16 Q. B. 493 ; *cf. Beverley* v. *Lincoln Gas Co.* (1837), 6 A. & E. 829 ; *Elphick* v. *Barnes* (1880), 5 C. P. D. 321 (death of horse delivered on sale or return) ; *Re Ferrier, ex p. the Trustee* v. *Donald*, [1944] Ch. 295 (goods taken in execution after delivery on sale or return) ; *Poole* v. *Smith's Car Sales (Balham) Ltd.*, [1962] 2 All E. R. 482 ; [1962] 1 W. L. R. 744, C. A.

Rule 5.—(1) Where there is a contract for the sale of unascertained or future goods by description, and goods of that description (*c*) and in a deliverable state (*d*) are unconditionally (*e*) appropriated to the contract, either by the seller with the assent of the buyer, or by the buyer with the assent of the seller, the property in the goods thereupon passes to the buyer (*f*). Such assent may be express or implied, and may be given either before or after the appropriation is made (*g*).

(2) Where, in pursuance of the contract, the seller delivers the goods to the buyer or to a carrier or other bailee [or custodier] (whether named by the buyer or not) for the purpose of transmission to the buyer, and does not reserve the right of disposal (*h*), he is deemed to have unconditionally appropriated the goods to the contract (*i*).

ILLUSTRATIONS

Rule 1.—(1) Contract for the sale of a stack of hay made on January 6; price to be paid on February 4, but hay not to be removed until May 1. On January 20 the stack is destroyed by fire. The property and risk passed when the contract was made, and the loss falls on the buyer (*k*).

(*c*) *Cf. Thornley* v. *Tuckwell* (*Butchers*), *Ltd.*, [1964] Crim. L. R. 127 (English lamb appropriated to contract for the sale of New Zealand lamb: no property passes).

(*d*) *Philip Head* v. *Showfronts, Ltd.*, [1970] 1 Lloyd's Rep. 140, *post*, illustration (10).

(*e*) So that where, as in the case of the usual c.i.f. contract, the seller retains the documents against payment of the price, his notice of appropriation does *not* pass the property : *Smyth* v. *Bailey*, [1940] 3 All E. R. 60, H. L.

(*f*) See, in illustration, *Rohde* v. *Thwaites* (1827), 6 B. & C. 388 ; *Aldridge* v. *Johnson* (1857), 7 El & Bl. 885 ; *Langton* v. *Higgins* (1859), 4 H. & N. 402 ; *Doak* v. *Bedford*, [1964] 1 All E. R. 311; [1964] 2 Q. B. 587; and see the principle fully discussed, and the authorities reviewed, in *Federspiel* v. *Twigg*, [1957] 1 Lloyd's Rep. 240, and in *Widenmeyer* v. *Burn Stewart & Co., Ltd.*, 1967 S. L. T. 129 (Ct. of Sess.).

(*g*) *Campbell* v. *Mersey Docks* (1863), 14 C. B., N.S., 412 ; *cf. Godts* v. *Rose* (1855), 17 C. B. 229, at p. 237 ; *Aldridge* v. *Johnson* (*supra*) ; *Jenner* v. *Smith* (1869), L. R. 4 C. P. 270.

(*h*) See *Smyth* v. *Bailey*, [1940] 3 All E. R. 60, H. L.

(*i*) For statement of principle, see *Mirabita* v. *Imperial Ottoman Bank* (1878), 3 Ex. D. 164, C. A. ; *Napier* v. *Dexters* (1926), 26 Ll. L. Rep. 62 ; and see Note C, Appendix II, *post*.

(*k*) *Tarling* v. *Baxter* (1827), 6 B. & C. 360.

Rule 2.—(2) Sale of condensing engine. It is to be severed from the realty and delivered free on rail at a specified price. It is damaged in transit before it reaches the railway. The property does not pass, as when it reaches the railway it is not in a deliverable state (*l*).

Rule 3.—(3) A sells 160 bags of cocoa to B at 59*s.* per load of 60 lbs., knowing that B will resell to X, and that the weights will be tested at X's place of business. This checkweighing is not suspensive of the contract of sale, nor a condition precedent to the property passing to B (*m*).

Rule 4.—(4) A delivers goods to B on sale or return. B pledges the goods. This adopts the sale, and the property passes to the buyer. A therefore cannot recover the goods from the pawnbroker (*n*).

(5) A delivers goods to B on sale for cash or return, the goods to remain the property of A till settled for or charged. C tells B he can find a customer for them, but, instead of doing so, he pawns them with D. A can recover the goods from D, the pawnbroker (*o*).

(6) A delivers diamonds to B on sale or return, and B delivers them to C on the like terms. C delivers them to D, and while they are in D's custody they are lost. As B cannot return the diamonds to A, he has by his dealing with them adopted the transaction and is liable for the price (*p*).

(7) Two days after goods are delivered by A to B on sale for cash or return within a week, execution is levied on B's goods on behalf of two creditors and the goods in question are seized. Once the goods have been seized, B no longer "retains" them within the meaning of s. 18, r. 4 (b); thus they have not become B's property, so that A is entitled to recover them (*q*).

Rule 5.—(8) A in England writes to B at Basle in Switzerland for a packet of patent dye, to be sent by parcel post. B posts the packet to A. The property passes to A as soon as the packet is posted in Basle (*r*).

(9) B orders 140 bags of rice from A, pays for them and asks for delivery. A sends him a delivery order for 125, and asks him

---

(*l*) *Underwood* v. *Burgh Castle Cement Syndicate*, [1922] 1 K. B. 343, C. A.

(*m*) *Nanka Bruce* v. *Commonwealth Trust*, [1926] A. C. 77, P. C.

(*n*) *Kirkham* v. *Attenborough*, [1897] 1 Q. B. 201, C. A.; *London Jewellers Ltd.* v. *Attenborough*, [1934] 2 K. B. 206, C. A.

(*o*) *Weiner* v. *Gill*, [1905] 2 K. B. 172, affirmed [1906] 2 K. B. 574, C. A.; *R.* v. *Eaton* (1966), 50 Cr. App. Rep. 189; *cf. Edwards* v. *Vaughan* (1910), 26 T. L. R. 545, C. A. The terms of the special contract took the case out of rule 4. Cozens-Hardy, M.R., at p. 546, said that "a transaction on sale or return was not a case where a person was in possession under an agreement to buy within s. 25 of the Sale of Goods Act, 1893." See also *Kempler* v. *Bravingtons, Ltd.* (1925), 133 L. T. 680, C. A. (reviewing previous cases).

(*p*) *Genn* v. *Winkel* (1912), 107 L. T. 434.

(*q*) *Re Ferrier, ex p. the Trustee* v. *Donald*, [1944] Ch. 295.

(*r*) *Badische Anilin Fabrik* v. *Basle Chemical Works*, [1898] A. C. 200, at pp. 203, 204.

to send for the remaining 15 at A's place of business. B waits a month before sending for them, and in the meantime they are stolen. The property in the 15 bags has passed to B, and he must bear the loss (*s*).

(10) A orders carpets from B to be laid by B on A's premises. B delivers carpets to A's premises, but before B can lay them they are stolen. The carpets are not in a deliverable state and the property in them has not passed to A (*t*).

Definitions.—For " contract of sale," see s. 1 (1), *ante,* and s. 62 (1), *post.* For " bailee," " buyer," " delivery," " future goods," " goods," " property," " seller " and " specific goods," see s. 62 (1), *post.* For " deliverable state," see s. 62 (4), *post.*

Cross-references.—*Cf.* p. 33, *ante* (shipping contracts); ss. 1 (1), 8 and 9, *ante* (price); s. 1 (2), *ante* (conditional and unconditional contract); s. 1 (3), *ante* (passing of property); s. 10, *ante* (time of payment); s. 13, *ante* (sale by description); s. 19, *post* (right of disposal); s. 28, *post* (payment and delivery); s. 35, *post* (acceptance and rejection); s. 55, *post* (" different intention "); s. 56, *post* (reasonable time a question of fact); and Note C, Appendix II, *post* (delivery to carrier).

COMMENT

As the English Courts have rejected the objective test of delivery for marking the time when the property is to pass, they have been forced to lay down more or less arbitrary rules for fixing the moment when the property is to be held to pass in cases where the parties have either formed no intention on the point, or have failed to express it.

The rules laid down by this section apply unless a contrary intention appears (*u*). It has, however, been said that very little is now needed to give rise to the inference that the property in specific goods is to pass only on delivery or payment (*a*).

Rule 1.—*Specific goods.*—See note to s. 17, *ante.* The first four rules deal only with specific goods.

Rule 2.—The final words, " and the buyer has notice thereof," were added in Committee on a suggestion from Scotland that it was unfair that the risk should be transferred to the buyer without notice. It is to be noted that this rule is negative. The case of an article which the seller is to manufacture for the buyer is sometimes treated as coming under this rule, but it generally comes

---

(*s*) *Pignataro* v. *Gilroy,* [1919] 1 K. B. 459. For converse case, see *Healy* v. *Howlett & Sons,* [1917] 1 K. B. 337.

(*t*) *Philip Head* v. *Showfronts, Ltd.*, [1970] 1 Lloyd's Rep. 140.

(*u*) See cases cited at p. 146, n. (*p*).

(*a*) *Ward* v. *Bignall,* [1967] 2 All E. R. 449, at p. 453; [1967] 1 Q. B. 534, at p. 545, *per* DIPLOCK, L.J.

under Rule 5. If a man orders a watch to be specially made for him, it is clear that the watchmaker may, if he likes, make two such watches, and that he keeps his contract by delivering either of them (*b*). PARKE, B., has pointed out that there may be an intermediate state of things. An article may be in course of manufacture, and the parties may have so far agreed upon it that the seller would break his contract if he delivered any other article, but there may be no intention that the property in it should pass before its completion (*c*). Unless a different intention be clearly shown, the rule is that the property in an article, which the seller is to make or complete for the buyer, does not pass until the article is delivered in a finished state, or until it is ready for delivery and is approved by the buyer in that state (*d*).

At one time the Courts seemed inclined to reverse the presumption in the case of shipbuilding contracts, where the ship was to be paid for by stated instalments as the work progressed (*e*); but in a subsequent case in the House of Lords it was held that there was no sound distinction between the case of a ship and any other manufactured article (*f*). These decisions depend upon the particular terms of the contracts with which they deal, and are not to be regarded as inconsistent (*g*).

If the specific goods sold are attached to, or form part of, realty at the time of the contract, and are to be severed by the buyer, the property will only pass when they are severed (*h*).

Rule 3.—As to the concluding words, " and the buyer has notice thereof," see note to last rule. Lord BLACKBURN, in his work on *Sale* (*i*), states this rule without confining its operation to acts to be done by the seller, and regards it as a rule arbitrarily adopted from the Roman law, where it was a logical deduction from the principle that there could be no sale until the price was fixed. But the Court of Exchequer in 1863 reviewed the cases, and came to

---

(*b*) *Cf. Atkinson* v. *Bell* (1828), 8 B. & C. 277, and *Xenos* v. *Wickham* (1866). L. R. 2 H. L. 296, at p. 316, *per* WILLES, J.

(*c*) *Laidler* v. *Burlinson* (1837), 2 M. & W. 602, at pp. 610, 611 (ship in course of building); *Wait* v. *Baker* (1848), 2 Exch. 1, at pp. 8, 9.

(*d*) *Clarke* v. *Spence* (1836), 4 A. & E. 448, at p. 466 (ship in course of building), reviewing the previous cases. As to an article begun by one person and finished by another, see *Oldfield* v. *Lowe* (1829), 9 B. & C. 73, at p. 78 (machinery); 34 Halsbury's Laws (3rd Edn.) 65.

(*e*) *Woods* v. *Russell* (1822), 5 B. & Ald. 942.

(*f*) *Seath* v. *Moore* (1886), 11 App. Cas. 350, at pp. 370, 380 (ship and its fittings); *Story on Sale*, § 316a, and notes to s. 17, *ante*.

(*g*) *Re Blyth Shipbuilding Co.*, [1926] Ch. 494, C. A. (reviewing all the earlier cases).

(*h*) *Jones* v. *Earl of Tankerville*, [1909] 2 Ch. 440, 442; *Kursell* v. *Timber Operators*, [1927] 1 K. B. 298, C. A.

(*i*) 1st ed., p. 152.

the conclusion that the rule should be qualified, as in the text, by confining it to acts to be done by the seller (*k*). The Act adopts this view, and the rule is now so confined. If the buyer were bound by the contract to do the weighing, it would be a matter of construction, having regard to s. 17 and s. 18, rule 1, when the property passed. A mere right of the buyer to weigh is immaterial (*l*). And if the parties have agreed on a provisional estimate of the total amount payable, to be more exactly calculated when the goods can be weighed, a common intention may be inferred that the transfer of the property shall not depend upon such weighing (*m*).

Rule 4.—*Sale or return, etc.*—When goods are sent on trial, or on approval, or on sale or return, the clear general rule is that the property remains in the seller till the buyer adopts the transaction, but it is quite competent to the parties to agree that the property shall pass to the buyer on delivery, and that, if he does not approve the goods, the property shall then revest in the seller (*n*). To use the language of continental lawyers, the condition on which the goods are delivered may be either suspensive or resolutive.

In some trades the usage is that when goods are delivered on fourteen days' approval, the property does not pass to the buyer on the expiration of that time, but the seller at any time after the fourteen days can call on the buyer either to take or to return the goods at once.

A transaction on sale or return is not a transaction where a person is in possession of the goods under an agreement to buy within s. 25 of the (Act *o*), nor does it become a sale because the goods perish or are damaged while in the possession of the bailee otherwise than by his act or default (*p*).

A delivery on sale or return within the meaning of the Rule may contemplate a sale to the transferee, or to a third party (*q*).

As to factors to be taken into account in determining what is a reasonable time within which to reject goods, see the comment to s. 35, *post*.

Rule 5.—*Generic goods.*—When there is a contract for the sale of unascertained goods, and the goods are afterwards selected by the buyer, or if selected by the seller are approved by the buyer,

---

(*k*) *Turley* v. *Bates* (1863), 2 H. & C., 200.

(*l*) *Nanka Bruce* v. *Commonwealth Trust*, [1926] A. C. 77, P. C., and *cf. Tansley* v. *Turner* (1835), 2 Bing. N.C. 151.

(*m*) *Martineau* v. *Kitching* (1872), L. R. 7 Q. B 436; *cf. Anderson* v. *Morice* (1874), L. R. 10 C. P. 58 ; and see also *Castle* v. *Playford*, (1872) L. R. 7 Ex. 98.

(*n*) *Cf. Head* v. *Tattersall* (1871), L. R. 7 Exch. 7.

(*o*) *Edwards* v. *Vaughan* (1910), 26 T. L. R. 545, at p. 546, C. A.

(*p*) *Elphick* v. *Barnes* (1880), 5 C. P. D. 321 (death of horse on sale or return); see also the remarks of WILLMER, L.J., in *Poole* v. *Smith's Car Sales* (*Balham*) *Ltd.*, [1962] 2 All E. R. 482 at p. 489; [1962] 1 W. L. R. 744 at p. 753, C. A.

(*q*) See *Poole* v. *Smith's Car Sales* (*Balham*), (*supra*).

no difficulty arises. The difficulty comes when the seller makes the selection pursuant to an authority derived from the buyer; and it is often a nice question of law whether the acts done by the seller merely express a revocable intention to appropriate certain goods to the contract, or whether they show an irrevocable exercise of a right of election.

> " The general rule seems to be that when, from the nature of an agreement, an election is to be made, the party who is by the agreement to do the first act, which from its nature cannot be done till the election is determined, has authority to make the choice in order that he may perform his part of the agreement: when once he has performed the act the choice has been made and the election irrevocably determined; till then he may change his mind as to what the choice shall be, for the agreement gives him till that time to make his choice " (*r*).

The expression that the property in the goods passes by their " appropriation to the contract," though consistently used in the modern cases, is not a fortunate one. In the first place, as PARKE, B., has pointed out, the term is used in two senses. It may mean that the goods are so far appropriated that the seller would break his contract by delivering any other goods, though they still remain his property, or it may, and usually does, mean that the goods are finally appropriated to the contract so as to pass the property in them to the buyer (*s*). In the second place, if the decisions be carefully examined, it will be found that in every case where the property has been held to pass, there has been an actual or constructive delivery of the goods to the buyer (*t*). If the term " delivery " had been substituted for " appropriation," probably less difficulty would have arisen; and it seems a pity that this was not done in the Act.

The difficulty in the shipbuilding cases usually arises in regard to appropriation.

> " For appropriation I think there must be some definite act, such as the affixing of the property to the vessel itself, or some definite agreement between the parties which amounts to an assent to the

---

(*r*) *Blackburn on Sale* (1st ed.), p. 128, citing *Heyward's Case* (1595), 2 Coke, 35a, where it is said, " the certainty and thereby the property begins by election."

(*s*) *Wait* v. *Baker* (1848), 2 Exch. 1, at p. 8.

(*t*) This passage was approved by PEARSON, J., in *Federspiel* v. *Twigg & Co.*, [1957] 1 Lloyd's Rep. 240 at pp. 255–256, subject to the possible qualification that there may be after such constructive delivery an actual delivery still to be made by the seller under the contract. *Cf. Wardar's* (*Import and Export*) *Co., Ltd.* v. *W. Norwood & Sons, Ltd.*, [1968] 3 All E. R. 602; [1968] 2 Q. B. 663, C. A., where the Court of Appeal left open the question whether there was an unconditional appropriation of frozen kidneys when these were removed from a cold store and placed on the pavement to await collection.

property in the materials passing from the builders to the purchasers" (*u*).

Under s. 20, *post*, risk normally passes with property. It is relevant, therefore, to consider whether in all the circumstances risk has passed (*a*).

Delivery to carrier.—The commonest method of appropriating goods to a contract is by delivering them to a carrier, and then, if there be authority so to deliver them, and the seller does not reserve the right of disposal (*b*),

> " the moment the goods which have been selected in pursuance of the contract are delivered to the carrier, the carrier becomes the agent of the vendee, and such a delivery amounts to a delivery to the vendee; and if there is a binding contract between the vendor and the vendee . . . then there is no doubt that the property passes by such delivery to the carrier. It is necessary, of course, that the goods should agree with the contract" (*c*).

19. Reservation of right of disposal.—(1) Where there is a contract for the sale of specific goods or where goods are subsequently appropriated to the contract, the seller may, by the terms of the contract or appropriation (*d*), reserve the right of disposal of the goods until certain conditions are fulfilled. In such case, notwithstanding the delivery of the goods to the buyer, or to a carrier or other bailee [or custodier] for the purpose of transmission to the buyer, the property in the goods does not pass to the buyer until the conditions imposed by the seller are fulfilled (*e*).

(2) Where goods are shipped, and by the bill of lading the goods are deliverable to the order of the

---

(*u*) *Re Blyth Shipbuilding Co.*, [1926] Ch. 494; *per* SARGANT, L.J., at p. 518.

(*a*) *Federspiel* v. *Twigg & Co.*, *supra*.

(*b*) As he does in the ordinary c.i.f. contract, by holding the documents against the price: *Smyth* v. *Bailey*, [1940] 3 All E. R. 60, H. L.

(*c*) *Wait* v. *Baker* (1848), 2 Exch. 1, at p. 8 (unindorsed bill of lading); *Napier* v. *Dexters, Ltd.* (1926), 26 Ll. L. Rep. 62; *Wardar's (Import and Export) Co., Ltd.* v. *W. Norwood & Sons, Ltd.*, [1968] 2 All E. R. 602; [1968] 2 Q. B. 663, C. A.; and see Note C, Appendix II, *post*.

(*d*) See, *e.g.*, *Wait* v. *Baker* (1848), 2 Exch. 1; *Ogg* v. *Shuter* (1875), 1 C. P. D. 47 (property reserved by terms of contract); *Turner* v. *Liverpool Docks* (1851), 6 Exch. 543 (property reserved by terms of appropriation).

(*e*) *Mirabita* v. *Imperial Ottoman Bank* (1878), 3 Ex. D. 164, *per* COTTON, L.J. See also *The Annie Johnson*, [1918] p. 154, at p. 163.

seller or his agent, the seller is primâ facie deemed to reserve the right of disposal (*f*).

(3) Where the seller of goods draws on the buyer for the price, and transmits the bill of exchange and bill of lading to the buyer together to secure acceptance or payment of the bill of exchange, the buyer is bound to return the bill of lading if he does not honour the bill of exchange, and if he wrongfully retains the bill of lading the property in the goods does not pass to him (*g*).

ILLUSTRATIONS

(1) A consigns goods to B by ship, and draws on him for the price. He discounts the bill with a bank, indorses the bill of lading in blank, and authorises the bank to hand the bill of lading to B when he accepts the bill of exchange. Apart from any special terms in the contract, the property in the goods is transferred to B when he accepts the bill of exchange (*h*).

(2) A contracts to sell goods to B, and ships them under bills of lading made out to order of A upon a ship chartered by A for B. A then draws a bill of exchange on B. He discounts it with a bank, lodging the bill of lading with the bank. B is notified by the bank to which he tenders the price. The bank refuses the tender and sells elsewhere. Held: A's appropriation was subject only to payment on tender. Property therefore passed to B when he tendered the price, and the bank is liable in conversion (*i*).

(*f*) *Ogg* v. *Shuter* (1875), 1 C. P. D. 47, C. A.; *Mirabita* v. *Imperial Ottoman Bank* (1878), 3 Ex. D. 164, at p. 172, C. A. But see *The Parchim*, [1918] A. C. 157. In *Joyce* v. *Swann* (1864), 17 C. B., N.S., 84, and in *Browne* v. *Hare* (1853), 3 H. & N. 484; (1859) 4 H. & N. 822, the inference was negatived.

(*g*) *Shepherd* v. *Harrison* (1871), L. R. 5 H. L. 116, see at p. 133, *per* Lord CAIRNS; *Cahn* v. *Pockett's Bristol Channel Co.*, [1899] 1 Q. B. 643, at p. 656, C. A. The same principle underlies *Saks* v. *Tilley* (1915), 32 T. L. R. 148, C. A., where diamonds were posted to a buyer, together with a bill of exchange for acceptance and an invoice marked "settled by acceptance". For the position where the seller's agent credits the seller and then sends the bill of lading and a bill of exchange for acceptance to the buyer, see *Jordesen & Co. and Kahn* v. *London Hardwood Co., Ltd.* (1913), 19 Com. Cas. 164; *Barton Thompson* v. *Vigers Brothers* (1906), 19 Com. Cas. 175; *Churchill and Sim* v. *Goddard*, [1936] 1 All E. R. 675; [1937] 1 K. B. 92.

(*h*) *The Prinz Adelbert*, [1917] A. C. 586, P. C.; *cf. The Derfflinger No.* 2 (1918), 87 L. J. P. C.; 195; *The Miramichi*, [1915] P. 71; *The Orteric*, [1920] A. C. 724, 733, P. C. See also *The Vesta*, [1921] A. C. 774, 783 where the property was held to have passed although the seller was obliged to take goods back if "found unsuited."

(*i*) *Mirabita* v. *Imperial Ottoman Bank* (1878), 3 Ex. D. 164. Had the seller's intention throughout been to deprive the buyer of the goods, rather than to pass them to him against payment, the result would have been different; see *per* COTTON, L.J., especially at pp. 172–173 and the cases cited at n. (*s*) on p. 35, *supra*.

(3) A contracts to sell goods to B, and arranges the shipment of the goods to B under a bill of lading naming A as consignee. A and B are associated companies and payment is to be made on credit terms. A has no intention of securing the price, but intends to reserve the right, if necessary, to divert the goods to another customer. Property only passes when the prospect of diverting the goods disappears, when A posts the bill of lading to B (*k*).

Definitions.—For "contract of sale," see s. 1 (1), *ante*, and s. 62 (1), *post*. For "bailee," "buyer," "delivery," "goods," "property," "seller" and "specific goods," see s. 62 (1), *post*.

Cross-references.—*Cf.* ss. 8 and 9, *ante* (price); s. 25, *post* (buyer in possession); s. 43 (1), *post* (seller's lien); and s. 32, *post*, and Note C, *post*, Appendix II (delivery to carrier).

COMMENT

This section, like the preceding sections, deals only with the transfer of the property in goods as between seller and buyer, and does not affect the protection afforded to innocent third parties by s. 25, *post*, and the Factors Acts (*l*), all of which prevail over the present section. COTTON, L.J.'s summary of the law upon which this section was largely based was as follows:

"In the case of such a contract [*i.e.*, a contract for the sale of unascertained goods], the delivery by the vendor to a common carrier, or, (unless the effect of the shipment is restricted by the terms of the bill of lading), shipment on board a ship of, or chartered for, the purchaser is an appropriation sufficient to pass the property. If, however, the vendor, when shipping the articles which he intends to deliver under the contract, takes the bill of lading to his own order, and does so not as agent, or on behalf of the purchaser, but on his own behalf, it is held that he thereby reserves to himself a power of disposing of the property, and that consequently there is no final appropriation, and the property does not on shipment pass to the purchasers. . . . If the vendor deals with, or claims to retain the bill of lading, in order to secure the contract price, as when he sends forward the bill of lading with a bill of exchange attached, with directions that the bill of lading is not to be delivered to the purchaser till acceptance or payment of the bill of exchange, the appropriation is not absolute, but until acceptance of the draft, or payment or tender of the price, is conditional only, and until such acceptance or payment or tender, the property in the goods does not pass to the purchaser" (*m*).

(*k*) *The Albazero*, [1974] 2 All E. R. 906, which has since been affirmed on appeal, as yet unreported, where the presumption was applied in the light of evidence of the seller's actual intention.

(*l*) *Cahn* v. *Pockett's Bristol Channel Co.*, [1899] 1 Q. B. 643, C. A.

(*m*) *Mirabita* v. *Imperial Ottoman Bank* (1878), 3 Ex. D. 164, at p. 172; see also at p. 170, *per* BRAMWELL, L.J. *Cf. Ex p. Banner* (1876), 2 Ch. D. 278; 34 Halsbury's Laws (3rd Edn.) 73.

This passage is however open to one observation: it does not cover the intermediate situation where the seller intends to reserve not his title, but simply a right to possession as security for the price, by taking the bill of lading in his own name (*n*).

The question whether any and if so what right of disposal is reserved is a question of fact (*o*). Relevant factors would include the seller's wishes (i) to secure payment or (ii) to raise money by pledge of shipping documents or (iii) to reserve the right to deliver or direct the cargo to another customer (*p*).

After loss of both title and the right to possession (*e.g.* after transfer of the shipping documents to the buyer), the unpaid seller may still be able to exercise the right of stoppage in transitu if the buyer becomes insolvent (*q*) but subject to the rights of any bonâ fide sub-buyer or pledgee from the buyer (*r*).

20. Risk *primâ facie* passes with property.—Unless otherwise agreed (*s*), the goods remain at the seller's risk until the property therein is transferred to the buyer, but when the property therein is transferred to the buyer, the goods are at the buyer's risk whether delivery has been made or not (*t*).

Provided that where delivery has been delayed through the fault of either buyer or seller the goods (*u*) are at the risk of the party in fault as regards any loss

---

(*n*) *The Parchim*, [1918] A. C. 157, H. L., and other cases considered, p. 35, *ante*.

(*o*) See p. 35, *ante*.

(*p*) See *The Albazero*, Illustration (3), *ante*.

(*q*) *Schotsman* v. *Lancs and Yorkshire Rail Co.* (1867), L. R. 2 Ch. App. 332, at p. 337; *Lyons* v. *Hoffnung* (1890), 15 App. Cas. 391, H. L. See further, s. 45, *post*.

(*r*) See the proviso to s. 47, *post*.

(*s*) *Martineau* v. *Kitching* (1872), L. R. 7 Q. B. 436; *Castle* v. *Playford* (1872), L. R. 7 Exch. 98; *Inglis* v. *Stock* (1885), 10 App. Cas. 263 (f.o.b. contract).

(*t*) For examples of seller's risk, see *Head* v. *Tattersall* (1871), L. R. 7 Exch. 7, see at p. 14; *Elphick* v. *Barnes* (1880), 5 C. P. D. 321, at p. 326 (death of horse delivered on sale or return). For examples of buyer's risk, see *Rugg* v. *Minett* (1809), 11 East, 210 (sale of goods in warehouse, sellers agreeing to pay rent for 30 days; goods destroyed by fire within this period); *Tarling* v. *Baxter* (1827), 6 B. &. C 360 (sale of standing hay, not to be cut by buyer before 1st May. Had seller undertaken to cut hay, property and risk would not have passed); *Sweeting* v. *Turner* (1871), L. R. 7 Q. B. 310 (risk of landlord distraining on goods sold); *cf. Widenmayer* v. *Burn Stewart & Co., Ltd.*, 1967 S. L. T. 129 (Ct. of Sess.) (a case of barter); 34 Halsbury's Laws (3rd Edn.) 79.

(*u*) " the goods " = contractual goods assembled by the seller for the purpose of fulfilling the contract and making delivery: *Demby Hamilton & Co.* v. *Barden*, [1949] 1 All E. R. 435.

which might not have occurred but for such fault (*v*).

Provided also that nothing in this section shall affect the duties or liabilities of either seller or buyer as a bailee [or custodier] of the goods of the other party (*w*).

ILLUSTRATIONS

(1) Furs delivered " on approval " with invoice. They are stolen by burglars. By the custom of the fur trade the goods are at the risk of the person ordering them on approval. The seller can recover the invoice price from the person to whom he delivered them (*a*).

(2) Sale of potatoes in clamp, delivery instructions to be given by buyer in 6 months, property to pass and payment to be completed on delivery. Seller undertakes to take reasonable care of the potatoes until delivery. The potatoes rot, although the seller takes reasonable care. Held: risk on buyer, who must pay for the potatoes as the risk and property are separated by the agreement (*b*).

(3) Sale of 120,000 gallons of white spirit out of 200,000 in a tank on wharf. No appropriation to the contract is made, but a delivery warrant is issued to the buyer, and the wharfingers attorn to the buyer. The warrant is not acted on for some months, and in the meantime the spirit deteriorates. The loss falls on the buyer (*c*).

(4) Agreement for sale of 30 tons of apple juice to be delivered weekly on buyer's instructions. Buyer fails to give instructions for last 10 tons, which seller has put in casks ready for delivery. The juice goes putrid as a result. Although property remains in the seller, the loss was the fault of the buyer and is at his risk (*d*).

Definitions.—For " bailee," " buyer," " delivery," " fault," " goods," " property " and " seller," see s. 62 (1), *post*.

Cross-references.—*Cf.* ss. 6 and 7, *ante* (goods perishing before sale); ss. 32 (2), (3) and 33, *post* (goods sent by carriers); s. 37,

(*v*) *Martineau* v. *Kitching* (1872), L. R. 7 Q. B. 436, at p. 456; *per* BLACKBURN, J., *Demby Hamilton & Co.* v. *Barden* (*supra*).

(*w*) *Cf. Head* v. *Tattersall* and *Elphick* v. *Barnes* (*supra*); *Poole* v. *Smith's Car Sales, Ltd.*, [1962] 2 All E. R. 482, at p. 489; [1962] 1 W. L. R. 744, at p. 753, *per* WILLMER, L.J.; *Wiehe* v. *Dennis Brothers* (1913), 29 T. L. R. 250 (pony left in possession of seller for a few days; seller liable as gratuitous bailee even if property and risk had passed).

(*a*) *Bevington* v. *Dale* (1902), 7 Com. Cas. 112.

(*b*) *Horn* v. *Minister of Food*, [1948] 2 All E. R. 1036.

(*c*) *Sterns* v. *Vickers*, [1923] 1 K. B. 78, C. A. The Court of Appeal left open the question whether property had passed, but it would seem that no legal property is acquired by a purchaser in such circumstances, although it is just possible that he might acquire an interest in equity in special circumstances; see further the comment to s. 16, *ante*. The finding that risk passed was based on the consideration that the seller had done all that he agreed to do, the wharfinger had attorned, and the buyers were free to take delivery when they wished. *Cf. J. J. Cunningham, Ltd.* v. *Robert Munro, Ltd.* (1922), 28 Com. Cas. 42 (bran deteriorates while waiting in lighters for buyer's ship; seller's risk, as buyer's contractual obligation to take delivery has not arrived); *Comptoir etc. du Boerenbond Belge* v. *Luis de Ridder*, [1949] 1 All E. R. 269; [1949] A. C. 293.

(*d*) *Demby Hamilton & Co.* v. *Barden*, [1949] 1 All E. R. 435.

post (neglect or refusal to take delivery); and s. 55 (1), *post* ("unless otherwise agreed").

History.—

The expression "*might* not have occurred" was substituted for "*would* not have occurred" in the first proviso at the instance of Lord WATSON. Although it is probably not true to say that it shifts the onus of proof on to the party in fault, it lessens the onus of proof on the innocent party (*d*).

COMMENT

"As a general rule," says BLACKBURN, J.,

> "*res perit domino*, the old Civil law maxim, is a maxim of our law, and, when you can show that the property passed, the risk of the loss *primâ facie* is in the person in whom the property is. If, on the other hand, you go beyond that and show that the risk attached to one person or the other, it is a very strong argument for showing that the property was meant to be in him. But the two are not inseparable. . . ." (*e*).

The parties may expressly agree that the risk shall pass to the buyer before the property in the goods passes or that the property shall pass before the risk.

Such an agreement may also be implied from the terms of the contract and the surrounding circumstances. Thus, risk may pass to the buyer before the property passes and even before the goods are delivered where the contract provides that the seller should take reasonable care of the goods until delivery (*f*), where the goods are an unascertained part of a larger bulk and the buyer has every opportunity to take delivery but elects not to (*g*), in f.o.b. contracts where property is reserved after shipment (*h*), and in c.i.f. contracts (*i*).

Risk will also frequently pass to the buyer once the goods are delivered to him even though the property in the goods is to remain in the seller until the price is paid (*k*). But the risk does not normally pass to a potential buyer where goods are delivered on sale or return (*l*).

(*d*) *Denby Hamilton & Co.* v. *Barden*, [1949] 1 All E. R. 435.

(*e*) *Martineau* v. *Kitching* (1872), L. R. 7 Q. B. 436, at pp. 454, 456; discussed and approved, *The Parchim*, [1918] A. C. 157, at p. 168, P. C. BLACKBURN, J.'s citation of the maxim *res perit domino* is misleading as to Roman Law, under which, as a general rule, risk passed as soon as the parties were agreed on a specific article and the price, but property passed only on delivery, and sometimes not even then: see Moyle's *Justinian* (5th Edn.) p. 434.

(*f*) *Horn* v. *Minister of Food*, Illustration (2), *ante*.

(*g*) *Stern* v. *Vickers*, Illustration (3), *ante*.

(*h*) P. 34-35, *ante*, and see the comment to s. 19, *ante*.

(*i*) P. 37, *ante*.

(*k*) *Cf. Comptoir d'Achat et de Vente du Boerenbond Belge S.A.* v. *Luis de Ridder*, [1949] 1 All E. R. 269; [1949] A. C. 293, H. L.; and see *Benjamin's Sale of Goods*, para. 405. Even where risk does not pass, the buyer will owe a duty of care as bailee of the goods—see *post*.

(*l*) *Cf. Bevington* v. *Dale*, Illustration (1), *ante*, where a trade custom to the contrary was proved.

In the absence of an express agreement to that effect, risk will rarely remain with the seller once the property in the goods has passed to the buyer except to a limited extent under the proviso to this section, where the seller is in default in delivering the goods, and under the special provisions of s. 32 (3), *post,* where transit by sea is involved (*m*).

Where the buyer is entitled to reject goods because of a breach of contract by the seller, and does reject them, so that the property in them revests in the seller, it would seem that the seller must bear any loss arising from damage to or deterioration of the goods even while the goods were temporarily the property of the buyer (*n*), provided at least that the loss was not the fault of the buyer.

Duties of bailee.—When the seller remains in possession of the goods after the property in them has passed to the buyer, or when the buyer gets possession of the goods before the property passes, as in the case of goods on trial, the party in possession is in each case a bailee (*o*).

The duty of a bailee is to take reasonable care of the goods, and to have them available undamaged for delivery up when agreed or required unless he is prevented from doing so without any fault on his or his servants' part (*p*). The onus is upon the bailee to prove that any loss or damage to goods occured without any failure on his part to take reasonable care (*q*). Further, in the case of a bailment under contract (as where a buyer is in possession of the goods, but the seller retains the property in them or *vice versâ*) the bailee's promise is that reasonable care will be taken by himself, his servants or agents (*r*).

The absence of any reward is only one of the issues taken into account in assessing the degree of care required. Reward in this sense exists where by an agreement of sale one or other of the parties is to have possession without property even though no specific sum is to be paid for such custody.

A buyer who rightfully rejects goods is an involuntary bailee if the seller refuses to retake, and is therefore perhaps only under a duty to avoid wilful or reckless damage (*s*).

A seller in possession under s. 37 following a buyer's failure to

---

(*m*) There is no reported authority, but it may be that a similar principle to that in s. 32 (3) will also apply where transit by land or air is involved.

(*n*) *Head* v. *Tattersall* (1871), L. R. 7 Exch. 7.

(*o*) See *Wiehe* v. *Dennis Bros.* (1913), 29 T. L. R. 250.

(*p*) *Houghland* v. *R. R. Low* (*Luxury Coaches*), *Ltd.*, [1962] 2 All E. R. 159; [1962] 1 Q. B. 694, C. A.; *Morris* v. *C. W. Martin & Sons, Ltd.*, [1965] 2 All E. R. 725; [1966] 1 Q. B. 716, C. A.

(*q*) *Ibid.*; *Coldman* v. *Hill*, [1919] 1 K. B. 443.

(*r*) *B. R. S.* v. *Arthur Crutchley & Co., Ltd.*, [1968] 1 All E. R. 811, C. A.; *Morris* v. *C. W. Martin & Sons, Ltd.*, *supra*, at pp. 731 and 728 respectively.

(*s*) *Howard* v. *Harris* (1884), 1 Cab. & El. 253; but see *Newman* v. *Bourne and Hollingsworth* (1915), 31 T. L. R. 209.

collect within a reasonable time, however, probably owes a duty of reasonable care since he has a right to charge for his care and custody.

Passing of risk but not property.—A purchaser to whom the property in the goods has not yet passed and who is not entitled to possession of them has been held to have no right of action in negligence against a third party who carelessly damages or destroys them even though the risk has passed to him (*t*).

As property and risk are separable, it follows that property and insurable interest may be separable. If goods are at a person's risk, he has an insurable interest, whether he has any property in the goods or not (*u*). And if he has the property in the goods, he has an insurable interest whether they are are at his risk or not (*v*). A provision in a contract that the risk shall not pass until a time necessarily later than the time when property would normally pass, provides a powerful indication that the property in the goods was not intended by the parties to pass until the later date (*w*).

Accessories or accretion.—The converse of the rule *res perit domino* also holds good, and any fruits or increase of the thing sold belong *primâ facie* to the party who has the property in it.

> " Any calamity befalling the goods after the sale is completed must be borne by the purchaser, and, by parity of reasoning, any benefit to them is his benefit, and not that of the vendor " (*x*).

Transfer of Title

21. Sale by person not the owner.—(1) Subject to the provisions of this Act (*a*), where goods are sold by a person who is not the owner thereof, and who does not sell them under the authority or with the consent of the owner, the buyer acquires no better title to the goods than the seller had (*b*), unless the

---

(*t*) *Margarine Union G.m.b.H.* v. *Cambay Prince S.S. Co., Ltd.*, [1967] 3 All E. R. 775; [1969] 1 Q. B. 219. Such a purchaser will commonly have contractual rights see s. 1 of the Bills of Lading Act, 1855, *post,* Appendix I, and " *The Dona Mari* ", [1973] 2 Lloyd's Rep. 366.

(*u*) As to the insurable interest of seller and buyer respectively, see *Anderson* v. *Morice* (1876), 1 App. Cas. 713; *Colonial Ins. Co.* v. *Adelaide Marine Insurance Co.* (1886), 12 App. Cas. 128, at p. 135; Chalmers' *Marine Insurance Act 1906* (5th ed.), pp. 11 *et seq.*

(*v*) *Inglis* v. *Stock* (1885) 10 App. Cas. 263, at p. 270.

(*w*) *President of India* v. *Metcalfe Shipping Co., Ltd.*, [1969] 1 All E. R. 861; [1969] 2 Q. B. 123; on appeal, [1969] 3 All E. R. 1549; [1970] 1 Q. B. 289, C. A.

(*x*) *Sweeting* v. *Turner* (1871), L. R. 7 Q. B. 310, at p. 313, *per* BLACKBURN, J.

(*a*) See ss. 22 to 25, *post.*

(*b*) For principle, see *Colonial Bank* v. *Whinney* (1886), 11 App. Cas. 426, at pp. 435, 436, *per* Lord BLACKBURN. For illustrations, see *Cooper* v. *Willomatt* (1845), 1 C. B. 672 (fraudulent sale by bailee) ; *Lee* v. *Bayes* (1856), 18 C. B. 599 (stolen goods sold by auction) ; *Consolidated Co.* v. *Curtis & Son* (1892), 1 Q. B. 495, 498 (goods included in bill of sale sold by auction) ; *The*

owner of the goods is by his conduct precluded from denying the seller's authority to sell.

(2) Provided also that nothing in this Act shall affect—

(a) The provisions of the Factors Acts, or any enactment enabling the apparent owner of goods to dispose of them as if he were the true owner thereof (*c*);

(b) The validity of any contract of sale under any special common law or statutory power of sale or under the order of a court of competent jurisdiction (*d*).

Telegrafo (1871), L. R. 3 P. C. 673, at p. 685 (goods taken by piracy); *Cundy* v. *Lindsay* (1878), 3 App. Cas. 459 (goods obtained by fraud and resold); *Helby* v. *Matthews*, [1895] A. C. 471 (wrongful disposition by hirer under hire-purchase agreement). But if a person acquires goods in good faith and spends money in improving them, the true owner will not be able to recover them without giving credit for the improvements; *Greenwood* v. *Bennett*, [1972] 3 All E. R. 586; [1973] 1 Q. B. 195, C. A. For a criticism of this case, see 1973 J. B. L. 64.

(*c*) See the Factors Act, 1889, the Factors (Scotland) Act, 1890, and the Bills of Lading Act, 1855, *post*; the Bankruptcy Act, 1914, s. 38 and 38A (reputed ownership); 3 Halsbury's Statutes (3rd Edn.) 87; as limited by the Hire-Purchase Act, 1965, s. 53; 30 Halsbury's Statutes (3rd Edn.) 61; see also the Hire-Purchase Act, 1964, ss. 27–29, which enable a *bonâ fide* private purchaser of a car which is the subject of a hire-purchase agreement to obtain a good title; and for certain purposes the Bills of Sale Act, 1878, *post*, Appendix I; *Lamb* v. *Wright & Co.*, [1924] 1 K. B. 857 (reputed ownership in case of hire-purchase agreement).

(*d*) As to pawnee, see *Martin* v. *Reid* (1862), 11 C. B., N.S., 730, at p. 736, *per* WILLES, J.; *Pigot* v. *Cubley* (1864), 15 C. B., N.S., 701. As to the power of a carrier to sell as agent of necessity, see *Sims & Co.* v. *Midland Railway*, [1913] 1 K. B. 103 (perishable goods). As to distrainor, see Woodfall's *Landlord and Tenant* (26th ed.), 436 *et seq.* As to sheriff, see s. 15, Bankruptcy and Deeds of Arrangement Act, 1913, as explained in *Curtis* v. *Maloney* [1950] 2 All E. R. 982; [1951] 1 K. B. 736, C. A., and *Dyal Singh* v. *Kenyan Insurance, Ltd.*, [1954] A. C. 287, P. C. (sales under execution); *Doe* v. *Donston* (1818), 1 B. & Ald. 230 (sale after expiration of office); *cf. Batchelor* v. *Vyse* (1834), 4 M. & Sc. 552 (excessive sale): *Manders* v. *Williams* (1849), 4 Exch. 339, (sale by sheriff of goods on sale or return); *Jones Brothers* v. *Woodhouse*, [1923] 2 K. B. 117 (sale by sheriff of goods on hire-purchase). As to master of ship, see *The Gratitudine* (1801), 3 Ch. Rob. 240, at p. 259, and *Kaltenbach* v. *Mackenzie* (1878), 3 C. P. D. 467, at p. 473. As to order of Court, see R. S. C., O. 29, r. 4. As to goods left with innkeeper, see the Innkeepers Act, 1878; 17 Halsbury's Statutes (3rd Edn.) 815; the Hotel Proprietors Act, 1956; 17 Halsbury's Statutes (3rd Edn.) 816; and *Chesham* v. *Beresford Hotel* (1913), 29 T. L. R. 584; *Marsh* v. *Commissioner of Police*, [1944] 2 All E. R. 393; [1945] K. B. 43. As to executor or administrator, see Williams and Mortimer: *Executors Administrators and Probate*, pp. 615–629. As to trustee in bankruptcy, see ss. 55, 56 of the Bankruptcy Act, 1914; 3 Halsbury's Statutes (3rd Edn.) 106, 108. As to sale by warehouseman for non-payment of his charges, see *Willets* v. *Chaplin & Co.* (1923), 39 T. L. R. 222.

ILLUSTRATIONS

Sub-section (1).—(1) A, a timber merchant, instructs the dock company with whom his timber is warehoused to accept delivery orders signed by his clerk. The clerk has a limited authority to sell to known customers. The clerk in an assumed name sells some of the timber to B, who knows nothing of A, or of the clerk under his real name. The clerk carries out the fraud by giving the dock company orders for the transfer of timber into his assumed name, and then in that name giving delivery orders to B. A can recover the value of the timber from B since the clerk, having no title to the timber, could give none to B. A has not held the clerk out to B as having his authority to sell, so that no question of estoppel can arise (*e*).

(2) A cotton broker innocently buys certain bales of cotton from a person who has fraudulently obtained them and who has himself no title to them, thinking they may suit an old customer. He afterwards sells and delivers the cotton to that customer. He is liable to the true owner for the conversion of the cotton (*f*).

(3) X wishes to obtain an advance on the security of his car. He goes to a car dealer and signs a form by which he makes an application to a finance company to hire the car on H.P. terms. The form contains a statement that the car is the sole unencumbered property of the car dealer. The car dealer then sells the car to the finance company. X is estopped from denying the car dealer's title to sell the car as its owner (*g*).

(4) M, a car dealer, displays a number of cars for sale. To the knowledge of P, another car dealer, some of these cars are the property of a finance company which has authorised M to sell them. Unknown to P this authority is withdrawn. M sells cars to P otherwise than in the ordinary course of business, representing them to be his own unencumbered property. They are in fact the property of the finance company. The finance company is not estopped from denying M's title to the cars (*h*).

Sub-section (2).—(3) The high bailiff of a County Court seizes certain goods under a warrant of execution. X, who is the true owner, claims them, but does not make the required deposit. The high bailiff sells them, and the price is duly paid into court. The purchaser acquires a good title under s. 135 of the County Courts Act, 1959 (*i*).

(*e*) *Farquharson Brothers* v. *King & Co.*, [1902] A. C. 325.

(*f*) *Hollins* v. *Fowler* (1875), L. R. 7 H. L. 757; see s. 23, *post*, as to voidable titles.

(*g*) *Stoneleigh Finance* v. *Phillips*, [1965] 1 All E. R. 513; [1965] 2 Q. B. 537, C. A.; *cf. Mercantile Credit* v. *Hamblin*, [1964] 3 All E. R. 592; [1965] 2 Q. B. 242, C. A.

(*h*) *Motor Credits* (*Hire Finance*) *Ltd.* v. *Pacific Motors Pty., Ltd.* (1963), 109 C. L. R. 87; reversed on another ground; *sub nom. Pacific Motor Auctions Proprietary, Ltd.* v. *Motor Credits* (*Hire Finance*), *Ltd.*, [1965] 2 All E. R. 105; [1965] A. C. 867.

(*i*) *Goodlock* v. *Cousins*, [1897] 1 Q. B. 558, C. A. (a decision under s. 156 of the County Courts Act, 1888, now replaced by the Act above referred to: 7 Halsbury's Statutes (3rd Edn.) 389).

(4) Consignment of perishable goods (tomatoes) from Jersey to London. Delivery is interrupted at Weymouth by a railway strike. The Railway Company sells the goods at Weymouth without communicating with the London buyers. The company is liable in damages, as in these circumstances it is not an agent of necessity (*k*).

Definitions.—For "contract of sale," see s. 1 (1), *ante*, and s. 62 (1), *post*. For "buyer," "Factors Acts," "goods," and "seller," see s. 62 (1), *post*.

Cross-references.—*Cf.* s. 12, *ante* (seller's warranty of title, etc.); and ss. 22 to 25, *post* (transfer of title).

COMMENT

Sub-section (1).—"The general rule of law," says WILLES, J., "is undoubted, that no one can transfer a better title than he himself possesses. *Nemo dat quod non habet*" (*l*).

In a case under the Factors Act, 1842, BLACKBURN, J., says: "At common law a person in possession of goods could not confer on another, either by sale or by pledge, any better title to the goods than he himself had. To this general rule there was an exception of sales in market overt [*see now* s. 22, *post*], and an apparent exception where the person in possession had a title defeasible on account of fraud [*see now* s. 23, *post*]. But the general rule was that, to make either a sale or a pledge valid against the owner of the goods sold or pledged, it must be shown that the seller or pledger had authority from the owner to sell or pledge, as the case might be. If the owner of the goods had so acted as to clothe the seller or pledger with apparent authority to sell or pledge, he was at common law precluded, as against those who were induced *bonâ fide* to act on the faith of that apparent authority, from denying that he had given such an authority, and the result as to them was the same as if he had really given it. But there was no such preclusion as against those who had notice that the real authority was limited" (*m*).

No title by finding.—The finder of lost goods acquires no title as against the true owner, and can give none except by sale in market overt.

"If a person leaves a watch or a ring on a seat in the park or on a table at a café, and it ultimately gets into the hands of a *bonâ fide* purchaser, it is no answer to the true owner to say that it was his carelessness, and nothing else, that enabled the finder to pass it off as his own" (*n*).

(*k*) *Springer* v. *Great Western Rail Co.*, [1921] 1 K. B. 257, C. A. See note on agent of necessity, *post*.

(*l*) *Whistler* v. *Forster* (1863), 14 C. B., N.S., 248, at p. 257; *Banque Belge* v. *Hambrouck*, [1921] 1 K. B. 321, at p. 329, C. A.

(*m*) *Cole* v. *North Western Bank* (1875), L. R. 10 C. P. 354, at p. 362; approved *Colonial Bank* v. *Whinney* (1886), 11 App. Cas. 426, at pp. 435, 436 (reputed ownership); *cf. City Bank* v. *Barrow* (1880), 5 App. Cas. 664, at p. 677, as to Roman and old French law, and Canadian law.

(*n*) *Farquharson Brothers* v. *King & Co.*, [1902] A. C. 325, at p. 336. But a finder has a good title against everyone except the true owner: *Armory* v. *Delamirie* (1721), 1 Str. 505. See the law fully discussed in *Hannah* v. *Peel*, [1945] 2 All E. R. 288; [1945] K. B. 509.

The saying that " findings are keepings " is not a maxim of the law. A person who finds goods and converts them to his own use, having reason to believe, at the time of finding, that the owner can be discovered by taking reasonable steps, may be guilty of theft (*o*). Presumably if goods are voluntarily abandoned, anyone taking possession becomes the owner in the full sense of the term, as was the case in the Civil Law (*p*).

Estoppel.—It sometimes happens that a man is precluded by his conduct from denying the truth of a particular state of affairs.

> " Where one man by his words or conduct wilfully causes another to believe in the existence of a certain state of things, and induces him to act on that belief, so as to alter his own previous position, the former is concluded from averring against the latter a different state of things as existing at the same time " (*q*).

Moreover, as PARKE, B., said in 1848,

> " By the term wilfully . . . we must understand . . . at least that he means his representation to be acted upon, and that it is acted upon accordingly; and if, whatever a man's real meaning may be, he so conducts himself that a reasonable man would take the representation to be true, the party making the representation would be equally precluded from contesting its truth; and conduct by negligence or omission when there is a duty cast upon a person by usage of trade or otherwise to disclose the truth may often have the same effect " (*r*).

Applying this rule to the circumstances envisaged by the present sub-section, the question to be asked is:

> " whether the true owner of the goods has so invested the person dealing with them with the indicia of property as that when an innocent person enters into a negotiation with the person to whom these things have been entrusted with the indicia of property the true owner of the goods cannot afterwards complain that there was no authority to make such a bargain " (*s*).

(*o*) Under the Theft Act, 1968; 8 Halsbury's Statutes (3rd Edn.) 782; as to what amounts to conversion by finder, see *Hollins* v. *Fowler* (1875), L. R. 7 H. L. 757, at p. 766.

(*p*) *Inst. II*, 1, 47 ; but see *Hibbert* v. *McKiernan*, [1948] 1 All E. R. 860, D. C. ; [1948] 2 K. B. 142.

(*q*) *Pickard* v. *Sears* (1837), 6 Ad. & El. 469, at p. 474.

(*r*) *Freeman* v. *Cooke* (1848), 18 L. J. Ex. 114, 119; and see *Gregg* v. *Wells* (1839), 10 Ad. & El. 90; *Cornish* v. *Abington* (1859), 4 H. & N. 549; *Wester Moffat Colliery Co.* v. *Jeffrey & Co.*, 1911, S. C. 346 (Ct. of Sess.); *Seton* v. *Lafone* (1887), 19 Q. B. D. 68, at p. 72, C. A.

(*s*) *Henderson & Co.* v. *Williams*, [1895] 1 Q. B. 521, at p. 525, *per* Lord HALSBURY; *Knights* v. *Wiffen* (1870), L. R. 5 Q. B. 660; but *cf. Farquharson Brothers* v. *King & Co.*, [1902] A. C. 325; *Mercantile Bank of India Ltd.* v. *Central Bank of India, Ltd.*, [1938] 1 All E. R. 52; [1938] A. C. 287, P. C., not following *Commonwealth Trust* v. *Akotey*, [1926] A. C. 72, P. C.; *Jerome* v. *Bentley*, [1952] 2 All E. R. 114. And see *Wah Tat Bank, and Overseas Chinese Banking Corporation, Ltd.* v. *Chan Chen Kum and Hua Siang Steamship Co.*, [1967] 2 Lloyd's Rep. 437; on appeal, [1971] 1 Lloyd's Rep. 439, P. C.

It is a question of fact whether the true owner has so invested the seller with the indicia of property. It is not enough that the true owner has carelessly allowed them to come into the hands of the seller (*t*). There must normally be a representation by the owner either that he has no interest in the goods (*u*), or that the seller has authority to sell the goods on his behalf in the manner in which he purports to (*v*), or that the seller is the owner of the goods (*a*) or is otherwise entitled to dispose of them, and that representation must be acted upon by the buyer to his detriment. In *Mercantile Credit Co., Ltd.* v. *Hamblin* (*b*) the Court of Appeal went further and held that in certain circumstances the owner of goods may owe a duty to at least some potential buyers not to put the seller into a position where he can hold himself out as having authority to sell. On the peculiar facts of that case, however, the Court held that the owner had not been negligent and that in any event the seller's fraud was not a reasonable foreseeable consequence of her acts (*c*).

The nature of the estoppel may be important. If the true owner has represented that the seller is owner of the goods, then the buyer will be able to rely on this representation whatever the circumstances of the sale, provided, of course, that he acts in good faith (*d*). If, on the other hand, the true owner has only represented that the seller

See also *Gillett* v. *Hill* (1834), 2 Cr. & M. 530 (wharfinger accepting delivery order estopped from denying he had the goods); *Laurie and Morewood* v. *Dudlin & Sons*, [1926] 1 K. B. 223 (warehouseman not estopped by mere receipt of delivery order); *Coventry* v. *Gt. Eastern Rail Co.* (1883), 11 Q. B. D. 776 (two delivery orders for same goods); *Cohen* v. *Mitchell* (1890), 25 Q. B. D. 262, C. A. (sale by bankrupt, trustee not intervening); *Waller* v. *Drakeford* (1853), 1 E. & B. 749; *Patterson Motors* v. *Riley* (1966), 56 D. L. R. (2d.) 278; 18 L. Q. R. 159; Atiyah, *Sale of Goods* (4th ed.), p. 181; Spencer Bower & Turner, *Estoppel by Representation* (2nd ed.), p. 69.

(*t*) *Farquharson Brothers* v. *King & Co.*, [1902] A. C. 325, P. C.; *Jerome* v. *Bentley*, [1952] 2 All E. R. 114.

(*u*) Thus, if Hire Purchase Information, Ltd. informs an inquirer that none of the finance companies who are members of it have any hire purchase interest in a particular car, it has been held to give that information on behalf of all its member finance companies, who are bound by it; *Moorgate Mercantile Co. Ltd.* v. *Twitchings* (1975), *Times*, 19th June, C. A. Leave to appeal was given.

(*v*) *Motor Credit* (*Hire Finance*), *Ltd.* v. *Pacific Motors Pty. Ltd.* (1963), 109 C. L. R. 87 (*ante*, illustration (4)); reversed on another ground, *sub nom. Pacific Motor Auctions Proprietary, Ltd.* v. *Motor Credits* (*Hire Finance*), *Ltd.*, [1965] 2 All E. R. 105; [1965] A. C. 867, P. C.; *Spencer* v. *North Country Finance Co.*, [1963] C. L. Y. 212.

(*a*) *Eastern Distributors, Ltd.* v. *Goldring*, [1957] 2 All E. R. 525; [1957] 2 Q. B. 600, C. A.; *Stoneleigh Finance, Ltd.* v. *Phillips*, [1965] 1 All E. R. 513; [1965] 2 Q. B. 537, C. A. (*ante*, illustration (3)); *Lloyds and Scottish Finance, Ltd.* v. *Williamson*, [1965] 1 All E. R. 641; [1965] 1 W. L. R. 404, C. A.; *cf. Mercantile Credit Co., Ltd.* v. *Hamblin*, [1964] 3 All E. R. 592; [1965] 2 Q. B. 242, C. A. If A gives B his car together with his registration book he does not thereby hold out B as having authority to sell the car: *Central Newbury Car Auctions* v. *Unity Finance*, [1956] 3 All E. R. 905; [1957] 1 Q. B. 371, C. A.

(*b*) [1964] 3 All E. R. 592; [1965] 2 Q. B. 242, C. A.

(*c*) But see *General & Finance Facilities* v. *Hughes* (1966), 110 Sol. Jo. 847.

(*d*) *Lloyds and Scottish Finance, Ltd.* v. *Williamson*, [1965] 1 All E. R. 641; [1965] 1 W. L. R. 404, C. A.

had authority to sell the goods as his agent then he will be able to recover them if they are sold in a manner not covered by the seller's ostensible authority as agent (*e*).

Moreover, in *Spiro* v. *Lintern* (*f*), the Court of Appeal held that if the true owner becomes aware, even after a contract has been entered into, that some other person has contracted to sell his property, he owes a duty of care to the purchaser to disclose to him the fact that that other person was not entitled so to act. If he does not disclose the true position to the purchaser, then he may be held to have represented to the purchaser by his conduct that the other person was entitled to act as he did, and if as a result the purchaser acts in a manner which would be to his disadvantage if the true owner were thereafter to assert his rights, then the true owner will be estopped from so asserting his rights. *Spiro* v. *Lintern* was a case concerned with the sale of a house, but the same principle would seem to apply in the case of a sale of goods.

By virtue of this sub-section, where the owner is precluded by his conduct from denying the seller's authority to sell, the buyer acquires a good title to the goods and not merely a right to plead an estoppel (*g*).

In *Snook* v. *London and West Riding Investments, Ltd.* (*h*), the defendants were allowed to rely on an estoppel although it was not expressly pleaded, all necessary facts being proved.

Sub-section (2).—*Special power.*—One person is sometimes invested by law with a special power to dispose of another person's property. For instance, a pawnbroker may sell unredeemed pledges; and a landlord who has duly distrained for rent may sell the goods so distrained. See the authorities collected in the footnote to sub-s. (2) (b), *ante*.

Agent of necessity.—It has always been held that the master of a ship may, in case of necessity, dispose of the ship and cargo and it has now been held that

> "the law as to the power of sale and the duty to take care of the goods which is laid down in the case of a carrier by sea applies to a carrier by land where the necessary conditions giving rise to such a power and duty exist,"

and

(*e*) *Motor Credit* (*Hire Finance*) *Ltd.* v. *Pacific Motors Pty. Ltd.* (1963), 109 C. L. R. 87 (*ante*, illustration (4)); reversed on another ground, *sub nom. Pacific Motor Auctions Proprietary, Ltd.* v. *Motor Credits* (*Hire Finance*), *Ltd.*, [1965] 2 All E. R. 105; [1965] A. C. 867, P. C.; *Davey* v. *Paine Brothers* (*Motors*), *Ltd.*, [1954] N. Z. L. R. 1122. *Raffoul* v. *Essanda, Ltd.* (1970), s. 72 S. R. (N. S. W.) 633.

(*f*) [1973] 3 All E. R. 319; [1973] 1 W. L. R. 1002, C. A.

(*g*) *Eastern Distributors Ltd.* v. *Goldring*, [1957] 2 All E. R. 525; [1957] 2 Q. B. 600, C. A.; *Mercantile Credit Co., Ltd.* v. *Hamblin*, [1964] 3 All E. R. 592; [1965] 2 Q. B. 242, C. A.

(*h*) [1967] 1 All E. R. 518; [1967] 2 Q. B. 786, C. A.

"the conditions necessary . . . are—(1) that a real necessity must exist for the sale, and (2) that it must be practically impossible to get the owner's instructions in time as to what shall be done" (*i*).

The exact limits of the doctrine are uncertain, but the only cases in which agency of necessity has been held to arise are cases in which there has been a previously existing contractual relationship between the parties giving rise to some duty on the one to take care of the other's goods.

Dicta favouring an extension of the doctrine to cases in which no such duty could be implied have been disapproved (*k*).

Co-owners.—The law relating to co-owners, who are not partners, is rather obscure. Probably a co-owner, in the absence of estoppel or authority from the other co-owners, could only transfer his own share (*l*). If one co-owner sells and retains the whole price, the remedy of the others at law against him is doubtful, unless the sale confers a good title to the whole. If co-owners cannot agree as to the possession or use of the goods owned in common the only remedy in equity was to apply for an injunction or for a receiver and sale (*m*), but now by s. 188 of the Law of Property Act, 1925 (*n*) the court has power to order a division on the application of persons interested in a moiety or upwards of such chattels. In a case in 1892, A sold a half-share in a gold snuff-box to B, on the terms that A was to retain possession till resale on joint account, and afterwards handed the box to B to sell it at Christie's. B, instead of selling, deposited the box with H, to whom he owed money. It was held that A could recover the box from H (*o*).

It seems that if two or more persons agree to purchase goods on joint account, notice to the one who effects the purchase of any defect in title of the seller affects the others also (*p*). Further,

"Any partner, co-adventurer, agent, or any other person in any position of trust must make no profit out of his position without the knowledge of his principals, and in particular cannot sell his own

(*i*) *Sims & Co.* v. *Midland Railway*, [1913] 1 K. B. 103, at p. 112; approved in *Springer* v. *Great Western Rail. Co.*, [1921] 1 K. B. 257. See also *Sachs* v. *Miklos*, [1948] 1 All E. R. 67, C. A.; [1948] 2 K. B. 23; *Munro* v. *Willmott*, [1948] 2 All E. R. 983; [1949] 1 K. B. 295.

(*k*) *Jebara* v. *Ottoman Bank*, [1927] 2 K. B. 254, at pp. 270, 271, commenting on *Prager* v. *Blatspiel*, [1924] 1 K. B. 566, where McCardie, J., reviewed the cases and favoured an extension of the doctrine.

(*l*) See *re Tamplin, Ex p. Barnett* (1890), 7 Morr. 70. As to partners, who *primâ facie* are agents for each other, see the Partnership Act, 1890, ss. 5 and 6; 24 Halsbury's Statutes (3rd Edn.) 505, 506. See also *Mann* v. *D'Arcy*, [1968] 2 All E. R. 172; [1968] 1 W. L. R. 893; *R.* v. *Bonner*, [1970] 2 All E. R. 97; [1970] 1 W. L. R. 838, C. A.

(*m*) *Lindley on Partnership* (9th ed.), pp. 38–41. There are very special conditions as to ships; see *ibid.*, p. 37; *cf. Nicol* v. *Cooper* (1896), 1 Com. Cas. 410 (sale of ship by managing owner, rights of dissenting shareholder).

(*n*) 27 Halsbury's Statutes (3rd Edn.) 605.

(*o*) *Nyberg* v. *Handelaar*, [1892] 2 Q. B. 202, C. A.

(*p*) *Oppenheimer* v. *Frazer*, [1907] 2 K. B. 50, 76, C. A.

goods to his principals without fully disclosing his own personal interest in the matter" (*q*).

As to co-adventurers in Scotland, see Green's *Encyclopædia of Scots Law*, tit. " Co-adventurers."

22. Market overt.—(1) Where goods are sold in market overt, according to the usage of the market, the buyer acquires a good title to the goods, provided he buys them in good faith and without notice of any defect or want of title on the part of the seller (*r*).

[(2) *Nothing in this section shall affect the law relating to the sale of horses.—Repealed.*](*s*).

(3) The provisions of this section do not apply to Scotland (*t*).

ILLUSTRATIONS

(1) Stolen sheep are brought to an auctioneer to sell, and are sold by him in market overt. The buyer is protected, but the auctioneer is liable for conversion (*u*).

(2) A motor-car is put up for auction in a statutory market by its bailee under a hire-purchase agreement. It remains unsold, but the bailee later sells it by private treaty in the same market, as is customary there. The buyer gets a good title, as the rules of market overt can apply in a market established by statutory powers and the sale need not be by a trader (*a*).

(3) Two stolen antique chandeliers are offered for sale in market overt on 21st October. One is sold at 4.30 p.m. on the same afternoon to a *bona fide* purchaser, having been exposed on the stall for only a few minutes. The other is exposed daily until 21st December when it is sold to another *bona fide* purchaser. The first sale was before sunset.

(*q*) *Kuhlirz* v. *Lambert Bros., Ltd.* (1913), 108 L. T. 565, at p. 567 (co-adventure in sale of coal to Austrian Government).

(*r*) *The Case of Market Overt* (1596), 5 Co. Rep. 83 b ; Tudor's *Merc. Cases* (3rd ed.), p. 274, and notes ; distinguished, *Bishopsgate Motor Finance Co.* v. *Transport Brakes, Ltd.*, [1949] 1 All E. R. 37, C. A. ; [1949] 1 K. B. 322 ; and see *Crane* v. *London Dock Co.* (1864), 5 B. & S. 313 ; *per* BLACKBURN, J., at p. 320, as to the usage of the market ; *cf. Vilmont* v. *Bentley* (1886), 18 Q. B. D. 322, at p. 331. See also 31 L. Q. R. 270 as to the City of London.

(*s*) Criminal Law Act, 1967, Sch. 3, Pt. III; 8 Halsbury's Statutes (3rd Edn.) 567.

(*t*) See *Todd* v. *Armsur*, 1882 9 R. (Ct. of Sess.) 901, as to conflict of laws, when it was held that a title acquired as the result of a sale in market overt in England will be recognised in Scotland. For Scottish cases as to horses, see Green's *Encyclopædia of Scots Law*, tit. " Horses."

(*u*) *Delaney* v. *Wallis* (1884), 15 Cox C. C. 525.

(*a*) *Bishopsgate Motor Finance Co.* v. *Transport Brakes, Ltd.*, [1949] 1 All E. R. 37, C. A. ; [1949] 1 K. B. 322.

The second sale was after sunset. The first buyer gets a good title; the second buyer does not (*b*).

Definitions.—For "buyer," "goods," and "seller", see s. 62 (1), *post*; for "good faith," see s. 62 (2), *post*.

Cross-references.—*Cf*. s. 12, *ante* (seller's warranty of title, etc.); and s. 21, *ante*, and ss. 23 to 25, *post* (transfer of title).

COMMENT

Sub-section (1).—The rules of market overt do not apply in Scotland or Wales (*c*), or the United States, and in England they only apply to a limited class of retail transactions.

All shops in the city of London are market overt, for the purposes of their own trade, provided that the goods are sold in the ordinary hours of business on business days between sunrise and sunset (*d*), and the whole transaction takes place in the open part of the shop. Thus a wharf in the city is not market overt (*e*), and a sale by sample is not within the custom, because the whole transaction must take place in the open market, and not merely the formation of the contract (*f*). So, too, a sale of jewellery to a tradesman in his showroom is not within the custom (*g*). Nor is the sale of a stolen watch in a first floor auction room where unredeemed pledges are sold periodically (*h*).

Outside the city of London, markets with the custom of market overt may exist by grant or prescription, and it seemed at one

---

(*b*) See *Reid* v. *Metropolitan Police Commissioner*, [1972] 2 All E. R. 97; [1973] Q. B. 551, C. A. The Court of Appeal appears to have arrived at this decision because it felt that despite the finding of the County Court Judge, the buyer had not been *bona fide*. The Court applied an ambiguous statement in *Co. Inst.*, Vol. II, pp. 713–714. In so doing it appears to have disregarded the wording of this section, and the principles of construing a codifying statute, set out at page 73, *ante*. The decision also runs contrary to the recommendations of the Law Reform Committee in its Report on the Transfer of Title to Chattels (Cmnd. 2958), that the law of market overt should be extended to all shops. It would also seem to lead to further anomalies in applying the law of market overt to many important markets, such as Covent Garden, which operates wholly or partly during the night. Such markets are, of course, generally well lit by electric lighting.

The other exceptions set out in *Co. Inst.* need not be referred to here, as they are embodied either in the text of this section or in the decisions based upon it, or are clearly no longer law.

(*c*) See The Laws in Wales Act, 1542, s. 47; 6 Halsbury's Statutes (3rd Edn.) 460.

(*d*) *Reid* v. *Metropolitan Police Commissioner*, [1973] 2 All E. R. 97; [1973] Q. B. 551, C. A.; see illustration (3), *ante*.

(*e*) *Wilkinson* v. *King* (1809), 2 Camp. 335.

(*f*) *Crane* v. *London Dock Co.* (1864), B. & S. 313.

(*g*) *Hargreave* v. *Spink*, [1892] 1 Q. B. 25; *cf. Ardath Tobacco Co.* v. *Ocker* (1931), 47 T. L. R. 177.

(*h*) *Clayton* v. *Le Roy*, [1911] 2 K. B. 1031; reversed on another ground [1911] 2 K. B. 1046, C. A.

time that the custom did not apply to a market established by a local Act (*i*). However, the Court of Appeal have now held that a sale between private persons in a statutory market can be a sale in market overt if made in accordance with the usage of the market (*k*).

Sub-section (2).—The Sale of Horses Act, 1555, and the Sale of Horses Act, 1588, laid down complicated regulations for the sale of horses in fairs or markets. These regulations were never observed in practice. These Acts have now been repealed together with sub-section (2) of this section (*l*), with the result that the sale of horses is now governed by the ordinary rules relating to market overt (*m*).

23. Sale under voidable title.—When the seller of goods has a voidable title thereto, but his title has not been avoided (*n*) at the time of the sale, the buyer acquires a good title to the goods, provided he buys them in good faith and without notice of the seller's defect of title (*o*).

ILLUSTRATIONS

(1) B has a room at 37 Wood Street. There is a reputable firm in the same street trading as "W. Blenkiron & Co." B orders goods in the name of "Blenkiron & Co." and A, thinking he is dealing with "W. Blenkiron & Co.," supplies them. There is no

(*i*) *Moyce* v. *Newington* (1878), 4 Q. B. D. 32, at p. 34, *per* COCKBURN, C.J.; and see *Lee* v. *Bayes* (1856), 18 C. B. 599 (sale by auction at horse repository).

(*k*) *Bishopsgate Motor Finance Co.* v. *Transport Brakes, Ltd.* (*supra*), where *Moyce* v. *Newington* (*supra*) was not followed.

(*l*) Criminal Law Act, 1967, Sch. 3, Pts. I, III; 8 Halsbury's Statutes (3rd Edn.) 565, 567.

(*m*) The repeals implemented one of the recommendations of the Law Reform Committee in its Report on the Transfer of Title to Chattels (Cmd. 2958), which also recommended that the law of market overt should be extended to all shops. See also [1956] Current Legal Problems 113.

(*n*) For avoidance, see *Clough* v. *London and Northwestern Rail. Co.* (1871), L. R. 7 Ex. 26; see also illustration (6) to this section and the footnote thereto.

(*o*) *Cundy* v. *Lindsay* (1878), 3 App. Cas. 459, at p. 464, *per* Lord CAIRNS; *cf. Robin and Rambler Coaches* v. *Turner*, [1947] 2 All E. R. 284; *Car and Universal Finance Co., Ltd.* v. *Caldwell*, [1964] 1 All E. R. 290; [1965] 1 Q. B. 525, C. A. (sale by dealer to finance company in course of hire-purchase transaction); see also *Stoneleigh Finance Ltd.* v. *Phillips*, [1965] 1 All E. R. 513; [1965] 2 Q. B. 537, C. A.; *Pollock on Possession*, pp. 203, 204; O. W. Holmes' *The Common Law*, lecture 9 (void and voidable contracts). The onus of proving notice or want of good faith is on the former owner: *Whitehorn Brothers* v. *Davison*, [1911] 1 K. B. 463, C. A.; *secus* as to the proviso to s. 2(1) of the Factors Act, 1889, *post*. The Law Reform Committee in its report on the Transfer of Title to Chattels (Cmd. 2958) has recommended that it should be for the purchaser to prove that he bought in good faith.

contract between A and B, and B cannot give any title to a sub-purchaser (*p*).

(2) B induces A to send him jewellery on approval by falsely representing that he has a good customer for it. He then pledges the goods with C. B then induces A to sell him the goods on credit, saying that he cannot ask his customer for cash before delivery. A cannot avoid the sale to B and recover the goods from C unless he can affect C with notice of the fraud (*q*).

(3) X goes to a jeweller and agrees to buy a ring. Pretending to be Sir G. B., he draws a cheque in that name and gives it in payment for the ring, which he is allowed to take away. X then pledges the ring with a pawnbroker, who takes it in good faith. The pawnbroker gets a good title, for there was a *de facto* contract between the jeweller and X (*r*).

(4) A motor-car obtained by fraud is sold to B, who buys it in good faith and without notice. B sells it to C, who had notice of the fraud, but was not a party to it. C gets a good title to the car (*s*).

(5) P advertises a car for sale. X comes to see it and says he likes it. He introduces himself as G, a well-known actor. P and X agree a price of £450. X writes a cheque and says he wants the car at once. P asks for proof of identity. X shows him a pass to Pinewood Studios with a photo of X, the name G, and an address. P then lets X have the car and log book. The cheque is not met but the car is resold to D. There is a voidable contract between P and X. The mistake as to identity was here merely a mistake as to an attribute of the buyer. X was therefore able to give a good title to D (*t*).

(6) A rogue obtains a motor car by fraud, and deliberately absconds. On discovering the fraud the seller immediately does everything possible to discover the rogue, and informs the police and the Automobile Association. Subsequently the rogue sells the car to another rogue who sells it to a bona fide purchaser. The first sale has been avoided and the bona fide purchaser acquires no title to the car under this section (*u*).

(*p*) *Cundy* v. *Lindsay* (1878), 3 App. Cas. 459.

(*q*) *Whitehorn Brothers* v. *Davison* [1911], 1 K. B. 463, C. A.; discussed and approved, *Folkes* v. *King*, [1923] 1 K. B. 282, C. A.; distinguished, *Heap* v. *Motorists Advisory Agency Ltd.*, [1923] 1 K. B. 577 (larceny by trick). As to seller's right to avoid a fraudulent purchase as against the buyer's trustee in bankruptcy, see *Ex p. Ward*, [1905] 1 K. B. 465; *Tilley* v. *Bowman, Ltd.*, [1910] 1 K. B. 745.

(*r*) *Phillips* v. *Brooks*, [1919] 2 K. B. 243; as explained in *Lake* v. *Simmons*, [1927] A. C. 487, at p. 501; and see *Barclays Bank* v. *Okenarhe*, [1966] 2 Lloyd's Rep. 87.

(*s*) *Peirce* v. *London Horse and Carriage Repository*, [1922] W. N. 170, C. A.

(*t*) *Lewis* v. *Averay*, [1971] 3 All E. R. 907; [1972] 1 Q. B. 198, C. A.; distinguishing *Ingram* v. *Little*, [1960] 3 All E. R. 332; [1961] 1 Q. B. 31, C. A., where on very similar facts it was held that there was no contract between the seller and the rogue, and that the seller could recover the car from the person to whom the rogue sold it. See further 1972 J. B. L. 151.

(*u*) *Car and Universal Finance Co., Ltd.* v. *Caldwell*, [1964] 1 All E. R. 290; [1965] 1 Q. B. 525, C. A.; but *cf. Newtons of Wembley* v. *Williams*, [1964] 3 All E. R. 532; [1965] 1 Q. B. 560, C. A., where the rogue conveyed a good title

Definitions.—For "buyer," "goods," "sale" and "seller," see s. 62 (1), *post*. For "good faith," see s. 62 (2), *post*.

Cross-references.—*Cf.* s. 12, *ante* (seller's warranty of title, etc.); and ss. 21 and 22, *ante*, and ss. 24 and 25, *post* (transfer of title).

COMMENT

Many cases to which this section applies will also fall within s. 25 (2), *post*, and s. 9 of the Factors Act, 1889. But the present section will also operate where the seller is not in possession, where his voidable title is not derived from a sale, or where he is a buyer in possession and the sale is not one made in the ordinary course of the business of a mercantile agent.

The section in terms deals with a seller, but a precisely similar rule applies to a pledgor (*v*), and to the creator of an equitable right by an incomplete pledge (*a*).

> "If the chattel has come into the hands of a person who professed to sell it, by a *de facto* contract, that is to say, a contract which has purported to pass the property to him from the owner of the property, there the purchaser will obtain a good title, even although afterwards it should appear that there were circumstances connected with that contract which would enable the original owner of the goods to reduce it and to set it aside" (*b*).

Such circumstances may include fraud, duress or undue influence; but where in fact no property has passed to the seller under the first transaction, the section has no application (*c*). Thus if the original seller believed that his purchaser was (*d*), or was acting on behalf of (*e*), some other person, then if the identity of the other contracting party was in all the circumstances so important that the proper inference was that there was no intention to contract with anybody else (*f*), then the purchaser will get no title. But if, as will normally be the case where the parties meet, the mistake as to identity was in all the circumstances merely a mistake as to an attribute of the purchaser, then the purchaser will get at least a voidable title to the

---

by virtue of the operation of ss. 2 and 9 of the Factors Act, 1889, *post*. For a conflicting decision of the Court of Session, see *Macleod* v. *Kerr*, 1965 S. L. T. 358; 29 M. L. R. 442. The Law Reform Committee in its report on the transfer of Title to Chattels (Cmd. 2958) recommended that a seller should not be able to avoid a sale until he has informed the buyer of his decision.

(*v*) *Whitehorn Brothers* v. *Davison*, [1911] 1 K. B. 463, C. A.

(*a*) *Attenborough* v. *St. Katherine's Dock Co.* (1878), 3 C. P. D. 450, C. A.

(*b*) *Cundy* v. *Lindsay* (1878), 3 App. Cas. 459, at p. 464.

(*c*) *Cundy* v. *Lindsay* (1878), 3 App. Cas. 459.

(*d*) *Ingram* v. *Little*, [1960] 3 All E. R. 332; [1961] 1 Q. B. 31, C. A.; *Boulton* v. *Jones* (1857), 2 H. & N. 564 (purchaser mistaken as to identity of seller); *Cundy* v. *Lindsay* (*supra*); *cf. Macleod* v. *Kerr*, 1965 S. L. T. 358 (Ct. of Sess.); 29 M. L. R. 442; *Porter* v. *Latec Finance Proprietary, Ltd.* (1964), 38 A. L. J. R. 184.

(*e*) *Hardman* v. *Booth* (1862), 32 L. J. Ex. 105.

(*f*) *Cf. Lewis* v. *Averay*, [1971] 3 All E. R. 907; [1972] 1 Q. B. 198, C. A.

goods (*g*), for where a fraudulent misrepresentation, even as to an attribute of the purchaser (*h*), has induced a contract of sale, or the creation of a power to pass the property (*i*), a third party taking before avoidance of the transaction will get a good title.

If the original seller only intended to pass possession without intending to create any power to pass a title (*k*), or intended to deal with some other property (*l*), no title will pass.

It follows that, broadly speaking, where goods are obtained by conduct which would before the Theft Act, 1968 came into force have amounted to larceny by a trick (*m*) no property passes, but where the obtaining is by false pretences a voidable title is created (*n*). This distinction was formerly of importance in connection with the Factors Acts and s. 25, *post.* If a mercantile agent gets possession of goods, intending at the time to misappropriate them, he has clearly committed larceny by a trick, for the common law judges held that in such circumstances the owner's apparent consent to the agent's possession was negatived (*o*). It was then much doubted (*p*) whether, that being so, the owner could be held to have consented to possession by the agent for the purposes of the Factors Acts. However, the better view is that Parliament did not have "any intention of applying the artificial distinctions of the criminal law to a commercial transaction," and that the consent necessary for the operation of the Factors Acts is only consent in fact (*q*).

If the transaction which passes the voidable title is subsequent in

---

(*g*) *Lewis* v. *Averay, supra.*

(*h*) *King's Norton Metal Co.* v. *Edridge Merrett & Co.* (1897), 14 T. L. R. 98, C. A. But compare *Newborne* v. *Sensolid (Great Britain), Ltd.*, [1953] 1 All E. R. 708; [1954] 1 Q. B. 45, C. A.; followed: *Black* v. *Smallwood* (1966), 117 C. L. R. 52 (H. C. of Austr.). See also *Wickberg* v. *Shatsky* (1969), 4 D. L. R. (3d) 540.

(*i*) *Whitehorn Brothers* v. *Davison* (*supra*).

(*k*) *Whitehorn Brothers* v. *Davison* (*supra*); *Folkes* v. *King*, [1923] 1 K. B. 282, C. A.

(*l*) *Morrisson* v. *Robertson*, 1908 S. C. 332.

(*m*) See Pollock and Wright, *Possession in the Common Law*, pp. 218–219; *Russell on Crime* (12th edn.) pp. 941 *et seq.*; *Ingram* v. *Little*, (*supra*) at pp. 349 and 69 respectively.

(*n*) In *Whitehorn Brothers* v. *Davison* (*supra*) and *Folkes* v. *King* (*supra*), the Court decided in each case that there was no larceny by a trick and accordingly a voidable contract.

(*o*) *Oppenheimer* v. *Frazer & Wyatt*, [1907] 2 K. B. 50, at pp. 71–72, C. A.

(*p*) *Cahn* v. *Pockett's Bristol Steam Packet Co.*, [1899] 1 Q. B. 643 at p. 659, C. A.; *Oppenheimer* v. *Frazer*, [1907] 2 K. B. 50, C. A.; *per* ATKIN L.J., in *Lake* v. *Simmons*, [1926] 2 K. B. 51 at pp. 70–71, C. A.

(*q*) *Folkes* v. *King* (*supra*); *Pearson* v. *Rose and Young*, [1950] 2 All E. R. 1027; [1951] 1 K. B. 275, C. A.; *Du Jardin* v. *Beadman*, [1952] 2 All E. R. 160; [1952] 2 Q. B. 712; and *per* DEVLIN, L.J., in *Ingram* v. *Little* (*supra*) at pp. 349 and 70 respectively. This view has been followed in New Zealand: *Davey* v. *Paine Brothers (Motors), Ltd.*, [1954] N. Z. L. R. 1122. See also [1957] Crim. L. R. pp. 28 and 96; the question was left open by Lord SUMNER in *Lake* v. *Simmons*, [1927] A. C. 487 at pp. 510, 511.

time to the disposition to the innocent third party, this section will still apply, for there is then a feeding of the third party's title (*r*).

[24. Revesting of property in stolen goods on conviction of offender.—*Repealed.*]

COMMENT

This section was repealed by Schedule 3, Part III, of the Theft Act, 1968. It is now provided by s. 31(2) of that Act, *post*, Appendix I, that where property has been stolen or obtained by fraud or other wrongful means, the title to that or any other property shall not be affected by reason only of the conviction of the offender.

Section 28 of the Theft Act, 1968, *post*, Appendix I, enables the convicting court to make an order for restitution in certain circumstances, but such an order does not affect the title to the goods in respect of which it is made. See further the comment to that section.

25. Seller or buyer in possession after sale.—(1) Where a person having sold goods continues or is in possession of the goods, or of the documents of title to the goods, the delivery or transfer by that person, or by a mercantile agent acting for him, of the goods or documents of title under any sale, pledge, or other disposition thereof, to any person receiving the same in good faith and without notice of the previous sale, shall have the same effect as if the person making the delivery or transfer were expressly authorised by the owner of the goods to make the same.

(2) Where a person having bought or agreed to buy goods obtains, with the consent of the seller, possession of the goods or the documents of title to the goods, the delivery or transfer by that person, or by a mercantile agent acting for him, of the goods or documents of title (*s*), under any sale, pledge, or other disposition thereof, to any person receiving the same in good faith and without notice of any lien or other right of the

---

(*r*) *Whitehorn Brothers* v. *Davison*, (*supra*) ; on the question of feeding title, see also *Butterworth* v. *Kingsway Motors*, [1954] 2 All E. R. 694.

(*s*) It seems that these need not be the same documents of title as those of which the buyer has obtained possession with the consent of the seller : see *per* SALMON, J. in *Mount* v. *Jay & Jay*, [1959] 3 All E. R. 307, at p. 311 ; [1960] 1 Q. B. 159 at pp. 168–169.

original seller in respect of the goods, shall have the same effect as if the person making the delivery or transfer were a mercantile agent in possession of the goods or documents of title with the consent of the owner.

For the purposes of this section—

(i) the buyer under a conditional sale agreement shall be deemed not to be a person who has bought or agreed to buy goods, and

(ii) " conditional sale agreement " means an agreement for the sale of goods which is a consumer credit agreement within the meaning of the Consumer Credit Act 1974 under which the purchase price or part of it is payable by instalments, and the property in the goods is to remain in the seller (notwithstanding that the buyer is to be in possession of the goods) until such conditions as to the payment of instalments or otherwise as may be specified in the agreement are fulfilled (*t*).

(3) In this section the term " mercantile agent " has the same meaning as in the Factors Acts.

ILLUSTRATIONS

Sub-section (1).—(1) A owns furs stored in a warehouse, where he owes for the storage charges. He sells the furs to the plaintiffs, promising to use the proceeds to pay the charges and then deliver the furs. Instead, he pledges the furs to the defendants, and with the proceeds of that transaction pays the storage charges and has the furs delivered to the defendants. The plaintiffs cannot recover the furs from the defendants without paying off their advance (*u*).

Sub-section (2).—(2) A sells certain copper to B, forwarding bill of lading indorsed in blank and bill of exchange for acceptance. B, who is insolvent, does not accept the bill of exchange, but transfers

---

(*t*) The part of this sub-section beginning " For the purposes of this section " was introduced by Sch. 4 to the Consumer Credit Act, 1974 with effect from a date to be appointed under s. 192 (4) of that Act. As yet no date has been appointed, but a similar deeming provision in s. 54 of the Hire-Purchase Act, 1965 (which is to be repealed from a date to be appointed under s. 192 (4)) excludes conditional sale agreements from the operation of this section.

(*u*) *City Fur Manufacturing Co.* v. *Fureenbond*, [1937] 1 All E. R. 799. *Cf. Eastern Distributors* v. *Goldring*, [1957] 2 All E. R. 525; [1957] 2 Q. B. 600, C. A.

the bill of lading to X in fulfilment of a contract to supply him with copper. X in good faith pays the price. A cannot stop the copper *in transitu* (*v*).

(3) A delivers a motor cab to B under a hire-purchase agreement. The hire amounts to £374 payable by 24 monthly instalments, and B may purchase the cab at any time within the two years by paying a further sum of £100. B pledges the cab with C, being at that time £58 in arrears with his instalments. A can sue C for the conversion of the cab, for B is not a person who has "agreed to buy" the cab; he has merely an option to purchase (*w*).

(4) B agrees to buy a plot of land from A, and also a motor-car if his solicitor approves the title to the land. B gets the motor-car, does not pay for it, but sells it to a *bonâ fide* purchaser. Subsequently B's solicitor disapproves the conditions of sale of the land. The purchaser of the motor-car gets a good title thereto, for B "agreed to buy" it (*x*).

Definitions.—For "delivery," "documents of title," "Factors Acts," "goods," "lien," "sale" and "seller," see s. 62 (1), *post*. For "good faith," see s. 62 (2), *post*. For "possession," see s. 1 (2) of the Factors Act, 1889, *post*.

Cross-reference.—*Cf*. s. 48 (2), *post* (re-sale by unpaid seller).

COMMENT

This section reproduces ss. 8 and 9 of the Factors Act, 1889, *post*, but omitting after "sale, pledge, or other disposition thereof" in sub-s. (1) and (2) the further words, "or under any agreement for sale, pledge, or other disposition thereof." For further notes on this section, see ss. 8 and 9 of the Factors Act, *post*. This section may override s. 19, *ante*.

26. Effect of writs of execution.—(1) A writ of fieri facias or other writ of execution against goods shall bind the property in the goods of the execution debtor as from the time when the writ is delivered to the sheriff to be executed; and, for the better manifestation of such time, it shall be the duty of the

(*v*) *Cahn* v. *Pockett's Bristol Channel Co.*, [1899] 1 Q. B. 643, C. A.

(*w*) *Belsize Motor Supply Co.* v. *Cox*, [1914] 1 K. B. 244. As C has an interest in the cab, the measure of damages is not the full value, but only the amount which A has lost: *Wickham Holdings, Ltd.* v. *Brooke House Motors, Ltd.*, [1967] 1 All E. R. 117; [1967] 1 W. L. R. 295, C. A., so that payments made by the hirer after the action is brought will be taken into account in assessing damages: *Western Credit Pty., Ltd.* v. *Dragan Motors Pty., Ltd.*, [1973] W. A. R. 184. *Cf. Whiteley* v. *Hilt*, [1918] 2 K. B. 808, C. A., where a hire-purchase agreement, before default, was held to be assignable, so that the assignee succeeded to the rights of the assignor.

(*x*) *Marten* v. *Whale*, [1917] 2 K. B. 480, C. A.

sheriff, without fee, upon the receipt of any such writ to endorse upon the back thereof the hour, day, month, and year when he received the same (*a*).

Provided that no such writ shall prejudice the title to such goods acquired by any person in good faith and for valuable consideration (*b*), unless such person had at the time when he acquired his title notice that such writ or any other writ by virtue of which the goods of the execution debtor might be seized or attached had been delivered to and remained unexecuted in the hands of the sheriff (*c*).

(2) In this section the term " sheriff " includes any officer charged with the enforcement of a writ of execution (*d*).

(3) The provisions of this section do not apply to Scotland.

Definitions.—For " goods " and " property," see s. 62 (1), *post*. For " good faith," see s. 62 (2), *post*.

Cross-references.—*Cf.* s. 12, *ante* (seller's undertaking as to title, etc.) ; and ss. 21 to 25, *ante* (transfer of title).

History.—The first part of sub-section (1) reproduces s. 15 (*e*) of the Statute of Frauds, with the addition that the sheriff is required to indorse the *hour* on the writ, but this accords with the practice. The proviso reproduces s. 1 of the Mercantile Law Amendment (Scotland) Act, 1856. Both these enactments were repealed by this Act.

COMMENT

It has been held that where goods are sold, otherwise than in market overt, after delivery of the writ, the words " shall bind the pro-

---

(*a*) See further, s. 10 of the Sheriffs Act, 1887; 30 Halsbury's Statutes (3rd Edn.) 582, as to giving a receipt.

(*b*) Thus a trustee in bankruptcy, who gives no value but derives his title by operation of law, will not be protected by this proviso: *Re Cooper*, [1958] 2 All E. R. 97, at p. 100; [1958] Ch. 922, at p. 928.

(*c*) As to the origin of the provision, see 2nd Report of the Mercantile Law Commission, 1855, p. 8.

(*d*) *Cf.* the definition of "sheriff" in s. 167 of the Bankruptcy Act, 1914; 3 Halsbury's Statutes (3rd Edn.) 159 ; and s. 29 of the Sheriff's Act, 1887 ; 30 Halsbury's Statutes (3rd Edn.) 592.

(*e*) Section 15 of the Revised Edition is commonly cited as s. 16.

perty in the goods" do not prevent the property from passing by the sale, but constitute the execution a charge upon the goods (*f*).

The effect of "shall bind the property" is that on delivery of the writ the sheriff acquires a legal right to seize enough of the debtor's goods to satisfy the amount specified in the writ (*g*). The proviso protects a purchaser against that right of seizure if the stated conditions are fulfilled, but it has no scope for operation where an actual seizure has already been effected (*g*).

The registration of a *lis pendens* does not affect goods (*h*).

As to practice, see R.S.C., O. 46 and 47, and notes in the Supreme Court Practice.

A creditor trustee under a deed of assignment is a purchaser for valuable consideration (*i*).

County Courts.—In the application of this section to County Courts, the Registrar now corresponds with the sheriff, and "the time when the writ is delivered to the sheriff" must be construed as meaning the time when application is made to the Registrar for the writ (*k*).

Part III

Performance of the Contract

27. Duties of seller and buyer.—It is the duty of the seller to deliver the goods, and of the buyer to accept and pay for them, in accordance with the terms of the contract of sale (*l*).

Definitions.—For "contract of sale," see s. 1 (1), *ante*, and s. 62 (1), *post*. For "buyer," "delivery," "goods," and "seller," see s. 62 (1), *post*.

Cross-references.—Cf. ss. 29 to 32, *post* (rules as to delivery), and s. 35, *post* (acceptance).

(*f*) *Woodland* v. *Fuller* (1840), 11 Ad. & El. 859, at p. 867. See also *Re Ferrier, Ex p. the Trustee* v. *Donald*, [1944] Ch. 295, at p. 297, where the effect of the seizure in execution of goods which were held on sale or return was discussed.

(*g*) *Lloyds and Scottish Finance Ltd.* v. *Modern Cars and Caravans* (*Kingston*), *Ltd.*, [1964] 2 All E. R. 732; [1966] 1 Q. B. 764.

(*h*) *Wigram* v. *Buckley*, [1894] 3 Ch. 483, at p. 492, C. A.

(*i*) *Beebee & Co.* v. *Turner's Successors* (1931), 48 T. L. R. 61.

(*k*) *Murgatroyd* v. *Wright*, [1907] 2 K. B. 333; as to execution issued from one County Court to another or from a local court as defined by s. 140 (3) of the County Courts Act, 1959, to a County Court, see the Administration of Justice Act, 1965, s. 22; 7 Halsbury's Statutes (3rd Edn.) 760.

(*l*) *Buddle* v. *Green* (1857), 27 L. J. Ex. 33; *Woolfe* v. *Horne* (1877), 2 Q. B. D. 355. See also 222 L. T. 129.

COMMENT

" In every contract of sale," says WATSON, B.,

" there is involved a contract on the one side to accept, and on the other to deliver."

" If," says MARTIN, B., in the same case,

" one buys goods of another in the possession of a third party, the vendor undertakes that they shall be delivered in a reasonable time. . . . If I buy a horse of you in another man's field, it is part of the contract that if I go for the horse I shall have it " (*m*).

The general obligation to deliver may, however, be modified by the terms of the contract. As BLACKBURN, J. says, there is no rule of law to prevent the parties from making whatever bargain they please (*n*). Thus, when the seller gives the buyer a delivery order for the goods, it may be a condition that the order should be given up to the warehouseman before the buyer can get the goods (*o*). Again, a man with his eyes open may buy the chance of obtaining goods and not the goods themselves ; see ss. 5 (2) and 12, *ante*.

As sale is a consensual contract, the parties may by agreement make the price payable how, when, and where they please (*p*) ; and when the time of payment arrives, the parties may agree that the debt shall be discharged by any means which amount to an accord and satisfaction. In one case the contract provided for payment of the price by instalments, the buyer to have possession, but the property not to pass until payment of the last instalment. The buyer made default in payment and the seller retook possession and sued for the instalments which ought to have been paid. It was held that the remedy was not for the price but for damages (*q*). See further, note to s. 49, *post* (action for price).

28. Payment and delivery are concurrent conditions.—Unless otherwise agreed (*r*), delivery of the

(*m*) *Buddle* v. *Green* (*supra*), at p. 34 ; *cf. Wood* v. *Baxter* (1883), 49 L. T. 45.

(*n*) *Calcutta Co.* v. *De Mattos* (1863), 32 L. J. Q. B. 322, at p. 328. See the passage cited at length, *post*, Note C, Appendix II ; and see *per* BRETT, L.J., in *Honck* v. *Muller* (1881), 7 Q. B. D. 92, at p. 103, C. A.

(*o*) *Bartlett* v. *Holmes* (1853), 13 C. B. 630 ; see, too, *Salter* v. *Woollams* (1841), 2 M. & Gr. 650, as explained in *Benjamin on Sale* (8th ed.), pp. 698, 699.

(*p*) But see the provisions as to credit-sale and conditional sale agreements in the Hire-Purchase Act, 1965; 30 Halsbury's Statutes (3rd Edn.) 61, and in the Consumer Credit Act, 1974, which repeals and replaces the Hire-Purchase Act, 1965 from a date to be appointed under s. 192 (4) of and Schs. 3 and 5 to the 1974 Act.

(*q*) *Att.-Gen.* v. *Pritchard* (1928), 97 L. J. K. B. 561.

(*r*) As, for instance, in a c.i.f. contract : see *E. Clemens Horst Co.* v. *Biddell Brothers*, [1912] A. C. 18, 22 ; *Orient Co.* v. *Brekke*, [1913] 1 K. B. 531. See further as to c.i.f. contracts, p. 36 *ante*,. See generally *post*, Appendix II, Note B (stipulations judicially construed).

goods and payment of the price are concurrent conditions, that is to say, the seller must be ready and willing to give possession of the goods to the buyer in exchange for the price and the buyer must be ready and willing to pay the price in exchange for possession of the goods (*s*).

ILLUSTRATION

Sale of hops under c.i.f. contract to be shipped to Hull, " terms net cash." The seller tenders the shipping documents to the buyer while the goods are at sea. This operates as tender of the goods, and the buyer must pay the price without waiting for the actual arrival and examination of the goods (*t*).

Definitions.—For " buyer," " delivery," " goods " and " seller," see s. 62 (1), *post.*

Cross-references.—*Cf.* ss. 1 (1) and 8, *ante* (price) ; s. 55, *post* (" unless otherwise agreed "). For a general note on conditions and warranties, see, *ante* p. 7. For their history, see Note A, *post,* Appendix II. For their effect, see s. 11, *ante. Cf.* also ss. 29–32, *post* (rules as to delivery), ss. 49 and 50, *post* (remedies of seller), and ss. 51 to 53, *post* (remedies of buyer).

COMMENT

" Where goods are sold," says BAYLEY, J.,

" and nothing is said as to the time of delivery or the time of payment . . . the seller is liable to deliver them whenever they are demanded upon payment of the price ; but the buyer has no right to have possession of the goods till he pays the price. . . . If goods are sold upon credit, and nothing is agreed upon as to the time of delivering the goods, the vendee is immediately entitled to the possession, and the right of possession and the right of property vest at once in him ; but his right of possession is not absolute, it is liable to be defeated if he becomes insolvent before he obtains possession " (*u*).

The language of BAYLEY, J., might be taken to imply that in cash sales payment was a condition precedent to delivery, but a reference to the cases cited in the second footnote to the section shows that payment and delivery have always been considered concurrent

(*s*) *Morton* v. *Lamb* (1797), 7 Term Rep. 125; *Rawson* v. *Johnson* (1801), 1 East, 203 ; *Wilks* v. *Atkinson* (1815), 1 Marshall, 412 ; *Pickford* v. *Grand Junction Railway* (1841), 8 M. & W. 372, at p. 378 ; *cf. Bussey* v. *Barnett* (1842), 9 M. & W. 312 ; *Bankart* v. *Bowers* (1866), L. R. 1 C. P. 484 ; *Paynter* v. *James* (1867), L. R. 2 C. P. 348.

(*t*) *E. Clemens Horst Co.* v. *Biddell Brothers,* [1912] A. C. 18 ; approving the judgment of KENNEDY, L.J., [1911] 1 K. B. 934, C. A.

(*u*) *Bloxam* v. *Sanders* (1825), 4 B. & C. 941, at p. 948 ; *cf. Chinery* v. *Viall* (1860), 5 H. & N. 288, at p. 293, as to credit sales.

conditions. It is not now necessary, as it was when the above cases were decided, for the plaintiff to allege that he had performed, or was ready and willing to perform, his part of the contract.

Where a man went into a restaurant and ordered dinner, and, after dining, said he could not pay for it, having only a halfpenny upon him, it was held that he could be convicted of obtaining credit by fraud but not of obtaining goods by false pretences (*a*). It seems therefore that under such circumstances there is an implied agreement for credit until the dinner is finished.

Evidence.—In an action for non-delivery, it seems that the buyer need not give evidence that he was ready and willing to pay, till the seller shows he was ready to deliver (*b*). By the same token, in an action for non-acceptance, the seller need not prove any tender of delivery. It is enough to show that he was ready and willing to deliver (*c*).

Where shares were sold, under a written contract, to be paid for at a future day, it was held that evidence might be received of a trade usage not to deliver till payment (*d*). On the other hand, where there was a contract in writing for the sale of hops at so much per cwt., evidence of a course of dealing between the parties to allow six months' credit was rejected (*e*). It is easier to draw imaginary distinctions between these cases than to harmonise the principles on which they rest.

29. Rules as to delivery.—(1) Whether it is for the buyer to take possession of the goods or for the seller to send them to the buyer is a question depending in each case on the contract, express or implied,

---

(*a*) *R.* v. *Jones* [1898] 1 Q. B. 119.

(*b*) *Wilks* v. *Atkinson* (1815), 1 Marsh., 412. " The averment of the plaintiff's readiness and willingness to perform his part of the contract will be proved by showing that he called on the defendant to accomplish his part " : notes to *Cutter* v. *Powell*, (1795) 2 Smith's L.C. (13th ed.), p. 1, at p. 15. See also notes to *Pordage* v. *Cole*, (1669) 1 Williams' *Notes to Saunders' Reports*, 319. But *cf. Longbottom & Co.* v. *Bass Walker & Co.*, [1922] W. N. 245, C. A. where the sellers were entitled to withhold delivery until they were paid for goods already delivered, and it was held to be immaterial that they were not ready to deliver at the contract date unless this showed that they would have been unable to perform the contract when the obligation to deliver arose.

(*c*) *Jackson* v. *Allaway* (1844), 6 M. & Gr. 942 ; *Baker* v. *Firminger* (1859), 28 L. J. Ex. 130 ; *Levey* v. *Goldberg*, [1922] 1 K. B. 688, at p. 692, *per* McCardie, J. " Whichever party," says Lord Halsbury, " was the actor, and is complaining of a breach of contract, is bound to show as a matter of law that he has performed all that was incident to his part of the concurrent obligations," *Forrestt* v. *Aramayo* (1900), 9 Asp. M. L. C. 134, C. A.

(*d*) *Field* v. *Lelean* (1861), 6 H. & N. 617, Ex. Ch. ; overruling as to usage, *Spartali* v. *Benecke* (1850), 10 C. B. 212.

(*e*) *Ford* v. *Yates* (1841), 2 M. & Gr. 549 ; as explained, *Lockett* v. *Nicklin* (1848), 2 Exch. 93.

between the parties (*f*). Apart from any such contract, express or implied, the place of delivery is the seller's place of business, if he have one, and if not, his residence: Provided that, if the contract be for the sale of specific goods, which to the knowledge of the parties when the contract is made are in some other place, then that place is the place of delivery.

(2) Where under the contract of sale the seller is bound to send the goods to the buyer, but no time for sending them is fixed, the seller is bound to send them within a reasonable time (*g*).

(3) Where the goods at the time of sale are in the possession of a third person, there is no delivery by seller to buyer unless and until such third person acknowledges to the buyer that he holds the goods on his behalf (*h*); provided that nothing in this section shall affect the operation of the issue or transfer of any document of title to goods (*i*).

(4) Demand or tender of delivery may be treated as ineffectual unless made at a reasonable hour. What is a reasonable hour is a question of fact.

(5) Unless otherwise agreed, the expenses of and incidental to putting the goods into a deliverable state must be borne by the seller.

ILLUSTRATIONS

(1) Sale of cotton seed to be shipped from Bombay in August or September. The goods are shipped in August, but the ship is stranded and cannot be got off for three months, and then has to be repaired. Unless the commercial object of the contract has been wholly defeated, the buyer cannot reject the goods on the ground that they have not been delivered within a reasonable time, but he may have a claim in damages for the delay (*k*).

(2) Sale of a cask of wine to be delivered at buyer's house. The carrier delivers the wine there to an apparently respectable person,

(*f*) As to f.o.b. contracts, see p. 34, *ante*.

(*g*) *Ellis* v. *Thompson* (1838), 3 M. & W. 445; see at p. 456 *per* ALDERSON, B.

(*h*) *Farina* v. *Home* (1846), 16 M. & W. 119, at p. 123; *Godts* v. *Rose* (1855), 17 C. B. 229; *Buddle* v. *Green* (1857), 27 L. J. Ex. 33; Pollock and Wright, *Possession in the Common Law*, p. 73.

(*i*) See the Bills of Lading Act, 1855, *post*, Appendix I, and the Factors Act, 1889, *post*; *cf. Hayman* v. *M'Lintock*, 1907 S. C. 936.

(*k*) *Re Carver & Co.* (1911), 17 Com. Cas. 59, at pp. 67, 70.

who signs for it, and then makes off with it. This is a good delivery to the buyer (*l*).

Definitions.—For "contract of sale," see s. 1 (1), *ante,* and s. 62 (1), *post.* For "business," "buyer," "delivery," "document of title," "goods," "seller" and "specific goods," see s. 62 (1), *post.* For "deliverable state," see s. 62 (4) *post.*

Cross-references.—*Cf.* s. 10, *ante* (stipulations as to time); ss. 25, *ante,* and 47, *post* (effect of documents of title); s. 32, *post,* (delivery to carrier); s. 55, *post* ("express or implied"); and s. 56, *post* (reasonable time a question of fact).

COMMENT

The delivery of the key of the place where the goods are may, by agreement, operate as a delivery of the goods (*m*). But where the goods are kept in a room or container on the seller's premises, then in the absence of any express agreement the seller is not to be taken to have delivered the goods by handing over the key to the room or container (*n*) unless the buyer is also licensed to enter on the seller's premises to take possession of them (*o*).

Sub-section (1).—*Place of delivery.*—This sub-section was much considered and several times altered in Committee. The first part deals incidentally with the mode of delivery, and the second part with the place of delivery. As regards mode of delivery there was very little authority, but the assumed rule was that it was for the buyer to take delivery and that, in the absence of any different agreement, the duty of the seller to deliver was satisfied by his affording to the buyer reasonable facilities for taking possession of the goods at the agreed place of delivery (*p*). It seems a pity that a more definite *primâ facie* rule has not been laid down by the Act.

As regards place of delivery, there was no authority in point, and text-book writers seem to have thought that the place of delivery was where the goods were to be found. The Act adopts a rule which is more in accordance with ordinary practice.

(*l*) *Galbraith and Grant* v. *Block,* [1922] 2 K. B. 155. *Cf. Thomas* v. *Alper,* [1953] C. L. Y. 3277.

(*m*) *Ellis* v. *Hunt* (1789), 3 Term Rep. 464; *Chaplin* v. *Rogers* (1800), 1 East, 192, at p. 195; *Elmore* v. *Stone* (1809), 1 Taunt, 458; *Ancona* v. *Rogers* (1876), 1 Ex. D. 285, at p. 290, C. A.; and see the question of so-called symbolic delivery discussed in Pollock and Wright, *Possession in the Common Law,* pp. 61–70; *cf.* note, p. 276, *post.*

(*n*) *Milgate* v. *Kebble* (1841), 3 M. & Gr. 100; *Wrightson* v. *McArihur and Hutchinson,* [1921] 1 K. B. 807, at p. 816.

(*o*) *Wrightson* v. *McArthur and Hutchinson, supra.*

(*p*) *Cf. Wood* v. *Tassell* (1844), 6 Q. B. 234; *Smith* v. *Chance* (1819), 2 B. & Ald. 753, at p. 755; *Salter* v. *Woollams* (1841), 2 M. & Gr. 650, as explained in *Benjamin on Sale* (8th ed.), pp. 698, 699.

Sub-section (2).—*Delivery " as required ".*—In a contract for goods to be delivered " as required," the buyer must require delivery within a reasonable time, but the seller cannot rescind the contract on the ground of delay without giving the buyer notice. " No doubt," says POLLOCK, C.B.,

> " where a contract is silent as to time, the law implies that it is to be performed within a reasonable time; but there is another maxim of law, viz., that *every reasonable condition is also implied,* and it seems to me reasonable that the party who seeks to put an end to a contract, because the other party has not, within a reasonable time, required him to deliver the goods, should in the first instance inquire of the latter whether he means to have them " (*q*).

A seller who is told to deliver goods at the purchaser's premises discharges his obligations if he delivers them without negligence to a person apparently having authority to receive them (*r*).

Sub-section (3).—*Goods in possession of third party.*—As regards documents of title, the common law drew a hard-and-fast distinction between bills of lading and other documents. The lawful transfer of a bill of lading was always held to operate as a delivery of the goods themselves because, while goods were at sea, they could not be otherwise dealt with (*s*). But the transfer of a delivery order or dock warrant operated only as a token of authority to take possession, and not as a transfer of possession (*t*); and, as between immediate parties, there is nothing to modify the common law rule. If, however, a buyer or mercantile agent, who is lawfully in possession of any document of title to goods, transfers it for value to a third person, the original seller's rights of lien and stoppage *in transitu* are thereby defeated.

Sub-section (4).—*Hours for delivery.*—This sub-section alters the law in so far as it makes the question of what is a reasonable hour a question of fact. It was formerly a question of law, and some highly technical rules for determining it were laid down by PARKE, B. (*u*).

Sub-section (5).—*Expenses of delivery.*—This is declaratory. " There is no implied contract," says Story,

---

(*q*) *Jones* v. *Gibbons* (1853), 8 Exch. 920, at p. 922. But this rule is not absolute—the facts may show a mutual intention to abandon the contract: *Pearl Mill Co.* v. *Ivy Tannery Co.*, [1919] 1 K. B. 78.

(*r*) *Galbraith* v. *Grant and Block*, [1922] 2 K. B. 155, *ante*, illustration 2.

(*s*) *Sanders* v. *Maclean* (1883), 11 Q. B. D. 327, at p. 341, *per* BOWEN, L.J. and *Biddell Brothers* v. *E. Clemens Horst Co.*, [1911] 1 K. B. 934, at p. 956, *per* KENNEDY, L.J.

(*t*) *Blackburn on Sale* (1st ed.), p. 302; *M'Ewan* v. *Smith* (1849), 2 H. L. Cas. 309.

(*u*) *Startup* v. *Macdonald* (1843), 6 M. & Gr. 593, Ex. Ch.

"that the vendee shall pay the vendor for any services in relation to the property rendered previous to the completion of the sale by delivery" (*a*).

For a list of express terms dealing with time and mode of delivery and judicially construed, see *post*, Appendix II, Note B.

30. Delivery of wrong quantity.—(1) Where the seller delivers to the buyer a quantity of goods less than he contracted to sell, the buyer may reject them, but if the buyer accepts the goods so delivered he must pay for them at the contract rate (*b*).

(2) Where the seller delivers to the buyer a quantity of goods larger than he contracted to sell, the buyer may accept the goods included in the contract and reject the rest, or he may reject the whole. If the buyer accepts the whole of the goods so delivered he must pay for them at the contract rate (*c*).

(3) Where the seller delivers to the buyer the goods he contracted to sell mixed with (*d*) goods of a different description not included in the contract, the buyer may accept the goods which are in accordance with the contract and reject the rest, or he may reject the whole (*e*).

(4) The provisions of this section are subject to any usage of trade, special agreement, or course of dealing between the parties (*f*).

(*a*) *Story on Sale*, § 297*a*.

(*b*) *Shipton* v. *Casson* (1826), 5 B. & C. 378, at p. 382; *Oxendale* v. *Wetherell* (1829), 4 Man. & Ry. 429; approved, *Colonial Ins. Co.* v. *Adelaide Ins. C.*, (1886), 12 App. Cas. 128, at pp. 138, 140; *Morgan* v. *Gath* (1865), 3 H. & C. 487; *Harland & Wolf* v. *Burstall* (1901), 84 L. T. 324 (470 loads of timber out of 500); *Behrend & Co.* v. *Produce Brothers Co.*, [1920] 3 K. B. 530.

(*c*) *Hart* v. *Mills* (1846), 15 M. & W. 85; *Cunliffe* v. *Harrison* (1851), 6 Exch. 903; *Tamvaco* v. *Lucas* (1859), 1 E. & E. 581; *cf. Dixon* v. *Fletcher* (1837), 3 M. & W. 146.

(*d*) Mixed with = "accompanied by" and does not connote physical confusion: *Moore & Co.* v. *Landauer & Co.*, [1921] 1 K. B. 73, at p. 76; affirmed, [1921] 2 K. B. 519, C. A.; *Ebrahim Dawood, Ltd.* v. *Heath*, [1961] 2 Lloyd's Rep. 512; but see the different view expressed *obiter* by SALTER, J., in *Barker & Co., Ltd.* v. *Agius Ltd.* (1927), 33 Com. Cas. 120 at pp. 131–132.

(*e*) *Levy* v. *Green* (1859), 8 E. & B. 575, Ex. Ch.; *Nicholson* v. *Bradfield Union* (1866), L. R. 1 Q. B. 620, at pp. 624, 625, per Lord BLACKBURN.

(*f*) See pp. 9–17, *ante*.

ILLUSTRATIONS

Sub-section (1).—(1) Contract for sale of two lots of cotton seed (200 and 500 tons) at different prices. Part of each lot is delivered, the whole of the contract price is paid, and the ship then goes on to another port. She returns two weeks later and tenders the rest. The buyer may keep what he has got, reject the rest, and recover back the price pre-paid for the rejected portion (*g*).

Sub-section (2).—(2) Sale of a cargo of wheat, which, with a limit of variation allowed by the contract, may amount to 4950 tons. The actual amount tendered is 4950 tons and 55 lbs., but the seller does not insist on payment for the extra 55 lbs. This is a good tender, and the buyer cannot reject it (*h*).

Sub-section (3).—(3) Contract for purchase of 3000 tins of canned fruit from Australia to be packed in cases, each containing 30 tins. When the goods are tendered in London a substantial part is tendered in cases containing 24. The buyer may reject the whole (*i*).

(4) Sale of 1500 tons " briquettes, size 2 inches." Only 25 tons correspond to this description, the remainder being larger. The buyer may accept the 25 tons and reject the remainder (*k*).

Definitions.—For " buyer," " delivery," " goods " and " seller," see s. 62 (1), *post*.

Cross-references.—*Cf.* s. 13, *ante* (description) ; ss. 29, *ante*, and 31 and 32, *post* (rules as to delivery) ; s. 35, *post* (acceptance and rejection) ; and s. 55, *post* (usage, etc.).

COMMENT

As the seller does not fulfil his contract by delivering a less quantity than he contracted to sell, so, conversely,

> " if a man contracts to buy 150 quarters of wheat, he is not at liberty to call for the delivery of a small portion without being prepared to receive the entire quantity " (*l*)

unless, of course, he has stipulated for the right to do so.

When the seller delivers a larger quantity of goods than was ordered, such delivery operates as a proposal for a new contract (*m*). This, presumably, is the effect of any tender of goods which are not in conformity with the contract.

(*g*) *Behrend & Co.* v. *Produce Brokers Co.*, [1920] 3 K. B. 530.

(*h*) *Shipton Anderson & Co.* v. *Weil Brothers*, [1912] 1 K. B. 574, 577: *de minimis non curat lex*. *Cf. Payne* v. *Lillico & Sons* (1920), 36 T. L. R. 569 (2 per cent. more or less); *Rapalli* v. *K. L. Take, Ltd.*, [1958] 2 Lloyd's Rep. 469, C. A.; *Margaronis Navigation Agency, Ltd.* v. *Henry W. Peabody & Co. of London, Ltd.*, [1964] 3 All E. R. 333; [1965] 2 Q. B. 430, C. A.

(*i*) *Moore & Co.* v. *Landauer & Co.*, [1921] 2 K. B. 519, C. A.

(*k*) *Barker* v. *Agius* (1927), 33 Com. Cas. 120.

(*l*) *Kingdom* v. *Cox* (1848), 5 C. B. 522, at p. 526, *per* WILDE, C.J.

(*m*) *Cunliffe* v. *Harrison* (1851), 6 Exch. 903, at p. 906, *per* PARKE, B.; *Gabriel Wade & English* v. *Arcos, Ltd.* (1929), 34 Ll. L. Rep. 306 (if buyer accepts the larger quantity he cannot then sue for delivery of wrong quantity).

When the seller is uncertain as to the exact amount he can deliver, he may protect himself by using such terms as so many tons "more or less," or "about" so many tons, and he is then allowed a reasonable margin (*n*). See list of terms judicially construed, *post*, Appendix II, Note B.

Frequently, in commercial contracts, a clause will be found providing that the "buyer shall not reject the goods herein specified, but shall accept or pay for the goods in terms of the contract against shipping documents," or like terms. The precise effect of such a clause may be hard to define, but it is clear that it can only operate where delivery of "the goods herein specified" is in fact made, and that if, for example, delivery of larger or smaller quantities is tendered, the buyer can reject since the "goods herein specified" have not, in the specified quantities, been tendered (*o*).

Sub-section (3) was amended in Committee. It has been held in Scotland not to apply to a case where goods are of the kind or description ordered, but a portion of them are of inferior quality (*p*), unless the contract is severable (*q*). Where in a f.o.b. contract, at the end of a clause dealing with over- and under-shipment, it was provided that "each item of this contract to be considered a separate interest," it was held that the buyer might reject the whole if part of the goods delivered did not comply with the contract specification as to quality (*r*).

The buyer's right to recover that part of the purchase price which relates to goods properly rejected is a right to recover money for a consideration which has wholly failed, at least for a rejection under s. 30 (1) and (3) (*s*), and *semble* under s. 30 (2) also.

31. Instalment deliveries.—(1) Unless otherwise agreed, the buyer of goods is not bound to accept delivery thereof by instalments (*t*).

(2) Where there is a contract for the sale of goods to be delivered by stated instalments, which are to be separately paid for, and the seller makes defective

(*n*) *McConnel* v. *Murphy* (1873), L. R. 5 P. C. 203 ; *Re Thornett & Fehr & Yuills*, [1921] 1 K. B. 219 (five per cent. more or less). As to importing such a term by usage, see *Moore* v. *Campbell* (1854), 10 Exch. 323.

(*o*) *Green* v. *Arcos, Ltd.* (1931), 47 T. L. R. 336, C. A. ; *Wilensko* v. *Fenwick*, [1938] 3 All E. R. 429.

(*p*) *Aitken & Co.* v. *Boullen*, 1908 S. C. 490 (rejection of part).

(*q*) *Cf. Jackson* v. *Rotax*, [1910] 2 K. B. 937, C. A. ; and see s. 11 (1) (c), *ante*.

(*r*) *Raahe O. Y. Osakeytio* v. *Goddard* (1935), 154 L. T. 124. See also *London Plywood & Timber Co.* v. *Nasic*, [1939] 2 K. B. 343.

(*s*) *Ebrahim Dawood, Ltd.* v. *Heath*, [1961] 2 Lloyd's Rep. 512.

(*t*) *Reuter* v. *Sala* (1879), 4 C. P. D. 239, C. A. Nor can he demand it: see note to last section ; 34 Halsbury's Laws (3rd Edn.) 101.

deliveries in respect of one or more instalments, or the buyer neglects or refuses to take delivery of or pay for one or more instalments, it is a question in each case depending on the terms of the contract and the circumstances of the case, whether the breach of contract is a repudiation of the whole contract or whether it is a severable breach giving rise to a claim for compensation but not to a right to treat the whole contract as repudiated (*u*).

ILLUSTRATIONS

Sub-section (1).—(1) Contract for sale of thirteen engraved plates "to be sent to me as published, the price of each plate £10 10s. 0d." This is an instalment contract and the buyer is bound to accept and pay for each plate separately (*v*).

Sub-section (2).—(2) Contract for sale of steel bars to be delivered over a period of three months, in about equal monthly quantities; payment cash in fourteen days after delivery, "all payments to be made on due date as a condition precedent to future deliveries." If the buyer does not pay according to contract, the seller may refuse unconditionally to make any further delivery (*a*).

(3) Contract for rosewood to be delivered by instalments during the year. The buyer, on what afterwards turns out to be erroneous grounds, refuses an offer to deliver the first instalment and repudiates the contract. The first instalment and the second instalment, which is according to contract, are subsequently tendered and refused. The buyer cannot afterwards set up, in mitigation of damages, that part of the first instalment was of slightly inferior character; *sed quaere* ? (*b*).

(4) Sale of 1100 pieces of blue gumwood to be delivered in two instalments. The first instalment of 750 pieces is of very inferior quality, and the buyer refuses to accept it. If an arbitrator finds

---

(*u*) *Mersey Steel & Iron Co.* v. *Naylor & Co.* (1884), 9 App. Cas. 434 (non-payment of one instalment in error): *Dominion Coal Co.* v. *Dominion Iron and Steel Co.*, [1909] A. C. 293, P. C.; *Steinberger* v. *Atkinson & Co.* (1914), 31 T. L. R. 110; *Payzu Ltd.* v. *Saunders*, [1919] 2 K. B. 581, C. A. (failure to pay punctually). See *Workman Clark* v. *Lloyd Brazileno*, [1908] 1 K. B. 968, 979, C. A., as to meaning of "compensation" in this section.

(*v*) *Howell* v. *Evans* (1926), 42 T. L. R. 310.

(*a*) *Ebbw Vale Steel Co.* v. *Blaina Iron Co.* (1901), 6 Com. Cas. 33, C. A.

(*b*) *Braithwaite* v. *Foreign Hardwood Co.*, [1905] 2 K. B. 543, C. A.; explained and followed, *Taylor* v. *Oakes, Roncoroni & Co.* (1922), 38 T. L. R. 349, 351; affd. 38 T. L. R. 517, C. A.; *Continental Contractors* v. *Medway* (1925), 23 Ll. L. Rep. 55, 124, C. A.; criticised, *British and Beningtons, Ltd.* v. *N. W. Cachar Tea Co.*, [1923] A. C. 48, at p. 70. The *Braithwaite* case is not to be taken as impugning the general rule that a buyer, who gives a wrong reason for refusing to perform his contract and afterwards discovers a sound reason, may then rely on the sound reason. See also *Universal Cargo Carriers Corporation* v. *Citati*, [1957] 2 All E. R. 70; [1957] 2 Q.B. 401.

that the quality of the instalment is of such a character as to amount to a repudiation of the contract, the Court will not disturb his finding (*c*).

(5) Contract for the sale of two ships. One ship is requisitioned by Government, and cannot be delivered. The buyer may refuse to accept the other ship (*d*).

Definitions.—For " contract of sale," see s. 1 (1), *ante*, and s. 62 (1), *post*. For " buyer," " delivery," " goods " and " seller," see s. 62 (1), *post*.

Cross-references.—*Cf.* s. 10, *ante* (stipulation as to time of payment not usually of the essence); s. 11, *ante* (repudiation, severable breach); s. 28, *ante* (payment and delivery usually concurrent conditions); ss. 29 *ante*, and 32, *post* (rules as to delivery); s. 35, *post* (acceptance); s. 41, *post* (seller's lien); and s. 55, *post* (" unless otherwise agreed "). *Cf.* also ss. 49 and 50, *post* (remedies of seller), and ss. 51 to 53, *post* (remedies of buyer).

COMMENT

Sub-section (1).—" Suppose," says BRAMWELL, L.J.,

> " a man orders a suit of clothes, the price being £7—£4 for the coat, £2 for the trousers, and £1 for the waistcoat; can he be made to take the coat only, whether they were all to be delivered together, or the trousers and waistcoat first ? "

and he then proceeds to show that this cannot be (*e*). On the other hand, the circumstances of a contract may be such that an agreement for delivery by instalments will be implied.

> " In many cases of contract to supply a quantity of goods to be delivered within a fixed period the whole quantity cannot, from the very nature of the case, be delivered at one time,"

as, for instance, in the case of contracts for the supply of provisions for the army and navy (*f*).

Sub-section (2).—It is very difficult to reconcile the older decisions in which it has been held that the refusal to deliver, accept, or pay for a particular instalment is a breach going to the root of the contract (*g*), with those in which the contrary has been held (*h*). " The rule of law," says Lord BLACKBURN,

---

(*c*) *Millars Karri & Co.* v. *Weddell & Co.* (1908), 100 L. T. 128. *Cf. Ballantine & Co.* v. *Cramp & Bosman* (1923), 129 L. T. 502 (sale of meat to be shipped by instalments).

(*d*) *Claddagh S.S. Co.* v. *Stevens & Co.*, 1919 S. C. (H. L.) 132.

(*e*) *Honck* v. *Muller* (1881), 7 Q. B. D. 92, at p. 99, C. A.

(*f*) *Colonial Ins. Co. of New Zealand* v. *Adelaide Ins. Co.* (1886), 12 App. Cas. 128, at pp. 138, 139, P. C.

(*g*) See *Withers* v. *Reynolds* (1831), 2 B. & Ad. 882; *Hoare* v. *Rennie* (1859), 5 H. & N. 19; *Honck* v. *Muller* (1881), 7 Q. B. D. 92, C. A.

(*h*) See *Jonassohn* v. *Young* (1863), 4 B. & S. 296; *Simpson* v. *Crippin* (1872), L. R. 8 Q. B. 14; *Freeth* v. *Burr* (1874), L. R. 9 C. P. 208; *Dumenil* v. *Ruddin*, [1953] 2 All E. R. 294; *Decro-Wall International S.A.* v. *Practitioners in Marketing, Ltd.*, [1971] 2 All E. R. 216; [1971] 1 W. L. R. 361, C. A.

"is that where there is a contract in which there are two parties, each side having to do something, if you see that the failure to perform one part of it goes to the root of the contract, it is a good defence to say, 'I am not going on to perform my part of it when that which is the root of the whole and the substantial consideration for my performance is defeated by your misconduct'" (*i*).

Each case must be judged on its own merits, and the main tests to apply are, first, the quantitative ratio which the breach bears to the contract as a whole, and, secondly, the degree of probability that such a breach will be repeated (*k*). The section in terms deals only with "stated instalments," but a similar principle applies where the instalments are not specified (*l*).

Where the passing of the property in specific goods is made dependent upon full payment of a price payable by instalments, the buyer may at any time pay the outstanding balance, even if it is not yet due, and the seller must then appropriate the payment to the goods (*m*).

Subject to s. 11 (1) (*c*), *ante*, under an instalment contract, the buyer may reject defective instalments while retaining the rest of the goods (*n*).

32. Delivery to carrier.—(1) Where, in pursuance of a contract of sale, the seller is authorised or required to send the goods to the buyer, delivery of the goods to a carrier, whether named by the buyer or not, for the purpose of transmission to the buyer is primâ facie deemed to be a delivery of the goods to the buyer (*o*).

(*i*) *Mersey Steel Co.* v. *Naylor & Co.* (1884), 9 App. Cas. 434, at p. 443; and see *per* JESSEL, M.R., in court below, 9 Q. B. D. 648, at p. 657. Steel to be delivered by instalments; non-payment of one instalment, under mistaken advice, held no repudiation; *cf. Munro & Co.* v. *Meyer*, [1930] 2 K. B. 312, at p. 332 ("each delivery to be treated as a separate contract and failure to give or take delivery shall not cancel the contract as to future deliveries"). See also *Fixby Engineering Co., Ltd.* v. *Auchlochan Sand and Gravel Co., Ltd.*, 1974 S. L. T. (Sh. Ct.) 58, where the authorities are reviewed.

(*k*) *Maple Flock Co. Ltd.* v. *Universal Furniture Products* (*Wembley*) *Ltd.*, [1934] 1 K. B. 148, C. A. But see *Decro-Wall International, S.A.* v. *Practitioners in Marketing, Ltd.*, [1971] 2 All E. R. 216; [1971] 1 W. L. R. 361, C. A.

(*l*) *Coddington* v. *Paleologo* (1867), L. R. 2 Exch. 193, 197; *Reuter* v. *Sala* (1879), 4 C. P. D. 239, C. A.; *Jackson* v. *Rotax Motor and Cycle Co.*, [1910] 2 K. B. 937, C. A.; 34 Halsbury's Laws (3rd Edn.) 101–105.

(*m*) *Lancs. Waggon Co.* v. *Nuttall* (1879), 42 L. T. 465, C. A.; *Croft* v. *Lumley* (1858), 6 H. L. Cas. 672.

(*n*) *The Hansa Nord*, [1974] 2 Lloyd's Rep. 216, at p. 226; reversed, apparently on other grounds, *sub. nom. Cehave N.V.* v. *Bremer Handelsgesellschaft* (1975), *Times*, 22nd July, C. A.

(*o*) For statement of principle, see *Wait* v. *Baker* (1848), 2 Exch. 1, at p. 7, *per* PARKE, B.; *Dunlop* v. *Lambert* (1839), 6 Cl. & F. 600, at p. 620, *per* Lord COTTENHAM; *Calcutta Co.* v. *De Mattos* (1863), 32 L. J. Q. B. 322, at p. 328, *per* BLACKBURN, J., cited in full, *post*, Appendix II, Note C; *Badische Anilin Fabrik* v. *Basle Chemical Works*, [1898] A. C. 200, at pp. 203, 204. For illustrations, see *Dutton* v. *Solomonson* (1803), 3 B. & P. 582 (carrier by land);

(2) Unless otherwise authorised by the buyer, the seller must make such contract with the carrier on behalf of the buyer as may be reasonable having regard to the nature of the goods and the other circumstances of the case. If the seller omit so to do, and the goods are lost or damaged in course of transit, the buyer may decline to treat the delivery to the carrier as a delivery to himself (*p*) or may hold the seller responsible in damages.

(3) Unless otherwise agreed, where goods are sent by the seller to the buyer by a route involving sea transit, under circumstances in which it is usual to insure, the seller must give such notice to the buyer as may enable him to insure them during their sea transit, and, if the seller fails to do so, the goods shall be deemed to be at his risk during such sea transit.

ILLUSTRATIONS

Sub-section (2).—(1) Contract for sale of electric engines, to be sent by rail. Sellers send them at owner's risk rate and they arrive damaged. Buyers can reject, as contract of carriage made by sellers not reasonable in the circumstances (*q*).

(2) C.i.f. contract for the sale of groundnuts Sudan/Hamburg with no term, express or implied, governing the route. The Suez Canal, which would be the normal route, then becomes closed to traffic. The sellers must ship *via* the Cape, which is a reasonable and practicable route in the circumstances (*r*).

Sub-section (3).—(3) Goods sold Antwerp, to be shipped as required, payment cash against bill of lading. Buyer directs seller to ship the goods to Odessa, leaving him to select the ship. Ship sails on 25th August, and is lost on the 26th. Buyer receives no notice of shipment till 29th August. He must pay for the goods, even

---

Bryans v. *Nix* (1839), 4 M. & W. 775 (canal-boat); *Alexander* v. *Gardner* (1835), 1 Bing. N. C. 671 (ship); *Ex p. Pearson* (1868), 3 Ch. App. 443 (railway); see also *Galbraith* v. *Grant & Block*, [1922] 2 K. B. 155; Bells' *Inquiries into the Contract of Sale of Goods and Merchandise*, p. 86. The same rules apply where goods are delivered to a carrier to convey to a pledgee; *Kum* v. *Wah Tat Bank*, [1971] 1 Lloyd's Rep. 439, P. C.

(*p*) *Clarke* v. *Hutchins* (1811), 14 East, 475; *Buckman* v. *Levi* (1813), 3 Camp. 414; *Young* v. *Hobson* (1949), 65 T. L. R. 365, C. A. *Story on*; *Sale*, § 305.

(*q*) *Young* v. *Hobson* (1949), 65 T. L. R. 365, C. A.

(*r*) *Tsakiroglou & Co. Ltd.* v. *Noblee & Thorl*, [1961] 2 All E. R. 179; [1962] A. C. 93, H. L.

though he has not insured, for he had sufficient information to enable him to do so before the goods were shipped (*s*).

Definitions.—For " contract of sale," see s. 1 (1), *ante*, and s. 62 (1), *post*. For " buyer," " delivery," " goods " and " seller," see s. 62 (1), *post*.

Cross-references.—*Cf.* s. 20, *ante* (risk) ; ss. 29 and 31, *ante* (rules as to delivery) ; ss. 51 to 53, *post* (remedies of the buyer) ; and s. 55, *post* (" unless otherwise agreed "). See also ss. 44 to 46, *post* (stoppage *in transitu*), and Note C, *post*, Appendix II (delivery to carrier). For the law relating to customary international contracts, see p. 33, *ante*.

COMMENT

Sub-section (1).—*Effect of delivery to carrier.*—The rule that delivery of goods to a carrier is *primâ facie* delivery to the buyer, passing to him the property and the risk, if they have not passed before, is the natural complement of the rule that *primâ facie* the proper place for delivery is the seller's abode, or the place where the goods are at the time of sale (s. 29, *ante*).

It is to be noted that, though the carrier is ordinarily the agent of the buyer to receive the goods, it has been held that he is not his agent to accept them. The cases on this point before the Act proceed on the ground that the carrier is not the buyer's agent to examine the goods, and that acceptance " means such an acceptance as precludes the purchaser from objecting to the quality of the goods " (*t*).

While the goods are in the hands of a carrier as such, they are liable to be stopped *in transitu* ; and of course they may be delivered to the carrier on such terms as to make him the seller's agent. When goods are sent " carriage forward " it is strong evidence that the delivery to the carrier was intended as a delivery to the buyer (*u*).

Sub-section (2).—*Seller's duty.*—" Delivery of goods to a carrier or wharfinger," says Lord ELLENBOROUGH,

> " with due care and diligence is sufficient to charge the purchaser, but he has a right to require that in making this delivery due care and diligence shall be exercised by the seller " (*v*).

Sub-section (3).—*Sea transit.*—As regards goods sent by sea, Bell, summing up the Scottish cases, says:

(*s*) *Wimble* v. *Rosenberg & Sons*, [1913] 3 K. B. 743, C. A., affirming BAILHACHE, J., on varying grounds—but see note (*a*), p. 194, *infra*.

(*t*) *Norman* v. *Phillips* (1845), 14 M. & W. 277, at p. 283 ; *Meredith* v. *Meigh* (1853), 2 E. & B. 364; *Hanson* v. *Armitage* (1822), 5 B. & Ald. 557; and see *Hammer and Barrow* v. *Coca-Cola*, [1962] N. Z. L. R. 723; 26 M. L. R. 194.

(*u*) See *post*, Appendix II, Note C.

(*v*) *Buckman* v. *Levi* (1813), 3 Camp. 414 at p. 415 ; *cf. Young* v. *Hobson* (1949), 65 T. L. R. 365, C. A.

"In delivering goods on ship-board, the seller is bound not only to charge the ship-master or shipping company with them effectually, but though not bound to insure, he must give such notice as to enable the buyer to insure" (*w*).

There appears to have been no English decision in point, but the Scottish rule has been adopted by the Act.

Sub-section (3) has no application to c.i.f. (*x*) and ex-ship contracts, but it applies to f.o.b. contracts unless the buyer waives notice or has from the contract itself or from other sources sufficient information to enable him to insure (*a*). It will also apply to c. and f. contracts and to ex-work contracts where the seller sends goods to the buyer by a route involving sea transit.

For a consideration of these various types of contract see pp. 33–41, *ante*.

33. Risk where goods are delivered at distant place.—Where the seller of goods agrees to deliver them at his own risk at a place other than that where they are when sold, the buyer must, nevertheless, unless otherwise agreed (*b*), take any risk of deterioration in the goods necessarily incident to the course of transit (*c*).

Definitions.—For "buyer," "delivery," "goods" and "seller," see s. 62 (1), *post*.

Cross-references.—*Cf.* ss. 6 and 7, *ante* (goods perishing before sale); s. 20, *ante* (risk generally); ss. 29 and 31, *ante* (rules as to delivery); s. 32 (2) and (3), *ante* (delivery to carrier); and s. 55 (1), *post* ("unless otherwise agreed").

(*w*) *Inquiries into the Contract of Sale of Goods and Merchandise*, p. 89. See a review of the Scottish cases by HAMILTON, L.J., in *Wimble* v. *Rosenberg*, [1913] 3 K. B. 743, at p. 762, C. A.

(*x*) *Law and Bonar, Ltd.* v. *British American Tobacco, Ltd.*, [1916] 2 K. B. 605 (leaving open the question whether the sub-section could apply to a c.i.f. contract made at a time when insurance other than that to be provided by the seller was usual).

(*a*) *Northern Steel and Hardware Co.* v. *Batt & Co.* (1917), 33 T. L. R. 516, C. A.; following the majority of the court in *Wimble* v. *Rosenberg*, [1913] 3 K. B. 743, C. A.; but note the judgment of HAMILTON, L.J., commenting on the Scottish law, and holding (contrary to the majority) that the sub-section can have no application to an f.o.b. contract, and the dissenting judgment of VAUGHAN WILLIAMS, L.J., at p. 751, holding that the seller's obligation to give notice is not dispensed with because the buyer has sufficient information from other sources to enable him to insure.

(*b*) See, *e.g.*, *Beaver Specialty, Ltd.* v. *Donald H. Bain, Ltd.* (1974), 39 D. L. R. (3d) 574 (Sup. Ct. of Can.) (sale f.o.b. the buyer's place of business).

(*c*) Compare the non-liability of the insurer in insurance law for *vice propre*, or inherent vice: Chalmers' *Marine Insurance Act*, 1906 (6th ed.), s 55, and notes.

COMMENT

" A manufacturer," says ALDERSON, B.,

> " who contracts to deliver a manufactured article at a distant place, must indeed stand the risk of any extraordinary or unusual deterioration ; but the vendee is bound to accept the article if only deteriorated to the extent that it is necessarily subject to in its course of transit from one place to the other " (*d*).

There appeared to be no reason for confining the rule to the case of a manufacturer, nor is the section inconsistent with the case of *Beer* v. *Walker* (*e*), where the buyer was held entitled to reject rabbits which arrived in Brighton in an unsaleable condition, though they were saleable when sent off from London. In the case of perishable goods, they are not really merchantable when sent off by the seller unless they are in such a condition as to continue to be saleable for a reasonable time (*f*).

34. Buyer's right of examining the goods.—(1) Where goods are delivered to the buyer, which he has not previously examined, he is not deemed to have accepted them unless and until he has had a reasonable opportunity of examining them for the purpose of ascertaining whether they are in conformity with the contract (*g*).

(2) Unless otherwise agreed (*h*), when the seller tenders delivery of goods to the buyer, he is bound,

---

(*d*) *Bull* v. *Robison* (1854), 10 Exch. 342, at p. 346 (hoop-iron sent by canal).

(*e*) (1877), 46 L. J. Q. B. 677. See also *Ollett* v. *Jordan*, [1918] 2 K. B. 41, 47 ; *Healy* v. *Howlett & Son*, [1917] 1 K. B. 337 ; *Broome* v. *Pardess Co-operative Society*, [1940] 1 All E. R. 603 ; *Mash & Murrell* v. *Joseph I. Emanuel Ltd.*, [1961] 1 All E. R. 485 ; [1961] 1 W. L. R. 862 ; reversed on an issue of fact, [1962] 1 All E. R. 77; [1962] 1 W. L. R. 16, C. A.; *Cordova Land Co., Ltd.* v. *Victor Brothers Incorporated*, [1966] 1 W. L. R. 793; *Oleificio Zucchi S.p.A.* v. *Northern Sales, Ltd.*, [1965] 2 Lloyd's Rep. 496, at pp. 517, 518.

(*f*) See comment to s. 14 (2), *ante*.

(*g*) *Lorymer* v. *Smith* (1822), 1 B. & C. 1 ; *Toulmin* v. *Hedley* (1845), 2 C. & K. 157, at p. 160 ; *Heilbutt* v. *Hickson* (1872), L. R. 7 C. P. 438, at p. 456, *per* BRETT, J.; *Bragg* v. *Villanova* (1923), 40 T. L. R. 154 (f.o.b. contract); *B. and P. Wholesale Distributors* v. *Marko*, [1953] C. L. Y. 3266 (opportunity for cursory examination not enough). As to waiver of inspection, see *Castle* v. *Sworder* (1861), 6 H. & N. 828, at p. 837; *Khan* v. *Duché* (1905), 10 Com. Cas. 87; *cf. van den Hurk* v. *Martens & Co.*, [1920] 1 K. B. 850 (ultimate destination); *Thornett* v. *Beers & Son*, [1919] 1 K. B. 486 (a case on s. 14 (2) of the Act).

(*h*) *Pettitt* v. *Mitchell* (1842), 4 M. & Gr. 819 ; *Polenghi Brothers* v. *Dried Milk Co.* (1904), 92 L. T. 64 ; *E. Clemens Horst Co.* v. *Biddell Brothers*, [1912] A. C. 18.

on request, to afford the buyer a reasonable opportunity of examining the goods for the purpose of ascertaining whether they are in conformity with the contract (*i*).

Definitions.—For " contract of sale," see s. 1 (1), *ante*, and s. 62 (1), *post*. For " buyer," " delivery," " goods " and " seller," see s. 62 (1), *post*.

Cross-references.—*Cf.* s. 35, *post* (acceptance) ; s. 14 (2), *ante* (proviso as to examination) ; s. 15 (2) (b), *ante* (sale by sample) ; and s. 55 (1), *post* (" unless otherwise agreed ").

COMMENT

Section 34 was formerly governed by, and had to be read in the light of, s. 35, *post*, so that a buyer who did any act amounting to an acceptance within the meaning of s. 35 could not reject the goods even though he had no reasonable opportunity of examining them (*k*).

Section 35 has been amended by s. 4 (2) of the Misrepresentation Act, 1967, *post*, which has inserted the words " (except where section 34 of this Act otherwise provides) " before the words " when the goods have been delivered to him, and he does any act in relation to them which is inconsistent with the ownership of the seller." A buyer will no longer therefore be deemed to have accepted the goods by doing an act inconsistent with the ownership of the seller unless and until he has had a reasonable opportunity of examining them in accordance with s. 34. What is a reasonable opportunity for examining the goods is a question of fact.

" Suppose," says BRAMWELL, B.,

> " I order a certain quantity of lime to be taken to a farm, and I am not there to object, and nobody else is there . . . to object to it, I shall not be at liberty afterwards to say : ' Those goods have not been accepted and received by me ' ; they have been, as much as it was possible, unless I had chosen to be there to make objection. So, on the other hand, if I go to a shop for an article I have previously ordered, and it is delivered to me, wrapped up, though I cannot see what it is, there cannot be the slightest question that I have received and accepted the goods, if they turn out to be in conformity with the order ; yet nobody can say that I shall not

(*i*) *Isherwood* v. *Whitmore* (1843), 11 M. & W. 347, at p. 350 (goods in closed casks).

(*k*) *Hardy* v. *Hillerns & Fowler*, [1923] 2 K. B. 490, C. A. ; followed in *E. & S. Ruben* v. *Faire Bros. & Co.*, [1949] 1 All E. R. 215 ; [1949] 1 K. B. 254 ; followed but doubted by ROXBURGH, J. in *Pelhams* v. *Mercantile Commodities Syndicate*, [1953] 2 Lloyds Rep. 281. It may be noted that it is only when the buyer was held to have accepted the goods in the second of the three ways contemplated by s. 35 that there was any question of a conflict between that section and s. 34 (1).

have a right to object to them afterwards, if they are not in conformity with the contract" (*l*).

Place for examination.—As a general rule, the place of delivery is *primâ facie* the place of examination; but the general rule is always liable to be displaced by the circumstances of the particular contract (*m*), and the parties may make terms of their own both as to place and kind of examination which may materially affect other terms in the contract (*n*).

35. Acceptance.—The buyer is deemed to have accepted the goods when he intimates to the seller that he has accepted them, or (except when section 34 of this Act otherwise provides) (*o*) when the goods have been delivered to him (*p*), and he does any act in relation to them which is inconsistent with the ownership of the seller (*q*), or when after the lapse of a reasonable time, he retains the goods without intimating to the seller that he has rejected them (*r*).

(*l*) *Castle* v. *Sworder* (1860), 29 L. J. Ex. 235, at p. 238; reversed, (1861) 30 L. J. Ex. 310, Ex. Ch.; see at p. 312.

(*m*) *Saunt* v. *Belcher & Gibbons* (1920), 90 L. J. K. B. 541; distinguishing *Van den Hurk* v. *Martens*, [1920] 1 K. B. 850; *Boks* v. *Rayner & Co.* (1921), 37 T. L. R. 519, affirmed by C. A., *ibid.* 800 (f.o.b. contract); *cf. Scalaris* v. *Ofverberg & Co.* (1921), 37 T. L. R. 307, C. A. (f.o.b. contract); *B. & P. Wholesale Distributors* v. *Marko*, [1953] C. L. Y. 3266; and compare the position where delivery is made to a carrier, p. 193, *ante*.

(*n*) *Potts & Co., Ltd.*, v. *Brown, Macfarlane & Co., Ltd.* (1925), 30 Com. Cas. 64 (where such a clause resulted in an extension of time for delivery); *Ruben* v. *Faire*, [1949] 1 All E. R. 215; [1949] 1 K. B. 254.

(*o*) The words in brackets were added by s. 4 (2) of the Misrepresentation Act, 1967, Appendix I, *post*. See further 30 M. L. R. 369.

(*p*) If the goods are delivered to a sub-purchaser at the request of the buyer, there is a constructive delivery to the buyer within the meaning of the section; *E. & S. Ruben* v. *Faire Bros. & Co.*, [1949] 1 All E. R. 215; [1949] 1 K. B. 254; and see *Hammer and Barrow* v. *Coca-Cola*, [1962] N. Z. L. R. 723; 26 M. L. R. 194. In a c.i.f. contract the goods are "delivered" within the meaning of this section when they are put on board a ship at the port of shipment: *Chao* v. *British Traders*, [1954] 1 All E. R. 779; [1954] 2 Q. B. 459.

(*q*) *Parker* v. *Palmer* (1821), 4 B. & Ald. 387; *Chapman* v. *Morton* (1843), 11 M. & W. 534; *Harnor* v. *Groves* (1855), 15 C. B. 667; *Heilbutt* v. *Hickson* (1872), L. R. 7 C. P. 438, at p. 451 *per* BOVILL, C.J.; *Mechan* v. *Bow McLachlan & Co.*, 1910 S. C. 758; *Wallis, Son and Wells* v. *Pratt*, [1911] A. C. 394, H. L., and see also *per* FLETCHER MOULTON, L.J. in the Court of Appeal, [1910] 2 K. B. at pp. 1013, 1015; *Hardy* v. *Hillerns & Fowler*, [1923] 2 K. B. 490 C. A.; *Libau Wood Co.* v. *H. Smith & Sons, Ltd.* (1930), 37 Lloyds Rep. 296; *E. & S. Ruben* v. *Faire Bros. & Co.*, [1949] 1 All E. R. 215; [1949] 1 K. B. 254; *Chao* v. *British Traders*, [1954] 1 All E. R. 779; [1954] 2 Q. B. 459; see also 108 L. J. 68.

(*r*) *Sanders* v. *Jameson* (1848), 2 Car. & Kir. 557; *Heilbutt* v. *Hickson* (1872), L. R. 7 C. P. 438, at p. 452 *per* BOVILL, C.J.; *Reeves* v. *Armour*, [1920] 3 K. B. 614; *Leaf* v. *International Galleries*, [1950] 1 All E. R. 693, C. A., *per* DENNING, L.J. at p. 695; [1950] 2 K. B. 86. See too the cases on "alse or return" noted to s. 18, rule (4), *ante*.

ILLUSTRATIONS

(1) C.i.f. contract for the sale of wheat. Buyers take delivery, and resell and deliver part to sub-purchasers. It is then discovered that the whole of the wheat is not in conformity with the contract. Prior to the coming into force of the Misrepresentation Act, 1967, the buyers would always have lost their right to reject (*s*). Now the buyers will not lose their right to reject if they have not examined the wheat prior to delivery unless and until they have had a reasonable opportunity of examining it to ascertain whether it is in conformity with the contract.

(2) Sale of 25 sacks of flour. Buyer, after discovering that the flour does not answer the contract description, uses two of the sacks and sells another. He has lost his right to reject (*t*).

(3) Contract for the supply of 2 feed-tanks to a firm of shipbuilders for a tug which they are building for the Admiralty. By the contract, the tanks are to be made " to British Admiralty latest tests and requirements ". The tanks are delivered without having been tested by the Admiralty inspector and the shipbuilders fit them into the tug in ignorance of this fact, although they could have ascertained it by examination of the tanks. Thereafter the Admiralty inspect and condemn the tanks. The shipbuilders have lost their right to reject (*u*).

(4) Contract for sale of 200,000 yo-yos by plaintiffs to C, 85,000 of which are to be delivered to N. Before deliveries begin, C re-sells the 85,000 yo-yos to N. The plaintiffs deliver them to N and they are found to be defective. Neither the delivery to N, which was in accordance with the terms of the contract, nor the re-sale, which was *before* delivery, disentitle C to reject the yo-yos (*v*).

Definitions.—For " buyer," " delivery," " goods " and " seller," see s. 62 (1), *post*.

Cross-references.—*Cf.* s. 11 (1) (c), *ante* (effect of acceptance); ss. 27 and 28, *ante* (buyer's duty to accept); s. 34, *ante* (buyer's right to examine); s. 36, *post* (rejected goods); s. 37, *post* (buyer's liability for non-acceptance); s. 50, *post* (seller's remedy for non-acceptance); and s. 56, *post* (reasonable time a question of fact).

COMMENT

Most of the numerous decisions relating to acceptance have arisen on the construction of the provisions of the Statute of Frauds relating to the sale of goods, or of s. 4 of this Act, all of which have now been repealed. For that reason they must be looked at critically. Before the repeal there could be an acceptance within the meaning of s. 4, which yet was not an acceptance in performance

(*s*) *Hardy & Co.* v. *Hillerns & Fowler,* [1923] 2 K. B. 490, C. A.
(*t*) *Harnor* v. *Groves* (1855), 15 C. B. 667.
(*u*) *Mechan* v. *Bow McLachlan & Co.*, 1910 S. C. 758.
(*v*) *Hammer and Barrow* v. *Coca-Cola,* [1962] N. Z. L. R. 723; 26 M. L. R. 194.

of the contract. For the purpose of determining whether there is an acceptance in this latter sense those cases may still be of use. At one time acceptance within the meaning of the Statute of Frauds was interpreted in the same sense as acceptance in performance of the contract, but in the course of time two different interpretations of the word were developed and there was much learning on the distinction between the two. All this is now happily obsolete.

Rules for presuming acceptance.—The question whether the buyer has accepted the goods is a question of fact.

Section 35 lays down three situations in which a buyer will be deemed to have accepted the goods. Only the second causes any difficulty. If goods are delivered to a buyer which he has not previously examined, he will not be deemed to have accepted them by doing an act inconsistent with the ownership of the seller unless and until he has had a reasonable opportunity of examining them for the purpose of ascertaining whether they conform with the contract (*a*). Subject to this, it is clear that if, after they are delivered to him, the buyer re-sells the whole of the goods he will be deemed to have accepted them (*b*), whether or not the defect or misdescription was discoverable on a reasonable examination (*c*). But a re-sale before delivery, even where the buyer has already examined the goods, but without finally accepting them, need not be an acceptance under s. 35, which covers acts inconsistent with the seller's ownership only if done after delivery (*d*). If, however, pursuant to such re-sale the buyer asks the seller to deliver the goods to the premises of the sub-purchaser, and the seller does so, then he may be held to have done so as agent for the buyer. In that case, the delivery will be an act, inconsistent with the seller's ownership, done by the buyer's agent, and, subject to s. 34 (1), will amount to an acceptance under s. 35 (*e*).

It might be thought that the rule as to re-sale after delivery would be the same when the buyer re-sells part only of the goods, but it seems that this is not always so. If none of the goods delivered are of the contract description, re-sale of part will

---

(*a*) See s. 34 (1), *ante*.

(*b*) *Wallis* v. *Pratt and Haynes*, [1911] A. C. 394. See also *Commercial Fibres* v. *Zabaida*, [1975] 1 Lloyd's Rep. 27 (defective packing apparent to agents who forwarded goods to sub-buyer).

(*c*) *Wallis* v. *Pratt and Haynes*, [1911] A. C. 394 (goods not complying with contract description); *Mechans, Ltd.* v. *Highland Marine Charters, Ltd.*, 1964 S. C. 48 (latent defect not discoverable when goods accepted).

(*d*) *Ruben* v. *Faire*, [1949] 1 All E. R. 215; [1949] 1 K. B. 254; *Hammer and Barrow* v. *Coca-Cola, supra*, illustration (4); 12 M. L. R. 368.

(*e*) *Ruben* v. *Faire, supra*; *cf. Hammer and Barrow* v. *Coca-Cola, supra*, illustration (4), where delivery to the sub-purchaser was a term of the main contract.

subject to s. 34 (1), amount to an acceptance of the whole (*f*); but it has been held, applying section 30 (3), that if the seller delivers goods part of which answer the contract description and part of which do not, a re-sale by the buyer of the part which conforms with the contract will not preclude him from rejecting the part which does not (*g*).

In a c.i.f. contract, where the transfer of the shipping documents passes the property in the goods to the buyer, it is not at first sight easy to see how the buyer can after that point do an act inconsistent with the owership of the seller when, *ex hypothesi*, the seller has no ownership. The problem was dealt with by DEVLIN, J. in *Chao* v. *British Traders* (*h*) where he said that the transfer of the shipping documents was effective only to pass a *conditional* property to the buyer. Thus if the buyer subsequently deals with the shipping documents, he is dealing only with his own conditional property in the goods and will not be taken thereby to have accepted them; but if he deals physically with the goods themselves after they have been landed, e.g. if he delivers them to a sub-buyer, he will be doing an act inconsistent with the seller's reversionary interest in the goods, and may thus be taken to have accepted them. Neither a pledge of the goods nor a sale of the documents, however, is inconsistent with the seller's reversionary interest (*i*).

Quite apart from the right to reject for breach of condition, which is controlled by section 11 (1) (c) of the Act, a buyer may also have a right to rescind the contract for material misrepresentation. In that case also the right will be lost if the buyer accepts the goods (*k*), not by reason of the operation of section 11 (1) (c), but as a result of the general limitations imposed by equity upon the right of rescission.

Rejection.—Rejection by the buyer must take place within a reasonable time. As SCRUTTON, L.J. has said,

> "When one party to a contract becomes aware of a breach of a condition precedent by the other, he is entitled to a reasonable time to consider what he will do, and failure to reject at once does not prejudice his right to reject if he exercises it within a reasonable time" (*l*).

(*f*) *Hardy* v. *Hillerns and Fowler*, [1923] 2 K. B. 490, C. A.

(*g*) *Barker* v. *Agius* (1927), 33 Com. Cas. 120.

(*h*) [1954] 1 All E. R. 779; [1954] 2 Q. B. 459; *J. Rosenthal & Sons, Ltd.* v. *Esmail*, [1965] 2 All E. R. 860, at p. 869; [1965] 1 W. L. R. 1117, at p. 1131, H. L.; see also 12 M. L. R. 368; *Hammer and Barrow* v. *Coca-Cola*, [1962) N. Z. L. R. 723; 26 M. L. R. 194.

(*i*) *A fortiori* where the buyer rejects the shipping documents but unloads and stacks the cargo as agent for the shipowner he is not deemed to have accepted the goods: *Libau Wood Co.* v. *H. Smith & Sons, Ltd.* (1930), 37 Ll. L. Rep. 296.

(*k*) See *Leaf* v. *International Galleries*, [1950] 1 All E. R. 693; [1950] 2 K. B. 86, C. A.; *Long* v. *Lloyd*, [1958] 2 All E. R. 402; [1958] 1 W. L. R. 753, C. A.; *Diamond* v. *B.C. Thoroughbred Breeders Society and Boyd* (1966), 52 D. L. R. (2d.) 146.

(*l*) *Fisher Reeves & Co.* v. *Armour & Co.*, [1920] 3 K. B. 614, at p. 624, C. A. *Cf. Chao* v. *British Traders*, [1954] 1 All E. R. 779; [1954] 2 Q. B. 459.

By s. 56, *post,* what is a reasonable time is a question of fact. In determining what is a reasonable time, the conduct of the seller may be taken into consideration, so that the time for rejection will be extended if the seller has induced the buyer to prolong the trial of the goods by a misrepresentation (*m*), or by a promise which he cannot fulfil (*n*), or by acquiescing by silence to an extended trial (*o*), or by encouraging the buyer to give the goods a fair trial (*p*). Nor does a buyer necessarily lose the right to reject because he tries to put goods sold into working order (*q*).

Where there are negotiations between the buyer and the seller with a view to settling the buyer's claim, these negotiations will also be taken into account (*r*).

A rejection must be clear and unequivocal or the buyer may be held to have accepted the goods on the strength of his subsequent conduct (*s*). Moreover, if the buyer continues to use the goods in a manner not consistent with a *bonâ fide* intention to reject them, he may not be entitled to rely on his rejection and may be treated as having accepted them (*t*). But once he has clearly rejected them and

---

(*m*) *Heilbutt* v. *Hickson* (1872), L. R. 7 C. P. 438; *Munro & Co.* v. *Bennet & Son,* 1911 S. C. 337; and see *Freeman* v. *Consolidated Motors, Ltd.* (1968), 69 D. L. R. (2d) 581 (right to reject not lost by accepting "forced alternative" imposed by seller); *Rafuse Motors* v. *Mardo Constructions* (1963), 41 D.L.R. (2d) 340 (settlement of claim to reject induced by misrepresentation of seller does not bind buyer); *Beldessi v. Island Equipment, Ltd.* (1974), 41 D. L. R. (3d) 147 (buyer showed " unusual patience " partly because of seller's representation that the fault was that of the operator of the machine sold).

(*n*) *Scholfield* v. *Emerson Brantingham Implements* (1918), 43 D. L. R. 509 (Sup. Ct. of Canada) (representation that the vehicle would be all right in time, or if not, then the seller would make it all right).

(*o*) *Lucy* v. *Mouflet* (1860), 5 H. & N. 229.

(*p*) *Rafuse Motors* v. *Mardo Constructions, supra.* See also *Barker* v. *Inland Truck Sales* (1970), 11 D. L. R. (3d) 469 (buyer retained possession of a truck for six months, giving seller every opportunity to correct defects in it and attempting to use it for the purpose for which he bought it: right to reject for breach of a fundamental term not lost). But this case is difficult to reconcile with *Long* v. *Lloyd,* [1958] 2 All E. R. 402; [1958] 1 W. L. R. 753, C. A., discussed in *Benjamin's Sale of Goods,* para. 903. See also *Burrough's Business Machines, Ltd.* v. *Feed Rite Mills* (1962), *Ltd.* (1974) 42 D. L. R. (3d) 303; *Public Utilities Commission* v. *Burrough's Business Machines, Ltd.* (1975), 52 D. L. R. (3d) 481 (Agreement for sale and installation of computer system: buyer entitled to reject after seller had tried unsuccessfully for several months to make it function satisfactorily).

(*q*) *Beldessi* v. *Island Equipment, Ltd.* (1974), 41 D. L. R. (3d) 147.

(*r*) *Polar Refrigeration Service, Ltd.* v. *Moldenhauer* (1967), 61 D. L. R. (2d) 462.

(*s*) *Chapman* v. *Morton* (1843), 11 M. & W. 534.

(*t*) *Cornwal* v. *Wilson* (1750), 1 Ves. Sen. 509 (re-sale in manner only consistent with there having been no rejection); *Electric Construction Co.* v. *Hurry and Young* (1897), 24 R. 312 (Ct. of Sess.) (continued user for 3 months after rejection); *Croom & Arthur* v. *Stewart & Co.* (1905), 7 F. (Ct. of Sess.) 563. See also *Central Farmers, Ltd.* v. *Smith,* 1974 S. L. T. (Sh. Ct.) 87. But compare *Public Utilities Commission* v. *Burrough's Business Machines, Ltd.* (1975), 52 D. L. R. (3d) 481 (buyer continued to use machine after rejection while obtaining a replacement because business could not function properly without it. Held: still entitled to reject).

rescinded the contract, a subsequent inadvertent act on his part inconsistent with the seller's ownership of them cannot amount to an acceptance and give new life to a contract already at an end (*u*). Where an effective examination of the goods is possible only by consuming or destroying a sample of them a buyer is not debarred from rejecting the remainder merely because he has consumed or destroyed such a sample, or because he has sold such a sample for consumption (*a*).

If the seller refuses to take the goods back when tendered, the buyer may still not deal with the goods as his own, for the property in them has revested in the seller. In exceptional cases where the goods are liable to deteriorate, he may be able to sell them as agent of necessity for the seller (*b*). Generally, however, he would probably be better advised to commence proceedings against the seller and apply to the court for an order for sale (*c*).

If the buyer deals with the goods without an order for sale and in circumstances in which he cannot claim to be an agent of necessity, then unless perhaps the seller's refusal is in such terms as to entitle the buyer to do as he wishes with the goods, the buyer may be liable to the seller for conversion. Probably, however, in such a case the seller could not claim that the buyer's conduct in selling the goods was inconsistent with a *bonâ fide* intention to reject them.

(*u*) *Breckwoldt* v. *Hanna* (1963), 5 W. I. R. 356, applying *Hardy & Co.* v. *Hillerns and Fowler*, [1923] 2 K. B. 490, C. A.

(*a*) *Lucy* v. *Mouflet* (1860), 5 H. & N. 229 (beer); *Winnipeg Fish Co.* v. *Whitman Fish Co.* (1909), 41 S.C.R. 453 (Sup. Ct. of Canada) (frozen fish).

(*b*) *Kemp* v. *Pryor* (1802), 7 Ves. Jun. 237, at p. 247. But *of. Bowstead on Agency* (13th ed.), p. 31. *Cornwal* v. *Wilson* (1750), 1 Ves. Sen. 509 turned upon a mercantile custom that where a factor acquired goods for a principal in excess of authority, the principal could reject them, so that they became the property of the factor, and in certain circumstances the principal could deal with them as agent for the factor. It does not establish any general proposition as to agents of necessity. In *Benjamin's Sale of Goods*, para. 875, it is suggested that *Caswell* v. *Coare* (1809), 1 Taunt. 566 and *Chesterman* v. *Lamb* (1834), 2 Ad. & El. 129 are possibly also examples of agency of necessity. But those cases are clearly concerned with damages recoverable for breach of warranty where the buyer has no right to treat the contract as terminated, and establish that although the property remains in the buyer, if he wishes to recover the expenses of keeping the goods until he can resell them, he should first give the seller the opportunity to rescind the contract by agreement: see Selwyn's *Law of Nisi Prius* (13th ed.), Vol. I, pp. 577–578: *M'Kenzie* v. *Hancock* (1826), Ry. & M. 436; *Watson* v. *Denton* (1835), 7 C. & P. 85, 90. *Cf.* p. 224, *post.*

(*c*) The application will be under R. S. C., Order 29, r. 4 in the High Court, or under C. C. R., Order 13, r. 12 in the County Court. In *The Hansa Nord*, [1974] 2 Lloyd's Rep. 216 (reversed, apparently on other grounds, *sub. nom. Cehave N.V.* v. *Bremer Handelgesellschaft* (1975), *Times*, 22nd July, C. A.) the buyer, fearing that the goods might become a total loss, obtained an order from a Dutch Court for their sale, the net proceeds to be deposited to the account of the legal owners. The goods were then sold by the Court's agent to the buyers, who used them for their own purpose. MOCATTA, J., held that this was not an act inconsistent with the seller's ownership for the purposes of this section.

Conditional acceptance.—Goods may, of course, by arrangement, be accepted conditionally and the acceptance may in such case be withdrawn on failure of the condition (*d*).

Scotland.—The right of rejecting goods as not being in conformity with the contract is larger in Scotland than in England. In Scotland a buyer may reject goods which he has accepted if he do so " timeously," whereas in England he could only do so if the contract contained what the continental lawyers call a " resolutive condition " (*e*).

36. Buyer not bound to return rejected goods.—Unless otherwise agreed, where goods are delivered to the buyer, and he refuses to accept them having the right so to do, he is not bound to return them to the seller, but it is sufficient if he intimates to the seller that he refuses to accept them (*f*).

Definitions.—For " buyer," " delivery," " goods " and " seller," see s. 62 (1), *post*.

Cross-references.—*Cf.* s. 35, *ante* (acceptance), and s. 55 (1), *post* (" unless otherwise agreed ").

COMMENT

" The buyer," says BRETT, J.,

> " may, in fact, return [the goods] or offer to return them, but it is sufficient I think, and the more usual course is to signify his rejection of them by stating that the goods are not according to contract, and that they are at the vendor's risk. No particular form is essential. It is sufficient if he does any unequivocal act showing that he rejects them " (*g*).

The buyer has no lien over the goods to secure repayment of the purchase price (*h*).

This section presupposes the relation of seller and buyer. It does not appear to touch the case of goods delivered to a man on the chance that he may buy them.

(*d*) *Lucy* v. *Mouflet* (1860), 5 H. & N. 229 ; *Heilbutt* v. *Hickson* (1872), L. R. 7 C. P. 438 ; *cf. Behrend & Co.* v. *Produce Brokers Co.*, [1920] 3 K. B. 530; *Adams* v. *Sprung, The Guardian*, 8th November 1961, C. A.

(*e*) See s. 11 (2), *ante*, and *Couston* v. *Chapman* (1872), L. R. 2 Sc. & Div. 250, at p. 254. For resolutive conditions in England, see *Lamond* v. *Davall* (1847), 9 Q. B. 1030 ; *Head* v. *Tattersall* (1871), L. R. 7 Exch. 7.

(*f*) *Grimoldby* v. *Wells* (1875), L. R. 10 C. P. 391; as to the place of rejection, see *Heilbutt* v. *Hickson* (1872), L. R. 7 C. P. 438, at p. 456, *per* BRETT, J. As to where the buyer properly rejects goods and the seller refuses to take them back, see *ante*, p. 202.

(*g*) *Grimoldby* v. *Wells* (1875), L. R. 10 C. P. 391, at p. 395.

(*h*) *Lyons & Co.* v. *May & Baker*, [1923] 1 K. B. 685.

The buyer is entitled to take into account in claiming damages any expenses incurred by him in looking after the goods until the seller recovers them in accordance with the ordinary rules of assessing damages for breach of contract (*i*).

37. Liability of buyer for neglecting or refusing delivery of goods.—When the seller is ready and willing to deliver the goods, and requests the buyer to take delivery, and the buyer does not within a reasonable time after such request take delivery of the goods, he is liable to the seller for any loss occasioned by his neglect or refusal to take delivery, and also for a reasonable charge for the care and custody of the goods (*k*) : Provided that nothing in this section shall affect the rights of the seller where the neglect or refusal of the buyer to take delivery amounts to a repudiation of the contract (*l*).

ILLUSTRATION

A contracts to build a steam launch for B by a fixed date, delivery to be made on a vessel found by B. A does not complete the launch by the contract time, but B does not notify A of a ship to receive the launch till A is ready to deliver. Neither party has any remedy against the other (*m*).

Definitions.—For "buyer," "delivery," "goods" and "seller," see s. 62 (1), *post*.

Cross-references.—*Cf.* s. 20, *ante* (risk) ; ss. 27 and 28, *ante* (buyer's duty to accept) ; ss. 29, 31 and 32, *ante* (rules as to delivery) ; s. 50, *post* (seller's remedy for non-acceptance) ; and s. 56, *post* (reasonable time a question of fact).

COMMENT

Conversely, if the seller is in default in making delivery, and the buyer, notwithstanding the delay, accepts the goods, he may recover damages for any loss occasioned by the delay ; see note to s. 51, *post*.

(*i*) See *post*, comment to ss. 53, 54. The authorities relied on in *MacGregor on Damages* (13th ed.), para. 594, and *Benjamin's Sale of Goods*, para. 875, are not in fact concerned with this situation, for the reasons given *ante*, n. (*b*) at p. 202.

(*k*) *Greaves* v. *Ashlin* (1813), 3 Camp. 426 ; *cf. Bloxam* v. *Sanders* (1825), 4 B. & C. 941, at p. 950; and see also *Penarth Dock Engineering Co.* v. *Pounds* [1963] 1 Lloyd's Rep. 359.

(*l*) *Cf. Mersey Steel Co.* v. *Naylor & Co.* (1884), 9 App. Cas. 434, at p. 443 ; *Braithwaite* v. *Foreign Hardwood Co.*, [1905] 2 K. B. 543, C. A., criticised, *British & Beningtons, Ltd.* v. *North Western Cachar Tea Co. Ltd.*, [1923] A. C. 48, at p. 70, *per* Lord SUMNER. See also s. 31, *ante*.

When the seller holds the goods in the exercise of his right of lien, he cannot charge for expenses of keeping them. See note to s. 41, *post.*

PART IV

RIGHTS OF UNPAID SELLER AGAINST THE GOODS

38. Unpaid seller defined.—(1) The seller of goods is deemed to be an " unpaid seller " within the meaning of this Act—

(a) When the whole of the price has not been paid or tendered (*n*);

(b) When a bill of exchange or other negotiable instrument has been received as conditional payment, and the condition on which it was received has not been fulfilled by reason of the dishonour of the instrument or otherwise (*o*).

(2) In this Part of this Act the term " seller " includes any person who is in the position of a seller, as, for instance, an agent of the seller to whom the bill of lading has been indorsed (*p*), or a consignor or agent who has himself paid, or is directly responsible for, the price (*q*).

Definitions.—For " goods," see s. 62 (1), *post.* The definition of " seller " in sub-s. (2) of this section extends, for this and the succeeding ten sections, the definition given in s. 62 (1), *post.*

Cross-references.—*Cf.* ss. 1 (1), 8 and 9, *ante* (price), and ss. 27 and 28, *ante* (buyer's duty to pay the price).

(*m*) *Forrestt* v. *Aramayo* (1900), 9 Asp. M. L. C. 134, C. A.

(*n*) *Hodgson* v. *Loy* (1797), 7 Term Rep. 440 ; *Feise* v. *Wray* (1802), 3 East, 93, at p. 102 ; *Van Casteel* v. *Booker* (1848), 2 Exch. 691, at pp. 702, 709 ; *Ex p. Chalmers* (1873), 8 Ch. App. 289 (severable contract). As to tender after the appointed day, see *Martindale* v. *Smith* (1841), 1 Q. B. 389

(*o*) *Feise* v. *Wray* (1802), 3 East, 93; *Griffiths* v. *Perry* (1859), 1 E. & E. 680; *Ex p. Lambton* (1875), 10 Ch. App. 405, at p. 415; *cf. Ex p. Stapleton* (1879), 10 Ch. D. 586, C. A. As to conditional payment, see p. 230, *post.*

(*p*) *Morison* v. *Gray* (1824), 2 Bing. 260. See, too, the Bills of Lading Act, 1855, *post,* Appendix I.

(*q*) *Feise* v. *Wray* (1802), 3 East, 93 ; *Tucker* v. *Humphrey* (1828), 4 Bing. 516 ; *cf. Ireland* v. *Livingston* (1872), L. R. 5 H. L. 395, at pp. 408, 409, *per* BLACKBURN, J. As to factors, see notes to *Kruger* v. *Wilcox* (1755), Tudor's *Merc. Cases* (3rd ed.), p. 370.

COMMENT

Sub-section (1).—In a case where the seller had discounted the buyer's acceptances, but the latter failed before the bills matured, it was held that the seller was unpaid, and MELLISH, L.J., says,

> "If the bill is dishonoured before delivery has been made, then the vendor's lien revives; or if the purchaser becomes openly insolvent before the delivery actually takes place, then the law does not compel the vendor to deliver to an insolvent purchaser" (*r*).

Sub-section (2).—The Courts show a strong inclination to give the rights of an unpaid seller against the goods to any one whose position can be shown to be substantially analogous to that of an ordinary seller (*s*).

But a buyer of goods who pays for them, and then on examination rejects them, has not the status of an unpaid seller, and has no lien on the goods for the price repayable to him (*t*).

39. Unpaid seller's rights.—(1) Subject to the provisions of this Act (*u*), and of any statute in that behalf (*a*), notwithstanding that the property in the goods may have passed to the buyer, the unpaid seller of goods, as such, has by implication of law—

(a) A lien on the goods [or right to retain them] for the price while he is in possession of them;

(b) In case of the insolvency of the buyer, a right of stopping the goods in transitu after he has parted with the possession of them;

(c) A right of re-sale as limited by this Act (*b*).

(2) Where the property in goods has not passed to the buyer, the unpaid seller has, in addition to his

---

(*r*) *Gunn* v. *Bolckow, Vaughan & Co.* (1875), 10 Ch. App. 491, at p. 501, overruling on this point, it seems, *Bunney* v. *Poyntz* (1833), 4 B. & Ad. 568; but see *Benjamin on Sale* (8th ed.), p. 844.

(*s*) *Cassaboglou* v. *Gibb* (1883), 11 Q. B. D. 797, at p. 804, *per* BRETT, M.R.; and, for examples, see *Jenkyns* v. *Usborne* (1844), 7 M. & Gr. 678, at p. 698 (re-sale by party who had contracted to buy goods); *Imperial Bank* v. *London & St. Katherine Docks Co.* (1877), 5 Ch. D. 195 (surety who has paid the price); *cf. Siffken* v. *Wray* (1805), 6 East, 371 (surety for price not entitled to stop goods in transit); *Longbottom & Co.* v. *Bass, Walker & Co.*, [1922] W. N. 245, C. A.

(*t*) *Lyons & Co.* v. *May & Baker*, [1923] 1 K. B. 685.

(*u*) See, as to lien, ss. 41 to 43 and ss. 47 and 48, *post*; as to stoppage *in transitu*, ss. 44 to 48, *post*.

(*a*) See the Factors Act, 1889, *post*.

(*b*) See s. 48, *post*.

other remedies, a right of withholding delivery similar to and co-extensive with his rights of lien and stoppage in transitu where the property has passed to the buyer.

Definitions.—For " buyer," " goods," " lien " and " property," see s. 62 (1), *post*. For " unpaid seller," see s. 38, *ante*.

Cross-references.—*Cf.* ss. 1 (1) and 8, *ante* (price) ; ss. 17 and 18, *ante* (passing of property) ; and ss. 27 and 28, *ante* (buyer's duty to pay price).

COMMENT

Sub-section (1).—The origin of the seller's lien in English law is doubtful. It is probably founded in the custom of merchants (*c*). The term " lien " is unfortunate, because the seller's rights, arising out of his original ownership, in all cases exceed a mere lien. They

> " perhaps come nearer to the rights of a pawnee with a power of sale than to any other common law rights " (*d*).

Many of the cases fail to distinguish the seller's right of lien from his right of stoppage *in transitu*. But it is important to keep them distinct, because, though the rights are analogous, they are in certain respects governed by different considerations (*e*). The seller's lien attaches when the buyer is in default, whether he be solvent or insolvent. The right of stoppage *in transitu* only arises when the buyer is insolvent. Moreover, it does not arise until the seller's lien is gone, for it presupposes that the seller has parted with the possession as well as with the property in the goods. It is peculiar to contracts of sale, and can only be exercised by a seller or person in the position of a seller (*f*).

The Courts look with great favour on the right of stoppage *in transitu* on account of its intrinsic justice (*g*). The decisions on the subject are very numerous, but as JESSEL, M.R., observes,

> " As to several of them there is great difficulty in reconciling them with principle ; as to others there is great difficulty in reconciling them with one another ; and as to the whole, the law on this subject is in a very unsatisfactory state " (*h*).

(*c*) *Blackburn on Sale* (1st ed.), p. 318.

(*d*) *Blackburn on Sale* (1st ed.), p. 325; *cf. Bloxam* v. *Sanders* (1825), 4 B. & C. 941, at p. 948; *Schotsmans* v. *Lancashire and Yorkshire Rail. Co.* (1867), 2 Ch. App. 332, at p. 340.

(*e*) *Blackburn on Sale* (1st ed.), p. 308; *cf. Bolton* v. *Lancashire and Yorkshire Rail. Co.* (1866), L. R. 1 C. P. 431, at p. 439, *per* WILLES, J.

(*f*) *Sweet* v. *Pym* (1800), 1 East. 4; *Siffken* v. *Wray* (1805), 6 East, 371. See the definition of " seller " for this purpose in s. 38 (2), *ante*.

(*g*) See *Cassaboglou* v. *Gibbs* (1883), 11 Q. B. D. 797, at p. 804 ; *Kemp* v. *Falk* (1882), 7 App. Cas. 573, at p. 590.

(*h*) *Merchant Banking Co.* v. *Phœnix Co.* (1877), 5 Ch. D. 205, at p. 220 (case of seller's lien).

These decisions must now of course be read subject to the Act.

Sub-section (2).—This sub-section was necessary because it would be a contradiction in terms to speak of a man having a lien upon his own goods. The enactment is declaratory (*a*).

Scotland.—The words "or right to retain them" were inserted when the bill was extended to Scotland.

The seller's "right of retention" in Scotland was more extensive than the seller's lien in England. Apart from statute the seller had the right to retain the goods not only for the price, but also for any other debt due from the buyer even if there had been a sub-sale (*b*). But the Mercantile Law Amendment (Scotland) Act, 1856, s. 2 (now repealed), altered the law in the case of sub-sales, and the Sale of Goods Act appears to apply a uniform rule to both countries.

The Scottish law as to stoppage *in transitu* appears to be similar to English law (*c*).

40. Attachment by seller in Scotland.—In Scotland a seller of goods may attach the same while in his own hands or possession by arrestment or poinding; and such arrestment or poinding shall have the same operation and effect in a competition or otherwise as an arrestment or poinding by a third party.

Definition.—For "seller," see s. 38 (2), *ante.*

COMMENT

This section is taken from s. 3 of the repealed Mercantile Law Amendment (Scotland) Act, 1856. It is probably restrained by the provisions of s. 47, *post.* *Cf.* Green's *Encyclopaedia of Scots Law*, tit. "Poinding," as to the procedure generally.

Unpaid Seller's Lien

41. Seller's lien.—(1) Subject to the provisions of this Act (*d*) the unpaid seller of goods who is in possession of them is entitled to retain possession of them until payment or tender of the price in the following cases, namely:—

(*a*) *Griffiths* v. *Perry* (1859), 1 E. & E. 680, at p. 688; *Ex p. Chalmers* (1873), 8 Ch. App. 289, at p. 292. But see 19 L. Q. R. 113 *per* A. COHEN, K.C.

(*b*) Mercantile Law Commission, 1855, 2nd Report, pp. 8, 9, 44; *Melrose* v. *Hastie*, 1851, 13 Dunl. (Ct. of Sess.) 880.

(*c*) *Allen* v. *Stein* (1790), M. 4949.

(*d*) See ss. 25 (2) and 39, *ante*, and ss. 42, 43, 47 and 48, *post.*

(a) Where the goods have been sold without any stipulation as to credit (*e*);

(b) Where the goods have been sold on credit, but the term of credit has expired (*f*);

(c) Where the buyer becomes insolvent (*g*).

(2) The seller may exercise his right of lien notwithstanding that he is in possession of the goods as agent or bailee [or custodier] for the buyer (*h*).

Definitions.—For "bailee," "buyer," "goods" and "lien," see s. 62 (1), *post*. For "unpaid seller," see s. 38, *ante*. For "insolvent," see s. 62 (3), *post*.

Cross-references.—*Cf.* ss. 1 (1) and 8, *ante* (price), and ss. 27 and 28, *ante* (buyer's duty to pay price).

COMMENT

Sub-section (1).—The lien is a lien for the price only, and not for charges for keeping the goods, for they are kept against the buyer's will (*i*).

A sale on credit excludes the lien during the currency of the credit (*k*), unless there be a trade usage to the contrary (*l*).

As regards instalment contracts, MELLISH, L.J. says that, once the buyer has become insolvent,

> "The seller, notwithstanding he may have agreed to allow credit for the goods, is not bound to deliver any more goods under the contract until the price of the goods not yet delivered is tendered to him; and that if a debt is due to him for goods already delivered, he is entitled to refuse to deliver any more till he is paid the debt due for those already delivered, as well as the price of those still to be delivered. . . . It would be strange if the right of a vendor

(*e*) *Bloxam* v. *Sanders* (1825), 4 B. & C. 941, at p. 948; *Miles* v. *Gorton* (1834), 2 Cr. & M. 504, at p. 511.

(*f*) *New* v. *Swain* (1828), 1 Dan. & Lloyd, 193, *per* BAYLEY, J.; *Bunney* v. *Poyntz* (1833), 4 B. & Ad. 568, at p. 569, *per* LITTLEDALE, J.

(*g*) *Bloxam* v. *Sanders* (1825), 4 B. & C. 941; *Bloxam* v. *Morley* (1825), 4 B. & C. 951; *Griffiths* v. *Perry* (1859), 1 E. & E. 680; *Ex. p. Lambton* (1875), 10 Ch. App., at p. 415; *Gunn* v. *Bolckow, Vaughan & Co.* (1875), 10 Ch. App. 491, at p. 501.

(*h*) *Townley* v. *Crump* (1835), 4 Ad. & El. 58; *Grice* v. *Richardson* (1877), 3 App. Cas. 319, P. C.; *Poulton* v. *Anglo-American Oil Co.* (1911), 27 T. L. R. 216, C. A.

(*i*) *Somes* v. *British Empire Shipping Co.* (1860), 8 H. L. Cas. 338, at p. 345 (shipwright's lien, but the rule was stated to apply to the seller's lien).

(*k*) *Spartali* v. *Benecke* (1850), 10 C. B. 212, at p. 223.

(*l*) *Field* v. *Lelean* (1861), 6 H. & N. 617, Ex. Ch.

who had agreed to deliver goods by instalments were less than that of a vendor who had sold specific goods" (*m*).

Even if the seller has broken his contract to deliver while the buyer is solvent, his lien revives on the buyer becoming insolvent, and the buyer's trustee is only entitled at most to nominal damages for the breach, unless the value of the goods at the time of breach was above the contract price (*n*).

When the seller exercises his right of lien, the buyer's trustee may affirm the contract and obtain the goods by tendering the price within a reasonable time (*o*), for it is clear law that the mere insolvency or bankruptcy of a party to a contract does not rescind it (*p*). But it seems that in the case of insolvency an agreement to rescind will be presumed on slight grounds (*q*).

A sub-purchaser also is probably entitled to obtain the goods by tendering the price to the original seller within a reasonable time (*r*).

Sub-section (2).—This sub-section was originally confined to the case where the buyer was insolvent (*s*). It was altered to its present form in Committee.

42. Part delivery.—Where an unpaid seller has made part delivery of the goods, he may exercise his right of lien [or retention] on the remainder, unless such part delivery has been made under such circumstances as to show an agreement to waive the lien (*t*) [or right of retention].

Definitions.—For "goods," "delivery" and "lien," see s. 62 (1), *post*. For "unpaid seller," see s. 38, *ante*.

Cross-references.—*Cf.* s. 29, *ante* (rules as to delivery); s. 31, *ante* (instalment deliveries); and ss. 39 and 41, *ante*, and s. 43, *post* (lien).

(*m*) *Ex p. Chalmers* (1873), 8 Ch. App. 289, at p. 291; *cf. Ex p. Stapleton* (1879), 10 Ch. D. 586, C. A.

(*n*) *Valpy* v. *Oakeley* (1851), 16 Q. B. 941; *Griffiths* v. *Perry* (1859), 1 E. & E. 680.

(*o*) *Ex p. Stapleton* (1879), 10 Ch. D. 586, C. A.

(*p*) *Mess* v. *Duffus* (1901), 6 Com. Cas. 165 (action for damages for non-acceptance).

(*q*) *Morgan* v. *Bain* (1874), L. R. 10 C. P. 15. As to trustee's right to disclaim onerous contracts, see s. 54 of the Bankruptcy Act, 1914.

(*r*) *Ex p. Stapleton*, (*supra*); and *cf. Kemp* v. *Falk* (1882), 7 App. Cas. 573, at p. 578, *per* Lord SELBORNE.

(*s*) Thus following BLACKBURN, J. in *Cusack* v. *Robinson* (1861), 1 B. & S. 299, at p. 308.

(*t*) *Dixon* v. *Yates* (1833), 5 B. & Ad. 313, at p. 341 (lien); *Miles* v. *Gorton* (1834), 2 Cr. & M. 504; *cf. Ex p. Cooper* (1879), 11 Ch. D. 68, C. A. (stoppage *in transitu*).

COMMENT

In a case where it was unsuccessfully contended that the delivery of part of a cargo to a sub-purchaser was a constructive delivery of the whole, Lord BLACKBURN says:

> " It is said that the delivery of a part is a delivery of the whole. It may be a delivery of the whole. In agreeing for the delivery of goods with a person, you are not bound to take an actual corporeal delivery of the whole in order to constitute such a delivery, and it may very well be that the delivery of a part of the goods is sufficient to afford strong evidence that it is intended as a delivery of the whole. If both parties intend it as a delivery of the whole, then it is a delivery of the whole; but if either of the parties does not intend it as a delivery of the whole, if either of them dissents, then it is not a delivery of the whole " (*u*).

The onus is on the party who relied upon the part delivery as a delivery of the whole to show that the part delivery took place under such circumstances as to make it a delivery of the whole (*v*).

Severable contract.—As regards severable contracts, if, for instance, delivery is to be made by three instalments, and the first instalment has been delivered and paid for, and the second has been delivered but not paid for, the seller cannot withhold delivery of the third instalment till he has been paid for both the second and third instalments, unless (1) the non-payment involves a repudiation of the contract under s. 31, *ante* (*a*), or (2) the buyer is insolvent (*b*). But any instalment which has been paid for must be delivered, even though the buyer be insolvent (*c*).

43. Termination of lien.—(1) The unpaid seller of goods loses his lien [or right of retention] thereon—

(a) When he delivers the goods to a carrier or other bailee [or custodier] for the purpose of transmission to the buyer (*d*) without reserving the right of disposal of the goods;

(*u*) *Kemp* v. *Falk* (1882), 7 App. Cas. 573, at p. 586; citing for the proposition *Dixon* v. *Yates* (*supra*).

(*v*) *Ex p. Cooper* (1879), 11 Ch. D. 68, C.A., at p. 73, *per* BRETT, L.J.

(*a*) *Steinberger* v. *Atkinson & Co.* (1914), 31 T. L. R. 110 (no repudiation).

(*b*) *Ex p. Chalmers* (1873), 8 Ch. App. 289.

(*c*) *Merchant Banking Co.* v. *Phœnix Bessemer Steel Co.* (1877), 5 Ch. D. 205; *Longbottom* v. *Bass Walker & Co.*, [1922] W. N. 245, C. A.

(*d*) *Bolton* v. *Lancashire and Yorkshire Rail. Co.* (1866), L. R. 1 C. P. 431, at p. 439, *per* WILLES, J.; Pollock and Wright, *Possession in the Common Law*, pp. 71, 72; and see the cases cited under s. 32, *ante*.

(b) When the buyer or his agent lawfully obtains possession of the goods (*e*);

(c) By waiver thereof.

(2) The unpaid seller of goods, having a lien [or right of retention] thereon, does not lose his lien [or right of retention] by reason only that he has obtained judgment [or decree] for the price of the goods (*f*).

Definitions.—For " bailee," " buyer," " delivery," " goods " and " lien," see s. 62 (1), *post*. For " unpaid seller," see s. 38, *ante*.

Cross-references.—*Cf.* ss. 1 (1) and 8, *ante* (price); s. 19, *ante* (reservation of right of disposal); s. 32, *ante* (delivery to carrier); and s. 49, *post* (action for price).

COMMENT

When goods are delivered to a carrier for transmission to the buyer the right of lien is lost, but a right of stoppage *in transitu* arises should the buyer become insolvent, and if this is exercised the right of lien revives. Since, in the case of the buyer's insolvency, the two rights are similar in their effects, they are sometimes confused in the cases.

For the most part, the cases on what constituted an actual receipt within the meaning of the repealed section of the Statute of Frauds and the repealed s. 4 of this Act appear to furnish the test for determining whether the seller's lien is gone or not. "The principle," says BLACKBURN, J.,

> " is that there cannot be an actual receipt by the vendee so long as the goods continue in the possession of the seller so as to preserve his lien " (*g*).

But the requirements of the two sections are not identical, for receipt of part of the goods by the buyer, coupled with acceptance, was enough to satisfy the old s. 4, though it would not wholly destroy the lien: s. 42, *ante*. And by s. 41 (2), *ante*, the seller may now exercise his right of lien notwithstanding that he is in possession of

---

(*e*) *Hawes* v. *Watson* (1824), 2 B. & C. 540; *Cooper* v. *Bill* (1865), 3 H. &C. 722; *Dodsley* v. *Varley* (1840), 12 A. and E. 632; *Valpy* v. *Gibson* (1847), 4 C. B. 837 (goods returned to seller for purpose of repacking); *cf. Schotsmans* v. *Lancashire and Yorkshire Railway* (1867), 2 Ch. App. 332, at p. 335, as to stoppage *in transitu*. The sub-section was amended in Committee. See also a Scottish case, *Paton's Trustees* v. *Finlayson*, 1923 S. C. 872 (potatoes lifted by buyer, but left on seller's land: right of retention upheld).

(*f*) *Houlditch* v. *Desanges* (1818), 2 Stark 337; *Scrivener* v. *Great Northern Rail. Co.* (1871), 19 W. R. 388. (*Quaere* whether the lien extends only to the price or also to costs on the judgment?)

(*g*) *Cusack* v. *Robinson* (1861), 1 B. & S. 299, at p. 308; *cf. Baldey* v. *Parker* (1823), 2 B. & C. 37, at p. 44, *per* HOLROYD, J.

the goods as agent or bailee for the buyer, so that the cases before the Act are no longer of authority on this latter point (*h*).

Subject to s. 47, *post*, when goods, at the time of sale, are in the possession of a third person, there is no delivery to the buyer, and the seller's lien therefore is not divested till such third person attorns to the buyer (*i*).

Again, the seller may deliver the goods to the buyer on terms such that the buyer holds them as bailee for the seller ; but in that case the seller has a special property in the goods arising out of the special agreement rather than a lien properly so called (*k*).

Waiver of lien.—The right of lien is given to the seller by implication of law : see s. 39, *ante*. It follows that it may be waived expressly. But it may also be waived by implication. The seller may reserve an express lien which excludes the implied one (*l*), or he may take a bill for the price which ordinarily would exclude his lien during its currency, though the lien would revive on its dishonour (*m*), or he may assent to a sub-sale (*n*), or he may part with the documents of title so as to exclude his lien under the provisions of the Factors Act, *post*, if the documents get into the hands of a holder for value. See, too, s. 55, *post*, as to negativing implied terms. The seller may also waive his lien by some wrongful act, such as dealing with the goods in a manner inconsistent with the mere right to have possession of them, as by wrongfully re-selling or consuming them (*o*), or by claiming to keep them on some ground other than his right of lien. The practical result of this is that if sued in trover by the buyer he cannot defeat the action by setting up his lien or pleading that the price was not tendered before action brought (*p*) : but the damages will only be the value of the buyer's actual interest in the goods, that is, their value less the price, or that part of it which is unpaid (*q*).

Stoppage in transitu

44. Right of stoppage in transitu.—Subject to the provisions of this Act (*r*), when the buyer of goods

(*h*) *Baldey* v. *Parker* (*supra*), at p. 44 ; *Bill* v. *Bament* (1841), 9 M. & W. 36, at p. 41 ; *Cusack* v. *Robinson* (*supra*).

(*i*) *McEwan* v. *Smith* (1849), 2 H. L. Cas. 309, and *ante*, note to s. 29 (3).

(*k*) *Dodsley* v. *Varley* (1840), 12 A. & E. 632, at p. 634, *per* Lord DENMAN.

(*l*) *Re Leith's Estate* (1866), L. R. 1 P. C. 296, at p. 305. As to effect of taking subsequent security, see *Angus* v. *McLachlan* (1883), 23 Ch. D. 330.

(*m*) *Valpy* v. *Oakeley* (1851), 16 Q. B. 941, at p. 951 ; *Griffiths* v. *Perry* (1859), 1 E. & E. 680, at p. 688.

(*n*) *Knights* v. *Wiffen* (1870), L. R. 5 Q. B. 660 ; and see, too, s. 47, *post*.

(*o*) *Gurr* v. *Cuthbert* (1843), 12 L. J. Ex. 309.

(*p*) See *Jones* v. *Tarleton* (1842), 9 M. & W. 675 (carrier) ; *Mulliner* v. *Florence* (1878), 3 Q. B. D. 484, C. A. (innkeeper) ; *Youngmann* v. *Briesemann* (1893), 67 L. T. 642, 644, C. A.

(*q*) *Chinery* v. *Viall* (1860), 5 H. & N. 288.

(*r*) See s. 39, *ante*, and ss. 45 to 47, *post*.

becomes insolvent, the unpaid seller who has parted with the possession of the goods has the right of stopping them in transitu, that is to say, he may resume possession of the goods as long as they are in course of transit, and may retain them until payment or tender of the price (*s*).

Definitions.—For " buyer " and " goods," see s. 62 (1), *post*. For " unpaid seller," see s. 38, *ante*. For " course of transit," see s. 45, *post*.

Cross-references.—*Cf*. ss. 1 (1) and 8, *ante* (price), and ss. 27 and 28, *ante* (buyer's duty to pay price).

COMMENT

" The vendors being unpaid," says Lord ESHER,

> " and the purchasers having become insolvent, according to the law merchant the vendors had a right to stop the goods while *in transitu*, although the property in such goods might have passed to the purchasers. The doctrine of stoppage *in transitu* has always been construed favourably to the unpaid vendor " (*t*).

The right of stoppage *in transitu* is a right against the goods themselves, and does not extend to policy moneys paid to the buyer by insurers for loss of or damage to the goods while in transit (*u*).

The term " stoppage *in transitu* " only applies in strictness to cases where the property in the goods has passed to the buyer (*v*). If the property has not passed, as when the seller has reserved the *jus disponendi*, his rights depend upon s. 39 (2), *ante*, which is declaratory, as it was clear before the Act that the seller's right of withholding or countermanding delivery extended to executory, as well as executed, contracts when the buyer was insolvent (*a*).

As to the carrier's lien and rights when goods are stopped *in transitu*, see s. 46 (2) *post*.

Since the right of stoppage *in transitu* arises by implication of law (s. 39, *ante*), it follows that it may be waived by the seller under the provisions of s. 55, *post*.

45. Duration of transit.—(1) Goods are deemed

(*s*) *Lickbarrow* v. *Mason* (1793), 6 East 22, *n*., H. L.; 1 Smith L. C. (13th ed.), p. 703; *Gibson* v. *Carruthers* (1841), 8 M. & W. 321; *Bolton* v. *Lancashire and Yorkshire Rail. Co.* (1866), L. R. 1 C. P. 431, at p. 439; *Bethell* v. *Clark* (1887), 19 Q. B. D. 553, at p. 561, affirmed, (1888), 20 Q. B. D. 615, C. A.; Pollock and Wright, *Possession in the Common Law*, pp. 72, 74, 214.

(*t*) *Bethell* v. *Clark* (1888), 20 Q. B. D. 615, at p. 617, C. A.

(*u*) *Berndtson* v. *Strang* (1868), 3 Ch. App. 588, at p. 591, *per* Lord CAIRNS; *cf*. *Phelps* v. *Comber* (1855), 29 Ch. D. 813, C. A.

(*v*) *Gibson* v. *Carruthers* (1841), 8 M. & W. 321.

(*a*) See *Griffiths* v. *Perry* (1859), 1 E. & E. 680, at p. 688; *Ex p. Chalmers* (1873), 8 Ch. App. 289, at p. 292.

to be in course of transit from the time when they are delivered to a carrier by land or water, or other bailee [or custodier] for the purpose of transmission to the buyer, until the buyer, or his agent in that behalf, takes delivery of them from such carrier or other bailee [or custodier] (*b*).

(2) If the buyer or his agent in that behalf obtains delivery of the goods before their arrival at the appointed destination, the transit is at an end (*c*).

(3) If, after the arrival of the goods at the appointed destination, the carrier or other bailee [or custodier] acknowledges to the buyer or his agent that he holds the goods on his behalf, and continues in possession of them as bailee [or custodier] for the buyer, or his agent, the transit is at an end, and it is immaterial that a further destination for the goods may have been indicated by the buyer (*d*).

(*b*) For principle, see *Bolton* v. *Lancashire and Yorkshire Rail. Co.* (1866), L. R. 1 C. P. 431, at p. 439, *per* WILLES, J. For illustrations, see *Whitehead* v. *Anderson* (1842), 9 M. & W. 518 (promise by captain to deliver when satisfied as to freight, transit not ended); *Dodson* v. *Wentworth* (1842), 4 M. & Gr. 1080 (goods delivered by carrier to warehouse to await orders, transit ended); *Valpy* v. *Gibson* (1847), 4 C. B. 837 (goods delivered to shipping agent of buyer, transit ended); *Schotsmans* v. *Lancashire and Yorkshire Rail. Co.* (1867), 2 Ch. App. 332 (goods delivered to general ship owned by buyer, transit ended); *Coventry* v. *Gladstone* (1868), L. R. 6 Eq. 44 (overside orders given by mate to holder of bill of lading, transit not ended); *Ex p. Gibbes* (1875), 1 Ch. D. 101 (goods shipped to Liverpool and then put on railway for buyer, transit ended); *Ex p. Watson* (1877), 5 Ch. D. 35 (ineffectual interruption of transit); *Ex. p. Barrow* (1877), 6 Ch. D. 783 (goods warehoused by carrier as forwarding agent, transit not ended); *Ex p. Rosevear China Clay Co.* (1879), 11 Ch. D. 560 (goods shipped on ship hired by buyer, destination not stated, transit not ended); *Kemp* v. *Falk* (1882), 7 App. Cas. 573, at p. 584 (goods on ship, cash receipts instead of delivery orders given to buyer, transit not ended); *Ex p. Francis* (1887), 4 Morr. 146 (goods shipped in vessel of buyer's agent, transit ended); *Bethell* v. *Clark* (1888), 20 Q. B. D. 615, C. A. (goods ordered to be delivered to the *Darling Downs* to Melbourne, transit not ended by shipment); followed *Lyons* v. *Hoffnung* (1890), 15 App. Cas. 391, P. C.; *Ex p. Hughes* (1892), 9 Morr. 294 (break in transit); *Kemp* v. *Ismay* (1909), 100 L. T. 996 (goods bought by agent in his own name for transmission abroad).

(*c*) *Whitehead* v. *Anderson* (1842), 9 M. & W. 518, at p. 534; *Blackburn on Sale* (1st ed.), p. 249; *London and North Western Rail Co.* v. *Bartlett* (1861), 7 H. & N. 400 (alteration of journey by agreement between carrier and consignee); see, too, dictum of BOWEN, L.J., in *Kendall* v. *Marshall, Stevens & Co.* (1883), 11 Q. B. D. 356, at p. 369.

(*d*) For principle, see *Kendall* v. *Marshall, Stevens & Co.* (1883), 11 Q. B D. 356, C. A., where the carrier attorned to buyer's agent. In illustration, see *Dixon* v. *Baldwen* (1804), 5 East, 175; *Valpy* v. *Gibson* (1847), 4 C. B. 837, where a re-delivery to seller for special purpose did not revive right of stoppage; *Ex p. Miles* (1885), 15 Q. B. D. 39, C. A.

(4) If the goods are rejected by the buyer, and the carrier or other bailee [or custodier] continues in possession of them, the transit is not deemed to be at an end, even if the seller has refused to receive them back (*e*).

(5) When goods are delivered to a ship chartered by the buyer it is a question depending on the circumstances of the particular case, whether they are in the possession of the master as a carrier, or as agent to the buyer (*f*).

(6) Where the carrier or other bailee [or custodier] wrongfully refuses to deliver the goods to the buyer, or his agent in that behalf, the transit is deemed to be at an end (*g*).

(7) Where part delivery of the goods has been made to the buyer, or his agent in that behalf, the remainder of the goods may be stopped in transitu, unless such part delivery has been made under such circumstances as to show an agreement to give up possession of the whole of the goods (*h*).

Definitions.—For " bailee," " buyer " and " goods," see s. 62 (1), *post*. For " seller," see s. 38 (2), *ante*, and s. 62 (1), *post*.

Cross-references.—*Cf.* ss. 39 and 44, *ante* (right of stoppage); s. 31, *ante* (part delivery); s. 32, *ante* (delivery to carrier); and s. 35, *ante* (rejection).

COMMENT

Duration of transit.—In order to form a clear notion of the meaning of the term " transit," two points should be noted: (1)

(*e*) *Bolton* v. *Lancashire and Yorkshire Rail. Co.* (1866), L. R. 1 C. P. 431; *James* v. *Griffin* (1837), 2 M. & W. 623.

(*f*) *Berndtson* v. *Strang* (1867), L. R. 4 Eq. 481, at p. 489; on appeal (1868), 3 Ch. App. 588, at p. 590, *per* Lord CAIRNS (the test is whether the master is the servant of the owner or the charterer); *Ex p. Rosevear China Clay Co.* (1879), 11 Ch. D. 560, C. A. (ship hired verbally); *cf. Schotsman* v. *Lancashire and Yorkshire Rail. Co.* (1867), 2 Ch. App. 332 (general ship owned by buyer, transit ended).

(*g*) *Bird* v. *Brown* (1850), 4 Exch. 786, at p. 790 (where carrier refused to deliver in consequence of an invalid notice to stop).

(*h*) *Bolton* v. *Lancashire and Yorkshire Railway* (1866), L. R. 1 C. P. 431, at p. 440, *per* WILLES, J.; *Ex p. Cooper* (1879), 11 Ch. D. 68, C. A.; *Kemp* v. *Falk* (1882), 7 App. Cas. 573, at p. 586, *per* Lord BLACKBURN; *cf.* s. 42, *ante*, as to seller's lien.

The goods may be *in transitu* although they have left the hands of the person to whom the seller intrusted them for transmission. It is immaterial how many agents' hands they have passed through if they have not reached their destination (*i*). (2) The term does not necessarily imply that the goods are in motion, for,

> " If the goods are deposited with one who holds them merely as an agent to forward and has the custody as such, they are as much *in transitu* as if they were actually moving " (*k*).

" The essential feature of a stoppage *in transitu*," says CAIRNS, L.J., " is that the goods should be . . . in the possession of a middleman . . ." (*l*).

When goods which have been sold are in the actual possession of a carrier or other bailee, three states of fact may exist with regard to them: First, the carrier or other bailee may hold them as agent for the seller; in that case the seller preserves his lien, and the right of stoppage *in transitu* does not arise. Secondly, the goods may be *in medio*. The carrier or other bailee may hold them in his character as such, and not exclusively as the agent of either the seller or buyer. In that case the right of stoppage *in transitu* exists. Thirdly, the carrier or other bailee may hold the goods, either originally or by subsequent attornment, solely as agent for the buyer. In that case there either has been no right of stoppage or it is determined. The difficulties that arise are rather difficulties of fact than of law.

Destination.—As regards the term " destination," BRETT, M.R. says that it means

> " sending [the goods] to a particular place to a particular person who is to receive them there, and not sending them to a particular place without saying to whom " (*m*);

and Lord FITZGERALD says,

> " Transit embraces not only the carriage of the goods to the place where delivery is to be made, but also delivery of the goods there according to the terms of the contract of conveyance " (*n*).

Termination of transit.—Where the attornment of the carrier is relied on, that attornment must be founded on mutual assent. If the carrier do not assent to hold the goods for the buyer, or if

---

(*i*) *Bethell* v. *Clark* (1888), 20 Q. B. D. 615, at p. 619, *per* FRY, L.J.; approved, *Lyons* v. *Hoffnung* (1890), 15 App. Cas. 391, P. C.

(*k*) *Blackburn on Sale* (1st ed.), p. 244.

(*l*) *Schotsmans* v. *Lancashire and Yorkshire Rail. Co.* (1867), 2 Ch. App. 332, at p. 338. *Cf.* the duration of transit for the purpose of insurance policies covering goods in transit: *Sadler Brothers Co.* v. *Meredith*, [1963] 2 Lloyd's Rep. 293; *Crows Transport, Ltd.* v. *Phoenix Assurance Co., Ltd.*, [1965] 1 All E. R. 596; [1965] 1 W. L. R. 383, C. A.

(*m*) *Ex p. Miles* (1885), 15 Q. B. D. 39, at p. 43, C. A.

(*n*) *Kemp* v. *Falk* (1882), 7 App. Cas. 573, at p. 588.

the buyer do not assent to his so holding them, there is no attornment (*o*).

The fact that the freight is unpaid is strong, though not conclusive, evidence that the carrier is in possession of the goods as carrier, and not as the buyer's agent (*p*).

In a case where goods were consigned to South Africa but were stopped by the buyer at Southampton, BAILHACHE, J., held that the *transitus* was at an end, saying,

> "Where the original *transitus* is interrupted by the buyers, I think the test is whether the goods will be set in motion again without further orders from the buyers; if not, the transit is ended, and the right to stop lost" (*q*).

So, too, where goods which, under the contract, were to be delivered "free house, London," arrived in London port, and the buyer directed the carrier to warehouse them and await further instructions, it was held that the *transitus* was at an end when the goods were collected from the steamer and warehoused in accordance with this instruction, and that in any case sub-s. (2) of the section would operate to terminate the transit (*r*).

Postal packets.—It has been held in New Zealand that packets containing goods sent by post by a seller to a buyer are not, while in the course of transit by post, delivered to or in the possession of "a carrier or other bailee" as no contract of bailment can arise. They cannot therefore be stopped in transit (*s*). The reasoning is difficult to follow, however, as a person may be a carrier without there being any contract of bailment (*t*). But the seller would seem to be prevented from exercising a right of stoppage in transitu by s. 64 of the Post Office Act, 1969, which protects postal packets from seizure under distress or in execution and from retention by virtue of a lien to the same extent as if they were the property of the Crown. Although there is no authority on the point, the right of stoppage in transitu would probably be held to be in the nature of a lien for the purpose of this section.

(*o*) See *James* v. *Griffin* (1837), 2 M. & W. 623 (offer to attorn not accepted by buyer); *Kemp* v. *Falk* (1882), 7 App. Cas. 573, at pp. 584, 586 (carrier not agreeing to change his character). See also *Blackburn on Sale* (1st ed.), p. 248.

(*p*) *Kemp* v. *Falk* (1882), 7 App. Cas. 573, at p. 584.

(*q*) *Reddall* v. *Union Castle Steamship Co.* (1914), 84 L. J. K. B. 360, 362.

(*r*) *Plischke* v. *Allison Bros. Ltd.*, [1936] 2 All E. R. 1009.

(*s*) *Postmaster General* v. *W. H. Jones & Co.*, [1957] N. Z. L. R. 829, applying *Triefus & Co.* v. *Post Office*, [1957] 2 All E. R. 387; [1957] 2 Q. B. 352, C. A.

(*t*) *Cf. Badische Anilin and Soda Fabrik* v. *Basle Chemical Works*, [1898] A. C. 200, at p. 204; *Schotsman* v. *Lancashire and Yorkshire Rail. Co.* (1867), L. R. 2 Ch. 332, at p. 338.

46. How stoppage in transitu is effected.— (1) The unpaid seller may exercise his right of stoppage in transitu either by taking actual possession of the goods (*u*), or by giving notice of his claim to the carrier or other bailee [or custodier] in whose possession the goods are (*v*). Such notice may be given either to the person in actual possession of the goods or to his principal. In the latter case the notice, to be effectual, must be given at such time and under such circumstances that the principal, by the exercise of reasonable diligence, may communicate it to his servant or agent in time to prevent a delivery to the buyer (*w*).

(2) When notice of stoppage in transitu is given by the seller to the carrier, or other bailee [or custodier] in possession of the goods, he must re-deliver the goods to, or accordind to the directions of, the seller (*a*). The expenses of such re-delivery must be borne by the seller.

ILLUSTRATIONS

(1) Goods are stopped in transit while in the hands of a railway company. The company has a lien for the charge of carrying these particular goods, but no general lien for other moneys due from the consignee, even though the consignment note purports to create such a lien (*b*).

(2) Goods sent by sea are stopped in transit at a port short of the destination. The unpaid seller declines to take any action with regard to the goods. He is liable for freight to destination and any landing charges that may have been incurred (*c*).

Definitions.—For " bailee," " buyer," " delivery " and " goods,"

(*u*) *Snee* v. *Prescot* (1743), 1 Atk. 245, at p. 250, *per* Lord HARDWICKE; *Whitehead* v. *Anderson* (1842), 9 M. & W. 518, at p. 534, *per* PARKE, B.

(*v*) *Litt* v. *Cowley* (1816), 7 Taunt. 169, at p. 170, *per* GIBBS, C.J.

(*w*) *Whitehead* v. *Anderson* (1842), 9 M. & W. 518 ; *Ex p. Watson* (1877), 5 Ch. D. 35, C. A. ; *Kemp* v. *Falk* (1882), 7 App. Cas. 573, at p. 585 ; *cf. Phelps* v. *Comber* (1885), 29 Ch. D. 813, C. A. (notice to consignee to hold proceeds ineffectual).

(*a*) *The Tigress* (1863), 32 L. J. P. M. & A. 97, at p. 102.

(*b*) *United States Steel Products Co.* v. *Great Western Rail. Co.*, [1916] 1 A. C., 189, 195, 916. It is not clear whether a forwarding agent in possession of goods has any lien over them. The point was raised in *Langley, Beldon and Gaunt, Ltd.* v. *Morley*, [1965] 1 Lloyd's Rep. 297, but was not decided as the forwarding agent was not in fact in possession of the goods.

(*c*) *Booth Steamship Co.* v. *Cargo Fleet Iron Co.*, [1916] 2 K. B. 570, C. A.

see s. 62 (1), *post.* For "unpaid seller," see s. 38, *ante.* For "seller," see s. 38 (2), *ante,* and s. 62 (1), *post.*

Cross-references.—*Cf.* ss. 39 and 44, *ante* (right of stoppage), and s. 45, *ante* (duration of transit).

COMMENT

"The seller," says Dr. LUSHINGTON,

> "exercises his right of stoppage *in transitu* at his own peril, and it is incumbent upon the master to give effect to a claim, as soon as he is satisfied it is made by the vendor, unless he is aware of a legal defeasance of the claim" (*d*).

If, after notice, lawfully given, the carrier delivers to the consignee or refuses to deliver to the seller, he is guilty of a conversion of the goods. In case of real doubt he should resort to an interpleader (*e*), for if the transit has ended and he wrongfully restores the goods to the seller, he is liable to the buyer in an action for conversion (*f*). The seller also has a remedy by injunction (*g*), or, if the goods be in the hands of the master of a ship, by arrest of the ship (*h*).

In a case in the Court of Appeal, BRAMWELL, L.J., doubted whether there was any obligation on the part of the principal to send on a notice of stoppage to his agent (*i*); but, when the case went to the House of Lords, Lord BLACKBURN expressly repudiated this doubt (*k*). Though, as between seller and carrier, the expenses of stoppage and re-delivery fall on the seller, it may be that the seller would be able to prove for them against the buyer's estate.

Re-sale by Buyer or Seller

47. Effect of sub-sale or pledge by buyer.—Subject to the provisions of this Act (*l*), the unpaid seller's right of lien [or retention] or stoppage in transitu is not affected by any sale, or other disposition

---

(*d*) *The Tigress* (1863), 32 L. J. P. M. & A. 97, at p. 101.

(*e*) *The Tigress* (*supra*), at p. 102; *cf. Litt* v. *Cowley* (1816), 7 Taunt. 169 at p. 170; *Pontifex* v. *Midland Rail. Co.* (1877), 3 Q. B. D. 23; discussed, *Booth Steamship Co.* v. *Cargo Fleet Iron Co.*, [1916] 2 K. B. 570, C. A.

(*f*) *Taylor* v. *Great Eastern Rail. Co.*, [1901] 1 K. B. 774.

(*g*) *Schotsmans* v. *Lancashire and Yorkshire Rail. Co.* (1867), 2 Ch. App. 332, at p. 340.

(*h*) *The Tigress* (1863), 32 L. J. P. M. & A. 97.

(*i*) *Ex p. Falk* (1880), 14 Ch. D. 446, C. A.

(*k*) *Kemp* v. *Falk* (1882), 7 App. Cas. 573, at p. 585.

(*l*) See s. 25 (2), *ante*, and *Cahn* v. *Pockett's Bristol Channel Co.*, [1899] 1 Q. B. 643, at p. 664, C. A.

of the goods which the buyer may have made (*m*), unless the seller has assented thereto (*n*).

Provided that where a document of title to goods has been lawfully transferred to any person as buyer or owner of the goods, and that person transfers the document (*o*) to a person who takes the document in good faith and for valuable consideration, then, if such last-mentioned transfer was by way of sale the unpaid seller's right of lien [or retention] or stoppage in transitu is defeated, and if such last-mentioned transfer was by way of pledge or other disposition for value, the unpaid seller's right of lien [or retention] or stoppage in transitu can only be exercised subject to the rights of the transferee.

ILLUSTRATIONS

(1) A, an oil merchant, sells oil to B without appropriating any particular oil to the contract. B sells 2 tons of the oil to C, and gives him a delivery order. C lodges the delivery order with A, indorsing it " please wait our orders." B falls into arrears with his payments, and A refuses to deliver the 2 tons to C. He has not lost his seller's lien, and is entitled to refuse, since nothing he has done was an assent to the re-sale (*p*).

(2) A having 6,000 bags of mowra seed sells 2,600 to B, which B pays for by cheque. A gives B delivery orders for the 2,600 bags, and B sells the seed to C, and indorses the delivery orders to him. If B's cheque is dishonoured, A has lost his right of lien and must deliver the seed to C (*q*).

(*m*) As to seller's lien, see *Dixon* v. *Yates* (1833), 5 B. & Ad. 313, at p. 339; *Farmeloe* v. *Bain* (1876), 1 C. P. D. 445. As to stoppage *in transitu*, see *Craven* v. *Ryder* (1816), 6 Taunt. 433; *Ex p. Golding Davis & Co.* (1880), 13 Ch. D. 628; *Kemp* v. *Falk* (1882), 7 App. Cas. 573. As to delivery orders before the Factors Act, 1877, see *McEwan* v. *Smith* (1849), 2 H. L. Cas. 309; and *Blackburn on Sale* (1st ed.), p. 302, which shows the common law effects of these documents.

(*n*) *Blackburn on Sale* (1st ed.), p. 271; *Stoveld* v. *Hughes* (1811), 14 East, 308; *Pearson* v. *Dawson* (1858), E. B. & E. 448; *Woodley* v. *Coventry* (1863), 2 H. & C. 164; *Knights* v. *Wiffen* (1870), L. R. 5 Q. B. 660; *Merchant Banking Co.* v. *Phoenix Bessemer Co.* (1877), 5 Ch. D. 205; *Mount* v. *Jay & Jay*, [1959] 3 All E. R. 307; [1960] 1 Q. B. 159.

(*o*) It seems that this means the identical document as that transferred to the buyer, even though such a construction may lead to very artificial results: see *per* SALMON, J. in *Mount* v. *Jay & Jay* (*supra*).

(*p*) *Mordaunt Brothers* v. *British Oil and Cake Mills*, [1910] 2 K. B. 502; *cf. Mount* v. *Jay & Jay*, [1959] 3 All E. R. 307; [1960] 1 Q. B. 159.

(*q*) *Ant. Jurgens Margarinefabrieken* v. *Louis Dreyfus & Co.*, [1914] 3 K. B. 40 (*issue* of delivery order by A = *transfer* of document of title).

Definitions.—For "buyer," "document of title," "goods," "lien" and "sale," see s. 62 (1), *post*. For "good faith," see s. 62 (2), *post*. For "unpaid seller," see s. 38, *ante*. For "seller," see s. 38 (2), *ante*, and s. 62 (1), *post*.

Cross-references.—*Cf.* ss. 39 and 41, *ante* (right of lien), and ss. 39 and 44, *ante* (right of stoppage).

COMMENT

The proviso reproduces and develops s. 10 of the Factors Act, 1889, *post*, the effect of which appears to be (i) to affirm the common law effect of the transfer of a bill of lading, and (ii) to put all the documents of title mentioned in s. 1 of that Act on the same footing as bills of lading for the purposes of that Act.

Transfer of bill of lading.—As regards bills of lading the law appears to be as follows :—

(1) As between buyer and seller, that is to say the immediate parties to the contract of sale, the indorsement of the bill of lading does not affect the right of stoppage, nor does a further indorsement by the buyer affect the right unless the indorsement be for value (*r*), but an antecedent debt may constitute such value (*s*).

(2) If the holder of the bill of lading re-sells the goods or otherwise disposes of them for value to a third person who pays the money, such third person acquires his interest in the goods subject to the original seller's right of stoppage *in transitu*, unless he gets a transfer of the bill of lading (*t*).

(3) Since the Bills of Lading Act, 1855, *post*, Appendix I, as well as before, a bill of lading may be indorsed by way of mortgage, pledge, or other security, and not only by way of absolute sale (*u*). Where a bill of lading is so transferred, the original seller retains his right of stoppage subject to the rights of the incumbrancer, and, further, he may compel the incumbrancer to resort to other goods pledged with him by his debtor, if such there be, before resorting to the goods covered by the bill of lading (*v*).

(4) The right of stoppage *in transitu* is wholly defeated when the bill of lading is assigned absolutely for a consideration which is wholly paid (*a*).

(*r*) *Lickbarrow* v. *Mason* (1793), 6 East, 22, *n*., H. L.; 1 Smith L. C (13th ed.), p. 703.

(*s*) *Leask* v. *Scott* (1877), 2 Q. B. D. 376, C. A.; dissenting from *Rodger* v. *Comptoir d'Escompte* (1869), L. R. 2 P. C. 393.

(*t*) *Kemp* v. *Falk* (1882), 7 App. Cas. 573, at p. 582, *per* Lord BLACKBURN.

(*u*) *Sewell* v. *Burdick* (1884), 10 App. Cas. 74. But as to Scotland, see *Hayman* v. *McLintock*, 1907 S. C. 936.

(*v*) *Re Westzinthus* (1833), 5 B. & Ad. 817; *Spalding* v. *Ruding* (1843), 6 Beav. 376; approved *Kemp* v. *Falk* (1882), 7 App. Cas. 573 at p. 585; *cf. Coventry* v. *Gladstone* (1868), L. R. 6 Eq. 44.

(*a*) *Lickbarrow* v. *Mason* (1793), 1 Smith L. C. (13th ed.), p. 703; *Leask* v. *Scott* (1877), 2 Q. B. D. 376, C. A.

(5) When the bill of lading is transferred to a sub-purchaser absolutely and for value, but that value is unpaid wholly or in part, there is probably no right to stop against the sub-purchaser for the original purchase price, even to the extent of what remains to be paid by the sub-purchaser (*b*).

See further the Bills of Lading Act, 1855, *post*, Appendix I.

Note that this section may override the operation of s. 19, *ante*, while according with the provisions of s. 25, *ante*.

48. Sale not generally rescinded by lien or stoppage in transitu.—(1) Subject to the provisions of this section, a contract of sale is not rescinded by the mere exercise by an unpaid seller of his right of lien [or retention] or stoppage in transitu (*c*).

(2) Where an unpaid seller who has exercised his right of lien [or retention] or stoppage in transitu re-sells the goods, the buyer acquires a good title thereto as against the original buyer (*d*).

(3) Where the goods are of a perishable nature (*e*), or where the unpaid seller gives notice to the buyer of his intention to re-sell, and the buyer does not within a reasonable time pay or tender the price, the unpaid seller may re-sell the goods and recover from the original buyer damages for any loss occasioned by his breach of contract (*f*).

(4) Where the seller expressly reserves a right of re-sale in case the buyer should make default, and on the buyer making default, re-sells the goods, the original contract of sale is thereby rescinded, but without prejudice to any claim the seller may have for damages (*g*).

(*b*) See *Ex p. Golding Davis & Co.* (1880), 13 Ch. D. 628, C. A.; *ex p. Falk* (1880), 14 Ch. D. 446, C. A.; on appeal, *sub nom. Kemp* v. *Falk* (1882), 7 App. Cas. 573; *cf. Phelps* v. *Comber* (1885), 29 Ch. D. 813, at p. 821; and see *Scrutton on Charterparties, etc.* (18th ed.), p. 185, discussing these cases.

(*c*) For an examination of the decisions before the Act, and the effect of this section on those decisions, see *Benjamin on Sale* (8th ed.), pp. 935–952.

(*d*) *Milgate* v. *Kebble* (1841), 3 M. & Gr. 100; *cf. Lord* v. *Price* (1874), L. R. 9 Exch. 54; *cf.* also s. 25 (1), *ante*, and 34 Halsbury's Laws (3rd Edn.) 141.

(*e*) *Cf. Maclean* v. *Dunn* (1828), 4 Bing. 722, at p. 728, where there had been a refusal to accept; 34 Halsbury's Laws (3rd Edn.) 142.

(*f*) *Page* v. *Cowasjee* (1866), L. R. 1 P. C. 127, at p. 145; *Lord* v. *Price* (1874), L. R. 9 Exch. 54, at p. 55; *ex p. Stapleton* (1879), 10 Ch. D. 586, C. A.

(*g*) *Lamond* v. *Davall* (1847), 9 Q. B. 1030.

Definitions.—For " contract of sale," see s. 1 (1), *ante*, and s. 62 (1), *post*. For " buyer " and " goods," see s. 62 (1), *post*. For " unpaid seller," see s. 38, *ante*.

Cross-references.—*Cf.* ss. 39 and 41, *ante* (right of lien) ; ss. 39 and 44, *ante* (right of stoppage) ; ss. 12 and 21 to 26, *ante* (transfer of title) ; ss. 1 (1) and 8, *ante* (price) ; ss. 27 and 28, *ante* (buyer's duty to pay price) ; ss. 49 and 50, *post* (remedies of seller) ; s. 10 (stipulations as to time); and s. 56, *post* (reasonable time a question of fact).

COMMENT

As long as the buyer is in default he is not entitled to the immediate possession of the goods, and therefore cannot maintain an action for conversion even against a wrongdoer in possession (*h*).

Re-sale by seller.—In *ex p. Stapleton* (*i*) it was said that when the buyer was insolvent, the seller might re-sell unless the trustee or a sub-purchaser tendered the price within a reasonable time, and nothing was said about notice. But as a fact the seller in that case gave fair notice of his intention to re-sell.

Sub-section (3) enables the seller to make the time of payment of the essence of the contract, so that when he later re-sells the seller is treating the buyer's failure to pay as a repudiation of the contract, and is re-selling as full owner (*k*). It would seem to follow that the seller is entitled to keep any extra profit, and that the law relating to deposits is that which generally governs deposits when a sale goes off through the fault of the buyer (*l*).

As to sale of perishable goods by order of the court, see R. S. C. O. 29, r. 4.

Quære whether in case of urgency, the seller in possession of the goods is justified in selling them on behalf of the buyer, acting as an " agent of necessity " (*m*).

Express reservation of title.—Contracts for the sale of goods commonly provide that although the buyer is entitled to immediate possession of the goods, property in the goods shall remain in the seller until the price is paid or some other condition performed.

Where there is such a provision, if the buyer repudiates the contract, for example by making it clear that he will not pay for the

---

(*h*) *Lord* v. *Price* (1874), L. R. 9 Exch. 54 (auction sale, goods neither removed nor paid for).

(*i*) (1879) 10 Ch. D. 586, C. A.

(*k*) *R. V. Ward, Ltd.* v. *Bignall*, [1967] 2 All E. R. 449; [1967] 1 Q. B. 534, C. A., overruling *Gallagher* v. *Shilcock*, [1949] 1 All E. R. 921; [1949] 2 K. B. 765. Generally stipulations as to time of payment are not deemed to be of the essence of a contract of sale—see s. 10, *ante*.

(*l*) p. 26, *ante*.

(*m*) *Prager* v. *Blatspiel*, [1924] 1 K. B. 566, *per* MCCARDIE, J. But on the facts the sale was held not to be *bonâ fide*, and the dicta of MCCARDIE, J., were disapproved by SCRUTTON, L.J., in *Jebara* v. *Ottoman Bank*, [1927] 2 K. B. 254, at pp. 271, 272. See note on agency of necessity, *ante*, notes to s. 21 (2).

goods, the seller may terminate the contract and recover the goods (*n*). Until such a termination the buyer has the right to possession and use of the goods. It will be a question of construction of each contract whether the buyer is also entitled to deal with the goods in any other way, as by disposing of them (*o*). Even if the contract forbids it, however, if a buyer does dispose of the goods, he may give a good title to a *bonâ fide* third party under s. 25 (2), *ante*.

A reservation of title clause gives the seller (so long at least as the buyer continues to hold the goods sold) a proprietary right which will prevent the goods forming part of the general assets of the buyer for distribution amongst his creditors in the event of his insolvency (*p*). Moreover, if the buyer has re-sold the goods, but the proceeds of sale are traceable, the seller may be able to recover those proceeds as held on trust for him (*q*).

Property and possession transferred to buyer.—It has commonly been assumed that if the buyer has acquired both title and possession under a contract, then in the absence of an express provision in the contract, or elsewhere, giving the seller rights over the goods, he will not be entitled to any such rights, even if he is entitled to treat the contract as at an end because it has been repudiated by the buyer (*r*).

But there is no clear authority in support of this proposition. Moreover, it has been decided that if property has passed but possession remains in the seller and the seller treats the contract as terminated by reason of the buyer's breach, the property in the goods automatically revests in the seller (*s*). There would seem to be no logical ground for holding that the question whether the property revests in the seller should turn on whether the seller or buyer is in possession at the material time (*t*).

(*n*) Subject to special provisions where the Hire-Purchase Act, 1965 or the Consumer Credit Act, 1974 applies.

(*o*) *Re Anchor Line (Henderson Brothers), Ltd.*, [1036] 2 All E. R. 941; [1937] Ch. 1, C. A.

(*p*) *McEntire* v. *Crossley Brothers, Ltd.*, [1895] A. C. 457, H. L. If the buyer is an individual and goes bankrupt, the reputed ownership provisions of s. 38 of the Bankruptcy Act, 1914 may, however, sometimes defeat the seller's claim. There is no equivalent statutory provision in the case of company liquidations.

(*q*) *Aluminium Ltd., Vaaseu B.V.* v. *Romalpa Aluminium* (1975, 119 Sol. J. 318, applying *re Hallett's Estate* (1880), 13 Ch. D. 696.

(*r*) See, *e.g.*, *Re Shackleton, Ex parte Whittaker* (1875), 10 Ch. App. 446, C. A.; *Re Wait*, [1927] 1 Ch. 606, at p. 640, *per* Atkin, L.J.

(*s*) *R. V. Ward, Ltd.* v. *Bignall*, [1967] 2 All E. R. 449; [1967] 1 Q. B 534, C. A. *Cf. Page* v. *Cowasjee* (1866), L. R. 1 P. C. 127, at p. 145.

(*t*) There is nothing in *Ward* v. *Bignall, supra*, to suggest any basis for such a distinction, and indeed Diplock, L.J., said, at pp. 457 and 550, " It is, of course, well established that where a contract for the sale of goods is rescinded after the property in the goods has passed to the buyer the rescission divests the buyer of his property in the goods ". In using the terms " rescind " and

If property does revest in the seller when he treats the contract as being at an end, he will have the same right to recover the property as if it had never passed to the buyer. This would seem to be so even if the buyer becomes bankrupt or, being a company goes into liquidation, or the trustee in bankruptcy takes the property subject to the rights of the seller (*u*).

Even if property does not remain or revest in the seller, in special circumstances, if an unpaid seller is entitled under the contract to give and does give special directions to the buyer as to the manner in which the goods are to be dealt with by the buyer, the seller may retain a beneficial interest in the goods until the price is paid (*v*).

PART V

ACTIONS FOR BREACH OF THE CONTRACT

Remedies of the Seller

49. Action for price.—(1) Where, under a contract of sale, the property in the goods has passed to the buyer, and the buyer wrongfully neglects or refuses to pay for the goods according to the terms of the contract, the seller may maintain an action against him for the price of the goods (*a*).

(2) Where, under a contract of sale, the price is payable on a day certain irrespective of delivery, and the buyer wrongfully neglects or refuses to pay such price, the seller may maintain an action for the price, although the property in the goods has not passed, and the goods have not been appropriated to the contract (*b*).

" rescission " in this context, DIPLOCK, L.J., made it clear that he was referring to the seller's " right to treat the contract as repudiated by the buyer ". While there does not appear to be any clear prior authority in support of this proposition, it would seem to be widely accepted where the contract is terminated before the buyer obtains possession.

(*u*) *Re Eastgate, Ex parte Ward,* [1905] 1 K. B. 465; *Tilley* v. *Bowman, Ltd.*, [1910] 1 K. B. 745, where goods were obtained by misrepresentation and the seller was allowed to rescind the contract and recover the goods from the trustees in bankruptcy.

(*v*) *Cf. Ex parte Banner* (1876), 2 Ch. D. 278, C. A.; *Tooke* v. *Hollingworth* (1793), 5 Term. Rep. 215; and compare the position where the seller is trustee of the purchase price, *post*, p. 258.

(*a*) *Scott* v. *England* (1844), 2 D. & L. 520 (re-sale of goods bought at auction, but not yet paid for); *cf. Kymer* v. *Suwercropp* (1807), 1 Camp. 109 (goods stopped *in transitu*); *Alexander* v. *Gardner* (1835), 1 Bing. N. C. 671 (goods lost at sea).

(*b*) *Dunlop* v. *Grote* (1845), 2 Car. & Kir. 153.

(3) Nothing in this section shall prejudice the right of the seller in Scotland to recover interest on the price from the date of tender of the goods, or from the date on which the price was payable, as the case may be.

ILLUSTRATIONS

(1) Contract for sale of goods to be paid for " net cash against documents on arrival of the steamer." If the buyer refuses the tender of the documents, the claim is for damages, and not for the price, for the price is not " payable on a day certain irrespective of delivery " (*c*).

(2) Goods sold and delivered, terms 6 months' credit, and then payment to be made by bills at 2 or 3 months at buyer's option. The cause of action does not arise till 9 months after date of the contract, and the Statute of Limitations begins to run from that time, and not from the time of delivery of the goods (*d*).

Definitions.—For "contract of sale," see s. 1 (1), *ante*, and s. 62 (1), *post*. For " action," " buyer," " delivery," " goods," " property " and " seller," see s. 62 (1), *post*.

Cross-references.—*Cf.* ss. 1 (1), 8 and 9, *ante* (price) ; s. 10 (1), *ante* (time for payment) ; ss. 17 and 18, *ante* (passing of property) ; ss. 27 and 28, *ante* (buyer's duty to pay price) ; and s. 54, *post* (interest and special damage).

History.—Before the Judicature Acts the price of goods sold could be recovered under the common *indebitatus* counts. The count for goods sold and delivered was applicable where the property had passed and the goods had been delivered to the buyer, and the price was payable at the time of action brought. The count for goods bargained and sold was applicable when the property had passed to the buyer and the contract had been completed in all respects except delivery, and the delivery was not a condition precedent to the payment of the price (*e*). Now it is sufficient to show facts disclosing either cause of action, but in practice the old counts are still largely used.

(*c*) *Stein Forbes & Co.* v. *County Tailoring Co.* (1916), 86 L. J. K. B. 448, citing the notes to *Pordage* v. *Cole*, Saunders' Reports (ed. E. V. Williams) (6th ed.), Vol. I, p. 319 l. *Cf. Colley* v. *Overseas Exporters*, [1921] 3 K. B. 302 (f.o.b. contract) ; *Napier* v. *Dexters, Ltd.* (1926), 26 Ll. L. Rep. 62 (considering previous cases).

(*d*) *Helps* v. *Winterbottom* (1831), 2 B. & Ad. 431 ; *cf. Schneider* v. *Foster* (1857), 2 H. & N. 4 (where the buyer has an option to take credit or pay cash and opts to pay cash, the cause of action accrues at once).

(*e*) Bullen & Leake's *Prec. of Pleading* (3rd ed.), pp. 38, 39 ; *Forbes* v. *Smith* (1863), 11 W. R. 574 ; *Colley* v. *Overseas Exporters*, [1921] 3 K. B. 302.

COMMENT

The general rule of English law is that, in the absence of any different agreement, when a debt becomes due, it is the duty of the debtor to go and tender the amount to his creditor, if in England, without waiting for any demand (*f*). Where goods are sold and delivered by a merchant in England to a buyer abroad, the price is payable in England, unless a different intention can be inferred from the contract (*g*).

Unless otherwise agreed (*h*), the price must be tendered in lawful money (*i*). Gold is legal tender up to any amount (*k*), silver or cupro-nickel coins of denominations of more than ten new pence are legal tender up to ten pounds; silver or cupronickel coins of up to ten new pence are legal tender up to five pounds; and bronze coins are legal tender up to twenty new pence (*l*). Bank of England notes payable to bearer on demand are legal tender for any amount notwithstanding that these are not now paid on demand in coin (*m*). When tender is pleaded as a defence, the sum tendered must be brought into Court (*n*).

Where the property in the goods has passed.—The seller frequently insists that property in the goods is to pass only when he is paid. It has been suggested that this is a provision which the seller may in some circumstances be entitled to waive, it being for his benefit alone, thereby causing the property to pass to the buyer, so as to entitle him to sue for the price (*o*). This may assist him where the price is otherwise payable on delivery and the buyer refuses to

---

(*f*) *Cf. Walton* v. *Mascall* (1844), 13 M. & W. 452, at p. 458; *Fessard* v. *Mugnier* (1865), 18 C. B. N. S. 286; *Startup* v. *Macdonald* (1843), 6 M. & Gr. 593, at pp. 623, 624; cases cited in *Charles Duval* v. *Gans*, [1904] 2 K. B. 685, C. A. (service out of jurisdiction); *Joachimson* v. *Swiss Bank*, [1921] 3 K. B. 110, at p. 116, C. A.

(*g*) *Rein* v. *Stein*, [1892] 1 Q. B. 753, C. A., approved *Johnson* v. *Taylor Bros.*, [1920] A. C. 144, at p. 155.

(*h*) There may be an express or implied agreement to take payment by bill or cheque. As to payment by bill, note, or cheque, see Chalmers' *Bills of Exchange* (13th ed.), pp. 338–344; as to price expressed in foreign currency, see *post*, notes to s. 54.

(*i*) A debtor has no right to demand change: *Betterbee* v. *Davis* (1811), 3 Camp. 70; Roscoe, *Evidence in Civil Actions* (20th ed.), pp. 682, 683. As to mode of tender, see *International Sponge Co.* v. *Andrew Watt & Sons*, [1911] A. C. 279 (payment to agent); *Mitchell-Henry* v. *Norwich Union*, [1918] 2 K. B. 67 (payment through post).

(*k*) Coinage Act, 1971, s. 2.

(*l*) Coinage Act, 1971, s. 2.

(*m*) Currency and Bank Notes Act, 1954, s. 1; 2 Halsbury's Statutes (3rd Edn.) 771.

(*n*) R. S. C., O. 18, r. 16; C. C. R., O. 9, r. 4 (9).

(*o*) *Napier* v. *Dexters, Ltd.* (1926), 26 Ll. L. Rep. at pp. 63–64, *per* ROCHE, J., and on appeal at pp. 187–188, *per* BANKES, L.J.; *Martin* v. *Hogan* (1917), 24 C. L. R. 231 (H. C. of Aus.).

take delivery. But in other circumstances, if the price is due, the seller may be entitled to claim it in any event at common law, apart from the provisions of sub-s. (2) of this s. (*p*), and if the price is paid the property will then pass under the terms of the contract.

According to the terms of the contract.—The seller cannot sue unless the neglect or refusal to pay is wrongful. It does not necessarily follow that because the property has passed the price is payable forthwith. The buyer is only bound to pay according to the terms of the contract. The sale may have been on credit, or payment may have been made to depend on some specific condition or contingency (*q*). If the contract is silent as to the time of payment an action for the price can be maintained, provided that (1) the property has passed to the buyer, and (2) the goods have either been delivered or the seller is ready and willing to deliver them in exchange for the price, for by s. 28, *ante*, payment of the price and delivery of the goods are concurrent conditions.

On a day certain irrespective of delivery.—In order to succeed in a claim under sub-s. (2) it is necessary to prove that the price payable on a day certain irrespective of delivery. "Day certain" has been held to mean a time specified in the contract not depending on a future or contingent event (*r*), and it has frequently been held that where payment is against shipping documents it is not on a day certain irrespective of delivery, so that no action for the price may be maintained (*s*).

But if this construction is correct, it would seem that this section does not cover all the situations in which an action for the price might have been maintained at common law where the property in the goods have not passed. In *Workman Clark & Co.* v. *Lloyd Brazileno* (*t*), the Court of Appeal held that a seller was entitled at common

(*p*) See *post*, "On a day certain irrespective of delivery".

(*q*) See, for example, *Calcutta Co.* v. *De Mattos* (1863), 32 L. J. Q. B. 322, p. 328.

(*r*) *Merchant Shipping Co.* v. *Armitage* (1873), L. R. 9 Q. B. 99; *Shell Mex, Ltd.* v. *Elton Cop. Dyeing Co.* (1928), 34 Com. Cas. 39; *Muller Maclean & Co.* v. *Lesley and Anderson*, [1921] W. N. 235. But in *Workman Clark & Co.* v. *Lloyd Brazileno*, [1908] 1 K. B. 968, the Court of Appeal all considered that s. 49 (2) applied in a case where an instalment of the price was payable when the keel of a ship was laid. The judgments as reported refer only to s. 49, but it is clear from the terms of the contract, under which the property had not passed, and from the arguments of counsel that only s. 49 (2) could apply. See also *Colley* v. *Overseas Exporters*, [1921] 3 K. B. 302, 306.

(*s*) *Muller Maclean & Co.* v. *Lesley and Anderson*, *supra*; *Stein Forbes & Co.* v. *County Tailoring Co.* (1916), 115 L. T. 215. But *cf. Polenghi* v. *Dried Milk Co., Ltd.* (1904), 10 Com. Cas. 42.

(*t*) [1908] 1 K. B. 968, C. A. In *Colley* v. *Overseas Exporters*, [1921] 3 K. B. 302, at pp. 309–310, McCardie, J., says that no action will lie for the price of goods until the property has passed except under the rule now embodied in s. 49 (2). But at p. 306, he clearly interprets the *Workman Clark* case, as coming within sub-s. (2).

law to summary judgment for an instalment of the price of a ship which under the terms of the contract was payable " when the keel . . . is laid ". In principle there seems no reason why the rules of common law should be excluded by this section, and there is some authority that where a contract clearly provides for the unconditional (*u*) payment of the price in given circumstances which arise, then the seller may claim the price (*v*). Thus, if the buyer has accepted delivery and not rejected the goods, payment being 90 days after delivery, and there being a provision that property should not pass until payment, the seller cannot claim damages for non-acceptance for the buyer has accepted the goods, and there seems no good reason why he should not claim the price.

Payment by bill.—Where there is an agreement for payment of the price by a bill payable at a future day and the bill is not given, the seller cannot sue for the price till the bill would have matured. His remedy before that time is by action for damages for breach of the agreement (*a*). Where a bill or cheque is given for the price, the general rule is that it operates as conditional payment (*b*). If the bill be dishonoured, the debt revives, and the buyer may be sued either on the bill or on the consideration (*c*). When the seller agrees to take the buyer's acceptance for the price, it is his duty to tender a bill to the buyer to get his acceptance (*d*).

(*u*) Payment " on delivery " or " against documents " is not unconditional in this sense.

(*v*) *Minister of Supply and Development* v. *Servicemen's Co-operative Joinery Manufacturers, Ltd.* (1951), 82 C. L. R. 621 (H. C. of Aus.) (" net cash before delivery "). See also *McEntire* v. *Crossley Brothers*, [1895] A. C. 497, *per* Lord HALSBURY; *Sandford* v. *Dairy Supplies, Ltd.*, [1941] N. Z. L. R. 141. *Polenghi* v. *Dried Milk Co., Ltd.* (1904), 10 Com. Cas. 42 (payment on arrival of ship against documents) also supports this proposition, but the case is inconsistent with *Stein Forbes & Co.* v. *County Tailoring Co.* (1916), 115 L. T. 215, where ATKIN, J., construed a similar provision as meaning payment upon delivery.

(*a*) *Paul* v. *Dod* (1846), 2 C. B. 800 ; distinguished, *Waynes Merthyr Steam Coal Co.* v. *Morewood* (1877), 46 L. J. Q. B. 746; but see *Bartholomew* v. *Markwick* (1864), 15 C. B. N.S. 711, where there was a repudiation of the contract. As to measure of damages, see *Gordon* v. *Whitehouse* (1846), 18 C. B. 747.

(*b*) *Goldshede* v. *Cottrell* (1836), 2 M. & W. 20; *Gunn* v. *Bolckow Vaughan & Co.* (1875), 10 Ch. App. 491, at p. 501; *Bolt and Nut Co.* (*Tipton*), *Ltd.* v. *Rowlands Nicholls & Co., Ltd.*, [1964] 1 All E. R. 137; [1964] 2 Q. B. 10, C. A.

(*c*) *Chalmers on Bills of Exchange* (13th ed.), p. 338; but a bill *may* be taken in absolute payment: see *ibid.*, p. 342; *Bolt and Nut Co.* (*Tipton*), *Ltd.* v. *Rowlands Nicholls & Co., Ltd.*, [1964] 1 All E. R. 137; [1964] 2 Q. B. 10, C. A. If the plaintiff sues on the consideration he must be the holder of the bill at the time of action brought: *Davis* v. *Reilly*, [1898] 1 Q. B. 1; *cf. Re a Debtor*, [1908] 1 K. B. 344, C. A. As to payment by cheque to agent authorised only to receive cash, see *Bradford & Sons* v. *Price* (1923), 92 L. J. K. B. 871; for the converse case, see *International Sponge Co.* v. *Watts*, [1911] A. C. 279. *Cf.* s. 38 (1) (b), *ante*.

(*d*) *Reed* v. *Mestaer* (1802), 2 *Comyn on Contract*, 229 ; Bullen & Leake's *Precedents of Pleading* (11th ed.), p. 927.

A bill of exchange is generally treated as cash not only as between remote parties, but also as between immediate parties. Where, therefore, the seller sues upon a bill given to him as payment for goods sold, he is entitled to judgment for the whole amount of the bill, without any stay of execution, even if the buyer has a cross-claim in damages against him in respect of the goods. The buyer must pay upon the bill and bring his cross-claim afterwards. But when a bill of exchange is given on the sale of goods, then, if the consideration wholly fails, as between the immediate parties the buyer who has accepted it is not liable upon it (*e*).

Payment by banker's irrevocable letter of credit.—This is considered at length p. 42, *ante*.

Interest.—For the English law, see the note to s. 3 of the Law Reform (Miscellaneous Provisions) Act, 1934, reproduced in Appendix I, *post*.

In Scotland it was said that

> " the seller may sue the purchaser for the price *and interest*, whether the goods sold are specific or not, provided goods according to the contract have been tendered to the purchaser " (*f*).

The Act preserves this rule.

50. Damages for non-acceptance.—(1) Where the buyer wrongfully neglects or refuses to accept and pay for the goods, the seller may maintain an action against him for damages for non-acceptance.

(2) The measure of damages is the estimated loss directly and naturally resulting, in the ordinary course of events, from the buyer's breach of contract (*g*).

(3) Where there is an available market for the goods in question the measure of damages is primâ facie (*h*) to be ascertained by the difference between the contract price and the market or current price at the time or times (*i*) when the goods ought to have been

---

(*e*) *All Trades Distributors, Ltd. v. Agencies Kaufman, Ltd.* (1969), 113 Sol. Jo. 995, C. A.

(*f*) Mercantile Law Commission, (1855) 2nd Report, p. 47. *Cf.* Green's *Encyclopædia of Scot's Law*, tit. "Interest."

(*g*) *Cort* v. *Ambergate Rail. Co.* (1851), 17 Q. B. 127; *Wayne's Merthyr Coal Co.* v. *Morewood* (1877), 46 L. J. Q. B. 746; *Re Vic. Mill Ltd.*, [1913] 1 Ch. 465, C. A.

(*h*) *Charter* v. *Sullivan*, [1957] 1 All E. R. 809; [1957] 2 Q. B. 117, C. A.

(*i*) See *Brown* v. *Muller* (1872), L. R. 7 Exch. 319 ; *Roper* v. *Johnson* (1873), L. R. 8 C. P. 167, as to non-delivery in instalment contracts.

accepted (*k*) or, if no time was fixed for acceptance, then at the time of the refusal to accept (*l*).

ILLUSTRATIONS

(1) V. Ltd. is in voluntary liquidation and in breach of an agreement, made before the winding-up, to buy and accept machines to be specially made by the sellers. The sellers can prove in the liquidation for the whole of the profit which they will lose on the machines which they have not yet begun to make (*m*).

(2) Buyers wrongfully refuse to accept a new motor car which they have agreed to buy from a dealer. The dealer returns the car to his suppliers, and can recover from the buyers the profit he would have made on the deal (*n*).

(3) Buyer wrongfully refuses to accept a new motor car which he has agreed to buy from a dealer. Cars of this kind are in short supply and the dealer can sell all he can get; he in fact re-sells this car within 10 days at the list price. He cannot recover any loss of profit from the defaulting buyer (*o*).

Definitions.—For " action," " buyer," " goods " and " seller," see s. 62 (1), *post.*

Cross-references.—*Cf.* ss. 1 (1), 8 and 9, *ante* (price); s. 10, *ante* (stipulations as to time); ss. 27 and 28, *ante* (buyer's duty to pay price); s. 35, *ante* (acceptance); s. 37, *ante* (delay in taking delivery); s. 48 (3), *ante* (right of re-sale); s. 51, *post* (non-delivery); and s. 54, *post* (interest and special damage).

COMMENT

This section deals only with general damages.

In general, where the property in the goods has not passed to the buyer, the seller's only remedy is an action for non-acceptance (*p*).

(*k*) Thus if the market price is lower than the contract price at the time when the goods ought to have been accepted, the seller is entitled to the difference even though the market price has subsequently risen and he ha re-sold for more than the contract price; *Campbell Mostyn* v. *Barnett*, [195 1 Lloyd's Rep. 65, C. A.

(*l*) *Phillpotts* v. *Evans* (1839), 5 M. & W. 475; *Barrow* v. *Arnaud* (1846), 8 Q. B. 595, at p. 609, Ex. Ch.; *cf. ex parte Stapleton* (1879), 10 Ch. D. 586, at p. 590, C. A. As to extension of time at buyer's request, see *Hickman* v. *Haynes* (1875), L. R. 10 C. P. 598.

(*m*) *Re Vic Mill*, [1913] 1 Ch. 465, C. A.

(*n*) *Thompson* v. *Robinson* (*Gunmakers*), [1955] 1 All E. R. 154; [1955] Ch. 177. See also *Victory Motors, Ltd.* v. *Bayda* (1973), 3 W. W. R. 747 (seller would have made loss selling trade-in car: loss taken into account in estimating seller's damages). But *cf. Jeremy's Garages, Ltd.* v. *MacAndrews Car Hire, Ltd.*, unreported as yet, 31st October, 1974, C. A. (losses actually made on re-selling trade-in cars disregarded in assessing seller's damages).

(*o*) *Charter* v. *Sullivan*, [1957] 1 All E. R. 809; (1957] 2 Q. B. 117, C. A. Nor will the seller be able to recover loss of profit if he delays selling the car until a time when supply exceeds demand: *Blythswood Motors, Ltd.* v. *Raeside*, 1966 S. L. T. 13 (Sheriff's Court).

(*p*) For cases in which he may sue for the price, see the comment to s. 49 (2), *ante.*

Where the property has passed he may sue either for the price (*q*) or for damages for non-acceptance. If the parties so choose, they can fix the damages by the contract itself (*r*).

The market price rule in sub-s. (3) is an obvious deduction from sub-s. (2), applicable in general to the ordinary goods of commerce for which there is more or less ready sale. The rule is so convenient that the Courts apply it whenever and so far as practicable. But it is only a *prima facie* rule depending on its conformity with the general principle. For example, if a contract be repudiated before the time fixed for delivery, and the seller accept the repudiation, he may not be justified in holding back the goods on a falling market and thus enhancing the damages. He must act reasonably (*s*). But " he is not bound to nurse the interests of the contract breaker, and so long as he acts reasonably at the time it ill lies in the mouth of the contract breaker to turn round afterwards, and complain, in order to reduce his own liability to a plaintiff, that the plaintiff has failed to do that which perhaps with hindsight he might have been wiser to do " (*t*).

Available market.—Various suggestions have been put forward as to what constitutes, or is required for there to be, an available market.

It was formerly suggested (*u*) that there had to be a particular place, such as an exchange where the goods could be sold, but this has been criticised (*v*) and expressly dissented from (*a*). An alternative view that " an available market merely means that the situation in the particular area was such that the particular goods could freely be sold, and that there was a demand sufficient to absorb readily all the goods that were thrust on it, so that if a purchaser defaulted, the goods in question could readily be disposed of " (*b*) was criticised in the Court of Appeal (*c*) where JENKINS, L.J., suggested that, since the sub-section contemplates a difference between contract price and

---

(*q*) Unless he has resold, in which case he must sue for damages : *Lamond* v. *Davall* (1847), 9 Q. B. 1030.

(*r*) *Diestal* v. *Stevenson & Co.*, [1906] 2 K. B. 345—subject of course to the general rules about penalties and, perhaps, to s. 55, *post*. See note (*e*) on p. 265.

(*s*) *Roth* v. *Taysen* (1895), 73 L. T. 628, and see note to s. 54, *post* ; for general principle, see *British Westinghouse Co.* v. *Underground Electric Railway*, [1912] A. C. 673, at p. 689.

(*t*) *Harlow and Jones* v. *Panex (International), Ltd.*, [1967] 2 Lloyd's Rep. 509, at p. 530, *per* ROSKILL, J.

(*u*) *Dunkirk Colliery* v. *Lever* (1878), 9 Ch. D. 20, at p. 25, C. A.

(*v*) *Thompson* v. *Robinson (Gunmakers)*, [1955] 1 All E. R. 154; [1955] Ch. 177; *Charter* v. *Sullivan*, [1957] 1 All E. R. 809; [1957] 2 Q. B. 117, C. A.

(*a*) *A. B. D. (Metals and Waste), Ltd.* v. *Anglo-Chemical and Ore Co., Ltd.*, [1955] 2 Lloyd's Rep. 456, at p. 466.

(*b*) *Thompson* v. *Robinson (Gunmakers)*, *supra;* followed in *A. B. D. (Metals and Waste), Ltd.* v. *Anglo-Chemical and Ore Co., Ltd.*, *supra*.

(*c*) *Charter* v. *Sullivan*, *supra*.

market price, there might only be an " available market " in the sense of the sub-section where the market price was regulated by supply and demand and thus subject to fluctuation, and not where the only " market " was an existing demand for goods which, for other reasons, could be sold only at a standard retail price.

Where there is only one alternative buyer available, who is willing to purchase part only of the contract goods, that buyer is not an " available market " (*d*).

In some cases where the seller has resold, the re-sale price has been assumed to furnish the correct measure of damages (*e*).

Where there is no available market, the measure of the seller's loss will normally be the difference between the sale price and the value of the goods to the seller at the time of the breach (*f*).

Where goods are sold at a fixed retail price, the " available market " test is inapplicable and the question is whether the seller has lost the opportunity of making an extra sale, so that he has lost the profit on that sale, or whether he has not in fact lost any profit, because he has been able quickly to resell the subject matter of the contract together with all other goods of that kind that he could obtain (*g*).

If goods are to be made specially for a buyer and the buyer repudiates before the seller begins to make the goods, the seller will be entitled to recover his loss of profit unless the buyer can prove that the seller acting reasonably and in the ordinary course of his business, has been able, as a result of the situation in which the buyer placed him, to take a course of action (such as taking on new orders) which diminished his loss. But if the buyer can prove that he has so diminished his loss then his damages would be reduced accordingly (*h*). The seller will not have to give credit for a profit arising from speculative work not in the ordinary course of his business (*i*).

(*d*) *Harlow and Jones* v. *Panex (International), Ltd.*, [1967] 2 Lloyd's Rep. 509.

(*e*) *Maclean* v. *Dunn* (1828), 4 Bing. 722 ; *Ex p. Stapleton* (1879), 10 Ch. D. 586, C. A. ; *Ryan* v. *Ridley* (1902), 8 Com. Cas. 105. But the seller must prove his actual loss ; it is not enough merely to prove the re-sale price if, for instance, the re-sale was on conditions different from those of the original sale : *Macklin* v. *Newbury Sanitary Laundry* (1919), 63 Sol. Jo. 337. Note also *Campbell Mostyn* v. *Barnett*, [1954] 1 Lloyd's Rep. 65, C. A. (wrongfully rejected goods resold by seller at more than contract price ; measure of damages nevertheless difference between contract price and market price at date of rejection).

(*f*) *Harlow & Jones* v. *Panex (International), Ltd.*, *supra*.

(*g*) *Thompson* v. *Robinson (Gunmakers)*, *ante*, illustration (2); *Charter* v. *Sullivan*, *ante*, illustration (3).

(*h*) *Hill & Sons* v. *Showell & Sons, Ltd.* (1918), 87 L. J. Q. B. 1106, H. L.

(*i*) *Ibid.*

Remedies of the Buyer

51. Damages for non-delivery.—(1) Where the seller wrongfully neglects or refuses to deliver the goods to the buyer, the buyer may maintain an action against the seller for damages for non-delivery (*k*).

(2) The measure of damages is the estimated loss directly and naturally resulting, in the ordinary course of events, from the seller's breach of contract (*l*).

(3) Where there is an available market for the goods in question (*m*) the measure of damages is primâ facie (*n*) to be ascertained by the difference between the contract price and the market or current price of the goods at the time (*o*) or times (*p*) when they ought to have been delivered, or, if no time was fixed, then at the time of the refusal to deliver (*q*).

ILLUSTRATIONS

(1) Sale of 10,000 tons of coal to be delivered at Lubeck, " penalty for non-execution of this contract, one shilling per ton." This is not a penalty, but liquidated damages ; and, if default is made in delivery, this measure must be applied to the instalments undelivered (*r*).

(2) A in June sells a cargo of coal to B, to be shipped in November. In October B sells the cargo to C, assigning his rights against A.

(*k*) *Ramsden* v. *Gray* (1849), 7 C. B. 961 ; *Jones* v. *Gibbons* (1853), 8 Exch. 920 (non-delivery of goods agreed to be delivered " as required ") ; *Lewis* v. *Clifton* (1854), 14 C. B. 245 (refusal to permit growing timber, which had been sold by auction, to be carried away).

(*l*) *Smeed* v. *Foord* (1859), 1 E. & E. 602 ; *Grébert-Borgnis* v. *Nugent* (1885), 15 Q. B. D. 85, C. A. (specially manufactured goods) ; *cf. Hammond* v. *Bussey* (1887), 20 Q. B. D. 79, at p. 93, C. A. ; and see note to s. 54, *post*.

(*m*) See *Marshall* v. *Nicoll*, 1919 S. C. (H. L.) 129, and *per* UPJOHN, J. in *Thompson* v. *Robinson* (*Gunmakers*), [1955] 1 All E. R. 154, at p. 159 ; [1955] Ch. 177, at p. 186; *Mouat* v. *Betts Motors, Ltd.*, [1958] 3 All E. R. 402; [1959] A. C. 71, P. C., *post*, illustration (4). See also the comment to the previous section.

(*n*) *Charter* v. *Sullivan*, [1957] 1 All E. R. 809; [1957] 2 Q. B. 117, C. A.

(*o*) *Leigh* v. *Paterson* (1818), 8 Taunt. 540 ; *Hinde* v. *Liddell* (1875), L. R. 10 Q. B. 265. As to c.i.f. contract, see *Sharpe & Co.* v. *Nosawa*, [1917] 2 K. B. 814 (time when the shipping documents, not the goods, should have been tendered).

(*p*) As to instalment deliveries, see *Brown* v. *Muller* (1872), L. R. 7 Exch. 319 ; *Roper* v. *Johnson* (1873), L. R. 8 C. P. 167 ; *cf. Bergheim* v. *Blaenavon Co.* (1875), L. R. 10 Q. B. 319.

(*q*) *Shaw* v. *Holland* (1846), 15 M. & W. 136, 146 ; *Josling* v. *Irvine* (1861), 6 H. & N. 512 ; *Ashmore* v. *Cox*, [1899] 1 Q. B. 436, 443 (date of repudiation of contract).

(*r*) *Diestal* v. *Stevenson & Co.*, [1906] 2 K. B. 345.

The cargo is not delivered. In an action by B the measure of damages is the difference between the contract price and the market price at the time of breach, i.e. in November. The sub-contract with C is to be disregarded, since it only affords evidence of the market value in October (*s*).

(3) Six contracts of sale of cotton waste on the terms that delivery is to be made within a reasonable time of the removal of a government embargo. The contracts are repudiated before the time has elapsed. Damages must be assessed in accordance with sub-s. (2). The market price at the time of repudiation is inapplicable (*t*).

(4) A sells a car to B for £1,207. B promises to offer it to A at, in the event, £1,157, if he wishes to re-sell it within two years. Within the two years he re-sells it to C for £1,700 without first offering it to A. Because of similar covenants given by him to his seller, A could only have re-sold the car for £1,207; had he re-purchased it. The measure of A's damages is £543, being the difference between the contract price and the market price, the only available market being the black market upon which B had re-sold the car (*u*).

Definitions.—For " action," " buyer," " delivery," " goods " and " seller," see s. 62 (1), *post.*

Cross-references.—*Cf.* ss. 1 (1), 8 and 9, *ante* (price) ; s. 10, *ante* (stipulations as to time) ; ss. 27 and 28, *ante* (seller's duty to deliver) ; ss. 29 to 31, *ante* (rules as to delivery) ; and s. 54, *post* (interest and special damage).

COMMENT

This section also, in terms, deals only with general damages. It is declaratory and is founded on *Hadley* v. *Baxendale* (*v*).

When the seller breaks his contract to deliver, one or more of four remedies may be open to the buyer, namely : (1) in all cases he may sue for damages for non-delivery ; (2) if the price has been prepaid by him, he may recover it back as money had and received for a consideration which has wholly failed (*a*) ; (3) if the goods be specified or ascertained he may sue for specific performance ; and (4) if the property in the goods has passed, and he is entitled to immediate possession, he has the ordinary remedies of an owner deprived of his goods, *viz.*, conversion or detinue as the case may be. The action for non-delivery lies when the goods are not delivered,

(*s*) *Williams* v. *Agius*, [1914] A. C. 510, 523 ; see also *Williams* v. *Reynolds* (1865), 6 B. & S. 495 ; *Thol* v. *Henderson* (1881), 8 Q. B. D. 457.

(*t*) *Millett* v. *Van Heek & Co.*, [1921] 2 K. B. 369, C. A.

(*u*) *Mouat* v. *Betts Motors, Ltd.*, [1958] 3 All E. R. 402; [1959] A. C. 71, P. C.

(*v*) (1854) 9 Exch. 341, at p. 354 ; 2 Smith L. C. (13th ed.), p. 529 ; and see now *Victoria Laundry* v. *Newman Industries*, [1949] 1 All E. R. 997, C. A.; [1949] 2 K. B. 528.

(*a*) But not if the contract itself was illegal ; see, *e.g.* *Berg* v. *Sadler & Moore*, [1937] 1 All E. R. 637; [1937] 2 K. B. 158, C. A.

or when goods which are not in conformity with the contract are tendered and rejected.

Section 51 does not in terms apply to actions for delay in delivery when the goods are ultimately accepted; and necessarily damages there are assessed on different lines from damages for non-delivery in order to give effect to the guiding principle stated in sub-s. (2), which applies throughout the law of contract.

Non-delivery.—The rule as to market price is clearly a deduction from the more general rule in sub-s. (2). " Where a contract to deliver goods at a certain price is broken," says TINDAL, C.J.,

> " the proper measure of damages in general is the difference between the contract price and the market price of such goods at the time when the contract is broken, because the purchaser having the money in his hands may go into the market and buy. So, if a contract to accept and pay for goods is broken, the same rule may be properly applied, for the seller may take his goods into the market and obtain the current price for them " (*b*).

Hence, if in an action for non-delivery no difference between the contract price and the market price is shown, the plaintiff in general is only entitled to nominal damages (*c*).

The rule is so convenient and obvious that the English Courts apply it whenever possible, even where it produces hardship in individual cases (*d*). In Scotland the rule is not so strictly applied (*e*).

But there are many cases in which the rule of market price is inapplicable. If it is partially applicable it will be applied with the necessary modifications, thus—

(1) The buyer may have prepaid the price. In that case he is entitled to recover back the money he has paid with interest (*f*), together with the difference between the price he paid and the full market price of the goods on the day when they ought to have been delivered (*g*).

(*b*) *Barrow* v. *Arnaud* (1846), 8 Q. B. 604, at p. 609, Ex. Ch.

(*c*) *Valpy* v. *Oakeley* (1851), 16 Q. B. 941; *Erie Natural Gas Co.* v. *Carroll*, [1911] A. C. 105, P. C. (substituted performance); *Aryeh* v. *Lawrence Kostoris & Son, Ltd.*, [1967] 1 Lloyd's Rep. 63.

(*d*) *Brady* v. *Oastler* (1864), 3 H. & C. 112 (special price for early delivery ignored); *Williams* v. *Reynolds* (1865), 6 B. & S. 495 (profit on re-sale excluded); *Thol* v. *Henderson* (1881), 8 Q. B. D. 457 (sub-contract by buyer disregarded); *Williams* v. *Agius*, [1914] A. C. 510 (sub-contract disregarded); *cf. Victoria Laundry* (*Windsor*) v. *Newman Industries, Ltd.*, [1949] 1 All E. R. 997; [1949] 2 K. B. 528, C. A.; *Mouat* v. *Betts Motors, Ltd.*, [1958] 3 All E. R. 402; [1959] A. C. 71, P. C. (*ante*, illustration (4)); and see *Charter* v. *Sullivan*, [1957] 1 All E. R. 809; [1957] 2 Q. B. 117, C. A.

(*e*) *Dunlop* v. *Higgins* (1848), 1 H. L. Cas. 381, at p. 403.

(*f*) As to interest, see s. 3 of The Law Reform (Miscellaneous Provisions) Act, 1934; reproduced in Appendix I, *post*.

(*g*) *Startup* v. *Cortazzi* (1835), 2 C. M. & R. 165; *cf. Barrow* v. *Arnaud* (1846), 8 Q. B. 604, at p. 610.

(2) The exact sort of goods the buyer has contracted for may not be obtainable, but if it is reasonable for him to buy in similar goods he may charge the seller with the difference in price (*h*).

(3) The seller may have repudiated his contract before the time for delivery arrives. In such a case the buyer may either accept the repudiation and sue at once, or he may refuse to accept it and wait for the time for performance before suing. If he refuses to accept the repudiation the measure of damages is calculated by reference to the value of the goods at the time when they ought to have been delivered and it seems that in such a case the buyer is under no obligation to mitigate his loss and may sit tight on a rising market (*i*). But since the contract is still open, the seller has the opportunity of changing his mind and eventually performing, and he may also take advantage of any subsequent events which operate to release him from his obligation to perform (*k*).

Where, however, the buyer accepts the seller's repudiation and sues forthwith, although the damages are still calculated by reference to the value of the goods at the time fixed for delivery (*l*), the buyer is under a duty to mitigate his loss if he can by buying in against the contract (*m*).

This is an application of the rule of *causa proxima* to damages. The *anticipatory* breach of a contract to deliver within a reasonable time comes under sub-s. (2), and not under sub-s. (3) (*n*).

(4) The time for delivery may have been extended at the seller's

---

(*h*) *Hinde* v. *Liddell* (1875), L. R. 10 Q. B. 265 ; *cf. Erie Natural Gas Co.* v. *Carroll,* [1911] A. C. 105, at p. 117, "the measure of damage is the cost of procuring the substituted article." This assumes that the article is a satisfactory and not an inferior substitute. *Cf.* also *Harbutt's Plasticine, Ltd.* v. *Wayne Tank and Pump Co., Ltd.*, [1970] 1 All E. R. 225; [1970] 1 Q. B. 447, C. A. (Old factory burnt down; measure of damages full cost of re-instating factory because not possible to acquire second hand factory. *Aliter* in case of second hand cars.)

(*i*) *Leigh* v. *Paterson* (1818), 8 Taunt. 540; *Roth* v. *Taysen* (1895), 73 L. T. 628, C. A.; *Michael* v. *Hart & Co.*, [1902] 1 K. B. 482, C. A.; *Tredegar Iron Co.* v. *Hawthorn* (1902), 18 T. L. R. 716, C. A. Dicta in *Nickoll* v. *Ashton* by MATHEW, J., at first instance, and VAUGHAN WILLIAMS, L.J., in the Court of Appeal, ([1900] 2 Q. B. 298 at p. 305 and [1901] 2 K. B. 126 at p. 138, respectively) might be taken to suggest the contrary, but it is not clear whether these remarks were based on accepted or non-accepted repudiation.

(*k*) *Avery* v. *Bowden* (1855), 5 E. & B. 714 ; (1856), 6 E. & B. 953, Ex. Ch.

(*l*) *Cf. Maredelanto Compania Naviera S.A.* v. *Bergbau-Handel G.m.b.H.*, [1970] 3 All E. R. 125; [1971] 1 Q. B. 164, C. A. (party in breach, had he waited, could lawfully have cancelled contract).

(*m*) *Roper* v. *Johnson* (1873), L. R. 8 C. P. 167; approved in *Garnac Grain Co. Inc.* v. *H. M. F. Faure and Fairclough, Ltd.*, [1967] 2 All E. R. 353, 360; [1968] A. C. 1130, 1140, H. L.

(*n*) *Millet* v. *Van Heek & Co.*, [1921] 2 K. B. 369, C. A. *Quaere* whether the final words of sub-s. (3) ever apply to such a contract.

request. In that case the extended time will be taken as the contract time (*o*).

(5) There may be no available market at the place of delivery, but regard may be had to the market price contemplated by the parties at the ultimate destination of the goods (*p*).

(6) The parties in their contract may have provided that in case of breach certain agreed damages should be paid by the party in default (*q*), and the question may then arise whether this is a penalty under the general law of contract (*r*).

(7) There may be a breach of a contract to deliver goods at a fixed date in a foreign country. In that case the damages must be calculated according to the market value at the time and place appointed for delivery and in the past, this normally involved the conversion of the foreign currency into English currency, according to the rate of exchange ruling at the time of breach (*s*), but it is not clear whether this is still the case (*t*).

Again the market price test may be wholly inapplicable, and then recourse must be had to the wider general principle of sub-s. (2). This is the case where there is no available market for the goods in question, as, for example, where the buyer has ordered some special article or articles to be expressly manufactured for him (*u*). In the case of non-delivery, where there is no market and the sale is for the purpose of re-sale, the measure of damages is the

---

(*o*) *Ogle* v. *Earl Vane* (1868), L. R. 3 Q. B. 272 (non-delivery), Ex. Ch.; *Hickman* v. *Haynes* (1875), L. R. 10 C. P. 598 (non-acceptance); *Tyers* v. *Rosedale Co.* (1875), L. R. 10 Exch. 195, Ex. Ch. See also *Rickards* v. *Oppenheim*, [1950] 1 All E. R. 420, C. A.; [1950] 1 K. B. 616, and see p. 23, *ante*:

(*p*) *Wertheim* v. *Chicoutimi Pulp Co.*, [1911] A. C. 301, P. C., applied in *Macauley & Culle* v. *Horgan*, [1925] 2 I. R. 1. *Cf. Sharpe & Co.* v. *Nosawa & Co.*, [1917] 2 K. B. 814, as to c.i.f. contract.

(*q*) *Dunlop* v. *New Garage*, [1915] A. C. 79; *Cellulose Acetate Silk Co., Ltd.* v. *Widnes Foundry* (1925), *Ltd.*, [1933] A. C. 20; *Imperial Tobacco Co., Ltd.* v. *Parslay*, [1936] 2 All E. R. 515; *Stewart* v. *Carapanayoti*, [1962] 1 All E. R. 418; [1962] 1 W. L. R. 34.

(*r*) See, e.g., *Dunlop* v. *New Garage* (*supra*); *Cooden Engineering* v. *Stanford*, [1952] 2 All E. R. 915; [1953] 1 Q. B. 86, C. A.; *Bridge* v. *Campbell Discount*, [1962] 1 All E. R. 385; [1962] A. C. 600; *Lombank Ltd.* v. *Excell*, [1963] 3 All E. R. 486; [1964] 1 Q. B. 415, C. A.; *Robophone Facilities Ltd.* v. *Blank*, [1966] 3 All E. R. 128; [1966] 1 W. L. R. 1428, C. A.

(*s*) *Di Ferdinando* v. *Simon Smits & Co.*, [1920] 3 K. B. 409, at pp. 414, 415, C. A.; approved, *S.S. Celia*, [1921] 2 A. C. 544. For the circumstances in which a loss caused by devaluation of a currency may be recovered, see *Aruna Mills, Ltd.* v. *Dhanrajmal Gobindram*, [1968] 1 All E. R. 113; [1968] 1 Q. B. 655; and see *post*, notes to s. 54. As to loss caused or aggravated by the injured party's impecuniosity, see p. 253, *post*.

(*t*) See p. 258, *post*.

(*u*) *Grébert-Borgnis* v. *Nugent* (1885), 15 Q. B. D. 85, C. A.; *Elbinger A. G.* v. *Armstrong* (1874), L. R. 9 Q. B. 473, at p. 476; *Leavey (J.) & Co., Ltd.* v. *Hirst (George) & Co. Ltd.*, [1943] 2 All E.R. 581; [1944] K. B. 24, C. A.

difference between the contract price and the re-sale price (*v*). Each case turns on its particular circumstances, and is usually complicated by questions of special damage (*w*). As to mitigation of damages, see *post*, notes to s. 54.

Goods lawfully rejected.—The buyer's duty to accept the goods arises only where tender of delivery is made in accordance with the terms of the contract of sale (see s. 27, *ante*), and if goods other than goods of the contract description are delivered, or if the delivery is otherwise not in accordance with the conditions of the contract, the buyer may rightfully reject the goods (provided he has not done any act amounting to acceptance within s. 35 of the Act) by notifying the seller within a reasonable time that he refuses to accept them: see ss. 35 and 36, *ante*.

When the buyer is entitled to reject the goods, and does so, he can recover back the price if he has paid it, for the consideration for its payment has wholly failed (*a*). Then arises the question: what further compensation, if any, is he entitled to? When he lawfully rejects the goods the position seems to be this: he has contracted for the supply of certain goods, and those goods have never been supplied to him. The seller, therefore, has failed in his obligation to deliver, and whatever damages would be recoverable in an action for non-delivery should on principle be recoverable in this case (*b*).

Where a contract for the supply of machinery contained a guarantee to replace defective parts, and concluded: "we do not give any other guarantee and we do not accept responsibility for consequential damages," it was held that the buyers were entitled to reject the goods delivered as not in accordance with the contract, and that the clause did not negative the seller's liability for extra expenses, etc., incurred in procuring other machinery, as this was damage arising naturally and directly from the failure to deliver according to contract (*c*).

(*v*) *Patrick* v. *Russo-British Grain Export Co.*, [1927] 2 K. B. 535; *Leavey* v. *Hirst*, [1943] 2 All E. R. 581; [1944] 1 K. B. 24; *Household Machines* v. *Cosmos Exporters*, [1946] 2 All E. R. 622; [1947] K. B. 217; *Kwei Tek Chao* v. *British Traders and Shippers*, [1954] 3 All E. R. 165; [1954] 2 Q. B. 459.

(*w*) *Hydraulic Co.* v. *McHaffie* (1878), 4 Q. B. D. 670, C. A. (special damage); *Grébert-Borgnis* v. *Nugent* (1885), 15 Q. B. D. 85, to C. A. (goods made to order); but see an attempt to apply the current price rule to specially manufactured goods: *Marshall* v. *Nicoll*, 1919 S. C. (H. L.) 129. See also another Scottish case, *British Motor Body Co.* v. *Shaw*, 1914 S. C. 922, 928 (motor-car made to order, undue delay).

(*a*) See p. 257, *post*.

(*b*) See *Bridge* v. *Wain* (1816), 1 Stark. 504, as commented on in *Elbinger A.G.* v. *Armstrong* (1874), L. R. 9 Q. B. 473, at p. 476, where this position seems to be assumed.

(*c*) *Millar's Machinery Co., Ltd.* v. *Way & Son* (1935), 40 Com. Cas. 204, C. A.

Delay in delivery.—A similar rule applies to damages for delay, when goods of a particular description are ordered and are ultimately accepted after the delay (*d*), there being a *primâ facie* rule that the damage is the difference between " the value of the article contracted for at the time when it ought to have been and at the time when it actually was delivered " (*e*). Accordingly, in the case of ordinary goods of commerce, the *primâ facie* measure of damages is the difference between the market price at the time when the goods ought to have been delivered and the time when they actually were delivered. But if that price has fallen, and yet the buyer has on delivery resold at a price above the market price, it has been held that the re-sale must be taken into account in fixing the damages (*f*). In this respect the action for delay would seem to differ from the action for non-delivery in which (special damages apart) circumstances " personal " or " accidental " to the buyer cannot be taken into account (*g*).

On the other hand the market price may have no bearing on the damages. Thus where the sellers had refused to make delivery, in the mistaken belief that the Control of Timber (No. 1) Order, 1939, made delivery illegal, the buyer, in addition to obtaining a declaration that he was entitled to delivery, claimed damages on the ground that if delivery had been made on the contract date he could have used the timber as he pleased, whereas now he would be subject to restrictions rendering the timber less valuable, and an inquiry as to damages was directed (*h*).

Trover or detinue.—Subject to the provisions of ss. 8 to 10 of the Factors Act, 1889 (now reproduced in ss. 25 and 47 of this Act, *ante*), where, under a contract of sale, the buyer is entitled to immediate possession of the goods, and the seller wrongfully neglects or refuses to deliver them, the buyer may maintain an action for damages or detention of the goods against the seller or any other person in possession of the goods, or an action for the conversion of

(*d*) *Fletcher* v. *Tayleur* (1855), 17 C. B. 21 ; *Smeed* v. *Foord* (1859), 1 E. & E. 602 ; *Cory* v. *Thames Ironworks Co.* (1868), L. R. 3 Q. B. 181 ; *Re Carver* (1911), 17 Com. Cas. 59. As to damages against a carrier for delay in delivering ordinary goods of commerce, see *Koufos* v. *Czarnikow, Ltd., Heron II*, [1967] 3 All E. R. 686; [1969] 1 A. C. 350, H. L.

(*e*) *Elbinger A. G.* v. *Armstrong* (1874), L. R. 9 Q. B. 473, at p. 477, *per* BLACKBURN, J.

(*f*) *Wertheim* v. *Chicoutimi Pulp. Co.*, [1911] A. C. 301, P. C. ; distinguished and criticised, *Slater* v. *Hoyle & Smith*, [1920] 2 K. B. 11, C. A. (action for breach of warranty); see s. 53 (2), *post*). See further, *Benjamin's Sale of Goods*, para. 1297.

(*g*) *Williams* v. *Agius*, [1914] A. C. 510, approving *Rodocanachi* v. *Milburn* (1886), 18 Q. B. D. 67, 77, C. A. See also *Victoria Laundry* v. *Newman Industries*, [1949] 1 All E. R. 997, C. A. ; [1949] 2 K. B. 528 ; *Brading* v. *McNeill & Co.*, [1946] 1 Ch. 145.

(*h*) *Rappaport* v. *London Plywood and Timber Co., Ltd.*, [1940] 1 All E. R. 576.

the goods against the seller or any other person who has dealt with the goods under such circumstances as to amount to a conversion thereof (*i*).

The measure of damages for conversion or, it would seem, detinue is the actual damage sustained by the buyer (*k*), provided, of course, that the damage is not too remote. The *primâ facie* measure of damages is the value of the goods at the time of the wrongful act (*l*), but this will not be so if the damage suffered by the plaintiff is less than the value of the goods.

As between seller and buyer, the buyer cannot recover larger damages by suing in tort rather than in contract. Thus if he has not paid the price and, instead of delivering the goods, the seller wrongfully sells them to a third party, the buyer can only recover the difference between the contract price and the value of the goods, even if he frames his action in tort (*m*).

As regards third parties, the *primâ facie* measure of damages is not normally applied where the plaintiff immediately prior to the conversion had only a limited interest in the goods (*n*).

When a man has sold goods to one person, a mere contract to sell them to another after the property has passed is not a conversion, but a delivery of them in pursuance of that contract is a conversion (*o*), unless at the time of re-sale the original buyer was in

---

(*i*) As to detinue, see *Langton* v. *Higgins* (1859), 4 H. & N. 402; *Whiteley* v. *Hilt*, [1918] 2 K. B. 808, 819. As to conversion or trover, see *Hollins* v. *Fowler* (1875), L. R. 7 H. L. 757; *Delaney* v. *Wallis* (1884), 13 L. R. Ir. 31, C. A. (sale by auctioneer in market overt).

(*k*) *Chinery* v. *Viall* (1860), 5 H. & N. 288; *Wickham Holdings* v. *Brooke House Motors, Ltd.*, [1967] 1 All E. R. 117; [1967] 1 W. L. R. 295, C. A.; and see *Whitely* v. *Hilt*, [1918] 2 K. B. 808, C. A.; *cf. Astley Industrial Trust* v. *Miller*, [1968] 2 All E. R. 36 (hire-purchase: detinue by third party; value of goods less than amount due to hirer at date of demand by him); *Aronson* v. *Mologa Holzindustrie A/G Leningrad* (1927), 32 Com. Cas. 276, C. A. (price paid by buyer; conversion by seller; fall in market price after conversion).

(*l*) *Chinery* v. *Viall, supra; France* v. *Gaudet* (1871), L. R. 6 Q. B. 199.

(*m*) This is because the seller, having failed to deliver, cannot sue for the price: *Chinery* v. *Viall, supra*; *cf. Johnson* v. *Stear* (1863), 15 C. B. N. S. 330; *Johnson* v. *Lancashire Rail. Co.* (1878), 3 C. P. D. 499, at p. 507; *Hiort* v. *London and North Western Rail. Co.* (1879), 4 Ex. D. 188, C. A. The position would seem to be different if the seller tortiously retakes the goods after delivery, since at that stage he has performed his part of the contract and may sue for the price notwithstanding his tort: *Stephens* v. *Wilkinson* (1831), 2 B. & Ad. 320; *Gillard* v. *Brittan* (1841), 8 M. & W. 575; *Page* v. *Cowasjee Edulgee* (1866), L. R. 1 P. C. 127, at pp. 146–147. This would seem to be so even though the tortious seizure is also a breach of the warranty of quiet possession implied by s. 12 (2) and, *semble*, could be set up under s. 53 (1) in diminution or extinction of the price: *Healing (Sales) Pty., Ltd.* v. *Inglis Electrix Pty., Ltd.* (1968), 121 C. L. R. 584 (High Ct. of Aus.).

(*n*) *Wickham Holdings* v. *Brooke House Motors, Ltd.*, [1967] 1 All E. R. 117; [1967] 1 W. L. R. 295, C. A.; and see *Whiteley* v. *Hilt*, [1918] 2 K. B. 808, C. A.; *cf. Astley Industrial Trust* v. *Miller*, [1968] 2 All E. R. 36.

(*o*) *Lancashire Wagon Co.* v. *Fitzhugh* (1861), 6 H. & N. 502.

default in paying the price (*p*). Ordinarily a person who buys and receives goods (otherwise than in market overt) which the seller had no right to sell is guilty of a conversion, however innocently he may have acted (*q*), but his liability has been much restricted by ss. 8 and 9 of the Factors Act, 1889, *post*, reproduced in s. 25 of this Act; and note also s. 23, *ante*.

52. Specific performance.—In any action for breach of contract to deliver specific or ascertained goods (*r*) the court may, if it thinks fit, on the application of the plaintiff, by its judgment [or decree] direct that the contract shall be performed specifically, without giving the defendant the option of retaining the goods on payment of damages. The judgment [or decree] may be unconditional, or upon such terms and conditions as to damages, payment of the price, and otherwise, as to the court may seem just, and the application by the plaintiff may be made at any time before judgment [or decree].

The provisions of this section shall be deemed to be supplementary to, and not in derogation of, the right of specific implement in Scotland.

Definitions—For "action," "defendant," "delivery," "goods," "plaintiff" and "specific goods," see s. 62 (1), *post*.

Cross-references.—*Cf.* ss. 1 (1), 8 and 9, *ante* (price), and s. 51, *ante* (action for non-delivery).

COMMENT

This section applies to all cases where the goods are specific or ascertained, whether the property therein has passed to the buyer or not. An arbitrator can award specific performance to the same extent as the High Court unless a contrary intention is expressed in

(*p*) *Milgate* v. *Kebble* (1841), 3 M. & Gr. 100, decided before the Act; "default" in this sense would now have to be construed in the light of s. 48 (3), *ante*.

(*q*) *Wilkinson* v. *King* (1809), 2 Camp. 335; *Cooper* v. *Willomatt* (1845), 1 C. B. 672.

(*r*) "Specific goods bear the meaning assigned to them in the definition clause, 'goods identified and agreed upon at the time a contract of sale is made.' 'Ascertained' probably means identified in accordance with the agreement after the time a contract of sale is made." *Per* ATKIN, L.J., *Re Wait*, [1927] 1 Ch. 606, at p. 630.

the arbitration agreement, and provided the contract does not relate to land or any interest in land (*s*).

If payment and delivery are concurrent, as they will be by virtue of s. 28, *ante*, unless otherwise agreed, it will be made a condition of an order for specific performance that the price is paid on delivery (*t*). So too, where all or part of the price is not payable immediately on delivery and the seller is entitled to a lien such as an unpaid seller's lien over the goods to secure payment of the price, the court will not order specific performance in a manner which will deprive the seller of his lien, but will seek to protect his rights while enforcing the contract against him (*u*).

In appropriate cases, a contract of sale may be rectified by the Court, and then in the same action specifically enforced as so amended (*v*).

The Court has power to order a distress to be levied on all the defendant's goods until he delivers the specific chattel (*a*).

The power of the court to order specific performance is a discretionary one. It is akin to the power of a court of equity to order specific performance of a contract and will be exercised on similar principles (*b*). In an action for detinue it was stated that:

> " The power vested in the Court to order the delivery up of a particular chattel is discretionary and ought not to be exercised when the chattel is an ordinary article of commerce, and of no special value or interest, and not alleged to be of any special value to the plaintiff, and where damages would fully compensate " (*c*).

The section deals only with claims for specific performance by a buyer against a seller. But a seller may also in a special case obtain specific performance against a buyer at least where the goods are specific or ascertained (*d*).

Non-specific or unascertained goods.—This section does not apply to non-specific or unascertained goods. In general, the courts

(*s*) Arbitration Act, 1950, s. 15 ; 2 Halsbury's Statutes (3rd Edn.) 446.

(*t*) *Cf. Hart* v. *Herwig* (1873), 8 Ch. App. 860, at p. 864.

(*u*) *Cf. Langen and Wind, Ltd.* v. *Bell*, [1972] 1 All E. R. 296; [1972] Ch. 685 (sale of shares).

(*v*) *U.S. of America* v. *Motor Trucks, Ltd.*, [1924] A. C. 196, P. C. ; but see *Rose* v. *Pim*, [1953] 2 All E. R. 739, C. A. ; [1953] 2 Q. B. 450.

(*a*) See R. S. C., O. 45, rr. 4, 13 (2); Appendix A, Forms 64 and 65.

(*b*) *Société des Industries Metallurgigues S.A.* v. *Bronx Engineering Co.*, [1975] 1 Lloyd's Rep. 465, *per* BUCKLEY and ORMROD, L.JJ.

(*c*) *Whiteley* v. *Hilt*, [1918] 2 K. B. 808, at p. 819, C. A. ; *Cohen* v. *Roche* [1927] 1 K. B. 169 (order for specific delivery of Hepplewhite chairs refused); *Behnke* v. *Bede Shipping Co.*, [1927] 1 K. B. 649 (specific performance of contract of sale of ship " of peculiar and perhaps unique value to the plaintiff granted) "; *Société des Industries Metallurgigues S.A.* v. *Bronx Engineering Co.*, *supra* (machinery specially made to order; obtainable elsewhere but would take 9–12 months; damages would be substantial and difficult to quantify. Held: these factors did not of themselves justify an order for specific performance).

(*d*) *Shell-Mex* v. *Elton Cop Dyeing Co.* (1928), 34 Com. Cas. 39, at p. 46.

will not order a seller to specifically perform such a contract, as the buyer can be adequately compensated for breach in damages (*e*).

It has recently been held, however, on an application for interlocutory relief, that where damages are not an adequate remedy, as where the goods are not or may not be obtainable elsewhere, then on general equitable principles, and subject to the normal limitations on such relief, the court may order specific performance of such a contract or grant an injunction restraining a seller from acting in breach of it (*f*).

Scotland.—In Scotland specific performance, or, as it is called, specific implement, is an ordinary and not an extraordinary remedy, and it can be demanded as of right wherever it is practicable (*g*).

53. Remedy for breach of warranty.—(1) Where there is a breach of warranty by the seller (*h*), or where the buyer elects, or is compelled, to treat any breach of a condition on the part of the seller as a breach of warranty (*i*), the buyer is not by reason only (*k*) of such breach of warranty entitled to reject the goods, but he may

(a) set up against the seller the breach of warranty in diminution or extinction of the price (*l*); or

---

(*e*) *Re Clarke* (1887), 36 Ch. D. 348, at p. 352, *per* COTTON, L.J.; *Fothergill* v. *Rowland* (1873), 43 L. J. Ch. 252, especially at p. 255, *per* JESSEL, M.R. See also *Dominion Coal Co., Ltd.* v. *Dominion Iron and Steel Co., Ltd. and National Trust Co., Ltd.* [1909] A. C. 293, P. C.

(*f*) *Sky Petroleum, Ltd.* v. *VIP Petroleum Ltd.*, [1974] 1 All E. R. 954; [1974] 1 W. L. R. 576. *Cf.* the conflicting opinions on this point of the members of the Court of Appeal in *Re Wait*, [1927] 1 Ch. 606., discussed pp. 142–143, *ante*, *Fry on Specific Performance* (6th ed.) 82–86; *Humboldt Flour Mills Co.* v. *Boscher* (1975), 50 D. L. R. (3d) 477 (specific performance held not to be available apart from this section).

(*g*) *Stewart* v. *Kennedy* (1890), 15 App. Cas. 75, at pp. 102, 105. *Cf.* *The S.S. Elorio*, 1916 S. C. 882 (discretion).

(*h*) *Syers* v. *Jonas* (1848), 2 Exch. 111, at p. 117; *Dawson* v. *Collis* (1851), 10 C. B. 523, at p. 523; *Behn* v. *Burness* (1863), 3 B. & S. 751, at pp. 755, 756, Ex. Ch.; *Heilbutt* v. *Hickson* (1872), L. R. 7 C. P. 438, at p. 451.

(*i*) See s. 11 (1) (c), *ante*, and *Street* v. *Blay* (1831), 2 B. & Ad. 456, at p. 463; *Gompertz* v. *Denton* (1832), 1 Cr. & M. 207; *Parsons* v. *Sexton* (1847), 4 C. B. 899; *Couston* v. *Chapman* (1872), L. R. 2 Sc. & Div. 250, at p. 254.

(*k*) *Bannerman* v. *White* (1861), 10 C. B. N. S. 844; *Behn* v. *Burness*, *supra*; *Heilbutt* v. *Hickson*, *supra*.

(*l*) *Mondel* v. *Steel* (1841), 8 M. & W. 858, 870, 871 (diminution); *Poulton* v. *Lattimore* (1829), 9 B. & C. 259 (extinction); *cf.* *Kinnear* v. *Brodie*, 1901 3 F. (Ct. of Sess.) 540, horse, warranted quiet, drowned as a result of its vice; buyer not liable for the price. But the warranty seems to have been treated as a condition).

(b) maintain an action against the seller for damages for the breach of warranty (*m*).

(2) The measure of damages for breach of warranty is the estimated loss directly and naturally resulting, in the ordinary course of events, from the breach of warranty (*n*).

(3) In the case of breach of warranty of quality such loss is primâ facie the difference between the value of the goods at the time of delivery to the buyer and the value they would have had if they had answered to the warranty (*o*).

(4) The fact that the buyer has set up the breach of warranty in diminution or extinction of the price does not prevent him from maintaining an action for the same breach of warranty if he has suffered further damage (*p*).

(5) Nothing in this section shall prejudice or affect the buyer's right of rejection in Scotland as declared by this Act (*q*).

ILLUSTRATIONS

(1) A sells sulphuric acid to B as commercially free from arsenic; B uses it for making glucose, which he sells to brewers, and the persons who drink the beer made by the brewers are poisoned. A does not know the purpose for which the acid is required. In an action for breach of warranty, B can recover the price of the acid

---

(*m*) *Davis* v. *Hedges* (1871), L. R. 6 Q. B. 687. The effect of the corresponding sub-section in the New South Wales Sale of Goods Act is discussed at length in *Healing* (*Sales*) *Pty., Ltd.* v. *Inglis Electrix Pty., Ltd.* (1968), 121 C. L. R. 584, 280, where differing conclusions are reached by the various members of the High Court of Australia.

(*n*) *Randall* v. *Raper* (1858), E. B. & E. 84 (seed barley of inferior quality sold with warranty. Re-sold with same warranty. Claim by sub-buyer. Seller liable to buyer for amount of sub-buyer's claim although buyer had not yet paid sub-buyer); *Smith* v. *Green* (1875), 1 C. P. D. 92 (cow with foot-and-mouth disease; sold to farmer and infected other cows); *Randall* v. *Newson* (1877), 2 Q. B. D. 102, C. A., at p. 111 (defective carriage-pole specially made for carriage); *Wilson* v. *Dunville* (1879), 4 L. R. Ir. 249 (brewers' grains which poisoned cattle); *Hammond* v. *Bussey* (1887). 20 Q. B. D. 79, C. A. (ship coal of particular quality; costs of reasonably defending actions by sub-buyer, re-sale to whom was contemplated by parties at time of sale).

(*o*) *Loder* v. *Kekulé* (1857), 3 C. B. N.S. 128; *Jones* v. *Just* (1868), L. R. 3 Q. B. 197; *cf. Heilbutt* v. *Hickson* (1872), L. R. 7 C. P. 438, at p. 453.

(*p*) *Mondel* v. *Steel* (1841), 8 M. & W. 858; *cf. Rigge* v. *Burbidge* (1846), 15 M. & W. 598.

(*q*) See s. 11 (2), *ante*, and s. 59, *post*.

and the value of other goods which have been spoiled by it, but not the damages he has had to pay to the brewers or damages for injury to the goodwill of his business (*r*).

(2) An orchid, warranted a *Cattelaya Alba*, is bought at a sale for £20. Two years afterwards it produces purple flowers. The purple variety is worth a few shillings, the white is worth £50. The buyer is entitled to £50 damages for breach of warranty (*s*).

(3) A, a grocer, sells tinned salmon to B which is unfit for food. B's wife is poisoned and dies. B may recover the reasonable expenses of medical attendance and the funeral, and also a reasonable sum for loss of his wife's services (*t*).

(4) Contract for the sale of cotton cloth of a certain quality. The cloth delivered is of inferior quality. The buyer can recover as damages the difference between the value of the cloth at the time of delivery, and the value it would have had if it had answered to the warranty. A sub-contract of which the seller had no notice must be disregarded (*u*).

(5) Contract for goods to be shipped in August. They are accepted and paid for. The buyer subsequently finds out that they were shipped in September. If this does not affect the market price of the goods at the time of delivery, the buyer can recover only nominal damages (*a*).

(6) A buys a fur coat from B for re-sale and sells it to X. By reason of a dye used on the collar, X contracts dermatitis and recovers damages from A. A is entitled to recover from B the damages and costs awarded against him in the action by X which he has reasonably defended, together with his own costs in that action as between solicitor and client (*b*).

Definitions.—For " action," " buyer," " delivery," " goods," " quality," " seller " and " warranty," see s. 62 (1), *post*.

Cross-references.—*Cf.* ss. 1 (1), 8 and 9, *ante* (price); ss. 10 to 15, *ante* (conditions and warranties); s. 62 (1) *post*, definition of " warranty "; Note A, *post*, Appendix II; s. 35, *ante* (acceptance and rejection); and s. 54 *post* (interest and special damage).

COMMENT

This section is the complement to s. 11, *ante*. Section 11 shows when goods may be rejected or when the buyer must resort to his remedy for breach of warranty under this section. Although

(*r*) *Bostock* v. *Nicholson*, [1904] 1 K. B. 725, C. A.; *cf. Holden* v. *Bostock* (1902), 50 W. R. 323, C. A., and see illustrations to s. 14, *ante*; and *Pinnock Brothers* v. *Lewis & Peat*, [1923] 1 K. B. 690, at p. 697.

(*s*) *Ashworth* v. *Wells* (1898), 78 L. T. 136, C. A.

(*t*) *Jackson* v. *Watson & Sons*, [1909] 2 K. B. 193, C. A.

(*u*) *Slater* v. *Hoyle & Smith*, [1920] 2 K. B. 11, C. A.

(*a*) *Taylor & Sons, Ltd.* v. *Bank of Athens* (1922), 27 Com. Cas. 142. *Cf. Kwei Tek Chao* v. *British Traders and Shippers*, [1954] 1 All E. R. 779; [1954] 2 Q. B. 459.

(*b*) *Bennett, Ltd.* v. *Kreeger* (1925), 41 T. L. R. 609; and see *Bowmaker (Commercial), Ltd.* v. *Day*, [1965] 2 All E. R. 856, n.; [1965] 1 W. L. R. 1396.

the buyer may not be able to reject the goods for simple breach of warranty, he may be entitled to reject them for fraud or some other invalidating cause. This conclusion is pointed to by the words "by reason *only* of such breach of warranty" in sub-s. (1), and by the provisions of s. 61 (2), *post*, as to savings of common law.

The seller's breach of warranty not only provides the buyer with a right of action if he has suffered damage, but also with a defence against an action for the price. Thus if the contract contains an arbitration clause limited to quality disputes, the seller's action for the price may still be stayed if the buyer insists on such an arbitration, notwithstanding that the buyer does not counterclaim for damages.

Sub-section (2).—The measure of damages for breach of warranty provided by this sub-section is the normal measure of damages for breach of contract. The general principles governing the award of damages for breach of contract are discussed in the comment to s. 54, *post*. Particular examples are considered later in the comment to this section.

Sub-section (3).—This sub-section lays down a *primâ facie* measure of damages in the case of a breach of warranty of quality. It does not apply to any other breach of warranty (*c*), but appears to be based on the same principle as where the seller fails to deliver the goods at all (*d*), namely that if the contract is broken, the buyer may sell the defective goods and purchase goods of the required quality (*e*). This principle may well be relevant in assessing damages for other breaches of warranty by the seller.

The difference in value of the goods.—The value of the goods will be the market value where there is an available market (*f*), for in such a case the buyer can go into the market, buy new goods, and sell those delivered at the relevant market price (*g*).

The buyer must prove his damage by adducing evidence as to the reduction in value of the goods (*h*). The contract price cannot be

---

(*c*) For damages for the breach of the condition and warranties implied by s. 12, see the comments to that section, *ante*.

(*d*) See s. 51 (3), *ante*.

(*e*) *Slater* v. *Hoyle and Smith*, [1920] 2 K. B. 11, C. A.

(*f*) *Ibid.*, at p. 22.

(*g*) See *Barrow* v. *Arnaud* (1846), 8 Q. B. 604, at p. 609, Ex. Ch., cited at length in the comment to s. 51, *ante*.

(*h*) *Aryeh* v. *Lawrence Kostoris & Son, Ltd.*, [1967] 1 Lloyd's Rep. 63. On the other hand, the Supreme Court of Canada has held that where goods are not fit for a purpose for which they are warranted, the *primâ facie* measure of damages is the price of the goods, and it is for the seller to show that they are still of some merchantable value: *Ford Motor Co. of Canada, Ltd.* v. *Haley*, [1967] S. C. R. 437.

assumed to represent the value which the goods would have had if they had answered to the warranty, both because the buyer may have been getting a bargain, or overpaying, and also, where delivery does not immediately follow the making of the contract, because the value may have changed in the interval. It may, however, be a factor to be taken into account where there is little other evidence (*i*).

So, too, the price at which the purchaser has agreed to re-sell will not be taken into account where other goods can be bought to enable the buyer to fulfil his sub-sale, but may be an element in estimating the value where alternative goods cannot be acquired (*k*). Indeed, where no substitute can be obtained for the goods, the price which the buyer is entitled to receive on the sub-sale will be very strong evidence of the value of the goods to the buyer, although account would have to be taken of the terms of the contract under which the goods are re-sold.

In appropriate cases, the difference in value may be assessed as the amount necessary to make the goods in question answer the warranty (*l*), or as an amount equal to a fair allowance against the price in respect of the damage (*m*).

The respective values should *primâ facie* be assessed at the date of delivery. But this rule is based on the proposition that the buyer can acquire alternative goods on that date to fulfil his requirements, and can sell the inferior goods on the same day. If the circumstances are such that the seller cannot reasonably expect the buyer to discover the defect until later, then the difference in value must be calculated at the time when the buyer can reasonably be expected to discover the defect (*n*). So, too, at least if the price falls, the relevant date may be postponed if the buyer is induced by the seller not to resell immediately (*o*).

In general, however, fluctuations in the market price of the inferior goods after the date of delivery, or where appropriate the date on which the defect is discovered, will be ignored (*p*).

Other recoverable heads of damage.—The measure of damages set out in sub-section (3) of this section is a *primâ facie* measure only.

(*i*) *Minster Trust, Ltd.* v. *Traps Tractors, Ltd.*, [1954] 3 All E. R. 136; [1954] 1 W. L. R. 963.

(*k*) *Slater* v. *Hoyle and Smith*, [1920] 2 K. B. 11, at pp. 17, 18.

(*l*) *Minster Trust, Ltd.* v. *Traps Tractors, Ltd.*, *supra*.

(*m*) *Biggin & Co., Ltd.* v. *Permanite, Ltd.*, [1950] 2 All E. R. 859, at p. 871; [1951] 1 K. B. 422, at p. 439, *per* Devlin, J.; reversed on other grounds, [1951] 2 All E. R. 191; [1951] 2 K. B. 314, C. A.

(*n*) *Van den Hurk* v. *Martens*, [1920] 1 K. B. 850 (packaged goods which seller knew buyer intended to re-sell and which it was not practical to examine before they were required for use); *Ashworth* v. *Wells*, illustration (2), *ante*.

(*o*) *Loder* v. *Kekule* (1857), 3 C. B. N. S. 128. *Cf.* the cases on a reasonable time for rejection, discussed in the comment to s. 35, *ante*.

(*p*) *Jones* v. *Just* (1868), L. R. 3 Q. B. 197; *Campbell Mostyn* (*Provisions*), *Ltd.* v. *Barnett Trading Co.*, [1954] 1 Lloyd's Rep. 65 C. A.; and see *Jamal* v. *Moolla Dawood*, [1916] 1 A. C. 175, at p. 179, P. C.

It is always open to a buyer to prove that he has suffered other loss which directly and naturally resulted in the ordinary course of events from the breach of warranty, and in such a case he will be entitled to damages to compensate him for such loss (*q*).

The buyer may claim such damages even where he has set up the breach of warranty in diminution or extinction of the price under sub-s. (4) for his loss may greatly exceed the price.

In considering what loss is within sub-s. (2) of this section, it is necessary first to consider whether and if so when, the buyer ought to have discovered that there had been a breach of warranty.

Where the defect is patent.—In general, a buyer cannot use an article at the risk of his seller if before he uses it he has notice that it is defective (*r*). So, too, if he buys it for re-sale, but before he re-sells it he becomes aware that it is defective, he will not normally be able to recover from his seller extra damages which he has had to pay the sub-buyer as a result of delivering defective goods over and above those he would have had to pay had he not delivered them (*s*).

If the buyer learns of the defects after the sub-sale, he ought to warn the sub-buyer if by so doing he can diminish his liability to him, and if he fails to do so, he will not be able to pass on the extra damages to his seller (*t*).

These propositions are probably illustrations of the general rule that a party to a contract should act reasonably even when the other party is in breach, and may not increase the liability of the other party by acting unreasonably. There may well be circumstances in which a buyer could reasonably use defective goods even with knowledge of the defect. Thus, if no alternative goods were available and there was a reasonable prospect that the defective goods would fulfil the purpose for which they were bought, it may sometimes be reasonable to use them for that purpose, and if in such a case the buyer in fact suffered additional loss, such loss might be recoverable from the seller (*u*).

(*q*) See further the distinction between general and special damages, considered in the comment to s. 54, *post*. For cases where the seller fails to deliver the goods and the buyer is able to recover more than the *primâ facie* measure of damages, see *ante*, comment to s. 51.

(*r*) *British Oil and Cake* v. *Burstall* (1923), 39 T. L. R. 406, at p. 407.

(*s*) *Biggin & Co., Ltd.* v. *Permanite, Ltd.*, [1951] 1 K. B. 422, at p. 435; reversed on another ground [1951] 2 K. B. 314, C. A.; see also *Dobell & Co., Ltd.* v. *Barber Garratt*, [1931] 1 K. B. 219, at p. 238 (a buyer cannot increase the damages recoverable from his seller by making further deliveries of cattle cake to a sub-buyer after he learns that the cattle cake is unfit for use as cattle food).

(*t*) *Biggin & Co., Ltd.* v. *Permanite, Ltd. supra.*

(*u*) In deciding what is reasonable, it would be necessary to take all factors into account, including the likely damage resulting from not using the goods, the prospect of successfully using them, and the extra loss which the buyer would suffer if he used them unsuccessfully. *Cf.* the rule that a party who acts reasonably in seeking to mitigate damages may recover additional loss which

Where the defect is discoverable.—A buyer will not normally be debarred from recovering damages just because the loss could have been avoided had he examined the goods, for he will normally be entitled to rely on the seller's warranty (*a*). It may be different if without any special examination the buyer is put on notice that something is wrong with the goods. The buyer must act reasonably. But he will not normally be held to have acted unreasonably just because, in the absence of any indication that there was anything wrong with the goods, he chooses to rely on the seller's warranty and does not subject the goods to a thorough examination (*a*).

Loss of profits.—If the seller should realise at the date of the contract that goods which are sold are likely to be used by the buyer with a view to profit, then damages for breach of warranty may include a sum in respect of any profits which the buyer has lost (*b*).

If the goods have more than one use, it would not seem to be necessary that the seller should know the precise use to which they are to be put by the buyer (*c*).

Expenses.—Subject to the same proviso as to foreseeability, a buyer may also recover damages in respect of wasted expenditure (*d*). Where there is no claim for loss of profit, this claim may even extend to cover expenditure incurred before the contract in anticipation of it (*e*).

It has sometimes been suggested that it is not possible to claim both for loss of profit and for expenses incurred. Such a general

---

he thereby incurs: *Lloyds and Scottish Finance, Ltd.* v. *Modern Cars and Caravans (Kingston), Ltd.*, [1964] 2 All E. R. 732; [1966] 1 Q. B. 764; *The World Beauty*, [1970] p. 144, at p. 156, C. A.; *MacGregor on Damages* (13th ed.), paras. 236–237.

(*a*) *Pinnock Brothers* v. *Lewis and Peat* [1923] 1 K. B. 690, 698 (where only a small parcel of goods was involved, and an expert examination would have been required to discover the defect); *Dobell & Co., Ltd.* v. *Barber and Garratt*, [1931] 1 K. B. 219, C. A. (where an analysis of cattle cake would have been required).

(*b*) See, *e.g.*, *Ashworth* v. *Wells*, *ante*, Illustration (2); *Holden* v. *Bostock* (1902), 18 T. L. R. 317 (sugar contaminated with arsenic sold to brewers; damages included market value of beer made with the sugar which the brewers had to destroy).

(*c*) *Bostock & Co., Ltd.* v. *Nicholson & Sons, Ltd.*, [1904] 1 K. B. 725; *Kendall* v. *Lillico*, [1968] 2 All E. R. 444; [1969] 2 A. C. 31, H. L.; *Bunting* v. *Tory* (1948), 64 T. L. R. 353, a decision of HILBERY, J., which is sometimes cited for the proposition that the seller should know the precise use, would seem to be inconsistent with these authorities. In any event, in that case, the only possible use for the bull which was bought was for breeding, and it is difficult to see why it should affect the measure of the damages that the buyer might have, but did not, sell the bull to somebody else instead of breeding from it himself.

(*d*) See, *e.g.*, *Molling & Co.* v. *Dean & Son, Ltd.* (1901), 18 T. L. R. 217 (cost of bringing goods back from sub-buyer who rejected them); *Smith* v. *Johnson* (1899), 15 T. L. R. 179) cost of pulling down and rebuilding wall).

(*e*) *Anglio Television* v. *Reed*, [1971] 3 All E. R. 690; [1972] 1 Q. B. 60, C. A.

proposition cannot be supported either by authority (*f*) or in principle. The object of damages, however assessed, is to put the injured party as far as possible in the same position as if the contract had been performed (*g*). Damages cannot be recovered twice over. Net profits are calculated by deducting from gross profits the expenses necessary to earn them. If as a result of the breach, the buyer makes a smaller net profit or succeeds only in breaking even, then he is put in the same position as if the contract had been performed if he receives by way of damages a sum equivalent to the net profits which he ought to have made but did not make (*h*). But if he incurs an overall loss, then he can only be recompensed if he receives in addition to the net profits a sum equivalent to that loss.

Thus if a man buys a machine for £5,000 which is warranted to produce him £10,000 profit during its lifetime of one year after allowing for the cost of the machine and all other expenses, but after he has spent £1,000 by way of necessary preliminary expenditure, and £500 reasonably trying to make the machine work, it finally proves to be valueless, the amount necessary to put him in the same position as if the contract had been performed must include £1,500 to cover him for his loss, in addition to damages in respect of the profit he should have made.

(*f*) Such a claim was allowed in *Molling* v. *Dean, supra, Foaminol Laboratories* v. *British Artid Plastics,* [1941] 2 All E. R. 393; *Hydraulic Engineering Co., Ltd.* v. *McHaffie Goslett and Co.* (1878), 4 Q. B. D. 670 (delay in delivery); *Steam Herring Fleet, Ltd.* v. *Richards & Co., Ltd.* (1901) 17 T. L. R. 731 (delay in delivery).

(*g*) *Cullinane* v. *British Rema,* [1953] 2 All E. R. 1257; [1954] 1 Q. B. 292, *per* JENKINS, L.J., at pp. 1264 and 308; *Monarch SS Co., Ltd.* v. *A. B. Karlshamns Oljefabriker,* [1949] 1 All E. R. 1, at p. 12; [1949] A. C. 196, at pp. 220–221. See also comment to s. 54, *post.*

(*h*) In *Cullinane* v. *British Rema, supra,* the buyer of a clay pulverising plant claimed loss of net profits for three years (deducting running expenses and an allowance for depreciation of the plant) plus the difference between the cost of the plant and its estimated residual value. The venture was a speculative one and the life expectancy of the plant was ten years. The plant failed to perform as warranted. The majority of the Court of Appeal held that in the absence of any claim for any loss of profit beyond three years and of any evidence or finding by the Official Referee as to what that profit would have been, the buyer was bound by his pleadings and they had to proceed on the assumption that no more profit would have been made after the three years—see *per* Lord EVERSHED at pp. 1263 and 305–306 and *per* JENKINS, L.J., at pp. 1266–1267 and 312. In those circumstances the buyer was put in the same position as if the contract had been performed, so far as was possible on the pleadings and the evidence. MORRIS, L.J., dissenting, considered that the evidence and the findings of the Official Referee indicated that further profit would have been made after the three years. The case is discussed at length by the High Court of Australia in *T. C. Industrial Plant Pty., Ltd.* v. *Roberts (Queensland), Pty., Ltd.* (1963), 37 A. L. J. R. 289. See also 1973 Can. Bar Rev. 490. See however *Anglia Television* v. *Reed,* [1971] 3 All E. R. 690, at p. 692; [1972] 1 Q. B. 60, at p. 64, where Lord DENNNING, M.R. accepts *Cullinane* v. *British Rema* as authority for the proposition that in some cases at least a plaintiff cannot claim both wasted expenditure and loss of profit.

Losses incurred by the buyer as a result of reselling the goods.—The question of re-sale has been considered in the House of Lords, where Lord PHILLIMORE said,

> " It is all a question of contract. . . Notice will not do of itself, nor will knowledge. But if the tribunal which tries the case comes to the conclusion that he contracted to sell . . . on terms that he should be responsible for damage which might accrue from his failure to provide for any one of certain objects, then he must be held liable " (*i*).

In order to recover damages in respect of such losses, a buyer must show that the seller ought to have contemplated the possibility of a sub-contract under which a claim might be made against the buyer (*k*). It is also essential that any difference between the first contract and the sub-contract should not be such as to make it impossible to say whether the injury which ultimately resulted to the sub-buyer and for which the buyer was liable was a natural and foreseeable result of the breach of the first contract (*l*).

Losses caused or aggravated by the buyer's impecuniosity.—Such losses will generally be irrecoverable (*m*), unless they can be shown to have been within the contemplation of the parties when the contract

---

(*i*) *Hall, Ltd.* v. *Pim Junior & Co., Ltd.* (1928), 33 Com. Cas. 324, at p. 373 (construing contract of London Corn Association). This decision is commented upon by SCRUTTON, L.J., in *Finlay & Co.* v. *Kwik Hoo Tong Handel Maatschappij*, [1929] 1 K. B. 400, at pp. 409–412, and by SANKEY, L.J., at pp. 417 and 418.

(*k*) *James Finlay* v. *N.V. Kwik Hoo Tong*, [1929] 1 K. B. 400, at p. 411, *per* SCRUTTON, L.J.; *Biggin & Co.* v. *Permanite*, [1951] 1 K. B. 422, reversed on another ground [1951] 2 All E. R. 191; [1951] 2 K. B. 314, C. A.

(*l*) *Biggin* v. *Permanite*, supra. See also *Kasler and Cohen* v. *Slavonski*, [1928] 1 K. B. 78, at p. 85; *Dexters, Ltd.* v. *Hill Crest Oil Co.* (*Bradford*), *Ltd.* [1926] 1 K. B. 348, at p. 359, C. A. For illustrations see *Hydraulic Engineering Co.* v. *McHaffie* (1878), 4 Q. B. D. 670, at p. 677, C. A. (part specially ordered to fulfil sub-contract); see also *Grébert-Bornis* v. *Nugent* (1885), 15 Q. B. D. 85, C. A. (goods ordered for French sub-contract); *Biggin & Co.* v. *Permanite*, [1951] 2 All E. R. 191; [1951] 2 K. B. 314, C. A. (buyers allowed to recover from sellers amount reasonably paid to sub-buyers in settlement of dispute). *Hammond* v. *Bussey* (1887), 20 Q. B. D. 79, C. A. (breach of warranty and sub-sale with similar warranty, costs of action reasonably defended), followed, *Agius* v. *Great Western Colliery Co.* (1899), 1 Q. B. 413, C. A.; *Bennett, Ltd.*, v. *Kreeger* (1925), 41 T. L. R. 609; *Alison & Co.* v. *Wallsend Slipway & Engineering Co.* (1927), 43 T. L. R. 323, C. A. (damages including costs of compromising claim); *Kasler & Cohen* v. *Slavonski*, [1928] 1 K. B. 78 (costs of actions between successive sellers); *Slavonski* v. *La Pelleterie de Roubain S. A.* (1927), 137 L. T. 645; *Dobell & Co.* v. *Barber & Garratt*, [1931] 1 K. B. 219 (damages paid under sub-contract recoverable); *Slade* v. *Sinclair & Wilcox* (1930), 74 Sol. Jo. 122; *Lloyds and Scottish Finance, Ltd.*, v. *Modern Cars and Caravans* (*Kingston*), *Ltd.*, [1964] 2 All E. R. 732; [1966] 1 Q. B. 764 (cost of interpleader proceedings reasonably incurred in an attempt to mitigate damages).

(*m*) *Trans Trust S. P. R. L.* v. *Danubian Trading Co., Ltd.*, [1952] 1 All E. R. 970; [1952] 2 Q. B. 297, C. A.; and see *Muhammad Tasa el Sheikh Ahmad* v. *Ali*, [1947] A. C. 414, P. C. Compare *Mrs. Eaton's Car Sales* v. *Thomesen*, [1973] N. Z. L. R. 686.

was made (*n*). But a seller will be unable to complain that a buyer has failed to take reasonable steps to mitigate his loss if he is unable to afford to take such steps (*o*).

Loss of goodwill.—Where goods are bought for re-sale or to enable the buyer to fulfil some other obligation, the buyer may suffer a loss of goodwill if as a result of a breach of contract by the seller, he is unable to fulfil his own obligations. He may recover damages in respect of any such loss if, but only if, he pleads and proves with sufficient certainty that he has suffered pecuniary loss, and that such pecuniary loss was within the contemplation of the seller when the contract was made (*p*).

Scotland.—In Scotland before the Act no distinction was drawn between warranties and conditions. Every material term was a condition, and the rule was that where the buyer could reject the goods, but had not done so, he could not sue for damages. The *actio quanti minoris* only applied to cases where the goods could not be returned; but now the buyer has a double remedy, guarded, however, by the court's power to order the consignation of the whole or part of the purchase price—see the comment to s. 59, *post*.

54. Interest and special damages.—Nothing in this Act shall affect the right of the buyer or the seller to recover interest or special damages in any case where by law interest or special damages may be recoverable, or to recover money paid where the consideration for the payment of it has failed.

Definitions.—For " buyer and " seller," see s. 62 (1), *post*.

(*n*) *Cf*. *Freedhoff* v. *Pomalift* (1971), 19 D. L. R. (3d.) 153, in which it was held by the Ontario Court of Appeal that where there is a contract for the sale of a revenue generating asset, the forseeable damage is the loss of revenue, not the loss occasioned by the buyer's inability to make payments using that revenue.

(*o*) *Cf*. *Robbins of Putney* v. *Meek*, [1971] R. T. R. 345; and see generally the passage on mitigation of damages, p. 255, *post*.

(*p*) *Foaminol Laboratories, Ltd*. v. *British Artid Plastics, Ltd*., [1941] 2 All E. R. 393, at p. 400; *cf*. *Simon* v. *Pawson and Leafs, Ltd*. (1932), 38 Com Cas. 151, C. A. In both these cases the claims failed on the grounds that the loss was not in the contemplation of the parties when the contract was made, and that in any event no pecuniary loss was proved. The buyer may show either that the seller should have contemplated that the breach of contract would naturally lead to a loss of the buyer's business or that the seller knew that the breach would probably lead to a loss of business: *Doe* v. *W. H. Bowater, Ltd*., [1916] W. N. 185. A claim for loss of goodwill was also disallowed in *Bostock* v. *Nicholson*, [1904] 1 K. B. 725. For a Canadian case where a claim for loss of custom was allowed, see *Lakelse Dairy Produce, Ltd*. v. *General Dairy Machinery, Ltd*. (1970), 10 D. L. R. (3d.) 277.

Cross-references.—*Cf.* ss. 37 and 49 to 53, *ante* (rights of action).

COMMENT

Interest.—See note to s. 3 of the Law Reform (Miscellaneous Provisions) Act, 1934, reproduced in Appendix I, *post*.

Object of damages.—In assessing damages for breach of contract the object aimed at is to put the injured party, so far as money can do it, into the same position as if the contract had not been broken, for this is the measure of the loss directly and naturally resulting from the breach (*q*).

Mitigation of damages.—But the injured party must act reasonably; and if it be open to him to take steps to mitigate the loss, he must do so. If he fails to take such steps, the defendant may set up the failure in mitigation of damages (*r*). This, in effect, is an application of the general rule of *causa proxima* to damages. If a man does not take reasonable steps to mitigate the loss arising from the other party's breach of contract, the resulting loss is proximately caused by his own conduct and not by the breach of contract. But a contracting party is not bound to take steps in mitigation of damages which would prove injurious to him in other ways (*s*). Provided he acts reasonably at the time, the other contracting party will not be permitted to complain that he failed to do that which, perhaps, with hindsight he might have been wiser to do (*t*). And if a

---

(*q*) *Wertheim* v. *Chicoutimi Pulp Co.*, [1911] A. C. 301, P. C.; *British Westinghouse Electric Co.* v. *Underground Railway*, [1912] A. C. 673, at p. 689. *Cf. Taylor & Sons, Ltd.* v. *Bank of Athens* (1922), 27 Com. Cas. 142, at p. 147 (nominal damages only given on facts). *Quære* how far this principle applies to breach of warranty in respect of dangerous articles, where the cause of action may be framed in tort?

(*r*) *Roth* v. *Taysen* (1895), 73 L. T. 628, C. A.; *Nickoll* v. *Ashton*, [1900 2 Q. B. 298; *Hill* v. *Showell* (1918), 87 L. J. K. B. 1106, H. L.; *Payzu, Ltd.* v. *Saunders*, [1919] 2 K. B. 581, C. A.; *Melachrino* v. *Nickoll and Knight*, [1920] 1 K. B. 693; *Houndsditch Warehouse Co., Ltd.* v. *Waltex, Ltd.*, [1944] K. B. 579.

(*s*) *Finlay & Co., Ltd.* v. *Kwik Hoo Tong Handel Maatschappij*, [1929] 1 K. B. 400 (buyer not bound to enforce sub-contracts when to do so would injure his commercial reputation); *Jewelowski* v. *Propp*, [1944] 1 All E. R. 483; [1944] 1 K. B. 510 (claimant not bound to expend money on a speculative venture in order to mitigate damages); *Lesters Leather and Skin Co.* v. *Home and Overseas Brokers* (1948), 64 T. L. R. 569, C. A. (no obligation to purchase from unfamiliar suppliers abroad goods which would only arrive long after contractual delivery date); *Heaven and Kesterten, Ltd.* v. *Etablissements Francois Albiac et cie*, [1956] 2 Lloyd's Rep. 316 (buyer who has rejected goods on ground of defective quality is not bound to accept such goods if offered in mitigation of damages; *aliter*, perhaps where the rejection was for reasons unconnected with the quality of the goods); *Robbins of Putney* v. *Meek*, [1971] R. T. R. 345 (impecunious seller forced to resell prematurely). *Cf.*, pp. 238, 254, *ante*.

(*t*) *Harlow & Jones* v. *Panex (International), Ltd.*, [1967] 2 Lloyd's Rep. 509.

contracting party takes a reasonable and prudent course to mitigate his loss, which naturally arises out of the circumstances in which he was placed by the breach, and which is not an independent or disconnected transaction, then any resulting diminution in the loss he has suffered will be taken into account although there was no duty on him so to act (*u*). Where a contracting party repudiates a wholly executory contract, the other contracting party is not bound to accept such repudiation, but can carry out his part of the contract and claim the full contract price, at least where he has a legitimate interest, financial or otherwise in so doing (*v*).

Special damages.—In pleading a care, a distintion is drawn, between general and special damages.

> " Special damage in such a context means the particular damage (beyond the general damage) which results from the particular circumstances of the case, and of the plaintiff's claim to be compensated, for which he ought to give warning in his pleadings in order that there may be no surprise at the trial " (*a*).

The Act deals only with general damages, and merely saves the law relating to special damages. Many of the cases fail to distinguish special from general damages, the reason being that both are governed by the same guiding rule. Given a particular contract, the measure of damages is the loss which naturally results from breach of a contract of the kind in question. Given a contract made under special circumstances to the knowledge of both parties, *e.g.*, a contract to fulfil a sub-contract, the measure of special damage is the loss which naturally results from the breach of a contract made under those particular circumstances. The underlying principles governing the award of damages for breach of contract were considered by the Court of Appeal in *Victoria Laundry* v. *Newman Industries* (*b*), where the following propositions were approved:

(1) The governing purpose of damages is to put the injured party in the same position, so far as money can do it, as if his rights had been observed, but this rule does not permit recovery of all loss resulting from a breach of contract, however improbable or unpredictable.

(2) Damages are recoverable only in respect of that part of the loss actually resulting as was the time of the contract reasonably foreseeable and liable to result from the breach.

(*u*) *British Westinghouse Electric and Manufacturing Co., Ltd.* v. *Underground Electric Rail. Co., Ltd.*, [1912] A. C. 673, at pp. 689–692, H. L.; *R. Pagnan and Frateli* v. *Corbisa Industrial Agropacuaria Limitdator*, [1971] 1 All E. R. 165; [1970] 1 W. L. R. 1306, C. A.; contrast *Aruna Mills, Ltd.* v. *Dhanrajmal Gobindram*, [1968] 1 All E. R. 113; [1968] 1 Q. B. 655 (independent transaction).

(*v*) *White and Carter* (*Councils*), *Ltd.* v. *Macgregor*, [1961] 3 All E. R. 1178; [1962] A. C. 413.

(*a*) *Ratcliffe* v. *Evans*, [1892] 2 Q. B. 524, at p. 528, *per* BOWEN, L.J.; see also *Anglo-Cyprian Trade Agencies* v. *Paphos Wine Industries*, [1951] 1 All E. R. 873

(*b*) [1949] 1 All E. R. 997; [1949] 2 K. B. 528, C. A.

(3) What was then reasonably foreseeable depends upon the knowledge then possessed by the parties, or, at all events, by the party who later commits the breach.

(4) For this purpose knowledge may be actual or imputed. A contract breaker is assumed to know what loss is liable to result in the ordinary course of things from his breach. He may also be shown to know of special circumstances outside the ordinary course of things of such a kind that a breach in those special circumstances would be liable to cause more loss. If so, the additional loss is also recoverable.

(5) For a contract breaker to be liable, it is not necessary that he should have asked himself what loss might result. As has often been pointed out, parties at the time of contracting contemplate performing the contract not breaking it. It is sufficient that, if he had considered the question, he would as a reasonable man have concluded that the loss in question was liable to result.

(6) It is not necessary to show that the loss was certain to result. It is enough if it is a " serious possibility " or " real danger " (*c*).

The same result will be arrived at if the supposed contemplation of the parties be wholly eliminated. Given a contract made without any special circumstances, then the measure of ordinary damage is the loss which naturally arises from the breach of such a contract. Given a contract made under special circumstances to the knowledge of both parties, then the special damages are those which naturally arise from a breach of such a contract under the particular circumstances.

Failure of consideration.—As to failure of consideration also, there is nothing peculiar to the contract of sale. Money paid on a consideration which has wholly failed can usually be recovered as money had and received (*d*). And where the plaintiff bought and paid for 175 tons of terra japonica, and only $156\frac{3}{4}$ tons were delivered, he was held entitled to recover a proportionate amount of the price under the common money counts (*e*).

Where the plaintiff bought a motor-car, used it for some months, and then found out that it had been stolen and had to be restored to its original owner, it was held that he was entitled to recover the full price he had paid. There was a total failure of consideration consequent on the breach of condition as to title (*f*).

(*c*) *Koufos* v. *Czarnikow, Ltd.*, [1967] 3 All E. R. 686; [1969] 1 A. C. 350, H. L., where the use of the words " on the cards " in the *Victoria Laundry* case was disapproved.

(*d*) See Bullen & Leake's *Precedents of Pleadings* (11th ed.), p. 321, and cases there collected. As to set-off of failure of consideration, see *Biggerstaff* v. *Rowatt's Wharf*, [1896] 2 Ch. 93, C. A.

(*e*) *Devaux* v. *Conolly* (1849), 8 C. B. 640; and see *Ebrahim Dawood Ltd.* v. *Heath*, [1961] 2 Lloyd's Rep. 512; but *cf. Covas* v. *Bingham* (1853), 2 E. & B. 836, where by the contract the bill of lading was made conclusive as to quantity.

(*f*) *Rowland* v. *Divall*, [1923] 2 K. B. 500, at p. 505, C. A. See further the comment to s. 12, *ante*.

If there is a total failure of consideration because the buyer has paid for the goods and the seller has then become bankrupt, or in the case of a company has gone into liquidation, before he supplies the goods, the buyer must normally prove in the bankruptcy or liquidation for the price. But in special circumstances, the seller may be held to be a trustee of the purchase price for the buyer, in which case the buyer will recover it in priority to the other creditors (*g*).

Rate of exchange.—Changes in the relative values of currencies are irrelevant if they occur after the date as at which damages fall to be assessed and are usually to be disregarded if they occur on or before that date, either because the loss flowing from the revaluation has no causal connection with the breach of contract, or because such a loss is not within the assumed contemplation of the parties (*h*). Until recently, it has also repeatedly been held that an English Court cannot give judgment in foreign currency. Where, therefore, a contract of sale was broken and the price was payable, or damages had to be calculated, in a foreign currency, the sum payable in English money had to be assessed according to the rate of exchange at the time of breach (*i*). Hence the necessity of a rule for conversion or " translation " (*k*).

But by virtue of art. 106 of the Treaty of Rome, this is no longer the case where the foreign currency is that of a Common Market country (*l*) and it may be that it is no longer the case in relation to other currencies also (*m*).

Where a foreign contract, *viz.* a hotel bill incurred in France,

---

(*g*) *Re Kayford, Ltd.*, [1975] 1 All E. R. 604; [1975] 1 W. L. R. 279; and see *Barclays Bank, Ltd.* v. *Quistclose Investments, Ltd.*, [1968] 3 All E. R. 651; [1970] A. C. 567, H. L.

(*h*) *Aruna Mills, Ltd.* v. *Dhanrajmal Gobindram*, [1968] 1 All E. R. 113; [1968 1 Q. B. 655 (where, on special facts, damages were awarded in respect of a loss flowing from a devaluation of the Indian rupee).

(*i*) *Barry* v *Van den Hurk*, [1920] 2 K. B. 709 (non-acceptance, price payable in New York in dollars) ; *Re Hodgson & Co.*, [1920] W. N. 198 (exchange on India) ; *Lebeaupin* v. *Crispin & Co.*, [1920] 2 K. B. 714 (breach of contract to deliver in Canada) ; *Di Ferdinando* v. *Simon Smits & Co.*, [1920] 3 K. B. 409, at p. 414, C. A. (action against carriers; rate of exchange ruling when goods should have been delivered) ; *Peyrae* v. *Wilkinson*, [1924] 2 K. B. 166 (the rule is the same for a debt payable in a foreign currency in England) ; *Madeleine Vionnet* v. *Wills*, [1940] 1 K. B. 72.

(*k*) *Di Ferdinando* v. *Simon Smits & Co.*, [1920] 3 K. B., at p. 415, *per* SCRUTTON, L.J. ; approved, *S.S. Celia*, [1921] 2 A. C. 544 ; and see *Re British-American Continental Bank*, [1923] 1 Ch. 276, C. A. ; *Ottoman Bank of Nicosia* v. *Chakarian*, [1938] A. C. 260, P. C. ; *Syndic in Bankruptcy of S. N. Khoury* v. *Khayat*, [1943] 2 All E. R. 406 ; [1943] A. C. 507, P. C.

(*l*) *Schorsch Meier G.m.b.H.* v. *Hennin*, [1975], 1 All E. R. 152; [1975] Q. B. 416, C. A.

(*m*) In *Schorsch Meier G.m.b.H.* v. *Hennin*, *supra*, the Court of Appeal held by a 2–1 majority that the rule was a procedural one which no longer applied quite apart from the effect of the Treaty of Rome. This decision was followed in *Miliangos* v. *George Frank* (*Textiles*), *Ltd.*, [1975] 1 All E. R. 1076; [1975] 2 W. L. R. 555, C. A., but an appeal in that case has now been heard by the House of Lords and judgment has been reserved.

was sought to be enforced in England, and the defendant before the hearing paid the nominal amount of the debt in francs to the plaintiffs, it was held that this was a good satisfaction notwithstanding that the rate of exchange had moved heavily against the plaintiffs after the debt was incurred (*n*).

PART VI

SUPPLEMENTARY

55. Exclusion of implied terms and conditions.

(1) Where any right, duty, or liability would arise under a contract of sale by implication of law, it may be negatived or varied by express agreement or by the course of dealing between the parties (*o*), or by usage, if the usage be such as to bind both parties to the contract (*p*); but the foregoing provision shall have effect subject to the following provisions of this section.

(2) An express condition or warranty does not negative a condition or warranty implied by this Act unless inconsistent therewith.

(3) In the case of a contract of sale of goods, any term of that or any other contract exempting from all or any of the provisions of section 12 of this Act shall be void.

(4) In the case of a contract of sale of goods, any term of that or any other contract exempting from all or any of the provisions of section 13, 14 or 15 of this Act shall be void in the case of a consumer sale and shall, in any other case, not be enforceable to the extent that it is shown that it would not be fair or reasonable to allow reliance on the term.

(5) In determining for the purposes of subsection (4) above whether or not reliance on any such term would be fair or reasonable regard shall be had to

---

(*n*) *Société des Hotels Le Touquet* v. *Cummings*, [1922] 1 K. B. 451, C. A.; see also *Cummings* v. *London Bullion Co.*, [1952] 1 All E. R. 383; [1952] 1 K. B. 327, C. A.

(*o*) See p. 15, *ante*.

(*p*) See p. 17, *ante*.

all the circumstances of the case and in particular to the following matters—

(*a*) the strength of the bargaining positions of the seller and buyer relative to each other, taking into account, among other things, the availability of suitable alternative products and sources of supply;

(*b*) whether the buyer received an inducement to agree to the term or in accepting it had an opportunity of buying the goods or suitable alternatives without it from any source of supply;

(*c*) whether the buyer knew or ought reasonably to have known of the existence and extent of the term (having regard, among other things, to any custom of the trade and any previous course of dealing between the parties);

(*d*) where the term exempts from all or any of the provisions of section 13, 14 or 15 of this Act if some condition is not complied with, whether it was reasonable at the time of the contract to expect that compliance with that condition would be practicable;

(*e*) whether the goods were manufactured, processed, or adapted to the special order of the buyer.

(6) Subsection (5) above shall not prevent the court from holding, in accordance with any rule of law, that a term which purports to exclude or restrict any of the provisions of section 13, 14 or 15 of this Act is not a term of the contract.

(7) In this section " consumer sale " means a sale of goods (other than a sale by auction or by competitive tender) by a seller in the course of a business where the goods—

(*a*) are of a type ordinarily bought for private use or consumption; and

(*b*) are sold to a person who does not buy or hold

himself out as buying them in the course of a business.

(8) The onus of proving that a sale falls to be treated for the purposes of this section as not being a consumer sale lie on the party so contending.

(9) Any reference in this section to a term exempting from all or any of the provisions of any section of this Act is a reference to a term which purports to exclude or restrict, or has the effect of excluding or restricting, the operation of all or any of the provisions of that section, or the exercise of a right conferred by any provision of that section, or any liability of the seller for breach of a condition or warranty implied by any provsion of that section.

(10) It is hereby declared that any reference in this section to a term of a contract includes a reference to a term which although not contained in a contract is incorporated in the contract by another term of the contract.

(11) This section is subject to section 61 (6) of this Act (*q*).

ILLUSTRATIONS

(1) "Seller gives no warranty express or implied as to description or any other matters": this clause would seem to be void whether or not the sale is a consumer sale as "any other matters" must include the warranties implied by s. 12 of the Act—see s. 55 (3) (*r*).

(2) "This guarantee expressly excludes any other guarantee or warranty statutory or otherwise": again, this term appears to be void as it excludes the warranties implied by s. 12 of the Act.

(3) "No allowance will be made for errors in description quantity weight or measurement": this term purports to restrict the liability of the seller for breach of s. 13 of the Act, and is therefore invalid in the case of a consumer sale and is unenforceable in the case of a non-consumer sale to the extent that it is shown that it would not be fair to allow reliance on it.

(4) "The purchaser is deemed to have examined the vehicle prior to this agreement and satisfied himself as to its condition": this would

(*q*) Sub-sections (2)–(11) were introduced by s. 4 of the Supply of Goods (Implied Terms) Act, 1973.

(*r*) Contrast s. 173 (1) of the Consumer Credit Act, 1974, under which terms in regulated agreements are invalid if, *and to the extent that,* they are inconsistent with certain provisions in that Act or in regulations made under it.

have the effect of excluding or limiting the operation of s. 14 (2) and is therefore invalid in the case of a consumer sale and may be unenforceable in the case of a non-consumer sale.

(5) " The purchaser acknowledges that he has not made known any particular purpose for which he requires the goods, and has fully examined them before entering into this agreement ": An acknowledgement in this form is generally treated as not being part of the contract (*s*). If it genuinely represents the true position, it does not seek to limit or exclude any liability to which the seller would be subject (or restrict any right of the buyer) for the position is no different, except as to evidence, than would have been the case if nothing had been written. If it is not true, there seems no reason why it should bind the purchaser quite apart from this section (*t*) but insofar as it might, it would be caught by this section.

(6) " Any claim must be made within 14 days from the final discharge of the goods ": this is invalid even in the case of a non-consumer sale, for it restricts the exercise of rights conferred by ss. 12–15 of the Act by limiting the time within which any such right must be exercised.

(7) " The buyer undertakes carefully to examine the goods immediately upon receipt and to notify the seller forthwith of any defect which such examination ought to reveal ": if the buyer does not examine the goods, his loss as the result of a discoverable defect may be increased. In this case, he will be able to recover the full amount of his loss but the seller will be able, if the clause is valid, to recoup the amount by which the loss has been increased as a result of the failure to examine promptly or properly. This appears to have the effect of restricting the liability of the seller, and if so is void in the case of a consumer sale. But if it is clearly brought to the attention of the buyer in a sale which is not a consumer sale, it would seem to be fair and reasonable, and in most cases the courts would probably allow. reliance on it, at least so long as such examination was practicable (*u*)

Definition.—For " Contract of sale," see s. 1 (1), *ante*, and s. 62 (1), *post*. For "warranty " and " business " see s. 62 (1), *post*.

Cross-references.—The words " unless otherwise agreed," or words to a similar effect, appear in ss. 10, 11 (1) (c), 18, 20, 28, 29, 30, 31 (1), 32 (2) and (3), 33, 34 (2) and 36, *ante*. For exemption clauses, see *ante*, p. 10. For examples of express terms judicially considered, see Appendix II, Note B. For questions as to conflict of laws see s. 55A, *post*. For contracts for the international sale of goods, see s 61 (6), *post*.

(*s*) *Lowe* v. *Lombank, Ltd.*, [1960] 1 All E. R. 611; [1960] 1 W. L. R. 196, C. A.

(*t*) There will generally be no estoppel, as in most cases where the representation is untrue, the seller must know that this is the case: *cf. Lowe* v. *Lombank, Ltd., supra.*

(*u*) Even in the case of a consumer sale, if it is clearly brought to the attention of the buyer, it may be arguable that even though the provision is void, extra loss suffered as a result of not examining the goods is too remote as it was in the contemplation of the parties that the buyer would examine the goods.

COMMENT

Sub-section (1) is declaratory. The various ways in which rights duties and liabilities which would otherwise be implied by law may be negatived or varied are considered, *ante*, pp. 9–17.

Sub-sections (2)–(11) were introduced by the Supply of Goods (Implied Terms) Act, 1973 implementing recommendations of the Law Commission in its Report on Amendments to the Sale of Goods Act, 1893 (Law Com. No. 24). Previously, the section was simply a declaratory section in the terms of what is now s. 55 (1). The new sub-s. (2) was taken from the former s. 14 (4) of the Act.

The combined effect of this section and s. 61 (6), *post*, means that it is now necessary to distinguish between three different types of contract for the sale of goods. These are contracts for the international sale of goods, as defined in s. 62 (1), *post;* contracts for the sale of goods which are not contracts for the international sale of goods but which are consumer sales as defined in s. 55 (7); and contracts for the sale of goods which are neither contracts for the international sale of goods nor consumer sales.

International sales.—By virtue of sub-s. (11), this section is subject to s. 61 (6). By virtue of s. 61 (6) the provisions of this section which restrict sellers from contracting out of the provisions of s. 12–15, *ante*, do not apply to contracts for the international sale of goods. The effect of an exemption clause in such a contract is therefore to be determined solely by general rules of construction.

Consumer sales.—In the case of other contracts which are consumer sales, by virtue of sub-ss. (3) and (4) of this section any term exempting from any of the provisions of ss. 12–15 is void, regardless of whether it is fair or not.

By the combined effect of sub-ss. (7) and (8) of this section, a party contending that a sale is not a consumer sale must show

(1) That the goods were not sold by a seller in the course of a business; or

(2) That the goods were sold by auction or by competitive tender; or

(3) That the goods are not of a type ordinarily bought for private use or consumption; or

(4) That the buyer bought the goods in the course of a business; or

(5) That the buyer held himself out as buying the goods in the course of a business (*a*).

Non-consumer sales.—Even if a sale is proved not to be a consumer sale, it is not possible to contract out of any of the

(*a*) Section s. 55 (7), *supra*. For the meaning of " business " see the comments to s. 62 (1), *post*.

provisions of s. 12. It is possible, however, to contract out of the provisions of ss. 13–15. But it is then open to the buyer to show that it would not be fair or reasonable to allow the seller to rely on the exemption term, and if and insofar as he can show this, that term will not be enforceable (*b*).

Sub-section (5) sets out some of the matters to be taken into account in determining whether it would be fair and reasonable to allow an exempting term to be relied on in any case. It may also be relevant to consider whether the buyer also deals in goods of the type sold, or whether he is buying them for a totally different business, and has no specialist knowledge of that type of goods. So, too, the seller's expertise or lack of expertise may be relevant at least to the extent that the seller makes it known to the buyer.

The question whether it is fair to allow reliance must be determined at the date of the hearing having regard to all the circumstances. The conduct of the seller and of the buyer after the contract was made may therefore also be relevant. The court may consider the steps taken by the seller to remedy any defect, and any steps taken by the buyer which may have reduced or aggravated his loss, or made it more difficult or expensive for the seller to remedy any defect. It would also seem to be possible to take into account an offer by the seller to make a voluntary payment.

The term is not enforceable only to the extent that it would not be fair or reasonable. It is therefore open to a court to allow partial reliance on a term in a proper case, or to allow reliance on it subject to undertakings from the seller.

Under the provisions of this section, the onus is on the buyer to show that it would not be fair or reasonable to allow reliance on the term. Under s. 3 of the Misrepresentation Act, 1967 (*c*), the onus is on the seller to show that it would be fair and reasonable to allow reliance on the exempting term.

Exempting terms.—For the purposes of this section, any term is an exempting term which purports to exclude or restrict, or has the effect of excluding or restricting, the operation of all or any of ss. 12–15 of the Act, or the exercise of a right conferred by any of those sections, or any liability of the seller for breach of any condition or warranty implied by any of those sections (*d*).

This would seem to include clauses limiting the time within which any claim must be made, clauses limiting damages, probably including genuine agreed damages clauses although these are not

(*b*) Section 55 (4), *supra*.
(*c*) Appendix I. *post*.
(*d*) Section 55 (9), *supra*.

normally treated as exemption clauses (*e*), and clauses limiting liability in any other way. It has been suggested that arbitration clauses are also caught by this provision (*f*) but it is difficult to see why this should be so.

The scope of this provision is best considered in relation to typical attempts in contracts for the sale of goods to exclude or restrict the liability of the seller. Some of these are considered in the Illustrations to this section, *ante*.

55A. **Conflict of laws.**—Where the proper law of a contract for the sale of goods would, apart from a term that it should be the law of some other country or a term to the like effect, be the law of any part of the United Kingdom, or where any such contract contains a term which purports to substitute, or has the effect of substituting, provisions of the law of some other country for all or any of the provisions of sections 12 to 15 and 55 of this Act, those sections shall, notwithstanding that term but subject to section 61 (6) of this Act, apply to the contract (*g*).

Cross-references.—For the proper law of contract, see p. 47, *ante*.

COMMENT

This section applies where a contract for the sale of goods contains (i) an express choice of foreign law; or (ii) a term " to like effect ", an expression which would seem to include a choice of forum clause, which commonly carries with it the inference that the proper law is the law of the forum (*h*); or (iii) a term which purports to substitute or has the effect of substituting provisions of foreign law for those of ss. 12 to 15 and 55, *ante*.

The section will not apply, however, in the case of a contract for the international sale of goods (*i*).

Where the contract is not one for the international sale of goods, in considering whether this section applies it is necessary to determine what would be the proper law of the contract ignoring any terms of type (i) or (ii). If, apart from such terms, the contract would be subject to

(*e*) *Suisse Atlantique Société D'Armement Maritime S.A.* v. *N.V. Rotterdamsche Kolen Centrale*, [1966] 2 All E. R. 61; [1967] 1 A. C. 361, H. L. Such a clause restricts the exercise of a right conferred by ss. 12–15. The fact that it does so in the interests of both parties would seem to be irrelevant. But *cf. Benjamin's Sale of Goods*, paras. 963, 972.

(*f*) *Benjamin's Sale of Goods*, para. 972.

(*g*) This section was introduced by s. 5 (1) of the Supply of Goods (Implied Terms) Act, 1973. See, generally, 90 L. Q. R. 42.

(*h*) P. 48, *ante*.

(*i*) Sections 61 (6), 62 (1), *post*.

English law, then ss. 12 to 15 and 55 apply. In relation to terms of type (iii), the section appears only to apply to contracts already subject to the law of some part of the United Kingdom, for if the proper law was that of some other country, ss. 12 to 15 and 55 would not apply in any event. It would thus seem to be unnecessary in view of s. 55 (3) and (4). The intention may have been that all terms not only of types (i) and (ii), but also of type (iii) should be ignored in deciding whether the proper law would be that of part of the United Kingdom but for such terms, but the section does not say this.

56. Reasonable time a question of fact.—Where, by this Act, any reference is made to a reasonable time the question what is a reasonable time is a question of fact.

Cross-references.—This definition is required by ss. 18 (4), 29 (2), 35, 37 and 48 (3), *ante*. *Cf.* s. 29 (4) (reasonable hours).

COMMENT

It is often difficult to say whether reasonable time is a question of law or a question of fact, or a mixed question of law and fact (*j*). The Act resolves the doubt as regards sale, by making it in all cases a question of fact.

57. Rights, &c., enforceable by action.—Where any right, duty, or liability is declared by this Act, it may, unless otherwise by this Act provided, be enforced by action.

Definition.—For " action," see s. 62 (1), *post*.

COMMENT

This section is possibly required in order to negative the rule of the common law that, when a statute is mandatory and provides no express penalty for disobedience to its provisions, any contravention of its provisions is punishable as a misdemeanour (*k*). The words " unless otherwise provided " refer to seller's lien and stoppage *in transitu* ; see ss. 38–48, *ante*.

58. Auction sales.—In the case of a sale by auction—

(1) Where goods are put up for sale by auction in

(*j*) *Phipson on Evidence* (11th ed.) § 33.
(*k*) Stephen's *Digest of Criminal Law* (9th ed.), art. 152.

lots, each lot is primâ facie deemed to be the subject of a separate contract of sale (*l*);

(2) A sale by auction is complete when the auctioneer announces its completion by the fall of the hammer (*m*), or in other customary manner. Until such announcement is made any bidder may retract his bid (*n*);

(3) Where a sale by auction is not notified to be subject to a right to bid on behalf of the seller, it shall not be lawful for the seller to bid himself or to employ any person to bid at such sale, or for the auctioneer knowingly (*o*) to take any bid from the seller or any such person: Any sale contravening this rule may be treated as fraudulent by the buyer (*p*);

(4) A sale by auction may be notified to be subject to a reserved [or upset] price, and a right to bid may also be reserved expressly by or on behalf of the seller (*q*).

Where a right to bid is expressly reserved, but not otherwise, the seller, or any one person on his behalf, may bid at the auction (*r*).

ILLUSTRATION

Sale by auction subject to a reserve price. The auctioneer by mistake knocks down a lot for less than the reserve. On finding

(*l*) *Emmerson* v. *Heelis* (1809), 2 Taunt. 38; *Roots* v. *Lord Dormer* (1832), 4 B. & Ad. 77; *cf. Couston* v. *Chapman* (1872), L. R. 2 Sc. & Div. 250, H. L. (a Scottish case).

(*m*) *Shankland* v. *Robinson & Co.*, 1920 S. C. (H. L.) 103.

(*n*) *Payne* v. *Cave* (1789), 3 Torm Rep. 148; *Warlow* v. *Harrison* (1859), 1 E. & E. 309. And so may the seller retract his offer: *Fenwick* v. *Macdonald*, 1904 6 F. (Ct. of Sess.) 850.

(*o*) *Mainprice* v. *Westley* (1865), 6 B. & S. 420; *cf.* Sale of Land by Auction Act, 1867, s. 5; 2 Halsbury's Statutes (3rd Edn.) 500.

(*p*) *Bexwell* v. *Christie* (1776), Cowp. 395, *per* Lord MANSFIELD; *Thornett* v. *Haines* (1846), 15 M. & W. 367; *Green* v. *Baverstock* (1863), 14 C. B. N.S., 204; *cf. Mortimer* v. *Bell* (1865), 1 Ch. App. 10, at p. 13. As to fictitious bids by person interested in the sale, but not the seller, see *Union Bank* v. *Munster* (1887), 37 Ch. D. 51; and the maxim of Roman Law, *alterius circumventio alii non præbet actionem.*

(*q*) See cases cited in previous note, and *Howard* v. *Castle* (1796), 6 Term Rep. 642, at p. 645, *per* GROSE, J.

(*r*) *Thornett* v. *Haines* (1846), 15 M. & W. 367, at p. 372; *Mortimer* v. *Bell* (1865), 1 Ch. App. 10 (where auctioneer and puffer both bid and sale was held void).

this out he refuses to complete the sale or sign the necessary memorandum. The buyer has no remedy against the auctioneer, and it is immaterial that he did not know what the actual reserve was (*s*).

Definitions.—For "contract of sale," see s. 1 (1), *ante*, and s. 62 (1) *post*. For "buyer," "seller," "goods" and "sale," see s. 62 (1), *post*.

Cross-references.—*Cf.* ss. 1 (1), 8 and 9, *ante* (price).

COMMENT

Sub-section (2).—If the contract is resolved into offer and acceptance, the bid constitutes the offer. As the offer may be retracted before acceptance, so, conversely, it has been held that if a sale be advertised, but the lots are afterwards withdrawn, an intending bidder has no right of action (*t*). An auctioneer who sells goods which he has no right to sell may or may not be guilty of conversion, according to the circumstances (*u*). He has no implied authority to give a warranty on behalf of the seller (*a*).

At common law there is an implied contract between an auctioneer and a purchaser that the purchaser should pay the price to the auctioneer. The auctioneer has a lien on the goods until he is paid, and is entitled to sue in his own name for the price (*b*).

Sub-section (3).—Formerly it seems to have been the rule in equity that, when a sale by auction was not expressly stated to be without reserve, the seller might employ one person to bid, so as to prevent the property going at an undervalue. The Sale of Land by Auction Act, 1867 (*c*), commonly called the Puffer's Act, was passed to abolish this rule. It first declares that any sale which would be invalid at common law, by reason of the employment of a puffer, shall be invalid in equity, and then proceeds to regulate sales at which a price is reserved or a right to bid is reserved, and in this it appears to go slightly further than the common law rule (*d*). That Act does not apply to the sale of goods by auction, but the present section is in substantial accordance with it (*e*).

(*s*) *McManus* v. *Fortescue*, [1907] 2 K. B. 1, C. A., criticising *Rainbow* v. *Howkins*, [1904] 2 K. B. 322. But it seems that an auctioneer warrants that he has his principal's authority to sell: *Anderson* v. *Croall*, 1904 6 F. (Ct. of Sess.) 153.

(*t*) *Harris* v. *Nickerson* (1873), L. R. 8 Q. B. 286.

(*u*) See *National Mercantile Bank* v. *Rymill* (1881), 44 L. T. 767; *Consolidated Co.* v. *Curtis & Son*, [1892] 1 Q. B. 495, 498; *Barker* v. *Furlong*, [1891] 2 Ch. 172; see, too, 36 Sol. Jo. 480. It seems he is liable unless he acts as "a mere conduit pipe".

(*a*) *Payne* v. *Leconfield* (1882), 51 L. J. Q. B. 642.

(*b*) *Chelmsford Auctions, Ltd.* v. *Poole*, [1973] 1 All E. R. 810; [1973] Q. B. 542, C. A.

(*c*) 2 Halsbury's Statutes (3rd Edn.) 499.

(*d*) *Parfitt* v. *Jepson* (1877), 46 L. J. Q. B. 529, at p. 533.

(*e*) See its provisions carefully discussed in Dart's *Vendors and Purchasers* (8th ed.), p. 188, and Fry's *Specific Performance* (6th ed.), pp. 341 *et seq.*

Sub-section (4).—There is no express decision on what amounts to notification under this subsection. But in *Hills and Grant Ltd.* v. *Hodson* (*f*), a case on s. 5 of the Sale of Land by Auction Act, 1867, which is *in pari materia*, it was held that that section does not require that the words "with reserve" or "without reserve" and no others shall be used, and that so long as it is made plain, by whatever words may be chosen, that the sale is subject to a reserve, that is sufficient. Presumably this applies also to the present subsection.

After much doubt it was settled that an agreement for a "knock-out" (*i.e.* a combination between intending bidders to refrain from bidding against each other) was not illegal. The seller could protect himself by fixing a reserve price (*g*). Conversely, an agreement between bidders to run up prices at an auction for a collateral purpose may not be illegal (*h*). But the common law has since been modified, and it is now provided by s. 3 of the Auctions (Bidding Agreements) Act, 1969 (*i*) that where a person who buys goods at an auction has been a party to an agreement between potential bidders not to bid for the goods (other than an agreement to buy the goods *bona fide* on joint account) and one of the parties to that agreement was a dealer as defined in s. 1 (2) of the Auctions (Bidding Agreements) Act, 1927 (*k*), then the seller may avoid the contract under which the goods were sold.

If the seller does avoid the contract and the goods are not restored to him, then s. 3 (2) of the 1969 Act provides that the parties to the agreement to abstain from bidding are jointly and severally liable to make good to the seller any loss he may have sustained because of the agreement.

As to an auctioneer's duty to put up his name, etc., during sale, see the Auctioneers Act, 1845, s. 7, the Auctions (Bidding Agreements) Act, 1927, s. 3, and the Auctions (Bidding Agreements) Act, 1969, s. 4 (*l*).

A by-law prohibiting sales by auction in an open market is *ultra vires* (*m*).

Scotland.—As regards Scotland, this section seems to be declaratory (*n*).

(*f*) [1934] Ch. 53.

(*g*) *Rawlings* v. *General Trading Co.*, [1921] 1 K. B. 635, C. A.; followed in *Cohen* v. *Roche*, [1927] 1 K. B. 169. *Cf. Fuller* v. *Abrahams* (1821), 6 Moore C. P. 316.

(*h*) *Doolubdass* v. *Ramloll* (1850), 5 Moo. Ind. App. 109, at p. 133, P. C.; but *cf. Scott* v. *Brown & Co.*, [1892], 2 Q. B. 724, C. A. (making market in shares).

(*i*) *Post*, Appendix I.

(*k*) "A person who in the normal course of his business attends sales by auction for the purpose of purchasing goods with a view to reselling them."

(*l*) 2 Halsbury's Statutes (3rd Edn.) 498.

(*m*) *Nicholls* v. *Tavistock Urban Council*, [1923] 2 Ch. 18.

(*n*) See Bell's *Principles of the Law of Scotland*, §§ 130–132; but *cf.* Green's *Encyclopaedia of Scots Law*, tit. "Auction," as to withdrawing bids.

59. Payment into court in Scotland when breach of warranty alleged.—In Scotland where a buyer has elected to accept goods which he might have rejected, and to treat a breach of contract as only giving rise to a claim for damages, he may, in an action by the seller for the price, be required, in the discretion of the Court before which the action depends, to consign or pay into Court the price of the goods, or part thereof, or to give other reasonable security for the due payment thereof.

Definitions.—For " action," " buyer," " goods " and " seller," see s. 62 (1), *post*.

Cross-references.—*Cf.* ss. 1 (1), 8 and 9, *ante* (price) ; s. 35, *ante* (acceptance and rejection) ; ss. 11 (2) and 53 (5) *ante*, (right of rejection).

COMMENT

The section is intended to guard against the abuse of the alternative remedy given to the buyer by s. 11 (2) but it has been said to have no application to either of the buyer's remedies under s. 53 (1) (*p*). Under the common law of Scotland, however, where an action is brought for the price of goods the Court has a discretionary power to order the consignation of the whole or part of the sum claimed where it appears from the pleadings that some sum at least will be due to the seller at the end of the day (*p*). In exercising its discretion both at common law (*p*) and under s. 59 (*q*) the Court will take into account the buyer's financial stability. At common law, and *semble* under s. 59, the Court will, *inter alia*, also take into account the fact that the buyer is continuing to use the goods and that he admits that they are of some use to him (*p*).

In England, unfortunately, the power of the Court is far less, and it is a commonly practised fraud to keep the goods and then set up against the price an alleged breach of warranty. In some such cases, of course, leave to defend under R. S. C., O. 14, r. 6 may in practice only be granted upon the condition that the defendant pay the amount claimed into Court, but there are substantial limitations on

---

(*o*) *George Cohen Sons & Co., Ltd.* v. *Jamieson and Paterson*, 1963 S. C. 289, 291; where s. 53(1)(b) was said to contemplate only a situation where the price had been paid: *sed quaere*; Brown, *Sale of Goods* (2nd ed.), p. 405. As to the *actio quanti minoris* and also the *actio redhibitoria* in Roman Law, see De Zulueta, *The Roman Law of Sale*, p. 50; and Arangio-Ruiz, *La compravendita in diritto romano*, pp. 361–399.

(*p*) *George Cohen Sons & Co., Ltd.* v. *Jamieson and Paterson*, *supra*; *Findlay Bannatyne & Co.* v. *Donaldson* (1842), 5 Bell, S. C. App. 105, H. L.

(*q*) *Porter Spiers* v. *Cameron*, [1950] C. L. Y. (Sc.), § 5324.

the Court's discretion in this respect (*r*). The right of rejection is larger in Scotland than in England.

[60. Repeals.—*Repealed.*]

COMMENT

This section and the Schedule, which effected certain repeals, were themselves repealed by the Statute Law Revision Act, 1908.

61. Savings.—(1) The rules in bankruptcy relating to contracts of sale shall continue to apply thereto, notwithstanding anything in this Act contained.

(2) The rules of the common law, including the law merchant (*s*), save in so far as they are inconsistent with the express provisions of this Act, and in particular the rules relating to the law of principal and agent (*t*) and the effect of fraud, misrepresentation, duress [or coercion], mistake, or other invalidating cause, shall continue to apply to contracts for the sale of goods (*a*).

(3) Nothing in this Act or in any repeal effected thereby shall affect the enactments relating to bills of sale, or any enactment relating to the sale of goods which is not expressly repealed by this Act.

(4) The provisions of this Act relating to contracts of sale do not apply to any transaction in the form

---

(*r*) See *Jacobs* v. *Booth's Distillery* (1901), 85 L. T. 262, H. L.

(*s*) As to the law merchant, and its power of expansion, see *Blackburn on Sale* (2nd ed.), p. 317; *Edelstein* v. *Schuler*, [1902] 2 K. B. 144, *per* BIGHAM, J.

(*t*) See, *e.g.*, *Keighley* v. *Durant*, [1901] A. C. 240 (contract made by one person in his own name, on behalf of another as undisclosed principal, but without the authority of the latter, cannot be ratified); *Universal Steam Navigation Co.* v. *McKelvie*, [1923] A. C. 492 (effect of signature "as agent"); *Prager* v. *Blatspiel*, [1924] 1 K. B. 566 (sale by "agent of necessity"); but see *Jebara* v. *Ottoman Bank*, [1927] 2 K. B. 254, at pp. 270, 271.

(*a*) *Clarke* v. *Army and Navy*, [1903] 1 K. B. 155, C. A.; *Wren* v. *Holt*, [1903] 1 K. B. 610, at p. 616 (liability for dangerous goods); *Leaf* v. *International Galleries*, [1950] 1 All E. R. 693, C. A.; [1950] 2 K. B. 86 (innocent misrepresentation); *Petrofina (Gt. Britain), Ltd.* v. *Martin*, [1966] 1 All E. R. 126; [1966] Ch. 146, C. A.; *Esso Petroleum Co., Ltd.* v. *Harper's Garage (Stourport), Ltd.*, [1967] 1 All E. R. 699; [1968] A. C. 269, H. L. (unreasonable restraint of trade).

of a contract of sale which is intended to operate by way of mortgage, pledge, charge, or other security (*b*).

(5) Nothing in this Act shall prejudice or affect the landlord's right of hypothec or sequestration for rent in Scotland (*c*).

(6) Nothing in ss. 55 or 55A of this Act shall prevent the parties to a contract for the international sale of goods from negativing or varying any right, duty or liability which would otherwise arise by implication of law under ss. 12 to 15 of this Act (*d*).

Definition.—For " contract of sale," see ss. 1 (1), *ante*, and 62 (1), *post*. For " contract for the international sale of goods " see s. 62 (1), *post*.

COMMENT

Sub-section (1).—The Act now in force in England is the Bankruptcy Act, 1914 (*e*). See in particular s. 34 (2), reputed ownership; s. 44, fraudulent preferences; s. 45, protected *bonâ fide* transactions; ss. 55, 56, power of trustee to sell; and s. 54, power of trustee to disclaim onerous contracts. As to Scotland, see the Bankruptcy (Scotland) Act, 1913.

Sub-section (2).—The word " coercion " was added when the Bill was extended to Scotland, as " duress " is not an indigenous term in Scottish law.

For the nature of fraud, see p. 286. *post*,

For the effect of misrepresentation, whether fraudulent, negligent or innocent, see p. 5. *ante*,

As to mistake, see p. 3. *ante*,

As to illegality, see p. 17. *ante*,

The object of this saving is (1) to fill up any *lacunæ* in the Act itself (*f*), and (2) to emphasise the fact that the law of sale is merely a chapter in the general law of contract. Rules of law common to the whole field of contract apply to contracts of sale unless they contravene some express provision of the Act embodying some special rule peculiar to sale.

(*b*) See *Maas* v. *Pepper*, [1905] A. C. 102; *The Orteric*, [1920] A. C. 724, 733, P. C.; and cases cited, *ante*, p. 51, note (*b*). As to mortgages and pledges, see *ante*, p. 57. In view of this express provision, some of the decisions before the Act must be looked at critically, *e.g.*, *McBain* v. *Wallace* (1881), 6 App. Cas. 588 (motive immaterial). See this sub-section discussed in *Gavin's Trustee* v. *Fraser*, 1920 S. C. 674.

(*c*) As to hypothec, see Green's *Encyclopædia of Scots Law*, tit. " Hypothec. "

(*d*) Introduced by s. 6 of the Supply of Goods (Implied Terms) Act, 1973.

(*e*) 3 Halsbury's Statutes (3rd Edn.) 33.

(*f*) See, *e.g.*, *Booth Steamship Co.* v. *Cargo Fleet Iron Co.* [1969] 2 K. B. 570, C. A.

Sub-section (3).—The Bills of Sale Acts at present in force in England are the Acts of 1878, 1882, 1890 and 1891 (*g*). The Act of 1878 alone effects sales as defined and dealt with by this Act. As to these Acts, see Appendix I, *post*. The bill of sale by which a British ship is transferred has nothing to do with bills of sale under the Bills of Sale Acts.

For examples of other Acts relating to sales, see p. 51, *ante*.

62. Interpretation of terms.—(1) In this Act, unless the context or subject matter otherwise requires—

"Action" includes counterclaim and set off, and in Scotland condescendence and claim and compensation:

Cross-references.—This definition is required by ss. 9, 49 to 53, 57 and 59, *ante*.

COMMENT

The definition is inclusive, not exhaustive. For the purposes of the Judicature Acts, "action" is defined as meaning a civil proceeding commenced by writ or in such other manner as may be prescribed by Rules of Court, and does not include a criminal proceeding by the Crown (*h*).

Scotland and foreign systems.—As to compensation in Scotland, which is equivalent to set-off in England, see Green's *Encyclopædia of Scots Law*, tit. "Compensation."

"Bailee" in Scotland includes custodier:

Cross-references.—Bailees are referred to in ss. 18, rule 5 (2), 19 (1), 20 (proviso), 41, 43, 45 and 46, *ante*.

COMMENT

As to bailments, see *Story on Bailments*, and notes to *Coggs* v. *Bernard* (1703), 1 Smith L. C. (13th ed.), p. 175.

Scotland.—As to pledge in Scotland, see Green's *Encyclopædia of Scots Law*, tit. "Pledge."

"Business" includes a profession and the activities of any government department (including a department of the Government of Northern Ireland), local authority or statutory undertaker;

(*g*) 3 Halsbury's Statutes (3rd Edn.) 245 *et seq.*

(*h*) Supreme Court of Judicature (Consolidation) Act, 1925, s. 225; 7. Halsbury's Statutes (3rd Edn.) 623: *cf. Cox* v. *Hoare* (1906), 95 L. T. 121.

Cross-references.—This definition was introduced by s. 7 (1) of the Supply of Goods (Implied Terms) Act, 1973, and is required by ss. 14, 29 and 55, *ante* and in the definition of " contract for the international sale of goods ", *post.*

COMMENT

The meaning of the word " business " must be construed in the context in which it appears (*i*), and although it has been said that " business " means almost anything which is an occupation as distinct from a pleasure (*k*), it is unlikely that what a man does with his spare time will be held to be a business unless it has some direct commercial involvement in it (*l*). An activity may be a business, however, even if there is no intention on the part of the person conducting it to make a profit (*m*), and this is recognised by the statutory definition above.

One or two isolated transactions will not by themselves be a business (*n*). But if they are proved to have been undertaken with the intent that they should be the first of several transactions, that is with the intent of carrying on a business, then they will be the first transactions in an existing business (*o*).

A sale of a capital asset of a business may in some circumstances at least be a sale in the course of a business (*p*).

" Buyer " means a person who buys or agrees to buy goods :

Cross-references.—This definition is required by ss. 1, 7 to 9. 11, 12, 14 to 23, 27 to 37, 39, 41, 43 to 51, 53, 54, 58 and 59, *ante*, See definition of " goods " in this section, *post.*

"Contract for the international sale of goods" means a contract of sale of goods made by parties whose

(*i*) *Abernethie* v. *Kleiman, Ltd.*, [1969] 2 All E. R. 790, at p. 792; [1970] 1 Q. B. 10, at p. 18, *per* HARMAN, L.J.

(*k*) *Rolls* v. *Miller* (1884), 27 Ch. D. 71, at p. 88, *per* LINDLEY, L.J.

(*l*) *Abernethie* v. *Kleiman, Ltd., supra, per* WIDGERY, L.J. at pp. 794 and 20. See also *Southwell* v. *Lewis* (1880), 45 J. P. 206 (farming for pleasure; occasional sale, not part of regular practice: not business).

(*m*) *South-West Suburban Water Co.* v. *St. Marylebone Guardians*, [1904] 2 K. B. 174; *Paddington Burial Board* v. *Inland Revenue Commissioners* (1884), 13 Q. B. D. 9.

(*n*) *Cf.* s. 189 (2) of the Consumer Credit Act, 1974.

(*o*) The business exists from the time of the commencement of the first transaction with the effect that it should be the first of its series: *Re Griffiths; ex parte Board of Trade* (1890), 60 L. J. Q. B. 235, at p. 237, *per* Lord ESHER, M.R.

(*p*) *London Borough of Havering* v. *Stevenson*, [1970] 3 All E. R. 609; [1970] 1 W. L. R. 1375, D. C.

places of business (or, if they have none, habitual residences) are in the territories of different States (the Channel Islands and the Isle of Man being treated for this purpose as different States from the United Kingdom) and in the case of which one of the following conditions is satisfied, that is to say—

(*a*) the contract involves the sale of goods which are at the time of the conclusion of the contract in the course of carriage or will be carried from the territory of one State to the territory of another; or

(*b*) the acts constituting the offer and acceptance have been effected in the territories of different States; or

(*c*) delivery of the goods is to be made in the territory of a State other than that within whose territory the acts constituting the offer and the acceptance have been effected:

Cross-references.—This definition is required by s. 61 (6), *ante,* and was introduced by s. 7 (1) of the Supply of Goods (Implied Terms) Act, 1973.

COMMENT

In order to show that a contract is one for the international sale of goods, it must be shown first that the contract was between parties whose places of business (or if none habitual residences) were in different states and secondly that one of the conditions (a), (b) or (c) applies.

In the ordinary meaning of the words a person may have more than one " place of business " (*q*). The words may refer to the principal place of business but it is more likely that they will be interpreted as meaning the place of business through or from which the contract was made.

The application of the conditions (a), (b) and (c) also gives rise to potential difficulties. As to (a) in particular, it is not clear whether it is sufficient that goods happen to be in the course of carriage when contracted to be sold or that one party happens to intend to take them to another state, or whether the contract must actually contemplate by its terms the goods being in the course of carriage or being carried from one to another state. An f.o.b. contract would normally involve this. A contract for sale by an international haulier of his stock of lorries (half of which are in the United Kingdom and

(*q*) *Davies* v. *British Geon, Ltd.,* [1956] 3 All E. R. 389; [1957] 1 Q. B. 1, C. A.

half of which are returning home from abroad) would not contemplate the goods being in the course of carriage, since it is pure chance where the lorries are when the contract is concluded.

The place where " the acts constituting the offer and acceptance have been effected " probably need not be the same as the place where the offer is made and the place where it is accepted (*r*).

" Contract of sale " includes an agreement to sell as well as a sale :

Cross-references.—*Cf.* the definition of " sale " in this section, *post,* and note the substantive definition of " contract of sale " in s. 1 (1), *ante.* The definition is required by ss. 3, 5 to 19, 21, 27, 29, 31, 32, 48, 49, 55, 58 and 61, *ante.*

COMMENT

The term " contract of sale " is used to include both executory and executed contracts of sale.

" Defendant " includes in Scotland defender, respondent, and claimant in a multiplepoinding :

Cross-reference.—This definition is required by s. 52, *ante.*

COMMENT

Multiplepoinding in Scotland corresponds with interpleader in England.

" Delivery " means voluntary transfer of possession from one person to another (*s*) :

Cross-references.—This definition is required by ss. 1, 9, 11, 18 to 20, 25, 27 to 37, 42, 43, 49, and 51 to 53, *ante.*

COMMENT

Delivery does not carry the same consequences in all circumstances. A delivery which is effectual for one purpose may be

---

(*r*) *Cf.* Article 1 (4) of the Uniform Law on the International Sale of Goods, which provides for this purpose that in the case of contracts by correspondence, offer and acceptance shall be considered to have been effected in the territory of the same State only if the letters and telegrams or other documentary communications which contain them have been sent and received in the territory of that State. The provisions of the Uniform Law may also be relevant in considering condition (c): see s. 1 (5) of the Uniform Law on International Sales Act 1967 and arts. 1 (5) and 19 of the Uniform Law on the International Sale of Goods. See also the definition of delivery, *post.*

(*s*) For a case where the "context of subject matter" was held to require a different meaning, see *W. E. Marshall & Co.* v. *Lewis and Peat (Rubber), Ltd.,* [1963] 1 Lloyd's Rep. 562.

ineffectual for other purposes. For instance, delivery to a carrier generally passes the property to the buyer, but does not defeat the right of stoppage *in transitu*, while delivery by the carrier to the consignee does defeat that right.

Pollock defines delivery as " voluntary dispossession in favour of another," and proceeds to say that

> " in all cases the essence of delivery is that the deliveror by some apt and manifest act puts the deliveree in the same position of control over the thing, either directly or through a custodian, which he himself held immediately before that act " (*t*).

Delivery may be actual or constructive. Delivery is constructive when it is effected without any change in the actual possession of the thing delivered, as in the case of delivery by attornment or symbolic delivery. Delivery by attornment may take place in three classes of cases (*u*): first, the seller may be in possession of the goods, but after sale he may attorn to the buyer, and continue to hold the goods as his bailee (*a*); secondly, the goods may be in the possession of the buyer before sale, but after sale he may hold them on his own account (*b*); thirdly, the goods may be in the possession of a third person as bailee for the seller, and after sale such third person may attorn to the buyer and continue to hold them as his bailee (*c*).

Pollock and Wright have carefully discussed the so-called " symbolic delivery " by giving the buyer the key of the place where the goods are stored. They show that the key is not the symbol of the goods, but that the transaction

> " consists of such a transfer of control in fact as the nature of the case admits, and as will practically suffice for causing the new possessor to be recognised as such " (*d*).

But the transfer of a bill of lading appears to afford a genuine instance of symbolic delivery (*e*). While goods are at sea, they can be dealt with on land only through the instrumentality of the bill of lading which represents them. The transfer of the bill of lading has the like effect as a delivery of the goods themselves. But a clause in a c.i.f. contract that " no claims shall be valid unless

---

(*t*) Pollock and Wright, *Possession in the Common Law*, pp. 43, 46.

(*u*) *Dublin City Distillery Co.* v. *Doherty*, [1914] A. C. 823, at p. 852, *per* Lord PARKER; and see cases reviewed by Lord ATKINSON, at p. 843.

(*a*) *Ibid.*; *Elmore* v. *Stone* (1809), 1 Taunt. 458; *Marvin* v. *Wallis* (1856), 6 E. & B. 726.

(*b*) *Story on Sale*, § 312*a*. *Cf. Cain* v. *Moon*, [1896] 2 Q. B. 283, 289; *Blundell-Leigh* v. *Attenborough*, [1921] 3 K. B. 235, C. A. (pledge).

(*c*) Pollock and Wright, *Possession in the Common Law*, p. 72; *Farina* v. *Home* (1846), 16 M. & W. 119, 123; and see s. 29 (3), *ante*.

(*d*) Pollock and Wright, *Possession in the Common Law*, p. 61.

(*e*) *Sanders* v. *Maclean* (1883), 11 Q. B. D. 327, at p. 341; *Biddell Bros.* v. *E. Clemens Horst Co.*, [1911] 1 K. B. 934, C. A., at p. 957, *per* KENNEDY, L.J. *The Prinz Adalbert*, [1917] A. C. 586, at p. 589, *per* Lord SUMNER.

made within two weeks after the goods are delivered" was held to mean delivery from the ship to the buyers, and not delivery of the documents by which the property passed (*f*).

Where goods are taken possession of by the buyer under a licence to seize, the transaction is equivalent to a delivery by the seller (*g*) and should perhaps be regarded as a case of actual delivery. Where possession is ambiguous, it must be attributed to the person having the title to the goods (*h*).

A delivery by mistake may be inoperative (*i*). For certain purposes, *e.g.*, the termination of the seller's lien, part delivery may operate as a delivery of the whole: see s. 42, *ante*.

It is to be noted that the Act makes no attempt to define "possession." The term is too elusive. Its meaning is always relative (*k*). For instance, when goods are under the control of an agent or servant, they are for some purposes in the possession of the principal or master, while for other purposes they are in the possession of the agent or servant. But a definition of possession for the purposes of the Factors Acts is given by s. 1 (2) of the Factors Act, 1889, *post*. The subject is exhaustively treated in Pollock and Wright, *Possession in the Common Law* (*l*), and O. W. Holmes' *The Common Law*, lectures 5 and 6. See also Harris: *Oxford Essays in Jurisprudence*, p. 61.

"Document of title to goods" has the same meaning as it has in the Factors Acts:

Cross-references.—This definition is required by ss. 25, 29 (3) and 47, *ante*.

COMMENT

See next definition. These Acts are reproduced with notes, *post*.

"Factors Acts" means the Factors Act, 1889, the Factors (Scotland) Act, 1890, and any enactment amending or substituted for the same:

(*f*) *Scriven Bros.* v. *Schmoll Fils & Co., Inc.* (1924), 40 T. L. R. 677.

(*g*) *Congreve* v. *Evetts* (1854), 10 Exch. 298, at p. 308, *per* PARKE, B.

(*h*) *French* v. *Gething*, [1921] 3 K. B. 280, at p. 290 (gift of furniture to wife).

(*i*) *Godts* v. *Rose* (1855), 17 C. B. 229; Pollock and Wright, *Possession in the Common Law*, pp. 100–114. *Cf. Bishop* v. *Shillito* (1819), 2 B. & Ald. 329 n., *per* BAYLEY, J. As to delivery by mistake as between two buyers, one lot of goods being lost, see *Denny* v. *Skelton* (1916), 115 L. T. 305.

(*k*) As to possession by wife when living with husband, see *French* v. *Gething*, [1922] 1 K. B. 236, C. A.

(*l*) Maitland, in an interesting article on the *Seisin of Chattels*, at 1 L. Q. R. 324, establishes that in early times the term "seisin" was applied to chattels as freely as the term "possession," and as its equivalent.

Cross-references.—This definition is required by ss. 21 (2) and 25 (3), *ante.*

COMMENT

These Acts are reproduced with notes, *post.*

" Fault " means wrongful act or default (*m*):

Cross-references.—This definition is required by ss. 7, 9 (2) and 20, *ante.*

COMMENT

This definition was inserted at the instance of Lord WATSON.

" The rule of law applicable to contracts is that neither of the parties can by his own act or default defeat the obligations he has undertaken to fulfil " (*n*).

" Future goods " means goods to be manufactured or acquired by the seller after the making of the contract of sale :

Cross-references.—This definition repeats that in s. 5 (1), *ante.* It is required by s. 18, rule (5), *ante.* *Cf.* definition of " goods " in this section, *post.*

COMMENT

The expression " future goods " has been criticised as awkward, but STIRLING, J., refers to it as " convenient " (*o*).

" Goods " include all chattels personal other than things in action and money, and in Scotland all corporeal moveables except money. The term includes emblements, [industrial growing crops], and things attached to or forming part of the land which are agreed to be severed before sale or under the contract of sale :

Cross-references.—This definition is required by ss. 1, 5 to 7, 9, 11, 12, 14 to 16, 18 to 51, 53, 58 and 59, *ante.* *Cf.* " contract of sale " defined in s. 1 (1) and in this section, *ante,* and " sale " defined in this section, *post.*

(*m*) *Cf. Asiatic Petroleum Co.* v. *Lennards Carrying Co.,* [1914] 1 K. B. 419.
(*n*) *Sailing Ship Blairmore Co.* v. *Macredie,* [1898] A. C. 593, at p. 607.
(*o*) *Ajello* v. *Worsley,* [1898] 1 Ch. 274, at p. 280.

COMMENT

A definition was needed because the expression " goods " is a term of uncertain extent, and is used in different Acts in different senses (*p*). Compare the definitions of " goods " given by s. 167 of the Bankruptcy Act, 1914 (*q*), by s. 10 of the Trading Stamps Act, 1964 and by s. 137 (2) of the Fair Trading Act, 1973, and contrast the definition of " personal chattels " given by s. 4 of the Bills of Sale Act, 1878, *post*, Appendix I. See also the definition of " property " in s. 4 and of " goods " in s. 34 (2) of the Theft Act, 1968, *post*, Appendix I.

It is to be noted (1) that this definition is *inclusive*, not exhaustive, and (2) that it is guarded by the opening words of the section, " unless the subject matter or context otherwise requires."

Chattels personal.—These include all tangible moveable property, except money. They are " properly and strictly speaking things moveable which may be annexed to or attendant on the person of the owner, and carried about with him from one part of the world to another " ; and they are commonly classified as chattels animate, chattels vegetable, and chattels inanimate (*r*). The property in, or right to, chattels personal is either *in possession* or *in action*. Property is in possession " where a man has the enjoyment, either actual or constructive, of the thing or chattel " (*s*).

Gas and water may be the subjects of theft (*t*), and there seems to be no reason why they should not be capable of being sold as goods (*u*). It is doubtful whether electricity can be goods for the purposes of this Act (*u*).

Things in action.—Property is in action where a man has not the enjoyment of the thing in question, but merely a right to recover it by legal process, whence the thing so recoverable is called a thing (or chose) *in action*. " Choses in action," says CHANNELL, J.—

(*p*) *Cf. The Noordam*, [1920] A. C. 904, at p. 908, *per* Lord SUMNER.

(*q*) 3 Halsbury's Statutes (3rd Edn.) 159.

(*r*) Williams & Mortimer, *Executors Administrators and Probate*, pp. 472 *et seq*.

(*s*) *Dicey on Parties to an Action*, p. 66.

(*t*) *Cf. Russell on Crimes* (12th ed.) pp. 887 *et seq.*; *Ferens* v. *O'Brien* (1883), 11 Q. B. D. 21. Electricity is the subject of a separate offence: see the Theft Act 1968, s. 13.

(*u*) See *County of Durham Electrical Power Co.* v. *Inland Revenue*, [1909] 2 K. B. 604, at p. 608, C. A., as to electricity (stamp question); *Bentley Brothers* v. *Metcalfe*, [1906] 2 K. B. 548, at pp. 552–553. In Canada, steam and gas have been held to be tangible personal property capable of being sold: *Re Social Services Tax Act, Re Central Heat Distribution, Ltd.* (1970), 74 W. W. R. 246 (steam); *Marleau* v. *People's Gas Supply Co., Ltd.*, [1940] 4 D. L. R. 433 (Sup. Ct. of Can.) (acetylene gas); *Bradshaw* v. *Booth's Marine, Ltd.*, [1973] 2 O.R. 646 (propane gas). But supplies of gas, water and electricity by statutory boards under a duty to supply are probably not sales: *cf. Pfizer* v. *Ministry of Health*, [1965] 1 All E. R. 450; [1965] A. C. 512, H. L.

" is a known legal expression used to describe all personal rights of property which can only be claimed or enforced by action, and not by taking physical possession " (*a*).

Scrip and shares are things in action, and so of course are bills, notes, and cheques (*b*).

Money.—Money (*i.e.*, current money) is necessarily excluded, because in sale the goods and the price are contrasted, and wholly different considerations apply to them. But a Jubilee five-pound gold piece, bought as a curiosity, may be treated as goods and not as money (*c*).

Emblements.—These are " such vegetable products as are the annual result of agricultural labour " (*d*).

The term " Industrial growing crops " was added when the Bill was extended to Scotland. It is probably a wider term than emblements or the Scottish equivalent, " way-going crops." Its possible effect may be to include in " goods " such crops as grass or clover, which did not come within the term of " emblements " (*e*).

Things attached to or forming part of land.—These words appear intended to give a rule for determining whether trees, fixtures and other things attached to or forming part of the soil are to be treated as goods or not.

Where an owner of land agrees to sell something attached to or forming part of his land, such as slate or slag, then unless the agreement obliges the buyer to sever the slate, slag, or other thing to be sold, the agreement will not normally be a contract for the sale of goods (*f*).

(*a*) *Torkington* v. *Magee*, [1902] 2 K. B. 427, at p. 430.

(*b*) *Humble* v. *Mitchell* (1839), 11 A. & E. 205 ; *Colonial Bank* v. *Whinney* (1886), 11 App. Cas. 426 (shares), approving the definition given by FRY, L.J., in the Court of Appeal, (1885), 30 Ch. D. 261 ; *Lang* v. *Smyth* (1831), 7 Bing. 284 (foreign bonds) ; *Frooman* v. *Appleyard* (1862), 32 L. J. Ex. 175 (certificate of railway stock).

(*c*) *Moss* v. *Hancock*, [1899] 2 Q. B. 111. As to price, see *ante*, s. 8 ; *cf. Banque Belge* v. *Hambrouck*, [1921] 1 K. B. 321, at p. 326, C. A., discussing the various meanings of " currency " and contrasting money and goods.

(*d*) *Williams on Personal Property*, (18th ed.), p. 162 ; *cf.* Williams & Mortimer, *Executors Administrators and Probate*, pp. 476 *et seq.*

(*e*) See *Kingsbury* v. *Collins* (1827), 4 Bing. 202 (teazles) ; *Graves* v. *Weld* (1833), 5 B. & Ad. 105 (clover) ; *Paton's Trustees* v. *Finlayson*, 1923 S. C. 872 (potatoes sold growing).

(*f*) *Morgan* v. *Russell & Sons*, [1909] 1 K. B. 354 (agreement to sell such cinders and slag as the buyer may desire to remove from the land); *Mills* v. *Stokman* (1967), 116 C. L. R. 61 (H. C. of Aus.) (licence to enter on land and remove slate); contrast *Amco* v. *Wade*, [1968] Qd. R. 445 (buyer agreed to enter land and remove sand and gravel within fixed time; sale of goods). See further, *Benjamin's Sale of Goods*, paras. 81 *et seq.*

Ships.—These are clearly chattels personal, but they are governed by so many special rules that it is doubtful how far they come under the denomination of "goods" for the purposes of the Act. "A ship," says TURNER, L.J.,

> "is not like an ordinary chattel. It does not pass by delivery, nor does the possession of it prove the title to it. There is no market overt for ships. In the case of American ships the laws of the United States provide the means of evidencing the title to them" (*g*);

as the Merchant Shipping Act, 1894, does for British ships (*h*). But unless and except in so far as there is some provision to the contrary, the Sale of Goods Act appears to apply (*i*).

"Lien" in Scotland includes right of retention:

Cross-references.—The word occurs in ss. 25, 39, 41 to 43, and 47, *ante*.

COMMENT

Cf. Factors (Scotland) Act, 1890, *post*. The common law extent of the right of retention in Scotland is cut down by the provisions of the Act. See Green's *Encyclopædia of Scots Law*, tit. "Retention."

"Plaintiff" includes [pursuer, complainer, claimant in a multiplepoinding] and defendant [or defender] counter-claiming:

Cross-reference.—This definition is required by s. 52, *ante*.

"Property" means the general property in goods, and not merely a special property:

Cross-references.—This definition is required by ss. 1, 2, 11, 12, 16 to 20, 24, 26, 39 and 49, *ante*. *Cf.* the definition of "goods" in this section, *ante*.

COMMENT

The essence of sale is the transfer of the ownership or general property in goods from seller to buyer for a price. See "the" property (that is, the general property), distinguished from "a"

(*g*) *Hooper* v. *Gumm* (1867), 2 Ch. App, 282, at p. 290; *cf. Lindley on Partnership* (12 ed.), p. 361.

(*h*) *Cf. Stapleton* v. *Haymen* (1864), 2 H & C. 918 (sale of ship to infant); *Manchester Ship Canal* v. *Horlock*, [1914] 2 Ch. 199, C. A. (sale of abandoned ship sunk in fairway).

(*i*) *Dloyd del Pacifico* v. *Board of Trade* (1930), 35 Ll. L. Rep. 217 (ship assumed to be goods for purpose of s. 14).

property (that is, merely a special property) by Lord Bowen (*k*). The use of the term " special property " is inveterate in English law. But it would have conduced to clearness if the term " special interest " had been used instead, while " property " was confined to its primary meaning as ownership or *dominium* (*l*).

The general property in certain goods may be in one person, while a special property in them is in another person: as in the case of a pledge where the pledgee has only a special property or interest, the general property remaining in the pledgor (*m*). The general property in goods may be transferred to one person subject to a special property or interest in another (*n*).

Again, the right of property in goods must be distinguished from the right to their present possession. The right of property may be in one person, while the right to possession may be in another, as in the case of a lien (*o*). So, too, property may be divided between owners, but the right to possession may be in one alone (*p*); and where there is a sale of specific goods for cash, the property passes by the contract, but the seller may (unless otherwise agreed) retain the goods till the price is paid. Again, goods may be sold which are in the possession of a third person, such as a carrier or warehouseman, who has no property in the goods but has a right to retain them till his charges are paid.

The Prize Court, when dealing with enemy goods, takes cognisance only of the general property in the goods, and disregards any special property or interest (*q*).

" Quality of goods " includes their state (*r*) or condition (*s*):

(*k*) *Burdick* v. *Sewell* (1884), 13 Q. B. D. 159, at p. 175, and (1884), 10 App. Cas. 74, at p. 93. This distinction is not recognised in Scotland: *Hayman* v. *M'Lintock*, 1907 S. C. 936, at p. 949; and see the term " special property " discussed, *Attenborough* v. *Solomon*, [1913] A. C. 76, at pp. 83, 84 (pledge).

(*l*) *The Odessa*, [1916] A. C. 145, at p. 159, P. C.

(*m*) *Halliday* v. *Holgate* (1868), L. R. 3 Exch. 299, Ex. Ch.

(*n*) *Franklin* v. *Neate* (1844), 13 M. & W. 481; *Jenkyns* v. *Brown* (1849), 14 Q. B. 496. See a lien distinguished from a pledge: *Donald* v. *Suckling* (1866), L. R. 1 Q. B. 585, at p. 612; *cf. Howes* v. *Ball* (1827), 7 B. & C. 481 (hypothecation); as to pledge, see p. 83, *ante*.

(*o*) *Mulliner* v. *Florence* (1878), 3 Q. B. D. 484, C. A.; *Blackburn on Sale* (1st ed.), pp. 198, 316; *Milgate* v. *Kebble* (1841), 3 M. & Gr. 100; Pollock and Wright, *Possession in the Common Law*, p. 120.

(*p*) *Nyberg* v. *Handelaar*, [1892] 2 Q. B. 202, C. A.

(*q*) *The Odessa*, [1916] A. C. 145; *The Ningchow*, [1916] P. 221 (enemy goods pledged with neutral).

(*r*) See *Tesco Stores, Ltd.* v. *Roberts*, [1974] 3 All E. R. 74; [1974] 1 W. L. R. 1253, D. C., where the meaning of " in the same state as when he purchased it " in s. 115 (1) of the Food and Drugs Act, 1955 is considered.

(*s*) See *Beer* v. *Walker* (1877), 46 L. J. Q. B. 677; *Niblett* v. *Confectioners' Materials Co.*, [1921] 3 K. B. 387, at pp. 396, 402; *Mash & Murrell* v. *Emanuel*, [1961] 1 All E. R. 485; [1961] 1 W. L. R. 862, reversed on an issue of fact, [1962] 1 All E. R. 77; [1962] 1 W. L. R. 16, C. A.

Cross-references.—This definition is required by ss. 14, 15 and 53 (3), *ante*. *Cf.* the definition of " goods " in this section, *ante*.

COMMENT

Flour or tobacco may be of excellent kind, but if it is sea-damaged it may not be of merchantable quality.

Lord DUNEDIN points out the distinction between the " description " of goods in the contract and their quality:

> " The tender of anything that does not tally with the specified description is not compliance with the contract. But when the article tendered does comply with the specified description, and the objection on the buyer's part is an objection to quality alone, then I think s. 14 (1) settles the standard, and the only standard by which the matter is to be judged " (*t*).

But, of course, in many cases, it will be the standard set by s. 14 (2) that has to be considered.

" Sale " includes a bargain and sale as well as a sale and delivery:

Cross-references. The word occurs in ss. 2, 23 to 25, 47 and 58, *ante*.

COMMENT

This definition follows from the definition of " contract of sale " given by s. 1 (1) and this section, *ante*. See notes to ss. 1 and 49, *ante*. See also headings " sale " and " seller " in Stroud's *Judicial Dictionary* (4th ed.).

" Seller " means a person who sells or agrees to sell goods:

Cross-references.—This definition is required by ss. 1, 5 to 7, 9, 11, 12, 14, 18 to 23, 25, 27 to 38, 40, 45 to 47, 49 to 51, 53, 54, 58 and 59, *ante*. *Cf.* the definition of " goods " in this section, *ante*.

COMMENT

As to seller selling to himself, see p. 75, *ante*. In other Acts the expression " seller " must, of course, be construed according to its context, and the subject-matter dealt with (*u*).

(*t*) *Manchester Liners, Ltd.* v. *Rea*, [1922] 2 A. C. 74, at p. 80.

(*u*) See, *e.g.*, *Pharmaceutical Society* v. *White*, [1901] 1 K. B. 601, C. A. As to who is a " hawker," see *O'Dea* v. *Crowhurst* (1899), 68 L. J. Q. B. 655. As to " seller " or " sale " in the Licensing Acts, see *Titmus* v. *Littlewood*, [1916] 1 K. B. 732, and the cases about clubs, p. 80, *ante*.

" Specific goods " means goods identified and agreed upon at the time a contract of sale is made :

Cross-references.—This definition is required by ss. 6, 7, 11 (1) (c), 17, 18, rules (1) to (3), 19 (1), 29 (1) and 52 (1), *ante*. *Cf.* definition of " contract of sale " in s. 1 (1) and this section, *ante*, and the definition of " goods " in this section, *ante*.

COMMENT

Specific goods must be distinguished from unascertained goods. Where there is a contract for the sale of specific goods, the seller would not fulfil his contract by delivering any other goods than those agreed upon. Where there is a contract for unascertained goods the seller fulfils his contract by delivering at the appointed time any goods which answer to the description in the contract. It is clear that " future goods," even though particularly described, do not come within the definition of specific goods, but for most purposes would be subject to the same considerations as unascertained goods ; see notes to ss. 17 and 18, *ante*.

" Warranty " as regards England and Ireland means an agreement with reference to goods which are the subject of a contract of sale, but collateral to the main purpose of such contract, the breach of which gives rise to a claim for damages, but not to a right to reject the goods and treat the contract as repudiated (*a*).

As regards Scotland a breach of warranty shall be deemed to be a failure to perform a material part of the contract (*b*).

Cross-references.—This definition is required by ss. 11, 12, 14 and 53, *ante*. *Cf.* the definition of " contract of sale " in s. 1 (1) and this section, *ante*, " goods " defined in this section, *ante*, and s. 35, *ante* (acceptance and rejection).

History.—Anson, in the 13th edition of his work on contracts, had collected six different senses in which the word warranty was used in the cases (*c*), but it is submitted that the definition selected by the Act is the most convenient. See Note A, *post*, Appendix II, where the history of the word is discussed at length.

(*a*) For the distribution between a condition and a warranty, see p. 8, *ante*.

(*b*) *Cf.* sections 11 (2) and 53 (5), *ante*, saving Scottish rules.

(*c*) *Anson on Contracts* (13th ed.), p. 368.

COMMENT

Conditions and warranties contrasted.—The Act draws throughout a distinction between the terms " condition " and " warranty," and defines the circumstances under which a condition may be treated as a warranty.

See further p. 8, *ante*.

(1A) Goods of any kind are of merchantable quality within the meaning of this Act if they are as fit for the purpose or purposes for which goods of that kind are commonly bought as it is reasonable to expect having regard to any description applied to them, the price (if relevant) and all the other relevant circumstances; and any reference in this Act to unmerchantable goods shall be construed accordingly.

Cross-references.—This definition is required by ss. 14 (2) and 15 (3), *ante*.

COMMENT

This sub-section was introduced by s. 7 (2) of the Supply of Goods (Implied Terms) Act, 1973. Its effect is discussed in the comment to s. 14 (2), *ante*.

---

(2) A thing is deemed to be done " in good faith " within the meaning of this Act when it is in fact done honestly, whether it be done negligently or not (*d*).

Cross-references.—This definition is required by ss. 22 (1), 23 25, 26 and 47, *ante*.

COMMENT

What constitutes fraud.—The House of Lords in *Derry* v. *Peek* (*e*) has exploded the notion of " legal fraud," and has established the principle that there is no *tertium quid* between good faith on the one hand and bad faith or fraud on the other. " First," says Lord HERSCHELL,

> " in order to sustain an action of deceit, there must be proof of fraud, and nothing short of that will suffice. Secondly, fraud is proved when it is shewn that a false representation has been made (1) knowingly, or (2) without belief in its truth, or (3) recklessly, careless whether it be true or false. Although I have treated the

---

(*d*) Taken from the Bills of Exchange Act, 1882, s. 90; 2 Halsbury's Statutes (3rd Edn.) 235; *cf. Jones* v. *Gordon* (1877), 2 App. Cas. 616, 629.

(*e*) (1889), 14 App. Cas. 337, at p. 374; and see *Weir* v. *Bell* (1878), 3 Ex. D. 238, at p. 243, *per* BRAMWELL, L.J. See fraud and illegality contrasted: *Ex p. Watson* (1888), 21 Q. B. D. 301, at p. 309.

second and third as distinct cases, I think the third is but an instance of the second; for one who makes a statement under such circumstances can have no real belief in the truth of what he states. To prevent a false statement being fraudulent, there must, I think, always be an honest belief in its truth. And this probably covers the whole ground; for one who knowingly alleges that which is false, has obviously no such honest belief. Thirdly, if fraud be proved, the motive of the person guilty of it is immaterial. It matters not that there was no intention to cheat or injure the person to whom the statement was made."

Gross negligence may be evidence of bad faith, but it is not the same thing, and does not entail the same consequences (*f*).

Though fraud avoids sale like any other contract it must be remembered, as Lord DUNEDIN says, that sale

"is a contract made at arm's length, not a contract *uberrimæ fidei* such as insurance" (*g*).

---

(3) A person is deemed to be insolvent within the meaning of this Act who either has ceased to pay his debts in the ordinary course of business, or cannot pay his debts as they become due, whether he has committed an act of bankruptcy or not, [and whether he has become a notour bankrupt or not] (*h*).

Cross-references.—This definition is required by ss. 39, 41 and 44, *ante*.

COMMENT

For the relevance of the words in brackets, see the Bankruptcy (Scotland) Act, 1913, ss. 5 and 6.

---

(4) Goods are in a "deliverable state" within the meaning of this Act when they are in such a state that

---

(*f*) *Jones* v. *Gordon* (1877), 2 App. Cas. 616, at p. 629, *per* Lord BLACKBURN. As to *dolus dans locum contractui*, see *Price & Pearce* v. *Bank of Scotland*, 1912 S. C. (H. L.) 19. In *Houghland* v. *Low, Ltd.*, [1962] 2 All E. R. 159; [1962] 1 Q. B. 694 C. A., ORMEROD, L.J., said: "I am not sure what is meant by the term 'gross negligence' which has been in use for a long time in cases of this kind. . . . I have always found some difficulty in understanding just what was 'gross negligence.'"

(*g*) *Shankland & Co.* v. *Robinson & Co.*, 1920 S. C. (H. L.) 103, at p. 111; *cf. Ward* v. *Hobbs* (1877), 3 Q. B. D. 150, at p. 157, *per* BRAMWELL, L.J.

(*h*) *Biddlecombe* v. *Bond* (1835), 4 A. & E. 332 (general inability to pay debts); *Ex p. Carnforth Co.* (1876), 4 Ch. D. 108, C. A., at p. 122 (inability to pay avowed either in act or word, and consequent intention not to pay debts); see *The Feliciana* (1915), 59 Sol. Jo. 546, as to an alien enemy dishonouring his acceptance in wartime.

the buyer would under the contract be bound to take delivery of them (*i*).

Cross-references.—This definition is required by s. 18, rules (1), (2), (3) and (5), and s. 29 (5) *ante*. *Cf.* the definitions of "buyer," "delivery," "goods" and "quality" in this section, *ante*.

[**63. Commencement.**—*Repealed*.]

COMMENT

This section was repealed as spent by the Statues Law Revision Act, 1908.

64. Short title.—This Act may be cited as the Sale of Goods Act, 1893.

[*Schedule repealed by S.L.R. Act*, 1908]

(*i*) See *Blackburn on Sale* (1st ed.), p. 152; *Pritchett & Co.* v. *Currie*, [1916] 2 Ch. 515, C. A.; *Underwood* v. *Burgh Castle Brick and Cement Synd. Co.*, [1922] 1 K. B. 343, C. A.; *Morison* v. *Lockhart*, 1912 S. C. 1017; *Kursell* v. *Timber Operators, Ltd.*, [1927] 1 K. B. 298 (timber to be cut by purchaser not in a deliverable state until so cut).

THE FACTORS ACT, 1889

(52 & 53 VICT. c. 45)

An Act to amend and consolidate the Factors Acts.

[26th August 1889.]

The Factors Act, 1889, which repeals the previous enactments dealing with similar subject-matter, is a partial application to English law of the French maxim, " en fait de meubles possession vaut titre." The present Act is the result of a long struggle between the mercantile community on the one hand and the principles of the common law on the other. The general rule of the common law was *nemo dat quod non habet* (*a*), and it was held that the mere fact that a person was in possession of goods or documents of title to goods did not enable him to dispose of those goods in contravention of his instructions about them. The merchants and bankers contended that, in the interests of commerce, if a person was put or left in the possession of goods or documents of title, he ought, as regards innocent third parties, to be treated as the owner of the goods. As BOWEN, L.J., has pointed out, the object of the Courts is to prevent fraud,

> " the object of mercantile usages is to prevent the risk of insolvency, not of fraud, and any one who attempts to follow and understand the law merchant will soon find himself lost if he begins by assuming that merchants conduct their business on the basis of attempting to insure themselves against fraudulent dealing. The contrary is the case. Credit, not distrust, is the basis of commercial dealings; mercantile genius consists principally in knowing whom to trust " (*b*).

To express the principle in other words, lawyers see only the pathology of commerce, and not its healthy physiological action, and their views are therefore apt to be warped and one-sided.

The first Factors Act was passed in 1823, the second in 1825 and the third in 1842. These enactments were a model of the art of saying few things in many words. They dealt with the powers of factors and other mercantile agents intrusted with the possession of goods or documents of title to goods, and their conjoint effect is carefully summed up by BLACKBURN, J., in *Cole* v.

(*a*) See s. 21 of the Sale of Goods Act, *ante*, and *Fuentes* v. *Montis* (1868), L. R. 3 C. P. 268, at p. 277, *per* WILLES, J.

(*b*) *Sanders* v. *Maclean* (1883), 11 Q. B. D. 327, at p. 343, C. A.; *cf. Speight* v. *Gaunt* (1883) 9 App. Cas. 1, at p. 20, *per* Lord BLACKBURN.

North Western Bank (*c*). After reviewing the history and policy of the Acts, he proceeds to say:

> " We do not think that the legislature wished to give to all sales and pledges in the ordinary course of business the effect which the Common Law gives to sales in market overt . . . The general rule of law is, that where a person is deceived by another into believing he may safely deal with property he bears the loss, unless he can show that he was misled by the act of the true owner. The legislature seems to us to have wished to make it the law that, where a third person has intrusted goods or the documents of title to goods to an agent who, in the course of such agency, sells or pledges the goods, he should be deemed by that act to have misled any one who *bonâ fide* deals with the agent, and makes a purchase from or an advance to him without notice that he was not authorised to sell or to procure the advance."

The Factors Act, 1877, dealt with a new subject-matter. After providing that a secret revocation of agency should be inoperative, it proceeded to deal with the case, not of agents, but of buyers or sellers left in possession of the documents of title to goods. The present Act reproduces and somewhat extends the effect of the four above-mentioned Acts.

" The Act," says Lord HERSCHELL,

> " is divided into parts. The first headed ' Preliminary ' consists of a definition clause. The last part headed ' Supplemental ' contains provisions as to the mode of transfer ' for the purposes of the Act ' and certain savings. The other two parts are headed respectively ' Dispositions by Mercantile Agents,' and ' Dispositions by Buyers and Sellers of Goods.' These headings are not, in my opinion, mere marginal notes, but the sections in the group to which they belong must be read in connection with them, and interpreted by the light of them. It appears to me that the Legislature has clearly indicated the intention that the provisions of s. 3 should not be treated as an enactment relating to all pledges of documents of title, but only to those effected by mercantile agents " (*d*).

By s. 21 (2) of the Sale of Goods Act, *ante*, nothing in that Act is to affect the provisions of the Factors Acts, or any enactment enabling the apparent owner of goods to dispose of them as if he were the true owner thereof.

Preliminary

1. Definitions.—(1) For the purposes of this Act—

The expression " mercantile agent " shall mean a

(*c*) *Cole* v. *North Western Bank* (1875), L. R. 10 C. P. 354, at p. 372, Ex. Ch.; *cf. Weiner* v. *Harris* (1910), 15 Com. Cas. 39, at p. 48, *per* FARWELL, L.J.

(*d*) *Inglis* v. *Robertson*, [1898] A. C. 616, at p. 630.

mercantile agent having in the customary course of his business as such agent authority either to sell goods, or to consign goods for the purpose of sale, or to buy goods, or to raise money on the security of goods :

Definition.—For " goods," see sub-s. (3), *post*.

Cross-references.—This definition is required by ss. 2 (1), (2) and (3), 4 to 6, 7 (2), 8, 9, 12 and 13, *post*. *Cf.* s. 6, *post* (agreements through clerks, etc.).

COMMENT

This definition is new, but is mainly declaratory. It extends the construction put on the repealed Acts in so far as it applies to agents " to buy goods," and perhaps also in so far as it applies to forwarding agents (*e*).

Under the repealed Acts, the terms used were simply " person " or " agent " intrusted with the possession of goods, but it was held that the Acts only applied to mercantile transactions, and that the term person or agent did not include a mere servant or caretaker, or one who had possession of goods for carriage, safe custody, or otherwise as an independent contracting party ; but only persons whose employment corresponded to that of some known kind of commercial agent like that class (factors) from which the Acts took their name (*f*). Thus, a person intrusted with furniture to keep in her own house for the plaintiff was held not to be an " agent " within the meaning of the Acts (*g*) ; and a wine merchant's clerk who, as such, was possessed of delivery orders, was held not to be an agent, within the meaning of the Acts, as to to be able to make a valid pledge in fraud of his master (*h*). It was further held that if a mercantile agent received goods in some other capacity the Act did not apply ; for instance, where goods were warehoused with a warehouseman who was also a broker, he could not pledge them in his capacity of broker (*i*). Under the present Act it was once held that a person employed to sell jewellery for a firm of jewellers at a small commission was not a mercantile agent (*k*), but that case has been expressly overruled (*l*).

(*e*) *Quære* how far *Hellings* v. *Russell* (1875), 33 L. T. 380, and *City Bank* v. *Barrow* (1880), 5 App. Cas. 664, are affected ?

(*f*) *Cole* v. *North Western Bank* (1875), L. R. 10 C. P. 354, at pp. 372, 373, *per* BLACKBURN, J. ; *City Bank* v. *Barrow* (1880), 5 App. Cas. 664, at p. 678.

(*g*) *Wood* v. *Rowcliffe* (1846), 6 Hare, 183.

(*h*) *Lamb* v. *Attenborough* (1862), 31 L. J. Q. B. 41 ; *cf. Oppenheimer* v. *Attenborough*, [1908] 1 K. B. 221, at p. 226, C. A.

(*i*) *Cole* v. *North Western Bank* (1875), L. R. 10 C. P. 354, Ex. Ch.

(*k*) *Hastings* v. *Pearson*, [1893] 1 Q. B. 62 ; see also *Tremoille* v. *Christie* (1893), 37 Sol. Jo. 650.

(*l*) *Weiner* v. *Harris*, [1910] 1 K. B. 285.

On the other hand, it was held that the repealed Acts applied to an isolated instance of employment, if the employment was such that persons who ordinarily carried on that kind of business would come within the Acts, and this is still the law (*m*).

The Court will look behind any document purporting to regulate the relationship of the parties to it, if in fact it is a cloak designed to hide the true nature of that relationship so as to deprive a third party of rights he would otherwise possess (*n*).

---

(2) A person shall be deemed to be in possession of goods or of the documents of title to goods, where the goods or documents are in his actual custody or are held by any other person subject to his control or for him or on his behalf:

Definitions.—For " goods," " documents of title " and " person," see sub-ss. (3), (4) and (6), *post*, respectively.

Cross-references.—This definition is required by ss. 1 (4), 2 (1), (2) and (3), 7 (1), 8 and 9, *post*.

COMMENT

This definition is taken from words used in s. 4 of the Factors Act, 1842, but it is generalised by the substitution of the word " person " for the word " agent." See note to definition of " delivery " in s. 62 (1) of the Sale of Goods Act, *ante*.

---

(3) The expression " goods " shall include wares and merchandise:

Cross-references.—This definition is required by ss. 1 (1) and (2), *ante*, 1 (4) and (5), 2 (1), (2) and (3), 3 to 5, 7 (1), 8 to 11, 12 (2) and (3), *post*.

(*m*) *Hayman* v. *Flewker* (1863), 32 L. J. C. P. 132 (pictures entrusted to insurance agent to sell on commission); *Lowther* v. *Harris*, [1927] 1 K. B. 393; *Budberg* v. *Jerwood and Ward* (1934), 51 T. L. R. 99; *Thoresen* v. *Capital Credit Corporation* (1962), 37 D. L. R. (2d.) 317; *R.* v. *Eaton* (1966), 50 Cr. App. Rep. 189. *Cf. Biggs* v. *Evans*, [1894] 1 Q. B. 88, decided on the repealed Acts, where it was held that an opal matrix table-top entrusted (but not for the purpose of sale) to a dealer in jewels and gems, and sold by him outside the ordinary course of his business, was not within the Act.

(*n*) *St. Margaret's Trust* v. *Castle*, [1964] C. L. Y. 1685, C. A.; and see generally pp. 75–76, *ante*.

COMMENT

The term used in the repealed s. 17 of the Statute of Frauds is " goods, wares, and merchandise." This definition, therefore, probably incorporates the numerous (and now otherwise obsolete) decisions on the meaning of those words in that Act. A new definition of " goods " is given by s. 62 (1) of the Sale of Goods Act for the purposes of that Act, *ante*.

---

(4) The expression " document of title " shall include any bill of lading, dock warrant, warehouse-keeper's certificate, and warrant or order for the delivery of goods (*o*), and any other document used in the ordinary course of business as proof of the possession or control of goods, or authorising or purporting to authorise, either by endorsement or by delivery, the possessor of the document to transfer or receive goods thereby represented:

Definitions.—For " goods " and " possession," see sub-ss. (3) and (2), *ante*, respectively.

Cross-references.—This definition is required by ss. 1 (2), *ante*, and 2 (1), (2) and (3), 3, 5, 8 to 11 and 12 (2), *post*. *Cf.* s. 11, *post* transfer of documents of title).

COMMENT

This definition is taken from s. 4 of the Act of 1842, with the addition of the " warehouse-keeper's certificate." It is incorporated by s. 62 (1) of the Sale of Goods Act, *ante*. The Act of 1825 included warehouse-keepers' certificates, but the Act of 1842 omitted them, and in a case in 1875 the Lords Justices held that these documents were not documents of title (*p*).

Cash receipts given in place of delivery orders are not documents of title (*q*), nor are the registration books of motor vehicles (*r*).

(*o*) As to delivery orders, see *Morgan* v. *Gath* (1865), 3 H. & C. 748, at p. 760 (definition) ; *Re Hall* (1884), 14 Q. B. D. 386 ; *Re Cunningham & Co.* (1885), 54 L. J. Ch. 448 (distinguished from bill of sale) ; *Union Credit Bank* v. *Mersey Docks*, [1899] 2 Q. B. 205.

(*p*) *Gunn* v. *Bolckow, Vaughan & Co.* (1875), 10 Ch. App. 491.

(*q*) *Kemp* v. *Falk* (1882), 7 App. Cas. 573, at p. 585, *per* Lord BLACKBURN. As to mate's receipts, see *Cowas-Jee* v. *Thompson* (1845), 5 Moo. P. C. C. 165 ; *Hathesing* v. *Laing* (1873), L. R. 17 Eq. 92; and compare *Kum* v. *Wah Tat Bank*, [1971] 1 Lloyd's Rep. 439, P. C.

(*r*) *Joblin* v. *Watkins and Roseveare*, [1949] 1 All E. R. 47 ; *Central Newbury Car Auctions* v. *Unity Finance*, [1956] 3 All E. R. 905, C. A.; [1957] 1 Q. B. 371; but the sale of a motor vehicle without its registration book is not normally a sale in the ordinary course of a mercantile agent's business: see p. 298, *post*,

Ordinarily, when the title to goods depends upon a written instrument, the document requires to be registered as a bill of sale for the purposes of the Bills of Sale Acts ; but most of the usual commercial documents of title are excluded by s. 4 of the Bills of Sale Act, 1878, reproduced *post*, Appendix I. And by the Bills of Sale Act, 1890 (*c*), as amended by the Bills of Sale Act, 1891 (*s*), certain mercantile letters of hypothecation are exempted from the provisions of the Bills of Sale Acts.

---

(5) The expression " pledge " shall include any contract pledging, or giving a lien or security on, goods, whether in consideration of an original advance or of any further or continuing advance or of any pecuniary liability :

Definition.—For " goods," see sub-s. (3), *ante*.

Cross-references.—This definition is required by ss. 2 (1), 3 to 6, 7 (2), 8, 9 and 12 (2), *post*.

COMMENT

This definition is new. Its terms seem wide enough to include a mortgage, that is, a contract transferring conditionally the general property in goods in consideration of a loan, and also perhaps a letter of hypothecation without possession. As to the common law definition of " pledge ", see *ante*.

The words " any pecuniary liability " are very wide, and are probably intended to meet cases such as the granting of a letter of credit to be operated by bills of exchange in consideration of the pledge of goods or documents.

---

(6) The expression " person " shall include any body of persons corporate or incorporate.

Cross-references.—This definition is required by ss. 1 (2), *ante*, and 2 (1), (2) and (3), 7 (1), 8 to 10 and 12 (2), *post*.

COMMENT

Cf. s. 19 of the Interpretation Act, 1889, which applies only to later Acts of Parliament.

In a contract of hire-purchase, and presumably also in a contract of sale, the provision of a car log book is a suspensive condition on which the existence of the contract depends, though the condition may be waived by taking delivery of the car and using it: *Bentworth Finance, Ltd.* v. *Lubert*, [1967] 2 All E. R. 810; [1968] 1 Q. B. 680, C. A. See also 99 L. J. 243 and 101 L. J. 481.

(*s*) 3 Halsbury's Statutes (3rd Edn.) 262.

Dispositions by Mercantile Agents

2. Powers of mercantile agent with respect to disposition of goods.—(1) Where a mercantile agent is, with the consent (*t*) of the owner, in possession of goods (*u*) or of the documents of title to goods, any sale, pledge, or other disposition of the goods, made by him when acting in the ordinary course of business of a mercantile agent, shall, subject to the provisions of this Act, be as valid as if he were expressly authorised by the owner of the goods to make the same; provided that the person taking under the disposition acts in good faith, and has not at the time of the disposition notice that the person making the disposition has not authority to make the same.

ILLUSTRATIONS

(1) A diamond broker obtains a parcel of diamonds from a diamond merchant on the pretence that he has a prospective customer in K. Instead of taking them to K, he pledges them with a pawnbroker. It is against the custom of the diamond trade for brokers to have authority to pledge, but the pawnbroker, who has had previous dealings with the broker, believes him to be a principal. The pledge stands good (*a*).

(2) A diamond broker obtains a parcel of diamonds from a diamond merchant on the pretence that he has a prospective customer in K. Instead of taking them to K, he hands them to B, requesting him to sell them. B arranges with another firm of diamond merchants to purchase the diamonds on joint account with him, and the price is paid to the broker. B, who shares an office with the broker, does not act in good faith, but the purchasing firm does. The purchasing firm is not protected, and they and B are jointly liable in an action for conversion (*b*).

(3) The owner of a car entrusts it to a motor-car agent to sell it on commission subject to a reserve price. The agent sells it to a *bonâ fide* purchaser below the reserve price, and misappropriates the money. The agent is a "mercantile agent," and the purchaser gets a good title to the car (*c*).

(*t*) As to "consent," see note to s. 23 of the Sale of Goods Act, *ante*.

(*u*) The word "goods" here bears its ordinary meaning; thus a car is still "goods" though it lack both ignition key and registration book and so cannot be used: *Stadium Finance* v. *Robbins*, [1962] 2 All E. R. 633; [1962] 2 Q. B. 664, C. A. But see *Pearson* v. *Rose & Young*, [1950] 2 All E. R. 1027; [1951] 1 K. B. 275.

(*a*) *Oppenheimer* v. *Attenborough*, [1908] 1 K. B. 221, C. A.

(*b*) *Oppenheimer* v. *Frazer & Wyatt*, [1907] 2 K. B. 50, C. A.; not followed in *Du Jardin* v. *Beadman*, [1952] 2 All E. R. 160; [1952] 2 Q. B. 712.

(*c*) *Folkes* v. *King*, [1923] 1 K. B. 282, C. A.; distinguished, *Heap* v. *Motorists Advisory Agency, Ltd.*, [1923] 1 K. B. 577 (larceny by trick, but the thief was not a mercantile agent).

(4) A hires a car to B on hire-purchase terms for use in B's self-drive hire business. B also sells cars as a mercantile agent. B sells this car to C. B is not in possession of the car in his capacity of a mercantile agent. C, therefore, does not get a good title to the car by virtue of this sub-section (*d*).

Definitions.—For " mercantile agent," " possession," " goods," " document of title," " pledge " and " person," see s. 1 (1), (2), (3), (4), (5) and (6), *ante*, respectively.

Cross-references.—*Cf.* s. 4, *post* (pledges for antecedent debt); s. 5, *post* (necessary consideration); s. 7, *post* (provisions as to consignors and consignees); Sale of Goods Act, 1893, s. 23, *ante* (sale under voidable title).

History.—This sub-section supersedes and reproduces, in altered language, ss. 2 and 4 of the Act of 1825 and s. 4 of the Act of 1842.

COMMENT

The consent of the owner: (1) the owner.—For the purposes of this sub-section, the owner must be a person in a position to give the mercantile agent express authority to dispose of the goods. Where rights of ownership are divided among two or more persons in such a way that the acts which the sub-section is contemplating could never be authorized save by both or all of them, those persons together constitute the owner (*e*). The mercantile agent himself may be part owner of the goods (*e*).

(2) The consent.—This sub-section no longer requires the goods or documents to be " intrusted " to the agent, but it suffices that they are in his possession with the owner's consent. How far this alteration of language extends the operation of the Act is not very clear; but it is now established that where a mercantile agent is intrusted with goods in some other capacity than that of mercantile agent, he cannot sell or pledge those goods contrary to instructions (*f*). Goods are in the possession of a mercantile agent in that capacity if his possession is in some way connected with his business

(*d*) *Astley Industrial Trust, Ltd.* v. *Miller*, [1968] 2 All E. R. 36. Note, however, that if C is a private purchaser as defined by s. 29 (2) of the Hire Purchase Act, 1964, he may now get a good title to the car by virtue of s. 27 of that Act.

(*e*) *Lloyds Bank, Ltd.* v. *Bank of America, National Trust and Savings Association*, [1938] 2 All E. R. 63; [1938] 2 K. B. 147, C. A.

(*f*) *Astley Industrial Trust, Ltd.* v. *Miller*, [1968] 2 All E. R. 36 (reviewing earlier cases) (*ante*, illustration (4)); *Belvoir Finance Co., Ltd.* v. *Harold G. Cole & Co., Ltd.*, [1969] 2 All E. R. 904. As to former rule, see *Monk* v. *Whittenbury* (1831), 2 B. & Ad. 484 (flour factor and wharfinger); *Cole* v. *North Western Bank* (1875), L. R. 10 C. P. 354 (warehouse-keeper and broker); *Biggs* v. *Evans*, [1894] 1 Q. B. 88 (dealer in jewels and gems intrusted with table-top to be sold on certain conditions only). As to meaning of "intrusted," see *Lake* v. *Simmons*, [1927] A. C. 487.

of mercantile agent even if he has no authority to dispose of the goods (*g*).

The consent is not vitiated by fraud on the part of the mercantile agent (*h*) even, it seems, where the fraud would before the Theft Act, 1968 have amounted to larceny by a trick (*i*). It is irrelevant that the consent has determined if the buyer has no notice that this is so: see s. 2 (2), *post*.

Sale pledge or other disposition.—Pledge is defined by s. 1 (5), *ante*, to include any contract pledging or giving a lien or security on goods. Such a contract would give only an equitable interest in the goods to the pledgee. A disposition must involve a transfer of property (*k*), but in the context it would seem that it would be sufficient to transfer an equitable interest in the goods. There are two cases in which the expression " sale pledge or other disposition " has been restrictively construed (*l*), but it is doubtful whether they would now be followed on this point (*m*).

Ordinary course of business of a mercantile agent.—It is a question of fact in each case whether a mercantile agent is acting in the ordinary course of his business. The situation may vary from place to place and from trade to trade. Even where it is in the ordinary course of a mercantile agent's business to sell or pledge goods, it may be necessary to consider whether the manner in which the sale or pledge is effected is within the ordinary course of such business (*n*), " that is to say, within business hours, at a proper place of business, and in other respects in the ordinary way in which a mercantile agent would act " (*o*).

There may be a general rule that agents should only sell for cash, cheque or bill (*p*), but there seems no inherent reason why an agent may not dispose of goods by way of exchange of for other valuable

(*g*) *Pearson* v. *Rose & Young*, [1950] 2 All.E. R. 1027, at p. 1032; [1951] 1 K. B. 275, at p. 288, *per* DENNING, L.J.; *Turner* v. *Sampson* (1911), 27 T. L. R. 200 (agent to obtain offers for picture and report back).

(*h*) *Lowther* v. *Harris*, [1927] 1 K. B. 393.

(*i*) See the comment to s. 23 of the Sale of Goods Act, *ante*.

(*k*) *Worcester Works Finance, Ltd.* v. *Cooden Engineering Co., Ltd.*, [1971] 3 All E. R. 708; [1972] 1 Q. B. 210, C. A.

(*l*) *Waddington & Sons* v. *Neale & Sons* (1907), 96 L. T. 786; *Jobling* v. *Watkins and Roseveare (Motors) Ltd.*, [1949] 1 All E. R. 47.

(*m*) They would seem to be inconsistent with *Worcester Works Finance, Ltd.* v. *Cooden Engineering Co., Ltd. supra*, and can be justified, if at all, only on the basis that the dispositions in question were not in the ordinary course of the agent's business. *Cf.* also *Stoneleigh Finance, Ltd.* v. *Phillips*, [1965] 1 All E. R. 513, C. A.; [1965] 2 Q. B. 537, C. A.

(*n*) *Janesich* v. *Attenborough* (1910), 102 L. T. 605 (not commercial rate of interest); *De Gorter* v. *Attenborough* (1904), 21 T. L. R. 19 (friend asked to pledge article for mercantile agent).

(*o*) *Oppenheimer* v. *Attenborough & Son*, [1907] 2 K. B. 221, at pp. 230–231, *per* BUCKLEY, L.J.

(*p*) *Biggs* v. *Evans*, [1894] 1 Q. B. 88, at p. 91.

consideration, or even by way of set-off (*q*), as long as it is, on the facts, within the ordinary course of business of such a mercantile agent.

A sale will not normally be in the ordinary course of business of a mercantile agent, where the mercantile agent sells the goods as agent and not as principal, and the buyer knows that the profits on the sale are to go to the agent and not the principal (*r*).

The sale of a second-hand car without a log book has been held not to be in the ordinary course of business of a mercantile agent (*s*), but the opposite conclusion was reached in another case where a new car was sold and a satisfactory explanation was given as to the absence of the log book (*t*).

The circumstances in which the vendor acquires possession of the goods has been held to affect the question of whether the mercantile agent is acting in the ordinary course of business (*u*). The better view, however, would seem to be that such circumstances are irrelevent in this context, although they may be important in deciding whether the goods have been given to the mercantile agent in his capacity as a mercantile agent (*a*).

Agent purporting to act as principal.—The third party cannot be in a worse position if the agent purports to be the owner of the goods than would have been the case if the agent had acted as agent (*b*). Accordingly, if he would have acquired a good title if the agent had acted as agent, he will still acquire a good title if the agent acts as owner (*c*). But it would seem that to come within this section, the transaction must still be one that would have been within the ordinary course of business of the agent had he acted as such (*d*).

A transaction would still be within the ordinary course of a mercantile agent's business, where the agent holds himself out as principal, although payment is made to the agent in a manner that would be outside his apparent or actual authority as agent (*e*).

(*q*) *Tingey & Co., Ltd.* v. *Chambers, Ltd.*, [1967] N. Z. L. R. 785.

(*r*) *Raffoul* v. *Essanda, Ltd.* (1970), 72 S. R. (N. S. W.) 633.

(*s*) *Pearson* v. *Rose and Young, Ltd.*, [1950] 2 All E. R. 1027; [1951] 1 K. B. 275, C. A.; *Lambert* v. *G. and C. Finance Corporation, Ltd.* (1963), 107 Sol. Jo. 666.

(*t*) *Astley Industrial Trust* v. *Miller*, [1968] 2 All E. R. 36. See further the comment to s. 1 (4), *ante*.

(*u*) *Stadium Finance* v. *Robbins*, [1962] 2 All E. R. 633, at p. 639; [1962] 2 Q. B. 664, at p. 675, *per* WILLMER, L.J.

(*a*) *Astley Industrial Trust* v. *Miller, supra*.

(*b*) *Oppenheimer* v. *Attenborough*, [1908] 1 K. B. 221, C. A. (Illustration (1), *ante*).

(*c*) *Oppenheimer* v. *Attenborough, supra; Janesich* v. *Attenborough* (1910), 102 L. T. 605.

(*d*) Thus in *Janesich* v. *Attenborough, supra*, a pledge was made by an agent acting as principal at 15 per cent interest, and the Court considered whether this was a commercial rate for interest such as an agent might agree to in the ordinary course of his business.

(*e*) *Oppenheimer* v. *Attenborough, supra; Janesich* v. *Attenborough, supra; Raffoul* v. *Essanda, Ltd.* (1970), 72 S. R. (N. S. W.) 633.

The powers of the mercantile agent.—It was held under the repealed Acts that the mercantile agent's powers were not exhausted by a single transaction. Thus, where the consignee of cotton pledged the bill of lading with a broker, authorising him to sell the cotton, and then with the broker's consent pledged the net proceeds to D, it was held that the latter transaction was valid as well as the former one (*f*). This case would seem still to be good law.

Under s. 1 of the Factors Act, 1842, no protection was given to a pledgee under a general pledge or lien, where goods which had been entrusted to a mercantile agent by their owner were pledged by him, together with other goods, as security for a loan (*g*). Section 2 (1) of the present Act is worded in more general terms. So far only the Supreme Court of British Columbia has considered the change of wording, holding that it has not altered the law (*h*).

Notice.—The term " notice " in this section probably means actual, though not formal, notice ; that is to say, either knowledge of the facts, or grounds for a suspicion of something wrong combined with a wilful disregard of the means of knowledge. The same construction would probably be put on it as upon the term " notice " in s. 29 of the Bills of Exchange Act, 1882, or in ss. 30 and 45 of the Bankruptcy Act, 1914 (*i*). Under this section, which creates an exception to the rule that no one can give a better title than he himself has, the person claiming the benefit of the proviso must prove his good faith and absence of notice ; while under s. 23 of the Sale of Goods Act, *ante*, which embodies an exception recognised by the common law to that rule, the onus is on the person who seeks to impeach the title of a buyer from a seller with a voidable title to prove in bad faith or notice (*k*).

(*f*) *Portalis* v. *Tetley* (1867), L. R. 5 Eq. 140.

(*g*) *Kaltenbach* v. *Lewis* (1883), 24 Ch. D. 54, C. A.; see especially at pp. 79–80. The point was not raised in the appeal to the House of Lords, (1885), 10 App. Cas. 617, but the decision of the Court of Appeal on this point received the approval of Lord BRAMWELL, at p. 635.

(*h*) *Thoresen* v. *Capital Credit Corporation, Ltd. and Active Bailiff* (1962) 37 D. L. R. (2d.) 317; and see *Re the Farmers' and Settlers' Co-operative Society, Ltd., City Bank of Sydney* v. *Barden* (1908), 9 S. R. (N. S. W.) 41.

(*i*) 3 Halsbury's Statutes (3rd Edn.) 204, 78, 96. See the term discussed, *Navulshaw* v. *Brownrigg* (1852), 21 L. J. Ch. 908, at p. 911 (Factors Act) ; *Raphael* v. *Bank of England* (1855), 17 C. B. 161, at p. 174, *per* WILLES, J. (bill of exchange) ; *Ex p. Snowball* (1872), 7 Ch. App. 534, at p. 549 (act of bankruptcy). See further, the comment to s. 8 of the Factors Act, *post*.

(*k*) *Cf. Heap* v. *Motorists Advisory Agency, Ltd.*, [1923] 1 K. B. 577 (Factors Act); *Halfway Garage (Nottingham)* v. *Lepley* (1964), *Guardian*, 8th February, C. A. (s. 25, Sale of Goods Act); and *Whitehorn Brothers* v. *Davison*, [1911] 1. K B. 463 (s. 23) Sale of Goods Act). The Law Reform Committee in its Report on the Transfer of Title to Chattels (Cmd.) 2958) has recommended that in a s. 23 situation it should be for the buyer to prove that he bought in good faith.

(2) Where a mercantile agent has, with the consent of the owner, been in possession of goods or of the documents of title to goods, any sale, pledge, or other disposition, which would have been valid if the consent had continued, shall be valid notwithstanding the determination of the consent: provided that the person taking under the disposition has not at the time thereof notice that the consent has been determined.

ILLUSTRATIONS

(1) A picture dealer in Paris sends pictures to an agent in London, some for sale, others for exhibition only. After revocation of his authority the agent pledges both lots of pictures with B, who takes them without notice of the revocation. B gets a good title (*l*).

(2) A, by fraud, induces plaintiff to sell him a car. On learning of the fraud, plaintiff rescinds the contract. Subsequently A sells the car in the Warren Street car market to B who acts in good faith and without notice of A's lack of title. By virtue of s. 9 of the Factors Act, A is in the same position as a mercantile agent. He obtained possession of the car initially with the consent of the owner, albeit the consent was obtained by fraud. The consent has since determined, but A may nevertheless give a good title to the car in a manner which would be within the ordinary course of a mercantile agent's business (*m*).

Definitions.—See note to sub-s. (1) of this section.

History.—See last note. This sub-section reproduces in altered language the provisions of s. 2 of the Factors Act, 1877, which was passed to override the decision in *Fuentes* v. *Montis* (*n*) where it was held that a mercantile agent was not intrusted with goods or documents within the meaning of the earlier Acts if his authority had been revoked.

---

(3) Where a mercantile agent has obtained possession of any documents of title to goods by reason of his being or having been, with the consent of the owner,

---

(*l*) *Moody* v. *Pall Mall Deposit Co.* (1917), 33 T. L. R. 306.
(*m*) *Newtons of Wembley, Ltd.* v. *Williams*, [1964] 3 All E. R. 532; [1965] 1 Q. B. 560, C. A.
(*n*) (1868), L. R. 3 C. P. 268; affirmed L. R. 4 C. P. 93, Ex. Ch.

in possession of the goods represented thereby, or of any other documents of title to the goods, his possession of the first-mentioned documents shall, for the purposes of this Act, be deemed to be with the consent of the owner.

Definitions.—See note to sub-s. (1) of this section.

History.—This sub-section reproduces in somewhat different language a provision in s. 4 of the Factors Act, 1842, which was intended to alter the law as laid down in *Phillips* v. *Huth* (*o*) and *Hatfield* v. *Phillips* (*p*). In the latter case it was held that a person intrusted with a bill of lading for the purpose of selling the goods mentioned in it was not, in consequence of being so intrusted, to be considered as intrusted with the dock warrant, notwithstanding that his possession of the bill of lading enabled him to obtain the dock warrant.

---

(4) For the purposes of this Act the consent of the owner shall be presumed in the absence of evidence to the contrary.

History.—This sub-section reproduces in somewhat different language the concluding paragraph of s. 4 of the Factors Act, 1842.

3. Effect of pledges of documents of title.—A pledge of the documents of title to goods shall be deemed to be a pledge of the goods.

Definitions.—For " goods," " documents of title " and " pledge," see s. 1 (3), (4) and (5), *ante*, respectively.

History.—This section is taken from a paragraph in s. 4 of the Factors Act, 1842.

COMMENT

The section, though general in its terms, is controlled by the heading " Dispositions by Mercantile Agents " of the part in which it occurs. It therefore applies only to pledges by " mercantile agents " as defined by s. 1 (1) (*q*).

(*o*) (1860), 6 M. & W. 572 ; see the comments on this case in *Cole* v. *North Western Bank* (1875), L. R. 10 C. P. 354, at p. 370.

(*p*) (1842), 9 M. & W. 647 ; affirmed (1845), 14 M. & W. 665.

(*q*) *Inglis* v. *Robertson*, [1898] A. C. 616.

4. Pledge for antecedent debt.—Where a mercantile agent pledges goods as security for a debt or liability due from the pledgor to the pledgee before the time of the pledge, the pledgee shall acquire no further right to the goods than could have been enforced by the pledgor at the time of the pledge.

Definitions.—For " mercantile agent," " goods " and " pledge," see s. 1 (1), (3) and (4), *ante*, respectively.

History.—This section reproduces in altered language the clumsily worded s. 3 of the Factors Act, 1825, as read with the proviso contained in s. 3 of the Factors Act, 1842. The substitution of the words " debt *or liability* " for " antecedent debt " is material (*r*).

COMMENT

The use of the word " due," though appropriate to the term " debt," seems inappropriate to the term " liability." The section should perhaps be read as if it ran " debt due from or liability incurred by."

The object of this section seems to be to draw a marked distinction between past consideration and present or future consideration (*s*). In terms it applies only to pledges of goods, but having regard to the language of s. 3, *ante*, it may be intended to apply also to pledges of documents.

5. Rights acquired by exchange of goods or documents.—The consideration necessary for the validity of a sale, pledge, or other disposition, of goods, in pursuance of this Act, may be either a payment in cash, or the delivery or transfer of other goods, or of a document of title to goods, or of a negotiable security, or any other valuable consideration; but where goods are pledged by a mercantile agent in consideration of the delivery or transfer of other goods, or of a document of title to goods, or of a negotiable security, the pledgee shall acquire no right

---

(*r*) For cases on the repealed sections, see *Jewan* v. *Whitworth* (1866) L. R. 2 Eq. 692; *Macnee* v. *Gorst* (1867), L. R. 4 Eq. 315; *Kaltenbach* v. *Lewis* (1885), 10 App. Cas. 617.

(*s*) But see *Industrial Acceptance Corpn.* v. *Whiteshall Finance Corporation* (1966), 57 D. L. R. (2d) 670, where the equivalent provision in the Manitoban Factors Act is considered in relation to a revolving credit given by a finance company to a car dealer.

or interest in the goods so pledged in excess of the value of the goods, documents, or security when so delivered or transferred in exchange.

Definitions.—For " mercantile agent," " goods," " document of title " and " pledge," see s. 1 (1), (3), (4) and (5), *ante*, respectively.

History.—By s. 4 of the Factors Act, 1842, it was provided, *inter alia*, that " any payment made, whether by money or bills of exchange, or other negotiable security, shall be deemed and taken to be *an advance* within the meaning of this Act." The first sentence of the present section considerably extends the scope of the old enactment, by substituting " valuable consideration " for " an advance " as above defined.

The second sentence of the section reproduces in somewhat different language the provisions of s. 2 of the Factors Act, 1842, which was intended to protect exchanges of goods and securities made in good faith, and to alter the law as laid down in *Taylor* v. *Kymer* and *Bonzi* v. *Stewart* (*t*).

Scotland.—As regards the application of this section to Scotland, see s. 1 (2) of the Factors (Scotland) Act, 1890, *post*.

6. Agreements through clerks, &c.—For the purposes of this Act an agreement made with a mercantile agent through a clerk or other person authorised in the ordinary course of business to make contracts of sale or pledge on his behalf shall be deemed to be an agreement with the agent.

Definitions.—For " mercantile agent " and " pledge," see s. 1 (1) and (5), *ante*, respectively.

History.—This section is taken from, and generalises, a paragraph in s. 4 of the Factors Act, 1842.

7. Provisions as to consignors and consignees.—(1) Where the owner of goods has given possession of the goods to another person for the purpose of consignment or sale, or has shipped the goods in the name of another person, and the consignee of the goods has not had notice that such person is not the

(*t*) (1832), 3 B. & Ad. 320 ; and (1842), 4 M. & Gr. 295 ; and see the comments on these cases in *Cole* v. *North Western Bank* (1875), L. R. 10 C. P. 354, at p. 370.

owner of the goods, the consignee shall, in respect of advances made to or for the use of such person, have the same lien on the goods as if such person were the owner of the goods, and may transfer any such lien to another person.

(2) Nothing in this section shall limit or affect the validity of any sale, pledge, or disposition, by a mercantile agent.

Definitions.—For " mercantile agent," " possession," " goods," " pledge " and " person," see s. 1 (1), (2), (3), (5) and (6), *ante*, respectively.

History.—The first sub-section reproduces in different language the provisions of s. 1 of the Factors Act, 1825 (*u*).

COMMENT

It is to be noted that the section applies only to goods and not to documents of title, and to cases where the consignee has not had notice that the consignor is not the owner. Lord BLACKBURN raised a doubt on the repealed enactment whether " notice " was co-extensive with knowledge (*a*). The term " advance " must probably be interpreted in the light of s. 5, *ante*.

The second sub-section shows that the present section is to be construed as amplifying, and not as derogating from, the powers of mercantile agents under s. 2, *ante*. Section 13, *post*, further saves the common law powers of factors and agents of that class.

Dispositions by Sellers and Buyers of Goods

8. Disposition by seller remaining in possession.—Where a person, having sold goods, continues, or is, in possession of the goods or of the documents of title to the goods, the delivery or transfer by that person, or by a mercantile agent acting for him, of the goods or documents of title under any sale, pledge, or other disposition thereof, or under any agreement for sale, pledge, or other disposition thereof, to any person

(*u*) See that enactment discussed in *Cole* v. *North Western Bank* (1875) L. R. 10 C. P. 354, at pp. 361–367 ; and *Johnson* v. *Crédit Lyonnais* (1877), 3 C. P. D. 32, at pp. 44, 45.

(*a*) *Mildred* v. *Maspons* (1883), 8 App. Cas. 874, at p. 885.

receiving the same in good faith and without notice of the previous sale, shall have the same effect as if the person making the delivery or transfer were expressly authorised by the owner of the goods to make the same.

Definitions.—For " mercantile agent," " possession," " goods," " document of title," " pledge " and " person," see s. 1 (1), (2), (3), (4), (5) and (6), *ante*, respectively.

Cross-references.—*Cf.* Sale of Goods Act, 1893, s. 25 (1), *ante* (sale, etc. by seller in possession), and s. 48 (3), *ante* (re-sale by unpaid seller).

History.—This section and the next are reproduced by s. 25 of the Sale of Goods Act, slightly modified by omitting the words " or under any agreement for sale, pledge, or other disposition thereof." The Sale of Goods Bill originally proposed to repeal these two sections, but the repeal was afterwards omitted, as the provisions of the Sale of Goods Act are slightly narrower (*b*).

COMMENT

This section alters the law as laid down about sellers in possession in *Johnson* v. *Crédit Lyonnais* (*c*). It was there held that if the buyer, for his own convenience, left the goods and documents of title in the hands of the seller, who fraudulently resold or pledged them, he could nevertheless recover the goods from the innocent purchaser or pledgee.

Continues, or is, in possession.—The words " continues . . . in possession " refer to the continuity of physical possession regardless of any private transaction between seller and buyer which might alter the legal title under which the seller is in possession (*d*). The words " or is " refer only to a case where the person who sold the goods had not got the goods when he sold them but they came into his possession afterwards (*e*).

(*b*) The repeal of these two sections has again been proposed recently by the Law Reform Committee in its Report on the Transfer of Title to Chattels (Cmd. 2958).

(*c*) (1877), 3 C. P. D 32, 40.

(*d*) *Worcester Works Finance, Ltd.* v. *Cooden Engineering Co., Ltd.*, [1971] 3 All E. R. 708; [1972] 1 Q. B. 182, C. A. *Pacific Motor Auctions Pty., Ltd.* v. *Motor Credit (Hire Finance), Ltd.*, [1965] 2 All E. R. 105, at p. 114; [1965] A. C. 867, at p. 888, P. C.; *Mercantile Credit Ltd.* v. *Upton & Son Pty., Ltd.* (1974), 48 A. L. J. R. 301 (H. C. of Aus.).

(*e*) *Worcester Works Finance, Ltd.* v. *Cooden Engineering Co., Ltd., supra.*

So, if the seller continues in possession in another capacity such as bailee or hirer by virtue of a term in the contract of sale itself (*f*) or of some other arrangement (*g*) he " continues in possession " within the meaning of the section even if he is in breach of such arrangement and his right to possession of the goods has determined so that he holds them as trespasser (*g*).

But there must be continuity of possession. If the seller delivers the goods to a purchaser who then leases them back to the seller, then at least as a general rule the section has no application (*h*).

By s. 1 (2) of the Factors Act, 1889, *ante*, a person is deemed to be in possession of the goods or of the documents of title to goods where the goods are held by any other person subject to his control or on his behalf.

Where the seller sold goods which were at the time of the sale deposited with a warehouseman, who had a lien thereon for his charges and promised to discharge the lien and to send the delivery orders to the buyer, but instead of so doing sold the goods to a second buyer who obtained delivery, it was held that the second sale was a sale by a seller in possession of the goods and the section accordingly operated in favour of the second buyer (*i*).

Delivery or transfer.—So that the section may apply there must be either delivery of the goods or a transfer of a document of title.

In a case where wine stored in a warehouse was sold, and the seller afterwards, by executing a memorandum of charge, pledged the wine to the warehouseman who had no notice of the sale, it was held that the warehouseman acquired no title, as there was no delivery of the goods and no transfer of a document of title to him (*k*).

In another case A bought a car from B but paid for it by a cheque which bounced. While in possession of the car A re-sold the car to C but did not deliver it to him. The car was re-possessed by B with the consent of A, and it was held by the Court of Appeal that this was a delivery or transfer of the car by A to B under a disposition of the car, so that B and not C was entitled to it (*l*).

(*f*) *Worcester Works Finance, Ltd.* v. *Cooden Engineering Co. Ltd., supra*, following *Pacific Motor Auctions Pty., Ltd.* v. *Motor Credit* (*Hire Finance*), *Ltd., supra,* and not following *Eastern Distributors* v. *Goldring*, [1957] 2 All E. R. 523; [1957] 2 Q. B. 600, C. A.

(*g*) *Worcester Works Finance, Ltd.* v. *Cooden Engineering Co., Ltd., supra.* See further 1972 J. B. L. 149; 88 L. Q. R 239.

(*h*) *Mitchell* v. *Jones* (1905), 24 N. Z. L. R. 932; approved in *Pacific Motor Auctions Proprietary, Ltd.* v. *Motor Credits* (*Hire Finance*), *Ltd.*, [1965] 2 All E. R. 105; [1965] A. C. 867, P. C.; *cf. Worcester Works Finance, Ltd.* v. *Cooden Engineering Co., Ltd.*, [1971] 3 All E. R. 708, at p. 712; [1972] 1 Q. B. 182, at pp. 217–218, where Lord DENNING, M. R., referred to a " substantial break in the continuity ".

(*i*) *City Fur Manufacturing Co., Ltd.* v. *Fureenbond* (*Brokers*), *London, Ltd.*, [1937] 1 All E. R. 799.

(*k*) *Nicholson* v. *Harper*, [1895] 2 Ch. 415.

(*l*) *Worcester Works Finance, Ltd.* v. *Cooden Engineering Co., Ltd., supra.*

Sale pledge or other disposition.—To constitute a disposition there must be more than a mere transfer of goods; there must be a disposal involving the transfer of an interest in property (*i*). As there is nothing in this section which requires the recipient to give value for the goods, a gift would seem to be sufficient for this purpose (*m*).

Notice.—Notice in this context would seem to mean actual notice so that a person receiving the goods need only show that he did not know of the sale and did not deliberately turn a blind eye (*n*).

The same effect . . . as if expressly authorised by the owner.—This is a wider provision than in s. 9, *post,* by virtue of which a disposition by a buyer in possession only has the same effect as if he was a mercantile agent in possession of the goods with the consent of the owner.

It is for the second purchaser to show that his case comes in all respects within this section or s. 25 of the Sale of Goods Act (*o*).

9. Disposition by buyer obtaining possession.—Where a person, having bought or agreed to buy goods, obtains with the consent of the seller possession of the goods or the documents of title to the goods, the delivery or transfer, by that person or by a mercantile agent acting for him, of the goods or documents of title (*p*), under any sale, pledge, or other disposition thereof, or under any agreement for sale, pledge, or other disposition thereof, to any person receiving the same in good faith and without notice of any lien or other right of the original seller in respect of the goods, shall have the same effect as if the person making the delivery or transfer were a mercantile agent in possession of the goods or documents of title with the consent of the owner (*q*).

For the purposes of this section—

(*m*) *Cf. Kitto* v. *Bilbie, Hobson & Co.* (1895), 72 L. T. 266; *Worcester Works Finance, Ltd.* v. *Cooden Engineering Co., Ltd., supra.*

(*n*) *Worcester Works Finance, Ltd.* v. *Cooden Engineering Co., Ltd., supra,* at pp. 712 and 219, *per* Lord DENNING, M.R. See also *Re Funduck and Horncastle* (1974), 39 D. L. R. (3d) 94. *Cf.* the comment to s. 2, *ante.*

(*o*) *Heap* v. *Motorists Advisory Agency, Ltd.,* [1923] 1 K. B. 577; *Halfway Garage (Nottingham)* v. *Lepley* (1964), *Guardian,* 8th February, C. A.

(*p*) It seems that these need not be the same documents as those of which the buyer has obtained possession. See *per* SALMON, J., in *Mount* v. *Jay & Jay,* [1959] 3 All E. R. 307 at p. 311; [1960] 1 Q. B. 159 at pp. 168–169.

(*q*) See illustration (2) to s. 2 (2), *ante.*

(i) the buyer under a conditional sale agreement shall be deemed not to be a person who has bought or agreed to buy goods, and

(ii) " conditional sale agreement " means an agreement for the sale of goods which is a consumer credit agreement within the meaning of the Consumer Credit Act 1974 under which the purchase price or part of it is payable by instalments, and the property in the goods is to remain in the seller (notwithstanding that the buyer is to be in possession of the goods) until such conditions as to the payment of instalments or otherwise as may be specified in the agreement are fulfilled (*r*).

Definitions.—For " mercantile agent," " possession," " goods," " document of title," " pledge " and " person," see s. 1 (1), (2), (3), (4), (5) and (6), *ante*, respectively.

Cross-references.—*Cf.* Sale of Goods Act, 1893, s. 25 (2), *ante* (sale by buyer in possession), and ss. 2 to 5, *ante* (dispositions by mercantile agents).

History.—As to the reason for not repealing this section by the Sale of Goods Act, 1893, see note to s. 8, *ante*.

COMMENT

This section alters the former law as regards buyers in possession, which was thus stated by BLACKBURN, J. (*a*):

" It has been repeatedly decided that a sale or pledge of a delivery order or other document of title (not being a bill of lading) by the vendee does not defeat the unpaid vendor's rights, because the vendee is not intrusted as an agent " (*b*).

(*r*) The provision relating to conditional sale agreements was introduced by Sch. 4 to the Consumer Credit Act, 1974 with effect from a date to be appointed under s. 192 (4) of that Act. At present no date has been appointed, but the provisions of this section and of s. 25 (2) of the Sale of Goods Act do not apply to conditional sale agreements by virtue of s. 54 of the Hire-Purchase Act, 1965. There are special provisions which apply to the sale of motor vehicles by a purchaser under a conditional sale agreement; see ss. 27–29 of the Hire-Purchase Act, 1964.

(*a*) *Cole* v. *North Western Bank* (1875), L. R. 10 C. P. 354, at p. 373. *Cf. Jenkyns* v. *Usborne* (1844), 7 M. & Gr. 678 ; *McEwan* v. *Smith* (1849), 2 H. L. Cas. 309.

(*b*) This refers to the distinction formerly existing between a buyer in possession and a mercantile agent to whom goods had been intrusted; see s. 2, *ante* (replacing in altered language ss. 2 and 4 of the Factors Act, 1842).

Bought or agreed to buy.—Possession under an agreement for sale or return is not within the Act (*c*): nor is possession under a hire-purchase agreement which confers a mere option to buy the goods at the end of the period of hiring on payment of a further, and usually nominal, sum, whether or not the hirer is obliged to continue the hiring for any specified period (*d*). This is so even though a promissory note is given as collateral security for, and not as payment of, the instalments to become due under the agreement (*e*). But possession under a conditional contract of sale (*f*), or under a so-called hire-purchase agreement in which there is no right to return the goods, but a mere provision that the goods shall not belong to the hirer until the last instalment has been paid, is within the Act (*g*). Such an agreement is a binding contract of sale, with a provision that the payment of the purchase price shall be deferred and that the property shall not pass until the full price is paid. It is a question of construction whether the contract is a contract of hire-purchase or a contract of sale (*h*).

It has also been held that if a buyer or mercantile agent is in possession of goods under an illegal contract, a third party acquiring the goods from that buyer or mercantile agent cannot prove that the possession is with the consent of the seller or owner because the consent is only under and subject to those agreements (*i*).

The consent of the seller.—Though the buyer must obtain possession with the consent of the seller, however fraudulent the buyer may have been in obtaining the possession (even, it seems, if he was guilty of larceny by a trick (*k*)) and however grossly he may abuse the confidence reposed in him or violate the mandate under which he got possession, he can by his disposition give a good title to the purchaser from him (*l*). Nor does it matter for this purpose that the

---

(*c*) *Edwards* v. *Vaughan* (1910), 26 T. L. R. 545, C. A.

(*d*) *Helby* v. *Matthews*, [1895] A. C. 471, and see *Belsize Motor Supply Co.* v. *Cox*, [1914] 1 K. B. 244. But see the exception to this rule for motor vehicles introduced by ss. 27–29 of the Hire-Purchase Act, 1964.

(*e*) *Modern Light Cars, Ltd.* v. *Seals*, [1934] 1 K. B. 32.

(*f*) *Marten* v. *Whale*, [1917] 2 K. B. 480, C. A.

(*g*) *Lee* v. *Butler*, [1893] 2 Q. B. 318, C. A. *Hull Rope Works Co.* v. *Adams* (1896), 65 L. J. Q. B. 114; *Thompson* v. *Veale* (1896), 74 L. T. 130, C. A.; *Wylde* v. *Legge* (1901), 84 L. T. 121.

(*h*) *McEntire* v. *Crossley Brothers*, [1895] A. C. 457, 467.

(*i*) *Belvoir Finance Co., Ltd.* v. *Harold G. Cole & Co., Ltd.*, [1969] 2 All E. R. 904; [1969] 1 W. L. R. 1877. *Sed quaere*, for the effect of such a principle is that a party to a contract derives an advantage over an innocent third party which he would not have obtained but for his own illegality.

(*k*) See the comment to s. 23 of the Sale of Goods Act, *ante*.

(*l*) *Cahn* v. *Pockett's Bristol Channel Co.*, [1899] 1 Q. B. 643, at p. 659. As to larceny by a trick, see *ante*, notes to s. 23 of the Sale of Goods Act.

seller's consent has since been withdrawn (*m*). To come within this section, the disposition must be one which would be within the ordinary course of a mercantile agent's business (*m*).

The first seller must probably be a person who is able to give a good title to the first buyer. A thief cannot be " the seller " within the meaning of this section nor can a person who has bought in good faith from a thief (*n*). The word " seller " would seem to be used here in contrast with the word " owner " later in the section because other persons than the owner are often able to confer a good title (*o*).

Delivery or transfer.—The delivery must be under a sale, pledge or other disposition or, in the case of s. 9 of the Factors Act, but not under s. 25 of the Sale of Goods Act, under any agreement for such sale, pledge or other disposition (*p*).

As if the person . . . were a mercantile agent in possession.—It would seem from this limitation that the buyer in possession will only give a good title if he disposes of the goods in a way which would be in the ordinary course of a mercantile agent's business (*q*), so as to satisfy the requirements of s. 2, *ante*. The buyer cannot therefore give a good title by making a gift of the goods, but it is not clear how literally this requirement is to be interpreted in the case of a private buyer—for example if a sale from a private house is excluded (*r*).

10. Effect of transfer of documents on vendor's lien or right of stoppage in transitu.—Where a document of title to goods has been lawfully transferred to a person as a buyer or owner of the goods, and that person transfers the document to a person who takes the document in good faith and for valuable consideration, the last-mentioned transfer shall have the same effect for defeating any vendor's lien or right of stoppage in transitu as the transfer of a bill of lading has for defeating the right of stoppage in transitu.

(*m*) *Newtons of Wembley, Ltd.* v. *Williams*, [1964] 3 All E. R. 532; [1965] 1 Q. B. 560, C. A.; *ante*, illustration 2 to s. 2 (2).

(*n*) *Brandon* v. *Leckie* (1972), 29 D. L. R. (3d) 633. There are several English cases in which a thief or a hirer sold goods and the buyer has subsequently resold them, but in no case has the ultimate purchaser sought to rely on this section. See also *Elwin* v. *O'Regan and Maxwell*, [1971] N. Z. L. R. 1124.

(*o*) See examples at p. 162, note (*d*), *ante*.

(*p*) See the comment to s. 8 of this Act, *ante*.

(*q*) *Newtons of Wembley, Ltd.* v. *Williams*, [1964] 3 All E. R. 532; [1965] 1 Q. B. 560, C. A.

(*r*) See further *Benjamin's Sale of Goods*, para. 538.

Definitions.—For " goods," " document of title " and " person," see s. 1 (3), (4) and (5), *ante*, respectively.

Cross-references.—*Cf.* Sale of Goods Act, 1893, s. 47, *ante* (effect of sub-sale or pledge by buyer).

History.—This section, which is now reproduced and developed by s. 47 of the Sale of Goods Act, was substituted for s. 5 of the Factors Act, 1877. It applies to all the documents of title mentioned in s. 1, *ante*, the common law rules relating to the effect of the transfer of a bill of lading on the seller's right of lien or stoppage *in transitu*, as to which see *ante*, notes to s. 47 of the Sale of Goods Act.

COMMENT

To some extent this section covers the same ground as the preceding section. But s. 9 requires that the transferee shall have no notice of the seller's lien or other rights, because it applies to cases where the buyer has obtained the goods or documents under a contract voidable on the ground of fraud. The present section omits the requirement as to absence of notice. The mere fact that the price is unpaid does not make it a fraud to transfer the goods or documents so as to defeat the seller's lien or right of stoppage *in transitu* (*s*).

Supplemental

11. Mode of transferring documents.—For the purposes of this Act, the transfer of a document may be by endorsement, or, where the document is by custom or by its express terms transferable by delivery, or makes the goods deliverable to the bearer, then by delivery.

Definitions.—For " goods " and " document of title," see s. 1 (3) and (4), *ante*, respectively.

History.—This section is taken from words used in s. 5 of the Factors Act, 1877, which are now generalised by being put into a separate section.

12. Saving for rights of true owner.—(1) Nothing in this Act shall authorise an agent to exceed or depart from his authority as between himself and his principal, or exempt him from any liability, civil or criminal, for so doing.

(*s*) *Cuming* v. *Brown* (1808) 9 East. 506 ; *Murdoch* v. *Wood*, [1921] W. N. 299.

Definition.—For " mercantile agent," see s. 1 (1), *ante.*

(2) Nothing in this Act shall prevent the owner of goods from recovering the goods from an agent or his trustee in bankruptcy at any time before the sale or pledge thereof, or shall prevent the owner of goods pledged by an agent from having the right to redeem the goods at any time before the sale thereof, on satisfying the claim for which the goods were pledged, and paying to the agent, if by him required, any money in respect of which the agent would by law be entitled to retain the goods or the documents of title thereto, or any of them, by way of lien as against the owner, or from recovering from any person with whom the goods have been pledged any balance of money remaining in his hands as the produce of the sale of the goods after deducting the amount of his lien (*t*).

Definitions.—For " mercantile agent," " goods," " document of title," " pledge " and " person," see s. 1 (1), (3), (4), (5) and (6), *ante.*

COMMENT

As a general rule, goods or documents of title, held by an agent for his principal, are considered as trust property, and do not pass to the agent's trustee in bankruptcy, though in some cases the reputed ownership provisions might apply: see Bankruptcy Act, 1914 s. 38 (*u*). Whether the proceeds of sale of goods entrusted to a factor and intermixed with his own moneys pass to his trustee in bankruptcy, or whether the principal can trace them as trust moneys will, it seems, depend on whether the relationship was a fiduciary one or merely one of debtor and creditor in account. Although it has been said that in the general course of business merchants voluntarily become the creditors, and not the *cestuis que trustent,* of their factors (*v*), each case must be decided on its own facts and a fiduciary relationship will be readily inferred (*w*). See *Williams on Bankruptcy* (18th ed.), pp. 300 *et seq.*

(*t*) See the effect of a similar section in the British Columbian Sale of Goods Act discussed in *Thoresen* v. *Capital Credit Corporation, Ltd. and Active Bailiff* (1962), 37 D. L. R. (2d.) 317, and see the comment to s. 2 (1), *ante.*

(*u*) 3 Halsbury's Statutes (3rd Edn.) 87. But see the limitation placed upon the operation of this section by s. 53 of the Hire-Purchase Act, 1965; 30 Halsbury's Statutes (3rd Edn.) 105, and by the new s. 38A of the Bankruptcy Act, 1914, introduced with effect from a date to be appointed by s. 192 and Sch. 4 to the Consumer Credit Act, 1974.

(*v*) *Pennell* v. *Deffell* (1853), 23 L. J. Ch. 115 ; see also *Henry* v. *Hammond,* [1913] 2 K. B. 515.

(*w*) *Re Hallet's Estate Knatchball* v. *Hallett* (1879), 13 Ch. D. 696, C. A.; *Ex p. Cooke, re Strachan* (1876), 4 Ch. D. 123; *Re Cotton* (1913), 108 L. T. 310.

(3) Nothing in this Act shall prevent the owner of goods sold by an agent from recovering from the buyer the price agreed to be paid for the same, or any part of that price, subject to any right of set-off on the part of the buyer against the agent (*x*).

Definitions.—For " mercantile agent " and " goods," see s. 1 (1) and (3), *ante*.

13. Saving for common law powers of agent.—The provisions of this Act shall be construed in amplification and not in derogation of the powers exerciseable by an agent independently of this Act.

Definition.—For " mercantile agent," see s. 1 (1), *ante*.

COMMENT

This section is new. It recognises what the judges have frequently pointed out, namely that the Factors Acts are partly declaratory and partly enacting (*a*). In dealing with the exceptions to the general rule, *nemo dat quod non habet*, WILLES, J., observes :

" A third case in which a man may convey a better title to goods than he himself had . . . is where an agent, who carries on a public business, deals with the goods in the ordinary course of it, though he has received secret instructions from his principal to deal with them contrary to the ordinary course of that trade. In that case he has . . . an ostensible authority to deal in such a way with the goods as agents ordinarily deal with them, and if he deals with them in the ordinary way of the trade he binds his principal " (*b*).

[**14. Repeals.**—*Repealed.*]

This section and the schedule, having done their work, are repealed by the Statute Law Revision Act, 1908. The Acts repealed by s. 14 were the Factors Act, 1823; the Factors Act, 1825; the Factors Act, 1842 and the Factors Act, 1877.

[**15. Commencement.**—*Repealed.*]

(*x*) As to the buyer's rights of set-off against an agent with whom he dealt, under the belief that he was a principal, see *Kaltenbach* v. *Lewis* (1885), 10 App. Cas. 617 ; *Cooke* v. *Eshelby* (1887), 12 App. Cas. 271.

(*a*) See *Cole* v. *North Western Bank* (1875), L. R. 10 C. P. 354, at pp. 360 *et seq.*

(*b*) *Fuentes* v. *Montis* (1868), L. R. 3 C. P. 268, at p. 277 ; *cf. Johnson* v. *Credit Lyonnais* (1877), 3 C. P. D. 32, at pp. 37–40; *Lloyds and Scottish Finance, Ltd.* v. *Williamson,* [1965] 1 All E. R. 641; [1965] 1 W. L. R. 404, C. A.; and see the comment to s. 21 of the Sale of Goods Act, *ante*.

[16. Extent.—*Repealed.*]

Sections 15 and 16 are repealed by the Statute Law Revision Act, 1908. Section 16 excluded Scotland, but the Factors Act, 1889, was applied to Scotland the next year by the Factors (Scotland) Act, 1890, *post.* It may be noted that the provisions of the Factors Acts are more nearly declaratory of Scottish common law than they were of English common law (*c*). See the subject discussed in Bell's *Principles of the Laws of Scotland* (9th ed.), pp. 824 *et seq.*

17. Short title.—This Act may be cited as the Factors Act, 1889.

Though the Act is called the Factors Act, the word " factor " does not occur in it. It applies to any mercantile agent within the meaning of s. 1. It might have been more appropriate to call it the " Mercantile Agents Act," as has been done in some colonies which have adopted it, but there was convenience in retaining the familiar name (*d*).

(*c*) *Vickers* v. *Hertz* (1871), L. R. 2 Sc. & Div. 113, at p. 119.

(*d*) For definitions of " factor," see *Story on Agency* (9th ed. revised), § 33, and *Ex p. Dixon* (1876), 4 Ch. D. 133, at p. 137.

THE FACTORS (SCOTLAND) ACT, 1890

(53 & 54 Vict. c. 40)

An Act to extend the Provisions of the Factors Act, 1889, to Scotland.

[14th August 1890.]

1. Application of 52 & 53 Vict. c. 45 to Scotland.—Subject to the following provisions, the Factors Act, 1889, shall apply to Scotland :—

(1) The expression " lien " shall mean and include right of retention ; the expression " vendor's lien " shall mean and include any right of retention competent to the original owner or vendor ; and the expression " set-off " shall mean and include compensation.

(2) In the application of section five of the recited Act, a sale, pledge, or other disposition of goods shall not be valid unless made for valuable consideration.

Comment

See s. 5 of the Factors Act, 1889, *ante*, and Green's *Encyclopædia of Scots Law*, tit. " Factors Acts."

2. Short title.—This Act may be cited as the Factors (Scotland) Act, 1890.

APPENDIX I—STATUTES

THE BILLS OF LADING ACT, 1855

(18 & 19 VICT., c. 111)

An Act to amend the Law relating to Bills of Lading

[4th August 1855.]

Whereas, by the custom of merchants, a bill of lading of goods being transferable by endorsement, the property in the goods may thereby pass to the endorsee, but nevertheless all rights in respect of the contract contained in the bill of lading continue in the original shipper or owner; and it is expedient that such rights should pass with the property: And whereas it frequently happens that the goods in respect of which bills of lading purport to be signed have not been laden on board, and it is proper that such bills of lading in the hands of a bonâ fide holder for value should not be questioned by the master or other person signing the same on the ground of the goods not having been laden as aforesaid:

1. Consignees, and endorsees of bills of lading empowered to sue.—Every consignee of goods named in a bill of lading, and every endorsee of a bill of lading, to whom the property in the goods therein mentioned shall pass upon or by reason of such consignment or endorsement, shall have transferred to and vested in him all rights of suit, and be subject to the same liabilities in respect of such goods as if the contract contained in the bill of lading had been made with himself.

COMMENT

" A bill of lading," says Lord BLACKBURN,

> " is a writing signed on behalf of the owner of the ship in which goods are embarked, acknowledging the receipt of the goods, and undertaking to deliver them at the end of the voyage, subject to such conditions as may be mentioned in the bill of lading " (*a*).

A bill of lading running " Received for shipment " instead of " Shipped " is, it seems, within the Act (*b*).

A " through bill of lading " is a bill of lading " made for the carriage of goods from one place to another by several shipowners or railway

---

(*a*) *Blackburn on Sale* (1st ed.), p. 275; see also *Sewell* v. *Burdick* (1884) 10 App. Cas. 74, at p. 105.

(*b*) *The Marlborough Hill*, [1921] 1 A. C. 444, P. C. But such a bill of lading would not comply with the requirements of a c.i.f. contract except in trades where it is shown to be usual: see p. 39, *ante*.

companies" (*c*). It seems doubtful how far the Act applies to these documents, which are of modern origin (*d*).

At common law the property in the goods could be transferred by the endorsement of the bill of lading, but the contract created by the bill of lading could not, therefore the endorsee could not sue on the contract in his own name (*e*). The Act confers this right while confirming the common law rights.

The property must pass "upon or by reason of the consignment or endorsement" (*f*). Where, therefore, a number of bills of lading are issued in respect of unseparated bulk cargoes, and are endorsed to different c.i.f. purchasers, so that no property (legal or equitable) passes to the buyer at least until the goods are appropriated to such contract and ascertained (*g*), this section would seem not to apply (*h*). A contractual relationship may however arise between the bill of lading holder and the ship by implication from presentation of the bill on delivery by the ship (*i*).

2. Saving as to stoppage in transitu, and claims for freight, &c.—Nothing herein contained shall prejudice or affect any right of stoppage in transitu, or any right to claim freight against the original shipper or owner, or any liability of the consignee or endorsee by reason or in consequence of his being such consignee or endorsee, or of his receipt of the goods by reason or in consequence of such consignment or endorsement.

COMMENT

As to non-liability of pledgee of bill of lading for freight, see *Sewell* v. *Burdick* (*k*). As to pledge of bill of lading and conversion before plaintiff's title accrued, see *Bristol Bank* v. *Midland Railway* (*l*).

3. Bill of lading in hands of consignee, etc., conclusive evidence of shipment as against master, etc.—Every bill of

---

(*c*) *Scrutton on Charter Parties, etc.* (18th ed.), art. 177. *Cf. Hansson* v. *Hamel and Horley*, [1922] 2 A. C. 36 (c.i.f. contract).

(*d*) See *Scrutton, supra,* and *British Shipping Laws,* Vol. 2, §§ 200–203, where their effect is fully discussed.

(*e*) *Thompson* v. *Dominy* (1845), 14 M. & W. 403.

(*f*) If property is reserved to the seller until delivery ex installation at the port of discharge, the section does not apply; *Gardano and Giamperi* v. *Greek Petroleum, George Mamidakis & Co.*, [1961] 3 All E. R. 919; [1962] 1 W. L. R. 40.

(*g*) See the comment to s. 16 of the Sale of Goods Act, *ante*. See also p. 37, *ante,* concerning c.i.f. contracts.

(*h*) It is clear that no property passes upon consignment or endorsement, and it would seem to be stretching the meaning of the words to say that it passes by reason of the consignment or endorsement.

(*i*) *Brandt* v. *Liverpool, Brazil and River Plate Steam Navigation Co.*, [1924] 1 K. B. 575.

(*k*) (1884), 10. App. Cas. 74.

(*l*) [1891] 2 Q. B. 653, C. A., commented upon in *Margarine Union G.m.b.H.* v. *Cambay Prince S.S. Co.*, [1967] 3 All E. R. 775; [1969] 1 Q. B. 219.

lading in the hands of a consignee or endorsee for valuable consideration representing goods to have been shipped on board a vessel, shall be conclusive evidence of such shipment (*m*) as against the master or other person signing the same, notwithstanding that such goods or some part thereof may not have been so shipped, unless such holder of the bill of lading shall have had actual notice at the time of receiving the same that the goods had not been in fact laden on board (*n*): Provided, that the master or other person so signing may exonerate himself in respect of such misrepresentation by showing that it was caused without any default on his part, and wholly by the fraud of the shipper, or of the holder, or some person under whom the holder claims.

The section is of limited effect only. Where a bill is signed by an agent the principal is not an " other person signing the same ", and so is not affected by the provisions of the section (*o*). Moreover, the bill is conclusive evidence only of shipment and by itself imposes no liability upon the signatory for non-delivery (*o*).

General Note.—"A cargo at sea," says BOWEN, L.J.,

> " while in the hands of the carrier, is necessarily incapable of physical delivery. During this period of transit and voyage, the bill of lading by the law merchant is universally recognised as its symbol, and the indorsement and delivery of the bill of lading operates as a symbolical delivery of the cargo. Property in the goods passes by such indorsement and delivery of the bill of lading whenever it is the intention of the parties that the property should pass, just as under similar circumstances the property would pass by an actual delivery of the goods " (*p*).

Bills of lading are commonly made out in several parts, one part being usually retained by the ship, the others being handed to the shipper. This practice has often given rise to frauds. The subject of bills of lading is fully dealt with in *Scrutton on Charter Parties, etc.* (18th ed.) and *Carver on Carriage by Sea, British Shipping Laws,* Vol. 2 and 3; but the following salient points may be noted:

(1) So long as the contract of carriage is not discharged, the bill of lading remains a document of title, by endorsement and delivery of which the property in the goods can be transferred, even though the goods have

---

(*m*) The estoppel relates only to the essentials of the goods as described in the bill (*i.e.*, those denoting or effecting the nature, quality or commercial value of the goods: *Parsons* v. *New Zealand Shipping Co.*, [1901] 1 Q. B. 548).

(*n*) But the shipowner is not liable if the master signs a bill of lading for goods not received on board: *Grant* v. *Norway* (1851), 10 C. B. 665. The rule is unsatisfactory and may yet be open to review—see *Scrutton on Charterparties, etc.* (18th ed.), pp. 112–113.

(*o*) *V/O Rasnoimport* v. *Guthrie & Co., Ltd.*, [1966] 1 Lloyd's Rep. 1, where the signatory to the bill signed as agent only and could not be regarded as contracting to carry or accept charge of the goods.

(*p*) *Sanders* v. *Maclean* (1883), 11 Q. B. D. 327, at p. 341 ; *cf. The Prinz Adalbert,* [1917] A. C. 586, at p. 589, *per* Lord SUMNER.

been landed (*q*), or wrongfully delivered to a third party without production of the bill (*r*).

(2) When two or more parts of a bill of lading are transferred to two or more different *bonâ fide* transferees for value, the property in the goods passes to the transferee who is first in point of time (*q*).

(3) Nevertheless, the person who has the custody of the goods may safely deliver them to the person who first presents the bill of lading (or a part thereof) to him, provided he acts in good faith and without notice of any prior claim (*s*).

(4) A contract to deliver a bill of lading is complied with by delivering one part, though the others are not accounted for (*t*).

(5) Except for the purposes of the Factors Act and of defeating the right of stoppage *in transitu*, the transferee of a bill of lading acquires no better title to the goods represented thereby than the transferor had. In this respect a bill of lading differs from a bill of exchange, or rather it resembles an overdue bill of exchange, which can only be negotiated subject to all equities attaching to it (*u*). As to the effect of the transfer of a bill of lading on the right of stoppage *in transitu*, see *ante*, notes to s. 47 of the Sale of Goods Act. See also the Factors Act, 1889, *ante*.

(6) Stipulations in a bill of lading must be construed according to the proper law of the contract; see note " conflict of laws," p. 47, *ante*.

(7) When a bill of exchange, with bill of lading attached, is presented for acceptance or payment, the person who in good faith presents it is not responsible for the authenticity of the bill of lading (*a*).

Foreign systems.—As to the United States and the Harter Act, see *Scrutton, loc. cit.*, App. V., and as to Australia, Canada, and New Zealand, *ibid.*, App. VI. The German Commercial Code, arts. 642–663, minutely regulates Bills of Lading.

THE BILLS OF SALE ACT, 1878 (*b*)
(41 & 42 VICT., c. 31)

An Act to consolidate and amend the Law for preventing Frauds upon Creditors by secret Bills of Sale of Personal Chattels.

[22nd July 1878.]

* * * * *

4. Interpretation of terms.—In this Act the following words and expressions shall have the meanings in this section assigned

(*q*) *Barclays Bank, Ltd.* v. *Customs and Excise Commissioners*, [1963] 1 Lloyd's Rep. 81; *Barber* v. *Meyerstein* (1870), L. R. 4 H. L. 317.

(*r*) See *Short* v. *Simpson* (1866), L. R. 1 C. P. 248, which appears to confirm this despite the doubt expressed in the headnote. See also *Bristol and West of England Bank* v. *Midland Railway*, [1891] 2 Q. B. 653, C. A., as explained in *Margarine Union G.m.b.H.* v. *Cambay Prince S.S. Co.*, [1967] 3 All E. R. 775; [1969] 1 Q. B. 219, for a case where a fresh pledge passed the right to possession after a misdelivery.

(*s*) *Glyn, Mills & Co.* v. *East & West India Dock* (1882), 7 App. Cas. 591.

(*t*) *Sanders* v. *Maclean* (1883), 11 Q. B. D. 327, C. A.

(*u*) *Gurney* v. *Behrend* (1854), 3 E. & B. 622. As to fraud, however, see *The Argentina* (1867), L. R. 1 A. & E. 370.

(*a*) *Guaranty Trust Co. of New York* v. *Hannay & Co.*, [1918] 2 K. B. 623, C. A.

(*b*) For the rest of this Act, see 3 Halsbury's Statutes (3rd Edn.) 245.

to them respectively, unless there be something in the subject or context repugnant to such construction; (that is to say),

The expression " bill of sale " shall include bills of sale, assignments, transfers, declarations of trust without transfer, inventories of goods with receipt thereto attached, or receipts for purchase-moneys of goods, and other assurances of personal chattels, and also powers of attorney, authorities, or licences to take possession of personal chattels as security for any debt, and also any agreement, whether intended or not to be followed by the execution of any other instrument, by which a right in equity to any personal chattels, or to any charge or security thereon, shall be conferred, but shall not include the following documents; that is to say, assignments for the benefit of the creditors of the person making or giving the same, marriage settlements, transfers or assignments of any ship or vessel or any share thereof, transfers of goods in the ordinary course of business of any trade or calling (*c*), bills of sale of goods in foreign parts or at sea, bills of lading, India warrants, warehouse-keepers' certificates, warrants or orders for the delivery of goods, or any other documents used in the ordinary course of business as proof of the possession or control of goods, or authorising or purporting to authorise, either by indorsement or by delivery, the possession of such document to transfer or receive goods thereby represented:

The expression " personal chattels " shall mean goods, furniture, and other articles capable of complete transfer by delivery, and (when separately assigned or charged) fixtures and growing crops, but shall not include chattel interests in real estate, nor fixtures (except trade machinery as hereinafter defined), when assigned together with a freehold or leasehold interest in any land or building to which they are affixed, nor growing crops when assigned together with any interest in the land on which they grow, nor shares or interest in the stock, funds, or securities of any government, or in the capital or property of incorporated or joint stock companies, nor choses in action, nor any stock or produce upon any farm or lands which by virtue of any covenant or agreement or of the custom of the country ought not to be removed from any farm where the same are at the time of making or giving of such bill of sale:

Personal chattels shall be deemed to be in the " apparent possession " of the person making or giving a bill of sale, so long as they remain or are in or upon any house, mill, warehouse, building, works, yard, land, or other premises occupied by him, or are used

(*c*) *Stephenson* v. *Thompson*, [1924] 2 K. B. 240, C. A. (sale of growing crops).

and enjoyed by him in any place whatsoever, notwithstanding that formal possession thereof may have been taken by or given to any other person (*d*):

" Prescribed " means prescribed by rules made under the provisions of this Act.

COMMENT

This Act has been amended by the Bills of Sale Act (1878) Amendment Act, 1882 (*e*), but the Act of 1882 only relates to bills of sale by way of security, and does not affect sales within the meaning of the Sale of Goods Act, 1893 ; see s. 61 (4) of that Act, *ante*. The Bills of Sale Acts, 1890 and 1891 (*f*), merely exempt certain mercantile letters of hypothecation from the definition of " bill of sale." These Acts apply only to England (*g*). There are corresponding enactments for Ireland, but there are no corresponding enactments for Scotland (*h*).

The Bills of Sale Acts strike at documents, and not at the transactions themselves (*i*). When the seller of goods remains in possession of them, and the buyer has to base his title or right to possession on some document which comes within the definition of a bill of sale, the document must be registered in accordance with the Act of 1878. If it be not so registered, the contract, though valid as between the parties, is void as against the seller's execution creditors, trustee in bankruptcy, or assignee for the benefit of creditors. See *Reed on the Bills of Sale Acts* (14th ed.), pp. 141 *et seq.*, where all the authorities are exhaustively reviewed. It may be noted that where the seller remains in possession, an entry of the sale by the auctioneer, or a note of the contract drawn up by the sheriff who has sold privately, constitutes a bill of sale (*k*). A delivery order for furniture is not a bill of sale (*l*), nor is an unregistered transfer of a ship or vessel (*m*), nor a letter of lien to bankers (*n*).

(*d*) As to the meaning of apparent possession, see *Koppel* v. *Koppel*, [1966] 2 All E. R. 187; [1966] 1 W. L. R. 802, C. A., reviewing the earlier cases. As to distinction between apparent possession and " possession, order, and disposition " in the reputed ownership provision in bankruptcy, see *Ancona* v. *Rogers* (1876), 1 Ex. D. 285, at p. 291, C. A.

(*e*) 3 Halsbury's Statutes (3rd Edn.) 261.

(*f*) 3 Halsbury's Statutes (3rd Edn.) 270, 271.

(*g*) *Cf. Coote* v. *Jecks* (1872), L. R. 13 Eq. 597 (English bill of sale cannot affect property in Scotland).

(*h*) Green's *Encyclopædia of Scots Law.* tit. " Bill of Sale."

(*i*) *North Central Wagon Co.* v. *Manchester, Sheffield and Lincolnshire Rail Co.* (1887), 35 Ch. D. 191, at p. 207.

(*k*) *Re Roberts* (1887), 36 Ch. D. 196 (auctioneer) ; *Ex p. Blandford* (1893), 10 Morr. 231 (sheriff).

(*l*) *Grigg* v. *National Guardian Assurance Co.*, [1891] 3 Ch. 206.

(*m*) *Gapp* v. *Bond* (1887), 19 Q. B. D. 200 (dumb-barge which did not require registration under Merchant Shipping Act).

(*n*) *Ex p. Carter*, [1905] 2 K. B. 772.

THE FINANCE ACT, 1901

(1 Edw. 7, c. 7)

An Act to grant certain duties of Customs and Inland Revenue, to alter other duties, and to amend the Law relating to Customs and Inland Revenue . . .

[26th July 1901.]

* * * * *

10. Addition or deduction of new or altered duties in the case of contract.—(1) Where any new customs import duty or new excise duty is imposed, or where any customs import duty or excise duty is increased, and any goods in respect of which the duty is payable are delivered after the day on which the new or increased duty takes effect in pursuance of a contract made before that day, the seller of the goods may, in the absence of agreement to the contrary, recover, as an addition to the contract price, a sum equal to any amount paid by him in respect of the goods on account of the new duty or the increase of duty, as the case may be.

(2) Where any customs import duty or excise duty is repealed or decreased, and any goods affected by the duty are delivered after the day on which the duty ceases or the decrease in the duty takes effect in pursuance of a contract made before that day, the purchaser of the goods, in the absence of agreement to the contrary, may, if the seller of the goods has had in respect of those goods the benefit of the repeal or decrease of the duty, deduct from the contract price a sum equal to the amount of the duty or decrease of duty, as the case may be.

(3) Where any addition to, or deduction from, the contract price may be made under this section on account of any new or repealed duty, such sum as may be agreed upon or in default of agreement determined by the Commissioners of Customs in the case of a customs duty, and by the Commissioners of Inland Revenue in the case of an excise duty, as representing in the case of a new duty any new expenses incurred, and in the case of a repealed duty any expenses saved, may be included in the addition to or deduction from the contract price, and may be recovered or deducted accordingly.

* * * * *

Comment

This section is amended by s. 7 of the Finance Act, 1902, which is as follows :—

"Section ten of the Finance Act, 1901, applies although the goods have undergone a process of manufacture or preparation, or have become a part or ingredient of other goods."

Under s. 4 of the Finance Act, 1908, now replaced by s. 1 (1) of the Customs and Excise Act, 1952, excise was transferred from the Inland Revenue to the Customs, and the Commissioners are now known as the Commissioners of Customs and Excise.

As to the construction of s. 10 of the Act of 1901, see generally *Corn Products Co.* v. *Fry* (*o*). Even though an action for the price of the goods is pending when the new duties are imposed, the seller can add the new duty to the price (*p*). It has been held that the words " duty paid " do not constitute an agreement to the contrary within the above section so as to disentitle the seller to recover the amount of an increase (*q*), but more recently an agreement to sell " free house London " was held to be an agreement to the contrary, and the foreign seller was held liable to pay duty imposed by the Abnormal Importations (Customs Duties) Act, 1931 (*r*). When a temporary customs duty was imposed under s. 3 of the Finance (No. 2) Act, 1964 these problems were avoided by s. 3 (8) which provided that for the purposes of s. 10 of the Act of 1901:

> " any terms in a contract expressly or impliedly making a seller liable for all customs duties (or all customs duties with exceptions not affecting duty under this section) shall not be deemed to amount to an agreement to the contrary (that is to say to an agreement that the said section 10 shall not apply)."

As to the adjustment of rights between seller and buyer on the commencement or variation of rates of charge of value added tax, see s. 42 of the Finance Act, 1972, *post.*

THE LAW REFORM (MISCELLANEOUS PROVISIONS) ACT, 1934

(24 & 25 Geo. 5, c. 41)

An Act to amend the law as to the effect of death in relation to causes of action and as to the awarding of interest in civil proceedings.

[25th July 1934.]

* * * * *

3. Power of courts of record to award interest on debts and damages.—(1) In any proceedings tried in any court of record for the recovery of any debt or damages, the court may, if it thinks fit, order that there shall be included in the sum for which judgment is given interest at such rate as it thinks fit on the whole or any part of the debt or damages for the whole or any part of the period between the date when the cause of action arose and the date of the judgment:

(*o*) [1917] W. N. 224.

(*p*) *Conway Brothers and Savage* v. *Mulheirn & Co., Ltd.* (1901), 17 T. L. R. 730. *Cf. Newbridge Rhondda Brewery Co.* v. *Evans* (1902), 86 L. T. 453 (a case of a conditional agreement for sale).

(*q*) *American Commerce Co., Ltd.* v. *Boehm* (*Frederick*), *Ltd.* (1919), 35 T. L. R. 224.

(*r*) *Lanificio Di Manerbio S.A.* v. *Gold* (1932), 76 Sol. Jo. 289.

Provided that nothing in this section—

(a) shall authorise the giving of interest upon interest (*s*); or

(b) shall apply in relation to any debt upon which interest is payable as of right whether by virtue of any enactment or otherwise; or

(c) shall affect the damages recoverable for the dishonour of a bill of exchange.

History.—At common law interest was not payable, even on a debt or other liquidated claim, *e.g.*, for the price of goods sold and delivered, unless there was a contract, express or implied, to pay it, though interest was apparently always payable when a bill of exchange was dishonoured (see now s. 57 of the Bills of Exchange Act, 1882 (*t*)).

Section 28 of the Civil Procedure Act, 1833 (*u*), allowed the jury to award interest as damages, if they thought fit, where a debt or sum of money was due by virtue of some written instrument at a certain time, or, if payable otherwise, then from the time when demand for payment had been made in writing, giving notice that interest would be claimed. Section 29 of the same Act (*v*) similarly allowed interest to be awarded as damages in actions of trover and trespass *de bonis asportatis* and in actions on policies of assurance.

Under neither of these two sections could there be any award of interest in an action for breach of a contract of sale where the claim was for damages for non-delivery, late delivery, non-acceptance, breach of warranty, or other unliquidated claim. And similarly, in actions for the price of goods, interest could only be claimed as of right if there was an agreement to pay it, though such an agreement might be inferred from the course of dealing between the parties or from trade usage (*a*). It could be awarded in the discretion of the Court under s. 28 of the Civil Procedure Act, 1833, *supra*, but only if the conditions specified by that section were satisfied, which in practice was not commonly the case.

COMMENT

The present section leaves untouched those cases where interest is payable as of right by virtue of contract or otherwise, and does not affect the measure of damages, which includes interest, on the dishonour of a bill of exchange. In all other actions for the recovery of debt or damages it gives the Court a discretion whether interest should be awarded or not; it is for the Court to specify the rate of interest and whether it is to be payable on the whole or part only of the sum in question, and the period during which, between the date the cause of action arose and judgment, the interest is to be allowed.

Interest is not awarded on the basis of anybody's fault but on the simple commercial basis that if the money had been paid at the appropriate commercial time, the other party would have had the use of it.

(*s*) But see *Bushwall Properties, Ltd.* v. *Vortex Properties, Ltd.*, [1975] 2 All E. R. 214. Court may give interest on damages awarded to cover interest lost by plaintiff through breach of contract.

(*t*) 3 Halsbury's Statutes (3rd Edn.) 220.

(*u*) Repealed by sub-s. (2) of this section.

(*v*) Also repealed by sub-s. (2) of this section.

(*a*) *Re Anglesey (Marquis), Willmot* v. *Gardner*, [1901] 2 Ch. 548, C. A.

It is a well-recognised rule of practice that *primâ facie* the losing party should be ordered to pay interest at a reasonable rate running from the date when the amount due should reasonably have been paid (*b*).

The onus is on the losing party to show sufficient reason why the usual practice should not apply in any particular case (*c*). Mere delay is not normally a ground for modifying the general rule as to interest, and an unexplained delay cannot be a basis for refusing to award interest (*d*). But if the plaintiff has been indemnified by an insurance company, he will not normally be awarded interest on the money when he has not been kept out of it (*e*).

It has been said (*f*) that the approach of the Commercial Court has been to compensate successful parties by awarding interest at a rate which broadly represents the rate at which they would have had to borrow the amount recovered over the period for which interest is awarded. But in practice, while the courts have taken account of the market rate, a borrower would have been fortunate to have been able to borrow at the rates awarded.

A rate of 1 per cent. over Bank Rate was commonly awarded (*g*). This principle was reviewed because of the replacement of Bank Rate by minimum lending rate, and what were then more realistic interest rates on judgment debts and on money in court placed on short term investment (*h*). Interest is now commonly awarded by reference to minimum lending rate or the base rate of the major banks.

By virtue of the new sub-section (1A) of the 1934 Act, introduced by s. 22 of the Administration of Justice Act, 1969, where damages are awarded in respect of personal injuries, interest will be awarded on those damages unless the defendant can show that there are special reasons why such interest should not be awarded.

(*b*) *Panchaud Freres S.A.* v. *R. Pagnan and Fratelli*, [1974] 1 Lloyd's Rep. 394, *per* KERR, J., at p. 409; *Harbutt's Plasticine, Ltd.* v. *Wayne Tank and Pump Co., Ltd.*, [1970] 1 All E. R. 225; [1970] 1 Q. B. 447, C. A.; *General Tire & Rubber Co.* v. *Firestone Tyre & Rubber Co., Ltd.*, [1975] 1 W. L. R. 819, at p. 836, H. L. See also *Kemp* v. *Tolland*, [1956] 2 Lloyd's Rep. 681.

(*c*) *Cremer* v. *General Carriers S.A.*, [1974] 1 All E. R. 1; [1974] 1 W. L. R. 341.

(*d*) *Panchaud Freres S.A.* v. *R. Pagnan and Fratelli*, [1974] 1 Lloyd's Rep. 394, *per* KERR, J., at p. 409; affirmed in the Court of Appeal, at pp. 411, 414. For a case where interest was reduced, see *United Fresh Meat Co.* v. *Charterhouse Cold Storage*, [1974] 2 Lloyd's Rep. 286.

(*e*) *Harbutt's Plasticine, Ltd.* v. *Wayne Tank and Pump Co., Ltd.*, *supra*.

(*f*) *Cremer* v. *General Carriers S.A.*, *supra*.

(*g*) *F.M.C. (Meat), Ltd.* v. *Fairfield Cold Stores, Ltd.*, [1971] 2 Lloyds' Rep. 221.

(*h*) *Cremer* v. *General Carriers S.A.*, *supra*. Judgment was delivered on 5th July, 1973, when interest on judgment debts was 7½ per cent. and interest on money in court of short-term investments was 8 per cent. The rate awarded by KERR, J., of 7½ per cent. was very close to the rate of 7·9 per cent. claimed by the plaintiff, whose calculations were based on average Bank Rate and Bank of England minimum lending rate. The subsequent rise in interest rates in general have not been reflected in interest rates on judgment debts or on money in court on short-term investment and they do not therefore at present seem to be a reliable basis for calculating the rate of interest to be awarded. In *Wallersteiner* v. *Moir* (*No. 2*), [1975] 1 All E. R. 849; [1975] 2 W. L. R. 389, the Court of Appeal awarded interest on general equitable principles at 1 per cent. over minimum lending rate, compounded with yearly rests, a rate which, both Lord DENNING, M.R. and SCARMAN, L.J. held, represented the value of the use of the sum of money in issue.

There is no obligation to claim interest under the Act on the indorsement of claim, the writ of summons, or in the pleadings (*i*). The claim for interest is not a cause of action and the defendant cannot and should not make a payment into Court in respect of it. If a case proceeds to trial after a payment into Court, any amount awarded at the trial on account of interest will be disregarded in determining the incidence of costs after the payment into Court (*k*). Where money is paid into Court, and the Court makes an order for payment out under R.S.C., O. 22, r. 5, in satisfaction of the cause of action, that order is not a judgment within s. 3 (1) of the 1934 Act, so that the plaintiff will not be entitled to interest on the sum paid in (*l*). It is nevertheless submitted that it is the better practice to claim interest. It is clearly desirable that the defendant should know what exactly is claimed, and if the plaintiff causes embarrassment by not telling him, it is conceivable that it might in certain circumstances influence the exercise by the Court of its discretion. It is therefore suggested that in all appropriate cases, such as, for example, claims by the buyer for the return of the price where no delivery has been made or claims by the seller for the price, etc., a claim for interest should be included in the prayer for relief on both writ and statement of claim. A county court is a court of record (s. 1 (2) of the County Courts Act, 1959); and what is said above applies also to the filling in of the *praecipe* and the settling of particulars of claim in county court actions.

Scotland.—The present section does not apply to Scotland, but s. 49 (3) of the Sale of Goods Act, 1893, *ante*, preserves the right which the seller has by Scottish law to receive interest from the date of tender of the goods pursuant to the contract, or the date of delivery, or the date the price otherwise becomes payable.

* * * * *

LAW REFORM (FRUSTRATED CONTRACTS) ACT, 1943

(6 & 7 Geo. 6, c. 40)

1. Adjustment of rights and liabilities of parties to frustrated contracts.—(1) Where a contract governed by English law

(*i*) *Riches* v. *Westminster Bank, Ltd.*, [1943] 2 All E. R. 725, C. A.; *Jefford* v. *Gee*, [1970] 1 All E. R. 1202; [1970] 2 Q. B. 130, C. A.

(*k*) *Jefford* v. *Gee*, *supra*.

(*l*) *Waite* v. *Redpath Dorman Long, Ltd.*, [1971] 1 All E. R. 513; [1971] 1 Q. B. 294. *Quaere* whether interest can be claimed under the Act if judgment is signed in default of appearance or pleadings, for there would not seem to be any "proceedings tried in any court of record". It is stated in the Supreme Court Practice 1973, Vol. I, para. 14/3–4/19 that interest may not be awarded under this section when judgment is given under R.S.C., O. 14 since the proceedings are not a trial, but that an issue may be directed to be tried as to whether the plaintiff ought to be awarded such interest and if so at what rate and for what period. No authority is cited for this proposition, however, which appears to put an unduly restrictive interpretation on the expression "proceedings tried", and would lead to unnecessary expense in many cases. See further *Wallersteiner* v. *Moir* (*No. 2*), [1975] 1 All E. R. 849, at p. 855; [1975] 2 W. L. R. 389, at p. 393, *per* Lord Denning, M.R.

(*m*) has become impossible of performance or been otherwise frustrated, and the parties thereto have for that reason been discharged from the further performance of the contract, the following provisions of this section shall, subject to the provisions of section two of this Act, have effect in relation thereto.

(2) All sums paid or payable to any party in pursuance of the contract before the time when the parties were so discharged (in this Act referred to as " the time of discharge ") shall, in the case of sums so paid, be recoverable from him as money received by him for the use of the party by whom the sums were paid, and, in the case of sums so payable, cease to be so payable:

Provided that, if the party to whom the sums were so paid or payable incurred expenses before the time of discharge in, or for the purpose of, the performance of the contract, the court may, if it considers it just to do so having regard to all the circumstances of the case, allow him to retain or, as the case may be, recover the whole or any part of the sums so paid or payable, not being an amount in excess of the expenses so incurred.

(3) Where any party to the contract has, by reason of anything done by any other party thereto in, or for the purpose of, the performance of the contract, obtained a valuable benefit (other than a payment of money to which the last foregoing sub-section applies) before the time of discharge, there shall be recoverable from him by the said other party such sum (if any), not exceeding the value of the said benefit to the party obtaining it, as the court considers just, having regard to all the circumstances of the case and, in particular,—

(a) the amount of any expenses incurred before the time of discharge by the benefited party in, or for the purpose of, the performance of the contract, including any sums paid or payable by him to any other party in pursuance of the contract and retained or recoverable by that party under the last foregoing sub-section, and

(b) the effect, in relation to the said benefit, of the circumstances giving rise to the frustration of the contract.

(4) In estimating, for the purposes of the foregoing provisions of this section, the amount of any expenses incurred by any party to the contract, the court may, without prejudice to the generality of the said provisions, include such sum as appears to be reasonable in respect of overhead expenses and in respect of any work or services performed personally by the said party.

(*m*) As to Scottish law, which is unaffected by this Act, see *Cantiare Sau Rocco* v. *Clyde Shipbuilding and Engineering Co., Ltd.*, [1924] A. C. 226.

(5) In considering whether any sum ought to be recovered or retained under the foregoing provisions of this section by any party to the contract, the court shall not take into account any sums which have, by reason of the circumstances giving rise to the frustration of the contract, become payable to that party under any contract of insurance unless there was an obligation to insure imposed by an express term of the frustrated contract or by or under any enactment.

(6) Where any person has assumed obligations under the contract in consideration of the conferring of a benefit by any other party to the contract upon any other person, whether a party to the contract or not, the court may, if in all the circumstances of the case it considers it just to do so, treat for the purposes of sub-section (3) of this section any benefit so conferred as a benefit obtained by the person who has assumed the obligations as aforesaid.

COMMENT

See generally, the note on frustration, p. 19, *ante*.

Sub-section (2).—See Sale of Goods Act, 1893, ss. 31 (2) and 49 (2), *ante*. It will be observed that it is not necessary that the consideration for the payment should have wholly failed, as was necessary at common law (*n*). This provision must be read in conjunction with sub-s. (3).

The proviso is an attempt to modify the hardship suffered by the seller at common law, and pointed out by Lord Simon, L.C. in the *Fibrosa Case* (*o*). It can only be invoked by the seller when the price or part of it was paid or payable under the contract of sale before the date of frustration. See further, sub-s. (4).

Sub-section (3).—This provision is new, but so far as contracts of sale of goods are concerned, it may be unnecessary to have recourse to it: for, if the buyer receives and keeps goods, he must as a general rule pay for them: in some cases he must pay a reasonable price, in others he must pay for them at the contract rate. See, *e.g.*, Sale of Goods Act, 1893, ss. 8 (2), 9 (1), 30 (1) (2) (3), *ante*. The same rule would no doubt apply if a case relating to a contract of sale arose under sub-s. (6).

2. Provision as to application of this Act.—

* * * * *

(2) This Act shall apply to contracts to which the Crown is a party in like manner as to contract between subjects.

(3) Where any contract to which this Act applies contains any provision which, upon the true construction of the contract, is

(*n*) See *Fibrosa S.A.* v. *Fairbairn Lawson Combe Barbour, Ltd.*, [1942] 2 All E. R. 122; [1943] A. C. 32.
(*o*) [1942] 2 All E. R. 122, at pp. 129, 130; [1943] A. C. 32, at p. 49.

intended to have effect in the event of circumstances arising which operate, or would but for the said provision operate, to frustrate the contract, or is intended to have effect whether such circumstances arise or not, the court shall give effect to the said provision and shall only give effect to the foregoing section of this Act to such extent, if any, as appears to the court to be consistent with the said provision.

(4) Where it appears to the court that a part of any contract to which this Act applies can properly be severed from the remainder of the contract, being a part wholly performed before the time of discharge, or so performed except for the payment in respect of that part of the contract of sums which are or can be ascertained under the contract, the court shall treat that part of the contract as if it were a separate contract and had not been frustrated and shall treat the foregoing section of this Act as only applicable to the remainder of that contract.

(5) This Act shall not apply—

(a) to any charterparty, except a time charterparty or a charterparty by way of demise, or to any contract (other than a charterparty) for the carriage of goods by sea; or

(b) to any contract of insurance, save as is provided by sub-section (5) of the foregoing section; or

(c) to any contract to which section seven of the Sale of Goods Act, 1893 (which avoids contracts for the sale of specific goods which perish before the risk has passed to the buyer) applies, or to any other contract for the sale, or for the sale and delivery, of specific goods (*p*), where the contract is frustrated by reason of the fact that the goods have perished.

COMMENT

See generally, the note on frustration, p.19 , *ante*.

Sub-section (3).—Parties to contracts are free to make whatever bargain they wish, provided that it is lawful. A clause, however, dealing with frustration may itself be or become illegal (*q*). It will be for the Court to decide whether, in a case where the contract is abrogated by the outbreak of war, an alien enemy will, on the conclusion of peace, be able to take advantage of the provisions of this Act.

Sub-section (4).—A contract for the sale of goods providing for delivery by instalments and payment for each instalment may be of such a nature that each delivery is equivalent to delivery under a separate contract to be paid for separately, and in such cases the contract is clearly severable, and no difficulty will arise. But even though the contract provides for delivery by instalments, it may still be an entire contract (*i.e.*, the whole of what is to be done on one side is

---

(*p*) For the definition of specific goods, see s. 62 (1) of the Sale of Goods Act, 1893, *ante*.

(*q*) *Ertel Bieber & Co.* v. *Rio Tinto Co.*, [1918] A. C. 260.

the consideration of the whole of what is to be done on the other (*r*), though divisible in performance) and

> " where there is an entire contract to deliver a large quantity of goods consisting of distinct parcels within a specified time, and the seller delivers part, he cannot, before the expiration of that time, bring an action to recover the price of that part delivered, because the purchaser may, if the vendor fail to complete his contract, return the part delivered. But if he retain the part delivered after the seller has failed in performing his contract, the latter may recover the vâlue of the goods which he has so delivered " (*s*).

Even if the buyer under the contract must pay for each instalment as delivered, the contract may still be entire, so that the seller's failure to complete may entitle the buyer to return the instalments received and recover the sums that he has paid ; for the consideration, being entire, fails entirely by failing partially (*t*). How far the buyer's rights in such cases are affected by this sub-section and other provisions of this Act is a question which must await judicial interpretation.

Sub-section (5) (c).—These words are inserted perhaps to make it clear that the Act does not apply to cases where the property in the goods has passed to the buyer, but have perished before delivery to him, or to cases falling within s. 6 of the Sale of Goods Act ; though it may be doubtful whether in either of these cases the contract can properly be said to be frustrated. As to what contracts fall within ss. 6 and 7 of the Sale of Goods Act, see p. 21 and the notes to those sections, *ante*, and also the note to s. 62 (1) of that Act, *ante*.

3. Short title and interpretation.—(1) This Act may be cited as the Law Reform (Frustrated Contracts) Act, 1943.

(2) In this Act the expression " court " means, in relation to any matter, the court or arbitrator by or before whom the matter falls to be determined.

THE EXCHANGE CONTROL ACT, 1947 (*u*)

(10 & 11 Geo. 6, c. 14)

An Act to confer powers, and impose duties and restrictions, in relation to gold, currency, payments, securities, debts, and the import, export, transfer and settlement of property, and for purposes connected with the matters aforesaid. [11th March 1947.]

* * * * *

33. Contracts, legal proceedings, etc.—(1) It shall be an implied condition in any contract that, where, by virtue of this Act, the permission or consent of the Treasury is at the time of the contract required for the performance of any term thereof, that term shall not be performed except in so far as the permission or consent is given or is not required:

(*r*) *Honck* v. *Muller* (1881), 7 Q. B. D. 92, 100.
(*s*) *Oxendale* v. *Wetherell* (1829), 9 B. & C. 386, at p. 387, *per* Parke, B.
(*t*) *Chanter* v *Leese* (1840), 5 M. & W. 698, 702.
(*u*) For the rest of this Act, see 22 Halsbury's Statutes (3rd Edn.) 900 *et seq.*

Provided that this sub-section shall not apply in so far as it is shown to be inconsistent with the intention of the parties that it should apply, whether by reason of their having contemplated the performance of that term in despite of the provisions of this Act or for any other reason.

* * * * *

SCHEDULE IV

* * * * *

4. (1) In any proceedings in a prescribed court and in any arbitration proceedings, a claim for the recovery of any debt shall not be defeated by reason only of the debt not being payable without the permission of the Treasury and of that permission not having been given or having been revoked.

* * * * *

COMMENT

This Act revised and perpetuated the Exchange Control system introduced, in the first instance as a war-time measure, by the Defence (Finance) Regulations, 1939.

Broadly speaking, the permission or consent of the Treasury is required for any transaction effected by a United Kingdom resident as a result of which a financial benefit accrues to a person resident outside the " sterling area." In practice consent is given as of course for certain types of transaction and altogether dispensed with for others; moreover, the provisions of the " Bretton Woods " Agreement (*a*) and the Anglo-American Loan Agreement (*b*) bind the U.K. Government not to refuse permission for " current transactions " (*c*), which include " payments due in connection with foreign trade " (*d*).

The effect of s. 33 (1) is to imply the condition therein set out into any contract in which the parties have not themselves made a provision of their own governing the subject-matter of the condition.

If the case falls within the proviso, the effect on the rights and liabilities of the parties of failure to obtain Treasury consent will be determined, according to the ordinary law governing the performance and discharge of contracts, by whatever arrangement the parties have come to. So if, for example, the statutory condition is not implied because the parties have agreed to make a payment in contravention of the provisions of the Act without obtaining the requisite consent, the Court would no doubt treat the contract as void for illegality, with the usual consequences to the parties.

If, however, the case does not fall within the proviso, it seems that the combined effect of the subsection and of paragraph 4 (1) of the Fourth Schedule is that the debtor need not (and indeed must not) pay so long as Treasury consent has not been given, but if the creditor sues him he cannot defend merely because there has been no Treasury

(*a*) Cmd. 6885; see Bretton Woods Agreements Act, 1945; 22 Halsbury's Statutes (3rd Edn.) 886; and S. R. & O. 1946, No. 36.

(*b*) Cmd. 6968.

(*c*) International Monetary Fund Agreement, Art. VIII (2) (a); Anglo-American Loan Agreement, Art. 7.

(*d*) I.M.F. Agreement, Art. XIX.; A.A.L. Agreement, Art. 11.

consent (*e*). His proper course then is to pay into Court under the appropriate Rules (*f*), and the creditor in his turn will require Treasury consent before he can take the money out of Court (*g*).

MISREPRESENTATION ACT, 1967

(1967, c. 7)

An Act to amend the law relating to innocent misrepresentation and to amend sections 11 *and* 35 *of the Sale of Goods Act*, 1893.

[22nd March 1967.]

1. Removal of certain bars to rescission for innocent misrepresentation.—Where a person has entered into a contract after a misrepresentation has been made to him, and—

(a) the misrepresentation has become a term of the contract; or

(b) the contract has been performed;

or both, then, if otherwise he would be allowed to rescind the contract without alleging fraud, he shall be so entitled, subject to the provisions of this Act, notwithstanding the matters mentioned in paragraphs (a) and (b) of this section.

COMMENT

Before this Act, it was doubtful whether a party could ever rescind a contract for non-fraudulent misrepresentation where the misrepresentation had become a term of the contract or where the contract had been performed. This section removes these possible bars to rescission but without affecting any other possible bar, as to which see *ante*, p. 6, note (*h*).

2. Damages for misrepresentation.—(1) Where a person has entered into a contract after a misrepresentation has been made to him by another party thereto and as a result thereof he has suffered loss, then, if the person making the misrepresentation would be liable to damages in respect thereof had the misrepresentation been made fraudulently, that person shall be so liable notwithstanding that the misrepresentation was not made fraudulently, unless he proves that he had reasonable ground to believe and did believe up to the time the contract was made that the facts represented were true.

(2) Where a person has entered into a contract after a misrepresentation has been made to him otherwise than fraudulently, and he

(*e*) *Cummings* v. *London Bullion Co.*, [1951] W. N. 102; reversed, [1952] 1 All E. R. 383; [1952] 1 K. B. 327, C. A.; applied, *Contract & Trading Co.* v. *Barbey*, [1959] 3 All E. R. 846; [1960] A. C. 244, H. L.

(*f*) R. S. C., O. 6, r. 2; O. 22, r. 9; O. 45, r. 2; O. 46, r. 6 (4), 7; O. 49, r. 7; O. 51, r. 1 (2).

(*g*) Fourth Schedule, paragraph 2.

would be entitled, by reason of the misrepresentation, to rescind the contract, then, if it is claimed, in any proceedings arising out of the contract, that the contract ought to be or has been rescinded, the court or arbitrator may declare the contract subsisting and award damages in lieu of rescission, if of opinion that it would be equitable to do so, having regard to the nature of the misrepresentation and the loss that would be caused by it if the contract were upheld, as well as to the loss that rescission would cause to the other party.

(3) Damages may be awarded against a person under subsection (2) of this section whether or not he is liable to damages under subsection (1) thereof, but where he is so liable any award under the said subsection (2) shall be taken into account in assessing his liability under the said subsection (1).

COMMENT

Subsection (1).— This subsection creates a statutory right to damages for non-fraudulent misrepresentation.

It is not clear whether the provision that a party is to be liable for misrepresentation under this section if he would have been liable had he been fraudulent equates such misrepresentation in all respects with fraudulent misrepresentation. Taken literally it seems to do so, but strange results would follow. Thus, the limitation period for fraud would have to be applied—as to which see s. 26 of the Limitation Act, 1939—so that an action under this sub-section may survive after any action on the contract has been barred. Moreover, no exclusion clause could exclude liability for damages under this subsection, for had the misrepresentation been fraudulent the party making it would have been liable as the exclusion clause could not have been construed as excluding liability for fraud (*h*). This point in no way conflicts with s. 3 of this Act, which is concerned with the effect to be given to an exclusion clause which, if valid, would be construed as excluding liability. To avoid these results, it is possible that the courts will construe the subsection restrictively, perhaps holding that " if the person making the misrepresentation would be liable to damages in respect thereof had the misrepresentation been made fraudulently " means that if damages would have been a possible remedy had the misrepresentation been fraudulent, then they shall be available under this subsection, but subject to the limitations appropriate in a case of negligence.

However, the right given by this sub-section is a statutory right. It is not based upon the tort of negligence despite being commonly described as negligent misrepresentation. Thus, there is no requirement that the person making the representation should owe any duty of care to the representee, or that he should voluntarily or otherwise take upon himself any liability for the accuracy of what he says. It seems more likely, therefore, that the person making the representation cannot exclude his liability by notice as he can exclude his liability for negligent mis-statement (*i*).

(*h*) *Pearson & Son, Ltd.* v. *Dublin Corporation*, [1907] A. C. 351.

(*i*) *Cf. Hedley Byrne & Co.* v. *Heller & Partners*, [1963] 2 All E. R. 575; [1964] A. C. 465, H. L.

To date, the courts appear to have assumed that the appropriate measure of damages is the same as that for breach of contract (*j*). But such a measure of damages cannot, it is submitted, be reconciled with the wording of the section. Damages for breach of contract are estimated to put the injured party in the same position as if the contract had been performed, for this is the measure of the loss directly and naturally resulting from the breach (*k*). It is not necessary that the injured party should have suffered any loss as a result of entering into the contract. It is sufficient that he has not made the profit that he should have made. But it is a necessary pre-condition to liability under this sub-section that the representee should have suffered loss as a result of entering into the contract, and if he has suffered no such loss, he should get no damages although he has not got the bargain he expected.

Nor can the measure of damages be the same as for negligence. Damages for negligence and for deceit are calculated in the same manner—so as to compensate the injured party for the loss he has suffered. But in negligence, damages are limited to such loss as may reasonably have been foreseen at the time of the negligent act or representation. Damages for deceit are not so limited (*l*). It is possible for a party to suffer loss as a result of a mis-representation which would be recoverable in an action for deceit, but not in an action for negligence. It is clear from the sub-section that in such a case the person making the representation is liable to damages.

It would seem, therefore, that the appropriate measure of damages must be the same as that for the tort of deceit, the measure suggested by the equation of the statutory liability with liability for fraud (*m*).

If the representor is to avoid liability under this subsection, he must show that he had reasonable grounds to believe, and did believe, the representation to be true not only when he made it, but also at all times up to the time the contract was made. Should new facts come to his notice before that time, he must disclose them to the other party.

Subsection (2).—It is open to a party who is entitled to rescind a contract for misrepresentation either to seek rescission from the court or to rescind without taking any proceedings. In either case, if the misrepresentation was fraudulent, even if it was negligent, and either party brings any proceedings in respect of the contract, then the court has a discretion to declare the contract subsisting and award damages instead. The court has no power to award damages under this sub-

(*j*) In *Gosling* v. *Anderson* (1972), *Times*, 8th February, the Court of Appeal appears to have assumed that the measure of damages was the same as in contract, but in that case the point was not argued and the Court of Appeal did not have to decide it, but simply directed an inquiry as to damages. A similar assumption was made, *obiter*, by Lord DENNING, M.R., in *Jarvis* v. *Swan Tours, Ltd.*, [1973] 1 All E. R. 71; [1973] Q. B. 233, C. A. Lord DENNING's dictum was applied by Judge FAY in *Davis & Co.* v. *Afa-Minerva, (E.M.I.), Ltd.*, [1974] 2 Lloyd's Rep. 27, but again the point does not seem to have been argued.

(*k*) See the comment to s. 54 of the Sale of Goods Act, *ante*.

(*l*) *Doyle* v. *Olby* (*Ironmongers*), *Ltd.*, [1969] 2 All E. R. 119; [1969] 2 Q. B. 158, C. A.

(*m*) *Watts* v. *Spence*, [1975] 2 W. L. R. 1039, where, however, GRAHAM, J. held, following *Gosling* v. *Anderson*, *supra*, and apparently overlooking *Doyle* v. *Olby*, *supra*, where the contrary was held, that damages for loss of bargain could be recovered in an action for fraud.

section as well as rescission, and it would seem that it cannot award damages if the right to rescind has been lost (*n*).

3. Avoidance of certain provisions excluding liability for misrepresentation.—If any agreement (whether made before or after the commencement of this Act) contains a provision which would exclude or restrict—

(a) any liability to which a party to a contract may be subject by reason of any misrepresentation made by him before the contract was made; or

(b) any remedy available to another party to the contract by reason of such a misrepresentation;

that provision shall be of no effect except to the extent (if any) that, in any proceedings arising out of the contract, the court or arbitrator may allow reliance on it as being fair and reasonable in the circumstances of the case.

COMMENT

This section only applies to a provision contained in an agreement. But such provisions are frequently contained in sales brochures, auction particulars and similar documents. It is not clear how far liability for misrepresentation may be limited or excluded by a provision in such a document. The position would seem, however, to be as follows:

(1). Liability for fraud cannot be excluded because no exemption clause will be construed as extending to cover a fraudulent misrepresentation (*o*).

(2). Liability for misrepresentation under section 2 (1) of this Act cannot be excluded by a provision other than one in an agreement, and arguably cannot be excluded even by a provision in an agreement (*p*). If a provision in an agreement should be held to be capable of excluding liability under s. 2 (1), by virtue of this section it will still only be valid to the extent that the court allows reliance on it.

(3). Liability for misrepresentation other than for fraud and for damages under s. 2 (1) may be excluded by a suitably worded clause in an agreement, but by virtue of this section, such a clause will be valid only to the extent that the court allows reliance on it. Liability for damages for negligent mis-statement under the law stated in *Hedley Byrne* v. *Heller Brothers* (*q*) may also be excluded by a suitably worded disclaimer independently of any contract, but that is because liability depends upon the assumption by the person making the statement of a duty of care, and he is therefore free to make it clear to the person to whom he makes the statement that he will not assume such a duty.

(4). The right to rescind for non-fraudulent misrepresentation is an equitable right, and while a court of equity may give some weight to an exclusion clause which is not in a contract, there seems no reason why, in doing what is equitable, it should allow reliance on such a provision unless it is fair and reasonable in all the circumstances to do so.

(*n*) See generally 30 M. L. R. 369; [1967] C. L. J. 239; 31 Conv. 234.

(*o*) *Pearson & Son Ltd.*, v. *Dublin Corporation*, [1907] A. C. 351, H. L.

(*p*) See the comment to that section, *ante*.

(*q*) [1963] 2 All E. R. 575; [1964] A. C. 465, H. L.

(5). Representors frequently seek to forestall any claim for misrepresentation either by warning the person to whom the representation is made that he should not rely on it, or by requiring him to sign an acknowledgment that he has not relied on any representation in entering into the contract. Such provisions are probably outside the scope of this section. But it does not follow that they will necessarily be effective. A warning not to rely on a representation will not normally assist the representor if he is fraudulent, and will therefore probably not assist him if he is sued for damages under s. 2 (1) (*r*). In other cases, the question will be whether he intended the representations to be relied on, and while the warning not to rely on them may be evidence that he did not so intend, it will not be conclusive. So too, an acknowledgment by a party to a contract that he has not relied on any representation may be evidence of that fact, and may sometimes give rise to an estoppel if the other party can show that the acknowledgment was intended to be acted upon, that he believed it, and that he did act upon it. But, if a person is required to sign such an acknowledgment as condition of entering into a contract, such an acknowledgement is unlikely to be given very much weight, particularly if the other evidence shows that the representor must have been aware that the representee was only entering the contract as a result of what he had been told.

(6). The section does not apply to clauses excluding or restricting liability or remedies for breach of contractual terms as such. However, if one provision excludes liability for both misrepresentation and breaches of contract, it would seem that the entire provision would be of no effect subject to the discretion of the court under this section, unless perhaps the part referring to misrepresentation can be severed from the rest (*s*). Moreover, if a misrepresentation is made in such circumstances that it becomes a term of the contract, it would seem that by reason of the misrepresentation the party making it is subjected to liability for breach of contract and the other party is entitled to certain remedies in respect of that breach. A provision excluding or restricting such liability or remedies would therefore seem to come within this section. If so, then a new distinction must be drawn between terms which are representations and terms which are not. The wording of the section is obscure, however, and it is unlikely that either of the interpretations suggested in this paragraph was considered or intended by the draftsman.

The section does not apply to clauses which prevent liability from arising, for example where a principal gives notice that an agent has no power to make representations on his behalf (*t*).

The section probably applies to agreed damage clauses, for such clauses, like other compromises, restrict the remedies available to the parties, but if the agreed damage clause is a fair one the court will generally allow reliance on it.

(*r*) See the comment to that section, *ante*.

(*s*) *Cf.* the rules relating to severance of covenants in restraint of trade.

(*t*) *Overbrooke Estates, Ltd.* v. *Glencombe Properties, Ltd.*, [1974] 3 All E. R. 511; [1974] 1 W. L. R. 1335.

Whatever the scope of the section, the court has the widest possible discretion in deciding whether it is fair and reasonable to give any effect to a provision which it holds to be covered by it. It may for example refuse to give any effect to the provision, or it may allow it to be relied on in respect of one misrepresentation but not another, or allow it to be relied on as excluding the right to rescind but not the right to claim damages, or as excluding liability for one loss but not liability for another.

In considering what is fair and reasonable, the court will probably take into account factors such as the relative negotiating position of the parties; whether the provision was clearly brought to the attention of the party against whom it operates (*u*); whether the liability which it seeks to exclude or limit is such as a fair minded party might reasonably seek to exclude or limit; the conduct of the parties generally; the nature of the misrepresentation or breach in respect of which a remedy is sought; and possibly the insurance position (*v*).

Under this section, it is for the representor to show that he ought fairly to be allowed to rely on the exemption clause, whereas under s. 55 (4) of the Sale of Goods Act, *ante*, it is for the buyer to show that it would not be fair for the seller to rely on the exemption clause.

4. Amendments of Sale of Goods Act, 1893.—(1) In paragraph (c) of section 11 (1) of the Sale of Goods Act 1893 (condition to be treated as warranty where the buyer has accepted the goods or where the property in specific goods has passed) the words " or where the contract is for specific goods, the property in which has passed to the buyer " shall be omitted.

(2) In section 35 of that Act (acceptance) before the words " when the goods have been delivered to him, and he does any act in relation to them which is inconsistent with the ownership of the seller " there shall be inserted the words " (except where section 34 of this Act otherwise provides) ".

COMMENT

See pp. 109 and 195–200, *ante* and articles cited in footnote.

5. Savings for past transactions.—Nothing in this Act shall apply in relation to any misrepresentation or contract of sale which is made before the commencement of this Act.

(*u*) *Cf.* the matter to be taken into account under s. 55 (5) of the Sale of Goods Act, *ante*. See also *Foley Motors, Ltd.* v. *McGhee*, [1970] N. Z. L. R. 649, where in applying a statute protecting hirers under hire-purchase agreements Richmond, J. commented that it might be argued that it is unfair and unreasonable to exclude an express warranty without at least drawing the attention of the purchaser to the clause in the contract and explaining its effect. He drew an analogy with moneylending contracts, where the failure by a lender to explain a clause in a loan agreement making the whole of the principal and interest due on default has been held to make the transaction harsh and unconscionable. See further *Meston on Moneylenders*, 5th ed., pp. 196-7.

(*v*) See generally 30 M. L. R. 369; [1967] C. L. J. 239; 31 Conv. 234.

COMMENT

Exclusion clauses agreed before the Act came into force are expressly affected by s. 3, but only when they apply to misrepresentations made after that date.

6. Short title, commencement and extent.—(1) This Act may be cited as the Misrepresentation Act 1967.

(2) This Act shall come into operation at the expiration of the period of one month beginning with the date on which it is passed.

(3) This Act, except section 4 (2), does not extend to Scotland.

(4) This Act does not extend to Northern Ireland.

COMMENT

The Act came into operation on 22nd April 1967.

The English law of misrepresentation and s. 11 (1) (c) of the Sale of Goods Act, 1893 do not apply to Scotland.

The Parliament of Northern Ireland has passed a similar Misrepresentation Act.

THEFT ACT 1968

(1968, c. 60)

An Act to revise the law of England and Wales as to theft and similar or associated offences . . . and for other purposes connected therewith.

[26th July 1968.]

1. Basic definition of theft.—(1) A person is guilty of theft if he dishonestly appropriates property belonging to another with the intention of permanently depriving the other of it; and " thief " and " steal " shall be construed accordingly.

(2) It is immaterial whether the appropriation is made with a view to gain, or is made for the thief's own benefit.

* * * * *

COMMENT

The terms " dishonestly ", " appropriates ", " property ", " belonging to another " and " permanently depriving " are defined in ss. 2–6 of the Act. " Gain " is defined in s. 34 (2).

* * * * *

24. Scope of offences relating to stolen goods.—(1) The provisions of this Act relating to goods which have been stolen shall apply whether the stealing occurred in England or Wales or elsewhere, and whether it occurred before or after the commencement of this Act, provided that the stealing (if not an offence under this Act) amounted to an offence where and at the time when the goods were stolen; and references to stolen goods shall be construed accordingly.

* * * * *

(4) For purposes of the provisions of this Act relating to goods which have been stolen (including subsections (1) to (3) above) goods obtained in England or Wales or elsewhere either by blackmail or in the circumstances described in section 15 (1) of this Act shall be regarded as stolen; and " steal ", " theft " and " thief " shall be construed accordingly.

* * * * *

28. Orders for restitution.—(1) Where goods have been stolen, and a person is convicted of any offence with reference to the theft (whether or not the stealing is the gist of his offence), the court by or before which the offender is convicted may on the conviction exercise any of the following powers:—

(a) the court may order anyone having possession or control of the goods to restore them to any person entitled to recover them from him; or

(b) on the application of a person entitled to recover from the person convicted any other goods directly or indirectly representing the first-mentioned goods (as being the proceeds of any disposal or realisation of the whole or part of them or of goods so representing them), the court may order those other goods to be delivered or transferred to the applicant; or

(c) on the application (*a*) of a person who, if the first-mentioned goods were in the possession of the person convicted, would be entitled to recover them from him, the court may order that a sum not exceeding the value of those goods shall be paid to the applicant out of any money of the person convicted which was taken out of his possession on his apprehension.

(2) Where under subsection (1) above the court has power on a person's conviction to make an order against him both under paragraph (b) and under paragraph (c) with reference to the stealing of the same goods, the court may make orders under both paragraphs provided that the applicant for the orders does not thereby recover more than the value of those goods.

(3) Where under subsection (1) above the court on a person's conviction makes an order under paragraph (a) for the restoration of any goods, and it appears to the court that the person convicted has sold the goods to a person acting in good faith, or has borrowed money on the security of them from a person so acting, then on the application (*a*) of the purchaser or lender the court may order that there shall be paid to the applicant, out of any money of the person convicted which was taken out of his possession on his apprehension, a sum not exceeding the amount paid for the purchase by the applicant or, as the case may be, the amount owed to the applicant in respect of the loan.

(4) The court shall not exercise the powers conferred by this section unless in the opinion of the court the relevant facts sufficiently appear from evidence given at the trial or the available documents, together with admissions made by or on behalf of any person in connection with any proposed exercise of the powers; and for this purpose " the available documents " means any written statements or admissions which were made for use, and would have been admissible, as evidence at the trial, the depositions taken at any committal proceedings and any written statements or admissions used as evidence in those proceedings.

(5) Any order under this section shall be treated as an order for the restitution of property within the meaning of sections 30 and 42

(*a*) An application is no longer essential: see s. 6 (2), Criminal Justice Act, 1972.

of the Criminal Appeal Act 1968 (which relate to the effect on such orders of appeals).

(6) References in this section to stealing are to be construed in accordance with section 24 (1) and (4) of this Act.

COMMENT

The court should apply the same principles in considering whether to exercise its discretion under this section as formerly were applied under s. 45 of the Larceny Act, 1916 (*b*).

If any doubt exists whether money or goods belong to a third party the criminal courts are not the correct forum in which that issue should be decided. It is only in the plainest cases, when there can be no doubt, that the court should make a restitution order (*c*).

The trial concludes when sentence is given. Evidence cannot be admitted after that time to assist the court in determining whether to make an order under this section (*d*).

By virtue of s. 6 (3) of the Criminal Justice Act, 1972, a court may now also make a restitution order under this section where on the conviction of a person of any other offence, it takes an offence with reference to the theft of the goods in question into consideration in determining sentence.

Moreover, s. 6 (2) of the 1972 Act enables the court to exercise its powers under subsections (1) (*c*) and (3) of this section without any application being made in that behalf.

An appeal lies from an order made under this section (*e*), and the Court of Appeal has power to vary or annul such order although the conviction is not quashed. (*f*).

Section 30 of the Criminal Appeal Act, 1968 and s. 6 (5) of the Criminal Justice Act, 1972 provide for the suspension of an order for the restitution of property (unless the court otherwise orders in a case where, in its opinion, the title to the property is not in dispute) for an initial period of 28 days and thereafter, in the event of an appeal, until the determination of the appeal. Section 42 of the Criminal Appeal Act, 1968, provides for a further suspension of the order in the event of an appeal to the House of Lords.

31. Effect on civil proceedings.—

* * * * *

(2) Notwithstanding any enactment to the contrary, where property has been stolen or obtained by fraud or other wrongful means, the title to that or any other property shall not be affected by reason only of the conviction of the offender.

COMMENT

Under s. 24 (1) of the Sale of Goods Act, 1893, where goods were stolen and the offender was prosecuted to conviction, the property in the stolen goods revested in the person who was the owner at the

(*b*) *R.* v. *Ferguson*, [1970] 2 All E. R. 820; [1970] 1 W. L. R. 1246, C. A.
(*c*) *Ibid.*, and see *Stamp* v. *United Dominions Trust* (*Commercial*), *Ltd.*, [1967] 1 All E. R. 251; [1967] 1 Q. B. 418, D. C.; *R.* v. *Church* (1970), 55 Cr. App. Rep. 65, D. C. (*d*) *R.* v. *Church, supra.*
(*e*) *R.* v. *Parker*, [1970] 2 All E. R. 458; [1970] 1 W. L. R. 1003, C. A.; Criminal Justice Act, 1972, s 6 (4). (*f*) Criminal Appeal Act 1968, s. 30 (4).

time of the theft notwithstanding any intermediate dealing with them, whether by sale in market overt or otherwise. Section 24 (2), however, limited the law as to revesting of property on conviction to cases of larceny, whereas prior to the enactment of the Sale of Goods Act, 1893 the rule had extended to cover cases where the goods had been obtained by false pretences, but under a *de facto* contract (*g*). Section 24 of the Sale of Goods Act is repealed by the Theft Act, 1968 (*h*), and the present sub-section is required in lieu of s. 24 (2).

(*g*) For the history of the rule, see p. 104 of the 15th edition of the present work.

(*h*) Sch. 3, Pt. III.

THE AUCTIONS (BIDDING AGREEMENTS) ACT 1969

(1969, c. 56)

An Act . . . to make fresh provision as to the rights of a seller of goods by auction where an agreement subsists that a person or persons shall abstain from bidding for the goods . . . [22nd October 1969.]

* * * * *

3. Rights of seller of goods by auction where agreement subsists that some person shall abstain from bidding for the goods.—(1) Where goods are purchased at an auction by a person who has entered into an agreement with another or others that the other or the others (or some of them) shall abstain from bidding for the goods (not being an agreement to purchase the goods bonâ fide on a joint account) and he or the other party, or one of the other parties, to the agreement is a dealer, the seller may avoid the contract under which the goods are purchased.

(2) Where a contract is avoided by virtue of the foregoing subsection, then, if the purchaser has obtained possession of the goods and restitution thereof is not made, the persons who were parties to the agreement that one or some of them should abstain from bidding for the goods the subject of the contract shall be jointly and severally liable to make good to the seller the loss (if any) he sustained by reason of the operation of the agreement.

(3) Subsection (1) above applies to a contract made after the commencement of this Act whether the agreement as to the abstention of a person or persons from bidding for the goods the subject of the contract was made before or after that commencement.

(4) Section 2 of the Auctions (Bidding Agreements) Act 1927 (right of vendors to treat certain sales as fraudulent) shall not apply to a sale the contract for which is made after the commencement of this Act.

(5) In this section, " dealer " has the meaning assigned to it by section 1 (2) of the Auctions (Bidding Agreements) Act 1927.

* * * * *

COMMENT

" Dealer " is defined by s. 1 (2) of the Auction (Bidding) Agreements) Act 1927 as " a person who in the normal course of his business attends sales by auction for the purpose of purchasing goods with a view to reselling them."

* * * * *

4. Copy of the Act to be exhibited at sale.—Section 3 of the Auctions (Bidding Agreements) Act 1927 (copy of Act to be exhibited at sale) shall have effect as if the reference to that Act included a reference to this Act.

* * * * *

FINANCE ACT 1972

(1972, c. 41)

An act to grant certain duties . . .

[27th July 1972.]

PART I

VALUE ADDED TAX

* * * * *

42. Adjustment of contracts on changes in tax.—(1) Where, after the making of a contract for the supply of goods or services and before the goods or services are supplied, there is a change in the tax charged on the supply, then, unless the contract otherwise provided, there shall be added to or deducted from the consideration for the supply an amount equal to the change.

(2) References in this section to a change in the tax charged on a supply include references to or from no tax being charged on the supply.

* * * * *

COMMENT

The Finance Act, 1972 replaced purchase tax, with value added tax. It is not within the scope of this book to deal with value added tax (*i*). In general, however, it is charged in the case of a contract for the sale of goods only where the supply is a taxable supply (*k*) by a taxable person in the course of a business carried on by him, and is payable by the person supplying the goods (*l*). But this does not mean that a seller is entitled to add the amount attributable to value added tax to the contract price at which he has agreed to sell the goods, unless there is a provision in the contract entitling him to do so. The price has still to be ascertained in accordance with s. 8 of the Sale of Goods Act, 1893, *ante*, though where a supplier will be liable for value added tax, he will naturally take this into account when making his quotation, or will quote a price of £x + value added tax.

The tax is chargeable in respect of any sale when the goods are treated as having been supplied in accordance with the rules set out in the Finance Act, 1972. If this is later than the date of the contract, the rate at which the tax charged may have been changed in the meantime, and in such a case, unless the parties have otherwise agreed, the section reproduced above provides for the price charged to be adjusted to take account of the increase or decrease in tax.

(*i*) See, *e.g.*, *De Voil on Value Added Tax*.

(*k*) "Taxable supply" is defined by s. 46 (1) of the Finance Act, 1972 to include any supply of goods other than an exempt supply. At present no sales of goods are exempt supplies. Some sale of goods are zero-rated, however, so that the supplier will not be liable for tax on the supply of those goods.

(*l*) Finance Act, 1972, s. 2 (2).

A similar provision in respect of customs and excise duties is contained in s. 1 of the Finance Act, 1901, *ante*, while a similar provision in respect of purchase tax was contained in s. 35 of the Purchase Tax Act, 1963.

POWERS OF CRIMINAL COURTS ACT 1973

(1973, c. 62)

An Act to consolidate certain enactments relating to the powers of courts to deal with offenders . . .

[25th October 1973.]

* * * * *

35. Compensation orders against convicted persons.—(1) Subject to the provisions of this Part of this Act, a court by or before which a person is convicted of an offence, in addition to dealing with him in any other way, may, on application or otherwise, make an order (in this Act referred to as " compensation order ") requiring him to pay compensation for any personal injury, loss or damage resulting from that offence or any other offence which is taken into consideration by the court in determining sentence.

(2) In the case of an offence under the Theft Act 1968, where the property in question is recovered, any damage to the property occurring while it was out of the owner's possession shall be treated for the purposes of subsection (1) above as having resulted from the offence, however and by whomsoever the damage was caused.

(3) No compensation order shall be made in respect of loss suffered by the dependants of a person in consequence of his death, and no such order shall be made in respect of injury, loss or damage due to an accident arising out of the presence of a motor vehicle on a road, except such damage as is treated by subsection (2) above as resulting from an offence under the Theft Act 1968.

(4) In determining whether to make a compensation order against any person, and in determining the amount to be paid by any person under such an order, the court shall have regard to his means so far as they appear or are known to the court.

(5) The compensation to be paid under a compensation order made by a magistrates' court in respect of any offence of which the court has convicted the offender shall not exceed £400; and the compensation or total compensation to be paid under a compensation order or compensation orders made by a magistrates' court in respect of any offence or offences taken into consideration in determining sentence shall not exceed the difference (if any) between the amount or total amount which under the foregoing provisions of this subsection is the maximum for the offence or offences of which the offender has been convicted and the amount or total amounts (if any) which are in fact ordered to be paid in respect of that offence or those offences.

COMMENT

This section, which re-enacts s. 1 of the Criminal Justice Act, 1972, gives a court before which a person is convicted of an offence (*m*) a new and wide discretionary power to order the person convicted to pay compensation to persons suffering injury, loss or damage as a result either of that offence or of any other offence which is taken into consideration by the court in determining sentence.

The 1972 Act made available such compensation for the first time to persons suffering loss as a result of an offence under the Trade Descriptions Act, 1968 and other statutes for the protection of consumers, thereby providing a simple way for consumers to obtain recompense in a clear case.

In most cases brought under the Trade Descriptions Act, 1968 and other consumer protection legislation, a person convicted of an offence is likely to be able to afford to pay compensation. In general, therefore, a person who seeks compensation in such a case, and who can show that he has suffered loss as a result of an offence, ought, if the facts are clear, to obtain compensation. But it is no longer the sole object of the court in making a compensation order to provide a quick and easy civil remedy for the victim (*n*). While compensation orders are not restricted to cases where the defendant can easily pay (*o*), an order will not normally be made which involves a large sum (*p*), or which will take the offender a long time to discharge (*q*).

There is no authority whether a court can award compensation in circumstances in which there is no civil liability, but the wording of sub-section (1) above seems wide enough to extend to such a case (*r*). Moreover, in *R.* v. *Thomson* (*s*) it was held that the concepts of causation which apply to the assessment of damages under the law of contract and tort do not apply in assessing compensation.

Cases where the facts are in dispute should normally be brought in the civil courts (*t*). But it does not follow that a criminal court cannot make a compensation order except where all the facts are before it. Both justices and judges in Crown Courts must do what they can to make a just order on such information as they have, even though they may not

---

(*m*) If magistrates commit an offender convicted by them to the Crown Court for sentence under s. 29 of the Magistrates Court Act, 1952, it is the Crown court and not the magistrates court which should consider whether to order compensation to be paid: *R.* v. *Brogan,* [1975] 1 All E. R. 879; [1975] 1 W. L. R. 393, C. A.

(*n*) *R.* v. *Oddy,* [1974] 2 All E. R. 666; [1974] 1 W. L. R. 1212, C. A.; *R.* v. *Inwood,* [1975] Cr. App. Rep. 70, C. A.

(*o*) *R.* v. *Bradburn* (1973), 57 Cr. App. Rep. 948, C. A.

(*p*) *R.* v. *Kneeshaw,* [1974] 1 All E. R. 896; [1975] Q. B. 57, C. A.

(*q*) *R.* v. *Bradburn, supra* (£400 at £2 per week reduced to £100 at £2 per week); *R.* v. *Daly,* [1974] 1 All E. R. 290; [1974] 1 W. L. R. 133, C. A. (£1,200 at £3·50 per week reduced to £600 at the same rate); *R.* v. *Wylie,* [1975] R. T. R. 94, C. A.

(*r*) In particular, sub-s. (2) specifically authorises the award of compensation which might not be recoverable as damages in a civil action. The absence of a civil remedy would presumably be a factor which would encourage the court to be more ready to award compensation, since one of the factors which influenced the Court of Appeal in *R.* v. *Bradburn* and *R.* v. *Daly, supra,* was the fact that the victims in those cases still had their civil remedies in damages even if an order was made: see s. 38 of the Powers of Criminal Courts Act 1973.

(*s*) [1974] 1 All E. R. 823; [1974] Q. B. 592, C. A. (*t*) *R.* v. *Kneeshaw, supra.*

be apprised of the detailed facts of an offence (*u*), and must ask themselves on the facts available whether loss or damage can fairly be said to have resulted to anyone from the offence for which the accused has been convicted or which has been taken into account (*v*).

Evidence is admissible to enable the court to decide whether to order compensation. Although, therefore, it is not necessary for an application to be made by the victim before the court can make a compensation order, there may be some advantage in doing so. In *R.* v. *Kneeshaw* (*a*), the Court of Appeal held that the court has a discretion as to the evidence which it will admit, even though it should hesitate to embark on a complicated investigation even at the suit of an applicant making a positive application (*b*).

Where there is more than one victim, or more than one injury or loss in respect of which the victim seeks compensation, the court in awarding compensation should specify how it is to be apportioned and in respect of what claims (*c*). Several compensation orders may be needed (*d*).

There is power to make a joint and several order against more than one offender, but this could lead to complications and should not normally be done (*e*).

Section 36 of this Act provides for the suspension of a compensation order pending an appeal and applies ss. 30 and 42 of the Criminal Appeal Act, 1968 to compensation orders as they apply to restitution orders under s. 28 of the Theft Act, 1968, *ante*. Section 37 enables a compensation order to be reviewed in certain circumstances, and s. 38 provides that damages subsequently awarded in civil proceedings should be reduced by the amount paid under the compensation order, and should not be enforced without the leave of the court to the extent that any sum remains to be paid under a compensation order.

(*u*) As in the case of many offences which are taken into consideration in determining sentence.

(*v*) *R.* v. *Thomson Holidays, Ltd.*, [1974] 1 All E. R. 823; [1974] Q. B. 592, C. A.

(*a*) [1974] 1 All E. R. 896; [1975] Q. B. 57, C. A.

(*b*) Under the law before 1972, the victim could instruct prosecuting counsel to make the application for compensation on his behalf, and a simple form which was in use in some police forces and which the victim signed authorising the prosecuting counsel to act on his behalf was approved and commended by the Court of Appeal for this purpose: *R.* v. *Salem Mohammed Monsoor Ali* (1972), 56 Cr. App. Rep. 301, C. A. The same procedure would seem to be equally effective under the new law at least so long as it is not necessary for the applicant to adduce any evidence in support of his application.

(*c*) *R.* v. *Oddy*, [1974] 2 All E. R. 666; [1974] 1 W. L. R. 1212, C. A.

(*d*) *R.* v. *Inwood*, [1975] Cr. App. Rep. 70, C. A.

(*e*) *R.* v. *Grundy*, [1974] 1 All E. R. 292; [1974] 1 W. L. R. 139, C. A. But the loss must result from an offence. No compensation order may be made, therefore, in respect of obligations arising or which might have arisen out of civil proceedings between the victim and the offender which determine before the offender is convicted: *Hammerton Cars* v. *Redbridge London Borough*, [1974] 2 All E. R. 216; [1974] 1 W. L. R. 484, D. C., where the court left open the question whether liability for costs in civil proceedings can be loss or damage resulting from the offence and whether, if it could be, it would be too remote. In principle, there seems no reason why such costs should not be capable of being a loss resulting from an offence at least in the case of an action between the victim and a third party over a state of affairs which resulted from the offence, provided that such litigation was reasonably undertaken by the victim. *Cf.* the rule on mitigation of damages discussed, p. 255, *ante*.

APPENDIX II—NOTES

NOTE A.—ON THE HISTORY OF THE TERMS CONDITION AND WARRANTY

I.—" CONDITION "

The term " condition " as applied to contracts of sale (*a*) appears to mean indifferently (1) an uncertain event on the happening of which the obligation of the contract is to depend, and (2) a term of the contract making its obligation depend on the happening of the event. Though the Act uses the term, it does not define it explicitly in the definition section (s. 62) : but s. 11 (1) (*b*) implicitly defines it as " a stipulation . . . the breach of which may give rise to a right to treat the contract as repudiated ".

The term seems to have been imported into the law of contract from the law of conveyancing. In conveyancing a distinction was drawn between " conditions " and " covenants " (*b*) which, in contracts, has now become obliterated.

The classification of conditions in English law is imperfect and unsatisfactory.

The division of conditions into positive and negative (*e.g.*, " if my horse wins the Derby "—" if my horse does not win the Derby ") is obvious and requires no comment.

JAMES, L.J., divides conditions into conditions precedent, subsequent and inherent, a classification which seems to involve a cross-division (*c*). Ordinarily they are divided into conditions precedent and conditions subsequent, that is to say, conditions which must be fulfilled before the obligation of the contract arises, and conditions on the happening of which an existent obligation is dissolved. This division corresponds generally, in sale at any rate, with the distinction drawn by Scottish law and the Continental codes between suspensive and resolutive conditions (*d*).

Conditions precedent have again (*e*) been divided into conditions precedent strictly so called and concurrent conditions. A condition is concurrent where the parties to a contract have reciprocally to perform certain acts at the same time. In the case of the failure

---

(*a*) For the modern view about contracts other than contracts of sale, see the judgment of DIPLOCK, L.J., in *Hong Kong Fir Shipping Co., Ltd.* v. *Kawasaki Kisen Kaisha, Ltd.*, [1962] 1 All E. R. 474 at pp. 485–489;[1962] 2 Q. B. 26 at pp. 65–73, C. A.

(*b*) *Bacon's Abridgement* (7th ed.), vol. ii, p. 116.

(*c*) *Ex p. Collins, Re Lees* (1875), 10 Ch. App. 367, at p. 372 (bill of sale case).

(*d*) Pothier's *Obligations*, by Evans, p. 112 ; French Civil Code, arts. 1181–1184 ; Italian Civil Code, arts. 1353, 1354 ; Bell's *Principles of the Law of Scotland* (9th ed.), pp. 47–50.

(*e*) And, in the editor's submission, confusingly.

of one party to perform his part of the contract, it is sufficient if the other party shows that he was ready and willing to perform his part, although he did not actually perform it.

Pothier's further division of condition precedent into potestative, casual, and mixed conditions, though followed in Scotland and by the Continental codes, is not recognised in England. But for accuracy some such subdivision is required.

The important distinction from the point of view of English law is between what may be called promissory conditions and contingent or casual conditions. Where there is a constitution of the latter kind, the obligations of both parties are suspended till the event takes place. Where there is a promissory condition, the non-performance of the condition by the promisor (unless excused by law), gives a right to the promisee to treat the contract as repudiated, that is to say he is discharged from his part of the contract, and, further, he has a claim for damages. In the one case the obligations of the contract never attach: in the other case they attach, but the contract is broken. If A says to B, " I will hire your horse and trap to-morrow if the day be fine," and B assents to this, the obligations of both parties depend on the agreed condition being fulfilled: but if A agrees with B to sell him a ton of hay and deliver it " on Monday for certain," there is a breach of contract by A if the hay be not so delivered.

In the older cases promissory conditions were referred to as " dependent covenants or promises " and were contrasted with independent covenants or promises, namely, stipulations the breach of which gives rise to a claim for damages, but not to a right to treat the contract as repudiated. Now the term " dependent promise " appears to be merged in the wider term " condition precedent."

The Indian Contract Act discards the term " condition," but seeks to preserve the distinction referred to above by dealing separately with " contingent contracts " and " reciprocal promises." The same result is arrived at by the distinction which is sometimes drawn between conditions *of* the contract and conditions *in* the contract.

II.—" WARRANTY "

The term " warranty " seems also to have been imported into the law of contract from the old law of conveyancing, where it signified an express or implied covenant by the grantor of real estate to indemnify the grantee if he should be evicted.

Its meaning was considerably widened in the law of contract, and before the Act was passed it was a term of very uncertain signification. It was sometimes used as strictly equivalent to " condition precedent," yet sometimes it was sought to contrast it with " condition precedent," or rather with a certain kind of condition precedent, namely, a promissory condition precedent. When used in the latter sense the distinction between " condition " and " warranty " corresponded with the distinction drawn by the older cases between what were known as " dependent " and " independent " covenants or promises.

In insurance law its use is curious. When it relates to any undertaking by the assured, it is used to denote a condition precedent of the strictest kind, the breach of which destroys the contract, as in the case of a sailing warranty. But the term is also used as a mere term of exclusion or limitation, as where goods are insured, warranted free from particular average under 3 per cent.

The chief controversy over the proper meaning of the term " warranty " arose in the law of sale, and the ambiguity of its use may have resulted from the want of a clear distinction in English law between sale—*i.e.*, the transfer of property in a thing—and the contract by which that transfer is effected. The term was used in two different senses, and judges and text-writers continually oscillated between them.

First, the term " warranty " was opposed to the term " condition precedent," and denoted a stipulation in a contract of sale, the breach of which gave rise to a claim for damages, but not to a right to reject the goods and treat the contract as repudiated. This is the meaning which, after much consideration, was adopted by the Act as regards England and Ireland (see ss. 11 and 62) (*f*). The objection to this use of the term appears to be that it does not cover the whole field of independent stipulations. For instance, where there is a contract for instalment deliveries, the obligation to pay for a particular instalment may be an independent promise, but it would not ordinarily be called a warranty.

Secondly, the term " warranty " was used to denote any auxiliary stipulation to a contract of sale, and in particular a stipulation relating to the title to, or the quality, condition, or fitness of, goods contracted to be sold. In this sense of the term a breach of warranty might give rise either to a mere claim for damages, or to a right to reject the goods and treat the contract as repudiated, according as the goods might have been accepted or not.

The weight of judicial authority before the Act was in favour of the first meaning, though etymologically and historically the second meaning appears more correct. [Warranty = Guarantee.] The objection to this use of the term is that it does not mark the distinction between a condition precedent and a collateral promise or undertaking. Using the term in the first sense, it is to be noted that many stipulations which in their inception are conditions (*e.g.*, the implied undertakings as to merchantableness and fitness for a particular purpose) may become contracted into warranties by virtue of subsequent events, and this fact doubtless explains much of the confusion of language to which the term has given rise.

(*f*) But note the saving for Scotland in those sections.

Citations

1.—" A *warranty* (concerning freeholds and inheritances) is a covenant real annexed to lands or tenements whereby a man and his heirs are bound to warrant the same, and either by voucher or by judgment in a writ of *warrantia chartæ* to yield other lands and tenements to the value of those that shall be evicted by a former title, else it may be used by way of rebutter " (*g*).

2.—" A *warranty* is an arrangement by which a seller assures to a buyer the existence of some fact affecting the transaction, whether past, present, or future " (*h*).

3.—" A *warranty*, properly so called, can only exist where the subject-matter of the sale is ascertained and existing, so as to be capable of being inspected at the time of the contract, and is a collateral engagement that the specific thing so sold possesses certain qualities, but the property passing by the contract of sale, a breach of the warranty cannot entitle the vendor to rescind the contract, and revest the property in the vendor without his consent. . . . But when the subject-matter of the sale is not in existence, or not ascertained at the time of the contract, an engagement that it shall, when existing or ascertained, possess certain qualities, is not a mere *warranty*, but a *condition*, the performance of which is precedent to any obligation on the vendee under the contract, because the existence of those qualities, being part of the description of the thing sold, becomes essential to its identity, and the vendee cannot be compelled to receive and pay for a thing different from that for what he contracted " (*i*).

4.—" An express *warranty* is a stipulation inserted in writing on the face of the policy, upon the literal truth or fulfilment of which the validity of the entire contract is dependent. These written stipulations either allege the existence of some fact or state of things at the time, or previous to the time, of making the policy, or they undertake for the happening of future events, or the performance of future acts. In the former case Mr. Marshall terms the stipulation an affirmative, and in the latter a promissory warranty " (*k*).

5.—" When it appears that the consideration has been executed in part, that which before was a warranty or condition precedent, loses the character of a condition, or, to speak more properly, ceases to be available as a condition and becomes a *warranty* in a narrower

(*g*) *Bacon's Abridgement* (7th ed.), pp. 356, 359, 361 ; and see *Williams's Real Property* (24th ed.), p. 714.

(*h*) New York Draft Code, § 877.

(*i*) Notes to *Cutter* v. *Powell* (1795), 2 Smith L. C. (7th ed.), p. 30.

(*k*) *Arnould's Marine Insurance* (6th ed.), p. 599; *Cranston* v. *Marshall* (1850), 5 Exch. 395, at p. 402 ; and see *Barnard* v. *Faber*, [1893] 1 Q. B. 340, at p. 343, *per* Bowen, L.J., as to fire insurance.

sense of the word, viz., a stipulation by way of agreement for the breach of which a compensation must be sought in damages" (*l*).

6.—"If upon a treaty about the buying of certain goods, the buyer should ask the seller if he would warrant them to be of such a value and his own goods, and the seller should warrant them, and then the buyer should demand the price and the seller should set the price, and then the buyer should take time for two or three days to consider, and then should come and give the seller his price, though the warranty here was before the sale yet this will be well, because the *warranty* is the ground of the treaty, and this is *warrantizando vendidit*" (*m*).

7.—"It was rightly held by HOLT, C.J., and has been uniformly adopted ever since, that an affirmation at the time of a sale is a *warranty*, provided it appear on evidence to have been so intended" (*n*).

8.—"Here, when F., a mutual acquaintance of the parties, introduced them to each other, he said, 'Mr. J. is in want of copper for sheathing a vessel,' and one of the defendants answered, 'We will supply him well.' As there was no subsequent communication, that constituted a contract and amounted to a *warranty*. But I wish to put the case on a broad principle. If a man sells an article he thereby *warrants* that it is merchantable—that is, fit for some purpose. If he sells it for a particular purpose he thereby *warrants* it fit for that purpose. . . . In every contract to furnish manufactured goods, however low the price, it is an implied *term* that the goods should be merchantable" (*o*).

9.—"Although the vendee of a specific chattel, delivered with a *warranty*, may not have a right to return it, the same reason does not apply to cases of executory contracts, where an article, for instance, is ordered from a manufacturer, who contracts that it shall be of a certain quality, or fit for a certain purpose, and the article sent is such as is never completely accepted by the party ordering it" (*p*).

10.—"A good deal of confusion has arisen in many of the cases on this subject from the unfortunate use made of the word '*warranty*.' Two things have been confounded together. A *warranty* is an express or implied statement of something which the party undertakes shall be part of a contract; and though part of the contract, yet collateral to the express object of it. But in many of the cases, some of which have been referred to, the circumstances of a party selling a

(*l*) Williams, *Notes to Saunders' Reports*, vol. i, p. 554, cited and approved *Heilbut* v. *Hickson* (1872), L. R. 7 C. P. 438, at p. 450; *cf. Stanton* v. *Richardson* (1872), L. R. 7 C. P. 421, at p. 436.

(*m*) *Lysney* v. *Selby* (1705), 2 Ld. Raym. 1118.

(*n*) *Per* BULLER, J., *Pasley* v. *Freeman* (1789), 3 Term. Rep. 51.

(*o*) *Per* BEST, C.J., *Jones* v. *Bright* (1829), 5 Bing. 533, at p. 543.

(*p*) *Per* Lord TENTERDEN, *Street* v. *Blay* (1831), 2 B. & Ad. 456, at p. 463 (horse case).

particular thing by its proper description has been called a *warranty*, and the breach of such contract a breach of warranty; but it would be better to distinguish such cases as a non-compliance with a contract which a party has engaged to fulfil, as if a man offers to buy peas of another, and he sends him beans, he does not perform his contract; but that is not a warranty; there is no warranty that he should sell him peas; the contract is to sell peas, and if he sends him anything else in their stead, it is a non-performance of it" (*q*).

11.—"We avoid the term '*warranty*,' because it is used in two senses, and the term 'condition,' because the question is whether that term is applicable. Then the effect is that the defendants required and the plaintiff gave his undertaking that no sulphur had been used. This undertaking was a preliminary stipulation; and if it had not been given, the defendants would not have gone on with the treaty which resulted in the sale. In this sense it was the condition upon which the defendants contracted, and it would be contrary to the intention expressed by this stipulation that the contract should remain valid if sulphur had been used. The intention of the parties governs in the making and in the construction of all contracts. If the parties so intend, the sale may be absolute, with a warranty superadded, or the sale may be conditional to be null if the warranty is broken; and upon this statement of facts we think that the intention appears to have been that the contract should be null if sulphur had been used" (*r*).

12.—"I agree with what MAULE, J., and CROWDER, J., say in *Hopkins* v. *Tanqueray*. CROWDER, J., says, in the plainest terms, in that case, that the conversation 'was a mere representation, and was evidently not made with an intention to warrant the horse. A representation to constitute a *warranty* must be shown to have been intended to form part of the contract.' It seems to me that that is perfectly correct" (*s*).

13.—"But with respect to statements in a contract descriptive of the subject-matter of it, or of some material incident thereof, the true doctrine established by principle, as well as authority, appears to be, generally speaking, that if such descriptive statement was intended to be a substantive part of the contract, it is to be regarded as a *warranty*, that is to say, a condition on the failure or non-performance of which the other party may, if he is so minded, repudiate the contract *in toto* and so be relieved from performing his part of it, provided it has not been partially executed in his favour. If, indeed, he has received the whole, or any substantial part, of the

(*q*) *Per* Lord ABINGER, *Chanter* v. *Hopkins* (1838), 4 M. & W. 399, at p. 404. *Cf. Kennedy* v. *Panama Mail Co.* (1867), L. R. 2 Q. B. 580, at p. 587; *Wallis* v. *Pratt*, [1910] 2 K. B. 1003, 1012, *per* FLETCHER MOULTON, L.J.

(*r*) *Per* ERLE, C.J., *Bannerman* v. *White* (1861), 10 C. B. N.S. 844, at p. 860.

(*s*) *Per* MARTIN, B., *Stucley* v. *Baily* (1862), 31 L. J. Ex. 483, at p. 489.

consideration from the promise on his part, the *warranty* loses the character of a *condition*, or, to speak perhaps more properly, perhaps, ceases to be available as a condition, and becomes a *warranty* in the narrower sense of the word, viz., a stipulation by way of agreement for the breach of which a compensation must be sought in damages " (*t*).

14.—" The wools are ' guaranteed about similar to samples.' Now such a clause may be a simple guarantee or *warranty*, or it may be a condition. Generally speaking when the contract is as to any goods, such a clause is a condition going to the essence of the contract; but when the contract is as to specific goods the clause is only collateral to the contract, and is the subject of a cross action or matter in reduction of damages. Here there is, I think, merely a *warranty* as distinguished from a *condition* " (*u*).

15.—" If the subject-matter of the contract is a specific existing chattel, a representation as to some quality attached to it or possessed by it is only a warranty unless the absence of that quality or the possession of it in a less degree makes the thing essentially different from that described in the contract (*v*).

16.—" As a matter of law I think every item in a description which constitutes a substantial ingredient in the identity of the thing sold is a *condition* " (*w*).

17.—" It has been said many times, and particularly in *Wallis, Son & Wallis* v. *Pratt & Haynes*, that whether any statement is to be regarded as a *condition* or a *warranty* must depend on the intention to be properly inferred from the particular statement made. A statement that an animal is sound in every respect would *primâ facie* be but a *warranty*, but in this case the learned judge found as a fact that the defendant went further and promised that he would take the animal back if she were no good . . . it seems to me plain that the language he used could not have been intended merely as a *warranty*, for a *warranty* would give no right of rejection to the purchaser " (*x*).

NOTE B.—STIPULATIONS JUDICIALLY CONSTRUED

The following terms and stipulations, among others, have been judicially construed, namely—

(*t*) *Per* WILLIAMS, J., *Behn* v. *Burness* (1863), 3 B. & S. 751, at p. 755.

(*u*) *Per* BLACKBURN, J., *Heyworth* v. *Hutchinson* (1867), L. R. 2 Q. B. 447, at p. 451.

(*v*) *Per* BAILHACHE, J., *Harrison* v. *Knowles and Foster*, [1917] 2 K. B. 606, at p. 610.

(*w*) *Per* SCOTT, L.J. (TUCKER and BUCKNILL, L.JJ., concurring) in *Couchman* v. *Hill*, [1947] 1 All E. R. 103, at p. 105; [1947] K. B. 554, at p. 559.

(*x*) *Per* EVERSHED, M.R., in *Harling* v. *Eddy*, [1951] 2 All E. R. 212, at p. 215; [1951] 2 K. B. 739, at pp. 741–742.

Terms as to shipment, etc.

" The names of the vessels to be declared as soon as the wools are shipped " (*y*).

" Provisional invoice with date of bill of lading to be sent by shipper's house to his buyer " (*a*).

" Shipped per *Diletta* as per bill of lading dated September or October " (*b*).

" For shipment in June and [or] July " (*c*).

" To be ready for shipment on June 15th, 1929 " (*d*).

" Shipment by steamer or steamers during February " (*e*).

" To be shipped during the months of March and/or April " (*f*).

" To be shipped by sailer or sailers from the Philippine Islands between May 1st and July 31st " (*g*).

" Clearance not later than 31st May " (*h*).

" To be shipped [from Egypt] during January, 1900, per steamship *Orlando*, and to be delivered in United Kingdom " (*i*).

" Delivered f.o.b. U.K. or Continental port " (*j*).

" F.o.b. stowed goods Danish port " (*k*).

" Shipment during October " (*l*).

" Expected to be ready to load late September " (*m*).

Shipment " per *Bristo City*, expected ready to load 3rd/5th January, or substitute " (*n*).

" On shipment " in c.i.f. contract (*o*).

" Shipment " (*p*).

" Shipment by Saturday's steamer " (*q*).

(*y*) *Graves* v. *Legg* (1854), 9 Exch. 709.

(*a*) *Berg & Sons* v. *Landauer* (1925), 42 T. L. R. 142 (buyers were held entitled to reject because provisional invoice stated the date of bill of lading to be February 1st, when in fact it was dated February 2nd).

(*b*) *Gattorno* v. *Adams* (1862), 12 C. B. N.S. 560.

(*c*) *Alexander* v. *Vanderzee* (1872), L. R. 7 C. P. 530, Ex. Ch.

(*d*) *Meyer, Ltd.* v. *Osakeyhtio Timber Co. Ltd.* (1930), 37 Ll. L. Rep. 212.

(*e*) *Brandt* v. *Lawrence* (1876), 1 Q. B. D. 344, C. A.; but see *Reuter* v. *Sala* (1879), 4 C. P. D. 239, C. A.

(*f*) *Bowes* v. *Shand* (1877), 2 App. Cas. 455.

(*g*) *Ashmore* v. *Cox*, [1899] 1 Q. B. 436.

(*h*) *Thelman Frères* v. *Texas Flour Mill Co.* (1900), 5 Com. Cas. 321, C. A.

(*i*) *Nickoll* v. *Ashton*, [1900] 2 Q. B. 298.

(*j*) *Muller Brothers* v. *G. M. Power Plant Co.* (1963), *Guardian*, 9th May.

(*k*) *Boyd & Co.* v. *Louca*, [1973] 1 Lloyd's Rep. 209.

(*l*) *Aron & Co.* v. *Comptoir Wegimont*, [1921] 3 K. B. 435; (c.i.f. contract).

(*m*) *S. Sanday & Co.* v. *Keighley Maxted & Co.* (1922), 27 Com. Cas. 296, C. A.; *Maredelanto Compania Naviera S.A.* v. *Bergbau-Handel G.m.b.H.*, [1970] 3 All E. R. 125; [1971] 1 Q. B. 164, C. A.

(*n*) *Thomas Borthwick (Glasgow), Ltd.* v. *Bunge & Co., Ltd.*, [1969] 1 Lloyd's Rep. 17.

(*o*) *Hansson* v. *Hamel & Horley*, [1922] 2 A. C. 36, at p. 47.

(*p*) *Mowbray, Robinson & Co.* v. *Rosser* (1922), 38 T. L. R. 413, C. A. (not covering put on rail); *Foreman & Ellams, Ltd.* v. *Blackburn*, [1928] 2 K. B. 60 (shipment to be construed as future shipment).

(*q*) *Wilson* v. *Wright*, [1937] 4 All E. R. 371, C. A.

"Shipment during 1934 navigation (*r*).

"The port of destination shall be declared by the last buyer to his seller not later than 21 days before commencement of shipment period" (*s*).

"Goods to be loaded under deck" (*t*).

"Demurrage" in f.o.b. contract (*u*).

"Invoicing back clause" if seller makes default in shipment (*v*).

"Should shipment be delayed by prohibition of export" (*w*).

"Shipment and destination: afloat per s.s. Merton Bay due London approximately June 8th" (*x*).

"For the shipping and unshipping of goods" (*a*).

As to arrival of ship or cargo, etc.

"On arrival" (*b*).

"150 tons of soda to arrive ex *Daniel Grant*" (*c*).

"100 hogsheads of oil expected to arrive by the ship *Resolute* from Madras" (*d*).

"100 bales of cotton now on passage from Singapore and expected to arrive at London per the *Ravenscraig*" (*e*).

"50 cases of tallow to be delivered on the safe arrival of the ship *Elgin*" (*f*).

"The cotton to be taken from the quay" (*g*).

"600 tons of nitrate of soda expected to arrive at port of call per *Precursor*" (*h*).

"To discharge at a safe port in United Kingdom, Manchester excepted" (*i*).

(*r*) *May & Hassell, Ltd.* v. *Exporters of Moscow* (1940), 45 Com. Cas. 128.

(*s*) *Carapanayoti* v. *Comptoir Commercial Andre and Cie S.A.*, [1972] 1 Lloyd's Rep. 139.

(*t*) *Montague L. Meyer, Ltd.* v. *Travaru A/B H. Cornelius* (1930), 46 T. L. R. 553; *Messers, Ltd.* v. *Morrison's Export Co., Ltd.*, [1939] 1 All E. R. 92.

(*u*) *Trading Society Kwik Hoo* v. *Sugar Commission* (1923), 129 L. T. 500.

(*v*) *Lancaster* v. *Turner & Co.*, [1924] 2 K. B. 222, C. A.; *Lang* v. *Crude Rubber Washing Co.* (1911), [1939] 2 K. B. 173 *n.*; *Adair & Co., Ltd.* v. *Birnbaum*, [1938] 4 All E. R. 775, C. A.; [1939] 2 K. B. 149.

(*w*) *Fairclough Dodd and Jones* v. *Vantol*, [1956] 3 All E. R. 921; [1957] 1 W. L. R. 136, H. L. See also *Koninklijke Bunge* v. *Compagnie Continentale D'Importation*, [1973] 2 Lloyd's Rep. 44.

(*x*) *Macpherson* v. *Howard Ross*, [1955] 2 All E. R. 445; [1955] 1 W. L. R. 640.

(*a*) *Pigott* v. *Docks and Inland Waterways Executive*, [1953] 1 All E. R. 22; [1953] 1 Q. B. 338.

(*b*) *Alewyn* v. *Pryor* (1826), R. & M. 406; *Boyd* v. *Siffkin* (1809), 2 Camp. 326.

(*c*) *Johnson* v. *Macdonald* (1842), 9 M. & W. 600.

(*d*) *Fischel* v. *Scott* (1854), 15 C. B. 69.

(*e*) *Gorrissen* v. *Perrin* (1857), 2 C. B. N.S. 681.

(*f*) *Hale* v. *Rawson* (1858), 4 C. B. N.S. 85; *cf. Simond* v. *Braddon* (1857), C. B. N.S. 324.

(*g*) *Neill* v. *Whitworth* (1866), L. R. 1 C. P. 684.

(*h*) *Smith* v. *Myers* (1871), L. R. 7 Q. B. 139, Ex. Ch.

(*i*) *Re Goodbody* (1899), 5 Com. Cas. 59, C. A.

Cargo expected to arrive by sailer from Bentos, " in case of non-arrival this contract to be void " (*k*).

" Subject to safe arrival " (*l*).

" Should the goods or any portion thereof be lost, this contract to be cancelled for the whole or each portion " (*m*).

" Subject to ' force majeure ' " (*n*).

" Subject to force majeure and shipment " (*o*).

" Unforeseen circumstances excepted " (*p*).

As to priority of delivery and payment.

" Delivered" (*q*).

" Payment, bill at two months from date of landing " (*r*).

" To be paid for by cash in one month " (*s*).

" Delivery forthwith ; payment, cash in 14 days from the making of the contract " (*t*).

Delivery order running : " we engage to deliver on presentation of this document " (*u*).

" To be free delivered and paid for in 14 days in cash " (*v*).

" Payment, cash in London in exchange for shipping documents " (*w*).

" The balance in cash on right delivery at Rangoon " (*a*).

" Freight to be payable on right delivery of the cargo " (*b*).

" The freight is payable in cash upon arrival of the vessel " (*c*).

" Payment to be made in net cash in London in exchange for bills of lading of each cargo or shipment " (*d*).

Payment " by cash in exchange for shipping documents " (*e*).

Payment " in cash on arrival in exchange for shipping or railway documents " (*f*).

(*k*) *Wyllie* v. *Povah* (1907), 12 Com. Cas. 317, 321.

(*l*) *Barnett* v. *Javeri & Co.*, [1916] 2 K. B. 390.

(*m*) *Clark* v. *Cox, McEuen & Co.*, [1921] 1 K. B. 139, C. A.

(*n*) *Re Comptoir Commercial Anversois*, [1920] 1 K. B. 868, C. A.; *Lebeaupin* v. *Crispin & Co.*, [1920] 2 K. B. 714.

(*o*) *Hong Guan & Co.* v. *Jumabhoy & Sons*, [1960] 2 All E. R. 100 ; [1960] A. C. 684

(*p*) *Wills & Sons* v. *Cunningham*, [1924] 2 K. B. 220.

(*q*) *Kwei Tek Chao* v. *British Traders and Shippers*, [1954] 1 All E. R. 779; [1954] 2 Q. B. 459 ; A/S *Tankexpress* v. *C.F. Belge des Pétroles*, [1948] 2 All E. R. 939 ; [1949] A. C. 76.

(*r*) *Alexander* v. *Gardner* (1835), 1 Bing. N. C. 671.

(*s*) *Spartali* v. *Benecke* (1850), 10 C. B. 212 ; but see *Field* v *Lelean* (1861), 6 H. & N. 617, Ex. Ch., as to usage.

(*t*) *Staunton* v. *Wood* (1851), 16 Q. B. 638.

(*u*) *Bartlett* v. *Holmes* (1853), 13 C. B. 630.

(*v*) *Godts* v. *Rose* (1855), 17 C. B. 229.

(*w*) *Tamvaco* v. *Lucas* (1859), 1 E. & E. 581.

(*a*) *Calcutta Co.* v. *De Mattos* (1863), 32 L. J. Q. B. 322.

(*b*) *Paynter* v. *James* (1867), L. R. 2 C. P. 348.

(*c*) *The Pantanassa*, [1970] 1 All E. R. 848; [1970] P. 187.

(*d*) *Sanders* v. *Maclean* (1883), 11 Q. B. D. 327, C. A.

(*e*) *Ryan* v. *Ridley* (1902), 8 Com. Cas. 105.

(*f*) *Polenghi* v. *Dried Milk Co.* (1904), 10 Com. Cas. 42

Payment " net cash after inspection " (*g*).
" Terms net cash " in c.i.f. contracts (*h*).
" Net cash against documents on arrival of the steamer " (*i*).
" Cash against documents " in ex-ship contract (*k*).

As to time of delivery, etc.

" Delivery at buyer's option in all April or sooner " (*l*).
" 10 tons of oil to be delivered within the last 14 days of March " (*m*).
" 5 tons of oilcakes to be put on board directly " (*n*).
" Delivery forthwith " (*o*).
Delivery " as required " (*p*).
" Delivery on April 17th, complete 8th May " (*q*).
" The lots to be cleared away within three days after the sale at the purchaser's expense " (*r*).
" To be finished as soon as possible " (*s*).
One thousand tons of iron " direct port specification to be given in the beginning of May " (*t*).
" Within a reasonable time after removal of embargo " (*u*).
" Subject to Brazilian export licence " (*a*).
" Timely fulfilment of contract rendered impossible owing to *force majeure* " (*b*).
" Delivery: prompt, as soon as export licence granted " (*c*).

As to cost of delivery

" Sellers shall not be responsible for delay in shipment . . . occasioned by . . . strike " (*d*).

(*g*) *Khan* v. *Duché* (1905), 10 Com. Cas. 87.
(*h*) *E. Clemens Horst Co.* v. *Biddell Brothers*, [1912] A. C. 18 (" net cash " =net cash against documents, and see note on c.i.f. contracts, p. 35, *ante*).
(*i*) *Stein Forbes & Co.* v. *County Tailoring Co.* (1916), 115 L. T. 215.
(*k*) *Yangtze Insurance Association* v. *Luckmanjee*, [1918] A. C. 585.
(*l*) *Cox* v. *Todd* (1825), 7 Dow. & Ry. K. B. 131.
(*m*) *Startup* v. *Macdonald* (1843), 6 Man. & G. 593, Ex. Ch. (tender at 8.30 on Saturday night).
(*n*) *Duncan* v. *Topham* (1849), 8 C. B. 225.
(*o*) *Staunton* v. *Wood* (1851), 16 Q. B. 638.
(*p*) *Jones* v. *Gibbons* (1853), 8 Exch. 920; *Jackson* v. *Rotax Motor and Cycle Co.*, [1910] 2 K. B. 937, C. A.; *Ross Brothers* v. *Shaw & Co.* [1917], 2 I. R. 367; *Pearl Mill Co.* v. *Ivy Tannery Co.*, [1919] 1 K. B. 78.
(*q*) *Coddington* v. *Paleologo* (1867), L. R. 2 Exch. 193.
(*r*) *Woolfe* v. *Horne* (1877), 2 Q. B. D. 355, C. A.; distinguishing *Attwood* v. *Emery* (1856), 1 C. B. N.S. 110.
(*s*) *Hydraulic Co.* v. *McHaffie* (1878), 4 Q. B. D. 670, C. A.
(*t*) *Kidston* v. *Monceau Iron Works* (1902), 7 Com. Cas. 82.
(*u*) *Millett* v. *Van Heek*, [1921] 2 K. B. 369, C. A.
(*a*) *Brauer* v. *Clark*, [1952] 2 All E. R. 497.
(*b*) *Podar* v. *Tagher*, [1949] 2 All E. R. 62; [1949] 2 K. B. 277.
(*c*) *Cassidy* v. *Osuusrukkukauppa*, [1957] 1 All E. R. 484; [1957] 1 W. L. R. 273.
(*d*) *European Grain and Shipping, Ltd.* v. *J. H. Rayner & Co., Ltd.*, [1970] 2 Lloyd's Rep. 239. *Cf. Koninklijke Bunge* v. *Compagnie Continentale D'Importation*, [1973] 2 Lloyd's Rep. 44.

" Free on board a foreign ship " (*e*).
" The cotton to be taken from the quay " (*f*).
" Goods to be taken from the deck " (*g*).
Delivery on payment of freight " and other conditions as per charterparty " (*h*).
" c. i. f. to buyer's wharf, Victoria Docks, London " (*i*).
" Cost of stevedoring to be paid by the Government" (buyer) (*k*).
" Free of customs formalities " (*l*).
" War risk for buyer's account " (*m*).
" Ex store, Rotterdam " (*n*).
" Cape surcharge buyer's account " (*o*).

As to price

" 2½ per cent. discount for cash, the duty to be deducted " (*p*).
" Market value " (*q*).
" Current Prices " (*r*).
" Terms—net cash, to be paid within six to eight weeks from date hereof " (*s*).
" Without reserve " (*t*).
" The highest bidder to be the purchaser " (*u*).
" Cash, or approved banker's bills " (*a*).
" Cash in a month less discount, or 4 months' bill at buyer's option " (*b*).
" Approved acceptance to buyer's draft " (*c*).
c. i. f. " insurance for 5 per cent. over net, invoice to be effected by sellers for account of buyers " (*d*).

(*e*) *Wackerbarth* v. *Masson* (1812), 3 Camp. 270.
(*f*) *Neill* v. *Whitworth* (1866), L. R. 1 C. P. 684, Ex. Ch.
(*g*) *Playford* v. *Mercer* (1870), 22 L. T. 41.
(*h*) *Steamship " County of Lancaster "* v. *Sharp* (1889), 24 Q. B. D. 158.
(*i*) *Acmê Wood Co.* v. *Sutherland, Innes & Co.* (1904), 9 Com. Cas. 170. (expenses under London clause).
(*k*) *White* v. *Williams* [1912] A. C. 814, P. C.
(*l*) *Jager* v. *Tolme & Runge*, [1916] 1 K. B. 939, C. A.
(*m*) *Groom* v. *Barber*, [1915] 1 K. B. 316.
(*n*) *Fisher Reeves & Co.* v. *Armour & Co.*, [1920] 3 K. B. 614, C. A.
(*o*) *Henry, Ltd.* v. *Wilhelm G. Clasen*, [1973] 1 Lloyd's Rep. 159, C. A.
(*p*) *Smith* v. *Blandy* (1825), Ry. & M. 257 at p. 260.
(*q*) *Orchard* v. *Simpson* (1857), 2 C. B. N.S. 299.
(*r*) *Jacks & Co.* v. *Palmer's Shipbuilding & Iron Co.* (1928), 98 L. J. K. B. 366, C. A.
(*s*) *Ashforth* v. *Redford* (1873), L. R. 9 C. P. 20.
(*t*) *Thornett* v. *Haines* (1846), 15 M. & W. 367.
(*u*) *Green* v. *Baverstock* (1863), 14 C. B. N.S. 204.
(*a*) *Smith* v. *Mercer* (1867), L. R. 3 Exch. 51.
(*b*) *Waynes Merthyr Steam Coal Co.* v. *Morewood* (1877), 46 L. J. Q. B. 746.
(*c*) *M'Dowall* v. *J. B. Snowball* 1904, 7 F. (Ct. of Sess.) 35.
(*d*) *Landauer* v. *Asser*, [1905] 2 K. B. 184 (over insurance, right of buyer to surplus) ; but as to increased value policies, see *Strass* v. *Spillers & Baker*, [1911] 2 K. B. 759.

c. i. f. contract, " duty paid " (*e*).

c. i. f. contract " payment cash (before delivery if required) against documents or delivery order " (*f*).

" Conditions—2½ per cent. discount, 1 month " (*g*).

Terms as to quantity, description and quality

(1) *Exemption clauses*

" Sellers give no warranty express or implied as to growth, description or any other matters " (*h*).

Nothing " is sold with a ' warranty ' unless specially mentioned at the time of offering " (*i*).

This guarantee " expressly excludes any other guarantee or warranty, statutory or otherwise " (*k*).

Sellers " give no warranty, express or implied, as to description, quality, productiveness, or any other matter . . . and will not be in any way responsible for the crop " (*l*).

" With all faults " (*m*).

" The goods to be taken with all faults and defects, damaged or inferior, if any, at valuation to be arranged mutually or by arbitration " (*n*).

" The lots are sold with all faults, imperfections and errors of description, the auctioneers not being responsible for the correct description, genuineness or authenticity of, or any fault or defect in any lot, and giving no warranty whatever " (*o*).

" The genuiness or authenticity of any lot is not guaranteed . . . No allowance whatever will be made for errors in description, quantity, weight or measurement " (*p*).

(*e*) *American Commerce Co.* v. *Boehm, Ltd.* (1919), 35 T. L. R. 224.

(*f*) *Re Denbigh Cowan & Co. & Atcherley & Co.* (1921), 90 L. J. K. B. 836, C. A.

(*g*) *Payzu, Ltd.* v. *Saunders*, [1919] 2 K. B. 581 C. A.

(*h*) *Wallis, Son and Wells* v. *Pratt*, [1911] A. C. 394.

(*i*) *Harling* v. *Eddy*, [1951] 2 All E. R. 212; [1951] 2 K. B. 739, C. A.

(*k*) *Baldry* v. *Marshall*, [1925] 1 K. B. 260; *William Barker (Junior) & Co., Ltd.* v. *Ed. T. Agius, Ltd.* (1927), 33 Com. Cas. 120.

(*l*) *Howcroft and Watkins* v. *Perkins* (1900), 16 T. L. R. 217.

(*m*) *Shepherd* v. *Kain* (1821), 5 B. & Ald. 240 (ship); *Taylor* v. *Bullen* (1850), 5 Exch. 779 (ship); *Ward* v. *Hobbs* (1878), 4 App. Cas. 13 (diseased pigs); *Lloyd del Pacifico* v. *Board of Trade* (1930), 35 Ll. L. R. 217, C. A. (" with all faults and errors of description "); *Munro & Co., Ltd.* v. *Meyer*, [1930] 2 K. B. 312, at p. 327 (clause applicable only to goods answering description); *Modiano Brothers & Son* v. *Pearson & Co., Ltd.* (1929), 34 Ll. L. R. 52 (" goods to be taken with all faults ", etc.); *Champanhac & Co., Ltd.* v. *Waller & Co., Ltd.*, [1948] 2 All E. R. 724 (" with all faults and imperfections ").

(*n*) *Ashington Piggeries, Ltd.* v. *Christopher Hill, Ltd.*, [1971] 1 All E. R. 847; [1972] A. C. 441, H. L.

(*o*) *Couchman* v. *Hill*, [1947] 1 All E. R. 103; [1947] K. B. 554, C. A.

(*p*) *Nicholson and Venn* v. *Smith-Marriott* (1947), 177 L. T. 189.

" Not accountable for errors in description " (*q*).

" The goods are not warranted free from defect rendering the same unmerchantable, which would be apparent on reasonable examination, any statement or rule of law notwithstanding " (*r*).

" All conditions, warranties and liabilities implied by statute, common law, or otherwise are excluded " (*s*).

" Any express or implied condition, statement or warranty, statutory or otherwise, not stated herein is hereby excluded " (*t*).

" No variation of these conditions will bind the garage proprietors unless made in writing signed by their authorised manager " (*u*).

" The vendors do not make or give and neither the Auctioneers nor any person in the employment of the Auctioneers has any authority to make or give any representation or warranty in relation to these properties " (*v*).

" The hirer is deemed to have examined . . . the vehicle prior to this agreement and satisfied himself as to its condition, and no warranty, condition, description or representation on the part of the owner as to the state or quality of the vehicle is given or implied " (*a*).

" Purchaser acknowledges that this agreement constitutes the entire contract and that there are no representations, warranties, or conditions, express or implied, statutory or otherwise, other than as contained herein " (*b*).

" This condition is in substitution for, and excludes all conditions warranties or liabilities of any kind relating to the goods sold whether as to fitness or otherwise and whether arising under the Sale of Goods Act 1893 or other statute or in tort or by implication of law or otherwise. In no event shall we be liable for any direct or indirect loss or damage (whether special consequential or otherwise) or any other claims except as provided for in these conditions " (*c*).

(*q*) *Harrison* v. *Knowles and Foster*, [1918] 1 K. B. 608, C. A.

(*r*) *Henry Kendall & Sons* v. *William Lillico & Sons, Ltd.*, [1968] 2 All E. R. 444; [1969] 2 A. C. 31, H. L.; contrast the different approach to a similar clause in *Lindsay & Co.* v. *European Grain and Shipping Agency*, [1963] 1 Lloyd's Rep. 437, C. A.; and see, *Pinnock Brothers* v. *Lewis and Peat*, [1923] 1 K. B. 690; *Canada Atlantic Grain Export Co.* v. *Eilers* (1929), 35 Ll. R. 206; *Munro & Co., Ltd.* v. *Meyer*, [1930] 2 K. B. 312.

(*s*) *Andrews Brothers, Ltd.* v. *Singer & Co., Ltd.*, [1934] 1 K. B. 17, C. A.

(*t*) *L'Estrange* v. *Graucob, Ltd.*, [1934] 2 K. B. 394.

(*u*) *Mendelssohn* v. *Normand, Ltd.*, [1969] 2 All E. R. 1215; [1970] 1 Q. B. 177, C. A.; but *cf. Jacobs* v. *Morris*, [1902] 1 Ch. 816, C. A.

(*v*) *Overbrooke Estates, Ltd.* v. *Glencombe Properties, Ltd.*, [1974] 3 All E. R. 511; [1974] 1 W. L. R. 1335; *cf. Heron* v. *Dilworth Equipment, Ltd.* (1963), 36 D. L. R. (2d.) 462; *Re J. I. Case Threshing and Machine Co.* v. *Mitten* (1919), 49 D. L. R. 30 (Sup. Ct. of Can.).

(*a*) *Webster* v. *Higgin*, [1948] 2 All E. R. 127, C. A.; and see *Docker* v. *Hyams*, [1969] 3 All E. R. 808; [1969] 1 W. L. R. 1060, C. A.

(*b*) *Francis* v. *Trans-Canada Trailer Sales, Ltd.* (1969), 6 D. L. R. (3d) 705; see also *Western Tractors, Ltd.* v. *Dyck* (1969), 7 D. L. R. (3d) 535.

(*c*) *R. G. MacLean, Ltd.* v. *Canadian Vickers, Ltd.* (1971), 15 D. L. R. (3d) 15.

"No warranty whatsoever is given by the owner as to the age, state or quality of the goods or as to their fitness for any purpose, and any implied warranties are hereby excluded" (*d*).

"There is no warranty as to the year model even if stated herein" (*e*).

"The goods delivered shall be deemed to be in all respects in accordance with the contract" unless complaint is made within 14 days (*f*).

"Any claim must be made within 14 days from the final discharge of the goods" (*g*).

"The products supplied shall be of good quality . . . provided that the company shall not be liable for any claim under this clause or in any other way whatsoever in relation to the products unless such claim is notified to the company within fourteen days of delivery of the products" (*h*).

"Horses not answering the description must be returned before 5 o'clock on Wednesday" (*i*).

"Horses warranted good workers, not answering such warranty, to be returned before 5 o'clock of the day after the sale, and shall then be tried by a person to be appointed by the auctioneer" (*k*).

"The buyer takes responsibility of any latent defects" (*l*).

"Differences of slight nature in quality, designs, shades, dimensions etc. is allowed" (*m*).

"Buyers shall not reject the goods herein specified, etc." (*n*).

"The works sell engines under the condition that they are free from any claims arising through the breakdown of any parts or stoppages of the engines, or from any consequential damage arising from the same direct or indirect" (*o*).

(*d*) *Yeoman Credit, Ltd.* v. *Apps*, [1961] 2 All E. R. 281; [1962] 2 Q. B. 508, C. A.

(*e*) *F. & B. Transport, Ltd.* v. *White Truck Sales Manitoba, Ltd.* (1965), 51 W. W. R. 124.

(*f*) *Beck & Co.* v. *Szymanowski & Co.*, [1924] A. C. 43.

(*g*) *Smeaton Hanscomb* v. *Sassoon I Setty Son & Co.*, [1953] 2 All E. R. 1471; [1953] 1 W. L. R. 1468.

(*h*) *Van der Sterren* v. *Cybernetics (Holdings) Pty. Ltd.* (1970), 44 A. L. J. R. 157 (H. C. of Austr.).

(*i*) *Head* v. *Tattersall* (1871), L. R. 7 Exch. 7; *cf. Chapman* v. *Withers* (1888) 20 Q. B. D. 824.

(*k*) *Hinchcliffe* v. *Barwick* (1880), 5 Ex. D. 177, C. A.

(*l*) *Henry Kendall & Sons* v. *William Lillico & Sons, Ltd.*, [1968] 2 All E. R. 444; [1969] 2 A. C. 31, H. L.

(*m*) *Esmail* v. *Rosenthal & Sons, Ltd.*, [1964] 2 Lloyd's Rep. 447, C. A.; on appeal, [1965] 2 All E. R. 860; [1965] 1 W. L. R. 1117.

(*n*) *Meyer, Ltd.* v. *Kivisto* (1929), 35 Ll. L. R. 265; *Meyer, Ltd.* v. *Travaru A/B H. Cornelius* (1930), 37 Ll. L. R. 204; *Meyer* v. *Osakeyhtio Carelia Timber Co.* (1930), 37 Ll. L. R. 212 (distinguishing *Meyer* v. *Kivisto, supra*); *Green* v. *Arcos, Ltd.* (1931), 47 T. L. R. 336; *White Sea Timber Trust, Ltd.* v. *W. W. North, Ltd.* (1932), 148 L. T. 263.

(*o*) *Pollock* v. *Macrae*, 1922 S. C. 192, H. L.

" We shall not be responsible in any event, for loss of profits, detention, personal injury, damage to property or for any other consequential damage, loss or expenses whatsoever arising from defects in the manufacture or erection of our goods " (*p*).

Defendants " will not accept responsibility for any loss . . . howsoever caused " (*q*).

No liability " for damage caused by water " (*r*).

" Nothing in this agreement shall render owners liable for any personal injury to the riders of the machines hired nor for any third party claims, nor losses of any goods belonging to the hirer in the machine " (*a*).

" I believe the mare to be sound, but I will not warrant her " (*b*).

(2) *Other terms as to quantity.*

" 18 pockets Kent hops " (*c*).

" 1,000 bales of gambier " (*d*).

" Cargo " (*e*).

" A full and complete cargo of sugar and molasses " (*f*).

" A small cargo of lath-wood (specifying lengths), in all about 60 cubic fathoms " (*g*).

" A cargo of from 2,500 to 3,000 barrels (seller's option) American petroleum " (*h*).

" The cargo " loaded on steamship R. (*i*).

" About 300 quarters more or less of foreign rye shipped at Hamburg " (*k*).

" 400 tons (approx.) " (*l*).

(*p*) *Mechans, Ltd.* v. *Highland Marine Charters, Ltd.*, 1964 S. C. 48 (Ct. of Sess.).

(*q*) *Mendelssohn* v. *Normand, Ltd.*, [1969] 2 All E. R. 1215; [1970] 1 Q. B. 177, C. A.

(*r*) *Akerib* v. *Booth*, [1961] 1 All E. R. 380; [1961] 1 W. L. R. 367, C. A.

(*a*) *White* v. *John Warwick & Co.*, [1953] 2 All E. R. 1021; [1953] 1 W. L. R. 1285, C. A.

(*b*) *Wood* v. *Smith* (1829), 5 M. & R. 124.

(*c*) *Spicer* v. *Cooper* (1841), 1 Q. B. 424.

(*d*) *Gorrissen* v. *Perrin* (1857), 2 C. B. N.S. 681.

(*e*) *Anderson* v. *Morice* (1876), 1 App. Cas. 713; *Colonial Ins. Co.* v. *Adelaide Ins. Co.* (1886), 12 App. Cas. 128, at pp. 129, 130; *Miller* v. *Borner*, [1900] 1 Q. B. 691, at 692; and see 34 Halsbury's Laws (3rd Edn.) 66.

(*f*) *Cuthbert* v. *Cumming* (1855), 11 Exch. 405, Ex. Ch.; *Margaronis Navigation Agency, Ltd.* v. *W. Peabody & Co. of London, Ltd.*, [1964] 3 All E. R. 333; [1965] 2 Q. B. 430, C. A.

(*g*) *Kreuger* v. *Blanck* (1870), L. R. 5 Exch. 179; but see *Ireland* v. *Livingston* (1872), L. R. 5 H. L. 395, at pp. 405, 410.

(*h*) *Borrowman* v. *Drayton* (1876), 2 Ex. D. 15, C. A.; *Re Harrison & Micks*, [1917] 1 K. B. 755 (remainder of cargo).

(*i*) *Paul* v. *Pim*, [1922] 2 K. B. 360.

(*k*) *Cross* v. *Eglin* (1831), 2 B. & Ad. 106.

(*l*) *Three Rivers Trading Co., Ltd.* v. *Gwinnear and District Farmers, Ltd.* (1967), 111 Sol. Jo. 831, C. A.

" Say from 1,000 to 1,200 gallons per month " (*m*).

" Say not less than 100 packs of combing skin at 7*d*. per lb." (*n*).

" The quantity to be taken from the bill of lading " (*o*).

"We hold to your order about 30 tons Saint Petersburg hemp " (*p*).

" 100 tons of Wallsend coals, more or less " (*q*).

" Say about 600 red pine spars averaging 16 inches " (*a*).

" 25 tons, more or less, Penang pepper; name of vessel or vessels to be declared within 60 days from the date of bill of lading " (*b*).

" About 150 tons of scrap iron " (*c*).

" The whole of the steel required for the Forth Bridge. The estimated quantity we understand to be 30,000 tons, more or less " (*d*).

" About 500 loads of timber " (*e*).

Nitric acid: " All our requirements during 12 months " (*f*).

" Whale oil. Entire production by a named steamer for the season 1930–1931 " (*g*).

" Buyer's total requirements up to 8,000 tons " (*h*).

" About 10,000 tons Newcastle coal in monthly shipments " (*i*).

" Quantity 1,750/2,500 tons " (coal contract) (*k*).

" Estimated 8/10 tons " (Australian sheep skins) (*l*).

" Remainder of the cargo, more or less about 5,400 quarters of wheat " (*m*).

Ship " dead weight capacity 450 tons—not accountable for errors in description " (*n*).

Eggs and butter " if and to the extent the same shall be required " (*o*).

(*m*) *Gwillim* v. *Daniell* (1835), 2 C. M. & R. 61; *Morris* v. *Levison* (1876), 1 C. P. D. 155, at p. 159.

(*n*) *Leeming* v. *Snaith* (1851), 16 Q. B. 275.

(*o*) *Covas* v. *Bingham* (1853), 2 E. & B. 836.

(*p*) *Moore* v. *Campbell* (1854), 10 Exch. 323.

(*q*) *Cockerell* v. *Aucompte* (1857), 2 C. B. N.S. 440; *cf. Bourne* v. *Seymour* (1855), 16 C. B. 337.

(*a*) *McConnel* v. *Murphy* (1873), L. R. 5 P. C. 203.

(*b*) *Reuter* v. *Sala* (1879), 4 C. P. D. 239, C. A.

(*c*) *McLay* v. *Perry* (1881), 44 L. T. 152.

(*d*) *Tancred* v. *Steel Co. of Scotland* (1890), 15 App. Cas. 125.

(*e*) *Harland* v. *Burstall* (1901), 6 Com. Cas. 113, 116 (preliminary negotiation excluded).

(*f*) *Berk* v. *International Explosives Co.* (1901), 7 Com. Cas. 20.

(*g*) *Hvalfangerselskabet Polaris Aktieselskap* v. *Unilever, Ltd.* (1933), 39 Com. Cas. 1, H. L.

(*h*) *Kier & Co., Ltd.* v. *Whitehead Iron & Steel Co., Ltd.*, [1938] 1 All E. R. 591.

(*i*) *Societé Anonyme* v. *Scholefield* (1902), 7 Com. Cas. 114 (custom of trade as to 5 per cent. margin).

(*k*) *Doe* v. *Bowater, Ltd.*, [1916] W. N. 185.

(*l*) *Tebbitts Bros.* v. *Smith* (1917), 33 T. L. R. 508, C. A.

(*m*) *Re Harrison & Micks*, [1917] 1 K. B. 755.

(*n*) *Harrison* v. *Knowles & Foster*, [1918] 1 K. B. 608, C. A.

(*o*) *Percival, Ltd.* v. *L. C. C. Asylums Committee* (1918), 87 L. J. K. B. 677.

4,000 tons of meal " 2 per cent. more or less " (*p*).

200 tons Australian tallow " five per cent. more or less " (*q*).

" Frozen meat carcasses " average not to exceed 60 lbs. (*r*).

" Requirements up to, but not exceeding 45,000 Imperial gallons per week " (*s*).

(3) *Other terms as to description and quality.*

Carriage to be built " to meet my convenience and taste " (*t*).

" To be to the entire satisfaction of the owner's representatives and ourselves " (*u*).

" Acceptance of the balance of the contract is subject to the sample parcel being satisfactorily received and approved by buyers " (*v*).

Sale of ship " subject to satisfactory survey " (*w*).

" After the completion of such inspection and/or survey, if any material defect or defects . . . shall have been found, the purchaser may give notice to the vendor . . . of his rejection of the yacht by indicating the nature of the defect or defects " (*a*).

" Scott and Co.'s mess pork " (*b*).

" Your wool at 16*s*. a stone " (*c*).

" Prime singed bacon " (*d*).

" Fine barley " and " good barley " (*e*).

" Ware potatoes " (*f*).

Potatoes " sound bags only " (*g*).

" 50 tons best palm oil; wet and inferior oil, if any, at a fair allowance " (*h*).

" Seed barley " (*i*).

(*p*) *Payne* v. *Lillico & Sons* (1920), 36 T. L. R. 569 ; *cf. Keighley* v. *Bryan* (1894), 70 L. T. 155 (3,000 tons of wheat, 10 per cent. more or less).

(*q*) *Re Thornett & Fehr & Yuills*, [1921] 1 K. B. 219 (seller's option).

(*r*) *Ballantine & Co.* v. *Cramp & Bosman* (1923), 129 L. T. 502.

(*s*) *Cory Bros. & Co., Ltd.* v. *Universe Petroleum Co., Ltd.* (1933), 46 Ll. L. R. 309.

(*t*) *Andrews* v. *Belfield* (1857), 2 C. B. N.S. 779.

(*u*) *Cammell Laird & Co., Ltd.* v. *The Manganese Bronze & Brass Co.*, [1934] A. C. 402 (contract for propellers for ship).

(*v*) *Wood Components of London* v. *James Webster, Ltd.*, [1959] 2 Lloyd's Rep. 200.

(*w*) *Astra Trust, Ltd.* v. *Adams and Williams*, [1969] 1 Lloyd's Rep. 81.

(*a*) *Docker* v. *Hyams*, [1969] 3 All E. R. 808; [1969] 1 W. L. R. 1060, C. A.

(*b*) *Powell* v. *Horton* (1836), 2 Bing. N. C. 668 ; *cf. Johnson* v. *Raylton* (1881), 7 Q. B. D. 438, C. A.

(*c*) *Macdonald* v. *Longbottom* (1860), 29 L. J. Q. B. 256, Ex. Ch. ; *cf. McCollin* v. *Gilpin* (1881), 6 Q. B. D. 516, C. A.

(*d*) *Yates* v. *Pym* (1816), 6 Taunt. 446.

(*e*) *Hutchinson* v. *Bowker* (1839), 5 M. & W. 535.

(*f*) *Smith* v. *Jeffryes* (1846), 15 M. & W. 561.

(*g*) *Glass's Fruit Markets, Ltd.* v. *A. Southwell & Son (Fruit), Ltd.*, [1969] 2 Lloyd's Rep. 398.

(*h*) *Lucas* v. *Bristow* (1858), E. B. & E. 907; *cf. Warde* v. *Stuart* (1856), 1 C. B. N.S. 88.

(*i*) *Carter* v. *Crick* (1859), 4 H. & N. 412.

Barley " about as per sample " (*j*).
" English sanfoin " (*k*).
" Copra cake " (*l*).
" Foreign refined rape oil " (*m*).
" 15 tons medium onions " (*n*).
" No. 3 hard amber durum wheat " (*o*).
Linen " authentic property of Charles I " (*p*).
" Four pictures, views in Venice. Canaletti " (*q*).
Timber to be shipped " under deck " (*r*).
Goods " afloat " (*s*).
Peaches " afloat per s.s. *Morton Bay,* due London approx. June 8th " (*t*).
" Briquettes, size 2 inches " (*a*).
50 tons galvanised steel sheets " assorted over 6, 7, 8, 9 and 10 feet long. Assorted tonnage per size " (*b*).
Rubber rolls of specified weight and size (*c*).
3,000 tins of canned fruit to be packed in cases each containing 30 tins " (*d*).
Moroccan canary seed " on a pure basis " (*e*).

(*j*) *Re Walkers,* [1904] 2 K. B. 152 (custom of London Corn Exchange).

(*k*) *Wallis* v. *Pratt,* [1911] A. C. 394; and see *Thornley* v. *Tuckwell* (*Butchers*), *Ltd.*, [1964] Crim. L. R. 127 (New Zealand lamb).

(*l*) *Pinnock Brothers* v. *Lewis and Peat,* [1923] 1 K. B. 690; see also *Henry Kendall & Son* v. *William Lillico & Sons, Ltd.*, [1968] 2 All E. R. 444; [1969] 2 A. C. 31, H. L.; *Ashington Piggeries, Ltd.* v. *Christopher Hill, Ltd.*, [1971] 1 All E. R. 847; [1972] A. C. 441, H. L.

(*m*) *Nichol* v. *Godts* (1854), 10 Exch. 191.

(*n*) *Rapalli* v. *K. L. Take,* [1958] 2 Lloyd's Rep. 469, C. A.

(*o*) *Toepfer* v. *Continental Grain Co.*, [1973] 1 Lloyd's Rep. 289.

(*p*) *Nicholson and Venn* v. *Smith-Marriott* (1947), 177 L. T. 189.

(*q*) *Power* v. *Barham* (1836), 4 Ad. & El. 473; *cf. Hyslop* v. *Shirlaw* 1905 7 F. (Ct. of Sess.) 875 (artist's name specified in receipt); *Leaf* v. *International Galleries,* [1950] 1 All E. R. 693; [1950] 2 K. B. 86, C. A. (picture " by John Constable").

(*r*) *Meyer, Ltd.* v. *Travaru A/B H. Cornelius* (1930), 37 Ll. L. R. 204; *White Sea Timber Trust, Ltd.* v. *W. W. North, Ltd.* (1932), 148 L. T. 263; and see *Messers, Ltd.* v. *Morrison's Export Co.*, [1939] 1 All E. R. 92 ("to be loaded on deck one-third at British Columbia ").

(*s*) *Benabu* v. *Produce Brokers Co., Ltd.* (1921), 37 T. L. R. 609; affd. *ibid.*, p. 851.

(*t*) *Macpherson Train & Co., Ltd.* v. *Howard Ross & Co., Ltd.*, [1955] 2 All E. R. 445; [1955] 1 W. L. R. 640.

(*a*) *William Barker* (*Junior*) *& Co., Ltd.* v. *Ed. T. Agius, Ltd.* (1927), 33 Com. Cas. 120; see also *Arcos* v. *E. and A. Ronaasen & Son,* [1933] A. C. 470 (staves ½ in. thickness); *Allard & Co.* (*Rubber*), *Ltd.* v. *R. J. Hawkins & Co.*, [1958] 1 Ll. R. 184 (steel rods ⅜ in. to ½ in. thickness).

(*b*) *Ebrahim Dawood, Ltd.* v. *Heath, Ltd.*, [1961] 2 Ll. R. 512.

(*c*) *E. and S. Ruben, Ltd.* v. *Faire Brothers & Co., Ltd.*, [1949] 1 All E. R. 215; [1949] 1 K. B. 254; contrast *Smeaton Hanscomb & Co., Ltd.* v. *Sassoon I. Setty Son & Co.*, [1953] 2 All E. R. 1471.

(*d*) *Re Moore & Co. and Landauer & Co.*, [1921] 2 K. B. 519, C. A.; *cf. Manbre Saccharine Co.* v. *Corn Products Co.*, [1919] 1 K. B. 198, 207.

(*e*) *Peter Darlington Partners, Ltd.* v. *Gosho Co., Ltd.*, [1964] 1 Lloyd's Rep. 149.

A reaping machine, " practically new ", used to cut only 50 acres (*f*).

Machinery described in catalogue as " new and unused " (*g*).

A new Singer car (*h*).

" A 1961 Triumph Herald 1200 car " (*i*).

A Holden Standard car not used as a taxi (*j*).

A new Bedford tipping lorry (*k*).

A 1958 truck. (*l*).

" All equipment reconditioned as necessary " (*m*).

Parcels of wood laths " about the specification mentioned below " (*n*).

" Quality as previously delivered " (*o*).

" 413 bales of wool guaranteed about similar to samples in selling broker's possession " (*p*).

Long staple salem cotton " guaranteed equal to sample; should the quality prove inferior to guarantee, a fair allowance to be made " (*q*).

Coal " reasonably free from stone and shale " (*r*).

" Norwegian Herring Meal, fair average quality of the season " (*s*).

" 300 tons fair usual quality Jelutong rubber " (*t*).

" Warranted sound " (*u*).

(*f*) *Varley* v. *Whipp*, [1900] 1 Q. B. 513.

(*g*) *Teheran-Europe Co.* v. *S. T. Belton (Tractors), Ltd.*, [1968] 2 All E. R. 886; [1968] 2 Q. B. 545, C. A.

(*h*) *Andrews Brothers, Ltd.* v. *Singer & Co., Ltd.*, [1934] 1 K. B. 17; and see *Morris Motors* v. *Lilley*, [1959] 3 All E. R. 737; [1959] 1 W. L. R. 1184; folld. in *Morris Motors* v. *Phelan*, [1960] 2 All E. R. 208 n.; [1960] 1 W. L. R. 352, 566 n.; *Taylor* v. *Combined Buyers, Ltd.*, [1924] N. Z. L. R. 627.

(*i*) *Beale* v. *Taylor*, [1967] 3 All E. R. 253; [1967] 1 W. L. R. 1193, C. A.

(*j*) *Armaghdown Motors, Ltd.* v. *Gray Motors, Ltd.*, [1963] N. Z. L. R. 5.

(*k*) *Astley Industrial Trust, Ltd.* v. *Grimley*, [1963] 2 All E. R. 33; [1963] 1 W. L. R. 584, C. A.

(*l*) *F. and B. Transport, Ltd.* v. *White Truck Sales Manitoba, Ltd.* (1965), 51 W. W. R. 124; contrast *Diamond* v. *B. C. Thoroughbred Breeders' Socy.* (1966), 52 D. L. R. (2d.) 146 (lineage of horse goes only to quality).

(*m*) *George Reid Inc.* v. *Nelson Machinery Co.*, [1973] 2 W. W. R. 597.

(*n*) *Vigers* v. *Sanderson*, [1901] 1 K. B. 608.

(*o*) *Steels and Busks* v. *Bleecker Bik & Co.*, [1956] 1 Lloyd's Rep. 228.

(*p*) *Heyworth* v. *Hutchinson* (1867), L. R. 2 Q. B. 447.

(*q*) *Azèmar* v. *Casella* (1867), L. R. 2 C. P. 677, Ex. Ch.; *Hookway* v. *Alfred Isaacs & Sons*, [1954] 1 Lloyd's Rep. 491 (" quality equal to London standard sample ").

(*r*) *Dominion Coal Co.* v. *Dominion Iron and Steel Co.*, [1909] A. C. 293, P. C.

(*s*) *Ashington Piggeries, Ltd.* v. *Christopher Hill, Ltd.*, [1971] 1 All E. R. 847; [1972] A. C. 441, H. L.

(*t*) *Re North Western Rubber Co.*, [1908] 2 K. B. 907 (usage) ; disapproved, *Produce Brokers Co.* v. *Olympia Oil & Cake Co.*, [1916] A. C. 314.

(*u*) *Kiddell* v. *Burnard* (1842), 9 M. & W. 668 ; *Holliday* v. *Morgan* (1858), 1 E. & E. 1; *Hinchcliffe* v. *Barwick* (1880), 5 Ex. D. 177, C. A. For lists of defects constituting unsoundness, see Oliphant's *Law of Horses* (6th ed.), chap. 4.

"Guaranteed" (*a*).
"Received £10 for a grey 4-year-old colt, warranted sound" (*b*).
"You need not look for anything, the horse is perfectly sound" (*c*).
Seller "agrees to protect the purchaser against all claims, losses, damages, costs and expenses which arise from or occur as a result of the sales against any item published" (*d*).

As to sale or return

"On sale or return" (*e*)
"On memorandum," *i.e.* on approval (*f*)
"On approbation, or sale for net cash or return" (*g*)
"If steamer not approved of, deposit to be returned immediately" (*h*)

NOTE C.—DELIVERY TO CARRIER

Frequent reference has been made to the rule that delivery of goods to a carrier is *primâ facie* a delivery to the buyer and a performance of the seller's contract which passes both the property and the risk to the buyer. It follows that, as a rule, if the goods are lost or destroyed, the buyer or consigneee is the proper person to sue the carrier. The most authoritative statement of the principle is in the judgment of the House of Lords in *Dunlop* v. *Lambert* (*i*), where it was held that if there was a special contract the consignor might sue the carrier though the goods might be the property of the consignee. Lord COTTENHAM there says (*k*):

> "It is no doubt true as a general rule that the delivery by the consignor to the carrier is a delivery to the consignee, and that the risk is after such delivery the risk of the consignee. This is so if, without designating the particular carrier, the consignee directs that the goods shall be sent by the ordinary conveyance: the delivery to the ordinary carrier is then a delivery to the consignee, and the consignee incurs all the risks of the carriage. And it is still more strongly so if the goods are sent by a carrier specially pointed out by the consignee himself, for such carrier then becomes his special agent.

(*a*) *Algoma Truck and Tractor Sales* v. *Bert's Auto Supply, Ltd.*, [1968] 2 O. R. 153.
(*b*) *Budd* v. *Fairmaner* (1831), 8 Bing. 48; *Anthony* v. *Halstead* (1877), 37 L. T. 433.
(*c*) *Schawel* v. *Reade*, [1913] 2 I. R. 64, H. L.
(*d*) *Helfand* v. *Royal Canadian Art Pottery* (1970), 11 D. L. R. (3d) 404.
(*e*) *Kirkham* v. *Attenborough*, [1897] 1 Q. B. 201, C. A.
(*f*) *Bevington* v. *Dale* (1902), 7 Com. Cas. 112.
(*g*) *Weiner* v. *Gill*, [1906] 2 K. B. 574, C. A.; followed *Kempler* v. *Bravingtons* (1925), 133 L. T. 680, C. A.; distinguished, *Weiner* v. *Harris*, [1910] 1 K. B. 285.
(*h*) *Haegerstrand* v. *Annie Thomas Steamship Co.* (1905), 10 Com. Cas. 67.
(*i*) (1839), 6 Cl. & Fin. 600.
(*k*) *Ibid.*, at pp. 620, 621.

But, though the authorities all establish the general inference I have stated, yet that general inference is capable of being varied by the circumstances of any special arrangement between the parties, or of any particular mode of dealing between them. If a particular contract be proved between the consignor and the consignee, and the circumstance of the payment of the freight and insurance is not alone a conclusive evidence of ownership—as where the party undertaking to consign undertakes to deliver at a particular place—the property, till it reaches that place and is delivered according to the terms of the contract, is at the risk of the consignor. And again, though in general the following the directions of the consignee, and delivering the goods to a particular carrier, will relieve the consignor from the risk, he may make such a special contract that, though delivering the goods to the carrier specially intimated to the consignee, the risk may remain with him; and the consignor may, by a contract with the carrier, make the carrier liable to himself. In an infinite variety of circumstances, the ordinary rule may turn out not to be that which regulates the liabilities of the parties" (*l*).

Delivery to carrier to pass property.—This passage is discussed by BLACKBURN, J., in an instructive judgment in *The Calcutta Co.* v. *De Mattos* (*m*), which has often been referred to in the text, but which was too long for insertion there. He says (*n*):

"What was the effect of the contract as regards the property in the goods and the right to the price, from the time of the handing over the shipping documents and paying half of the invoice value? There is no rule of law to prevent the parties, in cases like the present, from making whatever bargain they please. If they use words in the contract showing that they intend that the goods shall be shipped by the person who is to supply them, on the terms that when shipped they shall be the consignee's property, and at his risk, so that the vendor shall be paid for them whether delivered at the port of destination or not, this intention is effectual. Such is the common case where goods are ordered to be sent by a carrier to a port of destination. The vendor's duty is, in such cases, at an end when he has delivered the goods to the carrier, and, if the goods perish in the carrier's hands, the vendor is discharged and the purchaser is bound to pay him the price. If the parties intend that the vendor shall not merely deliver the goods to the carrier, but also undertake that they shall actually be delivered at their destination, and express such intention, this also is effectual; in such a case, if the goods perish in the hands of the carrier, the vendor is not only not entitled to the price, but he is liable for whatever damage may have been sustained by the purchaser in consequence of the breach of the vendor's contract to deliver at the place of destination. But the parties may intend an intermediate state of things; they may intend that the vendor shall deliver the goods to the carrier, and that, when he has done so, he shall have fulfilled his undertaking, so that he

(*l*) *Cf. Young* v. *Hobson* (1949), 65 T. L. R. 365.
(*m*) (1863), 32 L. J. Q. B. 322.
(*n*) *Ibid.*, at p. 328.

shall not be liable in damages for a breach of contract if the goods do not reach their destination; and yet they may intend that the whole or part of the price shall not be payable unless the goods do arrive. They may bargain that the property shall vest in the purchaser, as owner, as soon as the goods are shipped, that they shall then be both sold and delivered, and yet that the price (in whole or in part) shall be payable only on the contingency of the goods arriving, just as they might, if they pleased, contract that the price should not be payable unless a particular tree fall; but without any contract on the vendor's part in the one case to procure the goods to arrive, or in the other to cause the tree to fall. Where the contract is of this kind, the position of the vendor and purchaser, in case the goods do not arrive, is analogous to that of freighter and shipowner, in the ordinary contract of carriage on board a ship, in case the goods are prevented from arriving by one of the excepted perils. The shipowner is not bound to carry and deliver at all events; but, though he is excused if prevented by the excepted perils, yet no freight is earned or payable unless the goods are delivered. In the case of freight, also, the question often arises, whether a payment made at the port of shipment is an advance of part of the freight, returnable if the goods are not delivered and freight earned, or is an absolute payment, leaving only the balance contingent on the safe delivery of the goods—a question very analogous to the one that arises on the present contract" (*o*).

"Carriage forward."—The effect of the ordinary "carriage forward" contract is thus expressed by MELLOR, J.:

"There is evidence in the present case that these goods were, with the consent or by the authority of the purchaser, consigned by the vendors, as consignors, to be carried by the defendants as common carriers, to be delivered to the purchaser as consignee, and that the name of the consignee was made known to the defendants at the time of the delivery. Under such circumstances the ordinary inference is that the contract of carriage is between the carrier and the consignee, the consignor being the agent for the consignee to make it. It appears to us that there is evidence also that at the time of the delivery there was a specific mention that the freight was to be paid by the consignee. Under such circumstances the inference is almost irresistible that the contract for carriage in the present case was the ordinary contract for carriage at the cost and risk and under the control of the consignee" (*p*).

(*o*) See the cases as to prepayment of freight collected in *M'Lachlan on Shipping* (7th ed.), p. 434.

(*p*) *Cork Distilleries Co.* v. *Great Southern and Western Rail. Co.* (1874), L. R. H. L. 269, at p. 277.

INDEX

ABANDONMENT OF CONTRACT,
conduct evidencing, 22, 109

" ABOUT,"
quantity described by the word, 188, 363, 364

ACCEPTANCE,
breach of condition, after, 108, 109, 245
warranty, after, 108, 109, 245
c.i.f. contract, in case of, 200
conditional, 200
defined, 197
delivery after re-sale, as, 199
duty of buyer as to, 179
evidence of, 197
instalment contracts, in, 188
non-acceptance, damages for, 231
defective goods, of, 199
parol, of written offer, 93
part of goods, of, 109, 186–188
re-sale before delivery, need not be considered as, 199
rules for presuming, 199
wrong quantity delivered, where, 186

ACCESSORIES AND ACCRETIONS,
property and risk in, 161

ACKNOWLEDGEMENT,
third person holding, by, to buyer, 183, 277

ACT OF PARLIAMENT,
application of Interpretation Act, 1889 . . . 73
contract made illegal by, 20
See also SALE OF GOODS ACT, 1893, and Appendix I (STATUTES).

ACTION,
assignee of contract or price, by, 28, 29
available as remedy, when, 266
breach of warranty, for, 245
consideration, failure of, 257
consumer protection offences, 54
conversion or detinue, for, 30, 31, 220, 241–243
defined, 273
delayed delivery, for, 241
enemy, by or against, 89, 90
interest and special damages, 254
lex fori, application of, 49
loss on resale after buyer's default, for, 223
non-acceptance, for, 231
non-delivery, for, 235
price, for, 226
specific performance, for, 243
tort or contract, in, 7, 31, 241–243
See also DAMAGES.

ADVERTISEMENT,
trade description used in, 59

AFTER-ACQUIRED PROPERTY,
assignment of, 96

AGENT,
authority to contract distinguished from capacity, 86
carriers as, 191, 192, 368
confirming house, 46
del credere, 76
Factors Act, 1889, with reference to, 289 *et seq.*
See also FACTORS ACT.
mercantile. *See* MERCANTILE AGENT.
necessity, of, sale by, 167, 202, 224
obtaining advance on property of principal, 295
revocation of authority, sale after, 300
sale by unauthorised, 161, 164
saving for general law of principal and, 271
ship, master or captain of, 93, 167

AGREEMENT TO SELL,
breach of, remedies for, 84, 226 *et seq.*
contract of sale includes, 74, 84, 276
defined, 74, 75
distinguished from sale, 84
hire-purchase agreement as, 81
oral, written agreement modified or rescinded by, 21, 22
parol evidence as to, 4
passing into sale, 75, 84, 85
See also CONTRACT OF SALE; SALE.

ALIEN, ENEMY. *See* ENEMY.

" ALL FAULTS,"
sale with, 360

" ALL RISKS,"
insurance against, 40

AMBIGUITY
construction in case of, 75

ANTECEDENT DEBT,
pledge of document of title or goods for, 302
transfer of bill of lading for, 222, 320

APPARENT OWNER,
disposition by, 162, 290

APPARENT POSSESSION,
what amounts to, under Bills of Sale Act, 320

APPROPRIATION,
delivery to carrier as, 148
goods of, to contract, 140 *et seq.*

APPROVAL,
sale on, 147, 152, 368

ARBITRATION,
distinguished from valuation, 105

ARRIVAL,
sale of goods " to arrive," 356

ASCERTAINED GOODS,
passing of property, 143 *et seq.*
specific performance for, 243

ASSENT,
appropriation of goods to contract, to, 141, 148
attornment of carrier, to, 217, 218
fraud, induced by, 31, 172, 175
seller, of, to sub-sale by buyer, 220, 221

ASSIGNMENT,
after-acquired or future goods, of, 95, 96
contract of sale, of, 28–30
document of title, of, 185, 221, 295
equitable, 96
price, of, 28

ATTORNMENT,
carrier, by, to end stoppage *in transitu*, 217
delivery effected by, 183, 277

AUCTION,
abstention from bidding, 269
Bidding Agreements Act, 269, 343
exemption clause, protection concerning, judicially construed, 360, 361
rights of seller, 343
sale by, incidents of, 266–269, 343
not a consumer sale, 260, 263
under distress warrant, 112, 114

AVAILABLE MARKET,
meaning of term, 233

BAILEE,
buyer or seller as bailee for the other, 158, 160, 306
delivery by attornment of, 183, 277
delivery to, right of disposal reserved, 154
transmission to buyer, for, 148
includes custodier in Scotland, 273
seller's lien, when he holds as, 206
See also CARRIER; PLEDGE.

BAILMENT,
sale distinguished from, 76

BANKER,
contract between buyer and, 46
irrevocable letters of credit of, 42–46, 231

BANKRUPTCY,
agent, of, 312
insolvent, who is deemed to be, 287
See also INSOLVENCY.
saving for law of, 271
trustee in, affirmation of contract by, 210
disclaimer of contract by, 272
rights and duties of, 272
sale to bankrupt by, 75
saving in Factors Act as to, 312

BARGAIN AND SALE,
included in term " sale," 284

BARTER,
sale distinguished from, 79

BILL OF EXCHANGE,
accepted against bill of lading, 155
bill of lading compared with, 319
conditional payment, as, 205, 230
payment by, 230, 231

BILL OF LADING,
bill of exchange, compared with, 319
Bills of Lading Act, 1855, provisions of, 316
c.i.f. contract, with reference to, 37
defined, 316
document of title, as, 318
evidence of shipment, 317
foreign systems as to, 319
general note on, 318
liability under, 318
reservation of right of disposal by, 156
signature by agent, limitation upon, 318
symbolic delivery by, 277
"through bill of lading," 316
transfer of, effect of, 222, 316

BILL OF SALE,
defined, 320
general note on, 321
law as to, saving for, 271
letter of hypothecation, 83
mercantile documents not included in term, 320, 321
mortgage of goods as, 82
unregistered transfer of ship, 321

BREACH OF CONTRACT. *See* ACTION; DAMAGES.

BUSINESS,
defined, 128*n*., 273
place of, 275
purchase in the course of, 260, 261, 263
sale in the course of, 122–124, 128, 132

BUSINESS NAMES,
registration of, 91

BUYER,
acceptance of goods by, 197, 199
actions against, 226–234, 254
by, 235–259
bailee, as, 158, 160
capacity of, 85
conditional sale agreement, under, 176, 308
defined, 82, 274
delivery to, 179 *et seq.*
disposition by, after sale, 175, 220
before sale, 307, 310
duty to accept and pay, 179, 180
examination of goods by, 122, 131, 195
failure of, 131*n*.
export or import licence, obtaining, 41, 42
fetch goods away, when bound to, 182, 184
follow goods in hands of third person, power to, 84, 241
insolvent, when deemed to be, 287. *See* INSOLVENCY.
insurance, when to be effected by, 192
neglecting to take or refusing delivery, 204
obtaining goods or documents, 175, 307
payment and tender by, 179, 226 *et seq.*
pledge by, 220
possession, after sale, of, 175
consent of seller, with, 309
recovery by, 166, 188

BUYER – *continued*
rejection by, 200
reliance on seller's skill, 134
partial, 135
unreasonable, 136
representations, reliance upon, 166
repudiation of contract by, 186, 189
return of rejected goods by, 197, 203
risk of, when goods are at, 157, 191, 194
seller must be different from, 75
reliance on skill of, 123, 136
sub-sale by, 220, 307
title to goods, failure to acquire, 171–175

C.I.F. CONTRACT,
acceptance, rules as to, in case of, 200
delivery of goods, interpretation of, 197*n*.
general note on, 35, 36–40
risk passing to buyer, 159
stipulations in, judiciously construed, 359, 360
transfer of property under, 159

CAPACITY,
buy and sell, to, 85 *et seq*.
conflict of laws in relation to, 91
distinguished from authority, 86

CARGO,
sale of, 187, 363

" CARRIAGE FORWARD,"
effect of contract for, 370

CARRIER,
buyer's agent to receive goods, 193
" carriage forward " contract, effect of, 370
delivery to, 277
as appropriation, 148
duties of seller, 191 *et seq*.
goods in transit, 214, 215
passing property by, 368, 369
duty when goods stopped in transit, 219
refusal to deliver goods to buyer by, 216
reservation of right of disposal on delivery to, 154–156
risk of deterioration during transit, 194
seller's lien ended by delivery to, 211
ship chartered by buyer, 216
transit in hands of, duration of, 214–218

CAVEAT EMPTOR,
rule of, 126, 127

CHANCE,
sale of mere, 95

CHARGE. *See* BILL OF SALE; EXPENSES; MORTGAGE.

CHARTER-PARTY,
effect of, on duration of transit, 216

CHATTELS. *See* GOODS.

CHEQUE,
payment of price by, 230

CHOSE IN ACTION,
not included in term " goods," 279, 280

CIVIL LAW. *See* ROMAN LAW.

CLERK,
agreement by agency of, 303

CLUB,
liquor sold in, 80
member's liability, 93

COERCION,
saving for law as to, 271, 272

COGNATE CONTRACTS,
distinguish from sale, 75–77

COMMISSION,
sale on, 76

COMMIXTION,
goods mixed after sale, 143
mixed with goods not ordered, 186

COMMON LAW,
saving for, 271
special powers of sale under, 169

COMPENSATION,
orders against convicted persons, 27, 345

CONDITION OF GOODS,
caveat emptor and its exemptions, 126, 127
included in " quality," 283
sale by sample, in, 137

CONDITIONAL SALE AGREEMENT,
buyer, defined, 176, 308
under, cannot transfer title, 176, 308
defined, 111*n*., 176, 308
formalities, 4*n*.
nature of, 81, 82
Sale of Goods Act, application to, 81

CONDITIONS (IN CONTRACTS OF SALE),
breach of, effect of, 8, 10, 11–14, 108, 245
concurrent, 180
conditional acceptance, 203
contracts of sale, 80, 81
construction of, 105 *et seq*., 354
contingent, 349
delivery, as to, 357, 358
subject to, 146, 147, 151, 152
description, as to, 360–363, 365–368
on sale by, 117
discussion of definitions, 348
exclusion of implied, by agreement, etc., 259. *See* EXEMPTION CLAUSE.
express or implied, 259
fitness for particular purpose, of, 122 *et seq*.
fundamental breach, as to, 127
implied, exemption clauses, fair and reasonable, 259, 260, 264
void, 259, 263
generally, 105 *et seq*.
how negatived, 259
impossible of fulfilment, becoming, 97–100, 109
merchantable quality, as to, 122 *et seq*.
negatived or varied, how, 259
payment subject to, 205

CONDITIONS (IN CONTRACTS OF SALE)—*continued*
precedent or subsequent, 80, 97, 348, 349, 350
price, as to, 359
on re-sale, as to 52
promissory, 349
quality, as to, 360–363, 365–368
defined, 283
quantity, as to, 360–365
sale or return contracts, 147, 152, 368
sample, on sale by, 137, 366
stipulations judicially construed, 354 *et seq.*
subject to licences, 41
suspensive or resolutive, 80, 348
time, as to, 105, 358
title, as to, 111
waiver of, 31, 259
warranty, when to be treated as, 108, 245
See also ACTION; DAMAGES; WARRANTY.

CONDUCT,
sale implied from, 93
See also ESTOPPEL.

CONFIRMED CREDIT,
irrevocable letter of, 42–46, 231
when to be opened, 106

CONFIRMING HOUSE,
del credere agency distinguished, 47
functions of, 46

CONFLICT OF LAWS,
capacity in relation to, 91
contract of sale, 47–49, 265
evidence, law governing, 49
illegality, effect of, 48
lex fori, 49
lex loci contractus, 48, 91
lex situs, 49
market overt, title acquired in, 49
pledge, validity of, 49
procedure, law governing, 49
proper law of contract, 47, 48
sale of goods, as to, 50

CONSIDERATION,
disposition of goods, for, under Factors Act, 302
failure of, effect of, 188, 257, 258
price as, in contracts of sale, 74, 79. *See* PRICE.

CONSIGNEE,
bill of lading, rights under, 316
lien of, 303, 304

CONSTRUCTION,
ambiguous contract, of 75–77
canon of, as to codifying Act, 73
stipulations judicially construed, 354 *et seq.*
See also CONTRACT.

CONSTRUCTIVE DELIVERY,
how effected, 184, 210, 277

CONSUMER CREDIT,
Act of 1974 . . . 54, 63–65
enforcement, 71
powers of authorities, 72
offences, corporations, by, 65*n*.
defence of due diligence, 67
penalties, 69
prosecution, 71
agreement, conditional sale agreement, 176, 308
extortionate, 65
meaning, 63
ancillary credit business, 64
licensing, 64
business, defined, 64*n*.
provision of credit in course of, 64
credit, brokerage, 64
defined, 64
debtor-creditor-supplier agreement, 63, 64
Director-General of Fair Trading, functions, 63
notice of conviction, etc., to, 70
regulated agreement, 63, 64, 103*n*.
system of control, 63

CONSUMER PROTECTION,
Act of 1961 . . . 54, 55
offences, act or default of another, 66, 68
corporations, by, 65*n*.
defence of due diligence, 67, 68
regulations, power to make, 55
sale in breach of, 55
Act of 1971 . . . 54, 55
offences, corporations, by, 65*n*.
defence of due diligence, 67
regulations, power to make, 55
sale in breach of, 55
breach of statutory duty, remedies of buyer, 54, 55
compensation orders, 346
Consumer Credit Act 1974 . . . 54, 63–65,
Director-General of Fair Trading, functions, 61–63, 71
notice of conviction, etc., to, 70
enforcement of legislation, 70–72
Tair Trading Act 1973 . . . 54, 61–63
generally, 54
offences, accomplices, 56
act or default of another, 66
conviction, notice to Director-General of Fair Trading, 70
corporations, by, 65, 66
defences, act or default of another, 68
due diligence, 67, 68
reasonable precautions, 67, 68
obstruction of authorised officers, 72
penalties, 69
prosecution, 71
strict liability, of, 67
persons liable to prosecution, 65
powers of relevant authorities, 72
Restrictive Practices Court, 62
Trade Description Acts, 1968 and 1972 . . . 54, 56–61
Unsolicited Goods and Services Act, 1971 . . . 54, 61
Weights and Measures Act, 1963 . . . 54, 55

CONSUMER SALE,
defined, 111*n.*, 260
exemption clause, 120, 127, 138, 139, 259, 263
onus of proof that sale *not*, 261

CONTRACT,
abandonment of, evidence of, 22, 109
assignment of, 28–30
breach of, remedies for, 25–28
capacity to, 3
conditions and warranties contrasted, 8, 350
conflict of laws, 47–49, 265
consideration, 1
course of dealing, 15, 16
executed and executory, 84, 256
repudiation of, 256
exemption clauses, 10–14
express agreement, 9
formalities in, 3–5
formation of, 74 *et seq.*
frustrated, Law Reform, etc., Act, 1943 . . . 20, 21, 326–330
fundamental breach of, 11–14
illegality of, 17
supervening, 20, 21
implied terms, 14
impossibility, 19–21
intention to create legal relations, 1
intermediate term, 9
interpretation and construction in, 9
mistake, 3
offence by seller, compensation, 27, 345
offer and acceptance, 2
parol evidence, 4, 5
performance, 24
quasi, 30
relation of sale to, 1, 8
representations, 5–8
rescission, 21–23
saving for law of, 271
supervening illegality, 20, 21
terms of, 8–17
Treasury consent, 42, 330–332
Uniform Law on International Sales Act, 1967 . . . 50
usage, 17
variation, 21–24
waiver, 22–24

CONTRACT OF SALE,
absolute or conditional, 74, 80
actions on, 226 *et seq.* *See* ACTION.
assignment of, 28–30
auction sales, 266, 267, 343
buyer, duties of, 25
remedies of, 27, 28
capacity of parties, 85 *et seq.*
c.i.f. contract, 36–40
conditions and warranties in, 105 *et seq.* *See* CONDITION; WARRANTY.
conflict of laws, 265
consumer sale, 260. *See* CONSUMER SALE.
consideration for. *See* CONSIDERATION; PRICE.

CONTRACT OF SALE—*continued*
construction of, 121
conveyance, distinguished from, 84
customary international contracts, 33
definition of, 74, 75, 276
deposit in, 26
detinue, 30, 241
enforceable by action, 266. *See* ACTION.
examples of 76–78
exemption clause. *See* EXEMPTION CLAUSE.
ex-ship contract, 40
ex-store, 41
f.o.b. contracts, 34–36
foreign laws, and conflict of laws, 47–50
formalities of, 93
fraudulent representation, induced by, 173, 174
frustrated contracts, 20, 21, 100, 326–330
general nature of, 74 *et seq.*
implied terms. *See* IMPLIED TERMS.
includes sale and agreement to sell, 74, 276. *See* AGREEMENT TO SELL; SALE.
international, 33 *et seq.*
confirming houses, 46
conflict of laws, 47–49, 265
customary, 33 *et seq.*
defined, 274
exchange control, 42
exemption clauses, 120, 127, 138, 263, 272
foreign law, 49
licences, 41
payment by banker's irrevocable letters of credit, 42-46, 231
licences, 41, 42
mentally disordered person, with, 86
performance of, 24, 25
rules regulating, 179 *et seq.*
proper law of, 265
quasi-contracts of, 30, 86
remedies for breach of, 25–28
of unpaid seller against the goods, 206 *et seq.* *See* SELLER'S LIEN; STOPPAGE *IN TRANSITU*.
rescission of, 21–23
sale by estoppel, 32
seller, duties of, 24
rights of, 25, 26
severable, as, 108
stipulations in, judicially construed, 354 *et seq.*
subject-matter, existing or future goods, 94–100
terms of, division into conditions and warranties, 110
transfer of property and risk, 140 *et seq.* *See* TRANSFER.
title, 161 *et seq.* *See* TRANSFER.
trespass, 30
trover, 30, 241
usage, in relation to, 259. *See* USAGE OF TRADE.
variations of, 21–24
waiver of tort in, 31

CONVERSION,
agent or carrier by, 220
buyer's right of action for, 84, 241, 242
satisfied judgment for, effect of, 30

CONVEYANCE,
contract of sale distinguished from, 84
included in sale, 84

CO-OWNERS,
rights *inter se*, 168
sale by one to another, 74, 75

CORPORATION,
consumer protection offences, 65
defences, 69
contracts by, 93, 94

"COST, INSURANCE, AND FREIGHT,"
stipulations judicially construed, 359, 360. *See* C.I.F. CONTRACT.

COUNTER-INFLATION,
legislation, 51
statutory price control, 2

COUNTY COURT,
writ of execution issuing out of, 179

COURT,
sale by order, 162
specific performance ordered by, 243

CREDIT,
presumption against, 180–182
sale on, effect of, 181, 209
hire-purchase, nature of, 81
seller's lien with reference to, 208, 209
stoppage *in transitu* with reference to, 214

CREDIT-BROKER,
condition of fitness for purpose implied, 123, 136, 137
defined, 123

CREDIT-SALE AGREEMENT,
formalities, 4*n*.
law governing, 82

CROPS,
sale of, 281

CUSTODIER, 273. *See* BAILEE.

CUSTOM. *See* USAGE OF TRADE.

CUSTOMS DUTIES,
liability for, 322
price as affected by alteration of, 2, 323, 345

DAMAGE,
dangerous goods, from, 246, 247
transit, sustained during, 192, 194
See also LOSS; RISK.

DAMAGES,
agreed damage clause, 336
available market, 231, 235, 248
what constitutes, 233
carrier, where reasonable contract not made with, 192
causa proxima in relation to, 238, 255
conversion or detinue, in case of, 242
delay in delivering, for, 241
recovering rejected goods, 204
taking delivery, 204
full contract price, for, 256

DAMAGES—*continued*
general and special, compared, 256
goods lawfully rejected, 240
goodwill, loss of, 254
interest as, 227, 231, 254
loss caused by buyer's impecuniosity, 253
reasonably foreseeable, 253, 254, 256, 257, 334
measure of, 231, 334
misrepresentation, for, 332–334
mitigation of, duty as to, 234, 238, 250, 254, 255
nominal, 237, 247
non-acceptance of goods, for, 231
non-delivery, for, 235, 240
object of, 252, 255, 256, 334
quality, for breach of warranty of, 246, 248
rates of exchange as affecting, 239, 258
re-sale, defects discovered after, 250
goodwill, loss of, 254
losses incurred by buyer on, 253, 256
price on, 241
evidence of value, as, 249
rule in *Hadley* v. *Baxendale*, 236
special, recovery of, 254, 256
specific performance, in addition to, 243
statutory duty, breach of, 54
title, for breach of condition or warranty, 115, 116
tort, suing in, 242
warranty, for breach of, 54, 246–254, 285
defect, discoverable, 251
patent, 250
difference in value of goods, 248
expenses, 251
other recoverable heads, 249
profits, loss of, 251

DANGEROUS GOODS,
seller's duty as to, 246, 247

DEBT,
indebitatus counts, in action for, 227
pledge of goods for antecedent, 309

DECLARATION OF TRUST,
distinguished from gift, 79

DEED,
contract by, 93

DEFECTS,
buyer, actual or constructive knowledge of, 122, 127, 131
discoverable, 251
goods not warranted free from, 361
patent, 250
sale by sample, in case of, 137

DEFENDANT,
defined, 276

DEFINITIONS,
The Auctions (Bidding Agreements) Act, 1969 . . . 343
Bills of Sale Act, 1878 . . . 319
Factors Act, 1889 . . . 290
Factors (Scotland) Act, 1890 . . . 315
Interpretation Act, 1889 . . . 73

DEFINITIONS—*continued*
Sale of Goods Act, 1893 . . . 73, 273
Theft Act, 1968 . . . 339
Uniform Law on International Sales Act, 1967 . . . 50

DEL CREDERE AGENCY,
confirming house distinguished, 47
sale distinguished from, 76

DELAY,
acceptance inferred from, 197
making delivery, in, 241
risk shifted by, 157, 158
taking delivery, in, 204, 223

DELIVERY,
actual or constructive, 184, 211, 276
anticipation of, to end transit, 215
approval, on, 147, 152, 368
"as required," 185, 358
buyer in possession, by, 307, 310
buyer or bailee, to, 148
buyer's duty to accept and pay on, 179, 180
right to examine goods on, 122, 131, 195
carrier, to, 191 *et seq.*, 277, 368–370. *See* CARRIER.
conditions as to, 357, 358
defined, 276
delay in making, 157, 241
taking, 197, 223
deliverable state, defined, 287
distant place, at, risk in case of, 194
distress order in case of non-delivery, 244
ex-ship, 40
ex-store, 41
expenses of, 183, 185, 358, 359
instalments, in, 188
key of warehouse, of, 184, 277
mistake, by, 278
mode, of, 182, 277
non-delivery, actions for, 84, 235 *et seq.*
evidence in action for, 182
in case of rejection, 240
part, effect, of 188–191
seller's lien in case of, 210
stoppage *in transitu* in case of, 216
passing of property without, 84, 146
payment and, concurrent conditions, 180, 244, 357, 358
place of, 183, 184
priority of payment or, 180, 357, 358
refusal or neglect by buyer after request, 204
re-sale after, acceptance as, 199
reservation of right of disposal, 154
risk, in case of delay of, 157
rules as to, 182 *et seq.*
sale or return, on, 147, 152, 368
seller in possession, by, 304, 306
seller's duty as to, 179 *et seq.*
right to withhold, 206, 207. *See* SELLER'S LIEN.
stipulations as to, judicially construed, 357-359
symbolic, 184, 277
third person in possession of goods, where, 183, 185, 213, 277

DELIVERY—*continued*
time for, 183, 185, 358
transfer of document of title, 311
wrong quantity, of, 186

DELIVERY ORDER,
common law effect of, 185
document of title, is a, 293
transfer of, 185

DEPARTMENT OF TRADE AND INDUSTRY,
consumer protection powers, 55, 59, 71
powers, Trade Descriptions Act, 1968, under, 59
Weights and Measures Act, 1963, under, 55

DEPOSIT,
effect of, 26

DESCRIPTION,
conditions as to, 360–363, 365–368
errors in, 119–122, 306, 360, 361
exemption clause as to, 120, 360–363
goods not answering, 120–122
sale of goods by, 117, 128
exemption clause, 120, 259, 260, 264, 265
trade, meaning of, 56
words forming part of, 121

DESTINATION,
meaning of, 217

DESTRUCTION OF THING SOLD,
after agreement to sell, but before property passes, 99
before sale, 97
carrier, in hands of, 192, 194
general rule as to, 157–161. *See* RISK.

DETINUE,
appropriate remedy, when, 84, 241
effect of satisfied judgment in, 30

DISCOUNT,
payment within stated period, for, 106

DISPOSAL,
reservation of right of, 154, 211

DISPOSITION,
mercantile agent by, 175, 295 *et seq.*
seller or buyer in possession of goods or documents, by, 175, 221, 304, 307

DISTRESS,
sale of goods distrained, 75, 112, 114, 162, 167

DIVISIBLE CONTRACT,
instalment deliveries, 188, 189, 211
lots at an auction, 266, 267
warranties or conditions, 108. *See* CONDITION; WARRANTY.

DOCK WARRANT,
document of title, is a, 293
transfer of, 185

DOCUMENT OF TITLE,
bill of lading as, 318
Bills of Sale Acts, not within, 320
common law effect of, 185

DOCUMENT OF TITLE—*continued*
defined, 278, 293
derivative documents, 300
Factors Act, 1889, and, 295 *et seq.*
pledge of, effect, of 175, 221, 301
possession of, rights arising from, 175, 221, 300
seller's lien, effect on, 213
stoppage *in transitu*, effect on, 221, 310
transfer of, 175, 213, 221, 295 *et seq.*

DRUNKENNESS,
contract made in state of, 87

DURESS,
saving for law as to, 271, 272

E.E.C.,
resale price maintenance, condition as to, 52, 53

EMBLEMENTS,
included in "goods," 281

ENEMY,
contracts with, 20, 89, 90
definition of, 89
Prize Court rules as to goods of, 283
rights of action and appeal, 90

EQUITABLE ASSIGNMENT, 96. *See* ASSIGNMENT.

ESTOPPEL,
interpretation, 165
limitations upon, 166
nature of, 166
owner, against, where goods sold by another, 161, 162, 164
duty of care, 167
representation creating, 7, 166, 167
sale constituted by, 32

EVIDENCE,
compensation orders, in, 347
foreign law, of, 49
lex fori, governed by, 49
mitigation of damages, 255
non-acceptance, in action for, 182
non-delivery, in action for, 182
parol, admissibility of, 4
readiness to deliver or pay, of, 182

EX-SHIP CONTRACT,
incidents of, 40

EX-STORE CONTRACT,
what covered by, 41

EXAMINATION,
buyer's right of, 195
merchantable quality, implied condition excluded by, 122, 128, 131
on sale by sample, 137
patent defects, effects of, 131
place of, 140, 197
reasonable, defined, 140

EXCHANGE,
payment by bill of exchange, 230
price or damages payable in foreign currency, 239, 258
Treasury, when consent required, 42, 330

EXCHANGE (GOODS OR DOCUMENTS),
contract of, 79
goods or documents, of, under Factors Act, 302
sale distinguished from, 79

EXCISE DUTIES,
alteration of, effect on price, 2, 322, 345

EXECUTION,
bill of sale holder, rights of, 320
lex fori, governed by, 49
sheriff selling under, liability of, 114, 162*n.*, 178, 179
right of, 179
writ of, effect of, 177–179

EXEMPTION CLAUSE,
description, sale by, 120, 259, 260, 264, 265
fair or reasonable, determination whether, 259, 260, 264, 265, 335, 337
fitness for purpose, 127, 259, 260, 264, 265
fraudulent misrepresentation, invalid where, 335, 336
freedom from encumbrances, 116, 259, 263, 264
fundamental breach, for, 10, 11–14, 121
interpretation of, by court, 120, 121
limitations of, 120
misrepresentation, excluding liability for, 335–337
oral express term overriding, 5
quantity, 127, 259, 260, 264, 265
quiet possession, 116, 259, 263, 264
sale by sample, 138, 259, 260, 264, 265
terms with reference to, judicially construed, 360–363

EXPECTANCY,
sale of, 94

EXPENSES,
delay in recovering rejected goods, 204
delivery, of, 183, 185, 358, 359
seller's lien, in case of, 209
when buyer refuses or neglects to take delivery, 204
stoppage *in transitu*, of, 220
warranty of freedom from, 111

EXPORT LICENCE, 41, 42

F.O.B.,
contract, 33, 34–36, 194, 359
passing of risk under, 159

FACTORS ACT, 1889,
agent, saving for common law powers of, 313
agreements through clerks, 303
bankruptcy of agent, effect of, 312
buyer in possession before sale, 307
consignors and consignees, rights of, 303
definitions, 290
exchange of goods or documents, 302
history of legislation, 289
mercantile agent, dispositions by, 295
ordinary course of business of, 297
powers of, 299
mode of transferring documents, 311
pledge of documents or goods, 301, 302
possession defined, 292, 306
sale and exchange distinguished, 79
saving for, in Sale of Goods Act, 162, 271

FACTORS ACT, 1889—*continued*
seller in possession after sale, 304
seller's lien or stoppage *in transitu* under, 310, 315
true owner, etc., saving for, 311–313
See also DOCUMENT OF TITLE; MERCANTILE AGENT; PLEDGE.

FACTORS ACTS,
defined by Sale of Goods Act, 278
express saving for, 162

FACTORS (SCOTLAND) ACT, 1890,
provisions of, and application of English Act, 315

FAILURE OF CONSIDERATION,
effect of, 257

FAIR TRADING,
Act of 1973 . . . 54, 61–63
enforcement, 71
offences, act or default of another, 66
corporations, by, 68*n*.
defence of due diligence, 67
penalties, 69
powers of authorities, 72
Consumer Protection Advisory Committee, 62
Director-General of, appointment, 61
functions, 61, 62
credit control, system of, 63
notice of conviction, etc., to, 70
monopoly situations, 63
pyramid selling, 63
Restrictive Practices Court, 62

FALSE PRETENCES,
distinguished from larceny, 174
obtaining by, voidable title created by, 174

FAULT,
defined, 279

FI. FA. *See* EXECUTION.

FINANCE ACT, 1901,
provisions of, 322 *et seq*.

FINDING,
no title by, 164, 165

FIRM,
registration of business name, 91
sale to partner by, 75

FITNESS FOR PARTICULAR PURPOSE,
condition as to, 122 *et seq*.
exemption clause, 127, 259, 260, 264, 265
reasonable, 133–136
reliance on seller's skill etc., 134
partial, 135
unreasonable, 136
trade description as to, 56

FIXTURES
goods, as, 279

FOOD AND DRUGS,
Act of 1955, enforcement, 70
powers of authorities, 72
offences, act or default of another, 67, 68

FOOD AND DRUGS—*continued*
Act of 1955—*continued*
offences, corporations, by, 65
defence of due diligence, 67, 68
penalties, 69
prosecution, 70

FORCE MAJEURE,
effect of, 21, 358

FOREIGN CURRENCY,
price or damages payble in, 239, 258

FOREIGN LAW,
bill of lading, in relation to, 319
conflict of laws, 47–50
illegality according to, 48
presumption as to accord with English, 49
See also CONFLICT OF LAWS.

FRAUD,
auction sale in, 267
claim in court for price of goods, in, 270
contract induced by, 173, 174
effect of, generally, 286
exemption clause for, invalid, 335, 336
infant, on part of, 88
"legal," 286
mercantile agent, by, 297
misrepresentation, exclusion clause invalid, 335, 336
representation made in, 286
saving the law as to, 271
seller without title, by, 171, 173, 174, 300
what constitutes, 286

FREE ON BOARD, 33, 34–36, 194, 359

FREIGHT,
c.i.f. contracts, 33, 36–40, 194
claim for, saving, 317
pledge of bill of lading for, 317

FRUSTRATION. *See* IMPOSSIBILITY.

FUNDAMENTAL BREACH,
exemption clause to cover, 10, 11–14, 121
nature of, 127

FUNGIBLES,
what are, 147*n*.

FUTURE GOODS,
contract for sale of, 94
part, 96, 97
defined, 94, 279
when property passes in case of sale of, 148, 152

GAS,
whether included in "goods," 280

GENERIC GOODS,
destruction of, 97
genus numquam perit, application, 97
meaning of, 97
passing of property in, 141, 148, 153
sale of, 141, 152, 285
specific goods, distinguished from, 141, 285
unascertained part of ascertained whole, 142

GIFT,
sale distinguished from, 78

GOOD FAITH,
defined, 286

GOODS,
accessories or accretion, property in, 161
ascertained, must be, before property passes, 140
available market for, 231, 233, 235, 248
coin, when treated as, 281
dangerous, sale of, 246, 247
defined, 279
deliverable state, defined, 287
delivery. *See* DELIVERY.
emblements, growing crops, etc., 281
examination by buyer of, 122, 128, 131, 137, 138, 195
fungibles, 147*n*.
future. *See* FUTURE GOODS.
generic. *See* GENERIC GOODS.
"goods, wares, and merchandises," 292
land, things attached to or forming part of, 281
manufactured, to be, 94, 151
merchantability of, 122, 128–131
merchantable quality, defined, 286
mixed with others, not ordered, 143, 186
money, and things in action excepted from, 279, 280, 281
patent or trade name goods, 130, 132, 133
perishable, sale of, 223, 224
perishing before sale, 97–100
personal chattels, how far are, 279, 280
quality of, defined, 283
rejection by buyer of. *See* REJECTION.
ship, how far treated as, 282
specific. *See* SPECIFIC GOODS.
stolen. *See* STOLEN GOODS.
subject-matter of sale, as, 94
unascertained. *See* UNASCERTAINED GOODS.
unsolicited, Act of 1971 . . . 54, 61
wrong quantity delivered, 186

GROWING CROPS OR TREES,
sale of, 151, 281

GUARANTEE,
sale distinguished from, 76

HIRE-PURCHASE,
Act of 1965, repeal, 103*n*.
agreement for sale, 81, 309
contract, whether disposition by buyer valid, 309
possession under, 306, 309
rights under, 309

HORSE,
sale of, 171
warranty on sale of, construction of, 362, 363, 368

HOURS FOR DELIVERY,
reasonable, must be, 183, 185

HUSBAND,
liability for necessaries for wife, 91, 92
wife's right to pledge credit, presumption as to, 91

HYPOTHECATION,
letters of, not a bill of sale, 83, 321
meaning of, 83

ILLEGALITY,
consumer protection legislation, 55
general rules, 17–19
presumption as to, 19
saving for rules as to, 271
supervening, effect of, 20

IMPLIED TERMS,
course of dealing, from, 15
description, on sale by, 117
exclusion, restrictions on, 10
goods in existence, that, 97
how excluded, 259. *See* EXEMPTION CLAUSE.
quality and fitness, as to, 122 *et seq.*
sample, on sale by, 137
standard conditions of contract, 15, 16
title, as to, 111
usage, from, 124, 259

IMPORT LICENCE, 41, 42

IMPOSSIBILITY,
condition or warranty, effect on, 109
excuse for non-performance, as, 19–21, 98
goods which have perished, 97
under Law Reform (Frustrated Contracts) Act, 1943 . . . 20, 21, 100, 326–330
vis major, 21

INFANT,
capacity to buy or sell, 87–89
parents' liability for, 92

INNKEEPER,
power to sell goods left with, 162*n*.

INSOLVENCY,
buyer, of, reservation of title clause and, 225
revesting, effect on, 226
re-sale when buyer insolvent, 224
seller's lien when buyer insolvent, 207, 208, 209
stoppage *in transitu* on, 206, 207, 213 *et seq.*
what constitutes, 287
See also BANKRUPTCY; STOPPAGE *IN TRANSITU.*

INSPECTION, RIGHT OF. *See* EXAMINATION.

INSTALMENTS,
buyer not usually bound to accept, 188
contract for delivery by, 188–191
price payable by, 191
seller's lien and stoppage *in transitu* in case of delivery by, 210, 211, 216

INSURABLE INTEREST,
buyer or seller, of, 161*n*.
distinguished from property, 161

INSURANCE,
c.i.f. contract, in, 36–40
sea transit, in case of, 192, 194
seller's duty as to, 192

INTENTION,
pass property, to, 143
rules for ascertaining, 146 *et seq.*

INTEREST,
recoverable as damages, when, 227, 231, 254
saving for rules as to, 254
under Law Reform (Miscellaneous Provisions) Act, 1934 . . . 323

INTERMEDIATE TERM,
breach of, rights of other party, 6–9

INTERNATIONAL CONTRACT OF SALE.
See CONTRACT OF SALE.

INTERPRETATION.
See CONSTRUCTION; DEFINITIONS.

INTERPRETATION ACT, 1889,
construction of Sale of Goods Act subject to, 73

INTOXICATING LIQUORS,
sale of, 80

JOINT OWNERS,
rights *inter se*, 168
sale by one to another, 74, 75

JUS DISPONENDI,
reservation of, 154, 211, 223

KEY,
delivery by giving, 184, 277

KNOCK OUT,
auction sale, at, 269, 343

LAND,
sale of things attached to, 151, 279, 281

LARCENY,
false pretences, distinguished from, 174
trick, by, 174, 297, 309
See also STOLEN GOODS; THEFT.

LAW MERCHANT,
saving for, 271

LEGAL TENDER,
what is, 228

LEX FORI,
when applicable, 49

LEX LOCI CONTRACTUS,
when applicable, 48, 91

LEX SITUS,
when applicable, 49

LICENCE,
export or import, 41, 42

LICENCE TO SEIZE,
as bill of sale, 320
delivery by taking under, 96, 278

LIEN,
carrier, of, 214, 219
consignee, of, for advances, 303, 304
includes right of retention in Scotland, 282, 315
property, distinguished from, 283
unpaid seller, of, 206 *et seq.* *See* SELLER'S LIEN.

LIS PENDENS,
registration of, effect of, 179

LOSS,
c.i.f. contracts, 36–40
lost goods, no title to, by finding, 164, 165
seller's duty as to insurance, 192

MARKET OVERT,
rules, application of, 73*n*.
sale in, 169–171

MARKET PRICE,
measure of damages, as, 231 *et seq*.

MARRIED WOMAN,
divorce or judicial separation, 92
husband's liability for necessaries, 91, 92
incapacities, abolition of, 89
liability for child, 92
right to pledge husband's credit, 91, 92
separate estate, with, husband not liable, 92

MASTER OF SHIP,
owner's liability for necessaries supplied to, 93, 167
ship chartered by buyer, 216
special power of sale by, 162*n*.
stoppage, *in transitu*, duty in case of, 219
wrongful sale of cargo by, 49
See also CARRIER.

MAXIMS,
Caveat emptor, 126, 127
De minimis non curat lex, 187*n*.
every reasonable condition is implied, 185
Genus numquam perit, 97
Locus regit actum, 49
Nemo dat quod non habet, 164
Res perit domino, 159, 161
Ut res magis valeat quam pereat, 19

MEASURE,
sale by, 81, 147

MEASURE OF DAMAGES. *See* DAMAGES.

MENTALLY DISORDERED PERSON OR PATIENT,
contracting party, as, 86
necessaries, liability for, 85, 86

MERCANTILE AGENT,
acting in two capacities, 291, 292
agreement with clerk of, 303
buyer in possession as, 307, 310
defined, 290
disposition by, 175, 295, 297
fraudulent sale by, effect on title, 297, 299, 300
larceny by a trick by, 174
notice of lack of authority, 295, 299, 300
obtaining advance on property of principal, 295
pledge by, 297, 302
possession of, consent of owner, 296
determination of, 300
presumption as to, 301
derivative documents of title, of, 300
effect of, 175, 300

MERCANTILE AGENT—*continued*
powers of, 299
principal, purporting to act as, 298
sale by, 297, 298
Sale of Goods Act, 1893, under, 175
See also FACTORS ACT.

MERCHANTABLE, QUALITY,
defined, 129, 286
before 1973 . . . 128, 129
deterioration of goods in transit, 130, 194
exemption clause, 127, 259, 260, 264, 265
implied condition that goods are, 122 *et seq.*

MINOR,
necessaries for, 85, 87–89, 93

MISREPRESENTATION,
classification of representations, 5–8
effect of, 271, 286
exclusion clause, effect of, 335–337
trade description, 61
See also FRAUD.

MISREPRESENTATION ACT 1967,
avoidance of provisions excluding liability, 335
bars to rescission, removal of, 332
commencement of, 6, 338
damages, 332
extent of, 338
rescission, 333, 334, 335
Sale of Goods Act, 1893, amendments to, 337
savings, 337
short title of, 338

MISTAKE,
contract induced by, 3
delivery by, 278
identity of contracting party, as to, 3
rectification in case of, 244
saving for law as to, 271,
thing sold, in regard to, 3

MITIGATION OF DAMAGES,
duty as to, 234, 238, 250, 254, 255

MIXED GOODS,
mixed after sale, 143
rights of buyer in case of delivery of, 186

MONEY,
foreign currency, price or damages payable in, 239, 258
not included in "goods," 279, 281
what is legal tender of, 228

MONTH,
meaning of, 105

"MORE OR LESS,"
quantity, as description of, 188, 364

MORTGAGE,
defined, 82
pledge distinguished from, 83
sale distinguished from, 76
Sale of Goods Act, 1893, not within, 271, 272
See also PLEDGE.

MOTOR VEHICLE,
stipulations as to, judicially construed, 361
warranty as to condition of, 362

NECESSARIES,
defined, 85, 89
drunkard, supplied to, 85, 87
husband's credit, pledging for, rights of wife, 91
infant, supplied to, 85, 87–89
master of ship, supplied to, 93
mentally disordered person, supplied to, 85, 86
mistress, supplied to, 92
parent and child, 92

NEGOTIABLE INSTRUMENT.
See BILL OF EXCHANGE; BILL OF LADING.

NON-ACCEPTANCE, NON-DELIVERY. *See* ACCEPTANCE; ACTION; DAMAGES; DELIVERY.

NOTICE,
defective title, of, 169, 295, 299
insurance, with a view to, 192, 194
re-sale by seller, of, 223
sale by seller in possession, 305, 307
stoppage *in transitu*, to effect, 219
what is, under Factors Act, 295, 299

OFFER,
written, parol acceptance of, 93

OPTION,
buy, to, distinguished from conditional agreement, 81, 82

OWNER,
representations by, reliance upon, 166
sale by person who is not, 161, 169
saving for, under Factors Act, 311, 312

OWNERSHIP,
apparent, effect of, 162
special property, distinguished from, 282
transfer of, 74, 75, 84
See also PROPERTY; TRANSFER.

PARENT AND CHILD,
liability of parent for necessaries, 92

PAROL CONTRACT,
validity of, 93

PAROL EVIDENCE,
admissibility, 4, 93

PART ACCEPTANCE,
instalment contracts, 188
wrong quantity delivered, 186

PART DELIVERY,
goods, of, effect of, 186, 187, 188
seller's lien, effect on, 210
stoppage *in transitu*, effect on, 216

PART OWNERS,
rights *inter se*, 168
sale by one to another, 74, 75

PART PAYMENT,
instalment contracts, 188–191

PARTNER,
false trade description, liability for application, 58
sale by, to co-partner, 75

PATENT DEFECT,
sale of article with, 122 *et seq.*

PATENT OR TRADE NAME,
sale under, 130, 132, 133

PAWN. *See* PLEDGE.

PAWNBROKER,
sale by, 162*n.*, 167

PAYMENT,
according to terms, 229
banker's irrevocable letters of credit, 42–46, 231
bill of exchange or cheque, 230, 231
buyer's duty as to, 179, 228
confirmed credit, effect of, 42, 43, 231
when to be opened, 106
delivery and, *prima facie* concurrent conditions, 180, 244, 357, 358
deposit, 26
discount for prompt, 106
instalment, by, 191
contract, in case of, 188, 211
letter of credit, by, 42–46, 231
on day certain irrespective of delivery, 229
part payment. *See* PART PAYMENT.
place of, 228
prepayment of price, 237
stipulations as to, judicially construed, 357, 358
tender of. *See* TENDER.
time of, 105, 191, 229
unpaid seller's rights, 206
See also PRICE; SELLER'S LIEN; STOPPAGE *IN TRANSITU.*

PERFORMANCE,
buyer, by. *See* BUYER; SELLER.
delivery, by. *See* DELIVERY.
executory contract repudiated by other party, 256
general rules as to, 179 *et seq.*
impossibility of, 97, 100, 109
instalment contract, of, 188
mistake, etc., in connection with. *See* MISTAKE.
payment, by. *See* PAYMENT.
seller, by. *See* BUYER; SELLER.
specific, 243

PERISHABLE GOODS,
order of court, sale by, 224
re-sale on buyer's default, 223
when merchantable, 195

PERISHING OF GOODS,
effect of, before sale complete, 97–100
perish, meaning, 98
See also LOSS; RISK.

PERSON,
defined, 74, 294

PERSONAL CHATTELS,
apparent possession, when deemed in, 320
defined, Bills of Sale Act, 1878, under, 230
how far included in goods, 279, 280

PLACE,
delivery, of, 183, 184
examination, for, 140, 197
payment, of, 228

PLAINTIFF,
defined, 282

PLEDGE,
antecedent debt, for, 302
bill of lading, of, 221
buyer in possession of goods or documents, by, 175, 221, 307
conflict of laws as to, 49
consideration for, 302
defined, 83, 294
document of title, of, 175, 221, 301
exchange, rights acquired by, 302
lien, mortgage, and sale, distinguished from, 82, 83
mercantile agent, by, 297, 302
pledgee, rights of, 164
protection given to, 295, 297
Sale of Goods Act, 1893, non-application to, 271, 272
sale or return, of goods delivered on, 147
seller in possession of goods or documents, 175, 304, 307
stoppage *in transitu*, effect on, 221
unpaid seller's lien, effect on, 221
voidable title to goods under, 171

POSSESSION,
apparent, of maker of bill of sale, 320
buyer obtaining, 175, 307
defined, by Factors Act, 292, 306
property, and, transferred to buyer, revesting, 225, 226
quiet, implied warranty as to, 111, 112, 115
seller, of, after sale, 175
continues in possession, defined, 305, 306
third person, of, sale of goods in, 183, 185, 277
various meanings attributed to, 278
See also DELIVERY; PROPERTY.

POST,
delivery by, 218
stoppage *in transitu*, non-application to, 218

PRICE,
action for, 226 *et seq.*, 248
ascertainment of, 100–103, 141
bill of exchange or cheque for, 155, 205, 230, 231
breach of warranty defence in action for, 248
Code, 51
Commission, functions, 51
conditions as to, 359
control, statutory, 2, 51–53
currency, in what, 228, 239, 258
customs or excise duty, alteration of, 322, 323
defined, 74
deposit, payment of, 26
failure of consideration, 257
goods, of, false and misleading indications as to, 60, 103
indebitatus counts for, 227
instalments, payable by, 191
interest, power to award, 323
right to, 227, 231, 254

PRICE—*continued*
market, as measure of damages, 231 *et seq.*
payment of, mode and time of, 179–182
reasonable, what is, 101
recovery on rejection of goods, 240
re-sale price maintenance, 52, 53
reserved or upset, at auction sale, 267, 269
stipulations as to, judicially construed, 359, 360
sub-purchaser's payment of, 210, 224
tender of, 205, 228
to be agreed upon, 101
unpaid seller's rights, 206
valuation, fixed by, 103
value added tax, adjustment on change in, 2, 323, 344
wager, in nature of, 103
See also ACTION; PAYMENT.

PRINCIPAL AND AGENT. *See* AGENT.

PROPER LAW,
contract, of, application of 47
choice of, 48, 50

PROPERTY,
defined, 282
distinguished from right to possession, 283
estoppel, by, 32
insurable interest, distinct from, 161
possession, and, transferred to buyer, revesting, 225, 226
risk and benefit *prima facie* go with, 157
sale as a transfer of, 74, 84
special and general distinguished, 282
transfer of, 140 *et seq.* *See* TRANSFER.

PURCHASER. *See* BUYER.

QUALITY,
caveat emptor, and its exceptions, 126, 127
conditions as to, 360–363, 365–368
damages for breach of warranty of, 246
defined, 140
implied condition as to, 122 *et seq.*
includes state or condition of goods, 283
restrictions as to, 140
sample, sale by, in case of, 137
stipulation as to, judicially construed, 360 *et seq.*

QUANTITY,
stipulations judicially construed, 360 *et seq.*
trade description as to, 56
wrong, delivery of, 186–188

QUASI-CONTRACTS OF SALE,
note on, 30, 86, 102

QUIET POSSESSION,
enjoyment of, implied warranty of, 111, 112, 115

READINESS,
to pay or deliver, how proved, 182, 229

REASONABLE,
hours, 183, 185
price, 85, 101, 102
time, 266

RECTIFICATION,
of contract in case of mistake, 244

REGISTRATION,
business names, of, 91

REJECTION,
breach of condition or warranty, for, 106, 108–111
compensation, 240
defective goods of, after re-sale, 198
misdescription, for, 121
mode of, 201
non-delivery, as, 240
recovery of price after, 240
return of rejected goods, 203
rules as to, 201
Scotland, in, 109, 111, 203
when part not answering description, 200

REMEDIES,
by action, 266, 273
unpaid seller, of, against the goods, 205 *et seq*.
See also ACTION; BUYER; DAMAGES; SELLER.

REPRESENTATION,
classification of kinds of, 5–8
condition or warranty, as, 6
estoppel created by, 7
expression of opinion as, 5
false, 6
fraudulent, 7
incorporated in contract, 5
negligent misstatement, as, 7
term of contract, as, 5, 6
trade description as, 5, 61
See also ESTOPPEL; FRAUD; MISREPRESENTATION; WARRANTY

REPUDIATION. *See* PERFORMANCE.

RE-SALE,
auction, sale by, in case of, 81
buyer, by, 175, 220 *et seq*.
damages, defects discovered after, 250
goodwill, loss of, 254
losses incurred by buyer on, 253, 256
price on, 241
evidence of value, as, 249
evidence of acceptance, as, 197–200
price maintenance, 52
condition, notice of, 53
E.E.C., rules of, 52, 53
Resale Prices Act, 1964, under, 52
seller, by, 175, 220 *et seq*., 234
advantages of, 224
right where unpaid, 206, 223

RES PERIT DOMINO,
application of the maxim, 159

RESCISSION,
agreement of parties, by, 21, 22, 210
fraud, in case of, 269, 287
insolvency of buyer, on, 206
instalment contract, of, 188, 189
misrepresentation, for, 200, 332, 333, 334, 335
oral, when contract in writing, 21, 22
partial loss of goods, for, 99

RESCISSION—*continued*
resolutive condition, where there is, 80, 81, 152
seller's lien or stoppage *in transitu*, by, 223

RESERVATION,
right of disposal, of, 154, 211, 223
title, of, express, 224, 225

RETURN,
approval or sale or return, 147, 152, 159, 309, 368
breach of warranty, in case of, 81, 246
deposit, of, 27
rejection, in case of, 203

RISK,
" all risks " policies, 40
carrier in possession of goods, where, 192, 193, 368 *et seq*.
c.i.f. contract, in case of, 36
delivery at distant place, where, 194
destruction of goods before sale complete, 97–100
f.o.b. contract, in case of, 194
general incidence of, 157–161
insurance, seller's duty as to, 192
passing, but not property, 161
prima facie passes with property, 157
res perit domino, 159
sea transit, in case of, 192

ROMAN LAW,
future goods, sale of, 95
passing of property, as to, 145
price, how fixed, 101, 151
res perit domino, 159
See also MAXIMS.

SALE,
absolute or conditional, 74, 80
agreement to sell and, 74, 84. *See* TRANSFER OF PROPERTY.
auction sales, 112, 266–269, 343
induced by fraud, 269
buyer in possession, 175, 307
consumer. *See* CONSUMER SALE.
co-owner, by, 74, 75, 168
defined, 284
distress warrant, under, 112
exchange distinguished from, 70
formalities of, 93, 94
hire-purchase. *See* HIRE-PURCHASE AGREEMENT.
horse, of, 171, 362, 363, 368
included in " contract of sale," 74, 276
includes " bargain and sale," 284
mercantile agent, by, 297, 298
non-consumer, exemption clauses, 260, 263, 264
quasi-contracts of, 30, 86
relation to contract generally, 74 *et seq*.
seller in possession after, by, 175, 304, 307
statutory controls, 51 *et seq*.
trial or approval, on, 147, 152, 368
See also AGREEMENT TO SELL; CONTRACT OF SALE.

SALE OF GOODS ACT, 1893
code of law, as, 73
contracting out, restrictions on, 1

SALE OF GOODS ACT 1893—*continued*
construction of, 273 *et seq.*
corporations, saving for law of, 93, 94
general savings in, 271
special statutory or common law powers of sale, saved, 162
warranties under special Acts saved, 122

SALE OR RETURN,
delivery of goods on, 147, 152, 159, 309, 368

SAMPLE,
delivery of goods corresponding to, 120*n.*
sale by, 117, 137–140
exemption clause, 138, 259, 260, 264, 265

SCOTLAND,
arrestment or poinding by seller in, 208
auction sale in, 269
breach of warranty in, 109, 246, 254, 284
compensation in, 273, 315
consignation or payment into court in, 254, 270
discretion of court as to, 270
custodier defined, 273
equivalents of terms in Sale of Goods Act, 1893, in, 273 *et seq.*
Factors (Scotland) Act, 1890, provisions of, 315
insurance, duty of agent as to, 193, 194
interest, law as to, 227, 231, 326
landlord's hypothec in, 272
latent defects, law as to, 127
market overt rule not applicable to, 169
non-delivery, law as to damages for, 237
notour bankruptcy in, 287
passing of property, law as to, 145
risk in a contract, 80*n.*
price not material, conditions as to, 102*n.*
quanti minoris action in, 111, 254
rejection, law as to right of, 109, 111, 203
right of retention in, 208
specific implement in, 245
stoppage *in transitu* in, 208

SEA TRANSIT,
c.i.f., f.o.b. etc., contracts in case of, 33–41, 194
seller's duty in case of, 192
See also SHIP; SHIPMENT.

SEAL,
when required, 93

SELLER,
actions against, 235 *et seq.*
by, 226 *et seq.*
bailee, as, 158, 160, 277, 306
buyer, must be different person from, 75
capacity of, 85
customs and excise duties, effect of alteration of, 322, 323
liability for, 323
dangerous goods, duty as to, 247
defect in title of, 161 *et seq.* *See* TRANSFER OF TITLE.
defined, 284
delivery, duty as to, 179 *et seq.* *See* DELIVERY.
description, goods not answering, 120, 121
disposition by, after sale, 175, 304
duty to insure, 192

SELLER—*continued*
exemption clauses, protection by, 120, 360–363
export or import licence, obtaining, 41
judgment of, reliance upon, 123, 134, 135, 136
lien of. *See* SELLER'S LIEN.
possession, after sale, of, 175, 304–307
continues in, defined, 305, 306
rejection of goods by buyer, 200
re-sale by, advantages of, 224
rights of, 223
representations as to ownership by, 166
risk of, when goods are at, 157
skill of, reliance upon, 123, 134, 135, 136
stoppage *in transitu* by. *See* STOPPAGE *IN TRANSITU*.
title, failure to pass, 173, 174
voidable, of, 173
unpaid defined, and rights of, 205 *et seq.*

SELLER'S LIEN,
arises, in what cases, 206, 208
charge for keeping goods, 209
distinguished from stoppage *in transitu*, 207
document of title, effect of transfer of, 310
effect of part delivery, 210
origin and nature of, 207
rescission of contract by, 223
sub-sale by buyer, effect of, 220
termination of, 211
unpaid seller, definition and rights of, 205 *et seq.*
waiver of, 213

SEVERABLE CONTRACT. *See* DIVISIBLE CONTRACT.

SHARES,
things in action, are, 281

SHERIFF,
defined, 178
execution by, effect of, 177
sale by, 114

SHIP,
Bills of Sale Acts, outside, 273, 321
chartered by buyer, 216
ex-ship contract, effect of, 40
master of. *See* MASTER OF SHIP.
necessaries supplied to, 93
shipbuilding contracts, construction of, 144, 151, 153
transfer by bill of sale of, 93*n.*, 273
unregistered transfer of, 321

SHIPMENT,
bill of lading as evidence of, 316
effect of, 154, 155
f.o.b. contracts, 34
right of disposal reserved, 154, 155
seller's duty on making, 192
stipulations judicially construed, 355, 356
stoppage *in transitu*, 213 *et seq.*

SPECIAL DAMAGES,
saving for, and note on, 254, 256, 257
See also DAMAGES.

SPECIFIC GOODS,
caveat emptor in case of, 127, 128
defined, 285
generic goods, contrasted with, 141, 285
land, attached to, 151
non-existent, contract for sale of, 3
passing of property in case of, 143
patent or trade name, sold under, 131, 133
perishing of, before sale complete, 97–100
right to reject after property passes, 108–110
unascertained part of specific whole, 141

SPECIFIC PERFORMANCE,
failure, power to order distress in case of, 243
remedy by, 27, 243

SPIRITS,
intoxicating liquor, sale of, 80

STATUTE,
appendix of statutory provisions, 316 *et seq.*
sale under statutory power, 162
saving for Acts not expressly repealed, 271

STATUTE OF FRAUDS,
actual receipt within, 212

STOLEN GOODS,
auction, sold by, 161*n.*
Factors Act, effect of, 295
market overt, sold in, 169
restitution to owner of, on conviction of thief, 175, 339
sale of, generally, 295, 339, 340
title to, 295, 339
See also THEFT.

STOPPAGE *IN TRANSITU*,
contract of sale not usually rescinded by, 223
effect on executory contract, 214
goods in hands of carrier, 193
how effected, 219
part delivery, effect of, 216
postal packet sent to buyer, of, 218
re-sale by buyer, effect of, 220
right of, nature of, 206, 207, 214
seller's lien, distinguished from, 207
transfer of document of title, effect of, 157, 221, 310, 316
transit, duration of, 214
unpaid seller defined, 205

SUB-SALE,
damages, effect on, 234
effect on unpaid seller's rights, 220

SUSPENSIVE CONDITION,
incidents of, 80, 81

TENDER,
bill of exchange, of, 230
delivery, of, 183. *See* DELIVERY.
legal, what is, 228
payment into court where plea of, 228
price, of, 205, 228

TERMS AND STIPULATIONS,
intermediate, 6, 9
list of, judicially construed, 354 *et seq*,
See also CONDITIONS; WARRANTY.

THEFT. *See* STOLEN GOODS.
Theft Act, 1968 . . . 339
civil proceeding effect on, 341
compensation orders against convicted persons, 345
definitions of theft, 339
finding, by, 165
gas, water and electricity, of, 280
offences relating to stolen goods, 339
orders for restitution, 339 *et seq*.
thief, buyer from, compensation, 340
disposition by, 310

THINGS IN ACTION,
not included in " goods," 279, 280

THIRD PERSON,
attornment to buyer by, 183, 277
consignment by, rights of consignee in case of, 303
following goods in hands of, 84, 242
title of stolen goods, of, 171
transfer of title of, 112, 116

THROUGH BILL OF LADING,
c.i.f. contracts in, 37
defined, 316

TIMBER,
sale of, 279
shipment, requirement as to, 366

TIME,
construction of stipulations as to, 105, 358
delivery, for, 181, 185
demand of delivery, for, 183
essence of contract, when, 105–108
lex fori, governed by, 49
" month," meaning of, 105
payment, for, 180–182
reasonable, a question of fact, 266

TITLE,
bona fide purchaser, of, 171
document of. *See* DOCUMENT OF TITLE.
express reservation of, 224
implied undertakings as to, 111
purchaser under writ of execution, 162, 177
revocation of authority, effect on, 300
stolen goods, to, 169 *et seq*.
transfer of, 111, 116, 161 *et seq*. *See* TRANSFER.
voidable, 171, 299

TORT,
waiver of, 31. *See also* CONVERSION; DETINUE.

TRADE DESCRIPTION,
Act of 1968, enforcement, 71
powers of authorities, 72
generally, 54, 56–60
offences, 61
abroad, accomplices, 65

TRADE DESCRIPTION—*continued*
Act of 1968, offences,—*continued*
act or default of another, 66
compensation, 346
corporations, by, 65
defence of due diligence, 66
penalties, 69
prosecution, 71
Act of 1972, generally, 54, 56, 60
offences, 61
act or default of another, 66
advertisements, used in, 59
Department of Trade and Industry, powers of, 59
false, application of, 56, 57–59
meaning, 57
representation as, 5
course of business, made in, 7
statement of opinion as, 61*n.*
indication of certain matters, 56, 61
meaning, 56
misrepresentation, 61
origin of imported goods, indication of, 60
price, false or misleading indications as to, 60, 103
representation, as, 5, 61
supply or offers to supply any goods, 59, 99*n.*
application to, 56, 58
TRADE MARK,
infringement of, effect on sale of, 112
sale of goods bearing, 130
TRADE NAME,
sale of goods under, 130, 132, 133
TRADING STAMPS, 80
TRANSFER OF PROPERTY BETWEEN SELLER AND BUYER,
accessories or accretion, of, 161
appropriation of goods to contract, 141
articles specially made for buyer, in case of, 150, 151
ascertained, goods must be, 140
bill of lading, of, 222, 317
carrier, by delivery to, 148, 191, 368
conflict of laws as to, 47
document of title, of, 175, 221. *See* FACTORS ACT.
future goods, of, 94, 148, 150, 151
intention to, how ascertained, 146 *et seq.*
jus disponendi, reservation of, 154
price payable by instalments, 191
property, of, on sale, 74, 77, 84, 140 *et seq.*
risk, incidence in case of, 157, 191, 192, 193, 194, 195
sale on approval, or " on sale or return," 147, 152, 368
specific goods, of, 141, 146
unascertained or generic goods, of, 141, 148, 152
weighing or measuring, of goods requiring, 147, 152
TRANSFER OF TITLE,
buyer in possession, 175, 307
Factors Act, under, 162, 289 *et seq.*
general rule as to, 161 *et seq.*
market overt sales, 169
mistaken identity, effect of, 171–174
only such title as seller has, 111, 116
person, not the owner, by, 161 *et seq.*

TRANSFER OF TITLE—*continued*
seller remaining in possession, 175, 304, 306
special statutory or common law powers, 162
stolen goods, in case of, 169, 174, 339, 340
voidable title under, 171
writs of execution, under, 177
See also STOLEN GOODS; FACTORS ACT, 1889.

TRANSIT. *See* STOPPAGE *IN TRANSITU*.

TREASURY CONSENT, 42, 330–332

TROVER. *See* CONVERSION.

UBERRIMAE FIDEI,
sale not contract, of, 287

UNASCERTAINED GOODS,
destruction of, 97
genus numquam perit, application, 97
merning of, 141
passing of property in, 141, 148, 153
sale of, 141, 153
specific goods, distinguished from, 141, 285
unascertained part of ascertained whole, 142

UNDERSELLING,
resale price maintenance, 52, 53
restriction on, 52

UNPAID SELLER,
defined, 205
remedies of, against goods, 205 *et seq.*
See also RE-SALE; SELLER'S LIEN; STOPPAGE *IN TRANSITU*.

UNSOLICITED GOODS,
Act of 1971, offences, 61
corporations, by, 65*n*.
recipient, rights of, 61

USAGE OF TRADE,
conditions necessary for usage to have effect, 17
considered as part of agreement, 17
delivery of wrong quantity, effect of, 186
inconsistency with written term of agreement, 17
incorporation into contract of, 17, 124, 259
knowledge of both parties essential, 17
legality of, 17
reasonable, must be, 17
sale by sample evidenced by, 139
saving for law merchant, 271
warranty or condition implied by, 124, 259

VALUATION,
agreement to sell goods at, 103
distinguished from arbitration, 105

VALUE,
Factors Acts, in relation to, 302, 303
measure of damages, as, 247
price, in relation to, 101

VALUE ADDED TAX,
adjustment on changes of, 2, 323, 344

VARIATION,
contract of sale, of, 21, 22

VENDOR. *See* SELLER.

VIS MAJOR,
excuse, as, 21

WAGER,
avoidance of sale as being, 103
sale distinguished from, 76

WAIVER,
condition precedent, or warranty, of, 23, 108, 259
right to stop *in transitu*, of, 216
seller's lien, of, 211, 212
stipulation as to time, of, 107
tort, of, 31

WAR,
contract of sale, effect on, 326 *et seq.*
enemy, status of, 89

WARRANT FOR GOODS,
is a document of title, 293. *See* DOCUMENT OF TITLE.

WARRANTY,
action for breach of, 246, 247
citations, 351–354
condition, when to be treated as, 108, 245
consumer protection statutes, 54
defined, 349
by Sale of Goods Act, 285
discussion of definitions, 285, 348 *et seq.*
exemption clause. *See* EXEMPTION CLAUSE.
express or implied, 259
fitness for particular purpose, of, 122
freedom from charges or encumbrances, of, 111, 112, 115, 116
implied, exemption clause, fair and reasonable, 259, 260, 264
void, 259, 263
usage, annexed by, 123
merchantable quality, of, 121, 123, 124
motor vehicle, as to, judicially construed, 361, 362
negatived or varied, how, 259
quiet possession, of, 111, 112, 115, 116
reasonable fitness, of, 122
representation amounting to, 6
sale by sample, on, 139
without, 360, 361
stipulations judicially construed, 354, 360 *et seq.*
title, as to, 111
trading stamps, on redemption of, 80
See also CONDITION; REPRESENTATION; USAGE.

WATER,
damage by, liability as to, 363
whether included in " goods," 280

WEIGHING AND MEASURING,
Act of 1963, consumer protection, 54, 55, 56
enforcement, 70, 71
powers of authorities, 72
generally, 54
offences, act or default of another, 67, 68
corporations, by, 65*n*.
defence of due diligence, 67, 68
penalties, 69
prosecution, 70, 71
contract involving, 147

WHARFINGER,
delivery to, 193

WIFE. *See* Married Woman.

" WITHOUT RESERVE,"
in auction sale, 267

WORDS AND PHRASES,
list of, judicially construed, 354 *et seq.*

WORK AND MATERIALS,
sale, distinguished from, 76, 77

WRITING,
defined by Interpretation Act, 74
parol rescission of written contract, 4
variation of written contract, 4, 5

Printed and bound in Great Britain by R. J. Acford Ltd., Chichester, Sussex.